S0-CLY-654

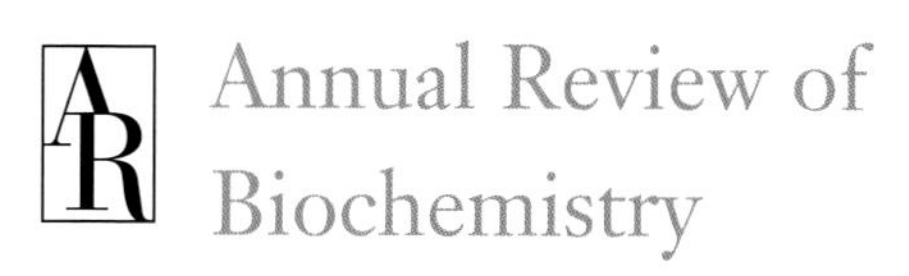
Annual Review of
Biochemistry

Production Editor: Jesslyn S. Holombo
Managing Editor: Jennifer L. Jongsma
Bibliographic Quality Control: Mary A. Glass
Electronic Content Coordinator: Suzanne K. Moses
Illustration Editor: Eliza K. Jewett-Hall

Annual Review of Biochemistry

Volume 79, 2010

Roger D. Kornberg, *Editor*
Stanford University School of Medicine

Christian R.H. Raetz, *Associate Editor*
Duke University Medical Center

James E. Rothman, *Associate Editor*
Yale University School of Medicine

Jeremy W. Thorner, *Associate Editor*
University of California, Berkeley

www.annualreviews.org • science@annualreviews.org • 650-493-4400

Annual Reviews
4139 El Camino Way • P.O. Box 10139 • Palo Alto, California 94303-0139

Annual Reviews
Palo Alto, California, USA

International Standard Serial Number: 0066-4154
International Standard Book Number: 978-0-8243-0879-7
Library of Congress Catalog Card Number: 32-25093

∞ The paper used in this publication meets the minimum requirements of American National Standards for Information Sciences—Permanence of Paper for Printed Library Materials, ANSI Z39.48-1992.

TYPESET BY APTARA
PRINTED AND BOUND BY FRIESENS CORPORATION, ALTONA, MANITOBA, CANADA

The Power of One

The brain is a correlating machine. If two events occur together frequently enough, we soon learn to expect one when the other occurs, and naturally take this as evidence that one causes the other. But it is no proof; rather it is distinguishing coincidence from causation that is the essence of proof in biochemistry, genetics, and all hard science. As biochemists, we have witnessed the relentless advance of our science resulting from the rigorous application of what is now frequently called the "reductionist" approach. We change only one variable (the concentration of an enzyme in a biochemical reconstitution or a single gene in a genetic experiment) at a time and compare the result with a control group, which is otherwise identical. Any change in outcome (rate of reaction, phenotype) must then be caused by the single change that was made. Critically, if even two variables were changed at once, we could no longer attribute the change in outcome to one or the other variable, making causal proof impossible with only the correlation remaining.

This is the "power of one." It accounts for every scientific fact in every textbook, and in this volume, we see its continuing power in article after article. But biochemistry is also at a turning point as the direct consequence of its success. We are now awash in complexity, drowning in a dizzying array of precise insights and three-letter abbreviations. Each has been shown by causal deduction to be vitally important in some fashion, but it is hard to avoid getting lost in all the details and avoid being driven into increasingly remote and specialized investigations. Adding to the dismay of those looking for simple principles, we have learned that genes, enzymes, and pathways often function in seeming duplicative and complex ways, and have guessed that this is because they may operate in a network whose output is only partly affected by one or the other component. If we could only understand the network—the principle of the circuit as distinct from detailed lists of its component genetic transistors, diodes, and resistors—we might gain a simpler, unified concept that permits the system (the physiology of the cell or the organism) to be better predicted and understood. This is the underlying aspiration of "systems biology."

The human mind cannot keep track of all the reductionist data on component genes and proteins, but computers can. Systems biologists use them to store and manipulate massive databases of causal results, to amalgamate them, and to correlate large numbers of compositional variables with outcomes (e.g., patterns of gene transcription, protein mass spectra, or single-nucleotide polymorphisms with physiology of the cell or the patient). This correlative science is done by statistics on an unprecedented scale, driven by the irresistible urge to use the mountains of data we have.

By definition, correlations can suggest hypotheses but cannot prove them, so there is a profound and irreconcilable difference between reductionist and systems science.

However, the latter constitutes a valid and complementary approach that needs to be understood and embraced by biochemists. The *Annual Review of Biochemistry* is facilitating this through articles in recent volumes, including this one, which feature high-throughput analyses of systems components (genomics, proteomics, metabolomics) and other such studies.

James E. Rothman, Associate Editor
February 5, 2010

Annual Review of Biochemistry

Volume 79, 2010

Contents

Preface

Prefatory Article

Recent Advances in Biochemistry

Indexes

Errata

An online log of corrections to *Annual Review of Biochemistry* articles may be found at http://biochem.annualreviews.org

Related Articles

From ***Annual Review of Biomedical Engineering,*** Volume 11 (2009)

Proteomics by Mass Spectrometry: Approaches, Advances, and Applications
John R. Yates, Cristian I. Ruse, and Aleksey Nakorchevsky

From ***Annual Review of Biophysics,*** Volume 39 (2010)

Global Dynamics of Proteins: Bridging Between Structure and Function
Ivet Bahar, Timothy R. Lezon, Lee-Wei Yang, and Eran Eyal

From ***Annual Review of Cell and Developmental Biology,*** Volume 25 (2009)

Membrane Traffic Within the Golgi Apparatus
Benjamin S. Glick and Akihiko Nakano

From ***Annual Review of Food Science and Technology,*** Volume 1 (2010)

Biochemistry and Genetics of Starch Synthesis
Peter L. Keeling and Alan M. Myers

From ***Annual Review of Genetics,*** Volume 43 (2009)

Active DNA Demethylation Mediated by DNA Glycosylases
Jian-Kang Zhu

From ***Annual Review of Genomics and Human Genetics,*** Volume 10 (2009)

Applications of New Sequencing Technologies for Transcriptome Analysis
Olena Morozova, Martin Hirst, and Marco A. Marra

From ***Annual Review of Immunology,*** Volume 28 (2010)

Molecular Basis of Calcium Signaling in Lymphocytes: STIM and ORAI
Patrick G. Hogan, Richard S. Lewis, and Anjana Rao

From ***Annual Review of Medicine,*** Volume 61 (2010)

Biological Mechanisms Linking Obesity and Cancer Risk: New Perspectives
Darren L. Roberts, Caroline Dive, and Andrew G. Renehan

From ***Annual Review of Microbiology,*** Volume 63 (2009)

Regulation of Translation Initiation by RNA Binding Proteins
Paul Babitzke, Carol S. Baker, and Tony Romeo

From ***Annual Review of Neuroscience,*** Volume 33 (2010)

The Genomic, Biochemical, and Cellular Responses of the Retina in Inherited Photoreceptor Degenerations and Prospects for the Treatment of These Disorders
Alexa N. Bramall, Alan F. Wright, Samuel G. Jacobson, and Roderick R. McInnes

From ***Annual Review of Nutrition,*** Volume 29 (2009)

Roles for Vitamin K Beyond Coagulation
Sarah L. Booth

From ***Annual Review of Pharmacology and Toxicology,*** Volume 50 (2010)

The RNA Polymerase I Transcription Machinery: An Emerging Target for the Treatment of Cancer
Denis Drygin, William G. Rice, and Ingrid Grummt

From ***Annual Review of Physiology,*** Volume 72 (2010)

Genomic Analyses of Hormone Signaling and Gene Regulation
Edwin Cheung and W. Lee Kraus

From ***Annual Review of Plant Biology,*** Volume 61 (2010)

Structure and Function of Plant Photoreceptors
Andreas Möglich, Xiaojing Yang, Rebecca A. Ayers, and Keith Moffat

From ***Annual Review of Public Health,*** Volume 31 (2010)

Genome-Wide Association Studies and Beyond
John S. Witte

Annual Reviews is a nonprofit scientific publisher established to promote the advancement of the sciences. Beginning in 1932 with the *Annual Review of Biochemistry*, the Company has pursued as its principal function the publication of high-quality, reasonably priced *Annual Review* volumes. The volumes are organized by Editors and Editorial Committees who invite qualified authors to contribute critical articles reviewing significant developments within each major discipline. The Editor-in-Chief invites those interested in serving as future Editorial Committee members to communicate directly with him. Annual Reviews is administered by a Board of Directors, whose members serve without compensation.

Annual Reviews of

Analytical Chemistry
Anthropology
Astronomy and Astrophysics
Biochemistry
Biomedical Engineering
Biophysics
Cell and Developmental Biology
Chemical and Biomolecular Engineering
Clinical Psychology
Condensed Matter Physics
Earth and Planetary Sciences
Ecology, Evolution, and Systematics
Economics
Entomology
Environment and Resources
Financial Economics
Fluid Mechanics
Food Science and Technology
Genetics
Genomics and Human Genetics
Immunology
Law and Social Science
Marine Science
Materials Research
Medicine
Microbiology
Neuroscience
Nuclear and Particle Science
Nutrition
Pathology: Mechanisms of Disease
Pharmacology and Toxicology
Physical Chemistry
Physiology
Phytopathology
Plant Biology
Political Science
Psychology
Public Health
Resource Economics
Sociology

SPECIAL PUBLICATIONS
Excitement and Fascination of Science, Vols. 1, 2, 3, and 4

Aaron Klug

From Virus Structure to Chromatin: X-ray Diffraction to Three-Dimensional Electron Microscopy

Aaron Klug

MRC Laboratory of Molecular Biology, Cambridge CB2 0QH, United Kingdom;
email: akl@mrc-lmb.cam.ac.uk

Annu. Rev. Biochem. 2010. 79:1–35

First published online as a Review in Advance on January 27, 2010

The *Annual Review of Biochemistry* is online at biochem.annualreviews.org

This article's doi:
10.1146/annurev.biochem.79.091407.093947

0066-4154/10/0707-0001$20.00

Key Words

protein disk, dislocation, nucleation and growth, chromatin, Royal Society

Abstract

Early influences led me first to medical school with a view to microbiology, but I felt the lack of a deeper foundation and changed to chemistry, which in turn led me to physics and mathematics. I moved to the University of Cape Town to work on the X-ray crystallography of some small organic compounds. I developed a new method of using molecular structure factors to solve the crystal structure, which won me a research studentship to Trinity College Cambridge and the Cavendish Laboratory. There I worked on the austenite-pearlite transition in steel. This is governed by the dissipation of latent heat, and I ended up numerically solving partial differential equations. I used the idea of nucleation and growth during the phase change, which had its echo when I later tackled the assembly of *Tobacco mosaic virus* (TMV) from its constituent RNA and protein subunits.

I wanted to move on to X-ray structure analysis of large biological molecules and obtained a Nuffield Fellowship to work in J.D. Bernal's department at Birkbeck College, London. There, I met Rosalind Franklin, who had taken up the study of TMV. I was able to interpret some of Franklin's beautiful X-ray diffraction patterns of the virus particle. From then on, my fate was sealed.

After Franklin's untimely death in 1958, I moved in 1962 to the newly built MRC Laboratory of Molecular Biology in Cambridge, which, under Max Perutz, housed the original MRC unit from the Cavendish Laboratory. I was thus privileged to join the Laboratory at an early stage in its expansion and consequently able to take advantage of, and to help build up, its then unique environment of intellectual and technological sophistication. There I have remained ever since.

Contents

EARLY LIFE AND EDUCATION

I was born in 1926 to Lazar and Bella (née Silin) Klug in Zelvas, Lithuania, but I remember nothing of the place because I was brought to South Africa as a child of two and grew up there. My father was trained as a saddler, but in fact as a young man, he worked in his father's business of rearing and selling cattle, so he grew up in the countryside. He had a traditional Jewish education and secular schooling, and though not a conventionally well-educated man, he had a gift for writing and had a number of articles published in the newspapers of the capital, for which he acted as what would now be called a stringer. Shortly after I was born, he emigrated to Durban, where members of my mother's family had settled at the turn of the century, and the rest of the family followed soon thereafter.

Durban was then a relatively sleepy town in subtropical surroundings. It was a fine place for a boy—there was the beach and the bush—and school was not too taxing. I went to a good school, Durban High School, which was run on traditional English lines, with a curriculum somewhat adapted to South African circumstances. Wc had some good masters, particularly in history and English. However, by the standards of today, there were few challenges other than Advanced Latin Prose Composition in the sixth Form. The philosophy of the school was quite simple: the bright boys specialized in Latin, the not so bright studied science, and the rest managed with geography or the like. There was a good library, but it was the playing fields that kept one out of mischief. I did not feel a particularly strong call to any one subject but read voraciously and widely and began to find science interesting. It was the book called *Microbe Hunters* by Paul de Kruif, well known in its time, which influenced me to begin medicine at the university as a way into microbiology.

At the University of Witwatersrand in Johannesburg, I took the premedical course, and in my second year, I took, among other subjects, biochemistry, or physiological chemistry as it was then called, which stood me in good stead in later years when I came to face biological material. However, I felt the lack of a deeper foundation and moved to chemistry, and this, in turn, led me to physics and mathematics. So finally, I took a science degree with chemistry, physics, and mathematics as major subjects.

I had by then decided that I wanted to do research in physics, and I went to the University of Cape Town, which was then offering scholarships that enabled one to do an MS degree in return for demonstrating in laboratory classes. The University lay in a beautiful site on the slopes of Table Mountain, which one climbed on the weekends. I was lucky to find as professor there R.W. James, the X-ray crystallographer who had brought to Cape Town the traditions of the Bragg school at Manchester. He was an excellent teacher, and I attended his undergraduate lectures as well as those in the MS course. From him, I acquired a feeling for optics and a knowledge of Fourier theory, and I remember particularly certain optical experiments on rather abstruse phenomena such as external and internal conical refraction, which fascinated me. After receiving my MS degree, I stayed on and worked on the X-ray analysis of some small organic compounds; in this way I developed a method of using molecular structure factors for solving crystal structures and taught myself some quantum chemistry to calculate bond lengths and other items of interest. During this time, I developed a strong interest, broadly speaking, in the structure of matter and how it was organized. I had now acquired a good knowledge of X-ray diffraction, not only through my own work, but also by having helped James check the proofs of his fine book—*The Optical Principles of the Diffraction of X-rays*—which is still a standard work. James wrote beautifully and fully and took great pains to make everything clear.

TMV: *Tobacco mosaic virus*

Cambridge

Supported by an 1851 Exhibition Scholarship and also by a Research Studentship from Trinity College, I went to Cambridge in 1949. Cambridge was the place for someone from the colonies or the dominions to study, and it was to the Cavendish Laboratory where one went to pursue physics. I wanted to work on some form of unorthodox X-ray crystallography, for example protein structure, but the MRC Unit where Perutz and Kendrew were working was full, and Bragg, then the Cavendish Professor, had closed down a project on order-disorder phenomena in alloys, which interested me. I finally found myself as a research student of D.R. Hartree, who had been a colleague of both Bragg and James at Manchester. He suggested to me a theoretical problem leftover from his work during the war on the cooling of steel through the austenite-pearlite transition, and I learned a fair amount of metallurgy in order to understand the physical basis of the phenomenon. It turned out, however, in the end that it was not special crystallographic insight that was required because the course of the transition was, in practice, governed by the diffusion of the latent heat engendered in the phase transition, and I ended up using numerical methods to solve the partial differential equations for heat flow in the presence of a phase transition. I learned a good deal during this time, particularly about computing and solid-state physics, and the idea of nucleation and growth during a phase change had its echo when I came later to think about the assembly of *Tobacco mosaic virus* (TMV) from its constituents.

After taking my PhD, I spent a year in the Colloid Science Department in Cambridge, working with F.J.W. Roughton, who had asked Hartree for someone to help him tackle the problem of simultaneous diffusion and chemical reaction, such as occurs when oxygen enters a red blood cell. The methods I had developed for the problem in steel were applicable here, and I was glad to put them to use on an interesting new problem. The quantitative data came from experiments in which thin layers of blood were exposed to oxygen or carbon monoxide. In the course of my stay there, I also showed how one could analyze the experimental kinetic curves for the reaction of hemoglobin with carbon dioxide or oxygen by simulations in the computer, which fit the rate constants.

Birkbeck College: Rosalind Franklin

This work made me more and more interested in biological matter, and I decided that I really wanted to work on the X-ray analysis of biological molecules. I obtained a Nuffield Fellowship to work in J.D. Bernal's department in Birkbeck College in London, and I moved there at the end of 1953. I joined a project on the protein ribonuclease, but shortly afterward, I met Rosalind Franklin, who had moved to Birkbeck earlier and had begun working on TMV. Her beautiful X-ray photographs fascinated me, and I was also able to interpret some that had apparently anomalous curved layer lines in terms of the splitting that occurs when the helical parameters are nonrational. From then on my fate was sealed. I took up the study of TMV, and in four short years, together with Kenneth Holmes and John Finch, who had joined us as research students, we were able to map out the general outline of the TMV structure. This work was done partly in parallel with that of Donald Caspar, then at Yale, but he spent 1955–1956 in Cambridge, and I formed an association with him, which continued across the Atlantic for many years. It was during this time that I met Francis Crick, and we published a paper together on diffraction by helical structures. I was fortunate to work with him again later and was able to learn, as he once wrote of Bragg, from watching the way he went about a problem.

Rosalind Franklin died in 1958, and supported by a National Institutes of Health grant, Finch, Holmes, and I continued the work on viruses, now extended to spherical viruses. We were joined soon after by Reuben Leberman, a biochemist. In 1962, we moved to the newly built MRC Laboratory of Molecular Biology in Cambridge, which, under the leadership of Perutz, was to house the original unit from the Cavendish Laboratory (Perutz, Kendrew, Crick, and, later, Brenner), enlarged by Sanger's group from the Biochemistry Department and Hugh Huxley from University College London. I was thus privileged to join the Laboratory at this stage in its expansion and was able to take advantage of, and to help build up, its then unique environment of intellectual and technological sophistication. I was Director of the Laboratory form 1976 to 1986 and since then have been a "retired worker" in the MRC unofficial jargon. The rest of my scientific career is largely a matter of record, and much of this is dealt with below.

However, I should perhaps add that, over the first 20 years back in Cambridge, I was actively involved in teaching undergraduates as well as, of course, in supervising research students. I was a director of studies in Natural Sciences at my college, Peterhouse, and under the tutorial, or—as it is called in Cambridge—supervision, system, I taught undergraduates myself. I liked teaching, and the contact with young minds kept one on one's toes, but increasing responsibilities in the Laboratory forced me to shed much of it in later years.

Before I went to Cambridge, I married Liebe Bobrow, whom I had met in Cape Town, where she was a music student at the University. She later trained in modern dance at the Jooss-Leeder School in London and became a choreographer and coordinator for the Cambridge Contemporary Dance Group when we returned to Cambridge in 1962. She also directed and acted in the theater. We had two sons, Adam and David, born in 1954 and 1963. Adam, after studying history and economics at Oxford and at the London School of Economics, went on to do research in economics at Princeton and emigrated to Israel to take up a professorship in Economic History at Ben-Gurion University. He died of prostate cancer in 2000. David studied physics at University

College London and is now a professor of chemistry at Imperial College London.

MACROMOLECULES IN BIOLOGICAL ASSEMBLIES

Within a living cell there occur a large number and variety of biochemical processes, almost all of which involve, or are controlled by, large molecules, the main examples of which are proteins and nucleic acids. These macromolecules do not, of course, function in isolation, but they often interact to form ordered aggregates or macromolecular complexes, sometimes so distinctive in form and function as to deserve the name of organelle. It is in such biological assemblies that the properties of individual macromolecules are often expressed in a cell. It is on some of these assemblies that I have worked for over 25 years, and these form the subject of this article.

The aim of our field of structural molecular biology was to describe the biological machinery in molecular, i.e., chemical, detail. The beginnings of this field were marked in 1962 when Max Perutz and John Kendrew received the Nobel Prize for the first solution of the structure of proteins. In the same year, Francis Crick, James Watson, and Maurice Wilkins were likewise honored for elucidating the structure of the double helix of DNA. In his Nobel Lecture, Perutz recalled how 40 years earlier, in 1922, Sir Lawrence Bragg, whose pupil he had been, came here to thank the Academy for the Nobel Prize awarded to himself and his father, Sir William, for having founded the new science of X-ray crystallography, by which the atomic structure of simple compounds and small molecules could be unraveled.

These men were not only my predecessors, but also some have been something like scientific elder brothers to me because the main subjects of my work have been both nucleic acids and proteins, the interactions between them, and the development of methods necessary to study the large macromolecular complexes arising from these interactions. In seeking to understand how proteins and nucleic acids interact, one has to begin with a particular problem, and I can claim no credit for the choice of my first subject, TMV. As described above, it was the late Rosalind Franklin who introduced me to the study of viruses and whom I was lucky to meet when I joined J.D. Bernal's department at Birkbeck College London in 1954. She had just switched from studying DNA to TMV, X-ray studies of which had been begun by Bernal in 1936. It was Rosalind Franklin who set for me the example of tackling large and difficult problems. Had her life not been cut tragically short, she might well have received a Nobel Prize for her work on DNA.

TOBACCO MOSAIC VIRUS

TMV is a simple virus consisting only of a single type of protein molecule and of RNA, the carrier of the genetic information. Its simple rod shape results from its design, namely a regular helical array of protein molecules, or subunits, in which is embedded a single molecule of RNA. This general picture was already complete by 1958 when Rosalind Franklin died (**Figure 1**). It is clear that the protein ultimately determines the architecture of the virus, an arrangement of $16^1/_3$ subunits per turn of a rather flat helix with adjacent turns in contact. The RNA is intercalated between these turns with three nucleotide residues per protein subunit, is situated at a radial distance of 40 Å from the central axis, and is therefore isolated from the outside world by the coat protein. The geometry of the protein arrangement forces the RNA backbone into a moderately extended single-strand configuration. Running up the central axis of the virus particle is a cylindrical hole with a 40-Å diameter, which we then thought to be a trivial consequence of the protein packing but later turned out to figure prominently in the story of the assembly.

At first sight, the growth of a helical structure like that of TMV presents no problem of comprehension. Each protein subunit makes identical contacts with its neighbors so that the bonding between them repeats over and over again. Subunits can have a precise built-in

Figure 1

Diagram summarizing the results of the first stage of the structure analysis of *Tobacco mosaic virus* (74). There are three nucleotides per protein subunit and $16^{1}/_{3}$ subunits per turn of the helix. Only about one-sixth of the length of a complete particle is shown.

geometry such that they can assemble themselves like steps in a spiral staircase in a unique way. Subunits would simply add one or a few at a time onto the step at the end of a growing helix, entrapping the RNA that would protrude there and generating a new step, and so on. It was in retrospect not too surprising when the classic experiments of Fraenkel-Conrat & Williams in 1955 (1) demonstrated that TMV could be reassembled from its isolated protein and nucleic acid components. They showed that, upon simple remixing, infectious virus particles were formed that were structurally indistinguishable from the original virus. Thus, all the information necessary to assemble the particle must be contained in its components; that is, the virus self-assembles, a term I introduced into the field. Later experiments (2) showed that the reassembly was fairly specific for the viral RNA, occurring most readily with the RNA homologous to the coat protein.

All this was very satisfactory, but there were yet some features which gave cause for doubt. First, other experiments (3) showed that foreign RNAs could be incorporated into virus-like rods, and these cast doubt on the belief that specificity in vivo was actually achieved during the assembly itself. Another feature about the reassembly that suggested to me that there were still missing elements in the story was its slow rate. Times of 8 to 24 h were required to give maximum yields of assembled particles. This seemed rather slow for the assembly of a virus in vivo because the nucleic acid is fully protected only on completion. These doubts, however, lay in the future, and before I describe their resolution, I return to the structural analysis of the virus and the virus protein.

X-ray Analysis of *Tobacco Mosaic Virus*: The Protein Disk

After Franklin's death, Holmes and I continued the X-ray analysis of the virus. Specimens for X-ray work can be prepared in the form of gels in which the particles are oriented parallel to each other, but randomly rotated about their own axes. These gels give good X-ray diffraction patterns, but because of their nature, the three-dimensional X-ray information is scrambled into two dimensions. Unscrambling these data to reconstruct the three-dimensional structure proved to be a major undertaking, and it was only in 1965 that Holmes and I obtained the first three-dimensional Fourier maps to a resolution of ~12 Å. In fact, only after another 10 years or so did the analysis by Holmes and his colleagues in Heidelberg (where he moved in 1968) reach a resolution approaching 4 Å in the best regions of the electron density map, but the resolution fell off significantly in other parts (4). At this resolution, it was not possible to identify individual amino acid residues with any certainty, and the ambiguities were too great to build unique atomic models. However, the map, taken together with the detailed map of the subunit we obtained in Cambridge (see below), yielded a considerable amount of information about the nature of the contacts with RNA (5).

These difficulties in the X-ray analysis of the virus were foreseen, and by the early 1960s, I came to realize that the way around this difficulty was to try to crystallize the isolated protein subunit of the virus, solve its structure at high resolution by X-ray crystallography, and then try to relate this to the virus structure solved to low resolution. We therefore began to try to crystallize the protein monomer. In order to frustrate the natural tendency of the protein to aggregate into a helix, Leberman introduced various chemical modifications in the hope of blocking the normal contact sites, but none of these modified proteins crystallized. The second approach was to try to crystallize small aggregates of the unmodified protein subunits. It had been known for some time, particularly from the work of Schramm & Zillig (6), that the protein on its own, free of RNA, could aggregate into a number of distinct forms in addition to that of the helix. I chose conditions under which the protein appeared to be mainly aggregated in a form with a sedimentation constant of about 4S, identified by Caspar as a trimer (7). We obtained crystals almost immediately but found (8) them to contain not the small oligomer expected, but a large one, corresponding to an aggregate with a sedimentation constant of 20S. The X-ray analysis showed that this was built from two juxtaposed layers, or rings, of 17 subunits each, and I named this form the two-layer disk (**Figures 2** and **3**). Our initial dismay in being faced with such a large structure, of molecular weight 600,000, was tempered by the fact that the geometry of the disk was clearly related to that of the virus particle. The cylindrical rings contained 17 subunits each, compared with $16^1/_3$ units per turn of the virus helix; therefore, the lateral bonding within the disks was likely to be closely related to that in the virus. We also showed, by analyzing electron micrographs, that the disk was polar, i.e., that its two rings faced in the same direction as do successive turns of the virus helix.

This was the first very large structure ever to be tackled in detail by X-ray crystallography, and it took us about 12 years to carry through the analysis to high resolution. The formidable technical problems were overcome only after the development in our laboratory of more powerful X-ray tubes and of special apparatus (cameras, computer-linked densitometers) for data collection from a structure of this magnitude. (In fact we had already begun building better X-ray tubes in London to use on weakly diffracting objects like viruses.) The 17-fold rotational symmetry of the disk also gave rise to redundant information in the X-ray data, which was exploited in the final analysis (9) to improve and extend the resolution of a map based originally on only one heavy-atom derivative. The map at 2.8-Å resolution (10) was interpreted in terms of a detailed atomic model for the protein (**Figures 2** and **3**), although the individual interactions upon RNA binding had yet to be deduced.

Protein Polymorphism

These results on the structure of the disk, which showed that it was fairly closely related to the virus helix, made me wonder whether the disk aggregate might not be fulfilling some vital biological role. It had been easy to dismiss it as perhaps an adventitious aggregate of a sticky protein or a storage form. The polymorphism of the TMV protein was first considered in some detail in 1963 by Caspar (7), who foresaw that some of the aggregation states might give insight into the way the protein functions. Quantitative studies of aggregation started by Lauffer et al. (11) in the 1950s concentrated on a rather narrow range of conditions, focusing mainly on understanding the forces driving aggregation (these are largely entropic). Because of the scattered nature of the earlier observations,

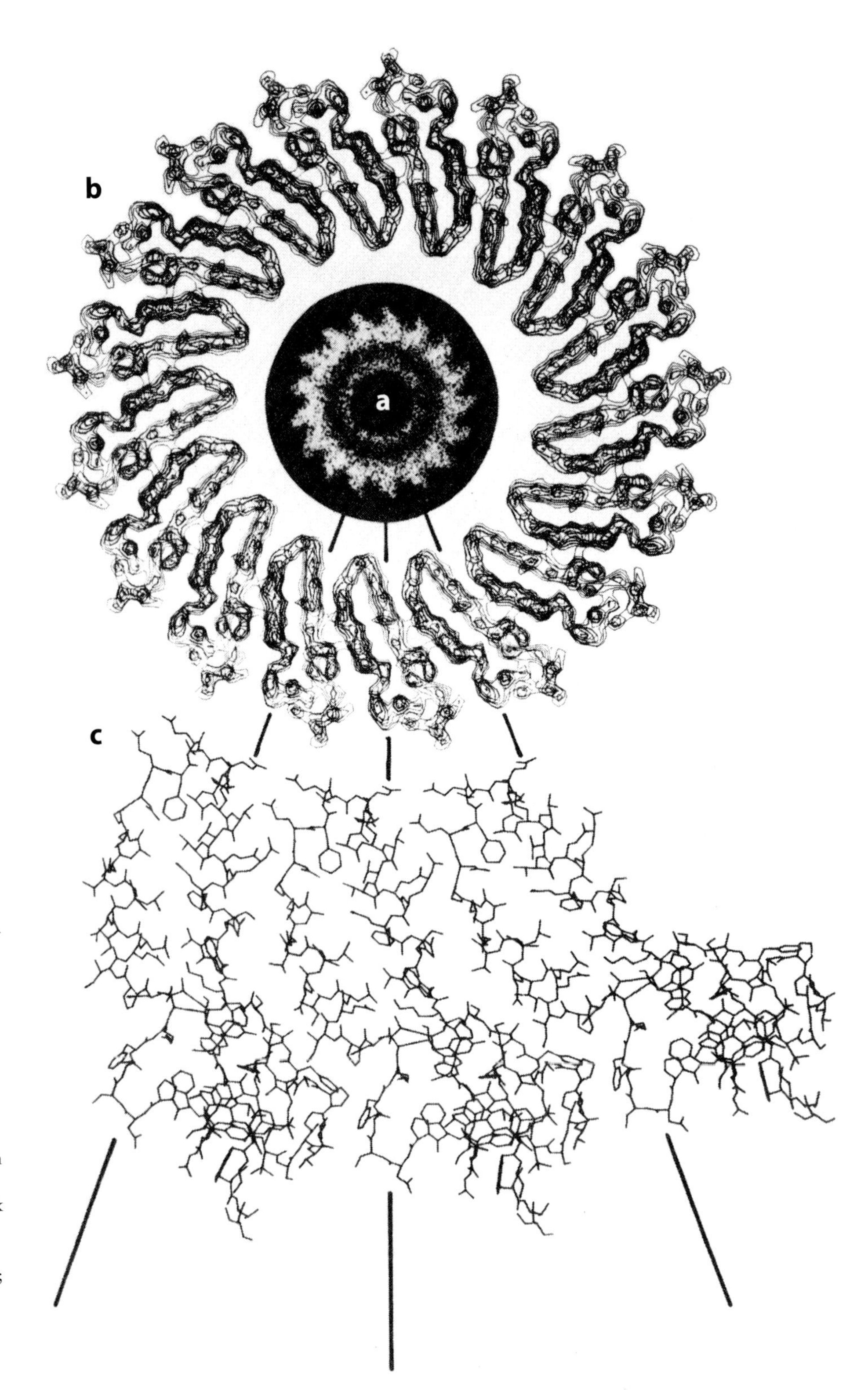

Figure 2

The disk viewed from above at successive stages of resolution. From the center outward there follow (*a*) a rotationally filtered electron microscope image at about 25-Å resolution (75); (*b*) a slice through the 5-Å electron density map of the disk obtained by X-ray analysis, showing rod-like α-helices (29); and (*c*) part of the atomic model built from the 2.8-Å map (Bloomer et al. 10).

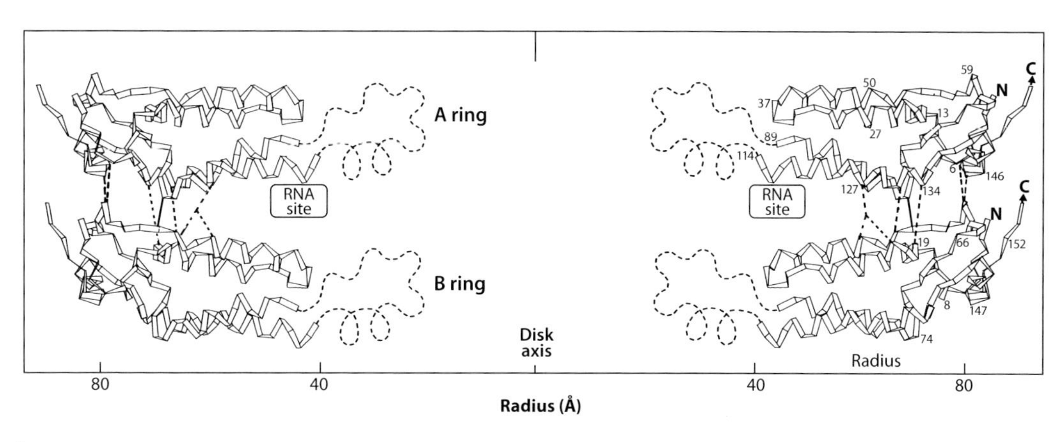

Figure 3

Section through a disk along its axis reconstructed from the results of X-ray analysis to a resolution of 2.8 Å (10). The ribbons show the path of the polypeptide chain of the protein subunits. Subunits of the two rings can be seen touching over a small area toward the outside of the disk but opening up into the "jaws" toward the center. The dashed lines at low radius indicate schematically the mobile portion of the protein in the disk, extending in from near the RNA-binding site to the edge of the central hole.

Durham, Finch, and I began a systematic survey of the aggregation states, which made their broad outline clear (12, 13). The results are summarized as a phase diagram (**Figure 4**).

At low or acid pH, the protein alone will form helices of indefinite lengths that are structurally very similar to the virus except for the lack of RNA. Above neutrality, the protein tends to exist as a mixture of smaller aggregates from about the trimer upward, and these aggregates are in rapid equilibrium with each other, commonly referred to as A-proteins. Near pH 7 and at about room temperature, the dominant form present is the disk, which is in a relatively slow equilibrium with the A-form in the ratio of about 4:1. The dominant factor controlling the state of aggregation of the coat protein is thus the pH. The control is mediated through groups, probably carboxylic acid residues as identified by Caspar (7), that bind protons abnormally in the helical state but not in the disk or A-form. Thus, the helical structure can be stabilized either in the virus by the interaction of the RNA with the protein or, in the case of the free protein, by protonating the acid groups. Consequently, these groups act as a "negative switch," ensuring that under physiological conditions the helix is not formed and thus that enough protein in the form of disks or A-protein is available to interact with the RNA during virus assembly.

A Role for the Disk

The disk aggregate of the protein therefore has a number of significant properties. It is not only closely related to the virus helix, but also is the dominant form of the protein under physiological conditions; moreover, disk forms had also been observed for other helical viruses. These strengthened my conviction that the disk form was not adventitious but might play a significant role in the assembly of the virus. What could this role be?

Assembly of any large aggregate of identical units, such as a crystal, can be considered from the physical point of view in two stages: first nucleation and then the subsequent growth or, in more biochemical language, as initiation and subsequent elongation. The process of nucleation—or, crudely, getting started—is frequently more difficult than growth. Thus, a simple mode of initiation, in which the free RNA interacts with individual protein subunits, does pose problems in getting started. At least 17 separate subunits would have to bind to the

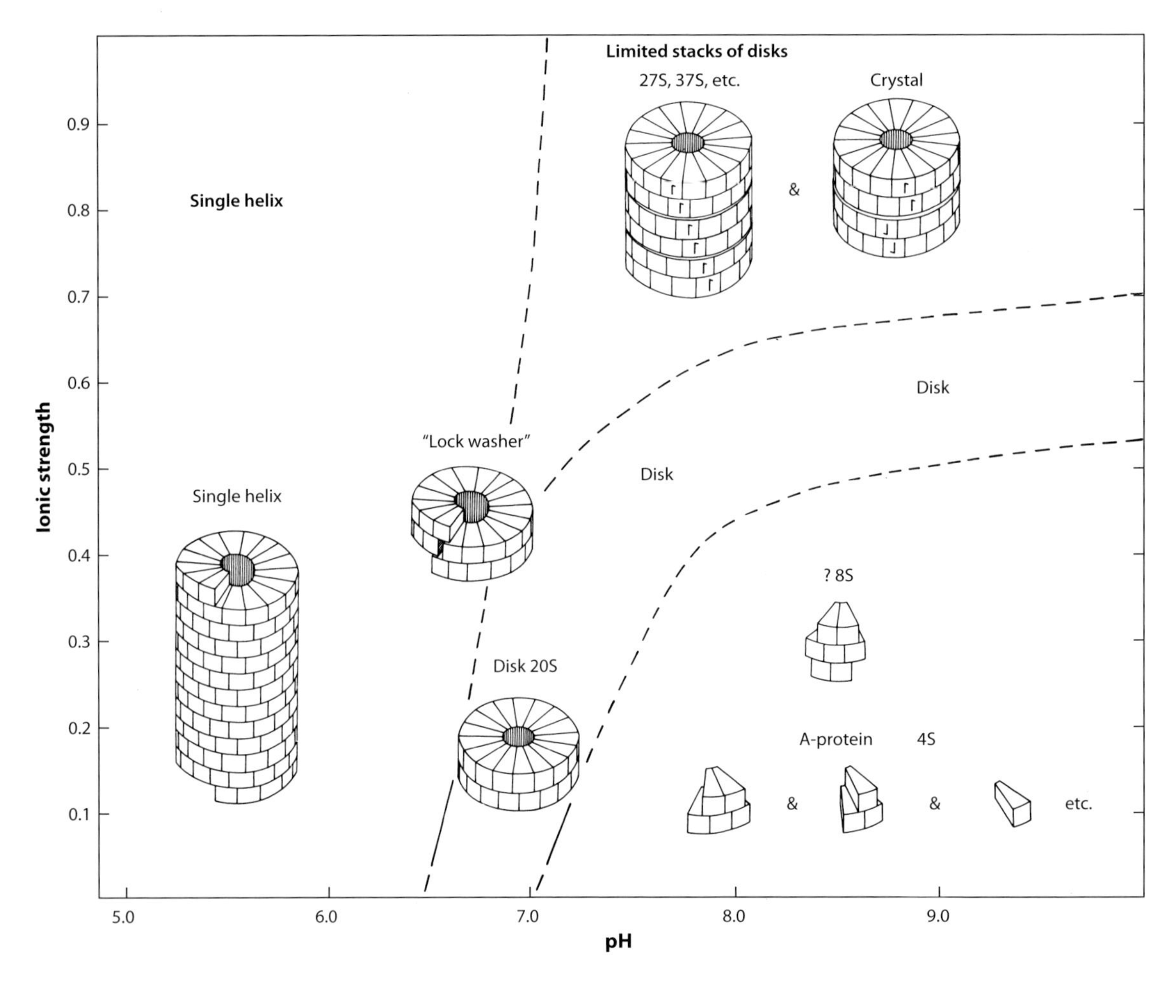

Figure 4

Diagram showing the ranges over which particular forms of *Tobacco mosaic virus* protein participate significantly in equilibrium (12). This is not a conventional phase diagram: A boundary is drawn where a larger species becomes detectable, but this does not imply that the smaller species disappears sharply. The "lock washer" indicated on the boundary between the 20S disk and the helix is not well defined and represents a metastable transitory state observed when disks are converted to helices by abrupt lowering of the pH.

flexible RNA molecule before the assembling linear structure could close round on itself to form the first turn of the virus helix. This difficulty could be avoided if a preformed disk were to serve as a jig upon which the first few turns of the viral helix could assemble to reach sufficient size to be stable. This mode of nucleation of helix assembly could also furnish a mechanism for recognition by the protein of its homologous RNA. The surface of the disk presents a set of 51 ($=17 \times 3$) nucleotide-binding sites, which could interact with a special long run of bases, resulting in an amplified discrimination that might not be possible with a few nucleotides. It thus seemed that the disk could solve both the physical and biological requirements for initiating virus growth and conferring specificity on the interaction. My working hypothesis is illustrated in **Figure 5**. It turned out that all the details in this diagram are wrong, but yet the spirit is correct. As A.N. Whitehead once observed, it is more important that an idea should be fruitful than that it should be (simply) true.

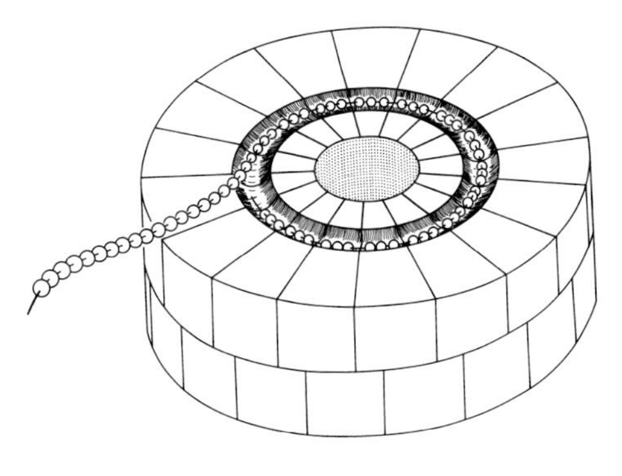
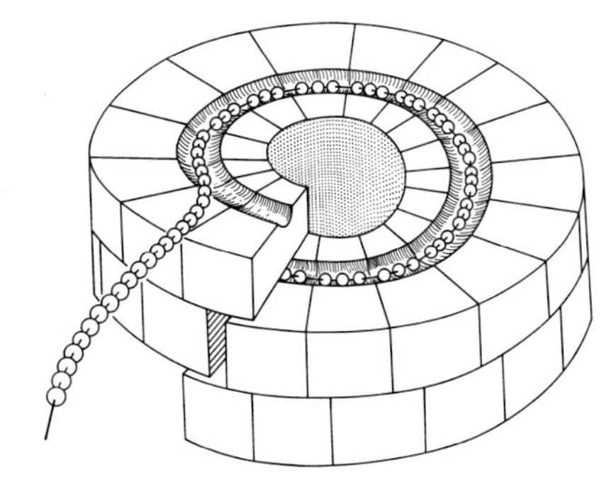

Figure 5

The role of the disk as originally conceived: The specific recognition of a special (terminal) sequence of *Tobacco mosaic virus* RNA initiates conversion of the disk form of the protein into two turns of helix. (See **Figure 7** for the mechanism that was finally established.)

This proposed mechanism of nucleation required the disk to be able to dislocate into a two-turn helix to form the beginning of the growing nucleoprotein rod. To test this, we carried out a very simple experiment, the pH drop experiment (14). This showed that an abrupt lowering of the pH would convert disks directly, within seconds, into short helices or lock washers (**Figure 4**), which stack on top of each other to give longer nicked helices, which in due course anneal to give more perfect helices. This conversion is an in situ one that does not require dissociation and then reassociation into a different form. The success of this experiment encouraged us to proceed to experiments with RNA itself, the natural substrate of the virus protein.

The first reconstitution experiments carried out by Butler and myself proved to be dramatic (15). When a mixture was made at pH 7 of the viral RNA and a disk preparation, complete virus particles were formed within 10 to 15 minutes, rather than over a period of hours, as was the case in the early reassembly experiments in which protein had been used in the disaggregated form (1). The notion that disks are involved in the natural biological process of initiation was strengthened by companion experiments (15) in which assembly was carried out with RNAs from different sources. These showed a preference, by several orders of magnitude, of disks for the viral RNA over foreign RNAs or synthetic polynucleotides of simple sequences. It is thus the disk state of the protein that is needed to achieve specificity in the interaction with the RNA. In the experiments cited earlier, in which virus-like rods were made containing TMV A-protein and foreign RNA (3), reactions were carried out at an acid pH, and under these artificial conditions, the protein alone would tend to form helical rods and so could entrap any RNA present.

In addition to this effect of disks on the rate of initiation, which had been predicted, we also found to our surprise that the disks appeared to enhance the rate of elongation, and therefore, we concluded that they must be actively involved in growth. This result was questioned by some other workers in the field and was the subject of argument (16, 17), but later discoveries on the configuration of RNA during incorporation into a growing particle, discussed below, have made the involvement of disks in the elongation, as well as in nucleation, much more intelligible.

The disk form of the protein therefore provided the elements that were missing from the simple reconstitution experiments using disaggregated protein, namely speed and specificity. We now knew what the disk did, so the next question was how did it do it?

The Interaction of the Protein Disk with the Initiation Sequence on RNA

Specificity in initiation ensures that only the viral RNA is picked out for coating by the viral protein. This must be brought about by the presence of a unique sequence on the viral RNA for interaction with the protein disk. Zimmern & Butler (18, 19) isolated the nucleation region containing this site by supplying limited quantities of disk protein, which were sufficient to allow nucleation to proceed but not subsequent growth, and then they digested away the uncoated RNA with nuclease. With the varying protein:RNA ratios and different digestion conditions, they found they could isolate a series of RNA fragments, all of which contained a unique common core sequence with variable extents of elongation at either end. These fragments could be rebound to the coat protein when it was in the form of disks. Among this population of fragments was a fragment only about 60 nucleotides long—just over the length necessary to bind round a single disk—and it appeared to represent the minimum protected core. Because of the strong rebinding of this fragment back to the disk, it seemed likely that it constituted the origin of assembly, where the normal nucleation reaction began.

However, the work on the RNA produced, in turn, another puzzle: the obvious expectation that the nucleation region would be near one end of the RNA turned out to be wrong. The nucleation occurs about one-sixth of the way along the RNA from the 3′ end (20), so that over 5000 nucleotides have to be coated in the major direction of elongation (3′–5′) and 1000 have to be coated in the opposite direction. Yet, growing nucleoprotein rods observed in the electron microscope (15) were always found to have all the uncoated RNA only at one end. Why were rods never seen with a tail at each end? The resolution of this conundrum came from considering the structure of the protein disk, to which I now return. Although the structure of the disk was solved in detail only in 1977, an earlier stage in the X-ray analysis gave the clue as to how it might interact with the RNA. At 5-Å resolution (21), the course of the polypeptide chain could be traced, and the basic design of the disk established (cf. **Figure 3**). The subunits of the upper ring of the disk lie in a plane perpendicular to the disk axis, whereas those of the lower ring are tilted downward toward the center, so that the two rings touch only toward the outside of the disk. In the neighborhood of the central hole, they are far apart, like an open pair of jaws, which could, as it were, "bite" a stretch of RNA entering through the central hole. Moreover, entry through the center would be facilitated because the inner region of the protein, from around the RNA-binding site inward, was found to be disordered and not packed into a regular structure.

It therefore looked very much as though the disk was designed to permit the RNA to enter through the central hole, effectively enlarged by the flexibility of the inner loop of protein, and intercalate between its two layers. The RNA, which would enter, thus would, of course, be the nucleation sequence that lies rather far from an end of the RNA molecule. This could, however, be achieved if the RNA doubled back on itself at a point near the origin of assembly and so entered as a hairpin loop. Indeed, the smallest RNA fragment that is protected during nucleation has a base sequence that can fold into a weakly paired double-helical stem with a loop at the top, that is a hairpin (**Figure 6**). This was proposed by Zimmern (19). The loop and top of the stem have an unusual sequence, containing a repeating motif of three nucleotides, with guanine G in one specific position, and usually A, or sometimes U, in the other two. Because there are three nucleotide-binding sites per protein subunit, such a triplet repeat pattern will place a specific base in a particular site on the protein molecule and could well lead to the recognition of the exposed RNA loop by the disk during the nucleation process.

Nucleation and Growth

The hypothesis for nucleation (22) then was that the special RNA hairpin would insert through the central hole of the disk into the jaws formed by the two layers of protein

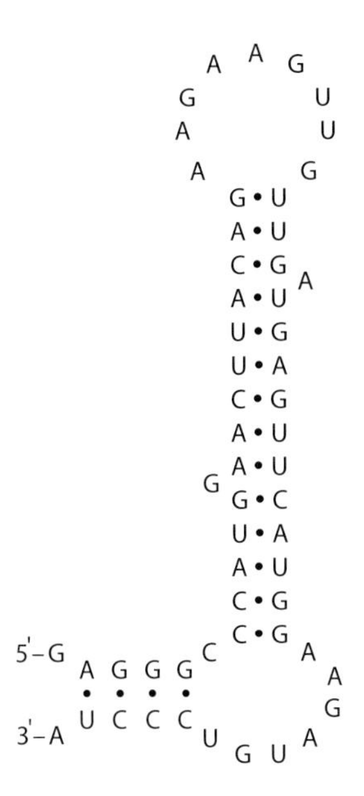

Figure 6

Postulated hairpin secondary structure of RNA in the nucleation region (19). This gives a weakly bonded double-helical stem and a loop at the top, probably the actual origin of assembly. The sequence at and near the top contains a repeating motif of three bases having G in the middle position and A or U in the outer positions.

subunits (**Figure 7**). The dimensions are quite suitable for this to occur, and the open loop could then bind to the RNA-binding sites on the protein. More of the rather unstable double helical stem would melt out and be opened as more of the RNA was bound within the jaws of the nucleating disk. Some, as yet unknown, feature of this interaction would cause the disk to dislocate into a short helical segment, entrapping the RNA and, after the rapid addition of a few more disks (18), would provide the first stable nucleoprotein particle. The subsequent events after nucleation can be called growth, and as stated above, there was a controversy about the particular way in which this proceeds. Our view was that elongation in the major direction of growth very likely takes place through the addition of more disks, as indeed our first reconstitution experiments drove us to conclude. The special configuration generated during the insertion of the loop into the center of the disk must be perpetuated as the rod grows, by pulling additional RNA up through the central hole. Thus, elongation could occur by a substantially similar mechanism to nucleation, only now, rather than requiring the specific nucleation loop of the RNA, it occurs by means of a "traveling loop," which can be inserted into the center of the next incoming disk. Therefore, this mechanism overcomes the main difficulty we envisaged in how a whole disk of protein subunits could interact with the RNA in the growing helix. Later, there was more evidence for growth by incorporation of blocks of subunits of roughly disk size (17), but the subject remained controversial for some time.

Later, we produced clear experimental confirmation of our hypothesis for the mechanism of nucleation. This predicted (23) that two tails of the RNA will be left at one end of the growing nucleoprotein rod formed and that one of these tails would project directly from one end but that the other would be doubled back all the way from the active growing point at the far end of the rod down the central hole of the growing rod (24). Both of these predictions were confirmed by Hirth's group (25) in Strasbourg, who obtained electron micrographs of growing rods in which the RNA is spread by partial denaturation; many particles show two tails protruding from the same end (25). In Cambridge, my colleagues used high-resolution electron microscopy, in which the two ends of the rods could be identified by their shapes to show that it is indeed the longer tail that is doubled back through the growing rod (26).

Design and Construction: Physical and Biological Requirements

These experiments showed that the formation of the protein disk is the key to the mechanism of the assembly of TMV. The protein subunit is not designed to form an endless helix, but a closed two-layer variant of it. This aggregate is stable and can be readily converted to the lock washer or helix-going form. The disk therefore represents an intermediate subassembly by means of which the entropically difficult

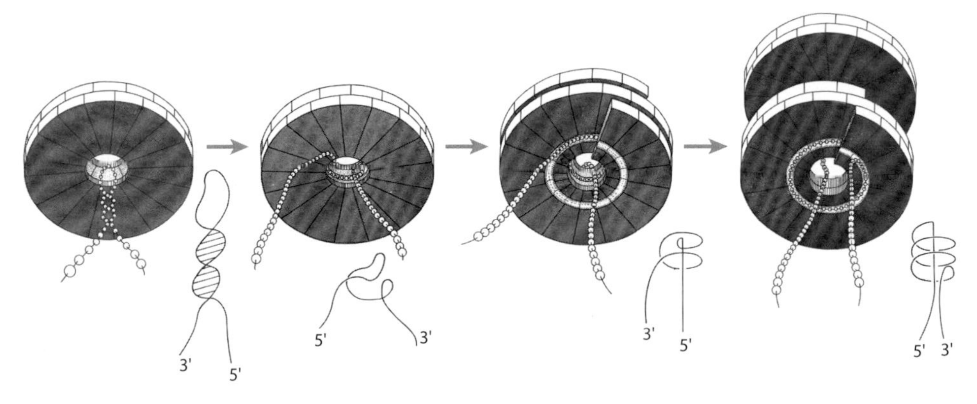

Figure 7

Nucleation of virus assembly occurs by the insertion of the hairpin of RNA (**Figure 6**) into the central hole of the protein disk and between the two layers of subunits. The loop at the top of the hairpin binds to form part of the first turn, opening up the base-paired stem as it does so, and causes the disk to dislocate into a short helix. This presumably "closes the jaws," entrapping the RNA between the turns of protein subunits and gives a start to the nucleoprotein helix (which can then elongate rapidly to some minimum stable size).

problem of nucleating helical growth is overcome. At the same time, the nucleation by the disk subassembly furnishes a mechanism for recognition of the homologous viral RNA (and rejection of foreign RNAs) by providing a long stretch of nucleotide-binding sites for interaction with the special sequence of bases on the RNA. The disk is thus an obligatory intermediate in the assembly of the virus, which simultaneously fulfills the physical requirement for nucleating the growth of the helical particle and the biological requirement for specific recognition of the viral RNA. TMV is self-assembling, self-nucleating, and self-checking.

There are a number of morals to be derived from the story of TMV assembly (23). The first is that one must distinguish between the design of a structure and the construction process used to achieve it. That is, although TMV looks like a helical crystal and its design lends itself to a picture of simple addition of subunits, its construction actually follows a more complex path that is highly controlled. It illustrates the point that function is inextricably linked with structure and shows how much can be done by one single protein. A most intricate structural mechanism has evolved to give the assembly an efficiency and purposefulness, and we now understand its basis. The general moral of all this is that not merely did nature once again confound our obvious preconceptions but that it also left enough clues for us to finally puzzle out what is happening. As Einstein once put it, "Raffmiert ist der Herr Gott, aber bösartig ist er nicht: The Lord is subtle, but he is not malicious."

CRYSTALLOGRAPHIC OR FOURIER ELECTRON MICROSCOPY

In 1955, Finch and I, in London, and Caspar, then in Cambridge, took up the X-ray analysis of crystals of spherical viruses. These had first been investigated by Bernal and his colleagues just before and after World War II, using "powder" and "still" photography. Finch and I worked on *Turnip yellow mosaic virus* and its associated empty shell, and Caspar on *Tomato bushy stunt virus*. Crick and Watson had predicted that spherical viruses ought to have one of the forms of cubic symmetry, and we showed that both viruses had icosahedral symmetry. Later, when Finch and I showed that poliovirus also had the same symmetry, we realized that there was some underlying principle at work, and this eventually led Caspar and me to formulate our theory of virus shell structure (27).

When my research group moved to Cambridge in 1962, we turned to electron microscopy because of the speed with which it enables one to tackle new subjects, and also because it produces a direct image, or so we thought. Armed with a theory of virus design and some X-ray data, we had some notion of how spherical shells of viruses might be constructed and thought we would be able to see the fine detail in electron micrographs. Thus, we knew what we were looking for, but we soon found that we did not understand what we were looking at: The micrographs did not present simple, direct images of the specimens. We soon discovered the limitations of electron microscopy. First, there were preparation artifacts and also radiation damage during observation. Second, artificial means of contrast enhancement had to be used as the majority of atoms in biological specimens have an atomic number too low to give sufficient contrast on their own. Third, the image formed depends on the operating conditions of the microscope and on the focusing conditions and aberrations present. Above all, because of the large depth of focus of the conventional microscope, all features along the direction of view are superimposed in the image. Finally, in the case of strongly scattering or thick specimens, there is multiple scattering within the specimen, which can destroy even the relation between object and image, but this was not the case in our investigations.

For these reasons, the detail one sees in a raw image is often unreliable and not easily interpretable without methods that correct for the operating conditions of the microscope and that can separate contributions to the image from different levels of the specimen. It is also important to be able to assess the degree of specimen preservation in each particular case. Over a period of about 10 years, these procedures for image processing of electron micrographs were developed by me and my colleagues. Their aim is to extract from electron micrographs the maximum amount of reliable information about the two- or three-dimensional structures that are being examined. Some applications of these methods to various problems studied in the MRC Laboratory over the first 15 years are given in **Table 1**. Electron microscopy combined with image reconstruction, supplemented wherever possible by X-ray studies on wet, intact material, provided what are now generally accepted models of the structural organization of a large number of biological systems such as those listed in the table. I describe below a limited number of examples that demonstrate the power of various techniques and the nature of the results they can give. Fuller accounts of the methods and the theory are available elsewhere (24, 28), but I would like to emphasize that these methods arose out of practical concerns and grew in the course of tackling concrete problems; nevertheless, they have proved to be of wide application.

Two-Dimensional Reconstruction: Digital Computer Processing

We began our studies on viruses, both spherical and helical, using the method of negative staining that had been recently introduced by Huxley, and by Brenner & Horne (29). In this method, the specimen is embedded in a thin amorphous layer of a heavy-metal salt, which simultaneously preserves and maps out the shape of the regions from which it is excluded. Much fine detail was to be seen, but one could not easily make sense of it in most cases. People simply thought that the specimens were being disordered because it was assumed that the negative stain gave, as it were, a footprint of the particle. I gradually came to realize that the confusion arose, not so much because of the disorder that the stain produced, as because there was a superposition of detail from the front and back of the particle, i.e., the stain was enveloping the whole particle, so forming a cast rather than a footprint. This interpretation was proved in two different ways that proceeded in parallel. First, in the case of the spherical viruses, one could build a model and compute or otherwise display it in projection, and we found that this could account for many if not all of the previously uninterpretable images (30). The

Table 1 Some applications of electron microscope image reconstruction at the MRC Laboratory of Molecular Biology, Cambridge, from 1964 through 1979

Viruses	Organelles	Enzymes
Helical		
Tobacco mosaic virus (TMV)	Microtubules from flagellar doublets and brain tubulin sheets	Haemocyanin
TMV protein disc		Glutamate dehydrogenase
Paramyxoviruses	Muscle filaments	Catalase
Icosahedral	Actin	Crystals
Human polyoma wart	Actin + tropomyosin	Tubes
Tomato bushy stunt virus	Actin + myosin + tropomyosin (inhibited and relaxed)	Sickle cell hemoglobin fibers
Turnip yellow mosaic virus		Purple membrane (bacteriohodopsin)
Bacterial virus R17	Bacterial flagella	Cytochrome oxidase
Nudaurelia capensis	Bacterial cell walls	
Cowpea mottle virus	Ribosome crystals	
Adenovirus hexon	Chromatin	
Aberrant hexon	Crystals of nucleosome cores	
Pentagonal tubes of polyoma	Tubes of histone octamers	
Phage T2 and T4	Gap junctions	
Head and its tubular variants (polyheads)		
Tail: sheath and core		
Baseplate		

uniqueness of the model was proved by tilting experiments in which the specimens on the grid and the model were tilted in the same manner through large angles (cf. figure 10 in Reference 31). The second approach was applied to helical structures, which are translationally periodic and therefore lend themselves to a direct image analysis, which I shall now describe.

Figure 8*a* shows an electron micrograph of a negatively stained specimen of a "polyhead," which is a variant of the head of T4 bacteriophage, consisting mainly of the major head protein. The particle has been flattened, and so its original tubular form is lost. The image clearly shows some structural periodicities, but these are difficult to discern, and such interpretations used to be left to subjective judgement. I realized that the optical (Fraunhofer) diffraction pattern produced from such an image would allow an objective analysis of all the periodicities present to be made (32). This is shown in **Figure 8*b***. Here, clear diffraction maxima can be seen; these fall into two sets, which can be accounted for as arising, respectively, from the near and far sides of the specimen. In this way, it was established that the negative stain was producing a complete cast of the particle rather than a one-sided footprint of it (32). Because this is a helically periodic structure, the diffraction maxima tend to lie on a lattice, and so they pick out genuine repeating features within the structure. In this case, the regular diffraction maxima extended to a spacing of about 20 Å, which demonstrated that the long-range order in the specimen was preserved to this resolution, and this is indeed sufficient to resolve individual protein molecules.

The confusion in the direct image is largely caused by the superposition of the near and far sides of the particle, and any one such side can be filtered out in an optical system by a suitably positioned mask that transmits only the desired diffracted rays (33). The filtered image, **Figure 8*c***, is immediately interpretable in terms of a particular arrangement of protein molecules (34). The fact that the background noise in the diffraction pattern has been filtered out also causes this clarity in a processed image. Background noise arises because of the individual variations between molecules, i.e., the

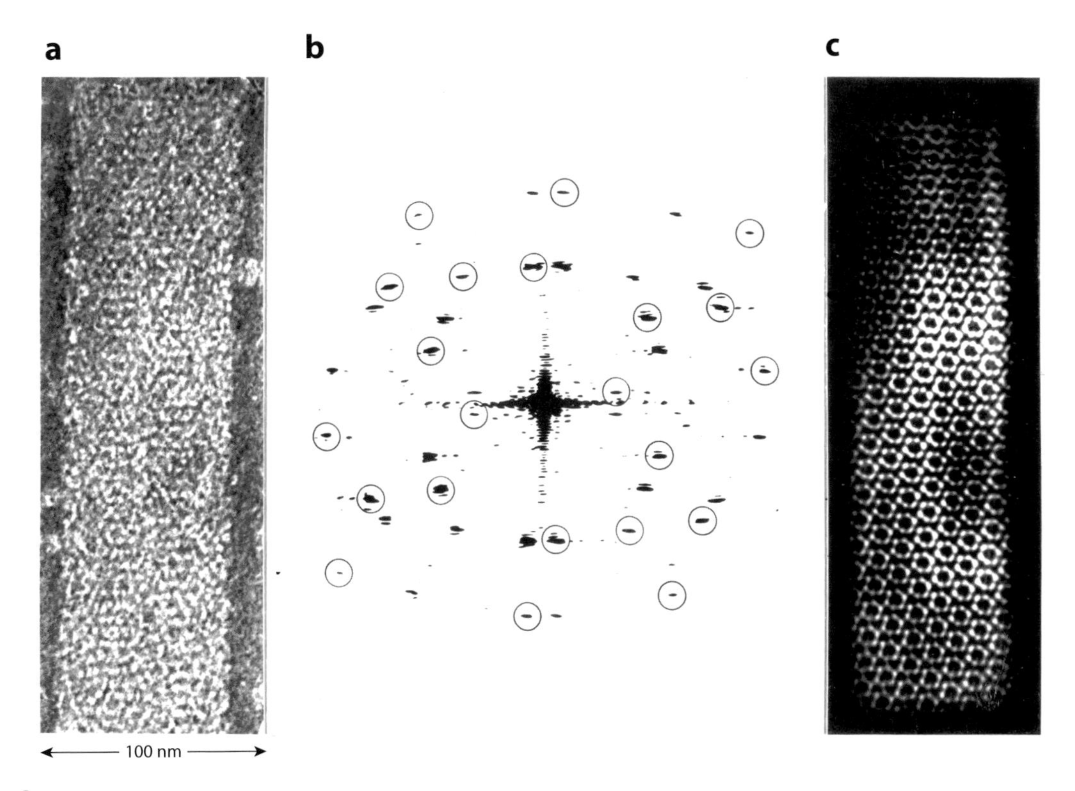

Figure 8

Optical diffraction and image filtering of the tubular structures known as "polyheads," consisting of the major head protein of T4 bacteriophage (34). (*a*) Electron micrograph of a negatively stained, flattened particle. (*b*) Optical diffraction pattern of (*a*), with circles drawn around one set of diffraction peaks that correspond to one layer of the structure. (*c*) Filtered image of one layer in (*a*) using the diffraction mask shown in (*b*). The apertures in the mask are chosen so that the averaging here extends locally only over a few unit cells. Individual molecules arranged in hexamers can be seen.

disorder, in the specimen, and these variations contribute randomly in all parts of the diffraction pattern. Indeed, the signal-to-noise ratio in the image has been enhanced by averaging over the copies of the molecules present in the arrangement. This idea of averaging over many copies of a repeated motif is central to the most powerful techniques developed so far to produce reliable images of biological specimens, and the three-dimensional procedures, which I will describe below, can also use this technique.

The essence of image processing of this type is that it is a two-step procedure after the first image has been obtained. First, the Fourier transform of the raw image is produced. Next, Fourier coefficients are manipulated, or otherwise corrected, and then transformed back again to reproduce the reconstructed image. These operations can be carried out most easily on a digital computer. Digital image processing, as first introduced by DeRosier and myself (35), allows a much greater flexibility than our original optical method and makes three-dimensional procedures possible.

Three-Dimensional Image Reconstruction

The first example (see **Figure 8**) is a relatively simple case where the problem is essentially that of separating contributions from two overlapping crystalline layers, and this figure shows how the method of Fourier analysis resolves the superposition in real space into separated sets of contributions in Fourier space. It was, however, already clear from the simple analysis of spherical viruses that to get a unique or reliable picture of a three-dimensional structure one must be able to view the specimen from many different directions (30). These different views were often provided by specimens lying in different orientations, but they can also be realized by tilting the specimen in the microscope, as mentioned above. Originally, the different views were interpreted by building models, but eventually, I saw that a set of transmission images taken in different views could be combined objectively to give a reconstruction of a three-dimensional object.

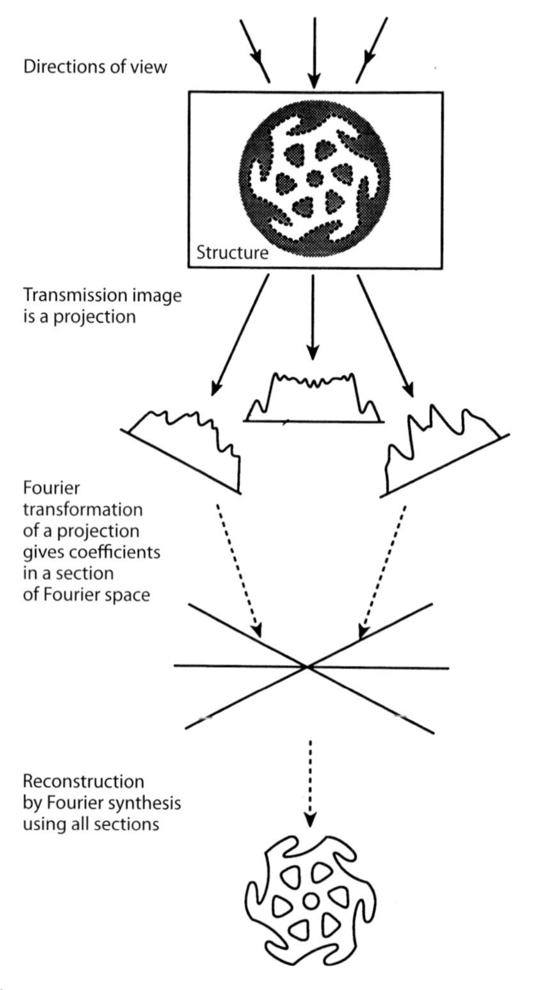

Figure 9

Scheme for the general process of three-dimensional reconstruction of an object from a set of two-dimensional projections (35).

This happened when DeRosier and I were studying the tail of bacteriophage T4, and our analysis showed that there were contributions to the image from the internal structure as well as from the front and back surfaces (35). To work in three dimensions, a generalized form of the two-dimensional filtering process had to be found, and—by making a connection with X-ray analysis—I realized that what is required is a three-dimensional Fourier synthesis. In the analysis of the X-ray diffraction patterns of TMV, I had used the idea that a helical structure could be built up mathematically out of a set of cylindrical harmonic functions; there is a relation between the number of functions that could be obtained and the number of different views available. Each new view gives additional harmonics of higher spatial frequency, and so, if one had enough views, one could build up the complete structure. Later, we came to see (35) that this synthesis was only a special case of a general theorem known to crystallographers as the projection theorem.

Thus, the general method of reconstruction, which we developed (see **Figure 9**), is based on the projection theorem, which states that the two-dimensional Fourier transform of a plane projection of a three-dimensional density distribution is identical to the corresponding central section of the three-dimensional Fourier transform that is normal to the direction of view. The three-dimensional transform can therefore be built up section by section using transforms of different views of the object, and the three-dimensional reconstruction is then produced by Fourier inversion. The important feature of the method is that it tells one how many different views are needed for a required resolution and how these are to be recombined into a three-dimensional map of the object (35, 36). The process is both quantitative and free from arbitrary assumptions. The approach is similar to conventional X-ray crystallography, except that the phases of the X-ray diffraction pattern cannot be measured directly, whereas here they can be computed from a digitized image. Different views could be collected from a single particle by using a tilting stage in the microscope if radiation damage could be prevented, but more realistically one must use several particles in different but identifiable orientations. In general, it is desirable to combine data from different particles so that imperfections can be averaged out. The Fourier method is only one way, out of several for solving the sets of mathematical equations, that relates the unknown three-dimensional density distribution with known projections in different directions (36). In fact, no other reliable method has been shown to be superior, and it is used in the X-ray CAT scanner for medical imaging. Moreover, the Fourier method has an advantage because it is carried out in steps (i.e., formation of the two-dimensional transforms and then recombination in three dimensions), so it is possible, as described above, to assess, select, and correct the data going into the final reconstruction.

Many applications were made in the following years. The first application was in fact to the phage tail of T4, the problem in which it arose. Particles with helical symmetry are the most straightforward to reconstruct because a reconstruction can be made from a single view of the whole particle, which is to a limited resolution and set by the helix symmetry. In physical terms, this is because a single image of a helical particle presents many different views of the repeating subunit, and it was this simplification that led us to use the phage tail as a first specimen for three-dimensional image reconstruction. Generally, more than one view is necessary, but any symmetry present will reduce the number required. Typically for small icosahedral viruses, three or four views are sufficient, but many more specimens must be investigated before the appropriate number can be found and averaging carried out (37). An example, from Crowther & Amos (38), is given in **Figure 10**.

Phase-Contrast Electron Microscopy

Electron microscopy, combined with some method of image analysis, when applied to negatively stained specimens has proved ideal for determining the arrangement and shape of small protein subunits within natural or artificial arrays, including two-dimensional crystals and macromolecular assemblies such as viruses and microtubules (24). The structural information obtainable has proved to be highly reliable with respect to detail down to about the 20-Å or 15-Å level. It became clear, however, that the degree of detail revealed was limited by the granularity of the negative stain and by the fidelity with which it follows the surface of the specimen (39). To obtain much higher-resolution information, better than about 10 Å, one should dispense with the stain and view the protein itself. At high resolution, there is a second problem: radiation damage. This can be reduced by cutting down the illuminating beam, but the statistical noise is then increased, and the raw image becomes less and less reliable. However, this difficulty can be overcome satisfactorily by imaging ordered arrays of molecules, so that the information from different molecules can be averaged, as described above, to give a statistically significant picture. The first problem, that of replacing the negative stain, yet avoiding dehydration, can be solved in two ways. One, developed by Dubochet, now in general use, is to use frozen hydrated specimens (40). The second, earlier method is that of Unwin & Henderson (41, 42), who, in their radical approach to determining the structure of unstained biological specimens by electron microscopy, used a dried-down solution of glucose to preserve the material.

The question then arose as to how this unstained specimen, effectively transparent to electrons, could be visualized. In the light microscopy of transparent specimens, the well-known Zernike phase-contrast method is used. Here the phase of the scattered beams relative to the unscattered beams is shifted by means of a phase plate, and then the scattered and unscattered beams are allowed to interfere in the image plane to produce an image. A successful electrostatic phase-contrast device for electron microscopy, quite analogous to the phase plate used in light microscopy, was constructed by Unwin (43), but it is not easy to make or use. A practical way of

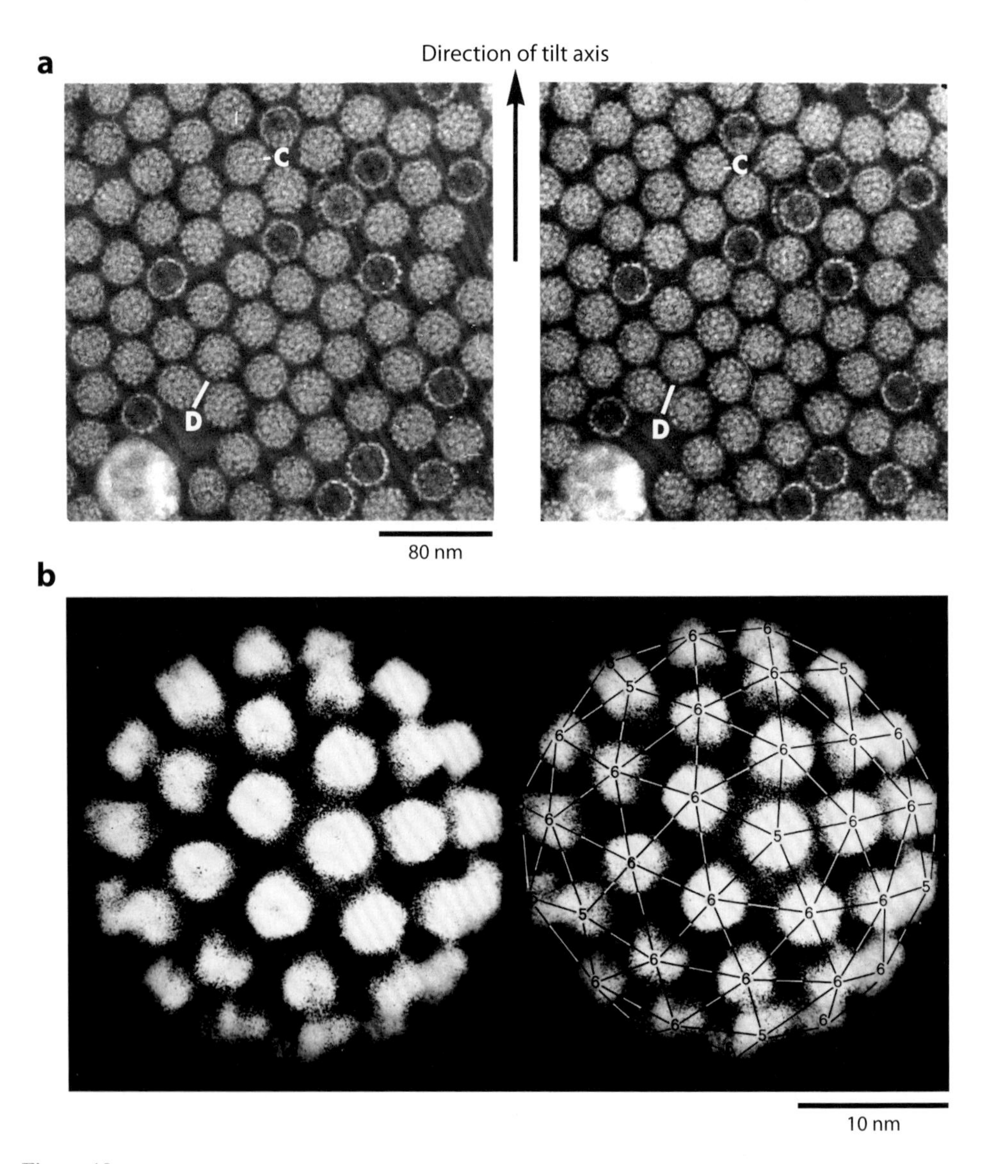

Figure 10

(*a*) Electron micrographs of the same field of negatively stained, closely packed particles of human wart virus before (*left*) and after (*right*) tilting the specimen grid through an angle close to 18° (31). (*b*) A three-dimensional reconstructed image of human wart virus (37, 38). Alongside is shown the underlying icosahedral surface lattice (27) with the fivefold and sixfold vertices marked.

producing phase contrast in the electron microscope is simply to record the image with the objective lens underfocused, and this was the method used by Unwin and Henderson. The defocusing phase-contrast method arose out of an academic study by Erickson and me of image formation in the electron microscope (44). This was undertaken because of a controversy that had developed concerning the nature of the raw image itself. When three-dimensional image reconstruction was introduced and applied to biological particles embedded in negative stain, objections were raised by various workers in the field of materials science, accustomed to dynamical effects in strongly scattering materials, to the premise that the image essentially represented the simple projection of the distribution of stain. It was asked whether multiple or dynamical scattering might not vitiate this assumption. To investigate this question, Erickson and I undertook an experimental study of negatively stained thin crystals of catalase as a function of the depth of focusing (44). We found that a linear or first-order theory of image formation would explain almost entirely the changes in the Fourier transform of the image. We concluded that the direct image, using a suitable value of underfocus dependent on the frequency range of interest, is a valid picture of the projection of the object density. When greater values of underfocus were used to enhance contrast, the image could be corrected to give a valid picture (44).

This study, although confined to the medium resolution range, included a practical demonstration that *a posteriori* digital image processing could be used to measure and compensate for the effects of defocusing, and we suggested that this approach could be directly extended to high resolution to compensate for the effects of spherical aberration as well as defocusing. It also provided a convenient way of producing phase contrast in the electron microscope in the case of unstained specimens. The image is recorded with the objective lens underfocused, which changes the phases of the scattered beams relative to the unscattered (or zero order) beam. Defocusing does not, however, act as a perfect phase plate analogous to that of Zernike because the phases are not all changed by the same amount, and successive bands of spatial frequencies contribute to the image with alternately positive and negative contrast. To produce a true image, the electron image must be processed to correct for the phase-contrast transfer of the microscope so that all spatial frequencies contribute with the same sign of contrast.

To produce their spectacular three-dimensional reconstructed image of the purple membrane of *Halobacterium* to a resolution of about 7 Å, Unwin & Henderson (41) took a series of very low-dose images of different pieces of membrane tilted at different angles. The final map represented an average over some 100,000 molecules. The small amount of contrast present in the individual micrographs was produced by underfocusing, which was then compensated for in the computer reconstruction by the method described above. For the first time, the internal structure of a protein molecule was seen by electron microscopy.

THE STRUCTURE OF CHROMATIN

The work on viruses not only gave results of intrinsic interest, but, as I indicated above, the difficulties in tackling large molecular aggregates led us to the development of methods and techniques that could be applied to other systems. A second example of this approach, and one that I think would not have gone so quickly without our earlier experience, is that of chromatin. Chromatin is the name given to the chromosomal material when extracted. It consists mainly of DNA, tightly associated with an equal weight of a small set of rather basic proteins called histones. I took up the study of chromatin in Cambridge in about 1972 when the protein chemists had shown that there were only five main types of histones, the apparent proliferation of species being the result of post-synthetic modifications, so that the structural problem appeared tractable.

DNA of the eukaryotic chromosome is probably a single molecule, amounting to several centimeters in length if laid out straight, and it must be highly folded to make the compact structure one can see in a chromosome. At the same time, it is organized into separate genetic or functional units, and the manner in which this folding is achieved, genes organized, and their expression controlled, became the subject of intense study throughout the world. The aim of my research group was to try to understand the structural organization of chromatin at various levels and to see what connections could be made with functional controls.

The large amounts in which histones occur suggested that their role was structural, and it was shown over the years 1972–1975 that the four histones H2A, H2B, H3, and H4 are responsible for the first level of structural organization in chromatin. As proposed by Roger Kornberg in 1974 (45), they fold successive segments of DNA [about 200 base pairs (bp) long] into compact bodies of ~100 Å in diameter, called nucleosomes. A string of nucleosomes, or repeating units, is thus created, and when these are closely packed, they form a filament about 100 Å in diameter. The role of the fifth histone, H1, was not clear at first. It is much more variable in sequence than the other four, being species and tissue specific. In the years 1975 and 1976, my colleagues and I showed that H1 is concerned with the folding of the nucleosome filament into the next higher level of organization and, later, how it performed this role.

This is not the place to tell in detail how this picture of the basic organization of chromatin emerged (46), but the idea of a nucleosome arose from the convergence of several different lines of work. The first indications for a regular structure came from X-ray diffraction studies on chromatin, which showed that there must be some sort of repeating unit, albeit not well ordered, on the scale of about 100 Å (48, 49). The first biochemical evidence for regularity came from the work of Hewish & Burgoyne (50), who showed that an endogenous nuclease in rat liver could cut DNA into multiples of a unit size, which was later shown by Noll using a different enzyme, micrococcal nuclease, to be about 200 bp (51). The fact that the nuclease cuts DNA of chromatin at regularly spaced sites, quite unlike its action on free DNA, was attributed to the fact that DNA is folded in such a way as to make only short stretches of free DNA, between these folded units, available to the enzyme. The third piece of evidence that led to the idea of a nucleosome was the observation by Kornberg & Thomas (52) that the two highly conserved histones, H3 and H4, existed in solution as a specific oligomer, the tetramer $(H3)_2(H4)_4$, which behaved rather like an ordinary multisubunit globular protein. On the basis of these different lines of evidence, Kornberg in 1974 (45) proposed a definite model for the basic unit of chromatin as a bead of about 100 Å in diameter, containing a stretch of DNA 200 bp long, condensed around the protein core made out of eight histone molecules, namely the H3H4 tetramer and two each of H2A and H2B. The fifth histone, H1, was somehow associated with the outside of each nucleosome. A quite unexpected feature of the model was that it was DNA that "coated" the histones, rather than the reverse.

However, in 1972, when Kornberg came to Cambridge, all this lay in the future. We began using X-ray diffraction to follow the reconstitution of histones and DNA because the X-ray pattern given by cell nuclei, or by chromatin isolated from them, limited as it was, was the only assay then available to follow the ordered packaging of DNA. These X-ray studies showed that almost 90% reconstitution could be achieved when DNA was simply mixed with an unfractionated total histone preparation, but all attempts to reconstitute chromatin by mixing DNA with a set of all four purified single species of histone failed, as if the process whereby the histones were being separated was denaturing them. We therefore looked for milder methods of histone extraction, and Kornberg found that the native structure could be reformed readily if the four histones were kept together in two pairs, H3 and H4 together, and H2A and H2B together, but not once they had been taken apart. It was this work that led Kornberg to investigate further the physicochemical properties of histones and to the discovery of the H3H4 histone tetramer (52), which in turn led him to the model of the nucleosome as described above.

The Structure of the Nucleosome

Approaches such as nuclease digestion and X-ray scattering on unoriented specimens of chromatin or nucleosomes in solution could reveal certain features of the nucleosome, but a full description of the structure can only come from crystallographic analysis, which gives complete three-dimensional structural information. In the summer of 1975, my colleagues and I therefore set about trying to prepare nucleosomes in forms suitable for crystallization. Nucleosomes purified from the products of micrococcal nuclease digestion contain an average of about 200 nucleotide pairs of DNA, but there is a rather wide distribution about the average, and such preparations are not homogeneous enough to crystallize. However, this variability in size can be eliminated by further digestion with micrococcal nuclease. Although the action of micrococcal nuclease on chromatin is first to cleave between nucleosomes, it subsequently acts as an exonuclease on the excised nucleosome, shortening DNA first to about 166 bp, where there is a brief pause in the digestion (53), and then to about 146 bp, where there is a clear plateau in the course of digestion before more degradation occurs. During this last stage the histone Hl is released (53), leaving as a major metastable intermediate a particle containing 146 bp of DNA complexed with a set of eight histone molecules. This enzymatically reduced form of the nucleosome is called the core particle, and its DNA content was found to be constant over many different species. DNA removed by the prolonged digestion, which had previously joined one nucleosome to the next, is called linker DNA.

A core particle therefore contains a well-defined length of DNA and is homogeneous in its protein composition. We naturally tried to crystallize preparations of core particles, but we were not at first successful, probably because of small traces of the fifth histone, Hl. Eventually, my colleague Leonard Lutter found a way to produce exceptionally homogeneous preparations of nucleosome core particles, and these formed good single crystals (54). The conditions for growing the crystals were based on our previous experience in crystallizing transfer RNA because we reasoned that a good part of the nucleosome core surface would consist of DNA. These experiments perhaps surprised biologists in showing dramatically that almost all DNA in the nucleus is organized in a highly regular manner.

The derivation of a three-dimensional structure from a crystal of a large molecular complex is, as for the TMV disk, a process that can take many years. We therefore concentrated on obtaining a picture of the nucleosome core particle at low resolution by a combination of X-ray diffraction and electron microscopy, supplemented where possible by biochemical and physicochemical studies. We first solved the packing in the crystals by analyzing electron micrographs of thin crystals and then obtained projections of the electron density along the three principal axes of the crystals, using X-ray diffraction amplitudes and electron microscope phases (54, 55). The nucleosome core particle turned out to be a flat disk-shaped object, about 110 Å by 110 Å by 57 Å, somewhat wedge-shaped, and strongly divided into two layers. We proposed a model in which DNA was wound into about $1^3/_4$ turns of a shallow superhelix of pitch about 27 Å around the histone octamer. There are thus about 80 nucleotides in each turn of the superhelix. This model for the organization of DNA in a nucleosome core also provided an explanation for the results of certain enzyme digestion studies on chromatin (54, 56), thus showing that what we had crystallized was essentially the native structure.

The first crystals we obtained were found to have the histone proteins within them partly proteolyzed, but their physicochemical properties remained very similar to those of the intact particle. We later grew crystals from intact nucleosome cores, which diffracted to a resolution of about 5 Å, and a detailed analysis was made in 1981 (57). Over the years Daniela Rhodes, Ray Brown, and Barbara Rushton grew crystals of core particles prepared from seven different organisms: All give essentially identical X-ray patterns, testifying to the universality of nucleosomes. There is a dyad axis of symmetry within the particle, which is not surprising because the eight histones occur in pairs and because DNA is studded with local dyad axes. High-angle diffuse X-ray scattering from the crystals shows that DNA of the core particle is in the B form. An electron density map of

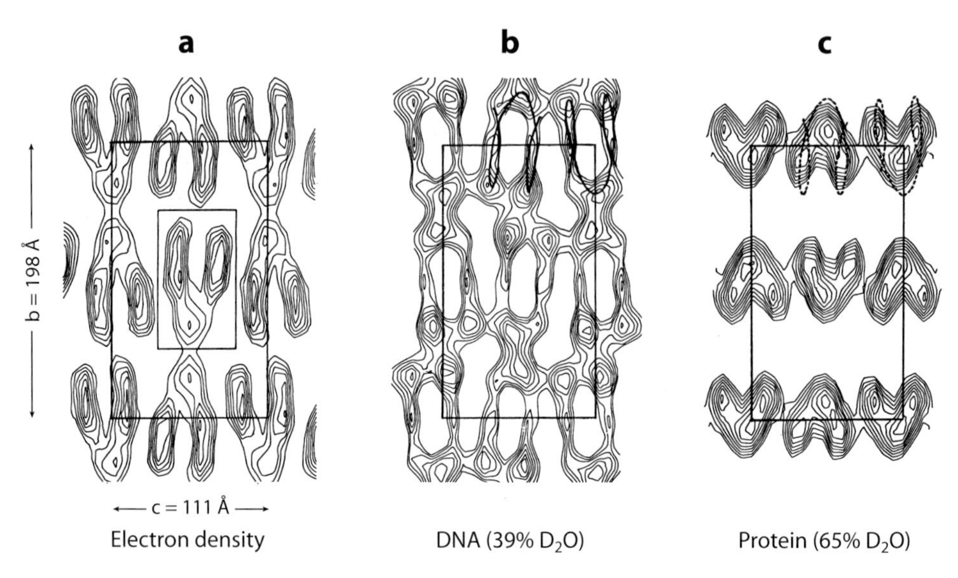

Figure 11

Fourier projection maps of the nucleosome core particle. (*a*) Map from X-ray data (57); (*b*) and (*c*) from neutron scattering data using contrast variation (58); (*b*) the DNA component with the path of the superhelix superimposed on the density; (*c*) the protein core component.

one of the principal projections of the crystal is shown in **Figure 11*a***. This map gives the total density in the nucleosome; the density of DNA is not distinguished from that of the protein. The contributions of protein and DNA can be distinguished by using neutron scattering combined with the method of contrast variation, and such a study was begun by John Finch and a group at the Institut Laue Langevin, Grenoble, when sufficiently large crystals were available (58). They obtained maps of DNA and protein along the three principal projections (see **Figure 11*b*,*c***). The map of DNA is consistent with the projection of about $1^3/_4$ superhelical turns as proposed earlier, and the map of the protein shows that the histone octamer itself is consistent with a wedge shape.

Three-Dimensional Image Reconstruction of the Histone Octamer and the Spatial Arrangement of the Inner Histones

An alternative to separating the contributions of DNA and the protein by neutron diffraction was to study the histone octamer directly. The histone octamer that forms the protein core of the nucleosome can exist in that form free in solution in high salt, which displaces DNA (59). In the course of attempts to crystallize it, we obtained ordered aggregates—hollow tubular structures—which were investigated by electron microscopy (60). The image reconstruction method, described above, was used to produce a low-resolution three-dimensional map and model of the octamer (**Figure 12*a***). As a check that the removal of DNA had not led to a change in the structure of the histone octamer, projections of this model were calculated and compared with the projections of the protein core of the nucleosome obtained from the neutron scattering study, mentioned above. There was a good agreement between the three maps, showing that the gross structure was not altered.

At the resolution of the analysis (20 Å), it was shown that the histone octamer possesses a twofold axis of symmetry, just as does the nucleosome core particle itself. Like the nucleosome core, the histone octamer is a flat wedge-shaped particle of bipartite character. Its periphery showed a system of ridges, which form a more or less continuous helical ramp

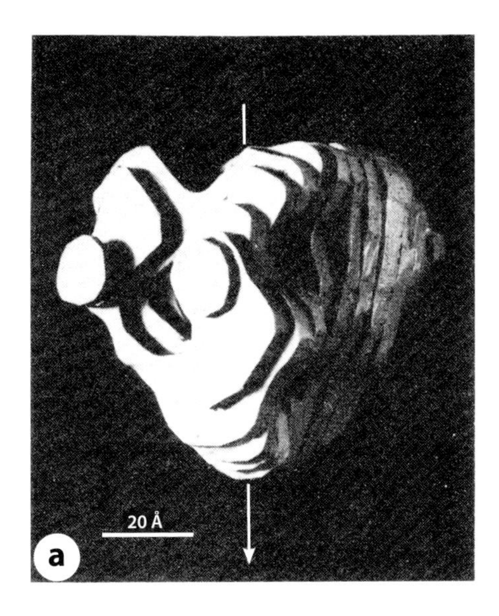

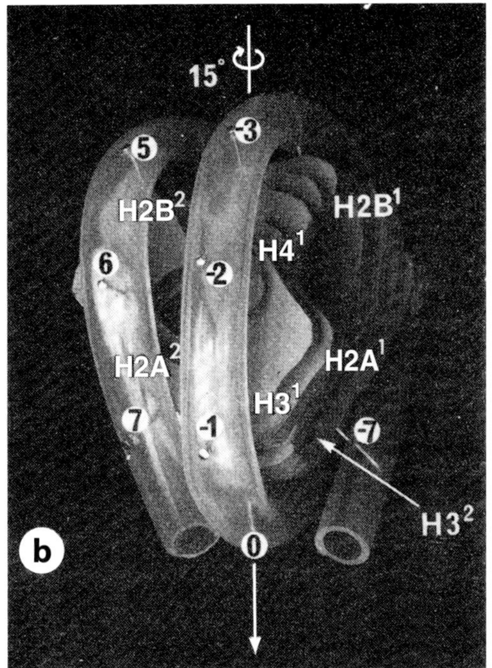

Figure 12

(*a*) Model of the histone octamer obtained by three-dimensional image reconstruction from electron micrographs (60). The dyad axis is marked. The ridges on the periphery of the model form a left-handed helical ramp on which $1^3/_4$ to 2 turns of a superhelix of DNA could be wound. (*b*) The histone octamer structure (*a*) with two turns of a DNA superhelix wound around it. (Note that, for clarity, the diameter of the plastic tube is smaller than the true scale for DNA.) Distances along DNA are indicated by the numbers −7 to +7, taking the dyad axis as the origin, to mark the 14 repeats of the double helix contained in the 146-bp nucleosome core. The assignment of the individual histones to various locations on the model is described in the text.

of a 70-Å external diameter and about a 27-Å pitch, exactly suitable for it to act as a spool on which could be wound about $1^3/_4$ turns of a superhelix of DNA in the appropriate dimensions (**Figure 12*b***). The resolution of the octamer map is too low to define individual histone molecules, but we have exploited the relation of the octamer to the superhelix of DNA to interpret them in terms of individual histones (60). This interpretation uses the results of Mirzabekov and his colleagues (61) on the chemical cross-linking of histones to nucleosomal DNA and also information on histone-histone proximities given by protein cross-linking. This data cannot be interpreted reliably without a three-dimensional model because a knowledge of the points of contact of histones along a strand of DNA is not sufficient to fix a spatial arrangement of the histones in the nucleosome core. Furthermore, because the two superhelical turns of DNA are close together, the pattern of histone-DNA cross-links need not directly reflect the linear order of histones along DNA. The three-dimensional density map restricts the number of possibilities and enables choices to be made.

In the spatial arrangement proposed, the helical density ramp in the octamer map is composed of a particular sequence of the eight histones, in the order H2A-H2B-H4-H3-H3-H4-H2B-H2A, with a dyad in the middle.

The $(H3)_2(H4)_2$ tetramer has the shape of a dislocated disk or single turn of a helicoid, which defines the central turn of a DNA superhelix. The structure of the histone tetramer explains the findings of many workers, expanding on the original observations of Felsenfeld and collegues (62), that H3 and H4 alone, in the absence of H2A and H2B, can confer nucleosome-like properties on DNA, in particular supercoiling and resistance to micrococcal nuclease digestion, whereas H2A and H2B alone cannot. It also explained the asymmetric dissociation of the histone octamer when the salt concentration is lowered: The octamer dissociates, through a hexameric intermediate, into a $(H3)_2(H4)_2$ tetramer and two H2A·H2B dimers (59, 63).

High-Resolution X-ray Studies of the Nucleosome

The next stage in the structural analysis of the nucleosome core particle in our laboratory reached 7-Å resolution, and a number of new features became evident (64). These studies were of material extracted from beef kidney nuclei and so contained mixed populations of DNA.

A major advance was made by Tim Richmond in studies of nucleosome core crystals fashioned by assembling particles from a single DNA sequence, with histone molecules created by protein synthesis (which were thus devoid of the chemical modification found on histones from natural populations). Richmond left the MRC Laboratory to set up his own laboratory at ETH, Zurich. After many years, an electron density map was obtained at 3-Å resolution (65), which revealed the interaction between the histone molecules and DNA in atomic detail.

The Role of H1 and Higher-Order Structures

These studies gave a fairly detailed picture of the internal structure of the nucleosome, but until 1975, there was still no clear idea of the relation of one nucleosome to another along the nucleosome chain or basic chromatin filament, nor of the next higher level of organization. It had been known for some time that the thickness of fibers observed in electron microscopic studies of whole-mount chromosome specimens varied from about 100 to 250 Å in diameter, depending on whether chelating agents had been used in the preparation. Taking this as a clue, Finch and I carried out some in vitro experiments on short lengths of chromatin prepared by brief micrococcal digestion of nuclei (66). In the presence of chelating agents, this native chromatin appeared as fairly uniform filaments of 100 Å in diameter. When Mg^{2+} ions were added, these coiled up into thicker, knobbly fibers about 250–300 Å in diameter, which were transversely striated at intervals of about 120–150 Å, corresponding apparently to the turns of an ordered, but not perfectly regular, helix or supercoil. Because the term supercoil had already been used in a different context, we called it a "solenoid" because the turns were spaced close together. On the basis of these micrographs and companion X-ray studies (67), we suggested that the second level of folding of chromatin was achieved by the winding of the nucleosome filament into a helical fiber with about six nucleosomes per turn. Moreover, we found that, when the same experiments were carried out on H1-depleted chromatin, only irregular clumps were formed, showing that the fifth histone, H1, is needed for the formation or stabilization of the ordered free structure.

Although these experiments told us the level at which H1 performs its function of condensing chromatin, the way in which the H1 molecule mediates the coiling of the 100-Å filament into the 300-Å fiber only became clear later by putting together evidence from its biochemistry, from crystallographic analysis, and from more refined electron microscope observations.

From observations on the course of nuclease digestion, taken in conjunction with the known X-ray structure of the nucleosome core, one can deduce where the Hl might be on the complete nucleosome. I have mentioned that there is an intermediate in the digestion of chromatin by micrococcal nuclease at about 166 bp of DNA, and it is during this step from 166 to 146 bp that Hl is released (52). Because the 146 bp of the particle correspond to $1^3/_4$ superhelical turns, we therefore suggested that the 166-bp particle contains two full turns of DNA (45). This brings the two ends of DNA on the nucleosome close together, so that both can be associated with the same single molecule of H1 (**Figure 13**). A particle consisting of the histone octamer and 166 bp has been called the chromatosome (68) and has been suggested by us and others to constitute the basic structural element of chromatin. In this particle, H1 would therefore be on the side of the nucleosome in the region of the entry and exit of the DNA superhelix.

This location follows in logic, but was histone H1 really there? Although H1 is too small a molecule to be seen directly by electron microscopy, its position in the nucleosome can be inferred from its effect on the appearance of chromatin, in the intermediate range of folding between the 100-Å nucleosome filament and the 300-Å solenoidal fiber. These intermediate stages were revealed in the course of a systematic study by Thoma et al. (69) of the folding of chromatin with increasing ionic strength. By employing monovalent salts rather than divalent ones, they exposed a range of structures showing increasing degrees of compaction as the ionic strength was raised. Thus, from the filament of nucleosomes around 1 mM, the extent of structure increased through a family of intermediate helical structures until, by 60 mM, the compact 300-Å fiber structure was formed, which was in all respects identical to that originally observed by Finch and me.

The location of H1 can be deduced by considering the difference between the structures observed in the range of ionic strength 1–5 mM in the presence or absence of H1 (**Figure 14**). In chromatin containing H1, an ordered structure is seen in which the nucleosomes are arranged in a regular zigzag with their flat faces down on the supporting grid. The zigzag form arises because DNA enters and leaves the nucleosome at sites close together, as one would expect from the combination of X-ray and biochemical evidence mentioned in the last paragraph (**Figure 13**). In chromatin depleted of H1, entrance and exit points are more or less on opposite sides and, in any case, randomly located. Indeed, at very low ionic strength, the nucleosomal structure

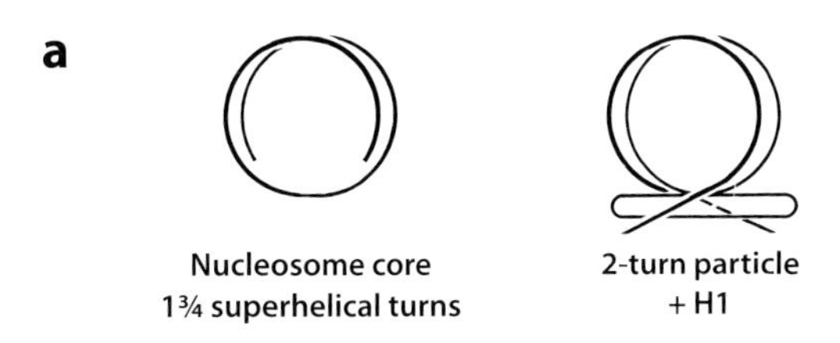

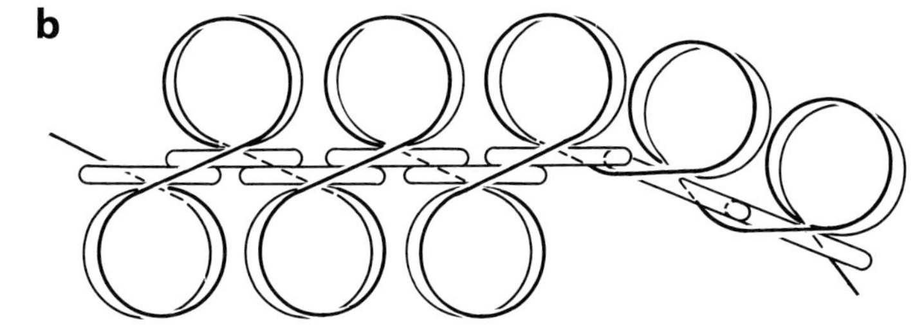

Figure 13

(*a*) If the 146 bp of DNA in the nucleosome core correspond to $1^3/_4$ superhelical turns, then the 166-bp particle corresponds to about two full superhelical turns. Because the 166-bp particle is the limit point for the retention of H1 (53), it must be located as shown. (*b*) Schematic diagram of the nucleosome filament, at low ionic strength, showing origin of the zigzag structure (**Figure 14**). The right side of the drawing shows a variant of the zigzag structure, which is often observed; this variant is formed by flipping a nucleosome by 180° about the filament axis.

unravels into a linearized form in which individual beads are no longer seen. When H1 is present, this is prevented from happening. We therefore concluded that H1, or part of it, must be located at and must stabilize the region where DNA enters and leaves the nucleosome, as was predicted.

In the zigzag intermediates, the H1 regions on adjacent nucleosomes appear to be close together or touching. We therefore suggested that, with increasing ionic strength, more of the H1 regions interact with one another, eventually aggregating into a helical polymer the along the center of the solenoid and thus accounting for its geometrical form (**Figure 15**). Polymers of H1 have indeed been shown to exist by chemical cross-linking experiments at both low and high ionic strength (62), but it remains to be shown that they are located in the center of the fiber. The important point, however, is that it appears to be the aggregation of H1 which accompanies, and indeed may control, the formation of the 300-Å fiber.

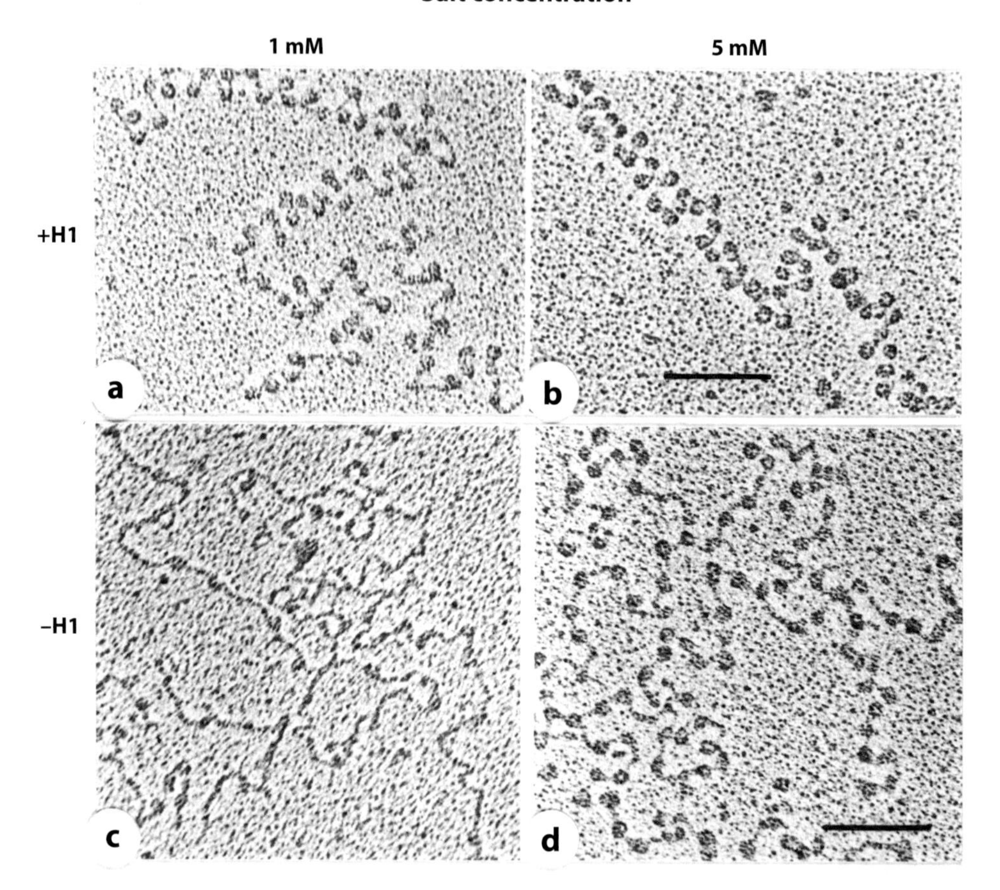

Figure 14

The appearance of chromatin with and without H1 at low ionic strength (69). When H1 is present, the first recognizable ordered structure is (*a*) a loose zigzag in which DNA enters and leaves the nucleosome at sites close together; at a somewhat higher salt concentration, (*b*) the zigzag is tighter. In the absence of H1, there is no order in the sense of a defined filament direction; (*c*) at the lower salt concentration, nucleosome beads are no longer visible, and the structure opened to produce a fiber of DNA coated with histones; (*d*) at a higher ionic strength, beads are again visible, but DNA enters and leaves the nucleosome more or less at random. The bar represents 100 nm. The open zigzag seen in electron micrographs arises because the nucleosomes fall with their flat faces on the electron microscope grid.

Refined Model of the 300-Å Chromatin Fiber

The solenoidal model for the 300-Å fiber (**Figure 15**) was a first-order model. To define the interal and external dimensions accurately, my colleague Rhodes and her coworkers (70) produced very long and regularly folded 300-Å fibers from in vitro reconstituted nucleosome arrays containing the linker histone H1 with increasing nucleosome repeat lengths (comprising 10 to 70 bp of linker DNA). They found that those containing the natural linker lengths of 10 to 40 bp produced fibers with a diameter of 33 nm and a repeat of 11 nucleosomes per 11 nm. Using the physical constraints imposed by these measurements, they built a model in which tight nucleosome packing is achieved through the interdigitation of nucleosomes from adjacent helical gyres of the solenoid (70). The model closely matches raw images of naturally folded chromatin arrays recorded in the solution state by using electron cryomicroscopy.

The Roles of the Histones

From the spatial arrangements of molecules proposed for the histone octamer and from the location deduced for histone H1, one can see the roles of the individual histones in folding DNA on the nucleosome (**Figure 16**) (60). The $(H3)_2(H4)_2$ tetramer has the shape of roughly a single turn of a helicoid, and this defines the central turn of the DNA superhelix. H2A and H2B add as two heterodimers, H2A·H2B, one on each face of the H3-H4 tetramer, each binding one extra half-turn of DNA, thereby completing the two-turn superhelix. Finally, H1 binds to the unique region at the side of the two-turn particle where three segments of DNA come together; this stabilizes and seals off the nucleosome, mediating folding to the next level of organization. Such a sequence of events, in time, would provide a structural rationale for the temporal order of assembly of histones onto newly replicated DNA (71–73).

We thus arrived at a moderately detailed model of the nucleosome and a description for the next higher level of folding. This provided a firm structural and chemical framework in which to consider the dynamic processes that take place in chromatin in the cell, that is, transcription, replication, and mitosis.

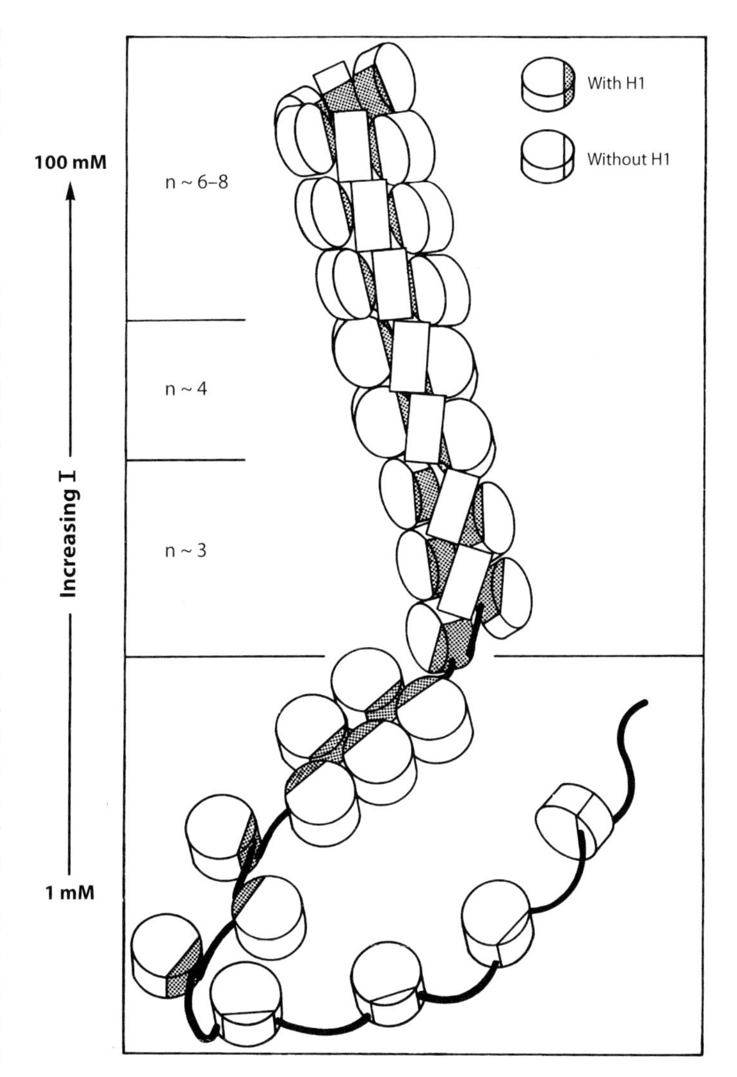

Figure 15

A model for chromatin condensation. An idealized drawing of helical superstructures formed by chromatin containing H1 with increasing ionic strength. The open zigzag of nucleosomes (*bottom left*) closes up to form helices with increasing numbers of nucleosomes per turn (n). The solenoid formed at high ionic strength probably has about six nucleosomes per turn (see text). When H1 is absent (*bottom right*), no zigzags or definite higher-order structures are found. (From Thoma et al. 69.)

THE ROYAL SOCIETY

The Royal Society for the Promotion of Natural Knowledge is the oldest scientific society, or organized academy of any kind, which has enjoyed a continuous existence. The older Italian one was short-lived. The Society represents British Commonwealth science; but it is not a government institution, like the Académie des Sciences of France or the Akademia Nauk of Russia.

The Society combines three different functions. First it is a learned society, established in 1660 to promote the then radical values of modern science, and particularly the notion that reliable knowledge about the natural world is best obtained by careful observation and controlled experiment. Second, it is in effect the academy of science for the United Kingdom (UK), leading the UK scientific community in its relation with UK society and government and with its counterparts in other countries. Third, the Society is a funding body using public and private monies to support the best individuals to undertake the most imaginative and far-reaching research in Britain and British Commonwealth countries. Finally, the Society promotes excellence in science by electing 40 Fellows each year and honoring scientific achievement by awarding prizes and medals.

I was elected president of the Society in 1995, which is a great honor and also a great responsibility. The work involved being in London two or three days a week as well as a good deal of international travel. I would, however, not have been able to undertake this without the selfless support of my wife, Liebe, and the initiatives she took to enliven our London home, a flat in the Royal Society building in St James.

My first surprise on becoming president was the scale of the Society's activities. I had been a Fellow for 27 years and had served on the Council and many committees, so I should have been prepared, but I was not. Moreover, I found myself from the very beginning plunged into the Society's fund-raising campaign, initiated by my predecessor.

Although the ongoing work of the Society involves mainly scientific advice, both formal and informal, to the founding bodies for research and (university) education, and is meant to be apolitical, my first brush in 1995 was with the ideological position of the then prime minister, who believed that research was better done

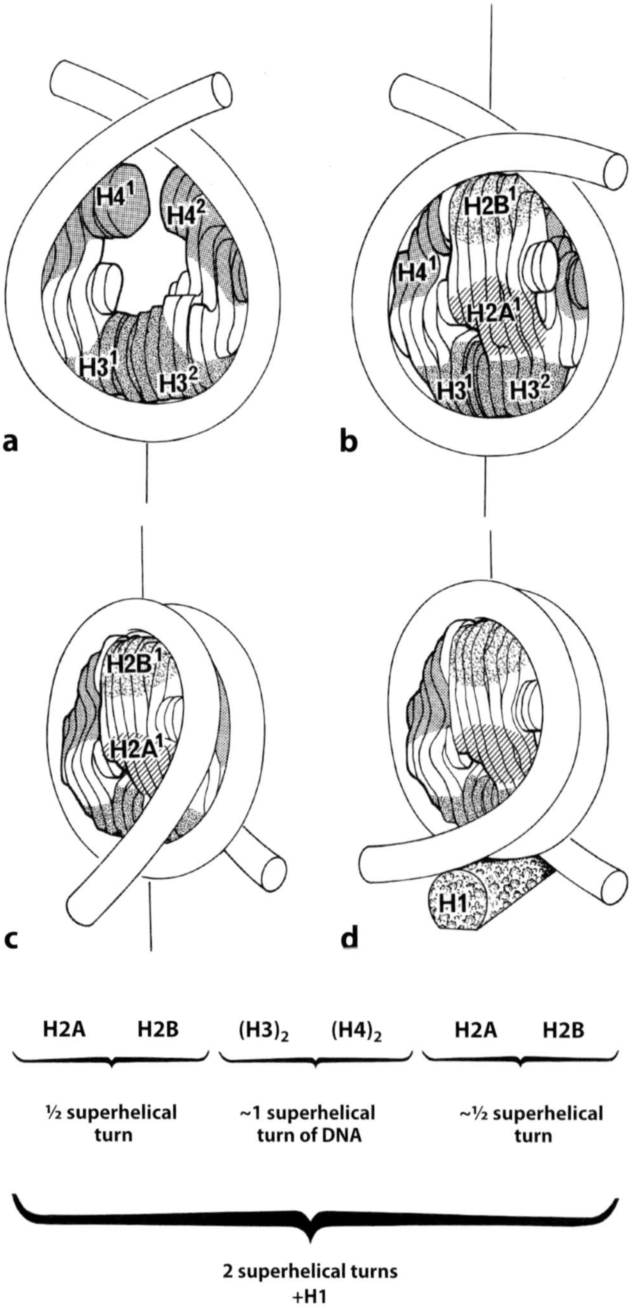

Figure 16

Exploded views of the nucleosome, showing the roles of the constituent histones. The patches on the histone core indicate locations of individual histone molecules, but the boundaries between them were then not known and are thus left unmarked. (*a*) The $(H3)_2(H4)_2$ tetramer has the shape of a lock washer and can act as a spool for 70–80 bp of DNA, forming about one superhelical turn. (*b*) An H2A·H2B dimer associates with one face of the tetramer. (*c*) H2A·H2B dimers on opposite faces each bind 30–40 bp DNA, or one-half a superhelical turn, to give a complete two-turn particle. (*d*) Histone H1 interacts with the unique configuration of DNA at the entry and exit points to seal off the nucleosome.

in the private sector and was intent on privatizing public sector research establishments, such as the MRC Laboratory, for example. From the Royal Society, we argued (ultimately successfully) that it was basic research that had the long-term potential for wealth creation by leading to totally new processes, products, and instruments. Second, there was strategic research, potentially relevant to a sector of the economy but without, as yet, identified customers. Thus, when the bovine spongiform encephalopathy (mad cow disease) and variant Creutzfeldt-Jakob disease (CJD) crisis broke out in March 1996, the little that was known about spongiform encephalopathies was largely the result of work by the Neuropathogenesis Unit (originally set up by the MRC and Agricultural Research Council). The normal CJD disease (a rare disease—one case per million per year) was hardly a main concern. Third, there was applied research, whereby principles were understood but had to be reduced to practice, a no less challenging task.

I have described this in some detail because it illustrates the Royal Society's role in scientific advice. It is given whether asked for or not, and the prestige of the Society ensures that it will get a hearing.

Another issue that I had to deal with during my presidency was global warming and the greenhouse effect. The physics is incontrovertible; a build up of carbon dioxide (and other such gases) must result in Earth's increased temperature. The only unknown, which might prevent this, was the possible absorption of carbon dioxide by the oceans, but this has proved to be illusory.

A closely related issue to this was the question of nuclear energy policy, which unfortunately is still not settled. I set up a Royal Society committee of experts in energy policy to see whether there are long-term alternatives. We found none. France had forged ahead with nuclear power stations: Over 78% of its energy supply now comes from these. Moreover, there have been no accidents or leaks, such as that in Russia at Chernobyl, which was not built under a thick concrete roof and this defect allowed the escape of radioactive material.

I have given these two examples of the policy and scientific issues to illustrate the work of the Society. There were several others, such as countering the opposition to genetically manipulated organisms, which were termed frankenfoods by the opponents. Fascinating as this all was, I was glad when my term as president ended in 2000, and I was able to get back to my zinc finger work. In the meantime, this had made progress thanks to the continuing work of my able research group at the MRC Lab.

DEVELOPING THE ZINC FINGER DESIGN INTO A NEW TECHNOLOGY

We had earlier shown in 1994 that a zinc finger construct could be used to switch off a deleterious gene in a mouse cell line or, conversely, to switch on a reporter gene. Now, there were additional examples from infectious virus diseases, wherein the virus titer could be reduced by 90% using a single application of a plasmid bearing the zinc finger construct. Moreover, there was now progress at our MRC spin-off company, Gendaq, in reaching a deeper understanding of the complexity of the zinc finger interactions with DNA. This resulted in our being able to make libraries or repertoires of highly specific zinc fingers and turn the promise of the zinc finger design into a robust technology.

CONCLUDING REMARKS

I outlined the chromatin work because it served as a contemporary paradigm for structural studies that try to connect the cellular and the molecular processes. One studies a complex system by dissecting it physically, chemically, or, in this case, enzymatically, and then tries to obtain a detailed picture of its parts by X-ray analysis and chemical studies as well as an overall picture of the intact assembly by electron microscopy. There is, however, a sense in which viruses and chromatin, which were the subject of my work, were still relatively simple systems. Much more complex systems, such as ribosomes and the mitotic apparatus, lay ahead, and later generations have taken on these formidable tasks, in some respects only just begun. I am glad to have had a hand in the beginnings of the foundation of structural molecular biology.

It was for my work on virus structures and chromatin that I received the undivided 1982 Nobel Prize in Chemistry, the citation for which reads "For the development of crystallographic electron microscopy and the determination of the structures of nucleic acid-protein complexes of biological importance." In a sense, the Nobel Committee acknowledged that the two parts of the citation were complementary; the technological advances would not have come about from studies on simpler macromolecular systems.

Envoi: Zinc Fingers

As explained in my review on zinc fingers, also in this volume (76), it was the work on the higher order of chromatin that prompted me to go on to consider what was then called "active chromatin," the opened up structure that was involved in transcription or was poised to do so. This led, serendipitously, through biochemical studies on the TFIIIA protein to the zinc finger motif, and hence to the possibility of using this design to synthesize DNA-binding proteins for control of gene expression.

DISCLOSURE STATEMENT

The author is not aware of any affiliations, memberships, funding, or financial holdings that might be perceived as affecting the objectivity of this review.

ACKNOWLEDGMENTS

It is obvious that I could not have accomplished all that has been summarized here without the help of many highly able and valued colleagues and collaborators. After Rosalind Franklin's death, I was able to continue and extend the virus work with John Finch and Kenneth Holmes, who were then students and who became colleagues. Over the years, I have had a transatlantic association with Donald Caspar and have benefited from his advice, criticism, and insights. I mention here only some of the names of my other collaborators in the several branches in which I have been involved: in the study of virus chemistry and assembly, Reuben Leberman, Tony Durham, Jo Butler, and David Zimmern; in virus crystallography, William Longley, Peter Gilbert, John Champness, Gerard Bricogne, and Anne Bloomer; in electron microscopy and image reconstruction, David DeRosier, Harold Erickson, Tony Crowther, Linda Amos, Jan Mellema, Nigel Unwin, and John Finch; in the structural studies on transfer RNA, Brian Clark (who provided the biochemical background without which the work could not have begun), Jon Robertus, Jane Ladner, and Tony Jack; in chromatin, Roger Kornberg (whose skill and insight transformed a messy project into a clear problem), Markus Noll, Len Lutter, Ray Brown and Daniela Rhodes (who fruitfully transferred their experience from tRNA to nucleosomes), and finally Tim Richmond and John Finch (who went on to higher-resolution X-ray studies on the nucleosome).

I dedicate this article to my wife Liebe, who has helped and supported me in many ways that have made my work possible.

LITERATURE CITED

1. Fraenkel-Conrat H, Williams RC. 1955. Reconstitution of active *Tobacco mosaic virus* from its inactive protein and nucleic acid components. *Proc. Natl. Acad. Sci. USA* 41:690–98

2. Fraenkel-Conrat H, Singer B. 1959. Reconstitution of *Tobacco mosaic virus*. III. Improved methods and the use of mixed nucleic acids. *Biochim. Biophys. Acta* 33:359–70
3. Matthews REF. 1966. Reconstitution of *Turnip yellow mosaic virus* RNA with TMV protein subunits. *Virology* 30:82–96
4. Stubbs G, Warren S, Holmes K. 1977. Structure of RNA and RNA binding site in *Tobacco mosaic virus* from 4-A map calculated from X-ray fibre diagrams. *Nature* 267:216–21
5. Holmes KCJ. 1979. Protein-RNA interactions during TMV assembly. *J. Supramol. Struct.* 12:305–20
6. Schramm G, Zillig WZ. 1955. Überdie Struktur des Tabakmosaikvirus IV. Die Reggregation des nucleinsäuve Freien. *Protein Z. Naturforschung Ser. B* 10:493–98
7. Caspar DLD. 1963. Assembly and stability of the *Tobacco mosaic virus* particle. *Adv. Protein Chem.* 18:37–121
8. Finch JT, Leberman R, Chang Y-S, Klug A. 1966. Rotational symmetry of the two turn disk aggregate of *Tobacco mosaic virus* protein. *Nature* 212:349–50
9. Bricogne G. 1976. Methods and programs for direct-space exploitation of geometric redundancies. *Acta Crystallogr. Sect. A* 32:832–47
10. Bloomer AC, Champness JN, Bricogne G, Staden R, Klug A. 1978. Protein disk of *Tobacco mosaic virus* at 2.8 angstrom resolution showing the interactions within and between subunits. *Nature* 276:362–68
11. Lauffer MA, Stevens CL. 1968. Structure of the *Tobacco mosaic virus* particle; polymerization of *Tobacco mosaic virus* protein. *Adv. Virus Res.* 13:1–63
12. Durham AC, Finch JT, Klug A. 1971. States of aggregation of *Tobacco mosaic virus* protein. *Nat. New Biol.* 299:37–42
13. Durham AC, Klug A. 1971. Polymerization of *Tobacco mosaic virus* protein and its control. *Nat. New Biol.* 299:42–46
14. Klug A, Durham ACH. 1971. The disk of TMV protein and its relations to the helical and other modes of aggregation. *Cold Spring Harb. Symp. Quant. Biol.* 36:449–60
15. Butler PJG, Klug A. 1971. Assembly of the particle of *Tobacco mosaic virus* from RNA and disks of protein. *Nat. New Biol.* 229:47–50
16. Fukuda M, Ohno T, Okada Y, Otsuki Y, Takebe I. 1978. Kinetics of biphasic reconstitution of *Tobacco mosaic virus* in vitro. *Proc. Natl. Acad. Sci. USA* 75:1727–30
17. Butler PJG, Lomonossoff GP. 1980. RNA-protein interactions in the assembly of *Tobacco mosaic virus*. *Biophys. J.* 32:295–312
18. Zimmern D, Butler PJG. 1977. The isolation of *Tobacco mosaic virus* RNA fragments containing the origin for viral assembly. *Cell* 11:455–62
19. Zimmern D. 1977. The nucleotide sequence at the origin for assembly on *Tobacco mosaic virus* RNA. *Cell* 11:463–82
20. Zimmern D, Wilson TMA. 1976. Location of the origin for viral reassembly on *Tobacco mosaic virus* RNA and its relation to stable fragment. *FEBS Lett.* 71:294–98
21. Champness JN, Bloomer AC, Bricogne G, Butler PJG, Klug A. 1976. The structure of the protein disk of *Tobacco mosaic virus* to 5 angstrom resolution. *Nature* 259:20–24
22. Butler PJG, Bloomer AC, Bricogne G, Champness JN, Graham J, et al. 1976. *Tobacco mosaic virus* assembly-specificity and the transition in protein structure during RNA packaging. In *Structure-Function Relationships of Proteins*, 3rd John Innes Symp., ed. R Markham, RW Home, pp. 101–10. Amsterdam: North-Holland/Elsevier
23. Klug A. 1979. The assembly of *Tobacco mosaic virus*: structure and specificity. *Harvey Lect.* 74:141–72
24. Crowther RA, Klug A. 1975. Structural analysis of macromolecular assemblies by image reconstruction from electron micrographs. *Annu. Rev. Biochem.* 44:161–82
25. Lebeurier G, Nicolaieff A, Richards KE. 1977. Inside-out model for self-assembly of *Tobacco mosaic virus*. *Proc. Natl. Acad. Sci. USA* 74:149–53
26. Butler PJG, Finch JT, Zimmern D. 1977. Configuration of *Tobacco mosaic virus* RNA during virus assembly. *Nature* 265:217–19
27. Caspar DLD, Klug A. 1962. Physical principles in the construction of regular viruses. *Cold Spring Harb. Symp. Quant. Biol.* 27:1–24
28. Klug A. 1979. Image analysis and reconstruction in the electron microscopy of biological macromolecules. *Chem. Scr.* 14:245–56

29. Brenner S, Horne RW. 1959. A negative staining method for high resolution electron microscopy of viruses. *Biochim. Biphys. Acta* 34:103–10
30. Klug A, Finch JT. 1965. Structure of viruses of the papilloma-polyoma type I. human wart virus. *J. Mol. Biol.* 11:403–23
31. Klug A, Finch JT. 1968. Structure of viruses of the papilloma-polyoma type IV. Analysis of tilting experiments in the electron microscope. Appendix: Symmetry relations between superposition patterns along different axes of view. *J. Mol. Biol.* 31:1–12
32. Klug A, Berger JE. 1964. An optical method for the analysis of periodicities in electron micrographs and some observations on the mechanism of negative staining. *J. Mol. Biol.* 10:565–69
33. Klug A, DeRosier DJ. 1966. Optical filtering of electron micrographs: reconstruction of one-sided images. *Nature* 212:29–32
34. DeRosier DJ, Klug A. 1972. Structure of the tubular variants of the head of bacteriophage T4 (polyheads). I. Arrangement of subunits in some classes of polyheads. *J. Mol. Biol.* 65:469–88
35. DeRosier DJ, Klug A. 1968. Reconstruction of three dimensional structure from electron micrographs. *Nature* 217:130–34
36. Crowther RA, DeRosier DJ, Klug A. 1970. The reconstruction of a three-dimensional structure from projections and its application to electron microscopy. *Proc. R. Soc. Lond. Ser. A* 317:319–40
37. Crowther RA, Amos LA, Finch JT, DeRosier DJ, Klug A. 1970. Three-dimensional reconstuctions of spherical viruses by Fourier synthesis from electron micrographs. *Nature* 226:421–25
38. Crowther RA, Amos LA. 1971. Three-dimensional image reconstructions of some small spherical viruses. *Cold Spring Harb. Symp. Quant. Biol.* 36:489–94
39. Unwin PNT. 1974. Electron microscopy of the stacked disk aggregate of *Tobacco mosaic virus* protein. I. Three-dimensional image reconstruction. *J. Mol. Biol.* 87:657–70
40. Taylor KA, Glaeser RM. 1976. Electron microscopy of frozen hydrated biological specimens. *J. Ultrastruct. Res.* 55:448–56
41. Unwin PNT, Henderson R. 1975. Molecular structure determination by electron microscopy of unstained crystalline specimens. *J. Mol. Biol.* 94:425–40
42. Henderson R, Unwin PNT. 1975. Three-dimensional model of purple membrane obtained by electron microscopy. *Nature* 257:28–32
43. Unwin PNT. 1972. Electron microscopy of biological specimens by means of an electrostatic phase plate. *Proc. R. Soc. Lond. Ser. A* 329:327–59
44. Erickson HP, Klug A. 1970. The fourier transform of an electron micrograph: effects of defocussing and aberrations and implications for the use of underfocus contrast enhancement. *Ber. Bunsen-Ges. Phys. Chem.* 74:1129–37
45. Kornberg RD. 1974. Chromatin structure: a repeating unit of histones and DNA. *Science* 184:868–71
46. Kornberg RD. 1977. Structure of chromatin. *Annu. Rev. Biochem.* 46:931–54
47. Kornberg RD, Klug A. 1981. The nucleosome. *Sci. Am.* 244:52–64
48. Wilkins MHF, Zubay G, Wilson HR. 1959. X-ray diffraction studies of the molecular structure of nucleohistone and chromosomes. *J. Mol. Biol.* 1:179–85
49. Luzzati V, Nicolaieff A. 1959. Etude par diffusion des rayons X aux petits angles des gels d'acide désoxyribonucléique et de nucléoprotéines. *J. Mol. Biol.* 1:127–33
50. Hewish DR, Burgoyne IA. 1973. The digestion of chromatin DNA at regularly spaced sites by a nuclear deoxyribonuclease. *Biochem. Biophys. Res. Commun.* 52:504–10
51. Noll M. 1974. Subunit structure of chromatin. *Nature* 251:249–51
52. Kornberg RD, Thomas JO. 1974. Chromatin structure: oligmers of the histone. *Science* 184:865–68
53. Noll M, Kornberg RD. 1977. Action of micrococcal nuclease on chromatin and the location of histone H1. *J. Mol. Biol.* 109:393–404
54. Finch JT, Lutter LC, Rhodes D, Brown RS, Rushton B, et al. 1977. Structure of nucleosome core particles of chromatin. *Nature* 269:29–36
55. Finch JT, Klug A. 1978. X-ray and electron microscope analysis of crystals of nucleosome cores. *Cold Spring Harb. Symp. Quant. Biol.* 42:1–15
56. Lutter LC. 1978. Kinetic analysis of deoxyribonuclease I cleavages in the nucleosome core: evidence for a DNA superhelix. *J. Mol. Biol.* 124:391–420

57. Finch JT, Brown RS, Rhodes D, Richmond T, Rushton B, et al. 1981. X-ray diffraction study of a new crystal form of the nucleosome core showing higher resolution. *J. Mol. Biol.* 145:757–69
58. Finch JT, Lewit-Bentley A, Bentley GA, Roth M, Timmins PA. 1980. Neutron diffraction from crystals of nucleosome core partcles. *Philos. Trans. R. Soc. Lond. B* 290:635–38
59. Thomas JO, Kornberg RD. 1975. An octamer of histones in chromatin and free in solution. *Proc. Natl. Acad. Sci. USA* 72:2626–30
60. Klug A, Rhodes D, Smith J, Finch JT, Thomas JO. 1980. A low resolution structure of the histone core of the nucleosome. *Nature* 287:509–16
61. Mirzabekov AD, Shick VV, Belyavsky AV, Bavykin SG. 1978. Primary organization of nucleosome core particle of chromatin: sequence of histone arrangement along DNA. *Proc. Natl. Acad. Sci. USA* 75:4184–88
62. Camerini-Otero RD, Felsenfeld G. 1977. Supercoiling energy and nucleosome formation: the role of the arginine-rich histone kernel. *Nucleic Acids Res.* 4:1159–81
63. Thomas JO, Kornberg RD. 1975. Cleavable cross-links in the analysis of histone-histone associations. *FEBS Lett.* 58:353–58
64. Richmond TJ, Finch JT, Rushton D, Rhodes D, Klug A. 1984. Structure of the nucleosome core particle at 7 angstrom resolution. *Nature* 311:532–37
65. Luger K, Mäder AW, Richmond RK, Sargent OF, Richmond TJ. 1997. Crystal structure of the nucleosome core particle at 2.8 A resolution. *Nature* 389:251–60
66. Finch JT, Klug A. 1976. Solenoidal model for superstructure in chromatin. *Proc. Natl. Acad. Sci. USA* 73:1897–901
67. Sperling L, Klug A. 1977. X-ray studies on "native" chromatin. *J. Mol. Biol.* 112:253–63
68. Simpson RT. 1978. Structure of the chromatosome, a chromatin particle containing 160 base pairs of DNA and all the histones. *Biochemistry* 17:5524–31
69. Thoma F, Koller T, Klug A. 1979. Involvement of histone H1 in the organization of the nucleosome and of the salt-dependent superstructures of chromatin. *J. Cell Biol.* 83:403–27
70. Robinson PJ, Fairall L, Huynh VA, Rhodes D. 2006. EM measurements define the dimensions of the "30-nm" chromatin fiber: evidence for a compact, interdigitated structure. *Proc. Natl. Acad. Sci. USA* 103:6506–11
71. Worcel A, Han S, Wong ML. 1978. Assembly of newly replicated chromatin. *Cell* 15:969–77
72. Senshu T, Fukuda M, Ohashi M. 1978. Preferential association of newly synthesized H3 and H4 histones with newly replicated DNA. *J. Biochem.* 84:985–88
73. Cremisi C, Yaniv M. 1980. Sequential assembly of newly synthesized histones on replication SV40 DNA. *Biochem. Biophys. Res. Commun.* 92:1117–23
74. Klug A, Caspar DLD. 1960. The structure of small viruses. *Adv. Virus Res.* 7:225–325
75. Crowther RC, Amos LA. 1971. Harmonic analysis of electron microscope images with rotational symmetry. *J. Mol. Biol.* 60:123–30
76. Klug A. 2010. The discovery of zinc fingers and their applications in gene regulation and genome manipulation. *Annu. Rev. Biochem.* 79:213–31

Genomic Screening with RNAi: Results and Challenges

Stephanie Mohr,[1,3] Chris Bakal,[1,2] and Norbert Perrimon[1,2]

[1]Department of Genetics, [2]Howard Hughes Medical Institute, and [3]Drosophila RNAi Screening Center, Harvard Medical School, Boston, Massachusetts 02115;
email: Stephanie_Mohr@hms.harvard.edu, chris.bakal@icr.ac.uk, perrimon@receptor.med.harvard.edu

Annu. Rev. Biochem. 2010. 79:37–64

First published online as a Review in Advance on April 1, 2010

The *Annual Review of Biochemistry* is online at biochem.annualreviews.org

This article's doi:
10.1146/annurev-biochem-060408-092949

0066-4154/10/0707-0037$20.00

Key Words

bioinformatics, cell biology, *Drosophila*, high-throughput screening

Abstract

RNA interference (RNAi) is an effective tool for genome-scale, high-throughput analysis of gene function. In the past five years, a number of genome-scale RNAi high-throughput screens (HTSs) have been done in both *Drosophila* and mammalian cultured cells to study diverse biological processes, including signal transduction, cancer biology, and host cell responses to infection. Results from these screens have led to the identification of new components of these processes and, importantly, have also provided insights into the complexity of biological systems, forcing new and innovative approaches to understanding functional networks in cells. Here, we review the main findings that have emerged from RNAi HTS and discuss technical issues that remain to be improved, in particular the verification of RNAi results and validation of their biological relevance. Furthermore, we discuss the importance of multiplexed and integrated experimental data analysis pipelines to RNAi HTS.

Contents

HIGH-THROUGHPUT RNAi SCREENING

RNA interference (RNAi) is an RNA-dependent gene-silencing process that is controlled by the RNA-induced silencing complex (RISC) and is initiated by short double-stranded RNA (dsRNA) molecules. In response to endogenous or exogenously introduced dsRNAs, the RNAi machinery knocks down (i.e., reduces but does not eliminate) the RNA targets of dsRNA in a sequence-specific manner (1, 2). The burgeoning of the RNAi field, recognized in its importance with a Nobel Prize to Andrew Fire and Craig Mello in 2006, has led to exciting new research in several areas. First, fundamental new biological insights have been obtained from the study of the genesis and function of small RNAs of 21–28 nucleotides (nt) in length that include microRNAs (miRNAs) and endogenous short interfering RNAs (endo-siRNAs) (recent reviews include References 3–6), as well as the cellular responses to RNA viruses (recent reviews include References 7 and 8). Second, much effort is ongoing regarding the potential use of RNAi-inducing reagents as therapeutics (see recent reviews in References 9–14). Third, RNAi is being harnessed as a molecular tool for gene- and transcript-specific knockdown of mRNA levels, facilitating large-scale study of gene function in a wide variety of cells, tissues, and organisms, including *Caenorhabditis elegans*, *Drosophila*, mammalian cells, the flatworm *Planaria*, and *Arabidopsis* (see early examples and reviews in References 1, 15–29).

RNA interference (RNAi): RNA-dependent gene silencing controlled and initiated by short double-stranded RNA molecules that can reduce levels of the mRNA target

HTS: high-throughput screen

The pairing of RNAi technologies with cDNA and genomic sequence data has made it possible to construct genome-scale libraries of RNAi reagents for performing RNAi high-throughput screens (HTSs) in a wide variety of cell types (30). As such, RNAi allows in many systems the type of systematic functional analyses that were previously practical for only a relatively small set of genetically tractable model organisms. Arguably, the most important impact in this regard has been the ability to perform genome-scale cell-based RNAi HTS in mammalian cells. Indeed, RNAi screening in mammalian cells has already led to a large number of results with important biomedical implications (see **Table 1** and below), including the identification of novel oncogenes and potential targets for the development of therapeutic treatments (recent reviews include References 11, 31–34).

Even in well-established genetic model systems, such as *C. elegans* and *Drosophila*, RNAi screening has had a profound impact, increasing the scope and pace at which gene interrogation can proceed. In *C. elegans*, RNAi screens are performed in vivo, usually following feeding of bacteria that express dsRNAs (29, 35, 36). In *Drosophila*, RNAi screening can be done either in vivo using transgenes that express RNAi reagents or in cell culture (reviews include References 28 and 29). Conveniently, the RNAi approach itself facilitates rapid transfer of information learned in model organisms to

Table 1 Results of genome-scale,[a] cell-based RNAi high-throughput screens in mammalian or *Drosophila* cells

Cell type	Screen type	Reagent	Primary hits	Secondary hits[b]	Field of study	References
Human cells						
HeLa	Plate reader & imaging	esiRNA	275	37	Cell division	120, 121
U2OS	Imaging	siRNA	1,152	18	Cell cycle	114
NCI-H1155	Plate reader	siRNA	87	6	Cancer biology	103
NIH3T3	Pooled	shRNA	15	3	Stress resistance	43
293T	Plate reader	siRNA	—	295	Host-pathogen interactions	79
293T, HeLa, MCF-7	Pooled	shRNA	30	8	Cell death	42
DLD1	Plate reader	siRNA	740	268	Signal transduction	50
HEK293	Pooled	shRNA	13,140	21	Cell adhesion	40
HeLa	Imaging	siRNA	305	124	Host-pathogen interactions	180
HeLa	Plate reader	siRNA	530	23	Signal transduction	156
HeLa-derived TZM-b1	Plate reader	siRNA	386	273	Host-pathogen interactions	78
HeLa P4/R5	Plate reader	siRNA	931	232	Host-pathogen interactions	80
Jurkat	Pooled	shRNA	11	5	Cancer biology	46
MCF-10A[c]	Pooled	shRNA	201	166	Cancer biology	45
MNT-1	Plate reader	siRNA	98	35	Pigmentation	142
RDG3	Imaging	siRNA	—	171	Stress resistance	133
BJtsLT	Pooled	shRNA	100	37	Cancer biology	126
DLD-1	Pooled	shRNA	368	83	Cancer biology	48
Huh7/Rep-Feo	Plate reader	siRNA	236	96	Host-pathogen interactions	181
Jurkat	Pooled	shRNA	252	7	Host-pathogen interactions	47
Huh 7.5.1	Imaging	siRNA	521	262	Host-pathogen interactions	182
Mouse cells						
NIH 3T3	Pooled	shRNA	—	28	Cancer biology	104
L929	Plate reader	siRNA	666	432	Cell death	122
B16-F0	Pooled	shRNA	78	22	Cancer biology	41
Oct4-Gip ESCs	FACS[d] & imaging	esiRNA	296	21	Stem cell biology	109
Oct4-Gip ESCs	FACS	siRNA	148	104	Stem cell biology	107
***Drosophila* cells**						
Kc167, S2R+	Imaging	dsRNA	438	—	Viability	97
S2	Imaging	dsRNA	—	121	Host-pathogen interactions	164
Clone 8	Plate reader	dsRNA	238	213	Signal transduction	152
Clone 8	Plate reader	dsRNA	509	96	Signal transduction	147
Kc167	Plate reader	dsRNA	—	90	Signal transduction	149
S2	Plate reader	dsRNA	474	121	Signal transduction	190
S2	Imaging	dsRNA	—	86	Host-pathogen interactions	167
S2	Imaging	dsRNA	305	~190	Host-pathogen interactions	169
S2	Imaging	dsRNA	210	112	Host-pathogen interactions	176
S2	Plate reader	dsRNA	—	14	Host-pathogen interactions	163, 191
S2	Plate reader	dsRNA	1,133	284	Protein secretion	139

(*Continued*)

Table 1 (*Continued*)

Cell type	Screen type	Reagent	Primary hits	Secondary hits[b]	Field of study	References
S2	FACS	dsRNA	488	—	Cell cycle and/or cell size	115
S2	Imaging	dsRNA	—	—	Signal transduction	130
S2	FACS	dsRNA	66	23	RNA biology	157
S2	Plate reader	dsRNA	75	4	Ion transport	129
S2R+	Plate reader	dsRNA	138	7	RNA biology	158
S2R+	Plate reader	dsRNA	1,168	331	Signal transduction	144
S2R+	Imaging	dsRNA	699	—	Signal transduction	131
S2R+	Imaging	dsRNA	1,500	27	Signal transduction	132
S2	Imaging	dsRNA	90	24	RNA biology	162
Kc167	Plate reader	dsRNA	81	47	Cell death	123
S2	Plate reader	dsRNA	47	1	RNA biology	159
S2*	Plate reader	dsRNA	18	5	Chromatin regulation	140
S2R+	Imaging	dsRNA	346	—	Signal transduction	145a
S2	Imaging	dsRNA	162	54	Host-pathogen interactions	173
DL1	Plate reader	dsRNA	176	110	Host-pathogen interactions	178
Kc167	Imaging	dsRNA	526	—	Lipids	82
Kc167	Plate reader	dsRNA	265	120	Transcription and/or translation	141
Primary neurons	Imaging	dsRNA	336	104	Neural outgrowth	136
S2	Plate reader	dsRNA	821	152	Mitochondria	138
S2	Plate reader	dsRNA	—	—	Phagocytosis	128
S2	Imaging	dsRNA	—	—	Centrioles and/or centrosomes	119
S2	Imaging	dsRNA	847	227	Lipids	83
S2	Imaging	dsRNA	292	133	Cancer biology	117
S2	Imaging	dsRNA	23	—	Transcription and/or translation	85
S2-derived RZ-14	Plate reader	dsRNA	177	—	RNA biology	160
S2R+	Imaging	dsRNA	119	39	Centrioles and/or centrosomes	118
S2R+	Imaging	dsRNA	133	72	RNA biology	161
S2R+	Imaging	dsRNA	~500	1	Mitochondria	137
S2R+	Plate reader	dsRNA	303	173	Circadian rhythms	125
Clone 8	Plate reader	dsRNA	~100	11	Signal transduction	155
S2	Plate reader	dsRNA	218	116	Host-pathogen interactions	37
Kc167	Plate reader	dsRNA	996	202	Cell cycle and/or cell size	116
S2R+	Plate reader	dsRNA	42	33	Cell death	124
S2R+	Imaging	dsRNA	—	—	Signal transduction	154
S2R+	Imaging	dsRNA	15	7	Cell cycle and/or cell size	134

[a]For this summary, we defined genome-scale with a cutoff of approximately 5000 genes (mammalian cell screens) or at least 70% of the genome (*Drosophila* cell screens).

[b]Here, we use secondary hits to refer to the largest set of primary hits that passed an additional test verifying the result at the reagent level (retest after re-synthesis or with another assay or cell type) or in most cases, at the gene level (retest with another reagent or single reagents from a pool, for example). In some cases, only a subset of primary hits were tested in secondary assays. For most reports, only a small number of genes (typically, one to five) were confirmed with a rigorous test, such as rescue of the RNAi effect with a cDNA, or were confirmed at the level of biological significance with another type of assay or an in vivo analysis.

[c]Additional cell types tested with smaller shRNA pools.

[d]FACS, fluorescence-activated cell sorter.

mammalian studies, by making it possible to design and execute RNAi knockdown of homologous genes in mouse or human cells (or in another relevant system, such as mosquito vectors of mammalian diseases; see, for example, Reference 37) subsequent to performing the large-scale screen in *Drosophila* or another model organism.

Because many aspects of RNAi screening have been reviewed previously, we have focused this review primarily on results of genome-scale cell-based screens in *Drosophila* and mammalian cells (**Table 1**). Following a discussion of the technical aspects of RNAi HTSs, we discuss in more detail what has been learned from the results of the large number of screens performed to date, including issues of false discovery, specific genes, and pathways newly implicated in various processes, and discuss how researchers are working toward systems-wide understandings of various biological processes. Where relevant, we refer to other sources for further reading on specific subtopics.

PERFORMING HIGH-THROUGHPUT, CELL-BASED RNAi SCREENS

The effects of RNAi can be compared with reduction-of-function (hypomorphic) genetic approaches. When the normal function of a gene is required for a given function, RNAi knockdown may lead to a phenotype detectable in an assay that tests that function, either directly or indirectly. As such, RNAi facilitates both small-scale studies and HTSs. With HTSs (see **Figure 1*a*,*b***), a large number of gene functions are interrogated concurrently, such that one can, at least in theory, begin to isolate multiple members of a functional pathway as well as implicate new genes in a given biological function, process, complex, or behavior. A cell-based RNAi HTS, as discussed here, is typically done in one of two formats: a pooled format, in which the library is introduced at random into cells (**Figure 1*a***), or an arrayed format, in which single genes are targeted by reagents in individual wells of a microtiter plate (**Figure 1*b***) (recently reviewed in References 32, 38, 39). Each of these approaches has significant advantages and disadvantages, and both have been successfully applied to the investigation of a number of different biological questions.

Pooled Format Screening

With a pooled screen (**Figure 1*a***), the RNAi reagent library [for mammalian cells, a viral-encoded short hairpin RNA (shRNA) library is typical] is introduced into cells en masse and at random, such that any given cell will contain approximately one gene-specific RNAi reagent. The screener may then perform a selection, in which only cells resistant to some treatment will survive [alternatively, a method such as fluorescence-activated cell sorter (FACS) can be used to isolate the specific subset of cells that are positive in the assay], followed by polymerase chain reaction (PCR) amplification of the RNAi reagents present in surviving cells and by sequencing to determine the identity of those reagents. The presence of a specific RNAi reagent after selection suggests that knockdown of the corresponding gene confers resistance to the treatment (see, for example, References 40–45).

Alternatively, the researcher may treat one or more subset of cells (or different cell types) differently, either before or after subjecting the cells to the pooled RNAi library (depending on the assay), creating a "reference set" and one or more "experimental sets" of cells. Subsequently, a molecular method, such as PCR amplification or microarray analysis, is used to detect which RNAi reagents are present in each set (via detection of the RNAi-inducing sequence itself or a unique molecular "barcode" that identifies each reagent). This makes it possible to determine which RNAi reagents are under- or over-represented in the experimental set(s) as compared with the reference set (see, for example, References 45–48).

In general, pooled approaches are more likely to be feasible in a standard lab setting than are arrayed screens, which require liquid handling automation and specialized

False discovery: experimental findings that cannot later be verified (i.e., false-positive results) or should be identified but are not (i.e., false-negative results)

RNAi reagent library: long or short dsRNAs designed to induce RNAi knockdown of specific genes (e.g., sets of dsRNAs, shRNAs, siRNAs, or esiRNAs)

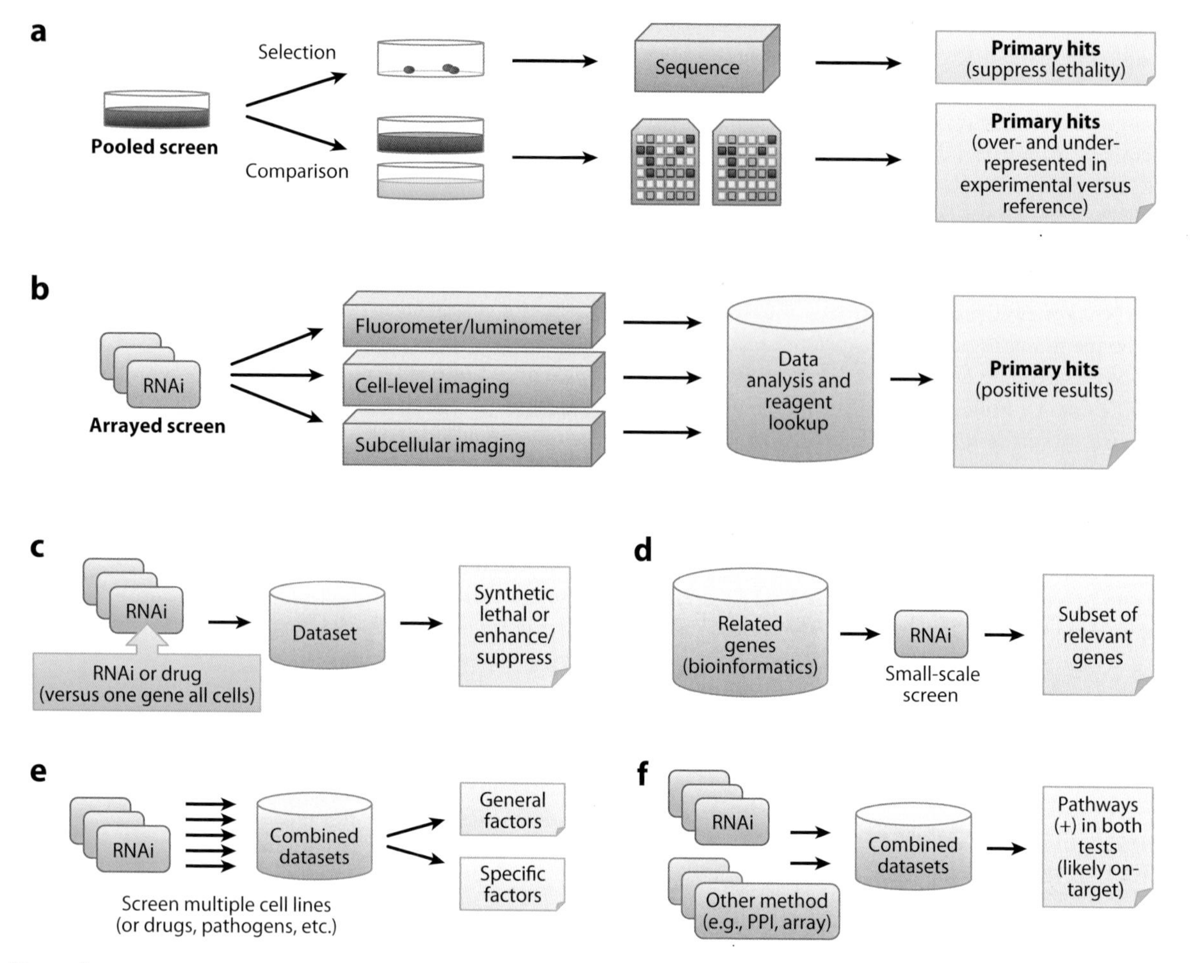

Figure 1

Approaches to high-throughput cell-based RNAi screening. (*a*) Pooled RNAi high-throughput screen (HTS) approach. (*b*) Arrayed RNAi HTS approach. (*c*) Modification of a pooled or arrayed approach via prior addition of a treatment, such as RNAi against a single gene in all cells or treatment with a small molecule. (*d*) Identification of related genes via informatics-based analysis (e.g., all kinases or genes previously implicated in a specific pathway or complex), followed by screening with reagents directed against the identified subset of genes. (*e*) A HTS with one assay type using multiple cell lines, the same cell line with multiple pathogens, or a similar multiplexed approach, followed by data integration to identify specific and general factors. (*f*) Parallel RNAi HTSs and an additional experimental high-throughput genomic or proteomic approach, followed by data integration to identify high-confidence hits. Abbreviations: PPI, protein-protein interaction; + sign, positive result.

assay readout instruments (see below). However, pooled approaches have the disadvantage that deconvolution of positive results can require a specialized and potentially costly approach (namely, microarrays to detect the RNAi reagents). Moreover, it is not currently feasible to use a pooled approach in conjunction with a high-content image-based cell assay (i.e., a microscopy screen), and there is at least some risk that representation of the library will not be uniform, creating a difference between the theoretical and the actual number and proportion of reagents tested in a given screen. Nevertheless, the approach has been successfully applied for a number of HTSs that have yielded interesting results, including a number of recent studies in human cells (**Table 1**) (see for example References 46, 48). Finally, it should be noted that, with a pooled approach, the time in cell culture after introduction of the library may be on the order of several days or weeks, whereas shorter incubations are more typical for arrayed format screens.

Arrayed Format Screening

The arrayed format is the more flexible format for RNAi HTSs as each unique RNAi reagent (or unique set of reagents, such as a small pool of independent siRNAs targeting a single gene) occupies a unique well in a microtiter plate, such as a 96- or 384-well plate, facilitating a wide variety of manipulations and readouts (**Figure 1*b***). Detection of the assay is typically done via measuring colorimetric, fluorescence, or luminescent readouts at the total well level (plate-reader screens) or via measuring fluorescent readouts at the cellular or subcellular level using imaging. Examples of cell-based assays performed in arrayed formats include detection of responses to an external stimulus (e.g., a stress, drug treatment, pathogen, signaling ligand, or metabolic substrate), such as via a transcriptional reporter; changes in the expression, modification, and/or subcellular localization of a protein; cell death, cell-cycle arrest, or other changes related to cell survival, metabolism, and/or division; changes in cellular or organelle size and/or morphology; and changes in transport and/or accumulation of an ion, metabolite, and/or biomolecule (30).

As a large number of individual assay plates must be screened to reach genome scale with an arrayed screen, screening in this format typically requires a fairly large total volume of assay reagents (e.g., media, antibodies, and dyes) and automated equipment (e.g., for liquid handling automation and assay readouts). However, arrayed screens have the advantage that, after the assay, one can easily determine which cells were treated with which specific RNAi reagents by simply looking up the identity of the reagent in a given well using a database or spreadsheet. Notably, arrayed screens also have the advantages that multiple, related phenotypes can be assayed in a single screen (e.g., via detection of multiple antibodies and/or fluorescent dyes) and one can have high confidence that all RNAi reagents in the library are tested in the screening assay (30).

Innovative and Multiplexed Screening Approaches

Researchers are increasing the complexity and usefulness of screening using the approaches outlined in **Figure 1*c-f***, such as via multiplexed and/or combinatorial approaches. The most accessible of these techniques for individual laboratories may be incorporation of bioinformatics analysis at genome scale to identify a subset of candidate genes, followed by experimental testing with a corresponding subset of reagents (**Figure 1*d***). For example, extensive screening has been done using sublibraries grouped by biochemical function (e.g., all human kinases). An increasing number of screens start with another type of bioinformatics-based approach to defining a candidate gene list, such as a literature-based analysis, followed by a small-scale screen. Moreover, integration of RNAi data with other "omics" approaches, such as protein-protein interaction maps, genetic interaction networks, and RNA-profiling experiments, can provide additional insight into relationships among the components of a network (49, 50).

RNAi Reagents

The specific RNAi reagent for knockdown is likely to be different for different types of cells, organisms, and assays, and reagent libraries are evolving as we learn more about RNAi mechanisms and rules for specific and effective design of RNAi reagents (recent reviews include those in References 12, 51, 52). The four types of RNAi reagent that are typically used for cell-based HTSs are dsRNAs, siRNAs, shRNAs, and endoribonuclease-prepared siRNAs (esiRNAs). In general, short dsRNA segments [~21 base pairs (bp), in the form of siRNA or shRNAs] are typical for mammalian systems, as longer segments can induce a nonspecific interferon response (30, 32, 38, 53–55). Longer segments (~500 bp) are typical for model systems that lack an interferon response, such as *C. elegans* and *Drosophila* (30, 36, 54–59). Once inside the cell, dsRNAs are processed by the endogenous RNAi machinery to generate small dsRNA segments (typically 20–22 bp) with a characteristic 2-bp 3′ overhang, the active agent for RNAi (recently reviewed in Reference 51).

Delivery to Cells

The appropriate delivery systems also differ for different cell types. Common delivery systems include viral transduction for shRNAs; lipid-mediated transfection or electroporation for shRNAs, siRNAs, esiRNAs, or dsRNAs (30, 32, 38, 52–54); or simply mixing cells with dsRNA in solution for most *Drosophila* cells, an approach referred to as "bathing" (29, 30, 54, 56–58).

Analysis and Follow-Up Studies

Subsequent to the primary screen, the resulting data are analyzed to identify positive results, "hits." As mentioned above, for pooled screens, this typically involves identifying the set of reagents that conferred resistance or those that are under- and/or overrepresented in the experimental set(s) as compared with the reference. Analysis of arrayed screens can involve application of specialized image analysis software or custom programs, as well as various methods of statistical analysis (60). RNAi screening has learned much from applying what was developed for statistical analysis of other methods, in particular for cell-based small-molecule screens, and much progress has been made. For example, several approaches to data normalization, establishment of appropriate thresholds for cutoffs, replicate tests, and other criteria have been established (60–68).

Important factors to consider in RNAi HTSs include (*a*) performing at least one replicate test in the primary screen, (*b*) including an appropriate type and number of "no treatment" and other controls, (*c*) a thoughtful array of the library and controls (e.g., randomized), (*d*) an early assessment of data quality to detect plate- or well-level problems such as dispensing errors, (*e*) data normalization, and (*f*) setting appropriate cutoff values for significant results (60–63, 66). Despite the recent improvements in addressing all of these factors during RNAi HTSs, subsequent data analysis, and follow-up tests, false discovery remains a significant and difficult problem to address. Statistical and experimental approaches can help to minimize the problem (60, 69). Sources contributing to false discovery are described in **Table 2**; methods of verification of RNAi HTS results at the level of the reagent, assay, gene, or biological process are described in **Table 3**.

HIGH-THROUGHPUT RNAi SCREEN RESULTS

Recognizing and Addressing False Discovery

Following the completion of the first full-genome screens in *Drosophila* and mammalian cells, it became apparent, from both comparative analysis of datasets and attempts to validate screen hits, that many primary screen hits were false positives attributable to off-target effects (OTEs) (70–73). Recognition of the problem, together with a better understanding of RNAi mechanisms, has prompted development of software tools for minimizing OTEs and better reagent libraries. Improved gene annotations, such the efforts of the ENCyclopedia Of DNA Elements (ENCODE) (74–76) and modENCODE (77) also contribute to improved reagent design. However, as we still lack a complete understanding of effective rules for reagent design and gene annotations continue to be revised, OTEs remain an issue. Intriguingly, even after improvements in library design (for a review, see Reference 52), overlap among screen hits from independent but related screens remains surprisingly low. For example, in multiple studies of human immunodeficiency virus (HIV) infection in mammalian

Table 2 Sources contributing to false discovery in RNAi high-throughput screens

Source	Contributes to	How to address during screening
Experimental noise inherent in large-scale studies	False positives and negatives	Use appropriate experimental controls, number of replicate tests, and statistical analyses for the specific screen performed
Bias inherent in the screen assay	False positives and negatives	Perform pilot tests to detect bias and flaws; correct for changes in cell number; include appropriate experimental controls; screen with multiple related assays
Off-target effects	False positives (and can obscure true positives)	Choose an optimized or verified RNAi reagent library; include more than one unique RNAi reagent per gene; learn from community annotation of RNAi reagents
Incomplete or incorrect gene models	False positives and negatives	Learn from community efforts to improve genome annotations
Potency of RNAi reagents	False negatives	Include more than one unique RNAi reagent per gene; choose an optimized or verified RNAi reagent library; work to improve reagent design
Knockdown causes weak phenotype not detected above a given cutoff	False negatives	Screen in a sensitized background; relax statistical cutoffs (at the cost of increasing false-positive discovery); increase the number of replicate tests to detect weak but repeatable results
The knockdown phenotype (e.g., cell death) obscures the screen assay phenotype	False negatives	Include all members of a pathway or complex in follow-up studies even if only a subset was identified in the screen; perform multiple screen assays; perform the screen assay in multiple cell lines.

cells, there is limited overlap among screen hits at the gene level (47, 78–81). Similarly, related screens performed in two different *Drosophila* cell types for components of the JAK/STAT signaling pathway resulted in only 25% overlap (82, 83). When analysis of either set of related screens was extended to the level of gene ontology or pathways, the results were more similar (81, 82).

Development of appropriate approaches to minimize false discovery rates remains a challenge as, to a large extent, minimizing false-positive results increases the number of false-negative results, and vice versa. The appropriate statistical cutoff applied to limit false discovery will vary depending on the tolerance for false discovery in one direction or the other, and the level of tolerance may be quite different for different screens. For example, if the ultimate goal is to identify the one gene (or a small set of genes) that confers a specific predicted gene activity, then a researcher might be fairly intolerant of false-negative results, for fear of tossing out the specific subset being sought. By contrast, for a screen aimed at studying a relatively understudied process, one might be willing to sacrifice a fair number of false negatives in the interest of working from a limited set of statistically high-confidence hits. It is worth noting that the false discovery tolerances of the researchers who analyze an initial study may be in conflict with the tolerances of those who analyze the data in subsequent studies (e.g., meta-analyses), emphasizing the need for data reporting standards, including facilitating easy access to raw data (for reanalysis), transparency of analysis methods that were used to determine the reported hits, and standardized reporting formats to facilitate data integration. Efforts at standardization include MIARE (for Minimum Information about an RNAi Experiment, **http://www.miare.org**), and information about reagents and data is being collected at the Probe (**http://www.ncbi.nlm.nih.gov/probe**) and PubChem (**http://pubchem.ncbi.nlm.nih.gov/**) databases at NCBI (**http://www.ncbi.nlm.nih.gov**).

Nevertheless, data analysis is neither the sole contributor nor the sole answer to the problem of false discovery. For example, the gene-level

Table 3 **Methods for experimental verification of RNAi screen results**

Method	Examples	Rationale
Retest the reagents with the same assay	Test several replicates (including a re-synthesized or new batch of reagent); test single reagents in arrayed format after a pooled approach	Reagent-level verification
Retest with a related assay and/or different cell type	Switch the reporters in a dual-reporter assay; test a different cell line, marker, or antibody; test in a different cell line	Reagent-level verification
Retest with unique reagents	Test reagents designed to target different regions of the gene	Gene-level verification (confidence increases when more than one works)
Assay small molecule(s)	Test a known inhibitor of the gene product in the assay; test small molecules in parallel with RNAi and compare pathways implicated in each	Gene-level verification (correlation is suggestive of an on-target effect)
Determine mRNA or protein levels in the presence of the RNAi reagent	Q-PCR or immunoblotting[a]	Gene-level verification (correlation between knockdown and phenotype is suggestive of an on-target effect)
Rescue in the presence of the RNAi reagent[b]	Test rescue with a genomic fragment, cDNA, or open reading frame construct that evades RNAi knockdown	Gene-level verification (rescue demonstrates an on-target effect)
Pattern of gene expression of mRNAs corresponding to hits	Q-PCR or microarray in specific cell types, stages, and/or tissues	Gene-level verification (expression in relevant tissues or stages is suggestive of a relevant finding)
Pattern of expression of the proteins corresponding to hits	Immunoblotting in specific cell types, stages, and/or tissues	Gene-level verification (expression in relevant tissues or stages is suggestive of a relevant finding)
Subcellular distribution of proteins corresponding to hits	GFP-tagged construct or immunofluorescence	Gene-level verification (expression in relevant subcellular compartments is indicative of a relevant finding)
RNAi-induced phenotype in another species	Test effect of knockdown of homologs in mammalian cells as a follow-up to a nonmammalian cell screen	Gene-level verification (similar phenotype provides compelling evidence of a biologically relevant finding)
Correlation with a related disease or disorder	Map disease-associated regions, mutations, and amplifications	Gene- and pathway-level verification (disease association is indicative of a relevant finding)
Protein-protein interactions	Coimmunoprecipitation, mass spectrometry, yeast two-hybrid screen	Gene- and pathway level verification (physical interactions among newly identified proteins or between new and established players are indicative of a relevant finding)
Genetic analysis in vivo	Test effects of mutations of gene hits in whole animals (same or different species than primary screen cells)	Gene and pathway-level verification [related phenotype provides compelling evidence of a relevant effect and can help refine the role(s) of the genes in specific pathways, events, or behaviors]

[a]Abbreviation: GFP, green fluorescent protein; Q-PCR, Quantitative Reverse Transcriptase PCR.

[b]The "gold standard" approach to verification of an RNAi result at the gene level; similar to a genetic test for complementation.

differences among HIV screen datasets might suggest that different subsets of reagents in the different libraries used for those screens may have resulted in robust knockdown, such that the datasets cannot be compared in a straightforward manner. Consistent with this, there seems to be significant variability in the robustness of specific RNAi reagents for mammalian RNAi HTSs, including both among and between sets of siRNAs, shRNAs, and bifunctional or miRNA-like reagents (12). The specific cell type, assay, and biological process being tested are also relevant. For example, another source of false discovery was recognized via analysis of two different screens in *Drosophila* cells that both interrogate the JAK/STAT pathway but had little overlap among screen hits. Comparison of the screen datasets revealed that there can be inherent bias in the assays and/or specific biological functions being addressed (84). Moreover, even when there is robust knockdown, cells may respond in a manner that makes it impossible to assay the process of interest. For example, in a screen for factors involved in hypoxia, several nonessential TOR pathway components were isolated; however, as knockdown of PTEN results in cell death, a role for PTEN in hypoxia could not be addressed (85).

As it seems likely that all of the above-mentioned factors (i.e., cell type, assay design, experimental noise, assay design, and the choice of reagent library) contribute to false discovery, it is important to continue to learn how to best address them. If robustness of reagents is a major contributing factor, then improving RNAi libraries such that each reagent is effective (and, ideally, knockdown occurs at comparable levels gene to gene) may in turn improve RNAi HTS results. There is variability in the effectiveness of RNAi reagents designed using a single set of tools, such that, in the absence of gene-by-gene testing (see, for example, Reference 86), this is currently difficult to accomplish. Recent evidence that miRNA-like approaches result in more robust knockdown, however, as well as using specific strategies designed to address difficult-to-target genes, may contribute to improved library design (12, 87). Moreover, problems related to pleiotropy might be overcome, at least in part, via screening in multiple cell lines, which are likely to have different essential requirements (see, for example, Reference 88).

Another approach to successfully overcome problems of false discovery is to combine RNAi with the results of other high-throughput methods, including overexpression screening, protein-protein interaction analysis, genomic analysis (e.g., mapping of disease-associated genomic regions or amplifications), and mRNA expression array data (see approaches outlined in **Figure 1*d–f***). These include projects in which new experimental studies were done in conjunction with a screen (see, for example, References 50, 89–94). In addition, bioinformatics-based analyses have been applied to preselect a set of genes to be tested or identify a high-confidence set of primary or verified hits (see, for example, References 89, 90, 95, 96).

Both genome-scale RNAi HTSs (**Figure 1*a,b***) and genome- or smaller-scale screens, combined with more integrated and/or multiplexed approaches (**Figure 1*c–f***), have led to important insights into a number of biological topics. Particular progress has been made in general cellular functions, signal transduction, cancer biology, and host cell responses to infection by bacterial, fungal, eukaryotic, or viral pathogens. Below, we highlight key findings from a large number of RNAi HTSs performed in *Drosophila* or mammalian cells (see also **Table 1**).

Biological Findings from Screens

Despite the relative infancy of cell-based RNAi HTSs, particularly as compared with more classic genetic screens, the approach has already resulted in a large number of studies with significant impact in a wide variety of fields. Indeed, many screens have investigated basic cellular processes, including how cells survive, proliferate, and divide, and more specialized cellular functions, such as responding to specific viral, bacterial, and fungal pathogens; surviving specific chemotherapeutic treatments; and generating pigment. Below and in **Table 1**, we summarize the results of many large-scale RNAi HTSs in mammalian and *Drosophila* screens published to date. Then, we discuss what has (and has not been) learned thus far about gene function and interactions at a system-wide level from the results of these studies.

Cell viability, proliferation, and cancer. Cell proliferation and survival are fundamental processes that are particularly relevant to human diseases such as cancer. Moreover, routine methods for detecting cell number, viability, and/or basic metabolic readouts in an automated fashion are well established. Thus, it is perhaps not surprising that screens aimed at identifying genes required for cell proliferation and survival were among the earliest cell-based RNAi HTS studies (97). More recently, a number of kinome-wide and other studies have identified factors essential for survival and/or proliferation of a number of mammalian cell types, with a particular emphasis on the requirements of cancer cells (44–46, 88, 92, 98–101). Each of these studies implicates several genes in cell survival and/or proliferation in various cell types. Taken together, the results of these studies highlight the different dependencies that various cell types have for survival, even when the cell lines differ only in expression of a specific factor such as the HPV16 E7 oncogene or mammalian von Hippel-Lindau (*VHL*) tumor suppressor gene (98, 99, 102).

At least one of the studies of cancer cell proliferation or survival looked at requirements for response to tumoricidal drugs (46). This and similar studies aimed at identifying factors required for death and/or survival in the presence of a specific chemical, hormone, or other treatment have yielded informative results. The findings include (*a*) identification of a requirement for mammalian PTPN1, NF1, SMARCB1, and SMARCE1 for the response of chronic myelogenous leukemia cells to imatinib (Gleevec) (46), a small-molecule inhibitor of BCR-ABL; (*b*) the sensitivity of MCF-10A breast cancer cells to disruption of DNA methyltransferase and proteasome activity (45); and (*c*) a role for *Drosophila* Sox14 as a positive regulator of hormone-induced cell death (95). A number of genes required for proliferation and/or viability of multiple cell lines were also identified in these studies, including more than 250 genes implicated as essential in 12 cancer cell lines (46), a set of 14 kinases implicated as essential across a diverse set of human cell types (88), and 19 genes essential in a set of three cancerous and one breast epithelial cell line (44).

Cancer biology. Extension of this type of approach to screening for drug sensitivity led to identification of factors that alter cellular responses to paclitaxel, a chemotherapeutic treatment used for breast cancer, and specifically implicated the genes *ACRBP*, *TUBGCP2*, and *MAD2* in various aspects of mitotic spindle assembly (103). In addition, a highly integrated approach that included not just cell-based RNAi HTSs to interrogate the kinome, but also overexpression and human genetic analyses, identified *IKBKE* as an oncogene in breast cancer (92). Furthermore, comparison of colorectal DLD-1 cancer cells with and without a mutant form of the oncogene *KRAS* point to sensitivity of Ras mutant cells to disruption of mitosis, including via disruption of ubiquitination and proteasome degradation of mitotic factors, in particular by way of perturbation of PLK1 (48). *KRAS* was also the subject of a genome-wide screen for epigenetic *Fas* silencing in K-*ras*-transformed mouse cells, which identified at least eight proteins (NPM2, TRIM66, ZFP354B, TSS, BMI1, DNMT1, SIRT6, and TRIM37) subsequently shown to bind *Fas* promoter regions (104). Another study identified genes that encode kinases such as CHK1 that, upon knockdown, can sensitize pancreatic cancer cells to treatment with the chemotherapeutic agent gemcitabine (105). Recently, RNAi screening of 30 patient samples (and 4 normal individuals for comparison) was applied to the study of leukemia and identified a number of genes associated with patient-specific sensitivities to downregulation of specific tyrosine kinases (106).

Stem cell biology. RNAi HTS assays for viability and/or proliferation, in addition to other assays, have also been applied to stem cells with the goal of identifying factors required for self-renewal (107–110). The results of these studies (reviewed in Reference 111) implicate the transcriptional regulators Cnot2 and Trim28 in self-renewal or differentiation of mouse stem cells (107) and implicate Paf1C in maintaining stem cell identity (109). The results of stem cell studies also suggest a role for human EXT1 in erythroid burst or colony formation (BFU-E or CFU-E) (108). Comparison of results from Hu et al. (107) and Ding et al. (109) further underscore the idea that even with similar assays and cells, screen results differ, a finding that Subramanian et al. (111) suggest may be attributable to the use of different libraries, or it may be that neither screen achieves saturation. A preliminary OCT4 regulatory network has been constructed on the basis of integrated analysis of these studies and other high-throughput datasets (112).

Cell division, cell death, and the cell cycle. Whereas many of the above studies used pooled screens or simple viability assays to detect changes in cell proliferation or viability, additional studies have used specific reporters, cell-cycle FACS analysis, or image-based assay readouts to uncover specific phenotypes related to cell division, cell death, or related processes. Among the findings of these studies are cell cycle roles for *Drosophila* kinases with known or putative roles in the cytoskeleton and signal transduction (113); a number of human kinases, phosphatases, and proteins involved in proteolysis (114); and the COP9 signalosome and the Wg/Wnt, MAPK, TOR and JAK/STAT signaling pathways (115). A number of signaling networks, including the TOR pathway (in *Drosophila*, Tor pathway), were implicated in a screen for survival of *Drosophila* cells after treatment with a DNA damaging agent (116). RNAi HTS studies in *Drosophila* cells also point to roles for a number of proteins in cell division and cell cycle events: HSET, a kinesin motor, was implicated in tumor-related multipolar mitoses (117); disruption of Polo or Centrosomin were found to block centrosome maturation (118). A number of proteins were implicated in centromere propagation, including at least three proteins shown to localize to centromeres, CAL1, CENP-C, and the mitotic cyclin CYCA (119); and the ubiquitin ligase SCF^{Slimb} was shown to regulate centriole duplication (96). Moreover, an early esiRNA-based screen in human HeLa cells identified several candidates involved in various aspects of mitosis and cytokinesis (120, 121).

Cell death and circadian rhythms. A study in mammalian cells has identified a number of genes required for RIP1 kinase-mediated necroptosis and apoptosis, including a role for the BH3-only Bcl-2 family member Bmf in cell death receptor-induced necroptosis (122). Cell death was also a specific focus of screens in *Drosophila* cells (123, 124), which have implicated metabolic regulators such as Charlatan and ARD1 in caspase activation (123) and identified *Tango7* and its mammalian homolog PCID1 (EIF3M) as effectors of cell death (124). In a cell-based assay for light-induced degradation of cryptochrome, three genes, subsequently validated in vivo, were identified as required for cryptochrome degradation: *Drosophila* CG17735, *ssh*, and *Bruce* (125). A study of circadian clock components in human cells linked circadian clocks to cancer and cell death via identification of ARNTL, a putative regulator of p53 (126). In addition, a screen for human kinases and phosphatases involved in the circadian clock identified casein kinase 2 as a circadian clock component that can phosphorylate PER2 (127). The role of calcium in cellular functions has also been the focus of several cell-based RNAi HTS studies in *Drosophila* cells. These studies shed light on calcium channels and homeostasis, including identification of a link between Draper-mediated phagocytosis and calcium homeostasis, and helped to identify the human immune deficiency-associated gene *Orai1* (128–132).

Cell death and stress responses. A number of screens have analyzed cellular responses to stress conditions. For example, a recent RNAi HTS study of conserved *Drosophila* genes assayed cellular responses to hypoxic conditions, leading to identification of Tor pathway components and *Protein tyrosine phosphatase 61F* as important for regulation of translation in response to hypoxia (85). In addition, a genome-wide study in mammalian cells that combined RNAi HTSs with expression analysis implicated TAF1 in regulation of apoptosis in response to genotoxic stress (42). In addition, a genome-wide pooled screen for mammalian genes, whose knockdown confers resistance to stress via the organic oxidant *tert*-butylhydroperoxide, linked retinol saturase, an enzyme that acts on vitamin A, to stress sensitivity (43). Finally, a "druggable genome"-wide image-based RNAi HTS allowed the simultaneous isolation of factors that disrupt formation of stress granules (SGs) and/or processing bodies (PBs) in response to arsenite treatment; components of the hexosamine biosynthetic pathway were implicated in SGs but not PB assembly, suggesting a role for *O*-linked *N*-acetylglucosamine modification in stress response (133).

Cell morphology, size, and adhesion. High-content image-based RNAi HTSs and other assays have also allowed a number of studies of cell morphology, cell size, and organelles, such as mitochondria. For example, an early screen in *Drosophila* cells identified several known and new genes with roles in cell shape and cytoskeletal regulation (27). A more recent report of factors involved in the control of cell size points to the Pvr, Ras, and Tor pathways as key growth regulators of cultured *Drosophila* cells (134). RNAi HTSs of the *Drosophila* kinome in multiple cell lines revealed general and cell type-specific genes required for cell morphology, including a role for *minibrain* (which encodes a homolog of the human DYRK1A protein) in a central nervous system-derived cell line (135). Coincidentally, DYRK-family kinases were also identified in a screen for components downstream of the NFAT signaling pathway (131). A screen in primary *Drosophila* neurons revealed genes with roles in outgrowth-related processes in *Drosophila* and mouse cells, including the vesicle trafficking genes *Sec61-alpha* and *Ran* (136). Huang et al. (40) used the pooled approach to identify cells that remain attached after induction of detachment via expression of a mutant form of c-Abl tyrosine kinase. Among the genes identified was *IL6ST*, which was previously implicated in cell-cell adhesion of another human cell type (40). A three-dimensional culture system was used with a pooled approach to identify mouse genes involved in metastasis. Among these genes was *Gas1*; the human homolog of *Gas1* is frequently downregulated in human metastatic melanoma cells and tumor samples (41).

Mitochondria and mitochondrial disease. Image-based RNAi HTSs in *Drosophila* cells identified a protein required for mitochondrial fission in *Drosophila* and mammalian cells, Mff (137). A luciferase reporter of citrate synthase (CS) activity was useful to identify genes subsequently confirmed to have in vivo relevance to mitochondrial CS activity: *barren*, *CG3249*, *HDAC6*, *klumpfuss*, *Rpd3 (HDAC1)*, *smt3*, *Src42A*, and *vimar* (138). Small-scale RNAi and overexpression screening with the subset of human genes that encode E3 ubiquitin ligases identified the mitochondrial MULAN (90). Another study that integrated multiple approaches (namely, combining mass spectrometry and bioinformatics to identify genes that were subsequently assayed via small-scale RNAi) identified mouse *C8orf38*, which is associated with an early lethal complex I–linked inherited human disease (93).

Other cell biological processes. A number of screens in *Drosophila* and mammalian cells have looked at additional aspects of cellular and molecular biology. For example, cells stably expressing a signal sequence–fused horseradish peroxidase were used to look for components required for protein secretion; four candidate proteins were shown to localize to the Golgi apparatus and seven to the endoplasmic reticulum (ER), and together the results led to identification of a number of previously uncharacterized "TANGO" genes (named for transport and Golgi organization) that should be interesting for further study (139). Image analysis was used in a genome-wide study of lipid droplets in *Drosophila* cells that implicated coat protein complex I (COPI) proteins also implicated in the Bard et al. study (139) and in host responses to infection (83). An RNAi HTS, with a set of 308 candidates identified in expression and literature analyses, led to identification of 20 genes implicated in homeostasis of cellular cholesterol levels, including *TMEM97* (89). Again in fly cells, full-genome RNAi HTS studies have identified genes involved in E2F repression, including *domino* (140), and genes required for SUMO-dependent transcriptional repression, which implicated a protein complex that includes MEP-1, Mi-2, and Sfmbt (141). Additionally, a recent genome-wide study in human melanocytes implicated a number of genes, including genes involved in tyrosinase expression and stability and in melanogenesis, which has relevance to a number of different human disorders (142).

Signal transduction. A number of screens that interrogated signal transduction have led to novel findings, including (*a*) identification of *Slpr2* in a screen for insulin signaling in adipocytes (143); (*b*) extensive analysis of pathways that intersect with JNK (49); (*c*) identification of novel components of the ERK (Map kinase) signaling pathway (144); (*d*) identification of kinase and phosphatase requirements in FOXO transcription factor regulation, which implicate protein kinase C (PKC) and glycogen synthase kinase 3β in insulin signaling (145); and (*e*) identification of *Drosophila moleskine (msk)* and implication of mammalian homologs of *msk* in TGF-β signaling (145a). Additionally, there have been at least three studies of Hedgehog (Hh) signaling, which implicate Dally-like protein (Dlp), PP2A, Cdc2l1, casein kinase 1α, and other kinases in aspects of Hh signaling (146–148); at least two studies of the JAK/STAT pathway (reviewed in References 84 and 149); and a large number of studies of the Wnt signal transduction pathway (reviewed in References 150 and 151).

Signal transduction: Wnt pathway. The Wnt pathway screen results are interesting in that they show that repeated RNAi HTS-based investigation of a pathway can continue to yield new and relevant findings. An early full-genome screen in *Drosophila* identified several candidates for Wnt regulation (152), and a screen focused on *Drosophila* transmembrane protein-encoding genes identified a conserved factor involved in Wnt secretion encoded by *evenness interrupted (evi)* (153). Several subsequent studies identified proteins required for subcellular localization of Wnt pathway components. A visual screen to detect disruption of the normal membrane localization of Dishevelled (Dsh) revealed the importance of pH and charge in recruitment of Dsh by Frizzled (154). And a screen for negative regulators of Wnt signaling identified Bili, which inhibits recruitment of Axin to Lrp6 during Wnt pathway activation (155). An arrayed screen for Wnt-related factors in mammalian cells revealed a link between transcription factor 7-like 2 and colorectal cancer (156). More recently, the results of concurrent small-molecule and RNAi screens in colorectal cancer cells identified Bruton's tyrosine kinase as an inhibitor of Wnt signaling that binds to the Wnt pathway component CDC73 (91). Additionally, a highly integrated and validated HTS RNAi screen in human DLD1 colon cancer cells led to identification of the nuclear chromatin-associated protein AGGF1, which has previously been implicated in human disease and was shown to be involved in β-catenin-mediated transcription in colon cancer cells (50).

RNA biology: RNAi. In an interesting overlap of experimental approach and biological topic, several cell-based RNAi HTSs have focused on RNA biology, including the study of RNAi mechanisms. Screens in vivo in *C. elegans* made early and notable contributions to this field, but screens in other systems have also contributed to our rapidly growing understanding of the mechanisms and control of RNAi, miRNAs, and other aspects of RNA biology. For example, studies in *Drosophila* cells include a screen for components required for uptake of dsRNA into cells, which revealed evolutionary conservation of the relevant genes in *C. elegans* (157), and a screen for factors that disrupt RNAi knockdown identified five previously described genes, including *AGO2* and *Hsc70–73* and two genes of unknown function (*CG17625* and *CG10883*) (158). A screen in *Drosophila* cells that focused on dissection of the miRNA pathways identified the conserved P-body component Ge-1 (159). Furthermore, multiple Argonaut-dependent pathways were interrogated in another screen, resulting in a new understanding of components shared among the pathways, with 54 genes in common for the siRNA, endo-siRNA, and miRNA pathways (160).

RNA biology: mRNA export and pre-mRNAs. Looking at different aspects of RNA biology, another *Drosophila* RNAi HTS addressed nuclear export of mRNA. The *Drosophila* PCI domain-containing protein (PCID2), which interacts with polysomes, was identified in the screen, and more generally, the list of genes identified in the screen emphasizes the links between mRNA export and other processes, including translation (161). A specific look at histone premRNAs in a *Drosophila* genome-wide RNAi HTSs using a low-content imaging approach identified factors previously known to be involved in mRNA cleavage and/or poyladenylation, i.e., zinc finger domain-containing and signaling genes as well as the histone variants H2Av and H3.3A/B (162).

Innate immunity and host-pathogen interactions. Innate immunity has been the focus of at least two RNAi HTSs in *Drosophila* screens, leading to the identification of *IAP2*, implicating the IAP family in nonapoptotic pathways (163); a new protein with IAP-like functions, Defense repressor 1; and the conserved gene *sickie*, required for Relish activation (164). Perhaps the largest category of RNAi HTSs in *Drosophila* and mammalian cells is for investigations of host cell responses to infection by viral, bacterial, fungal, or other pathogens (recently reviewed in References 8 and 165). In these screens, researchers typically treated a responsive cell type with a pathogen, followed by detection of a readout related to the pathogen and/or the host cell response, such as changes in the extent of localization, internalization, or proliferation of the pathogen; host cell death; or a specific host cell transcriptional or other response to infection. Screening in *Drosophila* cells has the advantage of being relevant not only to the conserved aspects of human host cell response but also to the responses of other dipterans (e.g., biting flies and mosquitoes) that act as disease vectors (reviewed in References 8, 165, 166). And screening directly in human cells is beginning to provide insights into significant threats to human health such as HIV (47, 78–80).

Responses to bacterial pathogens. Early studies in *Drosophila* cells looked at host-pathogen interactions relevant to bacterial infection. The findings included identification of Peste, a CD36 family protein involved in uptake into a *Drosophila* macrophage-like cell type; implicated the ESCRT (endosomal sorting complex required for transport) machinery in infection by mycobacteria (167, 168); and identified several candidates for interaction with *Listeria* (169). The latter study was followed up by RNAi in human cells, using a small library of RNAi reagents targeting putative vesicle trafficking genes; among the findings were 18 genes involved in vesicular trafficking of *Listeria* in a manner independent of listeriolysin O (170). RNAi interrogation of the human kinome implicated the AKT1 pathway in growth of the bacterial pathogens *Salmonella typhimurium* and *Mycobacterium* *tuberculosis* (171). Simultaneous RNAi knockdown of membrane trafficking components or knockdown of single components involved in ER-associated degradation affected *Legionella* replication in *Drosophila* cells (172). Genome-wide RNAi HTSs in *Drosophila* cells with another human disease-relevant infectious agent, *Chlamydia*, uncovered the importance of the multiprotein Tim-Tom complex, which is required for import of nuclear-encoded proteins into the mitochondria, in both *Drosophila* and mammalian cells (173).

Responses to other pathogens. Of course, RNAi HTSs have not been limited to bacterial pathogens. RNAi HTS interrogation of host cell-virus interactions has been the focus of several studies (see below). Additionally, RNAi HTSs in *Drosophila* cells demonstrated the feasibility of RNAi HTSs with the human fungal pathogen, *Candida albicans* (174). In addition, a kinome-wide RNAi HTS in human hepatoma cells was done to study the eukaryotic disease agent of malaria, *Plasmodium*. Following identification of candidates in the screen, candidates were tested in vivo by RNAi in mice, and the results support the idea that PKCζ is involved in invasion of hepatocytes by *Plasmodium* sporozoites (175).

Viral pathogen studies in *Drosophila* cells. RNAi HTSs in *Drosophila* have contributed to our general understanding of host cell responses to infection by viruses and have led to identification of specific factors. For example, an early screen utilizing an internal ribosome entry site (IRES)-containing virus, *Drosophila* C virus, implicated a large number of *Drosophila* ribosomal proteins in *Drosophila* C virus growth, a finding shown to have implications for treatment and for human infection by the IRES-containing poliovirus (176). Recently, *Ars2* (for arsenite resistance gene 2) was identified as a key component required for viral immunity in *Drosophila* cells (177). A modified influenza virus that can infect *Drosophila* cells was similarly used in a full-genome study to look at the influenza life cycle in host cells, and an extension of the results to human cells demonstrated the relevance of results with human homologs of several genes identified in the screen, including *ATP6V0D1*, *COX6A1*, and *NXF1* (178). Relevance to both human cells and a dipteran mosquito vector of human dengue virus (DENV), *Aedes aegypti*, was demonstrated after a full-genome screen with a DENV in *Drosophila* cells; specifically, 42 of the 82 genes identified in the *Drosophila* screen for which human homologs could be identified also acted as host factors for DENV in human cells (37). Additional studies have implicated the COPI coatmer in both viral replication and lipid homeostatis (82, 179). Thus, screening in *Drosophila* cells has rich potential for informing our understanding of general processes and specific genes involved in the interaction of viruses with their insect and human hosts, with likely implications for the development of therapeutic treatments.

Viral pathogen studies in mammalian cells. Recent studies in human cells have looked at infections with West Nile virus (180) and hepatitis C virus (181, 182), and at least four studies have focused on HIV (47, 78–80). For West Nile virus, researchers identified roles in various stages of infection or replication for the ubiquitin ligase CBLL1 and the monocarboxylic acid transporter MCT4, and this research implicated the ERAD pathway for transport of misfolded pathways previously implicated in the flavivirus life cycle (180). The COPI and PI4KA were implicated in a genome-wide RNAi HTS for factors involved in hepatitis C virus replication (181). The results of RNAi HTSs with HIV are at once encouraging and cautionary. On the one hand, taken together, the results of the screens implicate common activities and pathways with putative roles in various aspects of HIV infection, including genes involved in nuclear transport, GTP binding, and protein complex assembly, and point to a large number of potential therapeutic targets (47, 78–81). On the other hand, the degree of overlap among the screens is fairly modest at the gene level (81). This may reflect various sources of false discovery, including screen noise, timing of the assays, and methods of analysis (81), and shows that even large-scale screens fail to identify all genes required for a given process (i.e., the screens are not saturating). It also emphasizes the need to compare data among related screens and to combine standard RNAi HTSs with additional approaches, particularly when the goal of a study is to gain a systems-wide understanding of gene functions relevant to a given biological process or behavior.

COMBINATORIAL AND SENSITIZED SCREENS TO REVEAL FUNCTION

Combining multiple screening strategies provides an excellent means to identify genes that are relevant to particular cellular behaviors that may have been missed in single RNAi HTS assays (**Figure 1*d–f***). For example, the same assay can be performed iteratively in the same cell type in a sensitized background (e.g., in which another gene has been targeted by RNAi or overexpressed in all cells); under different environmental conditions (e.g., via exposure of cells to a small-molecule inhibitor); or across different cell types (which can be compared with screening in very different genetic backgrounds). Such screens are conceptually similar to studies in yeast, whereby the viability of gene deletions has been tested on backgrounds where other genes have been deleted (synthetic lethality) or overexpressed, and in the presence of various small molecules. Indeed, systematic analysis of gene pairs has revealed that the number of gene pairs required for viability far exceeds the number of single genes that are essential (reviewed in Reference 183), suggesting that taking a similar approach in RNAi HTSs will help identify many new pairs or sets of genes with related functions.

Many groups have begun to utilize combinatorial approaches in cell-based RNAi screens, not only to comprehensively describe the components of particular signaling networks but also to attempt to understand how the phenotypic output of a signaling molecule or network is dependent on the surrounding context. For example, how activation of molecules, such as Ras, ERK, JNK, and PI3K, can lead to proliferation in one cell type but to differentiation, migration, or apoptosis in others is unclear. To this end, Bakal et al. (49) recently implemented a combinatorial approach to describe genes involved in the regulation of JNK activity across different genetic backgrounds. In that study, JNK activity was monitored in live *Drosophila* cells using a fluorescence resonance energy transfer-based reporter. Systematic targeting of all *Drosophila* kinases and phosphatases individually by RNAi resulted in the isolation of 24 genes that regulate JNK activity in normal growth conditions (5% of genes tested). However, by performing the same screen 12 subsequent times in the presence of a second RNAi, which targeted a particular "query" gene, 55 more kinase and phosphatase regulators of JNK activity (17% of genes tested) were identified. A striking aspect to these screens is the difference in both the number and identity of the genes that were isolated in any given condition. For example, in that study, although 11 genes were identified as JNK suppressors following single RNAi, 17 enhancer genes were isolated in *Rac1*-deficient cells, 54 in *slpr/MLK* deficient cells, and 3 in *hippo* deficient cells. Similarly, combinatorial and integrative approaches to study of the ERBB network in cancer cells have been discussed recently by Sahin and colleagues (184, 185). The results of these studies dramatically illustrate how the results of a single RNAi screen are dependent on genetic context. Even inhibition of a single additional gene in the same cell line can significantly alter the final hit list of an RNAi screen. This suggests that the results from any screen in a specific cell type must be interpreted with the caveats that the genes identified could be highly specific to the cell type that was screened and that multiple cell types should be screened iteratively in order to comprehensively identify genes involved in specific cellular behaviors.

Genes that are repeatedly isolated in many screens are likely to correspond to the core regulators of a particular process. For example, the JNK phosphatase-encoding gene *puckered* (186) was repeatedly isolated in more than one *Drosophila* screen (including in *Rac1*-, *MLK*-, and *hippo*-deficient cells), reflecting the central role of JNK phosphatase as a JNK regulator in the majority of tissues and organisms studied to date. The isolation of genes only in specific backgrounds should provide insight into how genetic interactions modulate phenotypic outputs. Similar observations have been made in recent screens for regulators of ERK activation downstream of epidermal growth factor activation, where a genome-wide screen in the Kc cell line isolated 1405 genes, the identical screen in the S2R+ line isolated 1101 genes, and 422 genes were common to both screens (144). Another example is from a recent screen for genes required for the viability of four different cancer lines, which identified 1057 genes across the lines, revealing both a core of 23 genes required for viability across lines, as well as genes uniquely required for viability in each case (44). Thus, iterative screening is essential to understand how regulatory networks are capable of dynamically rewiring to maintain cellular function in the face of large genetic and environmental fluctuations and how subtle sequence variations lead to overt phenotypic differences.

CONCLUDING REMARKS AND PERSPECTIVES

As demonstrated by the various findings described above, the application of RNAi HTSs to gene discovery has been extremely successful in interrogating important biological questions. Collectively, the results of RNAi HTSs reaffirm the ideas that many gene networks are involved in a given process (e.g., cell viability, signal transduction, host-pathogen interactions) and that the various pathways and complexes represented in screen results are highly interlinked and interdependent. Both the issue of false discovery and the complexity of biological systems point to the importance of limiting false discovery rates. Improving RNAi reagent libraries, particularly for mammalian cell screens, is one important direction to follow toward limiting false discovery. However, there is no substitute for verification of results. Toward that goal, the development of improved, high-throughput methodologies for verification of RNAi results at the RNAi reagent and gene level, as well as for validation of RNAi results in vivo, will also be of significant benefit. Undoubtedly, the results of RNAi HTSs will continue to provide important contributions to biology and biomedicine in the future, as researchers both generate new screen data and perform reanalyses and meta-analyses of existing screen datasets, either alone or in conjunction with other studies. Although much of what has been learned is currently understood only at the level of individual genes or small functional networks or complexes, the scale and scope of studies made possible by RNAi HTSs, along with the emergence and refinement of other genomic, transcriptomic, and proteomic techniques, should also allow us to move toward system-wide understandings of gene networks in an increasing number of specific cell types, tissues, and organisms. In addition, RNAi screening may facilitate a shift toward genome-scale investigation in a larger number of different species, and in disease-related or similar studies, in the specific species of interest (or at the least, a closely related species) rather than in a distantly related model system. This is likely to be particularly true for organisms of specific industrial, agricultural, and biomedical importance. Thus, one can speculate that additional genome-scale collections may be generated in the future, such as for fungi of agricultural and industrial relevance (reviewed in Reference 18), parasitic nematodes (reviewed in Reference 187), and dipteran vectors of human disease [i.e., biting flies and mosquitoes (reviewed in References 165, 188, 189)].

SUMMARY POINTS

1. RNAi has emerged as a powerful method for genome-scale interrogation of gene function in a number of traditional and emerging model systems, including but not limited to *Drosophila* and mammalian cells.
2. The results of RNAi high-throughput screens (HTSs) are acutely sensitive to assay design and are subject to significant rates of false discovery, which can be addressed using various statistical, bioinformatics, and experimental approaches.
3. For high-throughput cell-based RNAi screens in *Drosophila* and human cells, primary hits (positive results) can be verified in high- or moderate-throughput modes to confirm the RNAi result; however, validation of the biological relevance of a given finding (i.e., via methods other than RNAi) remains principally a low-throughput process that would benefit from development of new technologies.
4. The results of high-throughput cell-based RNAi screens have led to discoveries in a wide variety of fields, with particular impacts in signal transduction, general cell biology, RNA biology, cancer biology, and host cell responses to infectious pathogens.

FUTURE ISSUES

1. Improved gene models, large-scale experimental verification of mRNA knockdown, new insights into effective RNAi reagent design, and the subsequent building of better genome-scale RNAi libraries are needed to support more effective RNAi screens.
2. Improved methods for RNAi reagent delivery, such as for difficult-to-transfect mammalian cell types, would open the door to screening in additional cellular contexts.
3. Continued efforts at standardization of data reporting, access to information about reagent design and efficacy, and availability of raw data from RNAi HTSs, along with the subsequent improved bioinformatics-based analyses would likely help inform our understanding of RNAi results.
4. New methods for large-scale verification and validation of screen results should be developed, and existing methods should be improved, made more affordable, and include development of genome-scale libraries for rescue experiments.
5. Screening pipelines should incorporate innovative, integrative, and multiplexed approaches, such as via concurrent RNAi and overexpression screens; screening of multiple cell lines, pathogens, drug treatments and/or assays; and capture of multiple assay or image readouts per screen.
6. Integration of high-quality results from other high-throughput datasets (e.g., microarray analysis, protein-protein interaction studies, genomic analyses) is needed to maximize the power of high-throughput approaches and gain a system-wide understanding of gene networks involved in various processes, events, and behaviors.

DISCLOSURE STATEMENT

The authors are not aware of any affiliations, memberships, funding, or financial holdings that might be perceived as affecting the objectivity of this review.

ACKNOWLEDGMENTS

We are grateful to Matthew Booker and Dr. Caroline Shamu for thoughtful comments on the manuscript. S.E.M. is supported by R01 GM067761 from the National Institute of General Medical Sciences at the National Institutes of Health. C.B. was a Fellow of the Leukemia and Lymphoma Society and is currently at the Institute of Cancer Research. N.P. is an investigator of the Howard Hughes Medical Institute.

LITERATURE CITED

1. Fire A, Xu S, Montgomery MK, Kostas SA, Driver SE, Mello CC. 1998. Potent and specific genetic interference by double-stranded RNA in *Caenorhabditis elegans*. *Nature* 391:806–11
2. Zamore PD, Tuschl T, Sharp PA, Bartel DP. 2000. RNAi: double-stranded RNA directs the ATP-dependent cleavage of mRNA at 21 to 23 nucleotide intervals. *Cell* 101:25–33
3. Carthew RW, Sontheimer EJ. 2009. Origins and mechanisms of miRNAs and siRNAs. *Cell* 136:642–55
4. Kim VN, Han J, Siomi MC. 2009. Biogenesis of small RNAs in animals. *Nat. Rev. Mol. Cell Biol.* 10:126–39
5. Bartel DP. 2009. MicroRNAs: target recognition and regulatory functions. *Cell* 136:215–33
6. Smibert P, Lai EC. 2008. Lessons from microRNA mutants in worms, flies and mice. *Cell Cycle* 7:2500–8
7. Umbach JL, Cullen BR. 2009. The role of RNAi and microRNAs in animal virus replication and antiviral immunity. *Genes Dev.* 23:1151–64
8. Cherry S. 2009. What have RNAi screens taught us about viral-host interactions? *Curr. Opin. Microbiol.* 12:446–52
9. Grimm D. 2009. Small silencing RNAs: state-of-the-art. *Adv. Drug Deliv. Rev.* 61:672–703
10. Lee SK, Kumar P. 2009. Conditional RNAi: towards a silent gene therapy. *Adv. Drug Deliv. Rev.* 61:650–64
11. Gondi CS, Rao JS. 2009. Concepts in in vivo siRNA delivery for cancer therapy. *J. Cell Physiol.* 220:285–91
12. Rao DD, Vorhies JS, Senzer N, Nemunaitis J. 2009. siRNA vs. shRNA: similarities and differences. *Adv. Drug Deliv. Rev.* 61:746–59
13. Singh SK, Hajeri PB. 2009. siRNAs: their potential as therapeutic agents—Part II. Methods of delivery. *Drug Discov. Today* 14:859–65
14. Hajeri PB, Singh SK. 2009. siRNAs: their potential as therapeutic agents—Part I. Designing of siRNAs. *Drug Discov. Today* 14:851–58
15. Brodersen P, Voinnet O. 2006. The diversity of RNA silencing pathways in plants. *Trends Genet.* 22:268–80
16. Small I. 2007. RNAi for revealing and engineering plant gene functions. *Curr. Opin. Biotechnol.* 18:148–53
17. Hamilton AJ, Baulcombe DC. 1999. A species of small antisense RNA in posttranscriptional gene silencing in plants. *Science* 286:950–52
18. Nakayashiki H, Nguyen QB. 2008. RNA interference: roles in fungal biology. *Curr. Opin. Microbiol.* 11:494–502
19. Romano N, Macino G. 1992. Quelling: transient inactivation of gene expression in *Neurospora crassa* by transformation with homologous sequences. *Mol. Microbiol.* 6:3343–53
20. Newmark PA, Reddien PW, Cebria F, Sanchez Alvarado A. 2003. Ingestion of bacterially expressed double-stranded RNA inhibits gene expression in planarians. *Proc. Natl. Acad. Sci. USA* 100(Suppl. 1):11861–65

21. Newmark PA. 2005. Opening a new can of worms: a large-scale RNAi screen in planarians. *Dev. Cell* 8:623–24
22. Kamath RS, Fraser AG, Dong Y, Poulin G, Durbin R, et al. 2003. Systematic functional analysis of the *Caenorhabditis elegans* genome using RNAi. *Nature* 421:231–37
23. Hammond SM, Bernstein E, Beach D, Hannon GJ. 2000. An RNA-directed nuclease mediates post-transcriptional gene silencing in *Drosophila* cells. *Nature* 404:293–96
24. Armknecht S, Boutros M, Kiger A, Nybakken K, Mathey-Prevot B, Perrimon N. 2005. High-throughput RNA interference screens in *Drosophila* tissue culture cells. *Methods Enzymol.* 392:55–73
25. Elbashir SM, Harborth J, Lendeckel W, Yalcin A, Weber K, Tuschl T. 2001. Duplexes of 21-nucleotide RNAs mediate RNA interference in cultured mammalian cells. *Nature* 411:494–98
26. Martin SE, Caplen NJ. 2007. Applications of RNA interference in mammalian systems. *Annu. Rev. Genomics Hum. Genet.* 8:81–108
27. Kiger AA, Baum B, Jones S, Jones MR, Coulson A, et al. 2003. A functional genomic analysis of cell morphology using RNA interference. *J. Biol.* 2:27
28. Perrimon N, Mathey-Prevot B. 2007. Applications of high-throughput RNA interference screens to problems in cell and developmental biology. *Genetics* 175:7–16
29. Boutros M, Ahringer J. 2008. The art and design of genetic screens: RNA interference. *Nat. Rev. Genet.* 9:554–66
30. Echeverri CJ, Perrimon N. 2006. High-throughput RNAi screening in cultured cells: a user's guide. *Nat. Rev. Genet.* 7:373–84
31. Kim SY, Hahn WC. 2007. Cancer genomics: integrating form and function. *Carcinogenesis* 28:1387–92
32. Iorns E, Lord CJ, Turner N, Ashworth A. 2007. Utilizing RNA interference to enhance cancer drug discovery. *Nat. Rev. Drug Discov.* 6:556–68
33. Wolters NM, MacKeigan JP. 2008. From sequence to function: using RNAi to elucidate mechanisms of human disease. *Cell Death Differ.* 15:809–19
34. Dasgupta R, Perrimon N. 2004. Using RNAi to catch *Drosophila* genes in a web of interactions: insights into cancer research. *Oncogene* 23:8359–65
35. Timmons L, Court DL, Fire A. 2001. Ingestion of bacterially expressed dsRNAs can produce specific and potent genetic interference in *Caenorhabditis elegans*. *Gene* 263:103–12
36. Wang J, Barr MM. 2005. RNA interference in *Caenorhabditis elegans*. *Methods Enzymol.* 392:36–55
37. Sessions OM, Barrows NJ, Souza-Neto JA, Robinson TJ, Hershey CL, et al. 2009. Discovery of insect and human dengue virus host factors. *Nature* 458:1047–50
38. Lord CJ, Martin SA, Ashworth A. 2009. RNA interference screening demystified. *J. Clin. Pathol.* 62:195–200
39. Sharma S, Rao A. 2009. RNAi screening: tips and techniques. *Nat. Immunol.* 10:799–804
40. Huang X, Wang JY, Lu X. 2008. Systems analysis of quantitative shRNA-library screens identifies regulators of cell adhesion. *BMC Syst. Biol.* 2:49
41. Gobeil S, Zhu X, Doillon CJ, Green MR. 2008. A genome-wide shRNA screen identifies *GAS1* as a novel melanoma metastasis suppressor gene. *Genes Dev.* 22:2932–40
42. Kimura J, Nguyen ST, Liu H, Taira N, Miki Y, Yoshida K. 2008. A functional genome-wide RNAi screen identifies TAF1 as a regulator for apoptosis in response to genotoxic stress. *Nucleic Acids Res.* 36:5250–59
43. Nagaoka-Yasuda R, Matsuo N, Perkins B, Limbaeck-Stokin K, Mayford M. 2007. An RNAi-based genetic screen for oxidative stress resistance reveals retinol saturase as a mediator of stress resistance. *Free Radic. Biol. Med.* 43:781–88
44. Schlabach MR, Luo J, Solimini NL, Hu G, Xu Q, et al. 2008. Cancer proliferation gene discovery through functional genomics. *Science* 319:620–24
45. Silva JM, Marran K, Parker JS, Silva J, Golding M, et al. 2008. Profiling essential genes in human mammary cells by multiplex RNAi screening. *Science* 319:617–20
46. Luo B, Cheung HW, Subramanian A, Sharifnia T, Okamoto M, et al. 2008. Highly parallel identification of essential genes in cancer cells. *Proc. Natl. Acad. Sci. USA* 105:20380–85

47. Yeung ML, Houzet L, Yedavalli VS, Jeang KT. 2009. A genome-wide short hairpin RNA screening of Jurkat T-cells for human proteins contributing to productive HIV-1 replication. *J. Biol. Chem.* 284:19463–73
48. Luo J, Emanuele MJ, Li D, Creighton CJ, Schlabach MR, et al. 2009. A genome-wide RNAi screen identifies multiple synthetic lethal interactions with the Ras oncogene. *Cell* 137:835–48
49. Bakal C, Linding R, Llense F, Heffern E, Martin-Blanco E, et al. 2008. Phosphorylation networks regulating JNK activity in diverse genetic backgrounds. *Science* 322:453–56
50. Major MB, Roberts BS, Berndt JD, Marine S, Anastas J, et al. 2008. New regulators of Wnt/β-catenin signaling revealed by integrative molecular screening. *Sci. Signal.* 1:ra12
51. Nowotny M, Yang W. 2009. Structural and functional modules in RNA interference. *Curr. Opin. Struct. Biol.* 19:286–93
52. Tilesi F, Fradiani P, Socci V, Willems D, Ascenzioni F. 2009. Design and validation of siRNAs and shRNAs. *Curr. Opin. Mol. Ther.* 11:156–64
53. Buchholz F, Kittler R, Slabicki M, Theis M. 2006. Enzymatically prepared RNAi libraries. *Nat. Methods* 3:696–700
54. Lehner B, Fraser AG, Sanderson CM. 2004. Technique review: how to use RNA interference. *Brief. Funct. Genomic Proteomics* 3:68–83
55. Clark J, Ding S. 2006. Generation of RNAi libraries for high-throughput screens. *J. Biomed. Biotechnol.* 2006:45716
56. Ramadan N, Flockhart I, Booker M, Perrimon N, Mathey-Prevot B. 2007. Design and implementation of high-throughput RNAi screens in cultured *Drosophila* cells. *Nat. Protoc.* 2:2245–64
57. DasGupta R, Gonsalves FC. 2008. High-throughput RNAi screen in *Drosophila*. *Methods Mol. Biol.* 469:163–84
58. Steinbrink S, Boutros M. 2008. RNAi screening in cultured *Drosophila* cells. *Methods Mol. Biol.* 420:139–53
59. O'Rourke EJ, Conery AL, Moy TI. 2009. Whole-animal high-throughput screens: the *C. elegans* model. *Methods Mol. Biol.* 486:57–75
60. Birmingham A, Selfors LM, Forster T, Wrobel D, Kennedy CJ, et al. 2009. Statistical methods for analysis of high-throughput RNA interference screens. *Nat. Methods* 6:569–75
61. Wiles AM, Ravi D, Bhavani S, Bishop AJ. 2008. An analysis of normalization methods for *Drosophila* RNAi genomic screens and development of a robust validation scheme. *J. Biomol. Screen.* 13:777–84
62. Boutros M, Bras LP, Huber W. 2006. Analysis of cell-based RNAi screens. *Genome Biol.* 7:R66
63. Stone DJ, Marine S, Majercak J, Ray WJ, Espeseth A, et al. 2007. High-throughput screening by RNA interference: control of two distinct types of variance. *Cell Cycle* 6:898–901
64. Zimmermann TS, Lee AC, Akinc A, Bramlage B, Bumcrot D, et al. 2006. RNAi-mediated gene silencing in non-human primates. *Nature* 441:111–14
65. Zhang XD, Espeseth AS, Johnson EN, Chin J, Gates A, et al. 2008. Integrating experimental and analytic approaches to improve data quality in genome-wide RNAi screens. *J. Biomol. Screen.* 13:378–89
66. Zhang XD, Heyse JF. 2009. Determination of sample size in genome-scale RNAi screens. *Bioinformatics* 25:841–44
67. Zhang XD, Kuan PF, Ferrer M, Shu X, Liu YC, et al. 2008. Hit selection with false discovery rate control in genome-scale RNAi screens. *Nucleic Acids Res.* 36:4667–79
68. Wang L, Tu Z, Sun F. 2009. A network-based integrative approach to prioritize reliable hits from multiple genome-wide RNAi screens in *Drosophila*. *BMC Genomics* 10:220
69. Echeverri CJ, Beachy PA, Baum B, Boutros M, Buchholz F, et al. 2006. Minimizing the risk of reporting false positives in large-scale RNAi screens. *Nat. Methods* 3:777–79
70. Jackson AL, Bartz SR, Schelter J, Kobayashi SV, Burchard J, et al. 2003. Expression profiling reveals off-target gene regulation by RNAi. *Nat. Biotechnol.* 21:635–37
71. Birmingham A, Anderson EM, Reynolds A, Ilsley-Tyree D, Leake D, et al. 2006. 3′ UTR seed matches, but not overall identity, are associated with RNAi off-targets. *Nat. Methods* 3:199–204
72. Kulkarni MM, Booker M, Silver SJ, Friedman A, Hong P, et al. 2006. Evidence of off-target effects associated with long dsRNAs in *Drosophila melanogaster* cell-based assays. *Nat. Methods* 3:833–38

73. Ma Y, Creanga A, Lum L, Beachy PA. 2006. Prevalence of off-target effects in *Drosophila* RNA interference screens. *Nature* 443:359–63
74. ENCODE Project Consortium. 2004. The ENCODE (ENCyclopedia of DNA elements) Project. *Science* 306:636–40
75. Birney E, Stamatoyannopoulos JA, Dutta A, Guigo R, Gingeras TR, et al. 2007. Identification and analysis of functional elements in 1% of the human genome by the ENCODE pilot project. *Nature* 447:799–816
76. Thomas DJ, Rosenbloom KR, Clawson H, Hinrichs AS, Trumbower H, et al. 2007. The ENCODE Project at UC Santa Cruz. *Nucleic Acids Res.* 35:D663–67
77. Celniker SE, Dillon LA, Gerstein MB, Gunsalus KC, Henikoff S, et al. 2009. Unlocking the secrets of the genome. *Nature* 459:927–30
78. Brass AL, Dykxhoorn DM, Benita Y, Yan N, Engelman A, et al. 2008. Identification of host proteins required for HIV infection through a functional genomic screen. *Science* 319:921–26
79. Konig R, Zhou Y, Elleder D, Diamond TL, Bonamy GM, et al. 2008. Global analysis of host-pathogen interactions that regulate early-stage HIV-1 replication. *Cell* 135:49–60
80. Zhou H, Xu M, Huang Q, Gates AT, Zhang XD, et al. 2008. Genome-scale RNAi screen for host factors required for HIV replication. *Cell Host Microbe* 4:495–504
81. Bushman FD, Malani N, Fernandes J, D'Orso I, Cagney G, et al. 2009. Host cell factors in HIV replication: meta-analysis of genome-wide studies. *PLoS Pathog.* 5:e1000437
82. Beller M, Sztalryd C, Southall N, Bell M, Jackle H, et al. 2008. COPI complex is a regulator of lipid homeostasis. *PLoS Biol.* 6:e292
83. Guo Y, Walther TC, Rao M, Stuurman N, Goshima G, et al. 2008. Functional genomic screen reveals genes involved in lipid-droplet formation and utilization. *Nature* 453:657–61
84. Muller P, Boutros M, Zeidler MP. 2008. Identification of JAK/STAT pathway regulators–insights from RNAi screens. *Semin. Cell Dev. Biol.* 19:360–69
85. Lee SJ, Feldman R, O'Farrell PH. 2008. An RNA interference screen identifies a novel regulator of target of rapamycin that mediates hypoxia suppression of translation in *Drosophila* S2 cells. *Mol. Biol. Cell* 19:4051–61
86. Lee G, Santat LA, Chang MS, Choi S. 2009. RNAi methodologies for the functional study of signaling molecules. *PLoS One* 4:e4559
87. Krueger U, Bergauer T, Kaufmann B, Wolter I, Pilk S, et al. 2007. Insights into effective RNAi gained from large-scale siRNA validation screening. *Oligonucleotides* 17:237–50
88. Grueneberg DA, Degot S, Pearlberg J, Li W, Davies JE, et al. 2008. Kinase requirements in human cells: I. Comparing kinase requirements across various cell types. *Proc. Natl. Acad. Sci. USA* 105:16472–77
89. Bartz F, Kern L, Erz D, Zhu M, Gilbert D, et al. 2009. Identification of cholesterol-regulating genes by targeted RNAi screening. *Cell Metab.* 10:63–75
90. Li W, Bengtson MH, Ulbrich A, Matsuda A, Reddy VA, et al. 2008. Genome-wide and functional annotation of human E3 ubiquitin ligases identifies MULAN, a mitochondrial E3 that regulates the organelle's dynamics and signaling. *PLoS One* 3:e1487
91. James RG, Biechele TL, Conrad WH, Camp ND, Fass DM, et al. 2009. Bruton's tyrosine kinase revealed as a negative regulator of Wnt-beta-catenin signaling. *Sci. Signal.* 2:ra25
92. Boehm JS, Zhao JJ, Yao J, Kim SY, Firestein R, et al. 2007. Integrative genomic approaches identify *IKBKE* as a breast cancer oncogene. *Cell* 129:1065–79
93. Pagliarini DJ, Calvo SE, Chang B, Sheth SA, Vafai SB, et al. 2008. A mitochondrial protein compendium elucidates complex I disease biology. *Cell* 134:112–23
94. Pearlberg J, Degot S, Endege W, Park J, Davies J, et al. 2005. Screens using RNAi and cDNA expression as surrogates for genetics in mammalian tissue culture cells. *Cold Spring Harb. Symp. Quant. Biol.* 70:449–59
95. Chittaranjan S, McConechy M, Hou YC, Freeman JD, Devorkin L, Gorski SM. 2009. Steroid hormone control of cell death and cell survival: molecular insights using RNAi. *PLoS Genet.* 5:e1000379
96. Rogers GC, Rusan NM, Roberts DM, Peifer M, Rogers SL. 2009. The SCF^{Slimb} ubiquitin ligase regulates Plk4/Sak levels to block centriole reduplication. *J. Cell Biol.* 184:225–39

97. Boutros M, Kiger AA, Armknecht S, Kerr K, Hild M, et al. 2004. Genome-wide RNAi analysis of growth and viability in *Drosophila* cells. *Science* 303:832–35
98. Baldwin A, Li W, Grace M, Pearlberg J, Harlow E, et al. 2008. Kinase requirements in human cells: II. Genetic interaction screens identify kinase requirements following HPV16 E7 expression in cancer cells. *Proc. Natl. Acad. Sci. USA* 105:16478–83
99. Bommi-Reddy A, Almeciga I, Sawyer J, Geisen C, Li W, et al. 2008. Kinase requirements in human cells: III. Altered kinase requirements in *VHL-/-* cancer cells detected in a pilot synthetic lethal screen. *Proc. Natl. Acad. Sci. USA* 105:16484–89
100. Grueneberg DA, Li W, Davies JE, Sawyer J, Pearlberg J, Harlow E. 2008. Kinase requirements in human cells: IV. Differential kinase requirements in cervical and renal human tumor cell lines. *Proc. Natl. Acad. Sci. USA* 105:16490–95
101. Ngo VN, Davis RE, Lamy L, Yu X, Zhao H, et al. 2006. A loss-of-function RNA interference screen for molecular targets in cancer. *Nature* 441:106–10
102. Manning BD. 2009. Challenges and opportunities in defining the essential cancer kinome. *Sci. Signal.* 2:pe15
103. Whitehurst AW, Bodemann BO, Cardenas J, Ferguson D, Girard L, et al. 2007. Synthetic lethal screen identification of chemosensitizer loci in cancer cells. *Nature* 446:815–19
104. Gazin C, Wajapeyee N, Gobeil S, Virbasius CM, Green MR. 2007. An elaborate pathway required for Ras-mediated epigenetic silencing. *Nature* 449:1073–77
105. Azorsa DO, Gonzales IM, Basu GD, Choudhary A, Arora S, et al. 2009. Synthetic lethal RNAi screening identifies sensitizing targets for gemcitabine therapy in pancreatic cancer. *J. Transl. Med.* 7:43
106. Tyner JW, Deininger MW, Loriaux MM, Chang BH, Gotlib JR, et al. 2009. RNAi screen for rapid therapeutic target identification in leukemia patients. *Proc. Natl. Acad. Sci. USA* 106:8695–700
107. Hu G, Kim J, Xu Q, Leng Y, Orkin SH, Elledge SJ. 2009. A genome-wide RNAi screen identifies a new transcriptional module required for self-renewal. *Genes Dev.* 23:837–48
108. Ali N, Karlsson C, Aspling M, Hu G, Hacohen N, et al. 2009. Forward RNAi screens in primary human hematopoietic stem/progenitor cells. *Blood* 113:3690–95
109. Ding L, Paszkowski-Rogacz M, Nitzsche A, Slabicki MM, Heninger AK, et al. 2009. A genome-scale RNAi screen for Oct4 modulators defines a role of the Paf1 complex for embryonic stem cell identity. *Cell Stem Cell* 4:403–15
110. Fazzio TG, Huff JT, Panning B. 2008. An RNAi screen of chromatin proteins identifies Tip60-p400 as a regulator of embryonic stem cell identity. *Cell* 134:162–74
111. Subramanian V, Klattenhoff CA, Boyer LA. 2009. Screening for novel regulators of embryonic stem cell identity. *Cell Stem Cell* 4:377–78
112. Chavez L, Bais AS, Vingron M, Lehrach H, Adjaye J, Herwig R. 2009. In silico identification of a core regulatory network of OCT4 in human embryonic stem cells using an integrated approach. *BMC Genomics* 10:314
113. Bettencourt-Dias M, Giet R, Sinka R, Mazumdar A, Lock WG, et al. 2004. Genome-wide survey of protein kinases required for cell cycle progression. *Nature* 432:980–87
114. Mukherji M, Bell R, Supekova L, Wang Y, Orth AP, et al. 2006. Genome-wide functional analysis of human cell-cycle regulators. *Proc. Natl. Acad. Sci. USA* 103:14819–24
115. Bjorklund M, Taipale M, Varjosalo M, Saharinen J, Lahdenpera J, Taipale J. 2006. Identification of pathways regulating cell size and cell-cycle progression by RNAi. *Nature* 439:1009–13
116. Ravi D, Wiles AM, Bhavani S, Ruan J, Leder P, Bishop AJ. 2009. A network of conserved damage survival pathways revealed by a genomic RNAi screen. *PLoS Genet.* 5:e1000527
117. Kwon M, Godinho SA, Chandhok NS, Ganem NJ, Azioune A, et al. 2008. Mechanisms to suppress multipolar divisions in cancer cells with extra centrosomes. *Genes Dev.* 22:2189–203
118. Dobbelaere J, Josue F, Suijkerbuijk S, Baum B, Tapon N, Raff J. 2008. A genome-wide RNAi screen to dissect centriole duplication and centrosome maturation in *Drosophila*. *PLoS Biol.* 6:e224
119. Erhardt S, Mellone BG, Betts CM, Zhang W, Karpen GH, Straight AF. 2008. Genome-wide analysis reveals a cell cycle-dependent mechanism controlling centromere propagation. *J. Cell Biol.* 183:805–18
120. Kittler R, Buchholz F. 2005. Functional genomic analysis of cell division by endoribonuclease-prepared siRNAs. *Cell Cycle* 4:564–67

121. Kittler R, Putz G, Pelletier L, Poser I, Heninger AK, et al. 2004. An endoribonuclease-prepared siRNA screen in human cells identifies genes essential for cell division. *Nature* 432:1036–40
122. Hitomi J, Christofferson DE, Ng A, Yao J, Degterev A, et al. 2008. Identification of a molecular signaling network that regulates a cellular necrotic cell death pathway. *Cell* 135:1311–23
123. Yi CH, Sogah DK, Boyce M, Degterev A, Christofferson DE, Yuan J. 2007. A genome-wide RNAi screen reveals multiple regulators of caspase activation. *J. Cell Biol.* 179:619–26
124. Chew SK, Chen P, Link N, Galindo KA, Pogue K, Abrams JM. 2009. Genome-wide silencing in *Drosophila* captures conserved apoptotic effectors. *Nature* 460:123–27
125. Sathyanarayanan S, Zheng X, Kumar S, Chen CH, Chen D, et al. 2008. Identification of novel genes involved in light-dependent CRY degradation through a genome-wide RNAi screen. *Genes Dev.* 22:1522–33
126. Mullenders J, Fabius AW, Madiredjo M, Bernards R, Beijersbergen RL. 2009. A large scale shRNA barcode screen identifies the circadian clock component ARNTL as putative regulator of the p53 tumor suppressor pathway. *PLoS One* 4:e4798
127. Maier B, Wendt S, Vanselow JT, Wallach T, Reischl S, et al. 2009. A large-scale functional RNAi screen reveals a role for CK2 in the mammalian circadian clock. *Genes Dev.* 23:708–18
128. Cuttell L, Vaughan A, Silva E, Escaron CJ, Lavine M, et al. 2008. Undertaker, a *Drosophila* Junctophilin, links Draper-mediated phagocytosis and calcium homeostasis. *Cell* 135:524–34
129. Zhang SL, Yeromin AV, Zhang XH, Yu Y, Safrina O, et al. 2006. Genome-wide RNAi screen of Ca^{2+} influx identifies genes that regulate Ca^{2+} release-activated Ca^{2+} channel activity. *Proc. Natl. Acad. Sci. USA* 103:9357–62
130. Feske S, Gwack Y, Prakriya M, Srikanth S, Puppel SH, et al. 2006. A mutation in Orai1 causes immune deficiency by abrogating CRAC channel function. *Nature* 441:179–85
131. Gwack Y, Sharma S, Nardone J, Tanasa B, Iuga A, et al. 2006. A genome-wide *Drosophila* RNAi screen identifies DYRK-family kinases as regulators of NFAT. *Nature* 441:646–50
132. Vig M, Peinelt C, Beck A, Koomoa DL, Rabah D, et al. 2006. CRACM1 is a plasma membrane protein essential for store-operated Ca^{2+} entry. *Science* 312:1220–23
133. Ohn T, Kedersha N, Hickman T, Tisdale S, Anderson P. 2008. A functional RNAi screen links *O*-GlcNAc modification of ribosomal proteins to stress granule and processing body assembly. *Nat. Cell Biol.* 10:1224–31
134. Sims D, Duchek P, Baum B. 2009. PDGF/VEGF signaling controls cell size in *Drosophila*. *Genome Biol.* 10:R20
135. Liu T, Sims D, Baum B. 2009. Parallel RNAi screens across different cell lines identify generic and cell type-specific regulators of actin organization and cell morphology. *Genome Biol.* 10:R26
136. Sepp KJ, Hong P, Lizarraga SB, Liu JS, Mejia LA, et al. 2008. Identification of neural outgrowth genes using genome-wide RNAi. *PLoS Genet.* 4:e1000111
137. Gandre-Babbe S, van der Bliek AM. 2008. The novel tail-anchored membrane protein Mff controls mitochondrial and peroxisomal fission in mammalian cells. *Mol. Biol. Cell* 19:2402–12
138. Chen J, Shi X, Padmanabhan R, Wang Q, Wu Z, et al. 2008. Identification of novel modulators of mitochondrial function by a genome-wide RNAi screen in *Drosophila melanogaster*. *Genome Res.* 18:123–36
139. Bard F, Casano L, Mallabiabarrena A, Wallace E, Saito K, et al. 2006. Functional genomics reveals genes involved in protein secretion and Golgi organization. *Nature* 439:604–7
140. Lu J, Ruhf ML, Perrimon N, Leder P. 2007. A genome-wide RNA interference screen identifies putative chromatin regulators essential for E2F repression. *Proc. Natl. Acad. Sci. USA* 104:9381–86
141. Stielow B, Sapetschnig A, Kruger I, Kunert N, Brehm A, et al. 2008. Identification of SUMO-dependent chromatin-associated transcriptional repression components by a genome-wide RNAi screen. *Mol. Cell* 29:742–54
142. Ganesan AK, Ho H, Bodemann B, Petersen S, Aruri J, et al. 2008. Genome-wide siRNA-based functional genomics of pigmentation identifies novel genes and pathways that impact melanogenesis in human cells. *PLoS Genet.* 4:e1000298
143. Tu Z, Argmann C, Wong KK, Mitnaul LJ, Edwards S, et al. 2009. Integrating siRNA and protein-protein interaction data to identify an expanded insulin signaling network. *Genome Res.* 19:1057–67

144. Friedman A, Perrimon N. 2006. A functional RNAi screen for regulators of receptor tyrosine kinase and ERK signalling. *Nature* 444:230–34
145. Mattila J, Kallijarvi J, Puig O. 2008. RNAi screening for kinases and phosphatases identifies FoxO regulators. *Proc. Natl. Acad. Sci. USA* 105:14873–78
145a. Xu L, Yao X, Chen X, Lu P, Zhang B, Ip YT. 2007. Msk is required for nuclear import of TGF-β/BMP-activated Smads. *J. Cell Biol.* 178:981–94
146. Evangelista M, Lim TY, Lee J, Parker L, Ashique A, et al. 2008. Kinome siRNA screen identifies regulators of ciliogenesis and hedgehog signal transduction. *Sci. Signal.* 1:ra7
147. Nybakken K, Vokes SA, Lin TY, McMahon AP, Perrimon N. 2005. A genome-wide RNA interference screen in *Drosophila melanogaster* cells for new components of the Hh signaling pathway. *Nat. Genet.* 37:1323–32
148. Lum L, Yao S, Mozer B, Rovescalli A, Von Kessler D, et al. 2003. Identification of Hedgehog pathway components by RNAi in *Drosophila* cultured cells. *Science* 299:2039–45
149. Muller P, Kuttenkeuler D, Gesellchen V, Zeidler MP, Boutros M. 2005. Identification of JAK/STAT signalling components by genome-wide RNA interference. *Nature* 436:871–75
150. Brown AM. 2005. Canonical Wnt signaling: High-throughput RNAi widens the path. *Genome Biol.* 6:231
151. Dasgupta R. 2009. Functional genomic approaches targeting the wnt signaling network. *Curr. Drug Targets* 10:620–31
152. DasGupta R, Kaykas A, Moon RT, Perrimon N. 2005. Functional genomic analysis of the Wnt-wingless signaling pathway. *Science* 308:826–33
153. Bartscherer K, Pelte N, Ingelfinger D, Boutros M. 2006. Secretion of Wnt ligands requires Evi, a conserved transmembrane protein. *Cell* 125:523–33
154. Simons M, Gault WJ, Gotthardt D, Rohatgi R, Klein TJ, et al. 2009. Electrochemical cues regulate assembly of the Frizzled/Dishevelled complex at the plasma membrane during planar epithelial polarization. *Nat. Cell Biol.* 11:286–94
155. Kategaya LS, Changkakoty B, Biechele T, Conrad WH, Kaykas A, et al. 2009. Bili inhibits Wnt/β-catenin signaling by regulating the recruitment of Axin to LRP6. *PLoS One* 4:e6129
156. Tang W, Dodge M, Gundapaneni D, Michnoff C, Roth M, Lum L. 2008. A genome-wide RNAi screen for Wnt/β-catenin pathway components identifies unexpected roles for TCF transcription factors in cancer. *Proc. Natl. Acad. Sci. USA* 105:9697–702
157. Saleh MC, van Rij RP, Hekele A, Gillis A, Foley E, et al. 2006. The endocytic pathway mediates cell entry of dsRNA to induce RNAi silencing. *Nat. Cell Biol.* 8:793–802
158. Dorner S, Lum L, Kim M, Paro R, Beachy PA, Green R. 2006. A genomewide screen for components of the RNAi pathway in *Drosophila* cultured cells. *Proc. Natl. Acad. Sci. USA* 103:11880–85
159. Eulalio A, Rehwinkel J, Stricker M, Huntzinger E, Yang SF, et al. 2007. Target-specific requirements for enhancers of decapping in miRNA-mediated gene silencing. *Genes Dev.* 21:2558–70
160. Zhou R, Hotta I, Denli AM, Hong P, Perrimon N, Hannon GJ. 2008. Comparative analysis of Argonaute-dependent small RNA pathways in *Drosophila*. *Mol. Cell* 32:592–99
161. Farny NG, Hurt JA, Silver PA. 2008. Definition of global and transcript-specific mRNA export pathways in metazoans. *Genes Dev.* 22:66–78
162. Wagner EJ, Burch BD, Godfrey AC, Salzler HR, Duronio RJ, Marzluff WF. 2007. A genome-wide RNA interference screen reveals that variant histones are necessary for replication-dependent histone pre-mRNA processing. *Mol. Cell* 28:692–99
163. Gesellchen V, Kuttenkeuler D, Steckel M, Pelte N, Boutros M. 2005. An RNA interference screen identifies *Inhibitor of Apoptosis Protein 2* as a regulator of innate immune signalling in *Drosophila*. *EMBO Rep.* 6:979–84
164. Foley E, O'Farrell PH. 2004. Functional dissection of an innate immune response by a genome-wide RNAi screen. *PLoS Biol.* 2:E203
165. Prudêncio M, Lehmann MJ. 2009. Illuminating the host—How RNAi screens shed light on host-pathogen interactions. *Biotechnol. J.* 4:826–37
166. Cherry S. 2008. Genomic RNAi screening in *Drosophila* S2 cells: What have we learned about host-pathogen interactions? *Curr. Opin. Microbiol.* 11:262–70

167. Philips JA, Rubin EJ, Perrimon N. 2005. *Drosophila* RNAi screen reveals CD36 family member required for mycobacterial infection. *Science* 309:1251–53
168. Philips JA, Porto MC, Wang H, Rubin EJ, Perrimon N. 2008. ESCRT factors restrict mycobacterial growth. *Proc. Natl. Acad. Sci. USA* 105:3070–75
169. Agaisse H, Burrack LS, Philips JA, Rubin EJ, Perrimon N, Higgins DE. 2005. Genome-wide RNAi screen for host factors required for intracellular bacterial infection. *Science* 309:1248–51
170. Burrack LS, Harper JW, Higgins DE. 2009. Perturbation of vacuolar maturation promotes listeriolysin O-independent vacuolar escape during *Listeria monocytogenes* infection of human cells. *Cell. Microbiol.* 11:1382–98
171. Kuijl C, Savage ND, Marsman M, Tuin AW, Janssen L, et al. 2007. Intracellular bacterial growth is controlled by a kinase network around PKB/AKT1. *Nature* 450:725–30
172. Dorer MS, Kirton D, Bader JS, Isberg RR. 2006. RNA interference analysis of *Legionella* in *Drosophila* cells: exploitation of early secretory apparatus dynamics. *PLoS Pathog.* 2:e34
173. Derré I, Pypaert M, Dautry-Varsat A, Agaisse H. 2007. RNAi screen in *Drosophila* cells reveals the involvement of the Tom complex in *Chlamydia* infection. *PLoS Pathog.* 3:1446–58
174. Stroschein-Stevenson SL, Foley E, O'Farrell PH, Johnson AD. 2009. Phagocytosis of *Candida albicans* by RNAi-treated *Drosophila* S2 cells. *Methods Mol. Biol.* 470:347–58
175. Prudêncio M, Rodrigues CD, Hannus M, Martin C, Real E, et al. 2008. Kinome-wide RNAi screen implicates at least 5 host hepatocyte kinases in *Plasmodium* sporozoite infection. *PLoS Pathog.* 4:e1000201
176. Cherry S, Doukas T, Armknecht S, Whelan S, Wang H, et al. 2005. Genome-wide RNAi screen reveals a specific sensitivity of IRES-containing RNA viruses to host translation inhibition. *Genes Dev.* 19:445–52
177. Sabin LR, Zhou R, Gruber JJ, Lukinova N, Bambina S, et al. 2009. Ars2 regulates both miRNA- and siRNA-dependent silencing and suppresses RNA virus infection in *Drosophila*. *Cell* 138:340–51
178. Hao L, Sakurai A, Watanabe T, Sorensen E, Nidom CA, et al. 2008. *Drosophila* RNAi screen identifies host genes important for influenza virus replication. *Nature* 454:890–93
179. Cherry S, Kunte A, Wang H, Coyne C, Rawson RB, Perrimon N. 2006. COPI activity coupled with fatty acid biosynthesis is required for viral replication. *PLoS Pathog.* 2:e102
180. Krishnan MN, Ng A, Sukumaran B, Gilfoy FD, Uchil PD, et al. 2008. RNA interference screen for human genes associated with West Nile virus infection. *Nature* 455:242–45
181. Tai AW, Benita Y, Peng LF, Kim SS, Sakamoto N, et al. 2009. A functional genomic screen identifies cellular cofactors of hepatitis C virus replication. *Cell Host Microbe* 5:298–307
182. Li Q, Brass AL, Ng A, Hu Z, Xavier RJ, et al. 2009. A genome-wide genetic screen for host factors required for hepatitis C virus propagation. *Proc. Natl. Acad. Sci. USA* 106:16410–15
183. Boone C, Bussey H, Andrews BJ. 2007. Exploring genetic interactions and networks with yeast. *Nat. Rev. Genet.* 8:437–49
184. Sahin O, Wiemann S. 2009. Functional genomics and proteomics approaches to study the ERBB network in cancer. *FEBS Lett.* 583:1766–71
185. Sahin O, Frohlich H, Lobke C, Korf U, Burmester S, et al. 2009. Modeling ERBB receptor-regulated G1/S transition to find novel targets for de novo trastuzumab resistance. *BMC Syst. Biol.* 3:1
186. Martin-Blanco E, Gampel A, Ring J, Virdee K, Kirov N, et al. 1998. Puckered encodes a phosphatase that mediates a feedback loop regulating JNK activity during dorsal closure in *Drosophila*. *Genes Dev.* 12:557–70
187. Rosso MN, Jones JT, Abad P. 2009. RNAi and functional genomics in plant parasitic nematodes. *Annu. Rev. Phytopathol.* 47:207–32
188. Brown AE, Catteruccia F. 2006. Toward silencing the burden of malaria: progress and prospects for RNAi-based approaches. *Biotechniques* Apr.(Suppl.):38–44
189. Catteruccia F, Levashina EA. 2009. RNAi in the malaria vector, *Anopheles gambiae*. *Methods Mol. Biol.* 555:63–75
190. Baeg GH, Zhou R, Perrimon N. 2005. Genome-wide RNAi analysis of JAK/STAT signaling components in *Drosophila*. *Genes Dev.* 19:1861–70
191. Goto A, Matsushita K, Gesellchen V, El Chamy L, Kuttenkeuler D, et al. 2008. Akirins are highly conserved nuclear proteins required for NF-κB-dependent gene expression in *Drosophila* and mice. *Nat. Immunol.* 9:97–104

Nanomaterials Based on DNA

Nadrian C. Seeman

Department of Chemistry, New York University, New York, New York 10003;
email: ned.seeman@nyu.edu

Annu. Rev. Biochem. 2010. 79:65–87

First published online as a Review in Advance on March 11, 2010

The *Annual Review of Biochemistry* is online at biochem.annualreviews.org

This article's doi:
10.1146/annurev-biochem-060308-102244

0066-4154/10/0707-0065$20.00

Key Words

branched DNA, chemical organization by DNA, designed DNA lattices, DNA-based nanomechanical devices, DNA objects, sticky-ended cohesion

Abstract

The combination of synthetic stable branched DNA and sticky-ended cohesion has led to the development of structural DNA nanotechnology over the past 30 years. The basis of this enterprise is that it is possible to construct novel DNA-based materials by combining these features in a self-assembly protocol. Thus, simple branched molecules lead directly to the construction of polyhedrons, whose edges consist of double helical DNA and whose vertices correspond to the branch points. Stiffer branched motifs can be used to produce self-assembled two-dimensional and three-dimensional periodic lattices of DNA (crystals). DNA has also been used to make a variety of nanomechanical devices, including molecules that change their shapes and molecules that can walk along a DNA sidewalk. Devices have been incorporated into two-dimensional DNA arrangements; sequence-dependent devices are driven by increases in nucleotide pairing at each step in their machine cycles.

Contents

1. INTRODUCTION AND PREREQUISITES FOR STRUCTURAL DNA NANOTECHNOLOGY

All the readers of this volume are aware that DNA is the molecule that nature uses as genetic material. The iconic antiparallel double helical structure assumed by its two strands facilitates high-fidelity recognition between the nucleotides of complementary molecules. Although every base can pair with every other base, including itself (1), the Watson-Crick pairing (2) of adenine (A) with thymine (T) and guanine (G) with cytosine (C) appears to be the favored type of interaction between polynucleotides if the sequences of the molecules permit it. Biology clearly exploits this form of interaction in the replication of genetic information and in its expression. Nevertheless, biology is no longer the only branch of science where DNA is finding a significant role: It is now possible to exploit DNA complementarity to control the structure of matter.

This article reviews the history and current status of using DNA to construct novel nanomaterials and to control their structures over time. We describe below that DNA is used today to build specific objects, periodic lattices in two and three dimensions, and nanomechanical devices. The dimensions of DNA are inherently on the nanoscale: The diameter of the double helix is about 2 nm, and the helical pitch is about 3.5 nm; hence, construction involving DNA is fundamentally an exercise in nanoscience and nanotechnology. Consequently, the area we discuss is called "structural DNA nanotechnology," and its goal is the finest possible level of control over the spatial and temporal structure of matter: Putting what you want where you want it in three dimensions (3D), when you want it there.

Structural DNA nanotechnology rests on three pillars: (*a*) hybridization, (*b*) stably branched DNA, and (*c*) convenient synthesis of designed sequences.

1.1. Hybridization

The self-association of complementary nucleic acid molecules, or parts of molecules (3), is implicit in all aspects of structural DNA nanotechnology. Individual motifs are formed by the hybridization of strands designed to produce particular topological species. A key aspect of hybridization is the use of sticky-ended cohesion to combine pieces of linear duplex DNA; this has been a fundamental component of genetic engineering for over 35 years (4). Sticky-ended cohesion is illustrated in **Figure 1*a***, where two double helical molecules are shown to cohere by hydrogen bonding. Not only is hybridization critical to the formation of structure, but it is deeply involved in almost all the sequence-dependent nanomechanical devices that have been constructed, and it is central to many attempts to build structural motifs in a sequential fashion (5). Among the various types of cohesion known between biological molecules, sticky-ended cohesion is very special: Not only do we know that two sticky ends will cohere with each other in a specific and programmable fashion (affinity), but we also know the structure that they form when they do cohere, a Watson-Crick double helix (6). This key point is illustrated in **Figure 1*b***. Thus, in contrast to other biologically based affinity interactions (e.g., an antigen and an antibody), we know on a predictive basis the local product structure formed when sticky ends cohere, without the need to determine the crystal structure first to establish the relative orientation of the two components.

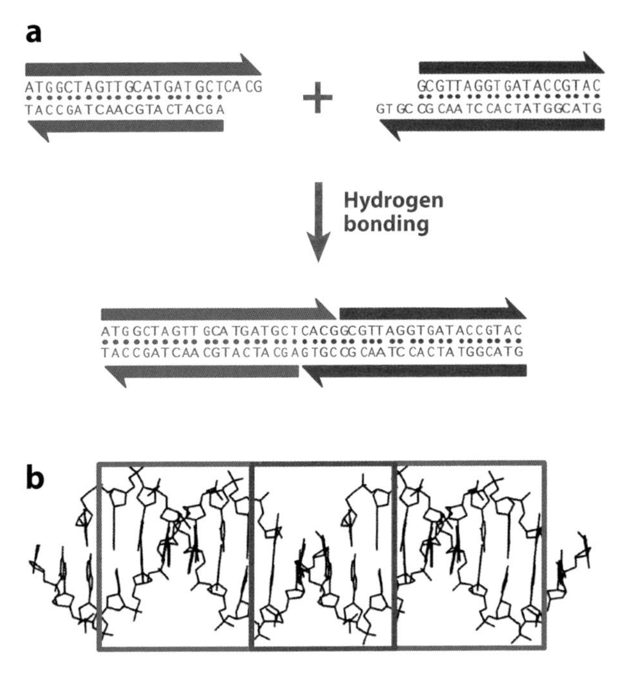

Figure 1

Sticky-ended cohesion. (*a*) Cohesion between two molecular overhangs. Two duplex molecules are shown, a red one and a blue one. Each has a single-stranded molecular overhang that is complementary to the overhang on the other molecule. When mixed, the two molecules can cohere in solution, as shown. (*b*) Structural features of stick-ended cohesion. A crystal structure (6) is shown that contains DNA decamers whose cohesion in the direction of the helix axis is directed by dinucleotide sticky ends. This interaction is seen readily in the red box, where the continuity of the chains is interrupted by gaps. The two blue boxes contain B-form duplex DNA. It is a half-turn away from the DNA in the red box, so it is upside down from it, but otherwise the structure is the same. Thus, sticky ends cohere to form B-DNA, and one can use this information in a predictive fashion to estimate the local structures of DNA constructs held together by sticky ends.

1.2. Stably Branched DNA

Although the DNA double helix is certainly the best-known structure in biology, little biology would occur if the DNA molecule were locked tightly into that structure with an unbranched helical axis. For example, triply branched replication forks occur during semiconservative replication (7), and four-arm branched Holliday junctions (8) are intermediates in genetic recombination. Likewise, branched DNA molecules are central to DNA nanotechnology. It is the combination of in vitro hybridization and synthetic branched DNA that leads to the ability to use DNA as a construction material (9). The fundamental notion behind DNA nanotechnology is illustrated in **Figure 2*a***, which shows the cohesion of four copies of a four-arm branched DNA molecule tailed in sticky ends associating to form a quadrilateral. In the example shown, only the inner sticky ends are used in forming the quadrilateral. Consequently, the structure can be extended to form an infinite lattice (9).

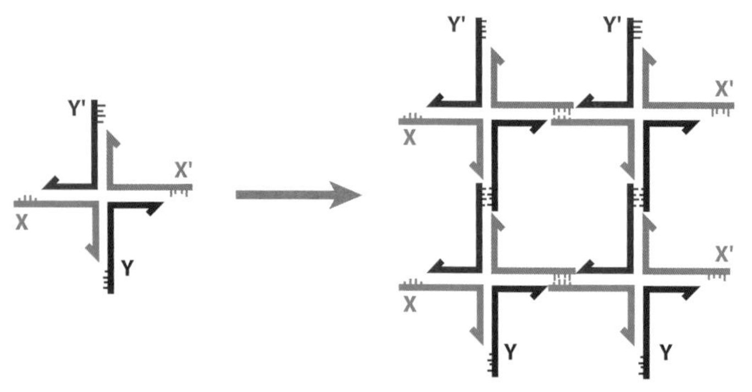

Figure 2

Self-assembly of branched DNA molecules to form larger arrangements. The image on the left shows a four-arm branched junction made from four differently colored strands. Its double helical domains are tailed in 5′ sticky ends labeled (clockwise from the left) X, Y′, X′, and Y; the sticky ends are indicated by small extensions from the main strand (our convention is to represent 3′ ends by arrowheads or, as here, by half arrowheads). The primed sticky ends complement the unprimed ones. The image on the right shows how four of these junctions can self-assemble through this complementarity to yield a quadrilateral. The sticky ends have come together in a complementary fashion. Note, this assembly does not use up all the available sticky ends, so that those that are left over could be used to generate a lattice in two dimensions (2D) and, indeed, in 3D.

1.3. Convenient Synthesis of Designed Sequences

Biologically derived branched DNA molecules, such as Holliday junctions, are inherently unstable because they exhibit sequence symmetry, i.e., the four strands actually consist of two pairs of strands with the same sequence. This symmetry enables an isomerization known as branch migration that allows the branch point to relocate (10). Branch migration can be eliminated if one chooses sequences that lack symmetry in the vicinity of the branch point. We discuss below different approaches to the use of symmetry in DNA nanotechnology, but the first approaches to DNA nanotechnology entailed sequence design that attempted to minimize sequence symmetry in every way possible. Such sequences are not readily obtained from natural sources, which leads to the third pillar supporting DNA nanotechnology, the synthesis of DNA molecules of arbitrary sequence (11). Fortunately, this is a capability that has existed for about as long as needed by this enterprise: Synthesis within laboratories or centralized facilities has been around since the 1980s. Today, it is possible to order all the DNA components needed for DNA nanotechnology, so long as they lack complex modifications, i.e., so-called "vanilla" DNA. In addition, the biotechnology enterprise has generated a demand for many variants on the theme of DNA (e.g., biotinylated molecules), and these molecules are also readily synthesized or purchased.

2. INITIAL STEPS IN THE PROCESS

There are two fundamental steps needed to perform projects in structural DNA nanotechnology: motif design and sequence design. In generating species more complex than the linear duplex DNA molecule, it is useful to have a protocol that leads to new DNA motifs in a convenient fashion; this protocol, based on reciprocal exchange, is presented in Section 2.1. Of course, whatever motif is designed, it must self-assemble from individual strands. Ultimately, it is necessary to assign sequences to the strands, sequences that will assemble into the designed motif, rather than some other structure. The sequence symmetry minimization procedure often used for sequence design is presented in Section 2.2.

2.1. Motif Design

Motif design relies on the operation of reciprocal exchange, the switching of the connections between DNA strands in two different double helices to produce a new connectivity. This notion is illustrated in **Figure 3*a*** where a red strand and a blue strand undergo reciprocal exchange to produce red-blue and blue-red strands. It is important to recognize that this is not an operation performed in the laboratory; it is done on paper or in the computer, and then the strands corresponding to the results of the operation are synthesized. Owing to the polar nature of DNA backbones, the operation can be performed between strands of the same polarity or between strands of opposite polarity. If only a single reciprocal exchange is performed, there is no difference, because the two products are just conformers of each other; however, if two or more operations are performed, different topologies result. Often a different ease of formation accompanies the two different topologies; empirically, the best-behaved molecules are those in which exchange takes place between strands of opposite polarity.

A number of important motifs generated this way are illustrated in **Figure 3*b***. The DX motif (12) with exchanges between strands of opposite polarity is shown at the upper left of that panel. This motif has been well characterized, and it is known that its persistence length is about twice that of a conventional linear duplex DNA (13). The DX+J motif is shown at the upper right of **Figure 3*b***. In this motif, the extra domain is usually oriented so as to be nearly perpendicular to the plane of the two helix axes; this orientation enables the domain to act as a topographic marker in the atomic force microscope (AFM). The motif shown at the bottom left of **Figure 3*b*** is the three-domain TX molecule. Below, we shall see that, by joining these motifs in a 1–3 fashion (the top helical domain of one molecule joins with the bottom domain of another molecule), two-dimensional arrays can be created that contain useful cavities (14, 15). By contrast with the DX and TX motifs shown, the PX motif and its topoisomer, the JX_2 motif (bottom right of **Figure 3*b***), result from reciprocal exchange between strands of the same polarity. In the case of the PX molecule, this happens everywhere that two double helices can

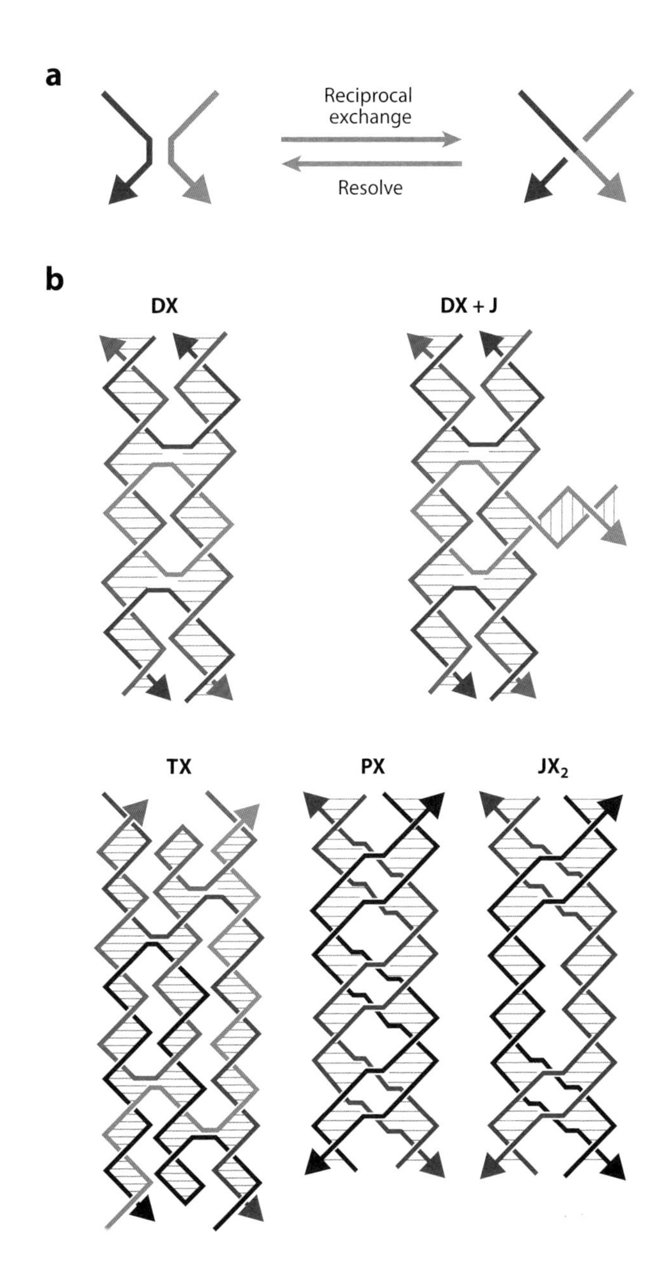

Figure 3

Motif generation by reciprocal exchange. (*a*) The fundamental operation. The basic operation of reciprocal exchange is shown: A red stand and a blue strand become a red-blue and a blue-red strand following the operation. (*b*) Motifs that can result from reciprocal exchange of DNA molecules. At the top of the panel are shown the DX motif and the DX+J motif. The DX motif results from two reciprocal exchanges between double helical motifs. The DX+J motif contains another DNA domain. Usually, this domain is oriented perpendicular to the plane of the two helix axes in the DX part of the motif. When this orientation is achieved, the extra domain can behave as a topographic marker for two-dimensional arrays containing the DX+J motif. The bottom row of the panel shows the TX motif at left, wherein a third domain has been added; again the exchanges take place between strands of opposite polarity. In the center and to the right are the PX motif and its topoisomer, the JX_2 motif. The PX molecule is formed by exchanges between strands of identical polarity at every possible position. The JX_2 molecule lacks two of these exchanges.

be juxtaposed; two exchanges are missing in the JX_2 molecule. These motifs are central to many of the robust nanomechanical devices that have been constructed (16).

2.2. Sequence Design and Symmetry Minimization

The design of DNA sequences that do not adhere to the strict linear duplex DNA paradigm is likely to result in molecules that correspond to excited states of some sort. Of course, the vast bulk of our knowledge about DNA structure and thermodynamics is predicated on a "ground state" linear duplex, rather than various excited states. It is evident that the goal of sequence design would naturally be to get the molecules to form the excited states that we seek to make. The free-energy cost of introducing a four-arm branch into a DNA molecule is not very high [+1.1 (±0.4) kcal mol^{-1} at 18°C in the presence of 10 mM Mg^{2+}] (17). However, before this was established, an effective method based on sequence symmetry minimization was worked out that has served well for the design of branched molecules (9, 18). This method is used for assigning small DNA motif sequences in many laboratories.

The basic approach to sequence symmetry minimization is illustrated in **Figure 4*a***, using the example of a four-arm branched junction built from four strands that each contain 16 nt (nucleotides). Each strand has been broken up into a series of 13 overlapping tetramers, so there are 52 tetramers in the entire molecule. The first two tetramers in strand one have been boxed, CGCA and GCAA. Sequence symmetry minimization would insist that each tetramer be unique. Furthermore, to ensure that the molecule cannot form linear duplex DNA at any point around the designed branch point, the linear complements to each of the 12 tetramers flanking the branch point are also forbidden. For example, the complement to the boxed CTGA (i.e., TCAG) is nowhere to be found in this molecule. Using these criteria, competition with the target structure (the four-octamer duplex components of the branch) will only come from trimers; thus, the boxed ATG segments could, in principle, be in the wrong place, but the free-energy differences between octamers and trimers win out, and the target is obtained without detectable impurities. In the case of four-arm branched molecules, branch migration may also be a problem (10), so the junction-flanking nucleotides are forbidden from having a twofold symmetric relationship among themselves. Other tricks used to avoid getting the wrong associations include: (*a*) the prevention of long stretches of Gs that could form other structures (particularly near crossover points) and (*b*) avoiding homopolymer tracts, polypurine or polypyrimidine tracts, alternating purine-pyrimidine tracts, and anything else that looks to the designer like it might be symmetric in the broadest sense of the word. Such precautions are clearly less important in the middle of a large stretch of linear duplex than they are near branch points.

In recent years, some issues involving symmetry minimization have been raised, either explicitly or implicitly. For example, in a 12-arm junction (**Figure 4*b***), it is not possible to flank the branch point with different base pairs, as it was with the four-arm junction. An entropic argument was used to design the junction in **Figure 4*b***, and it seems to have been successful (19). Mao and colleagues (20) have used physical restraints (e.g., strand length) and concentration, while maximizing symmetry, to obtain large arrangements of molecules designed to have as few as one strand. Both of these approaches consider nucleic acids as physical entities whose properties can be exploited; they represent advances from original considerations of nucleic acid sequences as arbitrary mathematical constructs. Arguably, the most dramatic example of ignoring sequence symmetry and just designing structures is Rothemund's DNA origami (21). We discuss DNA origami in more detail below, but basically, a long viral single strand, the sequence accepted as a given, is used to scaffold about 250 shorter strands to produce a two-dimensional (21) or three-dimensional shape, either with straight (22) or bent (23) features. Recently, Shih and coworkers (24) have

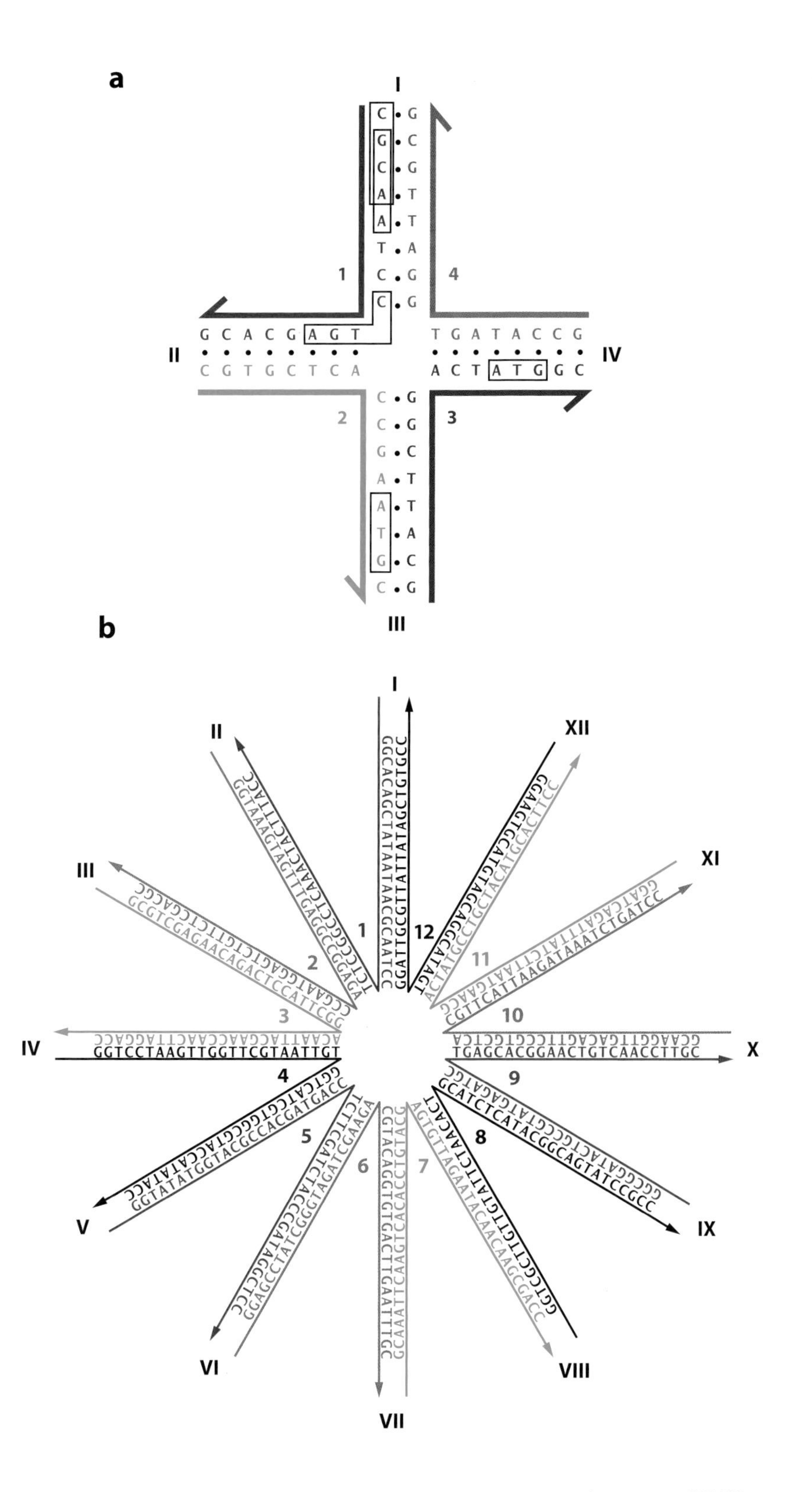

Figure 4

(*a*) Sequence design. The four-arm junction shown contains four 16 mers that are each broken up into 13 overlapping tetramers. Insisting that each tetramer be unique and that no tetramer complement those that flank the branch point leads to the formation of a stable branch, particularly if the twofold symmetry that enables branch migration is forbidden. Tetramers provide a "vocabulary" of 256 (less 16 self-complementary units) possible sequences to use, leading to competition from trimers. It would be difficult to design this molecule to have 56 unique trimers. Larger units clearly are to be used with larger constructs. (*b*) A 12-arm junction. It is not possible to eliminate symmetry around the center of this junction, so identical nucleotide pairs were spaced at four-step intervals around the junction.

demonstrated that two long complementary DNA molecules can be folded successfully into two different shapes before they rehybridize to form a linear duplex.

3. INDIVIDUAL CONSTRUCTS

DNA can be used to construct specific target structures, which are self-assembled and then may be ligated into closed species. These molecules may be characterized only by their topologies, but some have been characterized geometrically. These molecules are discussed in Section 3.1. Large constructs can be made by using a long scaffolding strand, which binds a large number of "helper" or "staple" strands that lock it into its final structure. This type of construction is known as DNA origami, and is described in Section 3.2.

3.1. Unscaffolded Targets

The earliest DNA constructs were best described as topological species, rather than geometrical species. This is because the earliest DNA motifs, i.e., branched junctions, were not robust but could be described as floppy if they were ligated (25, 26). Thus, both the linking and branching topologies of these molecules were well-defined because these features could be established by gels, but their detailed structures were not fixed. Idealized pictures of the first two structures, a cube (27) and a truncated octahedron (28), each with two double helical turns between vertices, are shown in **Figure 5**. All nicks in these molecules were sealed, and they were characterized topologically by denaturing gel electrophoresis. One of the issues with these structures is that they were not deltahedrons (polyhedrons whose faces are all triangles); deltahedrons, such as tetrahedrons (29), octahedrons (30), and icosahedrons (22), have all been produced during the current decade (22, 31). In addition, a protein has been encapsulated within a tetrahedron (32). The larger species have been characterized by electron microscopy. Other polyhedrons, such as DNA buckyballs (truncated icosahedrons), have been produced by carefully exploiting the interplay between junction flexibility and edge rigidity (31).

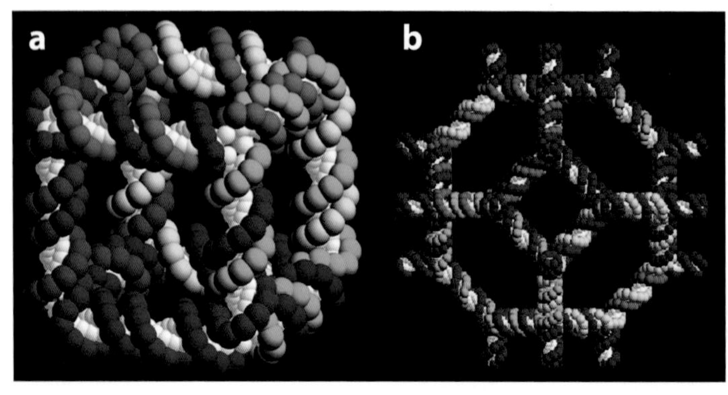

Figure 5

Early topological constructs built from DNA. (*a*) A cube-like molecule. This molecule is a hexacatenane; each edge corresponds to two double helical turns of DNA. Each backbone strand is drawn in a different color, and each one corresponds to a given face of the cube. Each is linked twice to each of the four strands that flank it, owing to the two-turn lengths of the edges. (*b*) A DNA-truncated octahedron. This molecule is a 14 catenane, again with each edge consisting of two turns of double helical DNA. Although the truncated octahedron has three edges flanking each vertex (i.e., it is "three connected"), it has been built using four-arm junctions.

The topological features of the early constructs also led to the development of single-stranded DNA topology. A crossover in a knot or a catenane can be regarded as being equivalent to a half-turn of DNA, which can be exploited accordingly (33). Thus, it has proved fairly simple to produce a variety of knotted molecules from single-stranded DNA (34), as well as a number of specifically linked catenanes. Some of these constructs have been used to characterize the topology of Holliday junctions (35). By using left-handed Z-DNA, it is possible to produce nodes of both signs in topological products. This aspect of DNA topology was exploited to produce the first Borromean rings from DNA (36). An RNA knot was used to discover that *Escherichia coli* DNA topoisomerase III (but not topoisomerase I) can act as an RNA topoisomerase (37).

3.2. DNA Origami

DNA origami entails the use of a scaffolding strand to which a series of smaller staple strands or helper strands are added. The staple strands are used to fold the scaffolding strand into a well-defined shape. The first example of a scaffolding strand in structural DNA nanotechnology was reported by Yan et al. (38), who used a scaffolding strand to make a one-dimensional barcode array. This application of scaffolding was followed quickly by Shih et al. (30), who used a long strand of DNA with five short helper strands to build an octahedron held together by PX cohesion (39). However, neither of these two advances had the dramatic impact of Rothemund's 2006 publication (21). He demonstrated that he was able to take single-stranded M13 viral DNA and get it to fold into a variety of shapes, including a smiley face (**Figure 6*a***). One of the great advantages of this achievement is that it creates an addressable surface area roughly 100 nm square. One can use the DX+J motif (**Figure 3*b***) developed for patterning two-dimensional arrays (40) (see below) to place patterns on DNA origami constructs. An example is seen in **Figure 6*b***, which shows a map of the Western Hemisphere. DNA origami has become widely used since it was introduced and has been employed for embedding nanomechanical devices (41), for making long six-helix bundles (42), for use as an aid to NMR structure determination (43), and for building three-dimensional objects (22), including a box that can be locked and unlocked, with potential uses in therapeutic delivery (44).

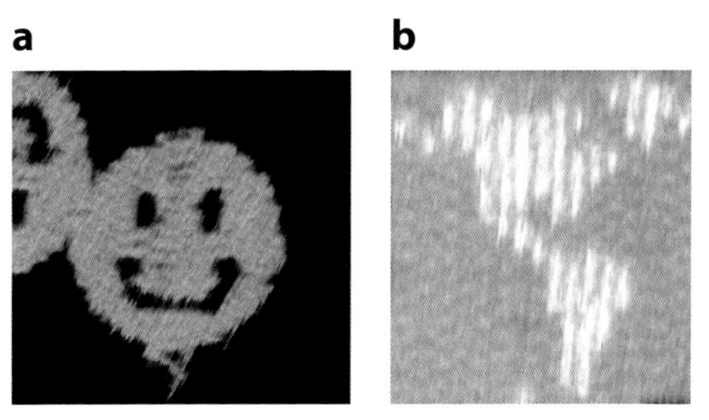

Figure 6

Atomic force micrographs of DNA origami constructs. (*a*) A smiley face. The scaffold strand, bound to helper strands, zigzags back and forth from left to right, yielding the structure seen. (*b*) A map of the Western Hemisphere. This image dramatically demonstrates the addressability of the DNA array. Each of the white pixels is made by a small DNA double helical domain, similar to the DX+J tile (**Figure 3**).

4. CRYSTALLINE ARRAYS

The original goal of structural DNA nanotechnology was to produce designed periodic matter (9). The first stage of this effort was the assembly of two-dimensional crystals from robust motifs. These two-dimensional crystals could be readily characterized by atomic force microscopy. Two-dimensional crystals are discussed in Section 4.1. Three-dimensional crystals have been assembled recently. They have been characterized by X-ray crystallographic methods, which require a more highy ordered sample than AFM. Three-dimensional crystals are discussed in Section 4.2.

4.1. Two-Dimensional Crystals

To generate periodic matter, it is necessary to have robust motifs that do not bend and flex readily; otherwise, a repeating pattern could fold up to form a cycle, poisoning the growth of the array. The DX molecule (see **Figure 3*b***) was the first molecule shown to be sufficiently rigid for this purpose (13, 45). The molecule was quickly exploited to produce periodic matter in two dimensions (40). The DX+J motif was used to impose patterns on these arrays. When DX+J molecules were included specifically in the pattern, deliberately striped features could be seen in the AFM, as shown in **Figure 7**. DNA motifs that form two-dimensional periodic (or aperiodic) arrays are often called "tiles" because they can tile the plane. There are numerous tiles that have been developed to tile the two-dimensional plane, including the three-domain TX tile (14), the six-helix hexagonal bundle (42), and the DX triangle (46). For reasons that are not well understood, two-dimensional crystals are relatively small, typically no more than a few micrometers in either dimension.

Two-dimensional crystals have proved to be an extremely good way to introduce students and other new investigators to structural DNA nanotechnology. The preparation of the AB* or ABCD* array (**Figure 7**), for example, is

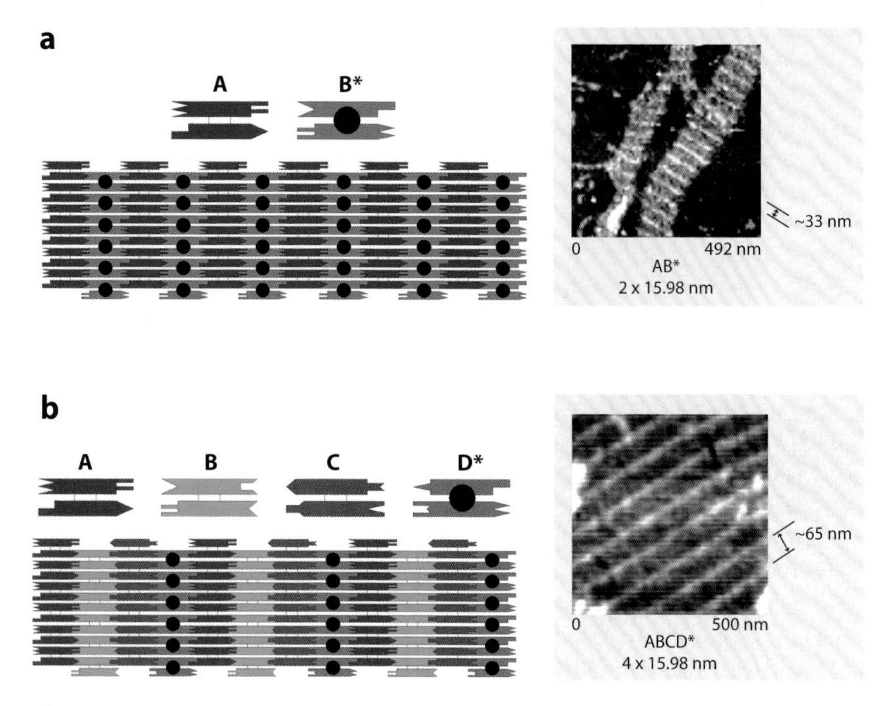

Figure 7

Two-dimensional arrays of DX and DX+J molecules. (*a*) An alternating array of DX and DX+J molecules. Two tiles are shown, a DX tile labeled A and a DX+J tile labeled B*. The black dot represents the extra domain of the DX+J. Sticky ends are shown as geometrically complementary shapes. The array that would be formed by these two tiles is shown below them, including a stripe-like feature formed by the extra domain of the DX+J tile. The horizontal direction of each tile is 16 nm. The atomic force microscope (AFM) image at right shows stripes separated by ~33 nm, near the predicted distance of 32 nm. (*b*) An array of three DX and one DX+J tiles. The A, B, and C tiles are DX molecules, and the D* tile is a DX+J tile. All tiles have the same dimensions as in (*a*). This leads to an ~65-nm separation of stripes, near the predicted distance of 64 nm, as seen in the AFM image at the right.

extremely simple: All the strands are mixed stoichiometrically, heated to 90°–95° and then cooled over 40 h in a styrofoam container (47); they are readily observed by atomic force microscopy when deposited on mica (47). Graduate students, undergraduates, and high school students are usually successful at producing beautiful AFM images on the first pass. There are some experiments that can discourage new investigators in structural DNA nanotechnology, but making two-dimensional periodic arrays is not among them.

4.2. Three-Dimensional Crystals

Control of the structure of matter would be incomplete without the ability to produce three-dimensional crystals of high quality. Although two-dimensional crystals are examined by AFM, the way to characterize the molecular structure of three-dimensional crystals is by X-ray crystallography. Under exceptionally good circumstances, one can resolve by AFM two double helices separated by 7 nm (two turns of DNA), but such resolution is not usually available (e.g., Reference 48). By contrast, X-ray crystallography is capable of resolutions of ~1 Å, the limitations being largely a function of crystal quality. All the early attempts to obtain high-quality crystals of DNA motifs resulted in crystals diffracting to no better than 10-Å resolution. If there are issues of molecular boundaries within the crystal to be determined, such a resolution is unlikely to provide an unambiguous answer (49). Recently, however, a tensegrity triangle (50), with two double helical turns per edge, has been crystallized to 4-Å resolution, and its structure has been determined by the single anomalous dispersion method (51). The tensegrity triangle has three double helical domains that point in three independent directions, so as to define a three-dimensional structure. An individual tensegrity triangle and its six neighbors are shown in stereoscopic projection in **Figure 8*a***. The three independent directions of the tensegrity triangle lead to a rhombohedral structure that flanks a cavity. This cavity, along with seven of the eight tensegrity triangles that flank it, is shown in stereoscopic projection in **Figure 8*b***; the volume of the cavity is ~100 nm^3.

It is evident from **Figure 8*a*** that the structure is held together in the directions of the helix axes. This is not happenstance. Nine other

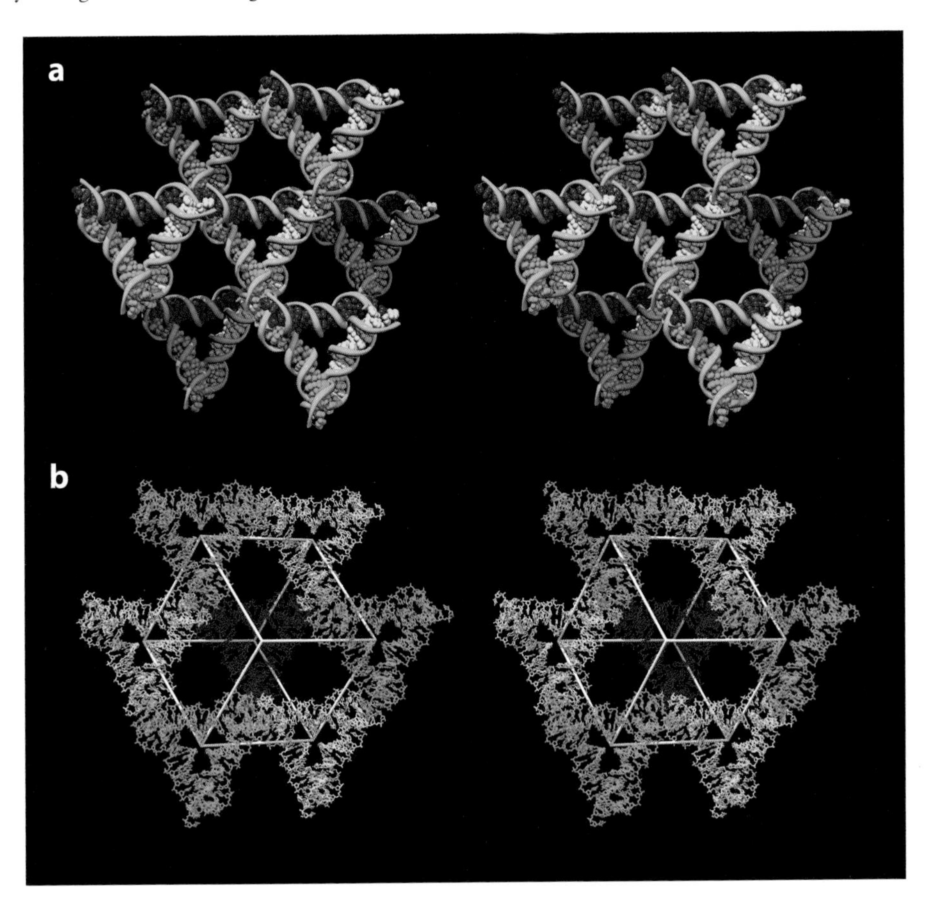

Figure 8

The three-dimensional lattice formed by the tensegrity triangles. (*a*) The surroundings of an individual triangle. This stereoscopic simplified image distinguishes the three independent directions in the colors (*red*, *green*, and *yellow*) of their base pairs. Thus, the central triangle is shown flanked by three other pairs of triangles in the three differently colored directions. (*b*) The rhombohedral cavity formed by the tensegrity triangles. This stereoscopic projection shows seven of the eight tensegrity triangles that comprise the corners of the rhombohedron. The outline of the cavity is shown in white. The red triangle at the back connects through one edge each to the three yellow triangles that lie in a plane somewhat closer to the viewer. The yellow triangles are connected through two edges each to three different green triangles, which are in a plane even nearer the viewer. A final triangle that would cap the structure has been omitted for clarity. That triangle would be directly above the red triangle and would be even closer to the viewer than the green triangles.

crystals have been characterized that are also based on the tensegrity triangle and are held together by 2-nt sticky ends (51). Some of the crystals (like the one shown in **Figure 8**) are threefold rotationally averaged by sticky-end selection, and some of the tensegrity triangles have longer edges, with either three turns or four turns. The resolution of the crystals is inversely related to edge length, ~6.5 Å for three turns and ~10–11 Å for four turns. It is unclear why this is the case, but it may be related to the fact that these are stick-like structures, rather than the ball-like structures usually examined by macromolecular crystallography. Since there are no molecular boundaries to establish for these structures, their resolution is not a key factor, so one of each of the larger structures has been determined by molecular replacement. The volume of the largest cavity is ~1100 nm^3 (roughly a zeptoliter). In contrast to two-dimensional crystals, the crystal sizes are macroscopic, ranging from 0.25 mm to 1 mm in linear dimensions. One of the strengths of the notion of holding molecules together by sticky ends is that one can design the number of molecules per crystallographic repeat. A crystal with two distinct two-turn motifs has been designed, and its structure determined by molecular replacement (T. Wang, R. Sha, J.J. Birktoft, J. Zheng, C. Mao & N.C. Seeman, in preparation).

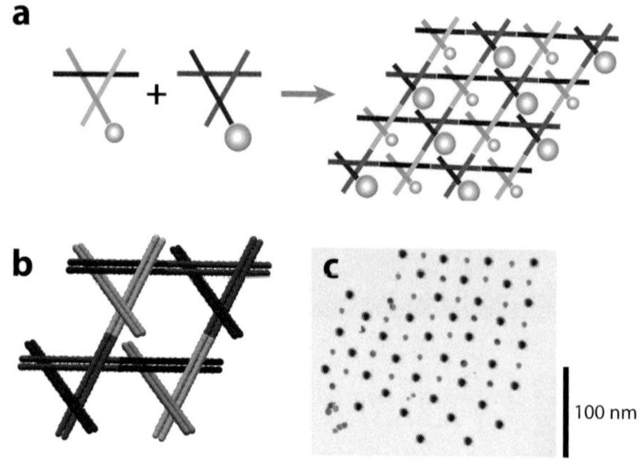

Figure 9

Organizing gold nanoparticles with a two-dimensional motif. (*a*) Two different three-dimensional-double crossover (3D-DX) motifs, containing 5-nm or 10-nm particles on one end of a propagation direction and yielding a checkerboard nanoparticle array. (*b*) The two 3D-DX motifs in greater detail assembled to form a two-dimensional array. (*c*) A transmission electron microscopy image showing a checkerboard array of gold nanoparticles.

5. OTHER TYPES OF DNA-BASED NANOMATERIALS

In addition to organizing conventional DNA species, there has been a lot of work with DNA-based systems. Section 5.1 discusses the use of DNA to organize proteins and nanoelectronic components. Section 5.2 discusses algorithmic assembly, derived from DNA-based computation, which differs from the periodic assembly described in the previous section. Section 5.3 describes efforts involving DNA analogs, and Section 5.4 describes the combination of DNA with coordination complexes.

5.1. Organization of Other Species

The use of DNA to organize other molecular species has been a central component of structural DNA nanotechnology since its inception (9). Originally, the notion was that three-dimensional crystals could organize biological macromolecules that were resistant to crystallization by traditional methods. It is clear from the last section that some progress is needed in resolution before such a goal can be realized. Nevertheless, structural DNA nanotechnology has done other things with proteins. Yan and his colleagues (52) have attached biotin groups to DNA arrays and used them to bind streptavidin. The same group has used aptamers on the surface of DNA origami to establish the cooperativity of protein-protein interactions (53).

Another target for structural DNA nanotechnology is the organization of nanoelectronic or nanophotonic components (54). In this regard, a number of studies have been devoted to organizing nanoparticles using DNA, sometimes within individual objects (55, 56), and sometimes within one-dimensional (57) or two-dimensional arrays containing multiple components (58, 59). **Figure 9** illustrates how multiple components, each carrying a different cargo, can be used to organize complex arrangements of metallic nanoparticles (59). The motif used is the three-dimensional-double crossover (3D-DX) motif, which is based on the tensegrity triangle, but contains eight-turn DX components in each direction of crystal propagation, rather than single helices. A 5-nm gold nanoparticle has been placed on the end of one of the propagation directions in one of the 3D-DX molecules, and a 10-nm gold nanoparticle has been put in the same site in the other 3D-DX molecule (**Figure 9*a***). Thus, one propagation direction is eliminated, but two-dimensional arrays can be formed. **Figure 9*b*** shows the 3D-DX motif and its lattice in greater detail, and **Figure 9*c*** shows a transmission electron microscopy image of the checkerboard pattern formed by the nanoparticles.

5.2. Algorithmic Assembly

The crystallographic background of the author has led to an emphasis on periodic matter in this article. However, it is also possible to organize aperiodic matter. Winfree and his colleagues (60–62) have focused on this aspect of assembly. The advantages of algorithmic assembly are that complex algorithmic patterns can be generated using only a few tiles. For example, Pascal's triangle modulo 2, in which all the even numbers are replaced by 0 and the odd numbers by 1, is known as a Sierpinski triangle. This fractal pattern can be generated by four tiles that correspond to the logical operations of exclusive OR (XOR), in addition to three tiles for the vertex, left and right sides of the triangle. The XOR gate yields a 0 if the two inputs are the same and a 1 if they are different. Currently, the main issue in algorithmic assembly is its extreme sensitivity to errors relative to periodic assembly. This problem derives from the fact that correct tiles in periodic assembly are competing only with incorrect tiles for a slot in the growing array, whereas in algorithmic assembly, correct tiles are competing with half-correct tiles (41, 64). Winfree and his colleagues have built Sierpinski triangles (60), most successfully when using an approach that "chaperones" the area of the aperiodic array (61). The same group has also built a set of tiles that can count binary groups (62). The great promise of counting is that two-dimensional crystals, and perhaps three-dimensional crystals, might be built to precise dimensions.

5.3. DNA Analogs

It is a natural question as to whether other nucleic acids are capable of being used as the basis of precise self-assembly. Jaeger's group (63, 65) has mined the loop-loop tertiary interactions in a variety of RNA crystal structures and has exploited them as the basis for self-assembled arrangements. They have taken robust motifs, such as tRNA molecules, and assembled them to form square-like species and small arrays (63, 65). Nobody seems to have tried to use RNA in the exclusively sticky-ended fashion that structural DNA nanotechnology has usually employed. Chaput and his colleagues (66) have built a branched junction from glycerol nucleic acid. Combining different types of nucleic acid molecules in the same construct is tricky because constructs like the DX molecule are very sensitive to double helical twist. Two-dimensional arrays mixing DNA and PNA were used to establish the helical repeat of the DNA/PNA double helix (67).

5.4. Seminatural Constructs

In addition to DNA analogs, it is possible to combine other chemical species with DNA in nanoconstructs. Sleiman's laboratory (68–71) has performed a number of studies in this vein. For example, they have placed metallo-organic complexes in junction sites (68), yielding different properties in the presence or absence of the metal; they have also placed metals at the corners of metal-nucleic acid cages (69). In a related development, Kieltyka et al. (70) have shown that it is possible to stimulate G-tetrad formation by use of an organometallic molecule that is square shaped. The same group has shown that DNA can pair with a molecule containing a conjugated backbone (71).

Although DNA is readily recognized as a special molecule, it is extremely valuable for the future of DNA-based materials in that it can be combined with the other molecular species that can be synthesized.

6. DNA-BASED NANOMECHANICAL DEVICES

The description above has emphasized the structure and the placement of atoms in space. It is natural to ask whether it is also possible to change their arrangement with time by exploiting some sort of transition that can be induced. Thus, a fourth dimension can be introduced in structural DNA nanotechnology. Section 6.1 describes changes in systems that are based on DNA structural transitions. However, such systems are largely limited to two states, regardless of the diversity of the components, and hence they ignore the programmability of DNA. Section 6.2 discusses transitions that are sequence dependent, so that a variety of individually addressable devices can coexist in the same solution.

6.1. Devices Based on Structural Transitions

There are a number of structural transitions of DNA that can be exploited to produce nanomechanical motion. The first such device was based on the transition from right-handed B-DNA to left-handed Z-DNA (72). There are two conditions necessary for the B-Z transition, a sequence that is capable of undergoing it readily and solution conditions that promote it (73). One can use the first requirement to limit how much of the DNA undergoes the transition (the proto-Z DNA) and use the second requirement to drive the transition one way or the other. The device consists of two DX components flanking a stretch of proto-Z DNA. **Figure 10*a*** shows how the two domains of the DX portions switch position when the device undergoes the transition. Other devices have been built that exploit the transition of oligo-dC sequences to the I form of DNA (74). The pH dependence of this device has been used to measure intracellular pH by combining it with fluorescent labels. The propensity of oligo-dG sequences to form tetraplex structures has also been used

Figure 10

DNA-based nanomechanical devices. (*a*) A DNA nanomechanical device based on the B-Z transition. The device consists of two DX molecules connected by a shaft containing 20 nt pairs (*yellow*) capable of undergoing the B-Z transition. Under B-promoting conditions, the short domains are on the same side of the shaft, but under Z-promoting conditions [added $Co(NH_3)_6^{3+}$], they are on opposite sides of the shaft. The pink and green FRET pairs are used to monitor this change. (*b*) The machine cycle of a PX-JX_2 device. Starting with the PX device on the left, the green set strands are removed by their complements (process I) to leave an unstructured frame. The addition of the yellow set strands (process II) converts the frame to the JX_2 structure, in which the top and bottom domains are rotated a half-turn relative to their arrangement in the PX conformation. Processes III and IV reverse this process to return to the PX structure. (*c*) AFM demonstration of the operation of the device. A series of DNA trapezoids are connected by devices. In the PX state, the trapezoids are in a parallel arrangement, but when the system is converted to the JX_2 state, they are in a zigzag arrangement. (*d*) Insertion of a device cassette into a two-dimensional array. The eight TX tiles that form the array are shown in different colored outlined tiles. For clarity, the cohesive ends are shown to be the same geometrical shape, although they all contain different sequences. The domain connecting the cassette to the lattice is not shown. The cassette and reporter helix are shown as red filled components; the marker tile is labeled "M" and is shown with a black filled rectangle, representing the domain of the tile that protrudes from the rest of the array. Both the cassette and the marker tile are rotated about 103° from the other components of the array (3-nt rotation). The PX arrangement is shown on the left, and the JX_2 arrangement is on the right. Note that the reporter hairpin points toward the marker tile in the PX state but points away from it in the JX_2 state.

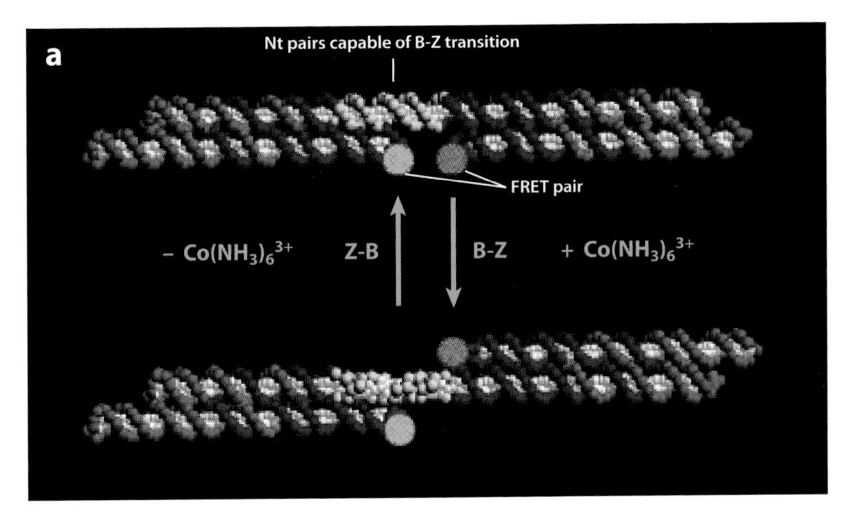
a
Nt pairs capable of B-Z transition
FRET pair
$- Co(NH_3)_6^{3+}$
Z-B
B-Z
$+ Co(NH_3)_6^{3+}$

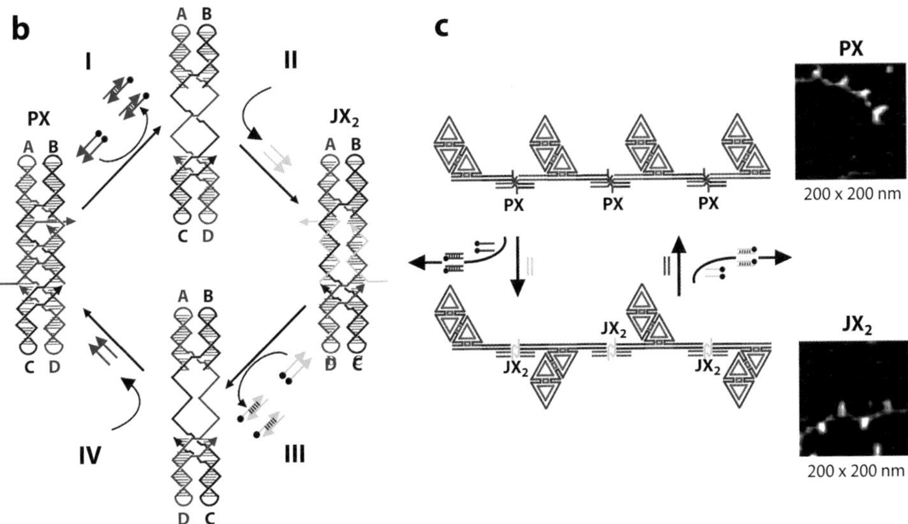
b
I
II
III
IV
PX
JX_2
A B
C D
D C
c
PX
PX PX PX
200 x 200 nm
JX_2
200 x 200 nm

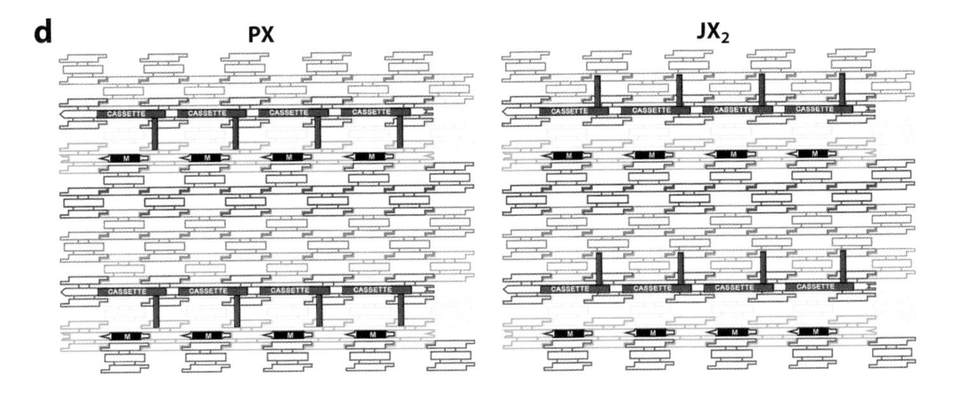
d
PX
JX_2
CASSETTE
M

to build potassium-responsive nanomechanical DNA-based devices (75).

6.2. Sequence-Dependent Devices

As intriguing and useful as DNA devices may be when based on structural transitions, the true power of using DNA is its programmability; single-trigger structural transitions can lead to only a few variants, even with nuanced DNA chemistry (34). The most effective way to construct multiple devices is to make them sequence dependent. Although the rate of response is limited by strand diffusion times, this seems to be the most robust way of achieving different responses from multiple devices in the same solution. The first sequence-dependent device was a molecular tweezers constructed by Yurke and his colleagues (76). The method that they used to change states was to add an 8-nt "toehold" to a controlling strand in the system. This toehold is designed to be unpaired when added to the device. It is removed by the addition of the complete complement to the strand containing the toehold. The complement binds to the toehold and removes the rest of the strand through branch migration. The reaction is downhill because more base pairs are formed by removing the toehold-containing stand than existed when it was bound to the device. Virtually every sequence-specific device utilizes this toehold principle.

A nanomechanical device is termed robust if it behaves like a macroscopic device, neither multimerizing nor dissociating when going through its machine cycle. The machine cycle of a robust sequence-dependent device controlled by "set" strands (16) is shown in **Figure 10*b***. This device has two states, called PX and JX_2, which differ from each other by a half-turn rotation between their tops and bottoms. Starting in the PX state at left, addition of the full complement to the green set strands (tailed in a biotin, represented by a black dot) removes the green strands, leaving a naked frame. Addition of the yellow set strands switches the device to the JX_2 state. The cycle is completed when the complements of the yellow strands are added to strip them from the frame and the green set strands are added. **Figure 10*c*** shows in a schematic that changing states can affect the structure of reporter trapezoids connected by the device. This panel also shows AFM images of these state changes.

The notion of combining devices and lattices enables one not only to place atoms at desired locations in a fixed fashion, but also to vary their positions with time in a programmed fashion. This goal has been achieved in 2D (15, 41) by developing a cassette with three components: One component is the PX-JX_2 device, the second is a domain that attaches the cassette to a two-dimensional array, and the third is a reporter arm whose motion can be detected by AFM. This system is illustrated in **Figure 10*d***, where the device has been incorporated into an eight-tile TX lattice. The PX state is shown on the left, and the JX_2 state is shown on the right. The reporter arm is seen to change its orientation relative to a marker tile (black) when the device state is switched.

The notion of exploiting toeholds that lead to systems with more base pairs dominates many activities in structural DNA nanotechnology. Clocked DNA walkers (77, 78) (devices that walk on sidewalks, where each step requires intervention) have been developed using this principle. Recently, the same notion has been used to produce an autonomous walker, which can move without intervention once initiated (79). Pierce and his colleagues (5) have recently suggested using this approach as a route to programming structural assembly. Likewise, Winfree and his colleagues (80) have used the same approach to control molecular circuitry. No purely DNA device based on any other reaction has been reported.

7. FURTHER CONSIDERATIONS

The topics discussed above do not cover all the applications of DNA to the construction of novel materials; others are discussed here. "Smart glue" applications are discussed in Section 7.1. The maximization of symmetry in DNA designs is discussed in Section 7.2. The cloning of DNA nanostructures in vivo is treated in Section 7.3.

7.1. Other Systems

DNA is a convenient material to use when building specific structural species. Space does not permit us to discuss the use of loosely organized DNA combined with proteins, a system that has great utility and has been pioneered by Niemeyer (81). Unstructured DNA as a smart glue has been utilized by Mirkin and his colleagues (82) when combined with gold nanoparticles. The difference between nanoparticles free in solution and those bound together by the presence of a DNA molecule is readily detected by a color change. Gang's (83) and Mirkin's (84) groups have also produced three-dimensional nanoparticle microcrystals using unstructured DNA.

7.2. Symmetry

We emphasized earlier that minimizing symmetry leads to greater control over any system, although Mao and coworkers (20) have indicated that the maximization of symmetry also leads to minimization of purification. This can lead to components of greater purity and possibly greater extent. For example, he and his colleagues have made large two-dimensional arrays using a single strand (20). In another take on symmetry, Yan and colleagues (85) have made finite symmetric arrangements of DNA with specific symmetries. As a consequence of the symmetry in these constructs, those investigators are able to use a relatively small number of DNA tiles for each of their constructs.

7.3. Cloning

As soon as one hears that DNA is being used to build materials, a question immediately presents itself: Can this material be reproduced biologically, rather than by complex laboratory procedures? The first suggestion for this approach is general but has not been attempted (86). Shih et al. (30) were able to clone a strand that folded into an octahedron, along with the aid of five helper strands. In 2007, Yan and his colleagues (87) reproduced a PX structure by in vitro rolling circle methods, a key initial step in the process. More recently, this group has managed to perform the same task in vivo for both a branched junction and a PX structure (88). However, it is important to point out that both of these cloned structures are topologically trivial: They are both equivalent to circles, and an open challenge is to produce a complex topology by cloning.

8. CONCLUSION AND PROSPECTS

It is hard to keep up with the advances that are coming rapidly from the field of DNA nanotechnology. In Section 8.1, I summarize what I feel are the key steps that have occurred in this field. In Section 8.2, I recap some important advances made by the growing number of laboratories working in the area.

8.1. A Toolbox Provided by Nature

The complementarity of DNA (2) and the ability of complementary strands to hybridize (3) facilitate the programmability of intermolecular affinity (4). In addition, the structural predictability of both sticky-ended cohesion products (6) and of specially designed branched DNA motifs (12, 14, 42, 45, 50) enable the programmability of structure. In this article, I have focused only on the aspects of DNA nanotechnology that derive from both of these features, emphasizing the capability of placing matter in particular spatial positions and the ability to change those structures at particular times through programmed molecular motion (16, 73, 76–79). There are many approaches to new materials, including new materials based on DNA (e.g., References 81 and 82), that are valuable and do not require this level of precision; however, the prejudices of this author and the limitations of space have prevented coverage of this aspect of DNA materials. The highest known precision of DNA constructs in 3D (250 μm in extent) is now 4-Å resolution, but other crystals (up to 1 mm in dimension) are readily available to 10-Å resolution (51). DNA origami has been estimated to provide a pixel size of about 60 Å (21). The quality of two-dimensional crystals (typical dimensions ~1 μm) is unclear because it is rare for the primary method of observing them, AFM, to give resolutions better than ~7 nm). It is clear that nature has provided a molecular basis for synthesizing molecules whose structures can be controlled with high precision outside the biological context. The properties of DNA may or may not be ideal for particular materials applications. However, there are large numbers of analogs (e.g., References 89 and 90) whose features may be more suitable and might be used instead.

8.2. A Rapidly Growing Field

The most exciting development in nucleic acid nanotechnology evolving in the past decade is the growth of the field. Around the turn of the millennium, the effort in this field was basically limited to the originating laboratory (9, 91). Today, 50–60 laboratories are involved in the effort. This article has a tilt that emphasizes my personal interests. Unfortunately, length restrictions have prevented me from mentioning much of the exciting work done by numerous investigators; I apologize to them for these omissions. The expansion of the field is arguably the most powerful development of the last decade: Different people have different perspectives on solutions to the problems that exist, leading to greater likelihoods of success. Furthermore, more investigators lead to more innovative approaches, exemplified by the developments of Rothemund's origami (21), Shih and coworkers' NMR system (43), Gothelf and colleagues' DNA box (44), Sleiman and coworkers' use of coordination chemistry (69), Yan and colleagues' cloning (88), Jaeger and coworkers' (63) and Paukstelis et al.'s (92) uses of tertiary interactions, and Chen & Mao's autonomous device (93). The increasing recognition of the power of directing molecular interactions by internally programmed molecular information is leading to an extremely bright future for this field. It is impossible to predict where this field is going, but with so many new and imaginative investigators involved, it will undoubtedly lead to increasingly exciting applications.

SUMMARY POINTS

1. The development of structural DNA nanotechnology relies on the control of DNA hybridization, on the ready availability of synthetic DNA, and on the ability to design and self-assemble stable branched DNA molecules.
2. Sequence design and motif design are key elements of this field.
3. It is possible to form molecules with the connectivities of various DNA polyhedrons by joining branched species via sticky ends; the shapes of molecules with only triangular faces will be correctly formed.
4. Stiff motifs are needed for periodic arrangements of DNA motifs. Using them, it is possible to self-assemble two-dimensional and three-dimensional crystalline arrangements. More complex arrangements, known as algorithmic assemblies, can also be self-organized within the context of a chaperoning border.
5. It is possible to use DNA to build nanomechanical devices. Some are based on DNA structural transitions, but others are sequence dependent, availing themselves of the full power of DNA programmability.

6. DNA can be used to organize other species in space, including proteins and nanoelectronic components.
7. DNA origami has resulted in facile construction of both two-dimensional and three-dimensional patterns and motifs. These range from a smiley face in two dimensions to a box with a keyed lock in three dimensions.
8. During the past decade, the field has expanded from a single laboratory to over 50 laboratories, signaling increasing interest in finding applications for programming the information in DNA beyond its genetic role, for structural and dynamic purposes.

FUTURE ISSUES

1. Can the resolution of three-dimensional DNA crystals be improved from 4 Å to 1–2 Å? If so, it will be possible to use this system as the basis for crystallizing biological macromolecules.
2. Can the sizes of two-dimensional DNA arrays be increased from ~1 μm to large areas? This would enable the use of DNA to organize nanoelectronics on a practical scale.
3. Can DNA origami tiles be self-assembled in the same way as small motifs? This would greatly increase the addressability of DNA-based surfaces.
4. Can nanoparticles and other nanoelectronic components be fit into three-dimensional DNA arrays? This would usher in a new era in the ability to organize nanoelectronics and nanophotonics from the bottom up, with an entirely new paradigm.
5. Can error-free algorithmic assembly be increased in extent and can it be extended to 3D? This will enable the construction of three-dimensional crystals of designated size.
6. Can nucleic acid analogs be used as effectively as DNA for nanoconstruction? If so, applications will not be limited to those compatible with an extensive polyanion with the structural characteristics of DNA.
7. Can some source of energy other than increased base pairing be employed for DNA-based nanodevices? If so, more sophisticated systems can be developed.
8. Can DNA-based nanodevices be made sufficiently sophisticated, autonomous, and inert to be used within a biological context? If so, cellular-level robotic therapy might be possible.

DISCLOSURE STATEMENT

The author is not aware of any affiliations, memberships, funding, or financial holdings that might be perceived as affecting the objectivity of this review.

ACKNOWLEDGMENTS

I wish to thank all of my students, postdocs, collaborators, and colleagues, who brought this field to its current state. I also wish to thank Dr. Paul W.K. Rothemund for the use of **Figure 6**. This work has been supported by grants GM-29544 from the National Institute of General Medical

Sciences, CTS-0608889 and CCF-0726378 from the National Science Foundation, 48681-EL and W911NF-07-1-0439 from the Army Research Office, N000140910181 from the Office of Naval Research, and a grant from the W.M. Keck Foundation.

LITERATURE CITED

1. Voet D, Rich A. 1970. The crystal structures of purines, pyrimidines and their intermolecular complexes. *Prog. Nucl. Acid Res. Mol. Biol.* 10:183–265
2. Watson JD, Crick FHC. 1953. Molecular structure of nucleic acids: a structure for deoxyribose nucleic acid. *Nature* 171:737–38
3. Rich A, Davies DR. 1956. A new two stranded helical structure: polyadenylic acid and polyuridylic acid. *J. Am. Chem. Soc.* 78:3548–49
4. Cohen SN, Chang ACY, Boyer HW, Helling RB. 1973. Construction of biologically functional bacterial plasmids in vitro. *Proc. Natl. Acad. Sci. USA* 70:3340–44
5. Yin P, Choi HMT, Calvert CR, Pierce NA. 2008. Programming biomolecular self-assembly pathways. *Nature* 451:318–23
6. Qiu H, Dewan JD, Seeman NC. 1997. A DNA decamer with a sticky end: the crystal structure of d-CGACGATCGT. *J. Mol. Biol.* 267:881–98
7. Watson JD, Crick FHC. 1953. Genetical implications of the structure of deoxyribonucleic acid. *Nature* 171:964–67
8. Holliday R. 1964. A mechanism for gene conversion in fungi. *Genet. Res.* 5:282–304
9. Seeman NC. 1982. Nucleic acid junctions and lattices. *J. Theor. Biol.* 99:237–47
10. Hsieh P, Panyutin IG. 1995. DNA branch migration. In *Nucleic Acids and Molecular Biology*, ed. F Eckstein, DMJ Lilley, 9:42–65. Berlin: Springer-Verlag
11. Caruthers MH. 1985. Gene synthesis machines: DNA chemistry and its uses. *Science* 230:281–85
12. Fu T-J, Seeman NC. 1993. DNA double crossover structures. *Biochemistry* 33:3311–20
13. Sa-Ardyen P, Vologodskii AV, Seeman NC. 2003. The flexibility of DNA double crossover molecules. *Biophys. J.* 84:3829–37
14. LaBean TH, Yan H, Kopatsch J, Liu F, Winfree E, et al. 2000. The construction, analysis, ligation and self-assembly of DNA triple crossover complexes. *J. Am. Chem. Soc.* 122:1848–60
15. Ding B, Seeman NC. 2006. Operation of a DNA robot arm inserted into a 2D DNA crystalline substrate. *Science* 31:1583–85
16. Yan H, Zhang X, Shen Z, Seeman NC. 2002. A robust DNA mechanical device controlled by hybridization topology. *Nature* 415:62–65
17. Lu M, Guo Q, Marky LA, Seeman NC, Kallenbach NR. 1992. Thermodynamics of DNA chain branching. *J. Mol. Biol.* 223:781–89
18. Seeman NC. 1990. De novo design of sequences for nucleic acid structure engineering. *J. Biomol. Struct. Dyn.* 8:573–81
19. Wang X, Seeman NC. 2007. Assembly and characterization of 8-arm and 12-arm DNA branched junctions. *J. Am. Chem. Soc.* 129:8169–76
20. Liu H, Chen Y, He Y, Ribbe AE, Mao C. 2006. Approaching the limit: Can one oligonucleotide assemble into large nanostructures? *Angew. Chem. Int. Ed. Engl.* 45:1942–45
21. Rothemund PWK. 2006. Scaffolded DNA origami for nanoscale shapes and patterns. *Nature* 440:297–302
22. Douglas SM, Dietz H, Liedl T, Högborg B, Graf F, Shih WM. 2009. Self-assembly of DNA into nanoscale three-dimensional shapes. *Nature* 459:414–18
23. Dietz H, Douglas SM, Shih WM. 2009. Folding DNA into twisted and curved nanoscale shapes. *Science* 325:725–30
24. Högborg B, Liedl T, Shih WM. 2009. Folding DNA origami from a double-stranded source of scaffold. *J. Am. Chem. Soc.* 131:9154–55
25. Ma R-I, Kallenbach NR, Sheardy RD, Petrillo ML, Seeman NC. 1986. Three-arm nucleic acid junctions are flexible. *Nucleic Acids Res.* 14:9745–53

26. Petrillo ML, Newton CJ, Cunningham RP, Ma R-I, Kallenbach NR, Seeman NC. 1988. The ligation and flexibility of four-arm DNA junctions. *Biopolymers* 27:1337–52
27. Chen J, Seeman NC. 1991. The synthesis from DNA of a molecule with the connectivity of a cube. *Nature* 350:631–33
28. Zhang Y, Seeman NC. 1994. The construction of a DNA truncated octahedron. *J. Am. Chem. Soc.* 116:1661–69
29. Goodman RP, Schaap IAT, Tardin CF, Erben CM, Berry RM, et al. 2005. Rapid chiral assembly of rigid DNA building blocks for molecular nanofabrication. *Science* 310:1661–65
30. Shih WM, Quispe JD, Joyce GF. 2004. DNA that folds into a nanoscale octahedron. *Nature* 427:618–21
31. He Y, Ye T, Su M, Zhang C, Ribbe AE, et al. 2008. Hierarchical self-assembly of DNA into symmetric supramolecular polyhedra. *Nature* 452:198–201
32. Erben CM, Goodman RP, Turberfield AJ. 2006. Single molecule protein encapsulation in a rigid DNA cage. *Angew. Chem. Int. Ed. Engl.* 45:7414–17
33. Seeman NC. 1992. The design of single-stranded nucleic acid knots. *Mol. Eng.* 2:297–307
34. Du SM, Stollar BD, Seeman NC. 1995. A synthetic DNA molecule in three knotted topologies. *J. Am. Chem. Soc.* 117:1194–1200
35. Fu T-J, Tse-Dinh Y-C, Seeman NC. 1994. Holliday junction crossover topology. *J. Mol. Biol.* 236:91–105
36. Mao C, Sun W, Seeman NC. 1997. Assembly of Borromean rings from DNA. *Nature* 386:137–38
37. Wang H, Di Gate RJ, Seeman NC. 1996. An RNA topoisomerase. *Proc. Natl. Acad. Sci. USA* 93:9477–82
38. Yan H, LaBean TH, Feng L, Reif JH. 2003. Directed nucleation assembly of DNA tile complexes for barcode-patterned lattices. *Proc. Natl. Acad. Sci. USA* 100:8103–8
39. Zhang X, Yan H, Shen Z, Seeman NC. 2002. Paranemic cohesion of topologically-closed DNA molecules. *J. Am. Chem. Soc.* 124:12940–41
40. Winfree E, Liu F, Wenzler LA, Seeman NC. 1998. Design and self-assembly of two-dimensional DNA crystals. *Nature* 394:539–44
41. Gu H, Chao J, Xiao SJ, Seeman NC. 2009. Dynamic patterning programmed by DNA tiles captured on a DNA origami substrate. *Nat. Nanotechnol.* 4:245–49
42. Mathieu F, Liao S, Mao C, Kopatsch J, Wang T, Seeman NC. 2005. Six-helix bundles designed from DNA. *Nano Lett.* 5:661–65
43. Douglas SM, Chou JJ, Shih WM. 2007. DNA-nanotube-induced alignment of membrane proteins for NMR structure determination. *Proc. Natl. Acad. Sci. USA* 104:6644–48
44. Andersen ES, Dong M, Nielsen MM, Jahn K, Subramani R, et al. 2009. Self-assembly of a nanoscale box with a controllable lid. *Nature* 459:73–76
45. Li X, Yang X, Qi J, Seeman NC. 1996. Antiparallel DNA double crossover molecules as components for nanoconstruction. *J. Am. Chem. Soc.* 118:6131–40
46. Ding B, Sha R, Seeman NC. 2004. Pseudohexagonal 2D DNA crystals from double crossover cohesion. *J. Am. Chem. Soc.* 126:10230–31
47. Seeman NC. 2002. Key experimental approaches in DNA nanotechnology. *Curr. Protoc. Nucleic Acid Chem.* Chapter 12:Unit 12.1
48. Mao C, Sun W, Seeman NC. 1999. Designed two-dimensional DNA Holliday junction arrays visualized by atomic force microscopy. *J. Am. Chem. Soc.* 121:5437–43
49. Kim SH, Quigley G, Suddath FL, McPherson A, Sneden D, et al. 1972. The three-dimensional structure of yeast phenylalanine transfer RNA: shape of the molecule at 5.5-Å resolution. *Proc. Natl. Acad. Sci. USA* 69:3746–50
50. Liu D, Wang W, Deng Z, Walulu R, Mao C. 2004. Tensegrity: construction of rigid DNA triangles with flexible four-arm junctions. *J. Am. Chem. Soc.* 126:2324–25
51. Zheng J, Birktoft JJ, Chen Y, Wang T, Sha R, et al. 2009. From molecular to macroscopic via the rational design of a self-assembled 3D DNA crystal. *Nature* 461:74–77
52. Park SH, Yin P, Liu Y, Reif JH, LaBean TH, Yan H. 2005. Programmable DNA self-assemblies for nanoscale organization of ligands and proteins. *Nano Lett.* 5:729–33
53. Rinker S, Ke YG, Liu Y, Chhabra R, Yan H. 2008. Self-assembled DNA nanostructures for distance-dependent multivalent ligand-protein binding. *Nat. Nanotechnol.* 7:418–22

54. Robinson BH, Seeman NC. 1987. The design of a biochip: a self-assembling molecular-scale memory device. *Protein Eng.* 1:295–300
55. Alivisatos AP, Johnson KP, Peng XG, Wilson TE, Loweth CJ, et al. 1996. Organization of 'nanocrystal molecules' using DNA. *Nature* 382:609–11
56. Mastroianni AJ, Claridge SA, Alivisatos AP. 2009. Pyramidal and chiral groupings of gold nanocrystals assembled using DNA scaffolds. *J. Am. Chem. Soc.* 131:8455–59
57. Li HY, Park SH, Reif JH, LaBean TH, Yan H. 2004. DNA-templated self-assembly of protein and nanoparticle linear arrays. *J. Am. Chem. Soc.* 126:418–19
58. Pinto YY, Le JD, Seeman NC, Musier-Forsyth K, Taton TA, Kiehl RA. 2005. Sequence-encoded self-assembly of multiple-nanocomponent arrays by 2D DNA scaffolding. *Nano Lett.* 5:2399–402
59. Zheng J, Constantinou PE, Micheel C, Alivisatos AP, Kiehl RA, Seeman NC. 2006. 2D nanoparticle arrays show the organizational power of robust DNA motifs. *Nano Lett.* 6:1502–4
60. Rothemund PWK, Papadakis N, Winfree E. 2004. Algorithmic self-assembly of DNA Sierpinski triangles. *PLoS Biol.* 2:2041–52
61. Fujibayashi K, Hariadi R, Park SH, Winfree E, Murata S. 2008. A fixed-width cellular automaton pattern. *Nano Lett.* 8:1791–97
62. Barish RD, Rothemund PWK, Winfree E. 2005. Two computational primitives for algorithmic self-assembly: copying and counting. *Nano Lett.* 5:2586–92
63. Chworos A, Severcan I, Koyfman AY, Weinkam P, Oroudjev E, et al. 2004. Building programmable jigsaw puzzles with RNA. *Science* 306:2068–72
64. Mao C, LaBean TH, Reif JH, Seeman NC. 2000. Logical computation using algorithmic self-assembly of DNA triple crossover molecules. *Nature* 407:493–96
65. Severcan I, Geary C, Venzemnieks A, Jaeger L. 2009. Square-shaped RNA particles from different RNA folds. *Nano Lett.* 9:1270–77
66. Zhang RS, McCullum EO, Chaput JC. 2008. Synthesis of two mirror-image 4-helix junctions derived from glycerol nucleic acid. *J. Am. Chem. Soc.* 130:5846–47
67. Lukeman PS, Mittal A, Seeman NC. 2004. Two dimensional PNA/DNA arrays: estimating the helicity of unusual nucleic acid polymers. *Chem. Commun.* 2004:1694–95
68. Yang H, Sleiman HF. 2008. Templated synthesis of highly stable, electroactive and dynamic metal-DNA branched junctions. *Angew. Chem. Int. Ed. Engl.* 47:2443–46
69. Yang H, McLaughlin CK, Aldaye FA, Hamblin GD, Rys AZ, et al. 2009. Metal-nucleic acid cages. *Nat. Chem.* 1:390–96
70. Kieltyka R, Engelbienne P, Fakhoury J, Autexier C, Moitessier N, Sleiman HF. 2008. A platinum supramolecular square as an effective G-quadruplex binder and telomerase inhibitor. *J. Am. Chem. Soc.* 130:10040–41
71. Lo PK, Sleiman HF. 2008. Synthesis and molecular recognition of conjugated polymer with DNA-mimetic properties. *Macromolecules* 41:5590–603
72. Rich A, Nordheim A, Wang AHJ. 1984. The chemistry and biology of left-handed Z-DNA. *Annu. Rev. Biochem.* 53:791–846
73. Mao C, Sun W, Shen Z, Seeman NC. 1999. A DNA nanomechanical device based on the B-Z transition. *Nature* 397:144–46
74. Modi S, Swetha MG, Goswami D, Gupta G, Mayor S, Krishnan Y. 2009. A DNA nanomachine maps spatiotemporal pH changes in living cells. *Nat. Nanotechnol.* 4:325–29
75. Hou X, Guo W, Xia F, Nie FQ, Dong H, et al. 2009. A biomimetic potassium-responsive nanochannel: G-quadruplex conformational switching in a synthetic nanopore. *J. Am. Chem. Soc.* 131:7800–5
76. Yurke B, Turberfield AJ, Mills AP Jr, Simmel FC, Newmann JL. 2000. A DNA-fuelled molecular machine made of DNA. *Nature* 406:605–8
77. Sherman WB, Seeman NC. 2004. A precisely controlled DNA bipedal walking device. *Nano Lett.* 4:1203–7
78. Shin JS, Pierce NA. 2004. A synthetic DNA walker for molecular transport. *J. Am. Chem. Soc.* 126:10834–35
79. Omabegho T, Sha R, Seeman NC. 2009. A bipedal DNA Brownian motor with coordinated legs. *Science* 324:67–71

80. Zhang DY, Turberfield AJ, Yurke B, Winfree E. 2007. Engineering entropy-driven reactions and networks catalyzed by DNA. *Science* 318:1121–25
81. Niemeyer CM. 2007. Functional devices from DNA and proteins. *Nano Today* 2:42–52
82. Mirkin CA, Letsinger RL, Mucic RC, Storhoff JJ. 1996. A DNA-based method for rationally assembling nanoparticles into macroscopic materials. *Science* 382:607–9
83. Nykypanchuk D, Maye MM, van der Lelie D, Gang O. 2008. DNA-guided crystallization of colloidal nanoparticles. *Nature* 451:549–52
84. Park SY, Lytton-Jean AKR, Lee B, Weigand S, Schatz GC, Mirkin CA. 2008. DNA-programmable nanoparticle crystallization. *Nature* 451:553–56
85. Liu Y, Ke Y, Yan H. 2005. Self-assembly of symmetric finite-size DNA arrays. *J. Am. Chem. Soc.* 127:17140–41
86. Seeman NC. 1991. The construction of three-dimensional stick figures from branched DNA. *DNA Cell Biol.* 10:475–86
87. Lin C, Wang X, Liu Y, Seeman NC, Yan H. 2007. Rolling circle enzymatic replication of a complex multi-crossover DNA nanostructure. *J. Am. Chem. Soc.* 129:14475–81
88. Lin C, Rinker S, Wang X, Liu Y, Seeman NC, Yan H. 2008. In vivo cloning of artificial DNA nanostructures. *Proc. Natl. Acad. Sci. USA* 105:17626–31
89. Freier SM, Altmann KH. 1997. The ups and downs of nucleic acid stability: structure-stability studies on chemically-modified DNA:RNA duplexes. *Nucleic Acids Res.* 25:4429–43
90. Egholm M, Buchardt O, Nielsen PE, Berg RH. 1992. Peptide nucleic-acids (PNA). Oligonucleotide analogs with an achiral peptide backbone. *J. Am. Chem. Soc.* 114:1895–97
91. Seeman NC. 1998. DNA nanotechnology: novel DNA constructions. *Annu. Rev. Biophys. Biomol. Struct.* 27:225–48
92. Paukstelis P, Nowakowski J, Birktoft JJ, Seeman NC. 2004. The crystal structure of a continuous three-dimensional DNA lattice. *Chem. Biol.* 11:1119–26
93. Chen Y, Mao C. 2004. Putting a brake on an autonomous DNA nanomotor. *J. Am. Chem. Soc.* 126:8626–27

Eukaryotic Chromosome DNA Replication: Where, When, and How?

Hisao Masai, Seiji Matsumoto, Zhiying You, Naoko Yoshizawa-Sugata, and Masako Oda

Genome Dynamics Project, Tokyo Metropolitan Institute of Medical Science, Tokyo 156-8506, Japan; email: masai-hs@igakuken.or.jp

Annu. Rev. Biochem. 2010. 79:89–130

First published online as a Review in Advance on April 7, 2010

The *Annual Review of Biochemistry* is online at biochem.annualreviews.org

This article's doi:
10.1146/annurev.biochem.052308.103205

0066-4154/10/0707-0089$20.00

Key Words

DNA helicase, prereplicative complex, replication fork, replication origin, replication stress, replication timing

Abstract

DNA replication is central to cell proliferation. Studies in the past six decades since the proposal of a semiconservative mode of DNA replication have confirmed the high degree of conservation of the basic machinery of DNA replication from prokaryotes to eukaryotes. However, the need for replication of a substantially longer segment of DNA in coordination with various internal and external signals in eukaryotic cells has led to more complex and versatile regulatory strategies. The replication program in higher eukaryotes is under a dynamic and plastic regulation within a single cell, or within the cell population, or during development. We review here various regulatory mechanisms that control the replication program in eukaryotes and discuss future directions in this dynamic field.

Contents

1. INTRODUCTION

Studies regarding the mechanisms of DNA replication for the past 57 years, since the discovery of double-stranded DNA by Watson and Crick, revealed that sequential assembly and reorganization of complex arrays of proteins are crucial for the coordinated execution of initiation, elongation, and termination processes of DNA replication. The progression of the replication fork is monitored strictly to ensure complete replication of the entire genome. The physiological significance of the execution of the entire process has been emphasized by the finding that defects in proteins for assembly and monitoring of the replication fork lead to genomic instability, resulting in carcinogenesis or a series of diseases collectively known as the "chromosome instability syndrome."

The extensive conservation of replication factors strongly suggests that the basic mechanism of DNA replication is evolutionally conserved. However, accumulating data also indicate that the modes of recognition and regulation of its firing are significantly more complex in eukaryotes than in prokaryotes and may be substantially different even between lower and higher eukaryotes. One of the most striking features of DNA replication in higher eukaryotes is its plasticity. A best-known example of plasticity is the very rapid replication of the entire genome of the fertilized eggs of amphibians (within a matter of 20–30 min), compared with the 8–10 h required for the genome replication in somatic cells. This robust process is achieved by very frequent initiation through utilization of virtually all the potential origins in contrast to the spatially and temporally regulated origin firing that occurs in differentiated cells. In spite of this, how this rather dramatic transition of the replication mode takes place is still under debate.

The challenges facing our understanding of eukaryotic DNA replication can be summarized by three questions—where, when, and how: Where in the genome and within the nuclei does DNA replication take place? When within the S phase does each replication origin start firing? And how is the entire process of DNA replication regulated?

In this review, we summarize recent progress in eukaryotic DNA replication and discuss how it contributes to our understanding of the basic mechanisms that enable highly strict and plastic regulation of this process. We also discuss how failures in the proper execution of replication program contribute to the etiology of various diseases including malignancy. Because the last review on eukaryotic DNA replication was published in the *Annual Review of Biochemistry* in 2002, we cite studies that were published mainly after that review. The readers are encouraged to refer to Reference 1 for the literature published before 2002.

2. WHERE?

Where on the genome and within the nuclei does DNA replication take place?

2.1. Assembly Sites of Replicative Complexes

2.1.1. Recognition by the initiator origin recognition complex. In bacteria, the initiator DnaA protein determines the site of initiation on the genome. Searches for the initiator of eukaryotic genomes led to the discovery of the origin recognition complex (ORC) (2). In the budding yeast, where ORC was first identified, it specifically recognizes the 11-bp or 17-bp conserved sequence present in the replicator sequences (3–5; see below). The ORC is evolutionally conserved and plays important roles in DNA replication across species (1). However, the ORC from other species exhibits DNA-binding specificity that is much more relaxed than that of budding yeast. For example, the fission yeast ORC recognizes DNA through the AT-hook motif present on the ORC4 subunit and preferentially binds adenine/thymine stretches (6), and the *Drosophila* ORC also shows some preference for AT-rich sequences. In contrast, the human ORC binds to any DNA without apparent sequence specificity (7, 8). However, the sites of initiation of DNA replication are not random, at least in differentiated somatic cells, suggesting that the ORC binds to genomic DNA with certain specificity. The characterization of sequence elements required for ectopic replication from mammalian replicators has identified AT-rich sequences (9–11), dinucleotide repeats (10), asymmetrical purine-pyrimidine sequences (11), and matrix attachment region (MAR) sequences (8, 12) as being required for efficient initiation of DNA replication. These sequences in the chromatin context may facilitate recognition by the ORC. It should be noted that even the ORC in budding yeast could exhibit a relaxed mode of DNA binding in vivo and in vitro (13–15), and thus, DNA binding without significant specificity may be a conserved feature of the ORC. This is quite different from bacterial initiators, which are highly specific to cognate replication origin sequences.

Other factors may facilitate the selection of specific segments on the genome. The potential factors include the topology of DNA. Replication in vitro at the chromosomal origins of DNA replication of *Escherichia coli* depends on the negative superhelicity of the template DNA (16). In *Drosophila melanogaster* as well as in fission yeast, the ORC exhibits a preference for supercoiled DNA (17, 18). Topoisomerases were reported to associate with a human replication origin (312). Transcription factors may also play significant roles in the specific localization of the ORC. At the *chorion* loci of *Drosophila* follicle cells, transcription factors that contain the Myb protein seem to facilitate the site-specific DNA replication of the ORC at ACE3 (amplification control element 3) and Ori-β (19). This locus also interacts with retinoblastoma (Rb) (20). At the rat aldolase B gene promoter origin, binding of the ORC1-interacting factor AIF-C to a specific sequence in the origin is required for efficient initiation of replication (21). Recruitment of the ORC to the Epstein-Barr virus (EBV) replicator, *oriP*, was shown to be dependent on the RNA that links *oriP*-bound EBV-coded nuclear antigen-1 (EBNA-1) and the ORC (22). Prereplicative complex (pre-RC) factors may also contribute to origin recognition. In budding yeast, Cdc6 ATPase activity contributes to stable and specific binding of the ORC-Cdc6 complex to origin, because its activation by nonorigin DNA promotes dissociation of Cdc6 (15). Binding of fission yeast ORC to the origin DNA is facilitated by Cdt1 and Cdc18 proteins (18).

Replication fork: the site of DNA replication where parental duplex DNA separates and DNA is synthesized by the replisome complex

Origin: a segment of DNA from which DNA replication is initiated

ORC: origin recognition complex

Prereplicative complex (pre-RC): a complex, composed of ORC, MCM, and loading factors, that is assembled at selective loci on the chromosome during early G1

Cell division cycle (Cdc): a mutation in yeasts, which causes arrest of cell growth at a specific cell cycle stage

SUMMARY OF GENOME-WIDE INVESTIGATION OF DNA REPLICATION USING MICROARRAY ANALYSES

In the eight years since the last review, several studies on genome-wide analyses of DNA replication have been published. They can be categorized into two groups: (*a*) mapping of replication origins and factor binding sites and (*b*) mapping of replication timing. The methods used in the first group include (*a*) analyses of replicated heavy-light and unreplicated light-light fractions [after labeling with bromodeoxyuridine (BrdU)] isolated at different time points during S phase and separated by a density gradient centrifugation, (*b*) analyses of copy number changes, (*c*) analyses of the single-strand DNAs generated, (*d*) analyses of the ORC-MCM-binding sites by ChIP-Chip assays, and (*e*) analyses of BrdU incorporation by ChIP-Chip assays. The methods for the second group include (*a*) fractionation of cells into early and late S-phase populations and analyses of BrdU-incorporating DNA, and (*b*) analyses of round mitotic cells collected after "baby machine"-based cell cycle synchronization [in which cells were pulse-labeled, washed, and incubated for different time intervals (126)].

The following are conclusions on replication origins drawn from the above studies:

1. In budding yeast, pre-RC sites (ORC and MCM binding) coincide with actual initiation sites (origins) (23, 81).
2. In fission yeast, the pre-RC is present with an average interorigin length of 10 kb (25). About 40% of the pre-RC assembly sites are utilized for initiation in the presence of hydroxyurea (on chromosomes I and II) (Y. Kanoh, personal communication).
3. No consensus sequence was identified for origins. AT-rich (>70%) intergenic regions serve as potential origins in fission yeast (24, 25, 86, 307).
4. In *Drosophila*, the consensus sequences for origins and for ORC binding have not yet been identified. The ORC-binding sites are overlapped by RNA polymerase II–binding sites, suggesting that transcription factors could contribute to selective binding of the ORC (26).
5. In human, origins are abundant in the CpG island promoter. Half of the origins are localized within or near CpG islands (27).
6. In human, origins are strongly associated with transcriptional regulatory elements (e.g., c-Jun and c-Fos) (27).
7. In mammals, origin sequences are evolutionarily conserved between different species (27).

2.1.2. Origin recognition complex and minichromosome maintenance complex-binding sites as potential origins. The minichromosome maintenance (MCM) complex, a heterohexameric complex of MCM 2, 3, 4, 5, 6, and 7 subunits, is loaded onto the ORC bound to chromatin at late mitosis (M) to early G1 to generate a pre-RC. Genome-wide chromatin-immunoprecipitation assays in budding yeast showed that the ORC- and/or MCM-binding sites include 95% of the known autonomously replicating sequence (ARS), suggesting that the ORC- and/or MCM-binding sites can predict the potential origins with high accuracy (23; **Supplemental Tables 1** and **2**. Follow the **Supplemental Material link** from the Annual Reviews home page at **http://www.annualreviews.org.**) See also the box titled Summary of Genome-Wide Investigation of DNA Replication Using Microarray Analyses. In fission yeast, the ORC- and MCM-binding sites also correlate well with the replication origins predicted on the basis of AT-richness (24, 25; Y. Kanoh, personal communication). These potential origins are present with an average of 10-kb intervals, mainly at the intergenic segments in fission yeast. Similar assays with the *Drosophila* chromosome identified distinct ORC-binding sites colocalized with the actual initiation sites (26). Although similar ChIP-Chip assays have not been reported in human, mapping of the initiation sites resulted in the identification of 283 origins in 1% of the human genome, which largely colocalized with transcriptional regulatory elements (27–29). These results indicate that potential origins are assembled at specific sites even on the genome of higher eukaryotes. Please refer to **Supplemental Table 3**. Follow the **Supplemental Material link** from the Annual Reviews home page at **http://www.annualreviews.org**.

2.2. Selection of Replication Initiation Sites of DNA Synthesis

There is general agreement that cells are equipped with excess numbers of pre-RCs on the genome (potential origins) and that only a subset of these pre-RCs is probably utilized for actual initiation (30, 31). Two of the outstanding questions are how this selection is regulated and how it is related to the plasticity of eukaryotic DNA replication, as discussed below (see Section 2.3.). Selection of the initiation sites may be regulated by various factors and may be achieved at the activation step as well as at the pre-RC assembly step.

8. Origin specification and firing efficiency are generally maintained across different cell types in mouse (28).
9. Origins for re-replication are not identical to those for normal replication in budding yeast (308).

The conclusions on replication timing are as follows:

1. Genomes can be divided into different replication timing domains; each domain contains one to tens of replication origins that fire in a coordinated fashion (26, 108–111).
2. In budding yeast, there is no correlation between replication timing and transcription or GC (guanine or cytosine) content (81).
3. In metazoan, generally significant correlation was detected between replication timing and transcription or GC content (**Supplemental Table 4**).
4. Replication timing domains undergo substantial reorganization during development or differentiation (111, 127).
5. Timing transition regions are associated with the frequent occurrence of mutations responsible for various diseases and single-nucleotide polymorphism (134). (Note: This conclusion was drawn by regular analyses, not by microarray analyses.)

Please refer to **Supplemental Tables 1–4**. Follow the **Supplemental Material link** from the Annual Reviews home page at **http://www.annualreviews.org**.

2.2.1. Distal elements. The distal DNA elements can have a profound effect on the initiation site. In the Chinese hamster ovary cells, DNA replication at the *DHFR* (dihydrofolate reductase) locus (which is amplified) initiates within the 55-kb intergenic region (initiation zone) (32). When the transcription reads through into the intergenic region as a result of deletion of the 3′ end of *DHFR*, the initiation is confined to the far end of the intergenic region. Introduction of a small DNA segment derived from the 3′ end of the gene (containing transcription terminator activity) generates the boundary defining the end of the originless segment (33). In the human *β-globin* locus, replication starts at the initiation region located at the 5′ end through the coding region of the β-like *β-globin* gene (34). This initiation event depends on the presence of the locus control region, which is located further upstream of the globin gene clusters and serves as a master control element of *β-globin* gene expression. In the cytokine cluster segment of mouse chromosome 11, initiation of replication at the 3′ side of the IL-13 gene requires the presence of the conserved noncoding sequence in the 10-kb intergenic region between IL-13 and IL-4 (35). The conserved noncoding sequence, containing the DNaseI-hypersensitive sites, is required for cell-type-specific expression of Th2 cytokines. At present, the mechanism by which these distant elements affect the initiation events is unknown.

2.2.2. Chromatin structures. Chromatin structures are important factors not only for transcription but also for origin selection. In yeasts, mutations in histone deacetylases (HDACs), such as Sir2, facilitate the initiation events. The *sir2* mutation rescues replication mutants involved in pre-RC assembly but not those involved in the following activation step, and in fact loading of the MCM complex onto origins is facilitated by the *sir2* mutation at some origins (36, 37). These sensitive origins contain a sequence called I^S, which mediates Sir2-dependent inhibition (38). In *Xenopus* egg extracts and human cells, HBO1, a histone acetyltransferase, is required for loading the MCM complex onto chromatin (39). HBO1 directly interacts with Cdt1 and enhances Cdt1-dependent rereplication (40). HBO1 interacts with the ORC and MCM2, suggesting that histone acetylation and deacetylation could affect the formation of the pre-RC. Thus, chromatin structures seem to affect the efficiency of pre-RC assembly, which may indirectly alter the initiation frequency at particular loci (41).

MCM: minichromosome maintenance

ARS: autonomously replicating sequence

Helicase: an enzyme capable of separating the duplex DNA using the energy obtained by hydrolysis of ATP

At the ACE3 and Ori-β origins of *Drosophila*, histone acetylation colocalizes with the ORC, and inhibition of HDAC results in hyperreplication concomitant with hyperacetylation and redistribution of the ORC (42). Site-specific DNA replication was observed in *Xenopus* egg extracts, where initiation is normally random, by introducing a transcriptional complex at a specific site (43). This initiation at the targeted site correlates with local hyperacetylation of histones. Because the distribution of the ORC is not altered, transcription factor-induced localized chromatin remodeling or histone acetylation may facilitate the assembly of the pre-RC and/or activation of the pre-RC (43). Thus, chromatin structure, more specifically histone acetylation, appears to facilitate initiation, affecting the site of initiation. By contrast, at the *β-globin* locus, histone modifications did not precisely correlate with the activity of the replication origin. Thus, chromatin structures may play a pivotal role in origin selection but are probably not essential for firing (44).

2.2.3. Transcription, topology, and nucleosome formation. Replication origins are abundant in the intergenic regions. Generally, origins present downstream of a transcriptional unit are less active compared to those present upstream (45, 46), and this was thought to be caused by interference of the origin function by read-through transcription (47–49). Topology is another important factor in initiation of DNA replication. Duplex unwinding is highly stimulated by the negative superhelicity. Therefore, we also consider the possibility that transcription-induced topological changes could affect initiation. It is known that transcription induces negative supercoiling behind the transcription bubble (50). Thus, two divergent transcription events will generate strong negative supercoiling in the intergenic region, facilitating strand opening. In fact, genome-wide determination of pre-RC assembly sites and actual initiation events in fission yeast revealed that the initiation frequency in the presence of hydroxyurea is 50% and 25% among the pre-RC assembled at the intergenic region between divergent transcription and that between convergent transcription, respectively (25; Y. Kanoh, personal communication). Furthermore, recent findings indicate that divergent transcription occurs in most of the active genes, including the ones regarded as single transcription units. This will maintain the promoter segment between the two divergently moving RNA polymerases nucleosome free (51), which will permit more efficient assembly of the pre-RC.

2.2.4. Other factors regulating the site of initiation. It was reported that the mammalian MCM4-6-7 helicase is specifically activated by T-rich single-stranded DNA (ssDNA), including the DNA derived from known replication origin sequences (52). The ability to activate the MCM helicase correlated with the presence of stretches of thymine residues that are often found in the vicinity of eukaryotic replication origins. Thus, the chance of origin activation may be determined by availability of single-stranded thymine-rich DNA, which may be regulated by the intrinsic nature of the sequence as well as by the conformational state of the pre-RC, or chromatin, and/or topological context.

The initiation of DNA replication can influence that from adjacent loci. In budding yeast, a propagating fork from a neighboring origin suppresses the firing of a potential origin, probably by disrupting the preformed pre-RC (53). UV-footprinting analyses showed that the pre-RC to post-RC conversion is observed at suppressed late origins after passage of the replication fork from neighboring origins (54). By analyses of single DNA molecules, it was demonstrated that this so-called origin interference does occur in human cells as well, maintaining a relatively constant interorigin distance (55). The precise mechanisms of origin interference are not known, although passive replication may well play a significant role in the phenomenon.

2.2.5. Regulation at the origin decision point. Analyses of the replication pattern of nuclei isolated from differentiated mammalian cells in *Xenopus* egg extracts resulted in the discovery that initiation takes place randomly in nuclei isolated from the early G1 phase. However, beyond a certain point during the G1 phase, called the origin decision point (ODP), nuclei exhibit site-specific initiation, characteristic of differentiated cells. The ODP is temporally distal to the timing decision point (TDP) yet occurs 2–3 h prior to the restriction point, and its execution requires kinase activity, proteolysis, and transcription (56). Transcription may eliminate subsets of the assembled pre-RC but is not sufficient for origin specification.

2.3. Plastic Regulation of Origin Selection

2.3.1. Highly plastic mode of eukaryotic DNA replication. Eukaryotic DNA replication exhibits a great deal of plasticity (**Figure 1**). In fact, studies by Taylor in 1977 indicated that the cellular replication program is dynamically regulated (57, 58). In those studies, Chinese hamster ovary cells were depleted of thymidine by treatment with the drug fluorodeoxyuridine (FdU). DNA synthesis was then analyzed by labeling of the newly synthesized DNA with tritiated thymidine and visualized by DNA fiber autoradiography. These experiments indicated that increased duration of exposure to FdU (more starved for thymidine) resulted in reduced distances between the labeled segments on DNA fibers and reduced rates of fork movement. Apparently, therefore, thymidine-starved cells compensate for the reduced fork movement by increasing the number of initiation sites to complete the S phase within a given time.

Nearly 30 years later, this finding was confirmed and extended by using a novel fluorescence microscopy technique (59). Using DNA segments spanning the adenosine deaminase 2 (*AMPD2*) gene, which was amplified in a Chinese hamster lung fibroblast cell line, it was clearly shown that cells with slower fork movement activate additional origins on this DNA segment. Conversely, accelerating the

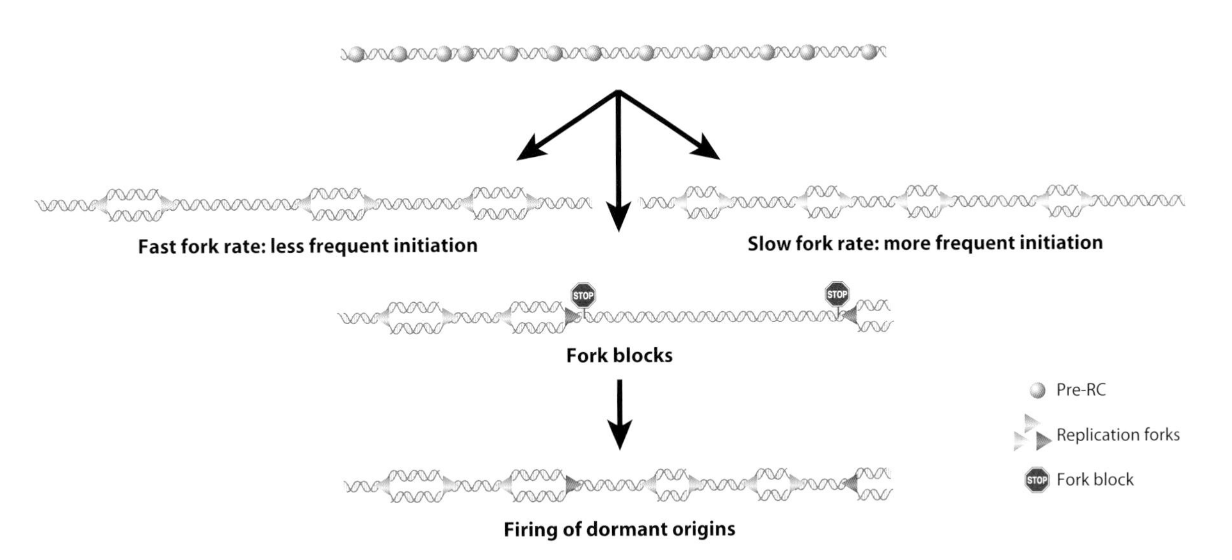

Figure 1

Plastic regulation of eukaryotic DNA replication. Eukaryotic genomes are equipped with potential replication origins (pre-RCs, *blue circles*) far exceeding in number what is actually required for the completion of genome replication. The initiation sites are selected from these potential origins. When the fork rate is fast, the initiation takes place less frequently (*left*), whereas initiation occurs more frequently when the fork rate is slower (*right*). When forks are stalled, the dormant (silent) origins fire (*center*).

Chromatin loop: looped DNA, whose root is juxtaposed and immobilized at a nuclear structure. Each loop contains coregulated genes or replicons

Replicon: a unit of DNA replication. On eukaryotic chromosomes, it is defined as a segment replicated from a single replicator (origin)

fork movement by adding adenine and uridine reduced the interorigin distance and restored the preferential firing at an efficient origin, resulting in a lower density of replication origins. Furthermore, fork rates are coordinately regulated with interorigin distances in the unperturbed normal cells (60). This highly plastic, reversible nature of the replication program permits the cells to respond to various internal and external interferences as well as to developmental requirements. Potential mechanisms by which plasticity of DNA replication programs may be achieved are considered below.

2.3.2. Chromatin loops and origin selection. It was reported that differentiated erythrocyte nuclei replicate only inefficiently (have fewer initiation events) in *Xenopus* egg extracts, but gain efficient replication ability (have high-frequency initiation) if they experience a single mitosis in egg extracts (61). This mitotic remodeling of chromatin is exemplified by the shortening of chromatin loops and recruitment of more initiator ORC proteins. Chromatin loops are known to regulate transcription by juxtaposing distal genomic loci and have been implicated in the determination of replicon sizes as well. The chromatin loop assays conducted with the above cells at different fork rates confirmed that loop sizes in G1 phase cells correlate with the interorigin distances in the preceding S phase. It was also shown that the loop sizes are established only after cells go though M phase. It was proposed that the "replication mark" left on the chromosome during the preceding S phase determines the loop sizes in the following M phase (62). In addition, the authors proposed that the mark may be related to the termination of replication forks present at the center of two active origins. Mitotic resetting in *Xenopus* egg extracts depends on topoisomerase II (61). Topoisomerase functions are essential for decatenation of two daughter strands at the location where two converging forks collide (63). Genome-wide mapping indicated the specific localization of topoisomerase II at the center of two origins (D. Fanchinetti & M. Foiani, personal communication). It is tempting to speculate that topoisomerase II involved in the decatenation of the daughter strands may contribute to the chromatin loop formation in the following M phase, keeping the memory of the origin density in the preceding S phase in the chromatin structures.

2.3.3. Dormant origins facilitate the maintenance of genomic integrity. The MCM complexes are present on chromatin in a number far in excess of the actual initiation sites (30, 31). What then are the roles of these seemingly surplus origins? It was reported in yeast that reduced numbers of pre-RCs can cause genomic instability (64, 65). Inhibition of pre-RC formation in human cancer cells by overexpression of the nondegradable form of geminin resulted in early or late S-phase arrest, eventually resulting in cell death (66). More recently, the effects of decreased pre-RC assembly on cell survival under normal and stress conditions were examined in *Xenopus* egg extracts and human cells (67–69). These studies concluded that these "dormant" origins are utilized when forks are stalled. This would facilitate the replication of those segments of the genome, which would otherwise stay unreplicated owing to flanking stalled forks on both sides. Cells with reduced levels of pre-RCs proceed through S phase without any apparent defect under an unperturbed condition but exhibit increased sensitivity to replication fork stress.

The adjustment of origin firing in response to fork progression rate appears to occur rapidly. The firing of additional origins takes place within a matter of minutes after the fork slows down, and the decrease in the overall number of active origins occurs within 2 h after a shift to "fast fork" conditions (62). The firing of dormant origins in response to fork stalling would also be a rapid cellular response. How does fork slowing or stalling induce activation of dormant or silent origins? One model hypothesizes that origin firing is a stochastic process and that the stalling fork increases the chance of dormant origins to be fired (see below). Under normal conditions, these origins would be replicated passively, and are inactivated (70). Other models hypothesize that the stalling fork alters the local chromatin structure or topology of the template DNA, thus creating a permissive structure for activation of the nearby pre-RCs (69). Indeed, the introduction of a site-specific double-strand break activates initiation at silent origins nearby (71). The effect appears to be local, and activation seems to take place earlier than expected from random activation, which may support the second model.

It has been generally assumed that stalling of the replication fork by hydroxyurea or aphidicolin, the condition used to slow the fork rate in the above experiments, induces checkpoint reactions and inhibits further origin firing (58, 72). It is unclear how the same treatment can induce opposite outcome. The conditions used for activation of dormant origins are generally much milder than those used for checkpoint activation (68, 69). Cells may differentially respond to the changing environment to adopt the most suitable cellular responses. Alternatively, as discussed above, origin recruitment may be a localized effect, affecting only the origins in the same cluster. The fork rate may dynamically affect the chromatin loop size during the S phase, activating additional origins (small loops) or suppressing potential origins (larger loops), when it is slowed or accelerated, respectively.

2.4. Localization of Chromatin within the Nucleus

2.4.1. Nuclear positioning is important for regulation of chromosome functions. Eukaryotic chromosomes are nonrandomly positioned within the nucleus. Each chromosome occupies a separate nonoverlapping position (interphase chromosome territory) in the nucleus (reviewed in References 73 and 74). Within each chromosome, the condensed heterochromatin tends to localize at the nuclear periphery (reviewed in Reference 75), and specific positioning within the nucleus is also observed for the chromatin regions where efficient transcription or DNA replication occurs. In budding yeast, proteins involved in transcriptional silencing, such as Sir4p, are abundantly present in discrete regions at the nuclear periphery, and positioning of chromosomal loci within these regions appears to facilitate transcription repression. Both the mating-type loci and telomeres of *Saccharomyces cerevisiae* tend to localize at the nuclear periphery (76). The nuclear periphery has been proposed to consist of a mosaic of microenvironments, some of which favor silencing and others transcriptional activation (77–79). It has been proposed that the DNA replication machinery is anchored at specific nuclear structures called the "replication factories." See Replication Factories: Site of DNA Replication? in the **Supplemental Material** for more details. Follow the **Supplemental Material link** from the Annual Reviews home page at **http://www.annualreviews.org**.

Checkpoint: a surveillance system that ensures ordered progression of the cell cycle

Replication timing: timing within the S phase when a segment of genome is replicated. Each chromosome consists of many segments of different replication timing

3. WHEN?

When within the S phase does each replication origin start firing?

3.1. Timing of DNA Replication within the S Phase

In normal human cells, the S phase lasts 6–8 h. The genomes are divided into distinct domains; some domains replicate early in the S phase, and others replicate later in the S phase (**Figure 1**). Replication timing appears to be predetermined in a given cell type. It has been known that transcriptionally inactive heterochromatin, contained in Giemsa dark chromosome bands or G bands, replicates late during the S phase, whereas Giemsa light or R bands containing euchromatin, where most transcription takes place, replicate early (80). However, as described in the following section, the timing seems to be related to the larger chromatin and nuclear architecture.

3.1.1. Determinants of timing regulation. In budding yeast, early- and late-firing origins have been clearly identified (81). Although the pre-RC is formed at all the potential origins at the late M/early G1 phase, activation of the pre-RC occurs in a genetically determined order on a chromosome (82). Telomeres are major elements that confer a late-replicating property (83). The replication timing of the origins near telomeres (within 10 kb) is rendered late as a result of the telomeric chromatin structure because the mutation in Sir3, required for spreading the silent domain, causes them to replicate early (84). The search for determinants of replication timing in late-firing origins distal to telomeres also led to identification of specific sequences that render early replicating origins to become late replicating (85).

In fission yeast, the definition of early- and late-firing origins is less clear. There is certainly a set of efficient origins that fire in the presence of hydroxyurea, and other origins are inefficient and normally do not fire under the same conditions (25, 86). It is not known whether these inefficient origins actually fire during the normal course of S phase or are passively replicated, although Cdc45 loading may take place in late S phase at some of these inefficient origins (87). However, the presence of sequences that inhibit early firing has been suggested; a GC-rich sequence was identified, which enforces late replication on ARS-driven plasmid replication in fission yeast (88). Relocation of an inefficient origin to the early replicating segment and that of an efficient origin to the late-replicating segment led to reversal of the replication timing in each case (25).

The mechanisms that regulate replication timing in metazoans remain elusive. Replication in *Xenopus* egg extracts occurs continuously throughout the S phase without apparent timing regulation in the entire genome. Incubation of nuclei from differentiated cells at various cell cycle stages in *Xenopus* egg extracts indicated that nuclei from M or early G1 phases replicate in an "embryonic" manner, whereas those from after the TDP within the G1 phase replicate with a unique timing regulation (in a differentiated manner) (89). However, the nature of the transition that occurs at the TDP is unknown. It was proposed that nuclear structures may play a role in determining the timing of DNA replication. Distinct patterns of DNA replication foci have been observed for mammalian cells in early, mid-, and late S phase. It was proposed also that anchoring of DNA to specific nuclear structures may determine the time at which the segment replicates within the S phase (80, 90). In fact, the transcriptionally inactive *β-globin* locus in Chinese hamster ovary cells becomes localized to the nuclear periphery between one and two hours after mitosis and then replicates late during the S phase within the same time frame as peripheral heterochromatin. This defines the TDP for the *β-globin* locus and suggests that establishment of the replication timing program coincides with its association with the nuclear periphery (90).

In budding yeast, late-replicating origins tend to localize close to the periphery of the nucleus specifically during the G1 phase, whereas early replicating origins tend to show random localization. Excision of one of the late-replicating origins (the ARS501-containing segment) as a circular plasmid during mitosis results in early replication, and excision during the G1 phase does not affect late replication of the excised ARS501-containing plasmid (91). Hence, interaction of the segment adjacent to ARS501 with nuclear periphery during G1 is required for establishing the late-replicating properties of ARS501. Furthermore, peripheral localization is required for its establishment but not for its maintenance. Indeed, ARS501 is abundantly found in nuclear periphery in the early G1 phase, but this distribution is not maintained in S phase, whereas early-firing origins are randomly localized within the nucleus throughout the cell cycle (92). These results suggest that the spatial arrangement of nuclear positions during the G1 phase may play a crucial role in determination of the replication timing in yeast, consistent with the above TDP proposal in mammalian cells.

3.1.2. Chromatin structures and replication timing. Deletion of Rpd3, a histone deacetylase, causes earlier firing of origins in budding yeast (93, 94). This is accompanied by advanced binding of Cdc45, an essential factor for generation of a replisome, to origins, suggesting that the firing step is accelerated under the condition of increased histone acetylation. A genome-wide search of initiation events indicates that more than 100 late-firing origins are deregulated in the absence of Rpd3L, one of the Rpd3 complexes (95). The Rpd3S complex (which is functionally distinct, but retains histone deacetylation activity) also affects replication timing, suggesting that histone deacetylation directly influences the initiation timing. Furthermore, a recent report indicated that not only histone acetylation but also methylation can accelerate S-phase progression and loading of Cdc45 at origins (96). Accelerated progression of S phase and Cdc45 loading at origins in the *rpd3* mutant are not observed in the histone H3 *K36/37R* mutant background or in the absence of Set2, a histone methyltransferase that methylates histone H3 K36. Further analyses indicate that the loading of Cdc45 at origins correlates with increased histone H3 K36 me1 and decrease of K36 me3, suggesting the opposing effects of these methylation events on initiation activities. These results indicate that the combination of multiple histone modifications is required for ordered activation of replication origins.

Special chromatin structures, called heterochromatin, constitute inactive chromatin. Generally, heterochromatin segments replicate late in the S phase. However, in fission yeast, two heterochromatic segments at centromere and mating-type switch loci replicate early in the S phase (97). Early replication of these heterochromatin segments may be achieved through interaction of the Swi6/HP1 heterochromatin-binding protein with the Dfp1/Him1 subunit (the fission yeast ortholog of Dbf4) of Hsk1 kinase (the fission yeast ortholog of Cdc7), which triggers advanced association of Sld3 (synthetically lethal with dpb11-1 mutant 3) with the centromeric pre-RC (98).

3.1.3. The *trans*-acting factors determining replication timing. Firing of replication origin depends on the phosphorylation events mediated by two kinases, the cyclin-dependent kinase (CDK) and the Dbf4-dependent Cdc7 kinase (Cdc7-Dbf4 or DDK) (99, 100). In budding yeast, firing of late origins is specifically abrogated in *clb5*Δ cells (one of the B-type cyclins), indicating that Clb5-dependent Cdc28 kinase plays a crucial role in activation of late origins (101). It was recently shown that Cdc2-cyclinA may regulate the timing of DNA replication. Induced activation of Cdc2-cyclinA resulted in premature activation of late replication origins in human cells (102).

The Cdc7-Dbf4 kinase complex is required for activation of each origin in budding yeast (103, 104). In fission yeast, origin firing efficiency is accelerated by an increased level of the Hsk1 catalytic subunit (homolog of Cdc7) or Dfp1 protein, its activation subunit (homolog of Dbf4) (87, 105), suggesting that Hsk1 kinase may also be a critical and limiting regulator. Both CDK and Cdc7 are required for the loading of Cdc45 protein at the origins (106), and overproduction of Cdc45 also increased origin efficiency (87). In budding and fission yeasts, loading of Cdc45 at origins is regulated under a temporal program, i.e., Cdc45 is loaded at early-firing origins at the onset of S phase, whereas it is loaded at late-firing origins late in S phase (87, 107). Thus, differential regulation of origins by CDK and/or Cdc7 may explain their early or late firing. How can these kinases differentially regulate the early and late-firing origins? In view of the significant roles played by the chromatin structure in timing control, histone modification near the origin may regulate access to these kinases. Alternatively, a specific factor that preferentially associates with the pre-RC at the early-firing origins may serve as a mark for early activation by these kinases. Furthermore, a factor that selectively associates with late-firing origins could

Replisome: a putative protein complex responsible for the coordinated synthesis of leading and lagging strands at a replication fork

CDK: cyclin-dependent kinase

ATR: ataxia-telangiectasia-mutated (ATM) and Rad3-related

negatively regulate the firing of these origins, and phosphorylation events may counteract this inhibition. In any case, the epigenetic marks on the protein complex at origins are likely to play crucial roles in differential activation.

3.1.4. Global genome organization and replication timing. Replication timing is generally regulated at a global genome level, and loss of a single origin does not affect global timing regulation. This is especially true in higher eukaryotes. Thus, each replication domain may be a functionally compartmentalized unit that coordinately regulates the origins contained within the domain. At first glance, the replication origins in budding yeast may appear to be regulated individually. However, large domains (>130 kb) contain multiple origins that initiate coordinately (85). Recent genome-wide analyses of origin firing in *clb5*Δ mutants indicated that the budding yeast genome can be divided into a Clb5-independent early-firing domain and a Clb5-dependent late-firing domain, suggesting that the replication domains may exist also in budding yeast (101).

In metazoan, replication timing is determined at the large chromosomal domain. A replication domain can be as long as several megabases, which may contain just a single replicon or more than 10 replicons (26, 108–111). Inactivation of any particular origin may not affect the overall timing of the segment (35). In fact, replication timing is independent of the site at which replication is initiated. The fact that the TDP occurs independent of the ODP supports this notion. Global genome organization may be coupled to subnuclear repositioning and may also be linked to specialized chromatin structures.

3.1.5. Checkpoint regulation of replication timing. In order to coordinate fork progression with progression of S phase and to maintain the integrity of the S-phase chromosome, eukaryotic cells monitor the progression of the replication fork and elicit the so-called intra-S-phase checkpoint, when they are threatened by replication stresses. This checkpoint response reduces the total rate of DNA replication, which can be achieved by inhibiting the firing of new origins and/or by slowing down ongoing replication forks (112, 113).

In budding yeast, inhibition of new origin firing appears to be a major pathway. The firing of late origins is restrained by checkpoint mechanisms, because a mutation in *mec1* or *rad53*, sensor or effector kinase, respectively, leads to early activation of the late origins in the presence of hydroxyurea, which inhibits the ongoing DNA synthesis owing to depletion of nucleotide precursors (114, 115). In mammalian cells, both the initiation and elongation phases of DNA replication are inhibited in a manner dependent on the Chk1 effector kinase (72, 112, 116).

Checkpoint kinases regulate origin firing even in the absence of fork stress. Ataxia-telangiectasia (AT) cells lacking Ataxia-telangiectasia mutate (ATM) proteins were originally observed to replicate more rapidly than the wild-type cells (117). Likewise, the rate of bulk DNA synthesis increased in cells exposed to caffeine, which inhibits both ATM and ATR (ataxia-telangiectasia-mutated and Rad3-related) protein kinases (118). Furthermore, reduction of Chk1 protein levels caused elevated origin firing and increased DNA synthesis during an unperturbed S phase (119, 120). Similarly, deletion of Cds1, a checkpoint effector kinase of fission yeast, exhibits unusual replication foci, potentially reflecting the perturbed temporal replication program (121). These results indicate that ordered firing of replication origins during a normal S phase is under the regulation of checkpoint functions.

3.1.6. Is the timing of replication stochastic or predetermined? Analyses of origin activation in *Xenopus* egg extracts and more recently in fission yeast led to the proposal that origins are selected from preformed pre-RCs in a stochastic manner. Mathematical calculation predicts that large gaps may remain unreplicated in the stochastic model (the "replication gap" problem) (122). This could be circumvented by the increased firing frequency in the late S phase (122, 123). In fission yeast, DNA fiber analyses of origin firing at specific loci during two consecutive rounds of S phase led to the conclusion that origin firing occurs stochastically at the preformed pre-RC sites (124). Chromosome-wide single-molecule analyses of replication kinetics in budding yeast showed that no two molecules have the same replication pattern, but averaging of all the patterns reproduced the results of bulk timing experiments (82), indicating that origin firing is probabilistic even in budding yeast. Firing of dormant origins in the presence of replication stress could be explained also by stochastic firing (70).

Genome-wide mapping of pre-RC assembly sites and initiation sites on the fission yeast genome indicates that replication is initiated only at the pre-RC sites (25). The location and extent of pre-RC assembly is highly reproducible, and pre-RCs are present on average once every 10 kb (25). Out of these origins, approximately 40% are utilized as efficient initiation sites in the presence of hydroxyurea (Y. Kanoh, personal communication). This selection is also highly reproducible and is not completely random. There is an obvious hierarchy in the selection of active initiation events. The data support the presence of preferred potential origins, which are selected for firing before other nonpreferred origins. Among the preferred origins that may exist in a cluster, the selection may be stochastic and exhibit variability among different cells or during the consecutive cell cycle. What then determines the hierarchy?

It was recently reported that loading of the ORC in fission yeast is cell cycle regulated, and it was proposed that the timing of ORC loading (thus pre-RC formation) during the late M phase may determine the initiation efficiency at the S phase (87). Prolongation of the M phase resulted in decreased initiation at "strong" origins and increased initiation at "weak" origins, and this may reflect decreased and increased ORC loading, respectively, at these origins during the M phase.

Overexpression of Hsk1, Dfp1, or Cdc45, but not Cdt1, resulted in a global increase of origin firing in fission yeast (87, 105). This suggests that the factors required for pre-RC activation are limiting. Because the levels of pre-RCs accumulating at early and late origins are roughly the same, it may not be the amount of a pre-RC that determines the preference. Advanced formation of the pre-RC at early origins may provide more time for the epigenetic or posttranslational modifications required for being preferentially targeted by Cdc7. Alternatively, there may be additional factors that are differentially recruited to early and late origins in a manner dependent on the timing of the pre-RC formation.

3.1.7. Transcription, chromatin, and replication timing: lessons from genome-wide analyses. A positive correlation between transcription and early replication has been well documented. Genome-wide analyses conducted in *Drosophila* and human cells support a positive correlation between transcription and replication timing (108, 110, 125, 126), whereas those in budding yeast revealed no significant correlation (81). The relationship in higher eukaryotes is indirect because 10% to 20% of late-replicating genes are expressed, and some genes change transcription without changes in replication timing and vice versa. The entire male X chromosome in C18 cells (isolated from wing imaginal discs) is early replicating, whereas the female X chromosome in Kc cells (derived from an embryo) contains late-replicating segments in spite of their similar transcription profiles. The enrichment of acetylated Lys16 of histone H4 may be responsible for early replication of the former chromosome (127). Thus, it would be accurate to state that replication timing is related to chromatin and nuclear architecture rather than transcription itself (109). (Refer to **Supplemental Table 4**.)

3.2. Developmental Regulation of Replication Timing

Differential regulation of replication timing in different cell types is most clearly illustrated at the *β-globin* locus, which replicates early in erythroid cells and late in nonerythroid cells (128–131). An approximately 1-Mb segment surrounding the *β-globin* gene constitutes the acetylated, DNase I-sensitive domain in erythroid cells, whereas it is within the hypoacetylated, DNase I-resistant domain in nonerythroid cells. The naturally occurring deletion of the locus control region, located 50–75-kb upstream of the *β-globin* locus, renders the transcription inactive and also makes the entire globin locus late replicating (130).

Another well-established example of the developmental regulation of replication timing is X chromosome inactivation in female mammals, which accompanies a shift from early to late replication of the inactive X chromosome in the epiblast of ~6.0 dpc (days postcoitum) mice (132). These facts suggest that replication timing is also under the control of development and differentiation. However, a limited number of studies in human transformed cells indicate that replication timing was unexpectedly similar between different cell types. In fact, a microarray-based comparison of the human chromosome 22 between fibroblast and lymphoblastoid cells revealed that only 1% of this chromosome differed in replication timing (see **Supplemental Table 4**) (108).

A dynamic change of replication timing was first demonstrated in timing analyses of mouse embryonic stem (ES) cells before and after induction of differentiation (111). Genome-wide replication timing in various ES cells as well as that in induced pluripotent stem cells from fibroblast cells was identical, indicating that replication timing could be viewed as a stable epigenetic mark specific to an individual cell type. However, upon induction of differentiation into neural precursor cells in vitro, a rather dramatic reorganization of replication domains (changes in up to 20% of the entire genome) was observed. Differentiation of ES cells lead to the consolidation of smaller differentially replicating domains into larger coordinately replicated domains (111). Extensive replication timing changes were also reported in *Drosophila* Kc (of embryonic origin) and C18 (from wing imaginal discs) cell lines or in human cell lines (127, 133). These results clearly indicate that replication timing undergoes major changes during the course of development.

Another determinant of replication timing is the boundary that defines the transition between early and late replication domains. The timing boundary segments may define the transition of the timing by two potential mechanisms. One may slow down the fork progression rate so that the early replication fork may not reach the late segment before the latter segment is fired for initiation. Alternatively, it may represent the chromosome domain, which might physically and spatially separate different replication timing domains within nuclei. A recent genome-wide microarray analysis has demonstrated the existence of molecular boundaries between the coordinately replicating units of chromosomes (109, 111). The results indicate that timing transition regions define long segments without origins of unidirectional replication.

It was reported that the timing transition seems to correlate with the junction between the syntenic segments. Accumulations of single-nucleotide polymorphisms and causative mutations for various diseases were reported in the timing transition segment (134). This may be due to the presence of replication fork blockages in these regions, impeding fork progression and eventually resulting in DNA damage. This would lead to an enhanced mutation rate owing to errors in the NHEJ (nonhomologous DNA end joining) or error-prone lesion bypass DNA synthesis. These fragile sites scattered on the genome have been implicated in chromosome breakage, leading to genome rearrangement (135). Fragility increases in a mutant of *mec1* from budding yeast or in the ATR-deficient cells of human (136, 137). These so-called fragile sites are now regarded as replication slow zones or fork pause sites of the eukaryotic genome. A potential link between the timing transition regions and fragile sites has also been reported (138). It is important to identify more precisely the timing transition segments and dissect the molecular events that take place in these regions.

4. HOW?

How is the entire process of DNA replication operated and regulated?

4.1. Initiation of Eukaryotic DNA Replication

4.1.1. Assembly of the prereplicative complex and its regulation. The initiation of eukaryotic DNA replication is a precisely regulated event that requires the ordered assembly of multiple protein complexes at replication origins (for reviews see References 1 and 139). The processes of replication initiation are divided into two steps (**Figure 2**). In the first step, the ORC, Cdc6, Cdt1, and MCM2~7 proteins are sequentially assembled on the chromosome at the late M to early G1 phase, generating a pre-RC. In the second step, pre-RCs are activated to generate active replication forks. At the G1-S transition, the activities of two kinases, the Cdc7-Dbf4 and the CDK, facilitate the loading of other essential replication proteins onto the pre-RC to activate the replicative DNA helicase and initiate chain elongation by DNA polymerases (99, 140–144).

Loading of the MCM complex onto DNA is referred to as DNA replication licensing. Upon entry into the S phase, multiple mechanisms ensure that no new pre-RC is formed, so that rereplication of any portion of the genome does not take place (139, 145). A major pathway for prevention of relicensing in metazoan is through regulation of Cdt1 activity by degradation or by its specific inhibitor, geminin. The degradation of Orc1 (146) and inhibition of reassociation of the MCM complex to chromatin may also be a part of the mechanisms for inhibition of relicensing. In yeast, inhibition of relicensing is achieved mostly through CDK-dependent mechanisms, including rapid CDK-dependent destruction of the Cdc6/Cdc18 protein during the S phase. After the loading of MCM2~7, the putative MCM loading machinery, comprising the ORC, Cdc6, and Cdt1, is no longer required for initiation of DNA replication, indicating that its primary function in DNA replication is to deliver MCM2~7 to the origins (147–149).

The ORC is an ATP-regulated DNA-binding protein that was first discovered in budding yeast (1, 2). The mammalian ORC consists of a stable core complex, composed of subunits ORC2 through ORC5, that interacts weakly with ORC1 and ORC6 (150). In human cell lines, ORC1 is tightly bound to the G1-phase chromatin and selectively degraded during S phase by ubiquitin-dependent proteolysis, and then it is resynthesized during the M-to-G1-phase transition (151).

The ATP-bound ORC binds to origin DNA, and then Cdc6, in a complex with ATP, binds to the ORC on DNA. The MCM2~7 and Cdt1, likely in a complex, join the ORC-Cdc6-DNA complex (the MCM complex is "associated" with chromatin and salt sensitive). ATP bound to Cdc6 is now hydrolyzed, and Cdc6 dissociates from the complex. At this step, dissociation of Cdt1 is also stimulated, which triggers stable binding of the MCM complex with DNA (the MCM complex is "loaded" onto chromatin and salt resistant) (152). ATP hydrolysis by the ORC is required for reiterative loading of the MCM complex, and ORC mutants defective in ATP hydrolysis cannot support pre-RC formation in vivo and are inviable (153). Electron microscopy studies showed that the ORC-Cdc6 complex forms a ring-shaped structure on DNA with dimensions similar to those of the ring-shaped MCM helicase (154). This implies that the ORC-Cdc6 ring binds the MCM ring to facilitate MCM helicase loading onto the origin (152).

Cdt1 plays a critical role in the licensing process. Loading of Cdt1 to chromatin is dependent on the ORC and Cdc6 (1, 155). Orc6 interacts with Cdt1 directly, thereby facilitating the loading of the MCM2~7 complex onto origin DNA (156). Cdt1 forms a complex with the MCM complex both in budding yeast and in mammalian cells (152, 157, 158) and plays a licensing role in the assembly

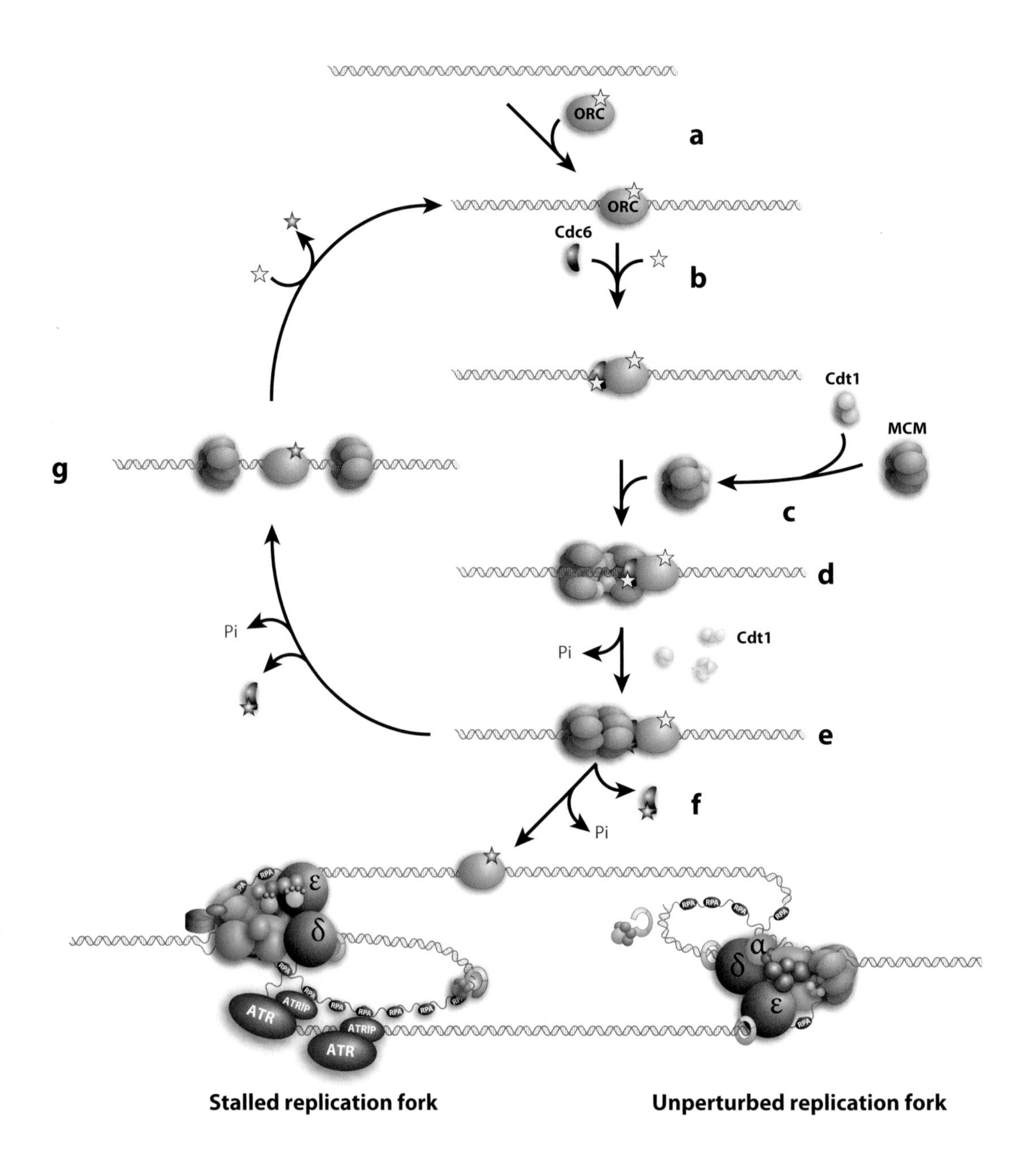
ORC
a
ORC
Cdc6
b
Cdt1
MCM
g
c
d
Pi
Cdt1
Pi
e
f
Pi
ε
δ
α
δ
ε
ATRIP
ATR
ATRIP
ATR
RPA
Stalled replication fork
Unperturbed replication fork

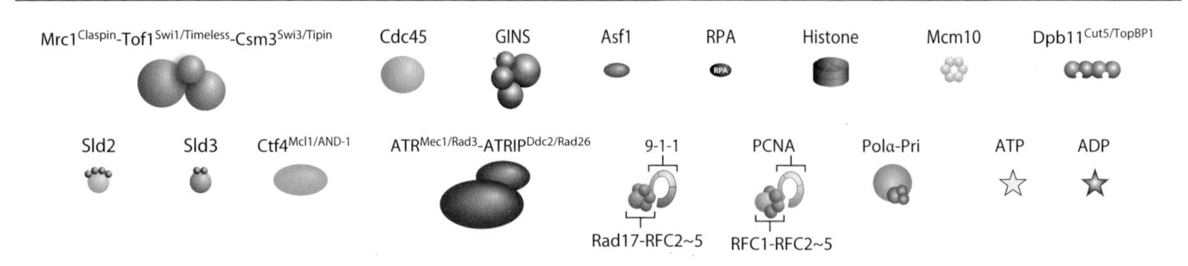
Mrc1Claspin-Tof1Swi1/Timeless-Csm3Swi3/Tipin
Cdc45
GINS
Asf1
RPA
Histone
Mcm10
Dpb11Cut5/TopBP1
Sld2
Sld3
Ctf4Mcl1/AND-1
ATRMec1/Rad3-ATRIPDdc2/Rad26
9-1-1
Rad17-RFC2~5
PCNA
RFC1-RFC2~5
Polα-Pri
ATP
ADP

of the pre-RC. The level of Cdt1 protein oscillates during the cell cycle, from being high in the G1 phase, low in S phase, and high again at the M-G1 transition (159). Geminin is destabilized during the G1 phase but accumulates during the S, G2, and M phases of the cell cycle. Geminin directly binds to Cdt1 and prevents the loading of the MCM2~7 complex onto DNA (160–165). Cdt1 is degraded during the S phase by ubiquitination-dependent proteolysis in a CDK-dependent manner or through interaction with proliferating cell nuclear antigen (PCNA), and the remaining Cdt1 is inhibited by geminin. This double regulation ensures that Cdt1 is not present during the S phase, resulting in strict prevention of relicensing of replication origins (155, 166–171). Recent reports indicate a possible regulation of stability or cellular localization of Cdc6 and Cdt1 proteins by acetylation (172, 173). c-Myc, a frequently deregulated proto-oncogene, is required for DNA replication during the G1 phase after pre-RC formation and before firing, and its deregulation induces replication-dependent DNA damage (174). The precise roles of c-Myc in regulation of the pre-RC remain to be investigated. Phosphorylation of the N-terminal segment of MCM2 by the Cdc7 kinase was reported to be important for pre-RC formation during the quiescence to S phase transition (175), although it is not known whether this regulation operates during the G1 phase of proliferating cells.

PCNA: proliferating cell nuclear antigen

Mechanistic studies regarding the initiation of DNA replication revealed striking conservation of the basic mechanisms through evolution (**Supplemental Figure 1**) (for more details, see the section Comparison of Initiation at Bacterial and Eukaryotic Origins in the **Supplemental Material**).

4.2. DNA Chain Elongation of Eukaryotic DNA Replication

4.2.1. Assembly of the replication fork and regulation. Conversion of the pre-RC to a replication fork complex (replisome) requires the association of additional factors. GINS [*Go-Ichi-Ni-San* (*Go-Ichi-Ni-San* represents the numbers 5, 1, 2, 3 in Japanese) for Sld5-Psf1-Psf2-Psf3] has been identified as a novel factor for the replisome (**Supplemental Figure 2**) (176–178). Its components were originally discovered through genetic

Figure 2

Assembly of the pre-RC and its activation. (*a*) The ATP-bound origin recognition complex (ORC) binds to potential origin sequences. (*b*) ATP-bound Cdc6 associates with the ORC. (*c*) The minichromosome maintenance (MCM) complex in association with Cdt1 joins the complex. (At this step, Cdt1 may not need to be in a complex with the MCM. Instead, it may bind to the ORC by itself.) (*d*) The MCM complex becomes "associated" with origin. (*e*) Hydrolysis of ATP on Cdc6 leads to "loading" of the MCM complex and release of Cdt1. (*f*) Dissociation of Cdc6 from origin and hydrolysis of ATP associated with the ORC completes the pre-RC assembly. (*g*) Alternatively, the MCM complex moves away and returns to the first step, repeating the loading of the MCM complex (reiterate loading). At the bottom of the figure (*right*), unperturbed replication fork; (*left*) stalled replication fork, which generates longer ssDNA segments that are coated by RPA and ATRIP. $ATR^{Mec1/Rad3}$-$ATRIP^{Ddc2/Rad26}$, a complex of ataxia-telangiectasia-mutated and Rad3-related (a phosphatidylinositol 3-kinase, which plays a major role in DNA replication stress checkpoint responses) and ATR interacting protein (superscripts represent the nomenclatures in budding and fission yeasts); Asf1, anti-silencing function 1 (histone H3-H4 chaperone); Cdc45, an essential protein for a functional helicase at the fork; $Ctf4^{Mc11/AND\text{-}1}$, a conserved replication fork factor whose functions are not precisely elucidated (superscripts represent the nomenclatures in fission yeast and higher eukaryotes); $Dpb11^{Cut5/TopBP1}$, DNA polymerase B 11 (required for initiation and checkpoint responses; superscripts represent the nomenclatures in fission yeast and higher eukaryotes); GINS, *Go-Ichi-Ni-San* complex (essential for a functional helicase at the fork); Mcm10, an essential factor for replication; $Mrc1^{Claspin}$-$Tof1^{Swi1/Timeless}$-$Csm3^{Swi3/Tipin}$, conserved factors required for stable maintenance of the replication fork (superscripts represent the nomenclatures in fission yeast and higher eukaryotes); PCNA-RFC-RFC2~5, proliferating nuclear antigen-replication factor C (clamp and clamp loader for DNA chain elongation); Polα-Pri, DNA polymerase α-primase; 9-1-1-Rad17-RFC2~5, Rad9-Rad1-Hus1 complex-Rad17-replication factor C2~5 (clamp and clamp loader for DNA damage responses); RPA, replication protein A; Sld2 or Sld3, synthetically lethal with *dpb11-1* mutant 2 or 3 (required for assembly of initiation complex and a target of CDK phosphorylation; *small orange dots* on Sld2 and Sld3 indicate phosphorylation).

RPC: replisome progression complex

CMG complex: Cdc45, MCM, and GINS complex

Mrc1: mediator of the replication checkpoint 1

screening for synthetic lethal mutations in combination with the *dpb11-1* mutant (*dpb11* is the Cut5-TopBP1 homolog) in budding yeast. They form a stable four-factor complex and are essential for DNA replication in both budding yeast and *Xenopus* egg extracts. GINS is necessary for stable engagement of Cdc45 with the nascent replisome and then associates stably with the MCM complex during S phase (179, 180). The crystal structure of a recombinant human GINS has been published, showing that it is a heterotetrameric complex (181–184). One report demonstrated that GINS does not contain a hole in the center and that the inner surface of the flat central cleft is unsuitable for DNA binding (181). Other reports showed that GINS has a ring-like structure with a small central channel (183, 184).

The replisome at the replication fork is assumed to be composed of many factors and to be involved not only in DNA replication but also in other chromosome transactions and chromatin regulation. This prediction was substantiated by the discovery of replisome progression complexes (RPCs) containing more than 20 replication-related proteins (185). In addition to the Cdc45, the MCM complex, and the GINS complex (referred to as CMG complex), the RPC contains Mrc1^{Claspin}, Tof1$^{\text{Swi1/Timeless}}$, and Csm3$^{\text{Swi3/Tipin}}$ (fork stabilizing factors). Swi1 and Swi3 from fission yeast were found to generate a stable complex termed FPC (fork protection complex) (186). Ctf1$^{\text{Mcl1/AND-1}}$, and components of the histone chaperones including FACT, were also identified in the RPC. Other replication factors such as topoisomerase I and MCM10 are associated weakly with the RPC (185).

The human GINS complex interacts physically with DNA polymerase-primase α and markedly stimulates its polymerase function but not its priming function (187). In contrast, the archaeal GINS homolog (from *Sulfolobus solfataricus*) is composed of two subunits and forms a complex with the MCM helicase. It directly interacts with the heterodimeric core primase but has no effect on the primase activity. These results suggest that the GINS complex may be important in coordinating the progression of the MCM helicase and priming events on the replication fork (188). Also, GINS in fission yeast is necessary for chromatin binding of DNA polymerase ε but is not required for that of DNA polymerase α (189). The *Pyrococcus furiosus* GINS complex interacts with cognate MCM complex and stimulates its ATPase and DNA helicase activities in vitro (190). Recent reports indicate that the budding yeast Ctf1$^{\text{Mcl1/AND-1}}$, required for efficient S-phase progression, interacts with GINS and DNA polymerase α and coordinates the progression of the MCM helicase and DNA polymerase α (191, 192). Thus, GINS-Ctf1$^{\text{Mcl1/AND-1}}$ coordinates duplex unwinding and DNA chain elongation at the replication fork. For more recent information on DNA polymerases, the reader is encouraged to read the most recent reviews (193–195).

In budding yeast, Sld2 and Sld3, which are targets of CDK phosphorylation, are required for the initiation of DNA replication (196–198). Sld2 and Sld3 were isolated as *sld* mutants (199). The Dpb11$^{\text{Cut5/TopBP1}}$ was identified as a multicopy suppressor of mutations in DNA polymerase ε (200). Dpb11$^{\text{Cut5/TopBP1}}$ has two pairs of tandem BRCT domains and is required for checkpoint reactions (200–202). The C-terminal pair of the BRCT domains in Dpb11$^{\text{Cut5/TopBP1}}$ binds to the phosphorylated Sld2 (Thr 84), and the N-terminal pair binds to the phosphorylated Sld3 (Thr 600 and Ser 622) (197, 203). The preformed DNA polymerase ε-Sld2-Dpb11$^{\text{Cut5/TopBP1}}$-GINS complex may bind to the Sld3-Cdc45 complex on the pre-RC in a manner dependent on CDK-mediated Sld3 phosphorylation (H. Araki, personal communication). Furthermore, the phosphomimetic mutant of Sld2-T84D, in combination with a mutant Sld3 (S600A + S622A) fused to Dpb11$^{\text{Cut5/TopBP1}}$ or in combination with the *cdc45/JET1* mutant, bypassed the CDK requirement for initiation of DNA replication, indicating that CDK-mediated phosphorylation of Sld2 and Sld3 is sufficient for initiation.

In budding yeast, Sld3 interacts with Cdc45, and both proteins associate with replication origins. Sld3 is required for interaction between the MCM complex and Cdc45 (204). Dpb11$^{Cut5/TopBP1}$, Sld3, Cdc45, and GINS assemble in a mutually dependent manner on replication origins to initiate DNA synthesis (176). After the establishment of DNA replication forks at early origins, Sld3 is no longer essential for the completion of chromosome replication, but Cdc45 and GINS associate stably with the MCM complex throughout the S phase (179). In contrast, in fission yeast, Sld3 functions in the initial phase of initiation complex assembly, followed by loading of GINS, Cut5/Dpb11, and then Cdc45 (142). The Hsk1 kinase, the fission yeast homolog of Cdc7, is required for chromatin loading of Sld3. In *Xenopus* extracts, Cdc7 must function before CDK for active DNA replication (205, 206). RecQ4L may be the metazoan homolog of Sld2, and it is also a target of CDK phosphorylation (207). Although the human ortholog of Sld3 has not been identified yet, the interaction of replication proteins dependent on CDK-mediated phosphorylation might be well conserved. Recently, a vertebrate factor with functions potentially related to Sld3 were reported (208, 209).

4.2.2. Activation of replicative helicase. Upon initiation of DNA replication in the S phase, the MCM complex moves away from replication origins as part of the DNA replication fork machinery (210–213). In the past decade, persuasive evidence has accumulated that suggests that the MCM2~7, a ring-shaped hexameric complex, is the replicative DNA helicase. Like most other replicative DNA helicases, the MCM2~7 complex consists of six subunits, containing highly conserved DNA-dependent ATPase motifs in their central regions (1, 214). The MCM proteins form several stable subcomplexes, including MCM2-3-4-5-6-7 (MCM2~7), MCM2-4-6-7, MCM4-6-7, and MCM3-5 (215–217). Biochemical characterization of the mouse MCM4-6-7 subcomplex revealed that it has intrinsic DNA-dependent ATPase and DNA helicase activities (215, 218). Later, helicase activity was reported in the MCM4-6-7 complex from other species as well (216, 217, 219). The activity and processivity of the MCM4-6-7 helicase can be highly stimulated by tailed substrate DNA (52, 216). The DNA helicase activity of the mouse MCM4-6-7 complex is highly enhanced also by the presence of thymine-rich ssDNA, which is often found near replication origins (52). Moreover, human FACT, a chromatin transcription factor, physically interacts with the MCM complex and promotes its DNA unwinding activity on forked nucleosomal templates in vitro (220).

Despite rather compelling genetic and biochemical evidence that the MCM complex is a replicative helicase at the fork, the isolated MCM2~7 complex does not show any detectable DNA helicase activity. Recently, however, such activity was detected in a purified recombinant budding yeast MCM2~7 complex, in a reaction mixture containing acetate and glutamate instead of chloride (221). It is yet to be determined whether the MCM2~7 complexes from other species exhibit DNA helicase activity under similar conditions.

As stated above, the CMG complex was identified (222), and evidence strongly suggests that both Cdc45 and GINS contribute to its helicase activity in vivo (210, 212, 213). Recently, the CMG helicase complex from *Drosophila* was reconstituted with purified components (223).

CDK and Cdc7-Dbf4 are the two conserved protein kinases involved in activation of the MCM helicase in the pre-RC (99, 100, 214). Evidence indicates that the MCM2~7 complex is a target of phosphorylation by Cdc7, which facilitates the loading of the Cdc45 protein onto origins (140, 141). Cdc7 phosphorylates mainly the N-terminal tail segments of MCM2, MCM4, and MCM6 proteins, and this phosphorylation is highly stimulated by prior phosphorylation by CDK in vitro (224). Cdc7 is an acidophilic kinase and favors the negatively charged segments generated by phosphorylation (225). Consistent with this finding, prior phosphorylation is required for Cdc7-mediated phosphorylation of in vitro assembled pre-RCs (144). The phosphorylated N-terminal tails of the MCM2, -4, or -6 proteins are recognized by Cdc45 or a complex containing Cdc45. These phosphorylation events appear to be redundant, and combinations of different N-terminal tail mutations lead to defective growth in fission yeast (141). It was reported recently that phosphorylation of S170 of budding yeast MCM2, a target site of Cdc7, is required for cell growth (226). The formation of the pre-RC on a yeast replication origin in vitro leads to the assembly of a head-to-head double-hexameric MCM connected through the N-terminal rings on a duplex DNA (227, 228). It is tempting to speculate that Cdc7-mediated phosphorylation of the N-terminal tail region may separate the two hexamers and convert them into active helicase complexes.

In budding yeast, the *mcm5-bob1* (carrying a proline to leucine substitution at position 83) bypasses Cdc7-Dbf4 (229) probably by conferring conformational change within the MCM complex, which would mimic the effect of Cdc7-mediated phosphorylation. The *bob1* mutation also reduces origin efficiency, and it was suggested that the mutant adopts several conformations, only one of which is active for origin activation (230–232). More recently, another bypass of Cdc7 was reported (233). Deletion of the N-terminal nonconserved segment of MCM4 containing the Cdc7 phosphorylation sites showed bypass of Cdc7 function. It was proposed that the function of Cdc7 is to antagonize the inhibitory effect of this segment on initiation. Several studies have provided information on the MCM structure, and several models for its mode of action have been proposed. Novel MCM-related proteins have been reported also. See the Structural Basis of MCM and Its Mode of Action and MCM-Related Proteins in the **Supplemental Material** for more details.

4.3. G1 Regulation and DNA Replication

Proliferation of mammalian cells is regulated by external growth signals, which induce a series of G1 signal transduction pathways ultimately leading to initiation of DNA replication. Key cell cycle regulators during G1 phase are Rb-E2F, Cdk4 and -6, and CDK inhibitors (**Figure 3**).

The best known link of these G1 regulators to DNA replication is the transcriptional induction of many replication factors, including licensing factors (Cdc6, Mcm7, and others) as well as Cdc45 by E2F at G1-S transition. Presumably E2F-mediated expression of these replication factors facilitates the licensing and initiation phases of DNA replication. These events are also subject to negative regulation. For example, induction of the CDK inhibitor $p16^{Ink4a}$ disrupts not only PCNA but also pre-RC assembly via a mechanism involving reduction of Cdc6 and Cdt1 expression. Interestingly, Cdc6 was reported to regulate the *INK4* locus, suggesting a feedback loop involving mutually coordinated regulation of INK4 and Cdc6 (234).

G1 CDKs also play a direct role in regulation of replication factors. Cdc6 is stabilized during the G1 phase by CDK-mediated phosphorylation of serine 54, allowing a time window for pre-RC assembly (235, 236). This pathway is subject to negative control by DNA damage acquired during the G1 phase, which results in p53 activation, induction of p21, and inhibition of Cdk2. The resulting decreases in Cdk2-mediated phosphorylation of serine 54 lead to destabilization of Cdc6 and inhibition of replication licensing (236).

It is well established that the Rb-CDK signaling axis regulates DNA replication via transcriptional control and posttranslational modification of licensing and initiation factors. However, several studies suggest that Rb may also regulate DNA synthesis directly, independently of its transcriptional repression activities. For example, Rb can interact directly with MCM7 (237), and the Rb-E2F complex colocalizes with replication foci in mammalian cells and with the ORC in *Drosophila* (20, 238). Ectopic activation of Rb leads to disruption of PCNA function but not other replication factors (239). This transient inhibition of replication may play an important role in response to DNA damage by facilitating the transfer of freed PCNA to repair foci (240).

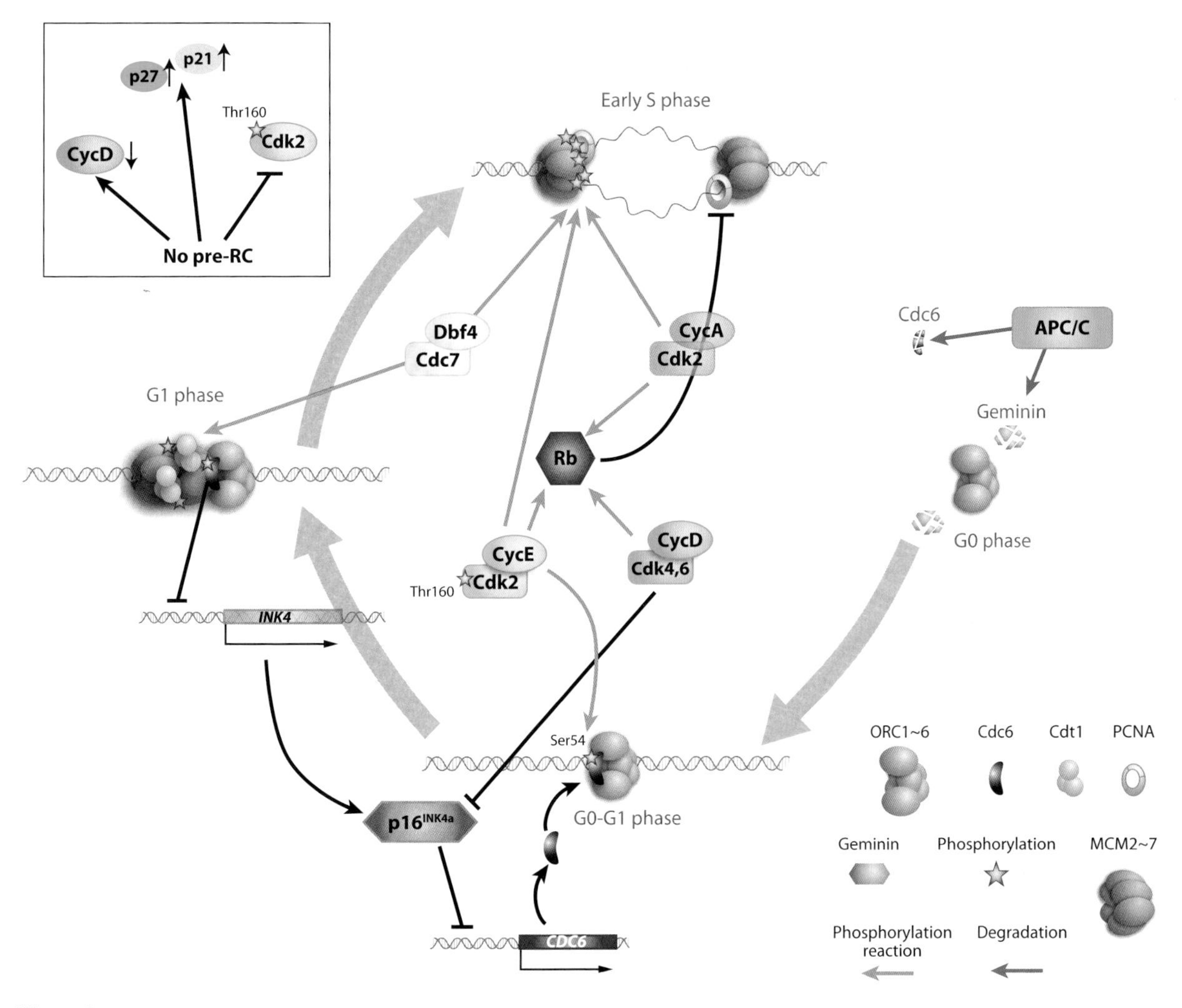

Figure 3

G1 regulation and DNA replication. In the transition from quiescence to growth, Geminin and Cdc6 are both degraded by APC/C. Cdc6, required for prereplicative complex (pre-RC) assembly, is transiently protected from degradation by Cdk2-CyclinE-mediated phosphorylation at Ser54, permitting pre-RC formation. Expression of Cdt1 and Cdc6, which is inhibited by $p16^{INK4a}$ (through Cdk4-CyclinD), as well as Cdc7-mediated phosphorylation of N-terminal segments of MCM2 are required for pre-RC formation. Cdc6 also regulates $p16^{INK4a}$ expression by interacting with its promoter region. DNA damage such as ionized irradiation induces p53, leading to inhibition of Cdk2-CyclinE activity by p21 (not shown in the figure). This leads to reduced phosphorylation of Ser54 in Cdc6 and reduced DNA replication. Pre-RC formation is required for expression of CyclinD and for activation phosphorylation of Thr160 of Cdk2. In the absence of a pre-RC, CDK activity is inhibited by p53-dependent loss of Thr160 phosphorylation and induction of p21 and p27 (inset). This is called a licensing checkpoint. Rb inhibits not only G1 regulators but also the S-phase replication machinery. Rb downregulates Cdk2-CyclinA. This in turn disrupts the proliferating cell nuclear antigen (PCNA) through an unknown mechanism (dissociation from chromatin). Cdc7 plays a crucial role in initiation of DNA replication by phosphorylating the N-terminal tails of MCM2, -4, and -6. Orange arrows indicate the phosphorylation reactions. Blue arrows indicate the degradation of the target proteins. Abbreviations: APC/C, anaphase promoting complex/cyclosome (E3 ubiquitin ligase); Cdc6, -7, cell division cycle 6, -7; Cdk2, cyclin-dependent kinase 2; Cdt1, cdc10-dependent transcripts 1; CycD and -E, CyclinD and -E; Dbf4, dumbbell former 4; G0-G1, quiescent to G1 phase; MCM2,-4,-6, minichromosome maintenance complex subunit 2, 4, 6; ORC1-6, origin recognition complex subunit 1-6; Rb, retinoblastoma.

Although G1 signaling via Rb and CDKs influences DNA replication, several recent studies also suggest that appropriate and timely activation of G1 CDKs and G1/S-phase progression is critically dependent on replication licensing. It was shown that impaired pre-RC assembly delays G1 phase progression via multiple mechanisms, including reduced transcriptional induction of Cyclin D1 (the activating binding partner of Cdk4/6), induction of CDK inhibitors p21 and p27, and p53-dependent loss of Thr160 phosphorylation of Cdk2 required for its activation (241–243). Further studies are necessary to determine the precise mechanisms of these novel licensing checkpoints.

These various mechanisms likely serve to integrate DNA replication with G1 to S-phase progression and may provide opportunities for dynamic and versatile regulation, thereby preventing progression into an aberrant S phase and/or facilitating the repair of damaged DNA. Interestingly, several studies have shown that licensing checkpoints are impaired in cancer cells, perhaps indicating that these novel checkpoints contribute to tumor suppression.

RECOGNITION OF STALLED REPLICATION FORKS IN BACTERIA

The stalled replication fork needs to be immediately recognized to elicit necessary cellular responses. In bacteria, a protein called PriA plays an essential role in recognition of fork stalls and replication restart (309). It specifically recognizes a stalled replication fork or processed stalled fork carrying the unelongated 3′ end of the nascent leading strand at the branch point (310). PriA contains a unique DNA-binding motif called a 3′ terminus-binding pocket (TT-pocket) structure in its N-terminal DNA-binding domain, and this domain, along with its C-terminal DEXH-type DNA helicase domain, is essential for recognition and high-affinity (Kd∼ <1 nM) binding of PriA at the stalled replication fork (311). The PriA protein generates a stable complex with the stalled forks and triggers reassembly of a replisome for restart of DNA replication. Although PriA is widely conserved in eubacterial species, structural and functional homologs of PriA in other kingdoms have not been identified yet, and it is not clear whether similar mechanisms operate in eukaryotes.

4.4. Detection of Fork Stalling and Cellular Responses to Stalled Replication Forks

The process of DNA replication, involving the separation of two DNA strands, inherently provokes a risky situation for the genomic DNA. Therefore, the chromosomes of S-phase cells are protected by layers of cellular strategies. Failure of these systems would immediately lead to chromosome instability and genetic alterations that could cause various diseases including malignancy (244).

It is known that oncogenic signals induce hyperproliferation, which may result in the generation of aberrant replication forks by misregulation of the replication licensing system or by direct hyperactivation of replication origins. This would lead to stalled replication forks, which may be dealt with by the cellular DNA damage response systems, inducing senescence or apoptosis in some occasions. If the DNA damage response system is compromised, cells would suffer from major genomic instability and would be converted to malignant tumors (245–247). Thus, the cellular mechanisms that deal with unexpected stalling of replication forks are crucial in protecting the genome from undergoing potentially lethal transformation.

4.4.1. Recognition of stalled replication forks. The progression of replication forks can be interfered with by external DNA insults and also by numerous internal factors. These include unusual secondary structures (repetitive sequences such as dinucleotide or trinucleotide repeats, specific G-rich nucleotide repeats, hairpins, quadruplexes, and triplexes), stalled transcription machinery, DNA-binding proteins or chromatin proteins blocking fork progression, or a reduced supply of nucleotides. In response to these obstacles, forks are stalled, or their rate of movement is reduced (136). The first step of the cellular response to stalled replication forks requires the recognition of such an event (see the sidebar titled Recognition of Stalled Replication Forks in Bacteria).

In eukaryotes, ssDNA generated at a stalled fork plays crucial roles in its cellular recognition. When forks are stalled, for example, by hydroxyurea, an inhibitor of ribonucleotide reductase, or by aphidicolin, an inhibitor of DNA polymerases, uncoupling of replicative helicase and DNA polymerases takes place, generating a ssDNA of sufficient length (248, 249). RPA binds to these ssDNA, and ATRIP in a complex with ATR brings this sensor/master kinase to the site of the fork stall (250).

However, the ssDNA-RPA complex is not sufficient for checkpoint activation. It was shown that the DNA synthesis function of DNA polymerase α is also required for checkpoint signaling. A primer-template structure (5′-recessed DNA), which may be synthesized on a lagging strand by DNA polymerase α, is required for loading of the Rad9-Rad1-Hus1 clamp, facilitated by the Rad17-RFC clamp loader (250–252). Rad17 and Rad9 are phosphorylated by ATR, and the phosphorylated Rad17 and Rad9 recruit Claspin and TopBP1, respectively (250). Subsequently, TopBP1 interacts with ATR and ATRIP, stimulating the kinase activity of ATR (253).

Using *Xenopus* egg extracts, various DNA structures were tested for their ability to activate an ATR-dependent checkpoint pathway. The results indicate that a primed ssDNA of sufficient length is required for efficient activation (254). These results are consistent with the requirement of DNA polymerase α, which generates DNA primer-template structures.

4.4.2. Regulation of the S-phase checkpoint. Replication fork stalling in response to various forms of DNA damage activates an S-phase checkpoint, which is mediated via the ATR-Chk1 signaling pathway. The S-phase checkpoint elicits the following reactions: (*a*) stabilization of stalled replication forks, (*b*) inhibition of initiation at unfired origins of replication, (*c*) inhibition of entry into G2 or M phase (through inactivation of Cdc25), and (*d*) slowing of ongoing fork progression.

mec1 (an ATR homolog) or *rad53* (a Chk2 homolog, functionally equivalent to Chk1) mutations in budding yeast repress late origins in the presence of hydroxyurea (114, 115). In mammalian cells, a known target of Chk1 is the tyrosine phosphatase Cdc25A, which activates Cdk2-CyclinE to facilitate the loading of Cdc45 onto origins (255). Chk1 phosphorylates serine 123 of Cdc25A, targeting the phosphatase for Skp1-Cullin-F-box/β-transducin repeat-containing protein ($SCF^{\beta TrCP}$)-mediated proteolysis, leading to inactivation of Cdk2 and unloading of Cdc45 in response to fork blocks. However, recent results indicate that Chk1 also induces dissociation of Cdc45 from chromatin in a Cdc25A- and Cdk2-independent manner (256).

RPA: replication protein A

Several studies have implicated Cdc7 as a target of the S-phase checkpoint. For example, it was shown using *Xenopus* egg extracts that etoposide induces dissociation of Cdc7 from Dbf4, thus inhibiting further firing of origins. Importantly, in those experiments, addition of Dbf4 counteracted etoposide-induced inhibition of origin firing, suggesting that Dbf4 is a target of the S-phase checkpoint (257). However, the putative mechanisms by which ATR/Chk1 signaling may negatively regulate Cdc7-Dbf4 in *Xenopus* are not known. Moreover, recent studies using cultured mammalian cells have shown that Cdc7 remains active in response to replication stress-inducing agents and is required for viability after genotoxin treatments (258). Therefore, it is not yet clear whether Cdc7 is a target of the intra-S checkpoint in all experimental systems (258–260).

Recent reports indicate that stalled replication forks not only block the initiation of late origins but also slow down the rate of fork movements (112, 261). How does the activated Chk1 slow down the fork? The $Tof1^{Swi1/Timeless}$ and $Csm3^{Swi3/Tipin}$, and $Mrc1^{Claspin}$ proteins, which are conserved replication fork factors, are required for checkpoint kinase activation (186, 262, 263, 264, 265). $Mrc1^{Claspin}$ is hyperphosphorylated in a manner dependent on the $Rad3^{Mec1/ATR}$ kinase in response to fork arrest (266, 267). In the mutant of *mrc1* or *tof1*, extensive unwinding of DNA occurs in the presence of hydroxyurea, which is mediated by the helicase components uncoupled from DNA synthesis (268). A similar extensively unwound DNA with a ssDNA tract of 500 to 800 bases was observed also in a *rad53* mutant (249). Thus, these proteins may antagonize the helicase action at the replication fork, contributing to the stabilization of stalled forks as well as to reduction of fork rate. Indeed, replication fork slowing in response to UV requires Tipin (261).

Another possible mechanism for fork pausing could involve recombination. Recombination may antagonize the slowing of replication forks by inducing template switching, causing a quick bypass reaction in yeasts. The *rqh1*Δ (encoding a RecQ-type DNA helicase) mutant, which displays hyperrecombination, does not slow forks in response to fork stresss (269). Thus, suppression of recombination is important for the slowing of replication forks in response to fork stress. Indeed, Mrc1^{Claspin} was recently reported to be required for suppression of homologous recombination (270). By contrast, homologous recombination functions are required for fork slowing in vertebrates (271).

4.4.3. Roles of conserved fork stabilization factors for normal fork progression. Mrc1^{Claspin} is required for efficient fork movement under normal growth conditions both in yeasts and mammalian cells (272–275). Slow S phase is not observed in *mrc1-AQ*, which is specifically defective in checkpoint responses due to the alanine substitutions of Rad3$^{\text{Mec1/ATR}}$-target SQ/TQ sequences, indicating that Mrc1^{Claspin} stimulates fork progression in a checkpoint-independent manner. Tof1$^{\text{Swi1/Timeless}}$ is also required for a normal DNA replication rate in human cells (261). The requirement of budding yeast Tof1$^{\text{Swi1/Timeless}}$ for normal DNA replication is controversial. In experiments using BrdU incorporation, Tof1$^{\text{Swi1/Timeless}}$ was shown to be required for a normal replication rate (276), whereas it is not required for a normal S-phase progression rate in density-shift experiments (273).

Mrc1^{Claspin} is required for recovery of DNA synthesis from hydroxyurea arrest but not from methylmethane sulfate arrest in budding yeast (276). This role also appears to be independent of checkpoint function. Tof1$^{\text{Swi1/Timeless}}$ is also required for fork recovery, but bulk replication is more efficiently restored in *tof1*Δ than in *mrc1*Δ, presumably because more efficient initiation/elongation takes place from late origins in *tof1*Δ during recovery. Tof1$^{\text{Swi1/Timeless}}$ but not Mrc1^{Claspin} is required for fork arrest at rDNA, at tRNA, or at centromeres (273). In fission yeast, Swi1 is required to pause at rDNA or at mating-type loci (277, 278). The current consensus is (*a*) Mrc1^{Claspin} and Tof1$^{\text{Swi1/Timeless}}$ proteins play distinct roles in the normal maintenance of the replication fork. (*b*) Mrc1^{Claspin} is required for efficient fork progression, and this function is independent of its checkpoint function. (*c*) Tof1$^{\text{Swi1/Timeless}}$ is required for fork arrest at specific loci. Fork stalling at artificial inverted repeat sequences was enhanced in *mrc1* or *tof1* mutants, suggesting that Mrc1^{Claspin} and Tof1$^{\text{Swi1/Timeless}}$ may be required to overcome the fork barrier owing to a secondary structure of DNA (279).

How does Mrc1^{Claspin} stimulate the progression of replication fork? Mrc1^{Claspin} interacts with fork factors, including Cdc45 and MCM. It can also bind preferentially to branched DNA in vitro (280, 281). These data suggest that Mrc1^{Claspin} is an integral fork factor linking MCM and other replication factors. Recently, Mrc1^{Claspin} was found to interact with DNA polymerase ε through both its N- and C-terminal sequences (282). This interaction occurs throughout the cell cycle and may serve to link a polymerase component to the helicase component. Indeed, Mrc1^{Claspin} was shown to interact with the C-terminal tail of MCM6 through its central domain. Thus, the physical link of a polymerase to the helicase may facilitate the progression rate of the replication fork (283). Stimulation of the fork movement by coupling of a replicative polymerase and helicase was reported previously in *E. coli* (284). Upon hyperphosphorylation of Mrc1^{Claspin} in response to hydroxyurea, the Mrc1N-DNA polymerase ε interaction is specifically lost. The loss of this interaction may cause a conformational change in DNA polymerase ε for checkpoint signaling. Alternatively, the released N-terminal portion of $Mrc1^{Claspin}$ may interact with other fork components to stabilize the stalled fork (282). Recently, a novel concept for intra-S-phase checkpoint was proposed (285). This model argues that the checkpoint slows down replication in all the origins. See the Novel Concept of Intra-S Checkpoint Regulation in the **Supplemental Material** for a more detailed discussion.

4.5. Chromatin Regulation of DNA Replication

Chromatin assembly is an integral part of eukaryotic DNA replication. See Chromatin Regulation of DNA Replication in the **Supplemental Material** for a more detailed discussion.

4.6. The Interplay between DNA Replication and other Chromosome Transactions

Chromosome dynamics during cell cycle progression is regulated in a coordinated manner. Recent reports emphasized the intricate coupling of the processes of DNA replication and other chromosome transactions. See Interplays Between DNA Replication and Other Chromosome Transactions in the **Supplemental Material** for additional discussion.

5. BEYOND DNA REPLICATION

5.1. DNA Replication and Diseases

Ample evidence indicates that defects in the process of DNA replication could result in genomic instability, which could lead to mutations and abnormal tissue growth as observed in cancer. It can also cause a group of diseases known as "chromosome instability syndromes." These diseases are in most cases predisposed to cancer because of various gene rearrangements or mutations. A hypomorphic mutation in MCM4 ($MCM4^{Chaos3}$) is known to cause severe genomic instability, and 80% of Chaos3 females die of mammary adenocarcinoma (286). Mutant mice expressing low levels of MCM2 exhibited severe deficiencies in proliferative cell compartments of various tissues and died at an early age owing to the development of lymphoma (287). The above defect may be due to a deficiency in a pre-RC formation. In contrast, deregulation of replication licensing also leads to G2/M checkpoint activation owing to head-to-tail collision of replication forks and generation of short-rereplicated DNA fragments (288). Cdt1 overproduction in rat or human fibroblasts induced chromosome aberrations without inducing rereplication (289, 290). Thus, perturbation of pre-RC formation, either inhibition or stimulation, can lead to DNA damage or a chromatin aberration, which could eventually lead to carcinogenesis. RecQL4, a potential mammalian homolog of Sld2, is responsible for Rothmund-Thomson syndrome, a rare autosomal recessive genodermatosis with achromosome fragility characterized by high incidences of skin and bone cancers (207). Thus, a defect or aberration in the processes of DNA replication can profoundly affect the maintenance of genomic integrity in mammals.

By contrast, normal and cancer cells respond to depletion of replication factors in significantly different manners (See **Supplemental Table 5**. Follow the **Supplemental Material link** from the Annual Reviews home page at **http://www.annualreviews.org**). Generally, cancer cells are very sensitive to depletion of replication factors and undergo cell death, which occurs in both p53-positive and -negative cells (291). For examples, Cdc7 depletion in cancer cells results in abortive S-phase progression and cell death after abnormal mitosis. In contrast, normal cells arrest in G1 phase in a p53-dependent manner (292, 293). Cdc6 depletion causes an S-phase defect and induces cell death in cancer cells (294). In normal cells, Cdc6 depletion induces an ATR-dependent checkpoint and blockade of DNA replication, thus escaping cell death (295). Downregulation of the ORC or depletion of an

MCM subunit may induce a licensing checkpoint, resulting in inhibition of CDK activity and delayed S-phase entry (see **Figure 3**) (241, 242, 296). In contrast, cancer cells progress into a lethal S phase. Thus, replication factors may be promising targets for cancer therapy. (Please refer to **Supplemental Table 5** for effects of depletion of replication factors in various mammalian cell lines.) The use of replication factors as biomarkers has also been explored because many of the replication factors are overexpressed in tumors and cancer cells (297).

Increasing evidence points to the role of replication timing in the regulation of chromatin structures and nuclear positioning, which may affect not only transcription but also other chromosome transactions. Aberrations in replication timing are associated with various types of human cancers (298, 299). A change in replication timing, presumably a switch from early to late S phase, may contribute to carcinogenesis by inducing silencing of tumor suppressor genes. In addition, certain chromosomes carrying translocations have replication timing defects that are accompanied by delays in mitotic chromosome condensation and an increased frequency of chromosomal instability. Mutations in *Drosophila* ORC subunits result in death at late larval stages, with defects in replication timing and chromosome condensation, as well as in sister chromatid cohesion, indicating a link between replication timing and chromosome condensation and segregation (300, 301). The regions of hypomethylation in ICF (immunodeficiency, centromeric heterochromatin, facial anomalies) syndrome (caused by mutations in the *DNMT3B* DNA methyltransferase gene) are associated with advanced replication timing and nuclease hypersensitivity (302). Some X chromosomes are hypomethylated but still silenced; escape from silencing is only seen with early replicating X chromosomes, suggesting that replication timing is a major determinant of gene silencing. Histone deacetylase 2 (HDAC2) is recruited to replication sites through DNMT1 (DNA methyltransferase 1) during late, but not early, S phase, thus raising the possibility that late replication maintains the heterochromatin state through HDAC2 (303). These findings suggest that modification of replication timing could be a potentially effective strategy in developing efficient therapies through reversal of the gene silencing. In fact, incorporation of an active replication origin in an expression vector prohibited the silencing frequently observed in the original vector (304).

5.2. The Cell Cycle and Other Biological Cycles

Recent reports suggest that cell cycle and checkpoint control may be coupled to the regulation of other biological cycles involved in setting the circadian rhythm and metabolism. See the Cell Cycle and Other Biological Cycles in the **Supplemental Material** for a more detailed discussion.

6. CONCLUDING REMARKS

During the past several years, several unexpected findings were reported on the where and when problems of eukaryotic DNA replication. The initiation sites are specific but not in the sense that applies to bacterial replicons. Each origin is much more inefficient, but its specificity is more relaxed, and initiation sites are selected from the list of equally active pre-RC sites. The choice may be made when the chromatin loops are generated in the preceding M phase. Alternatively, it may be determined during the ODP in the G1 phase when the majority of potential origins are rendered inactive. The timing may be determined on large chromosome domains. This may be closely linked to the nuclear repositioning that may occur during the early G1 phase. It is still not clear at this stage exactly what kind of mark is left on the chromatin or what kind of higher chromatin structures are formed to differentiate the initiation sites and timing of replication during the S phase.

A remarkable finding, reported in the past years, was the dynamic coregulation of fork rate and initiation at new origins. This significant degree of robustness of the replication program was indeed first reported in 1977, but it was largely unnoticed (57). It is now clear that cells respond to perturbation of DNA replication (e.g., reduced or enhanced fork rate or fork blocks) by altering the selection of potential origins to be used for initiation. When forks are stalled or significantly slowed down, potential origins (not used under normal conditions) or dormant origins are used to compensate for the deficiency. Obviously, cells are equipped with a system that monitors the fork rates and rapidly adjusting the locations of new initiation within the ongoing S phase. This could be achieved by changing the chromatin loop sizes so that replication is initiated with less or more intervals. Alternatively, it could be explained simply by the stochastic nature of the selection of active origins. Elucidation of the precise mechanism of coupling between fork rate and initiation frequency is an important issue that needs to be tackled in the near future.

Genome-wide analyses using microarrays provided the genomic landscape of replication origins and timing. In *Drosophila* and mammals, a strong correlation was observed between transcription and replication initiation sites. Most actively transcribed genes direct "divergent" transcripts, generating a nucleosome-free, negatively supercoiled state in the promoter region (305), which is favorable not only for transcription but also for DNA replication (pre-RC formation and initiation). This may be the reason why replication origins are abundant in the promoter segment. Localization of the pre-RC was examined genome wide in yeasts and *Drosophila*, and enrichment in the promoter region was observed, suggesting that the promoter regions provide a favorable chromatin environment for pre-RC assembly. It remains to be seen whether pre-RCs are assembled in a similar manner in mammals.

Investigation of replication timing with microarray analyses presented irrefutable evidence that replication timing is regulated during development. During the differentiation of ES cells, the timing changes (accompanied by nuclear repositioning) are observed within nearly 20% of the genome at the level of a large domain (600 kb) (111). Replication timing can be regarded at present as an epigenetic mark and might affect the expression profile (306).

Significant progress has been made in our understanding of the mechanisms of pre-RC formation. Currently, this process can be reconstituted with purified proteins both in *Xenopus* and budding yeast. The loading of the active MCM complex on chromatin requires the ORC, Cdc6, and Cdt1, and this process is under highly strict regulation to prevent reinitiation events within the single S phase. Efforts to reconstitute with purified proteins the steps following the pre-RC formation (i.e., loading of Cdc45 and initiation of DNA synthesis) on a defined template containing a specific origin sequence (e.g., budding or fission yeast origins or *oriP* from a model viral replicon) are in progress. The long-awaited origin-specific initiation of eukaryotic DNA replication in vitro is within our reach in the near future.

The replisome components are strikingly conserved across species. Newly conserved factors (GINS, Sld2, and Sld3) have been identified, and Sld2 and Sld3 have been shown to be critical targets of Cdk2 for assembly of the replisome. Isolation of the RPC indicates that many of these factors indeed generate a physically stable complex. Accumulating evidence suggests that the MCM complex is a central component of the replicative helicase. However, the MCM complex alone may not be sufficient to act as a replicative helicase. Identification of the helicase-active CMG complex suggests that the eukaryotic replicative helicase may be more complex than the prokaryotic counterpart. Additional biochemical studies of the helicase complex will not only clarify the mode of action of the MCM helicase at the fork but also eventually dissolve the MCM paradox. Other fork factors, such as Tof1$^{\mathrm{Swi1/Timeless}}$, Csm3$^{\mathrm{Swi3/Tipin}}$, and Mrc1^{Claspin}, may stabilize or slow down the fork upon encountering a replication stress signal by antagonizing the helicase action. Biochemical analyses of these proteins in the context of replication fork structures are needed.

A deficiency of replication factors can cause cancer or various other serious diseases collectively known as the chromosome instability syndrome. Yet, these factors could be cellular targets of novel cancer therapies. The unexpected link between cell cycle checkpoint and other biological cycles warrants further studies of replication factors at cellular and animal levels to elucidate the molecular basis of the potentially coordinated regulation of biological cycles.

SUMMARY POINTS

1. The ORC from metazoan (and even that from lower eukaryotes) binds to DNA with very little specificity, although it may generally prefer AT-rich sequences. The specificity may be provided by other factors, including transcription factors or chromatin structures.
2. Actual initiation sites are selected from many potential origins, determined by the pre-RC assembly present on the chromosomes, and the selection step is affected by many factors, including the distal regulatory elements, transcription state, topology, chromatin context, or the presence of nearby active origins.
3. The process of DNA replication shows high plasticity, with the fork rate and origin selection processes cross regulating each other, and dormant origins are activated when forks are stalled.
4. In metazoan, replication timing, although it correlates well with transcription, may be ultimately determined by the global chromatin context.
5. The two golden rules of eukaryotic DNA replication are "two-cycle engine" (pre-RC assembly and its activation occurring during mutually exclusive periods of cell cycle) and "once-per-cycle replication," both of which are independent of where and when during the S phase replication is initiated.
6. Thus, stochastic or opportunistic initiation within a given replication domain is compatible with most of the known experimental data.
7. Replication stress induces a checkpoint, which represses both the fork progression rate and new origin firing.
8. A huge RPC containing the helicase, fork stabilizers, and histone chaperones can be isolated. The MCM complex is the central helicase component of RPC and may team up with GINS and Cdc45 to assemble an active helicase (the CMG complex).

FUTURE ISSUES

1. How is the site-specific initiation achieved? In other words, how are origins differentially selected from the list of pre-RCs? Alternatively, is this selection truly stochastic?
2. What are the exact mechanisms that determine the timing of origin firing? Are there any novel factors involved in defining the timing pattern of potential origins or setting the replication timing domains?

3. What are the mechanisms that coordinate fork rate and new origin firing during the S phase, and how are the stalled fork signals transmitted to activate dormant origins?
4. What is the nature of the active replicative DNA helicase at the eukaryotic replication forks and how does it operate?
5. What is the molecular architecture of the eukaryotic replication fork? Could assembly of a replication fork and initiation of DNA replication be reconstituted with purified proteins?
6. How is the replication fork reorganized upon encounter with fork blocks, and what are the exact mechanisms for inhibition of new firing, slowing the replication fork, and the eventual reinitiation of DNA replication?
7. How do aberrations in replication and checkpoint factors lead to various diseases, and what is the molecular basis of cancer cell–specific cell death induced by inhibition of replication factors?
8. How are the replication cycle and checkpoint regulation coordinated with the circadian cycle or metabolic cycle? Are there potential "extra-DNA replication" functions of replication factors?

DISCLOSURE STATEMENT

The authors are not aware of any affiliations, memberships, funding, or financial holdings that might be perceived as affecting the objectivity of this review.

ACKNOWLEDGMENTS

The authors thank Dave Gilbert, Nick Rhind, Cyrus Vaziri, Anindya Dutta, and Hiroyuki Araki for critical reading of the manuscript. We also thank Yutaka Kanoh for the excellent artwork. We thank all the members of our laboratory for their hard work and helpful discussions. We also thank Ken-ichi Arai for guidance in the field of DNA replication and continued support throughout the course of the work conducted in the laboratory of H.M.

LITERATURE CITED

1. Bell SP, Dutta A. 2002. DNA replication in eukaryotic cells. *Annu. Rev. Biochem.* 71:333–74
2. Bell SP, Stillman B. 1992. ATP-dependent recognition of eukaryotic origins of DNA replication by a multiprotein complex. *Nature* 357:128–34
3. Palzkill TG, Newlon CS. 1988. A yeast replication origin consists of multiple copies of a small conserved sequence. *Cell* 53:441–50
4. Marahrens Y, Stillman B. 1992. A yeast chromosomal origin of DNA replication defined by multiple functional elements. *Science* 255:817–23
5. Theis JF, Newlon CS. 1997. The ARS309 chromosomal replicator of *Saccharomyces cerevisiae* depends on an exceptional ARS consensus sequence. *Proc. Natl. Acad. Sci. USA* 94:10786–91
6. Chuang RY, Kelly TJ. 1999. The fission yeast homologue of Orc4p binds to replication origin DNA via multiple AT-hooks. *Proc. Natl. Acad. Sci. USA* 96:2656–61
7. Vashee S, Cvetic C, Lu W, Simancek P, Kelly TJ, Walter JC. 2003. Sequence-independent DNA binding and replication initiation by the human origin recognition complex. *Genes Dev.* 17:1894–908

8. Schaarschmidt D, Baltin J, Stehle IM, Lipps HJ, Knippers R. 2004. An episomal mammalian replicon: sequence-independent binding of the origin recognition complex. *EMBO J.* 23:191–201
9. Paixao S, Colaluca IN, Cubells M, Peverali FA, Destro A, Giadrossi S, et al. 2004. Modular structure of the human lamin B2 replicator. *Mol. Cell. Biol.* 24:2958–67
10. Altman AL, Fanning E. 2004. Defined sequence modules and an architectural element cooperate to promote initiation at an ectopic mammalian chromosomal replication origin. *Mol. Cell. Biol.* 24:4138–50
11. Wang L, Lin C-M, Brooks S, Cimbora D, Groudine M, Aladjem MI. 2004. The human beta-globin replication initiation region consists of two modular independent replicators. *Mol. Cell. Biol.* 24:3373–86
12. Debatisse M, Toledo F, Anglana M. 2004. Replication initiation in mammalian cells: changing preferences. *Cell Cycle* 3:19–21
13. MacAlpine DM, Bell SP. 2005. A genomic view of eukaryotic DNA replication. *Chromosome Res.* 13:309–26
14. Nieduszynski CA, Knox Y, Donaldson AD. 2006. Genome-wide identification of replication origins in yeast by comparative genomics. *Genes Dev.* 20:1874–79
15. Speck C, Stillman B. 2007. Cdc6 ATPase activity regulates ORC x Cdc6 stability and the selection of specific DNA sequences as origins of DNA replication. *J. Biol. Chem.* 282:11705–14
16. Crooke E, Hwang DS, Skarstad K, Thony B, Kornberg A. 1991. *E. coli* minichromosome replication: regulation of initiation at oriC. *Res. Microbiol.* 142:127–30
17. Remus D, Beall EL, Botchan MR. 2004. DNA topology, not DNA sequence, is a critical determinant for *Drosophila* ORC-DNA binding. *EMBO J.* 23:897–907
18. Houchens CR, Lu W, Chuang RY, Frattini MG, Fuller A, Simancek P, et al. 2008. Multiple mechanisms contribute to *Schizosaccharomyces pombe* origin recognition complex-DNA interactions. *J. Biol. Chem.* 283:30216–24
19. Beall EL, Manak JR, Zhou S, Bell M, Lipsick JS, Botchan MR. 2002. Role for a *Drosophila* Myb-containing protein complex in site-specific DNA replication. *Nature* 420:833–37
20. Bosco G, Du W, Orr-Weaver TL. 2001. DNA replication control through interaction of E2F-RB and the origin recognition complex. *Nat. Cell Biol.* 3:289–95
21. Minami H, Takahashi J, Suto A, Saitoh Y, Tsutsumi K. 2006. Binding of AlF-C, an Orc1-binding transcriptional regulator, enhances replicator activity of the rat aldolase B origin. *Mol. Cell. Biol.* 26:8770–80
22. Norseen J, Thomae A, Sridharan V, Aiyar A, Schepers A, Lieberman PM. 2008. RNA-dependent recruitment of the origin recognition complex. *EMBO J.* 27:3024–35
23. Wyrick JJ, Aparicio JG, Chen T, Barnett JD, Jennings EG, et al. 2001. Genome-wide distribution of ORC and MCM proteins in *S. cerevisiae*: high-resolution mapping of replication origins. *Science* 294:2357–60
24. Segurado M, de Luis A, Antequera F. 2003. Genome-wide distribution of DNA replication origins at A+T-rich islands in *Schizosaccharomyces pombe*. *EMBO Rep.* 4:1048–53
25. Hayashi M, Katou Y, Itoh T, Tazumi A, Yamada Y, et al. 2007. Genome-wide localization of pre-RC sites and identification of replication origins in fission yeast. *EMBO J.* 26:1327–39
26. MacAlpine DM, Rodriguez HK, Bell SP. 2004. Coordination of replication and transcription along a *Drosophila* chromosome. *Genes Dev.* 18:3094–105
27. Cadoret J-C, Meisch F, Hassan-Zadeh V, Luyten I, Guillet C, et al. 2008. Genome-wide studies highlight indirect links between human replication origins and gene regulation. *Proc. Natl. Acad. Sci. USA* 105:15837–42
28. Sequeira-Mendes J, Diaz-Uriarte R, Apedaile A, Huntley D, Brockdorff N, Gomez M. 2009. Transcription initiation activity sets replication origin efficiency in mammalian cells. *PLoS Genet.* 5:e1000446
29. Karnani N, Taylor CM, Malhotra A, Dutta A. 2010. Genomic study of replication initiation in human chromosomes reveals the influence of transcription regulation and chromatin structure on origin selection. *Mol. Biol. Cell* 21:393–94
30. Edwards MC, Tutter AV, Cvetic C, Gilbert CH, Prokhorova TA, Walter JC. 2002. MCM2–7 complexes bind chromatin in a distributed pattern surrounding the origin recognition complex in *Xenopus* egg extracts. *J. Biol. Chem.* 277:33049–57

31. Hyrien O, Marheineke K, Goldar A. 2003. Paradoxes of eukaryotic DNA replication: MCM proteins and the random completion problem. *BioEssays* 25:116–25
32. Kalejta RF, Li X, Mesner LD, Dijkwel PA, Lin HB, Hamlin JL. 1998. Distal sequences, but not ori-β/OBR-1, are essential for initiation of DNA replication in the Chinese hamster *DHFR* origin. *Mol. Cell* 2:797–806
33. Mesner LD, Hamlin JL. 2005. Specific signals at the 3′ end of the *DHFR* gene define one boundary of the downstream origin of replication. *Genes Dev.* 19:1053–66
34. Aladjem MI, Groudine M, Brody LL, Dieken ES, Fournier RE, Wahl GM et al. 1995. Participation of the human beta-globin locus control region in initiation of DNA replication. *Science* 270:815–19
35. Hayashida T, Oda M, Ohsawa K, Yamaguchi A, Hosozawa T, et al. 2006. Replication initiation from a novel origin identified in the Th2 cytokine cluster locus requires a distant conserved noncoding sequence. *J. Immunol.* 176:5446–54
36. Pasero P, Bensimon A, Schwob E. 2002. Single-molecule analysis reveals clustering and epigenetic regulation of replication origins at the yeast rDNA locus. *Genes Dev.* 16:2479–84
37. Pappas DLJ, Frisch R, Weinreich M. 2004. The NAD^+-dependent Sir2p histone deacetylase is a negative regulator of chromosomal DNA replication. *Genes Dev.* 18:769–81
38. Crampton A, Chang F, Pappas DLJ, Frisch RL, Weinreich M. 2008. An ARS element inhibits DNA replication through a SIR2-dependent mechanism. *Mol. Cell* 30:156–66
39. Iizuka M, Matsui T, Takisawa H, Smith MM. 2006. Regulation of replication licensing by acetyltransferase Hbo1. *Mol. Cell. Biol.* 26:1098–108
40. Miotto B, Struhl K. 2008. HBO1 histone acetylase is a coactivator of the replication licensing factor Cdt1. *Genes Dev.* 22:2633–38
41. Burke TW, Cook JG, Asano M, Nevins JR. 2001. Replication factors MCM2 and ORC1 interact with the histone acetyltransferase HBO1. *J. Biol. Chem.* 276:15397–408
42. Aggarwal BD, Calvi BR. 2004. Chromatin regulates origin activity in *Drosophila* follicle cells. *Nature* 430:372–76
43. Danis E, Brodolin K, Menut S, Maiorano D, Girard-Reydet C, Mechali M. 2004. Specification of a DNA replication origin by a transcription complex. *Nat. Cell Biol.* 6:721–30
44. Prioleau MN, Gendron MC, Hyrien O. 2003. Replication of the chicken β-globin locus: early-firing origins at the 5′ HS4 insulator and the ρ- and β^A-globin genes show opposite epigenetic modifications. *Mol. Cell. Biol.* 23:3536–49
45. Nieduszynski CA, Blow JJ, Donaldson AD. 2005. The requirement of yeast replication origins for prereplication complex proteins is modulated by transcription. *Nucleic Acids Res.* 33:2410–20
46. Donato JJ, Chung SC, Tye BK. 2006. Genome-wide hierarchy of replication origin usage in *Saccharomyces cerevisiae*. *PLoS Genet.* 2:e141
47. Snyder M, Sapolsky RJ, Davis RW. 1988. Transcription interferes with elements important for chromosome maintenance in *Saccharomyces cerevisiae*. *Mol. Cell. Biol.* 8:2184–94
48. Haase SB, Heinzel SS, Calos MP. 1994. Transcription inhibits the replication of autonomously replicating plasmids in human cells. *Mol. Cell. Biol.* 14:2516–24
49. Saha S, Shan Y, Mesner LD, Hamlin JL. 2004. The promoter of the Chinese hamster ovary dihydrofolate reductase gene regulates the activity of the local origin and helps define its boundaries. *Genes Dev.* 18:397–410
50. Dayn A, Malkhosyan S, Mirkin SM. 1992. Transcriptionally driven cruciform formation in vivo. *Nucleic Acids Res.* 20:5991–97
51. Seila AC, Core LJ, Lis JT, Sharp PA. 2009. Divergent transcription: a new feature of active promoters. *Cell Cycle* 8:2557–64
52. You Z, Ishimi Y, Mizuno T, Sugasawa K, Hanaoka F, Masai H. 2003. Thymine-rich single-stranded DNA activates Mcm4/6/7 helicase on Y-fork and bubble-like substrates. *EMBO J.* 22:6148–60
53. Brewer BJ, Fangman WL. 1993. Initiation at closely spaced replication origins in a yeast chromosome. *Science* 262:1728–31
54. Tadokoro R, Fujita M, Miura H, Shirahige K, Yoshikawa H, et al. 2002. Scheduled conversion of replication complex architecture at replication origins of *Saccharomyces cerevisiae* during the cell cycle. *J. Biol. Chem.* 277:15881–89

55. Lebofsky R, Heilig R, Sonnleitner M, Weissenbach J, Bensimon A. 2006. DNA replication origin interference increases the spacing between initiation events in human cells. *Mol. Biol. Cell* 17:5337–45
56. Dimitrova DS. 2006. Nuclear transcription is essential for specification of mammalian replication origins. *Genes Cells* 11:829–44
57. Taylor JH. 1977. Increase in DNA replication sites in cells held at the beginning of S phase. *Chromosoma* 62:291–300
58. Gilbert DM. 2007. Replication origin plasticity, Taylor-made: inhibition vs recruitment of origins under conditions of replication stress. *Chromosoma* 116:341–47
59. Anglana M, Apiou F, Bensimon A, Debatisse M. 2003. Dynamics of DNA replication in mammalian somatic cells: nucleotide pool modulates origin choice and interorigin spacing. *Cell* 114:385–94
60. Conti C, Sacca B, Herrick J, Lalou C, Pommier Y, Bensimon A. 2007. Replication fork velocities at adjacent replication origins are coordinately modified during DNA replication in human cells. *Mol. Biol. Cell* 18:3059–67
61. Lemaitre J-M, Danis E, Pasero P, Vassetzky Y, Mechali M. 2005. Mitotic remodeling of the replicon and chromosome structure. *Cell* 123:787–801
62. Courbet S, Gay S, Arnoult N, Wronka G, Anglana M, et al. 2008. Replication fork movement sets chromatin loop size and origin choice in mammalian cells. *Nature* 455:557–60
63. Hiasa H, Marians KJ. 1996. Two distinct modes of strand unlinking during theta-type DNA replication. *J. Biol. Chem.* 271:21529–35
64. Lengronne A, Schwob E. 2002. The yeast CDK inhibitor Sic1 prevents genomic instability by promoting replication origin licensing in late G_1. *Mol. Cell* 9:1067–78
65. Tanaka S, Diffley JFX. 2002. Deregulated G_1-cyclin expression induces genomic instability by preventing efficient pre-RC formation. *Genes Dev.* 16:2639–49
66. Shreeram S, Sparks A, Lane DP, Blow JJ. 2002. Cell type-specific responses of human cells to inhibition of replication licensing. *Oncogene* 21:6624–32
67. Woodward AM, Gohler T, Luciani MG, Oehlmann M, Ge X, et al. 2006. Excess Mcm2–7 license dormant origins of replication that can be used under conditions of replicative stress. *J. Cell Biol.* 173:673–83
68. Ge XQ, Jackson DA, Blow JJ. 2007. Dormant origins licensed by excess Mcm2–7 are required for human cells to survive replicative stress. *Genes Dev.* 21:3331–41
69. Ibarra A, Schwob E, Mendez J. 2008. Excess MCM proteins protect human cells from replicative stress by licensing backup origins of replication. *Proc. Natl. Acad. Sci. USA* 105:8956–61
70. Blow JJ, Ge XQ. 2009. A model for DNA replication showing how dormant origins safeguard against replication fork failure. *EMBO Rep.* 10:406–12
71. Doksani Y, Bermejo R, Fiorani S, Haber JE, Foiani M. 2009. Replicon dynamics, dormant origin firing, and terminal fork integrity after double-strand break formation. *Cell* 137:247–58
72. Merrick CJ, Jackson D, Diffley JF. 2004. Visualization of altered replication dynamics after DNA damage in human cells. *J. Biol. Chem.* 279:20067–75
73. Cremer T, Cremer M, Dietzel S, Muller S, Solovei I, Fakan S. 2006. Chromosome territories—a functional nuclear landscape. *Curr. Opin. Cell Biol.* 18:307–16
74. Heard E, Bickmore W. 2007. The ins and outs of gene regulation and chromosome territory organisation. *Curr. Opin. Cell Biol.* 19:311–16
75. Shaklai S, Amariglio N, Rechavi G, Simon AJ. 2007. Gene silencing at the nuclear periphery. *FEBS J.* 274:1383–92
76. Hediger F, Neumann FR, Van Houwe G, Dubrana K, Gasser SM. 2002. Live imaging of telomeres: yKu and Sir proteins define redundant telomere-anchoring pathways in yeast. *Curr. Biol.* 12:2076–89
77. Taddei A, Hediger F, Neumann FR, Bauer C, Gasser SM. 2004. Separation of silencing from perinuclear anchoring functions in yeast Ku80, Sir4 and Esc1 proteins. *EMBO J.* 23:1301–12
78. Kumaran RI, Spector DL. 2008. A genetic locus targeted to the nuclear periphery in living cells maintains its transcriptional competence. *J. Cell Biol.* 180:51–65
79. Meaburn KJ, Misteli T. 2008. Locus-specific and activity-independent gene repositioning during early tumorigenesis. *J. Cell Biol.* 180:39–50

80. Zink D. 2006. The temporal program of DNA replication: new insights into old questions. *Chromosoma* 115:273–87
81. Raghuraman MK, Winzeler EA, Collingwood D, Hunt S, Wodicka L, et al. 2001. Replication dynamics of the yeast genome. *Science* 294:115–21
82. Czajkowsky DM, Liu J, Hamlin JL, Shao Z. 2008. DNA combing reveals intrinsic temporal disorder in the replication of yeast chromosome VI. *J. Mol. Biol.* 375:12–19
83. Ferguson BM, Fangman WL. 1992. A position effect on the time of replication origin activation in yeast. *Cell* 68:333–39
84. Stevenson JB, Gottschling DE. 1999. Telomeric chromatin modulates replication timing near chromosome ends. *Genes Dev.* 13:146–51
85. Friedman KL, Diller JD, Ferguson BM, Nyland SV, Brewer BJ, Fangman WL. 1996. Multiple determinants controlling activation of yeast replication origins late in S phase. *Genes Dev.* 10:1595–607
86. Feng W, Collingwood D, Boeck ME, Fox LA, Alvino GM, et al. 2006. Genomic mapping of single-stranded DNA in hydroxyurea-challenged yeasts identifies origins of replication. *Nat. Cell Biol.* 8:148–55
87. Wu P-YJ, Nurse P. 2009. Establishing the program of origin firing during S phase in fission yeast. *Cell* 136:852–64
88. Yompakdee C, Huberman JA. 2004. Enforcement of late replication origin firing by clusters of short G-rich DNA sequences. *J. Biol. Chem.* 279:42337–44
89. Dimitrova DS, Gilbert DM. 1999. The spatial position and replication timing of chromosomal domains are both established in early G1 phase. *Mol. Cell* 4:983–93
90. Li F, Chen J, Izumi M, Butler MC, Keezer SM, Gilbert DM. 2001. The replication timing program of the Chinese hamster beta-globin locus is established coincident with its repositioning near peripheral heterochromatin in early G1 phase. *J. Cell Biol.* 154:283–92
91. Raghuraman MK, Brewer BJ, Fangman WL. 1997. Cell cycle-dependent establishment of a late replication program. *Science* 276:806–9
92. Heun P, Laroche T, Raghuraman MK, Gasser SM. 2001. The positioning and dynamics of origins of replication in the budding yeast nucleus. *J. Cell Biol.* 152:385–400
93. Vogelauer M, Rubbi L, Lucas I, Brewer BJ, Grunstein M. 2002. Histone acetylation regulates the time of replication origin firing. *Mol. Cell* 10:1223–33
94. Aparicio JG, Viggiani CJ, Gibson DG, Aparicio OM. 2004. The Rpd3-Sin3 histone deacetylase regulates replication timing and enables intra-S origin control in *Saccharomyces cerevisiae*. *Mol. Cell. Biol.* 24:4769–80
95. Knott SR, Viggiani CJ, Tavare S, Aparicio OM. 2009. Genome-wide replication profiles indicate an expansive role for Rpd3L in regulating replication initiation timing or efficiency, and reveal genomic loci of Rpd3 function in *Saccharomyces cerevisiae*. *Genes Dev.* 23:1077–90
96. Pryde F, Jain D, Kerr A, Curley R, Mariotti FR, Vogelauer M. 2009. H3 k36 methylation helps determine the timing of cdc45 association with replication origins. *PLoS One* 4:e5882
97. Kim SM, Dubey DD, Huberman JA. 2003. Early-replicating heterochromatin. *Genes Dev.* 17:330–35
98. Hayashi MT, Takahashi T, Nakagawa T, Nakayama J, Masukata H. 2009. The heterochromatin protein Swi6/HP1 activates replication origins at the pericentromeric region and silent mating-type locus. *Nat. Cell Biol.* 11:357–62
99. Masai H, Arai K-I. 2002. Cdc7 kinase complex: a key regulator in the initiation of DNA replication. *J. Cell. Physiol.* 190:287–96
100. Sclafani RA. 2000. Cdc7p-Dbf4p becomes famous in the cell cycle. *J. Cell Sci.* 113:2111–17
101. McCune HJ, Danielson LS, Alvino GM, Collingwood D, Delrow JJ, et al. 2008. The temporal program of chromosome replication: genomewide replication in *clb5*Δ *Saccharomyces cerevisiae*. *Genetics* 180:1833–47
102. Katsuno Y, Suzuki A, Sugimura K, Okumura K, Zineldeen DH, et al. 2009. Cyclin A-Cdk1 regulates the origin firing program in mammalian cells. *Proc. Natl. Acad. Sci. USA* 106:3184–89
103. Donaldson AD, Fangman WL, Brewer BJ. 1998. Cdc7 is required throughout the yeast S phase to activate replication origins. *Genes Dev.* 12:491–501
104. Bousset K, Diffley JF. 1998. The Cdc7 protein kinase is required for origin firing during S phase. *Genes Dev.* 12:480–90
105. Patel PK, Kommajosyula N, Rosebrock A, Bensimon A, Leatherwood J, et al. 2008. The Hsk1(Cdc7) replication kinase regulates origin efficiency. *Mol. Biol. Cell* 19:5550–58

106. Zou L, Stillman B. 2000. Assembly of a complex containing Cdc45p, replication protein A, and Mcm2p at replication origins controlled by S-phase cyclin-dependent kinases and Cdc7p-Dbf4p kinase. *Mol. Cell. Biol.* 20:3086–96
107. Aparicio OM, Stout AM, Bell SP. 1999. Differential assembly of Cdc45p and DNA polymerases at early and late origins of DNA replication. *Proc. Natl. Acad. Sci. USA* 96:9130–35
108. White EJ, Emanuelsson O, Scalzo D, Royce T, Kosak S, et al. 2004. DNA replication-timing analysis of human chromosome 22 at high resolution and different developmental states. *Proc. Natl. Acad. Sci. USA* 101:17771–76
109. Birney E, Stamatoyannopoulos JA, Dutta A, Guigo R, Gingeras TR, et al. 2007. Identification and analysis of functional elements in 1% of the human genome by the ENCODE pilot project. *Nature* 447:799–816
110. Karnani N, Taylor C, Malhotra A, Dutta A. 2007. Pan-S replication patterns and chromosomal domains defined by genome-tiling arrays of ENCODE genomic areas. *Genome Res.* 17:865–76
111. Hiratani I, Ryba T, Itoh M, Yokochi T, Schwaiger M, et al. 2008. Global reorganization of replication domains during embryonic stem cell differentiation. *PLoS Biol.* 6:e245
112. Seiler JA, Conti C, Syed A, Aladjem MI, Pommier Y. 2007. The intra-S-phase checkpoint affects both DNA replication initiation and elongation: single-cell and -DNA fiber analyses. *Mol. Cell. Biol.* 27:5806–18
113. Kumar S, Huberman JA. 2009. Checkpoint-dependent regulation of origin firing and replication fork movement in response to DNA damage in fission yeast. *Mol. Cell. Biol.* 29:602–11
114. Santocanale C, Diffley JF. 1998. A Mec1- and Rad53-dependent checkpoint controls late-firing origins of DNA replication. *Nature* 395:615–18
115. Shirahige K, Hori Y, Shiraishi K, Yamashita M, Takahashi K, et al. 1998. Regulation of DNA-replication origins during cell-cycle progression. *Nature* 395:618–21
116. Chastain PD 2nd, Heffernan TP, Nevis KR, Lin L, Kaufmann WK, et al. 2006. Checkpoint regulation of replication dynamics in UV-irradiated human cells. *Cell Cycle* 5:2160–67
117. Painter RB, Young BR. 1976. Formation of nascent DNA molecules during inhibition of replicon initiation in mammalian cells. *Biochim. Biophys. Acta.* 418:146–53
118. Shechter D, Costanzo V, Gautier J. 2004. ATR and ATM regulate the timing of DNA replication origin firing. *Nat. Cell Biol.* 6:648–55
119. Maya-Mendoza A, Petermann E, Gillespie DAF, Caldecott KW, Jackson DA. 2007. Chk1 regulates the density of active replication origins during the vertebrate S phase. *EMBO J.* 26:2719–31
120. Petermann E, Maya-Mendoza A, Zachos G, Gillespie DAF, Jackson DA, Caldecott KW. 2006. Chk1 requirement for high global rates of replication fork progression during normal vertebrate S phase. *Mol. Cell. Biol.* 26:3319–26
121. Meister P, Taddei A, Ponti A, Baldacci G, Gasser SM. 2007. Replication foci dynamics: Replication patterns are modulated by S-phase checkpoint kinases in fission yeast. *EMBO J.* 26:1315–26
122. Rhind N. 2006. DNA replication timing: random thoughts about origin firing. *Nat. Cell Biol.* 8:1313–16
123. Goldar A, Labit H, Marheineke K, Hyrien O. 2008. A dynamic stochastic model for DNA replication initiation in early embryos. *PLoS One* 3:e2919
124. Patel PK, Arcangioli B, Baker SP, Bensimon A, Rhind N. 2006. DNA replication origins fire stochastically in fission yeast. *Mol. Biol. Cell* 17:308–16
125. Schubeler D, Scalzo D, Kooperberg C, van Steensel B, Delrow J, Groudine M. 2002. Genome-wide DNA replication profile for *Drosophila melanogaster*: a link between transcription and replication timing. *Nat. Genet.* 32:438–42
126. Farkash-Amar S, Lipson D, Polten A, Goren A, Helmstetter C, et al. 2008. Global organization of replication time zones of the mouse genome. *Genome Res.* 18:1562–70
127. Schwaiger M, Stadler MB, Bell O, Kohler H, Oakeley EJ, Schubeler D. 2009. Chromatin state marks cell-type- and gender-specific replication of the *Drosophila* genome. *Genes Dev.* 23:589–601
128. Gilbert DM. 2002. Replication timing and transcriptional control: beyond cause and effect. *Curr. Opin. Cell Biol.* 14:377–83

129. Reik A, Telling A, Zitnik G, Cimbora D, Epner E, Groudine M. 1998. The locus control region is necessary for gene expression in the human beta-globin locus but not the maintenance of an open chromatin structure in erythroid cells. *Mol. Cell. Biol.* 18:5992–6000
130. Cimbora DM, Schubeler D, Reik A, Hamilton J, Francastel C, et al. 2000. Long-distance control of origin choice and replication timing in the human beta-globin locus are independent of the locus control region. *Mol. Cell. Biol.* 20:5581–91
131. Simon I, Tenzen T, Mostoslavsky R, Fibach E, Lande L, et al. 2001. Developmental regulation of DNA replication timing at the human beta globin locus. *EMBO J.* 20:6150–57
132. Takagi N, Sugawara O, Sasaki M. 1982. Regional and temporal changes in the pattern of X-chromosome replication during the early post-implantation development of the female mouse. *Chromosoma* 85:275–86
133. Hansen RS, Thomas S, Sandstrom R, Canfield TK, Thurman RE, et al. 2010. Sequencing newly replicated DNA reveals widespread plasticity in human replication timing. *Proc. Natl. Acad. Sci. USA* 107:139–44
134. Watanabe Y, Fujiyama A, Ichiba Y, Hattori M, Yada T, et al. 2002. Chromosome-wide assessment of replication timing for human chromosomes 11q and 21q: disease-related genes in timing-switch regions. *Hum. Mol. Genet.* 11:13–21
135. Debatisse M, El AE, Dutrillaux B. 2006. Common fragile sites nested at the interfaces of early and late-replicating chromosome bands: cis acting components of the G_2/M checkpoint? *Cell Cycle* 5:578–81
136. Cha RS, Kleckner N. 2002. ATR homolog Mec1 promotes fork progression, thus averting breaks in replication slow zones. *Science* 297:602–6
137. Casper AM, Nghiem P, Arlt MF, Glover TW. 2002. ATR regulates fragile site stability. *Cell* 111:779–89
138. El Achkar E, Gerbault-Seureau M, Muleris M, Dutrillaux B, Debatisse M. 2005. Premature condensation induces breaks at the interface of early and late replicating chromosome bands bearing common fragile sites. *Proc. Natl. Acad. Sci. USA* 102:18069–74
139. Blow JJ, Dutta A. 2005. Preventing rereplication of chromosomal DNA. *Nat. Rev. Mol. Cell Biol.* 6:476–86
140. Sheu Y-J, Stillman B. 2006. Cdc7-Dbf4 phosphorylates MCM proteins via a docking site-mediated mechanism to promote S phase progression. *Mol. Cell* 24:101–13
141. Masai H, Taniyama C, Ogino K, Matsui E, Kakusho N, et al. 2006. Phosphorylation of MCM4 by Cdc7 kinase facilitates its interaction with Cdc45 on the chromatin. *J. Biol. Chem.* 281:39249–61
142. Yabuuchi H, Yamada Y, Uchida T, Sunathvanichkul T, Nakagawa T, Masukata H. 2006. Ordered assembly of Sld3, GINS and Cdc45 is distinctly regulated by DDK and CDK for activation of replication origins. *EMBO J.* 25:4663–74
143. Krasinska L, Besnard E, Cot E, Dohet C, Mechali M, et al. 2008. Cdk1 and Cdk2 activity levels determine the efficiency of replication origin firing in *Xenopus*. *EMBO J.* 27:758–69
144. Francis LI, Randell JCW, Takara TJ, Uchima L, Bell SP. 2009. Incorporation into the prereplicative complex activates the Mcm2–7 helicase for Cdc7-Dbf4 phosphorylation. *Genes Dev.* 23:643–54
145. Arias EE, Walter JC. 2007. Strength in numbers: preventing rereplication via multiple mechanisms in eukaryotic cells. *Genes Dev.* 21:497–518
146. Mendez J, Zou-Yang XH, Kim S-Y, Hidaka M, Tansey WP, Stillman B. 2002. Human origin recognition complex large subunit is degraded by ubiquitin-mediated proteolysis after initiation of DNA replication. *Mol. Cell* 9:481–91
147. Hua XH, Newport J. 1998. Identification of a preinitiation step in DNA replication that is independent of origin recognition complex and cdc6, but dependent on cdk2. *J. Cell Biol.* 140:271–81
148. Rowles A, Tada S, Blow JJ. 1999. Changes in association of the *Xenopus* origin recognition complex with chromatin on licensing of replication origins. *J. Cell Sci.* 112:2011–18
149. Maiorano D, Moreau J, Mechali M. 2000. XCDT1 is required for the assembly of prereplicative complexes in *Xenopus laevis*. *Nature* 404:622–25
150. Giordano-Coltart J, Ying CY, Gautier J, Hurwitz J. 2005. Studies of the properties of human origin recognition complex and its Walker A motif mutants. *Proc. Natl. Acad. Sci. USA* 102:69–74
151. Ohta S, Tatsumi Y, Fujita M, Tsurimoto T, Obuse C. 2003. The ORC1 cycle in human cells: II. Dynamic changes in the human ORC complex during the cell cycle. *J. Biol. Chem.* 278:41535–40
152. Randell JCW, Bowers JL, Rodriguez HK, Bell SP. 2006. Sequential ATP hydrolysis by Cdc6 and ORC directs loading of the Mcm2–7 helicase. *Mol. Cell* 21:29–39

153. Bowers JL, Randell JC, Chen S, Bell SP. 2004. ATP hydrolysis by ORC catalyzes reiterative Mcm2–7 assembly at a defined origin of replication. *Mol. Cell* 16:967–78
154. Speck C, Chen Z, Li H, Stillman B. 2005. ATPase-dependent cooperative binding of ORC and Cdc6 to origin DNA. *Nat. Struct. Mol. Biol.* 12:965–71
155. Fujita M. 2006. Cdt1 revisited: complex and tight regulation during the cell cycle and consequences of deregulation in mammalian cells. *Cell Div.* 1:22
156. Chen S, de Vries MA, Bell SP. 2007. Orc6 is required for dynamic recruitment of Cdt1 during repeated Mcm2–7 loading. *Genes Dev.* 21:2897–907
157. Tanaka S, Diffley JFX. 2002. Interdependent nuclear accumulation of budding yeast Cdt1 and Mcm2–7 during G1 phase. *Nat. Cell Biol.* 4:198–207
158. You Z, Masai H. 2008. Cdt1 forms a complex with the minichromosome maintenance protein (MCM) and activates its helicase activity. *J. Biol. Chem.* 283:24469–77
159. Nishitani H, Taraviras S, Lygerou Z, Nishimoto T. 2001. The human licensing factor for DNA replication Cdt1 accumulates in G1 and is destabilized after initiation of S-phase. *J. Biol. Chem.* 276:44905–11
160. McGarry TJ, Kirschner MW. 1998. Geminin, an inhibitor of DNA replication, is degraded during mitosis. *Cell* 93:1043–53
161. Wohlschlegel JA, Dwyer BT, Dhar SK, Cvetic C, Walter JC, Dutta A. 2000. Inhibition of eukaryotic DNA replication by geminin binding to Cdt1. *Science* 290:2309–12
162. Tada S, Li A, Maiorano D, Mechali M, Blow JJ. 2001. Repression of origin assembly in metaphase depends on inhibition of RLF-B/Cdt1 by geminin. *Nat. Cell Biol.* 3:107–13
163. Yanagi K, Mizuno T, You Z, Hanaoka F. 2002. Mouse geminin inhibits not only Cdt1-MCM6 interactions but also a novel intrinsic Cdt1 DNA binding activity. *J. Biol. Chem.* 277:40871–80
164. Lee C, Hong BS, Choi JM, Kim Y, Watanabe S, et al. 2004. Structural basis for inhibition of the replication licensing factor Cdt1 by geminin. *Nature* 430:913–17
165. Saxena S, Dutta A. 2005. Geminin-Cdt1 balance is critical for genetic stability. *Mutat. Res.* 569:111–21
166. Takeda DY, Parvin JD, Dutta A. 2005. Degradation of Cdt1 during S phase is Skp2-independent and is required for efficient progression of mammalian cells through S phase. *J. Biol. Chem.* 280:23416–23
167. Senga T, Sivaprasad U, Zhu W, Park JH, Arias EE, et al. 2006. PCNA is a cofactor for Cdt1 degradation by CUL4/DDB1-mediated N-terminal ubiquitination. *J. Biol. Chem.* 281:6246–52
168. Arias EE, Walter JC. 2006. PCNA functions as a molecular platform to trigger Cdt1 destruction and prevent rereplication. *Nat. Cell Biol.* 8:84–90
169. Xouri G, Squire A, Dimaki M, Geverts B, Verveer PJ, et al. 2007. Cdt1 associates dynamically with chromatin throughout G1 and recruits Geminin onto chromatin. *EMBO J.* 26:1303–14
170. Xouri G, Dimaki M, Bastiaens PIH, Lygerou Z. 2007. Cdt1 interactions in the licensing process: a model for dynamic spatiotemporal control of licensing. *Cell Cycle* 6:1549–52
171. Nishitani H, Sugimoto N, Roukos V, Nakanishi Y, Saijo M, et al. 2006. Two E3 ubiquitin ligases, SCF-Skp2 and DDB1-Cul4, target human Cdt1 for proteolysis. *EMBO J.* 25:1126–36
172. Paolinelli R, Mendoza-Maldonado R, Cereseto A, Giacca M. 2009. Acetylation by GCN5 regulates CDC6 phosphorylation in the S phase of the cell cycle. *Nat. Struct. Mol. Biol.* 16:412–20
173. Glozak MA, Seto E. 2009. Acetylation/deacetylation modulates the stability of DNA replication licensing factor Cdt1. *J. Biol. Chem.* 284:11446–53
174. Dominguez-Sola D, Ying CY, Grandori C, Ruggiero L, Chen B, et al. 2007. Non-transcriptional control of DNA replication by c-Myc. *Nature* 448:445–51
175. Chuang LC, Teixeira LK, Wohlschlegel JA, Henze M, Yates JR, et al. 2009. Phosphorylation of Mcm2 by Cdc7 promotes prereplication complex assembly during cell-cycle re-entry. *Mol. Cell* 35:206–16
176. Takayama Y, Kamimura Y, Okawa M, Muramatsu S, Sugino A, Araki H. 2003. GINS, a novel multiprotein complex required for chromosomal DNA replication in budding yeast. *Genes Dev.* 17:1153–65
177. Kubota Y, Takase Y, Komori Y, Hashimoto Y, Arata T, et al. 2003. A novel ring-like complex of *Xenopus* proteins essential for the initiation of DNA replication. *Genes Dev.* 17:1141–52
178. Kanemaki M, Sanchez-Diaz A, Gambus A, Labib K. 2003. Functional proteomic identification of DNA replication proteins by induced proteolysis in vivo. *Nature* 423:720–24
179. Kanemaki M, Labib K. 2006. Distinct roles for Sld3 and GINS during establishment and progression of eukaryotic DNA replication forks. *EMBO J.* 25:1753–63

180. Labib K, Gambus A. 2007. A key role for the GINS complex at DNA replication forks. *Trends Cell Biol.* 17:271–78
181. Kamada K, Kubota Y, Arata T, Shindo Y, Hanaoka F. 2007. Structure of the human GINS complex and its assembly and functional interface in replication initiation. *Nat. Struct. Mol. Biol.* 14:388–96
182. Boskovic J, Coloma J, Aparicio T, Zhou M, Robinson CV, et al. 2007. Molecular architecture of the human GINS complex. *EMBO Rep.* 8:678–84
183. Choi JM, Lim HS, Kim JJ, Song OK, Cho Y. 2007. Crystal structure of the human GINS complex. *Genes Dev.* 21:1316–21
184. Chang YP, Wang G, Bermudez V, Hurwitz J, Chen XS. 2007. Crystal structure of the GINS complex and functional insights into its role in DNA replication. *Proc. Natl. Acad. Sci. USA* 104:12685–90
185. Gambus A, Jones RC, Sanchez-Diaz A, Kanemaki M, van Deursen F, et al. 2006. GINS maintains association of Cdc45 with MCM in replisome progression complexes at eukaryotic DNA replication forks. *Nat. Cell Biol.* 8:358–66
186. Noguchi E, Noguchi C, McDonald WH, Yates JR 3rd, Russell P. 2004. Swi1 and Swi3 are components of a replication fork protection complex in fission yeast. *Mol. Cell. Biol.* 24:8342–55
187. De Falco M, Ferrari E, De Felice M, Rossi M, Hübscher U, Pisani FM. 2007. The human GINS complex binds to and specifically stimulates human DNA polymerase α-primase. *EMBO Rep.* 8:99–103
188. Marinsek N, Barry ER, Makarova KS, Dionne I, Koonin EV, Bell SD. 2006. GINS, a central nexus in the archaeal DNA replication fork. *EMBO Rep.* 7:539–45
189. Pai CC, Garcia I, Wang SW, Cotterill S, Macneill SA, Kearsey SE. 2009. GINS inactivation phenotypes reveal two pathways for chromatin association of replicative alpha and epsilon DNA polymerases in fission yeast. *Mol. Biol. Cell* 20:1213–22
190. Yoshimochi T, Fujikane R, Kawanami M, Matsunaga F, Ishino Y. 2008. The GINS complex from *Pyrococcus furiosus* stimulates the MCM helicase activity. *J. Biol. Chem.* 283:1601–9
191. Tanaka H, Katou Y, Yagura M, Saitoh K, Itoh T, et al. 2009. Ctf4 coordinates the progression of helicase and DNA polymerase alpha. *Genes Cells* 14:807–20
192. Gambus A, van Deursen F, Polychronopoulos D, Foltman M, Jones RC, et al. 2009. A key role for Ctf4 in coupling the MCM2–7 helicase to DNA polymerase α within the eukaryotic replisome. *EMBO J.* 28:2992–3004
193. Hübscher U, Maga G, Spadari S. 2002. Eukaryotic DNA polymerases. *Annu. Rev. Biochem.* 71:133–63
194. Johnson A, O'Donnell M. 2005. Cellular DNA replicases: components and dynamics at the replication fork. *Annu. Rev. Biochem.* 74:283–315
195. Burgers PM. 2009. Polymerase dynamics at the eukaryotic DNA replication fork. *J. Biol. Chem.* 284:4041–45
196. Masumoto H, Muramatsu S, Kamimura Y, Araki H. 2002. S-Cdk-dependent phosphorylation of Sld2 essential for chromosomal DNA replication in budding yeast. *Nature* 415:651–55
197. Tanaka S, Umemori T, Hirai K, Muramatsu S, Kamimura Y, Araki H. 2007. CDK-dependent phosphorylation of Sld2 and Sld3 initiates DNA replication in budding yeast. *Nature* 445:328–32
198. Zegerman P, Diffley JFX. 2007. Phosphorylation of Sld2 and Sld3 by cyclin-dependent kinases promotes DNA replication in budding yeast. *Nature* 445:281–85
199. Kamimura Y, Masumoto H, Sugino A, Araki H. 1998. Sld2, which interacts with Dpb11 in *Saccharomyces cerevisiae*, is required for chromosomal DNA replication. *Mol. Cell. Biol.* 18:6102–9
200. Araki H, Leem SH, Phongdara A, Sugino A. 1995. Dpb11, which interacts with DNA polymerase II(epsilon) in *Saccharomyces cerevisiae*, has a dual role in S-phase progression and at a cell cycle checkpoint. *Proc. Natl. Acad. Sci. USA* 92:11791–95
201. Myung K, Datta A, Kolodner RD. 2001. Suppression of spontaneous chromosomal rearrangements by S phase checkpoint functions in *Saccharomyces cerevisiae*. *Cell* 104:397–408
202. Mordes DA, Nam EA, Cortez D. 2008. Dpb11 activates the Mec1-Ddc2 complex. *Proc. Natl. Acad. Sci. USA* 105:18730–34
203. Tak Y-S, Tanaka Y, Endo S, Kamimura Y, Araki H. 2006. A CDK-catalysed regulatory phosphorylation for formation of the DNA replication complex Sld2-Dpb11. *EMBO J.* 25:1987–96
204. Kamimura Y, Tak YS, Sugino A, Araki H. 2001. Sld3, which interacts with Cdc45 (Sld4), functions for chromosomal DNA replication in *Saccharomyces cerevisiae*. *EMBO J.* 20:2097–107

205. Jares P, Blow JJ. 2000. *Xenopus* cdc7 function is dependent on licensing but not on XORC, XCdc6, or CDK activity and is required for XCdc45 loading. *Genes Dev.* 14:1528–40
206. Walter JC. 2000. Evidence for sequential action of cdc7 and cdk2 protein kinases during initiation of DNA replication in *Xenopus* egg extracts. *J. Biol. Chem.* 275:39773–78
207. Sangrithi MN, Bernal JA, Madine M, Philpott A, Lee J, et al. 2005. Initiation of DNA replication requires the RECQL4 protein mutated in Rothmund-Thomson syndrome. *Cell* 121:887–98
208. Kumagai A, Shevchenko A, Shevchenko A, Dunphy WG. 2010. Treslin collaborates with TopBP1 in triggering the initiation of DNA replication. *Cell* 140:349–59
209. Sansam CL, Cruz NM, Danielian PS, Amsterdam A, Lau ML, et al. 2010. A vertebrate gene, *ticrr*, is an essential checkpoint and replication regulator. *Genes Dev.* 24:183–94
210. Aparicio OM, Weinstein DM, Bell SP. 1997. Components and dynamics of DNA replication complexes in *S. cerevisiae*: redistribution of MCM proteins and Cdc45p during S phase. *Cell* 91:59–69
211. Labib K, Tercero JA, Diffley JF. 2000. Uninterrupted MCM2–7 function required for DNA replication fork progression. *Science* 288:1643–47
212. Pacek M, Walter JC. 2004. A requirement for MCM7 and Cdc45 in chromosome unwinding during eukaryotic DNA replication. *EMBO J.* 23:3667–76
213. Pacek M, Tutter AV, Kubota Y, Takisawa H, Walter JC. 2006. Localization of MCM2–7, Cdc45, and GINS to the site of DNA unwinding during eukaryotic DNA replication. *Mol. Cell* 21:581–87
214. Masai H, You Z, Arai K. 2005. Control of DNA replication: regulation and activation of eukaryotic replicative helicase, MCM. *IUBMB Life* 57:323–35
215. You Z, Komamura Y, Ishimi Y. 1999. Biochemical analysis of the intrinsic Mcm4-Mcm6-Mcm7 DNA helicase activity. *Mol. Cell. Biol.* 19:8003–15
216. Lee JK, Hurwitz J. 2000. Isolation and characterization of various complexes of the minichromosome maintenance proteins of *Schizosaccharomyces pombe*. *J. Biol. Chem.* 275:18871–78
217. Chong JP, Hayashi MK, Simon MN, Xu RM, Stillman B. 2000. A double-hexamer archaeal minichromosome maintenance protein is an ATP-dependent DNA helicase. *Proc. Natl. Acad. Sci. USA* 97:1530–35
218. Ishimi Y. 1997. A DNA helicase activity is associated with an MCM4, -6, and -7 protein complex. *J. Biol. Chem.* 272:24508–13
219. Kelman Z, Lee JK, Hurwitz J. 1999. The single minichromosome maintenance protein of *Methanobacterium thermoautotrophicum* ΔH contains DNA helicase activity. *Proc. Natl. Acad. Sci. USA* 96:14783–88
220. Tan BC-M, Chien C-T, Hirose S, Lee S-C. 2006. Functional cooperation between FACT and MCM helicase facilitates initiation of chromatin DNA replication. *EMBO J.* 25:3975–85
221. Bochman ML, Schwacha A. 2008. The Mcm2–7 complex has in vitro helicase activity. *Mol. Cell* 31:287–93
222. Moyer SE, Lewis PW, Botchan MR. 2006. Isolation of the Cdc45/Mcm2–7/GINS (CMG) complex, a candidate for the eukaryotic DNA replication fork helicase. *Proc. Natl. Acad. Sci. USA* 103:10236–41
223. Ilves I, Petojevic T, Pesavento JJ, Botchan MR. 2010. Activation of the MCM2-7 helicase by association with Cdc45 and GINS proteins. *Mol. Cell* 37:247–58
224. Masai H, Matsui E, You Z, Ishimi Y, Tamai K, Arai K. 2000. Human Cdc7-related kinase complex. In vitro phosphorylation of MCM by concerted actions of Cdks and Cdc7 and that of a criticial threonine residue of Cdc7 by Cdks. *J. Biol. Chem.* 275:29042–52
225. Kakusho N, Taniyama C, Masai H. 2008. Identification of stimulators and inhibitors of Cdc7 kinase in vitro. *J. Biol. Chem.* 283:19211–18
226. Bruck I, Kaplan DL. 2009. Dbf4-Cdc7 phosphorylation of Mcm2 is required for cell growth. *J. Biol. Chem.* 284:28823–31
227. Remus D, Beuron F, Tolun G, Griffith JD, Morris EP, Diffley JF. 2009. Concerted loading of Mcm2-7 double hexamers around DNA during DNA replication origin licensing. *Cell* 139:719–30
228. Evrin C, Clarke P, Zech J, Lurz R, Sun J, et al. 2009. A double-hexameric MCM2-7 complex is loaded onto origin DNA during licensing of eukaryotic DNA replication. *Proc. Natl. Acad. Sci. USA* 106:20240–45
229. Hardy CF, Dryga O, Seematter S, Pahl PM, Sclafani RA. 1997. mcm5/cdc46-bob1 bypasses the requirement for the S phase activator Cdc7p. *Proc. Natl. Acad. Sci. USA* 94:3151–55
230. Fletcher RJ, Bishop BE, Leon RP, Sclafani RA, Ogata CM, Chen XS. 2003. The structure and function of MCM from archaeal *M. thermoautotrophicum*. *Nat. Struct. Biol.* 10:160–67

231. Hoang ML, Leon RP, Pessoa-Brandao L, Hunt S, Raghuraman MK, et al. 2007. Structural changes in Mcm5 protein bypass Cdc7-Dbf4 function and reduce replication origin efficiency in *Saccharomyces cerevisiae*. *Mol. Cell. Biol.* 27:7594–602
232. Sclafani RA, Holzen TM. 2007. Cell cycle regulation of DNA replication. *Annu. Rev. Genet.* 41:237–80
233. Sheu YJ, Stillman B. 2010. The Dbf4-Cdc7 kinase promotes S phase by alleviating an inhibitory activity in Mcm4. *Nature* 463:113–17
234. Gonzalez S, Klatt P, Delgado S, Conde E, Lopez-Rios F, et al. 2006. Oncogenic activity of Cdc6 through repression of the INK4/ARF locus. *Nature* 440:702–6
235. Mailand N, Diffley JF. 2005. CDKs promote DNA replication origin licensing in human cells by protecting Cdc6 from APC/C-dependent proteolysis. *Cell* 122:915–26
236. Duursma A, Agami R. 2005. p53-Dependent regulation of Cdc6 protein stability controls cellular proliferation. *Mol. Cell. Biol.* 25:6937–47
237. Sterner JM, Dew-Knight S, Musahl C, Kornbluth S, Horowitz JM. 1998. Negative regulation of DNA replication by the retinoblastoma protein is mediated by its association with MCM7. *Mol. Cell. Biol.* 18:2748–57
238. Kennedy BK, Barbie DA, Classon M, Dyson N, Harlow E. 2000. Nuclear organization of DNA replication in primary mammalian cells. *Genes Dev.* 14:2855–68
239. Braden WA, Lenihan JM, Lan Z, Luce KS, Zagorski W, et al. 2006. Distinct action of the retinoblastoma pathway on the DNA replication machinery defines specific roles for cyclin-dependent kinase complexes in prereplication complex assembly and S-phase progression. *Mol. Cell. Biol.* 26:7667–81
240. Harrington EA, Bruce JL, Harlow E, Dyson N. 1998. pRB plays an essential role in cell cycle arrest induced by DNA damage. *Proc. Natl. Acad. Sci. USA* 95:11945–50
241. Machida YJ, Teer JK, Dutta A. 2005. Acute reduction of an origin recognition complex (ORC) subunit in human cells reveals a requirement of ORC for Cdk2 activation. *J. Biol. Chem.* 280:27624–30
242. Liu P, Slater DM, Lenburg M, Nevis K, Cook JG, Vaziri C. 2009. Replication licensing promotes cyclin D1 expression and G1 progression in untransformed human cells. *Cell Cycle*. 8:125–36
243. Nevis KR, Cordeiro-Stone M, Cook JG. 2009. Origin licensing and p53 status regulate Cdk2 activity during G(1). *Cell Cycle*. 8:1952–63
244. Blow JJ, Gillespie PJ. 2008. Replication licensing and cancer—a fatal entanglement? *Nat. Rev. Cancer* 8:799–806
245. Bartkova J, Rezaei N, Liontos M, Karakaidos P, Kletsas D, et al. 2006. Oncogene-induced senescence is part of the tumorigenesis barrier imposed by DNA damage checkpoints. *Nature* 444:633–37
246. Di Micco R, Fumagalli M, Cicalese A, Piccinin S, Gasparini P, et al. 2006. Oncogene-induced senescence is a DNA damage response triggered by DNA hyper-replication. *Nature* 444:638–42
247. Gorgoulis VG, Vassiliou LV, Karakaidos P, Zacharatos P, Kotsinas A, et al. 2005. Activation of the DNA damage checkpoint and genomic instability in human precancerous lesions. *Nature* 434:907–13
248. Byun TS, Pacek M, Yee M, Walter JC, Cimprich KA. 2005. Functional uncoupling of MCM helicase and DNA polymerase activities activates the ATR-dependent checkpoint. *Genes Dev.* 19:1040–52
249. Sogo JM, Lopes M, Foiani M. 2002. Fork reversal and ssDNA accumulation at stalled replication forks owing to checkpoint defects. *Science* 297:599–602
250. Zou L. 2007. Single- and double-stranded DNA: building a trigger of ATR-mediated DNA damage response. *Genes Dev.* 21:879–85
251. Zou L, Liu D, Elledge SJ. 2003. Replication protein A-mediated recruitment and activation of Rad17 complexes. *Proc. Natl. Acad. Sci. USA* 100:13827–32
252. Cortez D. 2005. Unwind and slow down: checkpoint activation by helicase and polymerase uncoupling. *Genes Dev.* 19:1007–12
253. Kumagai A, Lee J, Yoo HY, Dunphy WG. 2006. TopBP1 activates the ATR-ATRIP complex. *Cell* 124:943–55
254. MacDougall CA, Byun TS, Van C, Yee M, Cimprich KA. 2007. The structural determinants of checkpoint activation. *Genes Dev.* 21:898–903
255. Syljuåsen RG, Sørensen CS, Hansen LT, Fugger K, Lundin C, et al. 2005. Inhibition of human Chk1 causes increased initiation of DNA replication, phosphorylation of ATR targets, and DNA breakage. *Mol. Cell. Biol.* 25:3553 62

256. Liu P, Barkley LR, Day T, Bi X, Slater DM, et al. 2006. The Chk1-mediated S-phase checkpoint targets initiation factor Cdc45 via a Cdc25A/Cdk2-independent mechanism. *J. Biol. Chem.* 281:30631–44
257. Costanzo V, Shechter D, Lupardus PJ, Cimprich KA, Gottesman M, Gautier J. 2003. An ATR- and Cdc7-dependent DNA damage checkpoint that inhibits initiation of DNA replication. *Mol. Cell* 11:203–13
258. Tenca P, Brotherton D, Montagnoli A, Rainoldi S, Albanese C, Santocanale C. 2007. Cdc7 is an active kinase in human cancer cells undergoing replication stress. *J. Biol. Chem.* 282:208–15
259. Heffernan TP, Ünsal-Kaçmaz K, Heinloth AN, Simpson DA, Paules RS, et al. 2007. *J. Biol. Chem.* 282:9458–68
260. Tsuji T, Lau E, Chiang GG, Jiang W. 2008. The role of Dbf4/Drf1-dependent kinase Cdc7 in DNA-damage checkpoint control. *Mol. Cell* 32:862–69
261. Ünsal-Kaçmaz K, Chastain PD, Qu P-P, Minoo P, Cordeiro-Stone M, et al. 2007. The human Tim/Tipin complex coordinates an intra-S checkpoint response to UV that slows replication fork displacement. *Mol. Cell. Biol.* 27:3131–42
262. Chini CC, Chen J. 2003. Human claspin is required for replication checkpoint control. *J. Biol. Chem.* 278:30057–62
263. Chou DM, Elledge SJ. 2006. Tipin and Timeless form a mutually protective complex required for genotoxic stress resistance and checkpoint function. *Proc. Natl. Acad. Sci. USA* 103:18143–47
264. Alcasabas AA, Osborn AJ, Bachant J, Hu F, Werler PJ, et al. 2001. Mrc1 transduces signals of DNA replication stress to activate Rad53. *Nat. Cell Biol.* 3:958–65
265. Tanaka K, Russell P. 2001. Mrc1 channels the DNA replication arrest signal to checkpoint kinase Cds1. *Nat. Cell Biol.* 3:966–72
266. Yoo HY, Jeong SY, Dunphy WG. 2006. Site-specific phosphorylation of a checkpoint mediator protein controls its responses to different DNA structures. *Genes Dev.* 20:772–83
267. Kim JM, Kakusho N, Yamada M, Kanoh Y, Takemoto N, Masai H. 2008. Cdc7 kinase mediates Claspin phosphorylation in DNA replication checkpoint. *Oncogene* 27:3475–82
268. Katou Y, Kanoh Y, Bando M, Noguchi H, Tanaka H, et al. 2003. S-phase checkpoint proteins Tof1 and Mrc1 form a stable replication-pausing complex. *Nature* 424:1078–83
269. Willis N, Rhind N. 2009. Mus81, Rhp51(Rad51), and Rqh1 form an epistatic pathway required for the S-phase DNA damage checkpoint. *Mol. Biol. Cell* 20:819–33
270. Alabert C, Bianco JN, Pasero P. 2009. Differential regulation of homologous recombination at DNA breaks and replication forks by the Mrc1 branch of the S-phase checkpoint. *EMBO J.* 28:1131–41
271. Henry-Mowatt J, Jackson D, Masson JY, Johnson PA, Clements PM, et al. 2003. XRCC3 and Rad51 modulate replication fork progression on damaged vertebrate chromosomes. *Mol. Cell* 11:1109–17
272. Szyjka SJ, Viggiani CJ, Aparicio OM. 2005. Mrc1 is required for normal progression of replication forks throughout chromatin in *S. cerevisiae*. *Mol. Cell* 19:691–97
273. Hodgson B, Calzada A, Labib K. 2007. Mrc1 and Tof1 regulate DNA replication forks in different ways during normal S phase. *Mol. Biol. Cell* 18:3894–902
274. Lin S-Y, Li K, Stewart GS, Elledge SJ. 2004. Human Claspin works with BRCA1 to both positively and negatively regulate cell proliferation. *Proc. Natl. Acad. Sci. USA* 101:6484–89
275. Petermann E, Helleday T, Caldecott KW. 2008. Claspin promotes normal replication fork rates in human cells. *Mol. Biol. Cell* 19:2373–78
276. Tourriere H, Versini G, Cordon-Preciado V, Alabert C, Pasero P. 2005. Mrc1 and Tof1 promote replication fork progression and recovery independently of Rad53. *Mol. Cell* 19:699–706
277. Dalgaard JZ, Klar AJ. 2000. swi1 and swi3 perform imprinting, pausing, and termination of DNA replication in *S. pombe*. *Cell* 102:745–51
278. Mohanty BK, Bairwa NK, Bastia D. 2006. The Tof1p-Csm3p protein complex counteracts the Rrm3p helicase to control replication termination of *Saccharomyces cerevisiae*. *Proc. Natl. Acad. Sci. USA* 103:897–902
279. Voineagu I, Narayanan V, Lobachev KS, Mirkin SM. 2008. Replication stalling at unstable inverted repeats: interplay between DNA hairpins and fork stabilizing proteins. *Proc. Natl. Acad. Sci. USA* 105:9936–41

280. Zhao H, Russell P. 2004. DNA binding domain in the replication checkpoint protein Mrc1 of *Schizosaccharomyces pombe*. *J. Biol. Chem.* 279:53023–27
281. Sar F, Lindsey-Boltz LA, Subramanian D, Croteau DL, Hutsell SQ, et al. 2004. Human claspin is a ring-shaped DNA-binding protein with high affinity to branched DNA structures. *J. Biol. Chem.* 279:39289–95
282. Lou H, Komata M, Katou Y, Guan Z, Reis CC, et al. 2008. Mrc1 and DNA polymerase epsilon function together in linking DNA replication and the S phase checkpoint. *Mol. Cell* 32:106–17
283. Komata M, Bando M, Araki H, Shirahige K. 2009. The direct binding of Mrc1, a checkpoint mediator, to Mcm6, a replication helicase, is essential for replication checkpoint against methyl methanesulfonate-induced stress. *Mol. Cell. Biol.* 29:5008–19
284. Kim S, Dallmann HG, McHenry CS, Marians KJ. 1996. Coupling of a replicative polymerase and helicase: a tau-DnaB interaction mediates rapid replication fork movement. *Cell* 84:643–50
285. Alvino GM, Collingwood D, Murphy JM, Delrow J, Brewer BJ, Raghuraman MK. 2007. Replication in hydroxyurea: It's a matter of time. *Mol. Cell. Biol.* 27:6396–406
286. Shima N, Alcaraz A, Liachko I, Buske TR, Andrews CA, et al. 2007. A viable allele of Mcm4 causes chromosome instability and mammary adenocarcinomas in mice. *Nat. Genet.* 39:93–98
287. Pruitt SC, Bailey KJ, Freeland A. 2007. Reduced Mcm2 expression results in severe stem/progenitor cell deficiency and cancer. *Stem. Cells* 25:3121–32
288. Davidson IF, Li A, Blow JJ. 2006. Deregulated replication licensing causes DNA fragmentation consistent with head-to-tail fork collision. *Mol. Cell* 24:433–43
289. Tatsumi Y, Sugimoto N, Yugawa T, Narisawa-Saito M, Kiyono T, Fujita M. 2006. Deregulation of Cdt1 induces chromosomal damage without rereplication and leads to chromosomal instability. *J. Cell Sci.* 119:3128–40
290. Liu E, Lee AY, Chiba T, Olson E, Sun P, Wu X. 2007. The ATR-mediated S phase checkpoint prevents rereplication in mammalian cells when licensing control is disrupted. *J. Cell Biol.* 179:643–57
291. Feng D, Tu Z, Wu W, Liang C. 2003. Inhibiting the expression of DNA replication-initiation proteins induces apoptosis in human cancer cells. *Cancer Res.* 63:7356–64
292. Montagnoli A, Tenca P, Sola F, Carpani D, Brotherton D, et al. 2004. Cdc7 inhibition reveals a p53-dependent replication checkpoint that is defective in cancer cells. *Cancer Res.* 64:7110–16
293. Yoshizawa-Sugata N, Ishii A, Taniyama C, Matsui E, Arai K, Masai H. 2005. A second human Dbf4/ASK-related protein, Drf1/ASKL1, is required for efficient progression of S and M phases. *J. Biol. Chem.* 280:13062–70
294. Lau E, Zhu C, Abraham RT, Jiang W. 2006. The functional role of Cdc6 in S-G2/M in mammalian cells. *EMBO Rep.* 7:425–30
295. Lau E, Chiang GG, Abraham RT, Jiang W. 2009. Divergent S-phase checkpoint activation arising from pre-replicative complex deficiency controls cell survival. *Mol. Biol. Cell* 20:3953–64
296. Teer JK, Machida YJ, Labit H, Novac O, Hyrien O, et al. 2006. Proliferating human cells hypomorphic for origin recognition complex 2 and pre-replicative complex formation have a defect in p53 activation and Cdk2 kinase activation. *J. Biol. Chem.* 281:6253–60
297. Gonzalez MA, Tachibana KE, Laskey RA, Coleman N. 2005. Control of DNA replication and its potential clinical exploitation. *Nat. Rev. Cancer* 5:135–41
298. D'Antoni S, Mattina T, Di MP, Federico C, Motta S, Saccone S. 2004. Altered replication timing of the *HIRA/Tuple1* locus in the DiGeorge and Velocardiofacial syndromes. *Gene* 333:111–19
299. State MW, Greally JM, Cuker A, Bowers PN, Henegariu O, et al. 2003. Epigenetic abnormalities associated with a chromosome 18(q21-q22) inversion and a Gilles de la Tourette syndrome phenotype. *Proc. Natl. Acad. Sci. USA* 100:4684–89
300. Loupart ML, Krause SA, Heck MS. 2000. Aberrant replication timing induces defective chromosome condensation in *Drosophila* ORC2 mutants. *Curr. Biol.* 10:1547–56
301. Pflumm MF, Botchan MR. 2001. Orc mutants arrest in metaphase with abnormally condensed chromosomes. *Development* 128:1697–707
302. Hansen RS, Stoger R, Wijmenga C, Stanek AM, Canfield TK, et al. 2000. Escape from gene silencing in ICF syndrome: evidence for advanced replication time as a major determinant. *Hum. Mol. Genet.* 9:2575–87

303. Rountree MR, Bachman KE, Baylin SB. 2000. DNMT1 binds HDAC2 and a new corepressor, DMAP1, to form a complex at replication foci. *Nat. Genet.* 25:269–77

304. Fu H, Wang L, Lin C-M, Singhania S, Bouhassira EE, Aladjem MI. 2006. Preventing gene silencing with human replicators. *Nat. Biotechnol.* 24:572–76

305. Seila AC, Calabrese JM, Levine SS, Yeo GW, Rahl PB, et al. 2008. Divergent transcription from active promoters. *Science* 322:1849–51

306. Hiratani I, Gilbert DM. 2009. Replication timing as an epigenetic mark. *Epigenetics* 4:93–97

307. Dai J, Chuang R-Y, Kelly TJ. 2005. DNA replication origins in the *Schizosaccharomyces pombe* genome. *Proc. Natl. Acad. Sci. USA* 102:337–42

308. Tanny RE, MacAlpine DM, Blitzblau HG, Bell SP. 2006. Genome-wide analysis of rereplication reveals inhibitory controls that target multiple stages of replication initiation. *Mol. Biol. Cell* 17:2415–23

309. Marians KJ. 2000. PriA-directed replication fork restart in *Escherichia coli*. *Trends Biochem. Sci.* 25:185–89

310. Mizukoshi T, Tanaka T, Arai K, Kohda D, Masai H. 2003. A critical role of the 3′ terminus of nascent DNA chains in recognition of stalled replication forks. *J. Biol. Chem.* 278:42234–39

311. Tanaka T, Mizukoshi T, Sasaki K, Kohda D, Masai H. 2007. *Escherichia coli* PriA protein, two modes of DNA binding and activation of ATP hydrolysis. *J. Biol. Chem.* 282:19917–27

312. Abdurashidova G, Radulescu S, Sandoval O, Zahariev S, Danailov MB, et al. 2007. Functional interactions of DNA topoisomerases with a human replication origin. *EMBO J.* 26:998–1009

Regulators of the Cohesin Network

Bo Xiong[1] and Jennifer L. Gerton[1,2]

[1]Stowers Institute for Medical Research, Kansas City, Missouri 64110;
email: bxi@stowers.org, jeg@stowers.org

[2]Department of Biochemistry and Molecular Biology, University of Kansas Medical Center, Kansas City, Kansas 66160

Annu. Rev. Biochem. 2010. 79:131–53

First published online as a Review in Advance on March 23, 2010

The *Annual Review of Biochemistry* is online at biochem.annualreviews.org

This article's doi:
10.1146/annurev-biochem-061708-092640

0066-4154/10/0707-0131$20.00

Key Words

chromosome cohesion, cohesinopathy, posttranslational modification

Abstract

Chromosome cohesion is the term used to describe the cellular process in which sister chromatids are held together from the time of their replication until the time of separation at the metaphase to anaphase transition. In this capacity, chromosome cohesion, especially at centromeric regions, is essential for chromosome segregation. However, cohesion of noncentromeric DNA sequences has been shown to occur during double-strand break (DSB) repair and the transcriptional regulation of genes. Cohesion for the purposes of accurate chromosome segregation, DSB repair, and gene regulation are all achieved through a similar network of proteins, but cohesion for each purpose may be regulated differently. In this review, we focus on recent developments regarding the regulation of this multipurpose network for tying DNA sequences together. In particular, regulation via effectors and posttranslational modifications are reviewed. A picture is emerging in which complex regulatory networks are capable of differential regulation of cohesion in various contexts.

Contents

INTRODUCTION: THE COHESIN CYCLE

Sister chromatid cohesion is a cell cycle–regulated process and refers to the phenomenon that sister chromatids are held together from the time of their replication until they are pulled to opposite spindle poles at the metaphase to anaphase transition. This cohesion allows sister chromatids to orient correctly with respect to attachment to spindle microtubules, thus ensuring accurate segregation. Cohesion is mediated by a protein complex, known as the cohesin complex, which appears to form a ring that can encircle DNA. Additional factors assist in loading the complex onto DNA, establishing cohesion during S phase, maintaining cohesion in G2 phase, and dissolving cohesion at the metaphase to anaphase transition. Cohesin is also loaded at a double-strand break (DSB) site as part of the repair process. Mutations in the cohesin network manifest on a cellular level as chromosome segregation defects, DNA repair defects, and aberrant chromosome compaction. Hypomorphic mutations that preserve chromosome segregation can affect gene regulation. At the organismal level, hypomorphic mutations can cause developmental defects, which are likely related to molecular defects in gene regulation. We first review the proteins that make up the complex, followed by a discussion of how the complex associates with DNA, and then we review the accessory proteins (see **Table 1**). In the latter half of the review, we focus on mechanisms for regulating the cohesin network, including posttranslational modifications. We finish with a discussion of the association between the cohesin network and human disease.

Cohesion: the physical phenomenon of holding sister chromatids together

Cohesin: the complex of proteins that holds sister chromatids together

DSB: double-strand break

SMC family: structural maintenance of chromosomes family

THE COHESIN COMPLEX

Sister chromatids are held together by the cohesin complex. This complex consists of four subunits: Smc1, Smc3, Mcd1/Scc1/Rad21, and Scc3/SA1/SA2. Smc1 and Smc3 belong to the structural maintenance of chromosomes (SMC) superfamily. Members of this family are found in prokaryotes and have been conserved through evolution (1). Other members of this family include subunits of the condensin complex, which function in both chromosome condensation and dosage compensation (Smc2 and Smc4), and the Smc5-Smc6 complex and Rad50, which function in DNA damage repair (1). The molecular architecture of SMC proteins provides important clues to understand their function. These proteins contain globular domains at their N and C termini. In between is a long coiled-coil region that folds in half at another globular region known as the hinge domain. The hinge domain can interact with the hinge domain of another SMC. The result of the antiparallel coiled-coil structure is that

Table 1 Sister chromatid cohesion proteins (red indicates a meiosis-specific function)

Role in cohesion	*Saccharomyces cerevisiae*[a]	*Schizosaccharomyces pombe*	*Drosophila melanogaster*	Vertebrates
Cohesin complex subunits	Scc1/Mcd1	Rad21	Rad21	Rad21
	Scc3/Irr1	Psc3, Rec11	SA1	SA1, SA2, STAG3
	Smc1	Psm1	Smc1	Smc1α, Smc1β
	Smc3	Psm3	Smc3 (Cap)	Smc3
	Rec8	Rec8	c(2)M	Rec8
Loading	Scc2	Mis4	Nipped-B	NIPBL
	Scc4	Ssl3	Mau-2	Scc4/Mau-2
Establishment	Eco1/Ctf7	Eso1	Deco	Esco1, Esco2
Maintenance	Pds5	Pds5	Pds5	Pds5A, Pds5B
	Sgo1	Sgo1, Sgo2	Mei-S332	Sgo1, Sgo2
	PP2A	PP2A	PP2A	PP2A
	Rad61/Wpl1	Wpl1	Wapl	Wapl
	—	—	—	Sororin
Dissolution	Esp1	Cut1	Sse	Separase
	Pds1	Cut2	Pim, Thr	Securin
	Cdc5	Plo1	Polo	Plk1

the N and C termini come together to form a "head" domain with a functional ABC-type ATPase (**Figure 1**). Although Smc1 and Smc3 possess ATP-binding and hydrolysis domains, exactly how ATP hydrolysis fits into the function or regulation of the cohesin complex is unclear; ATP hydrolysis has been proposed to be coupled with DNA transport into the ring (2).

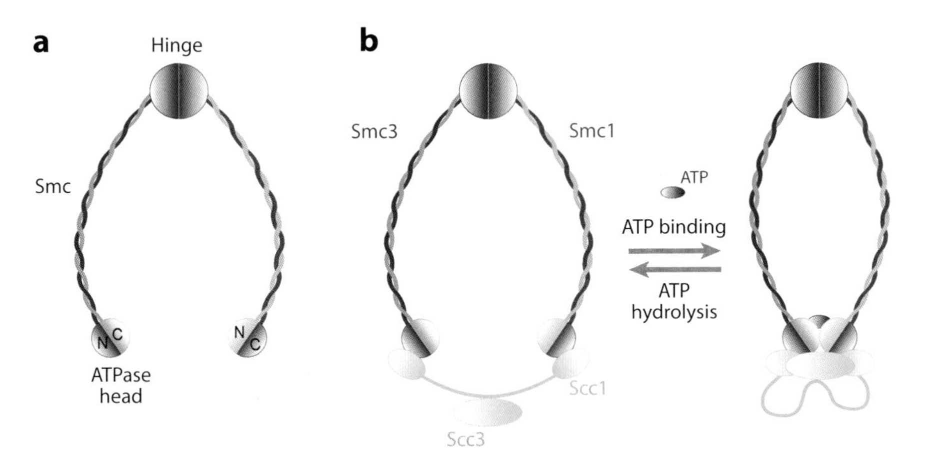

Figure 1

Structure of the Smc dimer and cohesin complex. (*a*) An Smc monomer possesses the hinge domain at one end and ATPase head domain at the other, connected by a long antiparallel coiled coil. The hinge domains of the Smc monomers can interact with each other to form an Smc dimer. The ATPase head domains, formed by N and C termini, play important roles in opening and closing of the ring structure. (*b*) Cohesin complex consists of a Smc1-Smc3 heterodimer and two non-Smc subunits Scc1/Mcd1/Rad21 and Scc3. The N terminus of Scc1 binds to the Smc3 head domain, and the C terminus binds to the Smc1 head domain. Scc1 also recruits Scc3 to the complex. ATP binding and hydrolysis regulate the engagement and disengagement of the Smc1-Smc3 head domains.

Mutations that block ATP binding or hydrolysis block cohesin binding to chromosomes (2, 3). For an excellent review of SMC protein structure and function, see Reference 4.

The Mcd1/Scc1/Rad21 subunit is known as the kleisin subunit. Kleisins are a superfamily of SMC protein partners evolutionarily conserved from prokaryotes to humans (5). The term kleisin is derived from the Greek word for closure. The kleisin component is the target of regulation by several different mechanisms. The N terminus of Mcd1 binds to the Smc3 head domain, and the C terminus binds to the Smc1 head domain (6), effectively bridging the heads of the SMC proteins. This subunit may also regulate the ATPase activity of the Smc1-Smc3 heterodimer (7). In electron micrographs, the Smc1-Smc3 heterodimer appears in a V shape (6, 8), and all together, the three proteins are thought to form a circular complex (6, 9, 10). Mcd1/Scc1/Rad21 also binds to the fourth subunit of the cohesin complex, Scc3/Irr1. Scc3/Irr1 contains HEAT repeats (protein-protein interaction motifs), but not much else is known about this subunit.

There are several meiosis-specific subunits for the cohesin complex. Rec8 is a meiosis-specific kleisin (11). Smc1β and STAG3 (an Scc3-like subunit) are additional meiosis-specific subunits present in vertebrates. Disruption of Smc1β causes complete infertility in mice of both sexes (12). Given this additional variety in subunits, there may be many different cohesin complexes present on vertebrate meiotic chromosomes. Although the exact number of complexes present and the different functionality of each complex are unclear, it is clear that cohesin influences meiotic chromosome structure (13–15) and recombination (12, 16).

COHESIN ASSOCIATION WITH CHROMATIN

The proposed ring structure of the cohesin complex lends itself to two models for cohesion. In one popular model, known as the embrace model, the cohesin ring holds within it two sister chromatids (6, 9). However, the diameter of the ring is estimated to be 45 nm, which would seem to preclude the embrace of two 30-nm chromatin fibers, although two 10-nm fibers (DNA in nucleosome form or "beads on a string") could be accomodated. In a second popular model, two cohesin rings each encircle a single chromatin fiber (17–19) and interact with each other, holding the sisters together. Additional models have also been proposed (20), and cohesin may be able to associate with DNA in multiple modes. Interestingly, interactions between the hinge domains of Smc1-Smc3 and the head domains may be required for the complex to function in cohesion (21–23). For a more detailed discussion of the structure of the complex, the reader is referred to excellent recent Annual Reviews (24, 25).

In any event, the cohesin complex can hold DNA sequences from sister chromatids together. However, it is important to note that, in several instances, cohesin has been shown to bind to DNA without actually mediating cohesion, and thus, binding of the subunits cannot be taken as an indicator of cohesion (10, 26). This result also suggests that cohesin can associate with DNA in at least two states: a cohesive state and a noncohesive state. Consistent with the idea that cohesin may be able to associate with DNA in more than one mode, Gerlich and colleagues (27) used photobleaching studies to show that there is a "fast" pool of cohesin throughout interphase and a "slow" pool in G2. These results have been interpreted as indicating that there is a fraction of cohesin that is "locked" in the cohesive state as the cell cycle proceeds (27, 28). This "stable" cohesin may be pericentromeric complexes involved in chromosome segregation (29). At least some significant fraction of cohesin appears to be mobile, and this fraction could be cohesin associated with chromosome arms, which may need to accommodate transcriptional changes. Although cohesin in chromosome arms provides cohesion of sister sequences, it may have additional or alternate roles to that of chromosome segregation, for instance, in gene regulation.

Thus, cohesin may have different dynamic properties and functionalities in different positions and chromatin contexts, and this may derive from different modes of binding.

The association of cohesin with DNA is a fascinating topic, and understanding this interaction may provide the key to understanding how cohesin appears to function in so many different processes. Despite intensive study over the past 15 years, our understanding of how this complex associates with DNA is still primitive. In recent years, ChIP in combination with microarrays has been used to examine the genome-wide pattern of binding of the complex in many organisms (30–36). One general feature of the pattern is that cohesin associates broadly across pericentromeric regions, and in this fashion, it functions in chromosome segregation (see box on Pericentromeric Cohesion). Another general feature that has emerged from these studies is that cohesin does not possess specificity for any particular DNA sequences on its own. Although cohesin colocalizes with CTCF (33–35), an insulator protein in mammalian cells that does have a cognate-binding sequence, knockdown of CTCF causes cohesin to delocalize from these sites, suggesting that cohesin does not bind this sequence specifically on its own (35). A third generality is that the binding in chromosome arms has some correlation with transcription, but the relationship varies in different organisms. In yeast, cohesin is located largely in intergenic regions, where transcription converges, but it can be located in transcriptionally inactive open reading frames, where it disappears upon transcriptional activation (30, 31). In flies and humans, cohesin associates with transcriptionally active regions and shows enrichment particularly at transcription start sites (32, 37). The underlying mechanistic basis for the correlations with transcriptional activity remains enigmatic.

Not only can cohesin association change with transcription, it may also play a role in the subnuclear localization of sequences. In

PERICENTROMERIC COHESION

Kinetochores are large protein structures that form at centromeres and attach chromosomes to microtubules. Microtubule-attached sister kinetochores can be separated by several microns in a variety of cell types. At pericentric regions, large domains of cohesin exist. It has long been a conundrum that cohesin binds to these regions, but these regions appear to transiently come apart during metaphase (150, 151). Thus, although pericentric chromatin appears to be highly elastic and can stretch or recoil in response to microtubule shortening or growth in mitosis, the embrace of a ring structure would seem to preclude this phenomenon. One proposal to explain this elasticity is that cohesin binds in a different mode at pericentric regions. Instead of mediating cohesion between sisters (interchromosomal cohesion), pericentric cohesin may organize chromatin into an intramolecular loop (152, 153). This could explain how cohesin is bound but the centromeres "breathe."

In mitosis, sister kinetochores are attached to microtubules from opposite spindle poles (bipolar attachment), but in meiosis I, sister kinetochores are attached to microtubules from the same spindle pole (monopolar attachment). Recently, the geometry of kinetochore orientation in meiosis and mitosis was examined in parallel and revealed that intermolecular cohesion at pericentric regions is present in meiosis I but absent in mitosis. An elegant system was developed in which the pericentric region could be excised by R recombinase and visualized using lacO/lacI-GFP within a single cell cycle (154). In mitosis, the excised circles did not remain together, indicating that, in fact, cohesion between sisters is not present at the core centromere in mitosis. In contrast, when the circles were excised from cells in meiosis I, the circles remained together. The kinetochores in mitosis have a back-to-back orientation that can accommodate biorientation, whereas kinetochores in meiosis I appear to have a side-by-side attachment that is conducive to monopolar attachment. The Moa1 protein helps to promote monopolar attachment in meiosis (155, 156). These results are consistent with the model that cohesin at pericentric regions in mitosis may be intramolecular. Extrapolating from these results, it appears cohesin can hold identical sequences together in one context (interchromosomal cohesion) and in pericentromeric regions may hold nonidentical sequences together (intrachromosomal cohesion). This model is consistent with the suggestion that cohesin can tether nonidentical sequences together.

ChIP: chromatin immunoprecipitation

CTCF: CCCTC-binding factor, a zinc finger protein that acts as an insulator protein

Insulator: a DNA sequence that can block the interaction of an enhancer and a promoter

budding yeast, cohesin associates with the open reading frame of *GAL2* when it is transcriptionally repressed but not when it is transcriptionally activated (26, 31). Furthermore, mutations in the cohesin network alter the induction profile and subnuclear localization of *GAL2* (38). Clustering of tRNA genes near the nucleolus and the clustering of telomeres are both disrupted by these same mutations. These results suggest that the cohesin network may contribute to gene regulation through subcellular positioning of sequences, an idea consistent with the observation that telomeres in Smc1$\beta^{-/-}$ mouse meiocytes fail to attach to the nuclear envelope (15). In contrast to yeast, cohesin associates with the *Abd-B* homeobox gene in a *Drosophila* cell line in which it is transcribed but not in a cell line in which it is silenced (32). Although the trend in yeast and flies appears to be diametrically opposed in terms of cohesin binding and transcription, when the underlying mechanistic connections between transcription, subnuclear localization, and cohesin association are better understood, the results may be reconcilable.

Because cohesin binds to DNA wrapped in nucleosomes, it is possible that it may recognize or prefer chromatin with particular modifications. In support of this idea, binding has been shown to be dependent on the chromatin remodeling complexes ISW (39) and RSC (19, 40–42). At the mating locus in budding yeast, binding appears to be dependent on the NAD-dependent histone deacetylase and silencing protein Sir2, a sirtuin (19). In *Schizosaccharomyces pombe*, binding of cohesin to the pericentric heterochromatic region depends on heterochromatin protein 1 (HP-1) (43). In flies, cohesin is generally absent from regions in which histone H3 is methylated by the Enhancer of zeste [E(z)] Polycomb group silencing protein. However, transcription is hypersensitive to cohesin levels in two exceptional cases where cohesin and the E(z)-mediated histone methylation simultaneously coat the entire *Enhancer of split* and *invected-engrailed* gene complexes in cells derived from the *Drosophila* central nervous system (44). Further studies to illuminate the colocalization of cohesin with various histone modifications may be informative with regard to both its chromatin-binding preferences and the correlation with transcriptional activity.

In addition to the relatively simple act of holding together sister chromatids, cohesin may influence the three-dimensional structure of chromosomes. One way cohesin might provide positional information has already been mentioned—it may help to tether specific sequences to discrete locations in the nuclear space. Another example of a unique chromosomal structure being stabilized by cohesin is presented in the Pericentromeric Cohesion box. Cohesin may help establish a unique intramolecular hairpin structure at pericentric domains (**Figure 2*a***). Furthermore, cohesin is necessary for chromosome compaction in fission and budding yeast (14, 45), and this can be separated from its role in chromosome segregation (38). It has been proposed that cohesin sites may define the boundary for the chromosome condensation machinery (45, 46). Cohesin-bound chromatin may interact to form a chromosome axis with loops emanating from the axis (**Figure 2*b***). Although speculative, these types of models of how cohesin contributes to chromatin organization help explain how mutations in the cohesin network can cause pleiotropic defects in a variety of DNA metabolic processes.

In addition to holding DNA sequences together, cohesin may also function in the cohesion of centrioles, which are part of the microtubule organizing center. This structure is duplicated once every cell cycle and nucleates the bipolar spindle for mitosis. The subunits of the cohesin complex have been shown to associate with centrioles (47, 48), and furthermore, cleavage of the kleisin subunit by the protease separase is necessary for the disengagement of centrioles during exit from mitosis (49). Thus, cohesin may be involved in additional duplication-separation events coupled to the cell cycle.

REGULATION OF THE COHESIN NETWORK: COHESIN ACCESSORY PROTEINS

On top of the four subunits of the cohesin complex proper, several accessory factors are needed throughout the cell cycle for chromosome cohesion to function correctly. We refer to the complex plus accessory factors as the cohesin network (**Figure 3**).

Loading Factors

Loading of the cohesin complex onto chromosomes requires a loading complex, Scc2-Scc4 (50–52). These proteins are evolutionarily conserved from yeast to man (53). Scc2 and Scc4 are required to load cohesin onto DNA during the G1 phase of the cell cycle in yeast, and telophase in vertebrates, prior to DNA replication. In yeast, cohesin appears to be at the same locations by ChIP in the G1 and G2/M phases. However, these locations were reported to be different from the locations of Scc2-Scc4, raising the possibility that cohesin is loaded at particular locations but quickly relocalizes to new locations (30, 54). Recently, this result has been challenged, with Kogut and colleagues (55) reporting that Scc2 and cohesin do in fact colocalize on chromosomes, similar to what has been reported in the fly genome (32). Scc2 also appears to be necessary for cohesion in the G2/M phase in response to a DNA DSB (56) and to load the condensin complex (54). Mutation of *SCC2/Nipbl* or *SCC4/MAU-2* has been shown to cause a variety of developmental defects, including impaired cell and axon migration (57–59). Overall, Scc2-Scc4 appears to promote cohesion by loading the cohesin complex onto chromatin.

Establishment Factors

ECO1/CTF7 is an essential gene in budding yeast (60). Eco1/Ctf7 is necessary to establish cohesion in yeast during S phase, but not for cohesin binding or maintenance of cohesion in G2/M phase (61). Eco1 contains a zinc

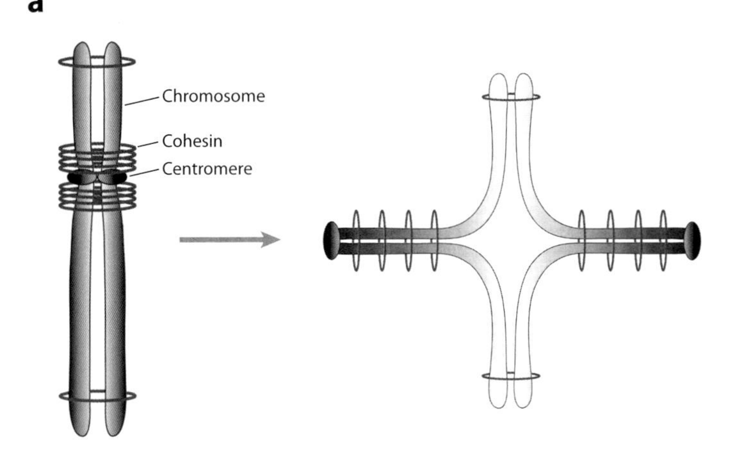

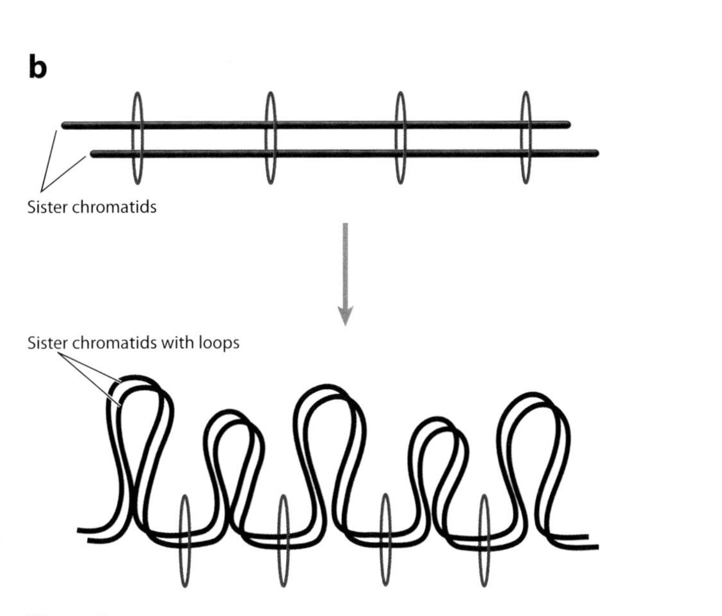

Figure 2

Linear maps of cohesin-binding sites may lend three-dimensional structure to chromosomes. (*a*) Cohesin binds to a wide region surrounding the budding yeast centromere (*black oval*). However, when pericentric chromatin on sisters is visualized, it is often separated. If cohesin in this region is intrachromosomal, rather than interchromosomal, and helps to form a hairpin structure, this could reconcile the high level of cohesin binding with the "breathing" of centromeres. The cartoon is schematic, and only one instance of cohesin binding in the arm is depicted for the sake of simplicity. The arrow is not meant to imply that one type of binding can convert to the other, only that the linear pattern could be interpreted in at least two ways. (*b*) Cohesin-binding sites within chromosome arms are found approximately every 11 kb in the budding yeast genome. The cohesin complex is depicted as a red circle. In this cartoon, a single cohesin complex is shown embracing two sisters. Cohesin sites may form boundaries for chromatin loops.

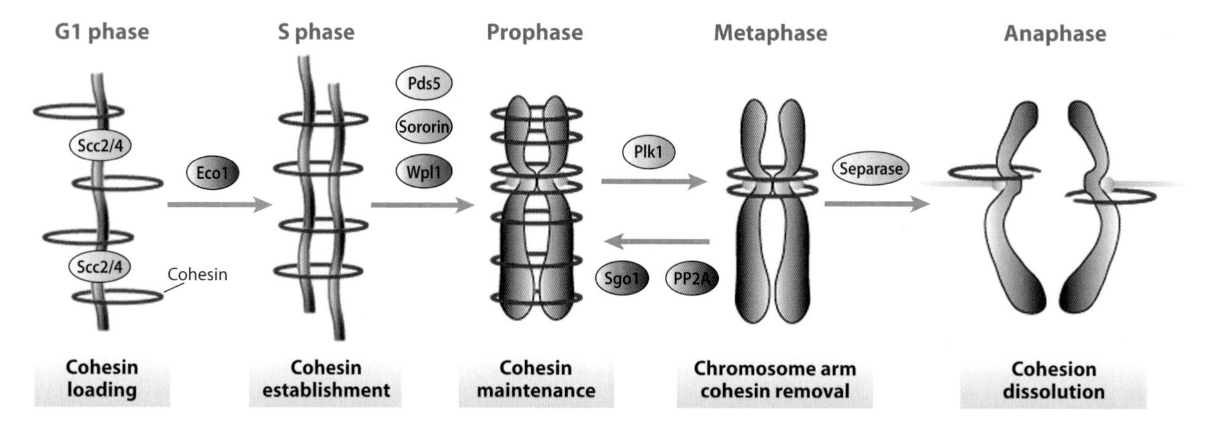

Figure 3

The cohesin network. The cohesin complex is initially loaded onto chromatin in G1 phase of the cell cycle, dependent on the activity of the loading complex Scc2-Scc4. Concomitant with DNA replication, cohesion is established between sister chromatids in the presence of the acetyltransferase Eco1. Upon entry into G2/M phase, cohesion is maintained by the proteins Pds5, sororin, and Wpl1. In mammalian cells, cohesin along chromosome arms is removed during prophase via phosphorylation of Rad21 by Polo-like kinase 1. In metaphase, cohesion is protected from removal by *shugoshin* and PP2A. At the onset of anaphase, a protease called separase cleaves the Mcd1/Scc1/Rad21 subunit, leading to the opening of the cohesin ring and the separation of sister chromatids.

finger in the N-terminal portion and an acetyltransferase domain in the C-terminal portion. Eco1 physically associates with components of the replication factor C complex (62). By piggybacking with the RFC complex, Eco1 may be able to take advantage of the proximity of sisters during DNA replication to promote cohesion. Eco1 appears to be a global regulator of cohesion (see below) and may itself be regulated. Deco is the name for the orthologous acetyltransferase needed for centromeric cohesion in flies (63–65). In humans, there are two paralogs of ECO1 known as ESCO1 and ESCO2.

Maintenance Factors

Several proteins have been shown to play a role in maintaining cohesion following S phase, including Pds5, Sororin, and Wapl. *PDS5* is an essential gene in budding yeast, which is required during G2/M phase for the maintenance of cohesion (66, 67). Metazoans have two *PDS5* genes, A and B, both of which contain HEAT repeats and are required for cohesion, but they differ in their expression (68). Both Pds5A- and Pds5B-deficient mice die at birth (69, 70). Pds5B$^{-/-}$ mice have multiple congenital abnormalities, but interestingly Pds5B$^{-/-}$ mouse embryonic fibroblasts lack sister cohesion defects (70).

Sororin is a vertebrate-specific component of the cohesin network (71, 72). Although sororin is dispensible for loading of cohesin onto chromosomes, it is needed for both establishment and maintenance of cohesion (72). Sororin is also needed for the repair of DSBs in G2 phase (72). Without sororin, the stably chromatin-bound population of cohesin decreases in G2 phase cells (72).

WPL1/RAD61/Wapl is an evolutionarily conserved gene involved in chromosome cohesion. *Wapl* (wings apart like) was first discovered in flies as a gene controlling heterochromatin organization (73). Wapl promotes the dissocation of cohesin from chromosome arms in mammalian cells during prophase and increases its turnover on chromatin during interphase (74, 75). In mammalian cells depleted of Wapl, cohesin remains associated with chromatids during mitosis, which results in a mitotic delay (74, 75). The characteristic X shape of bivalents is not observed because cohesin stays associated with chromosome arms, and they do not splay apart (74, 75). Overexpression of Wapl has the opposite effect, leading to premature separation of sisters. Somewhat similar to its role in mammalian cells, *WPL1* appears to be a cohesion inhibitor in yeast as deletion allows cohesion to be generated in G2/M phase (76). Deletion of *WPL1/RAD61* in budding yeast suppresses the temperature sensitivity of Eco1 acetyltransferase mutants (77–79), but cells have compromised sister chromatid cohesion (78–80).

The phenotypes associated with mutations in *WPL1* are complex. Although *WPL1* appears to promote cohesion because its deletion compromises cohesion, its main function in the cohesin network may be to act as a negative regulator of cohesion (**Figure 4**) (76, 79). A recent report demonstrated that a complex can be made in vitro that contains Pds5-Wpl1-Scc3, with Wpl1 binding to Scc3 and Pds5 (78). Another group demonstrated a Pds5-Wpl1 complex in yeast (79). This group proposed that Eco1 establishes cohesion in S phase by temporarily hindering the function of the Pds5-Wpl1 complex. Consistent with this, Wapl has been shown in a subcomplex with Pds5 in mammalian cells, where it could antagonize the cohesion-promoting function of Pds5 (75). However, mammalian Wapl has also been shown to be in a complex with Rad21 and SA1/Scc3 and has been proposed to prevent reassociation of the proteins with the cohesin complex following their release (74).

A recent study regarding the role of Wapl and Pds5 in the resolution of mitotic chromosome arms in *Xenopus* egg extracts illuminates the biochemical realities of these protein complexes (81). Careful cofractionation experiments reveal that these proteins can interact with each other and cohesin, but only at substoichiometric levels. Both Wapl and Pds5 appear to be needed for the dissolution of cohesion in chromosome arms upon mitotic entry. They appear to be able to act transiently but directly on the cohesin complex, and in particular on the kleisin component, over a very short window of time to cause a change from cohesive to released. The FGF motifs in the N terminus of human Wapl are necessary for its interaction with Pds5 and cohesin. Wapl is a relative newcomer to the cohesin network but promises to be a fascinating regulator. It will be interesting to study how this protein affects the posttranslational modifications discussed below.

Dissolution Factors

Cohesion is dissolved at the metaphase to anaphase transition. One of the major mechanisms of dissolution is proteolytic cleavage of the kleisin subunit by the protease known as separase/Esp1. Proteolysis of the kleisin allows the cohesin ring to come apart and chromosomes to separate (82, 83). Separase is normally inhibited by its binding to securin/Pds1 (84–87).

In mammalian cells, cohesin along chromosome arms is removed during prophase in a separase-independent mechanism (88, 89). Instead, this removal depends on Polo kinase, which phosphorylates the kleisin Rad21, allowing cohesion to dissolve. In meiosis, similar mechanisms allow cohesion along chromosome arms to dissolve prior to meiosis I, but centromeric cohesion is maintained until meiosis II when it is targeted by separase (90). Centromeric cohesion in both cycling and meiotic cells is protected from premature dissolution by *shugoshin* (Sgo1, Sgo2), Japanese for "guardian spirit," and the PP2A phosphatase (91–98).

REGULATION OF THE COHESIN NETWORK: POSTTRANSLATIONAL MODIFICATIONS

As the ability to identify posttranslational modifications by mass spectrometry technology has increased, the number of modifications identified has exploded. The new challenge is to discover the biological significance of these modifications. To date, there is evidence for the cell cycle-coupled regulation of the cohesin network by phosphorylation, acetylation, sumoylation, and ubiquitylation (for a summary see **Table 2**). Acetylation appears to specifically promote cohesion. There are examples of both positive and negative regulation by phosphorylation. Sumoylation and

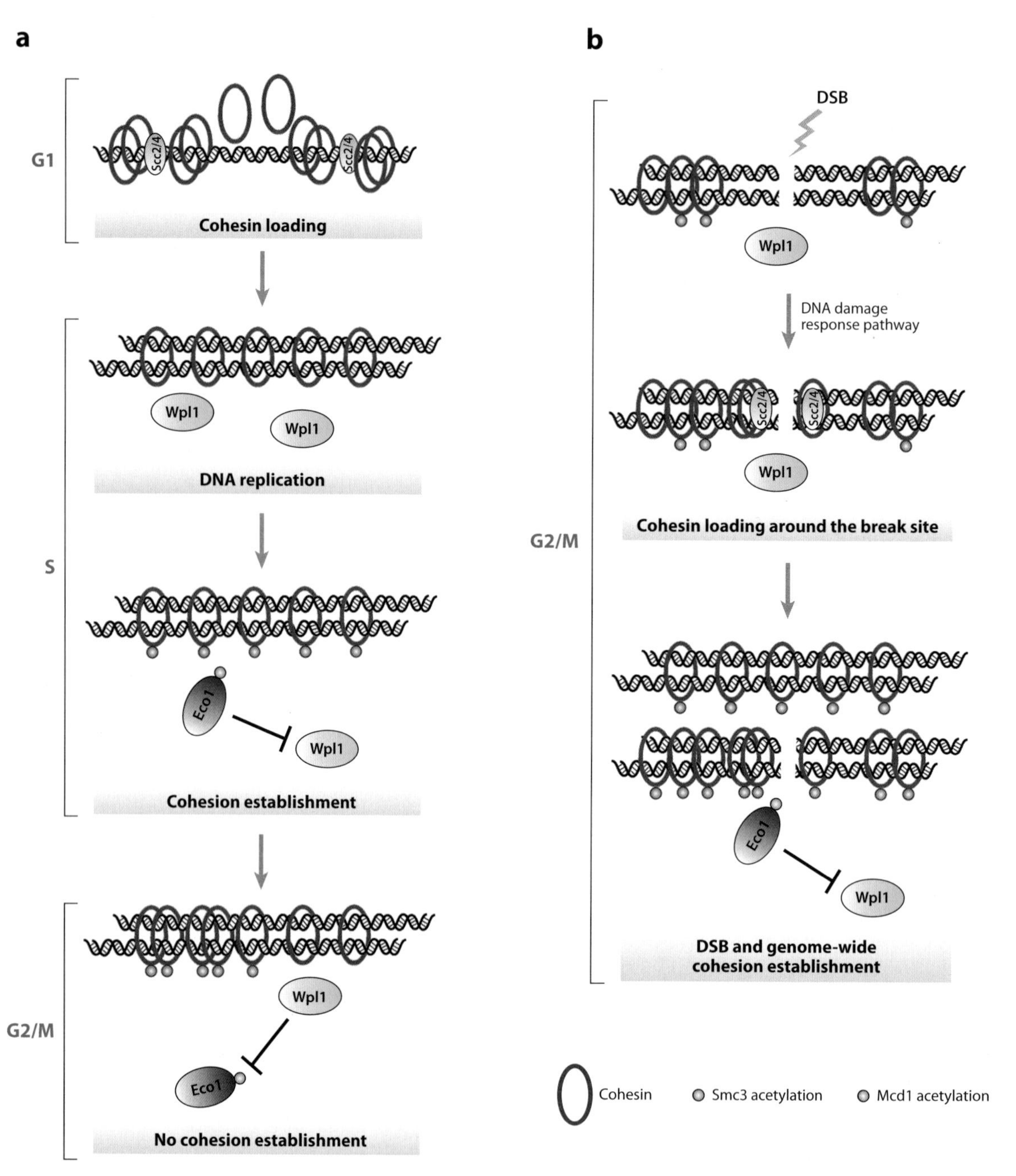

Figure 4

Model for establishment of cohesion in S phase and G2/M phase. (*a*) In G1/S phase, the Scc2-Scc4 complex loads cohesin onto the chromosomes. After DNA replication, Eco1-dependent Smc3 acetylation counteracts the activity of Wpl1 that hinders the establishment of cohesion. At this time, the chromatin-bound cohesin is converted to a cohesive state. Upon entering G2/M phase, cohesion cannot be newly generated because the activity of Eco1 decreases and is not high enough to counteract Wpl1. (*b*) When G2/M cells experience a DSB, the DNA damage response pathway induces DSB-proximal cohesin loading and then promotes Eco1-dependent Mcd1/Scc1 acetylation, which in turn allows cohesion establishment in G2/M phase by antagonizing Wpl1. Cohesion is established at the DSB site but is also reinforced genome wide.

Table 2 Catalog of posttranslational modifications associated with the cohesin network

Protein	Posttranslational modification	Amino acid	Addition dependent on protein	Removal dependent on protein	Biological function	References
yPds5	Sumoylated	—	Ubc9/Nfi1	Smt4	Dissolution of cohesion	133
yRec8	Phosphorylation	S136, T173, S179, S197, S199, S215, S386, S387, S410, S465, S466	Cdc5/Polo	PP2A	Loss of meiotic cohesin/cleavage of Rec8	126, 127
yMcd1	Phosphorylation	S175, S183, S194, S263, S273, S276, S325, S374, S389, S497	Cdc5/Polo	PP2A	Loss of mitotic cohesin/cleavage of Mcd1/Scc1	115, 116
yMcd1	Phosphorylation	S83	Mec1 Chk1	—	DSB-induced cohesion G2/M	120
yMcd1	Acetylation	K84, K210	Eco1	—	DSB-induced cohesion G2/M	76
ySmc3	Acetylation	K112, K113	Eco1	—	Promotes cohesion in S phase	77, 78, 103, 141
yPds1	Phosphorylation	S37, S71, S212, S213, S277, S292, T304	Cdc28/Cdk1	Cdc14?	Nuclear localization of separase	118, 119
yPds1	Ubiquitylation	—	APC/Cdc20	—	Degrades securin to release separase	117
hSmc1	Phosphorylation	S957, S966	ATM	—	DNA damage checkpoint	122, 123
hSmc3	Phosphorylation	S1067, S1083	ATM	—	DNA damage checkpoint	88, 121

ubiquitylation promote dissolution of cohesion. Thus, chromosome cohesion is regulated through multiple posttranslational modifications during the cell cycle. Furthermore, each posttranslational modification may depend in some manner on the others, but the cross talk between modifications is only beginning to be explored.

In addition to different posttranslational modifications contributing to the cell cycle-coupled regulation of cohesion, distinct posttranslational modifications appear to be required for different functions of the cohesin complex. For example, certain modifications are associated with the establishment of cohesion during S phase, and different modifications are associated with the loading of cohesin at a DSB site. In vertebrates, there is evidence that cohesin plays an additional role in DNA damage-induced checkpoints distinct from its role in cohesion at DSB sites, and this too is correlated with specific modifications.

Only recently has cohesin been demonstrated to promote the juxtaposition of DNA sequences for the outcome of gene regulation in human cells. Several recent reports have shown that CTCF, a major protein involved in insulation, and cohesin colocalize on mammalian chromosomes (33–35). At the developmentally regulated *IFNγ* locus, cohesin has been shown to be responsible for the interaction of distant *cis* sequences using the chromosome conformation capture method (99). However, although cohesin binding is necessary for the distant interactions, it is not sufficient. At the imprinted *H19/Igf2* locus, cohesin has been found to similarly help organize chromatin in three dimensions to direct gene expression by regulating the cross talk between the promoter and enhancer (35).

Interferon γ (IFNγ): a type II class interferon needed for innate and adaptive immunity

***H19/Igf2*:** a mammalian imprinted locus. H19 and Igf2 are expressed in a monoallelic manner from the maternal and paternal chromosomes, respectively

Earlier genetic experiments from the Dorsett lab (100, 101) had suggested Nipped-B and cohesin were involved in promoter-enhancer interactions in *Drosophila*. Although the location of some cohesin is dynamic, possibly to accommodate transcription (26), posttranslational modifications that might specifically facilitate this flexibility have yet to be identified.

Acetylation Associated with the Cohesin Network

Eco1/Ctf7 is an acetyltransferase that acetylates itself as well as proteins in the cohesin network. It is currently unclear whether autoacetylation is a mechanism of regulation. Even though null mutations of *ECO1* are lethal in yeast, mutations that decrease the acetyltransferase activity do not increase the rate of chromosome loss, arguing that the acetyltransferase activity of the protein is not essential for chromosome segregation (102). Recently, the cohesin subunit Smc3 has been shown to be acetylated by Eco1 on K112 and K113 (77, 78, 103, 141). In contrast to mutations in the acetyltransferase domain of Eco1, mutation of Smc3 on residues K112 and K113 leads to precocious sister chromatid separation (77). One potential explanation for this paradox is that the mutations that compromise the catalytic ability of Eco1 in vitro still retain enough catalytic activity in vivo to acetylate a sufficient fraction of Smc3 that chromosome segregation is not compromised. Alternately, the mutations in Smc3 that were designed to affect acetylation may have more pleiotropic effects on function. In fact, acetylation of Smc3 has been reported very recently to promote replication fork progression (104). Mutations in Smc3 that prevent acetylation increase the interaction with the regulators Pds5A and Wapl (104). The discovery of posttranslational modifications that modulate the interaction with regulatory proteins highlights the potential for additional layers of modulation.

The discovery of the regulation of the cohesin complex through Smc3 via acetylation opens up a host of new possibilities. SMC proteins are evolutionarily conserved from prokaryotes to humans. In eukaryotes, there are at least three protein complexes that contain SMC proteins, including cohesin, condensin, and the Smc5-6 complex, which plays a role in DSB repair and segregation of nucleolar sequences (105–107). Rad50, which is part of the Mre11-Rad50-Xrs2 (MRX) complex and plays an essential role in 5′ to 3′ resection of DNA breaks prior to repair and recombination, is also an SMC protein. In fact, loading of cohesin at a DSB is dependent on this complex (108). Perhaps acetylation by Eco1 or additional posttranslational modifications will be discovered that regulate this diverse group of SMC-containing complexes. This is an important area for future study because SMC complexes have such broad and profound impacts on genome architecture and organization.

For DSB repair, a different subunit of cohesin is targeted for acetylation. When an HO break is induced in budding yeast, cohesion occurs at the break site (108, 109) and also genome wide (56, 110). This cohesion depends on the acetyltransferase activity of Eco1 (56, 110). Mutation of residues K84 and K210 of the kleisin Mcd1/Scc1 causes a defect in genome-wide cohesion in G2/M phase in response to a DSB (76). An acetylation mimic mutant ($MCD1^{K84Q, K210Q}$) bypasses the need for Eco1, suggesting that Eco1 acetylates Mcd1/Scc1 on these residues (76). Although acetylation on K210 has been detected by mass spectrometry (76, 111), the evidence for the K84 modification is based solely on the genetic behavior of unacetylatable and acetylation mimics for this residue (76). The lysine residues that are postulated to be acetylated in Mcd1/Scc1 do not appear to be evolutionarily conserved in the meiotic kleisin Rec8, but this does not preclude the regulation of the meiotic form of the complex by acetylation. Even though it is clear that cohesin is part of a normal cellular response to a mitotic DSB, it is not clear what effect cohesin has on repair outcome. It has been assumed that cohesin is responsible for directing the choice of repair template, but experimental data to support this claim are lacking.

In addition to Smc3 and Mcd1/Scc1, Eco1 has been shown to have other targets in vitro, such as Pds5 and Scc3 (111). However, histones are not acetylated by Eco1 in vitro (111). It is still unknown if Pds5 and Scc3 are acetylated in vivo, and if so, what effect these acetylation events might have on the regulation of the cohesin network. Additional targets of Eco1 in vivo are likely to be discovered. It has been suggested that Wpl1 is antagonized by Eco1-dependent acetylation on Smc3 (79) and Mcd1/Scc1 (76). Studies in humans and yeast suggest that acetylation of Smc3 disrupts the association with Pds5-Wpl1, allowing cohesion to be established in S phase (77, 104). Overexpression of Eco1 also led to a reduced chromosomal association of Pds5 and Wpl1 (79). Thus, cohesion is likely to be regulated by acetylation at many levels.

In the field of chromatin research, there are histone acetyltransferases that transfer an acetyl group to specific lysine residues and histone deacetylases that remove them from specific lysine residues. For a review see Reference 112. By analogy then, it is easy to imagine that there may be a deacetylase that can counteract Eco1 that has yet to be discovered.

Phosphorylation Associated with the Cohesin Network

Phosphorylation alone of cohesin subunits can promote dissolution of cohesion. In mammalian cells, the cohesin subunit SA2 is phosphorylated in early mitosis by the Polo-like kinase, and this correlates with a loss of cohesion within chromosome arms, apparently without the action of the separase enzyme (88, 113, 114). Phosphorylation of Mcd1/Scc1 by Polo/Cdc5 in yeast at serine residues adjacent to the separase cleavage site appears to enhance both recognition and cleavage by separase (115, 116).

The cell cycle is driven in large part by a major kinase, Cdk1/Cdc28, which associates with different cyclins that control its substrate specificity. Because cohesion is tied to the cell cycle for chromosome segregation, it makes intuitive sense that it might be controlled by this system. Securin is a substrate for anaphase-promoting complex (APC)-dependent proteolysis (117). Phosphorylation of securin by Cdc28 allows efficient binding of securin to separase and promotes the interaction with securin that allows nuclear localization (118). Subsequent ubiquitylation of securin by the APC/cyclosome at its destruction box motif targets it to be degraded by the proteasome, thereby releasing separase. Interestingly, phosphorylation of securin in the neighborhood of its destruction box by Cdc28 inhibits ubiquitylation, and the mitotic exit-promoting phosphatase Cdc14 reverses phosphorylation and increases ubiquitylation, revealing the exquisite regulation possible for the exit from mitosis and the initiation of anaphase (119).

Phosphorylation of cohesin subunits also occurs in response to a DSB. In mitotic yeast cells, Mcd1/Scc1 is phosphorylated on S83, which may in turn increase its acetylation by Eco1 at S84 during DSB-induced cohesion (120). Although phosphorylation is dependent on the kinase Chk1, the kinase directly responsible for phosphorylation has not been identified (120). In vertebrate cells, Smc1 and Smc3 have been shown to be phosphorylated in response to DNA damage as targets of the ATM kinase (121–123). Thus, phosphorylation of cohesin subunits appears to occur as part of a DNA damage response. However, the response of the cohesin network to DNA damage is not limited to its ability to establish cohesion between sisters. Even though Smc1 and Smc3 are part of the cohesin complex, they also appear to be part of a distinct recombination complex 1 (RC-1) (124). And although the participation of Smc1 and Smc3 in a DNA damage-induced intra-S phase checkpoint is dependent on their ability to mediate cohesion, their role in a G2/M damage-induced checkpoint appears to be independent of their role in cohesion (125). Overall, phosphorylation appears to modulate the cohesin network to achieve an effective response to DNA damage.

As mentioned above, the kleisin component of the cohesin complex is often the target of regulation. Phosphorylation of the meiosis-specific kleisin Rec8 (11) helps to control cohesion during meiosis. In meiosis, centromeric cohesion must be preserved through meiosis I such that sisters are cohered until meiosis II. Given the role of phosphorylation in regulating cohesion and Mcd1/Scc1 during mitosis, Brar and colleagues (126) decided to investigate phosphorylation of Rec8 during meiosis in yeast. Rec8 is highly phosphorylated during meiosis, with several of the phosphorylation events dependent on Cdc5/Polo. In cells expressing a version of Rec8 with these sites mutated to nonphosphorylatable residues, Rec8 cleavage and anaphase I were delayed. Interestingly, elimination of recombination rescued the delay, indicating that recombination provides a signal that allows for loss of Rec8 cohesin from chromosome arms prior to anaphase I. In mice, hyperphosphorylation of Rec8 is also necessary for cleavage by separase, and mutations that prevent phosphorylation cause male sterility and meiotic delays in females (127). Because studies of the meiotic kleisin indicate that cleavage of Rec8 along chromosome arms is needed for the resolution of recombination intermediates, the regulation of this subunit by phosphorylation is crucial for a normal meiosis.

The protection of centromeric cohesion in meiosis I relies on preserving Rec8 in a hypophosphorylated state, which is dependent on MEI-S332/Sgo1 and the protein phosphatase PP2A. MEI-S332/Sgo1 localizes to the centromeres and to pericentric regions (96, 97, 128, 129). The pericentric localization depends on cohesin; Bub1, a kinase and a component of the spindle assembly checkpoint (129, 130); and the serine/threonine protein phosphatase PP2A, which is a heterotrimeric complex (92, 94, 95). Bub1 appears to phosphorylate MEI-S332/Sgo1, which helps to maintain its centromeric location. In fission yeast and budding yeast, PP2A may help to retain centromeric Rec8 in a hypophosphorylated state, which could counteract Polo kinase-dependent cleavage by separase. In human cells, there are two *shugoshins* that contribute to the hypophosphorylation of SA1/2 in mitosis (92, 94). Mutations in Bub1, MEI-S332/Sgo1, and PP2A cause precocious separation of sister chromatids at the first meiotic division, demonstrating that these factors are key components for the retention of pericentric cohesion. The evolutionarily conserved kinase AuroraB/Ipl1 is also needed to maintain the centromeric localization of PP2A and to protect centromeric cohesion during meiosis I (131, 132). Thus, there appears to be multilayered and dynamic phosphorylation at work at pericentric regions to preserve centromeric cohesion, and similar mechanisms can operate in mitosis and meiosis.

Sumoylation Associated with the Cohesin Network

Pds5 is required for the maintenance of cohesion in G2/M phase. The SUMO isopeptidase *SMT4* was identified as a high-copy suppressor of a temperature-sensitive *PDS5* mutant. *SMT4* did not suppress the temperature sensitivity of other cohesin mutants. Pds5 is conjugated to SUMO in a cell cycle–dependent manner, with the conjugated product peaking during mitosis. *SMT4* mutants had increased sumoylation on Pds5, whereas overexpression of *SMT4* decreased Pds5 sumoylation. On the basis of this evidence, it was suggested that Pds5 is sumoylated to promote the dissolution of cohesion (133). However, neither the sumoylated residues nor the enzyme responsible was identified. Sumoylation of Pds5 has not been demonstrated in any organism other than budding yeast to date.

Ubiquitylation Associated with the Cohesin Network

In addition to the ubiquitylation of securin already mentioned above, it has been suggested that ESCO2 may be targeted to the proteasome. Given the observation that ESCO2 expression peaked during S phase and was virtually undetectable during mitosis, van der Lelij and colleagues (134) tested whether ESCO2 was stabilized in mitotic human cells treated with the proteasome inhibitor MG132. ESCO2 was stabilized as measured by both immunofluorescence and Western blotting. siRNA targeting APC/C subunits did not have the same stabilizing effect, suggesting the stabilization was APC independent. The proteins involved in the ubiquitylation have not been identified, nor have the ubiquitylated residues. Nonetheless, these results suggest that ubiquitylation of ESCO2 may be a means of regulating a major regulator of the cohesin network.

MOLECULAR ETIOLOGY OF THE COHESINOPATHIES

The cohesinopathies are human developmental disorders caused by mutations in the cohesin network. Cornelia de Lange syndrome (CdLS) is a genetically dominant disorder caused by mutation of *SCC2/NIPBL*, *SMC1*, and *SMC3* (135–138). Mutations in *SMC1* and *SMC3* constitute only a few percent of cases and cause very mild versions of CdLS. Roberts syndrome (RBS) is a rare but severe genetically recessive disorder caused by mutation of *ESCO2* (139). Because a total loss of chromosome cohesion would lead to massive chromosome segregation and would not be compatible with life, these disorders are not thought to have chromosome missegregation as a predominant feature. Analysis of metaphase spreads from cells from CdLS patients and normal controls confirms only a small increase in precocious sister chromatid separation (140). When binding of cohesin was examined in CdLS patient fibroblasts, it was only modestly decreased (37).

Some of the predominant features associated with the cohesinopathies are growth and mental retardation, limb deformities, and craniofacial defects. The severity of CdLS can range from mild mental retardation to a multisystem disorder with physical, mental, and behavioral disabilities. The phenotypes of the disorders are more consistent with misregulation of genes during embryogenesis rather than defects in chromosome segregation. Interestingly, the most severe forms of the cohesinopathies are not caused by mutations in the subunits of the complex itself but by mutations in the regulators of the complexes *ESCO2* and *NIPBL*.

CdLS: Cornelia de Lange syndrome

The mutations in *ESCO2* associated with RBS are mostly null, but a point mutation that affects the acetyltransferase domain has been identified (142). This mutation suggests that a deficiency in the acetyltransferase activity is sufficient to cause disease. RBS cells display a unique heterochromatic repulsion or "puffing" chromosomal phenotype, which could be interpreted as a cohesion defect, although this defect does not appear to translate into precocious sister separation. Cells from these patients show sensitivity to genotoxic stress caused by mitomycin C, camptothecin, and etoposide (134, 142). These observations are consistent with a role for *ESCO2* in replication-coupled cohesion, and in fact, processivity of DNA replication forks in RBS cells is reduced (104). Although *ECO1* is essential in yeast, *ESCO2* does not appear to be essential in humans, but it is important to remember that humans possess a second *ECO1*-like gene, *ESCO1*.

A few patients with mild forms of CdLS have been found to have mutations in *SMC1* and *SMC3* (135, 136); however, the majority of patients with severe forms of the disease have mutations in *NIPBL*. No mutations in *SCC4/MAU-2* have been identified in association with CdLS. Approximately 40% of CdLS patients do not have an identifiable mutation. One possibility is that some of these patients may have mutations in the regulatory region of *NIPBL* (143). Discovering new gene mutations that are associated with CdLS will certainly promote our understanding of the molecular etiology of this disease. For a more thorough review of the human cohesinopathies, the reader is referred to Reference 144.

How do mutations in the cohesin network cause developmental disorders? Many ideas have been proposed, including cohesin contributing to various three-dimensional DNA conformations that either activate or repress transcription, depending on the context (145). It seems likely that certain gene expression programs during development may be dependent on regulation by the cohesin network, so that mutations can lead to changes in these programs, which result in developmental defects (37, 146–149). Although Scc2 is required to load cohesin in S phase and also promotes cohesion in response to a DNA DSB, Scc2 is also needed to load condensin, at least in budding yeast (54). Thus, the mutations in Scc2 may affect both cohesin and condensin, which might explain the severity of the phenotypes associated with mutations in this gene. Intriguingly, genocopies of a CdLS and of a Roberts mutation in budding yeast cause defects in chromosome condensation without causing defects in chromosome segregation (38). We speculate that some of the putative defects in gene regulation that lead to the cohesinopathies may stem in part from alterations in chromosome morphology.

CONCLUSION

The cohesin complex and its accessory factors contribute in many ways to genomic organization and stability. Proteins in this network have been shown to have roles in chromosome segregation, gene regulation, DSB repair, checkpoint function, and chromosome morphology. Owing to the pleiotropic effects of mutations in this network on cellular phenotypes, it will be important to understand how cohesin influences the genome. The regulation of the network by various posttranslational modifications has already been proven to influence various aspects of function, and the discovery of additional modulation will only add to the complexity. But in the end, a deep molecular understanding of the cohesin network is likely to contribute significantly to our models of higher-order chromosome organization and of its effect on processes such as DSB repair and gene regulation. A deeper understanding of the cohesin network will also be instrumental to unraveling the etiology of the human cohesinopathies.

SUMMARY POINTS

1. The cohesin complex holds sister chromatids together for chromosome segregation but is also capable of holding sequences together to affect gene regulation, DNA DSB repair, and chromosome morphology.
2. The cohesin network is made up of the four subunits in the cohesin complex as well as several proteins that act in different phases of the cell cycle to promote loading, establishment, maintenance, and dissolution of cohesion.
3. Although cohesin binding is necessary for cohesion and distant DNA-DNA interactions, it is not sufficient.
4. Cohesin may participate in tethering sequences to subnuclear positions.
5. Cohesin at chromosome arms and pericentric regions has different properties, is regulated differently, and may perform different biological functions.
6. Many protein components in the cohesin network are regulated by posttranslational modifications, and these modifications contribute to the diverse functions of the complex. The kleisin component of the complex, in particular, is often targeted for regulation.
7. The human cohesinopathy known as Roberts syndrome is a recessive disorder caused by mutations in the acetyltransferase *ESCO2*. The human cohesinopathy known as Cornelia de Lange syndrome (CdLS) is a dominant disorder caused by mutations in *SMC1*, *SMC3*, or *NIPBL/SCC2*.

FUTURE ISSUES

1. As new posttranslational modifications of cohesin proteins are identified, it will be important to discover how these modifications regulate the network.
2. New proteins involved in addition or removal of modifications to the cohesin network components are likely to be discovered.
3. The different ways in which cohesin might tether DNA sequences together are still enigmatic. Solving this puzzle is likely to lead to major insight into higher-order chromosome organization.
4. In only 60% of the CdLS cases has the mutation been identified. Discovery of additional mutations, especially new genes, will help us understand both the cohesin network and the disease etiology.

DISCLOSURE STATEMENT

The authors are not aware of any affiliations, memberships, funding, or financial holdings that might be perceived as affecting the objectivity of this review.

LITERATURE CITED

1. Losada A, Hirano T. 2005. Dynamic molecular linkers of the genome: the first decade of SMC proteins. *Genes Dev.* 19:1269–87
2. Weitzer S, Lehane C, Uhlmann F. 2003. A model for ATP hydrolysis-dependent binding of cohesin to DNA. *Curr. Biol.* 13:1930–40
3. Arumugam P, Gruber S, Tanaka K, Haering CH, Mechtler K, Nasmyth K. 2003. ATP hydrolysis is required for cohesin's association with chromosomes. *Curr. Biol.* 13:1941–53
4. Hirano T. 2006. At the heart of the chromosome: SMC proteins in action. *Nat. Rev. Mol. Cell Biol.* 7:311–22
5. Schleiffer A, Kaitna S, Maurer-Stroh S, Glotzer M, Nasmyth K, Eisenhaber F. 2003. Kleisins: a superfamily of bacterial and eukaryotic SMC protein partners. *Mol. Cell* 11:571–75
6. Haering CH, Lowe J, Hochwagen A, Nasmyth K. 2002. Molecular architecture of SMC proteins and the yeast cohesin complex. *Mol. Cell* 9:773–88
7. Arumugam P, Nishino T, Haering CH, Gruber S, Nasmyth K. 2006. Cohesin's ATPase activity is stimulated by the C-terminal Winged-Helix domain of its kleisin subunit. *Curr. Biol.* 16:1998–2008
8. Anderson DE, Losada A, Erickson HP, Hirano T. 2002. Condensin and cohesin display different arm conformations with characteristic hinge angles. *J. Cell Biol.* 156:419–24
9. Gruber S, Haering CH, Nasmyth K. 2003. Chromosomal cohesin forms a ring. *Cell* 112:765–77
10. Haering CH, Schoffnegger D, Nishino T, Helmhart W, Nasmyth K, Lowe J. 2004. Structure and stability of cohesin's Smc1-kleisin interaction. *Mol. Cell* 15:951–64
11. Klein F, Mahr P, Galova M, Buonomo SB, Michaelis C, et al. 1999. A central role for cohesins in sister chromatid cohesion, formation of axial elements, and recombination during yeast meiosis. *Cell* 98:91–103
12. Revenkova E, Eijpe M, Heyting C, Hodges CA, Hunt PA, et al. 2004. Cohesin SMC1 beta is required for meiotic chromosome dynamics, sister chromatid cohesion and DNA recombination. *Nat. Cell Biol.* 6:555–62
13. Novak I, Wang H, Revenkova E, Jessberger R, Scherthan H, Hoog C. 2008. Cohesin Smc1beta determines meiotic chromatin axis loop organization. *J. Cell Biol.* 180:83–90
14. Ding DQ, Sakurai N, Katou Y, Itoh T, Shirahige K, et al. 2006. Meiotic cohesins modulate chromosome compaction during meiotic prophase in fission yeast. *J. Cell Biol.* 174:499–508

15. Adelfalk C, Janschek J, Revenkova E, Blei C, Liebe B, et al. 2009. Cohesin SMC1beta protects telomeres in meiocytes. *J. Cell Biol.* 187:185–99
16. Hodges CA, Revenkova E, Jessberger R, Hassold TJ, Hunt PA. 2005. SMC1beta-deficient female mice provide evidence that cohesins are a missing link in age-related nondisjunction. *Nat. Genet.* 37:1351–55
17. Zhang N, Kuznetsov SG, Sharan SK, Li K, Rao PH, Pati D. 2008. A handcuff model for the cohesin complex. *J. Cell Biol.* 183:1019–31
18. Zhang N, Pati D. 2009. Handcuff for sisters: a new model for sister chromatid cohesion. *Cell Cycle* 8:399–402
19. Chang CR, Wu CS, Hom Y, Gartenberg MR. 2005. Targeting of cohesin by transcriptionally silent chromatin. *Genes Dev.* 19:3031–42
20. Huang CE, Milutinovich M, Koshland D. 2005. Rings, bracelet or snaps: fashionable alternatives for Smc complexes. *Philos. Trans. R. Soc. Lond. Ser. B* 360:537–42
21. Hirano M, Hirano T. 2006. Opening closed arms: long-distance activation of SMC ATPase by hinge-DNA interactions. *Mol. Cell* 21:175–86
22. Mc Intyre J, Muller EG, Weitzer S, Snydsman BE, Davis TN, Uhlmann F. 2007. In vivo analysis of cohesin architecture using FRET in the budding yeast *Saccharomyces cerevisiae*. *EMBO J.* 26:3783–93
23. Gruber S, Arumugam P, Katou Y, Kuglitsch D, Helmhart W, et al. 2006. Evidence that loading of cohesin onto chromosomes involves opening of its SMC hinge. *Cell* 127:523–37
24. Onn I, Heidinger-Pauli JM, Guacci V, Unal E, Koshland DE. 2008. Sister chromatid cohesion: a simple concept with a complex reality. *Annu. Rev. Cell Dev. Biol.* 24:105–29
25. Nasmyth K, Haering CH. 2009. Cohesin: its roles and mechanisms. *Annu. Rev. Genet.* 43:525–58
26. Bausch C, Noone S, Henry JM, Gaudenz K, Sanderson B, et al. 2007. Transcription alters chromosomal locations of cohesin in *Saccharomyces cerevisiae*. *Mol. Cell. Biol.* 27:8522–32
27. Gerlich D, Koch B, Dupeux F, Peters JM, Ellenberg J. 2006. Live-cell imaging reveals a stable cohesin-chromatin interaction after but not before DNA replication. *Curr. Biol.* 16:1571–78
28. McNairn AJ, Gerton JL. 2009. Intersection of ChIP and FLIP, genomic methods to study the dynamics of the cohesin proteins. *Chromosome Res.* 17:155–63
29. Yeh E, Haase J, Paliulis LV, Joglekar A, Bond L, et al. 2008. Pericentric chromatin is organized into an intramolecular loop in mitosis. *Curr. Biol.* 18:81–90
30. Lengronne A, Katou Y, Mori S, Yokobayashi S, Kelly GP, et al. 2004. Cohesin relocation from sites of chromosomal loading to places of convergent transcription. *Nature* 430:573–78
31. Glynn EF, Megee PC, Yu HG, Mistrot C, Unal E, et al. 2004. Genome-wide mapping of the cohesin complex in the yeast *Saccharomyces cerevisiae*. *PLoS Biol.* 2:E259
32. Misulovin Z, Schwartz YB, Li XY, Kahn TG, Gause M, et al. 2008. Association of cohesin and Nipped-B with transcriptionally active regions of the *Drosophila melanogaster* genome. *Chromosoma* 117:89–102
33. Parelho V, Hadjur S, Spivakov M, Leleu M, Sauer S, et al. 2008. Cohesins functionally associate with CTCF on mammalian chromosome arms. *Cell* 132:422–33
34. Stedman W, Kang H, Lin S, Kissil JL, Bartolomei MS, Lieberman PM. 2008. Cohesins localize with CTCF at the KSHV latency control region and at cellular c-myc and H19/Igf2 insulators. *EMBO J.* 27:654–66
35. Wendt KS, Yoshida K, Itoh T, Bando M, Koch B, et al. 2008. Cohesin mediates transcriptional insulation by CCCTC-binding factor. *Nature* 451:796–801
36. Rubio ED, Reiss DJ, Welcsh PL, Disteche CM, Filippova GN, et al. 2008. CTCF physically links cohesin to chromatin. *Proc. Natl. Acad. Sci. USA* 105:8309–14
37. Liu J, Zhang Z, Bando M, Itoh T, Deardorff MA, et al. 2009. Transcriptional dysregulation in NIPBL and cohesin mutant human cells. *PLoS Biol.* 7:e1000119
38. Gard S, Light W, Xiong B, Bose T, McNairn AJ, et al. 2009. Cohesinopathy mutations disrupt the subnuclear organization of chromatin. *J. Cell Biol.* 187:455–62
39. Hakimi MA, Bochar DA, Schmiesing JA, Dong Y, Barak OG, et al. 2002. A chromatin remodelling complex that loads cohesin onto human chromosomes. *Nature* 418:994–98
40. Baetz KK, Krogan NJ, Emili A, Greenblatt J, Hieter P. 2004. The ctf13-30/CTF13 genomic haploinsufficiency modifier screen identifies the yeast chromatin remodeling complex RSC, which is required for the establishment of sister chromatid cohesion. *Mol. Cell. Biol.* 24:1232–44

41. Huang J, Hsu JM, Laurent BC. 2004. The RSC nucleosome-remodeling complex is required for cohesin's association with chromosome arms. *Mol. Cell* 13:739–50
42. Yang XM, Mehta S, Uzri D, Jayaram M, Velmurugan S. 2004. Mutations in a partitioning protein and altered chromatin structure at the partitioning locus prevent cohesin recruitment by the *Saccharomyces cerevisiae* plasmid and cause plasmid missegregation. *Mol. Cell. Biol.* 24:5290–303
43. Nonaka N, Kitajima T, Yokobayashi S, Xiao G, Yamamoto M, et al. 2002. Recruitment of cohesin to heterochromatic regions by Swi6/HP1 in fission yeast. *Nat. Cell Biol.* 4:89–93
44. Schaaf CA, Misulovin Z, Sahota G, Siddiqui AM, Schwartz YB, et al. 2009. Regulation of the *Drosophila Enhancer of split* and *invected-engrailed* gene complexes by sister chromatid cohesion proteins. *PLoS One* 4:e6202
45. Guacci V, Koshland D, Strunnikov A. 1997. A direct link between sister chromatid cohesion and chromosome condensation revealed through the analysis of MCD1 in *S. cerevisiae*. *Cell* 91:47–57
46. Blat Y, Protacio RU, Hunter N, Kleckner N. 2002. Physical and functional interactions among basic chromosome organizational features govern early steps of meiotic chiasma formation. *Cell* 111:791–802
47. Guan J, Ekwurtzel E, Kvist U, Yuan L. 2008. Cohesin protein SMC1 is a centrosomal protein. *Biochem. Biophys. Res. Commun.* 372:761–64
48. Kong X, Ball AR Jr, Sonoda E, Feng J, Takeda S, et al. 2009. Cohesin associates with spindle poles in a mitosis-specific manner and functions in spindle assembly in vertebrate cells. *Mol. Biol. Cell* 20:1289–301
49. Nakamura A, Arai H, Fujita N. 2009. Centrosomal Aki1 and cohesin function in separase-regulated centriole disengagement. *J. Cell Biol.* 187:607–14
50. Ciosk R, Shirayama M, Shevchenko A, Tanaka T, Toth A, Nasmyth K. 2000. Cohesin's binding to chromosomes depends on a separate complex consisting of Scc2 and Scc4 proteins. *Mol. Cell* 5:243–54
51. Gillespie PJ, Hirano T. 2004. Scc2 couples replication licensing to sister chromatid cohesion in *Xenopus* egg extracts. *Curr. Biol.* 14:1598–603
52. Takahashi TS, Yiu P, Chou MF, Gygi S, Walter JC. 2004. Recruitment of *Xenopus* Scc2 and cohesin to chromatin requires the pre-replication complex. *Nat. Cell Biol.* 6:991–96
53. Watrin E, Schleiffer A, Tanaka K, Eisenhaber F, Nasmyth K, Peters JM. 2006. Human Scc4 is required for cohesin binding to chromatin, sister-chromatid cohesion, and mitotic progression. *Curr. Biol.* 16:863–74
54. D'Ambrosio C, Schmidt CK, Katou Y, Kelly G, Itoh T, et al. 2008. Identification of *cis*-acting sites for condensin loading onto budding yeast chromosomes. *Genes Dev.* 22:2215–27
55. Kogut I, Wang J, Guacci V, Mistry RK, Megee PC. 2009. The Scc2/Scc4 cohesin loader determines the distribution of cohesin on budding yeast chromosomes. *Genes Dev.* 23:2345–57
56. Strom L, Karlsson C, Lindroos HB, Wedahl S, Katou Y, et al. 2007. Postreplicative formation of cohesion is required for repair and induced by a single DNA break. *Science* 317:242–45
57. Bénard CY, Kebir H, Takagi S, Hekimi S. 2004. *mau-2* acts cell-autonomously to guide axonal migrations in *Caenorhabditis elegans*. *Development* 131:5947–58
58. Takagi S, Bénard C, Pak J, Livingstone D, Hekimi S. 1997. Cellular and axonal migrations are misguided along both body axes in the maternal-effect *mau-2* mutants of *Caenorhabditis elegans*. *Development* 124:5115–26
59. Seitan VC, Banks P, Laval S, Majid NA, Dorsett D, et al. 2006. Metazoan Scc4 homologs link sister chromatid cohesion to cell and axon migration guidance. *PLoS Biol.* 4:e242
60. Skibbens RV, Corson LB, Koshland D, Hieter P. 1999. Ctf7p is essential for sister chromatid cohesion and links mitotic chromosome structure to the DNA replication machinery. *Genes Dev.* 13:307–19
61. Toth A, Ciosk R, Uhlmann F, Galova M, Schleiffer A, Nasmyth K. 1999. Yeast cohesin complex requires a conserved protein, Eco1p(Ctf7), to establish cohesion between sister chromatids during DNA replication. *Genes Dev.* 13:320–33
62. Kenna MA, Skibbens RV. 2003. Mechanical link between cohesion establishment and DNA replication: Ctf7p/Eco1p, a cohesion establishment factor, associates with three different replication factor C complexes. *Mol. Cell. Biol.* 23:2999–3007
63. Williams BC, Garrett-Engele CM, Li Z, Williams EV, Rosenman ED, Goldberg ML. 2003. Two putative acetyltransferases, San and Deco, are required for establishing sister chromatid cohesion in *Drosophila*. *Curr. Biol.* 13:2025–36

64. Hou F, Zou H. 2005. Two human orthologues of Eco1/Ctf7 acetyltransferases are both required for proper sister-chromatid cohesion. *Mol. Biol. Cell* 16:3908–18
65. Hou F, Chu CW, Kong X, Yokomori K, Zou H. 2007. The acetyltransferase activity of San stabilizes the mitotic cohesin at the centromeres in a *shugoshin*-independent manner. *J. Cell Biol.* 177:587–97
66. Hartman T, Stead K, Koshland D, Guacci V. 2000. Pds5p is an essential chromosomal protein required for both sister chromatid cohesion and condensation in *Saccharomyces cerevisiae*. *J. Cell Biol.* 151:613–26
67. Panizza S, Tanaka T, Hochwagen A, Eisenhaber F, Nasmyth K. 2000. Pds5 cooperates with cohesin in maintaining sister chromatid cohesion. *Curr. Biol.* 10:1557–64
68. Losada A, Yokochi T, Hirano T. 2005. Functional contribution of Pds5 to cohesin-mediated cohesion in human cells and *Xenopus* egg extracts. *J. Cell Sci.* 118:2133–41
69. Zhang B, Chang J, Fu M, Huang J, Kashyap R, et al. 2009. Dosage effects of cohesin regulatory factor PDS5 on mammalian development: implications for cohesinopathies. *PLoS One* 4:e5232
70. Zhang B, Jain S, Song H, Fu M, Heuckeroth RO, et al. 2007. Mice lacking sister chromatid cohesion protein PDS5B exhibit developmental abnormalities reminiscent of Cornelia de Lange syndrome. *Development* 134:3191–201
71. Rankin S, Ayad NG, Kirschner MW. 2005. Sororin, a substrate of the anaphase-promoting complex, is required for sister chromatid cohesion in vertebrates. *Mol. Cell* 18:185–200
72. Schmitz J, Watrin E, Lenart P, Mechtler K, Peters JM. 2007. Sororin is required for stable binding of cohesin to chromatin and for sister chromatid cohesion in interphase. *Curr. Biol.* 17:630–36
73. Verni F, Gandhi R, Goldberg ML, Gatti M. 2000. Genetic and molecular analysis of wings apart-like (*wapl*), a gene controlling heterochromatin organization in *Drosophila melanogaster*. *Genetics* 154:1693–710
74. Gandhi R, Gillespie PJ, Hirano T. 2006. Human Wapl is a cohesin-binding protein that promotes sister-chromatid resolution in mitotic prophase. *Curr. Biol.* 16:2406–17
75. Kueng S, Hegemann B, Peters BH, Lipp JJ, Schleiffer A, et al. 2006. Wapl controls the dynamic association of cohesin with chromatin. *Cell* 127:955–67
76. Heidinger-Pauli JM, Unal E, Koshland D. 2009. Distinct targets of the Eco1 acetyltransferase modulate cohesion in S phase and in response to DNA damage. *Mol. Cell* 34:311–21
77. Ben-Shahar TR, Heeger S, Lehane C, East P, Flynn H, et al. 2008. Eco1-dependent cohesin acetylation during establishment of sister chromatid cohesion. *Science* 321:563–66
78. Rowland BD, Roig MB, Nishino T, Kurze A, Uluocak P, et al. 2009. Building sister chromatid cohesion: smc3 acetylation counteracts an antiestablishment activity. *Mol. Cell* 33:763–74
79. Sutani T, Kawaguchi T, Kanno R, Itoh T, Shirahige K. 2009. Budding yeast Wpl1(Rad61)-Pds5 complex counteracts sister chromatid cohesion-establishing reaction. *Curr. Biol.* 19:492–97
80. Warren CD, Eckley DM, Lee MS, Hanna JS, Hughes A, et al. 2004. S-phase checkpoint genes safeguard high-fidelity sister chromatid cohesion. *Mol. Biol. Cell* 15:1724–35
81. Shintomi K, Hirano T. 2009. Releasing cohesin from chromosome arms in early mitosis: opposing actions of Wapl-Pds5 and Sgo1. *Genes Dev.* 23:2224–36
82. Ciosk R, Zachariae W, Michaelis C, Shevchenko A, Mann M, Nasmyth K. 1998. An ESP1/PDS1 complex regulates loss of sister chromatid cohesion at the metaphase to anaphase transition in yeast. *Cell* 93:1067–76
83. Funabiki H, Kumada K, Yanagida M. 1996. Fission yeast Cut1 and Cut2 are essential for sister chromatid separation, concentrate along the metaphase spindle and form large complexes. *EMBO J.* 15:6617–28
84. Cohen-Fix O, Peters JM, Kirschner MW, Koshland D. 1996. Anaphase initiation in *Saccharomyces cerevisiae* is controlled by the APC-dependent degradation of the anaphase inhibitor Pds1p. *Genes Dev.* 10:3081–93
85. Yamamoto A, Guacci V, Koshland D. 1996. Pds1p, an inhibitor of anaphase in budding yeast, plays a critical role in the APC and checkpoint pathway(s). *J. Cell Biol.* 133:99–110
86. Yamamoto A, Guacci V, Koshland D. 1996. Pds1p is required for faithful execution of anaphase in the yeast, *Saccharomyces cerevisiae*. *J. Cell Biol.* 133:85–97
87. Funabiki H, Yamano H, Kumada K, Nagao K, Hunt T, Yanagida M. 1996. Cut2 proteolysis required for sister-chromatid separation in fission yeast. *Nature* 381:438–41

88. Hauf S, Roitinger E, Koch B, Dittrich CM, Mechtler K, Peters JM. 2005 . Dissociation of cohesin from chromosome arms and loss of arm cohesion during early mitosis depends on phosphorylation of SA2. *PLoS Biol.* 3:e69
89. Gimenez-Abian JF, Sumara I, Hirota T, Hauf S, Gerlich D, et al. 2004. Regulation of sister chromatid cohesion between chromosome arms. *Curr. Biol.* 14:1187–93
90. Buonomo SB, Clyne RK, Fuchs J, Loidl J, Uhlmann F, Nasmyth K. 2000. Disjunction of homologous chromosomes in meiosis I depends on proteolytic cleavage of the meiotic cohesin Rec8 by separin. *Cell* 103:387–98
91. McGuinness BE, Hirota T, Kudo NR, Peters JM, Nasmyth K. 2005. *Shugoshin* prevents dissociation of cohesin from centromeres during mitosis in vertebrate cells. *PLoS Biol.* 3:e86
92. Riedel CG, Katis VL, Katou Y, Mori S, Itoh T, et al. 2006. Protein phosphatase 2A protects centromeric sister chromatid cohesion during meiosis I. *Nature* 441:53–61
93. Lee J, Kitajima TS, Tanno Y, Yoshida K, Morita T, et al. 2008. Unified mode of centromeric protection by *shugoshin* in mammalian oocytes and somatic cells. *Nat. Cell Biol.* 10:42–52
94. Kitajima TS, Sakuno T, Ishiguro K, Iemura S, Natsume T, et al. 2006. *Shugoshin* collaborates with protein phosphatase 2A to protect cohesin. *Nature* 441:46–52
95. Tang Z, Shu H, Qi W, Mahmood NA, Mumby MC, Yu H. 2006. PP2A is required for centromeric localization of Sgo1 and proper chromosome segregation. *Dev. Cell* 10:575–85
96. Kitajima TS, Kawashima SA, Watanabe Y. 2004. The conserved kinetochore protein *shugoshin* protects centromeric cohesion during meiosis. *Nature* 427:510–17
97. Katis VL, Galova M, Rabitsch KP, Gregan J, Nasmyth K. 2004. Maintenance of cohesin at centromeres after meiosis I in budding yeast requires a kinetochore-associated protein related to MEI-S332. *Curr. Biol.* 14:560–72
98. Kerrebrock AW, Moore DP, Wu JS, Orr-Weaver TL. 1995. Mei-S332, a *Drosophila* protein required for sister-chromatid cohesion, can localize to meiotic centromere regions. *Cell* 83:247–56
99. Hadjur S, Williams LM, Ryan NK, Cobb BS, Sexton T, et al. 2009. Cohesins form chromosomal *cis*-interactions at the developmentally regulated IFNG locus. *Nature* 460:410–13
100. Rollins RA, Morcillo P, Dorsett D. 1999. Nipped-B, a *Drosophila* homologue of chromosomal adherins, participates in activation by remote enhancers in the *cut* and *Ultrabithorax* genes. *Genetics* 152:577–93
101. Rollins RA, Korom M, Aulner N, Martens A, Dorsett D. 2004. *Drosophila* Nipped-B protein supports sister chromatid cohesion and opposes the stromalin/Scc3 cohesion factor to facilitate long-range activation of the *cut* gene. *Mol. Cell. Biol.* 24:3100–11
102. Brands A, Skibbens RV. 2005. Ctf7p/Eco1p exhibits acetyltransferase activity–but does it matter? *Curr. Biol.* 15:R50–51
103. Zhang J, Shi X, Li Y, Kim BJ, Jia J, et al. 2008. Acetylation of Smc3 by Eco1 is required for S phase sister chromatid cohesion in both human and yeast. *Mol. Cell* 31:143–51
104. Terret M-E, Sherwood R, Rahman S, Qin J, Jallepalli PV. 2009. Cohesin acetylation speeds the replication fork. *Nature* 462:231–34
105. De Piccoli G, Cortes-Ledesma F, Ira G, Torres-Rosell J, Uhle S, et al. 2006. Smc5-Smc6 mediate DNA double-strand-break repair by promoting sister-chromatid recombination. *Nat. Cell Biol.* 8:1032–34
106. Torres-Rosell J, Machin F, Aragon L. 2005. Smc5-Smc6 complex preserves nucleolar integrity in *S. cerevisiae*. *Cell Cycle* 4:868–72
107. Torres-Rosell J, Machin F, Farmer S, Jarmuz A, Eydmann T, et al. 2005. *SMC5* and *SMC6* genes are required for the segregation of repetitive chromosome regions. *Nat. Cell Biol.* 7:412–19
108. Unal E, Arbel-Eden A, Sattler U, Shroff R, Lichten M, et al. 2004. DNA damage response pathway uses histone modification to assemble a double-strand break-specific cohesin domain. *Mol. Cell* 16:991–1002
109. Strom L, Lindroos HB, Shirahige K, Sjogren C. 2004. Postreplicative recruitment of cohesin to double-strand breaks is required for DNA repair. *Mol. Cell* 16:1003–15
110. Unal E, Heidinger-Pauli JM, Koshland D. 2007. DNA double-strand breaks trigger genome-wide sister-chromatid cohesion through Eco1 (Ctf7). *Science* 317:245–48
111. Ivanov D, Schleiffer A, Eisenhaber F, Mechtler K, Haering CH, Nasmyth K. 2002. Eco1 is a novel acetyltransferase that can acetylate proteins involved in cohesion. *Curr. Biol.* 12:323–28

112. Kouzarides T. 2007. Chromatin modifications and their function. *Cell* 128:693–705
113. Sumara I, Vorlaufer E, Stukenberg PT, Kelm O, Redemann N, et al. 2002. The dissociation of cohesin from chromosomes in prophase is regulated by Polo-like kinase. *Mol. Cell* 9:515–25
114. Losada A, Hirano M, Hirano T. 2002. Cohesin release is required for sister chromatid resolution, but not for condensin-mediated compaction, at the onset of mitosis. *Genes Dev.* 16:3004–16
115. Alexandru G, Uhlmann F, Mechtler K, Poupart MA, Nasmyth K. 2001. Phosphorylation of the cohesin subunit Scc1 by Polo/Cdc5 kinase regulates sister chromatid separation in yeast. *Cell* 105:459–72
116. Hornig NC, Uhlmann F. 2004. Preferential cleavage of chromatin-bound cohesin after targeted phosphorylation by Polo-like kinase. *EMBO J.* 23:3144–53
117. Visintin R, Prinz S, Amon A. 1997. CDC20 and CDH1: a family of substrate-specific activators of APC-dependent proteolysis. *Science* 278:460–63
118. Agarwal R, Cohen-Fix O. 2002. Phosphorylation of the mitotic regulator Pds1/securin by Cdc28 is required for efficient nuclear localization of Esp1/separase. *Genes Dev.* 16:1371–82
119. Holt LJ, Krutchinsky AN, Morgan DO. 2008. Positive feedback sharpens the anaphase switch. *Nature* 454:353–57
120. Heidinger-Pauli JM, Unal E, Guacci V, Koshland D. 2008. The kleisin subunit of cohesin dictates damage-induced cohesion. *Mol. Cell* 31:47–56
121. Luo H, Li Y, Mu JJ, Zhang J, Tonaka T, et al. 2008. Regulation of intra-S phase checkpoint by ionizing radiation (IR)-dependent and IR-independent phosphorylation of SMC3. *J. Biol. Chem.* 283:19176–83
122. Kim ST, Xu B, Kastan MB. 2002. Involvement of the cohesin protein, Smc1, in Atm-dependent and independent responses to DNA damage. *Genes Dev.* 16:560–70
123. Yazdi PT, Wang Y, Zhao S, Patel N, Lee EY, Qin J. 2002. SMC1 is a downstream effector in the ATM/NBS1 branch of the human S-phase checkpoint. *Genes Dev.* 16:571–82
124. Jessberger R, Riwar B, Baechtold H, Akhmedov AT. 1996. SMC proteins constitute two subunits of the mammalian recombination complex RC-1. *EMBO J.* 15:4061–68
125. Watrin E, Peters JM. 2009. The cohesin complex is required for the DNA damage-induced G2/M checkpoint in mammalian cells. *EMBO J.* 28:2625–35
126. Brar GA, Kiburz BM, Zhang Y, Kim JE, White F, Amon A. 2006. Rec8 phosphorylation and recombination promote the step-wise loss of cohesins in meiosis. *Nature* 441:532–36
127. Kudo NR, Anger M, Peters AH, Stemmann O, Theussl HC, et al. 2009. Role of cleavage by separase of the Rec8 kleisin subunit of cohesin during mammalian meiosis I. *J. Cell Sci.* 122:2686–98
128. Moore DP, Page AW, Tang TT, Kerrebrock AW, Orr-Weaver TL. 1998. The cohesion protein MEI-S332 localizes to condensed meiotic and mitotic centromeres until sister chromatids separate. *J. Cell Biol.* 140:1003–12
129. Kiburz BM, Reynolds DB, Megee PC, Marston AL, Lee BH, et al. 2005. The core centromere and Sgo1 establish a 50-kb cohesin-protected domain around centromeres during meiosis I. *Genes Dev.* 19:3017–30
130. Hamant O, Golubovskaya I, Meeley R, Fiume E, Timofejeva L, et al. 2005. A REC8-dependent plant *shugoshin* is required for maintenance of centromeric cohesion during meiosis and has no mitotic functions. *Curr. Biol.* 15:948–54
131. Resnick TD, Satinover DL, MacIsaac F, Stukenberg PT, Earnshaw WC, et al. 2006. INCENP and Aurora B promote meiotic sister chromatid cohesion through localization of the *shugoshin* MEI-S332 in *Drosophila*. *Dev. Cell* 11:57–68
132. Yu HG, Koshland D. 2007. The Aurora kinase Ipl1 maintains the centromeric localization of PP2A to protect cohesin during meiosis. *J. Cell Biol.* 176:911–18
133. Stead K, Aguilar C, Hartman T, Drexel M, Meluh P, Guacci V. 2003. Pds5p regulates the maintenance of sister chromatid cohesion and is sumoylated to promote the dissolution of cohesion. *J. Cell Biol.* 163:729–41
134. van der Lelij P, Godthelp BC, van Zon W, van Gosliga D, Oostra AB, et al. 2009. The cellular phenotype of Roberts syndrome fibroblasts as revealed by ectopic expression of ESCO2. *PLoS One* 4:e6936
135. Deardorff MA, Kaur M, Yaeger D, Rampuria A, Korolev S, et al. 2007. Mutations in cohesin complex members SMC3 and SMC1A cause a mild variant of Cornelia de Lange syndrome with predominant mental retardation. *Am. J. Hum. Genet.* 80:485–94

136. Musio A, Selicorni A, Focarelli ML, Gervasini C, Milani D, et al. 2006. X-linked Cornelia de Lange syndrome owing to SMC1L1 mutations. *Nat. Genet.* 38:528–30
137. Krantz ID, McCallum J, DeScipio C, Kaur M, Gillis LA, et al. 2004. Cornelia de Lange syndrome is caused by mutations in NIPBL, the human homolog of *Drosophila melanogaster* Nipped-B. *Nat. Genet.* 36:631–35
138. Tonkin ET, Wang TJ, Lisgo S, Bamshad MJ, Strachan T. 2004. NIPBL, encoding a homolog of fungal Scc2-type sister chromatid cohesion proteins and fly Nipped-B, is mutated in Cornelia de Lange syndrome. *Nat. Genet.* 36:636–41
139. Vega H, Waisfisz Q, Gordillo M, Sakai N, Yanagihara I, et al. 2005. Roberts syndrome is caused by mutations in ESCO2, a human homolog of yeast ECO1 that is essential for the establishment of sister chromatid cohesion. *Nat. Genet.* 37:468–70
140. Kaur M, DeScipio C, McCallum J, Yaeger D, Devoto M, et al. 2005. Precocious sister chromatid separation (PSCS) in Cornelia de Lange syndrome. *Am. J. Med. Genet. A* 138:27–31
141. Unal E, Heidinger-Pauli JM, Kim W, Guacci V, Onn I, et al. 2008. A molecular determinant for the establishment of sister chromatid cohesion. *Science* 321:566–69
142. Gordillo M, Vega H, Trainer AH, Hou F, Sakai N, et al. 2008. The molecular mechanism underlying Roberts syndrome involves loss of ESCO2 acetyltransferase activity. *Hum. Mol. Genet.* 17:2172–80
143. Kawauchi S, Calof AL, Santos R, Lopez-Burks ME, Young CM, et al. 2009. Multiple organ system defects and transcriptional dysregulation in the Nipbl(+/−) mouse, a model of Cornelia de Lange syndrome. *PLoS Genet.* 5:e1000650
144. Liu J, Krantz ID. 2008. Cohesin and human disease. *Annu. Rev. Genomics Hum. Genet.* 9:303–20
145. McNairn AJ, Gerton JL. 2008. Cohesinopathies: one ring, many obligations. *Mutat. Res.* 647:103–11
146. McNairn AJ, Gerton JL. 2008. The chromosome glue gets a little stickier. *Trends Genet.* 24:382–89
147. Horsfield JA, Anagnostou SH, Hu JK, Cho KH, Geisler R, et al. 2007. Cohesin-dependent regulation of Runx genes. *Development* 134:2639–49
148. Pauli A, Althoff F, Oliveira RA, Heidmann S, Schuldiner O, et al. 2008. Cell-type-specific TEV protease cleavage reveals cohesin functions in *Drosophila* neurons. *Dev. Cell* 14:239–51
149. Schuldiner O, Berdnik D, Levy JM, Wu JS, Luginbuhl D, et al. 2008. *piggyBac*-based mosaic screen identifies a postmitotic function for cohesin in regulating developmental axon pruning. *Dev. Cell* 14:227–38
150. Goshima G, Yanagida M. 2000. Establishing biorientation occurs with precocious separation of the sister kinetochores, but not the arms, in the early spindle of budding yeast. *Cell* 100:619–33
151. He X, Asthana S, Sorger PK. 2000. Transient sister chromatid separation and elastic deformation of chromosomes during mitosis in budding yeast. *Cell* 101:763–75
152. Yeh E, Haase J, Paliulis LV, Joglekar A, Bond L, et al. 2008. Pericentric chromatin is organized into an intramolecular loop in mitosis. *Curr. Biol.* 18:81–90
153. Bloom K, Sharma S, Dokholyan NV. 2006. The path of DNA in the kinetochore. *Curr. Biol.* 16:R276–78
154. Sakuno T, Tada K, Watanabe Y. 2009. Kinetochore geometry defined by cohesion within the centromere. *Nature* 458:852–58
155. Yokobayashi S, Watanabe Y. 2005. The kinetochore protein Moa1 enables cohesion-mediated monopolar attachment at meiosis I. *Cell* 123:803–17
156. Martin-Castellanos C, Blanco M, Rozalen AE, Perez-Hidalgo L, Garcia AI, et al. 2005. A large-scale screen in *S. pombe* identifies seven novel genes required for critical meiotic events. *Curr. Biol.* 15:2056–62

Reversal of Histone Methylation: Biochemical and Molecular Mechanisms of Histone Demethylases

Nima Mosammaparast[1,2] and Yang Shi[1,3]

[1]Department of Pathology, [2]Department of Laboratory Medicine, and [3]Division of Newborn Medicine, Department of Medicine, Children's Hospital, Harvard Medical School, Boston, Massachusetts 02115; email: yang_shi@hms.harvard.edu

Annu. Rev. Biochem. 2010. 79:155–79

First published online as a Review in Advance on April 7, 2010

The *Annual Review of Biochemistry* is online at biochem.annualreviews.org

This article's doi:
10.1146/annurev.biochem.78.070907.103946

0066-4154/10/0707-0155$20.00

Key Words

chromatin, demethylase, epigenetics, transcription

Abstract

The importance of histone methylation in gene regulation was suggested over 40 years ago. Yet, the dynamic nature of this histone modification was recognized only recently, with the discovery of the first histone demethylase nearly five years ago. Since then, our insight into the mechanisms, structures, and macromolecular complexes of these enzymes has grown exponentially. Overall, the evidence strongly supports a key role for histone demethylases in eukaryotic transcription and other chromatin-dependent processes. Here, we examine these and related facets of histone demethylases discovered to date, focusing on their biochemistry, structure, and enzymology.

Contents

THE ROLES OF HISTONE METHYLATION IN CHROMATIN REGULATION

Eukaryotic chromatin consists of DNA and its major protein component, histones. This group of highly conserved proteins includes the core histones H2A, H2B, H3, and H4, as well as the linker histone H1. A pair of each of the core histones is assembled with 146 base pairs of DNA to form a nucleosome, the basic unit of chromatin (1). Within the nucleosome, the core histones have a similar overall structure, with a globular, hydrophobic internal region that forms the histone fold, and amino-termini, which form flexible structures that emanate out of the nucleosome (2). These amino-termini (so-called N-terminal histone tails) contain a plethora of posttranslational modifications, which play key roles in the regulation of DNA replication, recombination, repair, and transcription (3). Histone modifications can alter the structure of the nucleosome and may also serve as a platform for the assembly of multiple protein complexes for covalent and noncovalent chromatin modifications, mediated by subunits that carry protein modules ("readers") that recognize specific histone modifications either singly or in combination (4).

Indeed, over 40 years ago, Allfrey and colleagues made the observation that two histone modifications, namely acetylation and methylation, may function to regulate RNA synthesis (5). It was unclear for some time after this initial discovery whether such modifications were simply a result of utilization of the chromatin template (i.e., for transcription) or whether such modifications regulated the chromatin landscape and its functional state. Strong evidence for the latter initially came from genetic studies in budding yeast, demonstrating the importance of the heavily modified H3 and H4 tails for transcriptional activation (6, 7). Conversely, regulation of transcriptional repression of specific genetic loci was also shown to require the same histone regions (8). Definitive evidence for the significance of histone modifications came from the first histone-modifying enzymes to be identified: the histone acetyltransferase Gcn5 (9) and the histone deacetylases Rpd3 (10) and Hda1 (11); the former two have known transcriptional coactivator and corepressor functions, respectively (12, 13). Although the function of histone acetylation is beyond the scope of this review (see Reference 14 for a review on this topic), it is accepted that this modification is, in general, associated with active transcription, regardless of the precise amino acid on the histone tails targeted for this modification. In contrast, histone methylation has proven to be a far more complex entity: Not only do distinct histone lysine residues have different, sometimes opposing, functions when methylated, but different degrees of methylation (mono-, di-, or trimethylation) on the same residue may have vastly differing functions (15, 16). In addition, histones can be methylated on arginine residues, a modification that has also been shown to play either positive or negative roles in transcription (17). To date, over 20 methyl marks on lysine and arginine residues have been identified, and the five lysine residues on the histone tails that have garnered the most attention are shown in **Figure 1**. However, additional sites of methylation are likely to be identified in the near future by mass spectrometry approaches, which will continue to add to the complexity of the epigenetic landscape.

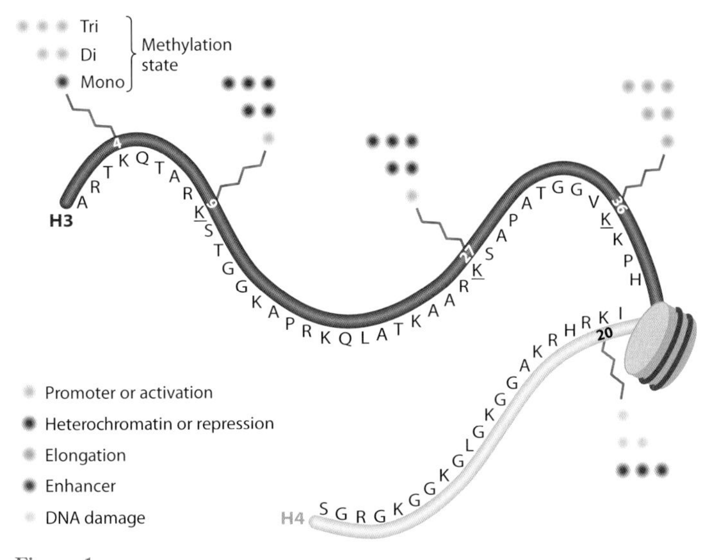

Figure 1

Major lysine methylation marks on the amino-termini of histones H3 (*purple*) and H4 (*blue*). The embedded numbers refer to the methylated amino acid residue on each histone. The general function of each mono-, di-, and trimethylation state is depicted in dots of distinct colors as shown in the figure key.

Transcriptional Activation and Euchromatin

One of the first methyl marks to be characterized in detail was methylation of histone H3 at lysine 4 (H3K4me). It was demonstrated that this modification is present in the transcriptionally active macronucleus of *Tetrahymena* but absent in the silent micronucleus (18). It is now appreciated that trimethylation of histone H3 at lysine 4 (H3K4me3) is strongly associated with and is important for transcriptional activation at promoter regions (**Figure 1**) (19–23). Early studies identified the yeast Set1 protein as the major methyltransferase responsible for this modification, and the *SET1* gene was shown to be important for maximal transcriptional activity of many genes (20, 24). A genome-wide analysis demonstrated that H3K4me3 is associated with promoters in human embryonic stem cells, although detectable full-length transcripts were lacking for most of these genes (25). In the same study, active transcriptional elongation was strongly linked to trimethylation at H3 lysine 36 (H3K36me3). Consistently, the methyltransferase responsible for this modification, Set2 in yeast, is associated with the elongating form of RNA polymerase II (26, 27). The role of H3K36me3 appears to be conserved in higher eukaryotes, as demonstrated by the requirement of Set2 homologs for trimethylation of H3K36 and their association with transcription elongation complexes (28, 29). Thus, trimethylation of H3K4 and H3K36 appears to play important roles in marking promoters and gene bodies during active transcription, respectively (**Figure 1**).

It should be noted that lower methylation states at these two H3 sites appear to have different functions than their trimethyl counterparts, particularly in higher eukaryotes. For example, in *Arabidposis*, H3K36me2/3 and H3K36me1 states are mediated by two different Set2 homologs, which have opposing effects on target gene transcription (30). Similarly, H3K4me1, but not H3K4me3, is associated with enhancer elements (15). Thus, it appears that eukaryotes have taken full advantage of the regulatory potential available in differing methyl states to mark the epigenetic landscape.

Plant homeodomain (PHD) finger: a protein domain characterized by a zinc finger containing seven cysteines and a histidine in a $C_4H_1C_3$ configuration (47)

Transcriptional Repression, Heterochromatin, and Bivalency

While H3K4me3 and H3K36me3 are strongly associated with transcriptional activation, trimethylation at H3K9, H3K27, and H4K20 are generally considered repressive modifications (**Figure 1**) (16). The first histone methyltransferase to be described, Su(var)3–9, promotes trimethylation of H3K9, which is a key mark of both constitutive and facultative heterochromatin in yeast as well as higher eukaryotes (31–34). Heterchromatin protein 1 (HP1) associates specifically with this modification via its chromodomain, providing a classic example of how a histone methyltransferase and its cognate modification reader protein cooperate in their functions (35–37). This repressive pathway on heterochromatin also involves trimethylation of H4K20, which is mediated by Su(var)4–20 (38). Recruitment of this enzyme requires both Su(var)3–9 and HP1, which puts H3K9me3 upstream of H4K20me3 and links the function of these two modifications (38). However, the precise molecular role of H4K20me3 in heterochromatin is still unclear.

Although H3K27me3 is also associated with repressed transcription, it is not a clear mark of constitutive heterochromatin. Instead, key developmental loci, such as the *HOX* genes, which may become active in specific tissues during mammalian development, are targeted for repression by enzyme complexes that methylate H3K27 (39, 40). The Polycomb repressive complex 2 (PRC2) has been shown to trimethylate H3K27 via the EZH2 (Enhancer of zeste homolog 2) subunit, which is critical for *HOX* gene repression (39, 40). It is important to note that repressive and active trimethyl marks are not always mutually exclusive. The prime example of this is the presence of both H3K4me3 and H3K27me3, the so-called bivalent domain, in developmentally important gene loci in embryonic stem (ES) cells (41), signifying an epigenetic state primed for either activation or repression. We predict that many other undiscovered bivalent (or multivalent) methylation states are likely to exist in the cell.

Cell Cycle, DNA Repair, and Recombination

Clearly the transcriptional roles of histone methylation have garnered the most attention, but other important nuclear functions are just as important. A catalytic mutation in the histone methyltransferase Set9, which targets H4K20 in *Schizosaccharomyces pombe*, mimics a point mutation phenotype in this histone residue, causing increased sensitivity to DNA damage and compromised cell cycle checkpoints (42). Dimethylation of H4K20 serves to recruit 53BP1 (Crb2 in *S. pombe*), a protein known to be involved in the DNA damage response, to sites of DNA damage in mammalian cells and yeast, respectively (42, 43). Methylation at H4K20 also directly recruits *S. pombe* Pdp1, a novel PWWP domain-containing protein involved in the DNA damage response pathway (44). H3K79 methylation has also been proposed to play roles in cell cycle checkpoints and DNA replication, although these roles are less clearly defined.

Like other DNA transactions in eukaryotes, recombination must also occur in a chromatin context. A specialized form of DNA recombination, V(D)J recombination, occurs during lymphocyte development and is essential for the diversification of the antigen repertoire. This process requires the RAG recombinase complex. RAG2 is targeted to chromatin by binding directly to H3K4me3, and this binding is mediated by its PHD finger (45, 46). Furthermore, a functional PHD finger is essential for recombinase activity in vivo, stressing the significance of recognizing this methylated state (45, 46). It is likely that hitherto unrecognized histone methylation marks play important roles in DNA recombination as well as in DNA damage repair in general.

CATALYTIC MECHANISMS OF HISTONE DEMETHYLATION

Unlike histone deacetylation, which involves simple hydrolysis of an amide bond, methylated histones were thought to be irreversible owing to the more stable nature of the C-N bond. Beyond the theoretical chemistry involved, the notion of methylation irreversibility was based on experiments demonstrating that the half-life of histone methyl marks was approximately equal to that of the histone itself (48, 49). Other mechanisms of histone demethylation, albeit indirect, were possible, such as active histone exchange (50) and proteolytic removal of histone amino-termini (51). Another possible mechanism involved the conversion of methylarginine to citrulline by a peptidylarginine deiminase, which works equally well on arginine and shows no preference for methylarginine (52, 53). However, all of these mechanisms would ultimately require either passive or active histone exchange to revert to the original unmethylated state.

A number of groups proposed mechanisms of direct histone demethylation primarily on the basis of bioinformatics and analogous chemical reactions. The first such proposal involved the use of *S*-adenosylmethionine (SAM) as the source of a reactive radical intermediate, which would target the *N*-methyl group, creating an unstable aminium cation radical that would spontaneously hydrolyze to form formaldehyde and a demethylated residue (54). Other possibilities involved oxidation of the methyl group coupled to reduction of a cofactor, releasing the methyl group as formaldehyde or as another higher oxidative state (55).

Although theoretically feasible, experimental evidence of enzymatic demethylation was lacking for several decades until the discovery of lysine-specific demethylase 1 [LSD1; also known as lysine (K)-specific demethylase 1A, or KDM1A]. Human LSD1 was originally identified as a component of the BRAF-histone deacetylase (HDAC) (or BHC) transcriptional corepressor complex containing the REST corepressor, CoREST, known to be important for repression of neuronal genes in nonneuronal cells (56–58). LSD1 itself could serve as a transcriptional repressor, and this function was reported to be dependent on its amine oxidase domain (59). Therefore, it was further hypothesized that because methylation of K4 on histone H3 was associated with transcriptional activation as described above, LSD1 may function as an H3K4 demethylase. Strikingly, LSD1 did indeed encode an H3K4 demethylase, revealing that histone methylation was a dynamic, reversible process (59). The demethylation reaction required the amine oxidase domain, released formaldehyde as a byproduct, reduced flavin adenine dinucleotide (FAD) as a cofactor, and was able to demethylate dimethyl but not trimethyl H3K4, presumably owing to the requirement of an iminium cation intermediate (**Figure 2*a***). As would be predicted for a key component of BHC, LSD1 was required for the downregulation of neuronal genes in nonneuronal cells, and knockdown of LSD1 increased H3K4 dimethylation at the promoters of these genes (59).

LSD1: lysine-specific demethylase 1

FAD: flavin adenine dinucleotide

The mechanism of histone demethylation by LSD1 appears to be highly conserved among most eukaryotes. Homologs of LSD1 have been characterized in *Arabidopsis*, *Drosophila*, *Caenorhabditis elegans*, and the fission yeast *S. pombe* (60–64). Notably, it appears that budding yeast does not encode an LSD1 homolog in its genome. In all of these examples, each organism contains at least two, if not more, LSD1 homologs. In humans, a recently characterized LSD1 homolog, LSD2, has also been shown to encode an H3K4me2/1 demethylase (65, 66). The LSD1 counterparts in plants, flies, and worms that have thus far been characterized also encode H3K4 demethylases. Surprisingly, *S. pombe* LSD1 homologs demethylate H3K9 instead of H3K4 (63, 64), suggesting that this mechanism of demethylation may also be used to target other sites of methylation. It remains to be seen if other nuclear amine oxidases exist that function as histone demethylases.

Subsequent to the discovery of LSD1, experimental evidence for an alternative oxidation-reduction mechanism for histone demethylation was first reported by Zhang and colleagues (67), immediately followed by reports from a number of groups that were independently pursuing new demethylases (68–70). This type of reaction was based on an Fe^{2+}- and oxygen-dependent catalytic center and was first

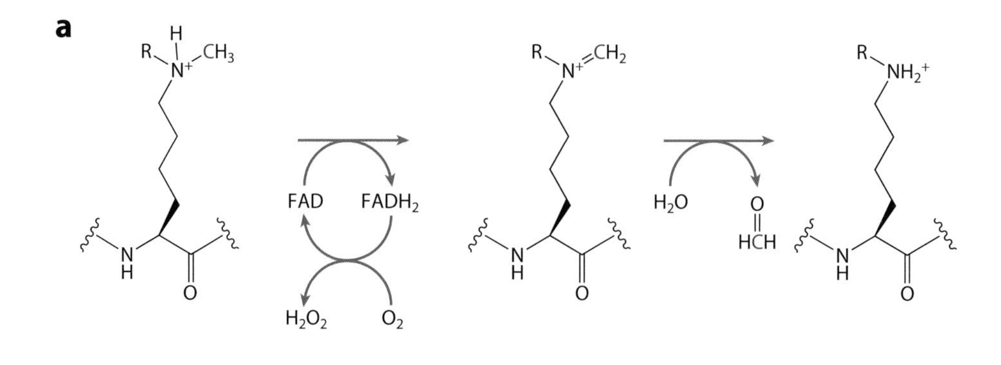

Figure 2

(*a*) Flavin adenine dinucleotide (FAD)-dependent amine oxidase mechanism of lysine demethylation mediated by lysine-specific demethylase 1 [LSD1; also known as lysine (K)-specific demethylase 1A, or KDM1A]. Red indicates carbons that will be demethylated in each reaction. (*b*) Fe^{2+}- and the 2-oxoglutarate-dependent dioxygenase mechanism of methyl-lysine demethylation mediated by the Jumonji C-terminal domain family of enzymes.

observed in *N*-dealkylation of DNA in the *Escherichia coli* AlkB enzyme (71, 72). It involved oxidative decarboxylation of α-ketoglutarate (also known as 2-oxoglutarate, or 2-OG), coupled to hydroxylation of the methyl group, creating an unstable hydroxymethyl ammonium intermediate, which was released as formaldehyde (**Figure 2*b***). A classical biochemical fractionation approach was used to successfully find such an activity in HeLa nuclear extracts, identified as FBXL11 (also known as JHDM1 and KDM2A) (67). This protein contained an Fe^{2+} dioxygenase Jumonji-C (JmjC) domain, similar to the active site of AlkB. Indeed the JmjC domain, as well as the indicated cofactors, was critical for the demethylase reaction, which was specific for H3K36me2. In theory, the JmjC-mediated catalysis could demethylate the trimethylated state, but H3K36me3 was not a substrate for FBXL11 (JHDM1 or KDM2A) (67). A subfamily of JmjC domain-containing proteins that demethylate trimethylated lysines on histone tails was reported shortly thereafter by a number of groups, independently pursuing new demethylases using candidate or affinity matrix-based pull-down approaches (68–70, 73). To date, many of the key methylated histone marks have a corresponding demethylase (**Figure 3**).

STRUCTURAL INSIGHTS INTO HISTONE DEMETHYLASES

A key question in understanding demethylase function is their substrate specificity. As discussed above, LSD1/KDM1 and JHDM1/KDM2A both exhibit exquisite lysine residue, as well as methyl state, specificity.

2-OG: 2-oxoglutarate

JmjC domain: Jumonji C-terminal domain

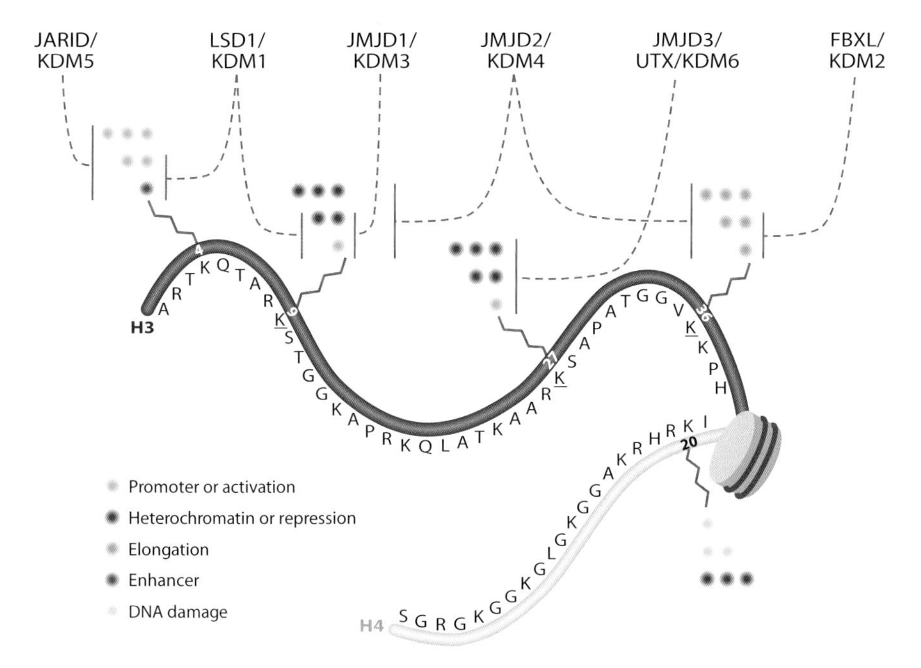

Figure 3

Substrate specificity of histone demethylases described to date. Dashed lines point to the methylated residue(s) that are demethylated by the indicated enzymes. The embedded numbers refer to the methylated amino acid residue on each histone. The general function of each mono-, di-, and trimethylation state is depicted in dots of distinct colors as shown in the figure key. Abbreviations: JARID, Jumonji and ARID domain protein (also known as KDM5); LSD1, Lysine-specific demethylase, also known as KDM1A; JMJD1, Jumonji domain protein 1, also known as KDM3; JMJD2, Jumonji domain protein 2, also known as KDM4; UTX, Ubiquitously transcribed tetratricopeptide repeat, X chromosome protein, also known as KDM6A; JMJD3, Jumonji domain protein 3, also known as KDM6B; FBXL, F-box and leucine-rich repeat protein, also known as KDM2.

Recent structural efforts have provided significant insights into some of these issues, as discussed below.

LSD1-Mediated Demethylation and Chemical Inhibitors

Analysis of the LSD1 structure has led to significantly greater insight into its function. Overall, the structure of LSD1 consists of an N-terminal SWIRM domain (74), and an amine oxidase domain split into two halves, consisting of a substrate-binding half and an FAD-binding half, which come together to form a globular domain (**Figure 4*a***) (75, 76). The active site of the enzyme is located in between these two halves. Two long, antiparallel α-helices divide and project away from the globular halves of the amine oxidase active site. This so-called tower domain serves as the binding interface between LSD1 and its protein cofactor, CoREST, and distinguishes it from other amine oxidases (75, 76). Although nucleosomal substrates are refractory to recombinant LSD1, addition of recombinant CoREST endows nucleosomal demethylation by LSD1, indicating that the primary function of CoREST is to enable LSD1 demethylation of nucleosomal substrates (77, 78). Consistently, mutations that inhibit the ability of CoREST to bind to DNA inhibit the ability of the complex to demethylate nucleosomes (76).

How does LSD1 achieve specificity for H3K4me2/me1? According to these structural studies, much of the specificity of LSD1 is intrinsic to its enzymatic pocket. The amine

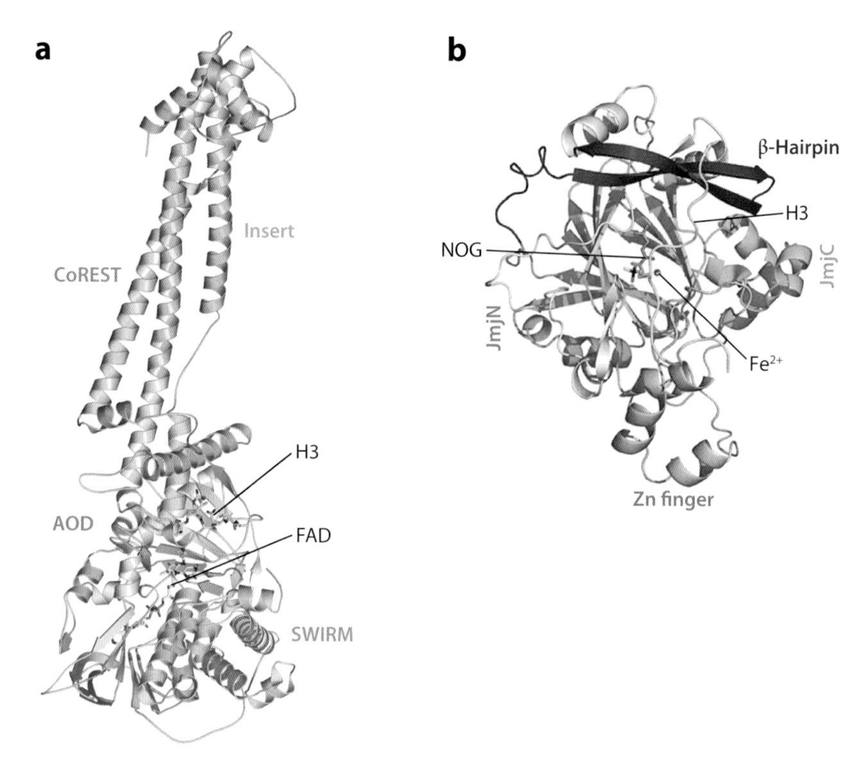

Figure 4

(*a*) Overall structure of LSD1 and CoREST bound to an H3 peptide mimetic and flavin adenine dinucleotide (FAD). Reprinted with permission from Reference 79. (*b*) Overall structure of JMJD2A bound to H3K36me3 and NOG. Reprinted with permission from Reference 90. Abbreviations: AOD, amine oxidase domain; CoREST, corepressor of RE1-silencing transcription factor; H3K36me3, H3 trimethylated at lysine 36; JmjC domain, Jumonji C-terminal domain; JMJD2A, Jumonji domain protein 2A; LSD1, lysine-specific demethylase 1; NOG, *N*-oxalylglycine; SWIRM, Swi3, Rsc8, and Moira domain.

oxidase domain of LSD1 is considerably larger than the related polyamine oxidases, presumably to accommodate the relatively bulkier histone tail substrate (75, 76). Molecular modeling of the H3 tail onto the structure of LSD1, as well as a cocrystal structure of LSD1 with an H3 peptide derivative, suggest that the first 12 residues of H3 make extensive contacts with the relatively acidic active site, whereas the more C-terminal portions of the H3 tail bind in a groove between the SWIRM domain and the amine oxidase domain (76, 79). Mutations within either region abrogate the catalytic activity of LSD1 (75, 76, 79). The importance of this extended network of contacts is reinforced by biochemical studies demonstrating that H3 peptides shorter than 21 residues are not efficiently demethylated by LSD1 (80). Within the catalytic pocket of LSD1, H3 adopts a serpentine structure, with three γ-turns, and becomes highly compressed such that, between Arg2 and Thr6 of H3, several hydrogen bonds are formed between backbone carbonyl and amide moieties (79). This binding mode places tight constraints on the H3 N terminus such that engagement of methylated K4 with the adjacent FAD cofactor at the active site does not permit more than three amino acids to be present on the N-terminal side of the methyllysine. This, combined with sequence-specific interactions between the active site and Arg2, Thr3, and Gln5, provides an explanation for the observed specificity of LSD1 toward H3K4me2/1 (79). Importantly, it appears that the active site of LSD1 can also accommodate H3K4me3, consistent with observations that a peptide containing this methyl state inhibits activity toward lower methyl states in a competitive fashion (75). Thus, the methyl state selectivity of LSD1 is not structurally inhibited but is chemically constrained, as predicted by its enzymatic mechanism.

Earlier studies of LSD1 demonstrated that certain modifications on the H3 tail have a significant impact on demethylase activity. For example, acetylation on lysine 9 increased the K_m of LSD1 for the H3 tail (which is normally ~4–40 μM, depending on reaction conditions) sixfold, whereas phosphorylation on serine 10 completely abolished activity (80), suggesting potential functional cross talk of these modifications with H3K4 demethylation. This is consistent with the abundance of electrostatic contacts between the LSD1 active site and the H3 amino terminus. Molecular modeling between LSD1 and H3 suggests that serine 10 is in close proximity to glutamate 559 on the LSD1 active site (76). Thus, it appears that demethylation of H3K4 by LSD1 would act downstream of the removal of other charge-altering modifications. The order of modification removal becomes significant when considering the mechanism of transcriptional repression by the LSD1 complex, which is discussed in greater detail below.

As discussed above, LSD1 belongs to a large family of FAD-utilizing amine oxidases. These include polyamine oxidases, such as spermine oxidase, which can be inhibited by polyamine analogs (81). Interestingly, certain polyamine analogs can also inhibit LSD1 at submicromolar concentrations, despite the significantly larger active-site cavity of LSD1 relative to maize polyamine oxidase (PAO) (82). Surprisingly, these compounds exhibited noncompetitive kinetics, suggesting that they do not compete with H3K4me2 for the LSD1 active site. Thus, the precise mechanism of action of these inhibitors remains to be determined. Monoamine oxidases (MAO-A and MAO-B) also belong to this family of enzymes and are responsible for the oxidative deamination of certain neurotransmitters, such as serotonin, norepinephrine, and dopamine (83). Monoamine oxidase inhibitors, or MAOIs, are a class of small-molecule inhibitors of the MAO family, which irreversibly inhibit various MAOs and have been used in the past as second- or third-line agents to treat depression and Parkinson's disease. Because LSD1 has a mechanism of action similar to MAOs, a panel of MAOIs were screened, and it was found that *trans*-2-phenylcyclopropylamine (also known as 2-PCPA or tranylcypromine) efficiently inhibits LSD1 (84). Kinetic analysis demonstrated that the IC_{50} of 2-PCPA for LSD1 was ~20 μM, which is approximately one order of magnitude greater than MAO-A or MAO-B (85). Spectroscopic and structural studies confirm that 2-PCPA forms a covalent adduct with FAD in the catalytic site, confirming its mechanism as an irreversible inhibitor (85, 86). The crystal structure of 2-PCPA-modified LSD1-CoREST complex demonstrates that the FAD-2-PCPA adduct is located in a hydrophobic pocket, although only two van der Waals contacts are seen between the phenyl ring of 2-PCPA and LSD1 (86). This is in contrast to the structure of 2-PCPA and MAO-B, where this phenyl ring forms several hydrophobic interactions, consistent with its higher potency as an inhibitor of MAO-B. However, the notion that this MAOI can function as an irreversible inhibitor of LSD1, coupled with structural insights, can serve as the basis for designing more specific and potent inhibitors of LSD1 for probing its function in various physiological as well as pathological contexts.

Jumonji C Domain-Mediated Demethylation and Chemical Inhibitors

As discussed above, the Fe^{2+}-dependent dioxygenase mechanism of demethylation by the JmjC family provided a potential framework for trimethyl lysine demethylation (**Figure 2*b***). The Jumonji domain protein 2 (JMJD2) family was the first of the JmjC proteins known to possess this activity, which targeted H3K9 and

H3K36 (**Figure 3**) (68, 70, 73). As of this writing, it is also the only JmjC family member with a published high-resolution structure (87–90). The structure of JMJD2A demonstrates that its catalytic core consists of a Jumonji N-terminal (JmjN) domain, a JmjC domain, an unexpected C-terminal zinc finger motif, as well as a β-hairpin and a mixed domain that serve to connect the JmjN and JmjC domains (**Figure 4*b***). The catalytic core consists of the JmjC domain and contains two histidine and one glutamate residue, which chelate the catalytic iron atom and hence are essential for activity. Not surprisingly, these residues are invariably present in other JmjC family members that are known to have demethylase activity. The JmjC domain of JMJD2A contains eight β-strands, which form a jelly-roll-like structure, a key feature of the cupin superfamily of metalloenzymes (89). Unlike the JmjC domain, the JmjN and C-terminal zinc finger are not present in all other Jmj demethylases. However, in JMJD2A, these two domains are important for maintaining overall stability of the JmjC catalytic core and hence are indispensible for activity (89). Domains adjacent to the JmjC domain in other JmjC family demethylases, such as the zinc finger domain of JMJD1A, have also been shown to be important for catalytic activity in vitro (91). Other domains, such as the tandem tudor domain of JMJD2A, which binds to methylated H3K4 and H4K20 (92), and the PHD finger of SMCX, which binds to H3K9me3 (93), likely function in chromatin targeting and may be important for demethylase activity in vivo. We speculate that these accessory domains may play important roles in the regulation of demethylase activity, either by interaction with other protein factors or as direct targets by posttranslational modification.

The structure of the catalytic core of JMJD2A with trimethylated peptide substrates has been solved by three independent groups and has provided considerable insight into how this enzyme recognizes its substrates (87, 88, 90). Binding of either H3K9me3 or H3K36me3 substrates does not result in large conformational changes relative to the apoenzyme. Surprisingly, unlike LSD1, relatively few side chain-to-side chain interactions are seen between JMJD2A and the peptide substrate; the vast majority of interactions are hydrogen bonds between the main chains of both the enzyme and the histone peptide. This mechanism of binding may explain the relative plasticity of JMJD2A in recognizing both methylated H3K9 and H3K36 substrates. Within the substrate, specificity seems to be conferred by the requirement of peptide bending between the +2 and +4 positions relative to the methylated lysine. A Gly → Pro substitution at position 12 within the context of the H3K9me3 peptide reduces substrate specificity nearly 30-fold (88). Conversely, substitution of the flexible amino acid glycine at positions 30 and 31 within the context of an H3K27me2 peptide permits catalytic activity on this substrate (90).

The catalytic center of JMJD2A is buried within a pocket that is lined by polar residues, which is quite distinct from other trimethyllysine-binding domains, such as the PHD fingers, which typically use a hydrophobic cage for methylated peptide binding (94). A number of these residues, namely Asn86, Gln88, Asp135, and Tyr175, are important for interaction with the peptide; mutation of these residues to alanine significantly reduces or completely abrogates activity (88, 90). Key residues within the substrate-binding pocket, particularly Ser288 and Thr289, appear to be important for methylation state selectivity of JMJD2A. Mutation of these residues to alanine and isoleucine, respectively, significantly shifts the specificity of JMJD2A toward dimethyl histone peptides (88, 90). The fact that members of the same subfamily of demethylases display differential methyl state specificities raises the interesting possibility of demethylases playing a fine-tuning role in regulating histone methylation dynamics (95). Interestingly, monomethyl groups do not function as substrates for JMJD2A, most likely because such a modified state is too distant from the catalytic Fe^{2+} or has reduced probability of contacting the catalytic center compared with the trimethyl or dimethyl states (87, 90). It is likely that the precise molecular makeup of analogous polar cavities within other JmjC family members will determine the methyl state selectivity of these enzymes. Future work on the structure of other JmjC proteins is needed to determine whether this or an alternative mechanism is used for substrate specificity.

As with LSD1, understanding the enzymatic mechanism of the JmjC family of demethylases has permitted the design of small molecular inhibitors of this enzyme family. *N*-oxalylglycine (NOG) is the amide analog of the JmjC cofactor 2-OG and can serve as a relatively weak competitive inhibitor for JmjC demethylases (87, 88, 90). Certain NOG mimetics, such as pyridine carboxylates, are more potent inhibitors; for example, one of these compounds has been shown to inhibit JMJD2E with IC_{50} in the low micromolar range (96). However, these small molecules may also inhibit other iron-dependent dioxygenases, such as HIF and collagen hydroxylases (96). Thus, as with MAOIs in the case of LSD1, these inhibitors need to be further refined to increase their specificity for greater utility.

HISTONE DEMETHYLASE COMPLEXES

Having discussed the two classes of histone demethylases in atomic detail, we now turn to the macromolecular complexes in which they function. As we discuss below, the biochemical activities of these complexes complement the biological context in which they function in vivo.

LSD1 Protein Complexes

Analysis of the macromolecular complexes that contain LSD1 has led to significant insight into the function of this enzyme. The LSD1 core complex contains LSD1, HDAC1-2, CoREST, BHC80, and BRAF35 (**Figure 5*a***) (77, 78); each and every subunit of the complex serves a distinct but collaborative role, and this has led to a model whereby the complex promotes transcriptional repression. Specifically, BRAF35 is believed to function as a targeting protein that directs the LSD1 complex to precise genomic locations (58), although other DNA-binding proteins, such as REST, may also play such a role in recruiting LSD1. HDAC1-2 deacetylate H3, which permits binding of CoREST, creates the preferred substrate for LSD1, which prefers H3 in the hypoacetylated state (78, 80, 97). The binding of CoREST to the nucleosome recruits LSD1, as described above, resulting in H3K4 demethylation. BHC80 contains a PHD finger that preferentially recognizes the H3K4me(0) state (98), thus preventing a futile cycle of H3K4 remethylation and promoting the repressed state.

As mentioned earlier, the LSD1-CoREST complex was originally found to function in repressing neuronal genes in nonneuronal tissues. However, its role may extend far beyond this function. An organ-specific knockout of LSD1 targeting the pituitary gland showed LSD1 functions to repress growth hormone expression in mature lactotropes, most likely via the CtBP (C-terminal binding protein)-CoREST complex (99). A study analyzing the mechanism of transcriptional repression by Gfi-1 (growth factor independent-1 transcription factor) demonstrated that this zinc finger protein recruited LSD1 and CoREST during hematopoietic differentiation (100). The majority of Gf1-1-occupied loci also contained LSD1 and CoREST, and Gfi-1-mediated repression, as well as hematopoietic differentiation, required both factors. A recent study has found that the CoREST complex functions with Nurr1, an orphan nuclear receptor, in astrocytes and microglia (101). This complex is recruited to repress expression of inflammatory mediators in these cells and prevent neurodegeneration. Thus, it is likely that the LSD1-CoREST complex is recruited by different factors to be utilized for repression in various contexts.

In addition to the CoREST complex, LSD1 has been very recently reported to be present in the NuRD chromatin-remodeling complex, where it also functions as an H3K4 demethylase (102). Therefore, the demethylase

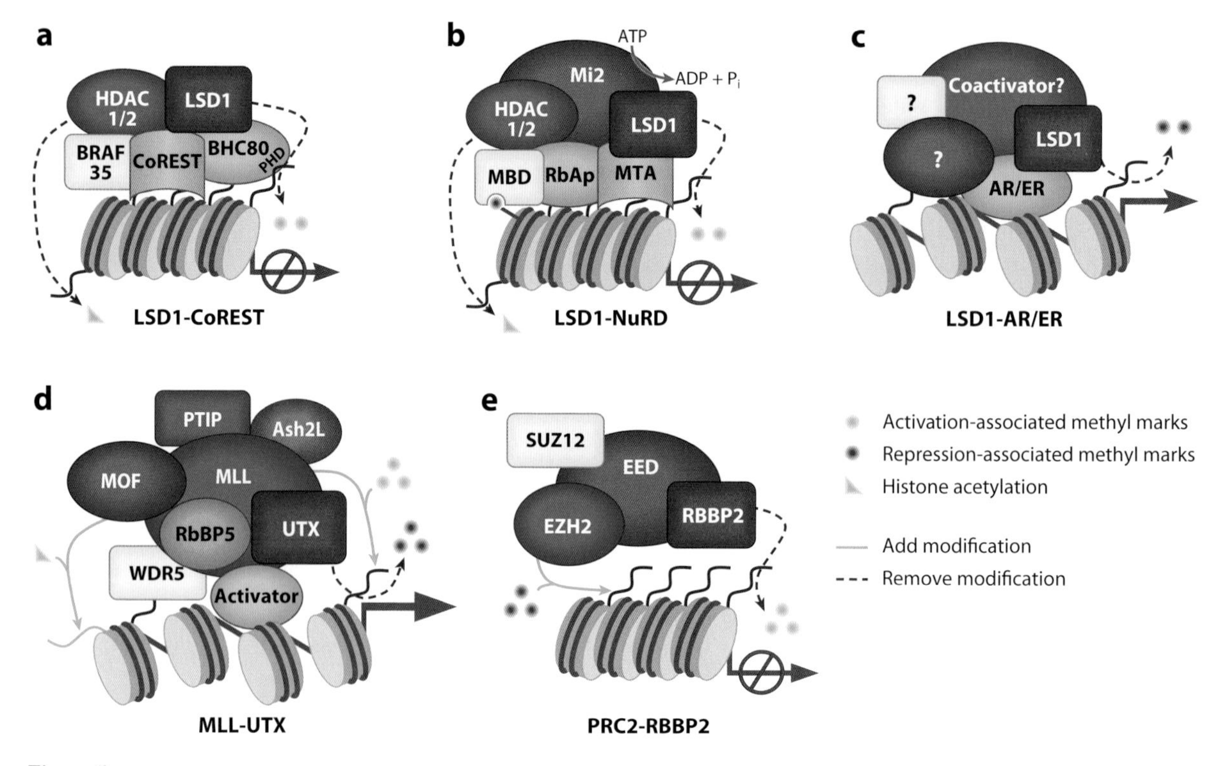

Figure 5

Key histone demethylase complexes and their activities described to date. Green and red circles represent activation- and repression-associated methyl marks, respectively, and green triangles represent histone acetylation. Solid green lines and dashed red lines represent enzymatic activities that add or remove specific modifications, respectively. (*a*) The lysine-specific demethylase 1 (LSD1)-corepressor of RE1-silencing transcription factor (CoREST) complex. (*b*) The LSD1-nucleosome remodeling and deacetylase (NuRD) complex. (*c*) The LSD1-nuclear hormone receptor (AR/ER) complex. (*d*) The mixed-lineage leukemia (MLL) protein-ubiquitously transcribed tetratricopeptide repeat, X chromosome protein (UTX) complex. (*e*) The Polycomb repressive complex 2 (PRC2)-retinoblastoma-binding protein 2 (RBBP2) complex. Abbreviations: AR/ER, androgen receptor/estrogen receptor; Ash2L, absent, small, or homeotic like 2; BHC80, BRAF and histone deacetylase complex 80; EED, embryonic ectoderm development; EZH2, enhancer of zeste homolog 2; HDAC1-2, histone deacetylase 1-2; MBD, methyl-CpG-binding domain protein 2; Mi2, myositis autoantigen 2; MOF, males absent on first; MTA, metastasis associated; PTIP, PAX interacting protein; RbAp, retinoblastoma-associated protein; RbBP, retinoblastoma-binding protein; SUZ12, suppressor of zeste homolog 12; WDR5, WD repeat domain protein 5.

activity of LSD1 can also be coupled to chromatin-remodeling Swi/Snf type ATPases important for transcriptional repression (**Figure 5*b***). In the NuRD complex, the metastasis-associated (MTA) proteins function in place of CoREST and play an analogous function to allow LSD1 demethylation of nucleosomal substrates (102). Like CoREST, the MTA proteins each contain a SANT domain, which serve as Myb-like DNA-binding domains. It is possible that other SANT domain-containing proteins may serve to target LSD1 to chromatin.

Although human LSD1 has a clear role in repression as an H3K4 demethylase, there is also evidence that it functions as an activator when present in other complexes, thus playing opposing roles in different physiological contexts. LSD1 has been demonstrated to associate directly with androgen receptor (AR), and in this molecular context, LSD1 can demethylate H3K9me2/me1 but not H3K4me2/1 (**Figure 5*c***) (103). As a result, transcription of AR-responsive genes, such as *PSA*, requires LSD1 for full activation. Interestingly, this nuclear hormone receptor-mediated gene activation event, which produces hydrogen peroxide as a result of LSD1-mediated demethylation, results in localized oxidative DNA damage. This DNA damage in turn recruits repair enzymes, such as 8-oxoguanine-DNA glycosylase, which, like LSD1, are important for the transcriptional activation process (104). Gene targeting studies have also supported an activation function for LSD1 in vivo. In contrast to pituitary lactotropes, LSD1 functions as an activator in somatotropes (99). Although the precise molecular complex in which LSD1 functions as an activator in this context is unclear, chromatin immunoprecipitation studies demonstrate direct recruitment of LSD1 to somatotrope-specific genes, which is concomitant with their activation during development (99). LSD1 was also shown to interact with estrogen receptor alpha (ERα) in a ligand-dependent fashion and was important for activation of a subset of ERα-dependent genes (105). Global analysis of LSD1 target genes demonstrated that a large portion of LSD1-occupied promoters are transcriptionally active and contain modifications consistent with gene activation. In agreement with its role in activation, specific H3K9 methyltransferases opposed LSD1 activity on these ERα gene targets (105). Evidence suggests that during estrogen-mediated gene activation, the chromosomal landscape changes significantly, promoting formation of interchromosomal granules. The formation of these granules, which are thought to be "hubs" for various transcription factors and splicing machinery, requires LSD1 (106). It is therefore possible that LSD1 plays an important role in transcriptional activation for a variety of nuclear receptors by demethylation of H3K9, although nonenzymatic roles of LSD1 in this context cannot be ruled out.

At present, it is unclear how LSD1 changes demethylase specificity from H3K4 to H3K9 when it engages different interacting partners. Although there are no in vitro results that demonstrate this activity for human LSD1, recombinant *S. pombe* LSD1 proteins have been shown to possess a low level of H3K9 demethylase activity (63, 64). As discussed above, the structure of LSD1 with an H3 peptide mimetic does not allow more than three amino acids N-terminal to the methylated lysine to be accommodated in the active site, thus precluding H3K9me2/1 as a substrate in this particular structural conformation (79). It is possible that the catalytic site of LSD1 is more plastic than its structure would suggest, and future studies are necessary to determine whether this is actually the case. For human LSD1 to demethylate H3K9me2/1, we anticipate that a whole new set of enzyme-substrate interactions would have to occur, which may be facilitated by the discovery of accessory factors associated with LSD1.

Jumonji C Protein Complexes

The discovery of the Jumonji family of histone demethylases increased the histone demethylase repertoire. Many known sites of histone methylation now have a corresponding demethylase (**Figure 3**). As with LSD1, the discovery of the biochemical complexes in which these enzymes exist has greatly increased our understanding of their functions. For instance, the JARID1 (KDM5) family of demethylases, which has four members, all act to antagonize H3K4me3 and hence are thought to be transcriptional corepressors (93, 107–110). However, the biochemical context in which each of these enzymes acts as a repressor is quite distinct. SMCX (also known as KDM5C), for example, has been reported to form a complex with the neuronal repressor REST and HDAC1-2, as well as the H3K9 methyltransferase G9a (110). By contrast, SMCY (KDM5D) associates with the Polycomb-like protein Ring6a, which promotes the demethylase activity of SMCY in vitro (109). Thus, identical demethylase activities encoded by different members of the same enzyme family may be recruited into various complexes for distinct functions. This may explain why mutations in SMCX cause X-linked mental retardation, even though at least three other JmjC family members possess precisely the same activity.

Many chromatin-modifying complexes, like the LSD1-CoREST complex, contain enzymatic activities that reinforce each other's functions, and the JmjC family is no exception. Purification of mixed-lineage leukemia (MLL)2-3 proteins, an H3K4 methyltransferase protein complex, identified its association with the H3K27me3/2 demethylase UTX (also known as KDM6A) (**Figure 5*d***) (111–113). It is well known that MLL proteins activate *HOX* gene transcription, corresponding with an increase in methylated H3K4 at these loci (114). Similarly, UTX is also present at these key developmental loci, promoting H3K27me3/2 demethylation, and like MLL, UTX is important for activation of *HOX* genes (113, 115, 116). These findings suggest that the MLL-UTX complex plays an important role in the coordinated regulation of H3K4 and H3K27 methylation states, with MLL mediating H3K4 methylation and UTX H3K27 demethylation, resulting in transcriptional activation of target genes. This strategy has also been used in another instance wherein the H3K4 trimethyl demethylase RBBP2 (KDM5A) is found to be associated with the PRC2 complex that mediates H3K27 trimethylation (**Figure 5*e***) (117). Together, this PRC2-RBBP2 complex coordinates H3K4 trimethyl demethylation and H3K27 trimethylation and functions antagonistically to that of the MLL-UTX complex to mediate Polycomb complex-mediated transcriptional repression. It therefore appears that cells have developed the strategy of assembling protein complexes containing functionally complementary enzymatic activities to coordinately regulate histone methylation at multiple lysines on histone tails to regulate chromatin and gene transcription.

Synergistic functions between histone demethylases and methylated histone reader proteins have also begun to emerge. The example of BHC80 recognizing H3K4me0 in the context of the LSD1-CoREST complex was discussed above (**Figure 5*a***). Demethylation by LSD1 generates a recognition surface for BHC80 binding, which in turn helps to keep LSD1 at target promoters to maintain the repressed state (98). Another example is HP1, which not only recruits Suv4–20h H4K20 methyltransferases, as alluded to previously in this review, but also binds directly to JMJD2A (KDM4A) in *Drosophila* (118). This association is important for the activation of the H3K36me3 demethylase activity of JMJD2A-KDM4A as well as for regulation of the global levels of H3K36me3. A number of JmjC proteins also contain PHD fingers (95, 119, 120), and it is likely that these domains have an intramolecular function similar to the one HP1 has for JMJD2A.

Clearly, most histone demethylases need to be considered in the context of large macromolecular complexes that modify the epigenetic landscape. Thus, analysis of the epigenomic landscape can also aid in predicting the presence of specific demethylase activities coupled with other activities or histone modification readers. For example, higher levels of H3K9me1, unlike H3K9me2/3, are detected at the transcriptional start site of highly active genes (16). This predicts that an H3K9me2/3 demethylase is coupled to a histone reader domain specific for H3K4me3. Whether this is mediated by one of the JMJD2-KDM4 family members, or another H3K9me2/3 demethylase, has yet to be determined.

HISTONE DEMETHYLASES IN BIOLOGY: KEY EXAMPLES

The biological functions of histone demethylases are tightly connected to the protein complexes in which they reside. However, because the biology of histone demethylases has been recently reviewed (119, 121, 122), we only highlight some of the key aspects of demethylase biology here. Studies of histone demethylases in the past few years uncovered their roles in many areas of biology, ranging from regulation of stem cell pluripotency and differentiation to cell proliferation and epigenetic memory.

As mentioned above, virtually every eukaryote studied to date, with the exception of budding yeast, encodes LSD1 homologs, and in all of these, LSD1 plays an important role in development. For example, in *Arabidopsis*, the LSD1 homolog *FLD* (*FLOWERING LOCUS D*) functions to repress the MADS box protein, *FLC* (*FLOWERING LOCUS C*). *FLC* acts to regulate the transition to flowering by repressing genes that promote the floral state. *FLD* promotes demethylation of H3K4me2 in vivo, both globally and at the *FLC* locus (61, 123). As a result, the loss of *FLD* function results in misregulation of *FLC* and causes a change in flowering timing. Functionally, *FLD* cooperates with *FCA* and *FY*, two proteins important for RNA processing and polyadenylation (61). Surprisingly, *DCL3*, a Dicer homolog in *Arabidopsis*, also functions to repress *FLC*, and *dcl3* mutations are partially epistatic to *fca* mutations, both of which harbor higher levels of H3K4me2 at the *FLC* locus (61). Thus, in *Arabidopsis*, the RNAi machinery intersects with a histone demethylase to regulate a key developmental transition. It will be of great interest to determine if this cooperative mechanism between histone demethylases and the RNAi machinery is broadly used in other organisms, particularly in mammalian cells.

Epigenetic states are thought to be reset in the germ line between each generation. Because histone demethylases erase an important epigenetic mark, demethylases may play a significant role in the phenomena of epigenetic reprogramming in the germ line. Mutations in *spr-5*, one of the homologs of LSD1 in *C. elegans*, cause progressive sterility and accumulation of H3K4me2 during successive generations (60). This sterility was shown to correlate with misregulation of spermatogenesis-expressed genes. It is likely that a number of demethylases play important roles in epigenetic reprogramming in the germ line as well as during early embryonic development. Indeed, a recent report has demonstrated that the LSD1 homolog LSD2 encodes an H3K4me2/1 demethylase and is important for DNA methylation of imprinted genes, specifically in oocytes (66). As a result, embryos derived from these oocytes die in utero, presumably owing to the loss of monoallelic expression of these genes.

Because epigenetic reprogramming also plays a significant role in stem cell function, it is perhaps no surprise that histone demethylases have also been shown to be important in this regard. JMJD1A (KDM3A) and JMJD2C (KDM4C), which demethylate H3K9me2 and H3K9me3, are important for maintaining ES cell self-renewal by regulating the expression of *Tcl1* and *Nanog*, respectively (124). Oct4, a well-characterized stem cell transcription factor, activates expression of these two demethylases, prominently placing these two enzymes in the midst of the genetic network responsible for maintaining the ES cell state. Studies of human epidermis, which serves as a model for mammalian tissue self-renewal, demonstrate that JMJD3 (KDM6B) may regulate the differentiation state of this tissue by counteracting H3K27 methylation (125). JMJD3 has also been shown to activate the tumor suppressor *INK4A-ARF* locus in response to oncogene- and stress-induced senescence (126), suggesting that this demethylase may play a role in tumor suppression. Interestingly, the JMJD3-related H3K27 demethylase UTX has recently been shown to be mutated in multiple tumors, strongly supporting a role for H3K27 demethylases in tumor suppression (127).

Two studies on a whole-animal knockout of JMJD1A (KDM3A) have revealed critical functions of this enzyme in animal development. This enzyme encodes an H3K9me2/1 demethylase and acts as a transcriptional activator for AR (91). Consistent with this, JMJD1A is highly expressed primarily in postmeiotic male germ cells, and male mice lacking the enzyme are infertile (128). These mutant mice have lower sperm counts, and their spermatids have a number of notable abnormalities, including loss of chromocenter organization and acrosomal structures that are not fully formed. Virtually all mature sperm from these mice are immotile and lack properly condensed chromatin. It appears that this demethylase is critical for direct activation of genes that promote spermatid maturation, including protamine 1 and transition nuclear protein 1. A parallel study of these mice also revealed that this demethylase is important for maintaining expression of genes involved in metabolism (129).

Specifically, JMJD1A is not only important for activating AR target genes but also acts in concert with other nuclear hormone receptors, including PPAR-γ and RXR-α, to promote metabolic gene activation in response to adrenergic signaling. Indeed, mice lacking this enzyme are markedly obese relative to their control littermates. Furthermore, these mutant mice are not capable of maintaining proper body temperature when exposed to the cold, which normally occurs at least in part by increasing metabolism in brown adipose tissue. The fact that these enzymes play significant roles in development and homeostasis further attests to the importance of histone methylation in gene regulation.

REGULATION OF HISTONE DEMETHYLASE ACTIVITY

How are all these histone demethylase activities regulated in the cell nucleus? As the field is still young, relatively little is known about the mechanisms that may regulate these enzymes. Thus far, two regulatory mechanisms have emerged: posttranslational modification of the demethylase and regulation by association with auxiliary factors. With regard to posttranslational modification, a recent study has shown that, in budding yeast, levels of Jhd2, an H3K4me3 demethylase, are tightly regulated by polyubiquitination, which targets this demethylase to the proteasome for degradation (130). Not4, a RING-finger protein that functions as an E3 ubiquitin ligase, and its cognate E2 ubiquitin-conjugating enzyme, Ubc4, mediate polyubiquitination of Jhd2. Cells that lack either Not4 or Ubc4 have globally decreased levels H3K4me3, owing to stabilization of Jhd2. Consequently, a Jhd2 target gene is repressed when Not4 is not present in vivo. As SMCX (JMJD5C), a mammalian counterpart of Jhd2, is also targeted for polyubiquitination by analogous ubiquination machinery, it is likely that this mechanism of demethylase regulation is conserved in higher eukaryotes (130). It will be interesting to determine whether other posttranslational modifications exist to activate, as opposed to downregulate, certain histone demethylases. Of note, the k_{cat} of the JmjC family of enzymes has been estimated to be $\sim$0.01 min^{-1}, over two orders of magnitude lower than LSD1 (80, 88). It is therefore possible that accessory factors or posttranslational modifications are important for JmjC activity in vivo. Activation of demethylases by auxiliary factors has already been discussed above, such as CoREST regulating LSD1 activity toward nucleosomal substrates (77, 78). We speculate that auxiliary factors are not limited to proteins but may also be noncoding RNAs as well.

DEMETHYLATION: BEYOND HISTONES

Like acetylation and phosphorylation, the nuclear proteins that are targeted for methylation likely number far more than histones, and not surprisingly, this modification also serves an important regulatory function for other proteins as well. A key example is p53, which is known to be modified by a plethora of posttranslational modifications, altering its function as a transcriptional activator. The notion of LSD1 substrate plasticity has been supported by recent findings that LSD1 can demethylate the tumor suppressor p53 (131). As with histones, methylation may activate or repress the function of p53 depending on the residue and degree of methylation. LSD1 appears to have preference for the demethylation of K370me2. This demethylation causes inactivation of p53 by inhibiting both the ability of p53 to bind to DNA and its association with 53BP1 (131). It is therefore possible that LSD1 may function as an oncoprotein by inactivating p53, and in fact, LSD1 is known to be overexpressed in various human tumors (132, 133). More recently, LSD1 has been demonstrated to demethylate DNA methyltransferase 1 (DNMT1), resulting in its stabilization in vivo (134). The notion that LSD1 can also target nonhistone proteins, such as p53 and DNMT1, suggests that the inherent substrate flexibility of LSD1 is likely to permit extension of its functions to other important methylated targets related to transcriptional regulation but, at the same time, adds another layer of complexity in interpreting the experimental findings given the expanded repertoire of target proteins.

HISTONES: BEYOND DEMETHYLATION

The discovery of LSD1 and the JmjC family of demethylases has revealed the existence of novel biochemical reactions occurring in the cell nucleus, which leads us to contemplate the existence of other analogous reactions that may be carried out by similar enzymes (**Figure 6**). It has been known for quite some time, for example, that hydroxylation of specific proteins, such as collagen in the extracellular matrix, is important for their structure and function (135). Many of these hydroxylation reactions are also carried out by

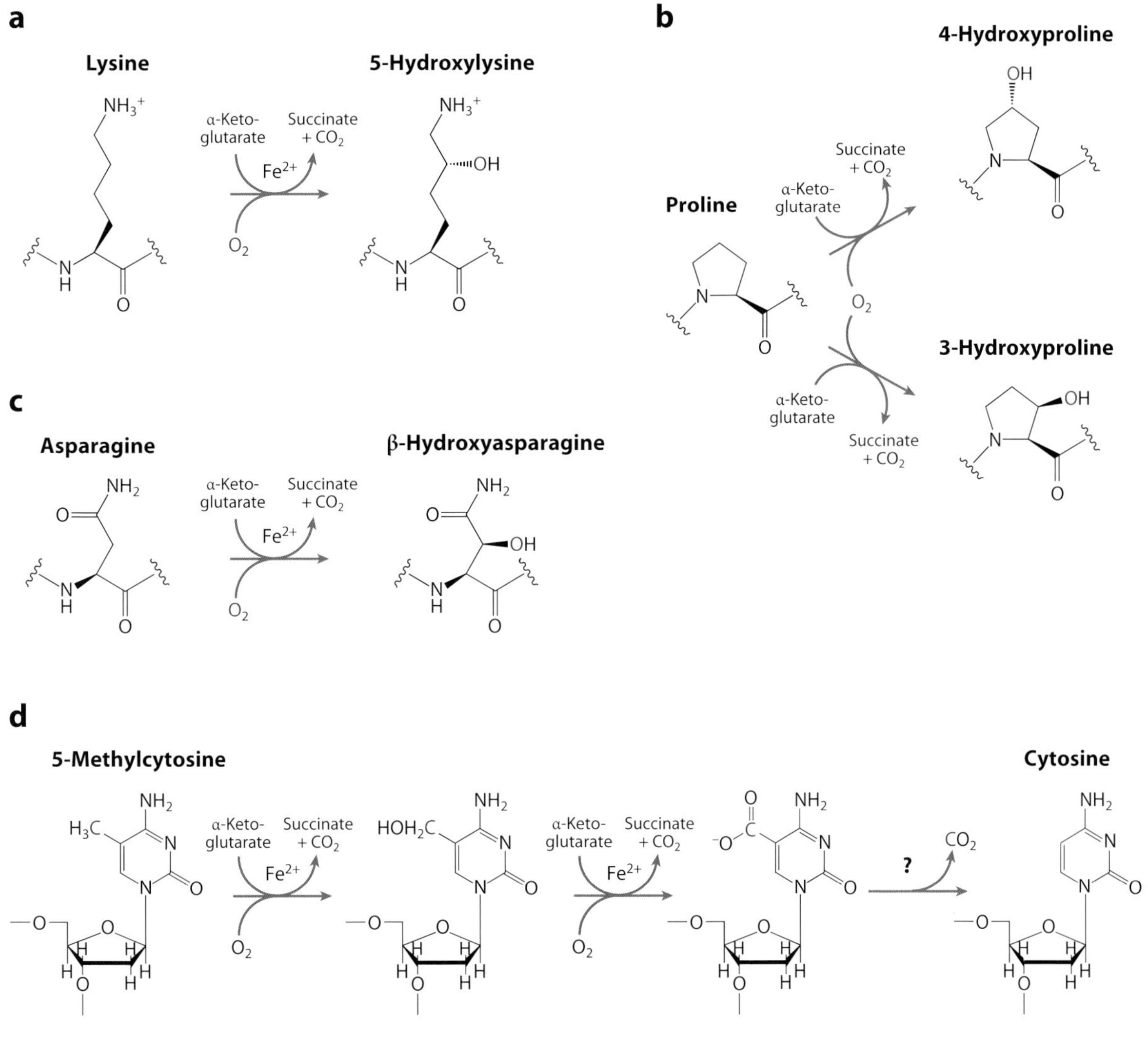

Figure 6

Potential reactions mediated by Fe^{2+} and 2-oxoglutarate-dependent dioxygenases. (*a*) Production of 5-hydroxylsine from lysine, (*b*) 4-hydroxyproline and 3-hydroxyproline from proline, and (*c*) β-hydroxyasparagine from asparagine. (*d*) Conversion of 5-methylcytosine to cytosine by successive oxidation followed by decarboxylation.

Fe^{2+}-dependent dioxygenases, and it is possible that this superfamily of enzymes may also hydroxylate histones. A recent study of JMJD6 has shown that this protein can hydroxylate lysine residues on U2AF65, a splicing factor, as well as H3 and H4 (136). Although its function on histone hydroxylation was not explored, hydroxylation of U2AF65 by JMJD6 was shown to regulate alternative splicing. JMJD6 was previously suggested to be an arginine demethylase (137), although the above new finding appears to be at odds with this suggestion. However, an alternative reaction, such as histone lysine hydroxylation, is possible for this enzyme. It is also tempting to consider that other residues, such as proline and asparagine (**Figure 6**), can also be hydroxylated on histones, as these modifications are known to exist on other nuclear proteins, including hypoxia-inducible factor (HIF) and factor-inhibiting HIF (FIH), respectively (138). The existence of a number of nuclear JmjC proteins, as well as other Fe^{2+}-dependent dioxygenases, which have yet to be reported to have any known activity, may imply that these enzymes are not histone demethylases but may function in alternative reactions, such as hydroxylation. The TET subfamily of Fe^{2+}-dependent dioxygenases has recently been reported to function as a DNA 5-methylcytosine hydroxylase (139), thus demonstrating the potential enzymatic repertoire of this large family of enzymes. We speculate that active DNA demethylation may be initiated by an enzyme, such as one of the TET family members, or another Fe^{2+}-dependent dioxygenase. However, further oxidation of the methyl group to form a carboxylate may be required prior to its release (**Figure 6*d***). This is analogous to the conversion of free thymine to uracil by sequential action of thymine hydroxylase, an Fe^{2+}-dependent dioxygenase, and isoorotate decarboxylase; both of these exist in certain fungi (140). Thus, from these various examples, it is clear that a whole host of enzymatic possibilities exist in this family, and we look forward to the elucidation of their activities and their respective functions in the future.

PERSPECTIVES

The flexible histone tails are heavily modified by a plethora of posttranslational modifications, which, in total, impact chromatin structure and function in various chromatin-templated reactions, such as transcription and DNA repair. Histone methylation is an integral part of this highly orchestrated process and therefore is essential for chromatin and epigenetic regulation. The discovery of histone demethylases adds an important missing layer to this complexity and provides an important conceptual framework for considering dynamic regulation of histone methylation in the cell. Importantly, these enzymes represent potentially exciting new drug targets for various disease indications, particularly cancer. The next few years will witness continued discovery of new histone-modifying enzymes and elucidation of exciting biology associated with histone demethylases.

SUMMARY POINTS

1. Histone methylation is an important histone posttranslational modification that serves to regulate many chromatin-dependent processes, including transcription, recombination, and DNA repair.
2. Demethylation of histones is an active process, mediated by a number of key enzymes discovered recently. Histone demethylases include LSD1, which uses an amine-oxidase mechanism, and the JmjC protein family, a group of Fe^{2+}-dependent dioxygenases. Known structures of LSD1 and JMJD2A have revealed very distinct mechanisms of histone recognition, thus providing an explanation for observed substrate specificity.

3. Histone demethylases exist in numerous macromolecular complexes, which can harbor other chromatin modifying enzymes, including histone deacetylases and methyltransferases, SWI/SNF remodeling factors, as well as modification readers, such as PHD and chromodomain-containing proteins. These associated factors, as well as other posttranslational modifications, may serve to regulate demethylase activity or potentially alter substrate specificity.
4. In vivo studies of histone demethylases have revealed important roles for these enzymes in stem cell biology, cellular proliferation and differentiation, and certain pathogenic states, particularly tumor development.
5. We predict that the amine oxidase family as well as the JmjC family of demethylases will have other undiscovered functions, including demethylation of nonhistone substrates, as well as the formation of novel histone modifications, such as hydroxylation.

FUTURE ISSUES

1. Although several methylated lysine residues on histones have a known demethylase, other lysine methylation sites, notably H3K79 and H4K20, lack a known demethylase. In addition to lysine, a number of arginine residues on histones are also methylated. Is methylation at the remaining lysine sites as well as arginine residues dynamically regulated by undiscovered demethylases?
2. Are there other novel chemical mechanisms of demethylation?
3. Recent studies suggest that histones may not be the only substrates for these newly identified demethylases. What are the nonhistone substrates for histone demethylases? What is the functional significance of such a "demethylome," and what is the interplay between methylation and other modifications? A proteomics approach may be warranted to identify methylated proteins and enzymes that mediate their methylation and demethylation.
4. JmjC domain–containing demethylases show differential specificities toward various lysine residues and also demonstrate methyl group number specificity, suggesting a fine-tuning role for demethylases. What are the structural mechanisms that explain this substrate selectivity?
5. How are demethylases recruited to their respective genomic locations? How is enzymatic recruitment and activity regulated during normal development and in different human disease states?
6. Current studies suggest that LSD1 is differentially recruited to distinct complexes for various cellular functions. Is this the paradigm for other histone demethylases as well?
7. Important cellular metabolites, such as FAD, Fe^{2+}, 2-OG, and O_2 serve as cofactors for histone demethylases. Is there a functional connection between metabolism and regulation of the activities of histone demethylases? For instance, do the cellular levels of these metabolites function to regulate demethylase activity?

DISCLOSURE STATEMENT

Y.S. is a cofounder and consultant for Constellation Pharma, a biotechnology company that develops epigenetic drugs. N.M. is not aware of any affiliations, memberships, funding, or financial holdings that might be perceived as affecting the objectivity of this review.

ACKNOWLEDGMENTS

We wish to thank members of the Shi lab for helpful discussions. We apologize to our colleagues whose work we were not able to cite in this review owing to space limitations. N.M. is supported by the Brigham and Women's Hospital Department of Pathology, Children's Hospital Boston Department of Laboratory Medicine, and a Ruth L. Kirschstein Institutional National Research Service Award (T32-HL007627). The work on histone demethylases in the Shi lab is supported by grants from the National Institutes of Health (GM058012, GM071004, and CA118487).

LITERATURE CITED

1. Kornberg RD. 1974. Chromatin structure: a repeating unit of histones and DNA. *Science* 184:868–71
2. Luger K, Mader AW, Richmond RK, Sargent DF, Richmond TJ. 1997. Crystal structure of the nucleosome core particle at 2.8 A resolution. *Nature* 389:251–60
3. Strahl BD, Allis CD. 2000. The language of covalent histone modifications. *Nature* 403:41–45
4. Ruthenburg AJ, Li H, Patel DJ, Allis CD. 2007. Multivalent engagement of chromatin modifications by linked binding modules. *Nat. Rev. Mol. Cell Biol.* 8:983–94
5. Allfrey VG, Faulkner R, Mirsky AE. 1964. Acetylation and methylation of histones and their possible role in the regulation of RNA synthesis. *Proc. Natl. Acad. Sci. USA* 51:786–94
6. Mann RK, Grunstein M. 1992. Histone H3 N-terminal mutations allow hyperactivation of the yeast *GAL1* gene in vivo. *EMBO J.* 11:3297–306
7. Durrin LK, Mann RK, Kayne PS, Grunstein M. 1991. Yeast histone H4 N-terminal sequence is required for promoter activation in vivo. *Cell* 65:1023–31
8. Lenfant F, Mann RK, Thomsen B, Ling X, Grunstein M. 1996. All four core histone N-termini contain sequences required for the repression of basal transcription in yeast. *EMBO J.* 15:3974–85
9. Brownell JE, Zhou J, Ranalli T, Kobayashi R, Edmondson DG, et al. 1996. Tetrahymena histone acetyltransferase A: a homolog to yeast Gcn5p linking histone acetylation to gene activation. *Cell* 84:843–51
10. Taunton J, Hassig CA, Schreiber SL. 1996. A mammalian histone deacetylase related to the yeast transcriptional regulator Rpd3p. *Science* 272:408–11
11. Carmen AA, Rundlett SE, Grunstein M. 1996. HDA1 and HDA3 are components of a yeast histone deacetylase (HDA) complex. *J. Biol. Chem.* 271:15837–44
12. Guarente L. 1995. Transcriptional coactivators in yeast and beyond. *Trends Biochem. Sci.* 20:517–21
13. Vidal M, Gaber RF. 1991. RPD3 encodes a second factor required to achieve maximum positive and negative transcriptional states in *Saccharomyces cerevisiae*. *Mol. Cell. Biol.* 11:6317–27
14. Shahbazian MD, Grunstein M. 2007. Functions of site-specific histone acetylation and deacetylation. *Annu. Rev. Biochem.* 76:75–100
15. Heintzman ND, Hon GC, Hawkins RD, Kheradpour P, Stark A, et al. 2009. Histone modifications at human enhancers reflect global cell-type-specific gene expression. *Nature* 459:108–12
16. Barski A, Cuddapah S, Cui K, Roh TY, Schones DE, et al. 2007. High-resolution profiling of histone methylations in the human genome. *Cell* 129:823–37
17. Bedford MT, Clarke SG. 2009. Protein arginine methylation in mammals: who, what, and why. *Mol. Cell* 33:1–13
18. Strahl BD, Ohba R, Cook RG, Allis CD. 1999. Methylation of histone H3 at lysine 4 is highly conserved and correlates with transcriptionally active nuclei in *Tetrahymena*. *Proc. Natl. Acad. Sci. USA* 96:14967–72
19. Schneider R, Bannister AJ, Myers FA, Thorne AW, Crane-Robinson C, Kouzarides T. 2004. Histone H3 lysine 4 methylation patterns in higher eukaryotic genes. *Nat. Cell Biol.* 6:73–77

20. Santos-Rosa H, Schneider R, Bannister AJ, Sherriff J, Bernstein BE, et al. 2002. Active genes are tri-methylated at K4 of histone H3. *Nature* 419:407–11
21. Noma K, Allis CD, Grewal SI. 2001. Transitions in distinct histone H3 methylation patterns at the heterochromatin domain boundaries. *Science* 293:1150–55
22. Litt MD, Simpson M, Gaszner M, Allis CD, Felsenfeld G. 2001. Correlation between histone lysine methylation and developmental changes at the chicken *β-globin* locus. *Science* 293:2453–55
23. Liang G, Lin JC, Wei V, Yoo C, Cheng JC, et al. 2004. Distinct localization of histone H3 acetylation and H3-K4 methylation to the transcription start sites in the human genome. *Proc. Natl. Acad. Sci. USA* 101:7357–62
24. Briggs SD, Bryk M, Strahl BD, Cheung WL, Davie JK, et al. 2001. Histone H3 lysine 4 methylation is mediated by Set1 and required for cell growth and rDNA silencing in *Saccharomyces cerevisiae*. *Genes Dev.* 15:3286–95
25. Guenther MG, Levine SS, Boyer LA, Jaenisch R, Young RA. 2007. A chromatin landmark and transcription initiation at most promoters in human cells. *Cell* 130:77–88
26. Li J, Moazed D, Gygi SP. 2002. Association of the histone methyltransferase Set2 with RNA polymerase II plays a role in transcription elongation. *J. Biol. Chem.* 277:49383–88
27. Xiao T, Hall H, Kizer KO, Shibata Y, Hall MC, et al. 2003. Phosphorylation of RNA polymerase II CTD regulates H3 methylation in yeast. *Genes Dev.* 17:654–63
28. Yoh SM, Lucas JS, Jones KA. 2008. The Iws1:Spt6:CTD complex controls cotranscriptional mRNA biosynthesis and HYPB/Setd2-mediated histone H3K36 methylation. *Genes Dev.* 22:3422–34
29. Nimura K, Ura K, Shiratori H, Ikawa M, Okabe M, et al. 2009. A histone H3 lysine 36 trimethyltransferase links Nkx2–5 to Wolf-Hirschhorn syndrome. *Nature* 460:287–91
30. Xu L, Zhao Z, Dong A, Soubigou-Taconnat L, Renou JP, et al. 2008. Di- and tri- but not monomethylation on histone H3 lysine 36 marks active transcription of genes involved in flowering time regulation and other processes in *Arabidopsis thaliana*. *Mol. Cell. Biol.* 28:1348–60
31. Rea S, Eisenhaber F, O'Carroll D, Strahl BD, Sun ZW, et al. 2000. Regulation of chromatin structure by site-specific histone H3 methyltransferases. *Nature* 406:593–99
32. Nakayama J, Rice JC, Strahl BD, Allis CD, Grewal SI. 2001. Role of histone H3 lysine 9 methylation in epigenetic control of heterochromatin assembly. *Science* 292:110–13
33. Boggs BA, Cheung P, Heard E, Spector DL, Chinault AC, Allis CD. 2002. Differentially methylated forms of histone H3 show unique association patterns with inactive human X chromosomes. *Nat. Genet.* 30:73–76
34. Peters AH, Mermoud JE, O'Carroll D, Pagani M, Schweizer D, et al. 2002. Histone H3 lysine 9 methylation is an epigenetic imprint of facultative heterochromatin. *Nat. Genet.* 30:77–80
35. Lachner M, O'Carroll D, Rea S, Mechtler K, Jenuwein T. 2001. Methylation of histone H3 lysine 9 creates a binding site for HP1 proteins. *Nature* 410:116–20
36. Bannister AJ, Zegerman P, Partridge JF, Miska EA, Thomas JO, et al. 2001. Selective recognition of methylated lysine 9 on histone H3 by the HP1 chromo domain. *Nature* 410:120–24
37. Eissenberg JC, Elgin SC. 2000. The HP1 protein family: getting a grip on chromatin. *Curr. Opin. Genet. Dev.* 10:204–10
38. Schotta G, Lachner M, Sarma K, Ebert A, Sengupta R, et al. 2004. A silencing pathway to induce H3-K9 and H4-K20 trimethylation at constitutive heterochromatin. *Genes Dev.* 18:1251–62
39. Cao R, Wang L, Wang H, Xia L, Erdjument-Bromage H, et al. 2002. Role of histone H3 lysine 27 methylation in Polycomb-group silencing. *Science* 298:1039–43
40. Muller J, Hart CM, Francis NJ, Vargas ML, Sengupta A, et al. 2002. Histone methyltransferase activity of a *Drosophila* Polycomb group repressor complex. *Cell* 111:197–208
41. Bernstein BE, Mikkelsen TS, Xie X, Kamal M, Huebert DJ, et al. 2006. A bivalent chromatin structure marks key developmental genes in embryonic stem cells. *Cell* 125:315–26
42. Sanders SL, Portoso M, Mata J, Bahler J, Allshire RC, Kouzarides T. 2004. Methylation of histone H4 lysine 20 controls recruitment of Crb2 to sites of DNA damage. *Cell* 119:603–14
43. Botuyan MV, Lee J, Ward IM, Kim JE, Thompson JR, et al. 2006. Structural basis for the methylation state-specific recognition of histone H4-K20 by 53BP1 and Crb2 in DNA repair. *Cell* 127:1361–73

44. Wang Y, Reddy B, Thompson J, Wang H, Noma K, et al. 2009. Regulation of Set9-mediated H4K20 methylation by a PWWP domain protein. *Mol. Cell* 33:428–37
45. Matthews AG, Kuo AJ, Ramon-Maiques S, Han S, Champagne KS, et al. 2007. RAG2 PHD finger couples histone H3 lysine 4 trimethylation with V(D)J recombination. *Nature* 450:1106–10
46. Liu Y, Subrahmanyam R, Chakraborty T, Sen R, Desiderio S. 2007. A plant homeodomain in RAG-2 that binds hypermethylated lysine 4 of histone H3 is necessary for efficient antigen-receptor-gene rearrangement. *Immunity* 27:561–71
47. Aasland R, Gibson TJ, Stewart AF. 1995. The PHD finger: implications for chromatin-mediated transcriptional regulation. *Trends Biochem. Sci.* 20:56–59
48. Byvoet P, Shepherd GR, Hardin JM, Noland BJ. 1972. The distribution and turnover of labeled methyl groups in histone fractions of cultured mammalian cells. *Arch. Biochem. Biophys.* 148:558–67
49. Thomas G, Lange HW, Hempel K. 1972. Relative stability of lysine-bound methyl groups in arginie-rich histones and their subfractions in Ehrlich ascites tumor cells in vitro. *Hoppe Seyler's Z. Physiol. Chem.* 353:1423–28
50. Ahmad K, Henikoff S. 2002. The histone variant H3.3 marks active chromatin by replication-independent nucleosome assembly. *Mol. Cell* 9:1191–200
51. Allis CD, Bowen JK, Abraham GN, Glover CV, Gorovsky MA. 1980. Proteolytic processing of histone H3 in chromatin: a physiologically regulated event in *Tetrahymena* micronuclei. *Cell* 20:55–64
52. Wang Y, Wysocka J, Sayegh J, Lee YH, Perlin JR, et al. 2004. Human PAD4 regulates histone arginine methylation levels via demethylimination. *Science* 306:279–83
53. Cuthbert GL, Daujat S, Snowden AW, Erdjument-Bromage H, Hagiwara T, et al. 2004. Histone deimination antagonizes arginine methylation. *Cell* 118:545–53
54. Chinenov Y. 2002. A second catalytic domain in the Elp3 histone acetyltransferases: a candidate for histone demethylase activity? *Trends Biochem. Sci.* 27:115–17
55. Bannister AJ, Schneider R, Kouzarides T. 2002. Histone methylation: dynamic or static? *Cell* 109:801–6
56. You A, Tong JK, Grozinger CM, Schreiber SL. 2001. CoREST is an integral component of the CoREST-human histone deacetylase complex. *Proc. Natl. Acad. Sci. USA* 98:1454–58
57. Hakimi MA, Dong Y, Lane WS, Speicher DW, Shiekhattar R. 2003. A candidate X-linked mental retardation gene is a component of a new family of histone deacetylase-containing complexes. *J. Biol. Chem.* 278:7234–39
58. Hakimi MA, Bochar DA, Chenoweth J, Lane WS, Mandel G, Shiekhattar R. 2002. A core-BRAF35 complex containing histone deacetylase mediates repression of neuronal-specifc genes. *Proc. Natl. Acad. Sci. USA* 99:7420–25
59. Shi Y, Lan F, Matson C, Mulligan P, Whetstine JR, et al. 2004. Histone demethylation mediated by the nuclear amine oxidase homolog LSD1. *Cell* 119:941–53
60. Katz DJ, Edwards TM, Reinke V, Kelly WG. 2009. A *C. elegans* LSD1 demethylase contributes to germline immortality by reprogramming epigenetic memory. *Cell* 137:308–20
61. Liu F, Quesada V, Crevillen P, Baurle I, Swiezewski S, Dean C. 2007. The *Arabidopsis* RNA-binding protein FCA requires a lysine-specific demethylase 1 homolog to downregulate FLC. *Mol. Cell* 28:398–407
62. Rudolph T, Yonezawa M, Lein S, Heidrich K, Kubicek S, et al. 2007. Heterochromatin formation in *Drosophila* is initiated through active removal of H3K4 methylation by the LSD1 homolog SU(VAR)3-3. *Mol. Cell* 26:103–15
63. Opel M, Lando D, Bonilla C, Trewick SC, Boukaba A, et al. 2007. Genome-wide studies of histone demethylation catalysed by the fission yeast homologues of mammalian LSD1. *PLoS ONE* 2:e386
64. Lan F, Zaratiegui M, Villen J, Vaughn MW, Verdel A, et al. 2007. *S. pombe* LSD1 homologs regulate heterochromatin propagation and euchromatic gene transcription. *Mol. Cell* 26:89–101
65. Karytinos A, Forneris F, Profumo A, Ciossani G, Battaglioli E, et al. 2009. A novel mammalian flavin-dependent histone demethylase. *J. Biol. Chem.* 284:17775–82
66. Ciccone DN, Su H, Hevi S, Gay F, Lei H et al. 2009. AOF1/KDM1A is a histone demethylase required to establish maternal genomic imprints. *Nature* 461:415–18
67. Tsukada Y, Fang J, Erdjument-Bromage H, Warren ME, Borchers CH, et al. 2006. Histone demethylation by a family of JmjC domain-containing proteins. *Nature* 439:811–16

68. Whetstine JR, Nottke A, Lan F, Huarte M, Smolikov S, et al. 2006. Reversal of histone lysine trimethylation by the JMJD2 family of histone demethylases. *Cell* 125:467–81
69. Fodor BD, Kubicek S, Yonezawa M, O'Sullivan RJ, Sengupta R, et al. 2006. Jmjd2b antagonizes H3K9 trimethylation at pericentric heterochromatin in mammalian cells. *Genes Dev.* 20:1557–62
70. Cloos PA, Christensen J, Agger K, Maiolica A, Rappsilber J, et al. 2006. The putative oncogene GASC1 demethylates tri- and dimethylated lysine 9 on histone H3. *Nature* 442:307–11
71. Trewick SC, Henshaw TF, Hausinger RP, Lindahl T, Sedgwick B. 2002. Oxidative demethylation by *Escherichia coli* AlkB directly reverts DNA base damage. *Nature* 419:174–78
72. Falnes PO, Johansen RF, Seeberg E. 2002. AlkB-mediated oxidative demethylation reverses DNA damage in *Escherichia coli*. *Nature* 419:178–82
73. Klose RJ, Yamane K, Bae Y, Zhang D, Erdjument-Bromage H, et al. 2006. The transcriptional repressor JHDM3A demethylates trimethyl histone H3 lysine 9 and lysine 36. *Nature* 442:312–16
74. Aravind L, Iyer LM. 2002. The SWIRM domain: a conserved module found in chromosomal proteins points to novel chromatin-modifying activities. *Genome Biol.* 3:research0039.1–39.7
75. Stavropoulos P, Blobel G, Hoelz A. 2006. Crystal structure and mechanism of human lysine-specific demethylase-1. *Nat. Struct. Mol. Biol.* 13:626–32
76. Yang M, Gocke CB, Luo X, Borek D, Tomchick DR, et al. 2006. Structural basis for CoREST-dependent demethylation of nucleosomes by the human LSD1 histone demethylase. *Mol. Cell* 23:377–87
77. Lee MG, Wynder C, Cooch N, Shiekhattar R. 2005. An essential role for CoREST in nucleosomal histone 3 lysine 4 demethylation. *Nature* 437:432–35
78. Shi YJ, Matson C, Lan F, Iwase S, Baba T, Shi Y. 2005. Regulation of LSD1 histone demethylase activity by its associated factors. *Mol. Cell* 19:857–64
79. Yang M, Culhane JC, Szewczuk LM, Gocke CB, Brautigam CA, et al. 2007. Structural basis of histone demethylation by LSD1 revealed by suicide inactivation. *Nat. Struct. Mol. Biol.* 14:535–39
80. Forneris F, Binda C, Vanoni MA, Battaglioli E, Mattevi A. 2005. Human histone demethylase LSD1 reads the histone code. *J. Biol. Chem.* 280:41360–65
81. Bianchi M, Polticelli F, Ascenzi P, Botta M, Federico R, et al. 2006. Inhibition of polyamine and spermine oxidases by polyamine analogues. *FEBS J.* 273:1115–23
82. Huang Y, Greene E, Murray Stewart T, Goodwin AC, Baylin SB, et al. 2007. Inhibition of lysine-specific demethylase 1 by polyamine analogues results in reexpression of aberrantly silenced genes. *Proc. Natl. Acad. Sci. USA* 104:8023–28
83. Shih JC, Chen K, Ridd MJ. 1999. Monoamine oxidase: from genes to behavior. *Annu. Rev. Neurosci.* 22:197–217
84. Lee MG, Wynder C, Schmidt DM, McCafferty DG, Shiekhattar R. 2006. Histone H3 lysine 4 demethylation is a target of nonselective antidepressive medications. *Chem. Biol.* 13:563–67
85. Schmidt DM, McCafferty DG. 2007. *trans*-2-Phenylcyclopropylamine is a mechanism-based inactivator of the histone demethylase LSD1. *Biochemistry* 46:4408–16
86. Yang M, Culhane JC, Szewczuk LM, Jalili P, Ball HL, et al. 2007. Structural basis for the inhibition of the LSD1 histone demethylase by the antidepressant *trans*-2-phenylcyclopropylamine. *Biochemistry* 46:8058–65
87. Ng SS, Kavanagh KL, McDonough MA, Butler D, Pilka ES, et al. 2007. Crystal structures of histone demethylase JMJD2A reveal basis for substrate specificity. *Nature* 448:87–91
88. Couture JF, Collazo E, Ortiz-Tello PA, Brunzelle JS, Trievel RC. 2007. Specificity and mechanism of JMJD2A, a trimethyllysine-specific histone demethylase. *Nat. Struct. Mol. Biol.* 14:689–95
89. Chen Z, Zang J, Whetstine J, Hong X, Davrazou F, et al. 2006. Structural insights into histone demethylation by JMJD2 family members. *Cell* 125:691–702
90. Chen Z, Zang J, Kappler J, Hong X, Crawford F, et al. 2007. Structural basis of the recognition of a methylated histone tail by JMJD2A. *Proc. Natl. Acad. Sci. USA* 104:10818–23
91. Yamane K, Toumazou C, Tsukada Y, Erdjument-Bromage H, Tempst P, et al. 2006. JHDM2A, a JmjC-containing H3K9 demethylase, facilitates transcription activation by androgen receptor. *Cell* 125:483–95
92. Huang Y, Fang J, Bedford MT, Zhang Y, Xu RM. 2006. Recognition of histone H3 lysine-4 methylation by the double tudor domain of JMJD2A. *Science* 312:748–51

93. Iwase S, Lan F, Bayliss P, de la Torre-Ubieta L, Huarte M, et al. 2007. The X-linked mental retardation gene SMCX/JARID1C defines a family of histone H3 lysine 4 demethylases. *Cell* 128:1077–88
94. Taverna SD, Li H, Ruthenburg AJ, Allis CD, Patel DJ. 2007. How chromatin-binding modules interpret histone modifications: lessons from professional pocket pickers. *Nat. Struct. Mol. Biol.* 14:1025–40
95. Shi Y, Whetstine JR. 2007. Dynamic regulation of histone lysine methylation by demethylases. *Mol. Cell* 25:1–14
96. Rose NR, Ng SS, Mecinovic J, Lienard BM, Bello SH, et al. 2008. Inhibitor scaffolds for 2-oxoglutarate-dependent histone lysine demethylases. *J. Med. Chem.* 51:7053–56
97. Lee MG, Wynder C, Bochar DA, Hakimi MA, Cooch N, Shiekhattar R. 2006. Functional interplay between histone demethylase and deacetylase enzymes. *Mol. Cell. Biol.* 26:6395–402
98. Lan F, Collins RE, De Cegli R, Alpatov R, Horton JR, et al. 2007. Recognition of unmethylated histone H3 lysine 4 links BHC80 to LSD1-mediated gene repression. *Nature* 448:718–22
99. Wang J, Scully K, Zhu X, Cai L, Zhang J, et al. 2007. Opposing LSD1 complexes function in developmental gene activation and repression programmes. *Nature* 446:882–87
100. Saleque S, Kim J, Rooke HM, Orkin SH. 2007. Epigenetic regulation of hematopoietic differentiation by Gfi-1 and Gfi-1b is mediated by the cofactors CoREST and LSD1. *Mol. Cell* 27:562–72
101. Saijo K, Winner B, Carson CT, Collier JG, Boyer L, et al. 2009. A Nurr1/CoREST pathway in microglia and astrocytes protects dopaminergic neurons from inflammation-induced death. *Cell* 137:47–59
102. Wang Y, Zhang H, Chen Y, Sun Y, Yang F, et al. 2009. LSD1 is a subunit of the NuRD complex and targets the metastasis programs in breast cancer. *Cell* 138:660–72
103. Metzger E, Wissmann M, Yin N, Muller JM, Schneider R, et al. 2005. LSD1 demethylates repressive histone marks to promote androgen-receptor-dependent transcription. *Nature* 437:436–39
104. Perillo B, Ombra MN, Bertoni A, Cuozzo C, Sacchetti S, et al. 2008. DNA oxidation as triggered by H3K9me2 demethylation drives estrogen-induced gene expression. *Science* 319:202–6
105. Garcia-Bassets I, Kwon YS, Telese F, Prefontaine GG, Hutt KR, et al. 2007. Histone methylation-dependent mechanisms impose ligand dependency for gene activation by nuclear receptors. *Cell* 128:505–18
106. Hu Q, Kwon YS, Nunez E, Cardamone MD, Hutt KR, et al. 2008. Enhancing nuclear receptor-induced transcription requires nuclear motor and LSD1-dependent gene networking in interchromatin granules. *Proc. Natl. Acad. Sci. USA* 105:19199–204
107. Christensen J, Agger K, Cloos PA, Pasini D, Rose S, et al. 2007. RBP2 belongs to a family of demethylases, specific for tri-and dimethylated lysine 4 on histone 3. *Cell* 128:1063–76
108. Klose RJ, Yan Q, Tothova Z, Yamane K, Erdjument-Bromage H, et al. 2007. The retinoblastoma binding protein RBP2 is an H3K4 demethylase. *Cell* 128:889–900
109. Lee MG, Norman J, Shilatifard A, Shiekhattar R. 2007. Physical and functional association of a trimethyl H3K4 demethylase and Ring6a/MBLR, a polycomb-like protein. *Cell* 128:877–87
110. Tahiliani M, Mei P, Fang R, Leonor T, Rutenberg M, et al. 2007. The histone H3K4 demethylase SMCX links REST target genes to X-linked mental retardation. *Nature* 447:601–5
111. Cho YW, Hong T, Hong S, Guo H, Yu H, et al. 2007. PTIP associates with MLL3- and MLL4-containing histone H3 lysine 4 methyltransferase complex. *J. Biol. Chem.* 282:20395–406
112. Issaeva I, Zonis Y, Rozovskaia T, Orlovsky K, Croce CM, et al. 2007. Knockdown of ALR (MLL2) reveals ALR target genes and leads to alterations in cell adhesion and growth. *Mol. Cell. Biol.* 27:1889–903
113. Lee MG, Villa R, Trojer P, Norman J, Yan KP, et al. 2007. Demethylation of H3K27 regulates polycomb recruitment and H2A ubiquitination. *Science* 318:447–50
114. Milne TA, Briggs SD, Brock HW, Martin ME, Gibbs D, et al. 2002. MLL targets SET domain methyltransferase activity to *Hox* gene promoters. *Mol. Cell* 10:1107–17
115. Agger K, Cloos PA, Christensen J, Pasini D, Rose S, et al. 2007. UTX and JMJD3 are histone H3K27 demethylases involved in *HOX* gene regulation and development. *Nature* 449:731–34
116. Lan F, Bayliss PE, Rinn JL, Whetstine JR, Wang JK, et al. 2007. A histone H3 lysine 27 demethylase regulates animal posterior development. *Nature* 449:689–94
117. Pasini D, Hansen KH, Christensen J, Agger K, Cloos PA, Helin K. 2008. Coordinated regulation of transcriptional repression by the RBP2 H3K4 demethylase and Polycomb-Repressive Complex 2. *Genes Dev.* 22:1345–55

118. Lin CH, Li B, Swanson S, Zhang Y, Florens L, et al. 2008. Heterochromatin protein 1a stimulates histone H3 lysine 36 demethylation by the *Drosophila* KDM4A demethylase. *Mol. Cell* 32:696–706
119. Cloos PA, Christensen J, Agger K, Helin K. 2008. Erasing the methyl mark: histone demethylases at the center of cellular differentiation and disease. *Genes Dev.* 22:1115–40
120. Klose RJ, Kallin EM, Zhang Y. 2006. JmjC-domain-containing proteins and histone demethylation. *Nat. Rev. Genet.* 7:715–27
121. Shi Y. 2007. Histone lysine demethylases: emerging roles in development, physiology and disease. *Nat. Rev. Genet.* 8:829–33
122. Nottke A, Colaiacovo MP, Shi Y. 2009. Developmental roles of the histone lysine demethylases. *Development* 136:879–89
123. Jiang D, Yang W, He Y, Amasino RM. 2007. *Arabidopsis* relatives of the human lysine-specific demethylase1 repress the expression of *FWA* and *FLOWERING LOCUS C* and thus promote the floral transition. *Plant Cell* 19:2975–87
124. Loh YH, Zhang W, Chen X, George J, Ng HH. 2007. Jmjd1a and Jmjd2c histone H3 Lys 9 demethylases regulate self-renewal in embryonic stem cells. *Genes Dev.* 21:2545–57
125. Sen GL, Webster DE, Barragan DI, Chang HY, Khavari PA. 2008. Control of differentiation in a self-renewing mammalian tissue by the histone demethylase JMJD3. *Genes Dev.* 22:1865–70
126. Agger K, Cloos PA, Rudkjaer L, Williams K, Andersen G, et al. 2009. The H3K27me3 demethylase JMJD3 contributes to the activation of the *INK4A-ARF* locus in response to oncogene- and stress-induced senescence. *Genes Dev.* 23:1171–76
127. van Haaften G, Dalgliesh GL, Davies H, Chen L, Bignell G, et al. 2009. Somatic mutations of the histone H3K27 demethylase gene *UTX* in human cancer. *Nat. Genet.* 41:521–23
128. Okada Y, Scott G, Ray MK, Mishina Y, Zhang Y. 2007. Histone demethylase JHDM2A is critical for Tnp1 and Prm1 transcription and spermatogenesis. *Nature* 450:119–23
129. Tateishi K, Okada Y, Kallin EM, Zhang Y. 2009. Role of Jhdm2a in regulating metabolic gene expression and obesity resistance. *Nature* 458:757–61
130. Mersman DP, Du HN, Fingerman IM, South PF, Briggs SD. 2009. Polyubiquitination of the demethylase Jhd2 controls histone methylation and gene expression. *Genes Dev.* 23:951–62
131. Huang J, Sengupta R, Espejo AB, Lee MG, Dorsey JA, et al. 2007. p53 is regulated by the lysine demethylase LSD1. *Nature* 449:105–8
132. Kahl P, Gullotti L, Heukamp LC, Wolf S, Friedrichs N, et al. 2006. Androgen receptor coactivators lysine-specific histone demethylase 1 and four and a half LIM domain protein 2 predict risk of prostate cancer recurrence. *Cancer Res.* 66:11341–47
133. Schulte JH, Lim S, Schramm A, Friedrichs N, Koster J, et al. 2009. Lysine-specific demethylase 1 is strongly expressed in poorly differentiated neuroblastoma: implications for therapy. *Cancer Res.* 69:2065–71
134. Wang J, Hevi S, Kurash JK, Lei H, Gay F, et al. 2009. The lysine demethylase LSD1 (KDM1) is required for maintenance of global DNA methylation. *Nat. Genet.* 41:125–29
135. Myllyharju J, Kivirikko KI. 2004. Collagens, modifying enzymes and their mutations in humans, flies and worms. *Trends Genet.* 20:33–43
136. Webby CJ, Wolf A, Gromak N, Dreger M, Kramer H, et al. 2009. Jmjd6 catalyses lysyl-hydroxylation of U2AF65, a protein associated with RNA splicing. *Science* 325:90–93
137. Chang B, Chen Y, Zhao Y, Bruick RK. 2007. JMJD6 is a histone arginine demethylase. *Science* 318:444–47
138. Kaelin WG Jr. 2005. Proline hydroxylation and gene expression. *Annu. Rev. Biochem.* 74:115–28
139. Tahiliani M, Koh KP, Shen Y, Pastor WA, Bandukwala H, et al. 2009. Conversion of 5-methylcytosine to 5-hydroxymethylcytosine in mammalian DNA by MLL partner TET1. *Science* 324:930–35
140. Smiley JA, Kundracik M, Landfried DA, Barnes VR Sr, Axhemi AA. 2005. Genes of the thymidine salvage pathway: thymine-7-hydroxylase from a *Rhodotorula glutinis* cDNA library and *iso*-orotate decarboxylase from *Neurospora crassa*. *Biochim. Biophys. Acta* 1723:256–64

The Mechanism of Double-Strand DNA Break Repair by the Nonhomologous DNA End-Joining Pathway

Michael R. Lieber

Norris Comprehensive Cancer Center, Departments of Pathology, Biochemistry and Molecular Biology, Molecular Microbiology and Immunology, and Biological Sciences, University of Southern California Keck School of Medicine, Los Angeles, California 90089; email: lieber@usc.edu

Annu. Rev. Biochem. 2010. 79:181–211

First published online as a Review in Advance on January 4, 2010

The *Annual Review of Biochemistry* is online at biochem.annualreviews.org

This article's doi:
10.1146/annurev.biochem.052308.093131

0066-4154/10/0707-0181$20.00

Key Words

Ku, DNA-PKcs, Artemis, XRCC4, DNA ligase IV

Abstract

Double-strand DNA breaks are common events in eukaryotic cells, and there are two major pathways for repairing them: homologous recombination (HR) and nonhomologous DNA end joining (NHEJ). The various causes of double-strand breaks (DSBs) result in a diverse chemistry of DNA ends that must be repaired. Across NHEJ evolution, the enzymes of the NHEJ pathway exhibit a remarkable degree of structural tolerance in the range of DNA end substrate configurations upon which they can act. In vertebrate cells, the nuclease, DNA polymerases, and ligase of NHEJ are the most mechanistically flexible and multifunctional enzymes in each of their classes. Unlike repair pathways for more defined lesions, NHEJ repair enzymes act iteratively, act in any order, and can function independently of one another at each of the two DNA ends being joined. NHEJ is critical not only for the repair of pathologic DSBs as in chromosomal translocations, but also for the repair of physiologic DSBs created during variable (diversity) joining [V(D)J] recombination and class switch recombination (CSR). Therefore, patients lacking normal NHEJ are not only sensitive to ionizing radiation (IR), but also severely immunodeficient.

Contents

THE BIOLOGICAL CONTEXT OF NONHOMOLOGOUS DNA END JOINING

NHEJ: nonhomologous DNA end joining

Unlike most other DNA repair and DNA recombination pathways, nonhomologous DNA end joining (NHEJ) in prokaryotes and eukaryotes evolved along themes of mechanistic flexibility, enzyme multifunctionality, and iterative processing to achieve repair of a diverse range of substrate DNA ends at double-strand breaks (DSBs) (1–3). Except for very limited protein homology for the Ku protein in prokaryotes and eukaryotes (2), the actual nuclease, DNA polymerase, and ligase components of NHEJ appear to have arisen independently but converged on these same mechanistic themes to handle the challenge of joining two freely diffusing ends of diverse DNA end configuration with a wide range of base or sugar oxidative damage (3).

Homology-Directed Repair versus Nonhomologous DNA End Joining

When DSBs arise in any organism, prokaryotic or eukaryotic, there are two major categories of DNA repair that can restore the duplex structure (**Figure 1**). If the organism is diploid (even if the diploidy is only transient, as in replicating bacteria or replicating haploid yeast), then homology-directed repair (HDR) can be used. The most common form of HDR is called homologous recombination (HR), which has the longest sequence homology requirements between the donor and acceptor DNA. Other forms of HDR include single-strand annealing and breakage-induced replication, and these require shorter sequence homology relative to HR (4, 5).

HR: homologous recombination

In nondividing haploid organisms or in diploid organisms that are not in S phase, a homology donor is not nearby. Hence, early in

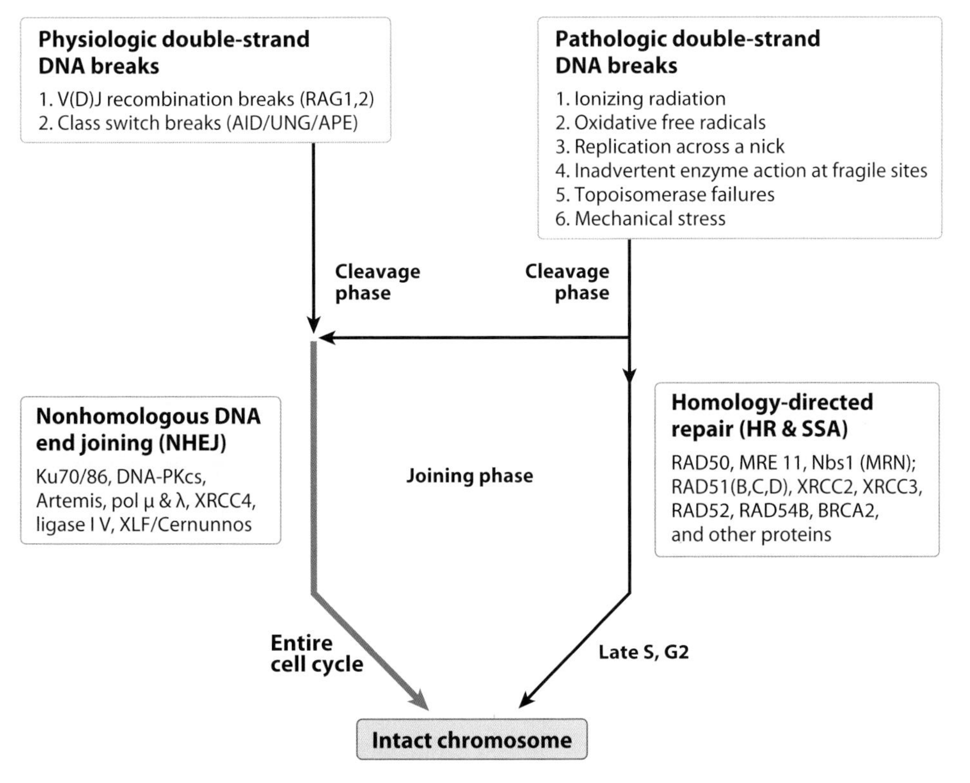

Figure 1

Causes and repair of double-strand DNA breaks. Physiologic and pathologic causes of double-strand breaks in mammalian somatic cells are listed at the top. During S and G2 phases of the cell cycle, homology-directed repair is common because the two sister chromatids are in close proximity, providing a nearby homology donor. Homology-directed repair includes homologous recombination (HR) and single-strand annealing (SSA). At any time in the cell cycle, double-strand breaks can be repaired by nonhomologous DNA end joining (NHEJ). Proteins involved in the repair pathways are listed. The NHEJ blue arrow is thicker to indicate its more frequent usage. Abbreviations for proteins listed in the figure include the following: AID, activation-induced deaminase; APE, apurinic/apyrimidinic endonuclease; pol μ, DNA polymerase μ; pol λ, DNA polymerase λ; and UNG, uracil N-glycosylase.

Table 1 Corresponding enzymes in prokaryotic and eukaryotic NHEJ (3)

Functional component	Prokaryotes	Eukaryotes	
		Saccharomyces cerevisiae	Multicellular eukaryotes
Tool belt protein	Ku (30–40 kDa)	Ku 70/80	Ku 70/80
Polymerase	Pol domain of LigD	Pol4	Pol μ and pol λ
Nuclease	Uncertain	Rad50:Mre11:Xrs2 (FEN-1)	Artemis:DNA-PKcs
Kinase/phosphatase	Phosphoesterase domain of LigD	Tpp1 and others	PNK and others
Ligase	Ligase domain of LigD	Nej1:Lif1:Dnl4	XLF:XRCC4: DNA ligase IV

evolution, another form of DSB repair had an opportunity to provide survival advantage, and nonhomologous DNA end joining (NHEJ) includes a set of DNA enzymes that have the mechanistic flexibility to provide such an advantage (**Table 1**) (6).

How the cell determines whether HR or NHEJ will be used to repair a break is still an active area of investigation. The HR versus NHEJ determination may be somewhat operational (7). If a homolog is not present near a DSB during the S/G2 phases, then HR cannot proceed, and NHEJ is the only option. During S phase, the sister chromatid is physically very close, thereby providing a homology donor for HR. Outside of the S/G2 phases, NHEJ is indeed the markedly preferred option. The precise molecular events, beyond issues of proximity and possible competition between Ku and RAD51 or -52, are yet to be deciphered (7–9). Recent data from *Saccharomyces cerevisiae* suggests that the DNA ligase IV complex may be key in suppressing the DNA end resection needed to initiate HR (10).

Causes and Frequencies of Double-Strand Breaks

There are an estimated 10 DSBs per day per cell; this estimate is based on metaphase chromosome and chromatid breaks in early passage primary human or mouse fibroblasts (11–13). Estimates of DSB frequency in nondividing cells are difficult to make because methods for assessing DSBs outside of metaphase are subject to even more caveats of interpretation.

IR: ionizing radiation

In mitotic cells of multicellular eukaryotes, DSBs are all pathologic (accidental) except the specialized subset of physiologic DSBs in early lymphocytes of the vertebrate immune system (**Figure 1**). Major pathologic causes of DSBs in wild-type cells include replication across a nick, giving rise to chromatid breaks during S phase. Such DSBs are ideally repaired by HR using the nearby sister chromatid.

All of the remaining pathologic forms of DSBs are repaired primarily by NHEJ because they usually occur when there is no nearby homology donor and/or because they occur outside of S phase. These causes include reactive oxygen species (ROS) from oxidative metabolism, ionizing radiation (IR), and inadvertent action of nuclear enzymes (14).

ROS are a second major cause of DSBs (**Figure 1**). During the course of normal oxidative respiration, mitochondria convert about ~0.1% to 1% of the oxygen to superoxide (O_2^-) (15). Superoxide dismutase in the mitochondrion (SOD2) or cytosol (SOD1) can convert this to hydroxyl free radicals, which may react with DNA to cause single-strand breaks. Two closely spaced lesions of this type on antiparallel strands can cause a DSB. About 10^{22} free radicals or ROS species are produced in the human body each hour, and this represents about 10^9 ROS per cell per hour. A subset of the longer-lived ROS may enter the nucleus via the nuclear pores.

A third cause of DSBs is natural IR of the environment. This includes gamma rays and X-rays. At sea level, ~300 million IR particles per hour pass through each person. As these traverse the body, they create free radicals along their path, primarily from water. When the particle comes close to a DNA duplex, clusters of free radicals damage DNA, generating one DSB in the genome for every 25 sites of single-strand damage (16). About half of the IR that strikes each of us comes from outside the earth. The other half arises from the decay of radioactive elements, primarily metals, within the earth.

A fourth cause of DSBs is inadvertent action by nuclear enzymes on DNA. These include failures of type II topoisomerases, which transiently break both strands of the duplex. If the topoisomerase fails to rejoin the strands, then a DSB results (17). Inadvertent action by nuclear enzymes of lymphoid cells, such as the RAG complex (composed of RAG1 and -2) and activation-induced deaminase (AID) are responsible for initiating physiologic breaks for antigen receptor gene rearrangement; however, they sometimes accidentally act at off-target sites outside the antigen receptor gene loci (18). In humans, these account for about half of all of the chromosomal translocations that result in lymphoma.

Finally, physical or mechanical stress on the DNA duplex is a relevant cause of DSBs. In prokaryotes, this arises in the context of desiccation, which is quite important in nature (19). In eukaryotes, telomere failures can result in chromosomal fusions that have two centromeres, and this results in physical stress by the mitotic spindle (breakage/fusion/bridge cycles) with DSBs (20).

In addition to the above for mitotic cells, meiotic cells have an additional source of DSBs, which is physiologic and is caused by an enzyme called Spo11, a topoisomerase II-like enzyme (21). Spo11 creates DSBs to generate crossovers between homologs during meiotic prophase I. These events are resolved by HR. Therefore, NHEJ is not relevant to Spo11 breaks. Interestingly, it is not clear that NHEJ occurs in vertebrate meiotic cells because one group reports the lack of Ku70 in spermatogonia (22). Human spermatogonia remain in meiotic prophase I for about 3 weeks, and human eggs remain in meiotic prophase I for 12 to 50 years; hence, these cells can rely on HR during these long periods. Given the error-prone nature of NHEJ (see below), reliance on HR may be one way to minimize alterations to the germ line at frequencies that might be deleterious to a population.

V(D)J: variable (diversity) joining

MECHANISM OF NONHOMOLOGOUS DNA END JOINING

The enzymes of NHEJ are able to function on a diverse range of DNA ends as substrates, as discussed in detail in the following sections.

Structural Diversity of Double-Strand Break DNA Ends

Perhaps the most intriguing aspect of NHEJ is the diversity of substrates that it can accept and convert to joined products. This demands a remarkable level of mechanistic flexibility at the level of protein-substrate interaction and is unparalleled in most other biochemical processes. Though we have substantial amounts of information on the DNA end configurations at DSBs, there are limitations to the information because of the diverse manner in which IR and ROS interact with DNA. Therefore, we know the most about the diversity of physiologic DSBs, specifically V(D)J recombination, because we know where the relevant enzymes initiate the cutting of the two DNA strands. In V(D)J recombination, we can examine many NHEJ outcomes from the same starting substrates. We can also vary the sequence of the two DNA ends being joined. All of these various overhangs are joined in vivo at about the same efficiency, regardless of sequence. Within this range of overhang variation then, NHEJ can accept a wide variety of overhang length, DNA end sequence, and DNA end chemistry.

Overview of the Proteins and Mechanism of Vertebrate Nonhomologous DNA End Joining

Like most DNA repair processes, NHEJ requires a nuclease to resect damaged DNA,

DNA polymerases to fill in new DNA, and a ligase to restore integrity to the DNA strands (**Figure 2**). Functional correspondence between the NHEJ proteins of prokaryotes, yeast (along with plants and invertebrates), and vertebrates can be inferred (**Table 1**). Prokaryotic NHEJ has been reviewed recently (23), and comparisons between prokaryotes and eukaryotes have been made (3). NHEJ in yeast, which appears to be similar for plants and

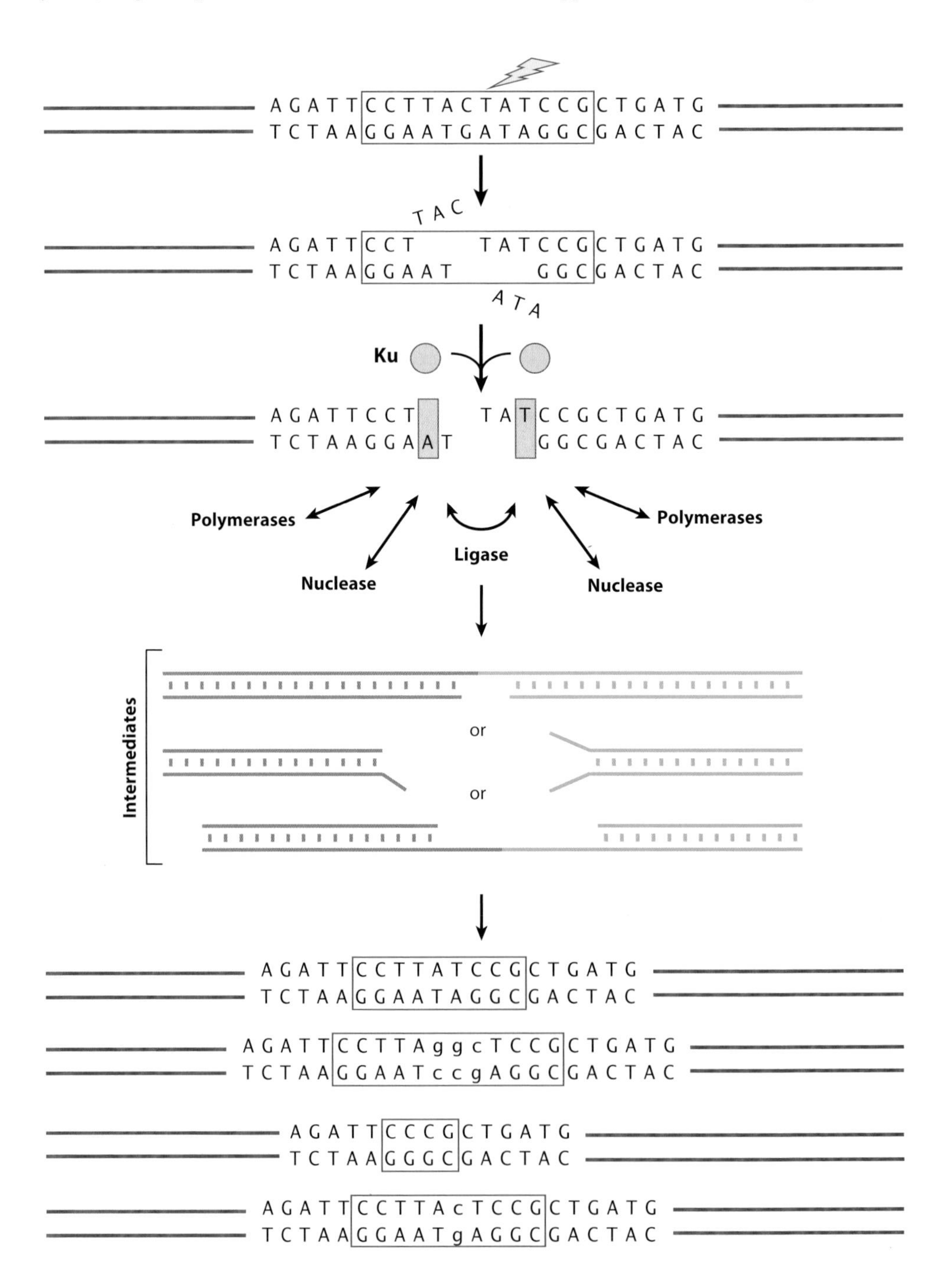

invertebrates, has been thoroughly reviewed as well (24, 25). Hence, the discussion here focuses on NHEJ in vertebrates, with appropriate comparisons to prokaryotic and yeast NHEJ.

When a DSB arises in vertebrates, it is thought that Ku is the first protein to bind on the basis of its abundance (estimated at ~400,000 molecules per cell) and its strong equilibrium dissociation constant (~10^{-9} M) for duplex DNA ends of any configuration (**Figures 3** and **4*a***) (26–29). Ku is a toroidal protein because of its crystal structure (30). Ku bound to a DNA end can be considered as a Ku:DNA complex, which serves as a node at which the nuclease, polymerases, and ligase of NHEJ can dock (31). One can think of Ku as a tool belt protein, similar to PCNA in DNA replication, where many proteins can dock. At a DSB, there are two DNA ends. Hence, it is presumed that there is a Ku:DNA complex at each of the two DNA ends being joined, thereby permitting each DNA end to be modified in preparation for joining.

Each Ku:DNA end complex can recruit the nuclease, polymerase, and ligase activities in any order (31, 32). This flexibility is the basis for the diverse array of outcomes that can arise from identical starting ends. The processing of the two DNA ends may transiently terminate when there is some small extent of annealing between the two DNA ends. The processing may permanently terminate when one or both strands of the left and right duplexes are ligated.

Ku likely changes conformation when bound to a DNA end versus when Ku is in free solution. The basis for this inference is that Ku does not form stable complexes with DNA-PKcs in the absence of DNA ends (33), and the same appears to apply for its interactions with DNA pols μ and λ and with XRCC4:DNA ligase IV (34, 35). The crystal structure for Ku lacks the C-terminal 19 kDa [167 amino acids (aa)] of Ku86 (an important region of interaction with DNA-PKcs and other proteins), and this may be a region for conformational change upon Ku binding to DNA (36).

The Artemis:DNA-PKcs complex has a diverse array of nuclease activities, including 5′ endonuclease activity, 3′ endonuclease

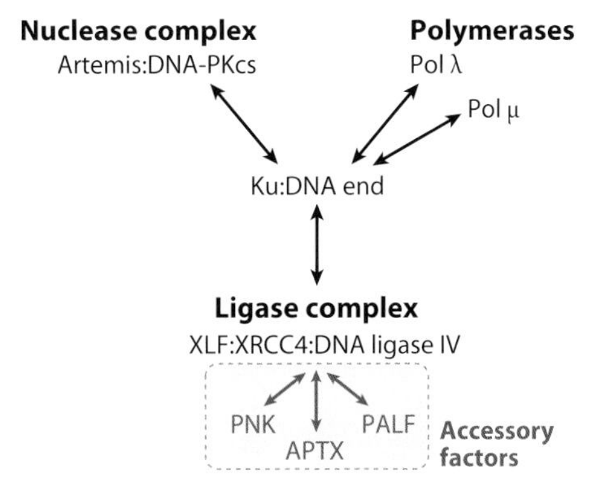

Figure 3

Interactions between nonhomologous DNA end-joining (NHEJ) proteins. Physical interactions between NHEJ components are summarized. In addition, interactions between XRCC4 and DNA-PKcs are discussed in the text, as are functional interactions. Abbreviations: APTX, aprataxin; PALF, PNK-APTX-like factor; PNK, polynucleotide kinase; XLF, XRCC4-like factor.

Figure 2

General steps of nonhomologous DNA end joining (NHEJ). The lightning arrow indicates ionizing radiation (IR), a reactive oxygen species (ROS), or an enzymatic cause of a DSB. Ku binding to the DNA ends at a double-strand breaks (DSBs) improves binding by nuclease, DNA polymerase, and ligase components. Note that Ku is thought to change conformation upon binding to the DNA end, as depicted by its shape change from a sphere to a rectangle. Flexibility in the loading of these enzymatic components, the option to load repeatedly (iteratively), and independent processing of the two DNA ends all permit mechanistic flexibility for the NHEJ process. This mechanistic flexibility is essential to permit NHEJ to handle a very diverse array of DSB end configurations and to join them. In addition to the overall mechanistic flexibility, each component exhibits enzymatic flexibility and multifunctionality, as discussed in the text. The figure shows that there are many alternative intermediates in the joining process (*middle*). These intermediates are reflected in a diverse DNA sequences at the junction of the joining process (*bottom*).

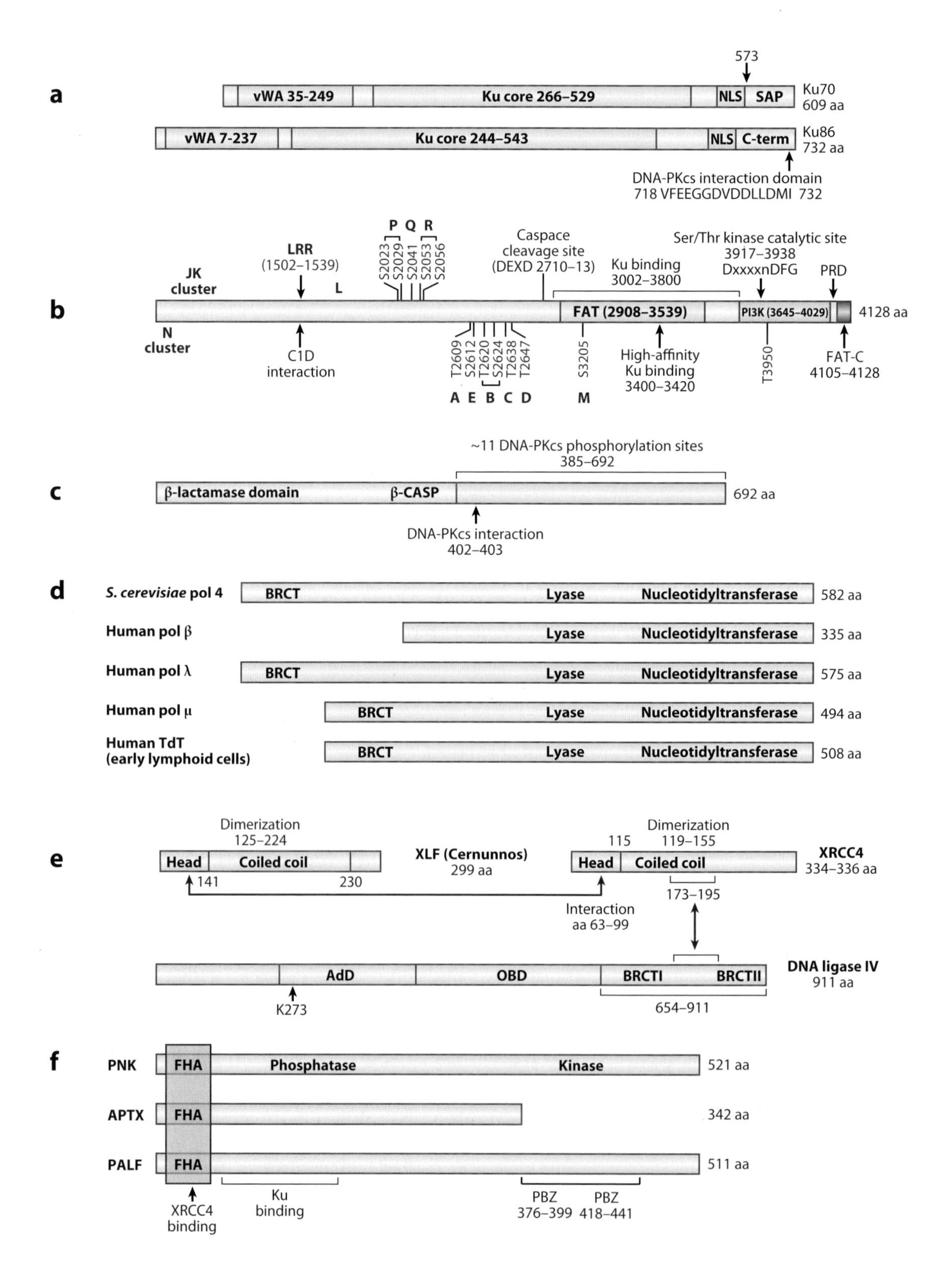

a
573
vWA 35-249
Ku core 266–529
NLS
SAP
Ku70
609 aa
vWA 7-237
Ku core 244–543
NLS
C-term
Ku86
732 aa
DNA-PKcs interaction domain
718 VFEEGGDVDDLLDMI 732
b
P Q R
S2023 S2029 S2041 S2053 S2056
LRR
(1502–1539)
JK cluster
L
Caspace cleavage site (DEXD 2710–13)
Ku binding 3002–3800
Ser/Thr kinase catalytic site 3917–3938 DxxxxnDFG
PRD
FAT (2908–3539)
PI3K (3645–4029)
4128 aa
N cluster
C1D interaction
T2609 S2612 T2620 S2624 T2638 T2647
A E B C D
S3205
M
High-affinity Ku binding 3400–3420
T3950
FAT-C 4105–4128
c
~11 DNA-PKcs phosphorylation sites
385–692
β-lactamase domain
β-CASP
692 aa
DNA-PKcs interaction
402–403
d
S. cerevisiae pol 4
BRCT
Lyase
Nucleotidyltransferase
582 aa
Human pol β
335 aa
Human pol λ
575 aa
Human pol μ
494 aa
Human TdT (early lymphoid cells)
508 aa
e
Dimerization 125–224
Head
Coiled coil
141
230
XLF (Cernunnos)
299 aa
115
Dimerization 119–155
XRCC4
334–336 aa
173–195
Interaction aa 63–99
AdD
OBD
BRCTI
BRCTII
DNA ligase IV
911 aa
K273
654–911
f
PNK
FHA
Phosphatase
Kinase
521 aa
APTX
342 aa
PALF
511 aa
XRCC4 binding
Ku binding
PBZ 376–399
PBZ 418–441

activity, and hairpin opening activity, in addition to an apparent 5′ exonuclease activity of Artemis alone (**Figure 4*b*,*c*** and **Supplemental Figure 1**. Follow the **Supplemental Material link** from the Annual Reviews home page at **http://www.annualreviews.org.**) (37). The Artemis:DNA-PKcs complex is able to endonucleolytically cut a variety of types of damaged DNA overhangs (38, 39). Hence, there is no obvious need for additional nucleases, although the 3′ exonuclease of PNK-APTX-like factor (PALF or APLF) and others are possibilities (see the Future Issues section, below).

Pols μ and λ are both able to bind to the Ku:DNA complexes by way of their BRCT domains located in the N-terminal portion of each polymerase (**Figure 4*d***) (32). Additional polymerases appear able to contribute when neither of these two polymerases is present (40, 41). As discussed below, pol μ is particularly well suited for functioning in NHEJ because it is capable of template-independent synthesis, in addition to template-dependent synthesis. Pol λ also has more flexibility than replicative polymerases.

A complex of XLF:XRCC4:DNA ligase IV is the most flexible ligase known, with the ability to ligate across gaps and ligate incompatible DNA ends (**Figure 4*e***) (42, 43). It can also ligate one strand when the other has a complex configuration (e.g., bearing flaps), and it can ligate single-stranded DNA (ssDNA), although with limited and substantial sequence preferences.

Therefore, the nuclease, polymerases, and ligase of NHEJ all have much greater mechanistic flexibility than their counterparts in other repair pathways. This flexibility permits these structure-specific proteins to act on a wider range of starting DNA end structures. One consequence of such flexibility in vertebrates may be the substantial diversity of junctional outcomes observed, even from identical starting ends, as discussed in the next section.

Variation in Products Even from Identical Starting Substrates

If we arbitrarily designate the two DNA ends as left and right, then the Ku bound at the left end could conceivably recruit the nuclease, and the Ku at the right DNA end might recruit the polymerase, or vice versa. It is likely that there are multiple rounds of action by the nuclease, polymerases, and ligase at the left and right ends of the DSB until the top or bottom strand is ligated. Therefore, the joining of the two ends

Figure 4

Diagrams of domains within nonhomologous DNA end-joining (NHEJ) proteins. (*a*) Ku is a heterodimer of Ku70 and 86. vWA designates von Willebrand domains. SAP designates a SAF-A/B, Acinus, and PIAS domain and may be involved in DNA binding. (*b*) DNA-PKcs autophosphorylation sites are shown in red (90, 95, 96). The function of each phosphorylation site (A-E, L, M, P-R) and the clusters (N and JK) are still under study. Adjacent phosphorylation sites that are linked by a bracket have not been functionally dissected from one another. LRR designates the leucine-rich region. The FAT-C domain is a FAT domain at the C terminus. PI3K designates the PI3 kinase domain. PRD designates the PI3K regulatory domain. (*c*) Artemis is phosphorylated by DNA-PKcs at 11 sites within the C-terminal portion (*green*) (161, 162). Amino acids 156 to 385 share conserved sequence with those metallo-β-lactamases that act on nucleic acids (163). This region has been called the β-CASP domain (metallo-β-lactamase-associated CPSF Artemis SNM1 PSO2) (164). (*d*) Polymerase X (Pol X) family. Pol μ and pol λ are generally involved in NHEJ in mammalian somatic cells. Terminal deoxynucleotidyl transferase (TdT) is only expressed in early lymphoid cells, where it participates in NHEJ primarily in the context of V(D)J recombination. (*e*) The NHEJ ligase complex consists of XLF (Cernunnos), XRCC4, and DNA ligase IV. The red arrows indicate the regions of physical interaction (118, 119). OBD in ligase IV is the oligo-binding domain, and AdB is the adenylation domain. (*f*) Polynucleotide kinase (PNK), aprataxin (APTX), and PNK-APTX-like factor (PALF, which is also called APLF) are ancillary components that bind to XRCC4 of the ligase complex. The PBZ domain appears to be important for poly-ADP ribose polymerase-1 (PARP-1) binding, for poly-ADP ribose binding, and for nuclease activity. FHA designates the forkhead-associated domain. Abbreviations: aa, amino acid; BRCTI and BRCTII, BRCA-1 C-terminal domain I and II; C-term, C terminus; DNA-PKcs, DNA-dependent protein kinase, N, N terminus, NLS, nuclear localization sequence.

is likely to be an iterative process with multiple possible routes all leading to a joining event, but with wide variation in the precise junctional sequence of the products.

Unlike pathologic causes of DSBs, which generally cannot generate predictable pairs of starting DNA ends, V(D)J recombination always generates two hairpinned coding ends (see the section on V(D)J recombination, below, for a full discussion, including signal ends). Though these two coding ends can be opened in a few different ways, the predominant hairpin opening position is 2 nucleotides (nt) 3′ of the hairpin tip in vivo (44) and in vitro (37). Hence, the two coding ends are often 3′ overhangs with a length of 4 nt each. Assuming that NHEJ within vertebrate B cells is representative of NHEJ in other tissue cell types (these junctions can be analyzed in cells that do not express terminal transferase), several inferences can be drawn from these relatively defined starting DNA ends.

First, the amount of nucleolytic resection (loss) from each DNA end varies, usually over a range of 0 to 14 base pairs (bp); but there are less frequent examples with resection up to ~25 bp (45). The rare instances where there is a loss greater than 25 bp may represent cases where the DNA end is released prematurely from whatever factors retain the two DNA ends in some proximity (see below). In vertebrate NHEJ, a complex of Artemis:DNA-PKcs is capable of endonucleolytically resecting a wide range of DNA end configurations. In yeast, plants, and invertebrates, the MRE11/RAD50/XRS2 (MRX) complex appears to be critical for some of the DNA end resection (24). The evolutionary inception of Artemis and DNA-PKcs coincides with the inception of V(D)J recombination (the vertebrate to invertebrate transition). The MRX nuclease system and the DNA-PKcs system both rely on the same conserved C-terminal tail for protein-protein interaction, also suggesting that the Artemis:DNA-PKcs complex may have evolved to replace the MRX complex for vertebrate NHEJ (46).

Second, nucleotide addition can occur at the DNA junction, even when terminal transferase is not present. In mammals, pol μ can add in a template-independent manner under physiologic conditions (7, 8). Mammalian pol λ does not appear to add in a template-independent manner except when Mg^{2+} is replaced with Mn^{2+} (47, 48). The precise biochemical properties of Pol X family members in other eukaryotes are not as clear. Interestingly, in bacteria, the polymerase activity intrinsic to the LigD protein is capable of adding 1 nt or ribonucleotide in an entirely template-independent manner (23), perhaps reflecting convergent evolution.

Pol μ and pol λ both seem to have much greater flexibility than most polymerases during template-dependent synthesis (47, 48). The template-independent addition by pol μ would sometimes be expected to fold back on itself (42), and the resulting stem-loop structure might function as a primer/template substrate, see step 1 in **Supplemental Figure 2**. Follow the Supplemental Material link from the Annual Reviews home page at **http://www.annualreviews.org** (48). This may account for the observed inverted repeats at many NHEJ junctions from chromosomal translocations in humans (49, 50). Both pol μ and pol λ can slip back on their template strand (51–53), and this may permit generation of direct repeats, accounting for such events seen in vivo (54–56). Direct repeats are also often seen at NHEJ junctions from human chromosomal translocations (49, 50). The direct and inverted repeats seen at these NHEJ junctions have been termed T nucleotides, where the T stands for templated (49, 50).

Therefore, even from a relatively homogeneous set of starting DNA ends as substrates, there is substantial variation in the nucleotide resection from each end and variation in the amount of template-independent addition to the two DNA ends. These two sources of variation are the basis for the heterogeneity at the joining site.

Mechanistic Flexibility, Iterative Processing, and Independent Enzymatic Functions as Conserved Themes

In the context of considering diverse substrates and diverse joining products, it is worth noting an additional facet of NHEJ flexibility: The proteins involved and the order of their action in NHEJ can vary at either of the two DNA ends. Each DNA end, especially when bound by Ku, is best considered as a node at which any of the NHEJ proteins can dock. If one of the polymerases arrives first at the left end, then this might enact the first step at that end. However, if the nuclease binds first at that end, then resection will occur first (32).

In addition to the theme of mechanistic flexibility, there is the theme of iterative processing of the junction (32, 38). For any given joining event, there might, for example, be only three steps involved: one each involving a nuclease, a polymerase, and then a ligase. However, in another joining scheme, there might be 10 steps with multiple appearances by each enzyme activity. Hence, each of the enzymatic components might be involved not at all, once, or many times.

Related but in addition to the theme of iterative processing is the independent function of the nuclease, polymerases, and ligase from one another and even from Ku. Each of the enzymatic activities has a substantial range and level of activity without any of the others, and even without Ku, when examined in purified biochemical systems. For example, Ku is entirely unnecessary in ligation of DNA ends by the ligase IV complex when those ends share 4 bp of terminal microhomology, but Ku is stimulatory for shorter microhomology lengths (42). Pols μ and λ are able to carry out fill-in synthesis, and pol μ does not always require Ku or XRCC4:DNA ligase IV to have terminal deoxynucleotidyl transferase (TdT)-like activity at a DNA end (42). The Artemis:DNA-PKcs complex does not require Ku or any other component to carry out its endonucleolytic functions (37). Therefore, independent function of each enzymatic activity and iterative processing and mechanistic flexibility are all noteworthy features of vertebrate NHEJ. *S. cerevisiae* NHEJ manifests mechanistic flexibility but within a narrower range of junctional outcomes than mammalian NHEJ (24, 57).

Enzymatic Revision of a Partially Completed Junction

In the context of iterative processing, the Artemis:DNA-PKcs nuclease complex is able to nick within the single-stranded portion of a gapped structure and within a bubble structure (38). Joinings where only one strand is ligated would often have a gapped configuration. Nicking of such a gap could permit nucleotides that were originally part of the arbitrarily designated left DNA end to become separated from that left end and become associated with the right DNA end. Then further nucleotide addition at the left end could separate these nucleotides from the left end. One can find potential examples of this at in vivo junctions. In other scenarios, with the flexibility of the ligase, there may be more nucleotides on the top strand than the bottom strand (42). The activated Artemis:DNA-PKcs complex can nick mismatched or bubble structures on either strand, thereby permitting additional rounds of junctional revision (38).

Terminal Microhomology Between the Initial Two DNA Ends Can Simplify the Protein Requirements

On the basis of both *S. cerevisiae* and mammalian in vivo NHEJ studies, the variation of the resulting junction is usually less when there is terminal microhomology at the ends (24, 57–60). This may reflect the involvement of fewer NHEJ proteins, and genetic studies support the view that not all of the NHEJ components are essential when the two DNA ends share terminal microhomology (60–66).

As mentioned above, when the two DNA ends happen to share 4-nt overhangs that are perfectly complementary, then the only component needed in purified biochemical systems is the XRCC4:DNA ligase IV (42). No nucleolytic resection or polymerase action is needed. Using purified proteins in vitro, XRCC4:ligase IV is adequate to join such a junction, and even ligase IV alone may be sufficient, all without Ku. Moreover, ligase I or III alone is sufficient for such joinings, though at a lower efficiency (42). In vivo generation of defined-DNA end configurations at DSBs is not simple, but there are two approaches that have been used. In *S. cerevisiae*, short oligonucleotide duplexes can be ligated onto the DNA ends of a linear plasmid, and then these can be transfected into cells (57). One can then harvest the joined circular molecules for analysis. On the basis of this result, it is clear that the joining dependence is simplified when there is terminal microhomology at the DNA ends. A second system for generating defined DNA ends is V(D)J recombination, as mentioned above. These ends are not as precisely defined as in the yeast system (because the precise DNA end configuration depends on how hairpin intermediates are opened), but there is the advantage that the ends are actually generated inside the nucleus. In V(D)J recombination, the coding ends (defined in **Supplemental Figure 3**. Follow the Supplemental Material link from the Annual Reviews home page at **http://www.annualreviews.org**) are usually configured with a 4-nt 3′ overhang. If the DNA ends are chosen to be complementary, then the dependence of the V(D)J recombination on Ku can be very minimal (60). However, NHEJ repairs such ends so as to align the microhomology in a disportionate fraction of the joins (58, 67).

Terminal Microhomology Can Bias the Diversity of Joining Outcomes, but Microhomology Is Not Essential

One of the strengths of NHEJ is that microhomology does not appear to be essential in mammalian cells (58, 67). The joining of incompatible DNA ends may be a key selective advantage that drove further evolutionary development of NHEJ in higher eukaryotes. The fact that some of this evolution was convergent rather than divergent further illustrates the strength of this selective advantage. Most natural DSBs generate incompatible ends with little or no microhomology within the first few nucleotides. [*S. cerevisiae* NHEJ shows mechanistically interesting differences from mammalian NHEJ insofar as yeast are very poor at blunt-end ligation and perhaps more reliant on at least one base pair of terminal microhomology (24, 25, 68).]

For mammalian NHEJ joins, the most common amount of observed terminal microhomology is 0 nt (45, 69). The next most common is 1 nt, and longer microhomologies are less common with increasing length. As mentioned above, when microhomology is present, then usage of that microhomology for a given pair of DNA ends can be dominant (57–60).

Overhangs with substantial terminal microhomology are uncommon in nature and are limited primarily to regions containing repetitive DNA. Like wild-type cells, in neoplastic cells arising in normal animals, the most common amount of terminal microhomology at NHEJ junctions is also zero. Increases in microhomology usage occur in two circumstances. First, if the experimental system being used specifically positions terminal microhomology at or near the DNA ends, then the high-microhomology outcome might be observed. Second, in animals lacking a complete wild-type NHEJ system, the NHEJ process may be slower, may show more resection, and may seek end alignments that are stabilized by more terminal microhomology (2 or 3 bp), as discussed in the next section.

Alternative Nonhomologous DNA End Joining

In experimental systems, when one or more proteins of NHEJ are mutated, the joining that occurs is said to be due to alternative NHEJ.

Ligase IV-independent joining. It has been elegantly demonstrated in murine and yeast genetic studies that end joining can occur in the absence of ligase IV (57, 69–71). Insofar as the only remaining ligase activity in the cells is due to ligase I in *S. cerevisiae* or ligase I or III in vertebrate cells, these joinings must be done by ligase I or III. Most of these joinings rely more heavily on the use of terminal microhomology than NHEJ in wild-type cells. In wild-type cells, plots of end-joining frequency versus microhomology length show a peak at 0 nt of microhomology and decline for increasing lengths (69). But for joinings without ligase IV, the peak is at 2.5 bp, and the frequency declines on both sides of this peak (69–71).

In biochemical systems using purified NHEJ proteins, it has been shown that human ligase I and III are able to join DNA ends that are not fully compatible (e.g., joining across gaps in the ligated strand), although this is still substantially less efficient than joining by the XRCC4:ligase IV complex (42, 43). Though relatively inefficient, this joining by ligase I or III is somewhat more efficient with two or more base pairs of terminal microhomology to stabilize the ends. Therefore, in the absence of the ligase IV complex, it may not be surprising that the peak microhomology usage changes from zero to between 2 and 3 bp.

The in vivo joining efficiency by mammalian ligase IV relative to ligase I and III is difficult to measure. Two measurements have been done in murine cells in which the joining occurred at DNA ends that have some increased opportunities for terminal microhomology, the class switch recombination (CSR) sequences (see below for explanation of CSR). In one study, cells lacking ligase IV are removed from mice and stimulated in culture to undergo CSR (70). Measurements of switch recombination can be done as early as 60 h after stimulation. In this case, end joining without ligase IV is reduced only 2.5-fold. In another case, a murine cell line was used to make the genetic knockout, and measurements could be done as early as 24 h, at which time the joining without ligase IV was reduced about ninefold (69). In both cases, the joining is almost certainly done by ligase I or III. The latter study suggests that the joining by ligase I or III is substantially less efficient at early times. In both studies, given sufficient time, the joining by ligase I or III improves to about half that of the wild-type cells (where nearly all joining is likely, though not proven, to be the result of ligase IV). Also, in both studies, the joining in wild-type cells was much less dependent on terminal microhomology than joining in the ligase IV knockout cells. One reasonable interpretation of these two studies is that ligase IV is more efficient (and perhaps faster) at joining incompatible DNA ends in vivo but that ligase I or III can join ends at a lower efficiency, especially when terminal microhomology can stabilize the DNA ends.

CSR: class switch recombination

In *S. cerevisiae*, end joining can occur in the absence of ligase IV, but it is at least 10-fold less efficient (24). Moreover, when the joining does occur, it tends to use microhomology (usually >4 bp) that is longer and more internal to the two DNA termini than is seen for wild-type yeast (57).

Ku-independent joining. Ku-independent end joining also occurs in both *S. cerevisiae* and mammalian cells. In yeast, such events can even be as efficient as end joining in the corresponding wild-type cells (24, 57). For ligase IV mutants in yeast, the joining relies on longer microhomology (usually >4 bp) that is more internal to the two DNA ends. Ku-independent end joining also can be seen in mammalian cells (60). Even in vertebrate V(D)J recombination, when the two DNA ends share 4 bp of terminal microhomology, the dependence on Ku for joining efficiency can be small (2.5-fold) (60). This indicates that terminal microhomology can substitute for the presence of Ku. We do not know with certainty what Ku-independent joining means mechanistically, but one possibility is that the ligase IV complex normally holds the DNA ends and that Ku stabilizes the ligase IV complex. But when Ku is absent, terminal microhomology may provide some of this stability, consistent with observations in biochemical systems (42).

Zero microhomology joins in the absence of ligase IV. For some ends joined in the absence of ligase IV, no microhomology is obvious (66). This raises the question whether ligase I or III can join ends with no terminal microhomology. XRCC4:ligase IV can ligate blunt ends (72), and ligase III is also able to do this at low efficiencies (35). All three ligases can do so when macromolecular volume excluders are present (35). Nevertheless, blunt-end ligation is much less efficient for all three ligases.

Another explanation of such events is that they involve template-independent synthesis by the Pol X family members (42). As discussed, in mammalian cells, pol μ and pol λ participate in NHEJ (42, 73, 74), as does Pol4 in *S. cerevisiae* (75). In Mn^{2+} buffers, both pol μ and pol λ can add nucleotides independently of a template, and in the more physiologic Mg^{2+} buffers, pol μ still shows robust template-independent addition. Such template-independent activity could permit additions to DNA ends that provide microhomology with another end; because that addition would be random, it would not have been scored as microhomology. One could consider such inapparent microhomology as polymerase-generated microhomology (also called invisible or occult microhomology, see below). This type of microhomology would not have been present in the two original DNA end sequences, and in that sense, it can be regarded as polymerase-generated microhomology.

In addition to template-independent synthesis by pol μ (which pol μ exhibits alone or in the context of other NHEJ proteins), pol μ together with Ku and XRCC4:ligase IV can synthesize across a discontinuous template, and this would also generate microhomology. However, this mechanism requires that the DNA end providing the template have a 3′ overhang to permit the polymerase to extend into that end. Hence, only a subset of DNA ends could be handled in this manner.

The ratio of template-independent versus template-dependent synthesis by pol μ at NHEJ junctions in such cases is not entirely clear, but both mechanisms occur under physiologic conditions in biochemical systems, and there is clearly some evidence for the template-independent pol μ addition within mammalian cells (41).

Nomenclature. In all organisms in which there is NHEJ, there are examples of DNA end joining in the absence of the major NHEJ ligase of that organism (76); even in mycobacteria, LigC can function in NHEJ when LigD is absent (77). Given that the ligase is often regarded as the signature enzymatic requirement of NHEJ, these joining events have been proposed to be the result of alternative NHEJ, or backup NHEJ. As mentioned above for most of these end joinings, there is substantial terminal microhomology. Hence, in *S. cerevisiae*, this joining has also been called microhomology-mediated end joining, but it is essentially an alternative NHEJ.

For eukaryotic end joining, one reasonable nomenclature is to use NHEJ as the general term and to simply note the exceptions, as for example, ligase IV-independent NHEJ, Ku-independent NHEJ, or DNA-PKcs-independent NHEJ (i.e., X-independent NHEJ, where X is the omitted protein). Until a specific pathway is delineated, this is a practical solution. It is quite conceivable (even likely) that ligase IV-independent NHEJ is merely NHEJ in which ligase I or III completes the ligation at a somewhat lower efficiency than ligase IV.

Evolutionary Comparisons of Nonhomologous DNA End Joining

Initially, NHEJ was thought to be restricted to eukaryotes because the best-studied prokaryote, *Escherichia coli*, cannot recircularize linear plasmids. However, when bioinformatists discovered a distantly diverged Ku-like gene in prokaryotic genomes, the existence of a similar NHEJ pathway in bacteria became clear (2, 78, 79). The bacterial Ku homolog appears to form a homodimer with a structure similar to the ring-shaped eukaryotic Ku heterodimer (80). The gene for an ATP-dependent ligase named LigD was typically found to be adjacent to the gene for Ku on the bacterial chromosome (2, 78). This linkage between Ku and an ATP-dependent ligase prompted extensive studies, which later defined a bacterial NHEJ pathway. In most bacterial species, unlike the eukaryotic

NHEJ ligase IV, LigD is a multidomain protein that contains three components within a single polypeptide: a polymerase domain, a phosphoesterase domain, and a ligase domain (23).

Why do not all bacteria have an NHEJ pathway? Bacterial NHEJ is nonessential under conditions of rapid proliferation because HR is active and a duplicate genome is present to provide homology donors (23, 81). Those bacteria that have the NHEJ pathway spend much of their life cycle in stationary phase, at which point HR is not available for DSB repair because of a lack of homology donors. In addition, desiccation and dry heat are two naturally occurring physical processes in nature that produce substantial numbers of DSBs in bacteria. Therefore, bacterial Ku and LigD are present in bacterial species that often form endospores because NHEJ is important for repair of DSB arising during long periods of sporulation.

INDIVIDUAL PROTEINS OF VERTEBRATE NONHOMOLOGOUS DNA END JOINING

Each of the individual NHEJ proteins carries an interesting detailed functional and structural literature, and more detailed reviews of each individual component are cited.

Ku

Ku was named on the basis of protein gel mobilities (actually 70 kDa and 83 kDa) of an autoantigenic protein from a scleroderma patient with the initials K.U. Ku86 is also called Ku80. The toroidal shape of Ku is consistent with studies showing that purified Ku can bind at DNA ends, and yet Ku can also slide internally at higher Ku concentrations (82). Ku can only load and unload at DNA ends. When linear molecules bearing Ku are circularized, the Ku proteins are trapped on the circular DNA. A minimal footprint size for Ku is ~14 bp at a DNA end (83). The key aspects of Ku in NHEJ have been discussed above, and the reader is referred to a detailed review about Ku for additional information (see Reference 84).

DNA-PKcs

DNA-PKcs has a molecular weight of 469 kDa and has 4128 aa. It is the largest protein kinase in biology, and the only one that is specifically activated by binding to duplex DNA ends of a wide variety of end configurations (33, 85–87). DNA-PKcs alone has an equilibrium dissociation constant of 3×10^{-9} M for blunt DNA ends, and this tightens to 3×10^{-11} M when Ku is also present at the end (88). Once bound, DNA-PKcs acquires serine and threonine kinase activity (89). But its initial phosphorylation target seems to be itself, with more than 15 autophosphorylation sites and probably with an equal number not yet defined (90). In addition to the relationship with Artemis discussed above, DNA-PKcs interacts with XRCC4 and phosphorylates a very long list of proteins in vitro (26, 91). In vivo evidence for functional effects of the additional protein phosphorylation targets is limited.

The best current structural information concerning DNA-PKcs alone is at 7-Å resolution by cryoelectron microscopy (cryo-EM) (92). At this resolution, α-helices are resolved, but this cryo-EM structure only contains a fraction of the total number of α-helical densities expected and therefore could not definitively reveal which portions of the structure are related to the primary amino acid sequence of DNA-PKcs. The "crown" in that structure is thought to contain the FAT domain and possibly parts of the kinase domain (**Figure 4*b***) (92). The base in that structure is the same as "proximal claws 1 and 2" in a cryo-EM structure of Ku:DNA-PKcs:DNA by another group and was shown to contain HEAT-like repeats at 7-Å resolution (93). In the Ku:DNA-PKcs:DNA and DNA-PKcs:DNA structures, the path of the duplex DNA is not entirely certain, and it is not clear which side of Ku is bound to DNA-PKcs (93, 94). The position of the C-terminal portion of Ku when bound to DNA-PKcs is also not determined, which is important because this interaction activates DNA-PKcs and is defined at the primary sequence level (46). Continued work using cryo-EM and other structural methods will undoubtedly be of great value.

It is not clear whether DNA-PKcs remains bound to the DNA ends throughout all processing steps of NHEJ (31, 95). Phosphorylation at the ABCDE cluster appears to increase the ability of other proteins, such as ligases, to gain access to the DNA ends, suggesting that DNA-PKcs may dissociate more readily after autophosphorylation at these sites (90, 95, 96).

DNA-PKcs interaction with other proteins is also important. As mentioned above, DNA-PKcs is critical for the endonucleolytic activities of Artemis (31, 37, 39, 97, 98). Activated DNA-PKcs stimulates the ligase activity of XRCC4:DNA ligase IV (90, 95, 96). Interestingly, the presence of XRCC4:DNA ligase IV stimulates the autophosphorylation activity of DNA-PKcs (96). Therefore, DNA-PKcs may be critical for the nucleolytic step, but also stimulatory for the ligation step.

Artemis

The Artemis:DNA-PKcs complex has 5′ endonuclease activity with a preference to nick a 5′ overhang so as to leave a blunt duplex end (37). The Artemis:DNA-PKcs complex also has 3′ endonuclease activity with a preference to nick a 3′ overhang so as to leave a 4-nt 3′ overhang. In addition, the Artemis:DNA-PKcs complex has the ability to nick perfect DNA hairpins at a position that is 2 nt past the tip. These three seemingly diverse endonucleolytic activities at single- to double-strand DNA transitions are similar to one another if one infers the following model for binding of the Artemis:DNA-PKcs complex to DNA (**Supplemental Figure 1**) (37). The complex appears to localize to a 4-nt stretch of ssDNA adjacent to a single-/double-strand transition and then nick on the 3′ side of that ssDNA 4-nt region. This would explain why 5′ overhangs are preferably removed to generate a blunt DNA end, but 3′ overhangs are nicked so as to preferably leave a 4-nt 3′ overhang. Moreover, it explains why a hairpin is nicked not at the tip but 2 nt 3′ of the tip (37). In perfect DNA hairpins, the last 2 bp do not form well, which means the tip is actually similar in many ways to a 4-nt single-stranded loop. Artemis nicks the hairpin on the 3′ side of that loop. The opened hairpin then becomes a 3′ overhang of 4 nt.

In V(D)J recombination, null mutants of DNA-PKcs and of Artemis are very similar (63, 99–101). Both fail to open the DNA hairpins, but the signal ends are joined. Biochemically, when a purified complex of Artemis:DNA-PKcs binds to an individual DNA hairpin molecule, that hairpin can activate the kinase activity of that DNA-PKcs protein (in *cis*) to phosphorylate itself and the bound Artemis within the C-terminal portion (96, 102). With respect to hairpin opening and its other endonucleolytic activities, Artemis:DNA-PKcs functions as if it were a heterodimer in which mutation of either subunit results in the failure of DNA end processing. A recent DNA-PKcs point mutation in a patient supports that view (103), as does a murine knockin model that recreates a truncation mutant of Artemis that removes the C-terminal portion where the sites of DNA-PKcs phosphorylation are located.

Some DNA ends containing oxidative damage to the bases or sugars require nuclease action to remove the damaged nucleotides. In these cases, involvement of Artemis:DNA-PKcs appears critical (39, 98).

Polymerase X Family

Three polymerases of the Pol X family can participate in NHEJ.

Polymerase μ. Pol μ has several remarkable activities as a polymerase under physiologic buffer conditions. First, it can carry out template-dependent synthesis with dNTP and rNTP, and it has substantial template-independent synthesis capability, like TdT (104). No other higher eukaryotic polymerase has this range of activities. The ability to add rNTP may be important for NHEJ during G1, when dNTP levels are low, but rNTP levels are high (74). Incorporation of U into the junction might then mark the junction for possible revision using uracil glycosylases at a later point in time. [The highly homologous Pol4 of *S. cerevisiae* also efficiently incorporates rNTPs (105). Interestingly, the bacterial polymerase for NHEJ (part of LigD) has the ability to incorporate ribonucleotides as well (23).] Second, like many error-prone polymerases, pol μ can slip on the template strand (48, 51, 52). Third, and as mentioned above, pol μ, when together with Ku and XRCC4:DNA ligase IV, can polymerize across a discontinuous template strand, essentially crossing from one DNA end to another (106, 107). Fourth, and also mentioned previously, pol μ has template-independent activity, which pol μ exhibits whether alone or together with Ku and XRCC4:DNA ligase IV (42).

Both the template-independent and the discontinous template polymerase activities are likely to be of great importance in the joining of two incompatible DNA ends. For example, in the case of two blunt DNA ends, the TdT-like activity of pol μ allows pol μ to add random nucleotides to each end. As soon as the resulting short 3′ overhangs share even 1 nt of complementarity (polymerase-generated microhomology), then ligation is much more efficient (42). In contrast, in the mechanism where pol μ (with Ku and XRCC4:ligase IV present) crosses from one DNA end to the other (template-dependent synthesis across a discontinuous template strand), the duplex end onto which the new synthesis extends must be a 3′ overhang to permit such extension by the polymerase (107). Hence, there are two mechanisms by which pol μ can create microhomology during the joining process (and these reaction intermediates would not be scored as microhomology events merely on basis of the final DNA sequence of the junctional product).

Structural studies of pol μ, TdT, and pol λ are defining the basis for the intriguing differences between these three highly related DNA polymerases (104). A region called loop 1 (and other positions, such as H329) is important for substituting for the template strand as TdT (always) and pol μ (sometimes) polymerize in their template-indepdendent mode (108, 109). Importantly, the crystal structures are on single-strand break DNA, and therefore, we do not know how these enzymes configure on DSBs.

Polymerase λ. Mouse in vivo systems, crude extract NHEJ studies, and purified NHEJ systems support a role for pol λ in NHEJ (41, 73, 110). Pol λ functions primarily in a standard template-dependent manner in Mg^{2+} buffers, but it has template-independent activity in Mn^{2+} (48, 104). The lyase domain in pol λ is functional, whereas the ones in pol μ and TdT do not appear to be functional. This permits pol λ to function after action by a glycosylase to remove a damaged base.

Terminal deoxynucleotidyl transferase. Terminal deoxynucleotidyl transferase (TdT or terminal transferase) is only expressed in pro-B/pre-B and pro-T/pre-T stages of lymphoid differentiation (111). Like the other two Pol Xs of NHEJ, TdT has an N-terminal BRCT domain. (Pol β is the only Pol X that is not involved in NHEJ, and it lacks any BRCT domain.) TdT only adds in a template-independent manner, consistent with a different loop 1 from pols μ and λ (104). TdT prefers to stack the incoming dNTP onto the base at the 3′ OH, accounting for its tendency to add runs of purines or runs of pyrimidines (45). TdT also has a lower K_m for dGTP, and this also biases its template-independent synthesis in vitro and in vivo (111, 112).

XLF, XRCC4, and DNA Ligase IV

DNA ligase IV (also called ligase IV or DNL4) is mechanistically flexible. In the absence of XRCC4, DNA ligase IV appears to still be capable of ligating not only nicks, but even compatible (4-nt overhang) ends of duplex DNA (72). With XRCC4, ligase IV is able to ligate ends that share 2 bp of microhomology and have 1-nt gaps, but addition of Ku improves this 10-fold (42). When Ku is present, XRCC4:DNA ligase IV is able to ligate even incompatible DNA ends at low efficiency (42). When XLF is also added, then XLF:XRCC4:DNA ligase IV, in the presence of Ku, can ligate incompatible DNA ends much more efficiently (43, 113).

Even 1 nt of terminal microhomology markedly increases the efficiency of ligation by Ku plus XRCC4:DNA ligase IV (43, 113), but some junctions formed within cells have no apparent microhomology (45). These could be cases where Ku plus XLF:XRCC4:DNA ligase IV ligate incompatible DNA ends or blunt ends. As mentioned, pol μ may add nucleotides either without a template or across a discontinuous template strand from the left to the right DNA end, and either of these mechanisms would not be scored as use of microhomology upon inspection of the sequence of the joined product junction (polymerase-generated microhomology). DNA ligase IV is predominantly preadenylated as it is purified from crude extracts. The reader is referred to Reference 114 for more details.

XRCC4 and XLF (Cernunnos). XRCC4 can tetramerize by itself, but it is unclear what function this serves (115). The crystal structure demonstrates a globular head domain and a coiled-coil C terminus when it forms a dimer (116, 117).

The crystal structure of XLF (Cernunnos) suggests a structure similar to XRCC4, with a globular head domain and a coiled-coil C terminus, where multimerization occurs (118, 119). When XLF is missing in humans, patients are IR sensitive and lack V(D)J recombination (120, 121). In mice, the IR defect is the same as in humans, but the V(D)J recombination defect is less severe in pre-B cells and yet is severe in mouse embryonic fibroblasts (when given exogenous RAGs) from the same mice (122). Considering the biochemical role of XLF in the joining of incompatible DNA ends, it has been suggested that TdT in the pre-B cells can provide "occult" or polymerase-generated microhomology, making joining less reliant on XLF, and this is a reasonable explanation of the data thus far (122).

Complexes of XLF, XRCC4, and DNA ligase IV and interactions with other NHEJ components. The interactions between XLF, XRCC4, and DNA ligase IV have been studied genetically and biochemically (120, 121, 123, 124). Gel filtration studies of XLF, XRCC4, and DNA ligase IV are most consistent with a stoichiometry of 2 XLF, 2 XRCC4, and 1 ligase IV (120). Complexes of XRCC4 and ligase IV are most consistent with a stoichiometry of 2 XRCC4 and 1 DNA ligase IV (115, 117). Further functional and structural work on the ligase complex will be of great value.

Both for *S. cerevisiae* and in mammalian purified proteins, Ku is able to improve the binding of XRCC4:DNA ligase IV at DNA ends. This interaction requires both Ku70 and 86 and the first BRCT domain within the C-terminal portion of ligase IV (aa 644 to 748) (91). The presence of DNA-PKcs enhances this complex formation, perhaps through interactions with XRCC4 (125–127). XRCC4:DNA ligase IV is able to stimulate DNA-PKcs kinase activity (96). The ligase complex also stimulates the pol μ and λ activities in the context of Ku (96). All of these findings suggest that the NHEJ components, although capable of acting independently, also evolved to function in a manner that is synergistic when in close proximity.

Polynucleotide Kinase, Aprataxin, and PNK-APTX-Like Factor

Polynucleotide kinase (PNK), aprataxin (APTX) and PALF (also called APLF) all interact with XRCC4 (**Figures 2** and **4*f***). PNK and XRCC4 form a complex via the PNK forkhead-associated (FHA) domain, but only after the CK2 kinase phosphorylates XRCC4 (127a). This same interaction occurs between PALF and XRCC4 as well as between APTX and XRCC4.

Polynucleotide kinase. For pathologic breaks caused by IR or free radicals, PNK plays an important role in several ways that illustrate a corollary theme of NHEJ: one of enzymatic multifunctionality (128–130). Mammalian PNK is both a kinase and a phosphatase. PNK has a kinase domain for adding a phosphate to a 5′ OH. PNK has a phosphatase domain that is important for removing 3′ phosphate groups, which can remain after some oxidative damage or partial processing (or after NIELS 1 or 2 remove an abasic sugar, leaving a 3′ phosphate group). Interestingly, the short 3′ overhang that the Artemis:DNA-PKcs complex prefers to leave at long 3′ overhangs represents an ideal substrate for PNK to add a 5′ phosphate at a recessed 5′ OH.

Removal of 3′ phosphoglycolate groups. Oxidative damage often causes breaks that leave a 3′ phosphoglycolate group, and these can be removed in either of two major ways. First, Artemis:DNA-PKcs can remove such groups using its 3′ endonucleolytic activity (39, 98). Second, 3′ phosphoglycolates can be converted to 3′ phosphate by tyrosyl DNA phosphodiesterase 1, whose major role is the removal of tyrosyl-phosphate linkages that arise when topoisomerases fail to religate transient DNA single-strand break reaction intermediates. Then PNK can remove the 3′ phosphate group.

Aprataxin. Aprataxin (APTX) is important in deadenylation of aborted ligation products in which an AMP group is left at the 5′ end of a nick or DSB owing to a failed ligation reaction (131, 132).

PNK-APTX-like factor. PALF and APLF are the same protein (511 aa, 57 kDa). The PALF designation stands for PNK and APTX-like FHA protein (133). APLF stands for aprataxin- and PNK-like factor (134, 135). Previously, it was also called C2orf13. PALF is an endonuclease and a 3′ exonuclease (133). This is interesting, given that Artemis lacks a 3′ exonuclease.

PHYSIOLOGIC DNA RECOMBINATION SYSTEMS

V(D)J and class switch recombination are two physiologic breakage and rejoining systems. NHEJ carries out the rejoining phase.

V(D)J Recombination

V(D)J recombination is one of the two physiologic systems for creating intentional DSBs in somatic cells, specifically in early B or T cells for the purpose of generating antigen receptor genes. RAG1 and RAG2 (both only expressed in early B and T cells) form a complex that can bind sequence specifically at recombination signal sequences (RSSs) that consist of a heptamer and nonamer consensus sequence, separated by either a 12- or 23-bp nonconserved spacer sequence (**Supplemental Figure 3**). [HMGB1 or -2 is thought to be part of this RAG complex on the basis of in vitro studies (136).] A given recombination reaction requires two such RSS sites, one 12-RSS site and one 23-RSS site (the 12/23 rule). The RAG complex initially nicks directly adjacent to each RSS and then uses that nick as a nucleophile to attack the antiparallel strand at each of the non-RSS ends (137). The two non-RSS ends are called coding ends because these regions join to encode a new antigen receptor exon. The nucleophilic attack generates a DNA hairpin at each of the two coding ends. The NHEJ proteins take over at this point, beginning with the opening of the two hairpins by Artemis:DNA-PKcs and followed by NHEJ joining (37). Like vertebrate NHEJ, most coding ends do not share significant terminal microhomology (45, 138). The NHEJ junctions formed in V(D)J recombination have proven to be useful for understanding NHEJ more generally.

The DSBs at the two RSS ends are called signal ends, and these are blunt and 5′ phosphorylated (139, 140). In cells that express terminal transferase, nucleotide addition can occur at these ends (141). But these ends only rarely suffer nucleolytic resection, presumably because of tight binding by the RAG complex (137). Joining of the two signal ends together to form a signal joint is also reliant on Ku and the ligase IV complex, but joining is not dependent on Artemis or DNA-PKcs (65, 137). [The fact that DNA-PKcs is required for coding joint formation (for Artemis:DNA-PKcs opening of hairpins), but not for signal joint formation, was a point of importance in the original description of scid mice (142). Scid mice have a mutant DNA-PKcs gene (143). Artemis-null mice behave similarly (63).]

Class Switch Recombination

CSR occurs only in B cells after they have already completed V(D)J recombination. It is the second of the two physiologic forms of DSB formation in somatic cells (64). CSR is necessary for mammalian B cells to change their immunoglobulin heavy chain gene from producing Igμ for IgM to Igγ, Igα, or Igε for making IgG, IgA, or IgE, respectively (**Supplemental Figure 4**. Follow the Supplemental Material link from the Annual Reviews home page at **http://www.annualreviews.org**). The process requires a B cell–specific cytidine deaminase, AID, which converts C to U within regions of ssDNA. In mammalian CSR, the single strandedness appears to be largely owing to formation of kilobase-length R loops that form at specialized CSR switch sequences because of the extremely (40% to 50%) G-rich RNA transcript that is generated at these specialized recombination zones (144, 145). This permits AID action on the nontemplate DNA strand. RNase H can resect portions of the RNA strand that pairs with the template strand, thereby exposing regions of ssDNA for AID action on that strand also. Once AID introduces C to U changes in the switch region, then UNG converts these to abasic sites, and apurinic/apyrimidinic endonuclease (APE1) can, in principle, nick at these abasic sites. Participation of other enzymes, such as Exo1, may assist in converting the nicks on the top and bottom strands into DSBs. NHEJ is largely responsible for joining these DSBs, but as mentioned above, elegant work has demonstrated the role of either ligase I or III, when ligase IV is missing (70).

CHROMOSOMAL TRANSLOCATIONS AND GENOME REARRANGEMENTS

Chromosomal translocations and genome rearrangements can occur in somatic cells, most notably in cancer. In addition, such genome rearrangements can occur in germ cells, giving rise to heritable genome rearrangements. Although the breakage mechanisms vary, the joining mechanism is usually via NHEJ.

Neoplastic Chromosomal Rearrangements

The vast majority of genome rearrangement-related DSBs (translocations and deletions) in neoplastic cells are joined by NHEJ, even though there is ample opportunity for participation of alternative ligases, if ligase IV is missing, as in experimental systems or extremely rare patients (66, 70). The breakage mechanisms in neoplastic cells include the following: random or near-random breakage mechanisms (owing to ROS, IR, or topoisomerase failures) in any cell type and V(D)J-type or CSR-type breaks in lymphoid cells (146). The lymphoid-specific breakage mechanisms can combine antigen-receptor loci with off-target loci at sequences that are similar to the RSS or CSR sequences (18, 147). In some lymphoid neoplasms, two off-target loci are recombined, and the breakage at each of the two sites can occur by any of the above mechanisms. Sequential action by AID followed by the RAG complex at CpG sites appears likely in some of the most common breakage events (called CpG-type events) in human lymphoma (146). In both CSR-type and CpG-type breaks, AID requires ssDNA to initiate C to U or meC to T changes, respectively. Departures from B-form DNA are relevant to such sites (148, 149).

Constitutional Chromosomal Rearrangements

The breakage mechanisms in germ cells are presumably primarily due to random causes (e.g., ROS, IR, or topoisomerase failures). Deviations from B-DNA are known to be relevant at long inverted repeats, where the most common constitutional translocations occur. The most common constitutional chromosomal rearrangement is the t(11;22) in the Emanuel syndrome (150). In this case, inverted repeats result in cruciform formation, creating a DNA structure that is vulnerable to DNA enzymes that can act on various portions of the cruciform. Once broken, the DNA ends are likely joined by NHEJ, on the basis of observed junctional sequence features.

During evolution, some of the chromosomal rearrangements that arise during speciation are almost certain to share themes with those discussed here, including breakage at sites of DNA structural variation and joining by NHEJ. Replication-based mechanisms are also likely to be very important for major genomic rearrangements (151, 152).

CHROMATIN AND NONHOMOLOGOUS DNA END JOINING

It is not yet clear how much disassembly of histone octamers must occur at a DSB for NHEJ proteins to function. In contrast to HR, where kilobases of DNA are involved and phosphorylated H2AX (γ-H2AX) alterations are important, NHEJ probably requires less than 30 bp of DNA on either side of a break.

If randomly distributed, 80% of DSBs would occur on DNA that is wrapped around histone octamers, and 20% would occur internucleosomally. For those breaks within a nucleosome, one study showed that Ku can bind, implying that the duplex DNA can separate from the surface of the nucleosome sufficiently to permit Ku to bind (153).

Several studies propose that γ-H2AX is important for NHEJ (16, 154). Much of the evidence is based on immunolocalization studies where the damage site may have contained a mixture of HR and NHEJ events within the 2000-Å confocal microscope section thickness. Differences in access within the euchromatic versus the heterochromatic regions are likely, but even early genetic insights concerning this are limited to yeast (155).

H2AX is only present, on average, in one of every ten human nucleosomes because H2A is the predominant species in histone octamers (16). Therefore, most DSBs would occur about 5 nucleosomes away (about 1 kb) from the nearest octamer containing an H2AX that is eligible for conversion to γ-H2AX via phosphorylation by ATM or DNA-PKcs at serine 139 of H2AX. Given this substantial distance from the site of the enzymatic repair, it is not clear that such H2AX phosphorylation events are critical for NHEJ.

When DNA-PKcs does phosphorylate H2AX, this increases the vulnerability of H2AX to the histone exchange factor called FACT (which consists of a heterodimer of Spt16 and SSRP1). Phosphorylated H2AX (γ-H2AX) is more easily exchanged out of the octamer, thereby leaving only a tetramer of $(H3)_2(H4)_2$ at the site, and this is more sterically flexible, thereby perhaps permitting DNA repair factors to carry out their work (156).

Poly-ADP ribose polymerase-1 (PARP-1) is able to downregulate the activity of FACT by ADP-ribosylation of the Spt16 subunit of FACT. This may be able to shift the equilibrium of γ-H2AX and H2AX in the nucleosomes. That is, PARP-1 activation at a site of damage might shift the equilibrium toward retention of γ-H2AX in the region, perhaps thereby aiding in recruitment or retention of repair proteins (156).

Hence, FACT may initially act proximally at the closest nucleosome to exchange γ-H2AX out and leave an $(H3)_2(H4)_2$ tetramer at the site of damage for purposes of flexibility of the DNA. FACT may act more regionally (distally) to favor the retention of γ-H2AX for purposes of integrating the repair process with repair protein recruitment, protein retention, and cell cycle aspects (156).

CONCLUDING COMMENTS

Mechanistic flexibility by multifunctional enzymes and iterative processing of each DNA end are themes that apply to all NHEJ across billions of years of prokaryotic and eukaryotic evolution. Because much of this evolution was convergent, it illustrates that these themes are important for solving this particular biological problem: the joining of heterogeneous DSBs.

SUMMARY POINTS

1. NHEJ evolved to directly repair DSBs. In haploid stationary phase organisms, there is no homology donor, and HR is not an option at all. Evolutionarily, assuming that many organisms were haploid, NHEJ likely represents a very early evolutionary DNA repair strategy.
2. In eukaryotes, most DSBs outside of S/G2 of the cell cycle are joined by NHEJ. Within S/G2 phases, homologous recombination is very active because the two sister chromatids are directly adjacent.
3. Key components of vertebrate NHEJ are Ku; DNA-PKcs; Artemis; Pol X members (pol μ and λ); and the ligase complex, consisting of XLF, XRCC4, and DNA ligase IV. Polynucleotide kinase (PNK) is important in a subset of NHEJ events.
4. Predominantly convergent evolution of NHEJ in prokaryotes and eukaryotes yielded mechanisms that reflect key themes for NHEJ and the repair of DSBs. These themes are
   a. mechanistic flexibility in handling diverse DNA end configurations by the nuclease, polymerase, and ligase activities; and
   b. iterative processing of each DNA end. Each DNA end, as well as incompletely ligated junctions, can undergo multiple rounds of revision by the nuclease, polymerases, and ligase.
5. When components of NHEJ are missing (e.g., genetically mutant yeast, or mice, or extremely rare human patients), the flexible nature of NHEJ permits substitutions by other enzymes. Rather than designate such substitutions as separate pathways (e.g., alternative NHEJ, backup NHEJ, microhomology-mediated NHEJ), one can include them as part of NHEJ but designate them as such (ligase IV-independent or Ku-independent NHEJ).
6. Terminal microhomology of one to a few nucleotides that are shared between the two DNA ends improves the efficiency of joining by NHEJ in vitro and can often, but not always, bias the outcome of the joining process toward using that microhomology in vivo. However, NHEJ does not require any microhomology in vitro or in vivo.
7. Many in vivo (even most, in vertebrate cells) NHEJ junctions have no apparent microhomology. Biochemical studies indicate that joining of fully incompatible ends can occur with absolutely no microhomology via Ku plus XLF:XRCC4:DNA ligase IV. For in vivo joins, one cannot rule out occult (inapparent) microhomology use, much of which may be polymerase generated.

FUTURE ISSUES

1. Are the two DNA ends held in proximity during NHEJ or is there synapsis? In biochemical systems, XRCC4:DNA ligase IV does not appear to require any additional

protein to help it bring two DNA ends together. This is especially clear when there are 4 bp of terminal microhomology; in which case, addition of Ku does not markedly stimulate joining. However, at 2 bp or less of terminal microhomology, Ku does improve XRCC4:DNA ligase IV ligation. This issue is relevant to whether the two DNA ends generated at a single DSB (proximal) are joined more readily than two DNA ends that arise far apart (as in a chromosomal translocation where two DSBs are involved). The issue of whether close DNA ends are joined more efficiently than ends that are far apart is a point of active study.

2. In what ways do the DNA damage response proteins mechanistically or functionally connect with the NHEJ enzymes? NHEJ at a single DSB may be so rapid and physically confined that the damage response pathways involving ATM, the RAD50:MRE11:NBS1 complex, γ-H2AX, and 53BP1 are not activated, but this is quite unclear and subject to speculation. Experimentally or with environmental extremes, a cell may be challenged with many DSBs, in which case, activation of the damage response pathways is increasingly likely. As these activate, the impact on the enzymology of NHEJ is not entirely clear.
3. Are there additional participants in NHEJ? The Werner's (WRN) 3′ exonuclease/helicase enzyme has been proposed as one candidate, but the IR-sensitivity data fail to show a large effect (157). WRN does interact with Ku and PARP-1, but it has been proposed that this may reflect a role in replication fork repair rather than NHEJ, and this seems reasonable (158). Metnase has been proposed as a possible NHEJ nuclease and helicase, but it also has decatenating activity (159, 160). Metnase is present in humans but not in apes, mice, or apparently any other vertebrates, and there is no yeast homolog. Moreover, there is no genetic knockout to demonstrate a role in NHEJ.

DISCLOSURE STATEMENT

The author is not aware of any affiliations, memberships, funding, or financial holdings that might be perceived as affecting the objectivity of this review.

ACKNOWLEDGMENTS

I thank T.E. Wilson, D. Williams, W. An, N. Adachi, and K. Schwarz for comments. I thank Jiafeng Gu and Xiaoping Cui, as well as other current members of my lab for comments. I apologize to those whose work was not cited because of length restrictions.

LITERATURE CITED

1. Aravind L, Koonin EV. 2000. SAP—a putative DNA-binding motif involved in chromosomal organization. *Trends Biochem. Sci.* 25:112–14
2. Aravind L, Koonin EV. 2001. Prokaryotic homologs of the eukaryotic DNA-end-binding protein Ku, novel domains in the Ku protein and prediction of a prokaryotic double-strand break repair system. *Genome Res.* 11:1365–74
3. Gu J, Lieber MR. 2008. Mechanistic flexibility as a conserved theme across 3 billion years of nonhomologous DNA end-joining. *Genes Dev.* 22:411–15
4. San Filippo J, Sung P, Klein H. 2008. Mechanism of eukaryotic homologous recombination. *Annu. Rev. Biochem.* 77:229–57

5. Sung P, Klein H. 2006. Mechanism of homologous recombination: mediators and helicases take on regulatory functions. *Nat. Rev. Mol. Cell Biol.* 7:739–50
6. Moore JK, Haber JE. 1996. Cell cycle and genetic requirements of two pathways of nonhomologous end-joining repair of double-strand breaks in *S. cerevisiae*. *Mol. Cell. Biol.* 16:2164–73
7. Sonoda E, Hochegger H, Saberi A, Taniguchi Y, Takeda S. 2006. Differential usage of non-homologous end-joining and homologous recombination in double strand break repair. *DNA Repair* 5:1021–29
8. Van Dyck E, Stasiak AZ, Stasiak A, West SC. 1999. Binding of double-strand breaks in DNA by human Rad52 protein. *Nature* 398:728–31
9. Hochegger H, Dejsuphong D, Fukushima T, Morrison C, Sonoda E, et al. 2006. Parp-1 protects homologous recombination from interference by Ku and ligase IV in vertebrate cells. *EMBO J.* 25:1305–14
10. Zhang Y, Hefferin ML, Chen L, Shim EY, Tseng HM, et al. 2007. Role of Dnl4-Lif1 in nonhomologous end-joining repair complex assembly and suppression of homologous recombination. *Nat. Struct. Mol. Biol.* 14:639–46
11. Lieber MR, Ma Y, Pannicke U, Schwarz K. 2003. Mechanism and regulation of human non-homologous DNA end-joining. *Nat. Rev. Mol. Cell. Biol.* 4:712–20
12. Lieber MR, Karanjawala ZE. 2004. Ageing, repetitive genomes and DNA damage. *Nat. Rev. Mol. Cell. Biol.* 5:69–75
13. Martin GM, Smith AC, Ketterer DJ, Ogburn CE, Disteche CM. 1985. Increased chromosomal aberrations in first metaphases of cells isolated from the kidneys of aged mice. *Isr. J. Med. Sci.* 21:296–301
14. Riballo E, Kuhne M, Rief N, Doherty A, Smith GCM, et al. 2004. A pathway of double-strand break rejoining dependent upon ATM, Artemis, and proteins locating to gamma-H2AX foci. *Mol. Cell* 16:715–24
15. Chance B, Sies H, Boveris A. 1979. Hydroperoxide metabolism in mammalian organs. *Physiol. Rev.* 59:527–603
16. Friedberg EC, Walker GC, Siede W, Wood RD, Schultz RA, Ellenberger T. 2006. *DNA Repair and Mutagenesis*. Washington, DC: ASM Press. 1118 pp.
17. Adachi N, Suzuki H, Iiizumi S, Koyama H. 2003. Hypersensitivity of nonhomologous DNA end-joining mutants to VP-16 and ICRF-193: implications for the repair of topoisomerase II-mediated DNA damage. *J. Biol. Chem.* 278:35897–902
18. Mahowald GK, Baron JM, Sleckman BP. 2008. Collateral damage from antigen receptor gene diversification. *Cell* 135:1009–12
19. Pitcher RS, Green AJ, Brzostek A, Korycka-Machala M, Dziadek J, Doherty AJ. 2007. NHEJ protects mycobacteria in stationary phase against the harmful effects of desiccation. *DNA Repair* 6:1271–76
20. Murnane JP. 2006. Telomeres, chromosome instability, and cancer. *DNA Repair* 5:1082–92
21. Zickler D, Kleckner N. 1999. Meiotic chromosomes: integrating structure and function. *Annu. Rev. Genet.* 33:603–754
22. Hamer G, Roepers-Gajadien HL, van Duyn-Goedhart A, Gademan IS, Kal HB, et al. 2003. Function of DNA-protein kinase catalytic subunit during the early meiotic prophase without Ku70 and Ku86. *Biol. Reprod.* 68:717–21
23. Shuman S, Glickman MS. 2007. Bacterial DNA repair by non-homologous end joining. *Nat. Rev. Microbiol.* 5:852–61
24. Daley JM, Palmbos PL, Wu D, Wilson TE. 2005. Nonhomologous end joining in yeast. *Annu. Rev. Genet.* 39:431–51
25. Wilson TE. 2009. Non-homologous recombination. In *Wiley Encyclopedia of Chemical Biology*, ed. TP Begley, 3:423–32. Hoboken, NJ: Wiley
26. Anderson CW, Carter TH. 1996. The DNA-activated protein kinase—DNA-PK. In *Molecular Analysis of DNA Rearrangements in the Immune System*, ed. R Jessberger, MR Lieber, pp. 91–112. Heidelberg: Springer-Verlag
27. Mimori T, Hardin JA. 1986. Mechanism of interaction between Ku protein and DNA. *J. Biol. Chem.* 261:10375–79
28. Blier PR, Griffith AJ, Craft J, Hardin JA. 1993. Binding of Ku protein to DNA. *J. Biol. Chem.* 268:7594–601

29. Falzon M, Fewell J, Kuff EL. 1993. EBP-80, a transcription factor closely resembling the human autoantigen Ku, recognizes single- to double-strand transitions in DNA. *J. Biol. Chem.* 268:10546–52
30. Walker EH, Pacold ME, Perisic O, Stephens L, Hawkins PT, et al. 2000. Structural determinants of phosphoinositide 3-kinase inhibition by wortmannin, LY294002, quercetin, myricetin, and staurosporine. *Mol. Cell* 6:909–13
31. Lieber MR. 2008. The mechanism of human nonhomologous DNA end joining. *J. Biol. Chem.* 283:1–5
32. Ma Y, Lu H, Tippin B, Goodman MF, Shimazaki N, et al. 2004. A biochemically defined system for mammalian nonhomologous DNA end joining. *Mol. Cell* 16:701–13
33. Yaneva M, Kowalewski T, Lieber MR. 1997. Interaction of DNA-dependent protein kinase with DNA and with Ku: biochemical and atomic-force microscopy. *EMBO J.* 16:5098–112
34. Nick McElhinny SA, Snowden CM, McCarville J, Ramsden DA. 2000. Ku recruits the XRCC4-ligase IV complex to DNA ends. *Mol. Cell. Biol.* 20:2996–3003
35. Chen L, Trujillo K, Sung P, Tomkinson AE. 2000. Interactions of the DNA ligase IV-XRCC4 complex with DNA ends and the DNA-dependent protein kinase. *J. Biol. Chem.* 275:26196–205
36. Walker JR, Corpina RA, Goldberg J. 2001. Structure of the Ku heterodimer bound to DNA and its implications for double-strand break repair. *Nature* 412:607–14
37. Ma Y, Pannicke U, Schwarz K, Lieber MR. 2002. Hairpin opening and overhang processing by an Artemis:DNA-PKcs complex in V(D)J recombination and in nonhomologous end joining. *Cell* 108:781–94
38. Ma Y, Schwarz K, Lieber MR. 2005. The Artemis:DNA-PKcs endonuclease can cleave gaps, flaps, and loops. *DNA Repair* 4:845–51
39. Yannone SM, Khan IS, Zhou RZ, Zhou T, Valerie K, Povirk LF. 2008. Coordinate 5′ and 3′ endonucleolytic trimming of terminally blocked blunt DNA double-strand break ends by Artemis nuclease and DNA-dependent protein kinase. *Nucleic Acids Res.* 36:3354–65
40. Wilson TE, Lieber MR. 1999. Efficient processing of DNA ends during yeast nonhomologous end joining: evidence for a DNA polymerase beta (Pol4)-dependent pathway. *J. Biol. Chem.* 274:23599–609
41. Bertocci B, De Smet A, Weill J-C, Reynaud CA. 2006. Nonoverlapping functions of DNA polymerases mu, lambda, and terminal deoxynucleotidyltransferase during immunoglobulin V(D)J recombination in vivo. *Immunity* 25:31–41
42. Gu J, Lu H, Tippin B, Shimazaki N, Goodman MF, Lieber MR. 2007. XRCC4:DNA ligase IV can ligate incompatible DNA ends and can ligate across gaps. *EMBO J.* 26:1010–23
43. Gu J, Lu H, Tsai AG, Schwarz K, Lieber MR. 2007. Single-stranded DNA ligation and XLF-stimulated incompatible DNA end ligation by the XRCC4-DNA ligase IV complex: influence of terminal DNA sequence. *Nucleic Acids Res.* 35:5755–62
44. Schlissel MS. 1998. Structure of nonhairpin coding-end DNA breaks in cells undergoing V(D)J recombination. *Mol. Cell. Biol.* 18:2029–37
45. Gauss GH, Lieber MR. 1996. Mechanistic constraints on diversity in human V(D)J recombination. *Mol. Cell. Biol.* 16:258–69
46. Falck J, Coates J, Jackson SP. 2005. Conserved modes of recruitment of ATM, ATR and DNA-PKcs to sites of DNA damage. *Nature* 434:605–11
47. Ramadan K, Maga G, Shevelev IV, Villani G, Blanco L, Hubscher U. 2003. Human DNA polymerase lambda possesses terminal deoxyribonucleotidyl transferase activity and can elongate RNA primers: implications for novel functions. *J. Mol. Biol.* 328:63–72
48. Ramadan K, Shevelev IV, Maga G, Hubscher U. 2004. De novo DNA synthesis by human DNA polymerase lambda, DNA polymerase mu, and terminal deoxynucleotidyl transferase. *J. Mol. Biol.* 339:395–404
49. Jäger U, Böcskör S, Le T, Mitterbauer G, Bolz I, et al. 2000. Follicular lymphomas' BCL-2/IgH junctions contain templated nucleotide insertions: novel insights into the mechanism of t(14;18) translocation. *Blood* 95:3520–29
50. Welzel N, T TL, Marculescu R, Mitterbauer G, Chott A, et al. 2001. Templated nucleotide addition and immunoglobulin JH-gene utilization in t(11;14) junctions: implications for the mechanism of translocation and the origin of mantle cell lymphoma. *Cancer Res.* 61:1629–36

51. Tippin B, Kobayashi S, Bertram JG, Goodman MF. 2004. To slip or skip, visualizing frameshift mutation dynamics for error-prone DNA polymerases. *J. Biol. Chem.* 279:5360–68
52. Domínguez O, Ruiz JF, de Lera T, García-Díaz M, González MA, et al. 2000. DNA polymerase mu (Pol μ), homologous to TdT, could act as a DNA mutator in eukaryotic cells. *EMBO J.* 19:1731–42
53. Bebenek B, Garcia-Diaz M, Blanco L, Kunkel TA. 2003. The frameshift infidelity of human DNA polymerase lambda: implications for function. *J. Biol. Chem.* 278:34685–90
54. Roth D, Wilson J. 1988. Illegitimate recombination in mammalian cells. In *Genetic Recombination*, ed. R Kucherlapapti, GR Smith, pp. 621–53. Washington, DC: Am. Soc. Microbiol.
55. Roth DB, Chang XB, Wilson JH. 1989. Comparison of filler DNA at immune, nonimmune, and oncogenic rearrangements suggests multiple mechanisms of formation. *Mol. Cell. Biol.* 9:3049–57
56. Roth DB, Wilson JH. 1986. Nonhomologous recombination in mammalian cells: role for short sequence homologies in the joining reaction. *Mol. Cell. Biol.* 6:4295–304
57. Daley JM, Laan RLV, Suresh A, Wilson TE. 2005. DNA joint dependence of Pol X family polymerase action in nonhomologous end joining. *J. Biol. Chem.* 280:29030–37
58. Gerstein RM, Lieber MR. 1993. Extent to which homology can constrain coding exon junctional diversity in V(D)J recombination. *Nature* 363:625–27
59. Gerstein RM, Lieber MR. 1993. Coding end sequence can markedly affect the initiation of V(D)J recombination. *Genes Dev.* 7:1459–69
60. Weinstock DM, Brunet E, Jasin M. 2007. Formation of NHEJ-derived reciprocal chromosomal translocations does not require Ku70. *Nat. Cell Biol.* 9:978–81
61. Gao Y, Chaudhurri J, Zhu C, Davidson L, Weaver DT, Alt FW. 1998. A targeted DNA-PKcs-null mutation reveals DNA-PK independent for Ku in V(D)J recombination. *Immunity* 9:367–76
62. Bassing CH, Swat W, Alt FW. 2002. The mechanism and regulation of chromosomal V(D)J recombination. *Cell* 109(Suppl.):S45–55
63. Rooney S, Sekiguchi J, Zhu C, Cheng H-L, Manis J, et al. 2002. Leaky scid phenotype associated with defective V(D)J coding end processing in Artemis-deficient mice. *Mol. Cell* 10:65–74
64. Dudley DD, Chaudhuri J, Bassing CH, Alt FW. 2005. Mechanism and control of V(D)J recombination versus class switch recombination: similarities and differences. *Adv. Immunol.* 86:43–112
65. Jung D, Giallourakis C, Mostoslavsky R, Alt FW. 2006. Mechanism and control of V(D)J recombination at the immunoglobulin heavy chain locus. *Annu. Rev. Immunol.* 24:541–70
66. Wang JH, Alt FW, Gostissa M, Datta A, Murphy M, et al. 2008. Oncogenic transformation in the absence of Xrcc4 targets peripheral B cells that have undergone editing and switching. *J. Exp. Med.* 205:3079–90
67. Boubnov NV, Wills ZP, Weaver DT. 1993. V(D)J recombination coding junction formation without DNA homology: processing of coding end termini. *Mol. Cell. Biol.* 13:6957–68
68. Daley JM, Wilson TE. 2005. Rejoining of DNA double-strand breaks as a function of overhang length. *Mol. Cell. Biol.* 25:896–906
69. Han L, Yu K. 2008. Altered kinetics of nonhomologous end joining and class switch recombination in ligase IV–deficient B cells. *J. Exp. Med.* 205:2745–53
70. Yan CT, Boboila C, Souza EK, Franco S, Hickernell TR, et al. 2007. IgH class switching and translocations use a robust non-classical end-joining pathway. *Nature* 449:478–82
71. Soulas-Sprauel P, Le Guyader G, Rivera-Munoz P, Abramowski V, Olivier-Martin C, et al. 2007. Role for DNA repair factor XRCC4 in immunoglobulin class switch recombination. *J. Exp. Med.* 204:1717–27
72. Grawunder U, Wilm M, Wu X, Kulesza P, Wilson TE, et al. 1997. Activity of DNA ligase IV stimulated by complex formation with XRCC4 protein in mammalian cells. *Nature* 388:492–95
73. Lee JW, Blanco L, Zhou T, Garcia-Diaz M, Bebenek K, et al. 2003. Implication of DNA polymerase lambda in alignment-based gap filling for nonhomologous DNA end joining in human nuclear extracts. *J. Biol. Chem.* 279:805–11
74. Nick McElhinny SA, Ramsden DA. 2003. Polymerase mu is a DNA-directed DNA/RNA polymerase. *Mol. Cell. Biol.* 23:2309–15

75. Tseng HM, Tomkinson AE. 2002. A physical and functional interaction between yeast Pol4 and Dnl4-Lif1 links DNA synthesis and ligation in nonhomologous end joining. *J. Biol. Chem.* 277:45630–37
76. Ferguson DO, Alt FW. 2001. DNA double-strand break repair and chromosomal translocations: lessons from animal models. *Oncogene* 20:5572–79
77. Aniukwu J, Glickman MS, Shuman S. 2008. The pathways and outcomes of mycobacterial NHEJ depend on the structure of the broken DNA ends. *Genes Dev.* 22:512–27
78. Doherty AJ, Jackson SP, Weller GR. 2001. Identification of bacterial homologues of the Ku DNA repair proteins. *FEBS Lett.* 500:186–88
79. d'Adda di Fagagna F, Weller GR, Doherty AJ, Jackson SP. 2003. The Gam protein of bacteriophage Mu is an orthologue of eukaryotic Ku. *EMBO Rep.* 4:47–52
80. Weller GR, Kysela B, Roy R, Tonkin LM, Scanlan E, et al. 2002. Identification of a DNA nonhomologous end-joining complex in bacteria. *Science* 297:1686–89
81. Pitcher RS, Brissett NC, Doherty AJ. 2007. Nonhomologous end-joining in bacteria: a microbial perspective. *Annu. Rev. Microbiol.* 61:259–82
82. de Vries E, van Driel W, Bergsma WG, Arnberg AC, van der Vliet PC. 1989. HeLa nuclear protein recognizing DNA termini and translocating on DNA forming a regular DNA-multimeric protein complex. *J. Mol. Biol.* 208:65–78
83. Yoo S, Kimzey A, Dynan WS. 1999. Photocross-linking of an oriented DNA repair complex. Ku bound at a single DNA end. *J. Biol. Chem.* 274:20034–39
84. Downs JA, Jackson SP. 2004. A means to a DNA end: the many roles of Ku. *Nat. Rev. Mol. Cell. Biol.* 5:367–78
85. Leuther KK, Hammarsten O, Kornberg RD, Chu G. 1999. Structure of the DNA-dependent protein kinase: implications for its regulation by DNA. *EMBO J.* 18:1114–23
86. Hammarsten O, Chu G. 1998. DNA-dependent protein kinase: DNA binding and activation in the absence of Ku. *Proc. Natl. Acad. Sci. USA* 95:525–30
87. Hammarsten O, DeFazio LG, Chu G. 2000. Activation of DNA-dependent protein kinase by single-stranded DNA ends. *J. Biol. Chem.* 275:1541–50
88. West RB, Yaneva M, Lieber MR. 1998. Productive and nonproductive complexes of Ku and DNA-PK at DNA termini. *Mol. Cell. Biol.* 18:5908–20
89. Hartley KO, Gell D, Smith GC, Zhang H, Divecha N, et al. 1995. DNA-dependent protein kinase catalytic subunit: a relative of phosphatidylinositol 3-kinase and the ataxia telangiectasia gene product. *Cell* 82:849–56
90. Meek K, Dang V, Lees-Miller SP. 2008. DNA-PK: the means to justify the ends? *Adv. Immunol.* 99:33–58
91. Costantini S, Woodbine L, Andreoli L, Jeggo PA, Vindigni A. 2007. Interaction of the Ku heterodimer with the DNA ligase IV/Xrcc4 complex and its regulation by DNA-PK. *DNA Repair* 6:712–22
92. Williams DR, Lee K-J, Shi J, Chen DJ, Stewart PL. 2008. Cryoelectron microscopy structure of the DNA-dependent protein kinase catalytic subunit (DNA-PKcs) at subnanometer resolution reveals α-helices and insight into DNA binding. *Structure* 16:468–77
93. Spagnolo L, Rivera-Calzada A, Pearl LH, Llorca O. 2006. Three-dimensional structure of the human DNA-PKcs/Ku70/Ku80 complex assembled on DNA and its implications for DNA DSB repair. *Mol. Cell* 22:511–19
94. Jin S, Kharbanda S, Mayer B, Kufe D, Weaver DT. 1997. Binding of Ku and c-Abl at the kinase homology region of DNA-dependent protein kinase catalytic subunit. *J. Biol. Chem.* 272:24763–66
95. Mahaney BL, Meek K, Lees-Miller SP. 2009. Repair of ionizing radiation-induced DNA double-strand breaks by non-homologous end-joining. *Biochem. J.* 417:639–50
96. Lu H, Shimazaki N, Raval P, Gu J, Watanabe G, et al. 2008. A biochemically defined system for coding joint formation in human V(D)J recombination. *Mol. Cell* 31:485–97
97. Goodarzi AA, Yu Y, Riballo E, Douglas P, Walker SA, et al. 2006. DNA-PK autophosphorylation facilitates Artemis endonuclease activity. *EMBO J.* 25:3880–89

98. Povirk LF, Zhou T, Zhou R, Cowan MJ, Yannone SM. 2007. Processing of 3′-phosphoglycolate-terminated DNA double strand breaks by Artemis nuclease. *J. Biol. Chem.* 282:3547–58
99. Rooney S, Alt FW, Lombard D, Whitlow S, Eckersdorff M, et al. 2003. Defective DNA repair and increased genomic instability in Artemis-deficient murine cells. *J. Exp. Med.* 197:553–65
100. Li L, Salido E, Zhou Y, Bhattacharyya S, Yannone SM, et al. 2005. Targeted disruption of the Artemis murine counterpart results in SCID and defective V(D)J recombination that is partially corrected with bone marrow transplantation. *J. Immunol.* 174:2420–28
101. Xiao Z, Dunn E, Singh K, Khan IS, Yannone SM, Cowan MJ. 2009. A non-leaky Artemis-deficient mouse that accurately models the human severe combined immune deficiency phenotype, including resistance to hematopoietic stem cell transplantation. *Biol. Blood Marrow Transpl.* 15:1–11
102. Jovanovic M, Dynan WS. 2006. Terminal DNA structure and ATP influence binding parameters of the DNA-dependent protein kinase at an early step prior to DNA synapsis. *Nucleic Acids Res.* 34:1112–20
103. van der Burg M, Ijspeert H, Verkaik N, Turul T, Wiegant W, et al. 2009. A DNA-PKcs mutation in a radiosensitive T-B- SCID patient inhibits Artemis activation and nonhomologous end-joining. *J. Clin. Investig.* 119:91–98
104. Moon AF, Garcia-Diaz M, Batra VK, Beard WA, Bebenek K, et al. 2007. The X family portrait: structural insights into biological functions of X family polymerases. *DNA Repair* 6:1709–25
105. Bebenek K, Garcia-Diaz M, Patishall SR, Kunkel TA. 2005. Biochemical properties of *Saccharomyces cerevisiae* DNA polymerase IV. *J. Biol. Chem.* 280:20051–58
106. Nick McElhinny SA, Havener JM, Garcia-Diaz M, Juarez R, Bebenek K, et al. 2005. A gradient of template dependence defines distinct biological roles for family X polymerases in nonhomologous end joining. *Mol. Cell* 19:357–66
107. Davis BJ, Havener JM, Ramsden DA. 2008. End-bridging is required for Pol mu to efficiently promote repair of noncomplementary ends by nonhomologous end joining. *Nucleic Acids Res.* 36:3085–94
108. Moon AF, Garcia-Diaz M, Bebenek K, Davis BJ, Zhong X, et al. 2007. Structural insight into the substrate specificity of DNA polymerase mu. *Nat. Struct. Mol. Biol.* 14:45–53
109. Juarez R, Ruiz JF, Nick McElhinny SA, Ramsden D, Blanco L. 2006. A specific loop in human DNA polymerase mu allows switching between creative and DNA-instructed synthesis. *Nucleic Acids Res.* 34:4572–82
110. Povirk LF. 2006. Biochemical mechanisms of chromosomal translocations resulting from DNA double-strand breaks. *DNA Repair* 5:1199–212
111. Chang LM, Bollum FJ. 1986. Molecular biology of terminal transferase. *CRC Crit. Rev. Biochem.* 21:27–52
112. Kornberg A, Baker T. 1992. *DNA Replication*. New York: Freeman
113. Tsai CJ, Kim SA, Chu G. 2007. Cernunnos/XLF promotes the ligation of mismatched and noncohesive DNA ends. *Proc. Natl. Acad. Sci. USA* 104:7851–56
114. Tomkinson AE, Vijayakumar S, Pascal JM, Ellenberger T. 2006. DNA ligases: structure, reaction mechanism, and function. *Chem. Rev.* 106:687–99
115. Modesti M, Junop MS, Ghirlando R, van de Rakt M, Gellert M, et al. 2003. Tetramerization and DNA ligase IV interaction of the DNA double-strand break protein XRCC4 are mutually exclusive. *J. Mol. Biol.* 334:215–28
116. Junop MS, Modesti M, Guarne A, Ghirlando R, Gellert M, Yang W. 2000. Crystal structure of the XRCC4 DNA repair protein and implications for end joining. *EMBO J.* 19:5962–70
117. Sibanda BL, Critchlow SE, Begun J, Pei XY, Jackson SP, et al. 2001. Crystal structure of an Xrcc4-DNA ligase IV complex. *Nat. Struct. Biol.* 8:1015–19
118. Andres SN, Modesti M, Tsai CJ, Chu G, Junop MS. 2007. Crystal structure of human XLF: a twist in nonhomologous DNA end-joining. *Mol. Cell* 28:1093–101
119. Li Y, Chirgadze DY, Bolanos-Garcia VM, Sibanda BL, Davies OR, et al. 2008. Crystal structure of human XLF/Cernunnos reveals unexpected differences from XRCC4 with implications for NHEJ. *EMBO J.* 27:290–300

120. Ahnesorg P, Smith P, Jackson SP. 2006. XLF interacts with the XRCC4-DNA ligase IV complex to promote nonhomologous end-joining. *Cell* 124:301–13
121. Buck D, Malivert L, de Chasseval R, Barraud A, Fondanèche MC, et al. 2006. Cernunnos, a novel nonhomologous end-joining factor, is mutated in human immunodeficiency with microcephaly. *Cell* 124:287–99
122. Li G, Alt FW, Cheng HL, Brush JW, Goff PH, et al. 2008. Lymphocyte-specific compensation for XLF/Cernunnos end-joining functions in V(D)J recombination. *Mol. Cell* 31:631–40
123. Deshpande RA, Wilson TE. 2007. Modes of interaction among yeast Nej1, Lif1 and Dnl4 proteins and comparison to human XLF, XRCC4 and Lig4. *DNA Repair* 6:1507–16
124. Callebaut I, Malivert L, Fischer A, Mornon JP, Revy P, de Villartay JP. 2006. Cernunnos interacts with the XRCC4/DNA-ligase IV complex and is homologous to the yeast nonhomologous end-joining factor Nej1. *J. Biol. Chem.* 281:13857–60
125. Hsu HL, Yannone SM, Chen DJ. 2002. Defining interactions between DNA-PK and ligase IV/XRCC4. *DNA Repair* 1:225–35
126. Wang YG, Nnakwe C, Lane WS, Modesti M, Frank KM. 2004. Phosphorylation and regulation of DNA ligase IV stability by DNA-dependent protein kinase. *J. Biol. Chem.* 279:37282–90
127. Leber R, Wise TW, Mizuta R, Meek K. 1998. The XRCC4 gene product is a target for and interacts with the DNA-dependent protein kinase. *J. Biol. Chem.* 273:1794–801
127a. Koch CA, Agyei R, Galicia S, Metalnikov P, O'Donnell P, et al. 2004. Xrcc4 physically links DNA end processing by polynucleotide kinase to DNA ligation by DNA ligase IV. *EMBO J.* 23:3874–85
128. Bernstein NK, Karimi-Busheri F, Rasouli-Nia A, Mani R, Dianov G, et al. 2008. Polynucleotide kinase as a potential target for enhancing cytotoxicity by ionizing radiation and topoisomerase I inhibitors. *Anticancer Agents Med. Chem.* 8:358–67
129. Bernstein NK, Williams RS, Rakovszky ML, Cui D, Green R, et al. 2005. The molecular architecture of the mammalian DNA repair enzyme, polynucleotide kinase. *Mol. Cell* 17:657–70
130. Williams RS, Bernstein N, Lee MS, Rakovszky ML, Cui D, et al. 2005. Structural basis for phosphorylation-dependent signaling in the DNA-damage response. *Biochem. Cell Biol.* 83:721–27
131. Ahel I, Rass U, El-Khamisy SF, Katyal S, Clements PM, et al. 2006. The neurodegenerative disease protein aprataxin resolves abortive DNA ligation intermediates. *Nature* 443:713–16
132. Rass U, Ahel I, West SC. 2008. Molecular mechanism of DNA deadenylation by the neurological disease protein aprataxin. *J. Biol. Chem.* 283:33994–4001
133. Kanno S, Kuzuoka H, Sasao S, Hong Z, Lan L, et al. 2007. A novel human AP endonuclease with conserved zinc-finger-like motifs involved in DNA strand break responses. *EMBO J.* 26:2094–103
134. Iles N, Rulten S, El-Khamisy SF, Caldecott KW. 2007. APLF (C2orf13) is a novel human protein involved in the cellular response to chromosomal DNA strand breaks. *Mol. Cell. Biol.* 27:3793–803
135. Macrae CJ, McCulloch RD, Ylanko J, Durocher D, Koch CA. 2008. APLF (C2orf13) facilitates nonhomologous end-joining and undergoes ATM-dependent hyperphosphorylation following ionizing radiation. *DNA Repair* 7:292–302
136. Swanson PC. 2004. The bounty of RAGs: recombination signal complexes and reaction outcomes. *Immunolog. Rev.* 200:90–114
137. Gellert M. 2002. V(D)J recombination: RAG proteins, repair factors, and regulation. *Annu. Rev. Biochem.* 71:101–32
138. Feeney AJ. 1992. Predominance of VH-D-JH junctions occurring at sites of short sequence homology results in limited junctional diversity in neonatal antibodies. *J. Immunol.* 149:222–29
139. Schlissel M, Constantinescu A, Morrow T, Peng A. 1993. Double-strand signal sequence breaks in V(D)J recombination are blunt, 5′-phosphorylated, RAG-dependent, and cell cycle regulated. *Genes Dev.* 7:2520–32
140. Roth DB, Zhu C, Gellert M. 1993. Characterization of broken DNA molecules associated with V(D)J recombination. *Proc. Natl. Acad. Sci. USA* 90:10788–92
141. Lieber MR, Hesse JE, Mizuuchi K, Gellert M. 1988. Lymphoid V(D)J recombination: nucleotide insertion at signal joints as well as coding joints. *Proc. Natl. Acad. Sci. USA* 85:8588–92

142. Bosma MJ, Carroll AM. 1991. The scid mouse mutant: definition, characterization, and potential uses. *Annu. Rev. Immunol.* 9:323–50
143. Blunt T, Finnie NJ, Taccioli GE, Smith GCM, Demengeot J, et al. 1995. Defective DNA-dependent protein kinase activity is linked to V(D)J recombination and DNA repair defects associated with the murine scid mutation. *Cell* 80:813–23
144. Yu K, Chedin F, Hsieh C-L, Wilson TE, Lieber MR. 2003. R-loops at immunoglobulin class switch regions in the chromosomes of stimulated B cells. *Nat. Immunol.* 4:442–51
145. Shinkura R, Tian M, Khuong C, Chua K, Pinaud E, Alt FW. 2003. The influence of transcriptional orientation on endogenous switch region function. *Nat. Immunol.* 4:435–41
146. Tsai AG, Lu H, Raghavan SC, Muschen M, Hsieh CL, Lieber MR. 2008. Human chromosomal translocations at CpG sites and a theoretical basis for their lineage and stage specificity. *Cell* 135:1130–42
147. Shimazaki N, Tsai AG, Lieber MR. 2009. H3K4me3 stimulates V(D)J RAG complex for both nicking and hairpinning in *trans* in addition to tethering in *cis*: implications for translocations. *Mol. Cell* 34:535–44
148. Tsai AG, Engelhart AE, Hatmal MM, Houston SI, Hud NV, et al. 2008. Conformational variants of duplex DNA correlated with cytosine-rich chromosomal fragile sites. *J. Biol. Chem.* 284:7157–64
149. Yu K, Lieber MR. 2003. Nucleic acid structures and enzymes in the immunoglobulin class switch recombination mechanism. *DNA Repair* 2:1163–74
150. Kurahashi H, Inagaki H, Ohye T, Kogo H, Kato T, Emanuel BS. 2006. Palindrome-mediated chromosomal translocations in humans. *DNA Repair* 5:1136–45
151. Lee JA, Carvalho CM, Lupski JR. 2007. A DNA replication mechanism for generating nonrecurrent rearrangements associated with genomic disorders. *Cell* 131:1235–47
152. Hastings PJ, Ira G, Lupski JR. 2009. A microhomology-mediated break-induced replication model for the origin of human copy number variation. *PLoS Genet.* 5:e1000327
153. Roberts SA, Ramsden DA. 2007. Loading of the nonhomologous end joining factor, Ku, on protein-occluded DNA ends. *J. Biol. Chem.* 282:10605–13
154. Stiff T, O'Driscoll M, Rief N, Iwabuchi K, Lobrich M, Jeggo PA. 2004. ATM and DNA-PK function redundantly to phosphorylate H2AX after exposure to ionizing radiation. *Cancer Res.* 64:2390–96
155. Shim EY, Ma JL, Oum JH, Yanez Y, Lee SE. 2005. The yeast chromatin remodeler RSC complex facilitates end joining repair of DNA double-strand breaks. *Mol. Cell. Biol.* 25:3934–44
156. Heo K, Kim H, Choi SH, Choi J, Kim K, et al. 2008. FACT-mediated exchange of histone variant H2AX regulated by phosphorylation of H2AX and ADP-ribosylation of Spt16. *Mol. Cell* 30:86–97
157. Yannone SM, Roy S, Chan DW, Murphy MB, Huang S, et al. 2001. Werner syndrome protein is regulated and phosphorylated by DNA-dependent protein kinase. *J. Biol. Chem.* 276:38242–48
158. Li B, Comai L. 2002. Displacement of DNA-PKcs from DNA ends by the Werner syndrome protein. *Nucleic Acids Res.* 30:3653–61
159. Hromas R, Wray J, Lee SH, Martinez L, Farrington J, et al. 2008. The human set and transposase domain protein Metnase interacts with DNA ligase IV and enhances the efficiency and accuracy of non-homologous end-joining. *DNA Repair* 7:1927–37
160. Williamson EA, Rasila KK, Corwin LK, Wray J, Beck BD, et al. 2008. The SET and transposase domain protein Metnase enhances chromosome decatenation: regulation by automethylation. *Nucleic Acids Res.* 36:5822–31
161. Ma Y, Pannicke U, Lu H, Niewolik D, Schwarz K, Lieber MR. 2005. The DNA-PKcs phosphorylation sites of human Artemis. *J. Biol. Chem.* 280:33839–46
162. Pannicke U, Ma Y, Lieber MR, Schwarz K. 2004. Functional and biochemical dissection of the structure-specific nuclease Artemis. *EMBO J.* 23:1987–97
163. Dominski Z. 2007. Nucleases of the metallo-beta-lactamase family and their role in DNA and RNA metabolism. *Crit. Rev. Biochem. Mol. Biol.* 42:67–93
164. Callebaut I, Moshous D, Mornon JP, de Villartay JP. 2002. Metallo-beta-lactamase fold within nucleic acids processing enzymes: the beta-CASP family. *Nucleic Acids Res.* 30:3592–601

RELATED RESOURCES

Huret JL. 2009. *Atlas of genetics and cytogenetics in oncology and haematology*. **http://atlasgeneticsoncology.org//**

Tsai AG, Lieber MR. 2009. *Index of /data/2008_cell_135_1130* (chromosomal translocation database). **http://lieber.usc.edu/data/2008_cell_135_1130/**

Int. Agency Res. Cancer/WHO. 2009. *IARC p53 mutation database*. **http://www-p53.iarc.fr/**

The Discovery of Zinc Fingers and Their Applications in Gene Regulation and Genome Manipulation

Aaron Klug

MRC Laboratory of Molecular Biology, Cambridge CB2 0QH, United Kingdom; email: akl@mrc-lmb.cam.ac.uk

Annu. Rev. Biochem. 2010. 79:213–31

First published online as a Review in Advance on January 4, 2010

The *Annual Review of Biochemistry* is online at biochem.annualreviews.org

This article's doi: 10.1146/annurev-biochem-010909-095056

0066-4154/10/0707-0213$20.00

Key Words

gene correction, gene targeting, modular design, protein engineering, transcription activation, transcription inhibition

Abstract

An account is given of the discovery of the classical Cys_2His_2 zinc finger, arising from the interpretation of biochemical studies on the interaction of the *Xenopus* protein transcription factor IIIA with 5S RNA, and of structural studies on its structure and its interaction with DNA. The finger is a self-contained domain stabilized by a zinc ion ligated to a pair of cysteines and a pair of histidines, and by an inner hydrophobic core. This discovery showed not only a new protein fold but also a novel principle of DNA recognition. Whereas other DNA binding proteins generally make use of the two-fold symmetry of the double helix, zinc fingers can be linked linearly in tandem to recognize nucleic acid sequences of varying lengths. This modular design offers a large number of combinatorial possibilities for the specific recognition of DNA (or RNA). It is therefore not surprising that the zinc finger is found widespread in nature, including 3% of the genes of the human genome.

The zinc finger design is ideally suited for engineering proteins to target specific genes. In the first example of their application in 1994, a three-finger protein was constructed to block the expression of an oncogene transformed into a mouse cell line. In addition, a reporter gene was activated by targeting an inserted zinc finger promoter. Thus, by fusing zinc finger peptides to repression or activation domains, genes can be selectively switched off or on. It was also suggested that by combining zinc fingers with other effector domains, e.g., from nucleases or integrases, to form chimeric proteins, genomes could be manipulated or modified. Several applications of such engineered zinc finger proteins are described here, including some of therapeutic importance.

Contents

INTRODUCTION

Ten years of research on the structure of chromatin led to the discovery of the nucleosome and an outline of its structure, as well as the next level for the folding of DNA in the 300-Å chromatin fiber (1, 2). This resulted in an interest in what was then called "active chromatin," the chromatin that is involved in transcription or that was poised to do so, and in finding a tractable system, which offered the possibility of extracting relatively large amounts of material for biochemical and structural studies.

The work of Robert Roeder and Donald Brown on the 5S RNA genes of *Xenopus laevis*, which are transcribed by RNA polymerase III (reviewed in Reference 3) was intriguing. They discovered that the correct initiation of transcription requires the binding of a 40-kDa protein factor, variously called factor A or transcription factor IIIA (TFIIIA), which had been purified from oocyte extracts. By deletion mapping, it was found that this factor interacts with a region about 50 nucleotides long within the gene, called the internal control region. This was the first eukaryotic transcription factor to be described.

7S RNP: 7S ribonucleoprotein particle

Immature oocytes store 5S RNA molecules in the form of 7S ribonucleoprotein particles (7S RNPs) (4), each containing a single 40-kDa protein, which was later shown (5) to be identical with TFIIIA. TFIIIA therefore binds both 5S RNA and its cognate DNA, and it was consequently suggested that it may mediate autoregulation of 5S gene transcription (5). Whether this autoregulation occurred in vivo or not, the dual interaction provided an interesting structural problem that could be approached because of the presence of large quantities of the protein TFIIIA in immature *Xenopus* oocytes.

In the autumn of 1982, Miller, a new graduate student, began studies on TFIIIA. This led to the discovery of a remarkable repeating motif within the protein, which was later, in laboratory jargon, called a zinc finger because it contained zinc (Zn) and gripped or grasped the DNA (6). This repeating structure was discovered through biochemistry and not, as some reviews have stated, by computer sequence analysis.

PREPARATION AND CHARACTERIZATION OF TFIIIA FROM THE 7S RNP

When Miller repeated the published protocols for purifying the 7S RNP, he obtained very low yields, which were attributed to dissociation. Brown and Roeder had used buffers that contained dithiothreitol (DTT) because the protein had a high cysteine content and EDTA to remove any contamination by metals, which hydrolyze nucleic acids. The gel filtration of the complex in 0.1 mM DTT resulted in a separate elution of protein and 5S RNA. However, when the strong reducing agent sodium borohydride did not disrupt the complex, it was realized that the protein was not held together by disulfide bridges and that a metal might be involved. After the particle was incubated with a variety of chelating agents, particle dissociation could be prevented only by prior addition of Zn^{2+} and not by a variety of other metals. Analysis of a partially purified 7S preparation by atomic absorption spectroscopy also revealed a significant concentration of Zn, with at least 5 mol Zn per mol particle.

While these experiments were in progress, Hanas et al. (7) reported the presence of Zn in the 7S RNP at a ratio of two per particle. This seemed to be an underestimate because their buffers contained 0.5 mM or 1 mM DTT, which has a high binding constant for Zn of about 10^{10}. Miller et al. (6) repeated the analysis with pure and undissociated particle preparations, without DTT, and took great care to ensure no contamination. The conclusion was that the native 7S RNP contains between 7 and 11 Zn ions (6). This result was consistent with the fact that the protein was known to contain large numbers of histidine and cysteine residues, the most common ligands for Zn in enzymes and other proteins. Such a result hinted at some kind of internal substructure.

DTT: dithiothreitol

A natural step was then to see if any such substructure could be revealed by proteolytic digestion. Miller et al. (6), who had already begun such studies, had found two products, an intermediate 33-kDa fragment and a limit 23-kDa fragment. At about that time Brown's group (8) also showed that, on treatment with proteolytic enzymes, the 40-kDa TFIIIA protein breaks down to a 30-kDa product, which is then converted to a 20-kDa product. They proposed that TFIIIA consists of three structural domains, which they identified as binding to different parts of the 50-base pair (bp) internal control region of the 5S RNA gene.

Carrying proteolytic studies further, Miller et al. (6) found that on prolonged proteolysis the TFIIIA product breaks down further, finally to a limit digest of about 3 kDa. In the course of this breakdown, periodic intermediates differing in size by about 3 kDa were seen. The correspondence in size between these last two values suggested that the 30-kDa domain of TFIIIA might contain a periodic arrangement of small, compact domains each of 3 kDa. If each such domain contained one Zn atom, then the observed high Zn content was explained.

This novel idea of small Zn-stabilized domains was strengthened by the timely publication by Roeder's group (9) of the amino acid sequence of TFIIIA derived from a cDNA clone. Upon inspection, the large number of cysteines and histidines present in the protein appeared to occur in a more or less regular pattern. A rigorous computer analysis by McLachlan showed that, of the 344 amino acids of the TFIIIA sequence residues, numbers 13–276 form a continuous run of nine tandemly repeated, similar units of about 30 amino acids, each containing

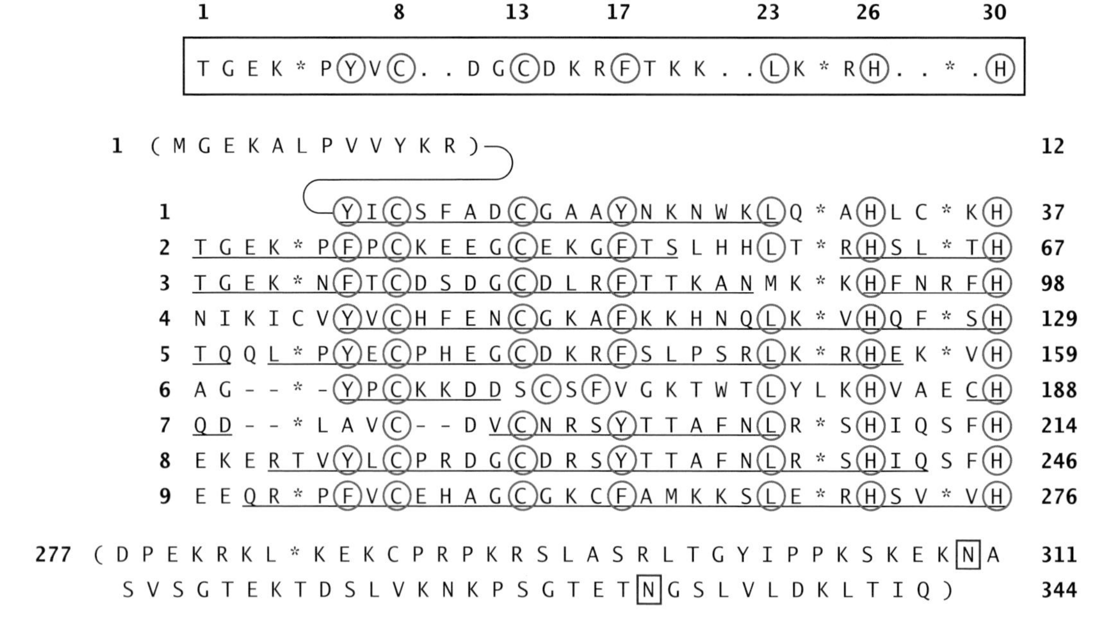

Figure 1

Amino acid sequence (9) of transcription factor IIIA from *X. laevis* oocytes aligned to show the repeating units (*underlined*) (6). The best-conserved residues are ringed in blue. Red rectangles show the most frequently occurring amino acid. The consensus sequences are shown in a box at the top, indexed on a repeat unit of 30 amino acids. The structural units or finger modules (**Figure 2**) begin at amino acid number 6, separated by a five-amino acid linker.

two invariant pairs of histidines and cysteines (**Figure 1**) (6). A repeating pattern in the sequence was also noticed by Brown et al. (10) who concluded, wrongly, that the whole protein was divided into 12 repeats, indexed on a 39-amino acid unit (although their abstract states "about 30").

A REPEATING STRUCTURE FOR TFIIIA

On the basis of the three different lines of evidence described above, namely (*a*) a 30-amino acid repeat in the sequence, which (*b*) corresponds in size to the observed periodic intermediates and the limit-digest product of 3 kDa, and (*c*) the measured Zn content of 7–11 atoms, Miller et al. (6) proposed that the 30-kDa region of the TFIIIA protein has a repeating structure consisting of nine 30-amino acid units (**Figure 2*a***). Twenty-five of the 30 amino acids in the repeat fold around a Zn ion to form a small independent structural domain or module, the "finger," and the five intervening amino acids provide the linkers between consecutive fingers (**Figure 2*b***). The Zn ion forms the basis of the folding by being tetrahedrally coordinated to the two invariant pairs of cysteines and histidines. In addition to this uniquely conserved pattern of Cys-Cys. . .His-His, each repeat also contains three other conserved amino acids, namely Tyr6 (or Phe6), Phe17, and Leu23, all of which are large hydrophobic residues (**Figure 1**). It was suggested that these might interact to form a hydrophobic cluster stabilizing the compact finger module. The 30-amino acid repeat is rich in basic and polar residues, but the largest number are found concentrated in the region between the second cysteine and the first histidine, implicating this region in particular in nucleic acid binding. This was later found to be the case (11).

Formally, when indexed on a 30-amino acid repeat, the repeating structure could be written as

1–5	6	8	13	17	23	26	30
linker	hX_1	$CX_{2,4}$	CX_3	$hX_{2,3}$	$HX_{3,4}H$	$HX_{3,4}$	H,

where h represents a conserved (large) hydrophobic residue. The proposal that each 25-amino acid module formed an independently folded, Zn-stabilized domain soon gained support from two lines of research. First, a study using extended X-ray absorption fine structure confirmed that the Zn ligands are two cysteines and two histidines (12). Second, Tso et al. (13) found that, in the DNA sequence of the gene for TFIIIA, the position of the intron-exon boundaries mark most of the proposed finger module domains.

In evolutionary terms, the multifingered TFIIIA may have arisen by gene duplication of an ancestral domain comprising ~30 amino acids. Because one such self-contained small domain would have had the ability to bind to nucleic acids and could be passed on by exon shuffling, Miller et al. (6) suggested that these domains might occur more widely in gene control proteins than in just this case of TFIIIA. The extent to which this prediction has been borne out [3% of the genes of the human genome, at the latest count (14)] is astonishing. Indeed, within months of the paper's publication, the investigators received word of sequences homologous to the zinc finger motif of TFIIIA. The first two were from *Drosophila*, the serendipity gene from Rosbash's group (15) and the *Krüppel* gene from Jäckle's group (16).

A NEW PRINCIPLE OF DNA RECOGNITION

The key points that emerged from Miller et al. (6) were that a new protein fold became known for nucleic acid binding and a novel principle of DNA recognition. The overall design for specific DNA recognition was distinctly different from that of the helix-turn-helix motif, found in the first DNA-binding proteins to be described. The latter binds to DNA as a symmetric dimer to a palindromic sequence on the DNA, thus making use of both the twofold symmetry of the DNA helix backbones and also the nucleotide sequence. Heterodimeric variations of this and other twofold symmetric designs were found later, but they still make use of the double-helix symmetry.

In contrast, the zinc finger is a DNA-binding module that can be linked tandemly in a linear, polar fashion to recognize DNA (or RNA) sequences of different lengths. Each finger domain has a similar structural framework but can achieve chemical distinctiveness through variations in a number of key amino

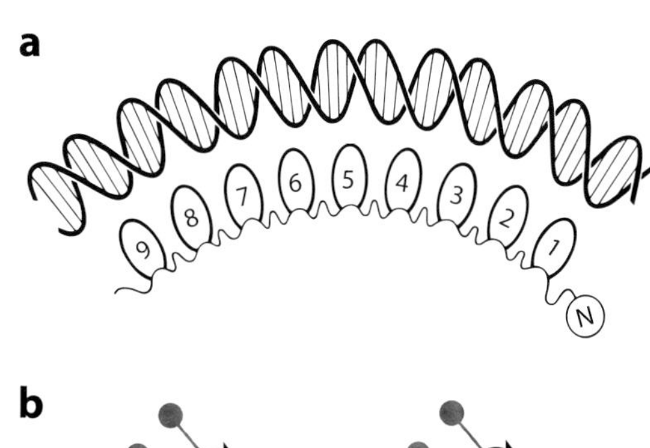

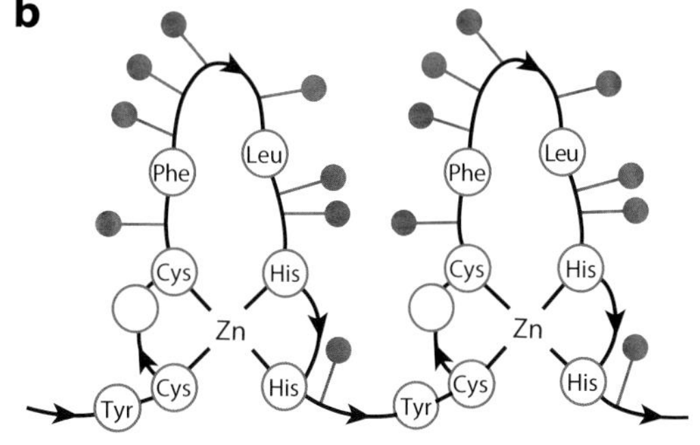

Figure 2

(*a*) Interpretation of the main structural feature of the protein TFIIIA and its interactions with 50 bp of DNA, showing combinatorial recognition by the modular design (6). (*b*) Folding scheme for a linear arrangement of repeating structural units ("zinc finger modules"), each centered on a tetrahedral arrangement of zinc ligands, Cys_2 and His_2. Gray dots indicate sites of amino acids capable of binding DNA. Also shown are the three hydrophilic groups, which were proposed to form a structural core, as confirmed later (see **Figure 4**, below) (11).

Table 1 Number of Cys_2 His_2 and Cys_4 genes in the genomes of various organisms (14)[a]

Organism	Total number of genes	Cys_2 His_2	Cys_4
Human	23,299	709 (3.0%)	48 (0.21%)
Mouse	24,948	573 (2.3%)	42 (0.17%)
Rat	21,276	466 (2.2%)	43 (0.2%)
Zebrafish	20,062	344 (1.7%)	53 (0.26%)
Drosophila	13,525	298 (2.2%)	21 (0.16%)
Anopheles	14,653	296 (2.0%)	20 (0.14%)
Caenorhabditis elegans	19,564	173 (0.88%)	270 (1.3%)
Caenorhabditis briggsae	11,884	115 (0.9%)	167 (1.4%)

[a]The Cys_4 genes refer to the steroid and thyroid hormone family of nuclear receptors, which were also later misleadingly named zinc fingers because of a similarity in the amino acid sequences, suggesting the presence of zinc tetrahedrally ligated to four cysteines. The structures are, however, quite different from the classical Cys_2 His_2 finger. Note how the proportion of Cys_2 His_2 genes increases with greater complexity on the evolutionary scale.

acid residues. This modular design thus offers a large number of combinational possibilities for the specific recognition of DNA (or RNA). It is not surprising that it is found widespread throughout so many different types of organisms (**Table 1**).

THE STRUCTURE OF THE ZINC FINGER AND ITS INTERACTION WITH DNA

Miller et al. noted (6) that, in addition to the characteristic arrangements of conserved cysteines and histidines that are fundamental in the folding of the finger by the coordinating Zn, there are three other conserved amino acids, notably Tyr6, Phe17, and Leu23, and suggested that they were likely to form a hydrophobic structural core of the folded structure. In other words, the seven conserved amino acids in each unit would provide the framework of tertiary folding, whereas some of the variable residues determined the specificity of each domain. Berg (17) built on these original observations by fitting known structural motifs from other metalloproteins to the consensus sequence of the TFIIIA finger motifs. His proposed model consisted of an antiparallel β-sheet, which contains a loop formed by the two cysteines, and an α-helix containing the His-His loop. The two structural units are held together by the Zn ion. In analogy with the way in which the bacterial helix-turn-helix motif binds DNA, DNA recognition was postulated to reside mostly in the helical region of the protein structure.

Berg's model was confirmed in outline by the NMR studies of Wright's group (18) on a single zinc finger in solution and by Neuhaus in the MRC laboratory (19, 20) on a two-finger peptide (**Figure 3**). Neuhaus' work took longer to solve the structure, but it had the merit of showing that adjacent zinc fingers are structurally independent in solution because they are joined by flexible linkers.

The precise pattern of amino acid interactions of zinc fingers with DNA remained unknown. The breakthrough came in 1991 when Pavletich & Pabo (11) solved the crystal structure of a complex of a DNA oligonucleotide specifically bound to the three-finger DNA-binding domain of the mouse transcription factor Zif268, an early response gene. The primary contacts are made by the α-helix, which binds in the DNA major groove through specific hydrogen-bond interactions from amino acids at helical positions –1, 3, and 6 to three successive bases (a triplet) on one strand of the DNA (**Figure 4*a***). Later, the second zinc finger-DNA complex solved, by Fairall & Finch (21) in Rhodes' group, revealed an important secondary interaction from helical position 2 to the other strand (**Figure 4*b*** and **Figure 5**). This is the canonical docking arrangement, but there are, however, some wide variations from

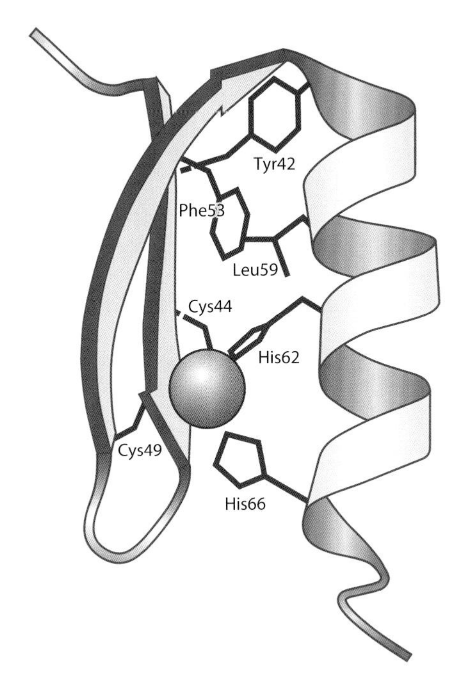

Figure 3

The structure of a zinc finger from a two-dimensional NMR study of a two-finger peptide in solution (20). The same study showed that the linker between the two modules is highly flexible (19).

this arrangement in the family of zinc finger-DNA complexes now known (22). There are also, of course, still other interactions, such as with the phosphates of the DNA backbones, but these do not play a direct part in specific recognition.

ZINC FINGER PEPTIDES FOR THE REGULATION OF GENE EXPRESSION

The mode of DNA recognition by a finger is thus principally a one-to-one interaction between individual amino acids from the recognition helix to individual DNA bases (11). This is quite unlike the case of other DNA-binding proteins, where one amino acid may contact two bases and vice versa. Moreover, because the fingers function as independent modules, fingers with different triplet specificities can be linked to give specific recognition of longer DNA sequences. For this reason, the zinc finger motifs are ideal natural building blocks for the de novo design of proteins for recognizing any given sequence of DNA. Indeed the first

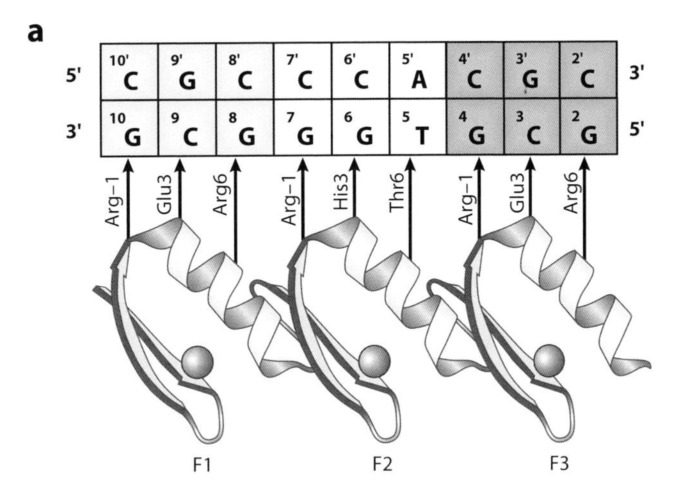

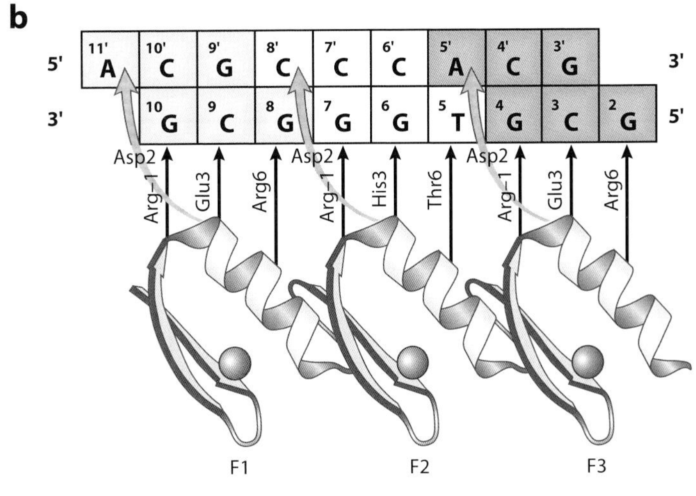

Figure 4

(*a*) Schematic diagram (32) of the first model of modular recognition of DNA by a three-zinc finger peptide, illustrating the results of the first crystal structure determination of the complex between the DNA-binding domain of the transcription factor Zif268 and an optimized DNA-binding site (11). Each finger interacts with a 3-bp subsite on one strand of the DNA, using amino acid residues in helical positions −1, 3 and 6. (*b*) Refined model of DNA recognition (32, 33). View of the potential hydrogen bonds to the second strand of the DNA, the so-called cross-strand interactions, emanating from position 2 on the recognition helix. This is based on the crystal structure of the tramtrack-DNA complex (21), the mutagenesis (32) and phage display selection studies of Isalan et al. (33), on the refined structure of the Zif268-DNA complex, and of variants by Pabo and his colleagues (22). The fingers ideally bind 4-bp overlapping subsites, so that adjacent fingers are functionally synergistic though structurally independent.

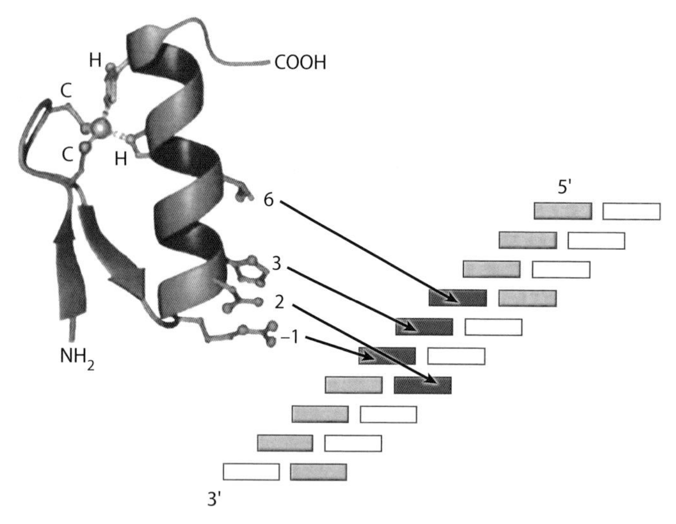

Figure 5

Another view of the refined model. Contacts with the DNA are made by amino acids at helical positions −1, 3, and 6 to the coding strand and from position 2 to the noncoding strand. Thus, the binding site for a finger is not simply a triplet of three successive bases, but a 4-bp site overlapping with that of the preceding finger (32). Illustration courtesy of Sangamo BioSciences, Inc.

experiments by Berg (17) and others, using site-directed mutagenesis, showed that it is possible to rationally alter the DNA-binding characteristics of individual zinc fingers when one or more of the amino acids in the α-helical positions are varied. As a small collection of these mutants accumulated, it became possible to find some regularities or rules relating amino acids on the recognition α-helix to corresponding bases in the bound DNA sequence.

The MRC laboratory adopted a different approach. The reason was that these rules did not take into account the fact that real DNA structures are not fixed in the canonical B form, but as was shown earlier, there are wide departures, depending on the DNA sequence (23, 24). This was further supported by the structure of the zinc finger-tramtrack DNA complex (21). Here the helical position used for the primary contact by the first finger with the 3′-most base of the cognate triplet (thymine) is not the canonical –1, but 2. The cause is that the DNA helix is deformed from the B form, with a thymine followed by an adenine at a helical rotation angle of 39°, rather than the canonical 36°, and preceded by another adenine at an angle of 33°. The interaction with the finger thus occurs at an ATA sequence, which has unusual flexibility, as noted long ago (23). DNA is not a rigid, passive participant in its interaction with proteins.

AFFINITY SELECTION FROM A LIBRARY OF ZINC FINGERS BY PHAGE DISPLAY

The alternative to this rational but biased design of proteins with new specificities was the isolation of desirable variants from a large pool or library. A powerful method, namely phage display, of selecting such proteins is the cloning of peptides (25) or protein domains (26) as fusions to the minor coat protein (pIII) of bacteriophage fd, which leads to their expression on the tip of the capsid. Phages displaying the peptides of interest can be affinity purified by binding to the target and then amplified in bacteria for use in further rounds of selection and for DNA sequencing of the cloned genes. This technology was applied to the study of zinc finger-DNA interactions after Choo demonstrated that functional zinc finger proteins could be displayed on the surface of fd phage and that such engineered phage could be captured on a solid support coated with the specific DNA (27, 28). The phage display method was also adopted by other groups working on zinc fingers, including those of Pabo and Barbas.

Phage display libraries comprising $\sim 10^7$ variants of the middle finger from the DNA-binding domain of Zif268 were created. A DNA oligonucleotide of fixed sequence was used to bind and hence purify phage from this library over several rounds of selection, returning a number of different but related zinc fingers which bind the given DNA. By comparing similarities in the amino acid sequence of functionally equivalent fingers, the likely mode of interaction of these fingers with DNA was deduced (27). Remarkably, most base contacts were found to occur from three primary positions on the α-helix of the zinc finger, correlating well with the implications of the crystal structure of Zif268 bound to DNA (11).

This demonstrated ability to select zinc fingers with desired specificity meant that, as the data from the selections accumulated, some wider general rules could be devised for a recognition code (28), and hence, DNA-binding peptides could be made to measure using the combinatorial strategy exemplified by TFIIIA. In other words, these general rules could be used for the rational design of a zinc finger peptide to recognize a short run of DNA sequence by mixing and matching individual specific fingers. Where the general rules for finger specificity led to an ambiguity, as in the case of closely related triplets, e.g., GCG and GTG, Choo & Klug (28) showed that zinc finger modules could nevertheless be selected to discriminate between them.

USE OF ENGINEERED ZINC FINGER PEPTIDES TO REPRESS GENE EXPRESSION IN A MOUSE CELL LINE

One interesting possibility for the use of such zinc finger peptides is to selectively target genetic differences in pathogens or transformed cells. In December 1994, Choo et al. (29) reported the first such application, which built a protein that recognized a specific DNA sequence both in vitro and in vivo. This was a crucial test of the understanding of the mechanism of zinc finger DNA recognition. This proof of the principle led to future zinc finger studies of potential applications in gene regulation for research purposes and for therapeutic correction. It also stimulated the creation of the first biotech companies (Sangamo BioSciences, Inc., in Richmond, California, and later Gendaq Ltd. in Cambridge, United Kingdom) to exploit the new technology.

In summary, a three-finger peptide was created that is able to bind site specifically to a unique 9-bp region of the p190 *bcr-abl* cDNA: This is a transforming oncogene, which arises by translocation between the tips of chromosomes 9 and 22, of which one product is the Philadelphia chromosome (29). Chromosome 22 contains a novel DNA sequence at the junction of two exons, one each from the two genomic parent *bcr* and *abl* genes. The engineered peptide discriminated in vitro against like regions of the parent *bcr* and *c-abl* genes, which differ in only a single base out of the 9-bp target, by factors greater than one order of magnitude (29).

This peptide also contained a nuclear localization signal (NLS) fused to the zinc finger domain so that the peptide could accumulate in the nucleus. Consequently, stably transformed mouse cells, made interleukin-3 independent by the action of the oncogene, were found to revert to IL-3 dependency on transient transfection with a vector expressing the peptide. This construct was also engineered to contain a c-myc epitope, which enabled investigators to follow by immunofluorescence the localization of the peptide to the nuclei of the transfected cells. When IL-3 is subsequently withdrawn from cell culture, over 90% of the transfected p190 cells become apoptotic (that is, showing chromosome degradation) within 24 h (**Figure 6**, left). These experiments were repeated on cells transformed by another related oncogene, p210 *bcr-abl*, which served as a control. All transfected p210 cells maintained their IL-3 dependency and remained intact on entry of the engineered peptide (**Figure 6**, right) (29).

Measurements of the levels of p190 *bcr-abl* mRNA extracted from cells treated with the peptide showed that the repression of oncogenic expression by the zinc finger peptide was due to a transcriptional block imposed by the sequence-specific binding of the peptide, which, with its highly basic NLS, presumably obstructed the path of the RNA polymerase. In later experiments to inhibit gene expression, a repression domain was added, such as the Kox domain from the *Xenopus* KRAB zinc finger family, and fused to the zinc finger construct (30, 31).

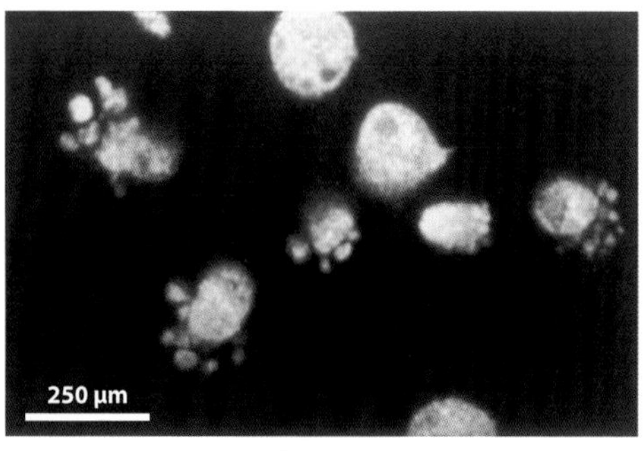

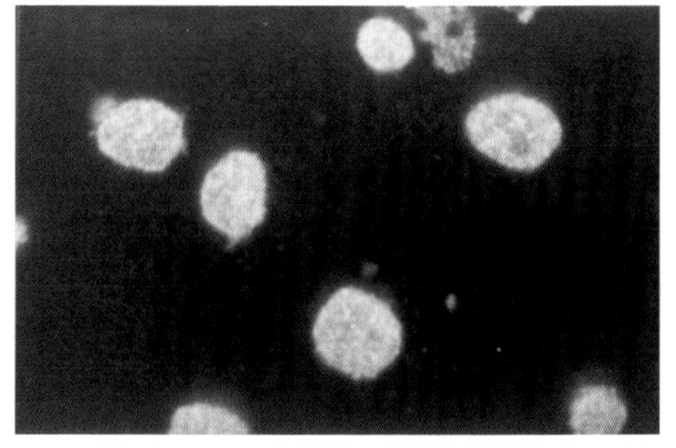

Figure 6

(*left*) An engineered site-specific zinc finger DNA-binding protein, designed de novo against an oncogenic *bcl-abl* fusion sequence (p190) transformed into a BaF3 mouse cell line, represses expression of the oncogene (29). Immunofluorescence image of cell nuclei 8 h after transfection with the zinc finger protein, showing apoptosis. After 24 h, 95% of the cells are destroyed. (*right*) The same on a control cell line, which has been transformed by a related but different *bcl-abl* oncogenic sequence (p210). The cells are not affected.

PROMOTER-SPECIFIC ACTIVATION BY A CHIMERIC ZINC FINGER PEPTIDE

These experiments showed that a zinc finger peptide could be engineered to switch off gene expression in vivo. The same paper (29) also described other experiments on a different cell system (cultured mouse fibroblasts) to show that a gene could also be switched on by a zinc finger construct. The same 9-bp sequence was used, but this time as a promoter for a CAT reporter gene contained in a plasmid. The peptide, which recognized the promoter sequence, was fused to a VP16 activation domain from herpes simplex virus and, on transient transfection, stimulated expression of the reporter gene by a factor of 30-fold above controls.

IMPROVING ZINC FINGER SPECIFICITY BY PROTEIN ENGINEERING

First having shown proof of the principle that engineered zinc finger peptides could be used to target DNA, improving the specificity of recognition became the focus of subsequent work. Although the main source of specificity lies in the amino acids at positions −1, 3, and 6 of the recognition α-helix of a zinc finger for successive bases lying on one strand of a DNA triplet, Isalan et al. (32) found that the "cross-strand" interaction, described above, from helical position 2 to the neighboring base pair on the adjacent triplet (**Figure 4*b***) can significantly influence specificity. Therefore, it has been necessary to revise the simple model that zinc fingers are essentially independent modules that bind 3-bp subsites to a model that considers functional synergy at the interface between adjacent independently folded zinc fingers. In this refined model, Zif268-like zinc fingers potentially bind 4-bp overlapping subsites (**Figures 4*b*** and **5**) (33).

Consequently, Isalan et al. (34) redesigned the method of phage library construction to allow for the optimization of the interaction that a finger makes with the DNA-binding site of the adjacent N-terminal finger. They adopted a bipartite selection strategy in which two halves of a three-finger DNA-binding domain are selected separately and then recombined in vitro to make a complete three-finger domain, which binds 9 bp and automatically allows the interface synergy between its constituent three fingers. These two separate (nonoverlapping) libraries are then used to perform the bipartite selections. Because all the steps are carried out in vitro, the method is rapid and easily adapted to a high-throughput automated format. This was applied commercially by the MRC spin-off company Gendaq (later acquired by Sangamo BioSciences) for up to $2 \times 4^5 = 2048$ binding sites to create a large archive of zinc finger peptides that recognize a vast number of DNA 9-bp sequences. Second, an important step forward was to increase the length of the DNA sequence targeted and hence its degree of rarity; at the same time, an increase was expected in the binding affinity of the longer zinc finger construct. Three zinc fingers recognize a 9-bp sequence, which would occur randomly many thousands of times in a large genome. Therefore, six fingers linked together would recognize a DNA sequence 18 bp in length, which is sufficiently long to constitute a rare address in the human genome. However, one cannot simply go on adding fingers with the conventional linkers because the periodicity of the packed fingers does not quite match the DNA periodicity. They thus tend to get out of register and are strained in doing so, leading to only a small increase in affinity. Investigators have, therefore, learned how to engineer longer runs of zinc fingers that can target longer DNA sequences and have affinities a thousand times greater than three-finger peptides (**Figure 7**) (35, 36).

An early design by Kim & Pabo (37) was to use a longer more flexible linker between two different preexisting three-finger domains to form a six-finger peptide, but this has not been used much in practical applications. The method of choice in current use is that developed in Cambridge by Moore et al. (36) (and later transferred to Sangamo via Gendaq). This uses two-finger binding domains, which can be obtained from the archive described above. Three of these are assembled into six-finger domains, using longer variants of the conserved six-amino acid linker TGEKP (**Figure 7*b***). The linkers contain an extra glycine residue or a glycine-serine-glycine tripeptide between the constituent two-finger modules. Such six-finger domains (denoted 3 × 2F) bind their 18-bp targets with picomolar affinity, as also do six fingers made by linking two three-finger domains (2 × 3F) made with an extended linker (36). However, the advantage of the 3 × 2F over the 2 × 3F strategy is that it is much more sensitive to a mutation or an insertion in the target sequence, with a loss of affinity of up to 100-fold. Thus, the 3 × 2F peptide discriminates more strongly than the 2 × 3F peptide between closely related DNA target sequences. The logic behind the 3 × 2F design was that the strain in binding longer DNA targets is more evenly distributed than in a 2 × 3F construct, and it indeed turned out to have the advantage expected. These six-finger peptides not only have picomolar affinities for their 18-bp targets but also give virtually single-gene recognition (38) when tested on DNA microarrays displaying 20 thousand different sequences. As mentioned above, a library of two-finger peptides begun in an MRC laboratory, Cambridge, was transferred via Gendaq to Sangamo.

Another strategy Moore et al. (35) developed was to target two noncontiguous 9-bp DNA sequences separated by up to 10 bp of unbound DNA. Using a nonspecific binding finger (in which all key amino acids had been mutated to serine) as a structured linker, investigators found that it could span a gap of 7 to 8 bp and maintain picomolar affinity. In contrast, the use of a flexible linker such as $(GSG)_n$ displayed no preference for a length of span, but the affinity was reduced to ~50 pM, probably attributable to the increased conformational entropy of the long peptide. These strategies have not yet been deployed.

SOME APPLICATIONS OF ENGINEERED ZINC FINGER PROTEINS

Zinc finger proteins (ZFPs) can be engineered with a variety of effector domains fused to polyzinc finger peptides, which can recognize virtually any desired DNA sequence with high affinity and specificity. They thus form the basis of a novel technology, which has increasing uses in research and medicine. An excellent summary of numerous such applications has been given by Pabo and his colleagues (39). A few of these are mentioned here.

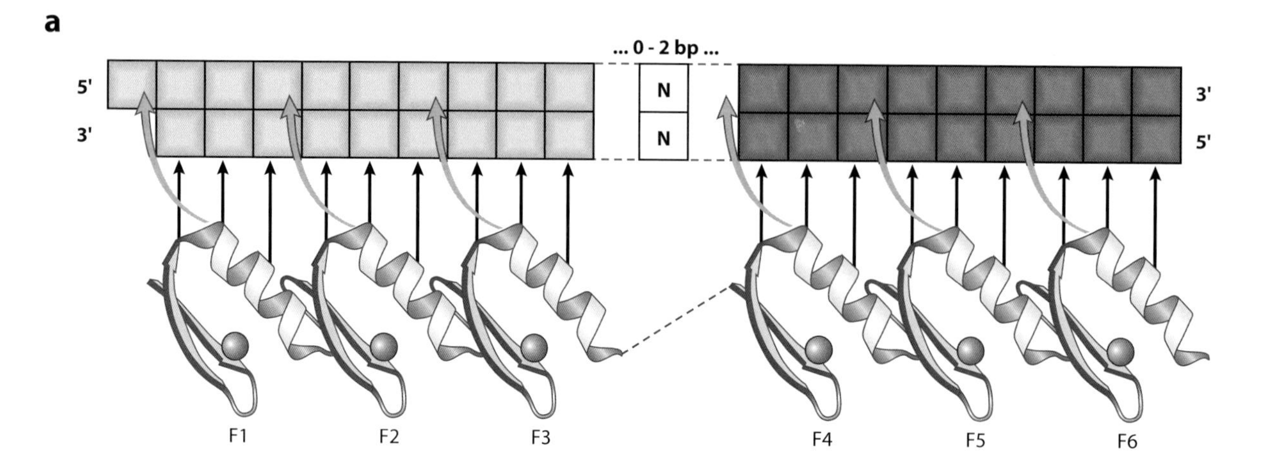

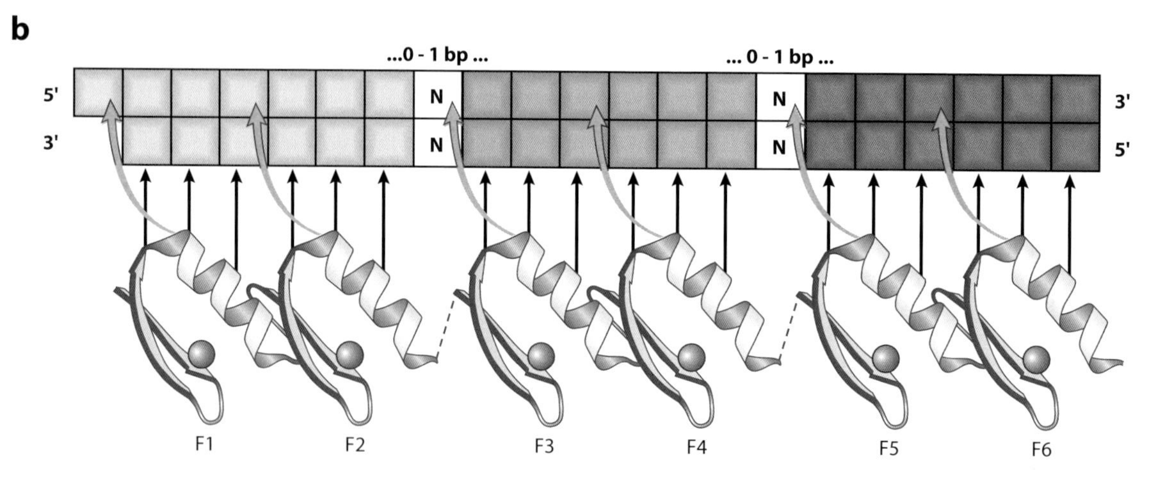

Figure 7

Two modes of generating six-zinc finger proteins for specific recognition of 18-bp sequences (35, 36). (*a*) Two three-finger peptides fused together using an extended canonical linker (2 × 3F scheme). (*b*) Three two-finger peptides linked using canonical linkers extended by an insertion of either a glycine residue or a glycine-serine-glycine sequence in the canonical linkers between fingers 2 and 3 and fingers 4 and 5, respectively.

VEGF: vascular endothelial growth factor

1. Inhibition of human immunodeficiency virus (HIV) expression (30): It was shown that ZFPs targeted to the HIV promoter long terminal repeat activated by the tat protein effectively repressed expression, and preliminary experiments in a cellular infection assay gave a threefold drop in infectivity.
2. Disruption of the effective cycle of infection of herpes simplex virus (31): A ZFP transcription factor designed to repress the promoter of a viral gene that is normally activated first in the replication cycle produced a tenfold reduction in the virus titer in an infected cell line (**Figure 8**). This was a good result, considering that there are five other "immediate early" genes that contribute to the infection. Several more of these would have to be repressed to reduce the titer further.
3. Activating the expression of vascular endothelial growth factor (VEGF)-A in a human cell line (40) and in an animal

model (41): These experiments have led to a therapeutic application (see below).

4. Regulation of the level of zinc finger expression by a small molecule (42): This can be used in controlling the dose and/or timing for a therapeutic application. An efficient way of achieving this is by fusing the ZFP to the ligand-binding domain of a steroid hormone nuclear receptor. In the absence of hormone, the ZFP transcription factor is retained in the cytoplasm, but after ligand binding, the ZFP translocates to the cell nucleus in active form.

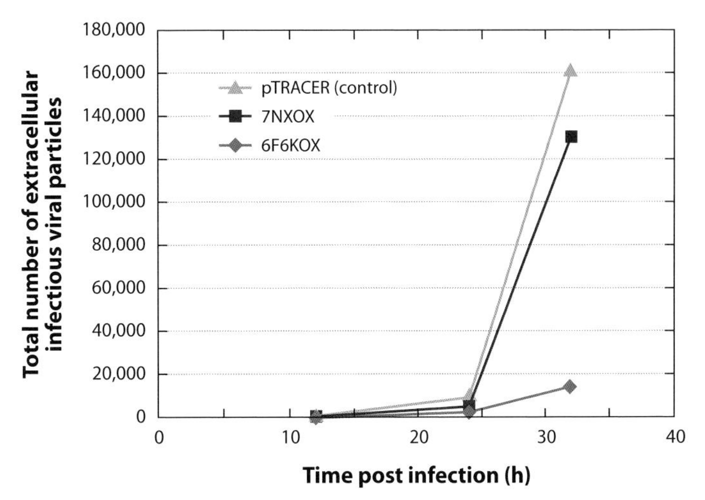

Figure 8

Repression of herpes simplex virus (HSV-1) infection (at 0.05 pfu/cell) of FACS-sorted HeLa cells (9 × 100,000) by zinc finger peptides fused to a KOX1 repression domain (31). The peptides are targeted to the viral gene *IE175 K*, the first of the six immediate early genes to be expressed by the virus. A six-finger recognition peptide, 6F6KOX, reduces the virus titer tenfold, whereas a three-finger peptide produces only a 20% reduction.

STIMULATION OF NEW VASCULATURE BY ENGINEERED ZINC FINGER PROTEINS

Following the work at Sangamo on the activation of the transcription factor VEGF by ZFPs in mouse and human cell lines (40), experiments showed that new blood vessels could be formed in a mouse ear (41). These did not leak, unlike the results that had been obtained by earlier workers who had been delivering various cDNA-spliced isoforms of the gene. The reason for the success with ZFP activation is that the latter acts on the promoter of the VEGF gene, and hence, all spliced isoforms are naturally produced when the gene is induced to promote angiogenesis.

Subsequent work by F. Giordano (unpublished results) at Yale and B. Annex (unpublished results) at Duke University showed increased blood flow in hind limbs of ischemic rabbits, with the ZFP delivered by a retrovirus or simply by injecting the DNA. Two years ago Sangamo began clinical trials to evaluate the ability of an appropriate ZFP to stimulate the natural growth of normal blood vessels in treating claudication, a symptom of peripheral arterial obstructive disease that causes poor blood flow in the legs. The phase 2 trials have shown great improvements and, indeed, also in the more serious condition of critical limb ischemia.

GENE CORRECTION BY HOMOLOGOUS RECOMBINATION USING SEQUENCE-SPECIFIC ZINC FINGER NUCLEASES

Gene correction is the process by which sequence alterations in defective or deleterious genes can be changed or corrected by homologous recombination (HR)-mediated gene conversion between the target locus and a donor construct encoding the corrective sequence (**Figure 9**, left). Monogenic disorders, such as X-linked severe combined immune deficiency (SCID), sickle-cell anemia, hemophilia, and Gaucher's disease, are caused by the inheritance of defective alleles of a single gene. The ability to replace this gene sequence via HR-mediated gene correction has the potential of fully restoring the gene function and providing a permanent cure for patients with these disorders. However, this process is highly inefficient in that the frequency of unaided HR at a specific locus occurs is only about 1 in 10^5 cells (43). This is far below a level that would be considered therapeutic. A double-stranded

ZFP: zinc finger protein

HR: homologous recombination

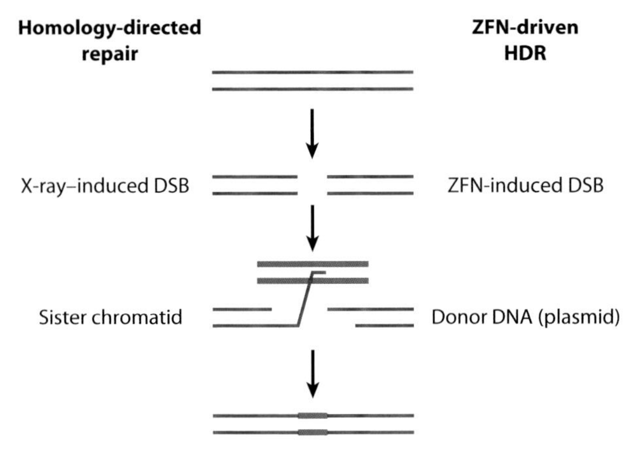

Figure 9

Gene modification or correction using homologous recombination (43) via "short-path gene conversion" stimulated by a double-stranded break (DSB) (44). The left side schematically depicts the repair of a random X-ray-induced DSB by homologous recombination using the sister chromatid as the repair donor. As shown on the right, a site-specific DSB is created by zinc finger protein nucleases (ZFNs). Abbreviation: HDR, homology-directed repair.

DSB: double-stranded break

ZFN: zinc finger nuclease

break (DSB) has been demonstrated by Jasin (44) to potentiate HR at a specific genetic locus by ~5000-fold. Therefore, the introduction of a corrective donor sequence together with a site-specific nuclease that would produce a DSB at or near the location of the mutation could stimulate gene correction to levels that would provide a therapeutic impact.

Jasin's demonstration was based on artificially introducing into an endogenous gene of a human cell line an 18-bp DNA sequence, which was the specific binding site for the homing endonuclease Sce1 and had only a small probability of occurring naturally elsewhere in the genome. However, to carry out gene correction in native cells requires the specific targeting of the mutated sequence, and a zinc finger peptide fused to a nuclease domain is the natural choice (**Figure 9**, right). A nuclease of this type has been developed by Chandrasegaran and coworkers (45) using an engineered ZFP fused to the nonspecific cleavage domain of the Fok1 type II restriction enzyme. This type of zinc finger nuclease (ZFN) has been used by Carroll and colleagues (46) to produce mutants in *Drosophila* by gene correction and by Porteus & Baltimore (47) in a green fluorescent protein model system to study gene modification in a human cell line. Following Chandrasegaran, the three-finger ZFP nucleases were introduced as pairs with tandem binding sites engineered in opposite orientations (**Figure 10**), with a 6-bp spacing separating the two half sites (45). In both cases, the efficiency of the targeting is not very high, but this was not crucial to those studies. Indeed, some of the *Drosophila* mutants were lethal, but they are normally selected on the phenotype. Both studies showed that a ZFN-produced DSB can markedly increase the rate of HR between a donor DNA construct and a reporter gene in their two different systems.

The results of Carroll's and of Porteus' laboratories stimulated work at Sangamo to apply the gene correction method to tackle monogenic disorders of the human genome. The workers set out to determine whether ZFNs could create a comparable increase in HR frequency at an endogenous human gene. They focused their efforts on the *IL2R*γ gene in which loss-of-function point mutations cause X-linked SCID. In the absence of bone marrow transplantation or gene therapy, this malady leads to death in early childhood. Treatment for the disease by gene therapy (48) has been performed by inserting one or more copies of the normal gene in the chromosomes of a number of affected children, that is, by gene addition rather than gene correction. After successful

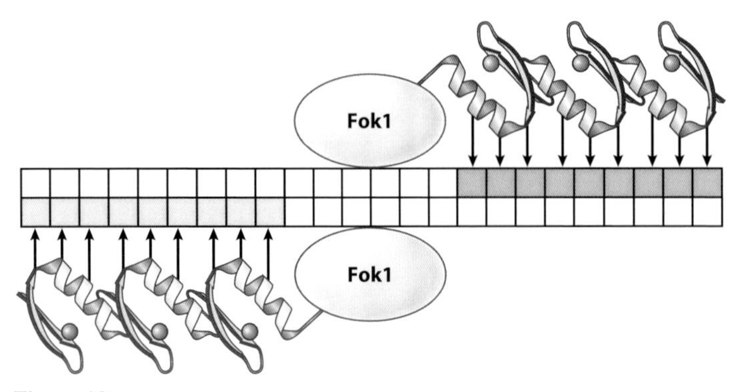

Figure 10

Gene correction using a pair of three-zinc finger protein nucleases (ZFNs) (45) to produce a double-stranded DNA break. The zinc finger peptides are linked to the nonspecific catalytic domain of the Fok1 endonuclease by a short amino acid linker. In the three-dimensional structure the two catalytic domains form a dimeric association.

treatment of the majority of the patients, two of them died some years later of leukemia, probably because the random insertion of the gene led to its coming under inappropriate control.

The middle exon of the *IL2Rγ* gene contains SCID mutation hot spots. Because maximal homology-driven recombination occurs when a DSB is evoked at or close to the mutated site, the Sangamo workers engineered a pair of ZFNs specific for the exact location of the mutation on the X chromosome. Two DNA-binding domains were assembled from the zinc finger archive, described above, each containing four highly specific zinc finger motifs, and thus simultaneously recognizing two different 12-bp sites, separated by a fixed distance between them. The chance of this particular pattern existing elsewhere in the genome is negligible. Having confirmed in vitro that the proteins bind as intended, they next improved them further by single amino acid substitutions in the zinc finger recognition helices, which gave an additional fivefold increase in potency. The results (49) showed an 18% to 20% rate of gene correction, which was stable after one month in culture, in the target cells (**Figure 11**).

This accomplishment is dramatic, with an increase by many orders of magnitude over anything achieved in the past by "gene targeting," particularly as no selection has been used. Moreover, measurements of both the mRNA and protein levels expressed by the corrected *IL2Rγ* gene showed that the mutation had been efficiently and stably corrected. The next steps in curing SCID disease are to isolate CD34 progenitor cells from the patient's bone marrow and correct them *ex vivo*. After this gene correction, the cells can be allowed to expand and then be reintroduced into the bone marrow to repopulate it with corrected cells. Of course, an extensive study of ZFN safety (e.g., verifying that no other DSBs are created outside the target and checking for immunogenicity) must first be undertaken, but there is every prospect of gene correction eventually becoming a reality for SCID and other monogenic diseases.

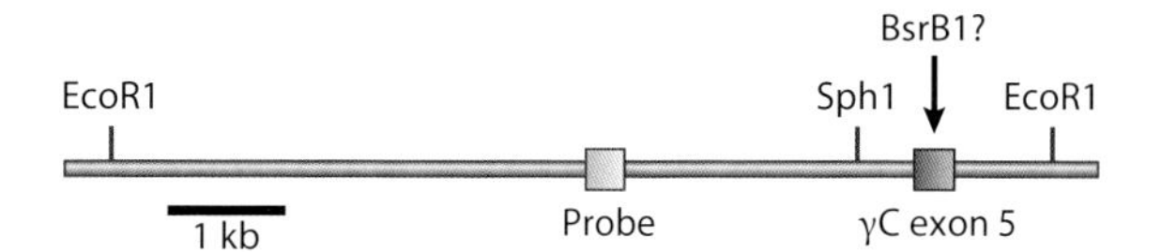

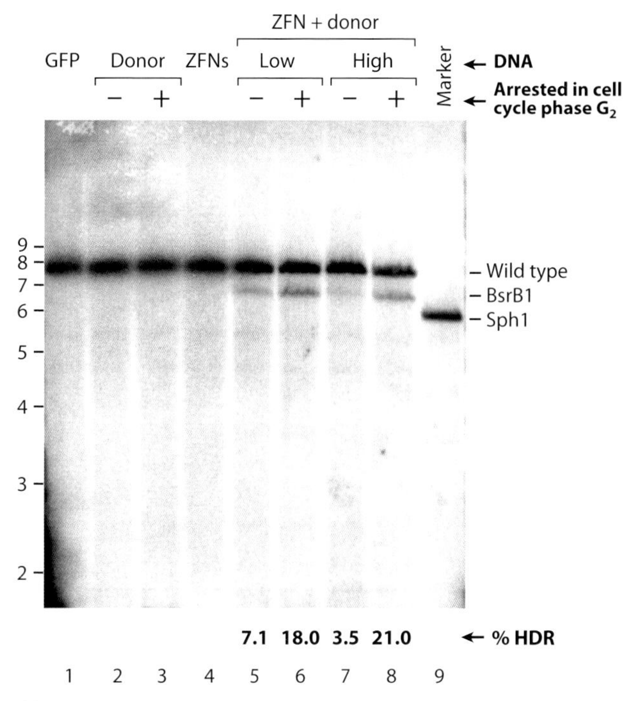

Figure 11

Results of gene correction for severe combined immune deficiency disease by the Sangamo group (49). High-frequency homology-driven repair (HDR) at the endogenous *IL2Rγ* locus of leukemia K562 cells produced by designed zinc finger nucleases (ZFNs). Cells were transfected with the indicated plasmids (at two different concentrations indicated by *low* and *high* to ensure that the exogenous DNA load was not limiting). The donor DNA was marked by a BsrB1 restriction site (not present in the K562 genome), which was created by replacing a single nucleotide of one codon so as to preserve the amino acid code. After one month, genomic DNA was isolated, digested to completion with EcoR1, BsrB1, and Dpn1, and the percentage of BsrB1-carrying chromosomes was measured by Southern blotting with the probe indicated. A Sph1 restriction fragment is also shown as a size marker. Abbreviation: GFP, green fluorescent protein.

SPECIFICITY OF ZINC FINGER CONSTRUCTS

Like any conventional small-molecule drug, a ZFN can produce secondary effects, but in this case, it is possible to assess this by measuring the binding to off-target sites on the DNA. This was done by Sangamo in their experiments on establishing HIV resistance in CD4 T cells by genome editing (50). Here the target was the HIV receptor CCR5, which was disrupted at 36% efficiency, and effects at the top 15 DNA sequence-related sites were read by 454 DNA pyrosequencing technology. The largest off-target effect (5.4%) was at the biologically related CCR2 site, and of the other 13 sites, one was at an intron of a gene *AβLIMZ* on chromosome 4, which showed a very low frequency of mutations in the two 38,000 sequences examined. Taken together with the "surveyer" Cel-1 nuclease data and the preservation of the biological properties of the cell, all the results support the conclusion that the ZFNs used in the CCR5 work are highly specific.

TARGETED GENE KNOCKOUT

Gene knockout is the most powerful tool for determining the function of a gene and also for permanently modifying the phenotype of a cell. Recent methods use HR (43) where recognition of the target is by the homology of the extra-chromosomal donor DNA (see above). The low efficiency of this process can necessitate screening thousands of clones. Rapid gene knockout can be achieved by simply using a ZFN to create a DSB at the target and, in the absence of a DNA donor, allowing it to be repaired by the natural process of nonhomologous end joining (NHEJ). This is an imperfect repair process and usually results in changes to the DNA sequence at the site of the break and hence to mutant (null) alleles of the protein products. In this way, investigators at Sangamo produced a disruption of the dihydrofolate reductase gene in CHO cells at frequencies greater than 1%, thus obviating the need for selection markers. This established a new method for gene knockouts (51), with applications to reverse genetics and recombinant protein productions in mammalian cells in a serum-free media.

As an unnamed reviewer remarked, "it is ironic that the NHEJ pathway, responsible for the annoying background of random integrants that has plagued gene targeting for so long, is the basis of a simple and efficient method for knockouts" (43). A spectacular demonstration of the power of NHEJ-targeted knockout has appeared recently. Meng et al. (52) and Doyon et al. (53) have shown that heritable mutations can be produced in zebrafish by generating mRNAs encoding ZFNs for the locus of interest and injecting them into embryos at the one-cell stage. Both groups also measured off-target effects: Meng et al. found a rate of ~1%, but Doyon et al. detected no off–target cleavage. The most likely (and comforting) explanation of the difference is that the latter used four-finger ZFNs and thus was more specific than the three-finger ZFNs used by Meng et al.

OTHER APPLICATIONS OF ENGINEERED ZINC FINGER PROTEINS

The Sangamo ZFP development programs for obstructive limb disease, diabetic neuropathy, and HIV/AIDS are the most advanced, but other programs underway include amyotrophic lateral sclerosis (ALS, a motor neuron disease) and nerve regeneration. These can be followed on the company's Web site at **http://www.sangamo.com**.

It should also be mentioned that customized zinc finger constructs for given DNA sequences can now be obtained commercially from Sigma-Aldrich, as laboratory reagents under the name Compo-Zr® technology.

Another application of zinc finger technology outside "the therapeutic space" is in the strategic partnership Sangamo has with Dow Agricultural Sciences for use of ZFNs in the breeding of enhanced plant crops. One example that has been announced is the layering of traits (quantitative trait loci or QTLs) in which particular alleles of a set of genes can be introduced to give a particular variant of a crop.

Thanks largely due to the development of a robust zinc finger technology by Sangamo BioSciences for therapeutic applications, the power of the new principle of gene recognition, discovered in 1985 (29), has been recognized by other groups. Thus, Sangamo's recent paper in *Nature* on genome modification in maize (54) was accompanied by a paper from another group on tobacco (55). A science correspondent for the *Daily Telegraph* recently called the technology "precision gene surgery." The availability of customized zinc finger reagents from Sigma-Aldrich, mentioned above, means that the technology is now accessible to all researchers without the means to make their own.

Frequent press releases from Sangamo BioSciences chart the progress of the applications, e.g., most recently to diabetic neuropathy, where nerve conduction has been re-established in what were essentially blocked nerves.

DISCLOSURE STATEMENT

I am on the Scientific Advisory Board of Sangamo BioSciences, Inc., a biotech company in Point Richmond, California, which is further developing the zinc finger technology for applications in therapeutics and plant crops engineering.

ACKNOWLEDGMENTS

I thank Dr. Philip Gregory and his colleagues at Sangamo BioSciences for many helpful discussions and for permission to use their figures (reproduced in **Figures 9**, **10**, and **11**). I would also like to acknowledge my former colleagues, Yen Choo, Mark Isalan, and Michael Moore at the MRC in Cambridge, and later at Gendaq, who helped develop zinc fingers into a robust technology. I am grateful to Jesslyn Holombo for her careful editing of the original manuscript.

LITERATURE CITED

1. Kornberg RD. 1977. Structure of chromatin. *Annu. Rev. Biochem.* 46:931–54
2. Klug A. 1983. From macromolecules to biological assemblies. In *Les Prix Nobel en 1982*, pp. 93–125. Stockholm: Nobel Found.
3. Brown DD. 1984. The role of stable complexes that repress and activate eukaryotic genes. *Cell* 37:359–65
4. Picard B, Wegnez M. 1979. Isolation of a 7S particle from *Xenopus laevis* oocytes: a 5S RNA-protein complex. *Proc. Natl. Acad. Sci. USA* 76:241–45
5. Pelham HRB, Brown DD. 1980. A specific transcription factor that can bind either the 5S RNA gene or 5S RNA. *Proc. Natl. Acad. Sci. USA* 77:4170–74
6. Miller J, McLachlan AD, Klug A. 1985. Repetitive zinc-binding domains in the protein transcription factor III A from *Xenopus* oocytes. *EMBO J.* 4:1609–14
7. Hanas JS, Hazuda DJ, Bogenhagen DF, Wu FY, Wu CW. 1983. *Xenopus* transcription factor A requires zinc for binding to the 5S RNA gene. *J. Biol. Chem.* 258:120–25
8. Smith DR, Jackson IJ, Brown DD. 1984. Domains of the positive transcription factor specific for the *Xenopus* 5S RNA gene. *Cell* 37:645–52
9. Ginsberg AM, King BO, Roeder RG. 1984. *Xenopus* 5S gene transcription factor, TFIIIA: characterization of a cDNA clone and measurement of RNA levels throughout development. *Cell* 39:479–89
10. Brown RS, Sander C, Argos P. 1985. The primary structure of transcription factor TFIIIA has 12 consecutive repeats. *FEBS Lett.* 186:271–74
11. Pavletich NP, Pabo CO. 1991. Zinc finger-DNA recognition: crystal structure of a Zif268-DNA complex at 2.1 Å. *Science* 252:809–17
12. Diakun GP, Fairall L, Klug A. 1986. EXAFS study of the zinc-binding sites in the protein transcription factor IIIA. *Nature* 324:698–99
13. Tso JY, Van Den Berg DJ, Korn LT. 1986. Structure of the gene for *Xenopus* transcription factor TFIIIA. *Nucleic Acids Res.* 14:2187–200
14. Bateman A, Birney E, Cerruti L, Durbin R, Etwiller L, et al. 2002. The Pfam protein families database. *Nucleic Acids Res.* 30:276–80

15. Vincent A, Colot HV, Rosbash M. 1985. Sequence and structure of the serendipity locus of *Drosophila melanogaster*: a densely transcribed region including a blastoderm-specific gene. *J. Mol. Biol.* 185:146–66
16. Rosenberg UB, Schröder C, Preiss A, Kienlin A, Côté S, et al. 1986. Structural homology of the product of the *Drosophila Krüppel* gene with *Xenopus* transcription factor IIIA. *Nature* 319:336–39
17. Berg JM. 1988. Proposed structure for the zinc binding domains from transcription factor IIIA and related proteins. *Proc. Natl. Acad. Sci. USA* 85:99–102
18. Lee MS, Gippert GP, Soman KV, Case DA, Wright PE. 1989. Three-dimensional solution structure of a single zinc finger DNA-binding domain. *Science* 245:635–37
19. Nakaseko Y, Neuhaus D, Klug A, Rhodes D. 1992. Adjacent zinc finger motifs in multiple zinc finger peptides from SW15 form structurally independent flexibly linked domains. *J. Mol. Biol.* 228:619–36
20. Neuhaus D, Nakaseko Y, Schwabe JW, Klug A. 1992. Solution structures of two zinc-finger domains from SWI5 obtained using two-dimensional ^{1}H nuclear magnetic resonance spectroscopy: a zinc-finger structure with a third strand of β-sheet. *J. Mol. Biol.* 228:637–51
21. Fairall L, Schwabe JW, Chapman L, Finch JT, Rhodes D. 1993. The crystal structure of a two zinc-finger peptide reveals an extension to the rules for zinc-finger/DNA recognition. *Nature* 366:483–87
22. Elrod-Erickson M, Benson TE, Pabo CO. 1998. High-resolution structures of variant Zif268-DNA complexes: implications for understanding zinc finger-DNA recognition. *Structure* 6:451–64
23. Klug A, Jack A, Viswamitra MA, Kennard O, Shakked Z, Steitz TAA. 1979. A hypothesis on a specific sequence-dependent conformation of DNA and its relation to the binding of the lac-repressor protein. *J. Mol. Biol.* 131:669–80
24. Rhodes D, Klug A. 1981. Sequence-dependent helical periodicity of DNA. *Nature* 292:378–80
25. Smith GP. 1985. Filamentous fusion phage: novel expression vectors that display cloned ontigers on the verion surface. *Science* 228:1315–17
26. McCafferty J, Griffiths AD, Winter G, Chiswell DJ. 1990. Phage antibodies: filamentous phage displaying antibody variable domains. *Nature* 348:552–54
27. Choo Y, Klug A. 1994. Toward a code for the interactions of zinc fingers with DNA: selection of randomized fingers displayed on phage. *Proc. Natl. Acad. Sci. USA* 91:11163–67
28. Choo Y, Klug A. 1994. Selection of DNA binding sites for zinc fingers using rationally randomized DNA reveals coded interactions. *Proc. Natl. Acad. Sci. USA* 91:11168–72
29. Choo Y, Sánchez-García I, Klug A. 1994. In vivo repression by a site-specific DNA-binding protein designed against an oncogenic sequence. *Nature* 372:642–45
30. Reynolds L, Ullman C, Moore M, Isalan M, West WJ, et al. 2003. Repression of the HIV-1 5′ LTR promoter and inhibition of HIV-1 replication by using engineered zinc-finger transcription factors. *Proc. Natl. Acad. Sci. USA* 100:1615–20
31. Papworth M, Moore M, Isalan M, Minczuk M, Choo Y, Klug A. 2003. Inhibition of herpes simplex virus 1 gene expression by designer zinc-finger transcription factors. *Proc. Natl. Acad. Sci. USA* 100:1621–26
32. Isalan M, Choo Y, Klug A. 1997. Synergy between adjacent zinc fingers in sequence-specific DNA recognition. *Proc. Natl. Acad. Sci. USA* 94:5617–21
33. Isalan M, Klug A, Choo Y. 1998. Comprehensive DNA recognition through concerted interactions from adjacent zinc fingers. *Biochemistry* 37:12026–33
34. Isalan M, Klug A, Choo Y. 2001. A rapid, generally applicable method to engineer zinc fingers illustrated by targeting the HIV-1 promoter. *Nat. Biotechnol.* 19:656–60
35. Moore M, Choo Y, Klug A. 2001. Design of polyzinc finger peptides with structured linkers. *Proc. Natl. Acad. Sci. USA* 98:1432–36
36. Moore M, Klug A, Choo Y. 2001. Improved DNA binding specificity from polyzinc finger peptides by using strings of two-finger units. *Proc. Natl. Acad. Sci. USA* 98:1437–41
37. Kim JS, Pabo CO. 1998. Getting a handhold on DNA: design of poly-zinc finger proteins with femtomolar dissociation constants. *Proc. Natl. Acad. Sci. USA* 95:2812–17
38. Tan S, Guschin D, Davalos A, Lee YL, Snowden AW, et al. 2003. Zinc-finger protein-targeted gene regulation: genome wide single-gene specifity. *Proc. Natl. Acad. Sci. USA* 100:11997–2002
39. Jamieson AC, Miller JC, Pabo CO. 2003. Drug discovery with engineered zinc-finger proteins. *Nat. Rev. Drug Discov.* 2:361–68

40. Liu PQ, Rebar EJ, Zhang L, Liu Q, Jamieson AC, et al. 2001. Regulation of an endogenous locus using a panel of designed zinc finger proteins targeted to accessible chromatin regions. Activation of vascular endothelial growth factor A. *J. Biol. Chem.* 276:11323–34
41. Rebar EJ, Huang Y, Hickey R, Nath AK, Meoli D, et al. 2002. Induction of angiogenesis in a mouse model using engineered transcription factors. *Nat. Med.* 8:1427–32
42. Beerli RR, Schopfer U, Dreier B, Barbas CF 3rd. 2000. Chemically regulated zinc finger transcription factors. *J. Biol. Chem.* 275:32617–27
43. Capecchi M. 1989. Altering the genome by homologous recombination. *Science* 244:1288–92
44. Jasin M. 1996. Genetic manipulation of genomes with rare-cutting endonucleases. *Trends Genet.* 12:224–27
45. Kim YG, Cha J, Chandrasegaran S. 1996. Hybrid restriction enzymes: zinc finger fusions to Fok I cleavage domain. *Proc. Natl. Acad. Sci. USA* 93:1156–60
46. Bibikova M, Beumer K, Trautman JK, Carroll D. 2003. Enhanced gene targeting with designed zinc finger nucleases. *Science* 300:764
47. Porteus MM, Baltimore D. 2003. Chimeric nucleases stimulate gene targeting in human cells. *Science* 300:763
48. Cavazzana-Calvo M, Hacein-Bey S, de Saint Basile G, Gross F, Yvon E, et al. 2000. Gene therapy of human severe combined immunodeficiency (SCID)-X-linked disease. *Science* 288:669–72
49. Urnov FD, Miller JC, Lee YL, Beausejour CM, Rock JM, et al. 2005. Highly efficient endogenous human gene correction using designed zinc-finger nucleases. *Nature* 435:646–51
50. Perez EE, Wang J, Miller JC, Jouvenot Y, Kim KA, et al. 2008. Establishment of HIV-1 resistance in CD4+ T cells by genome editing using zinc-finger nucleases. *Nat. Biotechnol.* 26:808–16
51. Santiago Y, Chan E, Liu P-Q, Orlando S, Zhang L, et al. 2008. Targeted gene knockout in mammalian cells by using engineered zinc-finger nucleases. *Proc. Natl. Acad. Sci. USA* 105:5809–14
52. Meng X, Noyes MB, Zhu LJ, Lawson ND, Wolfe SA. 2008. Targeted gene inactivation in zebrafish using engineered zinc-finger nucleases. *Nat. Biotechnol.* 26:695–701
53. Doyon Y, McCammon JM, Miller JC, Faraji F, Ngo C, et al. 2008. Heritable targeted gene disruption in zebrafish using designed zinc-finger nucleases. *Nat. Biotechnol.* 26:702–8
54. Shukla VK, Doyon Y, Miller JC, DeKelver RC, Moehle EA, et al. 2009. Precise genome modification in the crop species *Zea mays* using zinc-finger nucleases. *Nature* 459:437–41
55. Townsend JA, Wright DA, Winfrey RJ, Fu F, Maeder ML, et al. 2009. High-frequency modification of plant genes using engineered zinc-finger nucleases. *Nature* 459:442–45

Origins of Specificity in Protein-DNA Recognition

Remo Rohs,[1,2,*] Xiangshu Jin,[1,2,*] Sean M. West,[1,2] Rohit Joshi,[2] Barry Honig,[1,2] and Richard S. Mann[2]

[1]Howard Hughes Medical Institute, Center for Computational Biology and Bioinformatics, [2]Department of Biochemistry and Molecular Biophysics, Columbia University, New York, NY 10032; email: bh6@columbia.edu, rsm10@columbia.edu

Annu. Rev. Biochem. 2010. 79:233–69

First published online as a Review in Advance on March 24, 2010

The *Annual Review of Biochemistry* is online at biochem.annualreviews.org

This article's doi:
10.1146/annurev-biochem-060408-091030

0066-4154/10/0707-0233$20.00

*These authors contributed equally to this work.

Key Words

protein-DNA binding, direct readout, indirect readout, DNA base recognition, DNA shape recognition, narrow minor groove

Abstract

Specific interactions between proteins and DNA are fundamental to many biological processes. In this review, we provide a revised view of protein-DNA interactions that emphasizes the importance of the three-dimensional structures of both macromolecules. We divide protein-DNA interactions into two categories: those when the protein recognizes the unique chemical signatures of the DNA bases (base readout) and those when the protein recognizes a sequence-dependent DNA shape (shape readout). We further divide base readout into those interactions that occur in the major groove from those that occur in the minor groove. Analogously, the readout of the DNA shape is subdivided into global shape recognition (for example, when the DNA helix exhibits an overall bend) and local shape recognition (for example, when a base pair step is kinked or a region of the minor groove is narrow). Based on the >1500 structures of protein-DNA complexes now available in the Protein Data Bank, we argue that individual DNA-binding proteins combine multiple readout mechanisms to achieve DNA-binding specificity. Specificity that distinguishes between families frequently involves base readout in the major groove, whereas shape readout is often exploited for higher resolution specificity, to distinguish between members within the same DNA-binding protein family.

Contents

1. INTRODUCTION

Genomes are composed of both protein-coding and nonprotein-coding DNA sequences. Cells have the remarkable ability to decipher the information that is incorporated in both types of sequences. Biologists, on the other hand, are currently unable to do what the cell does—to interpret nonprotein-coding DNA sequences. An important step toward achieving this goal is to have a better understanding of protein-DNA recognition mechanisms. Traditionally, the analysis of noncoding DNA sequences has treated DNA as a linear string of nucleotides, which does not take into account the three-dimensional structure of DNA. In this review, we provide a new perspective on the problem of protein-DNA recognition, one that emphasizes the three-dimensional structures of both the DNA and the protein.

1.1. General Comments

More than 50 years after the structure of DNA was first proposed by Watson & Crick (1), biologists are still working to achieve a complete understanding of how proteins interact with genomes. One of the most important questions that remain is one of specificity—how do the large and diverse number of DNA-binding proteins encoded by eukaryotic genomes recognize their specific binding sites? Moreover, most DNA-binding proteins are part of large families that share DNA-binding domains with very similar biochemical properties. How do proteins with closely related DNA-binding domains carry out their unique functions in vivo? Providing answers to these questions is especially timely given the need to accurately annotate the many complete genome sequences that are now available, an endeavor that is still a major unsolved challenge.

The size and complexity of this problem has recently been underscored by several publications that use high-throughput approaches, such as protein-binding microarrays or the bacterial one-hybrid system, to generate an unprecedented database of the DNA sequence preferences for a large number of DNA-binding proteins (2–5). In one such recent report (6), the binding-site preferences for 104 mouse transcription factors, often including multiple members from the same transcription factor family, were described. To highlight just one example, the DNA-binding site preferences for 21 members of the Sox (SRY-related high-mobility group box)/TCF (T cell factor) family of transcriptional regulators were compared. Remarkably, although each factor executes unique functions, 14 of the 21 prefer to bind the sequence ACAAT. Moreover, although small differences in sequence preference were identified, these did not always correlate with the extent of sequence identity of the DNA-binding domains. For example, Sox1 preferred the sequence ATTTAAAT, whereas its two most closely related relatives (Sox14 and Sox21), as well as a much more distantly related family member, sex-determining region Y (SRY), preferred the sequence ACAAT. This study also revealed that many transcription factors have the capacity to recognize two distinct binding sites (so-called primary and secondary binding sites) and that there is a previously underappreciated interdependence between neighboring base pairs within a binding site.

Observations such as these raise a number of fundamental questions regarding protein-DNA recognition whose answers require a better understanding of the rules that govern how proteins bind to DNA sequences. We suggest that the linear sequence of base pairs in a binding site is only a small part of the story and that the three-dimensional structures of both macromolecules must be taken into account to fully understand protein-DNA recognition. In particular, local variations in DNA structure—DNA topography—may be as important as protein structure. A recent study that examined the evolutionary constraints on DNA topology strongly supports this point of view (7). Remarkably, the authors found that DNA topography of the human genome, as measured by hydroxyl radical cleavage patterns, is evolutionarily constrained. Moreover, these cleavage patterns, which are correlated with the solvent accessibility of the DNA helix (8), were found to be a much better predictor of functional DNA elements than the linear DNA sequence (7). Thus, to more fully understand the rules that govern protein-DNA recognition, we must consider both DNA structure and protein structure as equal partners.

1.2. Previous Definitions: Direct versus Indirect Readout Mechanisms

Understanding how proteins recognize their DNA-binding sites has a long history. Initially, on the basis of early low-resolution X-ray structures of nucleic acid duplexes (9), it was realized that the major groove of the DNA helix offered a set of base-specific hydrogen bond donors, acceptors, and nonpolar groups that could be recognized by a complementary set of donors and acceptors presented by amino acid side chains (10). Accordingly, the idea soon evolved that short DNA sequences could serve as binding sites that were specifically read by a complementary sequence of amino acids (11). This mechanism of protein-DNA recognition, now commonly referred to as direct readout, is evident in nearly all of the >1500 structures of protein-DNA complexes that have been solved and deposited in the Protein Data Bank (PDB). Nevertheless, as was realized many years ago (12), there is not a simple recognition code or one-to-one correspondence between DNA and protein sequences. Thus, direct readout, by itself, cannot be sufficient to account for the specificities of protein-DNA interactions.

Although elements of direct readout contribute to nearly all protein-DNA complexes, these structures also reveal that bound DNA frequently deviates from a standard B-form double helix. In some cases, deviations from a B-form helix are large and clearly contribute to DNA-binding specificity [e.g., the papillomavirus E2 protein and the TATA box-binding protein (TBP)] (13–15). In these cases, a bend or some other deformation of the DNA helix is required to establish a set of hydrogen bonds or nonpolar interactions between the protein and DNA that are much less likely to occur in the absence of the deformation. From such observations, the term indirect readout was coined (12). Indirect readout is defined as protein-DNA interactions that depend on base pairs that are not directly contacted by the protein (16). This broad definition includes situations where the DNA sequence creates or facilitates a DNA structure that is subsequently recognized by a protein, but also when the protein-DNA contact is mediated by a water molecule. In addition, over time, the term has been taken to mean any interaction between DNA and protein where the DNA is not a B-form helix. This even looser definition has limited value because it simply encompasses all interactions that are not direct.

1.3. Goals for this Review

In this review, we reevaluate the mechanisms that underlie protein-DNA recognition in light of new and previous structures of protein-DNA complexes. We suggest that the terms direct and indirect readout both describe idealized extremes that rarely exist in isolation in real protein-DNA complexes and therefore have limited value. For example, rarely are direct hydrogen bonds formed between protein side chains and DNA in the complete absence of any deviation from an ideal B-form helix. Conversely, rarely are protein-DNA interactions purely indirect. As detailed below, this reevaluation suggests that protein-DNA recognition utilizes a continuum of readout mechanisms that depend on the structural features and flexibility of both macromolecules, including the sequence-dependent propensity of DNA to assume conformations that deviate from ideal B-DNA. This more nuanced view suggests that protein-DNA and protein-protein recognition are in many ways analogous phenomena.

In order to reassess protein-DNA readout mechanisms, we divide this review into three main sections. In the first, we briefly discuss the range of protein structures that bind DNA. Because there are excellent recent reviews that already cover this topic (17–19), we simply summarize the major protein superfamiles that are observed in DNA-binding proteins. Second, because interactions between proteins and DNA depend on the interplay between both macromolecules, we review how DNA structures vary and the relationships between these structures and DNA sequence. Finally, with these structural considerations as a background, we review the range of interactions that are observed at protein-DNA interfaces, identifying common themes that are used both across and within individual families of DNA-binding proteins. We propose replacing the terms direct readout and indirect readout with the more informative terms, base readout and shape readout, which we further subdivide to reflect the way proteins recognize DNA sequences. Our goal is to present a richer and more subtle view of protein-DNA recognition that more accurately reflects the way in which evolution has fine-tuned these essential interactions.

Because the perspective offered here is structural in its origins, we do not review thermodynamic measurements of protein-DNA interactions nor do we summarize the many insights available from the application of simulation methodologies to the recognition problem (20). Rather, our goal is to review recent structural evidence regarding readout mechanisms of DNA sequences, recognizing that a deeper understanding of the underlying forces and their interactions requires the application of a variety of experimental and computational approaches to specific systems and on a genome-wide scale. It is our hope that the presentation and integration of structural data presented in this review serves to facilitate and to focus such studies.

2. STRUCTURE OF DNA-BINDING PROTEINS

The first protein-DNA complexes for which structural information was derived from X-ray crystallography were the catabolite gene activator protein (CAP) (21), Cro repressor (22), and λ repressor (23) bound to their binding sites. Since then, more than 1500 structures of protein-DNA complexes have been deposited in the Protein Data Bank.

Proteins utilize a wide range of DNA-binding structural motifs, such as the helix-turn-helix (HTH) motif of homeodomains, to recognize DNA. Many proteins also contain flexible segments outside a globular core that mediate important specific and nonspecific interactions. For example, λ repressor has an N-terminal arm that contacts bases in the major groove (24), the phage Φ29 transcriptional regulator p4 uses N-terminal β-turn substructures to make base-specific contacts in the major groove (25), and homeodomain proteins have N-terminal arms and linker regions that dock in the minor groove of the DNA (26–29). These flexible regions, which are sometimes not included in the strict definition of these DNA-binding domains, can have profound and essential roles in binding specificity.

According to the Structural Classification of Proteins (SCOP) database (30), DNA-binding proteins, whose structures are currently available in complexes with DNA, are grouped into more than 70 SCOP superfamilies (**Table 1**). Because of this large number, it is not possible to discuss each superfamily here, and thus, we focus only on a few representative examples. In **Table 1**, we group DNA-binding proteins into the following categories on the basis of the overall secondary structure content of the DNA-binding domains: mainly α, mainly β, mixed α/β, and multidomain proteins that have more than one of the aforementioned three domains. It is evident from the table that certain local motifs, such as the HTH motif, are used repeatedly and can be found within different global domain architectures. Moreover, depending on the protein and DNA-binding site, any one type of motif can be used in multiple ways to interact with DNA. These observations support one of the main points of this review: Protein-DNA interactions depend on the interplay between two equal partners, the DNA and the protein, and both macromolecules have their own characteristic three-dimensional structures that must accommodate the other to achieve specificity.

2.1. Mainly α

Proteins in 17 SCOP superfamilies have DNA-binding domains with mainly α-helical architecture, for example, homeodomains, leucine zipper proteins, and λ-repressor-like proteins. The α-helix is the most frequently used secondary structure element for specific DNA recognition in the major groove. The positioning of the helix in the major groove can vary between different protein families and also among different proteins within the same family, as reviewed previously (17). The Lac repressor (31, 32) and intron endonucleases (33–35) demonstrate that α-helices can also be used to interact with DNA in the minor groove. On the basis of the structural context in which the α-helices are found, the mainly α-class of proteins uses a number of local structural motifs for DNA binding.

2.1.1. Helix-turn-helix motif. The HTH motif is seen in many proteins in different SCOP superfamilies and is one of the most frequently represented structural motifs in DNA-binding proteins. The "recognition helix" of the HTH motif binds DNA through a series of hydrogen bonds and hydrophobic interactions with exposed bases, and the other helix stabilizes the interaction between the protein and DNA, but does not play a particularly strong role in recognition. Although the HTH motif is highly conserved, its structural context and precise orientation relative to the DNA-binding sites it recognizes can vary between different proteins, and the structures outside the HTH core region can differ greatly among various proteins. For example, in homeodomains, the second and third helices of the three-helix bundle comprise the HTH motif with the third helix (the recognition helix) contacting the major groove, in an orientation that is nearly parallel to the flanking DNA backbones. The motility gene repressor (MogR) DNA-binding domain contains seven α-helices connected by short loops: The first three helices form a three-helix bundle, the fourth helix forms a small dimerization interface, and helices 5–7 form a three-helix bundle DNA-binding domain that contains a HTH motif (α6 and α7), in which α7 is the recognition helix (36). Although the HTH motif is used most often in the major groove, some proteins use this motif to interact with the minor groove, for example, O6-alkylguanine-DNA alkyltransferase (AGT) (37).

A large class of HTH motif-containing proteins have an additional antiparallel β-sheet, hence its name "winged helix-turn-helix" (wHTH) motif (38). Proteins in many SCOP families contain the wHTH motif, including the hepatocyte nuclear factors-3 (HNF-3)/forkhead family of transcription factors (39),

Table 1 Architecture of DNA-binding proteins from the Structural Classification of Proteins (SCOP) database[a]

SCOP superfamily[b]	Number of PDB entries	Architecture of DNA-binding domains	DNA-binding motif
DNA/RNA polymerases	186	Multidomain, mixed α/β	
Nucleotidyltransferase	127	Multidomain, mixed α/β	
Ribonuclease H-like	104	Multidomain, mixed α/β	
Restriction endonuclease-like	89	Mixed α/β	
Homeodomain-like	75	Mainly α	Helix-turn-helix
Winged helix DNA-binding domain	75	Mainly α with a small β-ribbon (wing)	Winged helix-turn-helix
Lesion bypass DNA polymerase	60	Multidomain, mixed α/β	
Lambda repressor-like DNA-binding domains	57	Mainly α	Helix-turn-helix
Glucocorticoid receptor-like	53	Mixed α/β	Zinc finger
p53-like transcription factors	53	Mainly β	Immunoglobulin-like β-sandwich
DNA breaking-rejoining enzymes	45	Multidomain, mixed α/β	
DNA glycosylase	40	Mixed α/β	
S-adenosyl-L-methionine-dependent methyltransferases	40	Mixed α/β	
Histone fold	29	Mainly α	
Leucine zipper domain	27	Mainly α	Helix-loop-helix
TATA-box-binding protein-like	24	Mainly β	TBP β-sheet
Homing endonucleases	24	Mixed α/β	
C2H2 and C2HC zinc fingers	22	Mixed α/β	Zinc finger
E-set domains	21	Mainly β	Immunoglobulin-like β-sandwich
Chromo domain-like	19	Mainly β	β-barrel
DNA repair protein MutS	18	Multidomain, mixed α/β	
Ribbon-helix-helix	16	Mixed α/β	Ribbon-helix-helix
Uracil-DNA glycosylase-like	16	Mixed α/β	
His-Me finger endonucleases	14	Mixed α/β	
HMG box	13	Mainly α	Helix-turn-helix
Origin of replication-binding domain, RBD-like	13	Mixed α/β	
P-loop-containing nucleoside triphosphate hydrolases	12	Multidomain, mixed α/β	
Putative DNA-binding domain	12	Mainly α	
Zn2Cys6 DNA-binding domain	11	Mixed α/β	Zinc finger
IHF-like DNA-binding proteins	10	Mixed α/β	
RNase A-like	9	Mixed α/β	
Helix-loop-helix DNA-binding domain	8	Mainly α	Helix-loop-helix
SRF-like	8	Mixed α/β	
Zn2Cys4 DNA-binding domain	8	Mixed α/β	Zinc finger
C-terminal effector domain of the bipartite response regulators	7	Mainly α	Helix-turn-helix
DNase I-like	5	Mixed α/β	
Retrovirus zinc finger-like domains	5	Mixed α/β	Zinc finger
TrpR-like	5	Mainly α	Helix-turn-helix

(*Continued*)

Table 1 (*Continued*)

SCOP superfamily[b]	Number of PDB entries	Architecture of DNA-binding domains	DNA-binding motif
Viral DNA-binding domain	5	Mixed α/β	
PIN domain-like	5	Mixed α/β	Ribbon-helix-helix
Zinc finger design	4	Mixed α/β	Zinc finger
DNA-binding domains of HMG-I(Y)	4	Peptide	AT hook
Transcription factor IIA (TFIIA)	4	Mainly β	β-barrel
Replication terminator protein (Tus)	4	Multidomain, mixed α/β	
UDP/glycosyltransferase, glycogen phosphorylase	4	Mixed α/β	
Replication modulator SeqA, C-terminal DNA-binding domain	4	Mainly α	
DNA-binding domain	4	Mixed α/β	β-sheet
FMT C-terminal domain-like	4	Mixed α/β	
Sigma3 and sigma4 domains of RNA polymerase sigma factors	3	Mainly α	Helix-turn-helix
Methylated DNA-protein cysteine methyltransferase domain	3	Mixed α/β	
DNA-binding domain of intron-encoded endonucleases	3	Mixed α/β	
Cryptochrome/photolyase FAD-binding domain	3	Mixed α/β	
T4 endonuclease V	2	Mainly α	Helix-turn-helix
SMAD MH1 domain	2	Mixed α/β	
KorB DNA-binding domain-like	2	Mainly α	Helix-turn-helix
DNA topoisomerase IV, alpha subunit	2	Multidomain, mixed α/β	
SMAD MH1 domain	2	Mixed α/β	
5′ to 3′ exonuclease catalytic domain	2	Mixed α/β	
Metallo-dependent phosphatases	2	Multidomain, mixed α/β	
WD40 repeat-like	2	Mainly β	
Xylose isomerase-like	1	Mixed α/β	
RNA polymerase	1	Multidomain, mixed α/β	
GCM domain	1	Mixed α/β	β-sheet
ATP-dependent DNA ligase DNA-binding domain	1	Multidomain, mixed α/β	
Transposase IS200-like	1	Mixed α/β	
Thioredoxin-like	1	Multidomain, mixed α/β	
Holliday junction resolvase RusA	1	Mixed α/β	
Skn-1	1	Mainly α	
ARID-like	1	Mainly α	Helix-turn-helix
GCM domain	1	Mixed α/β	β-sheet
Phage replication organizer domain	1	Mainly α	
Bet v1-like	1	Mixed α/β	
AbrB/MazE/MraZ-like	1	Mainly β	

[a]This table lists DNA-binding protein domains in different SCOP superfamilies, whose structures in complexes with DNAs are available in the Protein Data Bank as of August 2009. When they are well defined, the DNA-binding motifs used by these SCOP superfamilies are listed in the fourth column.
[b]Abbreviations: Please see **http://supfam.mrc-lmb.cam.ac.uk/** for the nomenclature used in names of SCOP superfamilies.

Ets domain (40), and multiple antibiotic resistance (MarR)-like transcription factors (41). The "wing" typically sits over the minor groove to make additional DNA contacts. However, in some cases, the wings rather than the HTH motif contact the DNA in the major groove, as seen in regulatory factor X1 (RFX1) (42). Many proteins also contain a second wing, which makes additional DNA contacts.

2.1.2. Helix-loop-helix and leucine zipper motifs. The helix-loop-helix motif consists of a short α-helix connected by a loop to a longer α-helix. Part of this motif is a dimerization domain that interacts with other helix-loop-helix proteins to form homo- or heterodimers; the dimerization partner often determines DNA-binding affinity and specificity because two α-helices, one from each monomer, bind to the major groove of the target DNA (43–46).

2.2. Mainly β

Although less common than α-helices, β-strands and intervening loops embedded in the mainly β-domain structures are used by proteins in seven SCOP superfamilies to recognize specific DNA sequences.

2.2.1. TATA box-binding protein. TBPs use a large β-sheet surface to recognize DNA by binding in the minor groove (14, 15). Insertion of the concave, 10-stranded β-sheet of TBP into the groove requires profound DNA distortion. As discussed in the following sections, the TATA box DNA undergoes dramatic unwinding and bending that allows for contacts between the protein's concave surface and the edges of the base pairs in the otherwise recessed minor groove.

2.2.2. Immunoglobulin-like β-sandwich. Immunoglobulin-like structural domains are used for DNA binding in diverse families of proteins, such as p53-like transcription factors (47), E-set domains (48, 49), and Runt domains (50). The sequence conservation of the immunoglobulin-like domains in different families is low, and the structures outside the domain diverge significantly. Although the overall fold is a β-sandwich, DNA recognition is achieved mainly by intervening loops. Like the mainly α-helical DNA-binding domains, the orientation of the β-sandwich domains relative to the DNA varies among different proteins and different families of proteins.

2.2.3. β-trefoil. The β-trefoil is a capped β-barrel with an approximate threefold symmetry, i.e., four strands are repeated in a threefold arrangement, where strands 1 and 4 form the walls of the β-barrel and strands 2 and 3 contribute to the cap structure to give a 12-stranded structure. The β-trefoil domain of CSL [CBF-1, Su(H), Lag-1], the nuclear effector of Notch signaling, contacts DNA via the loop between strands βA1 and βA2 (51).

2.2.4. β-β-β-sandwich. The structure of $AgrA_C$ (52) reveals a novel topology of 10 β-strands arranged into three antiparallel β-sheets, which are arranged roughly parallel to each other in an elongated β-β-β-sandwich, and a small two-turn α-helix that is not involved in DNA binding. Base-specific contacts are made with residues from intervening loops at both the major and minor grooves.

2.3. Mixed α/β

A large number of proteins, which belong to 48 SCOP superfamilies, use mixed α/β domains to bind DNA, although the major secondary structure elements used for recognition can be any one or any combination of α-helix, β-strand, or loop.

2.3.1. Zinc finger proteins. The zinc finger is a compact ~30-amino acid DNA-binding domain. Zinc fingers are the most minimal of DNA-binding domains, with a relatively short α-helix, a two-stranded antiparallel β-sheet, and a Zn^{2+} ion coordinated by cysteine and histidine residues (53). Zinc fingers are classified by the type and order of the zinc coordinating residues, e.g., Cys_2His_2, Cys_4, and Cys_6. Zinc fingers often occur as tandem repeats with two, three, or more fingers that can bind in the major groove, typically spaced at 3-bp intervals. The α-helix of each domain (the recognition helix) makes sequence-specific contacts to DNA bases in the major groove; residues from a single recognition helix can contact four or more bases to yield an overlapping pattern of contacts with adjacent zinc fingers.

2.3.2. Ribbon-helix-helix motif. A family of transcription factors from bacteria contains the ribbon-helix-helix (RHH) motif (54) that consists of a two-stranded antiparallel β-ribbon followed by two α-helices. DNA recognition is achieved by insertion of the β-ribbon into the major groove, whereas the two helices comprise most of the hydrophobic core and are involved in dimerization. The prototypical examples are Met repressor MetJ (55) and Arc repressor (56).

2.3.3. Other mixed α/β domains. Structural studies of seemingly dissimilar restriction endonucleases with remarkable DNA sequence specificity demonstrated that they all share a common structural core with a mixed α/β architecture (57). A large amount of structural data also reveal that DNA polymerases, DNA lesion repair enzymes, and DNA-modifying enzymes all have mixed α/β domain structures (**Table 1**).

2.4. Multidomain Proteins

Many DNA-binding proteins contain multiple DNA-binding domains, which can work together to recognize different regions of a target sequence, achieving high affinity and recognition specificity. For example, POU domain proteins, such as Oct-1 (58) and Brn-5 (59), contain a homeodomain (POU_{HD}) and POU-specific domain (POU_S) that are connected by a flexible linker, and MarA (multiple antibiotic resistance A) consists of two HTH motifs that contact two successive major grooves (60). Other examples are the Rel-homology domain proteins, such as NF-κB p50, that have two immunoglobulin-like domains in each monomer: The N-terminal domain mediates DNA contacts primarily in the major groove, and the C-terminal domain mediates homo- and heterodimer interactions in addition to contacting DNA (48, 61). The side chains involved in dimer interactions lie along one face of the β-sandwich, leaving the loops free to contact the DNA. The *Escherichia coli* transcription factor Rob, which belongs to the AraC/XylS family, has two HTH domains: One binds specifically to DNA, whereas the other only forms a single salt bridge with the DNA backbone (62, 63). TCF (T cell factor) binds to specific DNA sequences through a high-mobility group (HMG) domain. Recent data suggest that DNA recognition by *Drosophila* TCF occurs through a bipartite mechanism, involving both the HMG domain and the C-clamp, which enables TCF to locate and activate wingless-regulated enhancers in the nucleus (64).

3. SEQUENCE-DEPENDENT VARIATIONS OF DNA STRUCTURE

Most current analyses of the information content in a nucleotide sequence view DNA as a one-dimensional string of letters based on an alphabet consisting of only four characters: A, C, G, and T. Yet these bases are chemical entities that, along with the inclusion of the backbone sugar and phosphate groups, create a three-dimensional double-stranded structure in which each base pair has a specific chemical and conformational signature (10). Although this textbook view of the double helix is well-known, what is much less appreciated is that DNA structures vary in a sequence-dependent manner (20, 65) and that structural variations are used by proteins to recognize DNA sequences (66).

In this section, we review the main ways in which DNA structures are known to deviate from idealized B-DNA. We distinguish between effects that vary the geometry of the helix in a localized manner (local shape, e.g., minor groove width and DNA kinks) from those that deform the overall cylindrical shape

Table 2 Tendency of DNA sequence elements to have specific structural characteristics[a]

Sequence element[b]	Structural characteristics	References
AT rich	B-DNA	72, 114
GC rich	A-DNA at low humidity	76, 77, 188
A-tract	B′-DNA, narrow minor groove, bending, rigid for $\geq$4 bp	81–83, 86, 217
TATA box	High deformability, A-DNA, TA-DNA upon TBP binding	78, 189
RY alternating (especially GC alternating)	Z-DNA at high salt concentration, upon cytosine methylation or supercoiling	79, 80, 197
YpR step (especially TpA step)	Compresses major groove, high deformability, hinge step, kinking	84, 88–90
RpY step	Compresses minor groove, low deformability	84, 88, 89

[a]The table reflects general tendencies for some sequences to have particular structural characteristics. It is important to stress, however, that DNA conformation depends on environmental conditions (e.g., humidity and salt concentration) and the larger sequence context (65, 76). For example, although AT-rich DNA is usually observed in B form, TATA box-containing oligonucleotides were crystallized in A form (189), which is the basis for TATA-binding protein specificity. In addition, owing to their high deformability, the structure of TATA boxes is affected by long-range sequence effects (218) and by supercoiling. A TATA box flanked by GC alternating regions can also assume a Z-DNA conformation (219).
[b]Abbreviations: A, adenine; C, cytosine; G, guanine; T, thymine; R, purine; Y, pyrimidine. The lower case "p" between nucleotides stands for phosphate to distinguish a base pair step from a base pair.

of the double helix (global shape, e.g., DNA bending, A-DNA, and Z-DNA). In addition, although some DNA sequences do not produce a well-defined structure per se, they may be highly flexible and therefore have a strong propensity to assume a non-B-like structure when bound to a protein. This property, commonly referred to as deformability, is another sequence-dependent feature that is used by proteins to recognize specific DNA sequences. To help make the connection between DNA sequence and DNA structure, **Table 2** lists DNA sequences that have a tendency to assume a particular DNA structure.

Differences in DNA shape can produce electrostatic potentials of varying magnitudes, a characteristic that can be read by proteins. For example, narrow minor grooves locally enhance the negative electrostatic potential of DNA through electrostatic focusing (66), which describes the deformation of field lines owing to the shape of the dielectric boundary between solute and solvent (67). This phenomenon was first described for a cavity of the superoxide dismutase protein (68) but has also been shown to play a role in codon-anticodon recognition in transfer RNAs (69), in shaping electrostatic potentials around diverse RNA structures (70), and in shifting pK_as in RNA catalytic sites (71). As discussed below, the effect appears to play an important role in protein-DNA recognition. In the following sections, we therefore refer to the electrostatic potential surfaces shown in **Figure 1**, which illustrate the close connection between shape and electrostatic potential in different DNA structures.

3.1. Global Shape Variations

In this section, we discuss the major ways in which DNA shape can vary in a global manner. These include different helical topologies and overall deformations of the DNA helix.

3.1.1. Polymorphisms of the double helix. Global shape variations include previously recognized polymorphisms of the double helix, B-DNA, A-DNA, TA-DNA, and Z-DNA, which we briefly discuss here.

3.1.1.1. B-DNA. The most common form of double-stranded DNA is B-DNA, which is generally favored in aqueous solution similar to the environment in cells (72). Most DNA-binding proteins recognize B-DNA and its structural variants. B-DNA is a right-handed double helix with base pairs oriented approximately perpendicular to the helix axis. Ideal

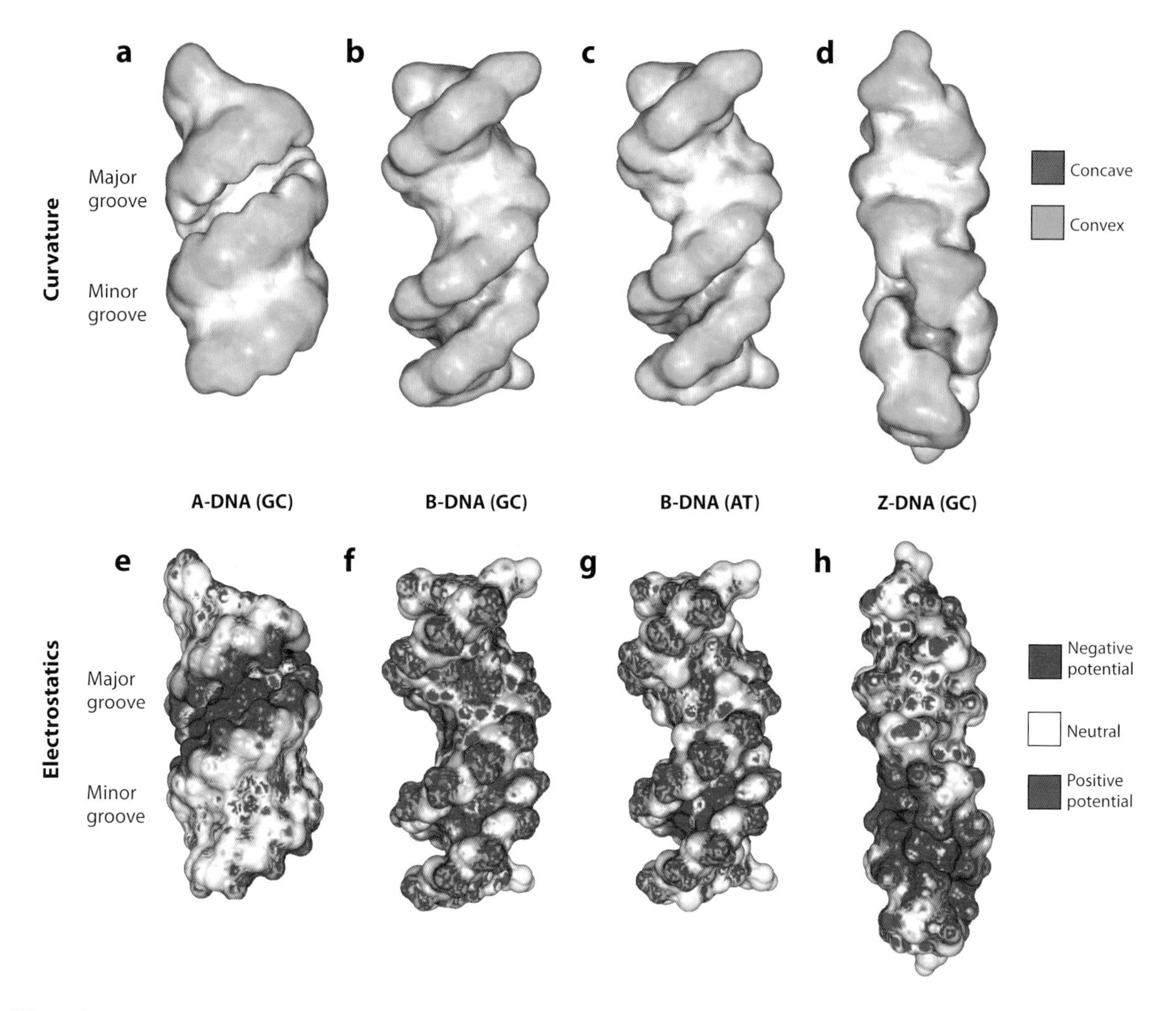

Figure 1

Molecular shape and electrostatic potential of A-DNA, B-DNA, and Z-DNA. The upper panels show the molecular shape in GRASP2 images (convex surfaces in *green* and concave surfaces in *dark gray*) (220) of the three helical forms of DNA constructed with the software tool, 3DNA (92) from fiber diffraction data (72, 80). Each DNA helix comprises 14 mers. The width and depth stated below were calculated with the software tool, Curves (221). The lower panels show how the electrostatic potential at the molecular surface varies owing to shape and atomic charges. The electrostatic potentials were calculated as described in (66) by solving the Poisson-Boltzmann equation with DelPhi (67, 222) at a salt concentration of 0.145 M (negative electrostatic potentials are shown in *red* and positive electrostatic potentials in *blue*). (*a*) A-DNA with a narrow, deep major groove (2.2-Å wide and 9.5-Å deep) and a wide, shallow minor groove (10.9-Å wide and no defined depth). The model is of the alternating sequence $d(GC)_7$. (*b*) B-DNA [alternating sequence $d(GC)_7$] with a wide, shallow major groove (11.4-Å wide and 4.0-Å deep) and a narrow, deep minor groove (5.9-Å wide and 5.5-Å deep). (*c*) B-DNA [alternating sequence $d(AT)_7$]. Because the models are built on the basis of fiber diffraction data, the shape of GC and AT alternating B-DNA does not reflect a sequence dependency. (*d*) Z-DNA lacks a major groove (13.2-Å wide and no defined depth), and the minor groove is narrow and deep (2.4-Å wide and 5.0-Å deep). The model is of the alternating sequence $d(GC)_7$. (*e*) A-DNA exhibits a strongly negative major groove but a hydrophobic minor groove surface, which is partially owing to its exposed C3′ endo sugar moieties. (*f*) B-DNA [alternating sequence $d(GC)_7$] exhibits a negative minor groove and less negative major groove. (*g*) B-DNA [alternating sequence $d(AT)_7$]. Variations in electrostatic potential between GC and AT alternating B-DNA reflect the different functional groups of the base pairs (e.g., positive guanine amino group in the GC minor groove and neutral thymine methyl group in the AT major groove). (*h*) Z-DNA exhibits a negative minor groove and a positive surface on opposing edges of the bases.

B-DNA exhibits a wide, shallow major groove and a narrow, deep minor groove (**Figure 1*b*,*c***) (65). As is evident from (**Figure 1*f*,*g***), the minor groove of B-DNA generally exhibits a more electronegative potential than the major groove. The differences in the potential in either groove between AT- and GC-rich sequences are due to the disposition of polar groups at the base edges; specifically AT-rich sequences display more negative electrostatic potentials in the minor groove than GC-rich sequences (**Figure 1*f*,*g***) (73, 74). These effects are further enhanced by sequence-dependent effects on groove width, as discussed below.

3.1.1.2. A-DNA. A-DNA is observed under dehydrated conditions and in some protein-DNA complexes (75). GC-rich sequences have an increased tendency to assume A-DNA or A/B intermediate conformations (76). This property is, in part, because GC base pairs have three hydrogen bonds, whereas AT base pairs have only two. This property makes GC base pairs more planar, allowing consecutive GC base pairs to slide relative to each other, which promotes the A/B transition (77). Although less pronounced, such a tendency is also observed for TATA boxes partly because the TpA step counters propeller twisting. A-DNA is also a right-handed double helix with the base pairs shifted toward the minor groove and, compared to B-DNA, tilted with respect to the helix axis by about 20°. This results in a narrow, very deep major groove and a wide, very shallow minor groove (**Figure 1*a***) (65). On the basis of this geometry, the A-DNA major groove resembles the shape of the B-DNA minor groove, which explains why, in contrast to B-DNA, the A-form major groove has a more negative electrostatic potential than its shallow minor groove (**Figure 1*e***) (70).

3.1.1.3. TA-DNA. TA-DNA is a variant of A-DNA observed in TATA boxes, which are specifically recognized by TBPs. It differs from A-DNA mainly by a larger base pair inclination of around 50° relative to the helix axis. This feature led to the description of TA-DNA as tilted A-DNA (78). The TA-DNA geometry exhibits a positive roll (rotation between adjacent base pairs with respect to the base pairing axis), which explains the opening of the TATA box minor groove observed in TATA-TBP complexes (14, 15).

3.1.1.4. Z-DNA. Alternating purine-pyrimidine sequences were observed to form a left-handed double helix under high salt concentrations (79, 80). Because of the zigzag conformation of its backbone, this topology was coined Z-DNA. Thought to form when B-DNA is deformed by supercoiling, Z-DNA does not have a pronounced major groove; instead, the base edges form a convex surface. The minor groove, however, resembles the dimensions of the B-DNA minor groove, but with a zigzag trajectory of the backbone (**Figure 1*d***) and a uniform negative electrostatic potential (**Figure 1*b***).

3.1.2. DNA bending. We define DNA bending as a curvature distributed over a stretch of several base pairs, leading to a different orientation of the regions on both sides of the curvature (**Figure 2*a***). Bending has frequently been observed for sequences that contain A-tracts, which are stretches of A:T base pairs that include ApA (TpT) and ApT, but not TpA, steps (81–83). Various models have been established to explain the molecular origin of bending (84, 85). These models either associate bending with wedge angles between adjacent base pairs, which can involve both roll and tilt, or with junctions between regions with negative base pair inclination (A-tracts) and regions with positive inclinations (83, 86).

It is likely that the phasing of wedge angles is the critical factor for overall curvature. If short A-tracts (regions with negative roll) are phased by half a helical turn, the overall curvature cancels owing to bending toward opposite sides of the helix. In a sequence where regions with negative roll are phased by a helical turn, the overall curvature is enhanced. The effect is further enhanced if regions with negative roll are in phase of half a helical turn with regions of positive roll

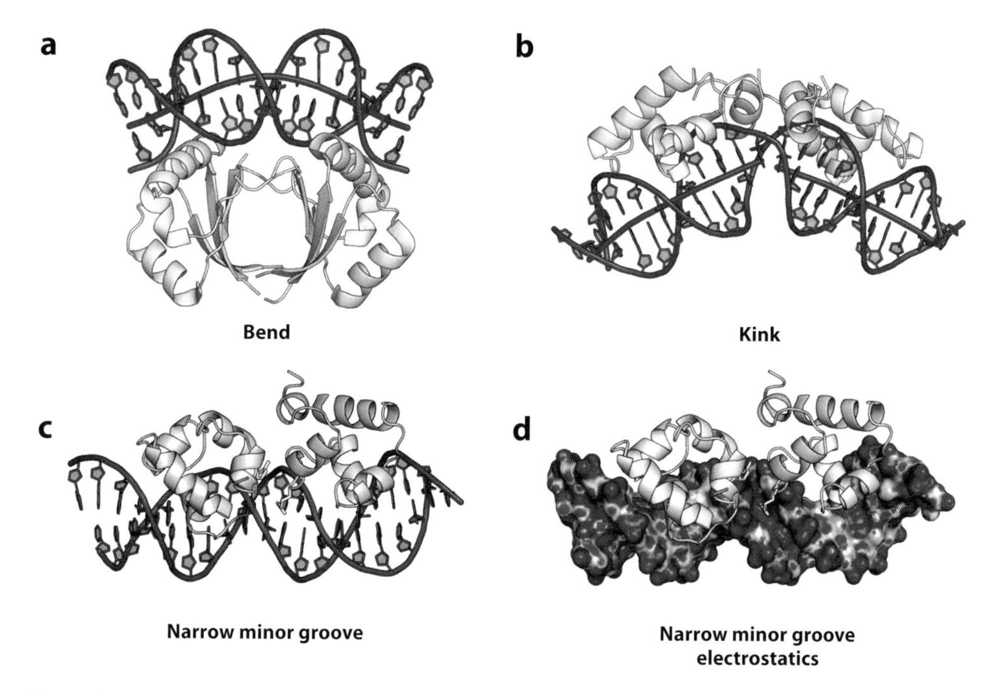

Figure 2

Illustration of DNA bending, kinking, and minor groove narrowing in protein-DNA complexes. (*a*) HPV-18 E2 bound to DNA (PDB ID 1jj4) shows bending over a large stretch of the helix. The smooth curvature is visualized by the helix axis (*blue*), calculated with Curves (221). (*b*) The Lac repressor kinks the DNA at a central CpG base pair step, stabilized by the partial intercalation of leucines (PDB ID 2kei). The helix axes calculated for both sides of the kink (*blue*) show an abrupt change in the helix trajectory caused by the kink. (*c*) Phage 434 repressor recognizes local shape deformations of its operator with arginine residues (PDB ID 2or1) (66). The narrow region of the minor groove that is contacted by arginines is highlighted in blue. (*d*) For the same structure shown in panel *c*, the electrostatic potential of the operator, calculated in the absence of the repressor, is plotted on the molecular surface. In comparison with **Figure 1*f*,*g***, the bottom of the minor groove is uniformly red, indicating enhanced negative electrostatic potential (66).

as both regions would bend the double helix in the same direction. Such a pattern has been reported for the nucleosome (84) and the papillomavirus E2-binding site (87). Ultimately, the source of sequence-dependent bending can be traced to the conformational properties of individual dinucleotide steps (88, 89), their tendency to form wedge angles, and the composition of these dinucleotide steps in a DNA sequence.

3.2. Local Shape Variations

Unlike global shape variations, we use the term local shape variations to refer to deviations from ideal B-DNA that originate from an individual base pair (e.g., a kink) or are localized in a small region of the double helix (e.g., minor groove narrowing).

3.2.1. DNA kinks. We distinguish a kink from a DNA bend by defining a kink as a local disruption of an otherwise linear helix (**Figure 2*b***). DNA kinks result from the complete or partial loss in stacking at a single base pair step. The pyrimidine-purine (YpR) steps TpA, CpA (TpG), and CpG are least stabilized through base stacking interactions, and of these, the TpA step has the weakest stacking interactions (**Table 2**) (65, 90). Therefore, it is the most flexible of the 10 unique dinucleotides and is referred to as a "hinge" step (86, 89). Because kinks occur at individual base pair steps, regions adjacent to a kink can remain in a straight B-form conformation or can be curved. Bending and kinking can enhance each other as is the case for CpA steps adjacent to an A-tract (91). Kinks are often stabilized by protein binding in cases where the loss of stacking interactions is compensated by the intercalation of hydrophobic side chains, which usually further deforms the kinked dinucleotide.

3.2.2. Minor groove narrowing. Minor groove width is another feature that varies locally in DNA structures (**Figure 2*c***) (66). Differences in minor groove width arise from differences in the hydrogen bonding pattern of each base pair and from differing stacking interactions for each dinucleotide step. To optimize both types of interactions, DNA structures vary with respect to three rotational parameters: *roll* (the relative rotation between adjacent base pairs with respect to the base pairing axis), *helix twist* (the relative rotation between adjacent base pairs with respect to the helix axis), and *propeller twist* (the relative rotation between bases within a base pair with respect to the base pairing axis) (92). ApT base pair steps usually have negative roll angles, which lead to a compression of the minor groove (**Table 2**) (84). In an A-tract sequence, ApT and ApA (TpT) steps exhibit a negative roll, and the bifurcated hydrogen bonds of A:T base pairs lead to propeller twisting, both enhancing minor groove narrowing (83, 87). In addition, several A:T base pairs in a row enhance propeller twisting by allowing the formation of interbase pair hydrogen bonds in the major groove (81). In contrast to ApA (TpT) and ApT, propeller-twisted TpA steps lead to a steric clash of the cross-strand adenines (86). Therefore, TpA steps tend to locally widen the minor groove and break rigid A-tract structures, and are thus referred to as hinge steps (**Table 2**) (89).

4. MECHANISMS OF PROTEIN-DNA RECOGNITION

Macromolecular interactions, whether they be protein-protein or protein-DNA in nature, depend on the three-dimensional structures of both interacting partners. In this section, we classify the types of readout mechanisms used by proteins to recognize DNA sequences in light of the types of DNA structures defined above.

4.1. General Comments

Protein-DNA interfaces involve on average 24 protein residues and 12 nucleotides (93), making it likely that each interface is composed of many different types of interactions. Although all interactions contribute to binding affinity, specificity can be viewed as resulting from a subset of interactions that are sequence specific. It is these specificity-determining contacts that we are most concerned with here.

Given our focus on specificity, it is important to define what we mean by this term and to point out that DNA-binding proteins generally exhibit multiple tiers of specificity. All homeodomains, for example, have an asparagine at position 51 (Asn51), which is important for the specific binding of these proteins to AT-rich sequences, such as TAAT (e.g., Engrailed and Antennapedia) (26, 94, 95). Thus, Asn51 can be considered to be a critical determinant of homeodomain DNA-binding specificity. However, as all homeodomains have Asn51, this residue cannot contribute to specificity within this superfamily. On a finer level, position 50 of the homeodomain partially fulfills this role: When it is a glutamine (Gln), the preferred binding sites are TAAT$\underline{\text{TG}}$ or TAAT$\underline{\text{TA}}$ (where the Gln contacts are underlined), but when it is a lysine, the preferred binding site is TAAT$\underline{\text{CC}}$ (96–99). However, the subset of homeodomain proteins that have a glutamine at position 50 is still very large and includes all of the Hox homeodomains, of which there are 39 in humans alone. Therefore, Gln50 cannot contribute to specificity within this subset of homeodomain proteins. In addition to Asn51 and Gln50, which are presented from a HTH recognition helix in the major groove, Hox proteins also bind to the minor groove, where DNA shape, in particular minor groove width, is read (29). As discussed below, this mode of protein-DNA recognition contributes to specificity within the Hox family. From this one example, we see that DNA-binding proteins use multiple readout mechanisms and that specificity is ultimately achieved by combinations of these mechanisms that successively fine-tune the selection of binding sites.

Although contacts between proteins and the DNA backbone are typically considered to have little impact on specificity (100), backbone contacts may play a role in specificity through the positioning of protein recognition elements in orientations that allow them to make other, more specific contacts, such as hydrogen bonds to the bases (101, 102). Indeed, protein families often contain conserved backbone-contacting residues that preserve the interface orientation for an entire family (102). In addition, specificity may depend on contacts to the DNA backbone if these contacts can only be made when the DNA assumes a sequence-dependent structure that deviates from ideal B-DNA (referred to below as nonideal B-DNA). An example is the readout of narrow minor groove regions, where the phosphates are located in positions that differ from ideal B-DNA. The Arg repressor from *Mycobacterium tuberculosis*, for instance, specifically recognizes a narrow minor groove region via extensive phosphate contacts from a four-stranded β-sheet that lies above the groove without inserting any side chain into the groove (103).

Protein-DNA recognition is also more complex than a simple docking process of two structurally preformed macromolecules. Some proteins fold only in the presence of DNA. For example, the leucine zippers of Fos and Jun are helical only when they form a heterodimer, and the basic regions are helical only when the dimer binds DNA (104, 105). Moreover, other domains in both proteins appear to be unstructured until bound by cofactors such as CREB-binding protein (CBP/p300) (106). Lymphoid enhancer factor-1 (LEF-1) also transitions from a relatively unstructured state to a well-folded domain upon DNA binding (107). The sequence-specific binding of Cys_2His_2 zinc finger proteins to DNA causes their linker regions to fold, cap, and thereby stabilize the preceding helix, which helps to orient the next zinc finger correctly for binding in the major groove (108). Finally, binding of the zinc finger domain of retinoid X receptor (RXR) to DNA leads to folding of the dimerization region, which is disordered in the unbound protein (109). DNA binding can also induce conformational changes in the bound protein that can change its properties. For example, the binding of the glucocorticoid receptor (GR) to its response elements induces conformational changes that expose transcriptional activation surfaces (110). Moreover, different GR-binding sites result in distinct GR activities, which, on the basis of X-ray data, could be explained by changes in the orientation of a GR loop induced by a modification of DNA backbone contacts (111).

DNA can also change conformation, and preexisting sequence-dependent conformations can be stabilized or enhanced upon protein binding (**Figure 3**). For example, in specifically designed noncognate GR complexes, the DNA is able to distort so as to maximize the number of cognate interface interactions, even if these are only maintained by a single strand (102, 112). Such effects make it difficult to unambiguously determine if nonideal B-DNA structures observed in protein-DNA complexes are intrinsic to the DNA sequence, induced by the protein, or some combination of the two. The relative impact of intrinsic versus induced effects on DNA structure can only be assessed with certainty by comparing the structure of the free DNA-binding site with its protein-bound form. Such structural information is currently restricted to the binding sites of only a handful of proteins, including the EcoRI endounclease (113, 114), Trp repressor (12, 115), Met repressor (55, 116), purine repressor (31, 116), NF-κB (48, 49, 117), Zif268 zinc fingers (**Figure 3*a***) (116, 118, 119), papillomavirus E2 protein (**Figure 3*b***) (13, 82, 120), and the Runt domain (50, 121, 122). The limited size of this group is largely because of the lack of free DNA structures (20). In their place, theoretical

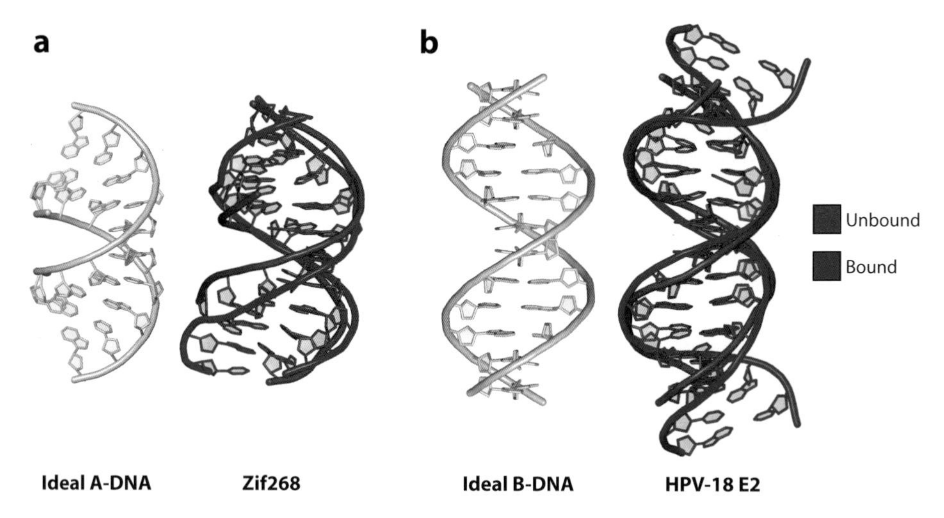

Figure 3

DNAs bound to proteins have features already present in unbound DNAs. (*a*) The structure of the unbound FIN-B sequence (PDB ID 2b1c) is similar to ideal A-DNA (*gray*), whereas the bound structure of the Zif268-DNA complex (PDB ID 1a1f) has some A-DNA characteristics, notably a wider minor groove than normally found in B-DNA. (*b*) The specific HPV-18 E2 site (PDB ID 1ilc) contains an A-tract AATT in the central region of the helix, which, although not contacted by the protein, bends the free-DNA structure (*red*) in a manner similar to that seen in the bound structure (*blue*) of the HPV-18 E2-DNA complex (PDB ID 1jj4). In comparison to ideal B-DNA (*gray*), the bending is reflected by a minor groove narrowing in the center of the free and bound DNA.

approaches have been developed to estimate the impact of intrinsic versus induced effects when only the bound form is available (123) or to predict the structure of the unbound DNA-binding site (20, 29, 87).

With this background in mind, below we discuss the various mechanisms proteins use to recognize their binding sites, attempting to organize them from a structure-based perspective (**Figure 4**). Note, we only have space in this review to support each readout mechanism with a small number of examples. Furthermore, because any one DNA-binding protein typically uses a variety of readout mechanisms, the same example may be used multiple times.

4.2. Base Readout

One well-established way for proteins to achieve DNA-binding specificity is through contacts with the bases in either the major or minor groove that recognize the chemical signature of the base or base pair. This type of recognition is generally mediated by the formation of hydrogen bonds between amino acids and bases, which convey the highest degree of specificity and, in some cases, by water-mediated hydrogen bonds or hydrophobic contacts (**Figure 4**).

4.2.1. Base-specific interactions in the major groove. In this section, we discuss the two main types of base readout mechanisms that occur in the major groove of the DNA, hydrogen bonds and hydrophobic interactions (**Figure 4**).

4.2.1.1. Hydrogen bonds with bases. Hydrogen bonds with bases can confer greater specificity in the major groove than in the minor groove because the four possible base pairs have a unique pattern of hydrogen bond donors and acceptors in the major but not in the minor groove (**Figure 5**) (10, 124). Proteins that form hydrogen bonds with bases in the major groove use HTH domains (e.g., homeodomains,

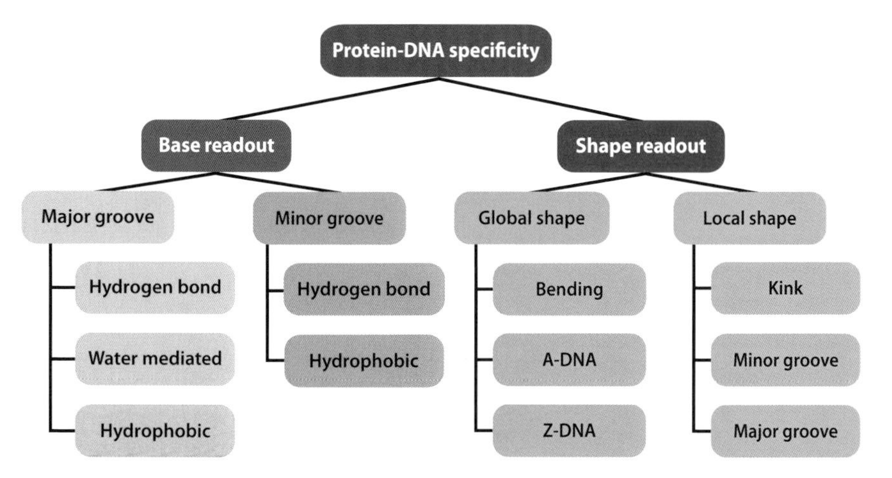

Figure 4

Types of protein-DNA recognition mechanisms used for specificity. We distinguish between two main classes of recognition: base readout and shape readout, which are further subdivided as illustrated.

434 repressor, λ repressor, Trp repressor, Myb), zinc finger domains (e.g., TFIIIA), immunoglobin fold domains (e.g., p53, NF-κB, STAT, and NFAT), and the N-terminal end of basic leucine zipper (bZip) domains or the Max transcription factor (17–19).

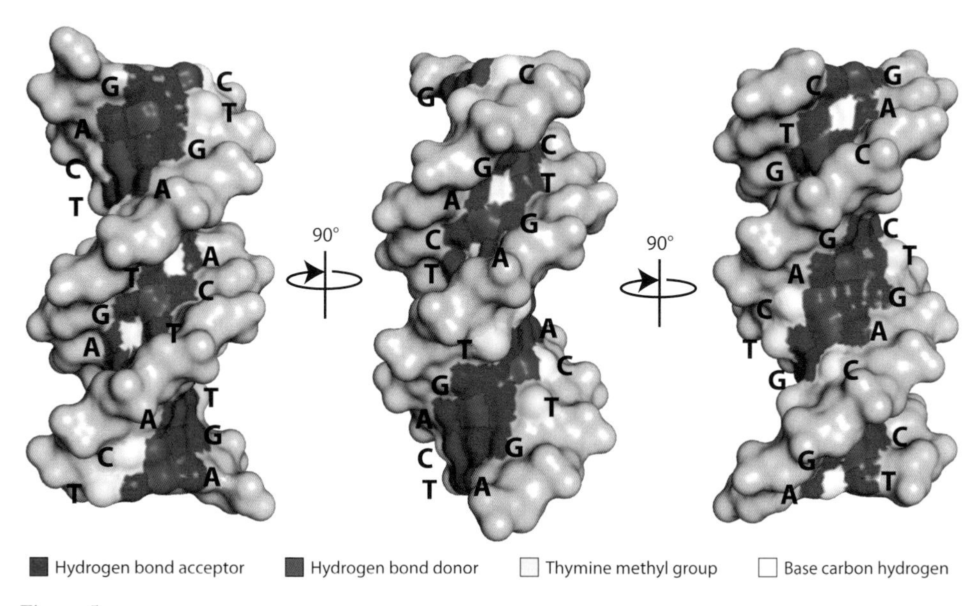

Figure 5

Base recognition in the major and minor groove. Sequence-specific patterns on the edges of the bases in the major groove underlie the ability of proteins to readout base pairs through hydrogen bonds and hydrophobic contacts (hydrogen bond acceptors in red, donors in blue, thymine methyl group in yellow, and base carbon hydrogens in white). In contrast, A:T versus T:A and C:G versus G:C are indistinguishable in the minor groove. The three panels show successive rotations of 90° around the helix axis. The dodecamer d(GACT)$_3$ was built on the basis of fiber diffraction data with 3DNA (92).

As noted above, the orientation of the recognition helix in the major groove is similar for homeodomain-DNA interfaces (125) but can vary among different families (17) and even within a given family, as between the Trp and λ repressors (100). In some cases, as observed for the KorA repressor, the recognition helix induces a widening of the major groove (126). In addition to α-helices, hydrogen bonds between β-sheets and bases can be used as well in specific recognition. Hydrogen bonds between bases in the major groove with the convex side of a β-sheet are observed in the binding of the MetJ and Arc repressors to DNA (127). The width of the major groove adjusts to the size of the β-sheet (widened in Arc repressor and narrowed in MetJ repressor), and the side of the β-sheet interacting with DNA generally exhibits more positive electrostatic potentials (127).

Specificity conveyed through hydrogen bonds in either groove depends on the number of contacts formed between protein residues and DNA bases but also on the uniqueness of the hydrogen bonding geometry. Bidentate hydrogen bonds (two hydrogen bonds with different donor and acceptor atoms) have the highest degree of specificity, followed by bifurcated hydrogen bonds (two hydrogen bonds that share the donor) and single hydrogen bonds. Whereas single hydrogen bonds usually do not contribute to specificity, bidentate hydrogen bonds are a source of remarkable selectivity (128). Bidentate hydrogen bonds can be formed with one base, two bases in a base pair, two adjacent bases in one strand, or two bases diagonally in different base pairs and opposite strands.

As discussed above, the specificity achieved through hydrogen bonds with bases depends on the pattern of donors and acceptors at the base edges in both grooves (**Figure 5**). Because DNA usually occurs in Watson-Crick geometry (1), this pattern is specific for each of the four base pairs in the major groove. However, base pair geometry can vary. For instance, Hoogsteen base pairs (129) have been observed in structures with deformed DNA sequences, such as the TBP/TATA box complex (130) and at the ends of oligonucleotides where the helical structure is preserved through stacking interactions [e.g., in a p53 tetramer complex (122)]. To date, a Hoogsteen base pair not present at the end of an oligonucleotide has only been observed in one complex with undistorted B-DNA, i.e., the MATα2 homeodomain bound to a specific binding site (131). Interestingly, the Hoogsteen base pair occurs in the center of the binding site CATGT$\underline{A}$ATT (underlined A) and was seen in crystals generated under various conditions (131). A transition from a Watson-Crick to Hoogsteen geometry alters the pattern of hydrogen bond donors and acceptors in both grooves and the conformation of the double helix. Although this single example should be interpreted with caution, it raises the possibility that non-Watson-Crick base pairs may contribute in important ways to binding specificity. As high-resolution structures are required to visualize such geometries, non-Watson-Crick base pairs may be present at a greater frequency than is evident in existing structures (see Note Added in Proof).

In many structures, the hydrogen bonds between protein and DNA are mediated by intervening water molecules. The bridging of hydrogen bonds by water molecules has frequently been observed for enzymes (132), and most hydrogen bonds in the Trp repressor-DNA interface are water mediated (12, 124). Mutagenesis experiments have shown that the CTAG tracts in both half sites of the Trp repressor's binding site are most critical for its sequence specificity (133). Highly ordered water molecules also mediate the specific readout of bases in the RXR-retinoid acid receptor (RAR)-DNA complex involving several arginine and lysine residues (134). Interestingly for the Lac repressor, the protein-DNA interface retains a significant portion of its hydration when it binds nonspecifically but not in the specific complex (135).

These data suggest that water-mediated hydrogen bonds in the major groove can be used for specific readout because they often reflect the positions of hydrogen bond donors and acceptors at the base edges. This is not the case for water molecules in the minor groove where the donor-acceptor patterns become unrecognizable.

4.2.1.2. Hydrophobic contacts with bases. Whereas hydrogen bonds with bases are highly specific in recognizing purines, hydrophobic contacts with bases are mainly used to read pyrimidines. Protein side chains employ hydrophobic interactions to differentiate thymine from cytosine (124) as in the bacteriophage 434 repressor and 434 Cro binding to their operator sites (136, 137). Four thymine methyl groups form a cleft that is specifically recognized by a valine in the lambdoid bacteriophage P22 c2 repressor-operator complex (138).

Hydrophobic contacts with bases also play a key role in the sequence-specific recognition of single-stranded DNA by bacterial cold-shock proteins, which recognize polythymine strands through stacking interactions with phenylalanines and histidines and distinguish thymine from cytosine through hydrogen bonding (139, 140).

4.2.2. Base-specific interactions in the minor groove. This section discusses the two forms of base readout observed in the minor groove: hydrogen bonds and hydrophobic contacts (**Figure 4**).

4.2.2.1. Hydrogen bonds with bases. Although, as discussed above, the pattern of donors and acceptors in the minor groove does not distinguish A:T from T:A or G:C from C:G base pairs (10) (**Figure 5**). Some proteins, such as zinc finger proteins with Cys_2Cys_2 GATA-like domains, that form hydrogen bonds in the major groove also bind in the minor groove (19). HMG proteins form hydrogen bonds in the minor groove (19) but rely on the recognition of DNA shape and flexibility, discussed below, to achieve specificity. This is also apparent for the binding of TBP to the minor groove as the six observed hydrogen bonds with the TATA box are not sufficient for the protein to attain specificity (14, 15, 141).

In some cases, a spine of hydration, a continuous string of water molecules, in narrow minor groove regions is contacted by proteins, as observed in the DNA complexes formed by the IFN-β enhanceosome (142, 143) and the integration host factor (IHF) (141). In other cases, only individual water molecules are displaced from narrow minor groove regions when amino acids intrude into the groove, e.g., α2-Arg7 in the MATa1-MATα2-DNA complex (144). The displacement of water molecules from the narrow minor groove has been shown to provide a strong thermodynamic driving force for DNA binding (145–147).

4.2.2.2. Hydrophobic contacts with bases. Architectural proteins only contact the minor groove, which is often associated with a dramatic widening and extensive hydrophobic contacts (141). This mechanism is employed by the TBP, SRY, and LEF-1. The TBP/TATA box interface is completely dehydrated, and the abundance of hydrophobic contacts in the interface (148) suggests that they contribute to specificity. Although 12 of the 16 hydrogen bond acceptors in the minor groove remain unsatisfied upon TBP binding, these base atoms mainly engage in hydrophobic contacts with nonpolar side chains (14, 15, 141).

4.3. Shape Readout

For most DNA-binding proteins, the readout of base pairs through hydrogen bonds or hydrophobic contacts is not sufficient to explain specificity. Other factors that have been proposed to contribute to specificity are sequence-dependent DNA structure and deformability (20, 149). These readouts, which all depend on deviations from ideal B-DNA, comprise a diverse set of mechanisms that all fall under the general heading of binding a nonideal B-DNA shape. As such, we collectively refer to them as shape readout (**Figure 4**). Furthermore, we distinguish between *local shape readout* mechanisms, in which the DNA helix deviates from ideal B-DNA in a localized manner, and *global shape readout* mechanisms, in which most of the DNA-binding site is either deformed or in a nonideal B-form conformation (**Figure 4**).

Both local and global shape readouts can contribute to DNA-binding specificity. For local shape readout, such as minor groove narrowing, recent results suggest that the shape of the minor groove within a binding site can be "read" by a complementary set of basic side chains, most typically arginines, when presented in the correct conformation (66). In contrast, global shape readout, such as a gradual bend in the DNA helix, may position elements of the DNA backbone such that these otherwise nonspecific contacts can become highly specific. Below, we discuss each of these types of readouts, providing specific examples to illustrate them.

4.3.1. Local shape readout. As described in the DNA structure section, the two predominant local shape deviations from ideal B-DNA are (*a*) small regions of 3–8 bp where the minor groove is narrow and (*b*) DNA kinks, which are caused by the unstacking of a single base pair step (**Figure 4**).

4.3.1.1. Minor groove shape. The N-terminal arms of homeodomain proteins have been observed in the minor grooves of several structures, but only recently have they been shown to play a role in DNA-binding specificity. In particular, the binding of the Hox protein Sex combs reduced (Scr) and its cofactor Extradenticle (Exd) to Scr-specific (*fkh250*) and Hox-Exd consensus (*fkh250^{con}*) binding sites shows how N-terminal arm arginines recognize a minor groove shape to achieve specificity (29). Whereas both Arg3 and Arg5 of Scr are ordered in the minor groove of the specific binding site (**Figure 6*c***), Arg3 is disordered when presented with the Hox-Exd consensus site (**Figure 6*d***). Arg3 does not form direct base contacts but instead forms a hydrogen bond with His -12, which, in turn, contacts the bases through a water-mediated hydrogen bond. Mutagenesis studies have shown that Arg3 plays a critical role in Scr in vivo specificity (29).

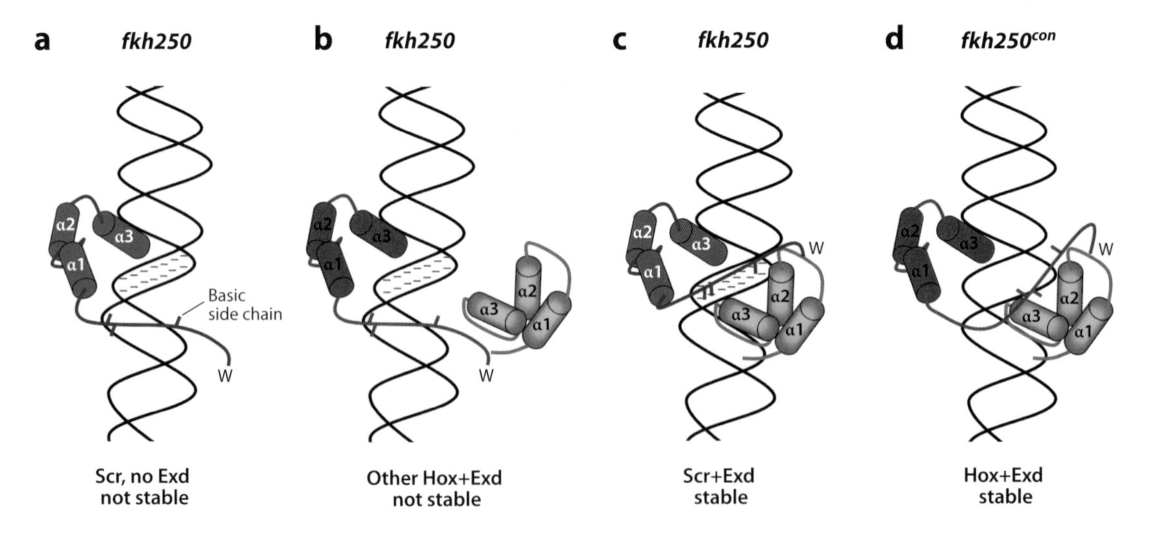

Figure 6

Hox DNA-binding specificity mediated by local shape recognition. All panels show either the *fkh250*-binding site or the *fkh250^{con}*-binding site. *fkh250*, but not *fkh250^{con}*, has two minor groove minima, which creates a more negative electrostatic potential (*minus signs*). The capital letter W refers to the Hox YPWM motif, which makes a direct contact with the cofactor Exd. See Reference 29 for details. (*a*) In the absence of Exd, Scr does not bind with high affinity to *fkh250* because basic side chains (*small bars*), in particular, arginines, on the N-terminal arm and linker of Scr are not positioned correctly. (*b*) Other Hox proteins do not bind well to *fkh250* even in the presence of Exd because their N-terminal arms and linker regions have different sequences. (*c*) The Scr-Exd heterodimer binds well to *fkh250* because the Scr N-terminal arm and linker region have the correct residues, and Exd positions them correctly by binding the YPWM motif (W). (*d*) Other Hox-Exd heterodimers bind well to *fkh250^{con}*. This binding site is not as selective because it has a less negative electrostatic potential. Thus, the sequences of the Hox N-terminal arms and linker regions are not as important for binding.

The Scr-specific and Hox-Exd consensus sites differ in minor groove shape, a structural feature that appears to be intrinsic to these sequences. These local variations in shape result in the enhancement of negative electrostatic potential at distinct positions that attract arginines into the minor groove (**Figure 2*c*,*d***) (20, 29). The Scr N-terminal arm uses these sequence-dependent variations in shape and electrostatic potential to achieve DNA-binding specificity (**Figures 6*c*,*d***) (150). Because narrow minor grooves are often associated with AT-rich sequences (**Table 2**), enhancement of negative electrostatic potential in the minor groove, which, in turn, is recognized by arginines, offers a general mechanism for sequence-specific recognition of DNA shape (66).

In addition to the results for Scr, mutagenesis studies on the Hox protein Ultrabithorax (Ubx) also suggest a role for linker and N-terminal arm residues in DNA-binding specificity, even when Ubx binds as a monomer (151, 152). Although no crystal structures are yet available to visualize these interactions, an intriguing possibility is that these residues may be reading differences in minor groove shape.

The use of arginines to bind to narrow regions of the minor groove is widespread among DNA-binding proteins (66). However, the manner in which the arginines are presented to the minor groove can differ (**Figure 7**). In the case of Scr-Exd, heterodimer formation between these two homeodomain proteins is necessary to position Arg3 and His-12, which are normally on an unstructured part of the Hox protein, so that these side chains can insert into the minor groove (**Figure 7*a***). In the case of the POU domain protein Brn-5 binding to its element *CRH-II*, the arginines that insert into a narrow region of the minor groove come from the linker region that separates the POU_{HD} from the POU_S domain (59). Thus, as with Scr-Exd, two DNA-binding domains are required to position the Brn-5 arginines, but in this case, both domains are in the same protein (**Figure 7*b***). Not all POU proteins use this method to position the relevant arginines (153). For example, the Oct-1/PORE complex uses the Arg2 and Arg5 side chains of two Oct-1 monomers to bind to two short A-tracts, ATTT and AAAT (154), and a Pit-1 dimer binds to DNA in a fashion similar to the Oct-1 dimer but uses Arg49 of the POU-specific domain to distinguish its ATAC site from the ATGC site of the Oct-1 dimer (153).

Proteins from families other than homeodomains also use the mechanism of local minor groove shape readout (66). The LEAFY gene regulator, for example, binds as a homodimer in which arginines present on the N-terminal arm of both monomers bind two distant narrow minor groove regions (155). In comparison, MogR uses arginines on a C-terminal extension from both monomers to contact a narrow minor groove composed of two antiparallel A-tracts that are separated by a TpA hinge step (**Figure 7*c***) (36). The γδ resolvase forms an arginine contact to a narrow minor groove with its N-terminal extension and uses another N-terminal arginine to contact the major groove (**Figure 7*d***) (156).

In all of the above examples, the arginines that insert into the minor groove come from otherwise unstructured strands that must be positioned owing to heterodimerization (Scr-Exd) or homodimerization (Oct-1, MogR), or via two adjacent DNA-binding domains in the same protein (Brn-5). Arginines that insert into minor grooves can also be integral to DNA-binding domains. For example, MEF2A, from the myocyte enhancer factor-2 family, uses its α1 helix, which is positioned on top of the minor groove, to contact the *MEF2A* minor groove (**Figure 7*e***) (157).

Minor groove-interacting arginines are often presented as part of short sequence motifs that include more than one arginine, such as RQR in Scr (29), RPR in Engrailed (26), RKKR in POU homeodomains (158), and RGHR in MATα2 (144). The observation that arginine-rich motifs bind to the minor groove was also made for the phage 434 repressor (KRPR) (**Figures 2*c*,*d***) and the Hin recombinase (GRPR), for which arginine mutations were shown to have a dramatic effect

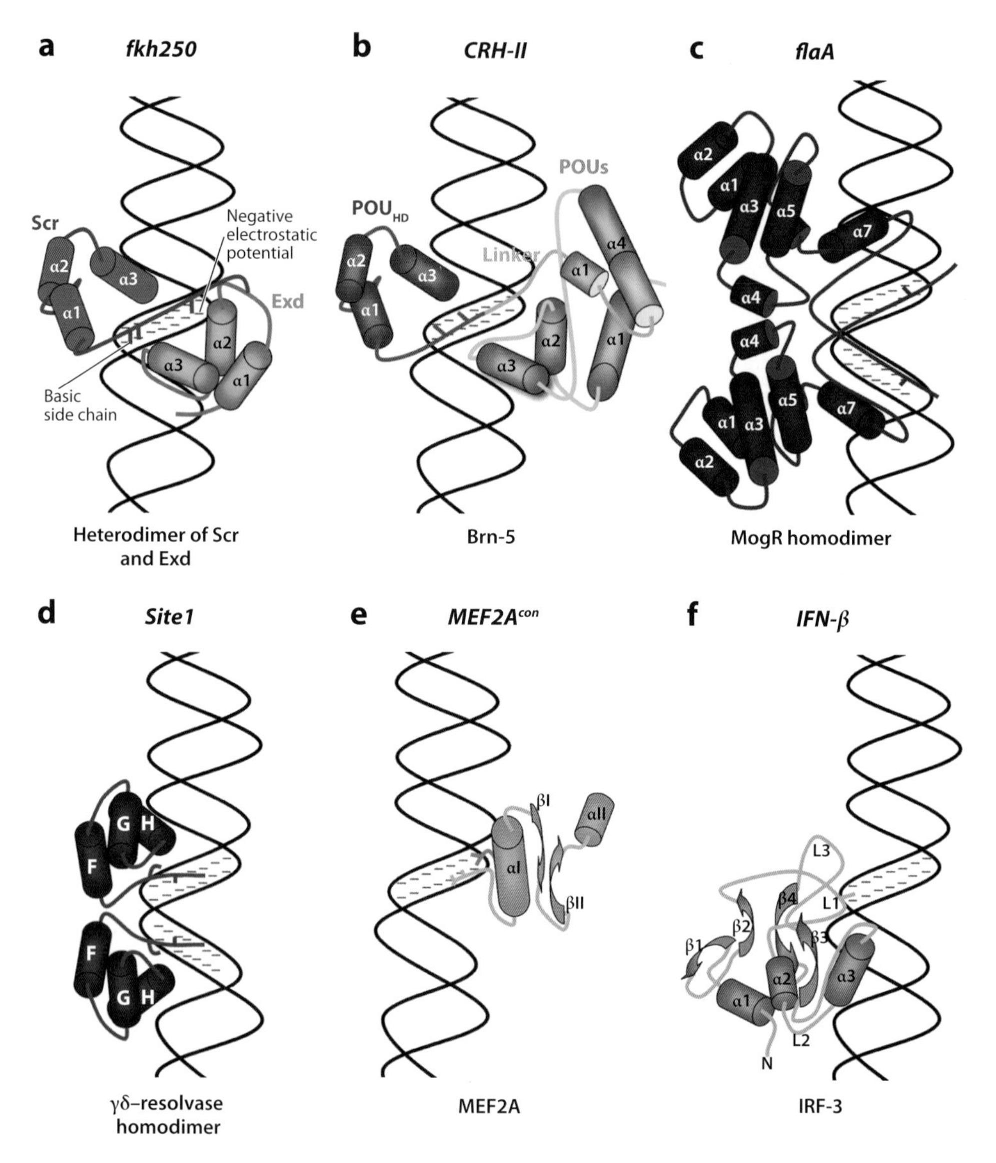

Figure 7

Examples of minor groove shape recognition. Each panel shows a different example in which basic side chains (*colored bars*) bind to minor grooves. (*a*) Arginine residues present on Scr's N-terminal arm and linker region require heterodimerization with Exd to be positioned correctly to insert into a narrow minor groove region of *fkh250* (PDB ID 2r5z). (*b*) Arginine residues present on the linker region that separates POU_{HD} from POU_S of Brn-5 insert into a narrow minor groove of the *CRH-II*-binding site (PDB ID 3d1n). (*c*) Arginine residues present on a C-terminal extension of a MogR homodimer insert into narrow regions of the *flaA*-binding site (PDB ID 3fdq). (*d*) An N-terminal extension from the $\gamma\delta$ resolvase has an arginine that inserts into a narrow minor groove and a second arginine that inserts into the major groove of its binding site (PDB ID 1gdt) (*e*) MEF2A recognizes a narrow minor groove of the *MEF2A*-binding site via an arginine and glycine present on an N-terminal strand and via a lysine present on α-helix αI (PDB ID 1egw). (*f*) A histidine residue of IRF-3 inserts into a narrow minor groove region of the *IFN-β* enhancer (PDB ID 1t2k).

on binding affinity (159). The RQR motif of Scr introduces its arginines like a fork into the minor grove, with the glutamine pointing away from the DNA like the fork's handle (29). Other arginine-rich motifs orient the arginine side chains differently, allowing them to recognize distinct minor groove shapes.

Unlike homeodomain proteins, which rely on both major and minor groove interactions to achieve specificity, the architectural proteins TBP, SRY, LEF-1, IHF, and HMG-I(Y) only contact the minor groove. For example, the N-terminal arm of IHF inserts two arginines deep into a narrow region of the minor groove complemented by a third arginine that contacts a different narrow region (141). HMG-I(Y) proteins bind to AT-rich minor grooves but, in contrast to IHF, stabilize essentially straight instead of deformed DNA (141).

Although arginine is the most abundant residue that inserts into minor grooves, lysines can also be observed in such regions, although at a much lower frequency (66). The difference between these two basic amino acids is due, at least in part, to the higher free energy associated with removing lysines, which have a less delocalized positive charge distribution, from the aqueous phase (66). The importance of solvation effects is illustrated by the IFN-β enhanceosome structure, which exhibits a number of lysines in the proximity of the minor groove, clearly solvated rather than intruding into the groove (142, 143, 160). However, the enhanceosome uses histidines (from IRF-3 and IRF-7) to penetrate narrow minor groove regions formed by A-tracts (142, 143, 160). His40 of IRF-1, which is conserved across the IRF family, also inserts into narrow minor groove regions (**Figure 7*f***) (161, 162). A histidine is also observed to insert into the minor groove in the Scr-Exd-*fkh250* structure (29).

4.3.1.2. Major groove shape. There are indications that sequence-dependent major groove shape is, like minor groove shape, also used as a readout mechanism. Indeed, minor and major groove geometries are correlated with each other (163). The human regulatory factor hRFX1 is a wHTH protein, which recognizes the DNA major groove with its β-hairpin wing in place of the recognition helix used by other wHTH proteins (42). In turn, hRFX1 protein places its H3 helix over the minor groove, from which a single lysine contacts the groove (42). The minor groove widens, resulting in a narrowing of the major groove that, in turn, improves major groove shape complementarity (38). In another example, domain 4 of the *E. coli* extracytoplasmic function σ factor, σ^E, specifically recognizes the GGAACTT element on the basis of major groove shape complementarity, which is achieved by narrowing the minor groove (164). The AT base pairs in the σ^E-binding site (underlined), which are highly conserved despite a lack of strong base contacts, are located in the center of a narrow minor groove (164) and were shown in genetic screening experiments to inhibit transcription when mutated (165).

4.3.1.3. Kinks. As discussed above, DNA kinks occur when the linearity of the helix is abruptly broken, often owing to the unstacking of a flexible base pair step, such as TpA (**Table 2**). Kinks can contribute to binding specificity by promoting conformations that optimize protein-DNA and protein-protein contacts. As an example, the conformational flexibility of the ATA region allows the Tramtrack-binding site to adjust to the contacting zinc finger (166). DNA recognition by endonuclease EcoRV also depends upon the deformability of a TpA step (167). The binding site of the γδ resolvase comprises a central TATA element and exhibits kinks at both TpA steps (156). The flexibility intrinsic to TpA steps also plays a role in the specific binding of the RevErb nuclear hormone receptor as it binds to a site that contains two TpA steps (168). Although neither of these steps engage in base-specific contacts with RevErb, they show different degrees of deformation, indicating the importance of their flexibility.

The DNA-binding site of the catabolite activator protein (CAP) shows dramatic kinks at two CpA (TpG) steps (16, 169), which cause, along with two additional smaller kinks, an overall bending of the DNA of about 90° around the protein (170, 171). The kink at the CpA (TpG) step makes it possible for an arginine residue to engage in partial stacking interactions with a thymine (124). The HincII endonuclease recognizes its cognate site GTYRAC on the basis of the deformability of its central YpR step and shows the highest affinity when this step is CpG (172). Similarly, the binding of the EcoRI endonuclease to the Dickerson dodecamer involves a kink at the center of its binding site (173).

4.3.1.4. Intercalation. Owing to weaker stacking interactions, kinks are often stabilized through the intercalation of protein side chains, which, in turn, causes further deformation of the DNA helix. The specific DNA-binding site of the Lac repressor adjusts to the protein by forming a kink of about 36° at its central CpG step, which widens the minor groove where two leucine residues interact with the kinked base pair step through partial intercalation (**Figure 2*b***) (135). By contrast, a nonspecific DNA sequence, which has been designed to be different in all positions from the Lac operator, does not form a kink upon binding to the Lac repressor, but the protein rearranges its backbone and side chain conformations to engage in phosphate contacts (174). When the purine repressor is bound to its cognate site GCAAACGTTTGC, a similar kink is observed at its central CpG step (underlined) and is stabilized by the partial intercalation of two leucine residues from the minor groove side (31). Although the conformations of the flanking A-tract regions are very similar in the structures of free and PurR-bound DNA, a kink is not observed in the unbound site (116). This observation argues that, in this case, it is not DNA structure per se but its deformability that is recognized by PurR.

The yeast TBP structure shows phenylalanine intercalations in the first and last base pair step (underlined) of its TATATAAA-binding site (14). Whereas the first intercalation site is a flexible TpA step, the second site is likely determined by spacing (141, 148). Architectural proteins that intercalate hydrophobic amino acids between base pairs from the minor groove are the HMG box proteins SRY and LEF-1 (141). These intercalating hydrophobic residues are conserved in HMG domains and are usually flanked by basic amino acids (175). SRY and LEF-1 both use Asn10 to convey specificity through tripartite polar contacts with base pairs preceding the intercalation pocket. Closely related to SRY, DNA-bending SOX domains represent another subgroup of HMG boxes (176). The SOX2-Oct-4-DNA ternary complex is characterized by the intercalation of methionine and phenylalanine residues into an ApA (TpT) step inducing a kink (154). The SOX17 protein also uses its HMG domain to cause a drastic kink of an ApA (TpT) step through the intercalation of a phenylalanine-methionine dipeptide (177).

4.3.2. Global shape readout. We include in this category the recognition of DNA sequences where the entire binding site is not in a classic B-form helix. Examples are the recognition of bent DNA, where the curvature is distributed along the entire helix, A-DNA, sequences that have elements of both A- and B-DNA, and Z-DNA (**Figure 4**).

4.3.2.1. Bent DNA. The papillomavirus E2 protein provides a clear example of DNA bending playing a role in protein-DNA recognition. The E2 protein binds as a dimer to two half sites separated by a linker of four base pairs (87, 178). Although only the underlined half sites of the ACCGN$_4$CGGT consensus-binding site are contacted by the protein, the variable linker optimizes these contacts through bending, which, in turn, enhances interactions between the protomers of the E2 dimer (13, 82). The DNA is similarly bent in complex with the E2 proteins of the bovine papillomavirus BPV-1 (13) and the human papillomavirus HPV-18 (**Figure 2*b***) (179). However, whereas the BPV-1 E2 protein binds with similar binding affinity to consensus sites with various linker sequences, the HPV-18 E2 protein shows a strong preference for AATT linkers (180), and the HPV-16

E2 protein shows a preference for AATT and AAAA linkers (178). X-ray crystallographic studies and Monte Carlo predictions stressed that the E2-binding site with AATT linker is also bent when not bound to the protein, whereas the site with ACGT linker is essentially straight (**Figure 3*b***) (82, 87, 120). A correlation of the structural data with binding studies suggests that high-affinity sites are prebent as seen in the E2-DNA complex, but low-affinity sites require the protein to induce the site to bend (178, 179).

Bending was also suggested to play a role in the specificity of homeodomains by facilitating contacts with the recognition helix (181). Specific DNA recognition by the phage 434 repressor is associated with the bending of its operator (149), which decreases with the number of G:C base pairs in its operator sequence (182). Long A-tracts are associated with bending and are present, for instance, in the binding sites of the MATa1-MATα2 heterodimer (144) and the NF-κB protein (48). The conformation of the NF-κB-binding site in its bound state is similar to the bending already present in its free state (117, 183). The RXR-RAR heterodimer recognizes the same half sites as the RXR homodimer. However, the smooth bending of the AAA region between both half sites in place of the kink induced by the RXR homodimer contributes to RXR-RAR specificity (134). The restriction endonucleases BglII and BamHI recognize DNA sites, AGATCT and GGATCC, with an identical core region (underlined), but bending differentiates both binding sites (184–186). In contrast, the similar binding sites of the endonucleases MunI and EcoRI, CAATTG and GAATTC, respectively, cannot be distinguished through bending and require an arginine contact to read the outer C:G base pair (186).

4.3.2.2. A-DNA. Whereas sugars are usually buried in the minor groove of B-DNA, they are exposed in A-DNA and provide about 50% of the protein-DNA interface in the TBP-DNA complex, where the DNA is in an A-form conformation (14). Although arginine and lysine frequently interact with nucleotides in B-DNA conformations, nonpolar amino acids, such as alanine, leucine, phenylalanine, and valine, contact nucleotides in A-DNA conformations (187). These types of contacts are thus associated with GC-rich sequences (76, 77, 188) and with TATA boxes (**Table 2**) (189). The higher accessibility of C3′-endo sugars of A-DNA in comparison to buried C2′-endo sugars of B-DNA (187) also contributes to the specificity of zinc finger proteins for GC-rich sequences (116) and of the TBPs for TATA boxes (78).

The transition from B-DNA to A-DNA that transforms the sugar conformations and widens the minor groove is often associated with the intrusion of hydrophobic residues into the minor groove (190). B- to A-transitions are often observed in complexes with endonucleases because A-DNA makes the phosphate oxygen of the bond that is cleaved more accessible (75). Other proteins that recognize A/B-intermediate conformations are the Trp repressor and the *Caenorhabditis elegans* Tc3 transposase (75). The transcription factor for polymerase IIIA (TFIIIA) also binds to an A-DNA-like binding site (191). In general, zinc finger proteins tend to bind A/B-intermediates in major grooves that are deep like A-DNA and wide like B-DNA (119) and that have the increased helix diameter typical for A-DNA (192). Zinc fingers from the human glioblastoma protein (GLI) show the base pair inclination that is distinct for A-DNA (193). In other complexes, only a limited number of base pairs exhibit A-DNA conformations, whereas the remaining site resembles B-DNA, as seen in two regions of the I-PolI-binding site (75).

Interestingly, binding sites of the mouse Cys_2His_2 zinc finger protein Zif268 crystallize in A-like conformations when both unbound and bound by the protein (**Figure 3*a***) (116, 118, 119). These observations suggest that this DNA sequence has an intrinsic tendency to assume an A-like conformation and that exposed hydrophobic surfaces of A-like sugars may be generally recognized by zinc fingers (191). Another example of the recognition of a DNA that has an A/B intermediate structure is the

Runt domain and its binding site (50). In this case, the unbound binding site was observed both in A-DNA (194) and B-DNA (121) conformations. Perhaps related to such observations is that some transcription factors, such as TFIIIA, Bicoid, and p53, bind to both DNA and RNA; the latter almost exclusively exhibits A-form topology (195).

4.3.2.3. Z-DNA. The zigzag positioning of phosphates along a left-handed Z-DNA helix is specifically recognized by the double-stranded RNA adenosine deaminase (ADAR1), which is an RNA-editing enzyme with a wHTH motif (196). Z-DNA structures have only been observed to form with purine-pyrimidine alternating sequences that can adopt a left-handed helix (79, 80, 197). The Zα-domain of ADAR1 has a conformation tailored to recognize a row of five phosphates in one zigzag-shaped backbone of Z-DNA. Because the tumor-associated DLM-1 protein also recognizes Z-DNA via five phosphates along a zigzag-shaped left-handed strand, phosphate positions seem to be the signature code recognized by Z-DNA-binding proteins (**Figures 1*d*,*h***) (198).

5. EXAMPLES OF HIGHER-ORDER PROTEIN-DNA INTERACTIONS

The above discussion highlights examples that illustrate specific readout mechanisms and thus provides a reductionist perspective on DNA recognition. However, individual DNA-binding proteins combine many, if not most, of these readout mechanisms to achieve the correct affinity and specificity required for function. To illustrate this, below we discuss a few examples of protein-DNA recognition in which combinations of readout mechanisms are clearly used.

5.1. The Nucleosome

The presence of nucleosomes in eukaryotic genomes profoundly affects the activity of transcription factors and other DNA-binding proteins (199–201). Although some factors can bind to nucleosomal DNA, others can only bind nucleosome-free DNA. For instance, the packaging of DNA in nucleosomes is expected to narrow the minor groove of TATA boxes, thus precluding TBP binding (148). In contrast, the bending of nucleosomal DNA was suggested to assist p53 binding at the DNA surface facing away from the histone core (91). Owing to the intimate relationship between protein-DNA recognition and nucleosome binding, attempts to predict nucleosome positions in genomic DNA have received a great deal of attention (202–205). Because DNA deformability (kinks), DNA bending, and local shape recognition all contribute to nucleosome positioning, these mechanisms need to be considered in any prediction algorithm.

The bendability of short sequences accommodates the wrapping of DNA around the histone core in the nucleosome (148, 149). The presence of short A-tracts of only three A:T base pairs stabilizes the deformation required for regions of the nucleosomal DNA facing the histones, where the minor groove is compressed (66, 206). Consequently, the distribution of short A-tracts in yeast in vivo sequences reflects the periodicity of a helical turn in congruence with the structural periodicity caused by the wrapping of nucleosomal DNA around the histone core (66). In addition, kinks caused by CpA steps adjacent to short A-tracts can enhance the overall curvature in regions where the minor groove faces the histones (91). And, because of their flexibility, the kinks resulting from TpA steps are also used to help wrap the DNA around the histone core. Taking both observations together, the deformability of short A-tracts and YpR steps provides more information about sequence periodicities than was originally observed for dinucleotides (202, 207–209).

The periodicity of short A-tracts in nucleosomal DNA also results in a periodic narrowing of the minor groove, which is, in turn, read by arginines present at the histone-DNA interface (66). Nucleosome-bound DNA contains, on average, 10 of these intrinsically narrow minor groove regions, most of which are likely to be contacted by arginines. Thus, in addition to DNA kinks and bends, nucleosome-DNA interactions also rely on the recognition of local variations in DNA shape (66).

5.2. *Escherichia coli* IHF

A combination of kinking, bending, and intercalation is used to achieve DNA-binding specificity for the *E. coli* nucleoid protein IHF, which also functions as a transcriptional activator (210). The IHF α/β heterodimer sharply bends DNA by about 160° to bring distant binding sites of the λ repressor into close proximity (211). IHF recognizes three DNA sites: TATCAA in the central region of its binding site, a 6-bp A-tract, and a TTG region at its flanks (210). The large bending is partially induced by the A-tract with its intrinsically narrow minor groove at one side of the IHF-DNA complex (212). On the other side of the complex, the TpG (CpA) step in the TTG element narrows the minor groove through kinking, which is recognized through the insertion of βArg46 (213). The TTG to TAG mutation, which shifts the YpR step 5′ by 1 bp, indicates that the IHF protein discriminates between A:T and T:A base pairs in this region due to the flexibility of the YpR step (213). The α-arm of the protein contacts the minor groove of the central consensus element with three arginine residues. Two large kinks at the ApA (TpT) steps caused by proline intercalations are the main contributors to the U-formed shape of the IHF-bound DNA (211).

5.3. Cooperativity

DNA-binding proteins often bind DNA cooperatively to create higher-order nucleoprotein complexes that reflect the combinatorial control of gene expression. DNA-binding cooperativity is most typically attributed to direct protein-protein interactions between adjacent DNA-binding factors that promote the assembly of higher-order complexes. Notable examples are Hox-Exd/Pbx heterodimers (28, 29, 214), the MATa1-MATα2 heterodimer (144), and the NFAT-Fos-Jun heterotrimer (215). Whereas cofactors in all of the previous examples directly bind to DNA, the cofactor CBFβ enhances the binding of the *Drosophila* Runt domain to DNA without forming any DNA contact (50).

In addition to this classical form of cooperativity, a sequence-dependent DNA structure may also promote the cooperative binding of multiple factors. One particularly striking example is the assembly of the IFN-β enhanceosome, which is composed of at least eight DNA-binding proteins: a heterodimer of ATF-2/c-Jun, a heterodimer of p50/Rel, and four IRF monomers, all bound to a highly conserved ~55-bp element (142, 143). In addition, the architectural protein HMGA1 binds, perhaps transiently, in the minor groove to at least two positions, inducing DNA bends that facilitate the assembly of the enhanceosome (216). Remarkably, despite the binding of eight transcription factors, a paucity of protein-protein interactions is observed, arguing that cooperativity is likely to be achieved in some other manner (142). One appealing suggestion is that the final DNA structure, which is optimized for enhanceosome assembly, depends on the intrinsic deformability of the DNA (160). According to this view, the binding of each factor improves the binding of the other factors through an effect on DNA structure. This idea follows logically from many of the other examples described above where DNA shape and deformability contribute to specificity on a smaller scale. Thus, if correct, a sequence-dependent DNA structure may be a critical component in the binding not only of individual factors to their binding sites, but also in the assembly of higher-order, multiprotein complexes. This idea fits well with another recent observation that was also pointed out at the beginning of this review, namely, that DNA shape is under evolutionary selection and provides a better indicator of functional elements than conservation of the linear DNA sequence (7).

SUMMARY POINTS

1. DNA-binding proteins use a wide range of mechanisms to bind specifically to binding sites.
2. The three-dimensional structure of the binding site must be taken into consideration when understanding binding specificity.
3. The main readout mechanisms are (*a*) the recognition of bases and (*b*) the recognition of DNA shape.
4. The recognition of bases can be further subdivided into those interactions that occur in the major groove, which provides the greatest potential for specificity, and those that occur in the minor groove.
5. The recognition of DNA shape can be further subdivided into the recognition of local shape variation (e.g., minor groove width) and the recognition of global shape variation (e.g., bent DNA).
6. Any one DNA-binding protein is likely to use a combination of readout mechanisms.
7. Readout mechanisms are often interrelated (e.g., bending toward the minor groove also narrows it).
8. The formation of higher-order protein-DNA complexes may depend on sequence-dependent DNA structures that are optimized to promote assembly.

FUTURE ISSUES

1. The annotation of genomes must take into account DNA structure.
2. The rules governing the relationships between DNA sequence and DNA structure need to be better understood.
3. Understanding intrinsic versus induced effects on DNA structure is an important goal and would benefit from additional structural analyses of free DNAs.
4. Understanding the rules governing binding specificity within a protein family would benefit from comparisons of structures of multiple family members, each bound to specific and nonspecific binding sites.

DISCLOSURE STATEMENT

The authors are not aware of any affiliations, memberships, funding, or financial holdings that might be perceived as affecting the objectivity of this review.

ACKNOWLEDGMENTS

This work was supported by National Institutes of Health (NIH) grants GM54510 (R.S.M.) and U54 CA121852 (B.H. and R.S.M.). The authors thank Z. Shakked, T. Tullius, T. Haran, and A. Aggarwal for helpful conversations.

LITERATURE CITED

1. Watson JD, Crick FH. 1953. Molecular structure of nucleic acids; a structure for deoxyribose nucleic acid. *Nature* 171:737–38
2. Berger MF, Badis G, Gehrke AR, Talukder S, Philippakis AA, et al. 2008. Variation in homeodomain DNA binding revealed by high-resolution analysis of sequence preferences. *Cell* 133:1266–76
3. Noyes MB, Christensen RG, Wakabayashi A, Stormo GD, Brodsky MH, Wolfe SA. 2008. Analysis of homeodomain specificities allows the family-wide prediction of preferred recognition sites. *Cell* 133:1277–89
4. Badis G, Chan ET, van Bakel H, Pena-Castillo L, Tillo D, et al. 2008. A library of yeast transcription factor motifs reveals a widespread function for Rsc3 in targeting nucleosome exclusion at promoters. *Mol. Cell* 32:878–87
5. Zhu C, Byers KJ, McCord RP, Shi Z, Berger MF, et al. 2009. High-resolution DNA-binding specificity analysis of yeast transcription factors. *Genome Res.* 19:556–66
6. Badis G, Berger MF, Philippakis AA, Talukder S, Gehrke AR, et al. 2009. Diversity and complexity in DNA recognition by transcription factors. *Science* 324:1720–23
7. Parker SC, Hansen L, Abaan HO, Tullius TD, Margulies EH. 2009. Local DNA topography correlates with functional noncoding regions of the human genome. *Science* 324:389–92
8. Greenbaum JA, Pang B, Tullius TD. 2007. Construction of a genome-scale structural map at single-nucleotide resolution. *Genome Res.* 17:947–53
9. Rosenberg JM, Seeman NC, Kim JJ, Suddath FL, Nicholas HB, Rich A. 1973. Double helix at atomic resolution. *Nature* 243:150–54
10. Seeman NC, Rosenberg JM, Rich A. 1976. Sequence-specific recognition of double helical nucleic acids by proteins. *Proc. Natl. Acad. Sci. USA* 73:804–8
11. Viswamitra MA, Kennard O, Jones PG, Sheldrick GM, Salisbury S, et al. 1978. DNA double helical fragment at atomic resolution. *Nature* 273:687–88
12. Otwinowski Z, Schevitz RW, Zhang RG, Lawson CL, Joachimiak A, et al. 1988. Crystal structure of trp repressor/operator complex at atomic resolution. *Nature* 335:321–29
13. Hegde RS, Grossman SR, Laimins LA, Sigler PB. 1992. Crystal structure at 1.7 A of the bovine papillomavirus-1 E2 DNA-binding domain bound to its DNA target. *Nature* 359:505–12
14. Kim Y, Geiger JH, Hahn S, Sigler PB. 1993. Crystal structure of a yeast TBP/TATA-box complex. *Nature* 365:512–20
15. Kim JL, Nikolov DB, Burley SK. 1993. Co-crystal structure of TBP recognizing the minor groove of a TATA element. *Nature* 365:520–27
16. Lawson CL, Berman HM. 2008. Indirect readout of DNA sequence by proteins. In *Protein-Nucleic Acid Interactions: Structural Biology*, ed. PA Rice, CC Correll, pp. 66–90. Cambridge, UK: R. Soc. Chem.
17. Garvie CW, Wolberger C. 2001. Recognition of specific DNA sequences. *Mol. Cell* 8:937–46
18. Luscombe NM, Austin SE, Berman HM, Thornton JM. 2000. An overview of the structures of protein-DNA complexes. *Genome Biol.* 1:REVIEWS001
19. Hong M, Marmorstein R. 2008. Structural basis for sequence-specific DNA recognition by transcription factors and their complexes. See Ref. 16, pp. 47–65
20. Rohs R, West SM, Liu P, Honig B. 2009. Nuance in the double-helix and its role in protein-DNA recognition. *Curr. Opin. Struct. Biol.* 19:171–77
21. McKay DB, Steitz TA. 1981. Structure of catabolite gene activator protein at 2.9 A resolution suggests binding to left-handed B-DNA. *Nature* 290:744–49
22. Anderson WF, Ohlendorf DH, Takeda Y, Matthews BW. 1981. Structure of the Cro repressor from bacteriophage lambda and its interaction with DNA. *Nature* 290:754–58
23. Pabo CO, Lewis M. 1982. The operator-binding domain of lambda repressor: structure and DNA recognition. *Nature* 298:443–47
24. Jordan SR, Pabo CO. 1988. Structure of the lambda complex at 2.5 A resolution: details of the repressor-operator interactions. *Science* 242:893–99
25. Badia D, Camacho A, Perez-Lago L, Escandon C, Salas M, Coll M. 2006. The structure of phage phi29 transcription regulator p4-DNA complex reveals an N-hook motif for DNA. *Mol. Cell* 22:73–81

26. Kissinger CR, Liu BS, Martin-Blanco E, Kornberg TB, Pabo CO. 1990. Crystal structure of an engrailed homeodomain-DNA complex at 2.8 A resolution: a framework for understanding homeodomain-DNA interactions. *Cell* 63:579–90
27. Wolberger C, Vershon AK, Liu B, Johnson AD, Pabo CO. 1991. Crystal structure of a MAT alpha 2 homeodomain-operator complex suggests a general model for homeodomain-DNA interactions. *Cell* 67:517–28
28. Passner JM, Ryoo HD, Shen L, Mann RS, Aggarwal AK. 1999. Structure of a DNA-bound Ultrabithorax-Extradenticle homeodomain complex. *Nature* 397:714–19
29. Joshi R, Passner JM, Rohs R, Jain R, Sosinsky A, et al. 2007. Functional specificity of a Hox protein mediated by the recognition of minor groove structure. *Cell* 131:530–43
30. Murzin AG, Brenner SE, Hubbard T, Chothia C. 1995. SCOP: a structural classification of proteins database for the investigation of sequences and structures. *J. Mol. Biol.* 247:536–40
31. Schumacher MA, Choi KY, Zalkin H, Brennan RG. 1994. Crystal structure of LacI member, PurR, bound to DNA: minor groove binding by alpha helices. *Science* 266:763–70
32. Lewis M, Chang G, Horton NC, Kercher MA, Pace HC, et al. 1996. Crystal structure of the lactose operon repressor and its complexes with DNA and inducer. *Science* 271:1247–54
33. Van Roey P, Waddling CA, Fox KM, Belfort M, Derbyshire V. 2001. Intertwined structure of the DNA-binding domain of intron endonuclease I-TevI with its substrate. *EMBO J.* 20:3631–37
34. Edgell DR, Derbyshire V, Van Roey P, LaBonne S, Stanger MJ, et al. 2004. Intron-encoded homing endonuclease I-TevI also functions as a transcriptional autorepressor. *Nat. Struct. Mol. Biol.* 11:936–44
35. Shen BW, Landthaler M, Shub DA, Stoddard BL. 2004. DNA binding and cleavage by the HNH homing endonuclease I-HmuI. *J. Mol. Biol.* 342:43–56
36. Shen A, Higgins DE, Panne D. 2009. Recognition of AT-rich DNA binding sites by the MogR repressor. *Structure* 17:769–77
37. Daniels DS, Woo TT, Luu KX, Noll DM, Clarke ND, et al. 2004. DNA binding and nucleotide flipping by the human DNA repair protein AGT. *Nat. Struct. Mol. Biol.* 11:714–20
38. Gajiwala KS, Burley SK. 2000. Winged helix proteins. *Curr. Opin. Struct. Biol.* 10:110–16
39. Clark KL, Halay ED, Lai E, Burley SK. 1993. Co-crystal structure of the HNF-3/fork head DNA-recognition motif resembles histone H5. *Nature* 364:412–20
40. Kodandapani R, Pio F, Ni CZ, Piccialli G, Klemsz M, et al. 1996. A new pattern for helix-turn-helix recognition revealed by the PU.1 ETS-domain-DNA complex. *Nature* 380:456–60
41. Hong M, Fuangthong M, Helmann JD, Brennan RG. 2005. Structure of an OhrR-*ohrA* operator complex reveals the DNA binding mechanism of the MarR family. *Mol. Cell* 20:131–41
42. Gajiwala KS, Chen H, Cornille F, Roques BP, Reith W, et al. 2000. Structure of the winged-helix protein hRFX1 reveals a new mode of DNA binding. *Nature* 403:916–21
43. Ferre-D'Amare AR, Pognonec P, Roeder RG, Burley SK. 1994. Structure and function of the b/HLH/Z domain of USF. *EMBO J.* 13:180–89
44. Ma PC, Rould MA, Weintraub H, Pabo CO. 1994. Crystal structure of MyoD bHLH domain-DNA complex: perspectives on DNA recognition and implications for transcriptional activation. *Cell* 77:451–59
45. Nair SK, Burley SK. 2003. X-ray structures of Myc-Max and Mad-Max recognizing DNA. Molecular bases of regulation by proto-oncogenic transcription factors. *Cell* 112:193–205
46. Parraga A, Bellsolell L, Ferre-D'Amare AR, Burley SK. 1998. Co-crystal structure of sterol regulatory element binding protein 1a at 2.3 A resolution. *Structure* 6:661–72
47. Cho Y, Gorina S, Jeffrey PD, Pavletich NP. 1994. Crystal structure of a p53 tumor suppressor-DNA complex: understanding tumorigenic mutations. *Science* 265:346–55
48. Ghosh G, van Duyne G, Ghosh S, Sigler PB. 1995. Structure of NF-$_{\kappa}$B p50 homodimer bound to a $_{\kappa}$B site. *Nature* 373:303–10
49. Muller CW, Rey FA, Sodeoka M, Verdine GL, Harrison SC. 1995. Structure of the NF-kappa B p50 homodiner bound to DNA. *Nature* 373:311–17
50. Tahirov TH, Inoue-Bungo T, Morii H, Fujikawa A, Sasaki M, et al. 2001. Structural analyses of DNA recognition by the AML1/Runx-1 Runt domain and its allosteric control by CBFbeta. *Cell* 104:755–67
51. Kovall RA, Hendrickson WA. 2004. Crystal structure of the nuclear effector of Notch signaling, CSL, bound to DNA. *EMBO J.* 23:3441–51

52. Sidote DJ, Barbieri CM, Wu T, Stock AM. 2008. Structure of the *Staphylococcus aureus* AgrA LytTR domain bound to DNA reveals a beta fold with an unusual mode of binding. *Structure* 16:727–35
53. Pavletich NP, Pabo CO. 1991. Zinc finger-DNA recognition: crystal structure of a Zif268-DNA complex at 2.1 A. *Science* 252:809–17
54. Schreiter ER, Drennan CL. 2007. Ribbon-helix-helix transcription factors: variations on a theme. *Nat. Rev. Microbiol.* 5:710–20
55. Somers WS, Phillips SE. 1992. Crystal structure of the met repressor-operator complex at 2.8 A resolution reveals DNA recognition by beta-strands. *Nature* 359:387–93
56. Raumann BE, Rould MA, Pabo CO, Sauer RT. 1994. DNA recognition by β-sheets in the Arc repressor-operator crystal structure. *Nature* 367:754–57
57. Pingoud A, Fuxreiter M, Pingoud V, Wende W. 2005. Type II restriction endonucleases: structure and mechanism. *Cell. Mol. Life Sci.* 62:685–707
58. Klemm JD, Rould MA, Aurora R, Herr W, Pabo CO. 1994. Crystal structure of the Oct-1 POU domain bound to an octamer site: DNA recognition with tethered DNA-binding modules. *Cell* 77:21–32
59. Pereira JH, Kim SH. 2009. Structure of human Brn-5 transcription factor in complex with CRH gene promoter. *J. Struct. Biol.* 167:159–65
60. Rhee S, Martin RG, Rosner JL, Davies DR. 1998. A novel DNA-binding motif in MarA: the first structure for an AraC family transcriptional activator. *Proc. Natl. Acad. Sci. USA* 95:10413–18
61. Chen FE, Huang DB, Chen YQ, Ghosh G. 1998. Crystal structure of p50/p65 heterodimer of transcription factor NF-κB bound to DNA. *Nature* 391:410–13
62. Kwon HJ, Bennik MH, Demple B, Ellenberger T. 2000. Crystal structure of the *Escherichia coli* Rob transcription factor in complex with DNA. *Nat. Struct. Biol.* 7:424–30
63. Muller CW. 2001. Transcription factors: global, detailed views. *Curr. Opin. Struct. Biol.* 11:26–32
64. Chang MV, Chang JL, Gangopadhyay A, Shearer A, Cadigan KM. 2008. Activation of wingless targets requires bipartite recognition of DNA by TCF. *Curr. Biol.* 18:1877–81
65. Shakked Z, Rabinovich D. 1986. The effect of the base sequence on the fine structure of the DNA double helix. *Prog. Biophys. Mol. Biol.* 47:159–95
66. Rohs R, West SM, Sosinsky A, Liu P, Mann RS, Honig B. 2009. The role of DNA shape in protein-DNA recognition. *Nature* 461:1248–53
67. Honig B, Nicholls A. 1995. Classical electrostatics in biology and chemistry. *Science* 268:1144–49
68. Klapper I, Hagstrom R, Fine R, Sharp K, Honig B. 1986. Focusing of electric fields in the active site of Cu-Zn superoxide dismutase: effects of ionic strength and amino-acid modification. *Proteins* 1:47–59
69. Sharp KA, Honig B, Harvey SC. 1990. Electrical potential of transfer RNAs: codon-anticodon recognition. *Biochemistry* 29:340–46
70. Chin K, Sharp KA, Honig B, Pyle AM. 1999. Calculating the electrostatic properties of RNA provides new insights into molecular interactions and function. *Nat. Struct. Biol.* 6:1055–61
71. Tang CL, Alexov E, Pyle AM, Honig B. 2007. Calculation of pK_as in RNA: on the structural origins and functional roles of protonated nucleotides. *J. Mol. Biol.* 366:1475–96
72. Leslie AG, Arnott S, Chandrasekaran R, Ratliff RL. 1980. Polymorphism of DNA double helices. *J. Mol. Biol.* 143:49–72
73. Jayaram B, Sharp KA, Honig B. 1989. The electrostatic potential of B-DNA. *Biopolymers* 28:975–93
74. Lavery R, Pullman B. 1981. The molecular electrostatic potential and steric accessibility of poly (dI.dC). Comparison with poly (dG.dC). *Nucleic Acids Res.* 9:7041–51
75. Lu XJ, Shakked Z, Olson WK. 2000. A-form conformational motifs in ligand-bound DNA structures. *J. Mol. Biol.* 300:819–40
76. Shakked Z, Guerstein-Guzikevich G, Eisenstein M, Frolow F, Rabinovich D. 1989. The conformation of the DNA double helix in the crystal is dependent on its environment. *Nature* 342:456–60
77. Ng HL, Kopka ML, Dickerson RE. 2000. The structure of a stable intermediate in the A ↔ B DNA helix transition. *Proc. Natl. Acad. Sci. USA* 97:2035–39
78. Guzikevich-Guerstein G, Shakked Z. 1996. A novel form of the DNA double helix imposed on the TATA-box by the TATA-binding protein. *Nat. Struct. Biol.* 3:32–37

79. Wang AH, Quigley GJ, Kolpak FJ, Crawford JL, van Boom JH, et al. 1979. Molecular structure of a left-handed double helical DNA fragment at atomic resolution. *Nature* 282:680–86
80. Arnott S, Chandrasekaran R, Birdsall DL, Leslie AG, Ratliff RL. 1980. Left-handed DNA helices. *Nature* 283:743–45
81. Nelson HC, Finch JT, Luisi BF, Klug A. 1987. The structure of an oligo(dA)·oligo(dT) tract and its biological implications. *Nature* 330:221–26
82. Hizver J, Rozenberg H, Frolow F, Rabinovich D, Shakked Z. 2001. DNA bending by an adenine-thymine tract and its role in gene regulation. *Proc. Natl. Acad. Sci. USA* 98:8490–95
83. Haran TE, Mohanty U. 2009. The unique structure of A-tracts and intrinsic DNA bending. *Q. Rev. Biophys.* 42:41–81
84. Zhurkin VB, Tolstorukov MY, Xu F, Colasanti AV, Olson WK. 2005. Sequence-dependent variality of B-DNA: an update on bending and curvature. In *DNA Conformation and Transcription*, ed. T Ohyama, pp. 18–34. Georgetown, TX/New York: Landes Biosci./Springer Sci. Bus. Media
85. Goodsell DS, Kaczor-Grzeskowiak M, Dickerson RE. 1994. The crystal structure of C-C-A-T-T-A-A-T-G-G. Implications for bending of B-DNA at T-A steps. *J. Mol. Biol.* 239:79–96
86. Crothers DM, Shakked Z. 1999. DNA bending by adenine-thymine tracts. In *Oxford Handbook of Nucleic Acid Structures*, ed. S Neidle, pp. 455–70. London: Oxford Univ. Press
87. Rohs R, Sklenar H, Shakked Z. 2005. Structural and energetic origins of sequence-specific DNA bending: Monte Carlo simulations of papillomavirus E2-DNA binding sites. *Structure* 13:1499–509
88. Gorin AA, Zhurkin VB, Olson WK. 1995. B-DNA twisting correlates with base-pair morphology. *J. Mol. Biol.* 247:34–48
89. Olson WK, Gorin AA, Lu XJ, Hock LM, Zhurkin VB. 1998. DNA sequence-dependent deformability deduced from protein-DNA crystal complexes. *Proc. Natl. Acad. Sci. USA* 95:11163–68
90. Mack DR, Chiu TK, Dickerson RE. 2001. Intrinsic bending and deformability at the T-A step of CCTTTAAAGG: a comparative analysis of T-A and A-T steps within A-tracts. *J. Mol. Biol.* 312:1037–49
91. Tolstorukov MY, Colasanti AV, McCandlish DM, Olson WK, Zhurkin VB. 2007. A novel roll-and-slide mechanism of DNA folding in chromatin: implications for nucleosome positioning. *J. Mol. Biol.* 371:725–38
92. Lu XJ, Olson WK. 2008. 3DNA: a versatile, integrated software system for the analysis, rebuilding and visualization of three-dimensional nucleic-acid structures. *Nat. Protoc.* 3:1213–27
93. Janin J, Rodier F, Chakrabarti P, Bahadur RP. 2007. Macromolecular recognition in the Protein Data Bank. *Acta Crystallogr. D* 63:1–8
94. Billeter M, Qian YQ, Otting G, Muller M, Gehring W, Wüthrich K. 1993. Determination of the nuclear magnetic resonance solution structure of an *Antennapedia* homeodomain-DNA complex. *J. Mol. Biol.* 234:1084–97
95. Ades SE, Sauer RT. 1995. Specificity of minor-groove and major-groove interactions in a homeodomain-DNA complex. *Biochemistry* 34:14601–8
96. Tucker-Kellogg L, Rould MA, Chambers KA, Ades SE, Sauer RT, Pabo CO. 1997. Engrailed (Gln50 → Lys) homeodomain-DNA complex at 1.9 A resolution: structural basis for enhanced affinity and altered specificity. *Structure* 5:1047–54
97. Grant RA, Rould MA, Klemm JD, Pabo CO. 2000. Exploring the role of glutamine 50 in the homeodomain-DNA interface: crystal structure of engrailed (Gln50 → ala) complex at 2.0 A. *Biochemistry* 39:8187–92
98. Hanes SD, Brent R. 1989. DNA specificity of the bicoid activator protein is determined by homeodomain recognition helix residue 9. *Cell* 57:1275–83
99. Treisman J, Gonczy P, Vashishtha M, Harris E, Desplan C. 1989. A single amino acid can determine the DNA binding specificity of homeodomain proteins. *Cell* 59:553–62
100. Pabo CO, Sauer RT. 1992. Transcription factors: structural families and principles of DNA recognition. *Annu. Rev. Biochem.* 61:1053–95
101. Luscombe NM, Thornton JM. 2002. Protein-DNA interactions: amino acid conservation and the effects of mutations on binding specificity. *J. Mol. Biol.* 320:991–1009

102. Siggers TW, Silkov A, Honig B. 2005. Structural alignment of protein-DNA interfaces: insights into the determinants of binding specificity. *J. Mol. Biol.* 345:1027–45
103. Cherney LT, Cherney MM, Garen CR, James MN. 2009. The structure of the arginine repressor from *Mycobacterium tuberculosis* bound with its DNA operator and co-repressor, L-arginine. *J. Mol. Biol.* 388:85–97
104. Ellenberger TE, Brandl CJ, Struhl K, Harrison SC. 1992. The GCN4 basic region leucine zipper binds DNA as a dimer of uninterrupted alpha helices: crystal structure of the protein-DNA complex. *Cell* 71:1223–37
105. Travers A. 1998. Transcription: activation by cooperating conformations. *Curr. Biol.* 8:R616–18
106. Weiss MA, Ellenberger T, Wobbe CR, Lee JP, Harrison SC, Struhl K. 1990. Folding transition in the DNA-binding domain of GCN4 on specific binding to DNA. *Nature* 347:575–78
107. Love JJ, Li X, Case DA, Giese K, Grosschedl R, Wright PE. 1995. Structural basis for DNA bending by the architectural transcription factor LEF-1. *Nature* 376:791–95
108. Laity JH, Dyson HJ, Wright PE. 2000. DNA-induced alpha-helix capping in conserved linker sequences is a determinant of binding affinity in Cys_2-His_2 zinc fingers. *J. Mol. Biol.* 295:719–27
109. Holmbeck SM, Dyson HJ, Wright PE. 1998. DNA-induced conformational changes are the basis for cooperative dimerization by the DNA binding domain of the retinoid X receptor. *J. Mol. Biol.* 284:533–39
110. Lefstin JA, Yamamoto KR. 1998. Allosteric effects of DNA on transcriptional regulators. *Nature* 392:885–88
111. Meijsing SH, Pufall MA, So AY, Bates DL, Chen L, Yamamoto KR. 2009. DNA binding site sequence directs glucocorticoid receptor structure and activity. *Science* 324:407–10
112. Luisi BF, Xu WX, Otwinowski Z, Freedman LP, Yamamoto KR, Sigler PB. 1991. Crystallographic analysis of the interaction of the glucocorticoid receptor with DNA. *Nature* 352:497–505
113. McClarin JA, Frederick CA, Wang BC, Greene P, Boyer HW, et al. 1986. Structure of the DNA-Eco RI endonuclease recognition complex at 3 A resolution. *Science* 234:1526–41
114. Drew HR, Wing RM, Takano T, Broka C, Tanaka S, et al. 1981. Structure of a B-DNA dodecamer: conformation and dynamics. *Proc. Natl. Acad. Sci. USA* 78:2179–83
115. Shakked Z, Guzikevich-Guerstein G, Frolow F, Rabinovich D, Joachimiak A, Sigler PB. 1994. Determinants of repressor/operator recognition from the structure of the *trp* operator binding site. *Nature* 368:469–73
116. Locasale JW, Napoli AA, Chen S, Berman HM, Lawson CL. 2009. Signatures of protein-DNA recognition in free DNA binding sites. *J. Mol. Biol.* 386:1054–65
117. Huang DB, Phelps CB, Fusco AJ, Ghosh G. 2005. Crystal structure of a free kappaB DNA: insights into DNA recognition by transcription factor NF-kappaB. *J. Mol. Biol.* 346:147–60
118. Elrod-Erickson M, Rould MA, Nekludova L, Pabo CO. 1996. Zif268 protein-DNA complex refined at 1.6 A: a model system for understanding zinc finger-DNA interactions. *Structure* 4:1171–80
119. Elrod-Erickson M, Benson TE, Pabo CO. 1998. High-resolution structures of variant Zif268-DNA complexes: implications for understanding zinc finger-DNA recognition. *Structure* 6:451–64
120. Rozenberg H, Rabinovich D, Frolow F, Hegde RS, Shakked Z. 1998. Structural code for DNA recognition revealed in crystal structures of papillomavirus E2-DNA targets. *Proc. Natl. Acad. Sci. USA* 95:15194–99
121. Bartfeld D, Shimon L, Couture GC, Rabinovich D, Frolow F, et al. 2002. DNA recognition by the RUNX1 transcription factor is mediated by an allosteric transition in the RUNT domain and by DNA bending. *Structure* 10:1395–407
122. Kitayner M, Rozenberg H, Kessler N, Rabinovich D, Shaulov L, et al. 2006. Structural basis of DNA recognition by p53 tetramers. *Mol. Cell* 22:741–53
123. Paillard G, Lavery R. 2004. Analyzing protein-DNA recognition mechanisms. *Structure* 12:113–22
124. Harrison SC, Aggarwal AK. 1990. DNA recognition by proteins with the helix-turn-helix motif. *Annu. Rev. Biochem.* 59:933–69
125. Billeter M. 1996. Homeodomain-type DNA recognition. *Prog. Biophys. Mol. Biol.* 66:211–25
126. Konig B, Muller JJ, Lanka E, Heinemann U. 2009. Crystal structure of KorA bound to operator DNA: insight into repressor cooperation in RP4 gene regulation. *Nucleic Acids Res.* 37:1915–24

127. Tateno M, Yamasaki K, Amano N, Kakinuma J, Koike H, et al. 1997. DNA recognition by beta-sheets. *Biopolymers* 44:335–59
128. Coulocheri SA, Pigis DG, Papavassiliou KA, Papavassiliou AG. 2007. Hydrogen bonds in protein-DNA complexes: where geometry meets plasticity. *Biochimie* 89:1291–303
129. Hoogsteen K. 1963. Crystal and molecular structure of a hydrogen-bonded complex between 1-methylthymine and 9-methyladenine. *Acta Crystallogr.* 16:907–916
130. Patikoglou GA, Kim JL, Sun L, Yang SH, Kodadek T, Burley SK. 1999. TATA element recognition by the TATA box–binding protein has been conserved throughout evolution. *Genes Dev.* 13:3217–30
131. Aishima J, Gitti RK, Noah JE, Gan HH, Schlick T, Wolberger C. 2002. A Hoogsteen base pair embedded in undistorted B-DNA. *Nucleic Acids Res.* 30:5244–52
132. Tainer JA, Cunningham RP. 1993. Molecular recognition in DNA-binding proteins and enzymes. *Curr. Opin. Biotechnol.* 4:474–83
133. Joachimiak A, Haran TE, Sigler PB. 1994. Mutagenesis supports water mediated recognition in the trp repressor-operator system. *EMBO J.* 13:367–72
134. Rastinejad F, Wagner T, Zhao Q, Khorasanizadeh S. 2000. Structure of the RXR-RAR DNA-binding complex on the retinoic acid response element DR1. *EMBO J.* 19:1045–54
135. Kalodimos CG, Biris N, Bonvin AM, Levandoski MM, Guennuegues M, et al. 2004. Structure and flexibility adaptation in nonspecific and specific protein-DNA complexes. *Science* 305:386–89
136. Aggarwal AK, Rodgers DW, Drottar M, Ptashne M, Harrison SC. 1988. Recognition of a DNA operator by the repressor of phage 434: a view at high resolution. *Science* 242:899–907
137. Wolberger C, Dong YC, Ptashne M, Harrison SC. 1988. Structure of a phage 434 Cro/DNA complex. *Nature* 335:789–95
138. Watkins D, Hsiao C, Woods KK, Koudelka GB, Williams LD. 2008. P22 c2 repressor-operator complex: mechanisms of direct and indirect readout. *Biochemistry* 47:2325–38
139. Max KE, Zeeb M, Bienert R, Balbach J, Heinemann U. 2007. Common mode of DNA binding to cold shock domains. Crystal structure of hexathymidine bound to the domain-swapped form of a major cold shock protein from *Bacillus caldolyticus*. *FEBS J.* 274:1265–79
140. Max KE, Zeeb M, Bienert R, Balbach J, Heinemann U. 2006. T-rich DNA single strands bind to a preformed site on the bacterial cold shock protein Bs-CspB. *J. Mol. Biol.* 360:702–14
141. Bewley CA, Gronenborn AM, Clore GM. 1998. Minor groove-binding architectural proteins: structure, function, and DNA recognition. *Annu. Rev. Biophys. Biomol. Struct.* 27:105–31
142. Panne D, Maniatis T, Harrison SC. 2007. An atomic model of the interferon-beta enhanceosome. *Cell* 129:1111–23
143. Escalante CR, Nistal-Villan E, Shen L, Garcia-Sastre A, Aggarwal AK. 2007. Structure of IRF-3 bound to the PRDIII-I regulatory element of the human interferon-beta enhancer. *Mol. Cell* 26:703–16
144. Li T, Jin Y, Vershon AK, Wolberger C. 1998. Crystal structure of the MATa1/MATalpha2 homeodomain heterodimer in complex with DNA containing an A-tract. *Nucleic Acids Res.* 26:5707–18
145. Crane-Robinson C, Dragan AI, Privalov PL. 2006. The extended arms of DNA-binding domains: a tale of tails. *Trends Biochem. Sci.* 31:547–52
146. Privalov PL, Dragan AI, Crane-Robinson C, Breslauer KJ, Remeta DP, Minetti CA. 2007. What drives proteins into the major or minor grooves of DNA? *J. Mol. Biol.* 365:1–9
147. Privalov PL, Dragan AI, Crane-Robinson C. 2009. The cost of DNA bending. *Trends Biochem. Sci.* 34:464–70
148. Patikoglou G, Burley SK. 1997. Eukaryotic transcription factor-DNA complexes. *Annu. Rev. Biophys. Biomol. Struct.* 26:289–325
149. Travers AA. 1989. DNA conformation and protein binding. *Annu. Rev. Biochem.* 58:427–52
150. Mann RS, Lelli KM, Joshi R. 2009. Hox specificity: unique roles for cofactors and collaborators. *Curr. Top. Dev. Biol.* 88:63–101
151. Liu Y, Matthews KS, Bondos SE. 2009. Internal regulatory interactions determine DNA binding specificity by a Hox transcription factor. *J. Mol. Biol.* 390:760–74
152. Liu Y, Matthews KS, Bondos SE. 2008. Multiple intrinsically disordered sequences alter DNA binding by the homeodomain of the *Drosophila* Hox protein Ultrabithorax. *J. Biol. Chem.* 283:20874–87

153. Phillips K, Luisi B. 2000. The virtuoso of versatility: POU proteins that flex to fit. *J. Mol. Biol.* 302:1023–39
154. Remenyi A, Lins K, Nissen LJ, Reinbold R, Scholer HR, Wilmanns M. 2003. Crystal structure of a POU/HMG/DNA ternary complex suggests differential assembly of Oct4 and Sox2 on two enhancers. *Genes Dev.* 17:2048–59
155. Hames C, Ptchelkine D, Grimm C, Thevenon E, Moyroud E, et al. 2008. Structural basis for LEAFY floral switch function and similarity with helix-turn-helix proteins. *EMBO J.* 27:2628–37
156. Yang W, Steitz TA. 1995. Crystal structure of the site-specific recombinase gamma delta resolvase complexed with a 34 bp cleavage site. *Cell* 82:193–207
157. Santelli E, Richmond TJ. 2000. Crystal structure of MEF2A core bound to DNA at 1.5 A resolution. *J. Mol. Biol.* 297:437–49
158. Remenyi A, Tomilin A, Pohl E, Lins K, Philippsen A, et al. 2001. Differential dimer activities of the transcription factor Oct-1 by DNA-induced interface swapping. *Mol. Cell* 8:569–80
159. Churchill ME, Travers AA. 1991. Protein motifs that recognize structural features of DNA. *Trends Biochem. Sci.* 16:92–97
160. Panne D. 2008. The enhanceosome. *Curr. Opin. Struct. Biol.* 18:236–42
161. Escalante CR, Yie J, Thanos D, Aggarwal AK. 1998. Structure of IRF-1 with bound DNA reveals determinants of interferon regulation. *Nature* 391:103–6
162. Fujii Y, Shimizu T, Kusumoto M, Kyogoku Y, Taniguchi T, Hakoshima T. 1999. Crystal structure of an IRF-DNA complex reveals novel DNA recognition and cooperative binding to a tandem repeat of core sequences. *EMBO J.* 18:5028–41
163. Boutonnet N, Hui X, Zakrzewska K. 1993. Looking into the grooves of DNA. *Biopolymers* 33:479–90
164. Lane WJ, Darst SA. 2006. The structural basis for promoter -35 element recognition by the group IV σ factors. *PLoS Biol.* 4:e269
165. Miticka H, Rezuchova B, Homerova D, Roberts M, Kormanec J. 2004. Identification of nucleotides critical for activity of the sigmaE-dependent rpoEp3 promoter in *Salmonella enterica* serovar Typhimurium. *FEMS Microbiol. Lett.* 238:227–33
166. Fairall L, Schwabe JW, Chapman L, Finch JT, Rhodes D. 1993. The crystal structure of a two zinc-finger peptide reveals an extension to the rules for zinc-finger/DNA recognition. *Nature* 366:483–87
167. Horton NC, Dorner LF, Perona JJ. 2002. Sequence selectivity and degeneracy of a restriction endonuclease mediated by DNA intercalation. *Nat. Struct. Biol.* 9:42–47
168. Sierk ML, Zhao Q, Rastinejad F. 2001. DNA deformability as a recognition feature in the reverb response element. *Biochemistry* 40:12833–43
169. Lawson CL, Swigon D, Murakami KS, Darst SA, Berman HM, Ebright RH. 2004. Catabolite activator protein: DNA binding and transcription activation. *Curr. Opin. Struct. Biol.* 14:10–20
170. Schultz SC, Shields GC, Steitz TA. 1991. Crystal structure of a CAP-DNA complex: the DNA is bent by 90°. *Science* 253:1001–7
171. Parkinson G, Wilson C, Gunasekera A, Ebright YW, Ebright RE, Berman HM. 1996. Structure of the CAP-DNA complex at 2.5 angstroms resolution: a complete picture of the protein-DNA interface. *J. Mol. Biol.* 260:395–408
172. Little EJ, Babic AC, Horton NC. 2008. Early interrogation and recognition of DNA sequence by indirect readout. *Structure* 16:1828–37
173. Kim YC, Grable JC, Love R, Greene PJ, Rosenberg JM. 1990. Refinement of Eco RI endonuclease crystal structure: a revised protein chain tracing. *Science* 249:1307–9
174. Kalodimos CG, Boelens R, Kaptein R. 2004. Toward an integrated model of protein-DNA recognition as inferred from NMR studies on the Lac repressor system. *Chem. Rev.* 104:3567–86
175. Travers A. 2000. Recognition of distorted DNA structures by HMG domains. *Curr. Opin. Struct. Biol.* 10:102–9
176. Weiss MA. 2001. Floppy SOX: mutual induced fit in HMG (high-mobility group) box-DNA recognition. *Mol. Endocrinol.* 15:353–62
177. Palasingam P, Jauch R, Ng CK, Kolatkar PR. 2009. The structure of Sox17 bound to DNA reveals a conserved bending topology but selective protein interaction platforms. *J. Mol. Biol.* 388:619–30

178. Hegde RS. 2002. The papillomavirus E2 proteins: structure, function, and biology. *Annu. Rev. Biophys. Biomol. Struct.* 31:343–60
179. Kim SS, Tam JK, Wang AF, Hegde RS. 2000. The structural basis of DNA target discrimination by papillomavirus E2 proteins. *J. Biol. Chem.* 275:31245–54
180. Hines CS, Meghoo C, Shetty S, Biburger M, Brenowitz M, Hegde RS. 1998. DNA structure and flexibility in the sequence-specific binding of papillomavirus E2 proteins. *J. Mol. Biol.* 276:809–18
181. Nelson HB, Laughon A. 1990. The DNA binding specificity of the *Drosophila* fushi tarazu protein: a possible role for DNA bending in homeodomain recognition. *New Biol.* 2:171–78
182. Koudelka GB, Carlson P. 1992. DNA twisting and the effects of noncontacted bases on affinity of 434 operator for 434 repressor. *Nature* 355:89–91
183. Edwards KJ, Brown DG, Spink N, Skelly JV, Neidle S. 1992. Molecular structure of the B-DNA dodecamer d(CGCAAATTTGCG)$_2$. An examination of propeller twist and minor-groove water structure at 2.2 Å resolution. *J. Mol. Biol.* 226:1161–73
184. Newman M, Strzelecka T, Dorner LF, Schildkraut I, Aggarwal AK. 1995. Structure of Bam HI endonuclease bound to DNA: partial folding and unfolding on DNA binding. *Science* 269:656–63
185. Viadiu H, Aggarwal AK. 2000. Structure of BamHI bound to nonspecific DNA: a model for DNA sliding. *Mol. Cell* 5:889–95
186. Lukacs CM, Aggarwal AK. 2001. BglII and MunI: what a difference a base makes. *Curr. Opin. Struct. Biol.* 11:14–18
187. Tolstorukov MY, Jernigan RL, Zhurkin VB. 2004. Protein-DNA hydrophobic recognition in the minor groove is facilitated by sugar switching. *J. Mol. Biol.* 337:65–76
188. Eisenstein M, Shakked Z. 1995. Hydration patterns and intermolecular interactions in A-DNA crystal structures. Implications for DNA recognition. *J. Mol. Biol.* 248:662–78
189. Shakked Z, Rabinovich D, Kennard O, Cruse WB, Salisbury SA, Viswamitra MA. 1983. Sequence-dependent conformation of an A-DNA double helix. The crystal structure of the octamer d(G-G-T-A-T-A-C-C). *J. Mol. Biol.* 166:183–201
190. Travers AA. 1995. Reading the minor groove. *Nat. Struct. Biol.* 2:615–18
191. Choo Y, Klug A. 1997. Physical basis of a protein-DNA recognition code. *Curr. Opin. Struct. Biol.* 7:117–25
192. Nekludova L, Pabo CO. 1994. Distinctive DNA conformation with enlarged major groove is found in Zn-finger-DNA and other protein-DNA complexes. *Proc. Natl. Acad. Sci. USA* 91:6948–52
193. Pavletich NP, Pabo CO. 1993. Crystal structure of a five-finger GLI-DNA complex: new perspectives on zinc fingers. *Science* 261:1701–7
194. Kitayner M, Rozenberg H, Rabinovich D, Shakked Z. 2005. Structures of the DNA-binding site of Runt-domain transcription regulators. *Acta Crystallogr. D* 61:236–46
195. Cassiday LA, Maher LJ 3rd. 2002. Having it both ways: transcription factors that bind DNA and RNA. *Nucleic Acids Res.* 30:4118–26
196. Schwartz T, Rould MA, Lowenhaupt K, Herbert A, Rich A. 1999. Crystal structure of the Zalpha domain of the human editing enzyme ADAR1 bound to left-handed Z-DNA. *Science* 284:1841–45
197. Herbert A, Rich A. 1999. Left-handed Z-DNA: structure and function. *Genetica* 106:37–47
198. Schwartz T, Behlke J, Lowenhaupt K, Heinemann U, Rich A. 2001. Structure of the DLM-1-Z-DNA complex reveals a conserved family of Z-DNA-binding proteins. *Nat. Struct. Biol.* 8:761–65
199. Li B, Carey M, Workman JL. 2007. The role of chromatin during transcription. *Cell* 128:707–19
200. Teytelman L, Ozaydin B, Zill O, Lefrancois P, Snyder M, et al. 2009. Impact of chromatin structures on DNA processing for genomic analyses. *PLoS One* 4:e6700
201. Segal E, Widom J. 2009. From DNA sequence to transcriptional behavior: a quantitative approach. *Nat. Rev. Genet.* 10:443–56
202. Segal E, Fondufe-Mittendorf Y, Chen L, Thastrom A, Field Y, et al. 2006. A genomic code for nucleosome positioning. *Nature* 442:772–78
203. Field Y, Kaplan N, Fondufe-Mittendorf Y, Moore IK, Sharon E, et al. 2008. Distinct modes of regulation by chromatin encoded through nucleosome positioning signals. *PLoS Comput. Biol.* 4:e1000216
204. Peckham HE, Thurman RE, Fu Y, Stamatoyannopoulos JA, Noble WS, et al. 2007. Nucleosome positioning signals in genomic DNA. *Genome Res.* 17:1170–77

205. Kaplan N, Moore IK, Fondufe-Mittendorf Y, Gossett AJ, Tillo D, et al. 2009. The DNA-encoded nucleosome organization of a eukaryotic genome. *Nature* 458:362–66
206. Satchwell SC, Drew HR, Travers AA. 1986. Sequence periodicities in chicken nucleosome core DNA. *J. Mol. Biol.* 191:659–75
207. Trifonov EN, Sussman JL. 1980. The pitch of chromatin DNA is reflected in its nucleotide sequence. *Proc. Natl. Acad. Sci. USA* 77:3816–20
208. Johnson SM, Tan FJ, McCullough HL, Riordan DP, Fire AZ. 2006. Flexibility and constraint in the nucleosome core landscape of *Caenorhabditis elegans* chromatin. *Genome Res.* 16:1505–16
209. Chung HR, Vingron M. 2009. Sequence-dependent nucleosome positioning. *J. Mol. Biol.* 386:1411–22
210. Swinger KK, Rice PA. 2004. IHF and HU: flexible architects of bent DNA. *Curr. Opin. Struct. Biol.* 14:28–35
211. Ellenberger T, Landy A. 1997. A good turn for DNA: the structure of integration host factor bound to DNA. *Structure* 5:153–57
212. Rice PA, Yang S, Mizuuchi K, Nash HA. 1996. Crystal structure of an IHF-DNA complex: a protein-induced DNA U-turn. *Cell* 87:1295–306
213. Lynch TW, Read EK, Mattis AN, Gardner JF, Rice PA. 2003. Integration host factor: putting a twist on protein-DNA recognition. *J. Mol. Biol.* 330:493–502
214. Piper DE, Batchelor AH, Chang CP, Cleary ML, Wolberger C. 1999. Structure of a HoxB1-Pbx1 heterodimer bound to DNA: role of the hexapeptide and a fourth homeodomain helix in complex formation. *Cell* 96:587–97
215. Chen L, Glover JN, Hogan PG, Rao A, Harrison SC. 1998. Structure of the DNA-binding domains from NFAT, Fos and Jun bound specifically to DNA. *Nature* 392:42–48
216. Yie J, Liang S, Merika M, Thanos D. 1997. Intra- and intermolecular cooperative binding of high-mobility-group protein I(Y) to the beta-interferon promoter. *Mol. Cell. Biol.* 17:3649–62
217. Stefl R, Wu H, Ravindranathan S, Sklenar V, Feigon J. 2004. DNA A-tract bending in three dimensions: solving the dA4T4 vs dT4A4 conundrum. *Proc. Natl. Acad. Sci. USA* 101:1177–82
218. Faiger H, Ivanchenko M, Haran TE. 2007. Nearest-neighbor non-additivity versus long-range non-additivity in TATA-box structure and its implications for TBP-binding mechanism. *Nucleic Acids Res.* 35:4409–19
219. Ellison MJ, Feigon J, Kelleher RJ 3rd, Wang AH, Habener JF, Rich A. 1986. An assessment of the Z-DNA forming potential of alternating dA-dT stretches in supercoiled plasmids. *Biochemistry* 25:3648–55
220. Petrey D, Honig B. 2003. GRASP2: visualization, surface properties, and electrostatics of macromolecular structures and sequences. *Methods Enzymol.* 374:492–509
221. Lavery R, Sklenar H. 1989. Defining the structure of irregular nucleic acids: conventions and principles. *J. Biomol. Struct. Dyn.* 6:655–67
222. Rocchia W, Sridharan S, Nicholls A, Alexov E, Chiabrera A, Honig B. 2002. Rapid grid-based construction of the molecular surface and the use of induced surface charge to calculate reaction field energies: applications to the molecular systems and geometric objects. *J. Comput. Chem.* 23:128–37
223. Kitayner M, Rozenberg H, Rohs R, Suad O, Rabinovich D, et al. 2010. Diversity in DNA recognition by p53 revealed by crystal structures with Hoogsteen base pairs. *Nat. Struct. Mol. Biol.* 17:423–29

NOTE ADDED IN PROOF

In section 4.2.1.1., we discuss the possibility of non-Watson-Crick base pairs playing a role in protein-DNA recognition. This hypothesis is supported by recent crystal structures of p53 tetramers bound to DNA-binding sites with contiguous half sites where the AT doublets of the CATG core regions exhibit Hoogsteen geometry (223). Although these Hoogsteen base pairs are embedded in essentially undistorted B-DNA, the alternate base pairing geometry affects local DNA shape. This observation expands the code of sequence readout.

Transcript Elongation by RNA Polymerase II

Luke A. Selth, Stefan Sigurdsson, and Jesper Q. Svejstrup

Mechanisms of Transcription Laboratory, Clare Hall Laboratories, Cancer Research UK London Research Institute, South Mimms, Hertfordshire EN6 3LD, United Kingdom; email: j.svejstrup@cancer.org.uk

Annu. Rev. Biochem. 2010. 79:271–93

First published online as a Review in Advance on April 1, 2010

The *Annual Review of Biochemistry* is online at biochem.annualreviews.org

This article's doi:
10.1146/annurev.biochem.78.062807.091425

0066-4154/10/0707-0271$20.00

Key Words

chromatin, elongation factors, gene traffic, transcription-associated recombination, transcription-repair coupling

Abstract

Until recently, it was generally assumed that essentially all regulation of transcription takes place via regions adjacent to the coding region of a gene—namely promoters and enhancers—and that, after recruitment to the promoter, the polymerase simply behaves like a machine, quickly "reading the gene." However, over the past decade a revolution in this thinking has occurred, culminating in the idea that transcript elongation is extremely complex and highly regulated and, moreover, that this process significantly affects both the organization and integrity of the genome. This review addresses basic aspects of transcript elongation by RNA polymerase II (RNAPII) and how it relates to other DNA-related processes.

Contents

1. INTRODUCTION

The RNA polymerase II (RNAPII) transcription cycle can be divided into several distinct steps. First, RNAPII is recruited to the promoter of a gene, where it forms a preinitiation complex with the general transcription factors. Initiation ensues, and RNAPII leaves the promoter behind in a process termed promoter clearance. Next, RNAPII enters processive transcript elongation, which comes to an end when the gene has been transcribed. Transcriptional termination then results in release and recycling of the polymerase so that it can transcribe again. During its journey across a gene, RNAPII faces significant challenges. First, the polymerase needs to escape the promoter, and production of the pre-mRNA transcript needs to be tightly coupled to RNA biogenesis (RNA capping, splicing, transcript cleavage, and polyadenylation). These events, and their coupling to the phosphorylation status of the C-terminal repeat domain (CTD) of the largest RNAPII subunit, have been the subject of excellent recent reviews (1, 2) and are not described here. Second, because transcript elongation occurs in the context of chromatin, RNAPII has to negotiate its way past nucleosomes and, very often, other obstacles to its progression, such as DNA damage. Third, transcription affects, and is affected by, several other DNA metabolic processes, such as DNA repair, recombination, and replication. Finally, transcript elongation in highly transcribed genes is carried out by several polymerase molecules simultaneously. Mechanisms regulating such gene "traffic" remain poorly understood.

This review is intended as a broad overview of the process of transcript elongation, but with particular emphasis on its interplay with other DNA-related processes.

2. BASIC MECHANISM OF TRANSCRIPT ELONGATION

The movement of RNA polymerases is not generated via the ATP-dependent "power strokes" that are typical of DNA helicases. Rather, transcript elongation seems to be governed by a Brownian ratchet mechanism, with one of the forward translocated states stabilized by the binding and hydrolysis of the correct incoming nucleotide (3–5). One consequence of this type of mechanism is that, although the polymerase moves rapidly forward on average, it can also move backward. So, even though the formation of a new phosphodiester bond is likely to be immediately followed by translocation of the new RNA 3′ terminus from the insertion site ($i+1$ site, or A site) to the i site of the polymerase catalytic center, the enzyme can also backtrack for one or several nucleotides so that the newly formed 3′ terminus comes out of alignment with the active site. Brownian motion can bring the end of the RNA back in register with the active site again, but the polymerase is also capable of endonucleolytic cleavage of the transcript, a process which is highly stimulated by elongation factor TFIIS (also called SII) (6, 7). This process may also be important for transcript proofreading, the process of removing incorrectly incorporated nucleotides, an idea that is discussed in more detail below. In general, the biochemistry of chain elongation, with its multiple equilibriums between different enzyme states, lends itself well to regulation by protein cofactors. Thus, general elongation factors such as TFIIF, Elongin, and ELL (**Table 1**) are likely to bring about their stimulatory effect by affecting these equilibriums, with the overall effect that pausing and stalling are reduced and forward translocation favored [these factors have previously been reviewed in detail (5, 8) and are not further described here].

2.1. Transcription Fidelity

The correct insertion of nucleotides into the nascent RNA transcript during transcript elongation is essential for accurate gene expression. RNAPII must balance the need for rapid transcription with the need for high fidelity, so that only the nucleoside triphosphate substrate specified by the DNA template is selected. An important structural feature of RNAPII called the trigger loop, a mobile element of the Rpb1 subunit, appears to play a key role in RNAPII fidelity (9, 10). The trigger loop is located beneath the active site and is involved in multiple interactions with the incoming nucleotide (11). These interactions ensure that the trigger loop and the incoming nucleotide are correctly aligned, which is required for nucleophilic attack and phosphodiester bond formation. Both nucleotide selection and phosphodiester bond formation may be mediated by the trigger loop and are likely to be coupled. Mismatched nucleotides in the active site are not aligned properly with the trigger loop and therefore result in a substantial reduction in the rate of phosphodiester bond formation (9, 11, 12).

Other factors also affect the fidelity of transcription. For example, by studying the canavinine sensitivity of yeast strains, it was shown that the Rpb9 subunit of RNAPII is involved in ensuring transcription fidelity (13). Recent steady-state kinetic studies suggest that one role of Rpb9 is to delay closure of the trigger loop on the incoming nucleotide, probably through interaction of the C-terminal domain of Rpb9 and the trigger loop (14). The fact that RNAPII can move backward as well as forward is also

Table 1 Examples of factors connected to RNA polymerase II transcript elongation

Basal elongation factors	**References[a]**
TFIIF	173
ELL	174
Elongin	175
TFIIS	176
Spt4/5 (DSIF in human)	177, 178
NELF	179
Chromatin elongation factors	
Swi-Snf	180
RSC	42, 44
ISWI	181
CHD	46, 182
Paf	183
Spt6	55, 184
FACT	119
Nap1	185
Vps75 (TAF-1/SET)	60, 186
Asf1	56
Elongator	78
SAGA	80
Nua3	96
Rpd3s	109
PCAF	97
COMPASS/Set1 (MLL)	109
Set2	109
Set3	109
Dot1	118
Rad6	118
Factors connecting transcription to DNA metabolism	
CSB (Rad26 in yeast)	129, 187
Def1	133
THO/TREX complex	152, 155
RECQL5	160

[a]Please note that the individual reference typically refers to a more recent study on the factor referenced, rather than to the study originally reporting the identification of the factor.

likely to play a significant role in maintaining transcription fidelity. In vitro, nucleotide misincorporation generally leads to slow addition of the next nucleotide and renders the transcript sensitive to the intrinsic transcript cleavage activity of the polymerase (15). Removal of a mispaired nucleotide is greatly enhanced by factors that stimulate transcript cleavage, such as TFIIS (16, 17). However, whether the proofreading activity of TFIIS is important for transcription fidelity in vivo is still unclear. Early data suggested that TFIIS did not have a significant role in this process (13, 18), but a more recent study suggests that TFIIS is an important contributor to transcription fidelity (19). The functions of Rpb9 and TFIIS in transcription are tightly coupled (20, 21), so Rpb9 might be involved in RNAPII fidelity both before and after nucleotide addition by affecting the function of the trigger loop and by mediating TFIIS function.

3. TRANSCRIPT ELONGATION THROUGH CHROMATIN

The structure of chromatin is extremely repressive to processes occurring on DNA, including transcription. Thus, although elongation through single nucleosomes was found to occur with reasonable efficiency (22), linear nucleosomal arrays present an extremely strong barrier to RNAPII passage (23, 24). How, then, are nucleosomes, one of the most stable nucleoprotein complexes, modified so that RNAPII elongation can proceed? One mechanism is nucleosome disassembly in front of elongating RNAPII (**Figure 1*a***). In vitro evidence for this was obtained in early experiments using mononucleosomes (25). The more recent advent of chromatin immunoprecipitation has been an important development in answering the question in vivo. Experiments measuring histone density at the yeast *GAL* genes showed that gene activation causes loss of nucleosomes, not only at the promoter but also within the coding region (26, 27). Such loss is indeed caused by elongating RNAPII, rather than by other events spreading from the activated promoter (28). More generally, genome-wide analyses of nucleosome occupancy in yeast revealed an inverse relationship between transcription rate and histone density in the coding region of genes, further suggesting that nucleosomes are disassembled during transcription (29, 30). In addition, in *Drosophila*, the histone H3.3 variant is more rapidly incorporated into active genes, and histone density is reduced at these loci, suggesting that higher organisms exhibit a similar pattern of chromatin plasticity (31).

Given the tripartite nature of the nucleosomal structure (H2A/H2B, H3/H4, DNA), it is perhaps not surprising that H2A/H2B and H3/H4 act as functionally distinct subcomplexes during transcription. H2A/H2B dimers, localized on the exterior of the nucleosome with fewer protein-DNA contacts, are exchanged rapidly in response to transcript elongation (**Figure 1*b***) (32–34). By contrast, histones H3 and H4 are less mobile, and their turnover is much more dependent on transcription than that of the H2A/H2B dimer, which is exchanged ubiquitously in both a replication- and transcription-independent manner (32). Collectively, these studies indicate that transcription elongation causes displacement of all core histones in a fashion that is often proportional to RNAPII density but that histones H2A/H2B are displaced more readily than H3/H4.

Whether displacement of nucleosomes in front of the polymerase is a prerequisite for transcription through chromatin is unclear. Indeed, compelling evidence for an ability of RNA polymerases to move through a single nucleosome in vitro without displacing it has been obtained (reviewed in Reference 35), and it is thus feasible that nucleosome removal does not always occur as RNAPII traverses the gene (**Figure 1*b***). This is consistent with in vivo data suggesting the existence of at least two distinct mechanisms to achieve efficient transcript elongation through chromatin: a pathway based on loss of nucleosomes and a histone acetylation-dependent mechanism correlating with little or no net loss of nucleosomes (26). The connection between histone modification and RNAPII transcript elongation is described in detail below. A strategy to allow RNAPII passage through chromatin involving histone modification rather than nucleosome disassembly is also supported by more recent analyses of genome-wide histone replacement patterns, which indicate that the rate of replication-independent H3 exchange varies widely between genes and is not invariably proportional to transcription rate (36). This may suggest that the level of "histone conservation," the proportion of histones that remain associated with DNA as polymerase transcribes through them, varies widely between genes.

Not surprisingly, the basic mechanism of RNAPII movement, with frequent backtracking as an integral element, is closely connected to transcript elongation through chromatin. For example, TFIIS is a major component of a chromatin transcription-enabling activity, which is required for efficient transcript elongation through chromatin (37). Thus, transcript elongation through nucleosomes is likely to exacerbate transcriptional backtracking and arrest, requiring transcript cleavage for efficient escape and resumed transcription. Mechanistic data to support such a model for TFIIS-stimulated transcript elongation through nucleosomes have been obtained (23, 38, 39).

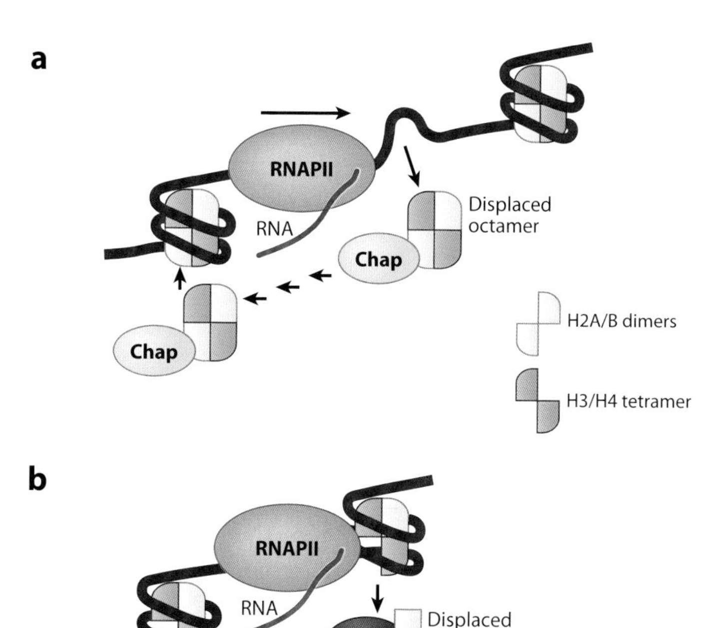

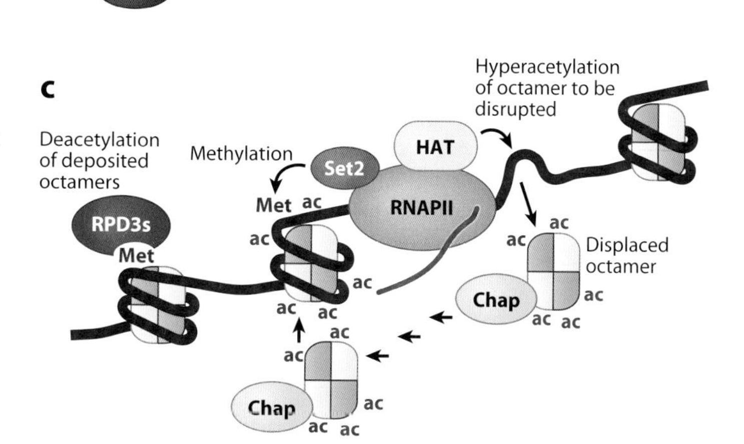

Figure 1

Transcript elongation through nucleosomes. (*a*) Transcription through a nucleosome can result in the release of histone proteins from DNA and their association with histone chaperones (Chaps). However, as indicated in (*b*), progression of RNA polymerase II (RNAPII) through a nucleosome may also occur without complete displacement of the histone proteins from DNA. In this case, a histone H2A/H2B dimer only is likely to be displaced. FACT, and possibly other histone chaperones, can receive this dimer and presumably reload it in the wake of the progressing polymerase. (*c*) Cotranscriptional histone modification. HATs, such as Elongator and Gcn5, acetylate histones as the polymerase travels through chromatin. This stimulates nucleosome dissociation and chaperoning of histone proteins. Upon reloading (or if RNAPII traverses the nucleosome without completely displacing it), the hyperacetylated nucleosome also becomes methylated by Set2 (bound to elongating RNAPII via the hyperphosphorylated C-terminal repeat domain). This, in turn, leads to recruitment of the histone deacetylase complex Rpd3S, which specifically associates with the methylated nucleosome and deacetylates it, so that the chromatin structure is reset. Abbreviations: ac, acetylation; Met, methylation.

3.1. Factors and Mechanisms That Help Remodel Chromatin During Transcript Elongation

To achieve the highly dynamic reconfiguration of chromatin that is required during transcript elongation, cells have evolved distinct groups of factors that are capable of modifying higher-order DNA structure. ATP-dependent chromatin remodelers and histone chaperones act primarily to disassemble and reassemble nucleosomes, while a multitude of enzymes catalyze covalent modification of histones. These factors are described briefly below, with particular emphasis on how their diverse mechanisms of action may regulate transcript elongation.

3.1.1. ATP-dependent chromatin remodelers. ATP-dependent chromatin remodeling complexes (remodelers) use the energy of ATP to modify the structure of chromatin. Four main families of remodelers have been identified: SWI-SNF, ISWI, CHD, and INO80/SWR (split ATPase domain) (reviewed in Reference 40). Research into the role of remodelers in transcription has focused primarily on their ability to modify promoter chromatin architecture (for reviews, see References 40 and 41), but accumulating evidence suggests that they can also play an important role in RNAPII transcript elongation. For example, Workman and colleagues (42) found that RSC, a SWI-SNF remodeler, can stimulate RNAPII transcript elongation through a mononucleosome in a simple reconstituted chromatin transcription system. This activity was enhanced by histone acetylation, which may increase the affinity of RSC for the nucleosome (43). Recent work showed that RSC is recruited to the coding regions of stress-activated genes in vivo, where it is required for the progress of RNAPII (44).

The remodeler Chd1 is also associated with chromatin at sites of active transcription (45, 46), and numerous lines of genetic evidence implicate this factor in transcript elongation as well. Moreover, Chd1 physically interacts with the elongation factors Paf, DSIF, and FACT (46–48). A group of Chd1-related factors found in *Drosophila* has also been implicated in transcription elongation. Kismet-L was first identified in a screen for suppressors of the transcriptional repressor Polycomb, and its absence leads to reduced levels of elongating RNAPII and the elongation factors Spt6 and Chd1 on polytene chromosomes (49). In general, it seems likely that different remodelers play redundant roles in transcript elongation, and indeed, numerous other examples of remodeler action in transcript elongation complement the few described above.

3.1.2. Histone chaperones. Histone chaperones are histone-binding proteins involved in intracellular histone dynamics, as well as histone storage and replication-associated chromatin assembly (for a review, see Reference 50). However, much recent evidence has revealed a role for these factors in regulating elongation-coupled changes to chromatin. FACT (facilitates chromatin transcription) was one of the first histone chaperones to be identified as having a role in transcript elongation (24, 51). FACT has intrinsic histone chaperone activity and appears to facilitate elongation at least partly by destabilizing nucleosome structure so that one histone H2A-H2B dimer is removed during RNAPII passage (**Figure 1*b***) (52). Subsequent work, mainly in yeast, has revealed that FACT plays an important role in transcript elongation through chromatin in vivo (for a review, see Reference 53).

Spt6 is a H3/H4 chaperone that is important for the maintenance of chromatin structure in yeast cells (54). In the absence of Spt6 activity, histones are depleted from the coding region of several yeast genes, unmasking cryptic transcription initiation sites (55), suggesting that Spt6 promotes the restoration of normal chromatin structure in the wake of RNAPII transcript elongation. Interestingly, defects in FACT and other histone chaperones gave rise to similar results, suggesting that histone chaperones in general are involved in both removing and redepositing histones during transcript elongation (**Figure 1*a*,*c*** and **Table 1**) (55–62).

The mechanism(s) of chaperone-mediated nucleosome disassembly and reassembly are yet to be fully elucidated, although we can envision at least three possibilities. First, as alluded to above in the case of FACT, these factors may directly alter nucleosome stability. Interestingly, this effect appears to be enhanced by histone acetylation (63, 64). Second, they may act as a sink for histones released during elongation and a source of histones during post-elongation chromatin reassembly. Finally, histone chaperones may act in combination with ATP-dependent chromatin remodeling machines to directly mediate nucleosome disassembly. For example, various histone chaperones, including Nap1, Vps75, and Asf1, can act in combination with RSC to promote full or partial nucleosome disassembly in vitro (60, 65). It must be noted that RNAPII itself is a strong force for directly displacing the nucleosome, which could then be chaperoned and reloaded by factors such as FACT, Spt6, and others.

Given that histone chaperones typically function in both chromatin disassembly and reassembly and that both activities are required for transcription-associated histone exchange, one prediction is that the absence of chaperone function would cause defects in new histone incorporation. Genome-wide and gene-specific measurement of histone exchange rates in cells lacking Nap1, Asf1, HIR, or Vps75 reveals that this is indeed the case (56, 60, 66–69).

3.1.3. Covalent histone modification. The above discussion has focused primarily on the most drastic change to the structure of nucleosomes, histone removal and replacement. A more subtle way of modifying chromatin structure, but by no means less important in terms of transcription elongation, is via covalent modification of histones. Histones are the substrates for a large and diverse array of modifying enzymes. The posttranslational modifications catalyzed by these factors are found primarily in the unstructured, amino terminal segments of histones that protrude from the nucleosome (70) and play an extremely important role in elongation via two main mechanisms. First, these modifications can alter the packaging of chromatin directly by affecting internucleosomal contacts or changing electrostatic charge. Second, the covalently attached moieties can act as a binding surface for elongation-associated effector complexes. These mechanisms are described in more detail below.

3.1.3.1. Histone acetylation. Transcription activity is typically correlated with dramatic histone hyperacetylation in the promoter of genes (71, 72). Interestingly, active transcription also correlates with histone acetylation in the coding region of genes, but here the observed increases are often surprisingly modest (73) or virtually nonexistent (72, 74, 75). This might argue that histone acetylation does not play an important role in RNAPII transcript elongation through chromatin. However, histone acetyltransferases (HATs) and histone deacetylases (HDACs) are enriched in the coding region of genes (59, 76–80), indicating that substantial turnover of acetylation occurs. Moreover, if the level of histone acetylation in chromatin is reduced by mutation of the GCN5 and ELP3 HATs, the dramatically reduced transcription level observed in several constitutively transcribed genes correlates remarkably well with lower acetylation levels in the coding region, but not in the promoter, of those genes (81). Together, these data thus argue that histone acetylation in the coding region of genes is important for gene activity. Indeed, different reconstituted chromatin transcription systems require HAT activity for productive elongation (37, 42). Moreover, nucleosomes assembled from full-length H3 and H4 inhibit transcript elongation more severely than nucleosomes formed from tailless histones, and acetylation of the tails overcomes this inhibitory effect (82).

A possible way of reconciling these apparently contradicting data would be if histone acetylation in coding regions of genes were dynamic and specifically coupled to the transcribing polymerase (83), with RNAPII-coupled histone acetylation being removed immediately in the wake of the elongation complex

(**Figure 1*c***). Although the HATs responsible for promoter nucleosome acetylation are well known (84), the enzyme(s) responsible for this function within coding regions has proved more difficult to identify. The Elongator HAT complex was isolated via its association with elongating RNAPII, suggesting a mechanism by which histone acetylation could be coupled to RNAPII movement through coding regions (85–87). Elongator stimulates chromatin transcription in vitro (88) and is associated with nascent mRNA molecules in vivo (77). Moreover, loss of Elongator function correlates with H3 hypoacetylation and impaired transcription in the coding regions, but not in the promoter, of genes in mammalian cells (78, 89). However, an Elongator mutation gives rise to pleiotrophic phenotypes, most of them not related to the transcription function of the complex, at least in yeast (90), and an involvement of additional RNAPII-associated HATs is therefore likely.

Gcn5 is another candidate for coupling RNAPII movement to histone acetylation. Deletion of this factor reduces H3 acetylation levels within coding regions of genes in yeast, resulting in defects in nucleosome eviction and elongation (26, 80, 81, 91). Gcn5 is a member of the SAGA (Spt-Ada-Gcn5-acetyltransferase) coactivator complex, whose members have been shown to interact physically and genetically with TFIIS and other elongation factors (92, 93). In addition, other members of the SAGA complex have recently been implicated in transcription elongation (94).

Other HATs have been indirectly implicated in the elongation phase of transcription. In human cells, HBO1-containing complexes are targeted to the coding regions of genes by combinatorial binding of three PHD fingers in two other subunits of the complex (95). Sas3, the catalytic subunit of the yeast NuA3 HAT complex, physically interacts with FACT (96). The p300/CBP-associated factor (PCAF) binds to hyperphosphorylated RNAPII and has been implicated in productive transcription by interaction with elongation factors (97, 98). As mentioned above, p300 is also required for efficient elongation through chromatin in vitro in an acetylation-dependent manner (37).

Histone acetylation is reversed by the action of HDACs in a process that also plays an important role during transcript elongation. Cotranscriptional H3K36 methylation by the RNAPII-associated Set2 complex recruits the Rp3dS HDAC complex, resulting in histone deacetylation in the wake of RNAPII passage (**Figure 1*c***). This pathway is crucial for reestablishing the chromatin structure that is perturbed by the movement of RNAPII through the gene. Thus, prevention of cryptic transcription initiation from within coding units in the wake of elongating RNAPII is achieved via two mechanisms: H3K36me-directed histone deacetylation and histone chaperone-mediated nucleosome reassembly (59, 79, 99).

3.1.3.2. Histone methylation. Whereas the effects of histone acetylation on nucleosome behavior and stability are increasingly well documented, the role of histone methylation is less clear. Nevertheless, considerable effort has been invested in mapping these modifications in genes and in identifying the histone methyltransferases responsible. Lysines can be modified by one, two, or three methyl groups, and these different methylation states can have distinct functions. In all likelihood, histone methylation does not have a significant structural role but rather acts as a tag for effector proteins containing methyl-binding domains, such as chromodomains, PHD finger domains, and tudor domains (100).

The elongating form of RNAPII is typically phosphorylated at serine 2 of the CTD, which is specifically recognized by the histone methyltransferase Set2. This enzyme methylates histone H3 lysine 36, resulting in accumulation of H3K36me in the coding region of genes (101–104). As mentioned above, this modification, in turn, recruits the Rpd3S HDAC complex in yeast, which recognizes di- and trimethylated H3K36 (105) via two binding modules, a chromodomain in Eaf3 and a PHD domain in Rco1 subunit, and catalyzes elongation-associated histone deacetylation

(59, 79, 99, 105, 106). The major role of H3K36 methylation may thus be to regulate the acetylation-deacetylation cycle connected with transcript elongation (**Figure 1*c***).

Another methylation mark, on histone H3 lysine 4, is also tightly coupled to transcript elongation. Phosphorylation of RNAPII at the CTD serine 5 by the general transcription factor TFIIH recruits the COMPASS complex to the 5′ end of genes, which, in turn, results in a peak of H3K4 trimethylation around promoters (107). In contrast, H3K4 dimethylation peaks immediately downstream of transcription start sites, whereas monomethylation is spread across the coding region of genes (72, 108). A recent study provided evidence for the idea that H3K4me2 recruits an HDAC complex via the PHD finger of the Set3 subunit of that complex (the Set3 complex), localizing the Hos2 and Hst1 subunits to deacetylate histones at the 5′ end of transcribed genes (109). In ways that are still poorly understood, this has a positive effect on transcription.

3.1.3.3. Histone ubiquitylation. Conjugation of ubiquitin to target proteins can lead to a diverse range of substrate fates. The best characterized of these is degradation by the 26S proteasome, which recognizes polyubiquitin chains and proteolytically degrades the tagged polypeptide (110). By contrast, monoubiquitylation serves as a signal to regulate function, transport, and/or processing of the substrate (111).

Monoubiquitylation of H2B at K123 (K120 in mammals) appears to be a general consequence of transcriptional activation and is found at both the promoter and coding regions of genes (112–114). The mechanism by which H2B is ubiquitylated in response to transcriptional cues has been described in a recent review (115). Briefly, the Rad6/Bre1/Lge1 complex ubiquitylates H2B at the promoter and coding regions of genes in a manner requiring the PAF complex and stimulated by other elongation factors and elongating RNAPII itself (115).

Two distinct mechanisms have been proposed to explain how monoubiquitylation of H2B affects movement of RNAPII through the coding region of genes. The first is to direct di- and trimethylation of lysines 4 and 79 on histone H3, modifications that regulate transcription elongation, as described above. In yeast, H3K4 and H3K79 methylation is carried out by Set1 [part of the COMPASS complex (MLL in humans)] and Dot1, respectively (116). In the absence of uH2B, monomethylation, but not di- or trimethylation, of the target residues can occur (117). A fascinating mechanism that is not yet fully understood has recently been shown to underlie this defect: When uH2B is lost, the COMPASS complex lacks an integral subunit, Cps35, that is required for the successive addition of methyl groups onto monomethylated K4 (118). Interestingly, Cps35 also interacts with Dot1 and is required for efficient H3K79 methylation, suggesting that it may also link uH2B to this modification. Exactly how uH2B mediates the association of Cps35 with methyltransferases to form catalytically active complexes is currently unknown, but this function clearly has an effect, albeit indirect, on transcript elongation.

The second way in which uH2B stimulates RNAPII elongation appears to be more direct. A recent study found that this modification enhances the movement of RNAPII through nucleosomes in vitro, possibly via FACT-dependent removal of an H2A/H2B dimer (119). In yeast, mutation of the H2B ubiquitylation site also affects RNAPII distribution in a manner that suggests elongation defects and that appears to be related to changes in chromatin structure (120, 121). A recent genome-wide study of the modification in human cells suggests that it is a widespread positive effector of elongation and further highlights its anticorrelation with H3K4 methylation patterns (122). Interestingly, uH2B also appears to be required to reform chromatin structure in the wake of elongating RNAPII in a manner that is coordinated with FACT function (123).

In summary, the mechanistic basis underlying the positive effect of H2B ubiquitylation on elongation through chromatin is still unclear but appears to relate to histone methylation and to structural changes to the nucleosome that might facilitate disassembly in front of elongating RNAPII, and reassembly in its wake.

In general, the problem of nucleosome disassembly and reassembly as elongating RNAPII travels through chromatin appears to have led to the evolution of many overlapping and/or redundant solutions, which invariably involve both factors required for the forward reaction (removal, making nucleosomes traversable) and those involved in the reverse reaction (resetting chromatin). Indeed, numerous histone-modifying enzymes, including HATs and HDACs, histone methyltransferases and demethylases, histone ubiquitylases and deubiquitylases, as well as histone chaperones, are required for efficient and correctly regulated transcript elongation. In contrast to transcriptional initiation, where a few general transcription factors are required for transcription at almost all genes, an absolute requirement for individual chromatin elongation factors is much less certain. Thus, even though some of the factors described here (RSC, SPT6, and FACT, for example) are encoded by essential genes, this does not necessarily indicate a general requirement for these factors in transcription: They might be required only for transcription of one, or a few, essential genes. Conversely, and more importantly, the fact that most of the individual processes and factors described above are not essential for viability in yeast certainly does not indicate that the ability to efficiently transcribe through a chromatin template is of little importance or that it is straightforward. Rather, it indicates that numerous redundant processes for enabling this crucial process have evolved.

4. INTERPLAY BETWEEN TRANSCRIPT ELONGATION AND OTHER DNA-RELATED PROCESSES

High levels of RNAPII transcription are associated with genome instability, resulting in increased rates of DNA recombination and mutagenesis. By contrast, certain DNA lesions in the transcribed strand of an active gene are removed much more rapidly than in the genome overall. Thus, transcription can have both positive and negative effects on genome integrity, but the interface between transcription and other processes occurring on DNA, such as replication and repair, remain poorly understood. However, a few recent studies have begun to elucidate the mechanisms and factors involved in this complex interplay. In the sections below, our current state of knowledge regarding the relationships between transcription elongation and DNA repair, recombination, and replication is described.

4.1. Transcription-Coupled DNA Repair and RNA Polymerase II Ubiquitylation

UV-induced DNA damage elicits a complex cellular response, including induction of genes required for repair, signaling to checkpoint proteins, and direct activation of various repair pathways. However, because they also block the progress of RNAPII, UV-induced DNA lesions in the coding strand of active genes pose a particular threat to genome integrity and cellular survival, and multiple mechanisms have evolved to minimize the detrimental effects of such obstacles. The best studied of these mechanisms is transcription-coupled nucleotide excision DNA repair (TC-NER), during which the transcribed strand of an active gene is repaired much faster than the nontranscribed strand and the genome in general (124, 125).

Another transcription-associated response to DNA damage is the polyubiquitylation of RNAPII (126), which ultimately results in proteasome-dependent polymerase degradation (127, 128). Like TC-NER, this process is triggered by the transcriptional arrest of RNAPII (and specifically targets this polymerase form), rather than by DNA damage per se (129, 130). The available data support the idea that TC-NER and RNAPII ubiquitylation and degradation represent interconnected, but distinct, alternative cellular pathways for contending with DNA damage in active genes (**Figure 2**) (126, 129–133). For example, yeast cells lacking the *DEF1* gene are unable to ubiquitylate and degrade RNAPII in response to DNA damage but have normal TC-NER (133). Conversely, cells lacking *RAD26* (or CSB in humans) have defective TC-NER but can still degrade RNAPII (129, 133). If both repair and RNAPII ubiquitylation are nonfunctional, cells become extremely UV sensitive (132, 133). In all likelihood, Rad26/CSB and Def1 associate with the damage-stalled polymerase and interact with each other to mediate a solution to the problem the stalled polymerase poses. First, Rad26/CSB attempts to direct repair of the transcription-obstructing lesion by recruiting and facilitating specific NER factors (134–136) and possibly by attempting to displace the polymerase or at least remodel the damaged DNA-polymerase interface (137), as is the case with the analogous bacterial enzyme Mfd (also called the transcription-repair coupling factor) (138). Rad26/CSB is a DNA-dependent ATPase in the Snf2/Swi2 family of DNA translocases (40, 134) and might use its intrinsic ATPase (translocase) activity for this latter purpose. Alternatively or additionally, Rad26/CSB could remodel chromatin to facilitate TC-NER (139) and enable transcription restart. However, if the DNA lesion cannot be removed or the polymerase cannot be displaced, Def1 (in yeast, and a related mechanism in humans) eventually overcomes the Rad26-mediated inhibition of RNAPII ubiquitylation (133) and works with the basic ubiquitylation machinery to mediate polyubiquitylation of RNAPII, which subsequently results in its degradation. Such a mechanism would allow access to the DNA lesion and potentially facilitate its removal by alternative DNA repair pathways, for example, general genome repair (see References 126 and 140 for details). Interestingly, RNAPII ubiquitylation is not confined to RNAPII arrested by DNA damage. Rather, it appears that any polymerase that stalls or arrests becomes a target for the ubiquitylation machinery. Ubiquitylation followed by degradation can thus be viewed as a last resort to clear genes of persistently stalled polymerases (126).

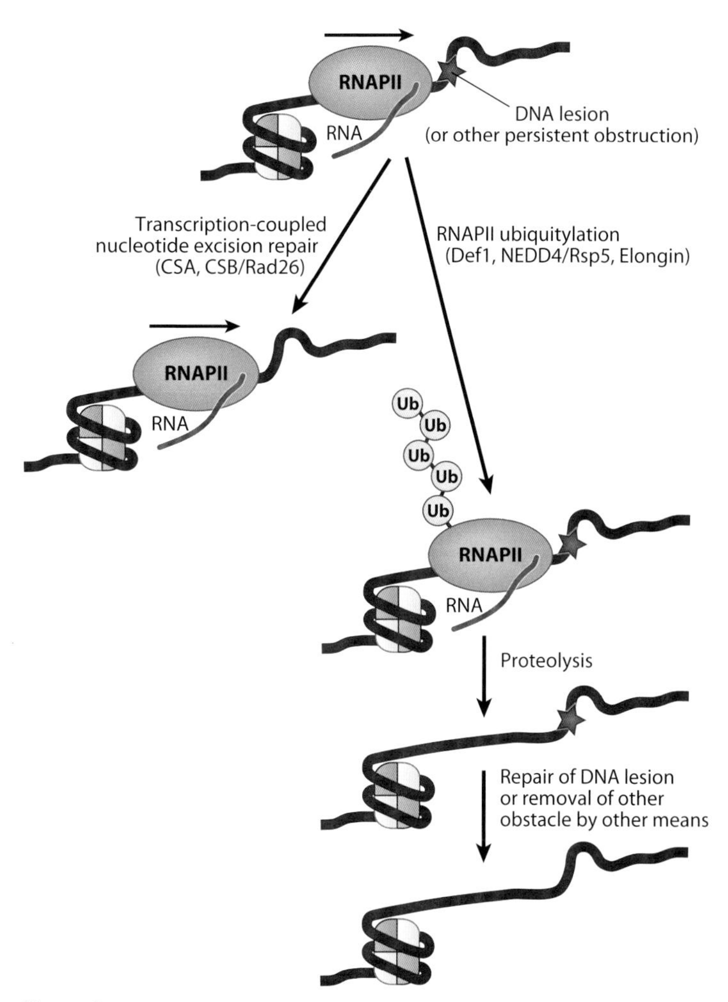

Figure 2

Alternative pathways for contending with transcript-obstructing DNA lesions. Bulky DNA lesions, such as those generated by UV light, are an absolute block to the progression of RNA polymerase II (RNAPII). Factors such as CSA and CSB enable such lesions to be preferentially repaired by transcription-coupled nucleotide excision repair, TC-NER (pathway on the left). This allows the polymerase to resume transcription. However, if such repair is not possible (or if the persistent block to transcription is caused by something else), the gene is unblocked by RNAPII ubiquitylation and proteolysis (pathway on the right). This presumably allows lesions and obstacles to be dealt with by other repair pathways [such as general (global) genome repair] or by the next polymerase reaching the lesion being able to trigger TC-NER.

Interestingly, recent data connect RNAPII ubiquitylation to the general elongation factor, Elongin, in an unexpected manner. Indeed, until recently, the mechanism regulating proteolysis of RNAPII had been controversial because two distinct ubiquitin ligases (E3s), Rsp5 (and its human homolog NEDD4) (129, 141) and the Elongin-Cullin complex (142–144), were both claimed to be absolutely required for RNAPII polyubiquitylation. This controversy was resolved when it was found that these E3s work sequentially in a two-step mechanism (145). Rsp5 first adds a monoubiquitin moiety to RNAPII, and the Elc1/Cul3 complex then extends this to produce a polyubiquitin chain, which signals for polymerase proteolysis. The relevance of Elongin complex having two distinct functions in RNAPII transcription remains unclear.

Other than DNA lesions repaired by NER, base damage or base loss [repaired by the base excision repair (BER) pathway] can also affect transcript elongation (see, for example, References 146 and 147), and it has been proposed that such lesions are subject to transcription-coupled repair as well. However, whether transcription-coupled BER is a physiologically relevant phenomenon is controversial after several key papers on the subject were retracted (reviewed in Reference 125).

4.2. Transcription-Associated Mutagenesis

Transcription also has a profound effect on DNA mutagenesis. In yeast, there is direct proportionality between transcription level and the mutation rate of a gene (148), and recent results show that the accumulation of apurinic and apyrimidinic sites is greatly elevated in highly transcribed DNA (149). Surprisingly, most of these sites stem from the removal of uracil, which is introduced erroneously during DNA replication in place of dTTP. This observation suggests an unexpected link between transcription and the fidelity of DNA replication, the biochemical basis of which remains to be uncovered.

4.3. Transcription-Associated DNA Recombination and Its Connection to DNA Replication

Over the past decade, it has become clear that the pathways leading to translatable mRNA, from transcription and RNA processing in the nucleus to RNA export into the cytoplasm, are very tightly controlled and that the individual steps are extensively interlinked. For example, factors required for pre-mRNA processing and nuclear export are recruited to RNA cotranscriptionally and also affect the efficiency of transcript elongation (2, 150). For a conceptual understanding of these complicated connections, it may be relevant to think of the entire process in terms of messenger ribonucleoprotein (mRNP) complexes, which need to be correctly assembled as the nascent RNA leaves transcribing RNAPII in order for the individual downstream steps to occur correctly. The TREX complex, which consists of proteins previously shown to reside in the THO complex (151) as well as the mRNA export factors Yra1 and Sub2 (152), may be important for correct function of such particles. THO complex is involved in suppressing recombination between direct repeats, but only when these are located in transcribed regions of active genes. Cells lacking normal THO function also have severe problems transcribing long and especially GC-rich genes, suggesting that this complex plays a role in transcript elongation and, by extension, that problems with elongation in these cells result in DNA hyperrecombination. Interestingly, defects in cells lacking the THO complex can be at least partially overcome if the occurrence of extended DNA:RNA hybrids in genes is suppressed, for example by a hammerhead ribozyme or overexpression of RNAse H (153). This implies that the RNA transcript can form hybrids with the transcribed region behind the advancing RNAPII in the absence of THO (perhaps because RNP particles are incorrectly formed or maintained) and that this is problematic for transcript elongation and induces DNA hyperrecombination.

Importantly, such transcription-associated recombination is dependent on replication, both in yeast and mammalian cells (154–156). Highly transcribed genes are generally an impediment for replication forks (157), and transcription-associated recombination is likely

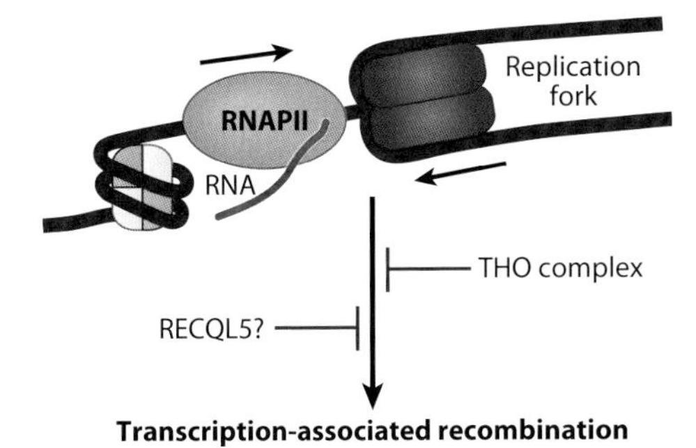

Figure 3

Transcription-associated recombination (TAR). TAR is a consequence of a clash between transcription and replication. In the example shown here, the replication fork meets the transcribing polymerase head-to-head. Transcription also reduces the fidelity of DNA replication, giving rise to elevated levels of mutagenesis.

the consequence of a collision between the DNA replication and transcription machineries (**Figure 3**) (154). In this connection, the RECQ family of helicases merits discussion. This enzyme family appears to be involved in aspects of replication-derived DNA recombination (158). Interestingly, RECQL5, a poorly understood member of the RECQ family, was recently found to interact directly with RNAPII and inhibit transcription in vitro (159, 160). RECQL5 is the only RECQ family member that associates with RNAPII (159). Mouse *recq5-/-* cells show elevated levels of chromosome crossovers and a higher incidence of cancer (161, 162) and also have phenotypes consistent with a role in certain aspects of DNA replication (163). The precise functional consequence of the intriguing interaction between RECQL5 and RNAPII is presently unknown, but like yeast cells lacking the THO complex, *recq5-/-* mouse cells have elevated levels of recombination between direct repeats. They are also prone to gross chromosomal rearrangements in response to replication stress (161). One possibility, therefore, is that RECQL5 somehow suppresses the recombinogenic effect of transcription during DNA replication (**Figure 3**), but this remains to be investigated.

5. GENE TRAFFIC

Several RNAPII molecules can transcribe the same gene simultaneously. The yeast *GAL* genes, for example, are thought to initiate a new round of transcription approximately every 6–8 s (164), meaning that the predicted spacing between polymerase active sites (assuming the transcription rate is 20–30 nucleotides/second on average) can be as little as 120–180 nucleotides. This may explain why *GAL*-driven genes (and other highly expressed genes) are essentially depleted of nucleosomes (26, 27, 29, 30). Traditionally, structural analysis, single-molecule approaches, and biochemical characterization of RNAPII activity in reconstituted transcription systems have focused on transcription at the level of individual polymerases (5, 12, 165), whereas techniques such as chromatin immunoprecipitation and live-cell imaging have been used to study the average behavior of polymerases on active genes in vivo (1, 166–168). However, regulation of gene traffic, i.e., the manner in which transcribing polymerases interfere with each other, remains very poorly understood. Interestingly, analysis of RNAPII dynamics by cell imaging has uncovered evidence for the idea that RNAPII frequently stalls or arrests, even during processive elongation in vivo (169). Indeed, although it is generally accepted that cellular transcript elongation occurs at an average rate of 20–30 nucleotides/second, evidence for a rate of up to 70 nucleotides/second, interrupted by relatively long periods of pausing, has been obtained (170).

In general, the speed of transcript elongation, combined with the high likelihood of transcriptional pausing or arrest, might be expected to result in frequent polymerase rear-end collisions. Moreover, as described above, RNAPII constantly encounters obstacles to its progression, such as DNA-binding proteins, nucleosomes, and DNA damage, which would be expected to enhance interaction between transcribing polymerases. Unfortunately, studies of RNAPII collision in vivo are difficult because conditions that lead to it occur randomly and unpredictably across the transcriptome. However, a recent study has addressed the consequence of RNAPII collision using highly purified polymerases (171). This study showed that rear-end collision by the trailing polymerase results in a transient state where the elongation complexes interact, followed by substantial backtracking of the trailing polymerase if progression of the leading polymerase is impossible. Importantly, the elongation complexes remain stable on DNA upon collision, with their activity and the integrity of transcription bubbles remaining intact. Subsequent TFIIS-stimulated transcript cleavage then allows resumed forward translocation, resulting in the trailing polymerase oscillating against the leading RNAPII if the obstruction cannot be traversed. However, if the leading polymerase is merely stalled at a pause site, collision by the trailing polymerase, combined with TFIIS action, can effectively drive it through the pause (171). These data suggest a basic biochemical mechanism that may underlie the extraordinary robustness of transcription, where individual polymerases are often required to traverse very long genes (in human cells, up to a million nucleotides long) without dissociation. In this model, conformational flexibility of RNAPII elongation complexes buffers the effect of running into objects on DNA, so that the polymerase in effect "bounces off" and retains full activity when obstacle collision occurs. The structural flexibility of RNAPII may be important to generally regulate the flow of polymerases on genes, with recent evidence supporting the idea that longer genes are more likely to experience traffic jams than shorter ones (172).

SUMMARY POINTS

1. Transcript elongation is governed by a Brownian ratchet mechanism, which allows backward as well as forward translocation of the polymerase. General elongation factors affect the multiple equilibriums between different enzyme states, thereby helping to drive the reaction toward forward translocation.
2. Transcription fidelity is ensured through multiple direct interactions of a structure called the trigger loop with the incoming nucleotide, as well as by transcript cleavage, which removes erroneously incorporated nucleotides.
3. Transcript elongation occurs on a chromatin template and involves temporary displacement and/or modification of nucleosomes. This can occur by a number of distinct mechanisms and requires histone chaperones, chromatin remodeling factors, and histone modifying enzymes. Resetting of the chromatin structure in the wake of RNAPII, in turn, requires a set of overlapping and distinct factors from the same groups.
4. There is a complex interplay between transcription and other DNA-related processes, such as DNA repair, recombination, and replication. High transcription levels increase mutagenesis and result in hyperrecombination.
5. By contrast, mechanisms that ensure that transcription is not hampered by RNAPII arrest-inducing DNA damage or other obstacles have evolved. DNA lesions in active genes are preferentially repaired, and persistently arrested RNAPII molecules are, as a last resort, removed by ubiquitylation and proteasome-mediated degradation.
6. Several polymerases often transcribe the same gene simultaneously. However, mechanisms that regulate gene traffic remain poorly understood.

FUTURE ISSUES

1. What are the biochemical mechanism and factor requirements for efficient transcript elongation through an array of nucleosomes?
2. What is the biochemical mechanism of transcription-coupled DNA repair? What is required for transcription to restart after damage removal?
3. What is the biochemical consequence of collision between RNAPII and replicating DNA polymerases? How does it lead to mutagenesis and DNA hyperrecombination?
4. How is gene traffic controlled? What are the consequences and physiological relevance of RNAPII collision?

DISCLOSURE STATEMENT

The authors are not aware of any affiliations, memberships, funding, or financial holdings that might be perceived as affecting the objectivity of this review.

ACKNOWLEDGMENTS

We apologize to all the investigators who could not be appropriately cited owing to space limitations. For example, recent work containing references to important prior studies is sometimes referred to rather than the prior studies themselves. Work in the Svejstrup lab is supported by a grant from the Association of International Cancer Research (AICR) and by an in-house grant from Cancer Research UK. Luke A. Selth is now at the Dame Roma Mitchell Cancer Research Laboratories, Hanson Institute, University of Adelaide, and Stefan Sigurdsson is now at the Biomedical Center, Faculty of Medicine, University of Iceland.

LITERATURE CITED

1. Margaritis T, Holstege FC. 2008. Poised RNA polymerase II gives pause for thought. *Cell* 133:581–84
2. Buratowski S. 2009. Progression through the RNA polymerase II CTD cycle. *Mol. Cell* 36:541–46
3. Bar-Nahum G, Epshtein V, Ruckenstein AE, Rafikov R, Mustaev A, Nudler E. 2005. A ratchet mechanism of transcription elongation and its control. *Cell* 120:183–93
4. Nudler E. 2009. RNA polymerase active center: the molecular engine of transcription. *Annu. Rev. Biochem.* 78:335–61
5. Herbert KM, Greenleaf WJ, Block SM. 2008. Single-molecule studies of RNA polymerase: motoring along. *Annu. Rev. Biochem.* 77:149–76
6. Izban MG, Luse DS. 1992. The RNA polymerase II ternary complex cleaves the nascent transcript in a 3′–5′ direction in the presence of elongation factor SII. *Genes Dev.* 6:1342–56
7. Reines D. 1992. Elongation factor-dependent transcript shortening by template-engaged RNA polymerase II. *J. Biol. Chem.* 267:3795–800
8. Shilatifard A, Conaway RC, Conaway JW. 2003. The RNA polymerase II elongation complex. *Annu. Rev. Biochem.* 72:693–715
9. Kaplan CD, Larsson KM, Kornberg RD. 2008. The RNA polymerase II trigger loop functions in substrate selection and is directly targeted by alpha-amanitin. *Mol. Cell* 30:547–56
10. Brueckner F, Cramer P. 2008. Structural basis of transcription inhibition by alpha-amanitin and implications for RNA polymerase II translocation. *Nat. Struct. Mol. Biol.* 15:811–18
11. Wang D, Bushnell DA, Westover KD, Kaplan CD, Kornberg RD. 2006. Structural basis of transcription: role of the trigger loop in substrate specificity and catalysis. *Cell* 127:941–54

12. Kornberg RD. 2007. The molecular basis of eukaryotic transcription. *Proc. Natl. Acad. Sci. USA* 104:12955–61
13. Nesser NK, Peterson DO, Hawley DK. 2006. RNA polymerase II subunit Rpb9 is important for transcriptional fidelity in vivo. *Proc. Natl. Acad. Sci. USA* 103:3268–73
14. Walmacq C, Kireeva ML, Irvin J, Nedialkov Y, Lubkowska L, et al. 2009. Rpb9 subunit controls transcription fidelity by delaying NTP sequestration in RNA polymerase II. *J. Biol. Chem.* 284:19601–12
15. Sydow JF, Brueckner F, Cheung AC, Damsma GE, Dengl S, et al. 2009. Structural basis of transcription: mismatch-specific fidelity mechanisms and paused RNA polymerase II with frayed RNA. *Mol. Cell* 34:710–21
16. Thomas MJ, Platas AA, Hawley DK. 1998. Transcriptional fidelity and proofreading by RNA polymerase II. *Cell* 93:627–37
17. Jeon C, Agarwal K. 1996. Fidelity of RNA polymerase II transcription controlled by elongation factor TFIIS. *Proc. Natl. Acad. Sci. USA* 93:13677–82
18. Shaw RJ, Bonawitz ND, Reines D. 2002. Use of an in vivo reporter assay to test for transcriptional and translational fidelity in yeast. *J. Biol. Chem.* 277:24420–26
19. Koyama H, Ito T, Nakanishi T, Sekimizu K. 2007. Stimulation of RNA polymerase II transcript cleavage activity contributes to maintain transcriptional fidelity in yeast. *Genes Cells* 12:547–59
20. Awrey DE, Weilbaecher RG, Hemming SA, Orlicky SM, Kane CM, Edwards AM. 1997. Transcription elongation through DNA arrest sites. A multistep process involving both RNA polymerase II subunit RPB9 and TFIIS. *J. Biol. Chem.* 272:14747–54
21. Hemming SA, Jansma DB, Macgregor PF, Goryachev A, Friesen JD, Edwards AM. 2000. RNA polymerase II subunit Rpb9 regulates transcription elongation in vivo. *J. Biol. Chem.* 275:35506–11
22. Lorch Y, LaPointe JW, Kornberg RD. 1987. Nucleosomes inhibit the initiation of transcription but allow chain elongation with the displacement of histones. *Cell* 49:203–10
23. Izban MG, Luse DS. 1991. Transcription on nucleosomal templates by RNA polymerase II in vitro: inhibition of elongation with enhancement of sequence-specific pausing. *Genes Dev.* 5:683–96
24. Orphanides G, LeRoy G, Chang CH, Luse DS, Reinberg D. 1998. FACT, a factor that facilitates transcript elongation through nucleosomes. *Cell* 92:105–16
25. Lorch Y, LaPointe JW, Kornberg RD. 1988. On the displacement of histones from DNA by transcription. *Cell* 55:743–44
26. Kristjuhan A, Svejstrup JQ. 2004. Evidence for distinct mechanisms facilitating transcript elongation through chromatin in vivo. *EMBO J.* 23:4243–52
27. Schwabish MA, Struhl K. 2004. Evidence for eviction and rapid deposition of histones upon transcriptional elongation by RNA polymerase II. *Mol. Cell. Biol.* 24:10111–17
28. Varv S, Kristjuhan K, Kristjuhan A. 2007. RNA polymerase II determines the area of nucleosome loss in transcribed gene loci. *Biochem. Biophys. Res. Commun.* 358:666–71
29. Lee CK, Shibata Y, Rao B, Strahl BD, Lieb JD. 2004. Evidence for nucleosome depletion at active regulatory regions genome-wide. *Nat. Genet.* 36:900–5
30. Bernstein BE, Liu CL, Humphrey EL, Perlstein EO, Schreiber SL. 2004. Global nucleosome occupancy in yeast. *Genome Biol.* 5:R62
31. Mito Y, Henikoff JG, Henikoff S. 2005. Genome-scale profiling of histone H3.3 replacement patterns. *Nat. Genet.* 37:1090–97
32. Jamai A, Imoberdorf RM, Strubin M. 2007. Continuous histone H2B and transcription-dependent histone H3 exchange in yeast cells outside of replication. *Mol. Cell* 25:345–55
33. Thiriet C, Hayes JJ. 2005. Replication-independent core histone dynamics at transcriptionally active loci in vivo. *Genes Dev.* 19:677–82
34. Kimura H, Cook PR. 2001. Kinetics of core histones in living human cells: little exchange of H3 and H4 and some rapid exchange of H2B. *J. Cell Biol.* 153:1341–53
35. Studitsky VM, Walter W, Kireeva M, Kashlev M, Felsenfeld G. 2004. Chromatin remodeling by RNA polymerases. *Trends Biochem. Sci.* 29:127–35
36. Gat-Viks I, Vingron M. 2009. Evidence for gene-specific rather than transcription rate-dependent histone H3 exchange in yeast coding regions. *PLoS Comput. Biol.* 5:e1000282

37. Guermah M, Palhan VB, Tackett AJ, Chait BT, Roeder RG. 2006. Synergistic functions of SII and p300 in productive activator-dependent transcription of chromatin templates. *Cell* 125:275–86
38. Izban MG, Luse DS. 1992. Factor-stimulated RNA polymerase II transcribes at physiological elongation rates on naked DNA but very poorly on chromatin templates. *J. Biol. Chem.* 267:13647–55
39. Kireeva ML, Hancock B, Cremona GH, Walter W, Studitsky VM, Kashlev M. 2005. Nature of the nucleosomal barrier to RNA polymerase II. *Mol. Cell* 18:97–108
40. Clapier CR, Cairns BR. 2009. The biology of chromatin remodeling complexes. *Annu. Rev. Biochem.* 78:273–304
41. Schnitzler GR. 2008. Control of nucleosome positions by DNA sequence and remodeling machines. *Cell Biochem. Biophys.* 51:67–80
42. Carey M, Li B, Workman JL. 2006. RSC exploits histone acetylation to abrogate the nucleosomal block to RNA polymerase II elongation. *Mol. Cell* 24:481–87
43. Kasten M, Szerlong H, Erdjument-Bromage H, Tempst P, Werner M, Cairns BR. 2004. Tandem bromodomains in the chromatin remodeler RSC recognize acetylated histone H3 Lys14. *EMBO J.* 23:1348–59
44. Mas G, de Nadal E, Dechant R, Rodriguez de la Concepcion ML, Logie C, et al. 2009. Recruitment of a chromatin remodelling complex by the Hog1 MAP kinase to stress genes. *EMBO J.* 28:326–36
45. Stokes DG, Tartof KD, Perry RP. 1996. CHD1 is concentrated in interbands and puffed regions of *Drosophila* polytene chromosomes. *Proc. Natl. Acad. Sci. USA* 93:7137–42
46. Simic R, Lindstrom DL, Tran HG, Roinick KL, Costa PJ, et al. 2003. Chromatin remodeling protein Chd1 interacts with transcription elongation factors and localizes to transcribed genes. *EMBO J.* 22:1846–56
47. Krogan NJ, Kim M, Ahn SH, Zhong G, Kobor MS, et al. 2002. RNA polymerase II elongation factors of *Saccharomyces cerevisiae*: a targeted proteomics approach. *Mol. Cell. Biol.* 22:6979–92
48. Kelley DE, Stokes DG, Perry RP. 1999. CHD1 interacts with SSRP1 and depends on both its chromodomain and its ATPase/helicase-like domain for proper association with chromatin. *Chromosoma* 108:10–25
49. Srinivasan S, Armstrong JA, Deuring R, Dahlsveen IK, McNeill H, Tamkun JW. 2005. The *Drosophila* trithorax group protein Kismet facilitates an early step in transcriptional elongation by RNA polymerase II. *Development* 132:1623–35
50. Park YJ, Luger K. 2006. Structure and function of nucleosome assembly proteins. *Biochem. Cell Biol.* 84:549–58
51. Orphanides G, Wu WH, Lane WS, Hampsey M, Reinberg D. 1999. The chromatin-specific transcription elongation factor FACT comprises human SPT16 and SSRP1 proteins. *Nature* 400:284–88
52. Belotserkovskaya R, Oh S, Bondarenko VA, Orphanides G, Studitsky VM, Reinberg D. 2003. FACT facilitates transcription-dependent nucleosome alteration. *Science* 301:1090–93
53. Reinberg D, Sims RJ 3rd. 2006. de FACTo nucleosome dynamics. *J. Biol. Chem.* 281:23297–301
54. Bortvin A, Winston F. 1996. Evidence that Spt6p controls chromatin structure by a direct interaction with histones. *Science* 272:1473–76
55. Kaplan CD, Laprade L, Winston F. 2003. Transcription elongation factors repress transcription initiation from cryptic sites. *Science* 301:1096–99
56. Schwabish MA, Struhl K. 2006. Asf1 mediates histone eviction and deposition during elongation by RNA polymerase II. *Mol. Cell* 22:415–22
57. Imbeault D, Gamar L, Rufiange A, Paquet E, Nourani A. 2008. The Rtt106 histone chaperone is functionally linked to transcription elongation and is involved in the regulation of spurious transcription from cryptic promoters in yeast. *J. Biol. Chem.* 283:27350–54
58. Mason PB, Struhl K. 2003. The FACT complex travels with elongating RNA polymerase II and is important for the fidelity of transcriptional initiation in vivo. *Mol. Cell. Biol.* 23:8323–33
59. Carrozza MJ, Li B, Florens L, Suganuma T, Swanson SK, et al. 2005. Histone H3 methylation by Set2 directs deacetylation of coding regions by Rpd3S to suppress spurious intragenic transcription. *Cell* 123:581–92
60. Selth LA, Lorch Y, Ocampo-Hafalla MT, Mitter R, Shales M, et al. 2009. An Rtt109-independent role for Vps75 in transcription-associated nucleosome dynamics. *Mol. Cell. Biol.* 29:4220–34

61. Formosa T, Ruone S, Adams MD, Olsen AE, Eriksson P, et al. 2002. Defects in SPT16 or POB3 (yFACT) in *Saccharomyces cerevisiae* cause dependence on the Hir/Hpc pathway. Polymerase passage may degrade chromatin structure. *Genetics* 162:1557–71
62. Nourani A, Robert F, Winston F. 2006. Evidence that Spt2/Sin1, an HMG-like factor, plays roles in transcription elongation, chromatin structure, and genome stability in *Saccharomyces cerevisiae*. *Mol. Cell. Biol.* 26:1496–509
63. Swaminathan V, Kishore AH, Febitha KK, Kundu TK. 2005. Human histone chaperone nucleophosmin enhances acetylation-dependent chromatin transcription. *Mol. Cell. Biol.* 25:7534–45
64. Ito T, Ikehara T, Nakagawa T, Kraus WL, Muramatsu M. 2000. p300-Mediated acetylation facilitates the transfer of histone H2A-H2B dimers from nucleosomes to a histone chaperone. *Genes Dev.* 14:1899–907
65. Lorch Y, Maier-Davis B, Kornberg RD. 2006. Chromatin remodeling by nucleosome disassembly in vitro. *Proc. Natl. Acad. Sci. USA* 103:3090–93
66. Walfridsson J, Khorosjutina O, Matikainen P, Gustafsson CM, Ekwall K. 2007. A genome-wide role for CHD remodelling factors and Nap1 in nucleosome disassembly. *EMBO J.* 26:2868–79
67. Kaplan T, Liu CL, Erkmann JA, Holik J, Grunstein M, et al. 2008. Cell cycle- and chaperone-mediated regulation of H3K56ac incorporation in yeast. *PLoS Genet.* 4:e1000270
68. Rufiange A, Jacques PE, Bhat W, Robert F, Nourani A. 2007. Genome-wide replication-independent histone H3 exchange occurs predominantly at promoters and implicates H3 K56 acetylation and Asf1. *Mol. Cell* 27:393–405
69. Kim HJ, Seol JH, Han JW, Youn HD, Cho EJ. 2007. Histone chaperones regulate histone exchange during transcription. *EMBO J.* 26:4467–74
70. Luger K, Mader AW, Richmond RK, Sargent DF, Richmond TJ. 1997. Crystal structure of the nucleosome core particle at 2.8 A resolution (see comments). *Nature* 389:251–60
71. Workman JL, Kingston RE. 1998. Alteration of nucleosome structure as a mechanism of transcriptional regulation. *Annu. Rev. Biochem.* 67:545–79
72. Pokholok DK, Harbison CT, Levine S, Cole M, Hannett NM, et al. 2005. Genome-wide map of nucleosome acetylation and methylation in yeast. *Cell* 122:517–27
73. Kouskouti A, Talianidis I. 2005. Histone modifications defining active genes persist after transcriptional and mitotic inactivation. *EMBO J.* 24:347–57
74. Clayton AL, Hebbes TR, Thorne AW, Crane-Robinson C. 1993. Histone acetylation and gene induction in human cells. *FEBS Lett.* 336:23–26
75. Hebbes TR, Thorne AW, Clayton AL, Crane-Robinson C. 1992. Histone acetylation and globin gene switching. *Nucleic Acids Res.* 20:1017–22
76. Wang A, Kurdistani SK, Grunstein M. 2002. Requirement of Hos2 histone deacetylase for gene activity in yeast. *Science* 298:1412–14
77. Gilbert C, Kristjuhan A, Winkler GS, Svejstrup JQ. 2004. Elongator interactions with nascent mRNA revealed by RNA immunoprecipitation. *Mol. Cell* 14:457–64
78. Close P, Hawkes N, Cornez I, Creppe C, Lambert CA, et al. 2006. Transcription impairment and cell migration defects in elongator-depleted cells: implication for familial dysautonomia. *Mol. Cell* 22:521–31
79. Keogh MC, Kurdistani SK, Morris SA, Ahn SH, Podolny V, et al. 2005. Cotranscriptional set2 methylation of histone h3 lysine 36 recruits a repressive rpd3 complex. *Cell* 123:593–605
80. Govind CK, Zhang F, Qiu H, Hofmeyer K, Hinnebusch AG. 2007. Gcn5 promotes acetylation, eviction, and methylation of nucleosomes in transcribed coding regions. *Mol. Cell* 25:31–42
81. Kristjuhan A, Walker J, Suka N, Grunstein M, Roberts D, et al. 2002. Transcriptional inhibition of genes with severe histone h3 hypoacetylation in the coding region. *Mol. Cell* 10:925–33
82. Protacio RU, Li G, Lowary PT, Widom J. 2000. Effects of histone tail domains on the rate of transcriptional elongation through a nucleosome. *Mol. Cell. Biol.* 20:8866–78
83. Svejstrup JQ. 2003. Transcription. Histones face the FACT. *Science* 301:1053–55
84. Kingston RE, Narlikar GJ. 1999. ATP-dependent remodeling and acetylation as regulators of chromatin fluidity. *Genes Dev.* 13:2339–52
85. Otero G, Fellows J, Li Y, de Bizemont T, Dirac AMG, et al. 1999. Elongator, a multisubunit component of a novel RNA polymerase II holoenzyme for transcriptional elongation. *Mol. Cell* 3:109–18

86. Wittschieben BO, Otero G, de Bizemont T, Fellows J, Erdjument-Bromage H, et al. 1999. A novel histone acetyltransferase is an integral subunit of elongating RNA polymerase II holoenzyme. *Mol. Cell* 4:123–28
87. Svejstrup JQ. 2007. Elongator complex: How many roles does it play? *Curr. Opin. Cell Biol.* 19:331–36
88. Kim JH, Lane WS, Reinberg D. 2002. Human elongator facilitates RNA polymerase II transcription through chromatin. *Proc. Natl. Acad. Sci. USA* 99:1241–46
89. Chen YT, Hims MM, Shetty RS, Mull J, Liu L, et al. 2009. Loss of mouse Ikbkap, a subunit of elongator, leads to transcriptional deficits and embryonic lethality that can be rescued by human IKBKAP. *Mol. Cell. Biol.* 29:736–44
90. Esberg A, Huang B, Johansson MJ, Bystrom AS. 2006. Elevated levels of two tRNA species bypass the requirement for elongator complex in transcription and exocytosis. *Mol. Cell* 24:139–48
91. Govind CK, Yoon S, Qiu H, Govind S, Hinnebusch AG. 2005. Simultaneous recruitment of coactivators by Gcn4p stimulates multiple steps of transcription in vivo. *Mol. Cell. Biol.* 25:5626–38
92. Wery M, Shematorova E, Van Driessche B, Vandenhaute J, Thuriaux P, Van Mullem V. 2004. Members of the SAGA and Mediator complexes are partners of the transcription elongation factor TFIIS. *EMBO J.* 23:4232–42
93. Milgrom E, West RW Jr, Gao C, Shen WC. 2005. TFIID and Spt-Ada-Gcn5-acetyltransferase functions probed by genome-wide synthetic genetic array analysis using a *Saccharomyces cerevisiae taf9-ts* allele. *Genetics* 171:959–73
94. Pascual-Garcia P, Govind CK, Queralt E, Cuenca-Bono B, Llopis A, et al. 2008. Sus1 is recruited to coding regions and functions during transcription elongation in association with SAGA and TREX2. *Genes Dev.* 22:2811–22
95. Saksouk N, Avvakumov N, Champagne KS, Hung T, Doyon Y, et al. 2009. HBO1 HAT complexes target chromatin throughout gene coding regions via multiple PHD finger interactions with histone H3 tail. *Mol. Cell* 33:257–65
96. John S, Howe L, Tafrov ST, Grant PA, Sternglanz R, Workman JL. 2000. The something about silencing protein, Sas3, is the catalytic subunit of NuA3, a yTAF(II)30-containing HAT complex that interacts with the Spt16 subunit of the yeast CP (Cdc68/Pob3)-FACT complex. *Genes Dev.* 14:1196–208
97. Cho H, Orphanides G, Sun X, Yang XJ, Ogryzko V, et al. 1998. A human RNA polymerase II complex containing factors that modify chromatin structure. *Mol. Cell. Biol.* 18:5355–63
98. Obrdlik A, Kukalev A, Louvet E, Farrants AK, Caputo L, Percipalle P. 2008. The histone acetyltransferase PCAF associates with actin and hnRNP U for RNA polymerase II transcription. *Mol. Cell. Biol.* 28:6342–57
99. Joshi AA, Struhl K. 2005. Eaf3 chromodomain interaction with methylated H3-K36 links histone deacetylation to Pol II elongation. *Mol. Cell* 20:971–78
100. Daniel JA, Pray-Grant MG, Grant PA. 2005. Effector proteins for methylated histones: an expanding family. *Cell Cycle* 4:919–26
101. Krogan NJ, Kim M, Tong A, Golshani A, Cagney G, et al. 2003. Methylation of histone H3 by Set2 in *Saccharomyces cerevisiae* is linked to transcriptional elongation by RNA polymerase II. *Mol. Cell. Biol.* 23:4207–18
102. Li B, Howe L, Anderson S, Yates JR 3rd, Workman JL. 2003. The Set2 histone methyltransferase functions through the phosphorylated carboxyl-terminal domain of RNA polymerase II. *J. Biol. Chem.* 278:8897–903
103. Schaft D, Roguev A, Kotovic KM, Shevchenko A, Sarov M, et al. 2003. The histone 3 lysine 36 methyltransferase, SET2, is involved in transcriptional elongation. *Nucleic Acids Res.* 31:2475–82
104. Xiao T, Hall H, Kizer KO, Shibata Y, Hall MC, et al. 2003. Phosphorylation of RNA polymerase II CTD regulates H3 methylation in yeast. *Genes Dev.* 17:654–63
105. Li B, Jackson J, Simon MD, Fleharty B, Gogol M, et al. 2009. Histone H3 lysine 36 dimethylation (H3K36me2) is sufficient to recruit the Rpd3s histone deacetylase complex and to repress spurious transcription. *J. Biol. Chem.* 284:7970–76
106. Li B, Gogol M, Carey M, Lee D, Seidel C, Workman JL. 2007. Combined action of PHD and chromo domains directs the Rpd3S HDAC to transcribed chromatin. *Science* 316:1050–54

107. Ng HH, Robert F, Young RA, Struhl K. 2003. Targeted recruitment of Set1 histone methylase by elongating Pol II provides a localized mark and memory of recent transcriptional activity. *Mol. Cell* 11:709–19
108. Liu CL, Kaplan T, Kim M, Buratowski S, Schreiber SL, et al. 2005. Single-nucleosome mapping of histone modifications in *S. cerevisiae*. *PLoS Biol.* 3:e328
109. Kim T, Buratowski S. 2009. Dimethylation of H3K4 by Set1 recruits the Set3 histone deacetylase complex to 5′ transcribed regions. *Cell* 137:259–72
110. Varshavsky A. 2005. Regulated protein degradation. *Trends Biochem. Sci.* 30:283–86
111. Pickart CM. 2001. Mechanisms underlying ubiquitination. *Annu. Rev. Biochem.* 70:503–33
112. Henry KW, Wyce A, Lo WS, Duggan LJ, Emre NC, et al. 2003. Transcriptional activation via sequential histone H2B ubiquitylation and deubiquitylation, mediated by SAGA-associated Ubp8. *Genes Dev.* 17:2648–63
113. Kao CF, Hillyer C, Tsukuda T, Henry K, Berger S, Osley MA. 2004. Rad6 plays a role in transcriptional activation through ubiquitylation of histone H2B. *Genes Dev.* 18:184–95
114. Xiao T, Kao CF, Krogan NJ, Sun ZW, Greenblatt JF, et al. 2005. Histone H2B ubiquitylation is associated with elongating RNA polymerase II. *Mol. Cell. Biol.* 25:637–51
115. Weake VM, Workman JL. 2008. Histone ubiquitination: triggering gene activity. *Mol. Cell* 29:653–63
116. Gerber M, Shilatifard A. 2003. Transcriptional elongation control and histone methylation. *J. Biol. Chem.* 278:26303–6
117. Osley MA. 2006. Regulation of histone H2A and H2B ubiquitylation. *Brief. Funct. Genomics Proteomics* 5:179–89
118. Lee JS, Shukla A, Schneider J, Swanson SK, Washburn MP, et al. 2007. Histone crosstalk between H2B monoubiquitination and H3 methylation mediated by COMPASS. *Cell* 131:1084–96
119. Pavri R, Zhu B, Li G, Trojer P, Mandal S, et al. 2006. Histone H2B monoubiquitination functions cooperatively with FACT to regulate elongation by RNA polymerase II. *Cell* 125:703–17
120. Tanny JC, Erdjument-Bromage H, Tempst P, Allis CD. 2007. Ubiquitylation of histone H2B controls RNA polymerase II transcription elongation independently of histone H3 methylation. *Genes Dev.* 21:835–47
121. Shukla A, Bhaumik SR. 2007. H2B–K123 ubiquitination stimulates RNAPII elongation independent of H3-K4 methylation. *Biochem. Biophys. Res. Commun.* 359:214–20
122. Minsky N, Shema E, Field Y, Schuster M, Segal E, Oren M. 2008. Monoubiquitinated H2B is associated with the transcribed region of highly expressed genes in human cells. *Nat. Cell Biol.* 10:483–88
123. Fleming AB, Kao CF, Hillyer C, Pikaart M, Osley MA. 2008. H2B ubiquitylation plays a role in nucleosome dynamics during transcription elongation. *Mol. Cell* 31:57–66
124. Svejstrup JQ. 2002. Mechanisms of transcription-coupled DNA repair. *Nat. Rev. Mol. Cell Biol.* 3:21–29
125. Hanawalt PC, Spivak G. 2008. Transcription-coupled DNA repair: two decades of progress and surprises. *Nat. Rev. Mol. Cell Biol.* 9:958–70
126. Svejstrup JQ. 2007. Contending with transcriptional arrest during RNAPII transcript elongation. *Trends Biochem. Sci.* 32:165–71
127. Bregman DB, Halaban R, van Gool AJ, Henning KA, Friedberg EC, Warren SL. 1996. UV-induced ubiquitination of RNA polymerase II: a novel modification deficient in Cockayne syndrome cells. *Proc. Natl. Acad. Sci. USA* 93:11586–90
128. Ratner JN, Balasubramanian B, Corden J, Warren SL, Bregman DB. 1998. Ultraviolet radiation-induced ubiquitination and proteasomal degradation of the large subunit of RNA polymerase II. Implications for transcription-coupled DNA repair. *J. Biol. Chem.* 273:5184–89
129. Anindya R, Aygun O, Svejstrup JQ. 2007. Damage-induced ubiquitylation of human RNA polymerase II by the ubiquitin ligase Nedd4, but not Cockayne syndrome proteins or BRCA1. *Mol. Cell* 28:386–97
130. Somesh BP, Reid J, Liu WF, Sogaard TM, Erdjument-Bromage H, et al. 2005. Multiple mechanisms confining RNA polymerase II ubiquitylation to polymerases undergoing transcriptional arrest. *Cell* 121:913–23
131. Kvint K, Uhler JP, Taschner MJ, Sigurdsson S, Erdjument-Bromage H, et al. 2008. Reversal of RNA polymerase II ubiquitylation by the ubiquitin protease Ubp3. *Mol. Cell* 30:498–506

132. Somesh BP, Sigurdsson S, Saeki H, Erdjument-Bromage H, Tempst P, Svejstrup JQ. 2007. Communication between distant sites in RNA polymerase II through ubiquitylation factors and the polymerase CTD. *Cell* 129:57–68
133. Woudstra EC, Gilbert C, Fellows J, Jansen L, Brouwer J, et al. 2002. A Rad26-Def1 complex coordinates repair and RNA Pol II proteolysis in response to DNA damage. *Nature* 415:929–33
134. Selby CP, Sancar A. 1997. Human transcription-repair coupling factor CSB/ERCC6 is a DNA-stimulated ATPase but is not a helicase and does not disrupt the ternary transcription complex of stalled RNA polymerase II. *J. Biol. Chem.* 272:1885–90
135. Tantin D, Kansal A, Carey M. 1997. Recruitment of the putative transcription-repair coupling factor CSB/ERCC6 to RNA polymerase II elongation complexes. *Mol. Cell. Biol.* 17:6803–14
136. Laine JP, Egly JM. 2006. Initiation of DNA repair mediated by a stalled RNA polymerase IIO. *EMBO J.* 25:387–97
137. Selby CP, Sancar A. 1997. Cockayne syndrome group B protein enhances elongation by RNA polymerase II. *Proc. Natl. Acad. Sci. USA* 94:11205–9
138. Park JS, Marr MT, Roberts JW. 2002. *E. coli* transcription repair coupling factor (Mfd protein) rescues arrested complexes by promoting forward translocation. *Cell* 109:757–67
139. Citterio E, Van Den Boom V, Schnitzler G, Kanaar R, Bonte E, et al. 2000. ATP-dependent chromatin remodeling by the Cockayne syndrome B DNA repair-transcription-coupling factor. *Mol. Cell. Biol.* 20:7643–53
140. Svejstrup JQ. 2003. Rescue of arrested RNA polymerase II complexes. *J. Cell Sci.* 116:447–51
141. Beaudenon SL, Huacani MR, Wang G, McDonnell DP, Huibregtse JM. 1999. Rsp5 ubiquitin-protein ligase mediates DNA damage-induced degradation of the large subunit of RNA polymerase II in *Saccharomyces cerevisiae*. *Mol. Cell. Biol.* 19:6972–79
142. Ribar B, Prakash L, Prakash S. 2006. Requirement of ELC1 for RNA polymerase II polyubiquitylation and degradation in response to DNA damage in *Saccharomyces cerevisiae*. *Mol. Cell. Biol.* 26:3999–4005
143. Ribar B, Prakash L, Prakash S. 2007. ELA1 and CUL3 are required along with ELC1 for RNA polymerase II polyubiquitylation and degradation in DNA-damaged yeast cells. *Mol. Cell. Biol.* 27:3211–16
144. Yasukawa T, Kamura T, Kitajima S, Conaway RC, Conaway JW, Aso T. 2008. Mammalian Elongin A complex mediates DNA-damage-induced ubiquitylation and degradation of Rpb1. *EMBO J.* 27:3256–66
145. Harreman M, Taschner M, Sigurdsson S, Anindya R, Reid J, et al. 2009. Distinct ubiquitin ligases act sequentially for RNA polymerase II polyubiquitylation. *Proc. Natl. Acad. Sci. USA* 106:20705–10
146. Charlet-Berguerand N, Feuerhahn S, Kong SE, Ziserman H, Conaway JW, et al. 2006. RNA polymerase II bypass of oxidative DNA damage is regulated by transcription elongation factors. *EMBO J.* 25:5481–91
147. Kuraoka I, Suzuki K, Ito S, Hayashida M, Kwei JS, et al. 2007. RNA polymerase II bypasses 8-oxoguanine in the presence of transcription elongation factor TFIIS. *DNA Repair* 6:841–51
148. Kim N, Abdulovic AL, Gealy R, Lippert MJ, Jinks-Robertson S. 2007. Transcription-associated mutagenesis in yeast is directly proportional to the level of gene expression and influenced by the direction of DNA replication. *DNA Repair* 6:1285–96
149. Kim N, Jinks-Robertson S. 2009. dUTP incorporation into genomic DNA is linked to transcription in yeast. *Nature* 459:1150–53
150. Luna R, Gaillard H, González-Aguilera C, Aguilera A. 2008. Biogenesis of mRNPs: integrating different processes in the eukaryotic nucleus. *Chromosoma* 117:319–31
151. Chavez S, Beilharz T, Rondon AG, Erdjument-Bromage H, Tempst P, et al. 2000. A protein complex containing Tho2, Hpr1, Mft1 and a novel protein, Thp2, connects transcription elongation with mitotic recombination in *Saccharomyces cerevisiae*. *EMBO J.* 19:5824–34
152. Strasser K, Masuda S, Mason P, Pfannstiel J, Oppizzi M, et al. 2002. TREX is a conserved complex coupling transcription with messenger RNA export. *Nature* 417:304–8
153. Huertas P, Aguilera A. 2003. Cotranscriptionally formed DNA:RNA hybrids mediate transcription elongation impairment and transcription-associated recombination. *Mol. Cell* 12:711–21
154. Prado F, Aguilera A. 2005. Impairment of replication fork progression mediates RNA polII transcription-associated recombination. *EMBO J.* 24:1267–76
155. Wellinger RE, Prado F, Aguilera A. 2006. Replication fork progression is impaired by transcription in hyperrecombinant yeast cells lacking a functional THO complex. *Mol. Cell. Biol.* 26:3327–34

156. Gottipati P, Cassel TN, Savolainen L, Helleday T. 2008. Transcription-associated recombination is dependent on replication in mammalian cells. *Mol. Cell. Biol.* 28:154–64
157. Azvolinsky A, Giresi PG, Lieb JD, Zakian VA. 2009. Highly transcribed RNA polymerase II genes are impediments to replication fork progression in *Saccharomyces cerevisiae*. *Mol. Cell* 34:722–34
158. Bachrati CZ, Hickson ID. 2008. RecQ helicases: guardian angels of the DNA replication fork. *Chromosoma* 117:219–33
159. Aygun O, Svejstrup J, Liu Y. 2008. A RECQ5-RNA polymerase II association identified by targeted proteomic analysis of human chromatin. *Proc. Natl. Acad. Sci. USA* 105:8580–84
160. Aygun O, Xu X, Liu Y, Takahashi H, Kong SE, et al. 2009. Direct inhibition of RNA polymerase II transcription by RECQL5. *J. Biol. Chem.* 284:23197–203
161. Hu Y, Raynard S, Sehorn MG, Lu X, Bussen W, et al. 2007. RECQL5/Recql5 helicase regulates homologous recombination and suppresses tumor formation via disruption of Rad51 presynaptic filaments. *Genes Dev.* 21:3073–84
162. Hu Y, Lu X, Barnes E, Yan M, Lou H, Luo G. 2005. Recql5 and Blm RecQ DNA helicases have nonredundant roles in suppressing crossovers. *Mol. Cell. Biol.* 25:3431–42
163. Hu Y, Lu X, Zhou G, Barnes EL, Luo G. 2009. Recql5 plays an important role in DNA replication and cell survival after camptothecin treatment. *Mol. Biol. Cell* 20:114–23
164. Iyer V, Struhl K. 1996. Absolute mRNA levels and transcriptional initiation rates in *Saccharomyces cerevisiae*. *Proc. Natl. Acad. Sci. USA* 93:5208–12
165. Arndt KM, Kane CM. 2003. Running with RNA polymerase: eukaryotic transcript elongation. *Trends Genet.* 19:543–50
166. Struhl K. 2007. Transcriptional noise and the fidelity of initiation by RNA polymerase II. *Nat. Struct. Mol. Biol.* 14:103–5
167. Saunders A, Core LJ, Lis JT. 2006. Breaking barriers to transcription elongation. *Nat. Rev. Mol. Cell. Biol.* 7:557–67
168. Singer RH, Lawrence DS, Ovryn B, Condeelis J. 2005. Imaging of gene expression in living cells and tissues. *J. Biomed. Opt.* 10:051406
169. Chubb JR, Trcek T, Shenoy SM, Singer RH. 2006. Transcriptional pulsing of a developmental gene. *Curr. Biol.* 16:1018–25
170. Darzacq X, Shav-Tal Y, de Turris V, Brody Y, Shenoy SM, et al. 2007. In vivo dynamics of RNA polymerase II transcription. *Nat. Struct. Mol. Biol.* 14:796–806
171. Saeki H, Svejstrup JQ. 2009. Stability, flexibility, and dynamic interactions of colliding RNA polymerase II elongation complexes. *Mol. Cell* 35:191–205
172. Swinburne IA, Miguez DG, Landgraf D, Silver PA. 2008. Intron length increases oscillatory periods of gene expression in animal cells. *Genes Dev.* 22:2342–46
173. Zhang C, Burton ZF. 2004. Transcription factors IIF and IIS and nucleoside triphosphate substrates as dynamic probes of the human RNA polymerase II mechanism. *J. Mol. Biol.* 342:1085–99
174. Kong SE, Banks CA, Shilatifard A, Conaway JW, Conaway RC. 2005. ELL-associated factors 1 and 2 are positive regulators of RNA polymerase II elongation factor ELL. *Proc. Natl. Acad. Sci. USA* 102:10094–98
175. Gerber M, Eissenberg JC, Kong S, Tenney K, Conaway JW, et al. 2004. In vivo requirement of the RNA polymerase II elongation factor Elongin A for proper gene expression and development. *Mol. Cell. Biol.* 24:9911–19
176. Kettenberger H, Armache KJ, Cramer P. 2003. Architecture of the RNA polymerase II-TFIIS complex and implications for mRNA cleavage. *Cell* 114:347–57
177. Rondon AG, Garcia-Rubio M, Gonzalez-Barrera S, Aguilera A. 2003. Molecular evidence for a positive role of Spt4 in transcription elongation. *EMBO J.* 22:612–20
178. Kim DK, Inukai N, Yamada T, Furuya A, Sato H, et al. 2003. Structure-function analysis of human Spt4: evidence that hSpt4 and hSpt5 exert their roles in transcriptional elongation as parts of the DSIF complex. *Genes Cells* 8:371–78
179. Narita T, Yamaguchi Y, Yano K, Sugimoto S, Chanarat S, et al. 2003. Human transcription elongation factor NELF: identification of novel subunits and reconstitution of the functionally active complex. *Mol. Cell. Biol.* 23:1863–73

180. Schwabish MA, Struhl K. 2007. The Swi/Snf complex is important for histone eviction during transcriptional activation and RNA polymerase II elongation in vivo. *Mol. Cell. Biol.* 27:6987–95
181. Mellor J, Morillon A. 2004. ISWI complexes in *Saccharomyces cerevisiae*. *Biochim. Biophys. Acta* 1677:100–12
182. Konev AY, Tribus M, Park SY, Podhraski V, Lim CY, et al. 2007. CHD1 motor protein is required for deposition of histone variant H3.3 into chromatin in vivo. *Science* 317:1087–90
183. Warner MH, Roinick KL, Arndt KM. 2007. Rtf1 is a multifunctional component of the Paf1 complex that regulates gene expression by directing cotranscriptional histone modification. *Mol. Cell. Biol.* 27:6103–15
184. Ardehali MB, Yao J, Adelman K, Fuda NJ, Petesch SJ, et al. 2009. Spt6 enhances the elongation rate of RNA polymerase II in vivo. *EMBO J.* 28:1067–77
185. Del Rosario BC, Pemberton LF. 2008. Nap1 links transcription elongation, chromatin assembly, and messenger RNP complex biogenesis. *Mol. Cell. Biol.* 28:2113–24
186. Gamble MJ, Erdjument-Bromage H, Tempst P, Freedman LP, Fisher RP. 2005. The histone chaperone TAF-I/SET/INHAT is required for transcription in vitro of chromatin templates. *Mol. Cell. Biol.* 25:797–807
187. Proietti-De-Santis L, Drané P, Egly JM. 2006. Cockayne syndrome B protein regulates the transcriptional program after UV irradiation. *EMBO J.* 25:1915–23

Biochemical Principles of Small RNA Pathways

Qinghua Liu and Zain Paroo

Department of Biochemistry, University of Texas Southwestern Medical Center, Dallas, Texas 75390; email: qinghua.liu@UTSouthwestern.edu

Annu. Rev. Biochem. 2010. 79:295–319

First published online as a Review in Advance on March 5, 2010

The *Annual Review of Biochemistry* is online at biochem.annualreviews.org

This article's doi:
10.1146/annurev.biochem.052208.151733

0066-4154/10/0707-0295$20.00

Key Words

assay, dsRNA, miRNA, piRNA, siRNA

Abstract

The discovery of RNA interference (RNAi) is among the most significant biomedical breakthroughs in recent history. Multiple classes of small RNA, including small-interfering RNA (siRNA), micro-RNA (miRNA), and piwi-interacting RNA (piRNA), play important roles in many fundamental biological and disease processes. Collective studies in multiple organisms, including plants, *Drosophila*, *Caenorhabditis elegans*, and mammals indicate that these pathways are highly conserved throughout evolution. Thus, scientists across disciplines have found novel pathways to unravel, new insights in probing pathology, and nascent technologies to develop. The field of RNAi also provides a clear framework for understanding fundamental principles of biochemistry. The current review highlights elegant, reason-based experimentation in discovering RNA-directed biological phenomena and the importance of robust assay development in translating these observations into mechanistic understanding. This biochemical template also provides a conceptual framework for overcoming emerging challenges in the field and for understanding an expanding small RNA world.

Contents

INTRODUCTION

Modern biological sciences encompass an expanding array of tools to probe the processes of life and disease. The rapid pace of discovery in the RNAi field has been the result of numerous outstanding contributions by investigators across biomedical disciplines. Although the importance of genetics, molecular biology, and cell biology cannot be overstated, advances in elucidating mechanisms of small RNA pathways have largely been achieved through traditional biochemical approaches. The purpose of the current review is to highlight the principles of classic biochemistry as applied to RNAi. Emphasis is placed on outlining the major achievements through which the field has matured, underscoring the rational basis for experimentation and key elements of assay development. For broader analyses, readers are directed to many wonderful recent reviews (1–9). In recounting the path the field has taken toward elucidating the biochemical mechanisms of RNAi, the present work may be particularly useful in providing a framework for understanding emerging regulatory RNA pathways.

Logical Flow of Classic Biochemistry

The purpose of biochemistry is to deconstruct and reconstruct biological events in a cell-free system. The entry point for biochemistry is identification of specific molecular events underlying a biological process (**Figure 1**). This molecular insight enables the design of a cell-free system that recapitulates the biological activity. Two critical elements are required for in vitro assay development: (*a*) Engineered substrates must harbor essential features for catalytic processing to defined products, and (*b*) biological extracts must contain robust enzymatic activity to execute the reaction. Biochemistry is susceptible to in vitro artifacts that are often caused by a deficiency in assay design, i.e., the biochemical readout does not accurately reflect the biological process. Successful reproduction of a biological process in vitro is a fundamental biochemical achievement, which enables direct and in-depth studies of a cellular event, minimizing the complexities that often confound cellular and in vivo studies.

Biochemical purification is a powerful and unbiased approach to identify the factors responsible for the biological activity (**Figure 1**). Two golden rules for biochemical purification are robust assay and unlimited material, i.e., a simple, rapid, and robust assay should be developed, often through repeated optimization of the original assay; and given the large-scale of purification, an abundant and economic source of material must be identified. Following this, a successful purification scheme can be developed, provided that the activity is resistant to salt exposure, exhibits excellent chromatographic behavior, and is sufficiently stable to withstand the arduous process of purification. A key indicator for success is the enrichment of specific activity (total activity relative to total protein) following sequential chromatography. The goal of purification is to approach homogeneity, i.e., a single factor(s) closely correlates with the activity, such that the candidate factor(s) responsible for the biochemical activity is (are) unequivocally identified by mass spectrometric analysis.

Functional validation of the candidate factor(s) typically involves three lines of experiments (**Figure 1**): (*a*) Immunodepletion of candidates should diminish, whereas immunoprecipitates concentrate, the catalytic activity from biological extract; (*b*) biochemical activity is reconstituted using recombinant proteins; and (*c*) biological function of candidates must be demonstrated in vivo. Genetic approaches are a powerful complement to biochemical studies and a gold standard for validation. However, the enormous amount of time and resources required for these studies, even in organisms accessible to genomic manipulation, often discourage such attempts. Fortunately, RNAi has significantly reduced these barriers for reverse genetic studies in most areas. Despite the high degree of difficulty, there is simply no substitute for traditional biochemistry in fundamental understanding of biological processes.

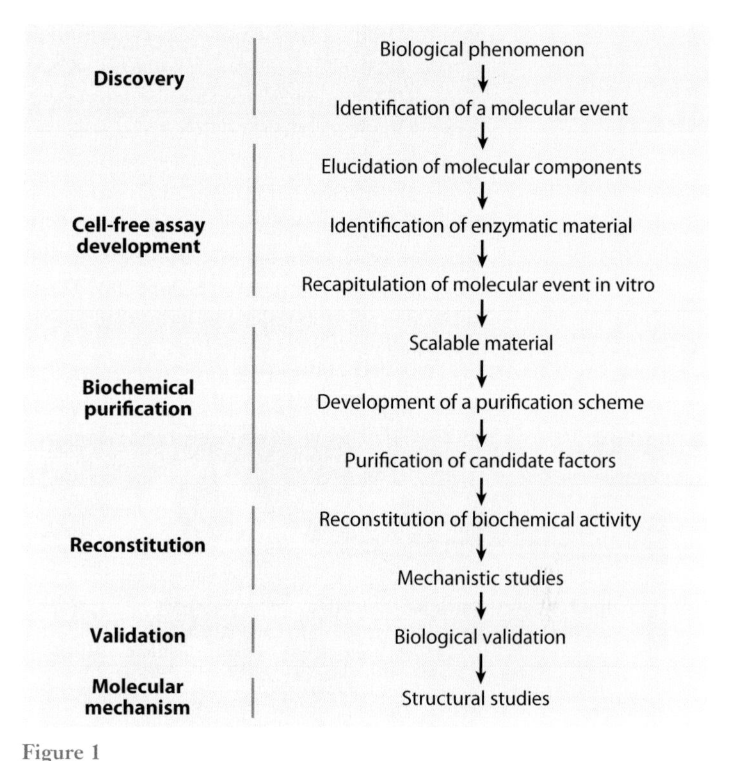

Figure 1

Logical flow of classic biochemistry.

HARNESSING THE PHENOMENON OF RNAi

The first documentation of RNA silencing was derived from attempts to alter aesthetic characteristics of petunia. Aiming to overexpress chalcone synthase to enrich flower pigmentation, Jorgensen and coworkers (10) unexpectedly noted transgene-induced silencing (**Figure 2*a***). Based on the premise of nucleic acid homology, antisense RNA and oligonucleotides were used as loss-of-function perturbations. However, the mechanisms by which these interventions effected gene silencing were poorly understood.

RNAi: RNA interference

Chromatography: a method of purifying molecules in solution by taking advantage of differences in binding properties to stationary phase surfaces

Immunodepletion: use of antibody-coated solid phase material to remove a particular protein from solution

Transgene: genetic material that is expressed through engineered methods

dsRNA: double-stranded RNA

Molecular Events Underlying RNAi

Andrew Fire and Craig Mello and their coworkers (11) systematically determined that injection of double-stranded RNA (dsRNA) homologous to *unc-22* messenger RNA (mRNA) was significantly more effective at mimicking the twitching phenotype of *unc-22* mutant worms than either sense or antisense RNA. This dsRNA-induced silencing phenomenon, termed RNA interference (RNAi), was shown to be systemic, heritable, and accompanied by reduction of target transcript. This work also established a green fluorescent protein (GFP) reporter assay such that RNAi silencing could be easily visualized in vivo (**Figure 2*b***). These seminal studies defined the molecular events underlying RNAi, i.e., dsRNA functions as a trigger for target transcript degradation.

Although the central dogma of molecular biology (DNA → RNA → protein) provided a useful framework to conceptualize gene expression, there were inherent constraints with this simplified picture. RNA was viewed primarily as a housekeeper simply transmitting genetic information from DNA to protein. Technically, RNA was traditionally studied as large polymers by agarose gel electrophoresis. Challenging these assumptions and reasoning that RNA silencing could be mediated by RNA smaller than typical transcripts, Hamilton & Baulcombe (12) employed polyacrylamide gel electrophoresis to connect RNA silencing with ~25-nucleotide (nt) RNA in plants. The simplicity of these experiments highlighted the sophistication of

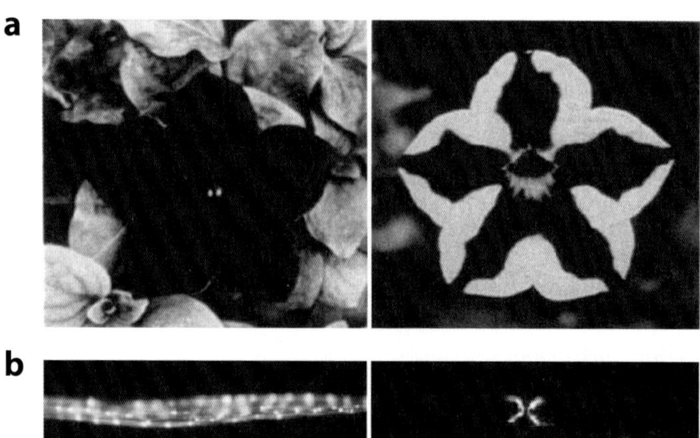

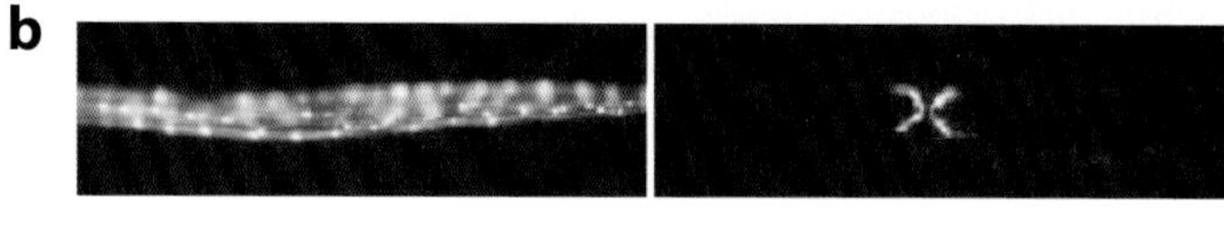

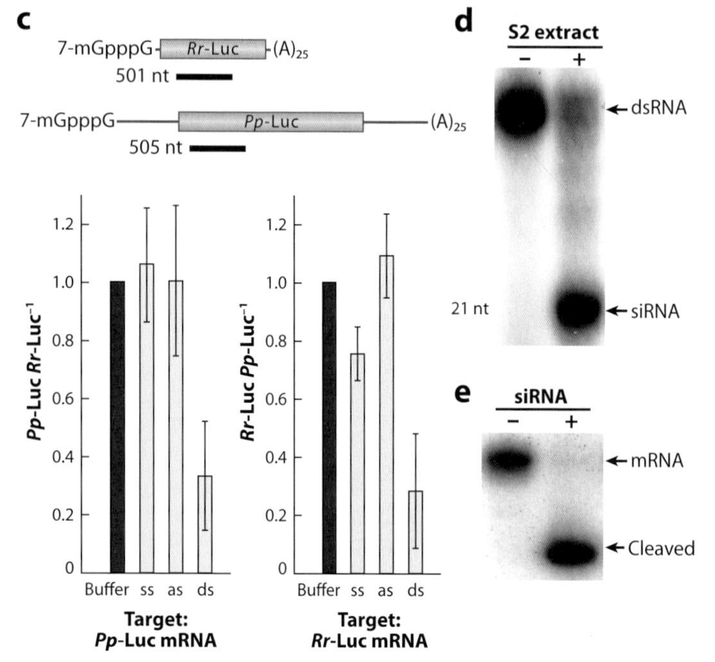

Figure 2

From biological phenomenon to cell-free assays. (*a*) Phenotypes of parental (*left*) and transgenic petunia expressing an ectopic chalcone synthase gene (*right*). Contrary to the enhanced pigmentation expected, Jorgensen and coworkers (10) observed reduced flower pigmentation. (*b*) Fire et al. (11) demonstrated dsRNA-induced silencing of a *gfp* transgene in adult worms. Animals were injected with control (*left*) or *gfp* dsRNA (*right*). (*c*) Tuschl et al. (13) established a cell-free system that recapitulated RNAi in *Drosophila* embryo extract. (*top*) Schematic indicates Renilla (*Rr*-Luc) and firefly luciferase (*Pp*-Luc) transcripts and homologous dsRNA. (*bottom*) Normalized reporter activity when using sense (ss), antisense (as), or dsRNA. (*d*) The Dicer assay showing that radiolabeled dsRNA is processed to 21–22-nt small-interfering RNA (siRNA) in *Drosophila* S2 cell extract. (*e*) The RNA-induced silencing complex (RISC) assay showing that duplex siRNA initiates RISC cleavage of 5′-radiolabeled target mRNA in S2 extract. Both assays were first established by Zamore (14) and Tuschl (16) and their coworkers in *Drosophila* embryo extract. **Figure 2*a*** was adapted by permission from the American Society of Plant Physiologists, *Plant Cell* © 1990 (10). **Figure 2*b*** was adapted by permission from Macmillan Publishers Ltd, *Nature* © 1998 (11). **Figure 2*c*** is reprinted by permission from Cold Spring Harbor Laboratory Press, *Genes and Development* © 1999 (13).

departing from the central dogma and facilitated the emergence of small regulatory RNAs.

Cell-Free Assay Development

Tuschl et al. (13) developed a cell-free system that recapitulated dsRNA-induced silencing and established a foundation for biochemical analysis of RNAi. Using dsRNA against luciferase transcripts, sequence-specific silencing of reporter mRNA translation was demonstrated in extract prepared from syncitial blastoderm *Drosophila* embryos (**Figure 2*c***). Similar to what was observed in worms (11), this in vitro silencing was also accompanied by degradation of the target transcript (13). These studies provided direct evidence for dsRNA and target mRNA as the molecular players involved and for mRNA degradation as the functional outcome of RNAi. To simplify this translation-based assay, Zamore et al. (14) employed radiolabeled mRNA to directly monitor target transcript degradation and indicated that RNAi was mediated by sequence-specific cleavage of target mRNA. Similarly, radiolabeled dsRNA was used to demonstrate that long dsRNA was processed to 21–22-nt small RNA fragments (**Figure 2*d***) (14). These findings provided a critical link between the long dsRNA silencing trigger used by Fire et al. (11) and the small RNA corresponding to RNAi observed by Hamilton & Baulcombe (12).

Scaling Up for Biochemical Purification

However, the limited quantities in which *Drosophila* embryos can be obtained presented a problem for biochemical purification. Hammond et al. (15) solved this problem by establishing RNAi activity in *Drosophila* Schneider 2 (S2) cells grown in large-scale suspension culture. In vitro target mRNA cleavage activity was achieved by "preloading" S2 cells with dsRNA prior to preparation of cell extract. Following chromatographic fractionation, an RNA-induced silencing complex (RISC) was isolated that harbored small RNA and catalyzed sequence-specific mRNA cleavage. Therefore, it was hypothesized that long dsRNA was processed to small RNA, which programmed the RISC to direct target transcript degradation (14, 15).

RISC: RNA-induced silencing complex

siRNA: small-interfering RNA

IDENTIFICATION OF SMALL-INTERFERING RNA

Careful mapping studies indicated that target transcript cleavage corresponded to regions complementary to dsRNA and occurred at 21–22-nt intervals similar to the size of dsRNA-derived small RNAs (16). To directly test if small RNAs mediate RISC activity, 21–22-nt RNA duplexes with symmetric 2-nt 3′ overhangs were synthesized to mimic dsRNA-processing products. Indeed, these synthetic oligonucleotides induced target mRNA cleavage at sites corresponding to the middle of small RNA (**Figure 2*e***) (16–18). Therefore, these dsRNA-derived small RNAs were designated as small-interfering RNAs (siRNAs). Furthermore, a most important contribution was made by demonstration of target transcript silencing through transfection of synthetic siRNAs into mammalian cells (19). This work provided the foundation for numerous RNAi-based applications, including powerful loss-of-function tools and a new class of potential therapeutics. As such, these landmark studies provided an excellent example of the importance of basic science, including traditional biochemistry, in generating new areas for biomedical applications.

A PARALLEL MICRO-RNA PATHWAY

These fundamental advances in the study of siRNA also provided new insight into a previously known regulatory RNA pathway. In *Caenorhabditis elegans*, LIN-14 protein is a master regulator of developmental gene expression (20). The *lin-4* locus is also required for proper timing of development, but this genomic region was not known to harbor a protein-coding sequence (21, 22). Rather, *lin-4* was found to encode ~22- and 61-nt RNAs that were

UTR: untranslated region

miRNA: micro-RNA

Epitope tagged: fusion of a protein-coding sequence with a short polypeptide marker to facilitate antibody recognition

complementary to multiple elements within the 3′-untranslated region (UTR) of *lin-14* mRNA (21, 22). Phenotypic and reporter assays indicated that both *lin-4* and 3′-UTR of *lin-14* were necessary and sufficient for developmental patterning (21, 22). Subsequent work revealed that lin-4 and the highly conserved 21-nt let-7 RNA govern a network of regulatory factors in directing temporal development (23). Similar to siRNA, these micro-RNAs (miRNAs), also known as small temporal RNAs (stRNA), were found to negatively regulate expression of target transcripts (24).

SMALL RNA BIOGENESIS

That siRNAs and miRNAs were established as functional guides in governing silencing of target transcripts raised the question of how these small RNAs were produced. Examination of dsRNA-processing products indicated an RNaseIII-like pattern of endonucleolytic cleavage (16, 25). Bernstein et al. (26) expressed candidate RNaseIII enzymes as epitope-tagged proteins in *Drosophila* S2 cells and found that CG4792 (named Dicer-1) immunoprecipitates processed dsRNA to siRNA in vitro. This work identified Dicer as the small RNA-generating enzyme and established the first RNAi factor with a defined biochemical activity. Subsequently, a six-step chromatographic purification of siRNA-generating activity from S2 cell extract revealed CG6493 (named Dicer-2) (27). dsRNA-mediated knockdown of Dicer-2, but not Dicer-1, in S2 cells decreased siRNA production and compromised RNAi silencing. Thus, two functional Dicer proteins are encoded in the *Drosophila* genome.

Genomic analysis indicated that the let-7 miRNA might be derived from a conserved stem-loop precursor (28), suggesting a similar dicing step may be required for miRNA biogenesis. By Northern blotting, Hutvagner et al. (29) verified the existence of this ~70-nt let-7 precursor (pre-let-7) in *Drosophila* pupae and HeLa cells. Furthermore, in vitro transcribed pre-let-7 was processed to mature let-7 in *Drosophila* embryo lysate. siRNA-mediated knockdown of Dicer in HeLa cells resulted in accumulation of pre-let-7 and attenuated expression of let-7. Thus, in addition to siRNA production, Dicer is also required for miRNA biogenesis.

To dissect small RNA pathways in *Drosophila*, Lee et al. (30) isolated *dicer-1* (*dcr-1*) and *dicer-2* (*dcr-2*) mutants from an elegant RNAi genetic screen. Analyses of these mutants indicated that Dicer-1 and Dicer-2 were responsible for miRNA and siRNA biogenesis, respectively (30). Reconstitution studies confirmed that recombinant Dicer-1 preferentially processed pre-miRNA independent of ATP, whereas Dicer-2 preferentially processed long dsRNA in an ATP-dependent manner (31). In contrast, mammalian genomes encode only one Dicer that generates both miRNA and siRNA (29, 32). Recombinant human Dicer preferentially processes pre-miRNA versus dsRNA (Q. Liu, unpublished information) and does not require ATP for catalysis (33, 34). Therefore, the biochemical characteristics of human Dicer more closely resemble *Drosophila* Dicer-1 versus Dicer-2.

Two hallmarks of RNaseIII catalysis are the discrete size of dsRNA products and the 2-nt 3′ overhang. Prokaryotic RNaseIII carries a single RNaseIII domain and functions as a homodimer (35), whereas eukaryotic Dicer contains two RNaseIII domains and functions as a monomer (34). Through mutagenesis studies of putative catalytic residues of *E. coli* RNaseIII and human Dicer, Zhang and coworkers (34) proposed that RNaseIII/Dicer contains a single catalytic center for processing dsRNA (**Figure 3*a*,*b***). The tandem RNaseIII domains of Dicer form an intramolecular dimer and cleave opposing strands of dsRNA in an offset manner to produce a 2-nt 3′ overhang. Similar results were obtained with corresponding experiments for *Drosophila* Dicer-1 (36). In addition, structural and modeling studies of a primitive Dicer from *Giardia intestinalis* suggested that the PAZ domain recognizes the dsRNA terminus and that the size of small RNA produced is determined by the physical distance between the PAZ and RNaseIII domains (**Figure 3*c***) (37).

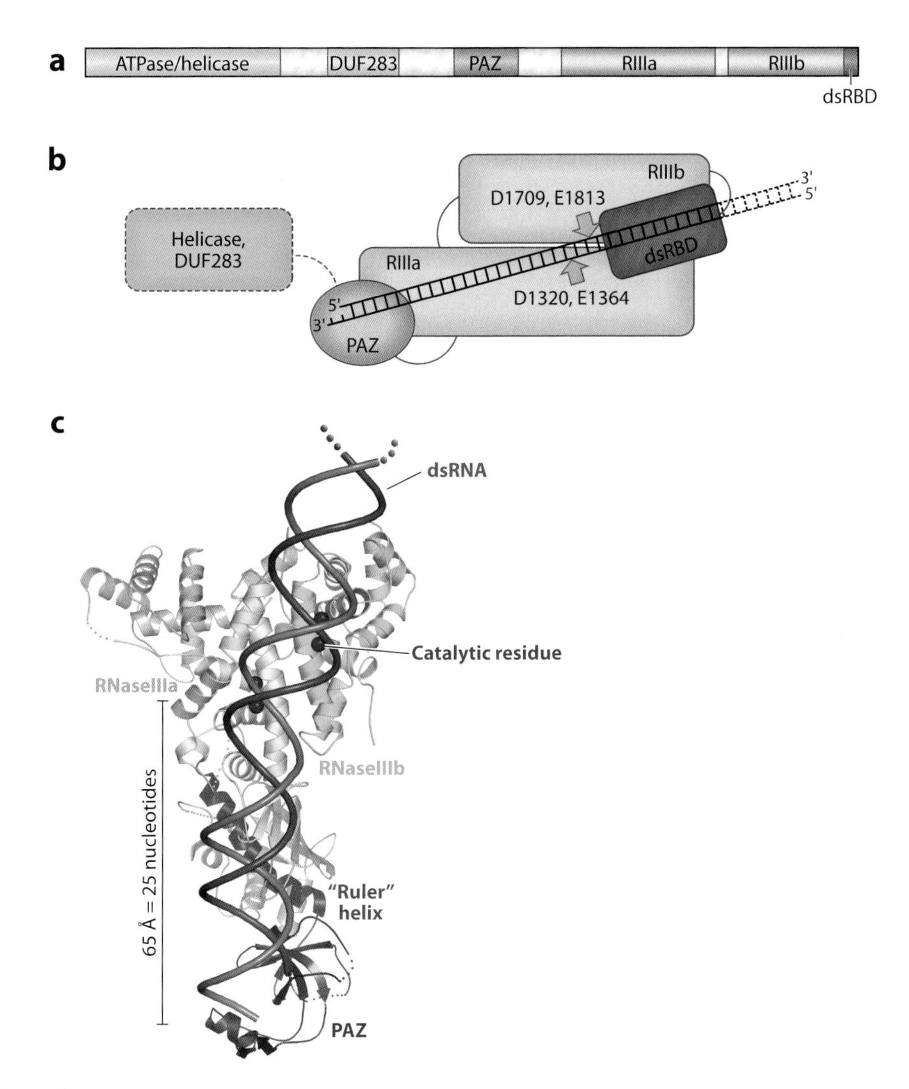

Figure 3

Molecular mechanism of Dicer function. (*a*) Domain structure of human Dicer, indicating ATPase/helicase, domain of unknown function (DUF)283, PAZ, RNaseIII (RIII), and dsRNA-binding domains (dsRBDs). (*b*) Single processing center model of Dicer catalysis. The PAZ domain recognizes the terminus of the dsRNA substrate. The two RNaseIII domains approximate to form one catalytic center and cleave the opposing strands of dsRNA in an offset manner, resulting in a characteristic 2-nt 3′ overhang. (*c*) Ribbon depiction of *Giardia* Dicer structure modeled with dsRNA substrate. The axe-like arrangement of the core enzyme is formed by a PAZ-dsRNA handle and RNaseIII domains forming the blade. The distance between PAZ and the catalytic residues (*balls*) is the ruler that determines the size of small RNA products. **Figure 3*a,b*** is from *Cell* © 2004 (34), reprinted by permission from Elsevier. **Figure 3*c*** provided courtesy of J. Doudna, University of California, Berkeley.

siRNA EFFECTOR FUNCTION

Identification of siRNA- and miRNA-generating enzymes intensified efforts to elucidate the mechanisms through which these small RNAs effected silencing. The work of Hammond et al. (15) indicated that the RNAi effector RISC contains a nuclease that targets mRNA through programming by siRNA. Chromatographic purification of preloaded RISC activity from S2 cell extract identified CG7439, named Argonaute2 (Ago2) (38). A similar independent purification scheme further suggested Ago2 involvement in *Drosophila* RISC (39). Human Argonautes were also identified as RISC constituents through affinity purification of biotinylated siRNA from HeLa cell extract (40). This family of proteins had previously been implicated in RNAi as *C. elegans*, *Neurospora*, and *Arabidopsis* mutants exhibited defects in silencing (41–43).

RISC was demonstrated as a magnesium-dependent endoribonuclease (44, 45). Affinity purification of epitope-tagged proteins indicated that all four human Agos associated with endogenous miRNA and transfected siRNA, but only Ago2-programmed RISC exhibited in vitro target mRNA cleavage activity (46, 47). Furthermore, Ago2-deficient murine embryonic fibroblast cells were defective for RNAi and could be rescued only by Ago2 but not by the Ago1 transgene (46). These findings strongly implicated Ago2 as a candidate effector nuclease for RISC.

Reasoning that catalytic residues may be surmised through sequence comparison between the cleaving Ago2 and noncleaving Ago proteins, Liu and coworkers (46) found that mutations of selected human Ago2 residues resulted in a loss of mRNA cleavage (slicer) activity. Song et al. (48) provided critical insight for Ago function by defining the crystal structure of a primitive Ago from *Pyrococcos furosus* (48). This analysis revealed that the PIWI domain of Ago has a fold similar to RNaseH, which degrades RNA in RNA-DNA hybrid. Thus, the PIWI domain of Ago2 could slice target mRNA in an mRNA-siRNA hybrid. Indeed, mutation of conserved RNaseH-based catalytic residues abolished in vitro slicer activity of immunoprecipitated Ago2 (46). A key biochemical achievement was the reconstitution of a minimal RISC enzyme using *E. coli*-produced recombinant human Ago2 and single-strand siRNA (49). Recombinant human Ago2-bearing mutations in catalytic residues D597, D669, and H807 did not exhibit slicer activity (46, 49). The lack of this "DDH" motif explains why human Ago1 and Ago4 do not exhibit slicer activity (1).

Much of what was learned from human Ago2 was also found for *Drosophila* Ago proteins. Immunoprecipitated siRNA-programmed dAgo2 exhibited in vitro slicer activity, whereas immunoprecipitated dAgo1 cleaved a target mRNA bearing a sequence complementary to bantam miRNA (50). Minimal RISC activity was reconstituted for recombinant dAgo1 and a truncated dAgo2 produced from *E. coli* (50). However, there appear to be key differences in how small RNAs program Argonautes in different organisms. In *Drosophila*, dAgo1 is programmed through miRNA, whereas dAgo2 is primarily programmed through siRNA (51). By contrast, human Ago1–4 were found to harbor both siRNA and miRNA (46, 47). The molecular basis for the difference between generalized functions of human Agos versus specialized roles of *Drosophila* Agos remains unknown.

Structural studies also provided great insight into the functional relationships among Ago, siRNA, and target mRNA (1). The PAZ domain of Ago contains an oligonucleotide-binding fold that interacts with the 2-nt 3′ overhang of siRNA (**Figure 4*a*,*b***) (1). In conjunction with PAZ, the PIWI, N-terminal, and middle (Mid) domains of Ago form a crescent base and a positively charged channel, wherein guide siRNA recognizes target mRNA, and the 5′ phosphate of siRNA is anchored within the binding pocket of the Mid domain (**Figure 4*b*,*c***) (1, 52, 53). Consistent with earlier biochemical studies (16, 44, 45), structural studies revealed that helical interaction between siRNA and target RNA positioned the scissile phosphate corresponding to the ninth and tenth positions of guide RNA at the catalytic center (1, 52, 53). Cleavage of target RNA results in a 5′ fragment with a 3′-hydroxyl terminus and a 3′ fragment with a 5′ phosphate (44, 45). Collectively, these biochemical and structural studies provided the foundation for understanding small RNA effector functions.

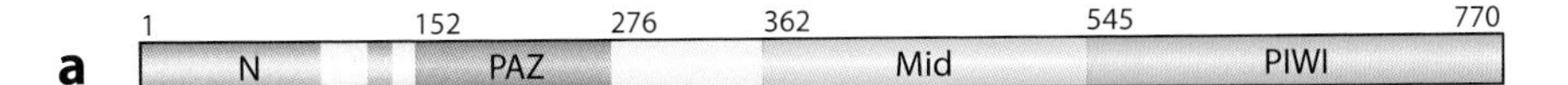

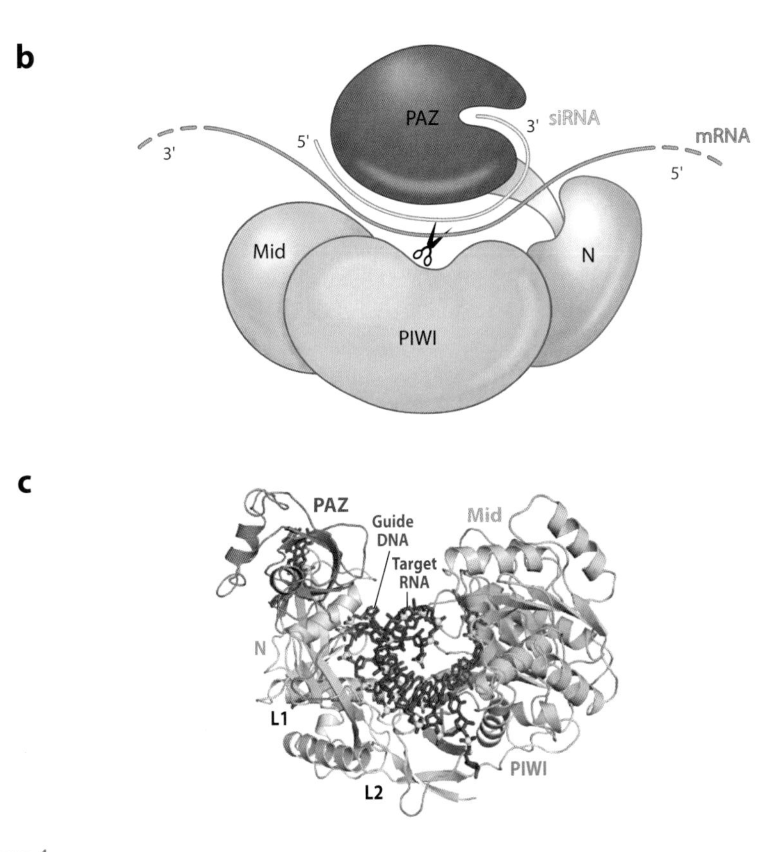

Figure 4

Molecular mechanism of slicer function. (*a*) Domain structure of argonaute from *Pyrococcus furiosus*. (*b*) A schematic representation of siRNA-directed mRNA cleavage. The 3′ end of siRNA is positioned in the cleft of the PAZ domain. The mRNA situates between the upper PAZ domain and the lower crescent-shaped base formed by the N-terminal, PIWI, and middle (Mid) domains. The catalytic site (*scissors*) slices mRNA at a position that corresponds to the ninth and tenth nucleotides of guide siRNA. (*c*) Crystal structure of *Thermus thermophilus* argonaute bound to 5′-phosphorylated 21-nt guide DNA and 20-nt target RNA. **Figure 4*a,b*** is from *Science* © 2004 (48), reprinted with permission from AAAS. **Figure 4*c*** was adapted by permission from Macmillan Publishers Ltd., *Nature* © 2008 (52).

miRNA EFFECTOR MECHANISMS

Despite the many similarities between siRNA and miRNA, where siRNA has a clearly defined effector mechanism, miRNA function is much more controversial [for a comprehensive review, see Sonenberg (53a) in the current issue]. A key reason for this disparity was the early establishment of a widely accepted in vitro assay for siRNA-induced RISC (siRISC) activity (14). The lack of a comparable, well-defined biochemical readout for miRNA-induced RISC (miRISC) activity has been a key obstacle in establishing biochemical assays to understand miRNA effector mechanisms.

Cross-linking: a technique used to induce covalent bond formation between proteins or between proteins and nucleic acids

Current thinking on animal miRNA function has been heavily influenced by the earliest report that lin-4-mediated repression of Lin-14 protein expression occurred in the absence of detectable changes in *lin-14* mRNA levels (22). Thus, it was proposed that miRNAs effect target silencing through inhibition of translation. However, these studies employed RNase protection assays to monitor transcript levels. Subsequent work using Northern blotting indicated reduction of full-length target transcripts (54). Moreover, cloning of *lin-41* 3′-UTR fragments indicated cleavage products upstream of let-7 target sites (54). Similar cleavage products were detected for 3′-UTR constructs bearing let-7 sites in human cells (55). Additionally, among the first miRNA-mRNA target relationships identified was miR-196-directed cleavage of *HOXB8* mRNA (56). Indeed, perturbation of miRNA expression followed by global transcript profiling has been widely successful in identifying miRNA targets (57). Thus, miRNAs appear to modulate protein expression through (*a*) a slicer mechanism for high homology mRNA targets, (*b*) slicer-independent mRNA cleavage or degradation, and (*c*) translational inhibition without changes in mRNA levels. Distinguishing among these outcomes may help resolve the stewing tension in the field between master miRNA regulators that govern expression of a few key targets versus fine-tuning miRNAs that moderately influence expression of a large number of transcripts (58).

Cell-free systems of miRNA-mediated silencing have been instrumental in identifying the molecular requirements for target transcript silencing (59, 60). Although these assays are not suited for biochemical purification, they may be useful in testing candidate factors that modulate miRNA-mediated silencing. Identification of miRNA-mediated deadenylation is a promising lead for more in-depth biochemical investigation (60). The lack of a widely accepted molecular readout, robust assay, and unlimited material remains an outstanding challenge for in-depth biochemical investigation.

ACTIVATING RISC

The establishment of robust enzymatic assays (**Figure 2*d*,*e***) enables biochemical identification of principal factors responsible for siRNA biogenesis and RISC function. However, understanding how these two processes connect has been a major challenge for the field. While newly synthesized siRNA is double stranded, only single-strand siRNA could directly program recombinant Ago2 into a minimal RISC (46, 49), suggesting that additional factors are required for incorporating nascent siRNA into RISC. Liu and coworkers (27) identified a biochemical link between dsRNA processing and RISC activation through biochemical purification of *Drosophila* siRNA-generating activity from S2 cell extract. Identified was a heterodimeric complex of Dicer-2 and CG7138, named R2D2 for its two dsRNA-binding domains (R2) and association with Dicer-2 (D2) (27). As recombinant R2D2 had no effect on dicing activity, it was hypothesized that R2D2 functions downstream of siRNA production. A partial reconstitution system was developed to determine that the Dicer-2-R2D2 complex coordinately recruits duplex siRNA to Ago2 to facilitate RISC activation (27). Genetic studies indicated that both Dicer-2 and R2D2 were required for efficient RISC activity and RNAi silencing (27, 30, 61). Native gel-shift analyses indicated that Dicer-2 and R2D2 interacted with duplex siRNA to assemble a RISC-loading complex that preceded formation of an active RISC (27, 62, 63). Moreover, cross-linking studies suggested that the Dicer-2-R2D2 complex senses thermodynamic asymmetry of siRNA with R2D2 preferentially binding the more stable end and Dicer-2 binding the less stable end (64–66). This asymmetric binding of Dicer-2-R2D2 appears to facilitate selection of the guide strand for incorporation into RISC.

Helicase Model of RISC Activation

The central step of RISC activation is the separation of two siRNA strands and loading of one strand, the guide strand, onto Ago2. Using native polyacrylamide gel electrophoresis to resolve double-strand and single-strand siRNA, Nykanen et al. (67) demonstrated that ATP was required for single-stranded siRNA production. Moreover, this siRNA-unwinding activity correlated well with the RISC activity following gel filtration chromatography (67). Thus, RNA helicases were implicated in RISC activation.

Multiple factors with annotated helicase domains have been implicated in RNAi. For example, both Dicer-1 and Dicer-2, known to influence RNAi activity, contain a DExH helicase motif (27, 30, 62). Ovary extracts deficient in the ATP-dependent helicase Armitage were defective for RISC activity (63). Human homologs of Armitage, MOV10, coimmunoprecipitated with human Ago1 and Ago2, and knockdown of MOV10 resulted in attenuated gene silencing (68). Affinity purification of biotinylated siRNA transfected in human cells indicated an association between RNA helicase A (RHA) and active RISC complex (69). Knockdown of RHA resulted in reduced association of siRNA with Ago2, decreased RISC activity, and attenuated gene silencing (69). In addition to duplex siRNA unwinding, RNA helicases may serve other important roles in small RNA pathways, including dsRNA processing, target mRNA recognition, and release of cleavage products. Given the importance of helicases in viral biology and many reports of cross talk between virology and RNAi, understanding the roles of helicases in regulatory RNA pathways may be of direct clinical significance.

Slicer Model of RISC Activation

An alternative model for RISC activation involves the same mechanism by which Ago2 and guide siRNA cleave target mRNA. Here, duplex siRNA is recruited to Ago2, and the (passenger) strand to be excluded from RISC is the target of slicer activity. Ago2 cleaves the passenger strand into 9-nt and 12-nt fragments, leaving the guide strand behind with Ago2 to form an active RISC. In support of this model, passenger strand cleavage was detected in wild-type, but not in *dcr-2*, *r2d2*, or *ago2* mutant *Drosophila* embryo lysates (50, 69, 70). Phosphorothioate and 2′-O-methyl siRNA modifications blocked passenger strand cleavage and attenuated RISC activity (69, 70). In S2 extract deficient in Ago2 activity, the production of single-strand siRNA was rescued with recombinant wild-type, but not catalytic mutant, Ago2 (71). A similar in vivo requirement for the slicer activity of Ago2 was also observed in flies (72). In *Neurospora*, the catalytic mutant *qde-2* (Ago2) strains were defective for production of single-strand, but not double-strand, siRNA (73). These studies indicated that the slicer mechanism plays a prominent role in *Drosophila* and *Neurospora* RISC activation.

Although various factors and models have been proposed, there is a lack of consensus in defining the biochemical mechanisms of RISC activation. One reason for this is the possibility of multiple mechanisms for RISC activation, each playing more or less significant roles depending on the species, tissues, and cell types under study. As not all Ago proteins exhibit slicer activity, passenger strand cleavage cannot explain all effector complex programming. Additionally, miRNAs have central mismatches that preclude Ago-mediated passenger strand cleavage. Thus, understanding the mechanisms by which small RNA effector complexes are activated remains a clear challenge for the field.

Overcoming Key Challenges for Understanding RISC Activation

Although an immensely powerful approach, biochemical fractionation and reconstitution impose many significant challenges. A critical obstacle in the setting of RNAi has been the salt sensitivity of in vitro RISC activity. Specifically, exposure to salt or chromatographic fractionation could irreversibly damage the ability to form RISC in *Drosophila* embryo or S2 extract (39, 74). Hammond et al. (15, 38) circumvented this problem by treating S2 cells with dsRNA to preload RISC prior to making cell extract. Unlike *de novo* RISC activity, preloaded

RISC activity was resistant to salt treatment and could withstand chromatographic fractionation. This disparity may be the result of stabilization of Ago2 conformation upon siRNA binding, conferring resistance to salt-induced structural changes. However, a solution to the salt-sensitivity problem was needed to purify additional factors involved in RISC activation.

It was determined that *Drosophila* Ago2 was a key salt-sensitive factor, and hence, the problem of RISC reconstitution was solved by using recombinant Ago2 purified at a low-salt condition (71). The production of full-length *Drosophila* Ago2 had been confounded by its extensive N-terminal polyglutamine repeats. Thus, an active, truncated Ago2 recombinant protein was generated that could fully restore RISC activity to *ago2* mutant extract (71). Recombinant Dicer-2-R2D2 and Ago2 proteins successfully reconstituted long dsRNA- and duplex siRNA-initiated RISC activity (71), demonstrating that these factors compose the catalytic core of RNAi. Such experiments may provide important guidance for solving key problems in reconstituting human RISC.

Liu and coworkers (71) employed this core reconstitution system to biochemically purify a new RISC activator, named C3PO (component 3 promoter of RISC), which consists of two evolutionarily conserved subunits: Translin/TB-RBP and Translin-associated factor X (Trax). Recombinant C3PO enhanced core RISC activity of Dicer-2/R2D2/Ago2, and genetic depletion of C3PO attenuated RISC activity in vitro and in vivo (71). Biochemical studies indicated that C3PO is a Mg^{2+}-dependent endoribonuclease (71). Mutagenesis of putative catalytic residues on the Trax subunit resulted in loss of C3PO's nuclease activity and RISC-enhancing activity, suggesting a functional link between these processes (71). Specifically, C3PO was found to promote RISC activation by removing siRNA passenger strand fragments (71). A similar mechanism had also been observed in *Neurospora*, wherein the exonuclease QIP degraded QDE-2-nicked duplex siRNA to facilitate RISC activation (73). These studies collectively illustrated a three-step process for slicer-mediated RISC activation: (*a*) Duplex siRNA is recruited to Ago2 (e.g., by Dicer-2-R2D2), (*b*) Ago2 cleaves the passenger strand, and (*c*) passenger strand fragments are actively removed by C3PO or QIP.

DEFINING HOLO-RISC

A three-component model of human RISC was proposed through studies of immunopreciptiated complexes containing Dicer, Ago2, and the HIV transactivating response element-binding protein (TRBP, a homolog of R2D2) from human cells (75, 76). Direct support for this model was provided by reconstitution studies employing recombinant Dicer, TRBP, and Ago2 proteins (77). Following incubation of individually purified components, gel filtration chromatography revealed the formation of a trimeric complex exhibiting pre-miRNA-processing and pre-miRNA-initiated RISC activity. However, recent reports have shown that recombinant human Ago2 devoid of pre-miRNA-processing activity could be programmed by pre-miRNA encoding 5′-, but not 3′-, derived miRNA to cleave target mRNA (77a; J.A. Doudna & Q. Liu, unpublished information). Thus, an alternative interpretation of the data is that the intact pre-miRNA may serve as a guide for human Ago2 to direct target mRNA cleavage. Importantly, long dsRNA- and duplex siRNA-initiated RISC activities have yet to be reconstituted in the human system. The lack of this reconstitution system has hindered further mechanistic understanding of human RISC.

The formation of an 80S holo-RISC was detected by sedimentation analysis in *Drosophila* embryo extract (62). Recombinant Dicer-2-R2D2 and Ago2 could reconstitute core RISC activity, and addition of C3PO markedly enhanced this baseline activity (71), suggesting that the trimeric model may be an underrepresentation of holo-RISC function. Development of a robust reconstitution system should greatly facilitate in-depth studies of the assembly, function, and regulation of holo-RISC. A key objective of biochemical reconstitution is to identify all factors that are necessary and sufficient for holo-RISC activity. Other objectives are to assign a specific role to each factor and to define the functional relationships among different components. Moreover, the reconstitution system can be employed to purify regulators of holo-RISC activity and to study posttranslational regulation of the RNAi machinery.

GENOME-ENCODED SMALL RNA

Perhaps the biggest surprise of the Human Genome Project was the relatively small number of protein-coding genes relative to genome size (78, 79). Only five percent of the genome is believed to encode proteins. miRNA, endogenous siRNA (endo-siRNA), and piwi-interacting RNA (piRNA) are constituents of the remaining noncoding genome. There is growing understanding of the importance of these and other noncoding RNAs in regulating genome function.

miRNA

Public miRNA databases indicate hundreds of miRNAs from many species with hundreds or thousands more estimated (80). These miRNAs have been identified through three main approaches (81): (*a*) forward genetics by isolating miRNA mutants, (*b*) bioinformatic predictions based on a stem-loop pre-miRNA and phylogenetic conservation, and (*c*) direct cloning and sequencing of small cellular RNA, which has been greatly expanded by new pyrosequencing technologies.

Hypothesizing that ~70-nt pre-miRNA was derived from longer transcripts, Lee et al. (82) performed RT-PCR using primers of increasing distance from the genomic region of pre-miRNA. This approach indicated the existence of several hundred bases long primary-miRNA (pri-miRNA) transcripts, which were confirmed by Northern blotting (82). These findings were particularly important as small RNAs, such as transfer RNA and small nucleolar RNA, are derived from short RNA polymerase III transcripts (83). That pri-miRNA transcripts were larger than expected suggested a different mechanism of transcription.

To elucidate features of pri-miRNA transcripts, total RNA was affinity purified using the guanosine cap-binding protein eIF4E (84, 85). All pri-miRNAs examined exhibited eIF4E binding, indicating 5′-guanosine capping of pri-miRNA transcripts. Affinity purification of total RNA using oligo-dT indicated that pri-miRNA transcripts were polyadenylated (84, 85). These characteristics suggested an RNA polymerase II (Pol II)-mediated transcription. Pharmacological inhibition of Pol II reduced pri-miRNA expression, and chromatin immunoprecipitation indicated Pol II occupancy of a pri-miRNA promoter (85). These findings demonstrate that miRNAs are derived from Pol II transcripts bearing a 5′-guanosine cap and 3′-poly(A) tail.

Although there was good understanding of Dicer's processing of ~70-nt pre-miRNA to mature miRNA, it was unclear how pri-miRNA transcripts, of hundreds or thousands of bases in length, were converted to pre-miRNA. Subcellular fractionation studies indicated that pri-miRNA and pre-miRNA processing occurred separately in the nucleus and cytoplasm, respectively (82). Through a series of mutagenesis studies, Lee et al. (86) deduced that the dsRNA feature of a pri-miRNA is required for its processing into pre-miRNA. Similar to the rationale employed in the discovery of Dicer (26), the specificity of RNaseIII for dsRNA processing proved useful here. The nuclear localization of the RNaseIII Drosha (87) suggested an excellent candidate for the pri-miRNA-processing enzyme. Indeed, pri-miRNA-processing activity immunoprecipitated with epitope-tagged Drosha, but not Dicer (86). Furthermore, siRNA-mediated knockdown of Drosha resulted in accumulation of pri-miRNA and reduced levels of pre-miRNA and miRNA (86). These findings indicate that miRNA biogenesis requires two spatially regulated catalytic steps. One exception is a small number of "mirtrons," whose pre-miRNAs are embedded in the introns of other transcripts, which

piRNA: piwi-interacting RNA

Immunoprecipitation: use of antibody-coated solid phase material to concentrate a particular protein out of solution are converted to pre-miRNA by the splicing machinery and processed by Dicer into miRNA (88–90).

dsRBP: double-stranded RNA-binding protein

To determine how pre-miRNA produced in the nucleus connect with the pre-miRNA-processing enzyme in the cytoplasm, a search for candidate transporters was conducted. The structural similarities between pre-miRNA and adenovirus VA1 RNA suggested that the VA1 transporter Exportin-5 (91) might play a role in pre-miRNA export (92). In human cells, siRNA-mediated knockdown of Exportin-5 decreased cytoplasmic pre-miRNA levels and reduced miRNA expression (93). In addition, the transport cofactor RanGTP was required for interaction between Exportin-5 and pre-miRNA in vitro (92). Export of radiolabeled pre-miRNA following nuclear injection of *Xenopus* oocyte was specifically enhanced by coinjection of recombinant Exportin-5 (94). Depletion of RanGTP or coinjection of Exportin-5 antibodies decreased pre-miRNA export and miRNA expression (94, 95). These results provide an important spatial link between the two distinct RNaseIII-mediated events required for miRNA biogenesis.

Dicer partners. Identification of Dicer and Drosha only partially represented the factors required for small RNA maturation. The discovery of RDE-4 and R2D2 (27, 96) suggested that Dicer and Drosha might function in tandem with dsRNA-binding proteins (dsRBPs). Bioinformatic analysis of the *Drosophila* genome revealed an R2D2-like CG6866, named R3D1/Loquacious (hereafter Loqs) for its three dsRNA-binding domains (R3) and association with Dicer-1 (D1) (31, 97, 98). Chromatographic fractionation of miRNA-generating activity showed excellent correlation with Dicer-1 and Loqs-PB, one of multiple Loqs isoforms (31). Unlike R2D2 that was required for siRISC assembly but not for siRNA production, Loqs is required for miRNA production (31, 97, 98) but not for miRISC assembly (99). Two roles for Loqs on miRNA production have been reported (31, 97, 98): (*a*) Loqs-PB facilitated Dicer-1-pre-miRNA interaction to enable efficient processing; and (*b*) expression of Loqs and Dicer-1 were interdependent, wherein loss of one attenuates expression of the other.

Computational analysis also indicated potential dsRBPs encoded in the human genome (97, 98). Human Dicer coimmunoprecipitated with TRBP and chromatographic fractionation of miRNA-generating activity showed close correlation with Dicer and TRBP (100–101a). Immunoprecipitation of epitope-tagged Dicer also revealed association with another dsRBP, protein activator of protein kinase R (PACT) (102). Knockdown of TRBP or PACT resulted in attenuated miRNA production and miRNA-mediated silencing (100–102). Similar to Loqs, TRBP was required for miRNA production through stabilization of Dicer expression and by facilitating Dicer-pre-miRNA interaction (100–102).

Drosha partner. A dsRBP partner (CG1800/Pasha) for *Drosophila* Drosha was identified by analyzing a yeast two-hybrid interaction map (103). Pasha coimmunoprecipitated with Drosha, and immunoprecipitates of both proteins exhibited pri-miRNA-processing activity in vitro (104). Moreover, chromatographic fractionation of pri-miRNA-processing activity exhibited good correlation with Drosha and Pasha (104). Immunoprecipitation of epitope-tagged Drosha from human cells revealed a putative dsRBP DGCR8 (105–107). The pri-miRNA-processing activity was observed for immunoprecipitates of Drosha and DGCR8 (105–107). Both recombinant Drosha and DGCR8 were required for in vitro reconstitution of pri-miRNA-processing activity (105, 108). Knockdown of Drosha or Pasha/DGCR8 resulted in accumulation of pri-miRNA and decreased miRNA expression (104–107). Thus, as a general rule, RNaseIII enzymes function in tandem with dsRBPs in small RNA pathways.

An interesting observation was made involving the autoregulation of Drosha and DGCR8. Following knockdown of Drosha in cultured cells, the levels of DGCR8/Pasha mRNA and protein were unexpectedly increased (109). Overexpression of Drosha reduced, whereas expression of a dominant-negative Drosha increased DGCR8 expression (109). Predicted stem-loop structures in the 3′-UTR of *dgcr8* mRNA were processed to ~70-nt products by Drosha, thereby negatively regulating DGCR8 expression (109). Thus, Drosha also governs mRNA expression through miRNA-independent mechanisms.

dsRBPs may also modulate the conformation of partner RNaseIII enzymes. Enhanced catalytic activity was observed for human Dicer in the absence of the N-terminal helicase domain (110). Because Dicer interacts with dsRBP partners through this region (36, 102), it has been proposed that TRBP contributes to Dicer function by relieving autoinhibition of Dicer (110). Another example of dsRBP function was provided through clinical studies that found two frameshift mutations in TRBP gene in cancer cell lines (110a). Both mutations introduced premature stop codons, resulting in reduced TRBP expression. Consistent with the role of TRBP in maintaining Dicer stability, these mutations resulted in reduced Dicer expression and lower miRNA production, and were associated with accelerated cell proliferation. These findings offer a possible genetic explanation for the reduced global miRNA expression in cancer cells (111, 112).

Endogenous siRNA

Endo-siRNAs have been shown to play important roles in regulating genome functions in diverse species. In *C. elegans*, exogenous dsRNAs are processed to rare primary siRNAs, which are amplified to more abundant secondary siRNAs (113). Specifically, primary siRNAs are loaded onto RDE-1 (homolog of Ago2) to cleave target mRNA, and RNA-dependent RNA polymerases (Rdrp) use cleaved mRNA as a template to prime synthesis of secondary siRNAs (114). Unlike primary siRNAs with 5′ monophosphate, secondary siRNAs have 5′ triphosphate (113) and are produced through a Dicer-independent mechanism (115). Secondary siRNAs are loaded onto multiple nonslicing Agos, which further contribute to target silencing (114). The production of secondary siRNAs corresponds to 5′ spreading of RNAi along the target mRNA and is linked to systemic and heritable silencing in worms (113, 114).

In *Schizosaccharomyces pombe*, a positive feedback loop of centromeric silencing is effected through the RNA-induced transcriptional silencing (RITS) complex (116). Centromeric repeats-derived dsRNAs are amplified by an Rdrp complex and processed by Dicer into siRNAs (116). Similar to RISC, these endo-siRNAs program RITS via Ago1, which recruits histone modifiers to establish and maintain heterochromatin and genomic silencing (116). In *Neurospora*, DNA damage induces the production of another class of QDE-2-interacting siRNAs, called qiRNAs (117). Biogenesis of qiRNAs requires Dicers, the DNA helicase QDE-3, and a DNA- and RNA-dependent RNA polymerase QDE-1 (117). Most qiRNAs are derived from ribosomal DNA repeats and are believed to contribute to the DNA damage response by inhibiting protein translation (117).

Recent studies have found diverse sources of endo-siRNAs in *Drosophila* and mammals (9). In general, dsRNA precursors are produced through (*a*) duplex formation of pseudogene and protein coding transcripts, (*b*) inverted repeat transcripts, (*c*) self-complementary mRNA, (*d*) retrotransposons, and (*e*) bidirectional sense and antisense transcripts (9). Although some endo-siRNAs are mapped to mRNA, others correspond to transposons and may contribute to transposon silencing in somatic cells (9). In *Drosophila*, though Dicer-1-Loqs-PB and Dicer-2-R2D2 complexes generate miRNAs and exo-siRNAs (2), respectively, the production of endo-siRNAs requires a noncanonical partnering of Dicer-2 and Loqs-PD (118, 119). These findings indicate functional cross talk among factors previously assigned to distinct small RNA pathways.

Rdrp: RNA-dependent RNA polymerase

Transposon: noncoding regions of DNA that exhibit chromosome mobility

piRNA

The recent history of small RNA pathways has yielded a powerful template by which to dissect other noncoding RNA pathways. Considering the possibility of small RNA independent of miRNA and siRNA, Aravin and colleagues (120) cloned 16–29-nt RNA from *Drosophila*. Among the RNAs identified were ~24-nt repeat-associated siRNAs (rasiRNAs), which correspond to transposable and repetitive elements within the genome. To identify the mammalian equivalent of rasiRNAs, multiple labs independently cloned a class of 26–30-nt RNAs that were highly abundant in mammalian testes (3). Like *Drosophila* rasiRNAs, these small RNAs coimmunoprecipitated with Piwi proteins and were designated as piRNAs. Both *Drosophila* and mammalian piRNAs contain 3′-terminal 2′-O-methyl modification that is catalyzed by a HEN1-like methyltransferase (3). Unlike *Drosophila* piRNA, most mammalian piRNAs are mapped uniquely in the genome and cluster to a small number of loci ranging from 10 to 83 kb (3). Genomic mapping of piRNAs does not indicate potential dsRNA precursors, and piRNA biogenesis appears to involve a Dicer-independent mechanism (3).

piRNA amplification. Sequencing of *Drosophila* piRNAs associated with different Piwi proteins revealed interesting characteristics (121, 122). Both Aubergine- and Piwi-associated piRNAs were mainly derived from the antisense strand of retrotransposons and showed a preference for uridine (U) at the 5′ end. In contrast, Ago3-associated piRNAs were primarily derived from the sense strand and exhibited a preference for adenosine (A) at the tenth nucleotide position. Furthermore, frequent complementarity was detected between the first ten bases of Aubergine- and Ago3-associated piRNAs. These findings, together with known slicer activity of piwi proteins (122), led to a "Ping-Pong" model for piRNA amplification: Sense piRNA guides Ago3 to cleave an antisense transcript to form the 5′ end of antisense piRNA bound to Aubergine, with each successive round of cleavage generating a new piRNA (121, 122). Although details of this model remain vague, this piRNA amplification loop appeared to be disrupted in *ago3* mutant flies (123). These studies indicated that different Piwi proteins conduct piRNA functions both cooperatively and independently of one another (123).

piRNA function. Studies in flies, fish, and mammals suggest that piRNAs play important roles in germ line development and maintenance of genomic integrity (3). *Drosophila* piRNAs are also derived from discrete genomic loci; several of these correspond to master regulatory regions of transposons (121). Mutations of the *flamenco* locus, a regulator of *gypsy*, *ZAM*, and *Idfix*, negated silencing of these retrotransposons and diminished production of piRNAs mapped to this region (121). In mammals, a significant portion of MILI-associated prepachytene piRNAs correspond to repeat elements and play a similar role in transposon silencing (3). piRNAs also direct silencing of specific genes. In *Drosophila*, expression of the *stellate* gene is suppressed by piRNAs derived from the *Suppressor of Stellate* locus, and loss of silencing results in overexpression of Stellate protein and male sterility (3). How piRNAs effect silencing remains a subject of debate. However, recent studies suggested that piRNAs may silence transcription by regulating DNA methylation (3).

REGULATION OF SMALL RNA PATHWAYS

Because of the importance of small RNA pathways in governing cellular activities, it has become increasingly important to understand how these pathways themselves are regulated. Spatiotemporal regulation of small RNA production and pathway components suggests tight coordination of gene silencing with wider biological processes. Defining the relationships between small RNA pathways and other cell regulatory systems represents an important and emerging area of investigation.

Endogenous Inhibitors of RNAi

The endogenous inhibitors of RNAi were first characterized in *C. elegans*. Noting that the worm nervous system was resistant to RNAi, several groups conducted genetic screens to isolate mutants exhibiting enhanced dsRNA-induced silencing in neurons. Disruption of *rrf-3*, encoding an Rdrp-like protein, enhanced RNAi phenotypes for neuronal and nonneuronal gene targets (124). Mutants of *eri-1*, encoding a protein carrying a 3′–5′ exonuclease domain, accumulated more siRNA following dsRNA feeding or injection. Recombinant ERI-1 specifically degraded siRNA, but not other RNA substrates in vitro (125). Trans-splicing has been shown to produce the helicase Eri-6/7, which inhibits exogenous, but is required for endogenous, RNAi (126). By contrast, little is known about the regulators and regulatory mechanisms of RNAi in *Drosophila* and humans.

Posttranscriptional Control of miRNA Biogenesis

Transcriptional control is a general mechanism for regulating miRNA expression. However, there were widespread reports of differential expression of miRNAs in the absence of apparent changes in pri- or pre-miRNA levels, suggesting that miRNA biogenesis is also regulated posttranscriptionally (127–130). For example, despite expression of pri-let-7, mature let-7 is not detected in mouse embryonic stem cells (131). Lin28 was identified as a candidate inhibitor of pri-let-7 processing through nucleic acid affinity purification (131, 132). Recombinant Lin28 inhibited in vitro pri-let-7 processing through recognition of the conserved loop region (132, 133). Overexpression of Lin28 resulted in accumulation of pri-let-7 and reduced expression of mature let-7 (131).

Lin28 has also been shown to govern pre-let-7 processing (134). Overexpression of Lin28 in human cells resulted in reduction of mature let-7 and accumulation of slower-migrating pre-let-7 species that were the result of 3′ polyuridylation (135). In vitro polyuridylation of pre-let-7 required recombinant Lin28 and the uridylyl transferase TUT4 (135, 135a). Collectively, these studies indicate that Lin28 negatively regulates let-7 biogenesis by steric inhibition of pri- or pre-let-7 processing and by recruiting uridyl transferases to mark pre-let-7 for degradation. The importance of this posttranscriptional regulatory mechanism is reflected through studies showing that knockdown of Lin28 unleashes tumor suppressor activity of let-7 and that Lin28 is a potential negative prognostic indicator for hepatocellular carcinoma (131, 136).

A number of other factors have also been identified as modulators of miRNA biogenesis. Photocross-linking and immunoprecipitation of the RNA-binding protein hnRNP A1 indicated association with the stem-loop region of pri-miR-18 (137). Unlike Lin28, which inhibits let-7 biogenesis, hnRNP A1 promoted in vitro processing of pri-miR-18 and was required for cellular expression of miR-18 (137). Following growth factor stimulation in human smooth muscle cells, Drosha processing of pre-miR-21 was enhanced by association of Smad and the RNA helicase p68 (138). The KH-type splicing regulatory protein (KSRP) coimmunoprecipitated with human Dicer and Drosha (139). KSRP binding to pri- and pre-miRNA loops facilitated maturation of a subset of miRNAs, including let-7 (139). Other factors, particularly those with known functions in RNA metabolism, are likely to emerge as regulators of small RNA biogenesis.

Modulating an miRISC-Target mRNA Relationship

Reasoning that 3′-UTR-binding factors may influence miRISC-target mRNA interactions, experiments were conducted that showed the AU-rich element-binding protein HuR reversed miR-122 suppression of cationic amino acid transporter-1 mRNA during nutrient stress (140). Noting that target transcripts often exhibit a conserved sequence surrounding miRNA target sites, Kedde et al. (141)

conducted immunoprecipitation experiments that indicated dead end 1 (Dnd1)–bound uridine-rich elements surrounding miRNA target sites, thereby precluding miRISC binding and inhibition of target transcripts. Similarly, Importin8 was shown to modulate interaction between miRISC and target mRNA (142). These findings suggest that miRNA-mediated silencing of target transcripts occurs in combination or in competition with other RNA-binding factors.

Cellular Regulation of Small RNA Pathways

siRNA, miRNA and piRNA have emerged as important regulators of biological processes. However, regulation of these pathways and their relationships with other cellular systems is only beginning to emerge. Human Ago2 coimmunoprecipitated with the prolyl-4-hydroxylase, C-P4H(I), and was hydroxylated at proline 700 (143). This modification was required for the stability of Ago2 proteins and siRNA-induced silencing. The Piwi family of proteins contain symmetric dimethyl arginine modifications that are catalyzed by the methyltransferase PRMT5 and are subsequently recognized by Tudor proteins (144–146). Both Piwi methylation and Tudor binding were required for proper stability, localization, and function of Piwi proteins in germ line development and transposon silencing (144–146).

The human miRNA-generating complex comprises Dicer and phospho-TRBP isoforms (101a). Mitogen activated protein kinase (MAPK)/Erk-mediated phosphorylation of TRBP stabilized the miRNA-generating complex and enhanced the capacity for miRNA production. Phosphorylation of TRBP resulted in a progrowth miRNA response, including upregulation of growth-promoting miRNAs and downregulation of the let-7 tumor suppressor miRNA, and was required for mitogenic signaling. MAPK/Erk also mediated oncogenic effects of Raf through another miRNA regulatory mechanism. By attenuating Myc-mediated transcription of Lin28, the Raf kinase inhibitory protein increased let-7 expression and reduced tumor cell invasion and metastasis (147). In addition, MAPK/Erk-mediated phosphorylation of serine-387 on human Ago2 has been shown to localize RISC to processing bodies (148). Thus, the MAPK/Erk pathway appears to target multiple RNAi factors to effect cell signaling. Collectively, these early studies indicate that small RNA pathways are governed by layers of regulatory mechanisms and are intimately connected with broader cellular systems.

CONCLUDING REMARKS

The early years following the birth of RNAi were marked by outstanding biochemical achievements providing a strong foundation for understanding small RNA function. The seminal work of Andrew Fire and Craig Mello (11) outlined dsRNA-induced target transcript silencing. Identification of these molecular events enabled development of a cell-free RNAi system using *Drosophila* embryo extract (13). Demonstration of dsRNA-induced mRNA cleavage activity in *Drosophila* S2 cells established scalable source material for biochemical purification of RISC (15). These milestone achievements provided the framework for subsequent studies elucidating biochemical mechanisms of regulatory RNA pathways.

In recent years, these advances have significantly slowed. Many of the current voids in understanding small RNA pathways stem from poorly defined molecular events. The controversy as to the mechanisms by which miRNAs exert target silencing is a clear example of this. Moreover, the lack of information on piRNA precursors and piRNA targets is a major limitation that precludes development of cell-free assays, which, in turn, severely limits understanding of the biochemical mechanisms of piRNA biogenesis and function. As genome analysis indicates expansive RNA networks far beyond current understanding, the biochemical principles of regulatory RNA functions outlined herein provide a framework for projected discoveries. Because of the importance of small RNAs in directing biological and pathological events, it has become increasingly important to understand the relationships between noncoding RNA pathways and wider cellular systems. A tantalizing prospect for small RNA function is the possibility of RNA-induced gene expression (149). Robust demonstration of de facto induction of gene expression would be a landmark achievement. Understanding the biochemical principles of known small RNA pathways will facilitate discovery of the many other regulatory RNA systems on the horizon.

SUMMARY POINTS

1. Demonstration of dsRNA-induced silencing represented a monumental shift away from the central dogma of molecular biology, revealing regulatory roles for RNA.
2. Identification of dsRNA-induced mRNA degradation indicated the molecular players of RNAi, enabling development of cell-free assays.
3. Demonstration of dsRNA-induced mRNA cleavage activity in *Drosophila* S2 cells provided scalable source material for biochemical purification of RISC.
4. These fundamental advances provided a conceptual framework for elucidating the biochemical mechanisms of an expanding number of small RNA pathways.

FUTURE ISSUES

1. What are the biochemical functions of nonslicing Agos?
2. How are small RNA effector complexes programmed?
3. What are the biochemical mechanisms by which miRNAs effect target silencing?
4. What defines a piRNA precursor, and how are primary piRNA generated?
5. What are the biochemcial mechanisms by which piRNAs effect target silencing?
6. What are the relationships between small RNA pathways and wider cellular systems?

DISCLOSURE STATEMENT

The authors are not aware of any affiliations, memberships, funding, or financial holdings that might be perceived as affecting the objectivity of this review.

ACKNOWLEDGMENTS

The perspective of classic biochemistry outlined in this article is a brief overview of teachings from Dr. Xiaodong Wang, whose exemplary studies of the biochemical mechanisms of apoptosis have inspired our endeavors. We apologize in advance to all the investigators whose outstanding research could not be appropriately cited owing to space limitations. Q.L. is supported by grants from the Welch Foundation and National Institutes of Health.

LITERATURE CITED

1. Tolia NH, Joshua-Tor L. 2007. Slicer and the argonautes. *Nat. Chem. Biol.* 3:36–43
2. Carthew RW, Sontheimer EJ. 2009. Origins and mechanisms of miRNAs and siRNAs. *Cell* 136:642–55

3. Siomi H, Siomi MC. 2009. On the road to reading the RNA-interference code. *Nature* 457:396–404
4. Ghildiyal M, Zamore PD. 2009. Small silencing RNAs: an expanding universe. *Nat. Rev. Genet.* 10:94–108
5. Mattick JS. 2009. The genetic signatures of noncoding RNAs. *PLoS Genet.* 5:e1000459
6. van Rij RP, Berezikov E. 2009. Small RNAs and the control of transposons and viruses in *Drosophila*. *Trends Microbiol.* 17:163–71
7. Malone CD, Hannon GJ. 2009. Small RNAs as guardians of the genome. *Cell* 136:656–68
8. Kim VN, Han J, Siomi MC. 2009. Biogenesis of small RNAs in animals. *Nat. Rev. Mol. Cell Biol.* 10:126–39
9. Okamura K, Lai EC. 2008. Endogenous small interfering RNAs in animals. *Nat. Rev. Mol. Cell Biol.* 9:673–78
10. Napoli C, Lemieux C, Jorgensen R. 1990. Introduction of a chimeric chalcone synthase gene into petunia results in reversible co-suppression of homologous genes in *trans*. *Plant Cell* 2:279–89
11. Fire A, Xu S, Montgomery MK, Kostas SA, Driver SE, Mello CC. 1998. Potent and specific genetic interference by double-stranded RNA in *Caenorhabditis elegans*. *Nature* 391:806–11
12. Hamilton AJ, Baulcombe DC. 1999. A species of small antisense RNA in posttranscriptional gene silencing in plants. *Science* 286:950–52
13. Tuschl T, Zamore PD, Lehmann R, Bartel DP, Sharp PA. 1999. Targeted mRNA degradation by double-stranded RNA in vitro. *Genes Dev.* 13:3191–97
14. Zamore PD, Tuschl T, Sharp PA, Bartel DP. 2000. RNAi: Double-stranded RNA directs the ATP-dependent cleavage of mRNA at 21 to 23 nucleotide intervals. *Cell* 101:25–33
15. Hammond SM, Bernstein E, Beach D, Hannon GJ. 2000. An RNA-directed nuclease mediates post-transcriptional gene silencing in *Drosophila* cells. *Nature* 404:293–96
16. Elbashir SM, Lendeckel W, Tuschl T. 2001. RNA interference is mediated by 21- and 22-nucleotide RNAs. *Genes Dev.* 15:188–200
17. Elbashir SM, Martinez J, Patkaniowska A, Lendeckel W, Tuschl T. 2001. Functional anatomy of siRNAs for mediating efficient RNAi in *Drosophila melanogaster* embryo lysate. *EMBO J.* 20:6877–88
18. Schwarz DS, Hutvagner G, Haley B, Zamore PD. 2002. Evidence that siRNAs function as guides, not primers, in the *Drosophila* and human RNAi pathways. *Mol. Cell* 10:537–48
19. Elbashir SM, Harborth J, Lendeckel W, Yalcin A, Weber K, Tuschl T. 2001. Duplexes of 21-nucleotide RNAs mediate RNA interference in cultured mammalian cells. *Nature* 411:494–98
20. Ambros V, Horvitz HR. 1987. The *lin-14* locus of *Caenorhabditis elegans* controls the time of expression of specific postembryonic developmental events. *Genes Dev.* 1:398–414
21. Lee RC, Feinbaum RL, Ambros V. 1993. The *C. elegans* heterochronic gene *lin-4* encodes small RNAs with antisense complementarity to *lin-14*. *Cell* 75:843–54
22. Wightman B, Ha I, Ruvkun G. 1993. Posttranscriptional regulation of the heterochronic gene *lin-14* by *lin-4* mediates temporal pattern formation in *C. elegans*. *Cell* 75:855–62
23. Reinhart BJ, Slack FJ, Basson M, Pasquinelli AE, Bettinger JC, et al. 2000. The 21-nucleotide *let-7* RNA regulates developmental timing in *Caenorhabditis elegans*. *Nature* 403:901–6
24. Olsen PH, Ambros V. 1999. The *lin-4* regulatory RNA controls developmental timing in *Caenorhabditis elegans* by blocking LIN-14 protein synthesis after the initiation of translation. *Dev. Biol.* 216:671–80
25. Bass BL. 2000. Double-stranded RNA as a template for gene silencing. *Cell* 101:235–38
26. Bernstein E, Caudy AA, Hammond SM, Hannon GJ. 2001. Role for a bidentate ribonuclease in the initiation step of RNA interference. *Nature* 409:363–66
27. Liu Q, Rand TA, Kalidas S, Du F, Kim HE, et al. 2003. R2D2, a bridge between the initiation and effector steps of the *Drosophila* RNAi pathway. *Science* 301:1921–25
28. Pasquinelli AE, Reinhart BJ, Slack F, Martindale MQ, Kuroda MI, et al. 2000. Conservation of the sequence and temporal expression of *let-7* heterochronic regulatory RNA. *Nature* 408:86–89
29. Hutvagner G, McLachlan J, Pasquinelli AE, Balint E, Tuschl T, Zamore PD. 2001. A cellular function for the RNA-interference enzyme Dicer in the maturation of the *let-7* small temporal RNA. *Science* 293:834–38
30. Lee YS, Nakahara K, Pham JW, Kim K, He Z, et al. 2004. Distinct roles for *Drosophila* Dicer-1 and Dicer-2 in the siRNA/miRNA silencing pathways. *Cell* 117:69–81

31. Jiang F, Ye X, Liu X, Fincher L, McKearin D, Liu Q. 2005. Dicer-1 and R3D1-L catalyze microRNA maturation in *Drosophila*. *Genes Dev.* 19:1674–79
32. Billy E, Brondani V, Zhang H, Muller U, Filipowicz W. 2001. Specific interference with gene expression induced by long, double-stranded RNA in mouse embryonal teratocarcinoma cell lines. *Proc. Natl. Acad. Sci. USA* 98:14428–33
33. Provost P, Dishart D, Doucet J, Frendewey D, Samuelsson B, Radmark O. 2002. Ribonuclease activity and RNA binding of recombinant human Dicer. *EMBO J.* 21:5864–74
34. Zhang H, Kolb FA, Jaskiewicz L, Westhof E, Filipowicz W. 2004. Single processing center models for human Dicer and bacterial RNase III. *Cell* 118:57–68
35. Blaszczyk J, Tropea JE, Bubunenko M, Routzahn KM, Waugh DS, et al. 2001. Crystallographic and modeling studies of RNase III suggest a mechanism for double-stranded RNA cleavage. *Structure* 9:1225–36
36. Ye X, Paroo Z, Liu Q. 2007. Functional anatomy of the *Drosophila* microRNA-generating enzyme. *J. Biol. Chem.* 282:28373–78
37. Macrae IJ, Zhou K, Li F, Repic A, Brooks AN, et al. 2006. Structural basis for double-stranded RNA processing by Dicer. *Science* 311:195–98
38. Hammond SM, Boettcher S, Caudy AA, Kobayashi R, Hannon GJ. 2001. Argonaute2, a link between genetic and biochemical analyses of RNAi. *Science* 293:1146–50
39. Rand TA, Ginalski K, Grishin NV, Wang X. 2004. Biochemical identification of Argonaute 2 as the sole protein required for RNA-induced silencing complex activity. *Proc. Natl. Acad. Sci. USA* 101:14385–89
40. Martinez J, Patkaniowska A, Urlaub H, Luhrmann R, Tuschl T. 2002. Single-stranded antisense siRNAs guide target RNA cleavage in RNAi. *Cell* 110:563–74
41. Tabara H, Sarkissian M, Kelly WG, Fleenor J, Grishok A, et al. 1999. The *rde-1* gene, RNA interference, and transposon silencing in *C. elegans*. *Cell* 99:123–32
42. Cogoni C, Macino G. 1997. Isolation of quelling-defective (*qde*) mutants impaired in posttranscriptional transgene-induced gene silencing in *Neurospora crassa*. *Proc. Natl. Acad. Sci. USA* 94:10233–38
43. Fagard M, Boutet S, Morel JB, Bellini C, Vaucheret H. 2000. AGO1, QDE-2, and RDE-1 are related proteins required for post-transcriptional gene silencing in plants, quelling in fungi, and RNA interference in animals. *Proc. Natl. Acad. Sci. USA* 97:11650–54
44. Schwarz DS, Tomari Y, Zamore PD. 2004. The RNA-induced silencing complex is a Mg^{2+}-dependent endonuclease. *Curr. Biol.* 14:787–91
45. Martinez J, Tuschl T. 2004. RISC is a 5′ phosphomonoester-producing RNA endonuclease. *Genes Dev.* 18:975–80
46. Liu J, Carmell MA, Rivas FV, Marsden CG, Thomson JM, et al. 2004. Argonaute2 is the catalytic engine of mammalian RNAi. *Science* 305:1437–41
47. Meister G, Landthaler M, Patkaniowska A, Dorsett Y, Teng G, Tuschl T. 2004. Human Argonaute2 mediates RNA cleavage targeted by miRNAs and siRNAs. *Mol. Cell* 15:185–97
48. Song JJ, Smith SK, Hannon GJ, Joshua-Tor L. 2004. Crystal structure of Argonaute and its implications for RISC slicer activity. *Science* 305:1434–37
49. Rivas FV, Tolia NH, Song JJ, Aragon JP, Liu J, et al. 2005. Purified Argonaute2 and an siRNA form recombinant human RISC. *Nat. Struct. Mol. Biol.* 12:340–49
50. Miyoshi K, Tsukumo H, Nagami T, Siomi H, Siomi MC. 2005. Slicer function of *Drosophila* Argonautes and its involvement in RISC formation. *Genes Dev.* 19:2837–48
51. Okamura K, Ishizuka A, Siomi H, Siomi MC. 2004. Distinct roles for Argonaute proteins in small RNA-directed RNA cleavage pathways. *Genes Dev.* 18:1655–66
52. Wang Y, Juranek S, Li H, Sheng G, Tuschl T, Patel DJ. 2008. Structure of an argonaute silencing complex with a seed-containing guide DNA and target RNA duplex. *Nature* 456:921–26
53. Wang Y, Sheng G, Juranek S, Tuschl T, Patel DJ. 2008. Structure of the guide-strand-containing argonaute silencing complex. *Nature* 456:209–13
53a. Fabian MR, Sonenberg N, Filipowicz W. 2010. Regulation of mRNA translation and stability by microRNAs. *Annu. Rev. Biochem.* 79:351–79
54. Bagga S, Bracht J, Hunter S, Massirer K, Holtz J, et al. 2005. Regulation by *let*-7 and *lin-4* miRNAs results in target mRNA degradation. *Cell* 122:553–63

55. Schmitter D, Filkowski J, Sewer A, Pillai RS, Oakeley EJ, et al. 2006. Effects of Dicer and Argonaute down-regulation on mRNA levels in human HEK293 cells. *Nucleic Acids Res.* 34:4801–15
56. Yekta S, Shih IH, Bartel DP. 2004. MicroRNA-directed cleavage of *HOXB8* mRNA. *Science* 304:594–96
57. Selbach M, Schwanhausser B, Thierfelder N, Fang Z, Khanin R, Rajewsky N. 2008. Widespread changes in protein synthesis induced by microRNAs. *Nature* 455:58–63
58. Paroo Z, Pertsemlidis A. 2009. microRNAs mature with help from cancer biology. *Genome Biol.* 10:310
59. Thermann R, Hentze MW. 2007. *Drosophila* miR2 induces pseudopolysomes and inhibits translation initiation. *Nature* 447:875–78
60. Wakiyama M, Takimoto K, Ohara O, Yokoyama S. 2007. *Let*-7 microRNA-mediated mRNA deadenylation and translational repression in a mammalian cell-free system. *Genes Dev.* 21:1857–62
61. Liu X, Jiang F, Kalidas S, Smith D, Liu Q. 2006. Dicer-2 and R2D2 coordinately bind siRNA to promote assembly of the siRISC complexes. *RNA* 12:1514–20
62. Pham JW, Pellino JL, Lee YS, Carthew RW, Sontheimer EJ. 2004. A Dicer-2-dependent 80S complex cleaves targeted mRNAs during RNAi in *Drosophila*. *Cell* 117:83–94
63. Tomari Y, Du T, Haley B, Schwarz DS, Bennett R, et al. 2004. RISC assembly defects in the *Drosophila* RNAi mutant armitage. *Cell* 116:831–41
64. Schwarz DS, Hutvagner G, Du T, Xu Z, Aronin N, Zamore PD. 2003. Asymmetry in the assembly of the RNAi enzyme complex. *Cell* 115:199–208
65. Khvorova A, Reynolds A, Jayasena SD. 2003. Functional siRNAs and miRNAs exhibit strand bias. *Cell* 115:209–16
66. Tomari Y, Matranga C, Haley B, Martinez N, Zamore PD. 2004. A protein sensor for siRNA asymmetry. *Science* 306:1377–80
67. Nykanen A, Haley B, Zamore PD. 2001. ATP requirements and small interfering RNA structure in the RNA interference pathway. *Cell* 107:309–21
68. Meister G, Landthaler M, Peters L, Chen PY, Urlaub H, et al. 2005. Identification of novel argonaute-associated proteins. *Curr. Biol.* 15:2149–55
69. Robb GB, Rana TM. 2007. RNA helicase A interacts with RISC in human cells and functions in RISC loading. *Mol. Cell* 26:523–37
70. Rand TA, Petersen S, Du F, Wang X. 2005. Argonaute2 cleaves the antiguide strand of siRNA during RISC activation. *Cell* 123:621–29
71. Liu Y, Ye X, Jiang F, Liang C, Chen D, et al. 2009. C3PO, an endoribonuclease that promotes RNAi by facilitating RISC activation. *Science* 325:750–53
72. Kim K, Lee YS, Carthew RW. 2007. Conversion of pre-RISC to holo-RISC by Ago2 during assembly of RNAi complexes. *RNA* 13:22–29
73. Maiti M, Lee HC, Liu Y. 2007. QIP, a putative exonuclease, interacts with the *Neurospora* Argonaute protein and facilitates conversion of duplex siRNA into single strands. *Genes Dev.* 21:590–600
74. Paroo Z, Liu Q, Wang X. 2007. Biochemical mechanisms of the RNA-induced silencing complex. *Cell Res.* 17:187–94
75. Gregory RI, Chendrimada TP, Cooch N, Shiekhattar R. 2005. Human RISC couples microRNA biogenesis and posttranscriptional gene silencing. *Cell* 123:631–40
76. Maniataki E, Mourelatos Z. 2005. A human, ATP-independent, RISC assembly machine fueled by pre-miRNA. *Genes Dev.* 19:2979–90
77. MacRae IJ, Ma E, Zhou M, Robinson CV, Doudna JA. 2008. In vitro reconstitution of the human RISC-loading complex. *Proc. Natl. Acad. Sci. USA* 105:512–17
77a. Tan GS, Garchow BG, Liu X, Yeung J, Morris JP 4th, et al. 2009. Expanded RNA-binding activities of mammalian Argonaute 2. *Nucleic Acids Res.* 7:7533–45
78. Lander ES, Linton LM, Birren B, Nusbaum C, Zody MC, et al. 2001. Initial sequencing and analysis of the human genome. *Nature* 409:860–921
79. Venter JC, Adams MD, Myers EW, Li PW, Mural RJ, et al. 2001. The sequence of the human genome. *Science* 291:1304–51
80. Lewis BP, Shih IH, Jones-Rhoades MW, Bartel DP, Burge CB. 2003. Prediction of mammalian microRNA targets. *Cell* 115:787–98

81. Lai EC. 2005. miRNAs: whys and wherefores of miRNA-mediated regulation. *Curr. Biol.* 15:R458–60
82. Lee Y, Jeon K, Lee JT, Kim S, Kim VN. 2002. MicroRNA maturation: stepwise processing and subcellular localization. *EMBO J.* 21:4663–70
83. Dieci G, Fiorino G, Castelnuovo M, Teichmann M, Pagano A. 2007. The expanding RNA polymerase III transcriptome. *Trends Genet.* 23:614–22
84. Cai X, Hagedorn CH, Cullen BR. 2004. Human microRNAs are processed from capped, polyadenylated transcripts that can also function as mRNAs. *RNA* 10:1957–66
85. Lee Y, Kim M, Han J, Yeom KH, Lee S, et al. 2004. MicroRNA genes are transcribed by RNA polymerase II. *EMBO J.* 23:4051–60
86. Lee Y, Ahn C, Han J, Choi H, Kim J, et al. 2003. The nuclear RNase III Drosha initiates microRNA processing. *Nature* 425:415–19
87. Wu H, Xu H, Miraglia LJ, Crooke ST. 2000. Human RNase III is a 160-kDa protein involved in preribosomal RNA processing. *J. Biol. Chem.* 275:36957–65
88. Okamura K, Hagen JW, Duan H, Tyler DM, Lai EC. 2007. The mirtron pathway generates microRNA-class regulatory RNAs in *Drosophila*. *Cell* 130:89–100
89. Ruby JG, Jan CH, Bartel DP. 2007. Intronic microRNA precursors that bypass Drosha processing. *Nature* 448:83–86
90. Berezikov E, Chung WJ, Willis J, Cuppen E, Lai EC. 2007. Mammalian mirtron genes. *Mol. Cell* 28:328–36
91. Gwizdek C, Ossareh-Nazari B, Brownawell AM, Doglio A, Bertrand E, et al. 2003. Exportin-5 mediates nuclear export of minihelix-containing RNAs. *J. Biol. Chem.* 278:5505–8
92. Yi R, Qin Y, Macara IG, Cullen BR. 2003. Exportin-5 mediates the nuclear export of premicroRNAs and short hairpin RNAs. *Genes Dev.* 17:3011–16
93. Yi R, Doehle BP, Qin Y, Macara IG, Cullen BR. 2005. Overexpression of exportin 5 enhances RNA interference mediated by short hairpin RNAs and microRNAs. *RNA* 11:220–26
94. Lund E, Guttinger S, Calado A, Dahlberg JE, Kutay U. 2004. Nuclear export of microRNA precursors. *Science* 303:95–98
95. Bohnsack MT, Czaplinski K, Gorlich D. 2004. Exportin 5 is a RanGTP-dependent dsRNA-binding protein that mediates nuclear export of pre-miRNAs. *RNA* 10:185–91
96. Tabara H, Yigit E, Siomi H, Mello CC. 2002. The dsRNA binding protein RDE-4 interacts with RDE-1, DCR-1, and a DExH-box helicase to direct RNAi in *C. elegans*. *Cell* 109:861–71
97. Förstemann K, Tomari Y, Du T, Vagin VV, Denli AM, et al. 2005. Normal microRNA maturation and germ-line stem cell maintenance requires Loquacious, a double-stranded RNA-binding domain protein. *PLoS Biol.* 3:e236
98. Saito K, Ishizuka A, Siomi H, Siomi MC. 2005. Processing of premicroRNAs by the Dicer-1-Loquacious complex in *Drosophila* cells. *PLoS Biol.* 3:e235
99. Liu X, Park JK, Jiang F, Liu Y, McKearin D, Liu Q. 2007. Dicer-1, but not Loquacious, is critical for assembly of miRNA-induced silencing complexes. *RNA* 13:2324–29
100. Chendrimada TP, Gregory RI, Kumaraswamy E, Norman J, Cooch N, et al. 2005. TRBP recruits the Dicer complex to Ago2 for microRNA processing and gene silencing. *Nature* 436:740–44
101. Haase AD, Jaskiewicz L, Zhang H, Laine S, Sack R, et al. 2005. TRBP, a regulator of cellular PKR and HIV-1 virus expression, interacts with Dicer and functions in RNA silencing. *EMBO Rep.* 6:961–67

101a. Paroo Z, Ye X, Chen S, Liu Q. 2009. Phosphorylation of the human microRNA-generating complex mediates MAPK/Erk signaling. *Cell* 139:112–22

102. Lee Y, Hur I, Park SY, Kim YK, Suh MR, Kim VN. 2006. The role of PACT in the RNA silencing pathway. *EMBO J.* 25:522–32
103. Giot L, Bader JS, Brouwer C, Chaudhuri A, Kuang B, et al. 2003. A protein interaction map of *Drosophila melanogaster*. *Science* 302:1727–36
104. Denli AM, Tops BB, Plasterk RH, Ketting RF, Hannon GJ. 2004. Processing of primary microRNAs by the Microprocessor complex. *Nature* 432:231–35
105. Han J, Lee Y, Yeom KH, Kim YK, Jin H, Kim VN. 2004. The Drosha-DGCR8 complex in primary microRNA processing. *Genes Dev.* 18:3016–27

106. Gregory RI, Yan KP, Amuthan G, Chendrimada T, Doratotaj B, et al. 2004. The Microprocessor complex mediates the genesis of microRNAs. *Nature* 432:235–40
107. Landthaler M, Yalcin A, Tuschl T. 2004. The human DiGeorge syndrome critical region gene 8 and its *D. melanogaster* homolog are required for miRNA biogenesis. *Curr. Biol.* 14:2162–7
108. Han J, Lee Y, Yeom KH, Nam JW, Heo I, et al. 2006. Molecular basis for the recognition of primary microRNAs by the Drosha-DGCR8 complex. *Cell* 125:887–901
109. Han J, Pedersen JS, Kwon SC, Belair CD, Kim YK, et al. 2009. Posttranscriptional crossregulation between Drosha and DGCR8. *Cell* 136:75–84
110. Ma E, MacRae IJ, Kirsch JF, Doudna JA. 2008. Autoinhibition of human Dicer by its internal helicase domain. *J. Mol. Biol.* 380:237–43
110a. Melo SA, Ropero S, Moutinho C, Aaltonen LA, Yamamoto H, et al. 2009. A TARBP2 mutation in human cancer impairs microRNA processing and DICER1 function. *Nat Genet.* 41:365–70
111. Lu J, Getz G, Miska EA, Alvarez-Saavedra E, Lamb J, et al. 2005. MicroRNA expression profiles classify human cancers. *Nature* 435:834–38
112. Kumar MS, Lu J, Mercer KL, Golub TR, Jacks T. 2007. Impaired microRNA processing enhances cellular transformation and tumorigenesis. *Nat. Genet.* 39:673–77
113. Pak J, Fire A. 2007. Distinct populations of primary and secondary effectors during RNAi in *C. elegans*. *Science* 315:241–44
114. Yigit E, Batista PJ, Bei Y, Pang KM, Chen CC, et al. 2006. Analysis of the *C. elegans* Argonaute family reveals that distinct Argonautes act sequentially during RNAi. *Cell* 127:747–57
115. Aoki K, Moriguchi H, Yoshioka T, Okawa K, Tabara H. 2007. In vitro analyses of the production and activity of secondary small interfering RNAs in *C. elegans*. *EMBO J.* 26:5007–19
116. Moazed D. 2009. Small RNAs in transcriptional gene silencing and genome defense. *Nature* 457:413–20
117. Lee HC, Chang SS, Choudhary S, Aalto AP, Maiti M, et al. 2009. qiRNA is a new type of small interfering RNA induced by DNA damage. *Nature* 459:274–77
118. Zhou R, Czech B, Brennecke J, Sachidanandam R, Wohlschlegel JA, et al. 2009. Processing of *Drosophila* endo-siRNAs depends on a specific Loquacious isoform. *RNA* 15:1886–95
119. Hartig JV, Esslinger S, Böttcher R, Saito K, Förstemann K. 2009. Endo-siRNAs depend on a new isoform of *loquacious* and target artificially introduced, high-copy sequences. *EMBO J.* 28:2932–44
120. Aravin AA, Lagos-Quintana M, Yalcin A, Zavolan M, Marks D, et al. 2003. The small RNA profile during *Drosophila melanogaster* development. *Dev. Cell* 5:337–50
121. Brennecke J, Aravin AA, Stark A, Dus M, Kellis M, et al. 2007. Discrete small RNA-generating loci as master regulators of transposon activity in *Drosophila*. *Cell* 128:1089–103
122. Gunawardane LS, Saito K, Nishida KM, Miyoshi K, Kawamura Y, et al. 2007. A slicer-mediated mechanism for repeat-associated siRNA 5′ end formation in *Drosophila*. *Science* 315:1587–90
123. Li C, Vagin VV, Lee S, Xu J, Ma S, et al. 2009. Collapse of germline piRNAs in the absence of Argonaute3 reveals somatic piRNAs in flies. *Cell* 137:509–21
124. Simmer F, Tijsterman M, Parrish S, Koushika SP, Nonet ML, et al. 2002. Loss of the putative RNA-directed RNA polymerase RRF-3 makes *C. elegans* hypersensitive to RNAi. *Curr. Biol.* 12:1317–19
125. Kennedy S, Wang D, Ruvkun G. 2004. A conserved siRNA-degrading RNase negatively regulates RNA interference in *C. elegans*. *Nature* 427:645–49
126. Fischer SE, Butler MD, Pan Q, Ruvkun G. 2008. Trans-splicing in *C. elegans* generates the negative RNAi regulator ERI-6/7. *Nature* 455:491–96
127. Hwang HW, Wentzel EA, Mendell JT. 2009. Cell-cell contact globally activates microRNA biogenesis. *Proc. Natl. Acad. Sci. USA* 106:7016–21
128. Thomson JM, Newman M, Parker JS, Morin-Kensicki EM, Wright T, Hammond SM. 2006. Extensive post-transcriptional regulation of microRNAs and its implications for cancer. *Genes Dev.* 20:2202–7
129. Obernosterer G, Leuschner PJ, Alenius M, Martinez J. 2006. Post-transcriptional regulation of microRNA expression. *RNA* 12:1161–67
130. Lee EJ, Baek M, Gusev Y, Brackett DJ, Nuovo GJ, Schmittgen TD. 2008. Systematic evaluation of microRNA processing patterns in tissues, cell lines, and tumors. *RNA* 14:35–42
131. Viswanathan SR, Daley GQ, Gregory RI. 2008. Selective blockade of microRNA processing by Lin28. *Science* 320:97–100

132. Newman MA, Thomson JM, Hammond SM. 2008. Lin-28 interaction with the Let-7 precursor loop mediates regulated microRNA processing. *RNA* 14:1539–49
133. Piskounova E, Viswanathan SR, Janas M, LaPierre RJ, Daley GQ, et al. 2008. Determinants of microRNA processing inhibition by the developmentally regulated RNA-binding protein Lin28. *J. Biol. Chem.* 283:21310–14
134. Rybak A, Fuchs H, Smirnova L, Brandt C, Pohl EE, et al. 2008. A feedback loop comprising *lin-28* and *let-7* controls pre-*let-7* maturation during neural stem-cell commitment. *Nat. Cell Biol.* 10:987–93
135. Heo I, Joo C, Cho J, Ha M, Han J, Kim VN. 2008. Lin28 mediates the terminal uridylation of let-7 precursor microRNA. *Mol. Cell* 32:276–84
135a. Heo I, Joo C, Kim YK, Ha M, Yoon MJ, et al. 2009. TUT4 in concert with Lin28 suppresses microRNA biogenesis through pre-microRNA uridylation. *Cell* 138:696–708
136. Chang TC, Zeitels LR, Hwang HW, Chivukula RR, Wentzel EA, et al. 2009. Lin-28B transactivation is necessary for Myc-mediated let-7 repression and proliferation. *Proc. Natl. Acad. Sci. USA* 106:3384–89
137. Guil S, Cáceres JF. 2007. The multifunctional RNA-binding protein hnRNP A1 is required for processing of miR-18a. *Nat. Struct. Mol. Biol.* 14:591–96
138. Davis BN, Hilyard AC, Lagna G, Hata A. 2008. SMAD proteins control DROSHA-mediated microRNA maturation. *Nature* 454:56–61
139. Trabucchi M, Briata P, Garcia-Mayoral M, Haase AD, Filipowicz W, et al. 2009. The RNA-binding protein KSRP promotes the biogenesis of a subset of microRNAs. *Nature* 459:1010–14
140. Bhattacharyya SN, Habermacher R, Martine U, Closs EI, Filipowicz W. 2006. Stress-induced reversal of microRNA repression and mRNA P-body localization in human cells. *Cold Spring Harb. Symp. Quant. Biol.* 71:513–21
141. Kedde M, Strasser MJ, Boldajipour B, Oude Vrielink JA, Slanchev K, et al. 2007. RNA-binding protein Dnd1 inhibits microRNA access to target mRNA. *Cell* 131:1273–86
142. Weinmann L, Hock J, Ivacevic T, Ohrt T, Mutze J, et al. 2009. Importin 8 is a gene silencing factor that targets Argonaute proteins to distinct mRNAs. *Cell* 136:496–507
143. Qi HH, Ongusaha PP, Myllyharju J, Cheng D, Pakkanen O, et al. 2008. Prolyl 4-hydroxylation regulates Argonaute 2 stability. *Nature* 455:421–24
144. Kirino Y, Kim N, de Planell-Saguer M, Khandros E, Chiorean S, et al. 2009. Arginine methylation of Piwi proteins catalysed by dPRMT5 is required for Ago3 and Aub stability. *Nat. Cell Biol.* 11:652–58
145. Vagin VV, Wohlschlegel J, Qu J, Jonsson Z, Huang X, et al. 2009. Proteomic analysis of murine Piwi proteins reveals a role for arginine methylation in specifying interaction with Tudor family members. *Genes Dev.* 23:1749–62
146. Reuter M, Chuma S, Tanaka T, Franz T, Stark A, Pillai RS. 2009. Loss of the Mili-interacting Tudor domain–containing protein-1 activates transposons and alters the Mili-associated small RNA profile. *Nat. Struct. Mol. Biol.* 16:639–46
147. Dangi-Garimella S, Yun J, Eves EM, Newman M, Erkeland SJ, et al. 2009. Raf kinase inhibitory protein suppresses a metastasis signaling cascade involving LIN28 and *let-7*. *EMBO J.* 28:347–58
148. Zeng Y, Sankala H, Zhang X, Graves PR. 2008. Phosphorylation of Argonaute 2 at serine-387 facilitates its localization to processing bodies. *Biochem. J.* 413:429–36
149. Vasudevan S, Tong Y, Steitz JA. 2007. Switching from repression to activation: MicroRNAs can up-regulate translation. *Science* 318:1931–34

Functions and Regulation of RNA Editing by ADAR Deaminases

Kazuko Nishikura

Department of Gene Expression and Regulation, The Wistar Institute, Philadelphia, Pennsylvania 19104-4268; email: kazuko@wistar.org

Annu. Rev. Biochem. 2010. 79:321–49

First published online as a Review in Advance on December 8, 2009

The *Annual Review of Biochemistry* is online at biochem.annualreviews.org

This article's doi:
10.1146/annurev-biochem-060208-105251

0066-4154/10/0707-0321$20.00

Key Words

double-stranded RNA, A→I RNA editing, noncoding RNA, RNA interference, microRNA, esiRNA

Abstract

One type of RNA editing converts adenosines to inosines (A→I editing) in double-stranded RNA (dsRNA) substrates. A→I RNA editing is mediated by adenosine deaminase acting on RNA (ADAR) enzymes. A→I RNA editing of protein-coding sequences of a limited number of mammalian genes results in recoding and subsequent alterations of their functions. However, A→I RNA editing most frequently targets repetitive RNA sequences located within introns and 5′ and 3′ untranslated regions (UTRs). Although the biological significance of noncoding RNA editing remains largely unknown, several possibilities, including its role in the control of endogenous short interfering RNAs (esiRNAs), have been proposed. Furthermore, recent studies have revealed that the biogenesis and functions of certain microRNAs (miRNAs) are regulated by the editing of their precursors. Here, I review the recent findings that indicate new functions for A→I editing in the regulation of noncoding RNAs and for interactions between RNA editing and RNA interference mechanisms.

Contents

INTRODUCTION

With the number of genes much fewer than previously expected, the complexity of higher organisms largely depends on posttranscriptional and posttranslational mechanisms that create different gene products and the diversity required for complex structural, enzymatic, and regulatory functions (1). A primary RNA transcript of the gene undergoes various maturation processes, such as 5′ capping, splicing, 3′ processing, and polyadenylation (2). RNA editing is one of the posttranscriptional mechanisms that introduce changes in RNA sequences encoded by genome sequences.

The phenomenon of RNA editing was first discovered more than 20 years ago in kinetoplastid protozoa (3). In the mitochondrial mRNA of these trypanosomes, many uridine nucleotides were found to be inserted or deleted to generate functional proteins (3). Since then, many other types of RNA editing mechanisms have been identified (4). In the animal kingdom, the most prevalent type of RNA editing that alters one nucleotide to another is mediated by adenosine deaminase acting on RNA (ADAR) enzymes; ADAR converts adenosines to inosines (A→I editing) in double-stranded RNA (dsRNA) substrates. A→I RNA editing can lead to a codon change and consequent alterations of protein-coding sequences of selected genes, resulting in a diversification of their protein functions. However, the vast majority of A→I RNA editing sites are in noncoding sequences. 5′ and 3′ untranslated regions (UTRs) and intronic retrotransposon elements, such as Alu and long interspersed elements (LINEs), are frequently targeted. Although the biological significance of these repetitive RNAs remains largely unknown, the interesting possibility that they are involved in the control of endogenous short interfering RNAs (esiRNAs) has emerged.

Precursors of certain microRNAs (miRNAs) also undergo A→I RNA editing. Here, editing regulates processing of precursor miRNAs into mature miRNAs or leads to selection of new target genes for silencing by edited miRNAs. These recent studies reveal new functions for A→I editing in regulation of noncoding RNAs and modulation of RNA interference (RNAi) pathways (5).

In this review, I briefly introduce the A→I RNA editing system, ADARs, editing-site selectivity, and representative A→I RNA editing targets of some protein-coding genes and diversification of their functions. However, I put much emphasis on A→I RNA editing of noncoding RNAs and also on the interaction between RNA editing and RNAi pathways. I do not attempt to cover all aspects of A→I RNA editing. More comprehensive reviews are available (6–10).

A→I RNA EDITING AND ADAR GENES

A→I RNA editing is mediated by ADARs. In the following sections, the A→I deamination mechanism, different ADAR genes, domain structures of ADARs, editing-site selectivity, regulation of ADAR gene expression, and their localization are described.

Adenosine deaminase acting on RNA (ADAR): ADAR catalyzes an RNA editing reaction whereby an adenosine is converted to an inosine

Retrotransposons: genetic elements containing retroviruses and transposons are replicated through an intermediate RNA stage

Hydrolytic Deamination of Adenosine to Inosine by ADARs

During A→I editing, adenosine is converted to inosine by hydrolytic deamination of the adenine base (**Figure 1*a***) (11, 12). The A→I deamination reaction is catalyzed by ADARs

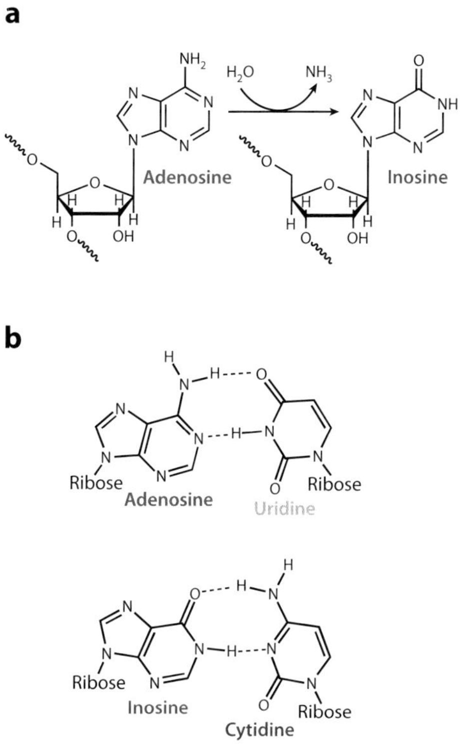

Figure 1

Deamination of adenosine to inosine by ADAR. (*a*) A hydrolytic deamination reaction converts adenosine to inosine. (*b*) Adenosine base pairs with uridine, whereas inosine base pairs, as if it were guanosine, in a Watson-Crick-bonding configuration with cytidine.

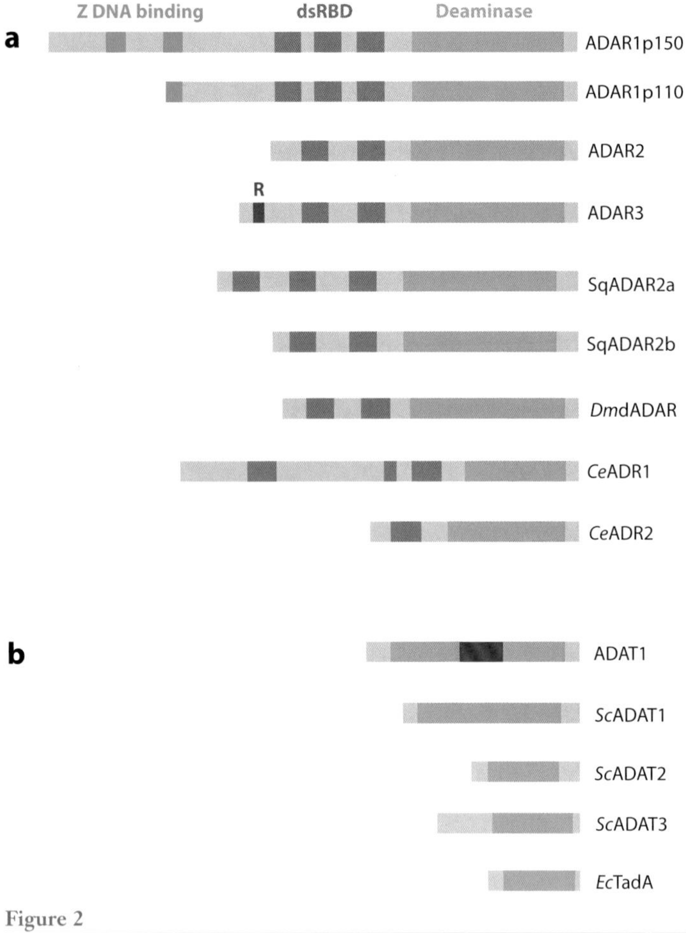

Figure 2

ADAR and ADAT family members. (*a*) The ADAR family is shown. Three vertebrate ADAR family members (ADAR1–3) are known. Two ADAR1 translation products (ADAR1p150 and ADAR1p110 isoforms) are known. Vertebrate ADARs, squid SqADAR2a and SqADAR2b (splicing isoforms), *Drosophila melanogaster* dADAR, and two *C. elegans* members (*Ce*ADR1 and *Ce*ADR2) share common functional domains: two to three repeats of the dsRNA-binding domain (dsRBD) and a catalytic deaminase domain. Certain structural features, such as Z-DNA-binding domains and the arginine-rich R domain, are unique to particular ADAR members. (*b*) The ADAT family is shown. Presented is the only known mammalian ADAT family member (ADAT1), its yeast homolog (*Sc*ADAT1), two additional yeast ADAT families (*Sc*ADAT2 and *Sc*ADAT3), and the only known bacterial ADAT family member (*Ec*TadA). The unique mammalian ADAT1 sequence is located within the deaminase domain (*gray bar*). ADARs target dsRNA. ADATs target tRNA, even though they lack any known RNA-binding motifs. Yeast *Sc*ADAT2 and *Sc*ADAT3 form active heterodimers, whereas ADAR1 and ADAR2 form active homodimers.

(**Figure 2*a***). ADARs were originally identified in *Xenopus laevis* eggs and embryos as a mysterious dsRNA-unwinding activity (13, 14). Soon after, however, it was revealed that this activity is a dsRNA-specific adenosine deaminase (11, 12). The first mammalian ADAR gene identified, human *ADAR1*, was cloned following biochemical purification (15, 16) and microsequencing of ADAR1 protein (17, 18), which led to identification of ADAR2 (19–21) and ADAR3 (**Figure 2*a***) (22, 23). The enzymatic activity of ADAR1 and ADAR2 has been demonstrated (17, 19–21). Although ADAR3 activity has not been demonstrated, its functional domain features are conserved (22, 23). These three ADARs are highly conserved in vertebrates (24, 25). A single ADAR2-like gene, *dADAR*, is present in *Drosophila melanogaster* (26), whereas two ADAR genes, *CeADR1* and *CeADR2*, exist in *Caenorhabditis elegans* (27). Two splicing isoforms of squid ADAR2, with high homology to human ADAR2, were also identified (**Figure 2*a***) (28). Interestingly, recent screening of invertebrate genome databases identified *ADAR1* and *ADAR2* but not *ADAR3* in sea urchin and sea anemones (29), indicating that ADAR1 and ADAR2 arose in early metazoan evolution, whereas *ADAR3* might have evolved more recently by possible duplication of *ADAR2* in vertebrates. Furthermore, *ADAR1* or *ADAR2* was lost in some species, such as insects and squid, during subsequent evolution. ADARs are absent in all protozoa, yeast, and plants (29).

In addition to the ADAR family, adenosine deaminases acting on tRNA (ADATs) have been identified because of their sequence homology to ADARs (**Figure 2*b***). ADATs are involved in A→I editing of tRNAs at or near the anticodon position. The ADAT members are conserved in eukaryotes from yeast to man (30). Moreover, a bacterial ortholog of the ADAT family, tRNA adenosine deaminase (TadA), also exists, indicating conservation of this A→I editing function between prokaryotes and eukaryotes (31). ADATs have been hypothesized to be the evolutionary ancestors of ADARs. Interestingly, it is not the adenosine deaminase acting on mononucleotides (ADAs) but the cytidine deaminases acting on mononucleotides (CDAs) that are the likely predecessors to ADATs and consequently ADAR (17, 30).

ADAR Domain Structure and the A→I Editing Mechanism

Common domain structures are found among members of the ADAR gene family (**Figure 2*a***). One to three repeats of the dsRNA-binding domain (dsRBD) (~65 amino acids), forming a highly conserved α-β-β-β-α configuration structure, are present among ADARs. The dsRBD makes direct contact with the dsRNA (32) and is required for dsRNA binding (33). Certain structural features are unique to particular ADAR members. For instance, ADAR1 contains two Z-DNA-binding domains, Zα and Zβ (34). The functional significance of Z domains remains largely unknown. ADAR3 contains an arginine-rich single-stranded RNA (ssRNA)-binding domain (R domain) in its N-terminal region (22). More recently, the presence of the R domain was detected also in a minor fraction of ADAR2 mRNAs, indicating the evolutionary conservation of this domain and thus its functional significance (35).

The C-terminal region of ADAR contains a catalytic domain consisting of amino acid residues that are conserved in several cytidine deaminases, including APOBEC1, that are involved in the C→U mRNA editing mechanism and are predicted to participate in the formation of the catalytic center containing a zinc ion (17, 36). The crystal structure of the catalytic domain of human ADAR2 reveals that histidine H394, glutamic acid E396, and two cysteine residues, C451 and C516, are involved in the coordination of a zinc atom and the formation of the catalytic center (37). Most interestingly, an inositol hexakisphosphate (IP_6) moiety is buried within the enzyme core and likely stabilizes multiple arginine and lysine residues present in the catalytic pocket. IP_6 is located very close to the catalytic center, strongly arguing that it plays a critical role during the hydrolytic deamination reaction (37).

Editing-Site Selectivity

A→I editing of dsRNA can be very specific, leading to deamination of select adenosine residues, or it can be almost random and lead to nonselective conversion of many adenosines. Both inter- and intramolecular dsRNAs of >20 base pair (bp) (two turns of the dsRNA helix) can serve as a substrate for ADAR (38). Many adenosine residues of long dsRNAs (>100 bp) are edited promiscuously, resulting in ~50% of all adenosine residues being converted to inosine. These are detected as so-called hypermutations of viral RNAs during replication and as a subsequent persistent infection with certain ss-RNA viruses, such as the measles virus (39), and they are also detected during extensive editing of sense-antisense RNA pairs made from transcripts of select genes, such as *Drosophila 4f-rnp* (40) and *C. elegans eri-6* and *eri-7* (41). By contrast, short dsRNAs (~20–30 bp) or long dsRNAs with mismatched bases, bulges, and loops (imperfect dsRNAs) are edited selectively; only a few adenosines are specifically chosen, indicating that the secondary structure within ADAR substrates dictates editing-site selectivity (42).

For site-selective A→I editing of protein-coding sequences, an imperfect fold-back dsRNA structure is formed between the exon sequence surrounding the editing site(s) and a downstream, usually intronic complementary sequence termed editing-site complementary sequence (ECS). This is demonstrated for the pre-mRNAs of the glutamate receptor B (GluR-B) Q/R site (43); the GluR-B,-C,-D R/G sites (44); or the A-E sites of serotonin (5-HT) receptor 2C (5-HT_{2C}R) (45, 46). However, the ECS of certain substrate RNAs is found within the exon sequence; for example, in the *Drosophila* K_v1.1 potassium ion channel (47); mammalian $GABA_A$ receptor α3 (Gabra-3) (48); and the self-editing site of dADAR (49). Furthermore, the dsRNA structure can be formed with the ECS through a complicated long-range pseudoknot, as shown for several recoding sites identified in *Drosophila* synaptotagmin I (50). Nonetheless, the ECS and the

Alu: a dispersed, middle-repetitive DNA sequence (~1.4 million copies) found in the human genome

LINE: a long interspersed element sequence that is typically used for non-long terminal repeat retrotransposons

Endogenous siRNA (esiRNA): repeat-associated siRNAs derived from endogenous repetitive sequences such as Alu or LINE retrotransposon elements

MicroRNA (miRNA): a small (19–23 nucleotides) single-stranded RNA that is processed from a precursor that consists of a short dsRNA region, bulges, and a loop

Noncoding RNA: RNA transcribed from DNA but not translated into protein

RNA interference (RNAi): a posttranscriptional gene-silencing process in which dsRNA triggers the degradation of homologous mRNA induced by siRNAs dsRNA structure are absolutely required for editing (8, 9).

Double-stranded RNA-binding domain (dsRBD): each dsRBD forms a domain (~65 amino acids) with α-β-β-β-α structures and makes direct contact with dsRNA

Z-DNA: left-handed DNA that differs from A- and B-DNA and that is believed to be involved in specific biological functions

Homodimerization of ADAR Is Required for Catalytic Activity

Some editing sites are preferentially edited only by ADAR1 or ADAR2, indicating a significant difference in their RNA-substrate interactions, possibly through their dsRBDs (51). ADAR1 or ADAR2 differ in the number of dsRBDs and in the spacing between the dsRBDs (51). In GluR-B pre-mRNA, the unique positioning of the second dsRBD of ADAR2, which is located close to two bulged bases adjacent to the R/G editing site, may contribute to selection of the adenine to be edited (51). The distinctive editing-site selectivity of ADAR1 and ADAR2 may also be mediated through functional interactions between the two monomers of ADAR1 and ADAR2, as such interactions possibly position specific adenosine residues relative to the catalytic center of ADAR (51, 52).

In vitro studies have revealed that the A→I editing activity of fly dADAR (53) and mammalian ADAR1 and ADAR2 (52) requires homodimerization. By contrast, ADAR3 does not dimerize at least in vitro, perhaps explaining the enzymatic inactivity of this ADAR family member (52). In vivo homodimerization of mammalian ADAR1 and ADAR2 was verified through studies using bioluminescence resonance energy transfer and fluorescence resonance energy transfer methods (54, 55). Whether ADAR dimerization requires RNA is the subject of debate (52–55). However, recent studies, using mutant ADAR1 and ADAR2 incapable of binding to dsRNA, indicate that dimerization is independent of RNA binding, suggesting that homodimer complex formation is mediated through protein-protein interactions between two monomers (33). Interestingly, a mutated ADAR subunit had a dominant-negative effect on dimer functions, indicating that the dsRBDs of the interacting monomers function cooperatively (33). The region involved in the dimerization of mammalian ADAR proteins remains to be established.

ADAR Regulation and Tissue and Cellular Distribution

Both ADAR1 and ADAR2 are present in many tissues, whereas ADAR3 is expressed only in brain (17, 19–23). Two isoforms of ADAR1, a full-length ADAR1p150 and a shorter, N-terminally truncated ADAR1p110, are known (**Figure 2*a***) (56). One of the three promoters that drive transcription of the *ADAR1* gene is interferon inducible, and the mRNA transcribed from this promoter directs translation of ADAR1p150 (57). Two other ADAR1 mRNAs, transcribed from constitutive promoters, direct the synthesis of ADAR1p110, which is initiated from a downstream methionine as the result of alternative splicing and skipping of the exon containing the upstream methionine (**Figure 2*a***). ADAR2 expression is regulated by the transcriptional activator CREB (cyclic adenosine monophosphate response element-binding) protein during experimental induction of ischemia in rat brain (58). The regulatory mechanism for ADAR3 is currently unknown.

The developmentally regulated expression of ADAR1 and ADAR2, starting around E10.0, has been reported in mice (59, 60). Finally, ADAR1 is regulated by the miRNA-mediated RNAi mechanism. The expression of ADAR1 is downregulated by miRNA-1 (miR-1) (61). Interestingly, miR-1 plays a critical role in the development of embryonic heart (62). Moreover, the earliest and highest expression of ADAR1 is detected in mouse embryonic heart at E10.0 (63), suggesting an important function of ADAR1 in mouse heart development and its tight regulation by miR-1.

ADAR1p150 is detected mainly in the cytoplasm (56, 64, 65). The cellular distribution of ADAR1p150 suggests that its targets, possibly including a different class of dsRNA substrates (e.g., esiRNAs, see below), may be localized to the cytoplasm (66). A nuclear localization signal has been identified in the third dsRBD of ADAR1 (65, 67). In addition, a functional nuclear export signal has been identified in the N-terminal $Z\alpha$ domain in full-length ADAR1p150, which is exported to the cytoplasm by the CRM1-RanGTP-mediated mechanism (65). Recently, however, the nuclear cytoplasmic shuttling of ADAR1p110 that lacks this $Z\alpha$ domain has been reported in certain cell lines (68). Interaction with the protein exportin-5, which is mediated via dsRNA binding, may be responsible for the CRM1-RanGTP-independent nuclear export of ADAR1. Although binding of dsRNA to the third dsRBD appears to facilitate nuclear export of ADAR1, it inhibits nuclear import of the complex. Transportin-1, which binds to the third dsRBD, has been identified as the import receptor for ADAR1 (68).

ADAR1, together with RanGTP and exportin-5, seems to act as a nuclear export carrier of certain dsRNAs. Interestingly, pre-miRNAs, processed from primary transcripts of miRNAs (pri-miRNAs) by Drosha/DGCR8, are exported into the cytoplasm by the RanGTP/exportin-5 complex (69). Thus, it would be interesting to determine whether ADAR1 is involved in the nuclear export of pre-miRNAs (possibly edited pre-miRNAs).

Nuclear import of ADAR2 and ADAR3 appears to be controlled by importin α family members: ADAR2 by importin $\alpha 4$ and $\alpha 5$, whereas ADAR3 by importin $\alpha 1$, which recognizes the N-terminal R domain (70). The presence of the factor ADBP-1, which binds to *Ce*ADR2 and regulates the nuclear localization and activity of *Ce*ADR2, has also been reported (71).

Finally, ADAR1p110 and ADAR2 accumulate in the nucleolus (64, 72). The localization of ADAR1p110 and ADAR2 in this compartment, which is dependent on functional dsRBDs, has been proposed to occur through their binding to rRNA or to small nucleolar RNA. They may be stored temporarily in the nucleolus but move out to the nucleoplasm as substrate dsRNAs appear (64, 72). However, the significance of the nucleolar localization of ADAR1p110 and ADAR2 remains largely unknown.

PHYSIOLOGY OF EDITING

Editing of coding sequences can result in dramatic alterations of the target gene functions. Deficiency in editing created in animal model systems revealed the importance of editing in vivo. Furthermore, the presence of human diseases related to A→I RNA editing has recently become known.

Diversification of Protein-Coding Gene Functions

The translation machinery reads an inosine as if it were guanosine (**Figure 1*b***), which could lead to codon changes. Despite the initial expectation, only a limited number of cellular genes (~30, mostly neurotransmitter receptors and ion channels) that are subjected to site-selective A→I RNA editing within their coding sequences have been identified. These include mammalian *GluR* (43), *5-HT$_{2C}$R* (46), and potassium channel *Kv1.1* (73), squid *Kv1.1A* (74), and *Drosophila Na$^+$ channel* (75) gene transcripts as well as the hepatitis delta virus antigen gene (76).

In most cases, RNA editing of protein-coding genes results in generation of protein isoforms and diversification of protein functions. For instance, the coding regions of receptors for several GluR ion channel subunits contain a total of eight A→I RNA editing sites (8, 43). One of these sites, the Q/R site, which is located within the channel pore-loop domain of the GluR-B subunit, plays a critical role in ion channel function. A→I editing (CAG glutamine → CIG arginine, thus Q/R site) of this site changes the tetrameric channel protein so that the channel becomes impermeable to Ca^{2+} (8, 43). Another example is the combinatorial editing of five sites (A-E sites) located within the second intracellular loop or G protein–coupling domain of 5-HT$_{2C}$R. Editing of these sites changes three codons—AUA (isoleucine), AAU (asparagine), and AUU (isoleucine)—to possibly six different amino acid residues (45, 46, 77). The result is that up to 24 receptor isoforms are expressed with substantially altered G protein–coupling functions, affecting 5-HT potency and ligand-binding affinity (45, 46, 77).

Among the 14 editing sites identified in the squid Kv1.1 channel, some sites affect the rate of deactivation, whereas other sites located within the N terminus inhibit tetramerization of individual subunits (74). However, the functional significance of most protein-recoding editing sites remains to be established (7). This includes sites recently found in mammalian Gabra-3 (48) as well as numerous sites in *Drosophila* neurotransmitters and ion channels identified by the comparative genomic method (73). The reader may refer to more comprehensive reviews on these protein-recoding type A→I editing targets and their biological significance (7, 8, 10, 73).

Advantage of RNA Editing over Gene Mutations

RNA editing has several advantages over gene mutations. As with alternative splicing, the extent of RNA editing can be differentially regulated (ranging from no editing to nearly 100% editing) and can be spatiotemporally controlled. By contrast, because of the lack of such control, gene mutations convey permanent, hardwired changes in the genome. This advantage of RNA editing is reflected in the developmental regulation of GluRs (8), 5-$HT_{2C}R$ (46), and Gabra-3 mRNA editing (48, 78, 79), as well as in editing targets containing multiple editing sites, such as GluR-B (8), 5-$HT_{2C}R$ (46), and fly Ca^{2+} channel α1-subunit (80). Combinatorial editing of the latter can generate numerous editing isoforms that are distributed differentially in various brain subregions, contributing to a complicated regulation of multiple, functionally diversified gene products. Interestingly, certain editing sites, such as the GluR-B Q/R site (81) and Gabra-3 (48), are hardwired in the fish and frog genomes (which contain guanosine instead of adenosine at sites subject to A→I editing in other species), indicating that the G→A mutation of this site and control of the codon through A→I editing must have occurred more recently for an evolutionary advantage.

RNA-Editing Deficiencies and Mutants with Altered Editing

Inactivation of *ADAR* gene family members has significant physiological consequences, as seen in phenotypic alterations of *ADAR* gene mutants created in various species. Mutant flies with a homozygous deletion in the *dADAR* gene exhibit brain-related changes, such as temperature-sensitive paralysis, uncoordinated locomotion, and age-dependent neurodegeneration, presumably resulting from a lack of editing of important dADAR target genes such as Na^+ (*para*), Ca^{2+} (*Dmca1A*), and glutamate-gated Cl^- channels (*DrosGluCl-α*) (26). *C. elegans* strains that contain homozygous deletions of both *CeADR1* and *CeADR2* display defective chemotaxis (27), although much reduced penetration of this phenotype has been reported by a separate group (71). Mice with a homozygous *ADAR2 null* mutation die several weeks after birth. These mice experience repeated episodes of epileptic seizures that originate from excess influx of Ca^{2+} and consequent neuronal death caused by underediting of the GluR-B Q/R site (82), which is a major target of ADAR2. However, *ADAR2*$^{-/-}$ mice can be rescued by introducing the genomic *GluR-B*^{R} mutation, which restores expression of GluR-B "edited" at the Q/R site (82). *ADAR3*$^{-/-}$ mice are viable and appear to be normal, although the possibility that ADAR1 and/or ADAR2 activity compensates for loss of ADAR3 function has not been excluded (M. Higuchi & P.H. Seeburg, personal communication). By contrast, the inactivation of *ADAR1* leads to an embryonic lethal phenotype because of widespread apoptosis (60, 63, 83). Thus, at least ADAR1 is absolutely required for life in mammals.

Recent studies using conditional inactivation of *ADAR1* revealed that ADAR1 also plays a critical role in the maintenance of hematopoietic stem cells, perhaps by preventing inappropriate induction of interferon signaling pathways and apoptosis of hematopoietic progenitor cells (84). Mutant mice overexpressing ADAR2 display adult-onset hyperphagia and consequent obesity (85). Interestingly, this obese phenotype was reproduced with a separate mutant mouse line overexpressing catalytically inactive ADAR2, indicating the possibility that hyperphagia and obese phenotype are caused by other ADAR2 functions, perhaps through its binding to a currently unknown dsRNA substrate (85).

Mutant animal model systems with permanently altered editing patterns of specific substrate RNAs have also been created. Phenotypes of these mutants generated important and often surprising insight into the in vivo significance of A→I editing of particular substrate RNAs. Self-editing of an *ADAR2* intronic site creates a proximal 3′ splice site containing a noncanonical adenosine-inosine dinucleoside. This leads to alternative splicing and consequent loss of functional ADAR2 protein expression owing to premature translation termination (86). The mutant mice, in which this *ADAR2* intronic self-editing is eliminated by removing the ECS sequence, lose ADAR2 autoediting and subsequent alternative splicing. ADAR2 protein expression is substantially increased in the mutant mice, confirming the function of this intronic editing site in a negative feedback regulatory mechanism (87).

Drosophila dADAR undergoes developmentally regulated self-editing of its own mRNA, which changes a conserved Ser (AGU) residue to Gly (IGU) in the catalytic domain and generates an isoform with much reduced editing activities (88). Ubiquitous expression of only the unedited isoform dADAR in embryos and in larvae of mutant flies is lethal because of the excess editing activities of the unedited enzyme (49).

The importance of GluR-B Q/R site editing was clearly demonstrated in heterozygous mice *GluR-B*$^{+/\Delta ECS}$ harboring an editing-incompetent *GluR-B* allele that lacks the ECS essential for editing (89). The unedited GluR-B subunit, although having relatively low expression levels (~25%), increased Ca^{2+} permeability in neurons and led to epileptic seizures and premature death by 3 weeks of age (89).

Mutant mouse lines harboring knockin *5-HT$_{2C}$R-VGV* or *5-HT$_{2C}$R-INI* alleles have been created, resulting in the sole expression of the fully edited 5-HT$_{2C}$R-VGV isoform or the unedited (editing-blocked) 5-HT$_{2C}$R-INI isoform (90). Although *INI* mice grew normally, *VGV* mice had drastically reduced fat mass, in spite of compensatory hyperphagia. Constitutive activation of the sympathetic nervous system and consequent increase in energy expenditure were detected in *VGV* mice. These studies revealed the presence of a previously unknown mechanism, mediated via 5-HT$_{2C}$R mRNA editing, that regulates energy expenditure and fat metabolism (90). This discovery raised the possibility that substantial variations in metabolic rate and obesity among individuals of different ages and ethnic backgrounds may be related, at least in part, to differences in the editing efficacy of 5-HT$_{2C}$R mRNAs (90).

RNA Editing in Human Diseases and Pathophysiology

Dysfunction of the A→I RNA editing mechanism can cause human diseases or pathophysiology (91, 92). Heterozygosity for functional null mutations in the ADAR1 gene—a total of 41 sites have been discovered—results in dyschromatosis symmetrica hereditaria, an autosomal dominant human pigmentary genodermatosis (93). Although the target dsRNA, as well as the mechanism underlying the pathophysiology, of this genetic disease is unknown, a longer isoform of ADAR1 (p150) has been given particular importance (91).

RNA-editing deficiencies also underlie disorders of the central nervous system. Underediting of the Q/R site of GluR-B pre-mRNA has been implicated in the death of motor neurons of sporadic amyotrophic lateral sclerosis (ALS) patients (94). Deficiency in the Q/R site of GluR-B pre-mRNA editing has also been proposed to be responsible for the apoptotic death of neurons during ischemia resulting from cardiac arrest and disruption of blood flow to the brain (58). RNA editing of 5-HT$_{2C}$R might be relevant to the cause of certain neuropsychiatric

Expressed sequence tag (EST): a single-pass, short read of complementary DNA that is generated from a transcribed region of the genome

disorders, such as depression and schizophrenia (92). Also, the editing pattern of 5-$HT_{2C}R$ mRNA is significantly altered in the prefrontal cortex of suicide victims (92, 95, 96).

Underediting of the GluR-B Q/R site (ADAR2 site) and reduced ADAR2 activities have been reported in human gliomas. Increased Ca^{2+} influx through GluRs with unedited GluR-B was proposed to underlie aggressive growth of tumor cells (97, 98). Furthermore, the loss of ADAR2 editing activity in pediatric astrocytomas has been reported (99). Results of a global bioinformatic survey of inverted Alu sequence editing also suggest that significant hypoediting occurs in various types of human cancers, due at least in part to simultaneous downregulation of ADAR1, ADAR2, and ADAR3 (100). This suggests that reduced A→I editing in general may be involved in the pathogenesis of cancer (100). However, the causative relevance of the reduction in A→I editing activities to the malignant transformation of cancers remains to be established.

A→I EDITING OF REPETITIVE NONCODING RNA

Physiologically important editing sites, such as those for GluR-B and 5-$HT_{2C}R$, and the consequent alterations of their protein functions, were discovered serendipitously. Identification of many more A→I RNA editing target genes was anticipated because a substantial amount of inosine was detected in rat brain $poly(A)^+$ RNA (101). Accordingly, bioinformatics strategies for globally searching for A→I editing sites have been developed, leading to the identification of a large number of new sites, which are mainly within noncoding regions.

Bioinformatics Screening and Identification of A→I Editing Sites in Repeat Sequences

Reverse transcriptase recognizes inosine as guanosine; and therefore, A→I editing is detected as an A→G change in the cDNA sequence or expressed sequence tag (EST). Accordingly, several groups have developed computational methods for genome-wide identification of A→I editing sites in cDNA (EST) sequencing databases (102–105). The screening strategy consists of an algorithm that aligns a cluster of A→G mismatches within cDNAs or ESTs to the genome sequence. A large number of human A→I RNA editing sites (~15,000 sites mapped in ~2,000 different genes) have been identified in noncoding regions that consist of inversely oriented repetitive elements (**Figure 3*a***), mostly within Alu elements and some LINE (102–105). It is predicted that more than 85% of pre-mRNAs may be edited, with the vast majority being targeted in introns and UTRs (102).

Editing depends on formation of RNA fold-back structures formed between intramolecular inverted Alu (LINE) repeats. It turns out that the distance between an Alu and the

Figure 3

Possible regulatory functions for noncoding RNA editing. (*a*) Extensive A→I editing of an RNA duplex structure that consists of inverted Alu or LINE repeats. The inverted Alu or LINE repeats in introns and untranslated regions (UTRs) form intramolecular RNA duplexes genome wide, which are then subjected to A→I RNA editing by ADAR. (*b*) Splicing machinery interprets an inosine as a guanosine. Therefore, splice sites might be created or deleted by A→I editing of intronic Alu fold-back dsRNAs, leading to the inclusion or exclusion of Alu exons (102). (*c*) A→I editing of a short interspersed element (SINE) fold-back dsRNA present within the 3′ UTR of cationic amino acid transporter 2 (CAT2) nuclear (CTN) RNA. Its binding to $p54^{nrb}$ might be involved in the regulatory mechanism that retains this RNA within nuclear speckles (118). When cells are placed under stress, CTN-RNA is cleaved and de novo polyadenylated at an alternative site to release the protein-coding mCAT2 mRNA, which is then translated for CAT2 proteins (118). (*d*) A→I editing of Alu or LINE fold-back dsRNAs might suppress the generation of esiRNAs and consequent RNAi-mediated silencing of retrotransposon activities and/or genes (mRNAs) harboring these sequences within UTRs in *trans*.

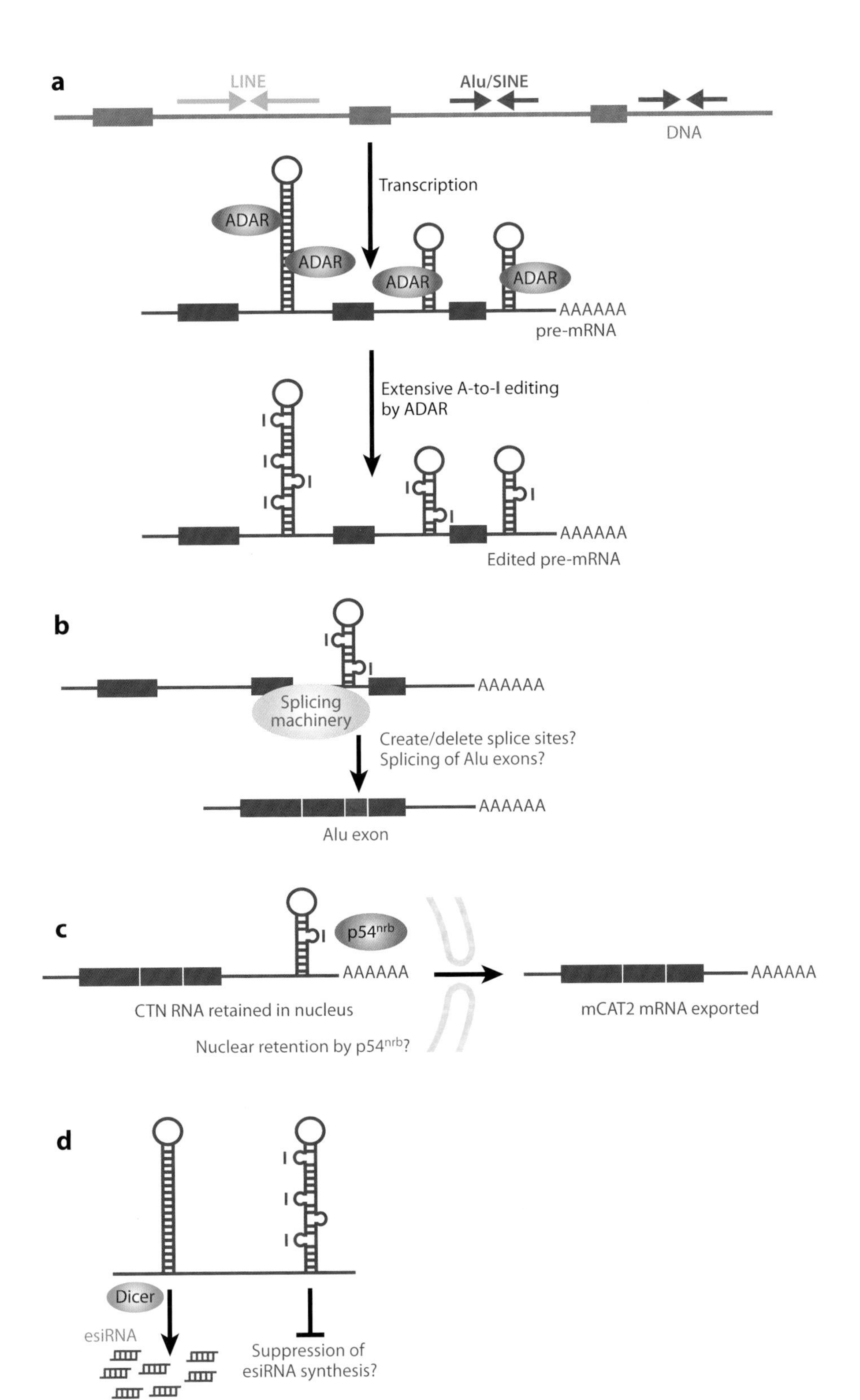
a
LINE
Alu/SINE
DNA
Transcription
ADAR
ADAR
ADAR
ADAR
AAAAAA
pre-mRNA
Extensive A-to-I editing
by ADAR
AAAAAA
Edited pre-mRNA
b
AAAAAA
Splicing
machinery
Create/delete splice sites?
Splicing of Alu exons?
AAAAAA
Alu exon
c
p54nrb
AAAAAA
CTN RNA retained in nucleus
Nuclear retention by p54nrb?
AAAAAA
mCAT2 mRNA exported
d
Dicer
esiRNA
Suppression of
esiRNA synthesis?

SINE: a short interspersed element of repetitive sequences, such as Alu elements, generated by retrotransposons

Single-nucleotide polymorphism (SNP): SNPs are base-pair substitutions that are the most common forms of genetic polymorphism; SNPs are typically biallelic

Wobble base pair: non-Watson-Crick pairing such as the thermodynamically less stable G·U and I·U pairing

closest inverted element is a critical determinant for the likelihood that a given element is targeted by the editing machinery (102, 103). Alu repeats are short interspersed elements (SINEs) that are unique to primates. Interestingly, bioinformatics screening for A→I editing sites in mouse, rat, and chicken transcriptomes revealed that noncoding repeat sequences are also major targets in these species, but the editing frequency is much lower than that observed in human transcriptomes (104, 106, 107). The substantial reduction in frequency is most likely a result of the differences in repeat length (e.g., ~300 bp versus ~150 bp for human Alu and mouse SINE, respectively) and of the higher sequence homogeneity among human Alus as compared to more divergent SINEs of other species (104, 106, 107).

Rare Editing of Protein-Coding Sequences

Combined bioinformatics and comparative genomics screening restricted to coding regions resulted in the identification of only a few editing target genes (108, 109). Application of a similar bioinformatics screening of *Drosophila* cDNA and EST databases identified 27 new protein-coding targets, raising the possibility that the insect editing system may more efficiently utilize recoding type A→I editing (110). Moreover, it became clear that some single-nucleotide polymorphism (SNP) databases included many A→I RNA editing sites. However, screening of a human SNP database identified only a few editing sites that result in recoding of protein sequences (111, 112). Finally, most recently, an unbiased screening method (the padlock approach) was developed that allows amplification and deep sequencing of a large number of human exons simultaneously. By focusing on the exons containing potential A→I RNA editing sites (A→G changes in cDNA and EST), 53 new recoding type editing sites were identified, increasing the repertoire of recoding type A→I RNA editing sites substantially (113). Nonetheless, these results together indicate that the most common targets of ADARs are the noncoding sequences of transcriptomes and also that protein recoding as a result of A→I RNA editing is relatively rare.

FUNCTIONAL IMPLICATIONS OF REPETITIVE RNA EDITING

The biological significance of widespread global A→I editing of noncoding, repetitive sequences remains largely a mystery (**Figure 3*a***). The editing of an A:U pair results in creation of an I·U wobble base pair and destabilization of the dsRNA structure. Some of the editing sites are at an A·C mismatched pair, with editing resulting in an I:C Watson-Crick pair and stabilization of the dsRNA structure. This change in the local and overall stability of the dsRNA structure and of inosines converted from adenosines seems to be recognized by several mechanisms. Accordingly, several possible functions of repeat element editing have been proposed.

Creation and Elimination of Splicing Sites?

The splicing machinery interprets an inosine as a guanosine. A→I editing could therefore create or delete splice donor and acceptor sites (**Figure 3*b***). For example, a highly conserved canonical 5′ splice site dinucleotide recognition sequence GU ($\underline{A}$U→$\underline{I}$U = $\underline{G}$U) or a 3′ splice acceptor site AG (A$\underline{A}$→ A$\underline{I}$ = A$\underline{G}$) can be created by editing. Similarly, editing of a 3′ splice site AG ($\underline{A}$G→$\underline{I}$G = $\underline{G}$G) can destroy the site (114). Indeed, several examples of inclusion and exclusion of the Alu exon due to editing of the Alu fold-back dsRNA sequence have been identified through analysis of human cDNA sequences (102). More recently, a series of exonizations of the intronic sequence of the human nuclear prelamin A recognition factor resulting in coding differences has been reported (115, 116). The A→I editing of the Alu sequence and exonization of the noncoding sequence may be one important evolutionary means for generating a new variant protein with a novel function (102, 115, 116).

Nuclear Retention of Inosine-Containing RNAs?

A nuclear-localized multifunctional protein called p54^{nrb} may play a role in the retention of extensively edited RNAs (I-RNA) within the nucleus (**Figure 3*c***) (117). It has been proposed that A→I editing of a dsRNA formed on inverted repeats of SINEs present within the 3′ UTR of mouse cationic amino acid transporter 2 (CAT2) nuclear RNA (CTN-RNA) and its binding to p54^{nrb} is involved in the mechanism that traps CTN-RNA within nuclear speckles (also known as interchromatin granule clusters) (118). During stress, CTN RNA is posttranscriptionally cleaved and de novo polyadenylated at an alternative site to produce protein-coding mCAT2 mRNA, which is then translated for CAT2 proteins (118). Whether the A→I editing of inverted Alu repeats in the 3′ UTR affects the nuclear retention mechanism was evaluated in human cells with an EGFP reporter system (119). Silencing of the reporter resulted from mRNA retention in the nucleus and was correlated with the extent of A→I editing of inverted Alu repeats as well as by its association with p54^{nrb}. However, more recent studies on reporter RNAs, as well as on endogenous *C. elegans* and human mRNAs with an inverted repeat dsRNA structure, revealed that the presence of the dsRNA structure or A→I editing of the dsRNA region plays no role in the nuclear retention of these mRNAs (120). In fact, these dsRNA-containing mRNAs are exported into the cytoplasm, integrated into polysomes, and translated efficiently, independent of their editing status (120). This raises doubt on whether A→I editing of inversely oriented repeat elements does indeed play a role in the nuclear retention mechanism (120).

Control of esiRNA Synthesis?

The concentration of editing sites in repeat elements brings up the question of whether A→I editing acts as an antitransposon mechanism by altering the element sequence and inhibiting the integration of transcribed Alu and LINE sequences back to the genome (102, 104). Alternatively, editing may protect dsRNAs derived from inverted repeats of these retrotransposons from entering into the RNAi pathway (5). Recently, generation of esiRNAs from long hairpin dsRNAs (hpRNAs) has been reported in insect somatic cells (121) and mouse oocytes (122, 123). Processing of founder dsRNAs into esiRNAs proceeds in a processive fashion, producing esiRNAs with a ~21-nucleotide periodicity. The vast majority of esiRNAs originate from long terminal repeats of transposons and retrotransposons, such as SINEs and LINEs, whereas a substantial number of the remaining esiRNAs arise from pseudogenes or protein-coding genes arranged in reverse orientations (124). These esiRNAs are generated by Dicer and are involved in the silencing of transposon activities and founder genes of the pseudogenes (122, 123). Most interestingly, some esiRNAs contain a limited number of A→G changes, indicating that the founder dsRNAs of these esiRNAs have undergone A→I RNA editing (125). dsRNA that is extensively edited in vitro by ADAR becomes resistant to Dicer (see the next section) (126). Therefore, generation of esiRNAs may be suppressed through extensive A→I editing of the fold-back dsRNA in cells and in tissues expressing high levels of ADARs (**Figure 3*d***).

INTERACTION OF RNA EDITING AND RNAi PATHWAYS

Accumulating evidence suggests that A→I RNA editing and RNAi pathways interact. In fact, the major function and evolutional reason for development of A→I RNA editing system may be to modulate the efficacy of RNAi. In following sections, the studies related to this particular aspect of A→I RNA editing are described.

Antagonistic Effects of A→I RNA Editing on RNAi

The A→I editing mechanism may interact with the RNAi pathway by competing for shared

Small interfering RNA (siRNA): a small (19–23 bp) noncoding dsRNA that is processed from a longer dsRNA by Dicer

dsRNA substrates and by antagonizing RNAi efficacy (5, 127). Dicer seems to distinguish dsRNAs that contain I·U wobble base pairs from dsRNAs that contain only Watson-Crick base pairs. Increasing deamination of long dsRNAs by ADAR progressively reduces the production of small interfering RNA (siRNA) by Dicer. Indeed, dsRNA that is extensively edited (~50% of the adenosines are converted to inosines) in vitro by ADAR becomes completely resistant to Dicer (126). By contrast, moderately deaminated dsRNA can be processed in vitro to siRNAs containing up to one I·U base pair per siRNA (126, 128). However, the substitution of even a single I·U for A:U base pair between the edited siRNAs and target mRNA could reduce the efficacy of RNAi (126). As already described in the previous section, a large proportion of esiRNAs containing a single A→G change has been identified recently in insect somatic cells (125). These results suggest that founder dsRNAs consisting of long hpRNAs of repetitive sequences are subject to A→I editing. Furthermore, they suggest that processing of the extensively edited founder dsRNAs by Dicer may be prohibited. Taken together, A→I RNA editing of long dsRNAs, such as those to be processed to esiRNAs, certainly can antagonize siRNA-mediated gene-silencing efficacy (**Figure 4*a***).

Degradation of Edited dsRNAs by Tudor Staphylococcal Nuclease

Extensive A→I editing of dsRNA may also result in degradation, consequently reducing the expression of esiRNAs. A ribonuclease activity that specifically cleaves inosine-containing dsRNA (I-dsRNA) has been reported (126). This ribonuclease preferentially cleaves dsRNA that contains multiple I·U base pairs (I-dsRNA) (126). Interestingly, Tudor staphylococcal nuclease (Tudor-SN), an RNA-induced silencing complex (RISC)-associated component that lacks an assigned function in the RNAi mechanism, has been identified as a potential I-dsRNA-specific ribonuclease or at least as an essential cofactor of ribonuclease activity (129). Extensive A→I editing might therefore lead to the degradation of dsRNAs by Tudor-SN, which in turn results in reduced esiRNA expression levels (**Figure 4*a***).

Sequestration of siRNA by ADAR1p150

In mammalian cells, ADAR1p150 might quench the function of siRNAs that have already been processed from long dsRNA by Dicer. In vitro studies revealed that cytoplasmic ADAR1p150 binds siRNA very tightly, and gene silencing by siRNA is significantly more effective in mouse fibroblasts that are homozygous for an ADAR1 null mutation than in wild-type cells (66). These findings implicate ADAR1p150 as a cellular factor that limits siRNA potency in mammalian cells by decreasing the effective siRNA concentration (66). Interestingly, induced ADAR1 gene expression is observed in mice injected with high doses of nonspecific siRNA (130), indicating that ADAR1 is involved in a cellular feedback mechanism in response to siRNA (**Figure 4*b***).

RNAi-Dependent Phenotypes of *ADAR null* Worms

C. elegans strains that contain homozygous deletions of both *CeADR1* and *CeADR2* genes have defective chemotaxis (**Figure 2*a***) (131). These phenotypic alterations, however, can be reverted in RNAi-defective worms, indicating that the *ADAR null* phenotype is RNAi dependent (131). Expression of a gene involved in the chemotaxis mechanism seems to be controlled by the balance between A→I editing and RNAi on dsRNA derived from the chemotaxis gene (**Figure 4*a***). Overly enhanced RNAi effects and suppression of the chemotaxis gene likely result in *ADAR null* worm phenotypes, but the identity of the chemotaxis gene as well as details of the interaction between RNA editing and the RNAi pathway remain unknown.

In addition, studies of *ADAR null* worms indicate that ADAR is involved in the mechanism that regulates the expression of transgenes. In

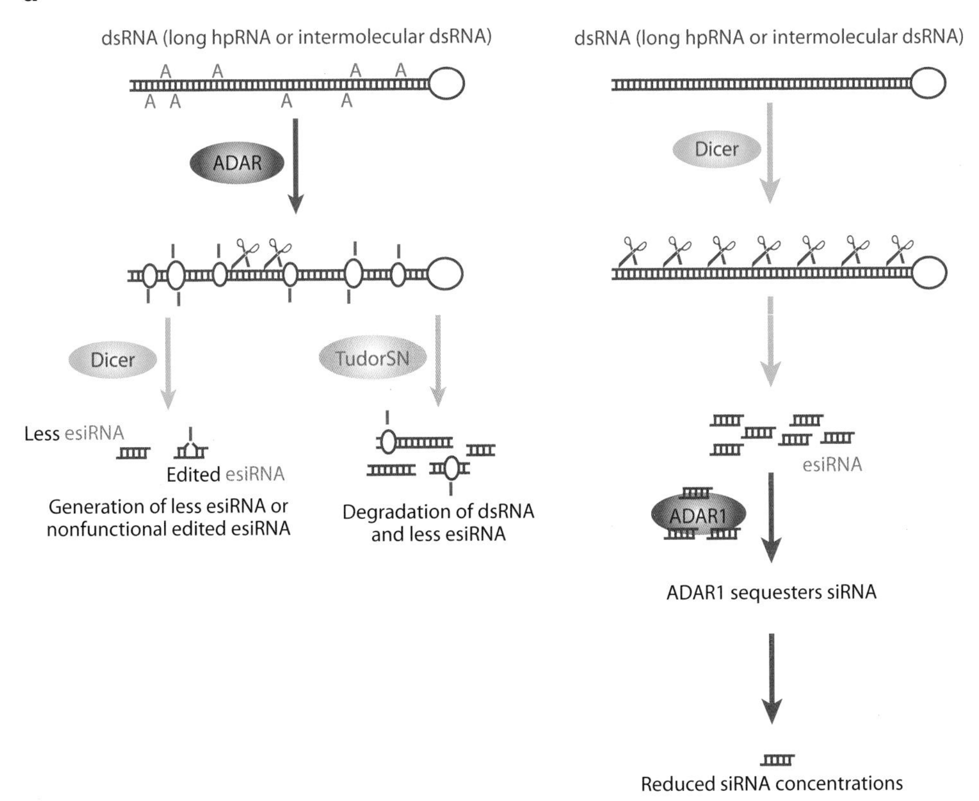

Figure 4

Interaction between RNA editing and RNAi pathways. Two possible ways in which RNA editing and RNAi pathways may interact. (*a*) The introduction of many I·U base pairs and the alteration of the dsRNA structure by ADAR leads to the generation of fewer esiRNAs by Dicer, because dsRNAs that contain many I·U base pairs become resistant to Dicer cleavage. Alternatively, extensively edited dsRNAs with multiple I·U base pairs may be rapidly degraded by Tudor staphylococcal nuclease (Tudor-SN), resulting in the generation of fewer esiRNAs. (*b*) In addition, a fraction of already processed siRNAs might be sequestered by certain ADAR gene family members, reducing the effective siRNA concentration. For instance, cytoplasmic ADAR1p150 binds siRNA tightly. Gene silencing by siRNA is significantly more effective in the absence of ADAR1, indicating that ADAR1p150 is a cellular factor that limits siRNA potency in mammalian cells by decreasing the effective siRNA concentration and its incorporation into RNA-induced silencing complex.

C. elegans, A→I RNA editing of dsRNAs derived from inverted repeats of transgenes seems to prevent silencing of the transgenes by RNAi (132). RNAi-mediated transgene or viral gene silencing (cosuppression) is very efficient in plants and fungi that lack ADAR genes and the A→I RNA editing system (133). Thus, ADAR gene families and the A→I RNA editing system might have evolved to counteract or regulate RNAi in the animal kingdom.

EDITING OF MIRNA AND ITS BIOLOGICAL SIGNIFICANCE

miRNA-mediated gene silencing controls many processes, such as development, differentiation, and apoptosis (133, 134). pri-miRNAs are processed sequentially by Drosha and Dicer (69). Nuclear Drosha, together with its partner, DGCR8, cleaves pri-miRNAs, releasing 60- to 70 nt pre-miRNAs. Correctly processed pre-miRNAs are recognized and exported from the nucleus by exportin-5 and RanGTP. Cytoplasmic Dicer together with the dsRNA-binding protein, transactivating response RNA-binding protein (TRBP), then cleaves pre-miRNAs into 20- to 22-nt siRNA-like duplexes. One or both strands of the duplex may serve as the mature miRNA. Following integration into miRNA-induced silencing complex (miRISC), miRNAs block the translation of partially complementary targets located in the 3′ UTR of specific mRNAs or guide the degradation of target mRNAs as do siRNAs (**Figure 5*a***) (69, 134).

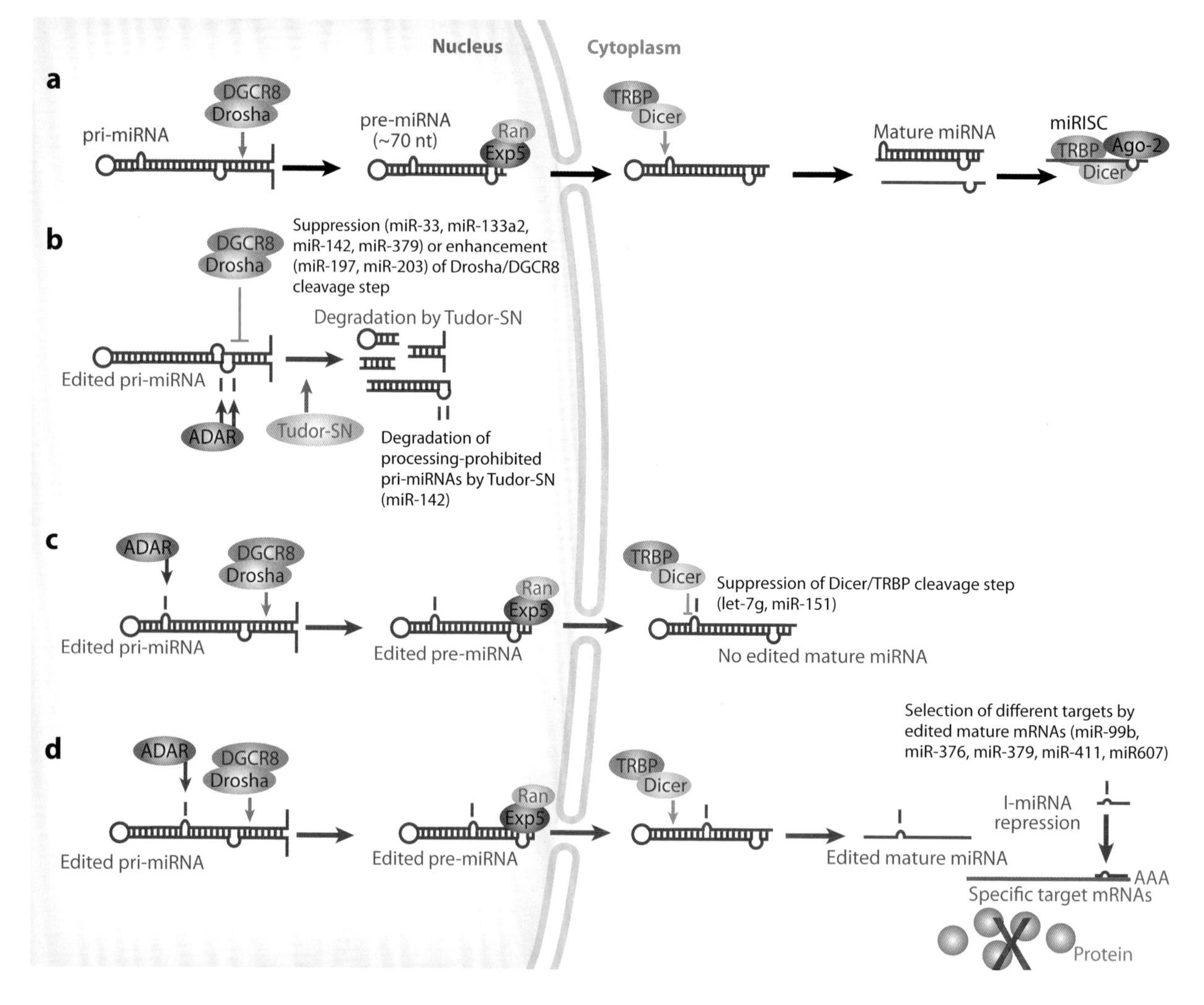

Figure 5

Regulation of miRNA processing and expression by RNA editing. (*a*) The Drosha/DGCR8 complex cleaves pri-miRNAs in the nucleus, producing ~70-nt pre-miRNA intermediates, which are exported by exportin-5 (Exp5) and RanGTP (Ran) into the cytoplasm. The Dicer/TRBP complex executes a second cleavage, generating mature miRNAs. (*b*) Drosha cleavage of pri- to pre-miRNA is suppressed by A→I editing of certain sites, such as the +4 and +5 sites of pri-miRNA-142 (miR-142). In addition, highly edited pri-miRNA-142 is degraded by Tudor staphylococcal nuclease (Tudor-SN) (138). (*c*) A→I editing of certain sites might lead to inhibition of the Dicer/TRBP cleavage step, as shown for pri-miR-151 editing. (*d*) Editing of pri-miRNAs at certain sites might lead to the expression of edited mature miRNAs and silencing of a different set of target genes owing to "seed" sequence alterations, as shown for miR-376.

Although certain pri-miRNAs have been reported to undergo A→I RNA editing in vivo (135–137), the biological significance of pri-miRNA remains unknown. Recent studies revealed that A→I RNA editing modulates processing and function of miRNAs.

Editing of Pri-miRNA Inhibits Drosha Cleavage

Several adenosine residues of pri-miR-142 are highly edited by ADAR1 and/or ADAR2. In vitro processing assays using Drosha/DGCR8 and Dicer/TRBP complexes revealed that editing inhibits Drosha cleavage of pri-miR-142 and that sites most inhibitory for processing of pri- to pre-miRNA-142 RNAs are the +4 and +5 sites located in the fold-back dsRNA stem near the Drosha cleavage site (**Figure 5*b***). As expected, miR-142-5p and miR-142-3p expression levels are significantly higher in spleens of B-cell-lineage-specific *ADAR1*$^{-/-}$ mice and *ADAR2*$^{-/-}$ mice than in spleens of wild-type control mice (138).

In view of the suppressive effects of A→I editing on the processing of pri-miR-142 and on the expression of mature miR-142 RNAs, detection of a substantial level of highly edited pri-miR-142 RNA sequences in wild-type mouse spleen is anticipated. However, only a relatively low level of edited pri-miR-142 RNAs can be detected. Highly edited pri-miR-142 RNAs (prohibited from cleavage by Drosha) appear to be rapidly degraded by Tudor-SN, a ribonuclease specific to I-dsRNAs (**Figure 5*b***) (129). Taken together, these findings suggest that steady-state levels of edited pri-miR-142 RNAs are regulated by editing frequency and Tudor-SN activity (138). However, some edited pri-miRNAs are stable, and perhaps only some edited pri-miRNAs with multiple I·U pairs are selected for degradation by Tudor-SN following A→I RNA editing.

Editing of Pri-miRNA Inhibits Dicer Cleavage

A major editing site (+3) and a minor site (−1) of pri-miR-151 were identified on its antisense strand and near its end loop (139). Analysis of *ADAR1*$^{-/-}$ embryos revealed that editing of the −1 and +3 sites is carried out by ADAR1 (139). No edited mature miR-151-3p RNAs were detected, but all pre-miR-151 molecules detected were completely edited at the +3 site. An in vitro pri-miRNA processing assay revealed that the A→I substitution at the −1 or +3 site had no effect on Drosha cleavage but inhibited cleavage of the edited pre-miR-151 by the Dicer/TRBP complex. These studies demonstrated that Dicer cleavage is the step inhibited by the editing of pri-miR-151 at the major +3 and minor −1 sites, resulting in accumulation of edited pre-miR-151 RNAs in the cytoplasm (139). Interestingly, in vitro studies showed that the editing efficiency at the +3 site is much higher with pre-miR-151 than with pri-miR-151. It is possible that certain pri-miRNAs, not edited in the nucleus, may be edited only after processing to pre-miRNAs in the cytoplasm, most likely by the only cytoplasmic ADAR, ADAR1p150 (139).

miRNA-induced silencing complex (miRISC): the miRNA-mediated RNAi machinery that contains miRNAs and other protein factors, such as Ago-2

Redirection of Silencing Target Genes by Edited miRNAs

A→I editing of miR-376 cluster transcripts can lead to the silencing of specific sets of genes, resulting in stringent and tissue-specific regulation of certain gene products. Six human and three mouse miR-376 family genes are included within the maternally imprinted loci. Most miR-376 cluster members undergo extensive and simultaneous A→I editing at two specific sites (+4 and +44 sites) in select human and mouse tissues and specific subregions of the brain (**Figure 6**) (140). The +4 site is edited by ADAR2, whereas the +44 site is selectively edited by ADAR1. Identification of the edited forms of mature miR-376 RNAs in certain tissues, such as brain, revealed that, unlike the case of pri-miR-142 and pri-miR-151, editing of pri-miR-376 RNAs at the two sites does not affect the Drosha and Dicer cleavage steps. Both of the two editing sites in pri-miR-376 RNAs (+4 and +44) are located within the functionally critical 5′ proximal "seed" sequences of

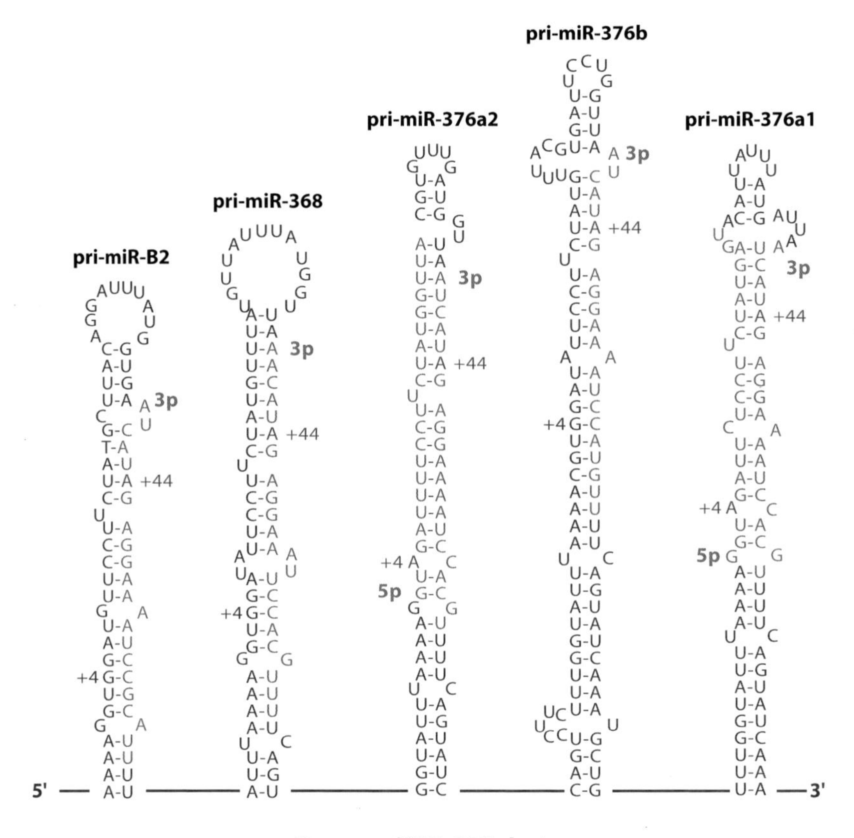

Figure 6

A→I RNA editing of pri-miR-376 RNAs. Hairpin structures of five human pri-miR-376 cluster member RNAs that are frequently edited in brain and other tissues. Regions processed into mature miRNAs are highlighted in blue. The two most highly edited adenosines (+4 and +44 sites) are highlighted in red. Editing sites that are numbered with the 5′ end of the human miR-376a1–5p sequence are counted as +1.

miRNA-376-5p and -3p strands. Two different sets of the potential target genes for unedited and edited miR-376a-5p (~80 each) were predicted *in silico*. Only a few overlapping genes were predicted to be targeted by both unedited and edited miR-376a-5p. In vitro reporter verification experiments for select unedited-version targets and edited-version targets confirmed that a single A→I base change is sufficient to redirect silencing miRNAs to a new set of targets.

Endogenous expression levels of one of the edited-version miR-376a-5p targets, pyrophosphate synthetase 1 (PRPS1), were assessed in wild-type and ADAR2$^{-/-}$ mouse cortices. The +4 site of miR-376a-5p is edited exclusively by ADAR2. No edited mature miR-376a-5p is expressed in *ADAR2*$^{-/-}$ mice, whereas both unedited and edited miR-376a-5p are expressed in brain cortex of wild-type mice. Only unedited mature miR-376a-5p RNAs were detected in liver of wild-type mice because of a nearly total lack of pri-miR376a RNA editing in this tissue. The levels of PRPS1 were almost twofold lower in wild-type mouse cortex than in ADAR2$^{-/-}$ mouse cortex. By contrast, no difference in PRPS1 expression was detected between wild-type and *ADAR2*$^{-/-}$ liver, revealing that the edited miR-376a-5p does repress this target gene in a tissue-specific manner. PRPS1 is involved in purine metabolism and the uric acid synthesis pathway. Analysis of uric acid levels confirmed that tissue-specific repression of PRPS1 is indeed reflected in a twofold increase in uric acid levels in *ADAR2*$^{-/-}$ cortex. Thus, editing of miR-376a appears to be one of the mechanisms that ensure tight regulation of uric acid levels in select tissues such as brain cortex (140).

Frequency and Fate of miRNA Editing

A large-scale survey of pri-miRNAs subject to A→I editing identified 86 A→I editing sites in 47 pri-miRNAs, which are highly edited in human brain (141). Statistical analysis revealed the most preferred sequence and structure surrounding pri-miRNA editing sites. The adenosine residue of the U<u>A</u>G triplet with a partner nucleotide of cytidine appears to be the most frequently and highly edited (141), consistent with the sequence preference of the highly edited adenosine residues found previously by analysis of synthetic dsRNAs (6). In vitro processing assays conducted on select pri-miRNAs (pri-let7-g, pri-miR-33, pri-miR-133a2, pri-miR-197, pri-miR-203, and pri-miR-379) suggest that the majority of pri-miRNA editing events suppresses or enhances miRNA processing steps (**Figure 5*b*,*c***). Additional edited mature miRNAs (miR-99b-3p, miR-379-5p, miR-411-5p, and miR-607) with altered seed sequences are also detected. These edited mature miRNAs are predicted to repress a set of genes that differ from those targeted by the unedited miRNAs (**Figure 5*d***). Approximately 16% of human pri-miRNAs are edited in brain, and thus the expression of a large number of genes is expected to be affected globally by A→I editing of miRNAs (141). Conversely, ADARs may have affected development of mammalian miRNAs during evolution because of a selective force imposed on miRNA sequences by A→I editing.

Although this is yet to be demonstrated, A→I editing of certain pri-miRNAs at specific sites may suppress their export from the nucleus by RanGTP/exportin-5 (**Figure 5*a***). A→I RNA editing is expected to affect the local stability of the duplex. The selection of the "effective" miRNA strand that is loaded onto miRISC and that guides the miRNA to the target mRNA depends on the local stability of the sense-antisense miRNA duplex (**Figure 5*a***) (133, 134, 142). Thus, editing may also affect the selection of the effective miRNA strand.

Modulation of the miRNA Target Sites by A→I Editing?

The possibility that editing of miRNA target sites may affect miRNA-mediated gene silencing has been proposed (143). Because A→I editing occurs frequently within 3′ UTRs, which are also common miRNA-binding (target) sites, A→I editing of 3′ UTRs may create or eliminate the miRNA target site. However, computational screening for the frequency of editing at sites that contain potential miRNA target sites revealed that RNA editing tends to avoid miRNA target sites (143).

INVOLVEMENT OF ADAR AND I-RNA IN OTHER FUNCTIONS

Several reports have implicated ADAR and/or inosine-containing RNAs (edited RNA molecules) in functions that are currently not thoroughly understood. Nonetheless, these reports may indicate important future directions of the field.

Function of I-dsRNA

dsRNAs containing tandem repeats of I·U base pairs have a unique geometry and reduced stability compared to those containing A:U or G·U pairs (129). Moreover, they appear to be recognized by certain cellular proteins, e.g., Tudor-SN (129). Hyperedited I-dsRNA may have a unique function (144). I-dsRNAs specifically bind a complex that consists of proteins previously characterized as components of stress granules. During stress, a subset of cellular mRNAs are translationally silenced by sequestration into stress granules. Apparently, I-dsRNAs complexed with stress granule component proteins downregulate gene expression by inhibiting the translation initiation step (144). One of the I-dsRNAs tested for its translation suppression activity was the edited pri-miR-142 (**Figure 5*b***) (138). The presence of edited pri-miR-142 substantially reduces the expression of a luciferase reporter, indicating that edited miRNAs in general may have the potential for regulating gene expression in *trans* (144).

Involvement of ADAR in Other RNA Processing Pathways?

A→I RNA editing machinery appears to interact with several posttranscriptional RNA processing pathways. Involvement of the intron sequence in the formation of the dsRNA structure essential for editing as well as identification of many intronic editing sites predicted that interactions occur between splicing machinery and editing enzymes. Both ADAR1 and ADAR2 associate with spliceosomal component proteins (Sm and SR) within large nuclear ribonucleoprotein particles (145). Indeed, coordination of editing and splicing has been noted for the editing of GluR-B pre-mRNA. Editing at the GluR-B Q/R site, 24 bases upstream of the intron 5′ splice donor site, stimulates splicing (146). Furthermore, RNA editing and the nonsense-mediated mRNA decay (NMD) mechanism also reportedly interact: The NMD-associated protein Upf1 associates with ADAR1, leading to the proposed degradation of pre-mRNAs containing edited Alu repeats by the NMD mechanism with the aid of Upf1 (147).

Heterochromatic Silencing?

The possible involvement of A→I RNA editing in the heterochromatic silencing mechanism was proposed following the identification of vigilin as a cellular factor that binds I-RNAs (148). Vigilin localized to heterochromatin is a component of a complex that contains ADAR1, the Ku86/Ku70 heterodimer (DNA-binding proteins involved in the DNA repair mechanism); RNA helicase A (RHA), the histone variant H2AX; and the heterochromatin protein HP1α (148). RHA has various proposed functions, such as unwinding a dsRNA structure formed around the exon-intron of the *D. melanogaster* Na^+ channel gene, which is also an A→I RNA editing target (75). The vigilin/ADAR1/Ku heterodimer/RHA complex recruits the DNA-protein kinase DNA-PKcs enzyme, which phosphorylates a set of target proteins including HP1α and H2AX and also plays a major role in the chromatin silencing mechanism (148). The presence of ADAR1 in this complex suggests its involvement and a role for A→I RNA editing or edited RNAs (I-RNAs) in the heterochromatic gene-silencing mechanism (148). Vigilin supposedly recognizes the genomic loci producing dsRNAs derived from repetitive sequences, such as retrotransposons, through binding to hyperedited I-RNAs and recruits other factors (such as histone methylase SUV39HZ) essential for transformation of the region into heterochromatin (149).

ADAR1 Function in Antiviral Mechanisms and Immunity?

Interferons induce the upregulation of ADAR1 (56), thus raising the possibility that ADAR1 functions in host defense mechanisms against viral infection and inflammation. For instance, ADAR1 edits the hepatitis C virus RNA genome and inhibits its replication (150). Inosine-containing mRNAs increase in T lymphocytes and macrophages stimulated with a variety of inflammatory mediators, including tumor necrosis factor-α and interferon-γ (151). Furthermore, ADAR1 interacts with nuclear factor 90 (NF90) family proteins, also known as interleukin enhancer-binding factor 3 (ILF3), and affects the NF90-mediated gene regulatory mechanism (152). NF90 is a known regulator that stimulates the activation of the cellular antiviral expression cascade and activates interferon-β (152). Suppression of the interferon activation pathway through the ADAR1-NF90 interaction is important for protecting hematopoietic progenitor cells from apoptosis (84). Finally, the inhibitory effects of ADAR1 on interferon-β induction of the cellular response to DNA treatment have been reported, supporting ADAR1's function in DNA-mediated immune responses (153).

CONCLUSIONS AND PERSPECTIVES

Nearly two decades ago, ADARs were originally discovered as a mysterious dsRNA-unwinding activity. Soon thereafter, ADARs were identified as dsRNA-specific adenosine deaminase enzymes that are essential for the recoding type A→I editing of important mammalian genes, such as *GluRs* and *5-HT*$_{2C}$*Rs*. Since then, great progress has been made with regard to identifying new RNA editing targets, the ADAR enzyme structure, and the site-selective editing mechanism. However, many questions remain to be answered.

One of the most important issues to address is the biological significance of A→I editing of noncoding RNAs, especially in relation to RNAi-mediated gene silencing. We are now beginning to realize that A→I editing and ADAR functions are more global, affecting expression of many more genes than previously anticipated. It appears that the primordial function of ADAR may be to deal with noncoding dsRNAs, such as those made from repetitive elements of transposons and retrotransposons as well as pri-miRNAs, and to counterbalance the efficacy of RNAi in the animal kingdom, possibly along with the expansion of repeat elements in the genome. Editing of protein-coding sites may have appeared more recently, perhaps by chance at the beginning, but were retained because of their extreme advantage. The Functional Annotation of Mouse (FANTOM) project has revealed that mouse retrotransposons are much more transcriptionally active than expected (154). The expression of these SINE, LINE, and LTR elements is tightly regulated in a tissue-specific manner (154). It would be interesting to determine whether A→I RNA editing is involved in the regulation of the retrotransposon transcriptome.

The inactivation of ADAR1 leads to an embryonic-lethal phenotype, which is caused by widespread apoptosis (60, 63, 83). Editing of currently unknown target dsRNA(s) appears to protect developing embryos from massive apoptosis. It is tempting to speculate that the critical function of ADAR1 may be to regulate the expression of noncoding RNAs, perhaps a select set of miRNAs or the founder dsRNAs of esiRNAs.

Apart from further investigation into RNAi and RNA editing interactions, studies in other areas are needed also. Studies on the interaction of ADAR with other proteins and factors (e.g., vigilin) revealed the involvement of A→I RNA editing in previously unexpected areas (e.g., heterochromatic gene silencing). Investigations into additional proteins and machinery that interact with ADAR are anticipated to reveal new functions of A→I RNA editing. ADAR expression levels in certain tissues and/or developing embryos do not necessarily correlate with A→I RNA editing activities, indicating the presence of a posttranslational regulatory mechanism (79). Almost no investigations exist on the posttranslational modification of ADARs, except for one report on sumoylation of ADAR1 and downregulation of its activity (155). Furthermore, a better understanding of the mechanism that controls the cellular distribution of ADARs (nuclear import and export) seems to be essential. Investigation into such mechanisms—e.g., identification of regulatory factors, control of ADAR cellular distribution, and posttranslational modifications—is needed to better understand the mechanism regulating ADAR activity.

SUMMARY POINTS

1. A→I RNA editing is catalyzed by adenosine deaminases acting on RNA (ADARs).
2. Three mammalian ADAR genes (*ADAR1–3*) with common functional domains have been identified.
3. ADARs edit protein-coding sequences of a limited number of genes, such as glutamate receptor GluR-B and serotonin receptor 2C, resulting in dramatic alterations of protein functions.
4. Deficiencies in the A→I RNA editing mechanism cause human diseases and pathophysiology.
5. Bioinformatics studies have identified numerous A→I RNA editing sites genome wide in Alu and LINE-repetitive RNA sequences located within introns and untranslated regions (UTRs).
6. Although the biological significance of A→I editing of repeat RNAs remains mostly unknown, one possibility is to control the synthesis of esiRNAs derived from retrotransposons.
7. A→I RNA editing and RNAi mechanisms appear to interact and compete for common substrate dsRNAs.
8. The biogenesis and function of certain miRNAs are regulated by the editing of their precursors.

FUTURE ISSUES

1. The physiological significance of recoding editing of newly discovered protein-coding genes needs to be identified.
2. Studies on mechanisms that regulate ADAR enzymatic activities, including posttranslational modification and nuclear cytoplasmic transport control, and that identify regulatory factors are necessary.
3. Structural studies on full-length ADAR proteins in complex with their RNA substrates are essential for understanding of the site-selective editing mechanism.
4. We need to better understand the significance of the editing of noncoding repetitive RNAs of transposons and retrotransposons.
5. Comprehensive identification of editing sites of pri-miRNAs and edited miRNAs and determination of their significance are required.
6. The interaction between RNAi and RNA editing pathways needs to be further investigated.
7. The RNA substrate(s) critical for apoptosis-prone $ADAR1^{-/-}$ mouse embryos must be identified.

DISCLOSURE STATEMENT

The author is not aware of any affiliations, memberships, funding, or financial holdings that might be perceived as affecting the objectivity of this review.

ACKNOWLEDGMENTS

I am grateful to members of my laboratory—Hiromitsu Ota, Boris Zinshteyn, and Bjørn-Erik Wulff—for their comments and suggestions. This work was supported in part by grants from the U.S. National Institutes of Health, the Ellison Medical Foundation, and the Commonwealth Universal Research Enhancement Program, Pennsylvania Department of Health.

LITERATURE CITED

1. Baltimore D. 2001. Our genome unveiled. *Nature* 409:814–16
2. Maniatis T, Reed R. 2002. An extensive network of coupling among gene expression machines. *Nature* 416:499–506
3. Benne R, Van den Burg J, Brakenhoff JP, Sloof P, Van Boom JH, Tromp MC. 1986. Major transcript of the frameshifted coxII gene from trypanosome mitochondria contains four nucleotides that are not encoded in the DNA. *Cell* 46:819–26
4. Gott JM, Emeson RB. 2000. Functions and mechanisms of RNA editing. *Annu. Rev. Genet.* 34:499–531
5. Nishikura K. 2006. Editor meets silencer: crosstalk between RNA editing and RNA interference. *Nat. Rev. Mol. Cell Biol.* 7:919–31
6. Bass BL. 2002. RNA editing by adenosine deaminases that act on RNA. *Annu. Rev. Biochem.* 71:817–46
7. Jepson JE, Reenan RA. 2008. RNA editing in regulating gene expression in the brain. *Biochim. Biophys. Acta* 1779:459–70
8. Seeburg PH, Hartner J. 2003. Regulation of ion channel/neurotransmitter receptor function by RNA editing. *Curr. Opin. Neurobiol.* 13:279–83
9. Reenan RA. 2001. The RNA world meets behavior: A→I pre-mRNA editing in animals. *Trends Genet.* 17:53–56
10. Keegan LP, Gallo A, O'Connell MA. 2001. The many roles of an RNA editor. *Nat. Rev. Genet.* 2:869–78
11. Bass BL, Weintraub H. 1988. An unwinding activity that covalently modifies its double-stranded RNA substrate. *Cell* 55:1089–98
12. Wagner RW, Smith JE, Cooperman BS, Nishikura K. 1989. A double-stranded RNA unwinding activity introduces structural alterations by means of adenosine to inosine conversions in mammalian cells and *Xenopus* eggs. *Proc. Natl. Acad. Sci. USA* 86:2647–51
13. Bass BL, Weintraub H. 1987. A developmentally regulated activity that unwinds RNA duplexes. *Cell* 48:607–13
14. Rebagliati MR, Melton DA. 1987. Antisense RNA injections in fertilized frog eggs reveal an RNA duplex unwinding activity. *Cell* 48:599–605
15. Kim U, Garner TL, Sanford T, Speicher D, Murray JM, Nishikura K. 1994. Purification and characterization of double-stranded RNA adenosine deaminase from bovine nuclear extracts. *J. Biol. Chem.* 269:13480–89
16. O'Connell MA, Keller W. 1994. Purification and properties of double-stranded RNA-specific adenosine deaminase from calf thymus. *Proc. Natl. Acad. Sci. USA* 91:10596–600
17. Kim U, Wang Y, Sanford T, Zeng Y, Nishikura K. 1994. Molecular cloning of cDNA for double-stranded RNA adenosine deaminase, a candidate enzyme for nuclear RNA editing. *Proc. Natl. Acad. Sci. USA* 91:11457–61
18. O'Connell MA, Krause S, Higuchi M, Hsuan JJ, Totty NF, et al. 1995. Cloning of cDNAs encoding mammalian double-stranded RNA-specific adenosine deaminase. *Mol. Cell. Biol.* 15:1389–97
19. Gerber A, O'Connell MA, Keller W. 1997. Two forms of human double-stranded RNA-specific editase 1 (hRED1) generated by the insertion of an Alu cassette. *RNA* 3:453–63

20. Lai F, Chen CX, Carter KC, Nishikura K. 1997. Editing of glutamate receptor B subunit ion channel RNAs by four alternatively spliced DRADA2 double-stranded RNA adenosine deaminases. *Mol. Cell. Biol.* 17:2413–24
21. Melcher T, Maas S, Herb A, Sprengel R, Seeburg PH, Higuchi M. 1996. A mammalian RNA editing enzyme. *Nature* 379:460–64
22. Chen CX, Cho DS, Wang Q, Lai F, Carter KC, Nishikura K. 2000. A third member of the RNA-specific adenosine deaminase gene family, ADAR3, contains both single- and double-stranded RNA binding domains. *RNA* 6:755–67
23. Melcher T, Maas S, Herb A, Sprengel R, Higuchi M, Seeburg PH. 1996. RED2, a brain-specific member of the RNA-specific adenosine deaminase family. *J. Biol. Chem.* 271:31795–98
24. Slavov D, Clark M, Gardiner K. 2000. Comparative analysis of the RED1 and RED2 A-to-I RNA editing genes from mammals, pufferfish and zebrafish. *Gene* 250:41–51
25. Slavov D, Crnogorac-Jurcevic T, Clark M, Gardiner K. 2000. Comparative analysis of the DRADA A-to-I RNA editing gene from mammals, pufferfish and zebrafish. *Gene* 250:53–60
26. Palladino MJ, Keegan LP, O'Connell MA, Reenan RA. 2000. A-to-I pre-mRNA editing in *Drosophila* is primarily involved in adult nervous system function and integrity. *Cell* 102:437–49
27. Tonkin LA, Saccomanno L, Morse DP, Brodigan T, Krause M, Bass BL. 2002. RNA editing by ADARs is important for normal behavior in *Caenorhabditis elegans*. *EMBO J.* 21:6025–35
28. Palavicini JP, O'Connell MA, Rosenthal JJ. 2009. An extra double-stranded RNA binding domain confers high activity to a squid RNA editing enzyme. *RNA* 15:1208–18
29. Jin Y, Zhang W, Li Q. 2009. Origins and evolution of ADAR-mediated RNA editing. *IUBMB Life* 61:572–78
30. Gerber AP, Keller W. 2001. RNA editing by base deamination: more enzymes, more targets, new mysteries. *Trends Biochem. Sci.* 26:376–84
31. Wolf J, Gerber AP, Keller W. 2002. tadA, an essential tRNA-specific adenosine deaminase from *Escherichia coli*. *EMBO J.* 21:3841–51
32. Ryter JM, Schultz SC. 1998. Molecular basis of double-stranded RNA-protein interactions: structure of a dsRNA-binding domain complexed with dsRNA. *EMBO J.* 17:7505–13
33. Valente L, Nishikura K. 2007. RNA binding-independent dimerization of adenosine deaminases acting on RNA and dominant negative effects of nonfunctional subunits on dimer functions. *J. Biol. Chem.* 282:16054–61
34. Herbert A, Alfken J, Kim YG, Mian IS, Nishikura K, Rich A. 1997. A Z-DNA binding domain present in the human editing enzyme, double-stranded RNA adenosine deaminase. *Proc. Natl. Acad. Sci. USA* 94:8421–26
35. Maas S, Gommans WM. 2009. Novel exon of mammalian ADAR2 extends open reading frame. *PLoS ONE* 4:e4225
36. Lai F, Drakas R, Nishikura K. 1995. Mutagenic analysis of double-stranded RNA adenosine deaminase, a candidate enzyme for RNA editing of glutamate-gated ion channel transcripts. *J. Biol. Chem.* 270:17098–105
37. Macbeth MR, Schubert HL, VanDemark AP, Lingam AT, Hill CP, Bass BL. 2005. Inositol hexakisphosphate is bound in the ADAR2 core and required for RNA editing. *Science* 309:1534–39
38. Nishikura K, Yoo C, Kim U, Murray JM, Estes PA, et al. 1991. Substrate specificity of the dsRNA unwinding/modifying activity. *EMBO J.* 10:3523–32
39. Cattaneo R, Schmid A, Eschle D, Baczko K, ter Meulen V, Billeter MA. 1988. Biased hypermutation and other genetic changes in defective measles viruses in human brain infections. *Cell* 55:255–65
40. Peters NT, Rohrbach JA, Zalewski BA, Byrkett CM, Vaughn JC. 2003. RNA editing and regulation of *Drosophila 4f-rnp* expression by *sas-10* antisense readthrough mRNA transcripts. *RNA* 9:698–710
41. Fischer SE, Butler MD, Pan Q, Ruvkun G. 2008. Trans-splicing in *C. elegans* generates the negative RNAi regulator ERI-6/7. *Nature* 455:491–96
42. Lehmann KA, Bass BL. 1999. The importance of internal loops within RNA substrates of ADAR1. *J. Mol. Biol.* 291:1–13

43. Higuchi M, Single FN, Kohler M, Sommer B, Sprengel R, Seeburg PH. 1993. RNA editing of AMPA receptor subunit GluR-B: A base-paired intron-exon structure determines position and efficiency. *Cell* 75:1361–70

44. Lomeli H, Mosbacher J, Melcher T, Hoger T, Geiger JR, et al. 1994. Control of kinetic properties of AMPA receptor channels by nuclear RNA editing. *Science* 266:1709–13

45. Wang Q, O'Brien PJ, Chen CX, Cho DS, Murray JM, Nishikura K. 2000. Altered G protein–coupling functions of RNA editing isoform and splicing variant serotonin$_{2C}$ receptors. *J. Neurochem.* 74:1290–300

46. Burns CM, Chu H, Rueter SM, Hutchinson LK, Canton H, et al. 1997. Regulation of serotonin-$_{2C}$ receptor G-protein coupling by RNA editing. *Nature* 387:303–8

47. Bhalla T, Rosenthal JJ, Holmgren M, Reenan R. 2004. Control of human potassium channel inactivation by editing of a small mRNA hairpin. *Nat. Struct. Mol. Biol.* 11:950–56

48. Ohlson J, Pedersen JS, Haussler D, Ohman M. 2007. Editing modifies the GABA$_A$ receptor subunit $\alpha 3$. *RNA* 13:698–703

49. Keegan LP, Brindle J, Gallo A, Leroy A, Reenan RA, O'Connell MA. 2005. Tuning of RNA editing by ADAR is required in *Drosophila*. *EMBO J.* 24:2183–93

50. Reenan RA. 2005. Molecular determinants and guided evolution of species-specific RNA editing. *Nature* 434:409–13

51. Stefl R, Xu M, Skrisovska L, Emeson RB, Allain FH. 2006. Structure and specific RNA binding of ADAR2 double-stranded RNA binding motifs. *Structure* 14:345–55

52. Cho DS, Yang W, Lee JT, Shiekhattar R, Murray JM, Nishikura K. 2003. Requirement of dimerization for RNA editing activity of adenosine deaminases acting on RNA. *J. Biol. Chem.* 278:17093–102

53. Gallo A, Keegan LP, Ring GM, O'Connell MA. 2003. An ADAR that edits transcripts encoding ion channel subunits functions as a dimer. *EMBO J.* 22:3421–30

54. Chilibeck KA, Wu T, Liang C, Schellenberg MJ, Gesner EM, et al. 2006. FRET analysis of in vivo dimerization by RNA-editing enzymes. *J. Biol. Chem.* 281:16530–35

55. Poulsen H, Jorgensen R, Heding A, Nielsen FC, Bonven B, Egebjerg J. 2006. Dimerization of ADAR2 is mediated by the double-stranded RNA binding domain. *RNA* 12:1350–60

56. Patterson JB, Samuel CE. 1995. Expression and regulation by interferon of a double-stranded-RNA-specific adenosine deaminase from human cells: evidence for two forms of the deaminase. *Mol. Cell. Biol.* 15:5376–88

57. Kawakubo K, Samuel CE. 2000. Human RNA-specific adenosine deaminase (ADAR1) gene specifies transcripts that initiate from a constitutively active alternative promoter. *Gene* 258:165–72

58. Peng PL, Zhong X, Tu W, Soundarapandian MM, Molner P, et al. 2006. ADAR2-dependent RNA editing of AMPA receptor subunit GluR2 determines vulnerability of neurons in forebrain ischemia. *Neuron* 49:719–33

59. Reymond A, Marigo V, Yaylaoglu MB, Leoni A, Ucla C, et al. 2002. Human chromosome 21 gene expression atlas in the mouse. *Nature* 420:582–86

60. Wang Q, Khillan J, Gadue P, Nishikura K. 2000. Requirement of the RNA editing deaminase ADAR1 gene for embryonic erythropoiesis. *Science* 290:1765–68

61. Lim LP, Lau NC, Garrett-Engele P, Grimson A, Schelter JM, et al. 2005. Microarray analysis shows that some microRNAs downregulate large numbers of target mRNAs. *Nature* 433:769–73

62. Srivastava D. 2006. Making or breaking the heart: from lineage determination to morphogenesis. *Cell* 126:1037–48

63. Wang Q, Miyakoda M, Yang W, Khillan J, Stachura DL, et al. 2004. Stress-induced apoptosis associated with null mutation of ADAR1 RNA editing deaminase gene. *J. Biol. Chem.* 279:4952–61

64. Desterro JM, Keegan LP, Lafarga M, Berciano MT, O'Connell M, Carmo-Fonseca M. 2003. Dynamic association of RNA-editing enzymes with the nucleolus. *J. Cell Sci.* 116:1805–18

65. Poulsen H, Nilsson J, Damgaard CK, Egebjerg J, Kjems J. 2001. CRM1 mediates the export of ADAR1 through a nuclear export signal within the Z-DNA binding domain. *Mol. Cell. Biol.* 21:7862–71

66. Yang W, Wang Q, Howell KL, Lee JT, Cho DS, et al. 2005. ADAR1 RNA deaminase limits short interfering RNA efficacy in mammalian cells. *J. Biol. Chem.* 280:3946–53

67. Eckmann CR, Neunteufl A, Pfaffstetter L, Jantsch MF. 2001. The human but not the *Xenopus* RNA-editing enzyme ADAR1 has an atypical nuclear localization signal and displays the characteristics of a shuttling protein. *Mol. Biol. Cell* 12:1911–24
68. Fritz J, Strehblow A, Taschner A, Schopoff S, Pasierbek P, Jantsch MF. 2009. RNA-regulated interaction of transportin-1 and exportin-5 with the double-stranded RNA-binding domain regulates nucleocytoplasmic shuttling of ADAR1. *Mol. Cell. Biol.* 29:1487–97
69. Kim VN. 2005. MicroRNA biogenesis: coordinated cropping and dicing. *Nat. Rev. Mol. Cell Biol.* 6:376–85
70. Maas S, Gommans WM. 2009. Identification of a selective nuclear import signal in adenosine deaminases acting on RNA. *Nucleic Acids Res.* 37:5822–29
71. Ohta H, Fujiwara M, Ohshima Y, Ishihara T. 2008. ADBP-1 regulates an ADAR RNA-editing enzyme to antagonize RNA-interference-mediated gene silencing in *Caenorhabditis elegans*. *Genetics* 180:785–96
72. Sansam CL, Wells KS, Emeson RB. 2003. Modulation of RNA editing by functional nucleolar sequestration of ADAR2. *Proc. Natl. Acad. Sci. USA* 100:14018–23
73. Hoopengardner B, Bhalla T, Staber C, Reenan R. 2003. Nervous system targets of RNA editing identified by comparative genomics. *Science* 301:832–36
74. Rosenthal JJ, Bezanilla F. 2002. Extensive editing of mRNAs for the squid delayed rectifier K^+ channel regulates subunit tetramerization. *Neuron* 34:743–57
75. Reenan RA, Hanrahan CJ, Ganetzky B. 2000. The mle^{napts} RNA helicase mutation in *Drosophila* results in a splicing catastrophe of the *para* Na^+ channel transcript in a region of RNA editing. *Neuron* 25:139–49
76. Polson AG, Bass BL, Casey JL. 1996. RNA editing of hepatitis delta virus antigenome by dsRNA-adenosine deaminase. *Nature* 380:454–56
77. Niswender CM, Copeland SC, Herrick-Davis K, Emeson RB, Sanders-Bush E. 1999. RNA editing of the human serotonin 5-hydroxytryptamine 2C receptor silences constitutive activity. *J. Biol. Chem.* 274:9472–78
78. Rula EY, Lagrange AH, Jacobs MM, Hu N, Macdonald RL, Emeson RB. 2008. Developmental modulation of $GABA_A$ receptor function by RNA editing. *J. Neurosci.* 28:6196–201
79. Wahlstedt H, Daniel C, Ensterö M, Öhman M. 2009. Large-scale mRNA sequencing determines global regulation of RNA editing during brain development. *Genome Res.* 19:978–86
80. Smith LA, Peixoto AA, Hall JC. 1998. RNA editing in the *Drosophila* Dmca1A calcium-channel alpha 1 subunit transcript. *J. Neurogenet.* 12:227–40
81. Wu YM, Kung SS, Chen J, Chow WY. 1996. Molecular analysis of cDNA molecules encoding glutamate receptor subunits, fGluR1 alpha and fGluR1 beta, of *Oreochromis* sp. *DNA Cell Biol.* 15:717–25
82. Higuchi M, Maas S, Single FN, Hartner J, Rozov A, et al. 2000. Point mutation in an AMPA receptor gene rescues lethality in mice deficient in the RNA-editing enzyme ADAR2. *Nature* 406:78–81
83. Hartner JC, Schmittwolf C, Kispert A, Muller AM, Higuchi M, Seeburg PH. 2004. Liver disintegration in the mouse embryo caused by deficiency in the RNA-editing enzyme ADAR1. *J. Biol. Chem.* 279:4894–902
84. Hartner JC, Walkley CR, Lu J, Orkin SH. 2009. ADAR1 is essential for the maintenance of hematopoiesis and suppression of interferon signaling. *Nat. Immunol.* 10:109–15
85. Singh M, Kesterson RA, Jacobs MM, Joers JM, Gore JC, Emeson RB. 2007. Hyperphagia-mediated obesity in transgenic mice misexpressing the RNA-editing enzyme ADAR2. *J. Biol. Chem.* 282:22448–59
86. Rueter SM, Dawson TR, Emeson RB. 1999. Regulation of alternative splicing by RNA editing. *Nature* 399:75–80
87. Feng Y, Sansam CL, Singh M, Emeson RB. 2006. Altered RNA editing in mice lacking ADAR2 autoregulation. *Mol. Cell. Biol.* 26:480–88
88. Palladino MJ, Keegan LP, O'Connell MA, Reenan RA. 2000. dADAR, a *Drosophila* double-stranded RNA-specific adenosine deaminase is highly developmentally regulated and is itself a target for RNA editing. *RNA* 6:1004–18
89. Brusa R, Zimmermann F, Koh DS, Feldmeyer D, Gass P, et al. 1995. Early-onset epilepsy and postnatal lethality associated with an editing-deficient *GluR-B* allele in mice. *Science* 270:1677–80
90. Kawahara Y, Grimberg A, Teegarden S, Mombereau C, Liu S, et al. 2008. Dysregulated editing of serotonin 2C receptor mRNAs results in energy dissipation and loss of fat mass. *J. Neurosci.* 28:12834–44

91. Maas S, Kawahara Y, Tamburro KM, Nishikura K. 2006. A-to-I RNA editing and human disease. *RNA Biol.* 3:1–9
92. Schmauss C. 2005. Regulation of serotonin 2C receptor pre-mRNA editing by serotonin. *Int. Rev. Neurobiol.* 63:83–100
93. Miyamura Y, Suzuki T, Kono M, Inagaki K, Ito S, et al. 2003. Mutations of the RNA-specific adenosine deaminase gene (*DSRAD*) are involved in dyschromatosis symmetrica hereditaria. *Am. J. Hum. Genet.* 73:693–99
94. Kawahara Y, Ito K, Sun H, Aizawa H, Kanazawa I, Kwak S. 2004. Glutamate receptors: RNA editing and death of motor neurons. *Nature* 427:801
95. Gurevich I, Tamir H, Arango V, Dwork AJ, Mann JJ, Schmauss C. 2002. Altered editing of serotonin 2C receptor pre-mRNA in the prefrontal cortex of depressed suicide victims. *Neuron* 34:349–56
96. Niswender CM, Herrick-Davis K, Dilley GE, Meltzer HY, Overholser JC, et al. 2001. RNA editing of the human serotonin 5-HT_{2C} receptor: alterations in suicide and implications for serotonergic pharmacotherapy. *Neuropsychopharmacology* 24:478–91
97. Ishiuchi S, Tsuzuki K, Yoshida Y, Yamada N, Hagimura N, et al. 2002. Blockage of Ca^{2+}-permeable AMPA receptors suppresses migration and induces apoptosis in human glioblastoma cells. *Nat. Med.* 8:971–78
98. Maas S, Patt S, Schrey M, Rich A. 2001. Underediting of glutamate receptor GluR-B mRNA in malignant gliomas. *Proc. Natl. Acad. Sci. USA* 98:14687–92
99. Cenci C, Barzotti R, Galeano F, Corbelli S, Rota R, et al. 2008. Down-regulation of RNA editing in pediatric astrocytomas: ADAR2 editing activity inhibits cell migration and proliferation. *J. Biol. Chem.* 283:7251–60
100. Paz N, Levanon EY, Amariglio N, Heimberger AB, Ram Z, et al. 2007. Altered adenosine-to-inosine RNA editing in human cancer. *Genome Res.* 17:1586–95
101. Paul MS, Bass BL. 1998. Inosine exists in mRNA at tissue-specific levels and is most abundant in brain mRNA. *EMBO J.* 17:1120–27
102. Athanasiadis A, Rich A, Maas S. 2004. Widespread A-to-I RNA editing of Alu-containing mRNAs in the human transcriptome. *PLoS Biol.* 2:e391
103. Blow M, Futreal PA, Wooster R, Stratton MR. 2004. A survey of RNA editing in human brain. *Genome Res.* 14:2379–87
104. Kim DDY, Kim TTY, Walsh T, Kobayashi Y, Matise TC, et al. 2004. Widespread RNA editing of embedded *Alu* elements in the human transcriptome. *Genome Res.* 14:1719–25
105. Levanon EY, Eisenberg E, Yelin R, Nemzer S, Hallegger M, et al. 2004. Systematic identification of abundant A-to-I editing sites in the human transcriptome. *Nat. Biotechnol.* 22:1001–5
106. Eisenberg E, Nemzer S, Kinar Y, Sorek R, Rechavi G, Levanon EY. 2005. Is abundant A-to-I RNA editing primate-specific? *Trends Genet.* 21:77–81
107. Neeman Y, Levanon EY, Jantsch MF, Eisenberg E. 2006. RNA editing level in the mouse is determined by the genomic repeat repertoire. *RNA* 12:1802–9
108. Clutterbuck DR, Leroy A, O'Connell MA, Semple CA. 2005. A bioinformatic screen for novel A-I RNA editing sites reveals recoding editing in BC10. *Bioinformatics* 21:2590–95
109. Levanon EY, Hallegger M, Kinar Y, Shemesh R, Djinovic-Carugo K, et al. 2005. Evolutionarily conserved human targets of adenosine to inosine RNA editing. *Nucleic Acids Res.* 33:1162–68
110. Stapleton M, Carlson JW, Celniker SE. 2006. RNA editing in *Drosophila melanogaster*: new targets and functional consequences. *RNA* 12:1922–32
111. Gommans WM, Tatalias NE, Sie CP, Dupuis D, Vendetti N, et al. 2008. Screening of human SNP database identifies recoding sites of A-to-I RNA editing. *RNA* 14:2074–85
112. Eisenberg E, Adamsky K, Cohen L, Amariglio N, Hirshberg A, et al. 2005. Identification of RNA editing sites in the SNP database. *Nucleic Acids Res.* 33:4612–17
113. Li JB, Levanon EY, Yoon JK, Aach J, Xie B, et al. 2009. Genome-wide identification of human RNA editing sites by parallel DNA capturing and sequencing. *Science* 324:1210–13
114. Valente L, Nishikura K. 2005. ADAR gene family and A-to-I RNA editing: diverse roles in posttranscriptional gene regulation. *Prog. Nucleic Acid Res. Mol. Biol.* 79:299–338

115. Lev-Maor G, Sorek R, Levanon EY, Paz N, Eisenberg E, Ast G. 2007. RNA-editing-mediated exon evolution. *Genome Biol.* 8:R29
116. Moller-Krull M, Zemann A, Roos C, Brosius J, Schmitz J. 2008. Beyond DNA: RNA editing and steps toward *Alu* exonization in primates. *J. Mol. Biol.* 382:601–9
117. Zhang Z, Carmichael GG. 2001. The fate of dsRNA in the nucleus: a p54^{nrb}-containing complex mediates the nuclear retention of promiscuously A-to-I edited RNAs. *Cell* 106:465–75
118. Prasanth KV, Prasanth SG, Xuan Z, Hearn S, Freier SM, et al. 2005. Regulating gene expression through RNA nuclear retention. *Cell* 123:249–63
119. Chen LL, DeCerbo JN, Carmichael GG. 2008. *Alu* element-mediated gene silencing. *EMBO J.* 27:1694–705
120. Hundley HA, Krauchuk AA, Bass BL. 2008. *C. elegans* and *H. sapiens* mRNAs with edited 3′ UTRs are present on polysomes. *RNA* 14:2050–60
121. Czech B, Malone CD, Zhou R, Stark A, Schlingeheyde C, et al. 2008. An endogenous small interfering RNA pathway in *Drosophila*. *Nature* 453:798–802
122. Tam OH, Aravin AA, Stein P, Girard A, Murchison EP, et al. 2008. Pseudogene-derived small interfering RNAs regulate gene expression in mouse oocytes. *Nature* 453:534–38
123. Watanabe T, Totoki Y, Toyoda A, Kaneda M, Kuramochi-Miyagawa S, et al. 2008. Endogenous siRNAs from naturally formed dsRNAs regulate transcripts in mouse oocytes. *Nature* 453:539–43
124. Okamura K, Chung WJ, Ruby JG, Guo H, Bartel DP, Lai EC. 2008. The *Drosophila* hairpin RNA pathway generates endogenous short interfering RNAs. *Nature* 453:803–6
125. Kawamura Y, Saito K, Kin T, Ono Y, Asai K, et al. 2008. *Drosophila* endogenous small RNAs bind to Argonaute 2 in somatic cells. *Nature* 453:793–97
126. Scadden AD, Smith CW. 2001. RNAi is antagonized by A→I hyper-editing. *EMBO Rep.* 2:1107–11
127. Bass BL. 2000. Double-stranded RNA as a template for gene silencing. *Cell* 101:235–38
128. Zamore PD, Tuschl T, Sharp PA, Bartel DP. 2000. RNAi: Double-stranded RNA directs the ATP-dependent cleavage of mRNA at 21 to 23 nucleotide intervals. *Cell* 101:25–33
129. Scadden AD. 2005. The RISC subunit Tudor-SN binds to hyper-edited double-stranded RNA and promotes its cleavage. *Nat. Struct. Mol. Biol.* 12:489–96
130. Hong J, Qian Z, Shen S, Min T, Tan C, et al. 2005. High doses of siRNAs induce *eri-1* and *adar-1* gene expression and reduce the efficiency of RNA interference in the mouse. *Biochem. J.* 390:675–79
131. Tonkin LA, Bass BL. 2003. Mutations in RNAi rescue aberrant chemotaxis of ADAR mutants. *Science* 302:1725
132. Knight SW, Bass BL. 2002. The role of RNA editing by ADARs in RNAi. *Mol. Cell* 10:809–17
133. Meister G, Tuschl T. 2004. Mechanisms of gene silencing by double-stranded RNA. *Nature* 431:343–49
134. Bartel DP. 2004. MicroRNAs: genomics, biogenesis, mechanism, and function. *Cell* 116:281–97
135. Blow MJ, Grocock RJ, van Dongen S, Enright AJ, Dicks E, et al. 2006. RNA editing of human microRNAs. *Genome Biol.* 7:R27.1–27.8
136. Luciano DJ, Mirsky H, Vendetti NJ, Maas S. 2004. RNA editing of a miRNA precursor. *RNA* 10:1174–77
137. Pfeffer S, Sewer A, Lagos-Quintana M, Sheridan R, Sander C, et al. 2005. Identification of microRNAs of the herpesvirus family. *Nat. Methods* 2:269–76
138. Yang W, Chendrimada TP, Wang Q, Higuchi M, Seeburg PH, et al. 2006. Modulation of microRNA processing and expression through RNA editing by ADAR deaminases. *Nat. Struct. Mol. Biol.* 13:13–21
139. Kawahara Y, Zinshteyn B, Chendrimada TP, Shiekhattar R, Nishikura K. 2007. RNA editing of the microRNA-151 precursor blocks cleavage by the Dicer-TRBP complex. *EMBO Rep.* 8:763–69
140. Kawahara Y, Zinshteyn B, Sethupathy P, Iizasa H, Hatzigeorgiou AG, Nishikura K. 2007. Redirection of silencing targets by adenosine-to-inosine editing of miRNAs. *Science* 315:1137–40
141. Kawahara Y, Megraw M, Kreider E, Iizasa H, Valente L, et al. 2008. Frequency and fate of microRNA editing in human brain. *Nucleic Acids Res.* 36:5270–80
142. Du T, Zamore PD. 2005. microPrimer: the biogenesis and function of microRNA. *Development* 132:4645–52
143. Liang H, Landweber LF. 2007. Hypothesis: RNA editing of microRNA target sites in humans? *RNA* 13:463–67

144. Scadden AD. 2007. Inosine-containing dsRNA binds a stress-granule-like complex and downregulates gene expression in *trans*. *Mol. Cell* 28:491–500
145. Raitskin O, Cho DS, Sperling J, Nishikura K, Sperling R. 2001. RNA editing activity is associated with splicing factors in lnRNP particles: the nuclear pre-mRNA processing machinery. *Proc. Natl. Acad. Sci. USA* 98:6571–76
146. Ryman K, Fong N, Bratt E, Bentley DL, Ohman M. 2007. The C-terminal domain of RNA Pol II helps ensure that editing precedes splicing of the GluR-B transcript. *RNA* 13:1071–78
147. Agranat L, Raitskin O, Sperling J, Sperling R. 2008. The editing enzyme ADAR1 and the mRNA surveillance protein hUpf1 interact in the cell nucleus. *Proc. Natl. Acad. Sci. USA* 105:5028–33
148. Wang Q, Zhang Z, Blackwell K, Carmichael GG. 2005. Vigilins bind to promiscuously A-to-I-edited RNAs and are involved in the formation of heterochromatin. *Curr. Biol.* 15:384–91
149. Zhou J, Wang Q, Chen LL, Carmichael GG. 2008. On the mechanism of induction of heterochromatin by the RNA-binding protein vigilin. *RNA* 14:1773–81
150. Taylor DR, Puig M, Darnell ME, Mihalik K, Feinstone SM. 2005. New antiviral pathway that mediates hepatitis C virus replicon interferon sensitivity through ADAR1. *J. Virol.* 79:6291–98
151. Yang JH, Luo X, Nie Y, Su Y, Zhao Q, et al. 2003. Widespread inosine-containing mRNA in lymphocytes regulated by ADAR1 in response to inflammation. *Immunology* 109:15–23
152. Nie Y, Ding L, Kao PN, Braun R, Yang JH. 2005. ADAR1 interacts with NF90 through double-stranded RNA and regulates NF90-mediated gene expression independently of RNA editing. *Mol. Cell. Biol.* 25:6956–63
153. Wang Z, Choi MK, Ban T, Yanai H, Negishi H, et al. 2008. Regulation of innate immune responses by DAI (DLM-1/ZBP1) and other DNA-sensing molecules. *Proc. Natl. Acad. Sci. USA* 105:5477–82
154. Faulkner GJ, Kimura Y, Daub CO, Wani S, Plessy C, et al. 2009. The regulated retrotransposon transcriptome of mammalian cells. *Nat. Genet.* 41:563–71
155. Desterro JM, Keegan LP, Jaffray E, Hay RT, O'Connell MA, Carmo-Fonseca M. 2005. SUMO-1 modification alters ADAR1 editing activity. *Mol. Biol. Cell* 16:5115–26

Regulation of mRNA Translation and Stability by microRNAs

Marc Robert Fabian,[1] Nahum Sonenberg,[1] and Witold Filipowicz[2]

[1]Department of Biochemistry and Goodman Cancer Research Center, McGill University, Montreal, Quebec, H3G 1Y6, Canada; email: marc.fabian@mail.mcgill.ca, nahum.sonenberg@mcgill.ca

[2]Friedrich Miescher Institute for Biomedical Research, 4002 Basel, Switzerland; email: Witold.Filipowicz@fmi.ch

Annu. Rev. Biochem. 2010. 79:351–79

The *Annual Review of Biochemistry* is online at biochem.annualreviews.org

This article's doi:
10.1146/annurev-biochem-060308-103103

0066-4154/10/0707-0351$20.00

Key Words

Argonaute, deadenylation, GW182, microRNA, repression

Abstract

MicroRNAs (miRNAs) are small noncoding RNAs that extensively regulate gene expression in animals, plants, and protozoa. miRNAs function posttranscriptionally by usually base-pairing to the mRNA 3′-untranslated regions to repress protein synthesis by mechanisms that are not fully understood. In this review, we describe principles of miRNA-mRNA interactions and proteins that interact with miRNAs and function in miRNA-mediated repression. We discuss the multiple, often contradictory, mechanisms that miRNAs have been reported to use, which cause translational repression and mRNA decay. We also address the issue of cellular localization of miRNA-mediated events and a role for RNA-binding proteins in activation or relief of miRNA repression.

Contents

INTRODUCTION

MicroRNAs (miRNAs) comprise a large family of small ~21-nucleotide-long noncoding RNAs that have emerged as key posttranscriptional regulators of gene expression in metazoan animals, plants, and protozoa. In mammals, miRNAs are predicted to control the activity of more than 60% of all protein-coding genes (1) and participate in the regulation of almost every cellular process investigated to date (reviewed in References 2–4). miRNAs regulate protein synthesis by base-pairing to target mRNAs. In animals, most studied miRNAs form imperfect hybrids with sequences in the mRNA 3′-untranslated region (3′ UTR), with the miRNA 5′-proximal "seed" region (positions 2–8) providing most of the pairing specificity (reviewed in References 2 and 5). Until very recently, it appeared that plant miRNAs generally base-pair to mRNAs with nearly perfect complementarity and trigger endonucleolytic mRNA cleavage by the RNA interference (RNAi) mechanism. However, new findings indicate that animal-like mechanisms also broadly operate in plants and that plant miRNAs can repress mRNA translation without a pronounced effect on mRNA stability (6, 7).

Generally, miRNAs inhibit protein synthesis either by repressing translation and/or by bringing about deadenylation and subsequent degradation of mRNA targets (reviewed in References 5, 8, and 9). More recently, however, some miRNAs were reported to activate mRNA translation (10–14). miRNAs function in the form of ribonucleoprotein complexes, miRISCs (miRNA-induced silencing complexes). Argonaute (AGO) and GW182 [glycine-tryptophan (GW) repeat-containing protein of 182 kDa] family proteins represent the best-characterized protein components (reviewed in References 8 and 9). Components of miRISC (including miRNAs as well as AGO and GW182 proteins) and repressed mRNAs are enriched in processing bodies (P bodies, also known as GW bodies), which are cytoplasmic structures thought to be involved in the storage or degradation of translationally repressed mRNAs (15, 16). Some P-body components are important for effective repression of protein synthesis by miRNAs (17–20). Recently, multivesicular bodies (MVBs) and endosomes were also identified as cellular organelles contributing to miRNA function or miRISC turnover (21, 22).

The mechanistic details of miRNA's function in repressing protein synthesis are not well understood. In addition, the results from studies conducted in different systems have often been contradictory. It is difficult to conclude whether the reported discrepancies are artifacts of different experimental approaches or whether miRNAs are indeed able to exert their repressive effects by disparate mechanisms (5, 8, 9, 23). This article reviews the current knowledge on mechanistic aspects of miRNA-induced repression of protein synthesis in animal and insect cells and discusses the disparities regarding different modes of miRNA function. We also highlight new findings indicating that miRNA-mediated repression is a regulated process. For example, under specific cellular conditions, miRNA-mediated repression can be prevented or reversed (24, 25). Moreover, factors have been identified that control repression by distinct subsets of miRNAs (26, 27). For recent reviews addressing mechanistic aspects of miRNA repression, see References 5, 8, 9, 23, 28, and 29. Other reviews discuss biogenesis (30, 31) and biological functions (2–4, 32) of miRNAs.

PRINCIPLES OF TARGET RECOGNITION BY miRNAS

miRNAs interact with their mRNA targets via base-pairing. With few exceptions, metazoan miRNAs base-pair with their targets imperfectly, following a set of rules which have been formulated based on experimental and bioinformatics analyses (2). The most stringent requirement is a contiguous and perfect Watson-Crick base-pairing of the miRNA 5′ nucleotides 2–8, representing the seed region nucleating the interaction. In addition, an A residue across position 1 of the miRNA and A or U across position 9 improve miRNA activity, although they do not need to base-pair with mRNA nucleotides. However, functional miRNA sites containing mismatches or even bulged nucleotides in the seed have also been identified as exemplified by the *Lin-41* mRNA targeted by *let-7* miRNA in *Caenorhabditis elegans* (33). Complementarity of the miRNA 3′ half is quite relaxed, though it stabilizes the interaction, particularly when the seed matching is suboptimal. Generally, miRNA-mRNA duplexes contain mismatches and bulges in the central region (miRNA positions 10–12) that prevent endonucleolytic cleavage of mRNA by an RNAi mechanism. AU-rich sequence context and structural accessibility of the sites may improve their efficacy (2). Usually, multiple sites, either for the same or different miRNAs, are required for effective repression, and when the sites are close to each other, they tend to act cooperatively (34, 35).

Most of the predicted and experimentally characterized miRNA sites are positioned in the mRNA 3′ UTR. However, animal miRNAs may also target 5′ UTR and coding regions of mRNAs, as documented by experiments involving both artificial and natural mRNAs and also by bioinformatic predictions (12, 36–39). Sites located in coding regions appear to be less robust than those in the 3′ UTR (36, 37), but inclusion of rare codons to slow down the ribosome transit through the miRNA site region can increase their potency, likely owing to the facilitated occupancy of the site by miRISC (37). Interestingly, in some instances, association of miRNAs with 5′-UTR target sites appears to activate rather than repress translation [(12, 13); and see below].

PROTEIN COMPONENTS OF miRNA RIBONUCLEOPROTEINS

The key components of miRISCs are proteins of the Argonaute family. These proteins contain three evolutionarily conserved domains, PAZ, MID, and PIWI, which interact with the

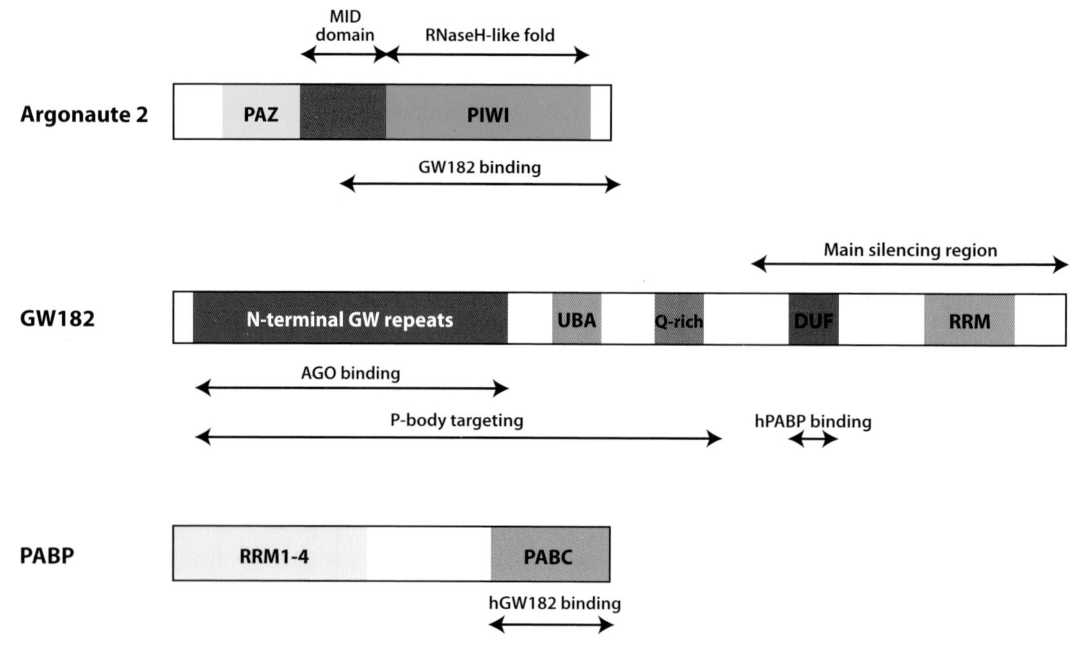

Figure 1

Schematic diagram of human Argonaute 2 (AGO2), GW182, and poly(A)-binding protein (PABP). Of the four human AGO proteins, only AGO2 functions in both miRNA repression and RNAi. It contains an enzymatically competent RNaseH-like PIWI domain, which endonucleolytically cleaves perfectly complementary RNA targets. There are three human GW182 paralogs (TNRC6A, -B, and -C), whereas *Drosophila* contains only one GW182 protein (dGW182, also known as Gawky), with a similar domain organization. *C. elegans* contains two proteins, AIN-1 and -2, which differ substantially from GW182s but perform analogous functions. The N-terminal region of GW182, containing glycine-tryptophan (GW) repeats, interacts with AGO proteins. The region, including GW-rich, ubiquitin-associated (UBA), and glutamine-rich (Q-rich) domains, is responsible for targeting GW182 proteins to P bodies. The C-terminal part of mammalian and *Drosophila* proteins (the main silencing region), containing DUF (domain of unknown function) motifs and RNA recognition motifs (RRMs), is a major effector domain, mediating translational repression and deadenylation of mRNA. Domains of PABP include four RRMs and a conserved C-terminal domain, PABC. Mammalian PABP binds directly to the silencing region of human GW182 proteins via PABC. Abbreviation: PAZ, Piwi-Argonaute-Zwilli domain.

3′ and 5′ ends of the miRNA, respectively (40, 41). The details of miRISC assembly are not well understood. The process may be coupled to miRNA processing by Dicer and to the selection of the mature miRNA strand from the complementary passenger strand (referred to as miRNA*).

Many Argonaute paralogs are encoded in metazoan and plant genomes but only some, known as AGO proteins, function in miRNA or both miRNA and siRNA pathways; others are dedicated to the function of piRNAs in germ cells or to other classes of small RNAs (40). In mammals, four AGO proteins, AGO1 through AGO4, function in miRNA repression, but only AGO2 (**Figure 1**), having an enzymatically competent RNaseH-like PIWI domain, which cleaves mRNA at the center of the siRNA-mRNA duplex, also functions in RNAi (42). Involvement of AGO1 through AGO4 in miRNA repression is demonstrated by their association with similar sets of miRNAs and proteins identified in immunoprecipitation experiments (42–45) and also by their ability to repress protein synthesis when artificially tethered to the mRNA 3′ UTR (46–48). Although these and some other data argue against paralog-specific functions of mammalian AGO proteins (49), there are indications that some AGO proteins are more potent repressors than others when tethered to reporter mRNAs (48). The cell- or tissue-specific differences in the relative abundance of individual AGOs suggest that the robustness of miRNA-mediated repression may differ between different types of cells (48). There are also indications that AGO2 in mammals may have some specific functions that cannot be complemented by the other Argonaute proteins. For example, knockdown of AGO2 in human HEK293 cells engenders a much more profound effect on miRNA-mediated repression than knockdowns of the other AGO proteins (50), and knockout of AGO2 but not the other Argonuates is embryonically lethal in mice (42). In addition, AGO2 is essential for hematopoiesis in mice (51). Because this requirement is independent of the endonucleolytic activity of AGO2, the specific role of AGO2 in hematopoiesis involves miRNA regulation rather than its potential role as an RNAi factor. In *Drosophila*, it was thought that AGO1 is exclusively dedicated to the miRNA pathway, whereas AGO2 functions in RNAi. However, recent data indicate that AGO2 also gets loaded with a subclass of miRNAs and represses protein synthesis via a mechanism that differs from the one induced by AGO1 [(52–54); and see below]. In *C. elegans*, which expresses 27 Argonaute proteins, only ALG (Argonaute-Like Gene)-1 and -2 function in the miRNA pathway (40).

Biochemical studies (55, 56) and X-ray structures of prokaryotic AGO-like proteins in complex with small RNAs (or their mimics), or in a ternary complex also including a target RNA (41, 57, 58), offer a molecular basis for some of the rules for mRNA recognition by miRNAs. The 5′-terminal miRNA nucleotide, in a monophosphorylated form, is anchored in a deep pocket at the junction of the MID and PIWI domains, with the terminal phosphate and the base interacting, either directly or via a magnesium ion, with conserved AGO amino acids. Nucleotides at miRNA positions 2–6 contact AGO through the phosphate-ribose RNA backbone and are displayed on the protein surface in a semihelical conformation, with the bases available for hydrogen bonding to the target mRNA. These properties explain why the nucleotide at the miRNA position 1 does not need to base-pair to the target and why perfect complementarity in the seed sequence is crucial for nucleating the miRNA-mRNA interaction (41, 57, 58). Association of the miRNA 3′ end with PAZ may be transiently disrupted to relieve the topological constraints during propagation of the miRNA-mRNA duplex over two helical turns (59).

GW182 proteins are another group of factors, which is crucial for the miRNA-induced repression (**Figure 1**) (60). They interact directly with and act downstream of AGOs. There are three mammalian GW182 proteins (known as TNRC6A, -B, and -C) (**Figure 1**) and a single *Drosophila* homolog (dGW182, also known as Gawky). GW182 proteins contain GW repeats in the N-terminal portion, followed by a glutamine (Q)-rich region, a domain of unknown function (DUF), and an RRM (RNA recognition motif) domain. Some GW182s contain in addition a putative ubiquitin-associated (UBA) domain. The *C. elegans* counterparts of GW182 proteins, AIN-1 and AIN-2, contain GW repeats but lack the DUF and RRM domains (61, 62). The GW repeats are responsible for the interaction of GW182 with the AGO proteins (63, 64). Disruption of the GW182-AGO interaction, by point mutations or a peptide competing with GW182 for AGO binding, abrogates miRNA-mediated repression (64, 65). The region extending from the N terminus to the Q-rich domain is responsible for targeting dGW182 to P bodies (17).

Direct tethering of GW182 to an mRNA in *Drosophila* cells leads to repression of protein synthesis even in the absence of AGO1, consistent with a mechanism whereby GW182 is the effector of AGO function (17). Recent mutagenesis analyses of mammalian GW182 proteins identified their C-terminal segment, encompassing the DUF and RRM domains as well as sequences C proximal to RRM as a minimal protein fragment, which effectively causes both translational repression and mRNA destabilization when tethered to the mRNA (66, 67). An equivalent fragment was also found to act as a repressive domain in dGW182

(63, 68), although one of these studies identified the Q-rich and N-terminal GW-rich regions of dGW182 as additional autonomous domains active in inducing repression (68). Structural and mutagenic analyses of the C-terminal inhibitory fragment of GW182 proteins indicate that regions flanking the RRM, rather than the DUF and RRM themselves, are important for repressing protein synthesis (63, 67, 69). In addition, the RRM appears not to exhibit RNA-binding activity (70; H. Mathys & W. Filipowicz, unpublished results).

miRISCs interact with several additional proteins that may function as regulatory factors that modulate miRNA function (reviewed in Reference 40). One example is the fragile X mental retardation protein, FMRP, and its *Drosophila* ortholog, dFXR, which are RNA-binding proteins known to act as modulators of translation (71, 72). Other examples include the RNA helicase RCK/p54, a P-body component that is essential for inducing repression (17, 20), and Importin 8 (Imp8), which, in addition to its role in transporting AGO2 to the nucleus, functions in miRNA repression in mammalian cells by enhancing the association of AGO2 complexes with target mRNAs (73). The TRIM-NHL family proteins, TRIM32 in mammalian cells and NHL-2 in *C. elegans*, were recently reported to enhance the activity of selected miRNAs by binding to core miRISC components (26, 27). Very little is known about how these proteins function in miRNA-mediated repression.

INTRODUCTION TO EUKARYOTIC TRANSLATION

The process of translation is divided into three steps: initiation, elongation, and termination. Initiation involves the assembly of an 80S ribosome complex positioned at the translation start site of the mRNA. This is followed by the elongation of the peptide chain. Termination entails the release of the newly synthesized protein and dissociation of ribosomal subunits from the mRNA. In eukaryotes, the rate-limiting step under most circumstances is initiation. Consequently, initiation is the most common target for translational control. All nuclear transcribed eukaryotic mRNAs contain at their 5′ end an m^7GpppN group (where N is any nucleotide) termed the 5′ cap, which facilitates ribosome recruitment to the mRNA. Some cellular and viral mRNAs are translated via alternative cap-independent mechanisms.

Cap-Dependent Translation

Cap-dependent translation requires the participation of at least 13 different eukaryotic initiation factors (eIFs). It is accomplished through a mechanism whereby the small (40S) ribosomal subunit, in a complex with a number of eIFs, binds the mRNA near the 5′ cap and scans the mRNA in a 5′→3′ direction until it encounters an AUG (or a near-cognate) codon in an optimal context (reviewed in Reference 74). Recruitment of ribosomes to the mRNA is facilitated by the 5′ cap and the 3′ poly(A) tail, via protein factors bound to these structures, the eIF4F complex, and the poly(A)-binding protein (PABP), respectively. The eIF4F contains three subunits (75, 76): (*a*) the eIF4A, an ATP-dependent RNA helicase that is thought to unwind the mRNA 5′-UTR secondary structure; (*b*) the eIF4E, a 24-kDa polypeptide that specifically interacts with the cap (77); and (*c*) the eIF4G, a large scaffolding protein that binds to both eIF4E and eIF4A and other proteins. The poly(A) tail functions as a translational enhancer whereby the PABP directly interacts with the eIF4G to effectively circularize the mRNA (78). The PABP-eIF4G interaction promotes mRNA circularization to stabilize the interaction of the eIF4E with the cap, thus enhancing the rate of translation initiation (79).

Cap-Independent Translation

The discovery of the internal ribosome entry sites (IRESs) in picornaviruses two decades ago (80, 81) has provided an alternative mechanism of translation initiation. IRESs provide an internal ribosome-binding site, thus bypassing the requirement for the cap. Subsequently, IRESs have been documented in a multitude of cellular mRNAs (82). Always (with the exception of hepatitis A virus), IRESs function independently of eIF4E. Certain IRESs [such as those of poliovirus and encephalomyocarditis virus (EMCV)] function via direct binding of the eIF4G subunit of the eIF4F complex to the IRES (83, 84). The hepatitis C virus (HCV) IRES bypasses the need for the entire eIF4 family of proteins and binds directly to eIF3 and the 40S ribosomal subunit (85). The CrPV (cricket paralysis virus) contains an intergenic IRES, which recruits the ribosome via a mechanism completely independent of initiation factors, whereby the IRES mimics an aminoacylated tRNA and positions itself within the P site of the ribosome (86, 87). This allows the CrPV IRES to initiate translation from a non-AUG codon.

miRNA-MEDIATED REPRESSION OF TRANSLATION

A large number of in vivo and in vitro studies addressed the mechanisms by which miRNAs suppress protein synthesis. These studies showed that miRNAs either inhibit translation of target mRNAs (**Figure 2**) or facilitate their deadenylation and subsequent degradation (**Figure 3**). In the following sections, we summarize these findings and analyze the molecular mechanisms involved. We discuss a complex relationship between mRNA translation and its deadenylation and decay, which could shed light on the source of diversity in the outcome of the miRNA mechanistic studies.

miRNA-Mediated Repression of Translation Initiation

Lin-4, the original miRNA, which was discovered in *C. elegans*, was initially shown to cause inhibition of translation of lin-14 without a reduction in mRNA levels or a shift in polysomes, leading to the conclusion that miRNAs inhibit mRNA translation at the elongation step (88–90). Although additional reports supported such a conclusion in other experimental systems, many other results pointed to defects in the control of translation initiation and mRNA stability.

miRNA-mediated repression of translation initiation was first observed in HeLa cells using both mono- and bicistronic reporter mRNAs whose 3′ UTRs were targeted by either endogenous (let-7) (46) or artificial (CXCR4) (91) miRNAs. Analysis of mRNA levels failed to detect pronounced degradation of miRNA-targeted mRNAs, demonstrating that translation was indeed inhibited (46, 91). Importantly, let-7 targeted mRNAs shifted to lighter fractions of polysomal density gradients, an event that is indicative of repressed translation at the initiation step caused by a defect in ribosome recruitment to the mRNA. This was not an isolated observation, as similar shifts were observed in Huh7 cells for the miR-122-targeted CAT-1 mRNA (24), in HEK293T cells for a miR-16-targeted reporter mRNA (92), and in *C. elegans* for multiple miRNA-targeted mRNAs, including the daf-12 and lin-41 mRNAs, which are regulated by the let-7 miRNA (93).

Targeting of cap-dependent translation. Several groups reported that mRNAs that lack a functional 5′-cap structure, or whose translation is cap-independent, are refractory to miRNA-mediated translational repression (46, 91, 94–97). mRNAs with a nonfunctional ApppG cap structure, targeted by the CXCR4 miRNA mimic, were not repressed as well (~twofold repression) as mRNAs bearing the m7G cap (~fivefold repression) in HeLa cells (91). mRNAs containing HCV (46), EMCV (46, 91) or CrPV (91) IRESs were refractory to miRNA-mediated repression in transfected HeLa cells. Moreover, tethering of either the eIF4E or the eIF4G to the intercistronic region of bicistronic mRNAs promoted translation of the second cistron regardless of whether let-7 target sites were present in its 3′ UTR (46). Collectively, these investigations in cultured cells pointed to the possibility that miRNAs interfere with either eIF4E function or eIF4E recruitment to the 5′-cap structure of miRNA-targeted mRNAs (**Figure 2*a***).

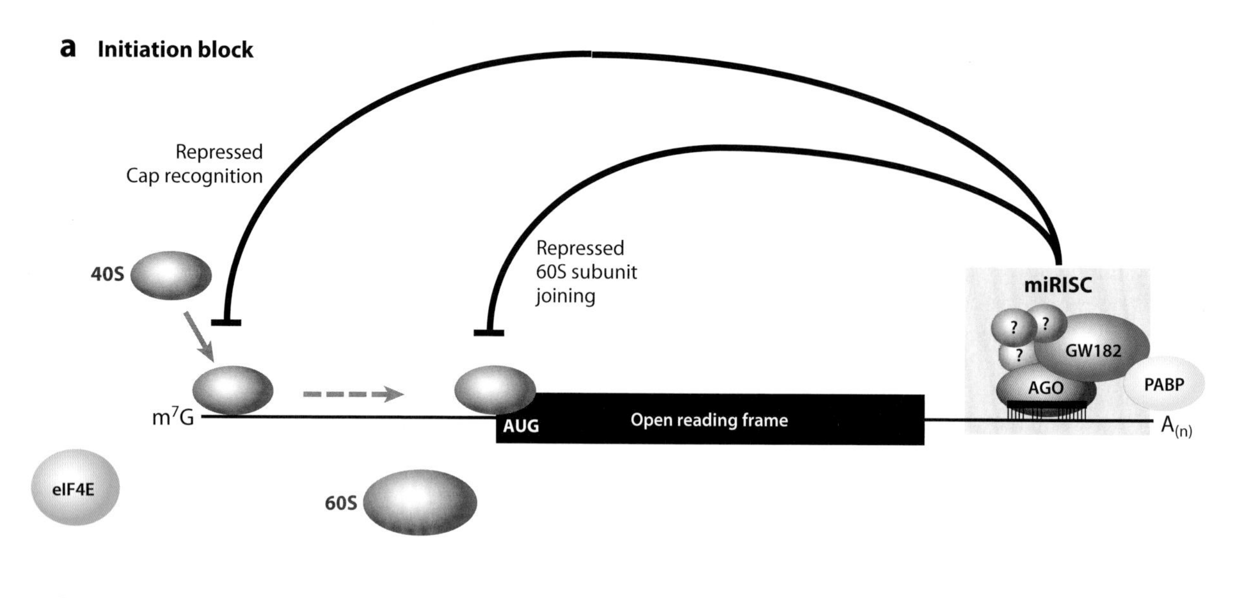

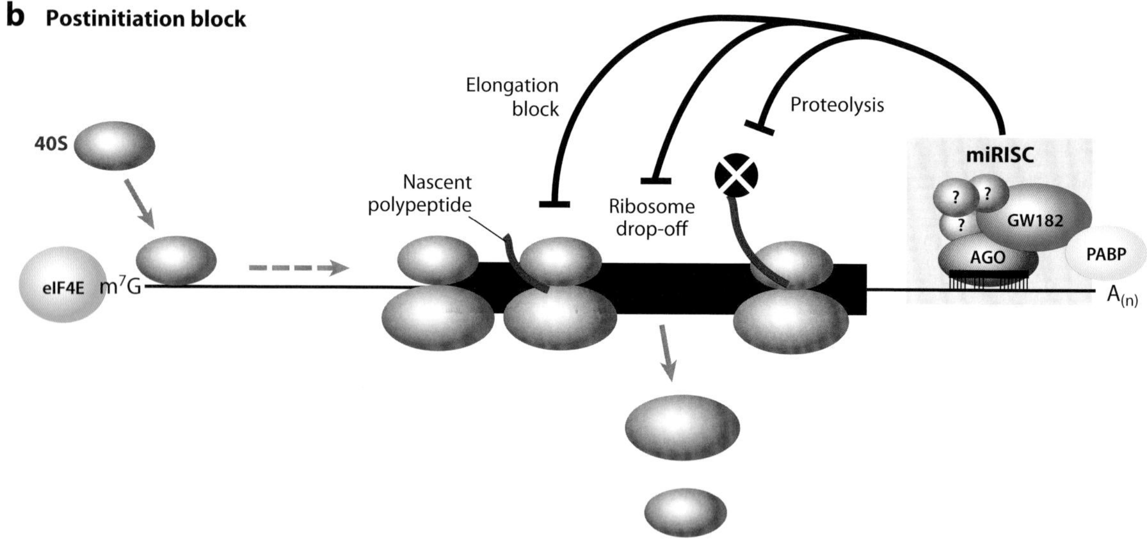

Figure 2

Schematic diagram of miRNA-mediated translational repression. (*a*) Initiation block: The miRISC inhibits translation initiation by interfering with eIF4F-cap recognition and 40S small ribosomal subunit recruitment or by antagonizing 60S subunit joining and preventing 80S ribosomal complex formation. The reported interaction of the GW182 protein with the poly(A)-binding protein (PABP) (106, 156) might interfere with the closed-loop formation mediated by the eIF4G-PABP interaction and thus contribute to the repression of translation initiation. (*b*) Postinitiation block: The miRISC might inhibit translation at postinitiation steps by inhibiting ribosome elongation, inducing ribosome drop-off, or facilitating proteolysis of nascent polypeptides. There is no mechanistic insight to any of these proposed "postinitiation" models. The 40S and 60S ribosomal subunits are represented by small and large gray spheres, respectively. Ovals with question marks represent potential additional uncharacterized miRISC proteins that might facilitate translational inhibition. Abbreviations: AGO, Argonaute; $A_{(n)}$, poly(A) tail; m^7G, the 5′-terminal cap.

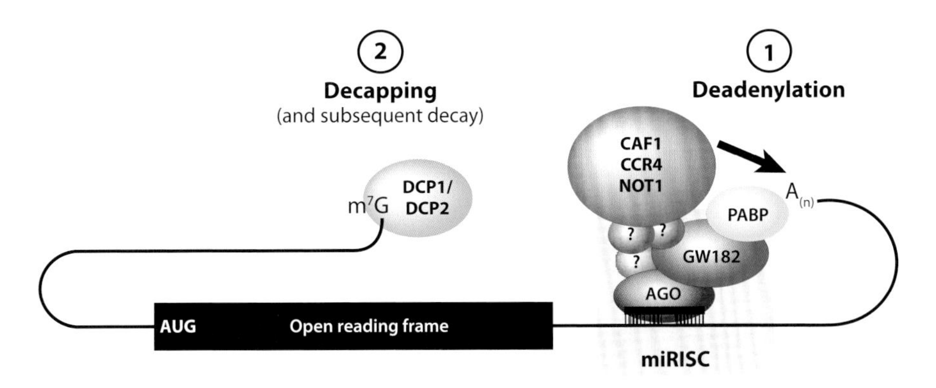

Figure 3

Schematic diagram of miRNA-mediated mRNA decay. The miRISC interacts with the CCR4-NOT1 deadenylase complex to facilitate deadenylation of the poly(A) tail [denoted by $A_{(n)}$]. Deadenylation requires the direct interaction of the GW182 protein with the poly(A)-binding protein (PABP) (see previous figures). Following deadenylation, the 5′-terminal cap (m^7G) is removed by the decapping DCP1-DCP2 complex. The open reading frame is denoted by a black rectangle. Abbreviations: AGO, Argonaute; CAF1, CCR4-associated factor; CCR4, carbon catabolite repression 4 protein; NOT1, negative on TATA-less.

The conclusions from the cell culture studies are strongly supported by in vitro experiments using cell-free systems that faithfully recapitulate the action of miRNAs in cells. In cell-free extracts from mouse Krebs II ascites cells (referred to as Krebs extracts) (97), *Drosophila* embryos (96), and HEK293 cells (95), inhibition of mRNA translation or deadenylation was dependent on the ability of the "seed sequence" of the miRNA to base-pair to the target sequence in the mRNA. Addition of oligonucleotides, which are complementary to miRNAs (antimiRs), to the extract prevented miRNA function. The three systems mentioned above made use of an endogenous miRNA targeting for in vitro synthesized mRNAs. All these aforementioned studies concluded that the miRNA-mediated translation inhibition occurs at the initiation step and is due to the interference with the cap recognition process. This is further supported by the findings that miRNAs failed to inhibit IRES-dependent translation or translation from ApppG-capped mRNAs (95–97). More detailed analyses revealed that miRNAs inhibited ribosome initiation complex formation; miR-2 inhibited both 40S ribosomal subunit recruitment and 80S initiation complex formation in fly embryo extract (96), and 80S initiation complex formation was impaired in mouse Krebs extract (97). A study by Zdanowicz et al. (98) points to the 5′-cap structure itself as being a direct target of miRNA-mediated translational repression. miR-2-targeted mRNAs bearing modifications to the triphosphate bridge of the 5′ cap demonstrated increased translational repression in both *Drosophila* embryonic extracts and S2 cells (98). Strong evidence for the notion that the cap recognition machinery is indeed the target for miRNA-mediated translational repression was the demonstration that adding a purified eIF4F complex to the Krebs extract alleviated translational repression of let-7-targeted mRNAs (97). In contrast to these results, Wang et al. (99) showed that in a rabbit reticulocyte lysate the CXCR4 artificial miRNA impairs translation by inhibiting the joining of the 60S subunit, even though the 5′ cap was required for the inhibition. It is possible that this inconsistence with former results stems from nuclease-treated rabbit reticulocyte lysate not displaying cap-poly(A) tail translational synergy (100, 101), which may, in turn, alter the outcome of miRNA-mediated repression.

Efforts to elucidate the mechanism by which miRNAs impede the cap recognition step of translation initiation have not been successful. In a promising study, Kiriakidou et al. (102) observed that AGO2 binds directly, albeit weakly, to the cap structure and suggested that this binding competes with eIF4E and results in inhibition of translation initiation. The authors reported that the AGO MID domain exhibits limited sequence homology to the cap-binding protein eIF4E and contains two aromatic residues that could function in a similar manner to those in eIF4E in sandwiching the cap structure. Mutating the two aromatics to valines abolished AGO2 interaction with m^7GTP-Sepharose and impaired its ability to repress translation when tethered to an mRNA 3′ UTR. Although this model is appealing, it was brought into question by Eulalio et al. (65), who demonstrated that mutation of the two aromatic residues interfered with the binding of AGO proteins to GW182, interacations that are required for miRNA-dependent repression. Moreover, structural modeling by Kinch & Grishin (103) indicated that AGO2 shares extremely limited, if any, structural similarity to eIF4E. Thus, it is questionable whether AGO proteins bind directly to the cap structure, and even if this is the case, the binding might not occur via the reported aromatic residues.

Another mechanism that the miRISC might utilize to inhibit cap-dependent translation is by interacting with a component of the cap-binding complex, eIF4F. Using a *Drosophila* embryo extract, Iwasaki et al. (52) demonstrated that both dAGO1, which associates with most miRNAs in *Drosophila*, and dAGO2, which is loaded only with a subclass of miRNAs (53, 54, 104, 105), can induce miRNA-mediated repression, albeit through different mechanisms of action. dAGO1 inhibits protein synthesis by repressing translation and inducing mRNA deadenylation and subsequent decay through its interaction with dGW182 (17, 65). In contrast, dAGO2 repression appears not to involve dGW182. Instead, dAGO2 was found to bind to eIF4E on targeted mRNAs, and the authors propose that dAGO2 represses cap-dependent translation by competing with eIF4G for binding to eIF4E. Notably, the dAGO2-eIF4E interaction mechanism is most likely not evolutionarily conserved as eIF4E has not been found to interact with mammalian AGO2 (106). Moreover, translation driven by tethered eIF4E was found to be refractory to let-7 repression in HeLa cells (46).

Repression by inhibiting the 80S complex assembly. As mentioned above, Wang et al. (99) reported the enrichment of 40S but not 60S ribosomal subunits in complexes formed by mRNA undergoing miRNA-mediated repression in reticulocyte lysates. The authors postulated that miRNAs may repress initiation by inhibiting 60S subunit joining (**Figure 2*a***), but the mechanism of the inhibition was not investigated. Another study also concluded that miRNAs might affect 60S joining (107). The 60S ribosomal subunit and its associated protein, eIF6, which prevents the 60S joining to 40S (108) and regulates translation (109), coimmunoprecipitated with the AGO2-Dicer-TRBP (TAR RNA-binding protein) complex (107). Depletion of eIF6 from either human cells or *C. elegans* partially alleviated the inhibition of let-7 or lin-4 miRNA targets, leading the authors to propose that miRISC association with eIF6 disrupts polysome formation by inhibiting 80S complex assembly. However, the validity of these results was brought into question by experiments showing that depletion of eIF6 from *Drosophila* S2 cells has had no noticeable effect on miRNA-mediated repression (65). Of note, knocking down eIF6 in *C. elegans* strongly interferes with the production of mature lin-4 miRNA (107), raising the possibility that it may have impacted the maturation and/or loading of miRNAs into the active miRISC. Also, Ding et al. (110) reported that knockdown of eIF6 in *C. elegans* enhanced rather than diminished let-7-mediated repression.

Translational repression of initiation and the poly(A) tail. It is well established that the PABP enhances cap-dependent translation of mRNAs, most probably by interacting with the eIF4G of the eIF4F complex (79, 111). Thus, miRNA-mediated mRNA deadenylation is expected to cause a decrease in translation initiation. Some discrepancies exist regarding the role of the poly(A) tail in miRNA-mediated translational repression. Both the 5′ cap and poly(A) tail were required for optimal translational repression of mRNA by a miRNA mimic in HeLa cells in one study (91), but no substantial difference in the repression between capped poly(A)$^+$ and poly(A)$^-$ mRNAs was noted by others (46). More recently, Beilharz et al. (112) reported that deadenylation of the miRNA-targeted mRNA promotes translational repression. Using a mammalian cell extract derived from HEK293 cells overexpressing AGO2 and GW182 proteins and let-7 miRNA, Wakiyama et al. (95) observed no miRNA repression of a nonadenylated mRNA reporter. However, as miRNA induced rapid mRNA deadenylation in the HEK293 cell extract, this result could reflect a potential mechanistic bias of the system favoring (because of the overexpression of miRISC components) the miRNA-mediated deadenylation rather than translational repression. Using an alternative strategy, several groups addressed the role of the poly(A) tail by replacing it with a stem loop, which acts in the 3′-end maturation of nonpolyadenylated histone mRNAs (65, 113), or by removing the poly(A) tail by the action of a ribozyme inserted in the reporter 3′ UTR (114). These experiments have revealed that in both mammalian and *Drosophila* cells, nonpolyadenylated mRNAs undergo miRNA-mediated repression, although the repression was not always as strong as for polyadenylated mRNAs (112–114). Taken together, these data demonstrate that miRNAs repress protein synthesis via both poly(A) tail–dependent and –independent mechanisms.

miRNA Repression at Postinitiation Steps

A number of studies concluded that miRNAs inhibit translation at postinitiation steps (**Figure 2**) (37, 88, 115–117). The most persuasive observations that have led to this conclusion originate from polysomal sedimentation analyses. Early investigations in *C. elegans* (88, 118) indicated that the lin-14 and lin-28 mRNAs, which are targets of lin-4 miRNA, remain associated with translating polysomes during larval development in spite of reduced protein levels. However, Ding & Grosshans (93) recently reexamined the polysome profiles of lin-14 and lin-28 mRNAs during *C. elegans* development and reported their shifting into lighter polysome fractions in response to lin-4 miRNA repression. Olsen & Ambros (88) analyzed lin-14 and lin-28 mRNAs at different developmental stages of *C. elegans* when lin-4 miRNA either is or is not expressed, but Ding & Grosshans compared wild-type and lin-4 mutant worms at the same stage of development. Thus, the experimental differences may account for the reported contradictory findings. Association of repressed mRNAs with functional polysomes was also observed in mammalian cells using reporter mRNAs targeted by both endogenous (37, 115, 116) and synthetic (117) miRNAs.

Several miRNAs, as well as AGO proteins, have also been reported to be associated with polysomal fractions in both mammalian (11, 115, 119, 120) and plant cells (7). This served as an argument in favor of the hypothesis that miRNAs inhibit translation elongation (**Figure 2b**) (116). However, the degree of translational repression is dependent on the number and possibly also positioning of miRNA target sites in the 3′ UTR (34, 35, 46). Thus, identification of miRNAs or other miRISC components in a polysome fraction of a sucrose gradient is not a definitive proof of the repression acting at the postinitiation translational steps, as cosedimenting miRNA-AGO complexes may not always be repressing their associated mRNAs.

Evidence supporting the postinitiation mechanism is not limited to association of miRNP components and repressed mRNA with translating polysomes. Several groups observed IRES-driven translation being repressed by the miRNA machinery (117, 121), in marked contrast to other studies (46, 91, 95, 97). In particular, Petersen et al. (117) observed cap-independent translation, driven by the HCV IRES and CrPV IRES, being repressed by the CXCR4 miRNA mimic. As IRES elements require fewer (e.g., EMCV and HCV) or any (e.g., CrPV) translation factors to initiate translation, these results are consistent with miRNAs inhibiting translation at a step other than initiation.

What could be the mechanism by which miRNAs inhibit translation at postinitiation steps? Unfortunately, there are no known molecular mechanisms to explain such inhibition. Conclusions drawn from metabolic labeling and ribosome run-off experiments led Petersen et al. (117) to propose that miRNAs may antagonize translation elongation by causing premature termination and subsequent ribosome drop-off (**Figure 2*b***). Interestingly, similar observations and conclusions were made for translationally repressed mammalian non-STOP mRNAs that lack in-frame termination codons (122).

On the basis of the demonstrated association of miRNA targets with polysomes, Nottrott et al. (115) proposed that the miRNA machinery recruits proteolytic enzymes to polysomes, which leads to the degradation of nascent polypeptides (**Figure 2*b***); a similar model had been previously put forward by Olsen & Ambros (88). Although not completely excluded, this model is highly improbable, as targeting polypeptides to the endoplasmic reticulum (ER) (using an ER signal recognition sequence), which should have made nascent proteins inaccessible to proteolysis, had no effect on the degree of the miRNA-mediated translational repression in HeLa cells (46). Moreover, high-throughput profiling of mammalian miRNA targets demonstrated an overrepresentation of mRNAs coding for membrane and ER proteins among translationally repressed mRNAs (123).

MODULATION OF miRNA-MEDIATED REPRESSION

miRNAs act preferentially by binding to the mRNA 3′ UTR. Hence, it comes as no surprise that 3′ UTR–binding proteins, such as HuR (24) or Dead-end 1 (Dnd1) (25), modulate miRNA-mediated repression. Regulation of miRNA repression by RNA-binding proteins is probably a widespread phenomenon. A comparative study of mRNAs interacting with Pumilio (PUF) proteins, which have been linked to let-7 repression of hbl-1 mRNA in *C. elegans* (124), showed a considerable enrichment of PUF-binding sites in the vicinity of predicted miRNA recognition sequences in human mRNAs (125). APOBEC3G (apolipoprotein B mRNA-editing enzyme catalytic polypeptide-like 3G) also appears to interfere with miRNA repression (92).

HuR is an AU-rich element (ARE)-binding protein, which counteracts the action of ARE-associating proteins known to destabilize mRNAs. It does so by competing with destabilizing proteins for binding to the mRNA 3′ UTR (reviewed in Reference 126). Bhattacharyya et al. (24) demonstrated that HuR relieves the miR-122-mediated repression of CAT-1 mRNA. In human hepatoma cells, CAT-1 mRNA is translationally repressed and localizes to P bodies in an miR-122-dependent manner. In response to cellular stress, such as amino acid starvation or ER stress, HuR translocates from the nucleus to the cytoplasm and, by binding to the CAT-1 mRNA 3′ UTR, causes the release of the mRNA from P bodies and into actively translating polysomes. In contrast to the situation with CAT-1 mRNA, repression of c-Myc mRNA by the let-7 miRNA is enhanced by the HuR binding to adjacent AREs (127). Consistently, depleting cells of HuR abrogates let-7-mediated inhibition of c-Myc. It is possible that HuR binding to the 3′ UTR modifies mRNA folding or accessibility of RISC to miRNA-binding sites. This scenario may apply to c-Myc, but is unlikely to operate during CAT-1 mRNA regulation. In the latter case,

HuR functions even when its binding site is positioned far away from the miR-122 sites. In addition, HuR derepresses mRNAs targeted not only by miR-122 but also by let-7 miRNA (24).

Dnd1 is an RNA-binding protein that is essential for primordial germ cell (PGC) survival in zebrafish (128) and mouse (129). Dnd1 prevents miR-221-mediated repression of the p27 mRNA in mammalian cells and miR-430-mediated repression of nanos1 and TDRD7 mRNAs in the PGCs of zebrafish (25). In zebrafish embryos, the nanos1 and TDRD7 mRNAs are deadenylated by miR-430 in somatic cells, but not in the PGCs regardless of the fact that miR-430 is present in both cell types (130). Depleting Dnd1, using an antisense strategy, led to a marked miR-430-dependent decrease in both nanos1 and TDRD7 mRNAs. It appears that Dnd1 prevents miRNA-mediated repression of nanos1 and TDRD7 mRNAs by binding to U-rich sequences adjacent to miRNA-binding sites. In this way, it interferes with the miRNA binding to the 3′ UTR of target mRNAs (25).

Two TRIM-NHL family proteins, NHL-2 in *C. elegans* (26) and TRIM32 in mice (27), were identified as positive regulators of miRNA activity. The *C. elegans* NHL-2 is required for full potency of let-7 and lsy-6 miRNAs. It interacts genetically and physically with the worm equivalents of AGO, GW182, and RCK/p54 proteins; the latter protein is implicated in miRNA repression in flies and mammals (20). The mouse TRIM32 enhances miRISC activity by interacting (via the NHL domain) with AGO1. Intriguingly, both TRIM-NHL proteins appear to enhance the activity of only some miRNAs. How these proteins enhance the repression of selected miRNAs and what specific features of miRISC or miRNA-mRNA complexes they recognize remain unknown.

miRNA-MEDIATED TRANSLATIONAL ACTIVATION

It was reported that miRNAs, in specific situations, activate rather than repress translation (11–13, 131). Vasudevan and coworkers (11, 131) found that miRNAs repress translation in proliferating cells but upregulate it in quiescent cells arrested in G0/G1. For example, under serum starvation conditions, the AGO2-miR369-3 complex bound to the 3′ UTR of TNFα mRNA was found to recruit the fragile X–related protein 1 (FXR1) and stimulate mRNA translation. Also, tethering of AGO2 or FXR1 to a reporter mRNA 3′ UTR activated translation in growth arrested cells. Broad translational activation by miRNAs or AGO2 in quiescent cells is rather unexpected. Of note, in G1-arrested cells in the *Drosophila* eye, miRNAs were found to repress translation of mRNAs (132). Thus, miRNA-mediated activation of translation is probably not a general mechanism in nonproliferating cells.

Examples of stimulatory effects of miRNAs interacting with the mRNA 5′ UTR were also reported. Orom et al. (12) found that miR-10a interacts with the 5′ UTR of many mRNAs encoding ribosomal proteins and is responsible for increased translation of these mRNAs in response to stress or nutrient shortage. The miRNA interaction was mapped to a region immediately downstream of the 5′ TOP (5′-terminal oligopyrimidine tract) motif characteristic of mRNAs encoding ribosomal proteins and some translation factors (133). Surprisingly, miR-10a binding to the 5′ UTR of ribosomal protein mRNAs does not seem to follow the classical miRNA-mRNA interaction rules for seed region base-pairing. Previously, a similar nonorthodox base-pairing was proposed for the interaction of miR-16 with the ARE-like element in mRNA 3′ UTR (134). It is important to establish what rules apply to these noncanonical miRNA-mRNA interactions.

miR-122, a liver-specific miRNA, stimulates replication of the HCV RNA in hepatoma cells by binding to the 5′ UTR upstream of the HCV RNA IRES (135). More recently, Henke et al. (13) found that miR-122 may also stimulate HCV RNA translation, possibly by increasing ribosome loading on the HCV IRES. As the HCV IRES can be translationally repressed by artificial tethering of multiple AGO2 molecules upstream of the IRES (121), it is possible that, in the experiments of Henke et al., miR-122 does not function by recruiting AGO and GW182 proteins, but rather acts as a chaperone modifying RNA structure and facilitating ribosome access to the HCV mRNA. Additional experimentation is required to understand why miR-122 enhances translation when binding to the HCV 5′ UTR but inhibits translation of cellular and reporter mRNAs bearing miR-122 sites in the 3′ UTR (24, 127, 136).

miRNA-MEDIATED mRNA DEADENYLATION AND DECAY

mRNA decay most often starts with the removal of the poly(A) tail by 3′–5′ exoribonucleases, which include (*a*) the CCR4 (carbon catabolite repression 4)-NOT1 (negative on TATA-less) complex, which contains, in addition to other proteins, the deadenylases CCR4/CNOT6 and CAF1/CNOT7 (CCR4-associated factor, an RNase D family deadenylase); (*b*) poly(A)-specific ribobonuclease (PARN); and (*c*) poly(A) nuclease (PAN) (137, 138). Either the mRNA is degraded in a 3′–5′ direction, or the 5′-terminal cap is first removed by the decapping enzyme (i.e., DCP1-DCP2 complex), and the body of the RNA is then degraded by Xrn1, a 5′–3′ exonuclease (139). mRNA stability is often under the control of *cis*-acting elements within the 3′ UTR, which recruit protein factors that, in turn, recruit deadenylation enzymes. Examples of these *cis*-acting elements include AU-rich elements, the *c-fos* RNA coding determinant, and miRNA target sites (28, 140, 141). miRNAs cause mRNA target degradation in human cells, *C. elegans*, *Drosophila* S2 cells, and zebrafish (17, 50, 142, 143). Many studies showed that perturbing the levels of specific miRNAs, or the activity of the miRNA machinery, has dramatic effects on the level of hundreds of miRNA targets. Several of these studies demonstrated that miRNA-mediated downregulation of target levels has important biological consequences (143–145).

Much evidence supports the idea that miRNAs destabilize target mRNAs through deadenylation and subsequent decapping and 5′–3′ exonucleolytic digestion. First, poly(A) length determination assays have demonstrated that miRNAs mediate deadenylation of a wide array of targets in a variety of systems. In zebrafish, miR-430 mediates the deadenylation of hundreds of maternal transcripts at the early stage of embryo development (143). Using mouse P19 embryonal carcinoma cells, Wu & Belasco (146) demonstrated that *lin-28* mRNA, whose levels decrease during retinoic acid-induced neuronal differentiation, is deadenylated through the activity of miR-125, a miRNA whose levels increase during differentiation (113). miRNA-mediated deadenylation has also been recapitulated in both mammalian and *Drosophila* cell-free extracts (52, 95, 106) (see below).

GW182 and miRNA-Mediated Deadenylation

Deadenylation and the subsequent decapping and decay of mRNAs targeted by miRNAs require the AGO and GW182 components of the miRISC (17). Knocking down or immunodepleting human AGO2 (106) or *Drosophila* AGO1 (17) abrogates miRNA-mediated deadenylation and stabilizes miRNA-targeted mRNAs. GW182 proteins, which interact with all mammalian AGO proteins (reviewed in References 43, 60, and 147), and *Drosophila* AGO1 (17), are also required for miRNA-mediated deadenylation and decapping (17, 18, 52, 65, 95, 106). The GW182-AGO interaction is mediated via GW repeats in the GW182 N terminus through binding to the AGO MID/PIWI domain (17, 64, 148–150). The relevance of this interaction for miRNA-mediated repression was tested using a GW-rich fragment of GW182 termed the AGO hook. A GW182 fragment encompassing the AGO hook region when expressed in *Drosophila* cells competes with GW182 for binding to AGO and interferes with miRNA-mediated repression (65). Moreover, adding the hook peptide blocked miRNA-mediated translational repression or deadenylation in vitro (64, 106, 148).

As mentioned above, tethering GW182 proteins to the mRNA represses translation and causes mRNA decay even in the absence of AGO proteins (17, 65, 68, 151), demonstrating that AGO proteins act as scaffolds to recruit GW182 to the mRNA. Thus, although AGO recruitment to mRNA can be circumvented, GW182 is indispensible. Knocking down GW182 in *Drosophila* S2 cells abrogates both translational repression and decay of miRNA-targeted mRNAs (17, 52, 68), demonstrating that GW182 is integral to both miRNA mechanisms of action. Similar results were observed in mammalian cells (67) and in *C. elegans* (62, 93). Thus, even though GW182 proteins interact with and recruit the mRNA decay machinery to miRNA-targeted mRNAs (see below), they likely also interact with translation factors and/or ribosomal subunits on target mRNAs to antagonize translation.

CCR4-NOT1 Complex and miRNA-Mediated Deadenylation

GW182 recruits the CCR4-NOT1 deadenylase complex to promote deadenylation of miRNA-targeted mRNAs (**Figure 3**). The complex can also be recruited to the mRNA by tethering GW182 to the 3′ UTR (17). CAF1 was identified as an AGO-interacting protein by MudPIT (multidimensional protein identification technology) analysis of human AGO1 and AGO2 immunoprecipitates from HEK293 cells (106). CAF1 was also pulled down from micrococcal nuclease-treated mouse Krebs ascites extracts using a biotin-labeled antilet-7 2′-*O*-methylated oligonucleotide, indicating that it interacts with let-7-loaded AGO proteins in an RNA-independent manner (106). Knocking down CAF1 or NOT1 blocked the majority of miRNA-mediated deadenylation and mRNA destabilization (114). In addition, immunodepleting CAF1 from a mammalian cell-free extract dramatically blocked let-7-mediated deadenylation (106). Thus, the requirement of the CCR4-NOT1 deadenylase complex for miRNA-mediated deadenylation is evolutionarily conserved.

Poly(A)-Binding Protein and miRNA-Mediated Deadenylation

The PABP is also required for miRNA-mediated deadenylation in vitro. Depleting the PABP from Krebs extracts blocked let-7-mediated deadenylation, which could be rescued by adding a recombinant PABP (106). How might a PABP function in miRNA-dependent deadenylation? Glutathione *S*-transferase pull-down experiments, in parallel with in vivo coimmunoprecipitations, revealed that the C terminus of GW182 directly interacts with the C terminus of the PABP in an RNA-independent manner. Moreover, blocking this interaction in Krebs extracts antagonized miRNA-mediated deadenylation in vitro (106). C-terminal fragments of human (67, 69) and *Drosophila* (63, 68) GW182 proteins, which contain both the DUF and RRM regions, mediate deadenylation and decay of mRNAs with an efficiency comparable to that of the full-length proteins when tethered to the reporter 3′ UTR. Of note, a sequence within the DUF region of both mammalian and *Drosophila* GW182 proteins shares similarity to a motif [termed a PAM2 motif (152–154)] in PABP-interacting proteins that binds the PABP C terminus (106). Moreover, the recently solved crystal structure of the mammalian DUF oligopeptide in a complex with the PABP C-terminal domain demonstrates that DUF exhibits a fold similar to the PAM2 motif (155). Thus, the DUF most likely functions as a PAM2-like motif to bind directly to the PABP C terminus.

Taken together, these data demonstrate that miRNA-mediated deadenylation requires the GW182 C terminus to interact with the PABP (**Figure 3**). Immunoprecipitated *Drosophila* GW182 protein interacts with the PABP (156), demonstrating that this interaction is evolutionarily conserved. A GW182-PABP interaction may have multiple roles in miRNA-mediated repression. It is conceivable that this interaction juxtaposes the PABP-associated poly(A) tail with the miRISC-associated deadenylase complex to facilitate initiation of the deadenylation reaction (106). The PABP is also a bona fide translation initiation factor that stimulates 40S ribosomal subunit recruitment and 80S complex formation (79). The PABP interacts with eIF4G (79, 157), Paip1 (a PABP-interacting protein that stimulates translation) (152, 158, 159), and the termination factor eRF3 (160). Thus, GW182 binding to the PABP may interfere with both translation initiation (by interfering with the mRNA "closed loop" conformation) and termination by blocking PABP binding to the factors listed above, through either competitive binding or steric hindrance. Consistent with this idea, adding a fragment of eIF4G that binds the PABP blocked miRNA-mediated deadenylation in vitro (106). In addition, overexpression of a fragment of GW182 in *Drosophila* S2 cells that binds PABP competed with PABP-eIF4G complexes (156). Using a dsRNA-mediated knockdown strategy to deplete various proteins in *Drosophila* S2 cells, Izaurralde and coworkers (17–19, 65) screened for factors involved in miRNA-mediated mRNA deadenylation, decapping, and decay. In addition to GW182, they identified the decapping DCP1-DCP2 complex and the decapping enhancer proteins Ge-1, EDC3, HPat, and Me31B. Knocking down key decapping factors led to the stabilization of the deadenylated miRNA targets. Thus, although miRNA-mediated deadenylation (**Figure 3**, step 1) is not sufficient for destabilization of target mRNAs, it mechanistically precedes the miRNA-mediated decapping (**Figure 3**, step 2) (18).

miRNA-mediated deadenylation is translation independent. It can proceed even when translation is blocked by translation inhibitors, such as cycloheximide (95, 106) or hippuristanol (106). It is also observed when the start codon of the mRNA is blocked by an antisense oligonucleotide (143) or by insertion of stable stem-loop structures in 5′ UTRs that block ribosome scanning (114). Likewise, it can occur on a model ApppN-capped mini-mRNA devoid of the coding region (106). These results demonstrate that a miRNA-targeted mRNA does not need to be translationally competent for deadenylation to proceed. Whether an actively translating mRNA first needs to be translationally repressed by the miRNA machinery to undergo deadenylation is an important question. Addition of a recombinant fragment of eIF4G that binds the PABP to Krebs extract blocks miRNA-mediated deadenylation (106), suggesting that the eIF4G-PABP interaction must be disrupted before deadenylation can commence. As the eIF4G-PABP interaction enhances cap-dependent translation, these data suggest that miRNA-mediated inhibition of translation may precede deadenylation. Indeed, kinetic analysis has revealed that let-7-mediated translational repression occurs prior to deadenylation in a mammalian cell-free extract (106).

miRNA-mediated deadenylation might also proceed on actively translating mRNAs. Using Xrn1 deletion yeast strains, Hu et al. (161) demonstrated that an actively translating mRNA can be cotranslationally deadenylated and decapped. In keeping with this result, Beilharz et al. (112) observed that the miRNA-mediated deadenylation precedes translational repression of let-7-target mRNAs in mammalian cells. Thus, miRNA-mediated deadenylation may be a cotranslational event as well. This might explain why miRNA-targeted mRNAs are frequently found in association with polysomes (115, 116, 119, 120). It would be interesting to investigate, using high-resolution poly(A) tail length determination assays, whether the polysome-associated miRNA targets are deadenylated.

Although miRNA-mediated deadenylation and subsequent mRNA decay appear to be widespread events (114, 123, 162), not all miRNA-targeted mRNAs are destabilized (for a review, see Reference 5). In addition, the Dicer mRNA in mammalian cells is translationally repressed by the let-7 miRNA, but its mRNA level remains for the most part unaffected (123). Moreover, even though deadenylation can contribute to miRNA-mediated repression, it is not absolutely required. Depletion of the CAF1 deadenylase, which abrogates let-7-mediated deadenylation, alleviates some but not all repression in Krebs extracts (106). Likewise, inhibiting deadenylation in mammalian cells, using an antisense blocking strategy, reduces but does not abolish miRNA-mediated repression (112). In addition, efficient miRNA-mediated deadenylation in vitro may require the PABP to contact the GW182 DUF motif (106), but the tethered C-terminal GW182 fragments lacking the DUF or containing mutations in its sequence are still able to efficiently repress protein synthesis in both *Drosophila* (63, 68) and mammalian (66, 67) cells. Thus, miRNA-mediated mRNA deadenylation and decay cannot account for the entirety of miRNA action.

A MULTITUDE OF INHIBITORY MECHANISMS?

On the basis of the data summarized above, it is clear that the molecular mechanisms by which miRNAs inhibit cap-dependent initiation of translation and mediate mRNA deadenylation and decay have begun to emerge. Nonetheless, there is considerable documentation to support alternative mechanisms in addition to translation initiation. One obvious and logical possibility is that miRNAs effect repression via several disparate or potentially overlapping mechanisms in a cell- or development-dependent manner. It is also possible that differences in the experimental design favor one mode of repression over another. For example, Lytle et al. (121) reported that the method of cell transfection (for example, cationic lipid versus electroporation) strongly influences the degree of miRNA-mediated repression. Differences in the transcriptional promoters used for driving expression of mRNA reporters might also account for some contradictory data (163). Although SV40 promotor-driven mRNAs shift into subpolysomal fractions upon repression by miRNA, the TK promotor-driven mRNAs do not. Therefore, promoter-dependent loading of specific RNA-binding proteins and/or differences in mRNA nuclear processing might dictate whether initiation or postinitiation repression dominates. Another attempt to explain the observed discrepancies is based on modeling of rate-limiting steps during translation. Nissan & Parker (164) postulated that some of the discrepancies could result from differences in rate-limiting steps in translation systems or in the mRNA reporters used by different investigators. For example, they argue that mRNAs containing IRESs may be refractory to miRNA inhibition because initiation is not the rate-limiting step during translation of these mRNAs.

CELLULAR COMPARTMENTALIZATION OF miRNA REPRESSION

Much evidence exists indicating that many components of the miRNA machinery and the repression process itself may not be localized in the cytosol but that they occur in association with different cellular organelles or structures.

Roles of P Bodies

Translationally repressed mRNAs can accumulate in discrete cytoplasmic foci known as P or GW bodies (15, 16) or in another class of cytoplasmic aggregates, stress granules (SGs), which form in response to various stress conditions (165). P bodies also seem to act as sites of the final steps of mRNA degradation (15, 16), although a recent report indicates that decapping and 5′–3′ exonucleolytic degradation of mRNA in yeast already occur on polysomes when mRNA still continues to be translated (161). P bodies are enriched in proteins involved in mRNA deadenylation, decapping, and degradation. For example, they contain the CCR4-NOT1 deadenylase complex, the decapping enzyme complex DCP1-DCP2, and the 5′–3′ exonuclease XRN1 (15, 16), key factors responsible for mRNA decay. P bodies are also enriched in a group of proteins referred to as decapping activators, including the helicase RCK/p54 (Me31B in *Drosophila*), HPat1, RAP55, EDC3, Ge-1/Hedls, and the heptameric LSm1-7 complex. As described above, some of the latter proteins (e.g., RCK/p54 and HPat1), or their homologs, act as translational repressors (15, 16, 165). P bodies lack ribosomes and most of the translation initiation factors.

Consistent with their role in translational repression and mRNA decay, P bodies in metazoans are also enriched in proteins participating in miRNA repression, such as AGOs and GW182s [the GW182 proteins are actually among the "founding" components of P bodies and this is why these structures are also known as GW bodies (166)], and in miRNAs themselves. Moreover, there are several reports describing a correlation between miRNA-mediated translational repression and accumulation of target mRNAs in P bodies and an inverse relationship between P-body localization and polysome association of target mRNAs (24, 46, 92, 167). For example, CAT-1 mRNA, a target of miR-122 in human hepatoma Huh7 cells, localizes to P bodies when translation is repressed but is redistributed outside of P bodies by stress. Moreover, transfecting miR-122 into cells that do not express miR-122 results in accumulation of CAT-1 mRNA in P bodies (24). Recently, Nathans et al. (168) demonstrated that miR-29a, by interacting with the 3′ UTR of the human immunodeficiency virus-1 (HIV-1), targets HIV-1 RNA to P bodies in human T lymphocytes and that disruption of P bodies by depletion of their components enhances HIV-1 viral production and infectivity. Consistent with their enrichment in P bodies, AGO and GW182 proteins and miRNAs interact with different P-body components (17, 20, 45, 46, 62, 167, 169), and the knockdown of some P-body components, e.g., RCK/p54 in mammalian cells (20) or Ge-1 and combinations of other decapping activators in *Drosophila* S2 cells (18), compromises miRNA-induced repression. Notably, a functional miRNA pathway is essential for formation of P-body aggregates. Global inhibition of miRNA biogenesis or depletion of proteins involved in miRNA repression results in dispersal of P bodies in mammalian and *Drosophila* S2 cells (169, 170).

Despite the aforementioned observations, which implicate P bodies in the miRNA-mediated silencing, important issues regarding the miRNA-P-body relationship remain unresolved. P bodies are highly dynamic structures, altering in both size and number during the cell cycle and in response to changes in the translational status of the cell (15, 16, 169–171). Depletion of certain P-body components has a strong effect on their integrity, as visualized by light microscopy. Interestingly, knockdowns of certain P-body components result in the dispersion of P bodies but do not prevent miRNA-mediated repression. These findings indicate that microscopically visible P bodies are not essential for the repression and that the presence of large P-body aggregates is a consequence rather than the cause of miRNA-induced silencing (170). However, these findings do not exclude the possibility that submicroscopic structures, possibly escaping elimination in knockdown experiments, contribute to the persisting repression. Indeed, studies of the interactions between core protein components during P-body formation in yeast indicate that assembly of submicroscopic complexes, consisting of P-body components, with individual mRNAs is sufficient to engender translational repression and/or decay (172).

The relative distribution of miRISC components between P bodies and the cytosol represents another issue of contention. Only ~1.3% of enhanced green fluorescent protein (EGFP)-tagged AGO2 localizes to P bodies in HeLa cells (173). Moreover, in FRAP (fluorescence recovery after photobleaching) experiments, the P-body-associated EGFP-AGO2 and also GFP-GW182 exchanged with the cytoplasm at a much slower rate than some P-body components involved in mRNA decay (173–175). In another study, it was found that only ~20% of ectopically expressed let-7 miRNA and repressed reporter mRNA localized to visible P bodies (46). Collectively, these data suggest that the repression either involves submicroscopic P bodies or occurs outside of them. Because many P-body components, including AGO proteins, are also found throughout the cytosol (16), a possible scenario is that repression by miRISCs is initiated in the cytosol and that the repressed mRNAs form P-body aggregates, either small or large, upon run off of ribosomes. P-body proteins having established inhibitory activity on translation (see above) might assist miRISCs in initiating the repression. The fact that miRNA repression can be recapitulated in cell-free extracts also argues against a primary role of P bodies in this process, inasmuch as these aggregates are unlikely to exist in cell-free extracts. However, pseudopolysomes that are formed in extracts from *Drosophila* embryos (96) might represent P-body-like aggregates, and it will be interesting to analyze them in some detail.

Role of Stress Granules

SGs form upon global repression of translation initiation (165). SGs share some protein components with P bodies, and SGs and P bodies are frequently located adjacent to each other, possibly exchanging their cargo material (175, 176). AGO proteins, artificial miRNA mimics, and repressed reporter mRNAs accumulate in SGs (173). Because SGs are known to form not only in response to stress, but also upon general inhibition of translation initiation (177, 178), SGs, like P bodies, may play a role in miRNA-mediated repression (173, 178). However, it is also possible that localization of miRISC components to SGs is solely due to pulling the mRNA-associated, but not necessarily inhibitory, miRISCs to SGs formed in response to stress. The latter possibility could explain why localization of AGO proteins to SGs, but not P bodies, is miRNA dependent (173). AGO proteins directly interact with core P-body components (17, 20, 64), but their localization to SGs might depend on association with miRNA to allow interaction with mRNA by base-pairing.

Role of Multivesicular Bodies and Endosomes

Association of a large fraction of AGO proteins with cellular membranes, such as the Golgi and ER, was noted some time ago, but its possible significance remained unexplored (179–181). Recent work carried out in *Drosophila* and in mammalian cells identified MVBs, specialized late endosomal compartments with a characteristic multivesicular morphology, as cellular organelles contributing to miRNA function or miRISC turnover (21, 22). MVBs sort endocytosed proteins into different compartments, including lysosomes (for proteolysis) and exosomes (for secretion). In both *Drosophila* and mammalian cells, blocking of MVB formation inhibits silencing by siRNAs and miRNAs, whereas blocking their turnover stimulates silencing. Dissection of the miRNA pathway, in vivo and in vitro, identified the loading of AGO proteins with small RNAs as a step that is enhanced when MVB turnover is impaired by inactivation of *HSP4* (Hermansky-Pudlak syndrome 4), a gene originally identified in a *Drosophila* screen, whose mutation enhances small RNA silencing (22). Gibbings et al. (21) reached similar conclusions by silencing different mammalian genes involved in MVB metabolism. Both studies investigating the role of MVBs/endosomes concluded that a large fraction of GW bodies [we use this nomenclature because GW182 proteins were mainly used to follow the localization of P/GW-body aggregates; in addition, Gibbings et al. (21) found that a considerable fraction of GW182-containing structures does not contain classical P-body markers] colocalizes with MVBs. Also, GW182 and some miRNAs, but not AGO proteins, were enriched in purified exosome-like vesicles secreted by MVBs. Whether the latter phenomenon represents a specific way of elimination of miRISC components from the cell or is indicative of the miRNA-mediated intercellular communication (182) remains to be established.

In summary, much still needs to be learned about the cellular localization of different steps in the assembly, function, and recycling of miRISCs. It will be interesting to establish how specific miRNAs are transported in neurons to get to dendritic spines, where they are implicated in regulation of local translation in response to synaptic stimulation (183, 184).

PERSPECTIVES

It is astonishing that miRNAs have evaded the radar of molecular biologists for so long, considering their paramount involvement and impact on organism and organ development, cellular differentiation, viral infection, and oncogenesis. What might we expect in the coming decade? One can anticipate solving the three-dimensional structures of the individual miRISC components and the complex itself. The knowledge generated from these structures, and possibly intermediates in miRISC assembly, should provide a comprehensive view of the molecular mechanism of miRISC formation and function.

An important challenge is to elucidate the interactions of miRISC with components of the translation and deadenylation machinery and to obtain three-dimensional structures of these supercomplexes. *Drosophila* has only one GW182 protein; however, there are several mammalian GW182 paralogs and isoforms. It is important for future studies to determine their tissue expression profiles during development and to establish whether they have redundant or unique functions. Another important challenge is to determine what dictates whether an mRNA-bound miRISC inhibits translation or initiates mRNA decay, or both. Indeed, 3′-UTR architecture, in combination with RNA-binding proteins, such as HuR and Dnd1, has already been shown to regulate miRNA accessibility and/or repressive function. Possibly, similar types of RNA-protein interactions may determine which mechanism, translational repression or deadenylation, is favored for miRNA-mediated repression.

Another important new field of miRNA research is identifying new posttranslational modifications to miRISC proteins and determining how these modifications impact miRISC function (reviewed in References 185 and 186). The Dicer-interacting protein TRBP was recently shown to be positively regulated through phosphorylation by the mitogen-activated protein kinase (MAPK) pathway (187). Moreover, human AGO2 undergoes several forms of posttranslational modification. AGO2 can be hydroxylated at proline 700 by the type I collagen prolyl-4-hydrolylase, a modification that stabilizes AGO2 and localizes it to P bodies (188). AGO and GW182 proteins are also known to be phosphorylated (166, 189). AGO2 phosphorylation at serine 387 facilitates its localization to P bodies (189). AGO2 may be negatively regulated by posttranslational modification, as phosphorylation of tyrosine 529 in the small RNA 5′ end-binding pocket interferes with small RNA loading (G. Meister, personal communication). Uncovering new posttranslational modifications and the signaling cascades that lead to these modifications is of paramount importance. As our understanding of the molecular biology of miRNA action increases, it will be possible to gain important insights into the role of miRNAs in health and disease.

SUMMARY POINTS

1. miRNAs inhibit protein synthesis by repressing translation and/or by bringing about deadenylation and subsequent degradation of mRNA targets. Generally, miRNAs function as part of ribonucleoprotein complexes, miRISCs (miRNA-induced silencing complexes), with miRNAs base-pairing to partially complementary sequences in the 3′ UTRs of target mRNAs. In certain instances, miRNAs have been also reported to activate translation of targeted mRNAs.
2. Core components of miRISCs include the AGO family of proteins, which directly anchor miRNAs in a deep pocket, and the GW182 family of proteins, which directly interact with AGO proteins via their GW repeats. GW182 proteins act downstream of AGO proteins to effect miRNA-mediated repression. AGO proteins function to bridge the miRNA to the silencing effectors, the GW182 proteins.

3. miRNAs have been found to repress translation at initiation, either by targeting the cap recognition step or by inhibiting ribosome 80S complex assembly, but repression at postinitiation steps has also been reported.
4. miRNA-mediated repression can be modulated by 3′ UTR-binding proteins such as HuR and Dnd1, and two AGO-interacting proteins of the TRIM-NHL protein family, the *C. elegans* NHL-2 protein and the mouse TRIM32 protein.
5. miRISC was shown to recruit the CCR4-NOT1 deadenylase complex to promote deadenylation of miRNA-targeted mRNAs. The PABP enhances miRNA-mediated deadenylation via its direct interaction with GW182.
6. miRNA-targeted translationally repressed mRNAs can accumulate in discrete cytoplasmic foci, such as P or GW bodies, or stress granules. A fraction of GW bodies colocalizes with multivesicular bodies (MVBs), membrane structures that play a role in miRNA-mediated repression.

FUTURE ISSUES

1. The detailed molecular mechanisms of how the miRISC represses translation are not known. Key issues include understanding how the miRISC directly contacts the deadenylation and translation machinery. In vitro reconstituted systems are likely to prove essential in addressing these issues.
2. AGO and GW182 proteins can undergo posttranslational modification. Future studies will eludicate the dynamics of these modifications and their importance for activity of the proteins. These studies will uncover the signaling pathways that posttranslationally regulate the activity of miRISC components.
3. While several RNA-binding proteins that modulate miRISC activity on specific target mRNAs have been discovered, many more are likely to be involved.

DISCLOSURE STATEMENT

The authors are not aware of any affiliations, memberships, funding, or financial holdings that might be perceived as affecting the objectivity of this review.

ACKNOWLEDGMENTS

This work was supported by a grant from the Canadian Institutes of Health Research (N.S.); the European Commission Framework Program 6 Project "Sirocco" and the Friedrich Miescher Institute, which is supported by the Novartis Research Foundation (W.F.); and by a postdoctoral fellowship from the Canadian Cancer Society (M.R.F.).

LITERATURE CITED

1. Friedman RC, Farh KK, Burge CB, Bartel DP. 2009. Most mammalian mRNAs are conserved targets of microRNAs. *Genome Res.* 19:92–105
2. Bartel DP. 2009. MicroRNAs: target recognition and regulatory functions. *Cell* 136:215–33

3. Bushati N, Cohen SM. 2007. microRNA functions. *Annu. Rev. Cell Dev. Biol.* 23:175–205
4. Ghildiyal M, Zamore PD. 2009. Small silencing RNAs: an expanding universe. *Nat. Rev. Genet.* 10:94–108
5. Filipowicz W, Bhattacharyya SN, Sonenberg N. 2008. Mechanisms of post-transcriptional regulation by microRNAs: Are the answers in sight? *Nat. Rev. Genet.* 9:102–14
6. Brodersen P, Sakvarelidze-Achard L, Bruun-Rasmussen M, Dunoyer P, Yamamoto YY, et al. 2008. Widespread translational inhibition by plant miRNAs and siRNAs. *Science* 320:1185–90
7. Lanet E, Delannoy E, Sormani R, Floris M, Brodersen P, et al. 2009. Biochemical evidence for translational repression by *Arabidopsis* microRNAs. *Plant Cell* 21:1762–68
8. Eulalio A, Huntzinger E, Izaurralde E. 2008. Getting to the root of miRNA-mediated gene silencing. *Cell* 132:9–14
9. Chekulaeva M, Filipowicz W. 2009. Mechanisms of miRNA-mediated post-transcriptional regulation in animal cells. *Curr. Opin. Cell Biol.* 21:452–60
10. Vasudevan S, Tong Y, Steitz JA. 2008. Cell-cycle control of microRNA-mediated translation regulation. *Cell Cycle* 7:1545–49
11. Vasudevan S, Tong Y, Steitz JA. 2007. Switching from repression to activation: microRNAs can up-regulate translation. *Science* 318:1931–34
12. Orom UA, Nielsen FC, Lund AH. 2008. MicroRNA-10a binds the 5′UTR of ribosomal protein mRNAs and enhances their translation. *Mol. Cell* 30:460–71
13. Henke JI, Goergen D, Zheng J, Song Y, Schuttler CG, et al. 2008. microRNA-122 stimulates translation of hepatitis C virus RNA. *EMBO J.* 27:3300–10
14. Niepmann M. 2009. Activation of hepatitis C virus translation by a liver-specific microRNA. *Cell Cycle* 8:1473–77
15. Parker R, Sheth U. 2007. P bodies and the control of mRNA translation and degradation. *Mol. Cell* 25:635–46
16. Eulalio A, Behm-Ansmant I, Izaurralde E. 2007. P bodies: at the crossroads of post-transcriptional pathways. *Nat. Rev. Mol. Cell Biol.* 8:9–22
17. Behm-Ansmant I, Rehwinkel J, Doerks T, Stark A, Bork P, Izaurralde E. 2006. mRNA degradation by miRNAs and GW182 requires both CCR4:NOT deadenylase and DCP1:DCP2 decapping complexes. *Genes Dev.* 20:1885–98
18. Eulalio A, Rehwinkel J, Stricker M, Huntzinger E, Yang SF, et al. 2007. Target-specific requirements for enhancers of decapping in miRNA-mediated gene silencing. *Genes Dev.* 21:2558–70
19. Rehwinkel J, Behm-Ansmant I, Gatfield D, Izaurralde E. 2005. A crucial role for GW182 and the DCP1:DCP2 decapping complex in miRNA-mediated gene silencing. *RNA* 11:1640–47
20. Chu CY, Rana TM. 2006. Translation repression in human cells by microRNA-induced gene silencing requires RCK/p54. *PLoS Biol.* 4:e210
21. Gibbings DJ, Ciaudo C, Erhardt M, Voinnet O. 2009. Multivesicular bodies associate with components of miRNA effector complexes and modulate miRNA activity. *Nat. Cell Biol.* 11:1143–49
22. Lee YS, Pressman S, Andress AP, Kim K, White JL, et al. 2009. Silencing by small RNAs is linked to endosomal trafficking. *Nat. Cell Biol.* 11:1150–56
23. Wu L, Belasco JG. 2008. Let me count the ways: mechanisms of gene regulation by miRNAs and siRNAs. *Mol. Cell* 29:1–7
24. Bhattacharyya SN, Habermacher R, Martine U, Closs EI, Filipowicz W. 2006. Relief of microRNA-mediated translational repression in human cells subjected to stress. *Cell* 125:1111–24
25. Kedde M, Strasser MJ, Boldajipour B, Oude Vrielink JA, Slanchev K, et al. 2007. RNA-binding protein Dnd1 inhibits microRNA access to target mRNA. *Cell* 131:1273–86
26. Hammell CM, Lubin I, Boag PR, Blackwell TK, Ambros V. 2009. *nhl-2* modulates microRNA activity in *Caenorhabditis elegans*. *Cell* 136:926–38
27. Schwamborn JC, Berezikov E, Knoblich JA. 2009. The TRIM-NHL protein TRIM32 activates microRNAs and prevents self-renewal in mouse neural progenitors. *Cell* 136:913–25
28. Standart N, Jackson RJ. 2007. MicroRNAs repress translation of m7Gppp-capped target mRNAs in vitro by inhibiting initiation and promoting deadenylation. *Genes Dev.* 21:1975–82
29. Nilsen TW. 2007. Mechanisms of microRNA-mediated gene regulation in animal cells. *Trends Genet.* 23:243–49

30. Winter J, Jung S, Keller S, Gregory RI, Diederichs S. 2009. Many roads to maturity: microRNA biogenesis pathways and their regulation. *Nat. Cell Biol.* 11:228–34
31. Kim VN, Han J, Siomi MC. 2009. Biogenesis of small RNAs in animals. *Nat. Rev. Mol. Cell Biol.* 10:126–39
32. Flynt AS, Lai EC. 2008. Biological principles of microRNA-mediated regulation: shared themes amid diversity. *Nat. Rev. Genet.* 9:831–42
33. Vella MC, Choi EY, Lin SY, Reinert K, Slack FJ. 2004. The *C. elegans* microRNA *let-7* binds to imperfect let-7 complementary sites from the *lin-41* 3′UTR. *Genes Dev.* 18:132–37
34. Grimson A, Farh KK, Johnston WK, Garrett-Engele P, Lim LP, Bartel DP. 2007. MicroRNA targeting specificity in mammals: determinants beyond seed pairing. *Mol. Cell* 27:91–105
35. Doench JG, Sharp PA. 2004. Specificity of microRNA target selection in translational repression. *Genes Dev.* 18:504–11
36. Easow G, Teleman AA, Cohen SM. 2007. Isolation of microRNA targets by miRNP immunopurification. *RNA* 13:1198–204
37. Gu S, Jin L, Zhang F, Sarnow P, Kay MA. 2009. Biological basis for restriction of microRNA targets to the 3′ untranslated region in mammalian mRNAs. *Nat. Struct. Mol. Biol.* 16:144–50
38. Rigoutsos I. 2009. New tricks for animal microRNAS: targeting of amino acid coding regions at conserved and nonconserved sites. *Cancer Res.* 69:3245–48
39. Kloosterman WP, Wienholds E, Ketting RF, Plasterk RH. 2004. Substrate requirements for let-7 function in the developing zebrafish embryo. *Nucleic Acids Res.* 32:6284–91
40. Peters L, Meister G. 2007. Argonaute proteins: mediators of RNA silencing. *Mol. Cell* 26:611–23
41. Jinek M, Doudna JA. 2009. A three-dimensional view of the molecular machinery of RNA interference. *Nature* 457:405–12
42. Liu J, Carmell MA, Rivas FV, Marsden CG, Thomson JM, et al. 2004. Argonaute2 is the catalytic engine of mammalian RNAi. *Science* 305:1437–41
43. Landthaler M, Gaidatzis D, Rothballer A, Chen PY, Soll SJ, et al. 2008. Molecular characterization of human Argonaute-containing ribonucleoprotein complexes and their bound target mRNAs. *RNA* 14:2580–96
44. Azuma-Mukai A, Oguri H, Mituyama T, Qian ZR, Asai K, et al. 2008. Characterization of endogenous human Argonautes and their miRNA partners in RNA silencing. *Proc. Natl. Acad. Sci. USA* 105:7964–69
45. Meister G, Landthaler M, Peters L, Chen PY, Urlaub H, et al. 2005. Identification of novel Argonaute-associated proteins. *Curr. Biol.* 15:2149–55
46. Pillai RS, Bhattacharyya SN, Artus CG, Zoller T, Cougot N, et al. 2005. Inhibition of translational initiation by let-7 microRNA in human cells. *Science* 309:1573–76
47. Pillai RS, Artus CG, Filipowicz W. 2004. Tethering of human Ago proteins to mRNA mimics the miRNA-mediated repression of protein synthesis. *RNA* 10:1518–25
48. Wu L, Fan J, Belasco JG. 2008. Importance of translation and nonnucleolytic Ago proteins for on-target RNA interference. *Curr. Biol.* 18:1327–32
49. Su H, Trombly MI, Chen J, Wang X. 2009. Essential and overlapping functions for mammalian Argonautes in microRNA silencing. *Genes Dev.* 23:304–17
50. Schmitter D, Filkowski J, Sewer A, Pillai RS, Oakeley EJ, et al. 2006. Effects of Dicer and Argonaute down-regulation on mRNA levels in human HEK293 cells. *Nucleic Acids Res.* 34:4801–15
51. O'Carroll D, Mecklenbrauker I, Das PP, Santana A, Koenig U, et al. 2007. A Slicer-independent role for Argonaute 2 in hematopoiesis and the microRNA pathway. *Genes Dev.* 21:1999–2004
52. Iwasaki S, Kawamata T, Tomari Y. 2009. *Drosophila* Argonaute1 and Argonaute2 employ distinct mechanisms for translational repression. *Mol. Cell* 34:58–67
53. Förstemann K, Horwich MD, Wee L, Tomari Y, Zamore PD. 2007. *Drosophila* microRNAs are sorted into functionally distinct Argonaute complexes after production by Dicer-1. *Cell* 130:287–97
54. Tomari Y, Du T, Zamore PD. 2007. Sorting of *Drosophila* small silencing RNAs. *Cell* 130:299–308
55. Parker JS, Parizotto EA, Wang M, Roe SM, Barford D. 2009. Enhancement of the seed-target recognition step in RNA silencing by a PIWI/MID domain protein. *Mol. Cell* 33:204–14
56. Ameres SL, Martinez J, Schroeder R. 2007. Molecular basis for target RNA recognition and cleavage by human RISC. *Cell* 130:101–12

57. Parker JS, Roe SM, Barford D. 2005. Structural insights into mRNA recognition from a PIWI domain-siRNA guide complex. *Nature* 434:663–66
58. Wang Y, Juranek S, Li H, Sheng G, Tuschl T, Patel DJ. 2008. Structure of an argonaute silencing complex with a seed-containing guide DNA and target RNA duplex. *Nature* 456:921–26
59. Wang Y, Juranek S, Li H, Sheng G, Wardle GS, et al. 2009. Nucleation, propagation and cleavage of target RNAs in Ago silencing complexes. *Nature* 461:754–61
60. Eulalio A, Tritschler F, Izaurralde E. 2009. The GW182 protein family in animal cells: new insights into domains required for miRNA-mediated gene silencing. *RNA* 15:1433–42
61. Zhang L, Ding L, Cheung TH, Dong MQ, Chen J, et al. 2007. Systematic identification of *C. elegans* miRISC proteins, miRNAs, and mRNA targets by their interactions with GW182 proteins AIN-1 and AIN-2. *Mol. Cell* 28:598–613
62. Ding L, Spencer A, Morita K, Han M. 2005. The developmental timing regulator AIN-1 interacts with miRISCs and may target the Argonaute protein ALG-1 to cytoplasmic P bodies in *C. elegans*. *Mol. Cell* 19:437–47
63. Eulalio A, Helms S, Fritzsch C, Fauser M, Izaurralde E. 2009. A C-terminal silencing domain in GW182 is essential for miRNA function. *RNA* 15:1067–77
64. Till S, Lejeune E, Thermann R, Bortfeld M, Hothorn M, et al. 2007. A conserved motif in Argonaute-interacting proteins mediates functional interactions through the Argonaute PIWI domain. *Nat. Struct. Mol. Biol.* 14:897–903
65. Eulalio A, Huntzinger E, Izaurralde E. 2008. GW182 interaction with Argonaute is essential for miRNA-mediated translational repression and mRNA decay. *Nat. Struct. Mol. Biol.* 15:346–53
66. Baillat D, Shiekhattar R. 2009. Functional dissection of the human TNRC6 (GW182-related) family of proteins. *Mol. Cell. Biol.* 29:4144–55
67. Zipprich JT, Bhattacharyya S, Mathys H, Filipowicz W. 2009. Importance of the C-terminal domain of the human GW182 protein TNRC6C for translational repression. *RNA* 15:781–93
68. Chekulaeva M, Filipowicz W, Parker R. 2009. Multiple independent domains of dGW182 function in miRNA-mediated repression in *Drosophila*. *RNA* 15:794–803
69. Lazzaretti D, Tournier I, Izaurralde E. 2009. The C-terminal domains of human TNRC6A, TNRC6B, and TNRC6C silence bound transcripts independently of Argonaute proteins. *RNA* 15:1059–66
70. Eulalio A, Tritschler F, Buttner R, Weichenrieder O, Izaurralde E, Truffault V. 2009. The RRM domain in GW182 proteins contributes to miRNA-mediated gene silencing. *Nucleic Acids Res.* 37:2974–83
71. Jin P, Alisch RS, Warren ST. 2004. RNA and microRNAs in fragile X mental retardation. *Nat. Cell Biol.* 6:1048–53
72. Ishizuka A, Siomi MC, Siomi H. 2002. A *Drosophila* fragile X protein interacts with components of RNAi and ribosomal proteins. *Genes Dev.* 16:2497–508
73. Weinmann L, Hock J, Ivacevic T, Ohrt T, Mutze J, et al. 2009. Importin 8 is a gene silencing factor that targets Argonaute proteins to distinct mRNAs. *Cell* 136:496–507
74. Pestova TV, Hellen CU. 2001. Functions of eukaryotic factors in initiation of translation. *Cold Spring Harb. Symp. Quant. Biol.* 66:389–96
75. Edery I, Humbelin M, Darveau A, Lee KA, Milburn S, et al. 1983. Involvement of eukaryotic initiation factor 4A in the cap recognition process. *J. Biol. Chem.* 258:11398–403
76. Grifo JA, Tahara SM, Morgan MA, Shatkin AJ, Merrick WC. 1983. New initiation factor activity required for globin mRNA translation. *J. Biol. Chem.* 258:5804–10
77. Sonenberg N, Rupprecht KM, Hecht SM, Shatkin AJ. 1979. Eukaryotic mRNA cap binding protein: purification by affinity chromatography on sepharose-coupled m7GDP. *Proc. Natl. Acad. Sci. USA* 76:4345–49
78. Sachs A. 2000. Physical and functional interactions between the mRNA cap structure and the poly(A) tail. In *Translational Control of Gene Expression*, ed. N Sonenberg, JWB Hershey, MB Mathews, pp. 447–66. Cold Spring Harbor, NY: Cold Spring Harb. Lab.
79. Kahvejian A, Svitkin YV, Sukarieh R, M'Boutchou MN, Sonenberg N. 2005. Mammalian poly(A)-binding protein is a eukaryotic translation initiation factor, which acts via multiple mechanisms. *Genes Dev.* 19:104–13

80. Pelletier J, Sonenberg N. 1988. Internal initiation of translation of eukaryotic mRNA directed by a sequence derived from poliovirus RNA. *Nature* 334:320–25
81. Jang SK, KraussIich HG, Nicklin MJ, Duke GM, Palmenberg AC, Wimmer E. 1988. A segment of the 5′ nontranslated region of encephalomyocarditis virus RNA directs internal entry of ribosomes during in vitro translation. *J. Virol.* 62:2636–43
82. Hellen CU, Sarnow P. 2001. Internal ribosome entry sites in eukaryotic mRNA molecules. *Genes Dev.* 15:1593–612
83. Hellen CU, Wimmer E. 1995. Translation of encephalomyocarditis virus RNA by internal ribosomal entry. *Curr. Top. Microbiol. Immunol.* 203:31–63
84. Kolupaeva VG, Pestova TV, Hellen CU, Shatsky IN. 1998. Translation eukaryotic initiation factor 4G recognizes a specific structural element within the internal ribosome entry site of encephalomyocarditis virus RNA. *J. Biol. Chem.* 273:18599–604
85. Pisarev AV, Shirokikh NE, Hellen CU. 2005. Translation initiation by factor-independent binding of eukaryotic ribosomes to internal ribosomal entry sites. *C. R. Biol.* 328:589–605
86. Spahn CM, Jan E, Mulder A, Grassucci RA, Sarnow P, Frank J. 2004. Cryo-EM visualization of a viral internal ribosome entry site bound to human ribosomes: The IRES functions as an RNA-based translation factor. *Cell* 118:465–75
87. Jan E, Sarnow P. 2002. Factorless ribosome assembly on the internal ribosome entry site of cricket paralysis virus. *J. Mol. Biol.* 324:889–902
88. Olsen PH, Ambros V. 1999. The *lin-4* regulatory RNA controls developmental timing in *Caenorhabditis elegans* by blocking LIN-14 protein synthesis after the initiation of translation. *Dev. Biol.* 216:671–80
89. Wightman B, Ha I, Ruvkun G. 1993. Posttranscriptional regulation of the heterochronic gene *lin-14* by *lin-4* mediates temporal pattern formation in *C. elegans*. *Cell* 75:855–62
90. Lee RC, Feinbaum RL, Ambros V. 1993. The *C. elegans* heterochronic gene *lin-4* encodes small RNAs with antisense complementarity to *lin-14*. *Cell* 75:843–54
91. Humphreys DT, Westman BJ, Martin DI, Preiss T. 2005. MicroRNAs control translation initiation by inhibiting eukaryotic initiation factor 4E/cap and poly(A) tail function. *Proc. Natl. Acad. Sci. USA* 102:16961–66
92. Huang J, Liang Z, Yang B, Tian H, Ma J, Zhang H. 2007. Derepression of microRNA-mediated protein translation inhibition by apolipoprotein B mRNA-editing enzyme catalytic polypeptide-like 3G (APOBEC3G) and its family members. *J. Biol. Chem.* 282:33632–40
93. Ding XC, Grosshans H. 2009. Repression of *C. elegans* microRNA targets at the initiation level of translation requires GW182 proteins. *EMBO J.* 28:213–22
94. Wang B, Love TM, Call ME, Doench JG, Novina CD. 2006. Recapitulation of short RNA-directed translational gene silencing in vitro. *Mol. Cell* 22:553–60
95. Wakiyama M, Takimoto K, Ohara O, Yokoyama S. 2007. Let-7 microRNA-mediated mRNA deadenylation and translational repression in a mammalian cell-free system. *Genes Dev.* 21:1857–62
96. Thermann R, Hentze MW. 2007. *Drosophila* miR2 induces pseudo-polysomes and inhibits translation initiation. *Nature* 447:875–78
97. Mathonnet G, Fabian MR, Svitkin YV, Parsyan A, Huck L, et al. 2007. MicroRNA inhibition of translation initiation in vitro by targeting the cap-binding complex eIF4F. *Science* 317:1764–67
98. Zdanowicz A, Thermann R, Kowalska J, Jemielity J, Duncan K, et al. 2009. *Drosophila* miR2 primarily targets the m^7GpppN cap structure for translational repression. *Mol. Cell* 35:881–88
99. Wang B, Yanez A, Novina CD. 2008. MicroRNA-repressed mRNAs contain 40S but not 60S components. *Proc. Natl. Acad. Sci. USA* 105:5343–48
100. Munroe D, Jacobson A. 1990. mRNA poly(A) tail, a 3′ enhancer of translational initiation. *Mol. Cell. Biol.* 10:3441–55
101. Soto Rifo R, Ricci EP, Décimo D, Moncorgé O, Ohlmann T. 2007. Back to basics: the untreated rabbit reticulocyte lysate as a competitive system to recapitulate cap/poly(A) synergy and the selective advantage of IRES-driven translation. *Nucleic Acids Res.* 35:e121
102. Kiriakidou M, Tan GS, Lamprinaki S, De Planell-Saguer M, Nelson PT, Mourelatos Z. 2007. An mRNA m7G cap binding-like motif within human Ago2 represses translation. *Cell* 129:1141–51

103. Kinch LN, Grishin NV. 2009. The human Ago2 MC region does not contain an eIF4E-like mRNA cap binding motif. *Biol. Direct* 4:2
104. Kawamura Y, Saito K, Kin T, Ono Y, Asai K, et al. 2008. *Drosophila* endogenous small RNAs bind to Argonaute 2 in somatic cells. *Nature* 453:793–97
105. Seitz H, Ghildiyal M, Zamore PD. 2008. Argonaute loading improves the 5′ precision of both microRNAs and their miRNA strands in flies. *Curr. Biol.* 18:147–51
106. Fabian MR, Mathonnet G, Sundermeier T, Mathys H, Zipprich JT, et al. 2009. Mammalian miRNA RISC recruits CAF1 and PABP to affect PABP-dependent deadenylation. *Mol. Cell* 35:868–80
107. Chendrimada TP, Finn KJ, Ji X, Baillat D, Gregory RI, et al. 2007. MicroRNA silencing through RISC recruitment of eIF6. *Nature* 447:823–28
108. Ceci M, Gaviraghi C, Gorrini C, Sala LA, Offenhauser N, et al. 2003. Release of eIF6 (p27BBP) from the 60S subunit allows 80S ribosome assembly. *Nature* 426:579–84
109. Gandin V, Miluzio A, Barbieri AM, Beugnet A, Kiyokawa H, et al. 2008. Eukaryotic initiation factor 6 is rate-limiting in translation, growth and transformation. *Nature* 455:684–88
110. Ding XC, Slack FJ, Grosshans H. 2008. The let-7 microRNA interfaces extensively with the translation machinery to regulate cell differentiation. *Cell Cycle* 7:3083–90
111. Tarun SZ Jr, Wells SE, Deardorff JA, Sachs AB. 1997. Translation initiation factor eIF4G mediates in vitro poly(A) tail-dependent translation. *Proc. Natl. Acad. Sci. USA* 94:9046–51
112. Beilharz TH, Humphreys DT, Clancy JL, Thermann R, Martin DI, et al. 2009. MicroRNA-mediated messenger RNA deadenylation contributes to translational repression in mammalian cells. *PLoS One* 4:e6783
113. Wu L, Fan J, Belasco JG. 2006. MicroRNAs direct rapid deadenylation of mRNA. *Proc. Natl. Acad. Sci. USA* 103:4034–39
114. Eulalio A, Huntzinger E, Nishihara T, Rehwinkel J, Fauser M, Izaurralde E. 2009. Deadenylation is a widespread effect of miRNA regulation. *RNA* 15:21–32
115. Nottrott S, Simard MJ, Richter JD. 2006. Human let-7a miRNA blocks protein production on actively translating polyribosomes. *Nat. Struct. Mol. Biol.* 13:1108–14
116. Maroney PA, Yu Y, Fisher J, Nilsen TW. 2006. Evidence that microRNAs are associated with translating messenger RNAs in human cells. *Nat. Struct. Mol. Biol.* 13:1102–7
117. Petersen CP, Bordeleau ME, Pelletier J, Sharp PA. 2006. Short RNAs repress translation after initiation in mammalian cells. *Mol. Cell* 21:533–42
118. Seggerson K, Tang L, Moss EG. 2002. Two genetic circuits repress the *Caenorhabditis elegans* heterochronic gene *lin-28* after translation initiation. *Dev. Biol.* 243:215–25
119. Kim J, Krichevsky A, Grad Y, Hayes GD, Kosik KS, et al. 2004. Identification of many microRNAs that copurify with polyribosomes in mammalian neurons. *Proc. Natl. Acad. Sci. USA* 101:360–65
120. Nelson PT, Hatzigeorgiou AG, Mourelatos Z. 2004. miRNP:mRNA association in polyribosomes in a human neuronal cell line. *RNA* 10:387–94
121. Lytle JR, Yario TA, Steitz JA. 2007. Target mRNAs are repressed as efficiently by microRNA-binding sites in the 5′ UTR as in the 3′ UTR. *Proc. Natl. Acad. Sci. USA* 104:9667–72
122. Akimitsu N, Tanaka J, Pelletier J. 2007. Translation of nonSTOP mRNA is repressed post-initiation in mammalian cells. *EMBO J.* 26:2327–38
123. Selbach M, Schwanhausser B, Thierfelder N, Fang Z, Khanin R, Rajewsky N. 2008. Widespread changes in protein synthesis induced by microRNAs. *Nature* 455:58–63
124. Nolde MJ, Saka N, Reinert KL, Slack FJ. 2007. The *Caenorhabditis elegans* pumilio homolog, *puf-9*, is required for the 3′UTR-mediated repression of the *let-7* microRNA target gene, *hbl-1*. *Dev. Biol.* 305:551–63
125. Galgano A, Forrer M, Jaskiewicz L, Kanitz A, Zavolan M, Gerber AP. 2008. Comparative analysis of mRNA targets for human PUF-family proteins suggests extensive interaction with the miRNA regulatory system. *PLoS One* 3:e3164
126. Hinman MN, Lou H. 2008. Diverse molecular functions of Hu proteins. *Cell Mol. Life Sci.* 65:3168–81
127. Kim HH, Kuwano Y, Srikantan S, Lee EK, Martindale JL, Gorospe M. 2009. HuR recruits let-7/RISC to repress c-Myc expression. *Genes Dev.* 23:1743–48

128. Weidinger G, Stebler J, Slanchev K, Dumstrei K, Wise C, et al. 2003. dead end, a novel vertebrate germ plasm component, is required for zebrafish primordial germ cell migration and survival. *Curr. Biol.* 13:1429–34
129. Youngren KK, Coveney D, Peng X, Bhattacharya C, Schmidt LS, et al. 2005. The *Ter* mutation in the dead end gene causes germ cell loss and testicular germ cell tumours. *Nature* 435:360–64
130. Mishima Y, Giraldez AJ, Takeda Y, Fujiwara T, Sakamoto H, et al. 2006. Differential regulation of germline mRNAs in soma and germ cells by zebrafish miR-430. *Curr. Biol.* 16:2135–42
131. Vasudevan S, Steitz JA. 2007. AU-rich-element-mediated upregulation of translation by FXR1 and Argonaute 2. *Cell* 128:1105–18
132. Lee YS, Nakahara K, Pham JW, Kim K, He Z, et al. 2004. Distinct roles for *Drosophila* Dicer-1 and Dicer-2 in the siRNA/miRNA silencing pathways. *Cell* 117:69–81
133. Hornstein E, Tang H, Meyuhas O. 2001. Mitogenic and nutritional signals are transduced into translational efficiency of TOP mRNAs. *Cold Spring Harb. Symp. Quant. Biol.* 66:477–84
134. Jing Q, Huang S, Guth S, Zarubin T, Motoyama A, et al. 2005. Involvement of microRNA in AU-rich element-mediated mRNA instability. *Cell* 120:623–34
135. Jopling CL, Yi M, Lancaster AM, Lemon SM, Sarnow P. 2005. Modulation of hepatitis C virus RNA abundance by a liver-specific microRNA. *Science* 309:1577–81
136. Jopling CL, Schutz S, Sarnow P. 2008. Position-dependent function for a tandem microRNA miR-122-binding site located in the hepatitis C virus RNA genome. *Cell Host Microbe* 4:77–85
137. Meyer S, Temme C, Wahle E. 2004. Messenger RNA turnover in eukaryotes: pathways and enzymes. *Crit. Rev. Biochem. Mol. Biol.* 39:197–216
138. Yamashita A, Chang TC, Yamashita Y, Zhu W, Zhong Z, et al. 2005. Concerted action of poly(A) nucleases and decapping enzyme in mammalian mRNA turnover. *Nat. Struct. Mol. Biol.* 12:1054–63
139. Coller J, Parker R. 2004. Eukaryotic mRNA decapping. *Annu. Rev. Biochem.* 73:861–90
140. Grosset C, Chen CY, Xu N, Sonenberg N, Jacquemin-Sablon H, Shyu AB. 2000. A mechanism for translationally coupled mRNA turnover: interaction between the poly(A) tail and a c-fos RNA coding determinant via a protein complex. *Cell* 103:29–40
141. Barreau C, Paillard L, Osborne HB. 2005. AU-rich elements and associated factors: Are there unifying principles? *Nucleic Acids Res.* 33:7138–50
142. Bagga S, Bracht J, Hunter S, Massirer K, Holtz J, et al. 2005. Regulation by *let-7* and *lin-4* miRNAs results in target mRNA degradation. *Cell* 122:553–63
143. Giraldez AJ, Mishima Y, Rihel J, Grocock RJ, Van Dongen S, et al. 2006. Zebrafish MiR-430 promotes deadenylation and clearance of maternal mRNAs. *Science* 312:75–79
144. Farh KK, Grimson A, Jan C, Lewis BP, Johnston WK, et al. 2005. The widespread impact of mammalian microRNAs on mRNA repression and evolution. *Science* 310:1817–21
145. Lim LP, Lau NC, Garrett-Engele P, Grimson A, Schelter JM, et al. 2005. Microarray analysis shows that some microRNAs downregulate large numbers of target mRNAs. *Nature* 433:769–73
146. Wu L, Belasco JG. 2005. Micro-RNA regulation of the mammalian *lin-28* gene during neuronal differentiation of embryonal carcinoma cells. *Mol. Cell. Biol.* 25:9198–208
147. Hock J, Weinmann L, Ender C, Rudel S, Kremmer E, et al. 2007. Proteomic and functional analysis of Argonaute-containing mRNA-protein complexes in human cells. *EMBO Rep.* 8:1052–60
148. Takimoto K, Wakiyama M, Yokoyama S. 2009. Mammalian GW182 contains multiple Argonaute-binding sites and functions in microRNA-mediated translational repression. *RNA* 15:1078–89
149. Lian SL, Li S, Abadal GX, Pauley BA, Fritzler MJ, Chan EK. 2009. The C-terminal half of human Ago2 binds to multiple GW-rich regions of GW182 and requires GW182 to mediate silencing. *RNA* 15:804–13
150. El-Shami M, Pontier D, Lahmy S, Braun L, Picart C, et al. 2007. Reiterated WG/GW motifs form functionally and evolutionarily conserved ARGONAUTE-binding platforms in RNAi-related components. *Genes Dev.* 21:2539–44
151. Li S, Lian SL, Moser JJ, Fritzler ML, Fritzler MJ, et al. 2008. Identification of GW182 and its novel isoform TNGW1 as translational repressors in Ago2-mediated silencing. *J. Cell Sci.* 121:4134–44

152. Roy G, De Crescenzo G, Khaleghpour K, Kahvejian A, O'Connor-McCourt M, Sonenberg N. 2002. Paip1 interacts with poly(A) binding protein through two independent binding motifs. *Mol. Cell. Biol.* 22:3769–82
153. Kozlov G, De Crescenzo G, Lim NS, Siddiqui N, Fantus D, et al. 2004. Structural basis of ligand recognition by PABC, a highly specific peptide-binding domain found in poly(A)-binding protein and a HECT ubiquitin ligase. *EMBO J.* 23:272–81
154. Khaleghpour K, Svitkin YV, Craig AW, DeMaria CT, Deo RC, et al. 2001. Translational repression by a novel partner of human poly(A) binding protein, Paip2. *Mol. Cell* 7:205–16
155. Jinek M, Fabian MR, Coyle S, Sonenberg N, Doudna JA. 2010. Structural insights into the human GW182-PABC interaction in microRNA-mediated deadenylation. *Nat. Struct. Mol. Biol.* 17:238–40
156. Zekri L, Huntzinger E, Heimstädt S, Izaurralde E. 2009. The silencing domain of GW182 interacts with PABPC1 to promote translational repression and degradation of microRNA targets and is required for target release. *Mol. Cell. Biol.* 29:6220–31
157. Tarun SZ Jr, Sachs AB. 1996. Association of the yeast poly(A) tail binding protein with translation initiation factor eIF-4G. *EMBO J.* 15:7168–77
158. Martineau Y, Derry MC, Wang X, Yanagiya A, Berlanga JJ, et al. 2008. Poly(A)-binding protein-interacting protein 1 binds to eukaryotic translation initiation factor 3 to stimulate translation. *Mol. Cell. Biol.* 28:6658–67
159. Craig AW, Haghighat A, Yu AT, Sonenberg N. 1998. Interaction of polyadenylate-binding protein with the eIF4G homologue PAIP enhances translation. *Nature* 392:520–23
160. Uchida N, Hoshino S, Imataka H, Sonenberg N, Katada T. 2002. A novel role of the mammalian GSPT/eRF3 associating with poly(A)-binding protein in cap/poly(A)-dependent translation. *J. Biol. Chem.* 277:50286–92
161. Hu W, Sweet TJ, Chamnongpol S, Baker KE, Coller J. 2009. Co-translational mRNA decay in *Saccharomyces cerevisiae*. *Nature* 461:225–29
162. Baek D, Villen J, Shin C, Camargo FD, Gygi SP, Bartel DP. 2008. The impact of microRNAs on protein output. *Nature* 455:64–71
163. Kong YW, Cannell IG, de Moor CH, Hill K, Garside PG, et al. 2008. The mechanism of micro-RNA-mediated translation repression is determined by the promoter of the target gene. *Proc. Natl. Acad. Sci. USA* 105:8866–71
164. Nissan T, Parker R. 2008. Computational analysis of miRNA-mediated repression of translation: implications for models of translation initiation inhibition. *RNA* 14:1480–91
165. Kedersha N, Anderson P. 2007. Mammalian stress granules and processing bodies. *Methods Enzymol.* 431:61–81
166. Eystathioy T, Chan EK, Tenenbaum SA, Keene JD, Griffith K, Fritzler MJ. 2002. A phosphorylated cytoplasmic autoantigen, GW182, associates with a unique population of human mRNAs within novel cytoplasmic speckles. *Mol. Biol. Cell* 13:1338–51
167. Liu J, Valencia-Sanchez MA, Hannon GJ, Parker R. 2005. MicroRNA-dependent localization of targeted mRNAs to mammalian P-bodies. *Nat. Cell Biol.* 7:719–23
168. Nathans R, Chu CY, Serquina AK, Lu CC, Cao H, Rana TM. 2009. Cellular microRNA and P bodies modulate host-HIV-1 interactions. *Mol. Cell* 34:696–709
169. Pauley KM, Eystathioy T, Jakymiw A, Hamel JC, Fritzler MJ, Chan EK. 2006. Formation of GW bodies is a consequence of microRNA genesis. *EMBO Rep.* 7:904–10
170. Eulalio A, Behm-Ansmant I, Schweizer D, Izaurralde E. 2007. P-body formation is a consequence, not the cause, of RNA-mediated gene silencing. *Mol. Cell. Biol.* 27:3970–81
171. Lian S, Jakymiw A, Eystathioy T, Hamel JC, Fritzler MJ, Chan EK. 2006. GW bodies, microRNAs and the cell cycle. *Cell Cycle* 5:242–45
172. Decker CJ, Teixeira D, Parker R. 2007. Edc3p and a glutamine/asparagine-rich domain of Lsm4p function in processing body assembly in *Saccharomyces cerevisiae*. *J. Cell Biol.* 179:437–49
173. Leung AK, Calabrese JM, Sharp PA. 2006. Quantitative analysis of Argonaute protein reveals microRNA-dependent localization to stress granules. *Proc. Natl. Acad. Sci. USA* 103:18125–30
174. Andrei MA, Ingelfinger D, Heintzmann R, Achsel T, Rivera-Pomar R, Luhrmann R. 2005. A role for eIF4E and eIF4E-transporter in targeting mRNPs to mammalian processing bodies. *RNA* 11:717–27

175. Kedersha N, Stoecklin G, Ayodele M, Yacono P, Lykke-Andersen J, et al. 2005. Stress granules and processing bodies are dynamically linked sites of mRNP remodeling. *J. Cell Biol.* 169:871–84
176. Wilczynska A, Aigueperse C, Kress M, Dautry F, Weil D. 2005. The translational regulator CPEB1 provides a link between dcp1 bodies and stress granules. *J. Cell Sci.* 118:981–92
177. Mazroui R, Sukarieh R, Bordeleau ME, Kaufman RJ, Northcote P, et al. 2006. Inhibition of ribosome recruitment induces stress granule formation independently of eukaryotic initiation factor 2alpha phosphorylation. *Mol. Biol. Cell* 17:4212–19
178. Leung AK, Sharp PA. 2006. Function and localization of microRNAs in mammalian cells. *Cold Spring Harb. Symp. Quant. Biol.* 71:29–38
179. Cikaluk DE, Tahbaz N, Hendricks LC, DiMattia GE, Hansen D, et al. 1999. GERp95, a membrane-associated protein that belongs to a family of proteins involved in stem cell differentiation. *Mol. Biol. Cell* 10:3357–72
180. Tahbaz N, Carmichael JB, Hobman TC. 2001. GERp95 belongs to a family of signal-transducing proteins and requires Hsp90 activity for stability and Golgi localization. *J. Biol. Chem.* 276:43294–99
181. Tahbaz N, Kolb FA, Zhang H, Jaronczyk K, Filipowicz W, Hobman TC. 2004. Characterization of the interactions between mammalian PAZ PIWI domain proteins and Dicer. *EMBO Rep.* 5:189–94
182. Valadi H, Ekstrom K, Bossios A, Sjostrand M, Lee JJ, Lotvall JO. 2007. Exosome-mediated transfer of mRNAs and microRNAs is a novel mechanism of genetic exchange between cells. *Nat. Cell Biol.* 9:654–59
183. Kosik KS. 2009. MicroRNAs tell an evo-devo story. *Nat. Rev. Neurosci.* 10:754–59
184. Fiore R, Siegel G, Schratt G. 2008. MicroRNA function in neuronal development, plasticity and disease. *Biochim. Biophys. Acta* 1779:471–78
185. Rudel S, Meister G. 2008. Phosphorylation of Argonaute proteins: regulating gene regulators. *Biochem. J.* 413:e7–9
186. Heo I, Kim VN. 2009. Regulating the regulators: posttranslational modifications of RNA silencing factors. *Cell* 139:28–31
187. Paroo Z, Ye X, Chen S, Liu Q. 2009. Phosphorylation of the human microRNA-generating complex mediates MAPK/Erk signaling. *Cell* 139:112–22
188. Qi HH, Ongusaha PP, Myllyharju J, Cheng D, Pakkanen O, et al. 2008. Prolyl 4-hydroxylation regulates Argonaute 2 stability. *Nature* 455:421–24
189. Zeng Y, Sankala H, Zhang X, Graves PR. 2008. Phosphorylation of Argonaute 2 at serine-387 facilitates its localization to processing bodies. *Biochem. J.* 413:429–36

Structure and Dynamics of a Processive Brownian Motor: The Translating Ribosome

Joachim Frank[1,2] and Ruben L. Gonzalez, Jr.[3]

[1]Howard Hughes Medical Institute, Department of Biochemistry and Molecular Biophysics, Columbia University, New York City, New York 10032

[2]Department of Biological Sciences and [3]Department of Chemistry, Columbia University, New York City, New York 10027; email: jf2192@columbia.edu, rlg2118@columbia.edu

Annu. Rev. Biochem. 2010. 79:381–412

First published online as a Review in Advance on March 17, 2010

The *Annual Review of Biochemistry* is online at biochem.annualreviews.org

This article's doi:
10.1146/annurev-biochem-060408-173330

0066-4154/10/0707-0381$20.00

Key Words

cryo-EM, FRET, protein synthesis, ribosome, X-ray

Abstract

There is mounting evidence indicating that protein synthesis is driven and regulated by mechanisms that direct stochastic, large-scale conformational fluctuations of the translational apparatus. This mechanistic paradigm implies that a free-energy landscape governs the conformational states that are accessible to and sampled by the translating ribosome. This scenario presents interdependent opportunities and challenges for structural and dynamic studies of protein synthesis. Indeed, the synergism between cryogenic electron microscopic and X-ray crystallographic structural studies, on the one hand, and single-molecule fluorescence resonance energy transfer (smFRET) dynamic studies, on the other, is emerging as a powerful means for investigating the complex free-energy landscape of the translating ribosome and uncovering the mechanisms that direct the stochastic conformational fluctuations of the translational machinery. In this review, we highlight the principal insights obtained from cryogenic electron microscopic, X-ray crystallographic, and smFRET studies of the elongation stage of protein synthesis and outline the emerging themes, questions, and challenges that lie ahead in mechanistic studies of translation.

Contents

1. INTRODUCTION

Translation: the process through which the ribosome synthesizes a protein by repeatedly incorporating aminoacyl-tRNAs as dictated by the messenger RNA

Cryogenic electron microscopy (cryo-EM): a technique for three-dimensional imaging of macromolecules in their native state with a transmission electron microscope

Protein synthesis, or translation, is universally catalyzed by the ribosome, a massive, two-subunit ribonucleoprotein molecular machine (**Figure 1*a***). New insights into the relationship between the conformational dynamics of the ribosome, its transfer RNA (tRNA) substrates, and its translation cofactors, as well as the mechanical, catalytic, and regulatory events that drive protein synthesis, are altering our mechanistic understanding of translation. This is particularly true for the translation elongation cycle (**Figure 1*b***), where a wealth of biochemical, structural, dynamic, and computational data have begun to advance the view of the elongating ribosome as a processive stochastic molecular machine (1–4). Synthesis of the data, toward an understanding of the role of conformational dynamics, benefits from concepts developed in recent years in studies of biomolecular motors (5–7).

1.1. Principles Underlying the Operation of Biomolecular Motors

Biomolecular motors harness the energy released from a chemical reaction, typically hydrolysis of a so-called high-energy phosphate compound, such as ATP or GTP, to perform mechanical work. The detailed mechanisms through which these systems transduce the energy of ATP or GTP hydrolysis into mechanical work remains an area of intense research. Static structures of biomolecular motors, such as those furnished by cryogenic electron microscopy (cryo-EM) or X-ray crystallography, often evoke the impression of smoothly running, deterministic machines, much like their macroscopic counterparts, captured in "snapshots" at certain time points along their respective reaction coordinates. In reality, however, nanometer-scale biomolecular motors operate in an environment where they are constantly subjected to the stochastic Brownian forces that arise from collisions of the surrounding media with the motor and its parts and where viscous drag easily dominates inertia—conditions that prohibit the smooth, deterministic motion usually associated with macroscopic machines. Thus, it has been proposed that many biomolecular motors employ, or at least partially employ, Brownian motor mechanisms of operation (5–7).

The principal idea underlying a Brownian motor mechanism is that force or motion is drawn from the stochastic thermal fluctuations to which these systems are constantly subjected. Because random thermal noise itself cannot confer processivity to a biomolecular motor, the directedness of the process is typically imparted by "biasing" or "rectifying" the stochastically fluctuating system through the intervention of (*a*) a substrate, cofactor, or allosteric effector binding event; (*b*) an irreversible chemical step; or (*c*) the release or diffusion of a reaction

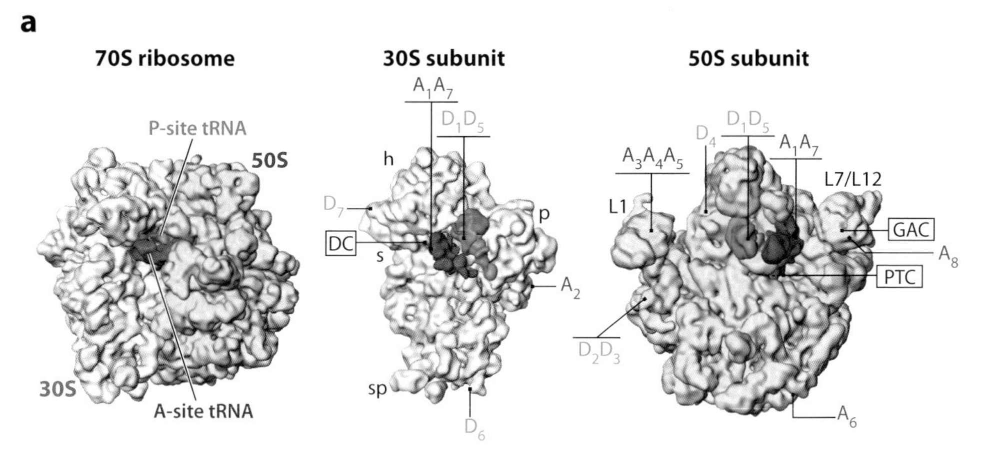

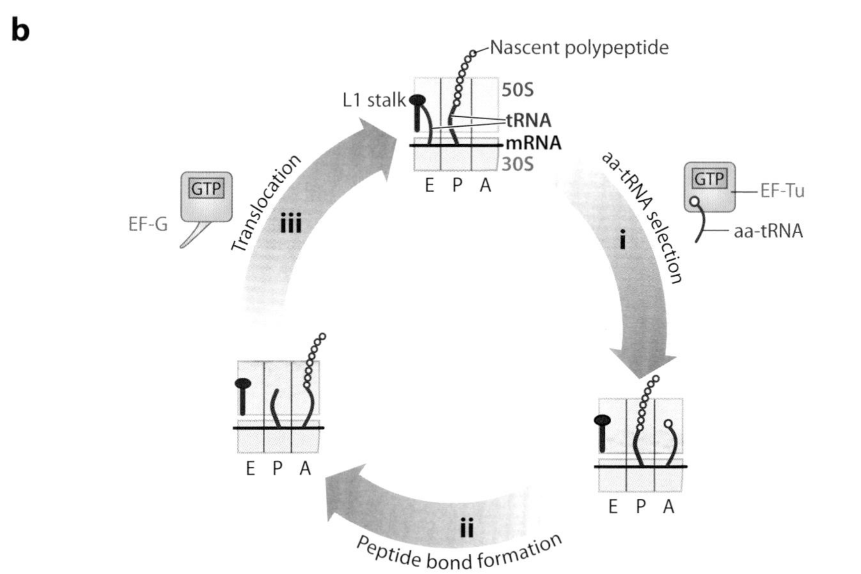

Figure 1

(*a*) Structure and dynamic features of the ribosome. Cryogenic electron microscopic map of the 70S ribosome, the 30S subunit, and the 50S subunit. The 30S and 50S subunits are shown with their intersubunit space facing the reader. The P- and A-site tRNAs are depicted in green and magenta, respectively, and their positions are denoted on the 70S ribosome. Major landmarks and mobile elements of the 30S subunit are the head (h), shoulder (s), platform (p), and spur (sp). The location of the decoding center (DC) active site is also denoted. Major landmarks and mobile elements of the 50S subunit are the L1 stalk (L1) and the L7/L12 stalk (L7/L12). The locations of the GTPase-associated center (GAC) and peptidyltransferase center (PTC) active sites are also denoted. The locations of all donor (D) and acceptor (A) fluorophore pairs that have thus far been used in single-molecule fluorescence resonance energy transfer investigations of translation elongation are labeled in green (D) and red (A). Details regarding each A-D pair are given in **Table 1**. (*b*) The elongation cycle of protein synthesis. The main steps of the translation elongation cycle, (*i*) aminoacyl-tRNA (aa-tRNA) selection, (*ii*) peptidyl transfer, and (*iii*) messenger RNA (mRNA)-tRNA translocation, are shown. The E, P, and A tRNA binding sites run vertically along both subunits. Further details regarding the mechanism of aa-tRNA selection and mRNA-tRNA translocation are provided in the captions for **Figures 2** and **3**, respectively. Abbreviations: EF, elongation factor.

Brownian motor: a biomolecular machine that rectifies or biases stochastic Brownian forces in the thermal bath to perform work

Translocation: movement of the messenger RNA and the A-site and P-site tRNAs through the ribosome by one codon

Free-energy landscape: plot of the free energy as a function of reaction and conformational coordinates

product away from a catalytic site. *Escherichia coli* RNA polymerase provides an excellent example of a well-studied molecular motor whose translocation is driven by a Brownian motor mechanism (8, 9). During the translocation step of the transcription elongation cycle, RNA polymerase has been observed to randomly oscillate between pre- and posttranslocation positions on the DNA template. Binding of nucleoside triphospate to the polymerase lowers the free energy of the forward position relative to the reverse position and thus imparts processivity to the polymerase.

An important aspect of Brownian motor function is the ability of the motor and its mechanical parts to undergo stochastic, thermally driven structural fluctuations. Indeed, it is the nanoscale dimensions of molecular motors and the energetically weak nature of the noncovalent interactions underlying their three-dimensional structures that permit biomolecular mechanical parts to operate at energies just above those available from the surrounding thermal bath. Brownian motors operate along a free-energy landscape in which fluctuations between two or more conformational states, such as the fluctuation of RNA polymerase between pre- and posttranslocation, are thermally accessible.

1.2. The Free-Energy Landscape of a Brownian Motor

Complex free-energy landscapes comprising numerous energy minima (valleys) and maxima (peaks) were originally introduced and developed in studies of protein (10, 11) and RNA folding (12–16). Viewed through this lens, an ensemble of protein or RNA molecules folds by navigating along a complex free-energy landscape, giving rise to multiple parallel folding pathways, locally stable folding intermediates, and kinetic trapping of the folding biopolymers (10–16). An excellent metaphor, provided by Dill & Chan (17), is that of water flowing along different routes down a collection of rugged hillsides that, despite experiencing different, trajectory-dependent physical obstacles to flow, ultimately collects at the same reservoir at the bottom of a deep valley. More recently, strong evidence has suggested that complex free-energy landscapes also underlie the catalytic cycles of various enzymes and ribozymes (18–28). In this view, individual enzymes or ribozymes within the ensemble can react via any one of numerous parallel reaction pathways. As the reaction proceeds, catalysis is guided by the differential stabilization (i.e., via ligand or substrate binding, product formation, and/or product release) of preexisting, thermally accessible, and on-pathway conformational intermediates. Although the role of enzyme or ribozyme conformational dynamics in guiding catalysis has been primarily developed using relatively simple model systems (18–28), it is quite likely that these ideas extend to catalysis in much more complex systems (29), including the mechanochemical cycles of Brownian motors.

1.3. Free-Energy Landscapes and the Concepts of States, Allosteric Regulation, and Induced Fit

The concept of a complex free-energy landscape forces a careful reconsideration of what is meant by the term state. The term state has often been colloquially used to refer to a single, relatively low-energy (i.e., stable) configuration of the molecule along the reaction trajectory. The corresponding picture is that of a linear progression of the entire system with defined points before and after. Contrasting with the concept of a state as a single, low-energy molecular configuration, increasing evidence supports the view that biomolecules (18–28), particularly complex biomolecular assemblies such as the ribosome (30, 31), are conformationally flexible and highly dynamic entities. Thus, states are much more adequately defined by reference to a complex free-energy landscape. Each valley in the landscape represents a free-energy minimum that is populated by an ensemble of conformations that collectively reflect a more-or-less stable state. The peaks separating the valleys represent energetic barriers between the various states, and depending on the heights of these barriers relative to the available average thermal energy [$RT = 2.5$ kJ mol^{-1} at 298 K, where R is the universal gas constant (8.314 J K^{-1} mol^{-1}) and T is temperature], transitions between states may be either thermally driven or may require the energy released by a chemical reaction.

The exact depths of the valleys and heights of the barriers in the free-energy landscape are a function of numerous variables and can be readily altered, for example, by changes in buffer conditions; by the binding of a substrate, cofactor, or allosteric effector; and by mutation. Thus, the proportion of molecules found in the valleys under any given condition and the ability of molecules to cross a barrier into a neighboring state are sensitive functions of these variables. It is this capability to redistribute the conformational ensemble and alter the rate of interconversion among the various conformers that allows ligands to allosterically regulate enzyme or ribozyme activity. Because binding of a cofactor (or allosteric effector) at a regulatory site can control the accessibility and population of conformations over the entire enzyme's or ribozyme's molecular surface, catalytic or binding site geometries can be very effectively regulated, regardless of their distance from the regulatory site.

The conformational dynamics and existence of multiple pathways implied by a complex free-energy landscape also require a reassessment of the concept of induced fit. Studies using dihydrofolate reductase (18, 21), ribonuclease A (23), and reverse transcriptase and its inhibition (24) as model systems have revealed that dynamics in the micro- to millisecond time regime allow the system to sample productive binding configurations, even in the absence of the ligand. Thus, the phenomenon of induced conformational changes in molecular interactions termed induced fit (32, 33) is perhaps better described as a conformational selection (34, 35) or selected fit (36) mechanism (28), in which the ligand simply binds to and stabilizes a productive binding configuration of the biomolecule during the time interval in which that configuration is sampled.

1.4. Single-Molecule Studies of Protein Synthesis

Recent studies relating to the structural dynamics of the translational machinery are providing compelling evidence that Brownian motor mechanisms operating along a complex free-energy landscape may underlie one of nature's most fundamental and complex multistep biochemical processes: protein synthesis (reviewed in References 1, 3, 4). It is in this context that single-molecule approaches to connect cryo-EM and X-ray snapshots in real time have emerged as a powerful tool for investigating the mechanisms through which the translational machinery couples chemical events, such as factor-dependent GTP hydrolysis, release of inorganic phosphate, and peptidyl transfer, to the mechanical steps of protein synthesis (37–60). Single-molecule fluorescence resonance energy transfer (smFRET) experiments, in particular, are uncovering the important role that large-scale, thermally driven conformational fluctuations of the translational machinery play in regulating mechanical events during translation elongation (38–44, 47, 52–56, 58–60). In this review, we integrate a rapidly evolving series of findings by cryo-EM (61–69), X-ray (70–86), and smFRET (38–47, 51–56, 58–60) studies of translation elongation that are beginning to define the complex free-energy landscape underlying translation elongation. Our analysis strongly suggests that Brownian motor mechanisms lie at the heart of at least two of the principal steps in the elongation cycle: aminoacyl-transfer RNA (aa-tRNA) selection and translocation. By using these two examples to describe the emerging mechanistic themes and identify the remaining questions and challenges, we hope to stimulate further investigation of the hypothesis that similar Brownian motor mechanisms underlie many, if not all, of the individual steps of protein synthesis (2–4).

Peptidyl transfer: ribosome-catalyzed transfer of the nascent polypeptide chain from the P site-bound peptidyl-tRNA to the A site-bound aa-tRNA

Single-molecule fluorescence resonance energy transfer (smFRET): measurement of the energy transfer efficiency between donor and acceptor fluorophores on single molecules

Aminoacyl-transfer RNA (aa-tRNA) selection: selection of an aa-tRNA cognate to the codon at the ribosomal A site by the mRNA-programmed ribosome

2. THE STRUCTURAL AND DYNAMIC TOOLKIT

At present, the tools of choice for characterizing ribosome structure and dynamics are cryo-EM, X-ray crystallography, and smFRET. These three techniques provide complementary and interdependent experimental information that are collectively driving our rapidly evolving view of the role that ribosome structural dynamics play in the mechanism and regulation of protein synthesis. Below, we describe these three approaches, discuss each of their advantages and disadvantages, and emphasize how the complementarity and interdependence of these three techniques overcome their individual disadvantages. Our intent is to present a persuasive argument that the synergistic application of these methods will ultimately provide a virtual movie of protein synthesis by the ribosome.

2.1. Cryogenic Electron Microscopy

The technique of cryo-EM combined with single-particle reconstruction (see References 87 and 88) produces a three-dimensional density map from thousands of projections of different molecules ideally having identical structure, trapped in random orientations within a thin layer of ice. In contrast to X-ray crystallography (see below), the molecule is frozen in its native state, without constraints from functionally meaningless intermolecular contacts. **Figure 1*a*** introduces the ribosome as seen by cryo-EM, with landmarks and important mobile elements denoted. In the application of this technique to the ribosome, the challenge is to find ways to trap the majority of the molecules in the same state, usually by the addition of small-molecule, ribosome-targeting antibiotics or nonhydrolyzable GTP analogs. Dynamics can be inferred by comparing ribosome complexes captured in successive states (see Reference 89). Any residual heterogeneity in a sample poses a problem. However, new and powerful classification methods have made it possible to divide the data into homogeneous subsets. Frequently, therefore, a single sample results in two or more reconstructions, each for a subpopulation of molecules in a defined state (e.g., References 62 and 68).

As an ensemble average, a cryo-EM reconstruction has features with varying resolution, reflecting local variability of conformation. Peripheral components sticking out into the solvent, or components that are functionally mobile, may therefore appear washed out. However, the most serious drawback is that presently, with a few exceptions (molecules with high symmetry), the density maps are of insufficient resolution to allow chain tracing. Much effort has therefore gone into the development of so-called hybrid techniques, i.e., tools for interpreting medium-resolution density maps of molecular complexes in terms of atomic-resolution X-ray structures of their components. Some of the novel flexible fitting tools (e.g., Reference 90) yield structures that are intact, are sterochemically correct, and are in optimal agreement with the density restraints, even though they are not uniquely determined by them owing to insufficient resolution.

2.2. X-Ray Crystallography

X-ray crystallography, whose stunning achievements form the basis of all current structural interpretations of translation, shows the molecule confined and packed in a crystal, in a conformation not necessarily related to its function. In some cases, a gallery of structures of the same molecule in different crystals or even in the same unit cell (e.g., Reference 91) may hint at the range of accessible or functionally important conformations. However, care must be taken when inferring dynamics from such comparisons of X-ray crystallographic structures, as it has been noted that the constraints imposed by the crystal lattice are likely to dampen or inhibit the range and extent of conformational changes that are observed (74, 92). Moreover, ligands frequently have to be modified or truncated to allow crystal formation; for instance, tRNA has often been substituted by an anticodon stem loop (ASL) (e.g., References 73 and 93), which limits the information one can gain about the way the intact tRNA interacts with the ribosome and how the intact tRNA might serve to transmit conformational events originating within the small 30S ribosomal subunit to the large 50S ribosomal subunit. Notably, X-ray crystallographic structures of elongation factor–bound ribosomes have recently emerged, allowing an atomic-resolution analysis of the interactions that these factors make with the ribosome during the elongation cycle (85, 86).

2.3. Single-Molecule Fluorescence Resonance Energy Transfer

Neither cryo-EM nor X-ray crystallography provides information on the evolving dynamic process itself. smFRET (94, 95) is uniquely suited to provide such information on a molecular motor in motion. smFRET draws directly from both cryo-EM and X-ray crystallography in the design of experiments (i.e., in the placement of donor and acceptor fluorophore pairs) for maximum information on dynamic distance changes associated with molecular function. In a rapidly evolving area of ribosome research, through careful positioning of donor-acceptor pairs, smFRET is providing real-time dynamic information on many of the structural rearrangements that have been inferred from comparisons of cryo-EM and X-ray structures (38–43, 44–46, 48, 51–60). **Figure 1*a*** and **Table 1** define the positions of all donor-acceptor pairs that have thus far been used in smFRET studies of translation.

Although smFRET provides time-resolved information on changes in molecular distances, typically only a single distance is monitored per donor-acceptor labeling scheme. Extension of

Table 1 Positions within the translational machinery that have been labeled with donor-acceptor fluorophore pairs in smFRET studies of translation elongation

Figure 1 designation	Donor fluorophore position	Acceptor fluorophore position	References
D_1/A_1	4-Thiouridine residue at position 8 of P-site $tRNA^{fMet}$	3-(3-Amino-3-carboxy-propyl) uridine residue at position 47 of A-site $tRNA^{Phe}$	38, 39, 44, 47, 52–56, 58, 59
	4-Thiouridine residue at position 8 of P-site $tRNA^{Phe}$	3-(3-Amino-3-carboxy-propyl) uridine residue at position 47 of A-site $tRNA^{Lys}$	52–54
D_2/A_2	C11 within an N11C single-cysteine mutant of ribosomal protein L9	C41 within a D41C single-cysteine mutant of ribosomal protein S6	43
D_3/A_3	C18 within a Q18C single-cysteine mutant of ribosomal protein L9	C202 within a T202C single-cysteine mutant of ribosomal protein L1	40, 60
D_4/A_4	C29 within a T29C single-cysteine mutant of ribosomal protein L33	C88 within an A88C single-cysteine mutant of ribosomal protein L1	42
D_5/A_5	3-(3-Amino-3-carboxy-propyl) uridine residue at position 47 of P-site $tRNA^{Phe}$	C202 within a T202C single-cysteine mutant of ribosomal protein L1	40, 41, 57, 60
	4-Thiouridine residue at position 8 of P-site $tRNA^{fMet}$	C55 within a S55C single-cysteine mutant of ribosomal protein L1	52, 53
	4-Thiouridine residue at position 8 of P-site $tRNA^{Phe}$	C55 within a S55C single-cysteine mutant of ribosomal protein L1	52, 53
D_6/A_6	Helix 44 of 16S rRNA (nucleotides 1450–1453)	Helix 101 of 23S rRNA (nucleotides 2853–2864)	45, 48
D_7/A_7	Helix 33a of 16S rRNA (nucleotides 1027–1034)	3-(3-Amino-3-carboxy-propyl) uridine residue at position 47 of A-site $tRNA^{Phe}$	51
D_8/A_8	C231 within an E231C single-cysteine mutant of EF-G	Unique native cysteine C38 within ribosomal protein L11	46

EF-Tu: elongation factor Tu

EF-G: elongation factor G

smFRET technology to simultaneously monitor two distances using three-color smFRET (i.e., using smFRET signals between one donor and two distinct acceptor fluorophores) within a single biomolecule has been recently reported (52, 53, 96, 97). However, the spectral properties required to generate appreciable energy transfer between the donor and both acceptors (and minimize the potentially complicating energy transfer between the two acceptors) are at odds with the spectral properties required to adequately detect the two resulting smFRET signals; thus, careful balancing of these opposing requirements (97) presents technical challenges [i.e., unacceptable amounts of donor or acceptor signal bleed through into the improper detection channel(s), very low signal-to-noise ratios, etc.] that often result in smFRET data that cannot be quantitatively analyzed (52, 53). Complicating matters further, the observed value of FRET efficiency is dependent on a number of spectroscopic and physical variables that must be carefully measured for each donor-acceptor labeling scheme in order to extract an accurate estimate of the distances between the donor and acceptor fluorophores (98). Therefore, relative to X-ray crystallography and cryo-EM, the structural resolution available from smFRET experiments is considerably limited, and the technique is best applied to obtain kinetic information on structural rearrangements that have already been well defined by comparisons of X-ray and/or cryo-EM structures. In the ideal smFRET experiment, donor and acceptor fluorophores are introduced into mobile and static structural elements of the ribosomal complex, respectively, such that the recorded smFRET data report on the intrinsic dynamics of the mobile element relative to a static landmark (39, 40, 42, 43, 45, 48, 51, 55, 56, 59, 60). smFRET studies where both fluorophores, or all three fluorophores in the case of three-color smFRET experiments, are positioned on mobile elements of the ribosomal complex are more difficult to interpret and often require prior knowledge, or a model, of the structural rearrangements that are being probed (38, 41, 44, 46, 52–54, 57, 58, 60).

Given the strengths and limitations of each of these techniques, the consensus is emerging that all three should be employed in the quest to understand the mechanism of protein synthesis (see for instance References 31 and 99). Indeed, the synthesis of cryo-EM, X-ray crystallography, and smFRET data has already yielded important insights into the mechanisms of aa-tRNA selection (39, 51, 55, 56, 59, 61, 64, 69, 73, 74, 85) and translocation (38, 40–43, 44–46, 52–54, 58, 60, 62, 63, 68, 86, 99, 100).

3. THE TRANSLATION ELONGATION CYCLE

The translation elongation cycle can be divided into three fundamental steps (**Figure 1*b***): (*a*) selection and incorporation of an aa-tRNA into the ribosomal A site, a step which is catalyzed by the essential GTPase elongation factor Tu (EF–Tu) (101); (*b*) peptidyl transfer of the nascent polypeptide from the peptidyl-tRNA at the ribosomal P site to the aa-tRNA at the A site, effectively deacylating the P-site tRNA and increasing the length of the nascent polypeptide now on the A-site tRNA by one amino acid (102); and (*c*) translocation of the ribosome along the messenger RNA (mRNA) template by precisely one codon and the accompanying joint movement of the newly deacylated tRNA from the P to the E site and the newly formed peptidyl-tRNA from the A to the P site, a step which is catalyzed by a second essential GTPase, elongation factor G (EF-G) (99, 103). At its conclusion, the translocation reaction places the subsequent mRNA codon at the A site such that the elongation cycle can be repeated for incorporation of the next mRNA-encoded amino acid.

3.1. Aminoacyl-tRNA Selection

aa-tRNA selection is a complex, multistep process during which an aa-tRNA, in the form of a ternary complex with elongation factor Tu (EF-Tu) and GTP, is selected by the ribosome from among at least 20 species of aa-tRNAs as dictated by the mRNA codon presented at the A site. Successful recognition of a cognate aa-tRNA results in transmission of a conformational signal to EF-Tu that triggers GTP hydrolysis and subsequent domain rearrangement of the factor resulting in its dissociation and the accommodation of the aa-tRNA into the ribosome.

3.1.1. Kinetic proofreading, induced fit, and the fidelity of protein synthesis. **Figure 2** presents the mechanism of aa-tRNA selection as deduced from comprehensive biochemical kinetic experiments (101, 104); where available, cryo-EM snapshots of ribosomal complexes that appear along the reaction pathway are also depicted. aa-tRNA selection is a process in which a series of selection steps control the ability of the incoming aa-tRNA to participate in the peptidyltransferase reaction. It is within this series of selection steps that kinetic proofreading (105, 106) and induced-fit (101, 106, 107) mechanisms discriminate aa-tRNAs that are cognate to the mRNA codon at the A site from those that are near- or noncognate.

aa-tRNA arrives at the ribosome as part of a ternary complex with EF-Tu and GTP (step $0 \rightarrow 1$, **Figure 2**). Selection of aa-tRNA is based on Watson-Crick base-pairing between the mRNA codon and aa-tRNA anticodon (steps $1 \rightarrow 3$, **Figure 2**). Recognition of a cognate or, with a much lower probability, a near-cognate codon-anticodon complex (39, 59, 104) triggers GTP hydrolysis and release of inorganic phosphate from EF-Tu (steps $3 \rightarrow 5$, **Figure 2**). This is an irreversible event which separates the initial selection stage of aa-tRNA selection from the subsequent proofreading stage (steps $5 \rightarrow 7$, **Figure 2**), thus providing two independent opportunities to reject near-cognate aa-tRNAs. An important proposal stemming from the kinetic data is that the rates of GTP hydrolysis (k_4, **Figure 2**) and peptidyl transfer (k_7, **Figure 2**) are limited not by active-site chemistry, but rather by two preceding structural rearrangements of the ribosomal elongation complex termed GTPase activation (k_3/k_{-3}) and aa-tRNA accommodation ($k_6/k_{6'}$, **Figure 2**). Notably, induced-fit mechanisms synergistically enhance the fidelity established by kinetic proofreading through selectively accelerating GTPase activation (k_3, **Figure 2**) and accommodation (k_6, **Figure 2**) in response to cognate versus near-cognate aa-tRNAs.

DC: decoding center

GAC: GTPase-associated center

3.1.2. Codon recognition and 30S subunit domain closure. In structural terms, aa-tRNA selection begins with binding of EF-Tu to the L7/L12 stalk near the GTPase-associated center (GAC) (**Figure 1*a***) of the 50S subunit and the formation and recognition of the codon-anticodon complex at the decoding center (DC) within the A site of the 30S subunit. NMR studies of an RNA oligonucleotide mimic of a portion of the DC (108, 109) suggested that in the absence of ligands the DC is conformationally dynamic (109), a finding that holds true within the authentic, ligand-free DC in an intact 30S subunit (74).

Details concerning the nature of the codon-anticodon interaction and its recognition and stabilization by the DC have emerged from X-ray studies of the 30S subunit. By soaking mRNA fragments and tRNA ASLs into crystals of the isolated 30S subunit, Ramakrishnan and coworkers (74) have demonstrated that binding of a cognate codon-anticodon complex into the DC stabilizes a specific local conformation of the otherwise disordered DC, in which the universally conserved nucleotides A1492, A1493, and G530 stably associate with the cognate codon-anticodon complex. In addition, binding of a cognate codon-anticodon complex into the DC induces a global conformational change of the 30S subunit, termed domain closure, in which the head and shoulder domains (**Figure 1*a***) rotate toward the center of the 30S subunit (73).

Based on comparisons of the X-ray crystal structures, the 30S subunit domain closure has been described as a conformational change that is induced upon recognition of a cognate codon-anticodon complex. However, it is likely that the head and shoulder domains within states 0, 1, and 2 in **Figure 2** are

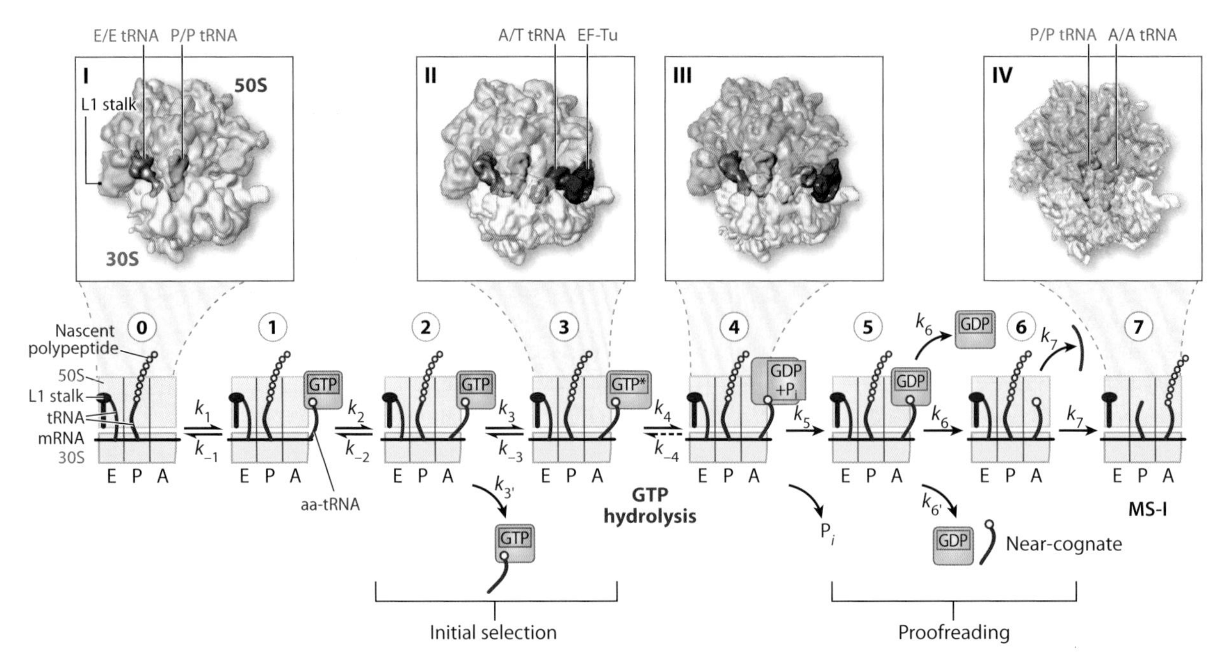

Figure 2

Distinct states and reversible/irreversible steps of the decoding and peptidyl transfer processes, and corresponding cryogenic electron microscopy (cryo-EM) maps, where available. In this schematic, tRNAs are colored according to their positions in the canonical (A, P, E) scheme, consistent with color choices in previous work [e.g. Reference (63)]. State 0: The posttranslocational state in which the A site is unoccupied, the P site contains a peptidyl-tRNA bound in the classical P/P (denoting the 30S P/50S P sites, respectively) configuration, and the E site contains a deacylated tRNA bound in the classical E/E configuration and in direct contact with the open L1 stalk. (Note: The E-site tRNA contacts the L1 stalk through its central fold, or elbow, domain.) Cryo-EM map *I* from Valle and coworkers (63). Step 0 → 1 (reversible; k_1/k_{-1}): The binding of elongation factior Tu (EF-Tu) in a ternary complex with aminoacyl-tRNA (aa-tRNA) and GTP to the ribosome via the L7/L12 stalk. State 1: The same as state 0 but with ternary complex bound to ribosome. Step 1 → 2 (reversible; k_2/k_{-2}): The probing of the mRNA codon by the aa-tRNA anticodon at the decoding center (DC). State 2: The same as state 1 but the aa-tRNA anticodon is engaged with the codon at the DC. Step 2 → 3 (reversible; k_3/k_{-3}): The cognate and a fraction of near-cognate ternary complexes are bound sufficiently long to induce GTPase activation of EF-Tu. Step 2 → 3 (irreversible; $k_{3'}$): The noncognate and a fraction of near-cognate ternary complexes are rejected as their binding to the ribosome fails to stabilize. State 3: The same as state 2 but with EF-Tu activated for GTP hydrolysis. Cryo-EM map *II*: The use of guanylyl iminodiphosphate prevents GTP hydrolysis (119; J. Sengupta, O. Kristensen, F. Fabiola, H. Gao, M. Valle, et al., in preparation). Step 3 → 4: (reversible; k_4/k_{-4}) GTP hydrolysis on EF-Tu. State 4: The same as state 3 but with EF-Tu bound in the GDP-P_i state. Cryo-EM map *III*: After GTP hydrolysis, kirromycin prevents conformational change of EF-Tu and locks the ternary complex in the A/T configuration (64). Step 4 → 5 (irreversible; k_5): The departure of P_i. State 5: The same as state 4 but with EF-Tu bound in the GDP only state. Step 5 → 6 (irreversible; k_6): The conformational change of EF-Tu, departure of EF-Tu·GDP, and accommodation of cognate aa-tRNA. (irreversible; $k_{6'}$): The conformational change of EF-Tu, departure of EF-Tu·GDP, and departure of near-cognate aa-tRNA. State 6: The same as state 5, but with aa-tRNA accommodated in the classical A/A configuration within the A site. Step 6 → 7 (irreversible; k_7): The departure of E-site tRNA and peptidyl transfer. (Note that the precise timing of the E-site tRNA departure has not been established, so this placement is tentative.) State 7: The macrostate I form of the pretranslocational complex is the same as state 6 but the nascent polypeptide is now covalently linked to the A-site tRNA, whereas the P-site tRNA is deacylated, and the E site is unoccupied. The ribosome is in its nonrotated conformation, the tRNAs are in their classical A/A and P/P positions, and the L1 stalk is in its open conformation. Cryo-EM map *IV* from Agirrezabala and coworkers (68).

inherently conformationally dynamic and that binding of a cognate codon-anticodon complex into the DC simply selectively stabilizes the global conformer that is observed in the X-ray studies. Indeed, the observation that, depending on experimental conditions, the ribosome has anywhere from a 1 in 1×10^3 (110, 111) through a 1 in 1×10^5 (112, 113) probability of misincorporating a near-cognate aa-tRNA suggests that binding of a near-cognate codon-anticodon complex into the DC has some probability of inducing or selectively stabilizing the domain-closed form of the 30S subunit.

Structural clues regarding how the domain closure event, inferred from X-ray studies of the isolated 30S subunit, might couple recognition of a cognate codon-anticodon complex at the DC to activation of GTP hydrolysis on EF-Tu at the GAC of the 50S subunit come from structural studies of ternary complex binding to ribosomes in the presence either of the nonhydrolyzable GTP analog, guanylyl iminodiphosphate (GDPNP), or of the antibiotic kirromycin. Kirromycin is a small-molecule antibiotic, which is known to block the aa-tRNA selection process at a step following GTP hydrolysis but prior to the conformational change of EF-Tu that normally results in its dissociation from the ribosome and the release of the aa-tRNA acceptor stem (114, 115). Recognition of a cognate or, with much lower probability, a near-cognate codon-anticodon complex at the DC leads to formation of the so-called A/T configuration of the ternary complex, a transient configuration visualized by cryo-EM (61, 64, 66, 69, 116–118) and, more recently, by X-ray crystallography (85), in which the aa-tRNA anticodon engages the codon at the DC while the aa-tRNA acceptor stem remains bound to EF-Tu (map *II*, **Figure 2**). The structures of the kirromycin-stabilized A/T configuration have not only pinned down the intermolecular contacts between EF-Tu and the ribosome but also demonstrated that, upon EF-Tu binding, the L11 stalk (**Figure 1*a***), which forms part of the GAC, curls inward toward the peptidyltransferase center (PTC) (**Figure 1*a***) of the 50S subunit (64, 85, 117, 119). Perhaps most importantly, the structure of the A/T configuration led to the discovery of a large conformational change of the tRNA body, as compared to its known X-ray structure (120). Apparently, this conformational change, characterized as a kink and twist in the anticodon stem, permits a geometry in which the aa-tRNA acceptor stem can remain bound to EF-Tu while the anticodon stem can continue probing the codon-anticodon interaction (61, 64, 69, 85). It has been suggested, as we point out below, that the unusual conformation of the aa-tRNA may have a role in the kinetic proofreading step.

Guanylyl iminodiphosphate (GDPNP): a GTP nonhydrolyzable analog

3.1.3. The frequency and rate of stably forming the A/T configuration depend on codon-dependent fluctuations of the ribosome-bound ternary complex. Dynamic information connecting the recognition of the codon-anticodon complex at the DC, and presumably the associated 30S subunit domain closure event, with formation of the A/T configuration comes from smFRET studies of aa-tRNA selection. Using an smFRET signal between a donor-labeled P-site peptidyl-tRNA and an acceptor-labeled incoming aa-tRNA (D_1/A_1, **Figure 1*a*** and **Table 1**), delivery of a cognate ternary complex to the ribosome generated a rapidly evolving, presteady-state smFRET signal that started at a 0-FRET value prior to ternary complex binding and concluded at a high-FRET value once aa-tRNA was fully accommodated into the A site (steps $0 \rightarrow 6$, **Figure 2**) (39). In contrast with this result, delivery of a near-cognate ternary complex yielded multiple reversible fluctuations between the 0-FRET value and a novel low-FRET value, which reports on the transient binding of the ternary complex to the ribosome (steps $0 \rightleftarrows 2$, **Figure 2**). Interestingly, delivery of a noncognate ternary complex failed to yield any detectable smFRET signal, demonstrating that transient binding of noncognate ternary complex either does not take place or is unobservable using this particular fluorophore labeling scheme and/or available time resolution. The transient binding configuration of the ternary complex that is characterized by the low-FRET value, therefore, is one in which the DC is able to inspect and recognize the codon-anticodon complex (state 2, **Figure 2**). Although the configuration of the ternary complex associated with the low-FRET value is a critical intermediate in the aa-tRNA selection pathway, the stochastic and transient nature with which it is sampled makes it very difficult to structurally characterize or biochemically investigate using ensemble experiments.

Steps along the smFRET trajectory that lie beyond the low-FRET, codon-dependent sampling of the DC can be investigated by substituting GTP with GDPNP in a manner analogous to the cryo-EM studies described above (39, 59). A cognate ternary complex delivered in the presence of GDPNP progresses through the low-FRET value and subsequently stabilizes at a mid-FRET value that directly corresponds to the A/T configuration visualized by cryo-EM (steps 0 → 3, **Figure 2**). Near-cognate ternary complexes, which are very effectively rejected from the configuration associated with the low-FRET value (steps 0 ⇄ 2, **Figure 2**), have a very low probability of stably achieving the A/T configuration associated with the mid-FRET value. Thus, the low- → mid-FRET transition reports on a codon-dependent structural rearrangement of the ribosome-bound ternary complex in which the aa-tRNA is brought closer to the peptidyl-tRNA (hence the increase in FRET) (step 2 → 3, **Figure 2**). This conformational change precedes GTP hydrolysis by EF-Tu, and its end point coincides with stable repositioning of EF-Tu at the GAC such that GTP hydrolysis by EF-Tu can be activated. It is therefore likely that the aa-tRNA itself participates in transmitting the codon-anticodon recognition signal from the DC to the GAC, inducing GTP hydrolysis by EF-Tu, in keeping with the results of tRNA cleavage (121) and mutational (122–124) studies.

Careful analysis of a large number of smFRET trajectories reveals that both cognate and near-cognate ternary complexes can fluctuate reversibly between the low- and mid-FRET states (i.e., low- ⇄ mid-FRET) in the presence of GDPNP (59). In addition, two subpopulations of low- → mid-FRET transitions were detected. One subpopulation remains at the mid-FRET value transiently, rapidly transiting back to the low-FRET value. This subpopulation likely represents unsuccessful attempts of the ternary complex to reposition at the GAC, an interpretation that is supported by similar smFRET experiments performed in the presence of the small-molecule, GAC-targeting antibiotic thiostrepton (56). The second population exhibits a stable and long-lived mid-FRET signal that indicates the successful repositioning of the ternary complex at the GAC. These data directly report on stochastic, thermally driven fluctuations of the ternary complex between configurations characterized by low- and mid-FRET values. Comparison of cognate and near-cognate ternary complex delivery in the presence of GDPNP demonstrates that near-cognate ternary complexes are ~sixfold less likely to undergo a low- → mid-FRET transition, exhibit a ~threefold decrease in the rate of low- → mid-FRET transitions, and are ~twofold less likely to undergo a successful versus an unsuccessful low- → mid-FRET transition. Thus, as part of the initial selection stage of aa-tRNA selection, thermally driven low- ⇄ mid-FRET fluctuations of the ribosome-bound ternary complex are biased in favor of cognate over near-cognate ternary complexes. If one assumes that low- → mid- and mid- → low-FRET transitions correspond to 30S domain closing and opening events, respectively, it is possible that rather than triggering a single, discrete 30S subunit domain closure event, recognition of a cognate codon-anticodon complex at the DC simply biases thermally driven open ⇄ closed fluctuations of the 30S head and shoulder domains. It would therefore be of great interest to design donor-acceptor pairs that directly report on the dynamics of the 30S subunit domain closure.

3.1.4. aa-tRNA distortion and fluctuations might drive and regulate the outcome of the accommodation reaction. Successful repositioning of a cognate and, with some probability, a near-cognate ternary complex at the GAC leads to GTPase activation (step 0 → 3, **Figure 2**) and GTP hydrolysis (step 3 → 4, **Figure 2**) on EF-Tu. The mechanism of GTPase activation of EF-Tu has been recently illuminated by two cryo-EM reconstructions of the A/T ternary complex configuration stabilized in the GDP-bound conformation using the antibiotic kirromycin (61, 69). These two recent cryo-EM reconstructions demonstrate that successful repositioning of EF-Tu at the GAC triggers opening of a hydrophobic gate within EF-Tu, which allows a crucial EF-Tu histidine residue to reorient toward GTP and activate a water molecule, subsequently leading to GTP hydrolysis. These findings have been recently confirmed at atomic resolution using X-ray crystallography (85). Upon GTP hydrolysis and release of the resulting inorganic phosphate (steps 4 → 5, **Figure 2**), EF-Tu undergoes a large conformational change, which results in the above-described dissociation of EF-Tu·GDP from the ribosome and the release of the aa-tRNA acceptor stem (steps 5 → 6, **Figure 2**). At this stage, a near-cognate aa-tRNA is preferentially released from the ribosome ($k_{6'}$) while a cognate aa-tRNA, on account of its optimal binding stability at the DC, is preferentially accommodated into the PTC (k_6). Assuming that aa-tRNA in its kinked, twisted conformation is in a high-energy configuration, the stability of its binding interactions at the DC may set the threshold for selection (125). The mid- → high-FRET transition observed in presteady-state smFRET studies of aa-tRNA selection reports directly on the dynamics of aa-tRNA as it is accommodated from its position within the A/T configuration of the ternary complex into the PTC (39). Interestingly, rapid mid- ⇄ high-FRET fluctuations are observed when the A/T configuration is stabilized either by GDPNP or by GDP·kirromycin (39). This observation suggests that even prior to GTP hydrolysis, release of inorganic phosphate, or the subsequent conformational change of EF-Tu, aa-tRNA can fluctuate and transiently sample the high-FRET, fully accommodated state but that stable accommodation of aa-tRNA into the PTC might require additional conformational processes at the PTC that are somehow coupled to the conformational change and/or dissociation of EF-Tu from the ribosome. Thus, in a manner analogous to the initial selection stage of aa-tRNA selection, these results suggest that the proofreading stage of aa-tRNA selection involves thermally driven fluctuations of the tRNA from its binding site on EF-Tu into the PTC that are biased to favor cognate over near-cognate tRNAs.

3.2. Peptidyl Transfer

Owing to their current spatial and time resolution limits, cryo-EM and smFRET have not significantly contributed to our structural and dynamic understanding of the peptidyltransferase reaction. There is, however, significant biochemical (126–128) and X-ray structural (70) evidence strongly suggesting that local structural rearrangements of the PTC regulate the conversion of the catalytic center of the ribosome from an inactive conformation to a conformation that supports the peptidyl-transfer reaction. X-ray crystallographic structures of the 50S subunit bound to various analogs of peptidyl- and aa-tRNA acceptor stems, as well as a peptidyltransferase transition state analog, suggest that docking of the acceptor end of aa-tRNA into the PTC triggers a coupled structural rearrangement of the PTC and the acceptor ends of the peptidyl- and aa-tRNAs. The resulting conformation of the substrate-bound PTC exposes the reactive carbonyl group at the C-terminal end of the peptidyl-tRNA, optimally positioning it and the α-NH_2 group of the incoming aa-tRNA for the in-line nucleophilic attack that transfers the nascent polypeptide from the P site–bound peptidyl-tRNA to the A site–bound aa-tRNA (70). Results of studies using analogs of peptidyl- and aa-tRNA acceptor stems within the isolated 50S subunit have recently been supported by similar studies using full-length peptidyl- and aa-tRNA analogs within intact, 70S ribosomes (129).

3.3. mRNA-tRNA Translocation

After peptidyl transfer, the A-site tRNA within the pretranslocation (PRE) ribosomal complex carries the nascent polypeptide chain while the P-site tRNA is deacylated. mRNA and tRNAs must now be advanced such that the next codon is placed into the DC of the posttranslocation

(POST) ribosomal complex (**Figure 3**). This is accomplished with the aid of EF-G·GTP, which binds to the PRE complex in a position similar to that of the ternary complex (63, 67, 85, 86, 130, 131). Evidently the need to transport mRNA and two tRNAs by the span of a codon (~13 Å) requires large, coordinated conformational changes of the PRE complex

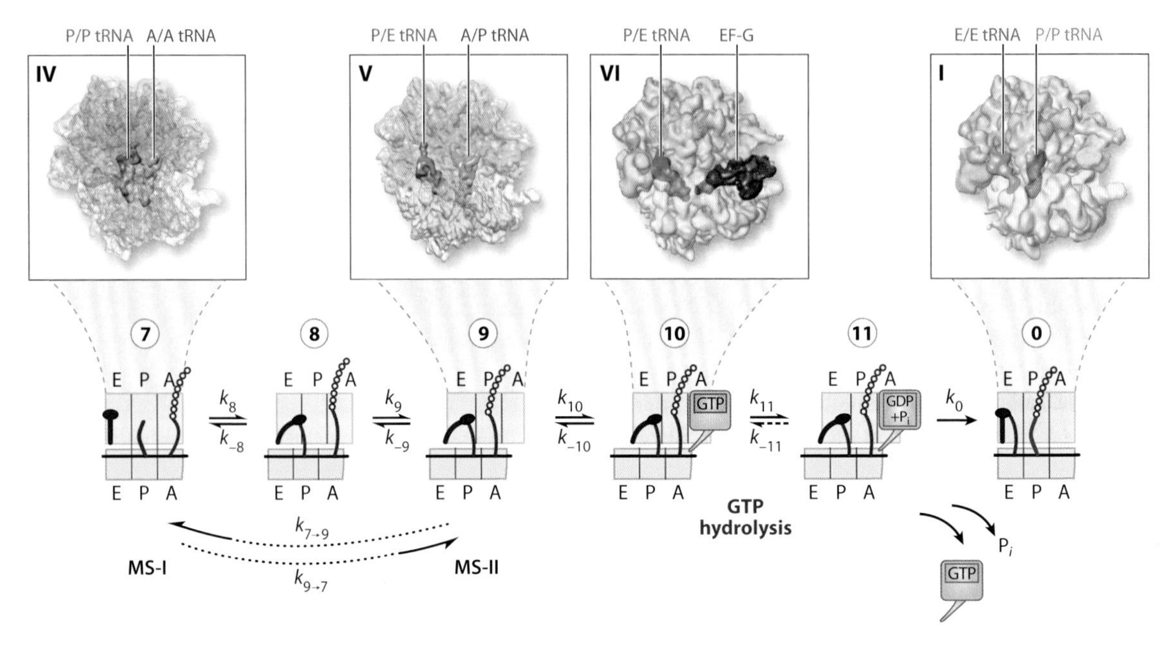

Figure 3

Distinct states and reversible/irreversible steps of the translocation process and corresponding cryo-EM maps, where available. In this schematic, unlike **Figure 2**, colors mark individual tRNAs, so that the steps of their translocation can be followed. State 7: The macrostate I (MS-I) form of the pretranslocational (PRE) complex is the same as state 7 in **Figure 2**. The A site contains the newly formed peptidyl-tRNA, the P site contains a deacylated tRNA, and the E site is unoccupied. The ribosome is in its nonrotated conformation, the tRNAs are in their classical A/A (denoting the 30S A/50S A sites, respectively) and P/P positions, and the L1 stalk is in its open conformation. Cryo-EM map *IV* from Agirrezabala and coworkers (68). Step 7 → 8 (reversible; k_8/k_{-8}): The rearrangement of MS-I into an intermediate state of ratcheting (44). State 8: This is the same as state 7 but with the PRE complex in an intermediate state of ratcheting consisting of a ribosome in a semirotated state and tRNAs in an intermediate classical A/A, hybrid P/E configuration that lies somewhere between the classical A/A and P/P configuration and the hybrid A/P and P/E configuration. The L1 stalk is in a "closed" position, where it forms a direct intermolecular contact with the hybrid P/E tRNA (44, 156). Step 8 → 9 (reversible; k_9/k_{-9}): The rearrangement of the intermediate state of ratcheting to macrostate II (MS-II) (44). State 9: The MS-II form of the PRE complex is the same as in state 8 but with the ribosome in the rotated state, tRNAs in hybrid A/P and P/E configurations, and the L1 stalk in a closed conformation where it directly contacts the hybrid P/E tRNA elbow. Cryo-EM map *V* from Agirrezabala and coworkers (68). Note: State 9 can alternatively be reached directly from state 7, bypassing the intermediate state 8. Step 9 → 10 (reversible; k_{10}/k_{-10}): The binding of EF-G in the GTP state. State 10: The same as in state 9 but with EF-G bound in GTP state, stabilizing MS-II. Cryo-EM map *VI* from Valle and coworkers (63). Note that EF-G in the presence of guanylyl iminodiphosphate stably binds only to ribosomes with an unoccupied A site (63, 68). Step 10 → 11 (reversible; k_{11}/k_{-11}): GTP hydrolysis on EF-G. State 11: The same as state 10 but with EF-G bound in the GDP-P$_i$ state. Step 11 → 0 (irreversible; k_0): The ribosome returns to the nonrotated position, the newly formed peptidyl-tRNA and the newly deacylated tRNA move into the classical P/P and E/E configurations, respectively, and the L1 stalk moves into the open position. EF-G·GDP and P$_i$ depart from the ribosome. State 0: The posttranslocation complex. The same as in state 11 but with the ribosome in the nonrotated position, the newly formed peptidyl-tRNA and the newly deacylated tRNA in classical P/P and E/E configurations, respectively, and the L1 stalk in the open position. Cryo-EM map *I* from Valle and coworkers (63). [Note that using fusidic acid, EF-G has also been trapped on the ribosome in the GDP state, with the ribosome in the nonrotated position, the newly formed peptidyl-tRNA and the newly deacylated tRNA in classical P/P and E/E configurations, respectively, and the L1 stalk in the open position as depicted in state 0 (see References 63 and 86).] For clarity, we have refrained from adding a panel depicting this configuration.

(steps 7 → 0, **Figure 3**). Indeed, the PRE complex transitions from one major conformation, termed macrostate-I (MS-I) (99) or global state 1 (GS-1) (41), to a second major conformation, termed macrostate-II (MS-II) (99) or global state 2 (GS-2) (steps 7 → 9, **Figure 3**) (41), prior to stable binding of EF-G (step 9 → 10, **Figure 3**); throughout this article, we use the term macrostates or MS. The MS-I → MS-II transition is characterized by, among other structural rearrangements, a counterclockwise ratchet-like rotation of the 30S subunit relative to the 50S subunit (denoted hereafter as the nonrotated → rotated ribosome transition), which was first observed by Frank & Agrawal (30) by comparing cryo-EM maps, confirming early proposals by Bretscher (132) and Spirin (133) that an intersubunit rotation is employed in mRNA-tRNA translocation.

Formation of the hybrid A/P (denoting the 30S A/50S P sites, respectively) and P/E configurations of the ribosome-bound tRNAs (denoted hereafter as the classical → hybrid tRNA transition), originally inferred from chemical modification studies by Moazed & Noller (134) and subsequently confirmed by Hardesty and coworkers (135) using ensemble FRET, is intricately linked with intersubunit rotation and is thus also observed to occur as part of the MS-I → MS-II transition (62, 68, 136, 137). In addition to intersubunit rotation and formation of the hybrid tRNA configurations, the MS-I → MS-II transition also encompasses a closing of the ribosomal L1 stalk (**Figure 1*a***), a highly mobile domain within the 50S subunit (denoted hereafter as the open → closed L1 stalk transition) (40–42, 63, 71, 72, 77, 83, 91, 138–142), and formation of an intermolecular contact between the closed L1 stalk and the hybrid P/E tRNA (denoted hereafter as the L1∘tRNA → L1•tRNA transition) (30, 63).

According to a general rule recognized by Zavialov & Ehrenberg (143) and Valle et al. (63), the MS-I → MS-II transition does not take place unless the P-site tRNA is deacylated, a feature of the conformational dynamics of the ribosomal elongation complex that can be rationalized by the need to stabilize the ribosome's conformation during aa-tRNA selection. This rule is appropriately described as a locking/unlocking mechanism—in the sense that the unlocking of a door is necessary but not sufficient for the door to open. In the present case, the deacylation of the P-site tRNA only provides the precondition for the MS-I → MS-II transition.

Macrostates (MS-I and MS-II): two states of the ribosome that are encountered in the process of translocation and are characterized by major conformational rearrangements

Ratchet-like rotation: counterclockwise rotation of the small subunit with respect to the large subunit, leading from macrostate I to macrostate II

Ribosome-stimulated GTP hydrolysis by EF-G and subsequent release of inorganic phosphate then causes EF-G to undergo a conformational change into the GDP form (86, 100), resulting in several interrelated events: decoupling of the mRNA-tRNAs complex from the DC (100), rotation of the 30S subunit head domain (91, 100, 131, 144), reversion of the rotation of the 30S subunit associated with intersubunit rotation (86, 99, 100), full advance of the next untranslated codon into the DC, and release of EF-G·GDP from the ribosome (steps 10 → 0, **Figure 3**). Head domain rotation and reverse rotation of the 30S subunit have been recognized as the steps leading to mRNA-tRNA translocation relative to the 30S subunit (99, 100).

3.3.1. Thermally driven, spontaneous fluctuations of pretranslocation complexes between macrostates I and II.

Initially, binding of EF-G to the PRE complex was thought to be required to bring about the MS-I → MS-II transition, including the associated classical → hybrid tRNA transition, as this was the condition under which the original cryo-EM observations were made (30, 63). This result, however, was at odds with the chemical modification studies of Moazed & Noller (134) and ensemble FRET studies of Hardesty and coworkers (135), which indicated that the classical → hybrid tRNA transition occurs spontaneously upon peptidyl transfer and in the absence of EF-G. The notion that the classical → hybrid tRNA transition within a PRE complex might be a reversible process was first suggested by Green and coworkers (145) as a possible resolution of the discrepancies between the chemical modification (134) and ensemble FRET (135) studies, on the one hand,

and the cryo-EM studies (30, 63), on the other. A few months later, the laboratories of Chu and Puglisi (38) reported the first smFRET investigation of tRNA dynamics within a PRE complex. Using the same tRNA-tRNA labeling scheme that was used in the smFRET studies of aa-tRNA selection (D_1/A_1, **Figure 1*a*** and **Table 1**), these studies directly demonstrated the reversible nature of the classical → hybrid tRNA transition, revealing that tRNAs within a PRE complex fluctuate stochastically between classical and hybrid configurations with free energies of activation, $\Delta G^{\ddagger}$, for the classical → hybrid transition of ~69 kJ mol^{-1} ($T = 296$ K) and for the hybrid → classical transition of ~70 kJ mol^{-1} ($T = 296$ K)—values that are, somewhat surprisingly, ~28-fold larger than the average thermal energy available at room temperature ($RT = 2.5$ kJ mol^{-1} at 296 K) (38).

The possibility that the tRNA fluctuations observed by Chu, Puglisi, and coworkers (38) might be accompanied by fluctuations of the entire PRE complex between the structurally observed MS-I and MS-II states in the absence of EF-G, as proposed by Kim et al. (58), suggested that a Brownian motor mechanism might underlie the translocation reaction, at least with respect to translocation of the tRNA acceptor ends within the 50S subunit. Initial experimental evidence suggesting that PRE complexes might spontaneously occupy the MS-II state in the absence of EF-G came from ensemble experiments in which an intersubunit FRET signal was used to demonstrate that PRE complexes could occupy the rotated ribosome conformation in the absence of EF-G (137). Direct experimental evidence suggesting that PRE complexes might fluctuate stochastically between the structurally observed MS-I and MS-II states in the absence of EF-G came from characterizing a variety of smFRET signals reporting directly on nonrotated and rotated ribosome conformations (D_2/A_2, **Figure 1*a***, and **Table 1**) (43), the open and closed L1 stalk conformations (D_3/A_3 and D_4/A_4, **Figure 1*a***, and **Table 1**) (40, 42, 60), and the formation and disruption of the L1 stalk-P/E tRNA intermolecular contact (D_5/A_5, **Figure 1*a***, and **Table 1**) (40, 41, 52, 60). Each of these individual smFRET signals has been found to stochastically fluctuate between two major states consistent with the MS-I and MS-II cryo-EM structures, thus defining individual nonrotated ⇄ rotated ribosome, open ⇄ closed L1 stalk, and L1∘tRNA ⇄ L1•tRNA dynamic equilibriums that, together with the classical ⇄ hybrid tRNA equilibrium, collectively define an MS-I ⇄ MS-II equilibrium.

Interestingly, data using a second intersubunit smFRET signal (D_6/A_6, **Figure 1*a*** and **Table 1**) were inconsistent with stochastic nonrotated ⇄ rotated fluctuations within a PRE complex (45). Instead, this study suggested that, during the elongation cycle, the energy of peptide bond formation drives the nonrotated → rotated transition, whereas the energy of ribosome-stimulated GTP hydrolysis on EF-G drives the reverse rotated → nonrotated transition. An alternative interpretation, however, that would reconcile the data obtained using the D_2/A_2 and D_6/A_6 intersubunit labeling schemes is that the D_6/A_6 labeling scheme simply reports on an as yet undefined conformational switch that is uniquely triggered upon each unlocking event (i.e., upon deacylation of the P-site tRNA) and uniquely reset upon each locking event (i.e., upon placement of the next peptidyl tRNA into the P site). Such a conformational switch need not directly correspond to, nor report on, the nonrotated ⇄ rotated fluctuations that are associated with the MS-I ⇄ MS-II equilibrium and that are reported on by the D_2/A_2 labeling scheme. Thus, the conformational cycle observed using the D_6/A_6 labeling scheme might report directly on the cycle of unlocking and locking that occurs during each round of the elongation cycle (146).

In complete agreement with the ensemble and smFRET investigations, two recent cryo-EM studies (62, 68), one of which was specifically performed under experimental conditions that alter the free-energy landscape of the MS-I ⇄ MS-II equilibrium (see Section 3.3.3. and **Figures 4** and **5**) in order to significantly populate MS-II (i.e., lower [Mg^{2+}], and **Figure 5**) (58, 68), were able

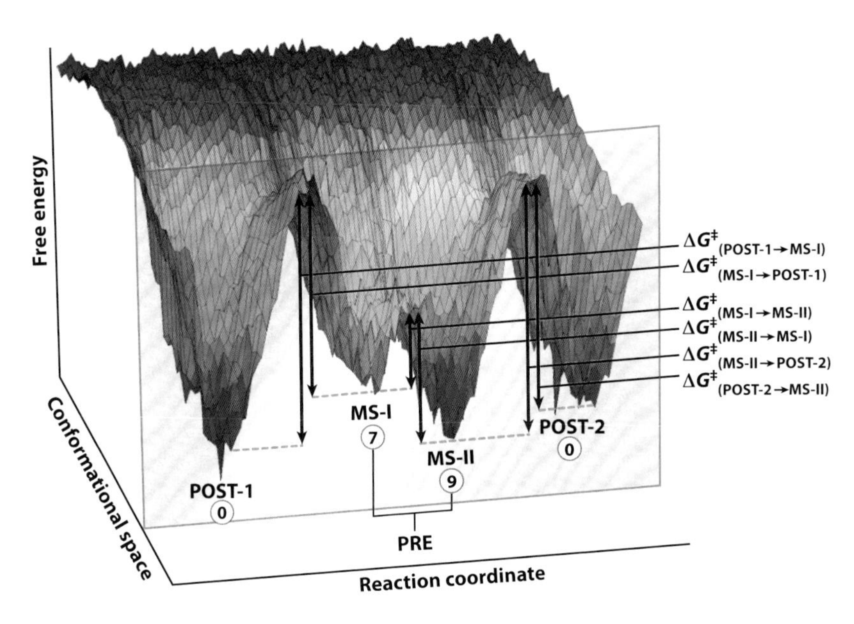

Figure 4

A heuristic schematic of the complex free-energy landscape of the elongation cycle, including macrostate (MS)-I and MS-II of the pretranslocational (PRE) ribosomal complex. Conformational changes of the ribosomal complex can occur along either the reaction coordinate or the conformational space axes. Conformational changes that take place along the reaction coordinate axis correspond to the rearrangements of the ribosomal complex that facilitate the elongation reaction that will ultimately transform posttranslocation (POST)-1 into POST-2. Conformational changes along the conformational space axis, by contrast, correspond to fluctuations among the ensemble of conformers that exist at all points along the reaction coordinate, leading to the availability of numerous parallel reaction pathways, which are the hallmark of a complex free-energy landscape. The energetic barriers separating POST-1 from the MS-I state of the PRE complex and the MS-II state of the PRE complex from POST-2 are large enough such that overcoming these barriers generally requires the energy released from GTP hydrolysis by elongation factor Tu and/or peptidyl transfer (for POST-1 → MS-I transitions) and GTP hydrolysis by EF-G (for MS-II → POST-2 transitions). The energetic barrier separating MS-I from MS-II, however, is small enough such that stochastic, thermally driven fluctuations between MS-I and MS-II are permitted. In addition, the ruggedness of the landscape strongly suggests that the valleys defining POST-1, MS-I, MS-II, and POST-2 are themselves composed of a multiplicity of smaller valleys separated by barriers even smaller than that separating MS-I from MS-II. Thus, POST-1, MS-I, MS-II, and POST-2 are each expected to be composed of an ensemble of conformations, with the population of any one member of the ensemble depending on the exact depth of its valley and heights of the barriers separating it from its neighbors. As experimentally demonstrated in **Figure 5**, the depth of the valleys within POST-1, MS-I, MS-II, and POST-2, as well as the depths of the POST-1, MS-I, MS-II, and POST-2 valleys themselves, are sensitive functions of environmental conditions (e.g., substrate, cofactor, or allosteric effector binding) or the dissociation of reaction products. The circled numbers listed underneath the POST-1, MS-I, MS-II, and POST-2 valleys refer to the equivalently labeled POST, MS-I, and MS-II complexes depicted in **Figure 3**. Abbreviation: $\Delta G^{\ddagger}$, free energy of activation.

to apply particle classification methods to single PRE complex samples in order to reveal the existence of two classes of particles with structures corresponding to MS-I and MS-II.

The structural rearrangements encompassed by MS-I ⇄ MS-II transitions are undoubtedly complex, involving substantial local as well as global reconfigurations of intra- and intersubunit ribosome-ribosome and ribosome-tRNA interactions (reviewed in Reference 31). Despite the complexity of these structural rearrangements, however, the

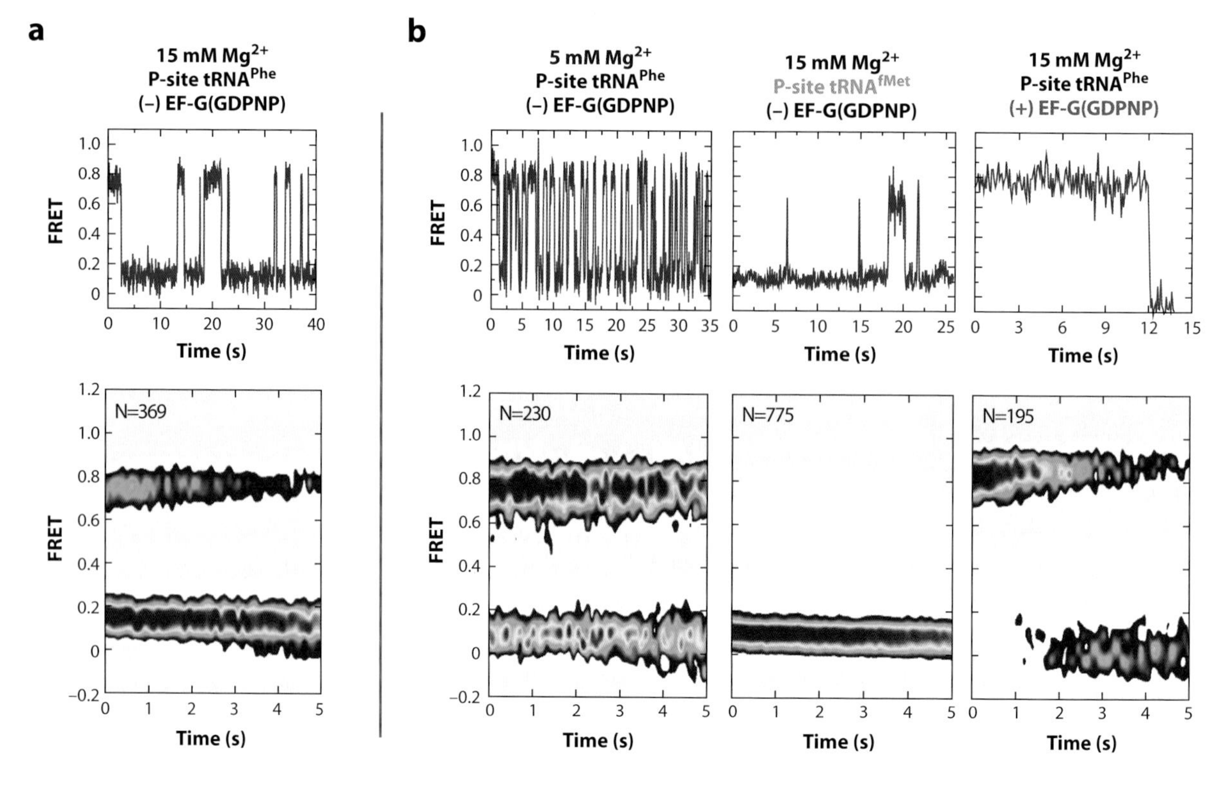
a
15 mM Mg2+
P-site tRNAPhe
(–) EF-G(GDPNP)
b
5 mM Mg2+
P-site tRNAPhe
(–) EF-G(GDPNP)
15 mM Mg2+
P-site tRNAfMet
(–) EF-G(GDPNP)
15 mM Mg2+
P-site tRNAPhe
(+) EF-G(GDPNP)
FRET
Time (s)
N=369
N=230
N=775
N=195

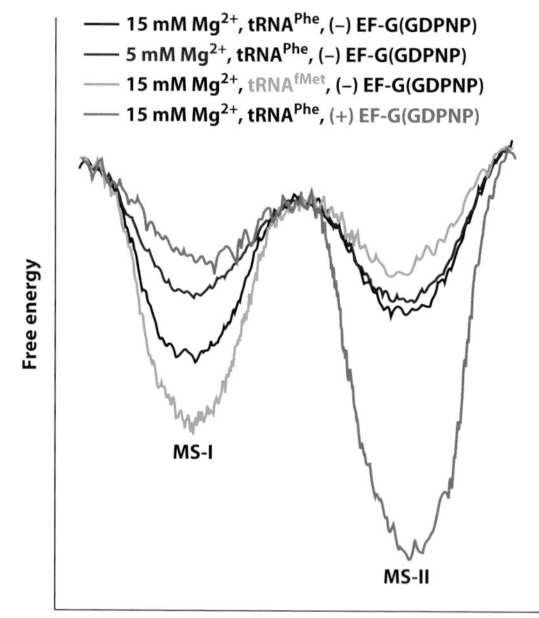
c
15 mM Mg2+, tRNAPhe, (–) EF-G(GDPNP)
5 mM Mg2+, tRNAPhe, (–) EF-G(GDPNP)
15 mM Mg2+, tRNAfMet, (–) EF-G(GDPNP)
15 mM Mg2+, tRNAPhe, (+) EF-G(GDPNP)
Free energy
MS-I
MS-II
Reaction coordinate

majority of smFRET studies reporting on the individual classical ⇄ hybrid tRNA (38, 58), nonrotated ⇄ rotated ribosome (43), open ⇄ closed L1 stalk (40, 42, 60), and L1∘tRNA ⇄ L1•tRNA (40, 41, 60) equilibriums within wild-type PRE complexes report fluctuations between just two major FRET states corresponding to the structures of MS-I or MS-II. Since individual MS-I ⇄ MS-II transitions must necessarily occur via some pathway (or, more likely, via any one of numerous parallel pathways), the failure of the majority of smFRET studies to detect any intermediate states connecting MS-I and MS-II is most likely due to either (*a*) the finite time resolution (typically 25–100 ms $frame^{-1}$ in studies of ribosome dynamics) with which smFRET studies can resolve energetically unstable, and thus transiently sampled, intermediate states or (*b*) the limited spatial resolution with which a specific donor-acceptor pair signal can detect the distance change associated with formation of a particular intermediate state (even an energetically stable intermediate state). Nevertheless, emerging cryo-EM (X. Agirrezabala, H. Liao, J. Fu, J.L. Brunelle, R. Ortiz-Meoz, et al., in preparation), X-ray crystallographic (93), and smFRET (44, 52) studies have reported observations of intermediate states connecting MS-I and MS-II. We expect that the reports of these new intermediate states will drive additional structural and dynamic studies that push the structural and time resolution limits of cryo-EM, X-ray crystallography, and smFRET in an effort to elucidate the physical basis of MS-I ⇄ MS-II transitions.

3.3.2. Origin of macrostates in the ribosomal architecture. An examination of the ribosome's architecture (77, 147) reveals the origin of the rotational instability that leads to the existence of the two macrostates (37): Two massive subunits are held together by a number of bridges with varying stability depending on their location across the intersubunit plane. As a general rule, tight RNA-RNA interactions occur in the center of the ribosome particle, whereas loose interactions involving at least one ribosomal protein occur at the periphery. An example of a tight interaction is that between

Figure 5

Modulating the free-energy landscape of the macrostate (MS)-I ⇄ MS-II equilibrium. (*a*) An equilibrium single-molecule fluorescence resonance energy (smFRET) versus time trajectory for a pretranslocation (PRE) ribosomal complex sample containing a donor-labeled P-site tRNA and an acceptor-labeled L1 stalk (41), (D_5/A_5, **Figure 1*a*** and **Table 1**) (*top*). The smFRET trajectory is calculated using $I_{Cy5}/(I_{Cy3}+I_{Cy5})$, where I_{Cy3} is the raw emission intensity of the Cy3 donor fluorophore, and I_{Cy5} is the raw emission intensity of the Cy5 acceptor fluorophore. In this labeling scheme, disruption of the L1 stalk-P/E (denoting the 30S P/50S E sites, respectively) tRNA contact (L1∘tRNA, MS-I) generates a FRET value centered at 0.16 FRET, whereas formation of the L1 stalk-P/E tRNA contact (L1•tRNA, MS-II) generates a FRET value centered at 0.76 FRET. Analysis of the dwell time spent in the MS-I state prior to transitioning to the MS-II state provides the average rate constant governing MS-I → MS-II transitions, and the analogous analysis for the dwell time spent in MS-II provides the average rate constant governing MS-II → MS-I transitions. These average rate constants can be converted to free energies of activation, $\Delta G^{\ddagger}$, for the two transitions using the equation $\Delta G^{\ddagger} = -RT\ln(hk/k_BT)$, where R is the universal gas constant (8.314 J K^{-1} mol^{-1}), T is the temperature (in K), h is Planck's constant (6.626×10^{-34} J s), k is the rate constant (in s), and k_B is Boltzmann's constant (1.381×10^{-23} J K^{-1}). The smFRET trajectory shown here, used as a point of reference, was recorded using a PRE complex containing a phenylalanine-specific tRNA ($tRNA^{Phe}$) in the P site, an unoccupied A site, and no addition of elongation factor G·guanylyl iminodiphosphate (EF-G·GDPNP) in a buffer containing 15 mM Mg^{2+}. A contour plot of the time evolution of population FRET (*bottom*) is generated by superimposing the first five of numerous individual smFRET trajectories, binning the data into 20 FRET bins and 30 time bins, and normalizing the resulting data to the most populated bin in the plot. N indicates the number of trajectories used to generate the contour plot. (*b*) Inspection of the smFRET trajectories (*top*) and contour plots of the time evolution of population FRET (*bottom*) reveal that lowering the [Mg^{2+}] from 15 mM to 5 mM (*left*), replacing the P-site $tRNA^{Phe}$ with formylmethionine specific tRNA ($tRNA^{fMet}$) (*center*), and binding of EF-G·GDPNP (*right*) all markedly alter the free-energy landscape of the MS-I ⇄ MS-II equilibrium, changing the average rates and corresponding $\Delta G^{\ddagger}$s for the MS-I → MS-II and MS-II → MS-I transitions and thereby modulating the observed MS-I and MS-II equilibrium populations. (*c*) Two-dimensional free-energy profile of the MS-I ⇄ MS-II equilibrium in which the conformational space coordinate has been averaged to a single average conformer. The plot summarizes how the $\Delta G^{\ddagger}$ for the MS-I → MS-II and MS-II → MS-I transitions is altered by changes in [Mg^{2+}], P site tRNA identity, and EF-G·GDPNP binding.

helix 44 on the 30S subunit and helix 69 on the 50S subunit (i.e., bridge B2a), and an example of a loose interaction is that between ribosomal protein S19 on the 30S subunit and helix 38 on the 50S subunit (i.e., bridge B1a).

The architectural properties of the ribosome are such that they give rise to a rotational motion when the molecule is in its thermal environment, as shown by normal mode analysis either of the X-ray structure (138, 148) or of an elastic network representation of the cryo-EM density map itself (149). These studies all indicated that intersubunit rotation is indeed a predominant mode of motion in this molecule. Normal-mode analysis also revealed a correlated motion of the L1 stalk; however, this motion goes in the opposite direction: As the 30S subunit rotates counterclockwise, the L1 stalk moves clockwise, toward the intersubunit space, again in agreement with experimental findings (40–43, 60, 62, 63, 68). Movement of the L1 stalk has been implicated in the transport of the deacylated tRNA from the P/P to the P/E position (41, 63). Thus, we have the important result that the ribosome is constructed in such a way that energy supplied from the ambient surroundings is harnessed toward productive work (99, 148).

Quite likely, this rationale of optimal energy harnessing is the reason why the intersubunit motion is encountered in other translational processes, as well (see tabulation in Reference 99).

1. In initiation, GTP hydrolysis and the release of initiation factor 2 in its GDP form from the ribosomal initiation complex are accompanied by a MS-II → MS-I transition (48, 150, 151), implying that the ribosomal initiation complex is initially assembled in MS-II, and then brought into MS-I, the proper state for acceptance of the first ternary complex of the elongation cycle.
2. In termination, the binding of GTP to ribosome-bound release factor 3 triggers an MS-I → MS-II transition of the ribosomal termination complex (57, 137, 152, 153); this conformational change is responsible for the dissociation of release factor 1 or 2 from a ribosomal termination complex (57, 153).
3. In ribosome recycling, binding of ribosome recycling factor to the posttermination ribosomal complex biases the ribosome toward MS-II (57, 154, 155).

3.3.3. Modulating the complex free-energy landscape of the MS-I ⇄ MS-II equilibrium. As expected for a complex free-energy landscape, the depths of the valleys and heights of the peaks underlying the MS-I ⇄ MS-II equilibrium are affected by a variety of factors. For example, using smFRET signals directly reporting on either the classical ⇄ hybrid tRNA, nonrotated ⇄ rotated ribosome, open ⇄ closed L1 stalk, or L1∘tRNA ⇄ L1•tRNA equilibriums as proxies for the MS-I ⇄ MS-II equilibrium, the presence of a nascent polypeptide chain versus an amino acid on the A-site tRNA decreases $\Delta G^{\ddagger}$ for the MS-I → MS-II transition by ~4 kJ mol^{-1} while having little to no effect on $\Delta G^{\ddagger}$ for the MS-II → MS-I transition (38, 41, 44). Likewise, lowering the [Mg^{2+}] from 15 mM to 3.5 mM decreases $\Delta G^{\ddagger}$ for the MS-I → MS-II transition by ~3 kJ mol^{-1} but has little to no effect on $\Delta G^{\ddagger}$ for the MS-II → MS-I transition (**Figure 5**) (58; J. Fei & R.L. Gonzalez, Jr., in preparation). Similarly, the identity of the P-site tRNA can modulate the free-energy landscape underlying the MS-I ⇄ MS-II equilibrium; for example, formylmethionine-specific tRNA ($tRNA^{fMet}$) increases $\Delta G^{\ddagger}$ for the MS-I → MS-II transition by ~1 kJ mol^{-1} and decreases $\Delta G^{\ddagger}$ for the MS-II → MS-I transition by ~2 kJ mol^{-1} relative to phenylalanine-specific tRNA ($tRNA^{Phe}$), thereby providing the mechanistic basis for the widely reported propensity of $tRNA^{fMet}$ to occupy the classical P/P configuration (40, 41), findings that have been subsequently confirmed (42, 43, 52, 53).

The ruggedness of a free-energy landscape typically speaks to the scale of the conformational change(s) that can be derived from a specified change in energy. In the case of a smooth free-energy landscape, thermal fluctuations, which are small in energy, can yield only correspondingly small changes in structure. In contrast, thermal fluctuations within a rugged free-energy landscape can lead to large conformational changes (17). Thus, the observation that thermal fluctuations can propel the large-scale conformational changes encompassing MS-I ⇄ MS-II transitions (38, 40–44, 52–54, 58, 60, 62, 68) reveals the rugged nature of the free-energy landscape underlying the translocation reaction. The ruggedness takes into account a property of MS-I ⇄ MS-II transitions that is both predicted and more recently observed. The complexity of the structural rearrangement encompassed by MS-I ⇄ MS-II transitions, shown by atomic modeling of cryo-EM density maps for MS-I and II (68), leads us to predict that individual transitions involve an entire cascade of individual small-scale rearrangements. This suggests the existence of subsidiary minima within the valleys characterizing MS-I and MS-II, thereby yielding substates that are individually populated. Interventions, such as tRNA or ribosome mutagenesis or modification (including the fluorescent labeling necessary for smFRET studies), depletion of functionally important ribosomal proteins from ribosomes, binding of antibiotic inhibitors of translocation, replacing tRNAs with ASLs, subjecting ribosomes to crystal packing forces, et cetera, may lead to the lowering of one of the subsidiary wells and thus bias the system toward the corresponding substate. Experimental evidence for the existence of such substates is now emerging from cryo-EM (156), X-ray crystallography (93), and smFRET (44, 52). For instance, using an smFRET signal reporting on classical ⇄ hybrid tRNA transitions, the transition from the classical A/A and P/P configuration to the hybrid A/P and P/E configuration within a PRE complex containing a ribosomal RNA mutation that destabilizes the A/P configuration was proposed to significantly populate an A/A and P/E intermediate configuration (44), an intermediate configuration that had been previously proposed on the basis of tRNA mutagenesis experiments (157) and ensemble kinetic measurements (158) using wild-type PRE complexes. Indeed, a cryo-EM reconstruction of the mutant PRE complex shows that the majority of imaged particles are in an intermediate state between MS-I and MS-II (state 8, **Figure 3**) in which the tRNAs occupy the A/A and P/E configuration (156). Although the same smFRET study proposed that the A/A and P/E intermediate configuration is also significantly populated in wild-type PRE complexes, two cryo-EM studies of wild-type PRE complexes failed to observe this intermediate (62, 68). This discrepancy between the smFRET and cryo-EM studies most likely arises from the present time and/or structural resolutions with which such intermediates can be defined by either smFRET or cryo-EM in wild-type PRE complexes (29). Similarly, recent X-ray crystal structures of a new crystal form of the 70S ribosome crystallized in (*a*) the absence of tRNAs, (*b*) in the presence of a P-site tRNA ASL, and (*c*) the presence of both A- and P-site tRNA ASLs have revealed an additional intermediate state between MS-I and MS-II that can be stabilized by confinement of a wild-type ribosome within a packed crystal in the absence of intact tRNAs (93). We expect that such experiments will continue to define the rugged free-energy landscape underlying the MS-I ⇄ MS-II equilibrium.

Perhaps the largest alteration of the rugged free-energy landscape underlying the MS-I ⇄ MS-II equilibrium, and certainly the most important in terms of a Brownian motor mechanism of translocation, is the effect of EF-G·GTP binding. As described above, MS-I ⇄ MS-II transitions are stochastic and thermally driven such that MS-II is transiently sampled in the absence of EF-G·GTP, one of the hallmarks of a Brownian motor mechanism. Thus, binding of EF-G·GTP to the PRE complex should serve to rectify these stochastic MS-I ⇄ MS-II fluctuations such that MS-II is conformationally selected and/or transiently stabilized en route to the second step of the translocation reaction, namely the EF-G-catalyzed translocation of the tRNA anticodon ends and the mRNA on the 30S subunit.

Indeed, this is exactly what has been observed using multiple smFRET signals reporting on the nonrotated ⇄ rotated ribosome (43), open ⇄ closed L1 stalk (40, 42), and L1∘tRNA ⇄ L1•tRNA (41, 53) equilibriums within analogs of PRE complexes in which the P site contains a deacylated tRNA and the A site is unoccupied (PRE^{-A}). In all cases, the binding of EF-G·GDPNP rectifies the system toward MS-II. Interestingly, EF-G·GDPNP-bound PRE^{-A} complexes can still exhibit spontaneous MS-I ⇄ MS-II fluctuations, indicating that such complexes are not necessarily statically trapped in MS-II (40). In fact, binding of EF-G·GDPNP to a PRE^{-A} complex can increase the frequency with which such fluctuations occur relative to that observed in the corresponding, EF-G·GDPNP-free PRE^{-A} complex. EF-G binding, therefore, can rectify MS-I ⇄ MS-II fluctuations toward MS-II either by lowering the $\Delta G^{\ddagger}$ for the MS-I → MS-II transition and/or by raising the $\Delta G^{\ddagger}$ for the MS-II → MS-I transition. Perhaps most interestingly, the identity of the P-site tRNA dictates the kinetic strategy used by EF-G. For example, in the presence of P site–bound $tRNA^{fMet}$, EF-G lowers $\Delta G^{\ddagger}$ for the MS-I → MS-II transition by ~5 kJ mol^{-1} while having little to no effect on $\Delta G^{\ddagger}$ for the MS-II → MS-I transition. Contrasting with this result, the presence of a P site–bound $tRNA^{Phe}$ raises $\Delta G^{\ddagger}$ for the MS-II → MS-I transition so high that this transition becomes virtually inaccessible, effectively trapping the system in MS-II. Satisfyingly, the dynamic nature of the EF-G·GDPNP-bound PRE^{-A} complex and the P-site tRNA-dependent regulation of these dynamics have been recently confirmed (53). It is particularly remarkable that the direct binding interactions that EF-G·GDPNP makes with the ribosome at the factor-binding site near the A site can allosterically modulate the kinetics of L1 stalk opening and closing at a hinge that is located ~170 Å away within the E site; this observation highlights the long-range allosteric coupling that links the various dynamic processes that operate within the PRE complex (40).

3.3.4. A Brownian motor mechanism may also underlie translocation of the tRNA anticodons and the mRNA on the 30S subunit. The observation that, in the absence of EF-G, the ribosome can efficiently reverse-translocate the entire tRNA-mRNA complex under certain experimental conditions (159, 160) strongly supports the hypothesis that Brownian motor mechanisms underlie the entire translocation reaction. Thus, in addition to the Brownian motor mechanism that drives translocation of the tRNA acceptor ends within the 50S subunit, a Brownian motor mechanism might also underlie translocation of the tRNA anticodons and the mRNA on the 30S subunit. In this scenario, binding of EF-G·GTP would rectify MS-I ⇄ MS-II fluctuations toward MS-II, and upon ribosome-stimulated GTP hydrolysis, EF-G·GDP would serve to rectify fluctuations of the tRNA anticodons and the mRNA in the forward direction (i.e., toward POST-2 in **Figure 4**). Indeed, LepA (EF-4), a recently discovered translational GTPase that catalyzes reverse translocation (161), may have the opposite function, namely to rectify fluctuations of the tRNA anticodons and the mRNA in the reverse direction (i.e., toward POST-1 in **Figure 4**). The use of smFRET studies to investigate reverse translocation and the roles of EF-G and LepA in promoting forward and reverse translocation, respectively, should provide the means for testing these possibilities.

4. FUTURE GOALS AND PERSPECTIVES

The current driving forces in the characterization of ribosomal dynamics and the way they affect translation are cryo-EM, X-ray crystallography, and smFRET. The emerging view of the ribosome as a processive Brownian motor, whose states are described by a rugged free-energy landscape that can be modulated by environmental conditions as well as by the ribosome's tRNA substrates and accessory translation factors, makes it clear how little we know from experiments to date. It is even more humbling to realize that the free-energy landscape describing elongation that is used as an example in **Figure 4** is just one of numerous such landscapes that will require detailed study. For example, the scope of this article has not allowed us to discuss further modifications of the elongation free-energy landscape by the interactions that the exiting polypeptide makes with the ribosomal exit tunnel as well as with the translocon, which are clearly required to round up the characterization of states. Beyond elongation, the initiation, termination, and ribosome recycling stages of protein synthesis all also involve an MS-I $\rightarrow$ MS-II transition (99, 150–155). Indeed, very recent smFRET studies have demonstrated how initiation factors (48), release factors (57), and ribosome recycling factors (57) rectify and thereby regulate the MS-I $\rightleftarrows$ MS-II equilibrium during the initiation, termination, and ribosome recycling stages of protein synthesis. A true understanding of the workings of the ribosome as a molecular motor will require a continued effort to sample the relevant landscapes systematically, covering all important substates and macrostates by cryo-EM or, when feasible, by X-ray crystallography, and all transitions among these substates and macrostates by smFRET.

Efforts to reach this goal still face a number of obstacles. Resolving the substates by cryo-EM with the necessary spatial resolution requires a substantial increase in data collection and a sharpening of classification tools. The prospect of digital online data capture with 8000 $\times$ 8000 pixel field size on the electron microscope becoming an affordable option offers hope that data collection in the millions of particles will be a reality. Regarding classification, the recent works by the Carazo (162, 163) and Penczek (164, 165) groups have given us valuable tools for unsupervised classification, i.e., without the need to use reference structures. Exciting progress is also being made in the development of time-resolved electron microscopy through the employment of flash photolysis or nanotechnology (166, 167). Thanks to these efforts, it is now possible to obtain a two-component mix of ribosome and substrate (such as EF-G) at a defined time point and freeze the mixture at some point several milliseconds later (167). By changing the length of reaction time in a systematic way, it should be possible to follow the emergence of newly occupied states.

Likewise, several limitations currently restrict the potential of smFRET, defining areas that would benefit from further technical development. Perhaps the major limitation involves the ever-present trade-offs between time resolution, signal-to-noise ratio, and the efficiency with which large, statistically relevant data sets can be collected. On the one hand, wide-field microscopies, such as total internal reflection fluorescence microscopy, allow data to be simultaneously collected on hundreds of molecules with a time resolution on the order of tens of milliseconds to hundreds of milliseconds (95, 168), limited primarily by the signal-to-noise ratio of the data but also by the ability of the computer to rapidly write the data to disk. Although these techniques allow a large, statistically relevant data set to be collected relatively rapidly, the time resolution may not be high enough to characterize short-lived, but mechanistically interesting, conformational substates. On the other hand, confocal fluorescence microscopies allow data to be collected on one molecule at a time with microseconds to tens of microseconds time resolution (95, 168). Here the time resolution promises data that will be mechanistically richer but will require longer periods of data collection to generate large, statistically relevant data sets. Finally, model-free inference of the number of states underlying smFRET versus time trajectories and the corresponding rates of transitions between these states remains a major goal of developing data analysis algorithms. Currently, most data analysis methods model the observed smFRET trajectories using a hidden Markov model and implement a maximum likelihood-based inference approach on individual trajectories that requires the user to either guess the number of states present in the data or overfit the data intentionally by asserting an excess number of states (169–172). Recently, however, a variational Bayesian approach has been introduced that allows inference of the number of states and transition rates between states from individual smFRET trajectories without requiring user guesses and/or overfitting (40, 60).

The fact that cryo-EM and smFRET can draw complementary pictures of the same process provides great opportunities in coming to an understanding of translation. However, future studies aiming to integrate data from cryo-EM with those of smFRET in a more quantitative way will require looking at the same sample with both techniques because there are many indications that even small changes in experimental conditions and molecular constructs may have pronounced effects on the free-energy landscape.

SUMMARY POINTS

1. The ribosome is a highly dynamic molecular machine, specifically a processive Brownian motor.
2. The conformational dynamics of the ribosome, and the way the ribosome interacts with its ligands, have been experimentally studied by three primary methods: cryo-EM and X-ray crystallography, yielding three-dimensional snapshots, and, more recently, smFRET, yielding the real-time conformational trajectories of single molecules.
3. Detailed insights into the molecular mechanisms underlying aa-tRNA selection and mRNA-tRNA translocation can be gained by combining results from cryo-EM, X-ray crystallography, and smFRET.
4. smFRET studies indicate that, during both aa-tRNA selection and translocation, ribosomes and tRNAs fluctuate among several conformational states and that the choices of pathways taken are sensitive to many factors, such as the exact experimental conditions, the identity of the tRNAs involved, the presence of small-molecule antibiotic inhibitors of aa-tRNA selection and/or translocation, and the action of translation factors.
5. Translocation is facilitated by architectural features of the ribosome that allow it to interconvert between two conformations, or macrostates, with little expenditure of energy. These states are linked to the transition of the tRNA configuration from classical A/A and P/P configurations to the hybrid A/P and P/E configurations.
6. Evidence from smFRET and cryo-EM for spontaneous transitions between the macrostates has led to a revision of the role of EF-G in translocation, from instrumental to ancillary, accelerating a process structurally ingrained in the ribosome.
7. A free-energy landscape depiction of the degrees of freedom of the complex formed by ribosome and its ligands during the elongation cycle is a useful heuristic tool, which suggests that concepts recently developed in the study of enzyme catalysis are applicable to this much larger system.

FUTURE ISSUES

1. In terms of the free-energy landscape of translation, our current knowledge is restricted predominantly to elongation and is limited to only a few states and pathways. A detailed, systematic mapping of the full landscape throughout all stages of protein synthesis is clearly required.

2. As far as cryo-EM is concerned, addressing this broad program will require two developments: (*a*) time-resolved methods allowing the study of a presteady-state system developing over time and (*b*) further advances in classification methods to sort heterogeneous molecular populations into homogeneous subsets.
3. Likewise, the use of smFRET for mapping out the free-energy landscapes of the translating ribosome will benefit greatly from several developments: (*a*) further advances in detector technologies to increase the signal-to-noise ratio and time resolution, (*b*) automated and high-throughput data collection schemes to improve the efficiency with which large, statistically relevant data sets can be collected, and (*c*) continued development of automated, high-throughput, and model-free data analysis algorithms.
4. As the experimental knowledge base on ribosomal dynamics from cryo-EM, X-ray crystallography, and smFRET increases, molecular dynamics simulations of selected pathways will lead to a fuller understanding of molecular mechanisms of translation.

DISCLOSURE STATEMENT

The authors are not aware of any affiliations, memberships, funding, or financial holdings that might be perceived as affecting the objectivity of this review.

ACKNOWLEDGMENTS

This work was supported by HHMI and NIH-NIGMS grants (R37 GM29169 and R01 GM55440) to J.F. and a Burroughs Wellcome Fund CABS Award (CABS 1004856), an NSF CAREER Award (MCB 0644262), an NIH-NIGMS grant (R01 GM084288), and an American Cancer Society Research Scholar Grant (RSG GMC-117152) to R.L.G. We thank Jingyi Fei, Daniel D. MacDougall, Joseph D. Puglisi, Colin E. Aitken, and Peter V. Cornish for insightful discussions, Li Wen, Bo Chen, Jingyi Fei, Daniel D. MacDougall, and Margaret M. Elvekrog for comments on the manuscript, and Lila Iino-Rubenstein, Jingyi Fei, and Jonathan E. Bronson for assistance with the preparation of the illustrations.

LITERATURE CITED

1. Spirin AS. 2002. Ribosome as a molecular machine. *FEBS Lett.* 514:2–10
2. Garai A, Chowdhury D, Ramakrishnan TV. 2009. Stochastic kinetics of ribosomes: single motor properties and collective behavior. *Phys. Rev. E.* 80:011908
3. Spirin AS. 2009. The ribosome as a conveying thermal ratchet machine. *J. Biol. Chem.* 284:21103–19
4. Spirin AS. 2004. The ribosome as an RNA-based molecular machine. *RNA Biol.* 1:3–9
5. Astumian RD. 1997. Thermodynamics and kinetics of a Brownian motor. *Science* 276:917–22
6. Peskin CS, Odell GM, Oster GF, Cordova NJ, Ermentrout B. 1993. Cellular motions and thermal fluctuations: the Brownian ratchet. *Biophys. J.* 65:316–24
7. Cordova NJ, Ermentrout B, Oster GF. 1992. Dynamics of single-motor molecules: the thermal ratchet model. *Proc. Natl. Acad. Sci. USA* 89:339–43
8. Abbondanzieri EA, Greenleaf WJ, Shaevitz JW, Landick R, Block SM. 2005. Direct observation of base-pair stepping by RNA polymerase. *Nature* 438:460–65
9. Bai L, Shundrovsky A, Wang MD. 2004. Sequence-dependent kinetic model for transcription elongation by RNA polymerase. *J. Mol. Biol.* 344:335–49
10. Bryngelson JD, Onuchic JN, Socci ND, Wolynes PG. 1995. Funnels, pathways, and the energy landscape of protein folding: a synthesis. *Proteins* 21:167–95

11. Onuchic JN, Wolynes PG. 2004. Theory of protein folding. *Curr. Opin. Struct. Biol.* 14:70–75
12. Shcherbakova I, Mitra S, Laederach A, Brenowitz M. 2008. Energy barriers, pathways, and dynamics during folding of large, multidomain RNAs. *Curr. Opin. Chem. Biol.* 12:655–66
13. Thirumalai D, Lee N, Woodson SA, Klimov D. 2001. Early events in RNA folding. *Annu. Rev. Phys. Chem.* 52:751–62
14. Bokinsky G, Zhuang XW. 2005. Single-molecule RNA folding. *Acc. Chem. Res.* 38:566–73
15. Chu VB, Herschlag D. 2008. Unwinding RNA's secrets: advances in the biology, physics, and modeling of complex RNAs. *Curr. Opin. Struct. Biol.* 18:305–14
16. Treiber DK, Williamson JR. 2001. Beyond kinetic traps in RNA folding. *Curr. Opin. Struct. Biol.* 11:309–14
17. Dill KA, Chan HS. 1997. From Levinthal to pathways to funnels. *Nat. Struct. Biol.* 4:10–19
18. Schnell JR, Dyson HJ, Wright PE. 2004. Structure, dynamics, and catalytic function of dihydrofolate reductase. *Annu. Rev. Biophys. Biomol. Struct.* 33:119–40
19. Zhuang X, Kim H, Pereira MJ, Babcock HP, Walter NG, Chu S. 2002. Correlating structural dynamics and function in single ribozyme molecules. *Science* 296:1473–76
20. Nagel ZD, Klinman JP. 2006. Tunneling and dynamics in enzymatic hydride transfer. *Chem. Rev.* 106:3095–118
21. Benkovic SJ, Hammes GG, Hammes-Schiffer S. 2008. Free-energy landscape of enzyme catalysis. *Biochemistry* 47:3317–21
22. Hammes-Schiffer S, Benkovic SJ. 2006. Relating protein motion to catalysis. *Annu. Rev. Biochem.* 75:519–41
23. Cole R, Loria JP. 2002. Evidence for flexibility in the function of ribonuclease A. *Biochemistry* 41:6072–81
24. Das K, Bauman JD, Clark AD Jr, Frenkel YV, Lewi PJ, et al. 2008. High-resolution structures of HIV-1 reverse transcriptase/TMC278 complexes: Strategic flexibility explains potency against resistance mutations. *Proc. Natl. Acad. Sci. USA* 105:1466–71
25. Zhuang XW. 2005. Single-molecule RNA science. *Annu. Rev. Biophys. Biomol. Struct.* 34:399–414
26. Zhuang X, Bartley LE, Babcock HP, Russell R, Ha T, et al. 2000. A single-molecule study of RNA catalysis and folding. *Science* 288:2048–51
27. English BP, Min W, van Oijen AM, Lee KT, Luo G, et al. 2006. Ever-fluctuating single enzyme molecules: Michaelis-Menten equation revisited. *Nat. Chem. Biol.* 2:87–94
28. Boehr DD, McElheny D, Dyson HJ, Wright PE. 2006. The dynamic energy landscape of dihydrofolate reductase catalysis. *Science* 313:1638–42
29. Munro JB, Sanbonmatsu KY, Spahn CM, Blanchard SC. 2009. Navigating the ribosome's metastable energy landscape. *Trends Biochem. Sci.* 34:390–400
30. Frank J, Agrawal RK. 2000. A ratchet-like inter-subunit reorganization of the ribosome during translocation. *Nature* 406:318–22
31. Korostelev A, Ermolenko DN, Noller HF. 2008. Structural dynamics of the ribosome. *Curr. Opin. Chem. Biol.* 12:674–83
32. Koshland DE Jr. 1958. Application of a theory of enzyme specificity to protein synthesis. *Proc. Natl. Acad. Sci. USA* 44:98–104
33. Williamson JR. 2000. Induced fit in RNA-protein recognition. *Nat. Struct. Biol.* 7:834–37
34. Tsai CJ, Kumar S, Ma B, Nussinov R. 1999. Folding funnels, binding funnels, and protein function. *Protein Sci.* 8:1181–90
35. Bosshard HR. 2001. Molecular recognition by induced fit: How fit is the concept? *News Physiol. Sci.* 16:171–73
36. Wang C, Karpowich N, Hunt JF, Rance M, Palmer AG. 2004. Dynamics of ATP-binding cassette contribute to allosteric control, nucleotide binding and energy transduction in ABC transporters. *J. Mol. Biol.* 342:525–37
37. Wen JD, Lancaster L, Hodges C, Zeri AC, Yoshimura SH, et al. 2008. Following translation by single ribosomes one codon at a time. *Nature* 452:598–603
38. Blanchard SC, Kim HD, Gonzalez RL Jr, Puglisi JD, Chu S. 2004. tRNA dynamics on the ribosome during translation. *Proc. Natl. Acad. Sci. USA* 101:12893–98

39. Blanchard SC, Gonzalez RL Jr, Kim HD, Chu S, Puglisi JD. 2004. tRNA selection and kinetic proofreading in translation. *Nat. Struct. Mol. Biol.* 11:1008–14
40. Fei J, Bronson JE, Hofman JM, Srinivas RL, Wiggins CH, Gonzalez RLJ. 2009. Allosteric collaboration between elongation factor G and the ribosomal L1 stalk directs tRNA movements during translation. *Proc. Natl. Acad. Sci. USA* 106:15702–7
41. Fei J, Kosuri P, MacDougall DD, Gonzalez RL Jr. 2008. Coupling of ribosomal L1 stalk and tRNA dynamics during translation elongation. *Mol. Cell* 30:348–59
42. Cornish PV, Ermolenko DN, Staple DW, Hoang L, Hickerson RP, et al. 2009. Following movement of the L1 stalk between three functional states in single ribosomes. *Proc. Natl. Acad. Sci. USA* 106:2571–76
43. Cornish PV, Ermolenko DN, Noller HF, Ha T. 2008. Spontaneous intersubunit rotation in single ribosomes. *Mol. Cell* 30:578–88
44. Munro JB, Altman RB, O'Connor N, Blanchard SC. 2007. Identification of two distinct hybrid state intermediates on the ribosome. *Mol. Cell* 25:505–17
45. Marshall RA, Dorywalska M, Puglisi JD. 2008. Irreversible chemical steps control intersubunit dynamics during translation. *Proc. Natl. Acad. Sci. USA* 105:15364–69
46. Wang Y, Qin H, Kudaravalli RD, Kirillov SV, Dempsey GT, et al. 2007. Single-molecule structural dynamics of EF-G–ribosome interaction during translocation. *Biochemistry* 46:10767–75
47. Uemura S, Dorywalska M, Lee TH, Kim HD, Puglisi JD, Chu S. 2007. Peptide bond formation destabilizes Shine-Dalgarno interaction on the ribosome. *Nature* 446:454–57
48. Marshall RA, Aitken CE, Puglisi JD. 2009. GTP hydrolysis by IF2 guides progression of the ribosome into elongation. *Mol. Cell* 35:37–47
49. Vanzi F, Vladimirov S, Knudsen CR, Goldman YE, Cooperman BS. 2003. Protein synthesis by single ribosomes. *RNA* 9:1174–79
50. Vanzi F, Takagi Y, Shuman H, Cooperman BS, Goldman YE. 2005. Mechanical studies of single ribosome/mRNA complexes. *Biophys. J.* 89:1909–19
51. Dorywalska M, Blanchard SC, Gonzalez RL, Kim HD, Chu S, Puglisi JD. 2005. Site-specific labeling of the ribosome for single-molecule spectroscopy. *Nucleic Acids Res.* 33:182–89
52. Munro JB, Altman RB, Tung CS, Cate JH, Sanbonmatsu KY, Blanchard SC. 2010. Spontaneous formation of the unlocked state of the ribosome is a multistep process. *Proc. Natl. Acad. Sci. USA.* 107:709–14
53. Munro JB, Altman RB, Tung CS, Sanbonmatsu KY, Blanchard SC. 2010. A fast dynamic mode of the EF-G-bound ribosome. *EMBO J.* 29:770–81
54. Feldman MB, Terry DS, Altman RB, Blanchard SC. 2009. Aminoglycoside activity observed on single pre-translocation ribosome complexes. *Nat. Chem. Biol.* 6:54–62
55. Effraim PR, Wang J, Englander MT, Avins J, Leyh TS, et al. 2009. Natural amino acids do not require their native tRNAs for efficient selection by the ribosome. *Nat. Chem. Biol.* 5:947–53
56. Gonzalez RL Jr, Chu S, Puglisi JD. 2007. Thiostrepton inhibition of tRNA delivery to the ribosome. *RNA* 13:2091–97
57. Sternberg SH, Fei J, Prywes N, McGrath KA, Gonzalez RL Jr. 2009. Translation factors direct intrinsic ribosome dynamics during translation termination and ribosome recycling. *Nat. Struct. Mol. Biol.* 16:861–68
58. Kim HD, Puglisi J, Chu S. 2007. Fluctuations of transfer RNAs between classical and hybrid states. *Biophys. J.* 93:3575–82
59. Lee TH, Blanchard SC, Kim HD, Puglisi JD, Chu S. 2007. The role of fluctuations in tRNA selection by the ribosome. *Proc. Natl. Acad. Sci. USA* 104:13661–65
60. Bronson JE, Fei J, Hofman JM, Gonzalez RL Jr, Wiggins CH. 2009. Learning rates and states from biophysical time series: a Bayesian approach to model selection and single-molecule FRET data. *Biophys. J.* 97:3196–205
61. Villa E, Sengupta J, Trabuco LG, LeBarron J, Baxter WT, et al. 2009. Ribosome-induced changes in elongation factor Tu conformation control GTP hydrolysis. *Proc. Natl. Acad. Sci. USA* 106:1063–68
62. Julian P, Konevega AL, Scheres SH, Lazaro M, Gil D, et al. 2008. Structure of ratcheted ribosomes with tRNAs in hybrid states. *Proc. Natl. Acad. Sci. USA* 105:16924–27
63. Valle M, Zavialov AV, Sengupta J, Rawat U, Ehrenberg M, Frank J. 2003. Locking and unlocking of ribosomal motions. *Cell* 114:123–34

64. Valle M, Zavialov A, Li W, Stagg SM, Sengupta J, et al. 2003. Incorporation of aminoacyl-tRNA into the ribosome as seen by cryo-electron microscopy. *Nat. Struct. Biol.* 10:899–906
65. Li W, Agirrezabala X, Lei J, Bouakaz L, Brunelle JL, et al. 2008. Recognition of aminoacyl-tRNA: a common molecular mechanism revealed by cryo-EM. *EMBO J.* 27:3322–31
66. Stark H, Rodnina MV, Wieden HJ, Zemlin F, Wintermeyer W, van Heel M. 2002. Ribosome interactions of aminoacyl-tRNA and elongation factor Tu in the codon-recognition complex. *Nat. Struct. Biol.* 9:849–54
67. Stark H, Rodnina MV, Wieden HJ, van Heel M, Wintermeyer W. 2000. Large-scale movement of elongation factor G and extensive conformational change of the ribosome during translocation. *Cell* 100:301–9
68. Agirrezabala X, Lei J, Brunelle JL, Ortiz-Meoz RF, Green R, Frank J. 2008. Visualization of the hybrid state of tRNA binding promoted by spontaneous ratcheting of the ribosome. *Mol. Cell* 32:190–97
69. Schuette JC, Murphy FV 4th, Kelley AC, Weir JR, Giesebrecht J, et al. 2009. GTPase activation of elongation factor EF-Tu by the ribosome during decoding. *EMBO J.* 28:755–65
70. Schmeing TM, Seila AC, Hansen JL, Freeborn B, Soukup JK, et al. 2002. A pre-translocational intermediate in protein synthesis observed in crystals of enzymatically active 50S subunits. *Nat. Struct. Biol.* 9:225–30
71. Korostelev A, Trakhanov S, Laurberg M, Noller HF. 2006. Crystal structure of a 70S ribosome-tRNA complex reveals functional interactions and rearrangements. *Cell* 126:1065–77
72. Selmer M, Dunham CM, Murphy FV 4th, Weixlbaumer A, Petry S, et al. 2006. Structure of the 70S ribosome complexed with mRNA and tRNA. *Science* 313:1935–42
73. Ogle JM, Murphy FV, Tarry MJ, Ramakrishnan V. 2002. Selection of tRNA by the ribosome requires a transition from an open to a closed form. *Cell* 111:721–32
74. Ogle JM, Brodersen DE, Clemons WM Jr, Tarry MJ, Carter AP, Ramakrishnan V. 2001. Recognition of cognate transfer RNA by the 30S ribosomal subunit. *Science* 292:897–902
75. Carter AP, Clemons WM, Brodersen DE, Morgan-Warren RJ, Wimberly BT, Ramakrishnan V. 2000. Functional insights from the structure of the 30S ribosomal subunit and its interactions with antibiotics (see comments). *Nature* 407:340–48
76. Wimberly BT, Brodersen DE, Clemons WM Jr, Morgan-Warren RJ, Carter AP, et al. 2000. Structure of the 30S ribosomal subunit. *Nature* 407:327–39 (also comment pp. 306–7)
77. Yusupov MM, Yusupova GZ, Baucom A, Lieberman K, Earnest TN, et al. 2001. Crystal structure of the ribosome at 5.5 A resolution. *Science* 292:883–96
78. Yusupova GZ, Yusupov MM, Cate JH, Noller HF. 2001. The path of messenger RNA through the ribosome. *Cell* 106:233–41
79. Cate JH, Yusupov MM, Yusupova GZ, Earnest TN, Noller HF. 1999. X-ray crystal structures of 70S ribosome functional complexes. *Science* 285:2095–104
80. Schmeing TM, Moore PB, Steitz TA. 2003. Structures of deacylated tRNA mimics bound to the E site of the large ribosomal subunit. *RNA* 9:1345–52
81. Ban N, Nissen P, Hansen J, Moore PB, Steitz TA. 2000. The complete atomic structure of the large ribosomal subunit at 2.4 A resolution. *Science* 289:905–20
82. Nissen P, Hansen J, Ban N, Moore PB, Steitz TA. 2000. The structural basis of ribosome activity in peptide bond synthesis. *Science* 289:920–30
83. Harms J, Schluenzen F, Zarivach R, Bashan A, Gat S, et al. 2001. High resolution structure of the large ribosomal subunit from a mesophilic eubacterium. *Cell* 107:679–88
84. Schluenzen F, Tocilj A, Zarivach R, Harms J, Gluehmann M, et al. 2000. Structure of functionally activated small ribosomal subunit at 3.3 angstroms resolution. *Cell* 102:615–23
85. Schmeing TM, Voorhees RM, Kelley AC, Gao YG, Murphy FV 4th, et al. 2009. The crystal structure of the ribosome bound to EF-Tu and aminoacyl-tRNA. *Science* 326:688–94
86. Gao YG, Selmer M, Dunham CM, Weixlbaumer A, Kelley AC, Ramakrishnan V. 2009. The structure of the ribosome with elongation factor G trapped in the posttranslocational state. *Science* 326:694–99
87. Frank J. 2006. *Three-Dimensional Electron Microscopy of Macromolecular Assemblies*. New York: Oxford Univ. Press

88. Glaeser R, Downing K, DeRosier D, Chiu W, Frank J. 2007. *Electron Crystallography of Biological Macromolecules*. New York: Oxford Univ. Press
89. Mitra K, Frank J. 2006. Ribosome dynamics: insights from atomic structure modeling into cryo-electron microscopy maps. *Annu. Rev. Biophys. Biomol. Struct.* 35:299–317
90. Trabuco LG, Villa E, Mitra K, Frank J, Schulten K. 2008. Flexible fitting of atomic structures into electron microscopy maps using molecular dynamics. *Structure* 16:673–83
91. Schuwirth BS, Borovinskaya MA, Hau CW, Zhang W, Vila-Sanjurjo A, et al. 2005. Structures of the bacterial ribosome at 3.5 A resolution. *Science* 310:827–34
92. Carter AP, Clemons WM Jr, Brodersen DE, Morgan-Warren RJ, Hartsch T, et al. 2001. Crystal structure of an initiation factor bound to the 30S ribosomal subunit. *Science* 291:498–501
93. Zhang W, Dunkle JA, Cate JH. 2009. Structures of the ribosome in intermediate states of ratcheting. *Science* 325:1014–17
94. Ha T. 2001. Single-molecule fluorescence resonance energy. *Methods* 25:78–86
95. Joo C, Balci H, Ishitsuka Y, Buranachai C, Ha T. 2008. Advances in single-molecule fluorescence methods for molecular biology. *Annu. Rev. Biochem.* 77:51–76
96. Clamme JP, Deniz AA. 2005. Three-color single-molecule fluorescence resonance energy transfer. *ChemPhysChem* 6:74–77
97. Hohng S, Joo C, Ha T. 2004. Single-molecule three-color FRET. *Biophys. J.* 87:1328–37
98. Clegg RM. 1992. Fluorescence resonance energy transfer and nucleic acids. *Methods Enzymol.* 211:353–88
99. Frank J, Gao H, Sengupta J, Gao N, Taylor DJ. 2007. The process of mRNA-tRNA translocation. *Proc. Natl. Acad. Sci. USA* 104:19671–78
100. Taylor DJ, Nilsson J, Merrill AR, Andersen GR, Nissen P, Frank J. 2007. Structures of modified eEF2 80S ribosome complexes reveal the role of GTP hydrolysis in translocation. *EMBO J.* 26:2421–31
101. Rodnina MV, Gromadski KB, Kothe U, Wieden HJ. 2005. Recognition and selection of tRNA in translation. *FEBS Lett.* 579:938–42
102. Beringer M, Rodnina MV. 2007. The ribosomal peptidyl transferase. *Mol. Cell* 26:311–21
103. Shoji S, Walker SE, Fredrick K. 2009. Ribosomal translocation: one step closer to the molecular mechanism. *ACS Chem. Biol.* 4:93–107
104. Daviter T, Gromadski KB, Rodnina MV. 2006. The ribosome's response to codon-anticodon mismatches. *Biochimie* 88:1001–11
105. Hopfield JJ. 1974. Kinetic proofreading: a new mechanism for reducing errors in biosynthetic processes requiring high specificity. *Proc. Natl. Acad. Sci. USA* 71:4135–39
106. Rodnina MV, Wintermeyer W. 2001. Ribosome fidelity: tRNA discrimination, proofreading and induced fit. *Trends Biochem. Sci.* 26:124–30
107. Zaher HS, Green R. 2009. Fidelity at the molecular level: lessons from protein synthesis. *Cell* 136:746–62
108. Fourmy D, Recht MI, Blanchard SC, Puglisi JD. 1996. Structure of the A site of *E. coli* 16S ribosomal RNA complexed with an aminoglycoside antibiotic. *Science* 274:1367–71
109. Fourmy D, Yoshizawa S, Puglisi JD. 1998. Paromomycin binding induces a local conformational change in the A site of 16S rRNA. *J. Mol. Biol.* 277:333–45
110. Gromadski KB, Rodnina MV. 2004. Kinetic determinants of high-fidelity tRNA discrimination on the ribosome. *Mol. Cell* 13:191–200
111. Gromadski KB, Daviter T, Rodnina MV. 2006. A uniform response to mismatches in codon-anticodon complexes ensures ribosomal fidelity. *Mol. Cell* 21:369–77
112. Thomas LK, Dix DB, Thompson RC. 1988. Codon choice and gene expression: Synonymous codons differ in their ability to direct aminoacylated-transfer RNA binding to ribosomes in vitro. *Proc. Natl. Acad. Sci. USA* 85:4242–46
113. Dix DB, Thompson RC. 1989. Codon choice and gene expression: Synonymous codons differ in translational accuracy. *Proc. Natl. Acad. Sci. USA* 86:6888–92
114. Wolf H, Chinali G, Parmeggiani A. 1977. Mechanism of the inhibition of protein synthesis by kirromycin. *Eur. J. Biochem.* 75:67–75
115. Parmeggiani A, Swart GW. 1985. Mechanism of action of kirromycin-like antibiotics. *Annu. Rev. Microbiol.* 39:557–77

116. Stark H, Rodnina MV, Rinke-Appel J, Brimacombe R, Wintermeyer W, van Heel M. 1997. Visualization of elongation factor Tu on the *Escherichia coli* ribosome. *Nature* 389:403–6
117. Valle M, Sengupta J, Swami NK, Grassucci RA, Burkhardt N, et al. 2002. Cryo-EM reveals an active role for aminoacyl-tRNA in the accommodation process. *EMBO J.* 21:3557–67
118. LeBarron J, Grassucci RA, Shaikh TR, Baxter WT, Sengupta J, Frank J. 2008. Exploration of parameters in cryo-EM leading to an improved density map of the *E. coli* ribosome. *J. Struct. Biol.* 164:24–32
119. Frank J, Sengupta J, Gao H, Li W, Valle M, et al. 2005. The role of tRNA as a molecular spring in decoding, accommodation, and peptidyl transfer. *FEBS Lett.* 579:959–62
120. Shi H, Moore PB. 2000. The crystal structure of yeast phenylalanine tRNA at 1.93 A resolution: a classic structure revisited. *RNA* 6:1091–105
121. Piepenburg O, Pape T, Pleiss JA, Wintermeyer W, Uhlenbeck OC, Rodnina MV. 2000. Intact aminoacyl-tRNA is required to trigger GTP hydrolysis by elongation factor Tu on the ribosome. *Biochemistry* 39:1734–38
122. Cochella L, Green R. 2005. An active role for tRNA in decoding beyond codon:anticodon pairing. *Science* 308:1178–80
123. Ledoux S, Olejniczak M, Uhlenbeck OC. 2009. A sequence element that tunes *Escherichia coli* $tRNA_{GGC}^{Ala}$ to ensure accurate decoding. *Nat. Struct. Mol. Biol.* 16:359–64
124. Murakami H, Ohta A, Suga H. 2009. Bases in the anticodon loop of $tRNA^{Ala}_{GGC}$ prevent misreading. *Nat. Struct. Mol. Biol.* 16:353–58
125. Yarus M, Valle M, Frank J. 2003. A twisted tRNA intermediate sets the threshold for decoding. *RNA* 9:384–85
126. Brunelle JL, Youngman EM, Sharma D, Green R. 2006. The interaction between C75 of tRNA and the A loop of the ribosome stimulates peptidyl transferase activity. *RNA* 12:33–39
127. Weinger JS, Parnell KM, Dorner S, Green R, Strobel SA. 2004. Substrate-assisted catalysis of peptide bond formation by the ribosome. *Nat. Struct. Mol. Biol.* 11:1101–6
128. Youngman EM, Brunelle JL, Kochaniak AB, Green R. 2004. The active site of the ribosome is composed of two layers of conserved nucleotides with distinct roles in peptide bond formation and peptide release. *Cell* 117:589–99
129. Voorhees RM, Weixlbaumer A, Loakes D, Kelley AC, Ramakrishnan V. 2009. Insights into substrate stabilization from snapshots of the peptidyl transferase center of the intact 70S ribosome. *Nat. Struct. Mol. Biol.* 16:528–33
130. Agrawal RK, Penczek P, Grassucci RA, Frank J. 1998. Visualization of elongation factor G on the *Escherichia coli* 70S ribosome: the mechanism of translocation. *Proc. Natl. Acad. Sci. USA* 95:6134–38
131. Agrawal RK, Heagle AB, Penczek P, Grassucci RA, Frank J. 1999. EF-G-dependent GTP hydrolysis induces translocation accompanied by large conformational changes in the 70S ribosome. *Nat. Struct. Biol.* 6:643–47
132. Bretscher MS. 1968. Translocation in protein synthesis: a hybrid structure model. *Nature* 218:675–77
133. Spirin AS. 1969. A model of the functioning ribosome: locking and unlocking of the ribosome subparticles. *Cold Spring Harb. Symp. Quant. Biol.* 34:197–207
134. Moazed D, Noller HF. 1989. Intermediate states in the movement of transfer RNA in the ribosome. *Nature* 342:142–48
135. Odom OW, Picking WD, Hardesty B. 1990. Movement of tRNA but not the nascent peptide during peptide bond formation on ribosomes. *Biochemistry* 29:10734–44
136. Ermolenko DN, Spiegel PC, Majumdar ZK, Hickerson RP, Clegg RM, Noller HF. 2007. The antibiotic viomycin traps the ribosome in an intermediate state of translocation. *Nat. Struct. Mol. Biol.* 14:493–97
137. Ermolenko DN, Majumdar ZK, Hickerson RP, Spiegel PC, Clegg RM, Noller HF. 2007. Observation of intersubunit movement of the ribosome in solution using FRET. *J. Mol. Biol.* 370:530–40
138. Wang Y, Rader AJ, Bahar I, Jernigan RL. 2004. Global ribosome motions revealed with elastic network model. *J. Struct. Biol.* 147:302–14
139. Berk V, Zhang W, Pai RD, Cate JH. 2006. Structural basis for mRNA and tRNA positioning on the ribosome. *Proc. Natl. Acad. Sci. USA* 103:15830–34
140. Connell SR, Takemoto C, Wilson DN, Wang H, Murayama K, et al. 2007. Structural basis for interaction of the ribosome with the switch regions of GTP-bound elongation factors. *Mol. Cell* 25:751–64

141. Tama F, Miyashita O, Brooks CL 3rd. 2004. Normal mode based flexible fitting of high-resolution structure into low-resolution experimental data from cryo-EM. *J. Struct. Biol.* 147:315–26
142. Blaha G, Stanley RE, Steitz TA. 2009. Formation of the first peptide bond: the structure of EF-P bound to the 70S ribosome. *Science* 325:966–70
143. Zavialov AV, Ehrenberg M. 2003. Peptidyl-tRNA regulates the GTPase activity of translation factors. *Cell* 114:113–22
144. Spahn CM, Gomez-Lorenzo MG, Grassucci RA, Jorgensen R, Andersen GR, et al. 2004. Domain movements of elongation factor eEF2 and the eukaryotic 80S ribosome facilitate tRNA translocation. *EMBO J.* 23:1008–19
145. Sharma D, Southworth DR, Green R. 2004. EF-G-independent reactivity of a pre-translocation-state ribosome complex with the aminoacyl tRNA substrate puromycin supports an intermediate (hybrid) state of tRNA binding. *RNA* 10:102–13
146. Aitken CE, Petrov A, Puglisi JD. 2010. Single ribosome dynamics and the mechanism of translation. *Annu. Rev. Biophys.* 39:491–513
147. Frank J, Zhu J, Penczek P, Li Y, Srivastava S, et al. 1995. A model of protein synthesis based on cryo-electron microscopy of the *E. coli* ribosome. *Nature* 376:441–44
148. Tama F, Valle M, Frank J, Brooks CL 3rd. 2003. Dynamic reorganization of the functionally active ribosome explored by normal mode analysis and cryo-electron microscopy. *Proc. Natl. Acad. Sci. USA* 100:9319–23
149. Wriggers W, Agrawal RK, Drew DL, McCammon A, Frank J. 2000. Domain motions of EF-G bound to the 70S ribosome: insights from a hand-shaking between multi-resolution structures. *Biophys. J.* 79:1670–78
150. Allen GS, Zavialov A, Gursky R, Ehrenberg M, Frank J. 2005. The cryo-EM structure of a translation initiation complex from *Escherichia coli*. *Cell* 121:703–12
151. Myasnikov AG, Marzi S, Simonetti A, Giuliodori AM, Gualerzi CO, et al. 2005. Conformational transition of initiation factor 2 from the GTP- to GDP-bound state visualized on the ribosome. *Nat. Struct. Mol. Biol.* 12:1145–49
152. Klaholz BP, Myasnikov AG, Van Heel M. 2004. Visualization of release factor 3 on the ribosome during termination of protein synthesis. *Nature* 427:862–65
153. Gao H, Zhou Z, Rawat U, Huang C, Bouakaz L, et al. 2007. RF3 induces ribosomal conformational changes responsible for dissociation of class I release factors. *Cell* 129:929–41
154. Agrawal RK, Sharma MR, Kiel MC, Hirokawa G, Booth TM, et al. 2004. Visualization of ribosome-recycling factor on the *Escherichia coli* 70S ribosome: functional implications. *Proc. Natl. Acad. Sci. USA* 101:8900–5
155. Gao N, Zavialov AV, Li W, Sengupta J, Valle M, et al. 2005. Mechanism for the disassembly of the posttermination complex inferred from cryo-EM studies. *Mol. Cell* 18:663–74
156. Fu J, Kenney D, Munro JB, Lei J, Blanchard SC, Frank J. 2009. The P-site tRNA reaches the P/E position through intermediate positions. *J. Biomol. Struct. Dyn.* 26:794–95
157. Pan D, Kirillov S, Zhang CM, Hou YM, Cooperman BS. 2006. Rapid ribosomal translocation depends on the conserved 18–55 base pair in P-site transfer RNA. *Nat. Struct. Mol. Biol.* 13:354–59
158. Pan D, Kirillov SV, Cooperman BS. 2007. Kinetically competent intermediates in the translocation step of protein synthesis. *Mol. Cell* 25:519–29
159. Shoji S, Walker SE, Fredrick K. 2006. Reverse translocation of tRNA in the ribosome. *Mol. Cell* 24:931–42
160. Konevega AL, Fischer N, Semenkov YP, Stark H, Wintermeyer W, Rodnina MV. 2007. Spontaneous reverse movement of mRNA-bound tRNA through the ribosome. *Nat. Struct. Mol. Biol.* 14:318–24
161. Qin Y, Polacek N, Vesper O, Staub E, Einfeldt E, et al. 2006. The highly conserved LepA is a ribosomal elongation factor that back-translocates the ribosome. *Cell* 127:721–33
162. Scheres SH, Valle M, Grob P, Nogales E, Carazo JM. 2009. Maximum likelihood refinement of electron microscopy data with normalization errors. *J. Struct. Biol.* 166:234–40
163. Scheres SH, Gao H, Valle M, Herman GT, Eggermont PP, et al. 2007. Disentangling conformational states of macromolecules in 3D-EM through likelihood optimization. *Nat. Methods* 4:27–29

164. Zhang W, Kimmel M, Spahn CM, Penczek PA. 2008. Heterogeneity of large macromolecular complexes revealed by 3D cryo-EM variance analysis. *Structure* 16:1770–76
165. Penczek PA, Yang C, Frank J, Spahn CM. 2006. Estimation of variance in single-particle reconstruction using the bootstrap technique. *J. Struct. Biol.* 154:168–83
166. Shaikh TR, Barnard D, Meng X, Wagenknecht T. 2009. Implementation of a flash-photolysis system for time-resolved cryo-electron microscopy. *J. Struct. Biol.* 165:184–89
167. Lu Z, Shaikh TR, Barnard D, Meng X, Mohamed H, et al. 2009. Monolithic microfluidic mixing-spraying devices for time-resolved cryo-electron microscopy. *J. Struct. Biol.* 168:388–95
168. Cornish PV, Ha T. 2007. A survey of single-molecule techniques in chemical biology. *ACS Chem. Biol.* 2:53–61
169. McKinney SA, Joo C, Ha T. 2006. Analysis of single-molecule FRET trajectories using hidden Markov modeling. *Biophys. J.* 91:1941–51
170. Qin F, Auerbach A, Sachs F. 2000. A direct optimization approach to hidden Markov modeling for single channel kinetics. *Biophys. J.* 79:1915–27
171. Andrec M, Levy RM, Talaga DS. 2003. Direct determination of kinetic rates from single-molecule photon arrival trajectories using hidden Markov models. *J. Phys. Chem. A* 107:7454–64
172. Qin F, Auerbach A, Sachs F. 2000. Hidden Markov modeling for single channel kinetics with filtering and correlated noise. *Biophys. J.* 79:1928–44

Adding New Chemistries to the Genetic Code

Chang C. Liu and Peter G. Schultz

The Scripps Research Institute, La Jolla, California 92037; email: ccliu@berkeley.edu, schultz@scripps.edu

Annu. Rev. Biochem. 2010. 79:413–44

First published online as a Review in Advance on March 18, 2010

The *Annual Review of Biochemistry* is online at biochem.annualreviews.org

This article's doi:
10.1146/annurev.biochem.052308.105824

0066-4154/10/0707-0413$20.00

Key Words

aminoacyl-tRNA synthetase, protein engineering, protein evolution, translation

Abstract

The development of new orthogonal aminoacyl-tRNA synthetase/tRNA pairs has led to the addition of approximately 70 unnatural amino acids (UAAs) to the genetic codes of *Escherichia coli*, yeast, and mammalian cells. These UAAs represent a wide range of structures and functions not found in the canonical 20 amino acids and thus provide new opportunities to generate proteins with enhanced or novel properties and probes of protein structure and function.

Contents

INTRODUCTION

Proteins carry out a remarkable range of functions—from photosynthesis to transcription and signal transduction—with only 20 amino acids. Nonetheless, proteins often require chemistries beyond those contained in the canonical 20 amino acids for function, including cofactors, such as pyridoxal, thiamine, flavins, and metal ions (1), and posttranslational modifications (PTMs) such as methylation, glycosylation, sulfation, and phosphorylation (2). Moreover, the modification of proteins by chemical methods can lead to new physicochemical, biological, or pharmacological properties (3–8). A number of archaea and eubacteria even encode the noncanonical amino acids selenocysteine or pyrrolysine for added function (9). Thus, the creation of organisms with expanded genetic codes that include amino acids beyond the common 20 building blocks might allow the design of peptides and proteins with enhanced or novel activities. Additional building blocks might also facilitate the incorporation of biophysical probes into proteins for the analysis or control of protein structure and function in vitro and in living cells. Finally, an expanded genetic code may provide an advantage in the evolution of new molecular or organismal function. Here, we describe efforts to augment the genetic codes of both prokaryotic and eukaryotic organisms with unnatural amino acids (UAAs) that have novel properties, and we illustrate the utility of this methodology in exploring protein structure and function and in generating proteins with new and/or enhanced activities.

PTM: posttranslational modification

Unnatural amino acid (UAA): an amino acid not specified by the existing genetic code, which encodes only 20 amino acids

BACKGROUND

Chemists have exquisite control over the structures of small molecules, but the control of macromolecular, especially protein, structure and function still represents a major challenge. A number of approaches are being pursued to address this limitation. The first of these is solid-phase peptide synthesis, which allows chemists to rationally manipulate polypeptide structure using natural and unnatural amino acids (5, 10). Though this method is subject to size limitations (peptides containing over 50–100 amino acids are difficult to synthesize in significant quantities), it can be used in conjunction with semisynthetic methods to produce larger full-length proteins containing UAAs. For example, in native chemical ligation, a peptide containing a C-terminal α-thioester is reacted with a second peptide containing an N-terminal cysteine. The resulting thioester linkage undergoes an acyl rearrangement to form a native peptide bond, joining the two smaller peptides (7). A C-terminal α-thioester leaving group can also be generated by an intein-mediated reaction arrested before the final splicing step. This method, termed expressed protein ligation (EPL), allows the linkage of a recombinantly expressed protein to a synthetic UAA-containing peptide (8). Solid-phase peptide synthesis and ligation techniques have been used to modify protein backbones (11), make polymer-modified erythropoietin analogs (12, 13), introduce fluorescent probes into peptides and proteins, and produce modified signaling proteins, ion channels, and histones (14–16). However, the general application of these techniques is limited by the need for protecting group chemistry, the restrictions on the sites of ligation, the constraints on protein folding, and the inherent extracellular nature of these synthetic techniques.

Biosynthetic approaches have also been developed for the in vitro synthesis of proteins containing UAAs. In these techniques, truncated tRNAs are enzymatically ligated to chemically aminoacylated nucleotides, effectively decoupling the identity of the tRNA from that of the attached amino acid (17–21). Cell-free translation systems then use these aminoacylated tRNAs for protein synthesis in response to either "blank" (nonsense or frameshift codons that can be used to specify an UAA) or coding codons. Though technically challenging, this method has been very useful in the study of protein structure and function. Examples include the incorporation of amino acids with modified backbones (22, 23), fluorophores (24–26), functional groups corresponding to PTMs (27), photo- and chemically reactive side chains (28), and altered pK_as (29). This method has also been combined with mRNA display to generate selectable peptide libraries that contain UAAs (30, 31). A variation of this approach involves the injection or transfection of chemically or otherwise aminoacylated tRNAs into living cells (32, 33). For example, by chemically acylating an amber suppressor tRNA, such as one derived from *Tetrahymena thermophilia* $tRNA^{Gln}$ (which affords high efficiency and fidelity of UAA incorporation when microinjected into *Xenopus* oocytes) (34), UAAs spanning a range of electronic, structural, and conformational properties have been incorporated into ion channels directly in cells (35–37). This has allowed for detailed structure-function studies by the systematic modification of individual protein residues (38), including the determination of the role of proline isomerization in channel gating (39). Nevertheless, these methods are limited by the accessibility and stability of the aminoacyl-tRNA adducts, the stoichiometric use of aminoacylated tRNAs that cannot be continuously delivered, and the disruptive nature of microinjection and transfection techniques.

Another approach that is applicable to living cells involves the use of wild-type aminoacyl-tRNA synthetases (aaRSs) to incorporate UAAs that are close structural analogs of canonical amino acids (40, 41). In this approach, a strain auxotrophic for one of the common 20 amino acids is used to substitute that amino acid with an UAA analog. Although the resulting wholesale replacement of a common amino acid by an UAA cannot sustain exponential growth, nondividing cells are still viable and are able to overexpress proteins that contain the UAA. In these newly synthesized proteins, the canonical amino acid is efficiently replaced by its UAA analog at all sites (42, 43). The diversity in the range of UAA analogs that can be incorporated using this approach has been increased by aaRS overexpression (44), active-site engineering (45, 46), editing domain mutations (47), and sorting by cell-surface display of the UAA whose incorporation was desired (48). The global incorporation of UAA analogs by this method has a number of useful applications. For example, substitution of methionine with selenomethionine introduces a heavy atom into proteins for crystallographic phasing experiments (49); replacement of methionine by noreleucine in cytochrome P450, leucine by 5′,5′,5′-trifluoroleucine in chloramphenicol acetyltransferase, and tryptophan by 4-aminotryptophan in barstar and green fluorescent protein (GFP) yields proteins with new activities and properties (50–53); and timed substitution of methionine or phenylalanine

aaRS: aminoacyl-tRNA synthetase

by alkyne-containing UAA analogs allows the tracking of newly synthesized proteins (54, 55). Such methods, however, are not site-specific (rather, they are residue-specific and result in global replacement), do not usually allow for continuous cell growth, and are only generally applicable to UAAs that are close analogs of canonical amino acids.

Orthogonal aminoacyl-tRNA synthetase/iso-tRNA (aaRS/iso-tRNA) pair: an aaRS/iso-tRNA pair that is specific for its cognate amino acid and does not exhibit cross-reactivity with other aaRS/iso-tRNA pairs

EF: elongation factor

RF: release factor

TRANSLATION WITH NEW AMINO ACIDS

To further enhance our ability to control the structure and properties of proteins, both in vitro and in the context of living cells, we sought to directly create organisms that genetically encode 21 or more amino acids. In this approach, the biological, chemical, or physical properties of new amino acids are precisely defined by the chemist at the bench, but because these UAAs are genetically encoded, their incorporation into proteins should occur with the same fidelity, efficiency, and genetic manipulability of natural protein synthesis. To realize this goal, we developed a general method for the engineering and direct integration of aaRS and tRNA components into the translational machinery.

Translation is a unique biological process in which mRNA templates the assembly of a distinct biopolymer, a polypeptide chain, through a tRNA adapter molecule. This process is different from the synthesis of DNA and RNA where the same basic recognition elements are required in the template and product. As a consequence, translation has a greater intrinsic capacity to be adapted to accommodate new building blocks. The relationship of template (mRNA) to product (polypeptide) is defined by the genetic code, which assigns a specific amino acid to each triplet codon and utilizes aminoacyl-tRNA adapters to establish the map between mRNA and protein sequence. The fidelity and efficiency of translation rely on numerous molecular recognition steps. First, the 20 amino acids are specifically loaded onto 20 isoacceptor tRNA (iso-tRNA) sets by aaRSs, each specific for its own unique amino acid substrate and its own unique set of iso-tRNAs (56–58). This results in a network of orthogonal aaRS/iso-tRNA pairs (that is, a given aaRS, which is specific for its cognate amino acid, will only recognize its cognate iso-tRNA set, which itself is only recognized by the given aaRS) (59). Next, aminoacyl-tRNAs enter the ribosome and recognize the correct mRNA codons. This is accomplished by elongation factor Tu (EF-Tu), which binds and transports the range of aminoacyl-tRNAs into the ribosome, where standard (and sometimes wobble) base pairing mediates the recognition between an incoming aminoacyl-tRNA's anticodon loop and the mRNA codon being read. Correct anticodon-codon pairing results in a peptidyl transfer reaction between the incoming aminoacyl-tRNA and the growing polypeptide chain, a reaction catalyzed by the ribosome's peptidyl transfer center. Finally, translocation and eventually release factor (RF) binding continue and ultimately terminate translation of the desired protein product.

One can expand the genetic code of an organism to include new amino acids by adding new components to this template-directed biosynthetic machinery. These include a cell-permeable or biosynthesized UAA, a unique codon, a corresponding iso-tRNA set (in this case, an iso-tRNA set with cardinality one is simplest), and a cognate aaRS. These components must satisfy a number of criteria: First, the UAA must be metabolically stable and have good cellular bioavailability; it must be tolerated by EF-Tu and the ribosome, but it must not be a substrate for any endogenous aaRSs. Second, the unique codon must be recognized by the new tRNA but not by any endogenous tRNAs. Third, the aaRS/tRNA pair must be specific for the UAA, functional in the host organism, and orthogonal in the context of all endogenous aaRS/tRNA pairs in the organism (60). In general, most UAAs added to the media are taken up by both prokaryotic and eukaryotic cells (exceptions include highly charged amino acids, which can be modified as metabolically labile derivatives or incorporated into dipeptides to increase permeability).

In addition, natural aaRSs have evolved high specificity for their cognate amino acids, and the aminoacyl-binding site of EF-Tu and the ribosome are highly promiscuous (in vitro and in vivo translation experiments have demonstrated that the range of acceptable substrates includes many noncanonical amino acid side chains as well as D-amino acids and even α-hydroxy acids where ester rather than amide backbone bonds are formed) (61). Thus, the first criterion is easily met for most UAAs. Because codon recognition is determined by simple base pairing rules, the second criterion is easily fulfilled by choosing a blank (nonsense, frameshift, or otherwise unused) codon and designing a tRNA with the corresponding anticodon. The real challenge is to fulfill the third criterion, aaRS/tRNA orthogonality and aminoacylation specificity.

Although several strategies to generate orthogonal aaRS/tRNA pairs have been explored, ultimately, the most straightforward solution involves the importation of a heterologous aaRS/tRNA pair from a different domain of life. This is because tRNA recognition by aaRSs can be domain or species specific (62), a feature that can serve as the basis for orthogonality. The anticodon loop of the imported tRNA is then mutated to create a blank codon ($codon_{BL}$) suppressor tRNA ($tRNA_{SB}$), and the orthogonality of the resulting aaRS/$tRNA_{SB}$ pair is assessed. If necessary, the orthogonality of this pair is improved by a two-step process, involving both positive and negative rounds of selection to identify functional optimized $tRNA_{SB}$s that exhibit no cross-reactivity with endogenous aaRSs (60, 63). Finally, structure-based mutagenesis and a similar two-step selection strategy are used to alter the specificity of the heterologous aaRS so that it uniquely recognizes the UAA of interest (60, 64, 65). This process has allowed for the systematic directed evolution of aaRS/tRNA pairs that are specific for a variety of UAAs (**Figure 1**) and are orthogonal in bacteria, yeast, and mammalian cells (**Table 1**). This approach should, in theory, make translation with UAAs accessible in any organism.

Encoding Unnatural Amino Acids in *Escherichia coli*

$Codon_{BL}$: any codon that does not encode a natural amino acid for protein synthesis

$tRNA_{SB}$: a tRNA that suppresses a blank codon

To genetically encode UAAs in *E. coli*, one first imports a heterologous aaRS/tRNA pair from archea or eukaryotes, mutates the anticodon loop to generate $tRNA_{SB}$, and if necessary, uses mutagenesis and selection to improve the orthogonality of the imported pair. Selection is achieved by transforming a library of mutant $tRNA_{SB}$s (which is based on the heterologous $tRNA_{SB}$) into *E. coli* cells that contain the toxic barnase gene with a $codon_{BL}$ at permissive sites. In this negative selection step, only clones containing mutant $tRNA_{SB}$s that are not substrates for endogenous aaRSs grow. The surviving $tRNA_{SB}$s are then transformed into *E. coli* cells that express the heterologous aaRS and have a β-lactamase gene with a $codon_{BL}$ at a permissive site. In this positive selection step, the presence of ampicillin in the growth media kills all clones that contain a nonfunctional $tRNA_{SB}$, leaving only $tRNA_{SB}$ mutants that are aminoacylated by the heterologous aaRS. The result is a $tRNA_{SB}$ that functions with its cognate aaRS as a highly orthogonal pair in *E. coli*.

A similar two-step selection scheme is then used to alter the specificity of the heterologous aaRS so that it uniquely recognizes the UAA of interest (**Figure 2*a***). First, a library of aaRS mutants, containing randomized residues in the amino acid–binding site, is constructed on the basis of available crystal structures. This library is transformed into *E. coli* cells that express $tRNA_{SB}$ and a gene encoding chloramphenicol acetyltransferase (CAT) with a $codon_{BL}$ mutation at a permissive site. In this positive selection step, these cells are grown in the presence of chloramphenicol and the UAA of interest so that only the aaRS mutants capable of aminoacylating $tRNA_{SB}$ with the UAA and/or endogenous amino acids live. Surviving mutants are then transformed into *E. coli* cells that express $tRNA_{SB}$ and the toxic barnase gene with $codon_{BL}$ mutations at permissive sites. In this negative selection step, cells are grown in the absence of the UAA so that all clones whose mutant aaRS aminoacylates endogenous

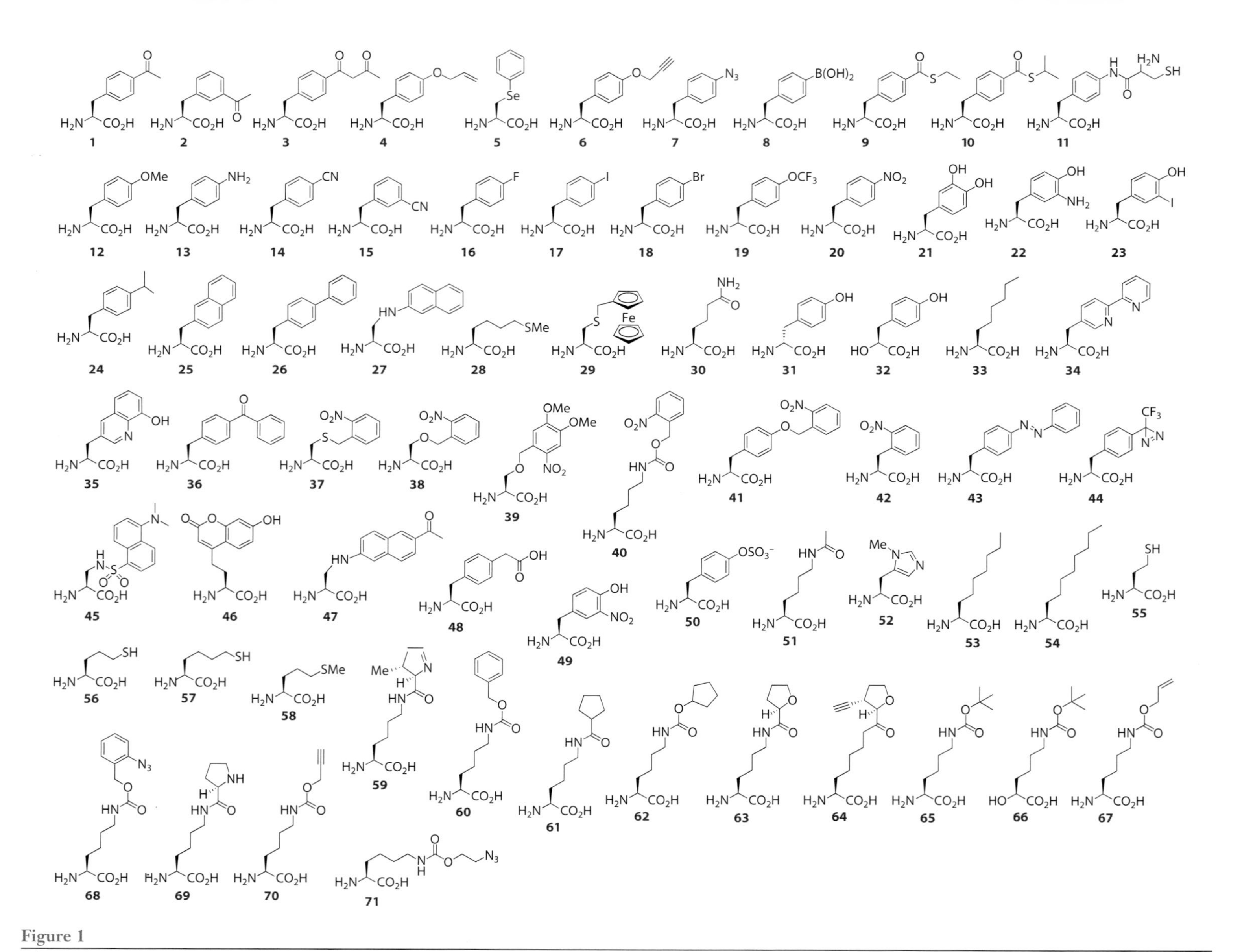

Figure 1

Chemical structures of genetically encoded unnatural amino acids (and hydroxy acids in the case of molecules 32 and 66).

Table 1 Summary of unnatural amino acids incorporated via expanded genetic codes

Unnatural amino acid (UAA)	Common name (if applicable)	Organism(s) in which UAA is encoded[a]	References[b,c] (and notes)
1	*p*-Acetylphenylalanine	*E. coli*, yeast, mammalian cells	66, 123, 131, 133
2	*m*-Acetylphenylalanine	*E. coli*	67
3		*E. coli*	68
4	*O*-allyltyrosine	*E. coli*	69
5	Phenylselenocysteine	*E. coli*	70 (precursor to dehydroalanine)
6	*p*-Propargyloxyphenylalanine	*E. coli*, yeast, mammalian cells	71, 125, 131, 133
7	*p*-Azidophenylalanine	*E. coli*, yeast, mammalian cells	72, 123, 125, 131, 133
8	*p*-Boronophenylalanine	*E. coli*	73
9		*E. coli*	H. Zeng & P.G. Schultz, unpublished
10		*E. coli*	H. Zeng & P.G. Schultz, unpublished
11		*E. coli*	M. Jahnz & P.G. Schultz, unpublished
12	*O*-methyltyrosine	*E. coli*, yeast, mammalian cells	74, 123, 124, 131, 133, 134
13	*p*-Aminophenylalanine	*E. coli*	64, 75
14	*p*-Cyanophenylalanine	*E. coli*	76
15	*m*-Cyanophenylalanine	*E. coli*	J. Chittuluru & P.G. Schultz, unpublished
16	*p*-Fluorophenylalanine	*E. coli*	100 (requires auxotrophic strain)
17	*p*-Iodophenylalanine	*E. coli*, yeast, mammalian cells	78, 123, 131, 133
18	*p*-Bromophenylalanine	*E. coli*	77, 78, 139
19		*E. coli*	79
20	*p*-Nitrophenylalanine	*E. coli*	80
21	L-DOPA	*E. coli*	81
22	3-Aminotyrosine	*E. coli*	82
23	3-Iodotyrosine	*E. coli*, yeast, mammalian cells	83, 132
24	*p*-Isopropylphenylalanine	*E. coli*	64
25	3-(2-Naphthyl)alanine	*E. coli*	84
26	Biphenylalanine	*E. coli*	85
27		Yeast, mammalian cells	126
28		Yeast, mammalian cells	127
29		Yeast, mammalian cells	130
30	Homoglutamine	*E. coli*	104
31	D-tyrosine	*E. coli*	J. Guo & P.G. Schultz, unpublished
32	*p*-Hydroxyphenyllactic acid	*E. coli*	86 (requires a strain with disrupted *tyrB* and *aspC*)
33	2-Aminocaprylic acid	Yeast, mammalian cells	89, 127
34	Bipyridylalanine	*E. coli*	85
35	HQ-alanine	*E. coli*	87
36	*p*-Benzoylphenylalanine	*E. coli*, yeast, mammalian cells	88, 123, 131, 133, 134, 169
37	*o*-Nitrobenzylcysteine	Yeast, mammalian cells	124

(*Continued*)

Table 1 (*Continued*)

Unnatural amino acid (UAA)	Common name (if applicable)	Organism(s) in which UAA is encoded[a]	References[b,c] (and notes)
38	*o*-Nitrobenzylserine	Yeast, mammalian cells	N. Wu & P.G. Schultz, unpublished
39	4,5-Dimethoxy-2-nitrobenzylserine	Yeast, mammalian cells	128
40	*o*-Nitrobenzyllysine	*E. coli*, yeast, mammalian cells	120
41	*o*-Nitrobenzyltyrosine	*E. coli*	89
42	2-Nitrophenylalanine	*E. coli*	90
43		*E. coli*	91
44		*E. coli*	92
45	Dansylalanine	Yeast, mammalian cells	129, 134
46		*E. coli*	93
47		Yeast, mammalian cells	126; J. Guo, H.S. Lee, E.A. Lemke, R.D. Dimla, & P.G. Schultz, unpublished results
48	*p*-Carboxymethylphenylalanine	*E. coli*	94
49	3-Nitrotyrosine	*E. coli*	95
50	Sulfotyrosine	*E. coli*	96
51	Acetyllysine	*E. coli*, yeast, mammalian cells	114, 121
52	Methylhistidine	Yeast, mammalian cells	F. Peters, J. Chittuluru, & P.G. Schultz, unpublished
53	2-Aminononanoic acid	Yeast, mammalian cells	127
54	2-Aminodecanoic acid	Yeast, mammalian cells	127
55		Yeast, mammalian cells	127
56		Yeast, mammalian cells	127
57		Yeast, mammalian cells	127
58		Yeast, mammalian cells	127
59	Pyrrolysine	*E. coli*, yeast, mammalian cells	111 (UAARS from a natural expanded genetic code)
60	Cbz-lysine	*E. coli*, yeast, mammalian cells	112, 114
61		*E. coli*, yeast, mammalian cells	115
62		*E. coli*, yeast, mammalian cells	116, 120
63		*E. coli*, yeast, mammalian cells	115, 117
64		*E. coli*, yeast, mammalian cells	117
65	Boc-lysine	*E. coli*, yeast, mammalian cells	112, 114
66		*E. coli*, yeast, mammalian cells	118
67	Allyloxycarbonyllysine	*E. coli*, yeast, mammalian cells	112
68		*E. coli*, yeast, mammalian cells	112
69		*E. coli*, yeast, mammalian cells	116
70		*E. coli*, yeast, mammalian cells	119
71		*E. coli*, yeast, mammalian cells	119

[a]Underlined font, functionality not experimentally demonstrated but based on parent aminoacyl-tRNA synthetase (aaRS)/tRNA pair orthogonality.
[b]References are those pertinent to the original encoding of the UAA.
[c]Abbreviation: UAARS, UAA-specific mutant aminoacyl-tRNA synthetase.

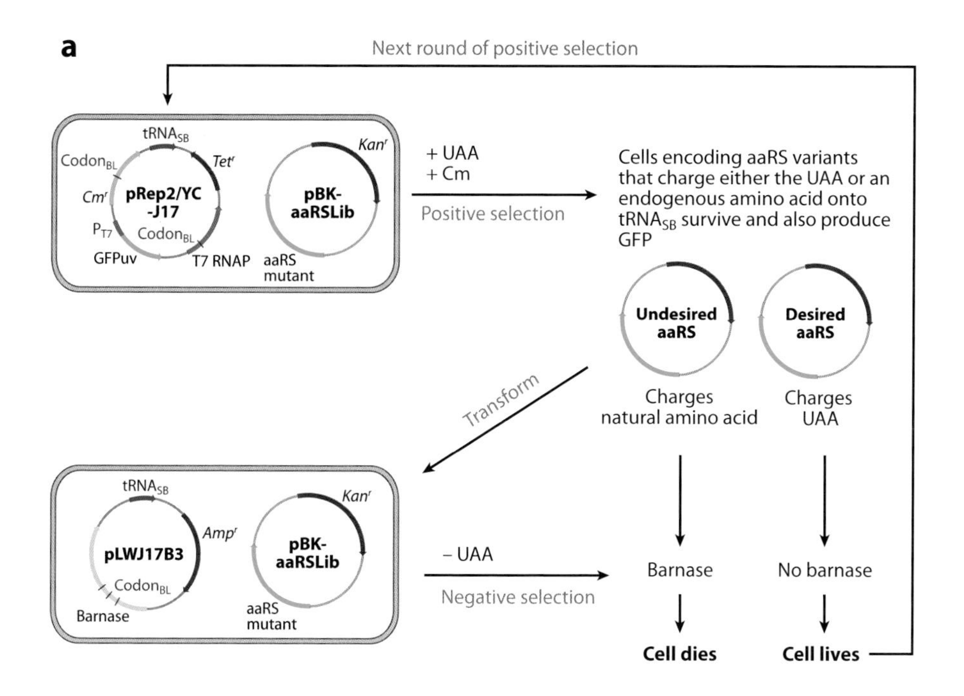

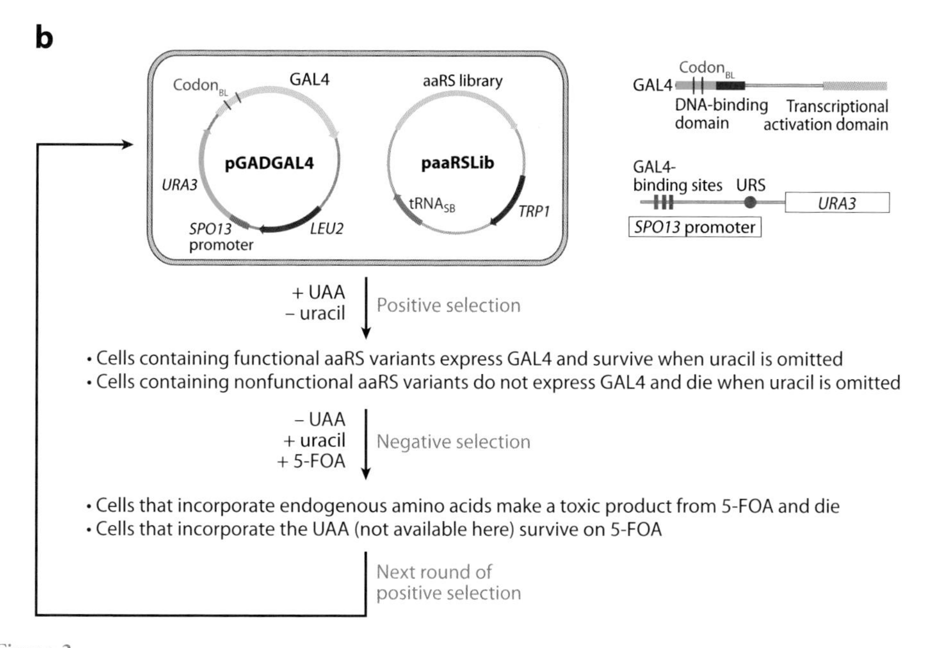

Figure 2

Selection schemes for genetically encoding amino acids in *E. coli* and yeast. (*a*) Two-step selection for the directed evolution of orthogonal UAARS/$tRNA_{SB}$ pairs in *E. coli*. (*b*) Two-step selection for the directed evolution of orthogonal UAARS/$tRNA_{SB}$ pairs in yeast. Abbreviations: aaRS, aminoacyl-tRNA synthetase; $codon_{BL}$, blank codon; 5-FOA, 5-fluoroorotic acid; GFP_{UV}, green fluorescent protein; P_{T7}, bacteriophage T7 promoter; T7 RNAP, bacteriophage T7 RNA polymerase; $tRNA_{SB}$, suppressor tRNA; UAA, unnatural amino acid; UAARS, UAA-specific mutant aaRS; URS, upstream repression sequence.

UAARS: an aminoacyl-tRNA synthetase specific for an unnatural amino acid substrate

amino acids die. This leaves only mutant aaRSs that aminoacylate $tRNA_{SB}$ with the UAA. The two-step selection is usually repeated for two to three additional rounds to yield an UAA-specific mutant aaRS (UAARS). We can assume that the UAARS does not aminoacylate endogenous tRNAs to any appreciable extent as this would lead to missense incorporation throughout the proteome, resulting in cell death. Through this selection scheme and its more facile variants (64, 65), many structurally distinct amino acids have been genetically encoded in *E. coli*, most successfully using the amber (TAG) codon as $codon_{BL}$ and importing the heterologous *Methanocaldococcus jannaschii* TyrRS/$tRNA^{Tyr}$ (*Mj*TyrRS/*Mj*$tRNA^{Tyr}$) pair from which ~38 orthogonal UAARS/$tRNA_{SB}$ pairs specific for chemically distinct UAAs (e.g., UAAs **1–15**, **17–26**, **31**, **32**, **34–36**, **41–44**, **46**, and **48–50**, shown in **Figure 1** and **Table 1**) have been derived (64, 66–96). In general, these UAAs are incorporated with excellent fidelity and yields. Indeed, shake-flask expression yields in the 1–100 mg/L range and/or optimized high-density fermentation yields in the g/L range have been achieved for almost all UAA-containing mutant proteins. However, mRNA context effects, protein folding and stability, and other factors can lead to low incorporation efficiency at some desired sites; these areas remain an opportunity for methodological improvements.

Although evolution of orthogonal UAARS/$tRNA_{SB}$ pairs from the *Mj*TyrRS/*Mj*$tRNA^{Tyr}$ pair has been very successful, the ability to evolve aaRSs specific for any given amino acid depends on structural constraints in the active site. Thus, it may be necessary to consider other aaRS/tRNA pairs with structurally diverse active sites as starting points. Toward this end, several additional orthogonal pairs have been adapted for use in *E. coli*. These include pairs from *Saccharomyces cerevisiae* (e.g., AspRS/$tRNA^{Asp}$, GlnRS/$tRNA^{Gln}$, TyrRS/$tRNA^{Tyr}$, and PheRS/$tRNA^{Phe}$) (97–100), pairs from archea (e.g., *Mj*TyrRS/*Mj*$tRNA^{Tyr}$) (101), and pairs with hybrid or consensus components (e.g., *S. cerevisiae* TyrRS/*E. coli* initiator $tRNA_f^{Met}$, *Methanobacterium thermoautotrophicum* LeuRS/*Halobacterium* sp. *NRC-1* $tRNA^{Leu}$, *Pyrococcus horikoshii* or *Methanosarcina mazei* GluRS/consensus tRNA derived from archael $tRNA^{Glu}$ sequences, and *P. horikoshii* LysRS/consensus tRNA from archael $tRNA^{Lys}$ sequences) (102–105). The utility of each pair is dependent on efficient expression in bacteria, orthogonality, ability to suppress the desired $codon_{BL}$ (once the tRNA's anticodon loop is changed), and ability to evolve the aaRS active site to accommodate novel amino acids. All of these characteristics can be optimized through promoter engineering, rational mutations, and selection. Recently, some pairs have also been imported, without loss of orthogonality, into *Mycobacteria smegmatis*, *bovis*, and *tuberculosis*, thus allowing the genetic incorporation of UAAs into these organisms to study their pathogenesis and produce vaccines (106).

An additional set of heterologous pairs has been added to this list with the recent characterization of naturally occurring 21 amino acid methanogens that genetically encode pyrrolysine (107, 108). These organisms contain a pyrrolysyl-tRNA synthetase (PylRS) that aminoacylates $tRNA^{Pyl}{}_{CUA}$ with pyrrolysine; pyrrolysyl-$tRNA^{Pyl}{}_{CUA}$ in turn incorporates this twenty-first amino acid in response to the amber codon. In its native contexts, the PylRS/$tRNA^{Pyl}{}_{CUA}$ pair functions alongside the 20 canonical aaRS/iso-tRNA set pairs without any cross-reactivity among pairs. Therefore, its importation into other organisms might also preserve orthogonality, especially given that the ancient pre-LUCA PylRS likely underwent significant horizontal gene transfer among other aaRS/tRNA pairs in early life (109, 110). Indeed, this has so far been the case. For instance, importation of the *Methanosarcina barkeri* PylRS/*M. barkeri* $tRNA^{Pyl}$, *M. mazei* PylRS/*M. mazei* $tRNA^{Pyl}$, and *Desulfitobacterium hafniense* PylRS/*D. hafniense* $tRNA^{Pyl}$ pairs (and certain hybrid pair combinations thereof) into *E. coli* and mammalian cells maintains orthogonality without requiring further modification (111–114). This has allowed pyrrolysine (**59**) and many of its analogs (e.g., UAAs **61–67** and **69–71**) to be genetically encoded in bacteria (111, 112, 114–119). In addition, rational mutagenesis/screening or the general two-step selection scheme has been used to alter the specificity of PylRS/$tRNA^{Pyl}{}_{CUA}$ pairs to encode structurally distinct amino acids (e.g., UAAs **40**, **51**, **60**, and **68**), including a photocaged lysine and acetyllysine (112, 114, 120, 121).

The identification of multiple heterologous aaRS/tRNA pairs orthogonal in *E. coli* also makes possible the simultaneous encoding of two or more UAAs, which requires two or more different blank codons recognized by two or more mutually orthogonal aaRS/tRNA pairs. An example is the use of both the AGGA frameshift and TAG nonsense codons with corresponding orthogonal UAARS/$tRNA_{SB}$ pairs derived from the *P. horikoshii* LysRS/tRNA (a consensus tRNA from archael $tRNA^{Lys}$ sequences) and the *Mj*TyrRS/*Mj*$tRNA^{Tyr}$ pairs to site specifically incorporate two distinct UAAs (**30** and **12**, respectively) into a single myoglobin protein in *E. coli* (104). In this case, it was not necessary to explicitly engineer mutual orthogonality, probably because the two UAARSs were derived from pairs that specify different natural amino acids. If necessary, one could evolve orthogonality of one pair in the context of the other and vice versa by using the general two-step selection scheme described above.

Encoding Unnatural Amino Acids in Yeast

The processing, expression, and recognition of tRNAs in yeast are distinct from those in *E. coli*. Therefore, new heterologous aaRS/tRNA pairs are needed to expand the genetic code of yeast. Toward this end, several aaRS/tRNA pairs, including the *E. coli* GluRS/human initiator tRNA, the *E. coli* TyrRS/*E. coli* $tRNA^{Tyr}$, the *E. coli* LeuRS/*E. coli* $tRNA^{Leu}$, and the *M. mazei* PylRS/*M. mazei* $tRNA^{Pyl}$ pairs (102, 114, 122–124), have been shown to be orthogonal in *S. cerevisiae*. To change the amino acid specificity of the aaRS/tRNA pairs for incorporation of UAAs in response to $codon_{BL}$, a two-step selection similar to that used in *E. coli* is applied (**Figure 2*b***) (123). First, a library of aaRS mutants is generated and transformed into a *S. cerevisiae* uracil auxotrophic strain containing a GAL4 transcriptional activator with $codon_{BL}$ at two permissive sites, a *URA3* gene (required for uracil synthesis) under the control of GAL4, and the heterologous tRNA with its anticodon mutated to suppress $codon_{BL}$ ($tRNA_{SB}$). In positive selection, these cells are grown in the presence of UAA and in the absence of uracil. Clones expressing aaRS mutants that can aminoacylate $tRNA_{SB}$ with the UAA and/or natural amino acids produce GAL4 by suppressing $codon_{BL}$, which results in uracil synthesis and survival. This is followed by negative selection, where surviving cells are grown in the absence of UAA but in the presence of uracil and 5-fluoroorotic acid (5-FOA). When GAL4 is produced, *URA3* is expressed, which converts 5-FOA into a toxic product, killing the cell. Therefore, those clones containing aaRS mutants that aminoacylate $tRNA_{SB}$ with natural amino acids express *URA3* and die. The clones that survive both steps of selection are ones that aminoacylate $tRNA_{SB}$ exclusively with the desired UAA. This process is often repeated for several rounds to yield orthogonal UAARS/$tRNA_{SB}$ pairs specific for the UAA of interest.

In this way, multiple UAARS/$tRNA_{SB}$ pairs based on both the *E. coli* TyrRS/*E. coli* $tRNA^{Tyr}$ and *E. coli* LeuRS/*E. coli* $tRNA^{Leu}$ pairs have been evolved, resulting in the addition of ~22 UAAs to the yeast genetic code (e.g., UAAs **1**, **6**, **7**, **12**, **17**, **27–29**, **33**, **36–39**, **45**, **47**, and **52–58**) (123–130). Furthermore, these evolved UAARS/$tRNA_{SB}$ pairs retain their activity and orthogonality when imported into the methylotrophic yeast *Pichia pastoris*, a host that facilitates large-scale recombinant protein expression and control over glycosylation patterns. Indeed, expression systems containing optimized transcriptional control elements have been created for the transfer of UAARS/$tRNA_{SB}$ pairs from *S. cerevisiae* to *P. pastoris*. In one case, these systems were used to produce human serum albumin containing the UAA **1** in shake-flask expression yields of >100 mg/L (131).

Encoding Unnatural Amino Acids in Mammalian Cells

In mammalian cells, directed evolution efforts are severely limited by technical challenges associated with transformation efficiency, slow doubling times, and growth conditions; therefore, the two-step selection strategy is not ideal for evolving aaRSs with altered specificities here. Instead, there are a number of aaRS/tRNA pairs orthogonal both in *E. coli* or *S. cerevisiae* and in mammalian cells. Thus, evolution of UAA specificity can be carried out in *E. coli* or *S. cerevisiae*, where large libraries can be constructed. The resulting UAARS/$tRNA_{SB}$ pairs can then be transferred into mammalian cells without loss of orthogonality. This shuttle approach has been used for several UAARS/$tRNA_{SB}$ pairs, including ones derived from the *M. mazei* PylRS/*M. mazei* $tRNA^{Pyl}$ pair, which is orthogonal in both *E. coli* and mammalian cells (114, 120); and ones derived from the *E. coli* TyrRS/*Bacillus stearothermophilus* $tRNA^{Tyr}$ pair, which is orthogonal in eukaryotes (132, 133). In the latter case, *B. stearothermophilus* $tRNA^{Tyr}$ is used rather than *E. coli* $tRNA^{Tyr}$ because it contains intact internal A- and B-box promoters that drive proper tRNA expression in mammalian cells (132). More recently, use of external H1 and U6 promoters and 3′ flanking regions derived from mammalian sequences has led to the general production of functional heterologous tRNAs (e.g., *E. coli* $tRNA^{Tyr}$ and *E. coli* $tRNA^{Leu}$) in mammalian cells, thus amending the list of shuttle UAARS/$tRNA_{SB}$ pairs with ones derived from the *E. coli* TyrRS/*E. coli* $tRNA^{Tyr}$ and the *E. coli* LeuRS/*E. coli* $tRNA^{Leu}$ pairs (134; W. Liu, J. Guo, & P.G. Schultz, unpublished results). In this way, the genetic codes of mammalian cells including CHO cells, 293T cells, and even primary neurons have been augmented with ~10 structurally diverse UAAs (e.g., UAAs **1**, **6**, **7**, **12**, **17**, **36**, **45**, **47**, and **52**) incorporated by UAARS/$tRNA_{SB}$ pairs evolved in *S. cerevisiae* (133, 134; J. Guo, H.S. Lee, E.A. Lemke, R.D. Dimla, & P.G. Schultz, unpublished results). In addition, several lysine and pyrrolysine derivatives (e.g., UAAs **40**, **51**, **60**, **62**, and **65**) have been incorporated into proteins in mammalian cells by UAARS/$tRNA_{SB}$ pairs derived from the PylRS/$tRNA^{Pyl}$ pairs (114, 120). For these UAAs, laboratory mammalian expression systems result in protein yields in the microgram per 10^7 cells range with no misincorporation detected. These orthogonal pairs are also being transferred to *Caenorhabditis elegans* and *Mus musculus* cells with the aim of creating whole multicellular organisms with expanded genetic codes that allow the cotranslational incorporation of UAAs for in vivo biological studies.

Other Methods for Genetically Encoding Unnatural Amino Acids

Alternative approaches have also been used to genetically encode UAAs. For example, rational mutagenesis of heterologous aaRS/$tRNA_{SB}$ pairs followed by their use in auxotrophic *E. coli* strains has allowed the site-specific incorporation of several UAAs (e.g., tryptophan analogs, UAA **16** and UAA **18**) (100, 135, 136). In this approach, a largely orthogonal yeast aaRS/tRNA pair, modified to suppress a blank codon, is imported into *E. coli*. The yeast aaRS recognizes the UAA of interest but also recognizes one or more of the common 20 amino acids, often owing to limitations on rational design. To favor incorporation of the UAA, an auxotrophic *E. coli* strain that does not biosynthesize the corresponding common amino acid is used so that its concentration can be strictly limited. Even though an auxotrophic strain is necessary, the UAA does not replace a common amino acid throughout the proteome but rather is separately encoded. Another approach uses a rationally designed orthogonal aaRS/tRNA pair that recognizes 3-iodotyrosine (**23**) and site specifically incorporates it into proteins in response to the amber codon in mammalian cells (132). To further favor incorporation of UAA **23** over tyrosine, the editing domain from PheRS was fused to the mutant aaRS (137). These methods can be quite useful, but they are often limited by moderate aaRS/tRNA pair orthogonality, modest UAA specificity, and difficulties inherent to rational design, especially in addressing UAAs that have structural features significantly different from the canonical amino acids. Therefore, they have mostly been improved or supplanted by directed evolution methods.

Recognition of Unnatural Amino Acids by Evolved Aminoacyl-tRNA Synthetases

To understand how the evolved UAA-specific synthetases recognize their substrates, the X-ray crystal structures of a number of aaRSs have been solved. For UAARSs based on TyrRS, these studies reveal that this class I synthetase contains an amino acid–binding pocket with a high degree of tolerance for substitutions. For the most part, substitutions that arise from the two-step selection scheme alter the side chains that line the amino acid–binding pocket to increase complementarity with UAAs while decreasing complementarity with endogenous amino acids. For example, the UAARS specific for *p*-benzoylphenylalanine (**36**), evolved in *E. coli* from the heterologous *Mj*TyrRS, accommodates the large side chain of UAA **36** primarily through the mutation Tyr32Gly, which deepens the substrate-binding pocket (138). In some cases, substitutions can also change the conformation of the protein backbone, suggesting a significant degree of structural plasticity. For example, in the UAARS specific for *p*-bromophenylalanine (**18**), evolved from *Mj*TyrRS, the Asp158Pro mutation terminates the α_8-helix, flipping residues 158, 159, and 162, which normally line the binding pocket, outward (139). Mutations that accommodate UAAs are often selected together with ones that disfavor binding of natural amino acids. For example, in the UAARS specific for *p*-acetylphenylalanine (**1**), evolved from *Mj*TyrRS, substitutions Tyr32Leu and Asp158Gly have the dual effect of removing hydrogen bonding with tyrosine's phenolic hydroxyl group and opening up the binding pocket to accommodate the acetyl group of UAA **1**. To further increase affinity for UAA **1**, the unmutated residue Gln109 forms a hydrogen bond with the acetyl group (140).

Perhaps the best illustration of aaRS tolerance to substitutions is the selection of an UAARS specific for bipyridylalanine (**34**) (85). In this case, a library of mutants based on *Mj*TyrRS did not directly yield a aaRS specific for bipyridylalanine but did yield one specific for the isostere biphenylalanine (**26**). The crystal structure of the biphenylalanine-specific UAARS bound to UAA **26** was solved, and the structure served as a guide for the design of a new library of aaRS mutants, which was subjected to further selection for incorporation of UAA **34**. Though one may expect that the accumulation of additional substitutions should reduce synthetase stability and activity, this second selection yielded an active bipyridylalanine-specific UAARS, suggesting that the thermophilic *Mj*TyrRS is highly tolerant to mutations in its amino acid–binding domain. This explains, at least in part, why a large number of UAARSs have been successfully evolved from the heterologous *Mj*TyrRS.

Taken together, these structural studies support the use of structure-based library design and selection and suggest that successful selection of new synthetases from libraries based on known structures of *Mj*TyrRS, UAARSs, and the class I *E. coli* LeuRS (from which several UAARSs have been evolved but for which crystal structures are few) (141) will continue. Furthermore, structural and phylogenetic studies on the class II PylRS suggest that its specificity for pyrrolysine is achieved by interactions localized to its side chain–binding pocket (109); therefore, aaRS libraries based on PylRS should also yield new amino acid specificities, a prediction that recent experiments have already begun to confirm. Nonetheless, despite the fact that aaRSs have been engineered to recognize rather large, structurally complex amino acids (including those with ferrocenyl, dimethoxynitrobenzyl seryl, dansyl, and bipyridyl side chains), the development of strategies and aaRS libraries to accept even greater structural diversity remains a challenge.

Blank Codons

Amber nonsense codons and, to a lesser extent, ochre, opal, or frameshift codons have been used to specify UAAs. In each case, suppression is in competition with other processes, such as RF binding to nonsense codons or recognition of frameshift codons by tRNAs with three-base anticodons, both of which lead to decreased cotranslational incorporation efficiency. Although a number of efforts are employed to mitigate these undesired effects—examples include the overproduction of $tRNA_{SB}$ suppressors, synonymous mutation of codons surrounding $codon_{BL}$, selection of efficient suppressors of frameshift codons that compete well with embedded rare three-base codons, and modulation of RF levels (142–145)—and other efforts have tried to circumvent them—for example, a sense codon was used to encode an UAA in an auxotrophic strain by exploiting wobble position stability differences and limiting endogenous amino acid concentrations (146)—a general solution that involves the conversion of sense codons into blank codons would be ideal.

In theory, such a strategy is feasible with genome synthesis, whereby certain sense codons in an organism are replaced in a wholesale manner through synonymous mutations. Although effects on mRNA folding and translation initiation rates would have to be considered (147, 148), the decreased degeneracy of the code would allow the reassignment of unused sense codons to UAAs. Such an approach is currently being carried out in *S. cerevisiae* and, if successful, would not only avoid the inefficiencies associated with nonsense and frameshift suppression, but would also allow several UAAs to be simultaneously encoded through mutually orthogonal UAARS/$tRNA_{SB}$ pairs. Such strains might allow completely unnatural polypeptides to be made ribosomally. An alternative approach that uses unnatural base pairs in addition to A-T and C-G could also provide new blank codons, in this case by extending the number of possible three-base codons beyond 64 (149–151). Although the inefficiency of unnatural base pairs in replication, transcription, and translation has kept this approach from moving beyond simple in vitro expression systems, the modularity of genetic code elements should allow for the direct integration of unnatural codons once these problems are solved.

Optimized Systems for Protein Expression

The practical utility of organisms with 21 (or greater) amino acid codes is dependent on the efficiency and fidelity with which proteins containing UAAs can be expressed. Therefore, several improvements to the components involved in cotranslational UAA incorporation have been made. The most basic of these involve optimization of expression vectors. For example, in one case, a 20-fold increase in the yield of an UAA-containing myoglobin protein was observed in *E. coli* when the original two-plasmid system encoding *M. jannaschii* $tRNA_{CUA}$ (*Mj*$tRNA_{CUA}$, an amber $tRNA_{SB}$) under the control of the *lpp* promoter and an UAARS derived from *Mj*TyrRS (*Mj*UAARS) under the control of the *glnS* promoter was replaced by a single-plasmid system encoding 3 or 6 copies of *Mj*$tRNA_{CUA}$ under the control of the *ProK* promoter (this promoter drives endogenous $tRNA^{Pro}$ expression in *E. coli*) and the *Mj*UAARS under the control of a strong mutant *glnS* promoter (142). This plasmid design (pSup) was further improved by encoding an additional copy of the *Mj*UAARS under the control of inducible *ara* and *T7* promoters (79). The resulting pSUPAR plasmids afford shake-flask expression yields ranging from ~40 mg/L to ~65 mg/L for UAA containing the FAS-TE (fatty acid synthase thioesterase domain) as well as high yields for many other proteins tested (although yields do vary based on the individual proteins and codon context effects) (152). Plasmid system improvements have also been made for UAA incorporation in *S. cerevisiae* by increasing promoter strengths, raising plasmid copy number, and using optimized codons (153). For example, a particularly effective optimization effort significantly increased the yield of an UAA-containing protein over earlier systems by placing *E. coli* $tRNA^{Leu}{}_{CUA}$ (an amber $tRNA_{SB}$ that is part of an orthogonal pair in yeast) under the control of an external promoter containing A- and B-box sequences (154). Expression was conducted in a nonsense-mediated decay-deficient *S. cerevisiae* strain, which prevents the degradation of mRNAs that contain premature stop codons.

Synthetase mutations and tRNA optimization have resulted in further improvements in these expression systems. For example, a T252A mutation in the editing domain of *E. coli* LeuRS variants increases the fidelity of UAA incorporation by reducing net aminoacylation with an undesired common amino acid (129), and a D286R substitution in *Mj*TyrRS variants increases the efficiency of *Mj*UAARS/*Mj*$tRNA_{CUA}$ pairs by mediating better recognition of the CUA anticodon (155). In addition, *Mj*$tRNA_{CUA}$ has been engineered for higher expression levels of UAA-containing proteins by randomizing tRNA T-stem nucleotides at the EF-Tu-binding interface and conducting stringent selection for increased amber suppression efficiency (156). The resulting selected tRNAs are in general less toxic to *E. coli* and in some cases enhance expression efficiency in an UAA-dependent manner. Some of these resulting tRNAs have been combined with the pSUPAR plasmid design to achieve UAA-containing protein overexpression yields that approach wild-type levels (wild-type refers to the analogous proteins where the UAA is replaced by a common amino acid) (157).

Though more challenging, elongation factors and ribosomes themselves can also be optimized for UAA incorporation. For example, the recent development of orthogonal ribosomes that only translate mRNAs containing artificial 5′ sequences allows for ribosome evolution by decoupling mutant and endogenous ribosome action (158). The use of this approach yielded the orthogonal ribosome ribo-X, which contains mutations that likely decrease its interaction with RF-1, thereby preventing early termination and increasing the efficiency of translation with UAAs (159). It is also possible to develop EF-Tu mutants that better tolerate UAAs, especially those with large side chains or altered backbones. This has been demonstrated in an in vitro translation system with rationally designed EF-Tu variants that have mutations in their aminoacyl-binding pockets (160). Although their activity for natural amino acids is lower than that of wild-type EF-Tu, these variants have increased activity for large aromatic UAAs, thus increasing the efficiency of their incorporation. Selection in cells with UAARS/$tRNA_{SB}$ pairs should allow tuning of EF-Tu-binding site variants to better accommodate UAAs. Other efforts, including the selection of optimal nucleotide contexts surrounding blank codons, random genome-wide mutagenesis and complementation to find other factors that affect UAA-incorporation efficiency, and enhancement of cellular uptake of charged UAAs by their delivery as dipeptides or precursors, are currently being explored.

GENETICALLY ENCODED UNNATURAL AMINO ACIDS AND THEIR APPLICATIONS

Clearly, the 20 common amino acids are sufficient for all known forms of life. However, if one assumes a substantial proteomic contribution to fitness during the solidification of the genetic code (Crick's "frozen accident" factor) (161, 162), then one must conclude that the identities of the 20 canonical amino acids are a consequence of factors present during early evolution (163, 164). Therefore, additional genetically encoded amino acids may offer evolutionary advantages to modern organisms and certainly extend our ability to manipulate the physicochemical and biological properties of proteins. To this end, we and others have genetically encoded UAAs representing an extensive range of structural and electronic properties not found in the common 20 amino acids (**Figure 1**).

UAAs with chemically reactive groups can be used as bioorthogonal handles for the site-specific in vitro, and in some cases intracellular, modification of proteins; they can also be used to introduce a new or enhanced catalytic function into proteins. Such amino acids include UAAs **1–11**, **17**, **18**, **64**, **67**, **68**, **70**, and **71**. Proteins containing these UAAs have been selectively modified with nonproteinogenic moieties using a variety of selective chemistries, including oxime condensation reactions, click chemistry, Michael addition reactions, and Suzuki couplings. Members of another class of UAAs represented by UAAs **12–33**, **36**, **44**, **45–47**, **53–63**, **65**, **66**, and **69** contain, in their side chains, in vitro or cellular probes of protein structure and function. These UAAs can be used as IR, NMR, and fluorescent probes, as well as redox-active reagents, heavy atoms for X-ray structure determination, and probes of hydrogen bonding and packing interactions in proteins. This class also includes photocross-linking amino acids, which can be used as in vitro or cellular probes of protein-protein and protein-nucleic acid interactions. UAAs can also be used to directly engineer into proteins new functions that may be challenging to generate with the common 20 amino acids. Such UAAs include metal ion-binding, photoisomerizable, photocaged, and photoreactive amino acids (**34–44**). Another class of UAAs is represented by UAAs **48–52**, which are all free amino acids that correspond to the product of a PTM. These UAAs allow the expression of proteins containing defined PTMs in simple hosts; lift the sequence restrictions to which PTMs must otherwise adhere; and have been used to express sulfated, nitrosylated, and methylated proteins. In addition, *p*-carboxymethylphenylalanine (**48**) was substituted in the transcription factor STAT1 as a stable mimetic of phosphotyrosine and overcame the complexities associated with removal of this PTM by phosphatases (94). Below, we overview a number of examples to illustrate the utility of genetically encoded UAAs in the study and manipulation of protein structure and function.

Probes of Protein Structure and Function

Many biophysical and mechanistic studies require significant quantities of proteins with a probe incorporated at a unique site in a protein. UAA mutagenesis methodology is well suited to many such problems. For example, in a recent application of UAAs as NMR probes, milligram quantities of site specifically labeled FAS-TE were produced from $\leq$25 mg of either $^{13}C/^{15}N$-labeled *O*-methyltyrosine (**12**), ^{15}N-labeled *o*-nitrobenzyltyrosine (**41**), or OCF_3Phe (**19**) (79). The resulting variants of FAS-TE (33 in total were generated) were then characterized by ^{1}H-^{13}C HSQC, ^{1}H-^{15}N HSQC, and ^{19}F NMR in the absence and presence of a small-molecule inhibitor. Using the isolated NMR signals of the UAA probes, these experiments identified the binding site of the inhibitor and revealed conformation changes of certain residues that occur upon binding. A particularly interesting aspect of these studies was the incorporation of a ^{15}N-labeled variant of the photoreactive UAA **41**, which upon irradiation decages to afford site-specific isotopic labeling at only one of the many tyrosines. When an active-site tyrosine was mutated to a ^{15}N-labeled UAA **41**, chemical shifts corresponding to ligand binding were observed only after the Tyr was decaged.

UAAs with unique IR and X-ray diffraction signatures have also been efficiently incorporated into proteins to study structure and dynamics. For example, *p*-cyanophenylalanine (**14**) contains a cyano group that absorbs in a clear spectral window at $\sim$2200 cm^{-1} and is sensitive to subtle changes in local environment. It was incorporated into myoglobin to probe metal ion and ligand binding (76) and is also used as a Stark effect probe of local protein dipoles. Similarly, *p*-azidophenylalanine (**7**), which absorbs in a clear spectral window at $\sim$2100 cm^{-1} and is sensitive to changes in electrostatic environment, was incorporated site specifically at various sites in rhodopsin. After expression and reconstitution into lipid membranes, changes in the stretch frequency of the azido group upon light-induced receptor activation were analyzed by Fourier transform infrared (FTIR) difference spectroscopy for several rhodopsin mutants, revealing changes in the local environments of specific amino acids (165). In addition, solution of the X-ray crystal structures of *O*-acetylserine sulfhydrylase, *N*-acetyltransferase, and T4 lysozyme was facilitated by the site-specific incorporation of UAAs containing heavy atoms (Zn^{2+}-bound UAA **35**, **23**, and **17**, respectively) (78, 83, 87). Several other recent studies have also exploited UAA mutagenesis to selectively introduce novel biophysical probes. These include the dissection of radical propagation in ribonucleotide reductase with a redox-active amino acid (82), the exploration of K^+ channel inactivation by perturbing the diameter of the inactivation peptide (134), the mutation of backbone amides to esters in order to evaluate the contribution of hydrogen bonding interactions to protein stability (86, 118), and the study of protein dynamics in dihydrofolate reductase with deuterated amino acids site specifically incorporated by genetically encoded photocaged variants (166). In this last example, FTIR characterization of specifically placed C-D bonds allowed the time-resolved characterization of the microenvironment of Tyr100 upon substrate binding.

Because the genetic encoding of UAAs exploits the cells' own translational machinery, one can use genetically encoded biophysical probes to study processes both in vitro and in living cells. For example, one can genetically encode small fluorescent UAAs at surface sites in a protein with minimal structural perturbation (this is in contrast to traditional fluorescent genetic tags such as GFP and its variants, which are limited by their large size). Recently, such a strategy was used to incorporate UAA **47** into histones, whose nuclear localization in yeast and mammalian cells could then be visualized through fluorescence microscopy (**Figure 3*a***) (J. Guo, H.S. Lee, E.A. Lemke, R.D. Dimla, & P.G. Schultz, unpublished results). Furthermore, because the fluorescence properties of UAA **47** and other fluorescent UAAs are highly sensitive to environmental changes, and these

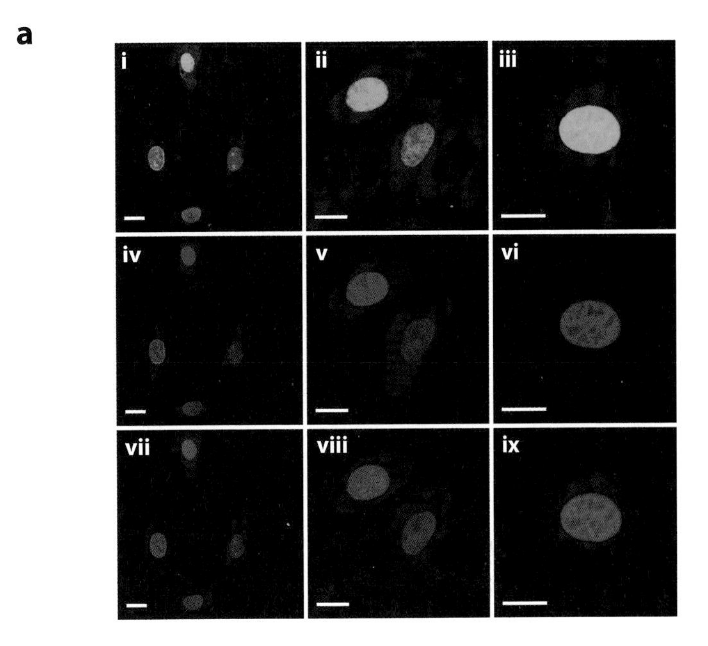

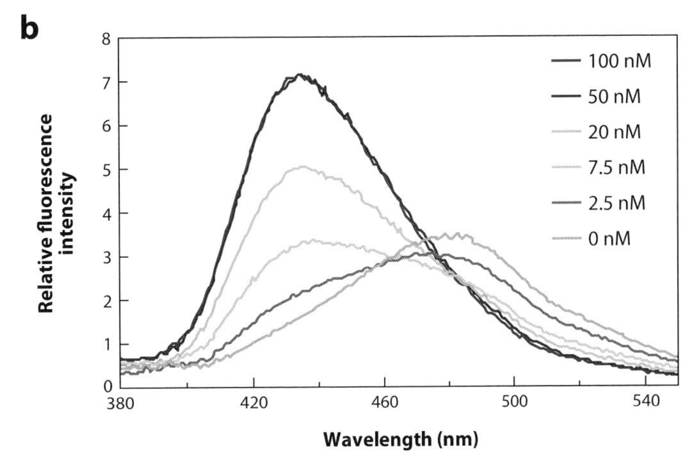

Figure 3

(*a*) Fluorescence images of labeled histone H3 in CHO cells acquired by confocal, two-photon microscopy. Each scale bar is 10 μm. (*i-iii*) CHO cells expressing an mCherry-labeled histone H3 with UAA **47** at position 59 (excitation, 730 nm, two photon; emission, 420–500 nm). (*iv-vi*) CHO cells expressing an mCherry-labeled histone H3 with UAA **47** at position 59 (excitation, 543 nm; emission, 600–700 nm). (*vii*) Composite of images *i* and *iv*. (*viii*) Composite images of *ii* and *v*. (*ix*) Composite of images *iii* and *vi*. (*b*) Fluorescence spectra of glutamine-binding protein with UAA **47** at position 160 and different concentrations of Gln (excitation, 350 nM).

amino acids can be selectively incorporated at or near a site of interest, this labeling method can be used for the characterization of local protein conformation changes, protein folding, and biomolecular interactions (87, 93, 126, 129, 167). Recent examples include the characterization of myoglobin unfolding using UAA **46** (e.g., urea-dependent unfolding of the amino terminus prior to global unfolding), and the detection of ligand binding by glutamine-binding protein (QBP) using UAA **47** (**Figure 3*b***). In the latter case, UAA **47** was placed in a deep cleft that lies between the two domains of QBP, which are connected by two peptide hinges. Upon binding of glutamine, significant conformational changes in this hinge region were directly revealed [without the use of a Förster resonance energy transfer (FRET) pair] by a significant shift in the emission λ_{max} and intensity of UAA **47**. The K_d of the mutant QBP for Gln was determined from these spectral changes and found to be close to that of wild-type protein.

In addition to fluorescent probes, one can use photocaged and photoisomerizable UAAs to study protein function. In vitro these UAAs have been used to photoregulate ligand-protein binding (91), site specifically cleave proteins (90), and probe protein-nucleic acid interactions (168). Such genetically encoded UAAs can also allow the study of cellular processes. For example, 4,5-dimethoxy-2-nitrobenzylserine (**39**), a photocaged serine, was encoded in the transcription factor Pho4 at either Ser114 (termed S2) or Ser128 (termed S3). These variants, fused to GFP, were localized in the nucleus when expressed in *S. cerevisiae* in low-phosphate media. Photolysis with a visible laser pulse (465 nm) decaged S2 or S3, leading to subsequent phosphorylation and export from the nucleus by Msn5 under P_i-rich conditions (**Figure 4*a***). The kinetics of this process were monitored in real time to better understand the activities of differentially phosphorylated Pho4 mutants (**Figure 4*b***) (128).

Another use of UAA incorporation is the characterization of protein-protein and protein-nucleic acid interactions in vitro or in living cells through photocross-linking. In many cases, biomolecular interactions are transient or unstable to in vitro isolation conditions, thus requiring covalent cross-linking of the interacting molecules to isolate relevant complexes. In a proof-of-principle study, *p*-benzoylphenylalanine (**36**), which upon irradiation inserts into nearby C-H bonds (or relaxes back to the ground state), was site specifically incorporated into the adapter protein Grb2. This protein mediates extracellular signaling to the Ras protein by binding various target molecules, including members of the epidermal growth factor (EGF) receptor family. When expressed in CHO cells together with the EGF receptor, in vivo light-induced cross-linking resulted in the identification of covalently linked Grb2-EGF receptor complexes (169). Furthermore, the efficiency of cross-linking varied with the position of *p*-benzoylphenylalanine incorporation. This methodology is now being used in many labs to map cell circuitry and identify orphan ligands and/or receptors.

Protein Therapeutics

UAA mutagenesis is beginning to find many applications in the generation of therapeutic proteins, where the production of large quantities of homogenously modified protein is desired. Three examples illustrate this best. First, is the use of immunogenic amino acids to break immunological tolerance and generate therapeutic vaccines against self-proteins associated with cancer or inflammation. In this approach, incorporation of *p*-nitrophenylalanine (**20**) into a target protein results in immunogenic epitopes that upon immunization induce high-titer, long-lived, and cross-reactive antibodies to native proteins. For example, murine TNF-α mutants containing UAA **20** at position 11 or position 86 induce, in mice, antibodies that are cross-reactive with native murine TNF-α and protect against lipopolysaccharide-induced death (**Figure 5**) (170, 171). This occurs through the stimulation of T cells by epitopes containing UAA **20**, which ultimately leads to activation of autoreactive B cells that produce

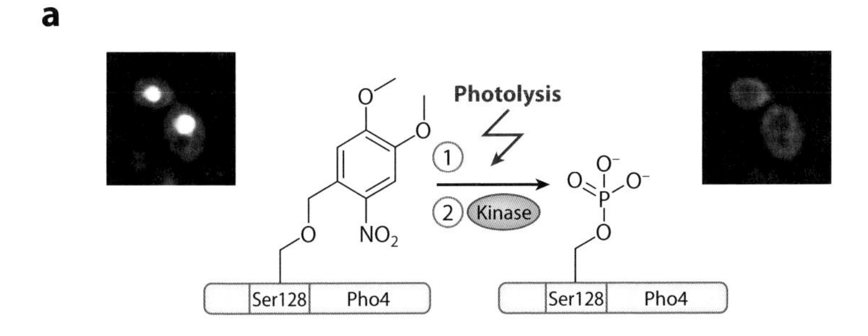

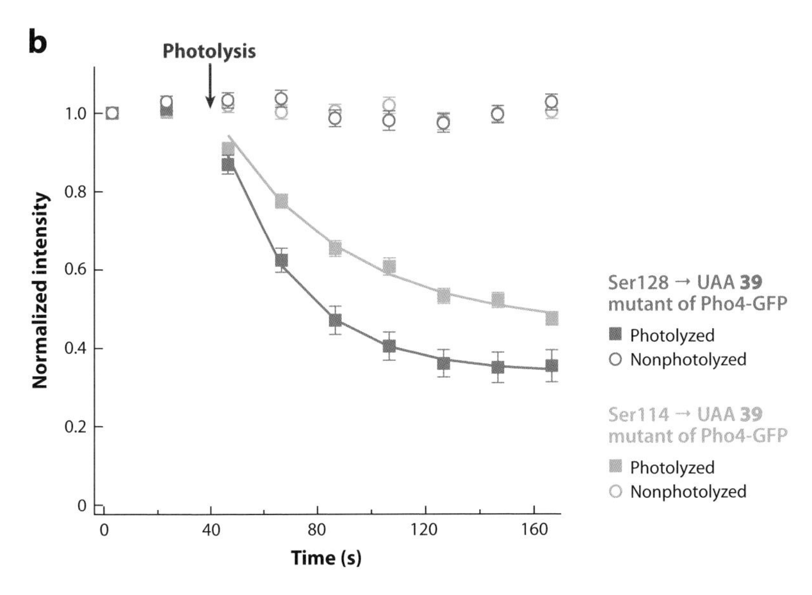

Figure 4

In vivo photolysis and nuclear export of Pho4 in yeast under high-phosphate conditions. (*a*) Magnification of two cells before and after photolysis (Ser128 → UAA **39** mutant of Pho4-GFP), and scheme depicting photolysis followed by serine phosphorylation. (*b*) Real-time analysis of in vivo photolysis of Pho4-GFP mutants. The normalized average fluorescence intensity is plotted as a function of time for the Ser128 → UAA **39** mutant of Pho4-GFP (*blue*) and the Ser114 → UAA **39** mutant of Pho4-GFP (*green*). Panel *b* reprinted with permission from Nature Publishing Group and adapted from Reference 128.

polyclonal antibodies against the native self-protein, a mechanism that should be general to many antigens and epitopes against which a strong immune response is desired. This approach is currently being applied to other self-antigens, such as EGF and HER2 (involved in cancer), PCSK9 (involved in cardiovascular disease), and C5a (involved in inflammation), as well as to conserved epitopes of human immunodeficiency virus (HIV) and malaria that are difficult to target with traditional vaccines. This approach should also prove useful as a tool to knock down levels of secreted proteins through targeted antibody responses. Moreover, the demonstration that a single *p*-nitrophenylalanine residue can break tolerance provides support to the intriguing notion that the enzymatic posttranslational generation of nitrotyrosine in proteins, stimulated by local inflammation and cytokine release, could create immunogenic epitopes that are the underlying initiating event in many autoimmune diseases.

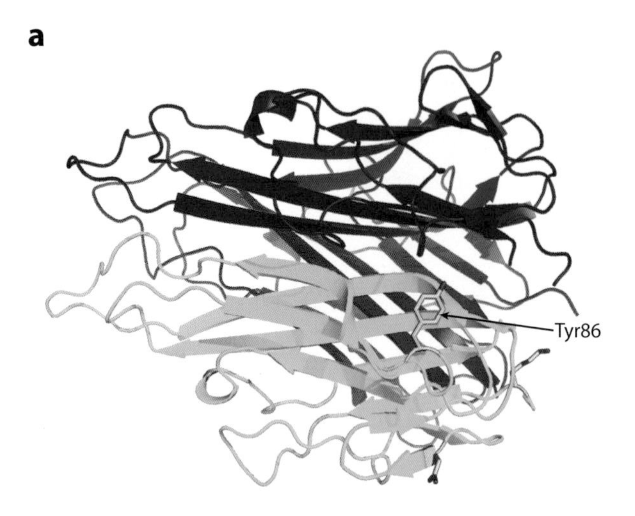

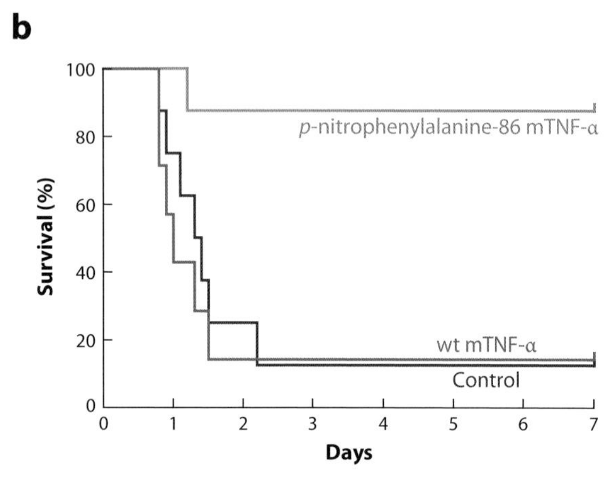

Figure 5

(*a*) X-ray crystal structure of murine tumor necrosis factor alpha (TNF-α) trimer with Tyr86 indicated (Protein Data Bank code 2TNF). (*b*) Substitution of Tyr86 in murine TNF-α with *p*-nitrophenylalanine and subsequent vaccination leads to a robust T cell–driven immune response, which cross-reacts with wild-type (wt) TNF-α and protects mice from lipopolysaccharide (LPS) challenge. Survival of mice vaccinated with the Tyr86 → UAA **20** TNF-α mutant is compared to survival of mice vaccinated with wild-type TNF-α. Survival is plotted as a function of days after LPS challenge. Reprinted with permission from the National Academy of Sciences, U.S.A., copyright © 2008 and adapted from Reference 170.

A second example is the production of the anticoagulant leech protein sulfohirudin. Because natural sulfohirudin is the product of posttranslational tyrosine sulfation, which occurs in higher eukaryotes, it can only be obtained in minute amounts through traditional methods. As such, it is not available in clinically useful quantities; instead, a recombinant hirudin that lacks sulfation is commercially expressed and used as a direct thrombin inhibitor (DTI) with a K_i of ~307 fM. However, by expanding the genetic code of *E. coli* to incorporate sulfotyrosine (**50**), one can access sulfohirudin directly (96). This allows for the expression of the higher-affinity DTI sulfohirudin (K_i = 26 fM) in excellent yields, which also facilitates the crystallization of the sulfohirudin/thrombin complex (**Figure 6**) (172), and should allow clinical application of sulfohirudin.

Third is the production of protein conjugates containing a site specifically attached toxin, radioisotope, polyethylene glycol, or even another protein (creating a bispecific therapeutic). In these cases, an UAA containing a bioorthogonal reactive functional group is incorporated at a specific position, which is then further functionalized in vitro with another

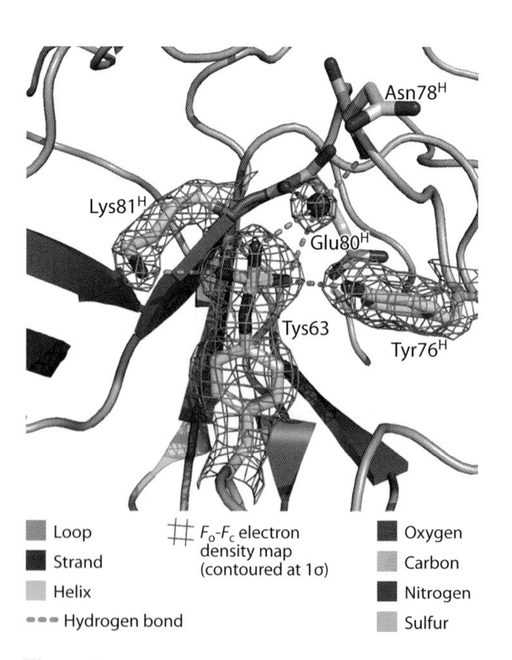

Figure 6

Sulfotyrosine at position 63 (Tys63) in the X-ray crystal structure of the sulfohirudin/thrombin complex. Interactions mediated by the sulfate of UAA **50** are shown. (^{H} refers to the heavy chain of thrombin.) Adapted with permission from Reference 172, copyright © 2007 American Chemical Society.

small molecule or macromolecule of interest. This method overcomes the challenges of nonspecific labeling typically associated with electrophilic reagents or the labeling of cysteine residues that may not be unique or may be involved in folding. In nontherapeutic contexts, this strategy has been used for glycan and lipid conjugation (70, 173), dual-fluorophore labeling for single-molecule FRET studies (174), labeling of G protein–coupled receptors with fluorophores (175), nonnative histone modification (176), and scarless protein purification (73). In therapeutic contexts, this strategy is especially valuable as it allows the creation of homogenous therapeutic agents with defined pharmacological activity and toxicity. Moreover, because the cotranslational incorporation of UAAs is highly efficient, clinically useful quantities of labeled proteins can be produced. Indeed UAA-containing therapeutic candidates can be generated in yields of grams per liter on a >1000 liter fermentation scale in bacteria (high-yielding *Pichia* and mammalian systems are in development). For example, human growth hormone (hGH) containing *p*-acetylphenylalanine (**1**) has been expressed commercially and site specifically PEGylated in kilogram quantities (177). The modified hGH has a significantly improved serum half-life while retaining biological activity and is currently in Phase II clinical trials. Similar approaches are being applied to cytokines, growth factors, and other proteins for clinical use. Various antibodies have also been efficiently expressed with keto amino acid **1** at specific sites for conjugation to therapeutically or diagnostically relevant molecules. For example, in a recent study, *p*-acetylphenylalanine mutants of the α-HER2 antibody trastuzumab were conjugated to the toxins saporin or auristatin, resulting in targeted killing of HER2-positive cells (B. Hutchins, S. Kazane, J. Yin, P.G. Schultz, & V.V. Smider, unpublished results). In addition, the conjugation of antibodies via amino acid **1** to oligonucleotides allows for polymerase chain reaction–based immunodiagnostics as well as the creation of spatially defined arrays of proteins and possibly cells (S. Kazane, B. Hutchins, P.G. Schultz, & V.V. Smider, unpublished results). Efforts are also under way to create structurally defined bispecific antibodies using the additional orthogonal chemistries afforded by UAAs to selectively couple proteins at uniquely defined sites and also to conjugate therapeutic peptides to carrier proteins, such as HSA and antibodies, to improve their half-lives. In all these experiments (and especially in their potential clinical use), large quantities of homogeneous high-molecular-weight UAA-containing proteins are required; thus, the ease and efficiency inherent to translation with expanded genetic codes are highly desirable.

Protein Evolution with an Expanded Genetic Code

It is quite possible that the ability to encode additional amino acids with novel properties would be evolutionarily advantageous, especially since nature's choice of 20 could have been arbitrarily fixed at the point of transition between communal and Darwinian evolution paradigms and subsequently sustained by the code's inertia (162). Furthermore, in the limited scope of laboratory-directed evolution, which concerns only one or few specific functions over a short time rather than general organismal fitness over thousands or millions of years, one can easily envision a selective advantage conferred by additional amino acids. Because the templated assembly of polypeptides from mRNA on the ribosome establishes a direct link between genes (information) and proteins (phenotype), UAA mutagenesis methodology can easily be adapted to the evolution of proteins with novel or enhanced function. As a first step in this direction, robust phage-based systems for use with 21 amino acid organisms were recently developed (178, 179). In one such system, phage-displayed protein libraries are generated in *E. coli* that encode 21 amino acids (X-*E. coli*) such that library members containing a TAG codon in a gene of interest will have a corresponding UAA in the displayed protein. Displayed proteins are then subjected to selection, and the surviving phage clones are used to reinfect X-*E. coli*. Like traditional phage-display with the canonical 20 amino acids, rounds of mutation, amplification, and selection result in the discovery of new proteins with desired properties that, in this case, may be achieved by UAAs.

After optimizing this system to minimize the expression bias against proteins containing UAAs, several in vitro molecular evolution experiments were conducted. For example, a phage-displayed library of naive antibodies containing six NNK-randomized residues (N = A, C, G, or T; K = G or T) in the V_H CDR3 was produced in X-*E. coli* that encodes the UAA sulfotyrosine (**50**). Selection experiments were then performed on the basis of affinity for the HIV coat protein gp120 (179). After several rounds of amplification and selection, the phage population converged on one antibody clone that contained a sulfotyrosine. The selection of this phage-displayed antibody was based on gp120-binding function, as determined by expression studies, and this function was dependent on UAA **50**, as confirmed by enzyme-linked immunosorbent assays. Another in vitro evolution experiment used X-*E. coli* that encoded the reactive UAA *p*-boronophenylalanine (**8**) to generate an unbiased phage library of naive antibodies containing NNK-randomized residues. The resulting phage particles were exposed to selection for binding to the acyclic glycan glucamine (180). After one round, over 50% of the phage population contained UAA **8**, and after three rounds, over 80% contained UAA **8**. This convergence on UAA-containing sequences was concurrent with functional enrichment for antibodies that bind glucamine, presumably by using the boronate functionality as a "chemical warhead" to react with diols. Furthermore, the function of the selected *p*-boronophenylalanine-containing clones cannot be attributed to UAA **8** alone but requires the contribution of the whole selected sequence, as demonstrated in a series of studies on both phage-displayed and soluble Fab format clones. Several other studies using phage-based evolution with 21 amino acids are currently being conducted, including the use of UAAs to overcome PTM sequence constraints (181), the use of reactive UAAs to evolve inhibitors of glycoproteins and proteases, and the incorporation of metal ion–binding UAAs to evolve metalloproteins with novel activities.

In addition to these phage-based evolution experiments, other efforts to include UAAs in directed evolution experiments are under way. For example, we are currently synthesizing and selecting bioactive cyclic peptides that contain UAAs, generated through the use of split inteins or a recently characterized ribosomal synthesis pathway (182). Furthermore, although there have been several recent successes of protein engineering with UAAs—for example, the generation of a redox-active, bipyridylalanine-containing catabolite activator protein mutant that site specifically cleaves DNA (183)—we expect that protein design alone will not be enough as we move toward the application of UAAs to more complex functions. Therefore, a combination of computational design and molecular evolution techniques are being employed in several new efforts (for example, to engineer metalloproteins that use a metal ion–binding UAA as a structural element or for catalysis). Future studies are also likely to include organismal evolution experiments, including those with an autonomous 21 amino acid *E. coli* that biosynthesizes its own UAA (75).

Taken together, these experiments demonstrate that the "chemical potential" contained in expanded amino acid sets can, through directed evolution, be realized as enhanced protein function and show that an expanded genetic code can confer a selective advantage in protein evolution. In all these cases, the integration of orthogonal UAARS/$tRNA_{SB}$ pairs into the translation paradigm provides the direct link between genotype and phenotype required for protein evolution with UAAs.

CONCLUSION

Genetically encoded UAAs seamlessly integrate new chemistry into biology and offer a number of advantages over other approaches for chemically manipulating protein structure and function. First, the generality of methods for developing orthogonal UAARS/$tRNA_{SB}$ pairs allows the systematic genetic encoding of new UAAs in vivo to meet new needs in chemistry and biology. Second, the modularity of components involved in this approach allows the combination of multiple blank codons with multiple UAARS/$tRNA_{SB}$ pairs to achieve the biosynthesis of proteins containing many or perhaps exclusively UAAs. Finally, the incorporation of UAAs directly through the translation machinery greatly simplifies the efficient production of homogenous mutant proteins, the development of cell biological probes, and the evolution of new protein function. Thus by removing the constraints imposed by nature on the number and properties of genetically encoded amino acids, exciting new opportunities in basic and applied science research have been (and will continue to be) realized.

SUMMARY POINTS

1. A general method based on the engineering of modular orthogonal unnatural amino acid–specific mutant aminoacyl-tRNA synthetase (UAARS)/tRNA pairs makes possible the genetic encoding of unnatural amino acids (UAAs) in *E. coli*, yeast, and mammalian cells.
2. Approximately 70 UAAs, representing various structures and chemistries distinct from those contained in the 20 natural amino acids, have been genetically encoded in response to blank codons, such as nonsense triplet codons and frameshift quadruplet codons.
3. Structural studies of engineered UAARSs suggest that their actives sites have considerable structural plasticity but that mutations in the amino acid–binding sites of synthetases result in small, local perturbations rather than unpredictable large-scale changes, underscoring the utility of structure-based synthetase library design and selection to manipulate aminoacyl-tRNA synthetase (aaRS) substrate specificity.
4. A two-step selection system, which uses a positive step to select for aaRSs that aminoacylate the cognate tRNA with the desired UAA and a negative step to remove aaRSs that recognize endogenous host amino acids, is highly effective in the evolution of aaRS specificity and orthogonality.
5. The integration of UAARS/tRNA pairs directly into the translation machinery allows for the efficient and selective introduction of UAAs into proteins in living cells.
6. Applications of genetically encoded UAAs are widespread and are already leading to novel protein therapeutics, new tools to study protein structure and function, and the directed evolution of proteins that use UAAs to achieve new function.

FUTURE ISSUES

1. The further development of novel UAARS/tRNA pairs for the incorporation of new, chemically distinct UAAs.
2. The further optimization of the translational machinery for more efficient expression of proteins with UAAs at any defined site.

3. The creation of new codon$_{BL}$s and mutually orthogonal UAARS/tRNA pairs to allow the in vivo expression of proteins containing multiple UAAs or even fully unnatural polypeptide polymers.
4. The further application of UAA mutagenesis methods to the generation of novel protein therapeutics (e.g., vaccines, bispecific proteins, immunotoxins) and to the development of tools (e.g., fluorophores, photoreactive amino acids, amino acids containing modifications normally added posttranslationally) to study protein structure and function in vitro and in living cells.
5. The further use of expanded genetic codes to evolve new proteins with novel physical, biochemical, and catalytic properties that are difficult to produce using only the 20 canonical amino acids.
6. The creation of multicellular organisms that have expanded genetic codes for the in vivo study of protein structure and function.

DISCLOSURE STATEMENT

Some of the work described has been licensed through the Scripps Research Institute. The institution and inventors, including C.C.L. and P.G.S., receive license fees.

ACKNOWLEDGMENTS

C.C.L. is supported by predoctoral fellowships from the Fannie and John Hertz Foundation and the National Science Foundation. P.G.S. is supported by the National Institutes of Health, the U.S. Department of Energy, and the Skaggs Institute of Chemical Biology.

LITERATURE CITED

1. Silverman RB. 2002. *Organic Chemistry of Enzyme-Catalyzed Reactions*. London/San Diego: Academic. 800 pp.
2. Walsh CT. 2005. *Posttranslational Modification of Proteins: Expanding Nature's Inventory*. Englewood, CO: Roberts. 490 pp.
3. Chalker JM, Bernardes GJ, Lin YA, Davis BG. 2009. Chemical modification of proteins at cysteine: opportunities in chemistry and biology. *Chem. Asian J.* 4:630–40
4. Bernardes GJ, Chalker JM, Errey JC, Davis BG. 2008. Facile conversion of cysteine and alkyl cysteines to dehydroalanine on protein surfaces: versatile and switchable access to functionalized proteins. *J. Am. Chem. Soc.* 130:5052–53
5. Merrifield RB. 1969. Solid-phase peptide synthesis. *Adv. Enzymol. Relat. Areas Mol. Biol.* 32:221–96
6. Marshall GR. 2003. Solid-phase synthesis: a paradigm shift. *J. Pept. Sci.* 9:534–44
7. Dawson PE, Kent SB. 2000. Synthesis of native proteins by chemical ligation. *Annu. Rev. Biochem.* 69:923–60
8. Muir TW. 2003. Semisynthesis of proteins by expressed protein ligation. *Annu. Rev. Biochem.* 72:249–89
9. Ambrogelly A, Palioura S, Söll D. 2007. Natural expansion of the genetic code. *Nat. Chem. Biol.* 3:29–35
10. Kimmerlin T, Seebach D. 2005. '100 years of peptide synthesis': ligation methods for peptide and protein synthesis with applications to beta-peptide assemblies. *J. Pept. Res.* 65:229–60
11. Baca M, Kent SB. 1993. Catalytic contribution of flap-substrate hydrogen bonds in "HIV-1 protease" explored by chemical synthesis. *Proc. Natl. Acad. Sci. USA* 90:11638–42

12. Kochendoerfer GG, Chen SY, Mao F, Cressman S, Traviglia S, et al. 2003. Design and chemical synthesis of a homogeneous polymer-modified erythropoiesis protein. *Science* 299:884–87
13. Chen SY, Cressman S, Mao F, Shao H, Low DW, et al. 2005. Synthetic erythropoietic proteins: tuning biological performance by site-specific polymer attachment. *Chem. Biol.* 12:371–83
14. Vazquez ME, Nitz M, Stehn J, Yaffe MB, Imperiali B. 2003. Fluorescent caged phosphoserine peptides as probes to investigate phosphorylation-dependent protein associations. *J. Am. Chem. Soc.* 125:10150–51
15. Flavell RR, Muir TW. 2009. Expressed protein ligation (EPL) in the study of signal transduction, ion conduction, and chromatin biology. *Acc. Chem. Res.* 42:107–16
16. Valiyaveetil FI, Leonetti M, Muir TW, MacKinnon R. 2006. Ion selectivity in a semisynthetic K^+ channel locked in the conductive conformation. *Science* 314:1004–7
17. Hecht SM, Alford BL, Kuroda Y, Kitano S. 1978. "Chemical aminoacylation" of tRNA's. *J. Biol. Chem.* 253:4517–20
18. Heckler TG, Chang LH, Zama Y, Naka T, Chorghade MS, Hecht SM. 1984. T4 RNA ligase mediated preparation of novel "chemically misacylated" $tRNA^{Phe}$ S. *Biochemistry* 23:1468–73
19. Roesser JR, Xu C, Payne RC, Surratt CK, Hecht SM. 1989. Preparation of misacylated aminoacyl-$tRNA^{Phe}$'s useful as probes of the ribosomal acceptor site. *Biochemistry* 28:5185–95
20. Robertson SA, Noren CJ, Anthony-Cahill SJ, Griffith MC, Schultz PG. 1989. The use of 5′-phospho-2 deoxyribocytidylylriboadenosine as a facile route to chemical aminoacylation of tRNA. *Nucleic Acids Res.* 17:9649–60
21. Noren CJ, Anthony-Cahill SJ, Griffith MC, Schultz PG. 1989. A general method for site-specific incorporation of unnatural amino acids into proteins. *Science* 244:182–88
22. Koh JT, Cornish VW, Schultz PG. 1997. An experimental approach to evaluating the role of backbone interactions in proteins using unnatural amino acid mutagenesis. *Biochemistry* 36:11314–22
23. Eisenhauer BM, Hecht SM. 2002. Site-specific incorporation of (aminooxy)acetic acid into proteins. *Biochemistry* 41:11472–78
24. Murakami H, Hohsaka T, Ashizuka Y, Hashimoto K, Sisido M. 2000. Site-directed incorporation of fluorescent nonnatural amino acids into streptavidin for highly sensitive detection of biotin. *Biomacromolecules* 1:118–25
25. Taki M, Hohsaka T, Murakami H, Taira K, Sisido M. 2001. A non-natural amino acid for efficient incorporation into proteins as a sensitive fluorescent probe. *FEBS Lett.* 507:35–38
26. Taki M, Hohsaka T, Murakami H, Taira K, Sisido M. 2002. Position-specific incorporation of a fluorophore-quencher pair into a single streptavidin through orthogonal four-base codon/anticodon pairs. *J. Am. Chem. Soc.* 124:14586–90
27. Fahmi NE, Dedkova L, Wang B, Golovine S, Hecht SM. 2007. Site-specific incorporation of glycosylated serine and tyrosine derivatives into proteins. *J. Am. Chem. Soc.* 129:3586–97
28. Kanamori T, Nishikawa S, Shin I, Schultz PG, Endo T. 1997. Probing the environment along the protein import pathways in yeast mitochondria by site-specific photocrosslinking. *Proc. Natl. Acad. Sci. USA* 94:485–90
29. Thorson J, Chapman E, Murphy E, Judice J, Schultz P. 1995. Linear free energy analysis of hydrogen bonding in proteins. *J. Am. Chem. Soc.* 117:1157–58
30. Li S, Millward S, Roberts R. 2002. In vitro selection of mRNA display libraries containing an unnatural amino acid. *J. Am. Chem. Soc.* 124:9972–73
31. Frankel A, Li S, Starck SR, Roberts RW. 2003. Unnatural RNA display libraries. *Curr. Opin. Struct. Biol.* 13:506–12
32. Kohrer C, Yoo JH, Bennett M, Schaack J, RajBhandary UL. 2003. A possible approach to site-specific insertion of two different unnatural amino acids into proteins in mammalian cells via nonsense suppression. *Chem. Biol.* 10:1095–102
33. Rodriguez EA, Lester HA, Dougherty DA. 2006. In vivo incorporation of multiple unnatural amino acids through nonsense and frameshift suppression. *Proc. Natl. Acad. Sci. USA* 103:8650–55
34. Saks ME, Sampson JR, Nowak MW, Kearney PC, Du F, et al. 1996. An engineered *Tetrahymena* $tRNA^{Gln}$ for in vivo incorporation of unnatural amino acids into proteins by nonsense suppression. *J. Biol. Chem.* 271:23169–75

35. Dougherty DA. 2000. Unnatural amino acids as probes of protein structure and function. *Curr. Opin. Chem. Biol.* 4:645–52
36. Beene DL, Dougherty DA, Lester HA. 2003. Unnatural amino acid mutagenesis in mapping ion channel function. *Curr. Opin. Neurobiol.* 13:264–70
37. England PM, Zhang Y, Dougherty DA, Lester HA. 1999. Backbone mutations in transmembrane domains of a ligand-gated ion channel: implications for the mechanism of gating. *Cell* 96:89–98
38. Dougherty DA. 2008. Physical organic chemistry on the brain. *J. Org. Chem.* 73:3667–73
39. Lummis SCR, Beene DL, Lee LW, Lester HA, Broadhurst RW, Dougherty DA. 2005. *Cis-trans* isomerization at a proline opens the pore of a neurotransmitter-gated ion channel. *Nature* 438:248–52
40. Hendrickson TL, de Crécy-Lagard V, Schimmel P. 2004. Incorporation of nonnatural amino acids into proteins. *Annu. Rev. Biochem.* 73:147–76
41. Link AJ, Mock ML, Tirrell DA. 2003. Non-canonical amino acids in protein engineering. *Curr. Opin. Biotechnol.* 14:603–9
42. van Hest JC, Tirrell DA. 1998. Efficient introduction of alkene functionality into proteins in vivo. *FEBS Lett.* 428:68–70
43. Link AJ, Tirrell DA. 2003. Cell surface labeling of *Escherichia coli* via copper(I)-catalyzed [3+2] cycloaddition. *J. Am. Chem. Soc.* 125:11164–65
44. Kiick KL, van Hest JC, Tirrell DA. 2000. Expanding the scope of protein biosynthesis by altering the methionyl-tRNA synthetase activity of a bacterial expression host. *Angew. Chem. Int. Ed. Engl.* 39:2148–52
45. Datta D, Wang P, Carrico IS, Mayo SL, Tirrell DA. 2002. A designed phenylalanyl-tRNA synthetase variant allows efficient in vivo incorporation of aryl ketone functionality into proteins. *J. Am. Chem. Soc.* 124:5652–53
46. Sharma N, Furter R, Kast P, Tirrell DA. 2000. Efficient introduction of aryl bromide functionality into proteins in vivo. *FEBS Lett.* 467:37–40
47. Döring V, Mootz HD, Nangle LA, Hendrickson TL, de Crécy-Lagard V, et al. 2001. Enlarging the amino acid set of *Escherichia coli* by infiltration of the valine coding pathway. *Science* 292:501–4
48. Link AJ, Vink MK, Agard NJ, Prescher JA, Bertozzi CR, Tirrell DA. 2006. Discovery of aminoacyl-tRNA synthetase activity through cell-surface display of noncanonical amino acids. *Proc. Natl. Acad. Sci. USA* 103:10180–85
49. Yang W, Hendrickson WA, Crouch RJ, Satow Y. 1990. Structure of ribonuclease H phased at 2 A resolution by MAD analysis of the selenomethionyl protein. *Science* 249:1398–405
50. Cirino PC, Tang Y, Takahashi K, Tirrell DA, Arnold FH. 2003. Global incorporation of norleucine in place of methionine in cytochrome P450 BM-3 heme domain increases peroxygenase activity. *Biotechnol. Bioeng.* 83:729–34
51. Montclare JK, Tirrell DA. 2006. Evolving proteins of novel composition. *Angew. Chem. Int. Ed. Engl.* 45:4518–21
52. Bae JH, Rubini M, Jung G, Wiegand G, Seifert MH, et al. 2003. Expansion of the genetic code enables design of a novel "gold" class of green fluorescent proteins. *J. Mol. Biol.* 328:1071–81
53. Budisa N, Rubini M, Bae JH, Weyher E, Wenger W, et al. 2002. Global replacement of tryptophan with aminotryptophans generates non-invasive protein-based optical pH sensors. *Angew. Chem. Int. Ed. Engl.* 41:4066–69
54. Beatty KE, Xie F, Wang Q, Tirrell DA. 2005. Selective dye-labeling of newly synthesized proteins in bacterial cells. *J. Am. Chem. Soc.* 127:14150–51
55. Beatty KE, Liu JC, Xie F, Dieterich DC, Schuman EM, et al. 2006. Fluorescence visualization of newly synthesized proteins in mammalian cells. *Angew. Chem. Int. Ed. Engl.* 45:7364–67
56. Ibba M, Söll D. 2000. Aminoacyl-tRNA synthesis. *Annu. Rev. Biochem.* 69:617–50
57. Ibba M, Söll D. 2004. Aminoacyl-tRNAs: setting the limits of the genetic code. *Genes Dev.* 18:731–38
58. Giege R, Sissler M, Florentz C. 1998. Universal rules and idiosyncratic features in tRNA identity. *Nucleic Acids Res.* 26:5017–35
59. Chin JW. 2006. Modular approaches to expanding the functions of living matter. *Nat. Chem. Biol.* 2:304–11

60. Liu DR, Schultz PG. 1999. Progress toward the evolution of an organism with an expanded genetic code. *Proc. Natl. Acad. Sci. USA* 96:4780–85
61. Wang L, Schultz PG. 2004. Expanding the genetic code. *Angew. Chem. Int. Ed. Engl.* 44:34–66
62. Kwok Y, Wong JT. 1980. Evolutionary relationship between *Halobacterium cutirubrum* and eukaryotes determined by use of aminoacyl-tRNA synthetases as phylogenetic probes. *Can. J. Biochem.* 58:213–18
63. Wang L, Schultz PG. 2001. A general approach for the generation of orthogonal tRNAs. *Chem. Biol.* 8:883–90
64. Santoro SW, Wang L, Herberich B, King DS, Schultz PG. 2002. An efficient system for the evolution of aminoacyl-tRNA synthetase specificity. *Nat. Biotechnol.* 20:1044–48
65. Melancon CE 3rd, Schultz PG. 2009. One plasmid selection system for the rapid evolution of aminoacyl-tRNA synthetases. *Bioorg. Med. Chem. Lett.* 19:3845–47
66. Wang L, Zhang Z, Brock A, Schultz PG. 2003. Addition of the keto functional group to the genetic code of *Escherichia coli*. *Proc. Natl. Acad. Sci. USA* 100:56–61
67. Zhang Z, Smith BA, Wang L, Brock A, Cho C, Schultz PG. 2003. A new strategy for the site-specific modification of proteins in vivo. *Biochemistry* 42:6735–46
68. Zeng H, Xie J, Schultz PG. 2006. Genetic introduction of a diketone-containing amino acid into proteins. *Bioorg. Med. Chem. Lett.* 16:5356–59
69. Zhang Z, Wang L, Brock A, Schultz PG. 2002. The selective incorporation of alkenes into proteins in *Escherichia coli*. *Angew. Chem. Int. Ed. Engl.* 41:2840–42
70. Wang J, Schiller SM, Schultz PG. 2007. A biosynthetic route to dehydroalanine-containing proteins. *Angew. Chem. Int. Ed. Engl.* 46:6849–51
71. Deiters A, Schultz PG. 2005. In vivo incorporation of an alkyne into proteins in *Escherichia coli*. *Bioorg. Med. Chem. Lett.* 15:1521–24
72. Chin JW, Santoro SW, Martin AB, King DS, Wang L, Schultz PG. 2002. Addition of *p*-azido-L-phenylalanine to the genetic code of *Escherichia coli*. *J. Am. Chem. Soc.* 124:9026–27
73. Brustad E, Bushey ML, Lee JW, Groff D, Liu W, Schultz PG. 2008. A genetically encoded boronate-containing amino acid. *Angew. Chem. Int. Ed. Engl.* 47:8220–23
74. Wang L, Brock A, Herberich B, Schultz PG. 2001. Expanding the genetic code of *Escherichia coli*. *Science* 292:498–500
75. Mehl RA, Anderson JC, Santoro SW, Wang L, Martin AB, et al. 2003. Generation of a bacterium with a 21 amino acid genetic code. *J. Am. Chem. Soc.* 125:935–39
76. Schultz KC, Supekova L, Ryu Y, Xie J, Perera R, Schultz PG. 2006. A genetically encoded infrared probe. *J. Am. Chem. Soc.* 128:13984–85
77. Wang L, Xie J, Deniz AA, Schultz PG. 2003. Unnatural amino acid mutagenesis of green fluorescent protein. *J. Org. Chem.* 68:174–76
78. Xie J, Wang L, Wu N, Brock A, Spraggon G, Schultz PG. 2004. The site-specific incorporation of *p*-iodo-L-phenylalanine into proteins for structure determination. *Nat. Biotechnol.* 22:1297–301
79. Cellitti SE, Jones DH, Lagpacan L, Hao X, Zhang Q, et al. 2008. In vivo incorporation of unnatural amino acids to probe structure, dynamics, and ligand binding in a large protein by nuclear magnetic resonance spectroscopy. *J. Am. Chem. Soc.* 130:9268–81
80. Tsao ML, Summerer D, Ryu Y, Schultz PG. 2006. The genetic incorporation of a distance probe into proteins in *Escherichia coli*. *J. Am. Chem. Soc.* 128:4572–73
81. Alfonta L, Zhang Z, Uryu S, Loo JA, Schultz PG. 2003. Site-specific incorporation of a redox-active amino acid into proteins. *J. Am. Chem. Soc.* 125:14662–63
82. Seyedsayamdost MR, Xie J, Chan CT, Schultz PG, Stubbe J. 2007. Site-specific insertion of 3-aminotyrosine into subunit $\alpha 2$ of *E. coli* ribonucleotide reductase: direct evidence for involvement of Y_{730} and Y_{731} in radical propagation. *J. Am. Chem. Soc.* 129:15060–71
83. Sakamoto K, Murayama K, Oki K, Iraha F, Kato-Murayama M, et al. 2009. Genetic encoding of 3-iodo-L-tyrosine in *Escherichia coli* for single-wavelength anomalous dispersion phasing in protein crystallography. *Structure* 17:335–44
84. Wang L, Brock A, Schultz PG. 2002. Adding L-3-(2-Naphthyl)alanine to the genetic code of *E. coli*. *J. Am. Chem. Soc.* 124:1836–37

85. Xie J, Liu W, Schultz PG. 2007. A genetically encoded bidentate, metal-binding amino acid. *Angew. Chem. Int. Ed. Engl.* 46:9239–42
86. Guo J, Wang J, Anderson JC, Schultz PG. 2008. Addition of an alpha-hydroxy acid to the genetic code of bacteria. *Angew. Chem. Int. Ed. Engl.* 47:722–25
87. Lee HS, Spraggon G, Schultz PG, Wang F. 2009. Genetic incorporation of a metal-ion chelating amino acid into proteins as a biophysical probe. *J. Am. Chem. Soc.* 131:2481–83
88. Chin JW, Martin AB, King DS, Wang L, Schultz PG. 2002. Addition of a photocrosslinking amino acid to the genetic code of *Escherichia coli*. *Proc. Natl. Acad. Sci. USA* 99:11020–24
89. Deiters A, Groff D, Ryu Y, Xie J, Schultz PG. 2006. A genetically encoded photocaged tyrosine. *Angew. Chem. Int. Ed. Engl.* 45:2728–31
90. Peters FB, Brock A, Wang J, Schultz PG. 2009. Photocleavage of the polypeptide backbone by 2-nitrophenylalanine. *Chem. Biol.* 16:148–52
91. Bose M, Groff D, Xie J, Brustad E, Schultz PG. 2006. The incorporation of a photoisomerizable amino acid into proteins in *E. coli*. *J. Am. Chem. Soc.* 128:388–89
92. Tippmann EM, Liu W, Summerer D, Mack AV, Schultz PG. 2007. A genetically encoded diazirine photocrosslinker in *Escherichia coli*. *ChemBioChem* 8:2210–14
93. Wang J, Xie J, Schultz PG. 2006. A genetically encoded fluorescent amino acid. *J. Am. Chem. Soc.* 128:8738–39
94. Xie J, Supekova L, Schultz PG. 2007. A genetically encoded metabolically stable analogue of phosphotyrosine in *Escherichia coli*. *ACS Chem. Biol.* 2:474–78
95. Neumann H, Hazen JL, Weinstein J, Mehl RA, Chin JW. 2008. Genetically encoding protein oxidative damage. *J. Am. Chem. Soc.* 130:4028–33
96. Liu CC, Schultz PG. 2006. Recombinant expression of selectively sulfated proteins in *Escherichia coli*. *Nat. Biotechnol.* 24:1436–40
97. Pastrnak M, Magliery T, Schultz PG. 2000. A new orthogonal suppressor tRNA/aminoacyl-tRNA synthetase pair for evolving an organism with an expanded genetic code. *Helv. Chim. Acta* 83:2277–86
98. Liu DR, Magliery TJ, Pastrnak M, Schultz PG. 1997. Engineering a tRNA and aminoacyl-tRNA synthetase for the site-specific incorporation of unnatural amino acids into proteins in vivo. *Proc. Natl. Acad. Sci. USA* 94:10092–97
99. Ohno S, Yokogawa T, Fujii I, Asahara H, Inokuchi H, Nishikawa K. 1998. Co-expression of yeast amber suppressor tRNATyr and tyrosyl-tRNA synthetase in *Escherichia coli*: possibility to expand the genetic code. *J. Biochem.* 124:1065–68
100. Furter R. 1998. Expansion of the genetic code: site-directed *p*-fluoro-phenylalanine incorporation in *Escherichia coli*. *Protein Sci.* 7:419–26
101. Wang L, Magliery TJ, Liu DR, Schultz PG. 2000. A new functional suppressor tRNA/aminoacyl-tRNA synthetase pair for the in vivo incorporation of unnatural amino acids into proteins. *J. Am. Chem. Soc.* 122:5010–11
102. Kowal AK, Kohrer C, RajBhandary UL. 2001. Twenty-first aminoacyl-tRNA synthetase-suppressor tRNA pairs for possible use in site-specific incorporation of amino acid analogues into proteins in eukaryotes and in eubacteria. *Proc. Natl. Acad. Sci. USA* 98:2268–73
103. Santoro SW, Anderson JC, Lakshman V, Schultz PG. 2003. An archaebacteria-derived glutamyl-tRNA synthetase and tRNA pair for unnatural amino acid mutagenesis of proteins in *Escherichia coli*. *Nucleic Acids Res.* 31:6700–9
104. Anderson JC, Wu N, Santoro SW, Lakshman V, King DS, Schultz PG. 2004. An expanded genetic code with a functional quadruplet codon. *Proc. Natl. Acad. Sci. USA* 101:7566–71
105. Anderson JC, Schultz PG. 2003. Adaptation of an orthogonal archaeal leucyl-tRNA and synthetase pair for four-base, amber, and opal suppression. *Biochemistry* 42:9598–608
106. Wang F, Robbins S, Shen W, Schultz PG. 2010. Genetic incorporation of unnatural amino acids into proteins in *Mycobacterium tuberculosis*. *PLoS ONE* 5(2):e9354
107. Krzycki JA. 2005. The direct genetic encoding of pyrrolysine. *Curr. Opin. Microbiol.* 8:706–12
108. Srinivasan G, James CM, Krzycki JA. 2002. Pyrrolysine encoded by UAG in Archaea: charging of a UAG-decoding specialized tRNA. *Science* 296:1459–62

109. Kavran JM, Gundllapalli S, O'Donoghue P, Englert M, Söll D, Steitz TA. 2007. Structure of pyrrolysyl-tRNA synthetase, an archaeal enzyme for genetic code innovation. *Proc. Natl. Acad. Sci. USA* 104:11268–73
110. Woese CR, Olsen GJ, Ibba M, Söll D. 2000. Aminoacyl-tRNA synthetases, the genetic code, and the evolutionary process. *Microbiol. Mol. Biol. Rev.* 64:202–36
111. Blight SK, Larue RC, Mahapatra A, Longstaff DG, Chang E, et al. 2004. Direct charging of tRNA(CUA) with pyrrolysine in vitro and in vivo. *Nature* 431:333–35
112. Yanagisawa T, Ishii R, Fukunaga R, Kobayashi T, Sakamoto K, Yokoyama S. 2008. Multistep engineering of pyrrolysyl-tRNA synthetase to genetically encode N(epsilon)-(o-azidobenzyloxycarbonyl) lysine for site-specific protein modification. *Chem. Biol.* 15:1187–97
113. Nozawa K, O'Donoghue P, Gundllapalli S, Araiso Y, Ishitani R, et al. 2009. Pyrrolysyl-tRNA synthetase-$tRNA^{Pyl}$ structure reveals the molecular basis of orthogonality. *Nature* 457:1163–67
114. Mukai T, Kobayashi T, Hino N, Yanagisawa T, Sakamoto K, Yokoyama S. 2008. Adding l-lysine derivatives to the genetic code of mammalian cells with engineered pyrrolysyl-tRNA synthetases. *Biochem. Biophys. Res. Commun.* 371:818–22
115. Li WT, Mahapatra A, Longstaff DG, Bechtel J, Zhao G, et al. 2009. Specificity of pyrrolysyl-tRNA synthetase for pyrrolysine and pyrrolysine analogs. *J. Mol. Biol.* 385:1156–64
116. Polycarpo CR, Herring S, Berube A, Wood JL, Söll D, Ambrogelly A. 2006. Pyrrolysine analogues as substrates for pyrrolysyl-tRNA synthetase. *FEBS Lett.* 580:6695–700
117. Fekner T, Li X, Lee MM, Chan MK. 2009. A pyrrolysine analogue for protein click chemistry. *Angew. Chem. Int. Ed. Engl.* 48:1633–35
118. Kobayashi T, Yanagisawa T, Sakamoto K, Yokoyama S. 2009. Recognition of non-alpha-amino substrates by pyrrolysyl-tRNA synthetase. *J. Mol. Biol.* 385:1352–60
119. Nguyen DP, Lusic H, Neumann H, Kapadnis PB, Deiters A, Chin JW. 2009. Genetic encoding and labeling of aliphatic azides and alkynes in recombinant proteins via a pyrrolysyl-tRNA synthetase/tRNA(CUA) pair and click chemistry. *J. Am. Chem. Soc.* 131:8720–21
120. Chen PR, Groff D, Guo J, Ou W, Cellitti S, et al. 2009. A facile system for encoding unnatural amino acids in mammalian cells. *Angew. Chem. Int. Ed. Engl.* 48:4052–55
121. Neumann H, Peak-Chew SY, Chin JW. 2008. Genetically encoding N(epsilon)-acetyllysine in recombinant proteins. *Nat. Chem. Biol.* 4:232–34
122. Edwards H, Schimmel P. 1990. A bacterial amber suppressor in *Saccharomyces cerevisiae* is selectively recognized by a bacterial aminoacyl-tRNA synthetase. *Mol. Cell. Biol.* 10:1633–41
123. Chin JW, Cropp TA, Anderson JC, Mukherji M, Zhang Z, Schultz PG. 2003. An expanded eukaryotic genetic code. *Science* 301:964–67
124. Wu N, Deiters A, Cropp TA, King D, Schultz PG. 2004. A genetically encoded photocaged amino acid. *J. Am. Chem. Soc.* 126:14306–7
125. Deiters A, Cropp TA, Mukherji M, Chin JW, Anderson JC, Schultz PG. 2003. Adding amino acids with novel reactivity to the genetic code of *Saccharomyces cerevisiae*. *J. Am. Chem. Soc.* 125:11782–83
126. Lee HS, Guo J, Lemke EA, Dimla RD, Schultz PG. 2009. Genetic incorporation of a small, environmentally sensitive, fluorescent probe into proteins in *Saccharomyces cerevisiae*. *J. Am. Chem. Soc.* 131:12921–23
127. Brustad E, Bushey ML, Brock A, Chittuluru J, Schultz PG. 2008. A promiscuous aminoacyl-tRNA synthetase that incorporates cysteine, methionine, and alanine homologs into proteins. *Bioorg. Med. Chem. Lett.* 18:6004–6
128. Lemke EA, Summerer D, Geierstanger BH, Brittain SM, Schultz PG. 2007. Control of protein phosphorylation with a genetically encoded photocaged amino acid. *Nat. Chem. Biol.* 3:769–72
129. Summerer D, Chen S, Wu N, Deiters A, Chin JW, Schultz PG. 2006. A genetically encoded fluorescent amino acid. *Proc. Natl. Acad. Sci. USA* 103:9785–89
130. Tippmann EM, Schultz PG. 2007. A genetically encoded metallocene containing amino acid. *Tetrahedron* 63:6182–84
131. Young TS, Ahmad I, Brock A, Schultz PG. 2009. Expanding the genetic repertoire of the methylotrophic yeast *Pichia pastoris*. *Biochemistry* 48:2643–53

132. Sakamoto K, Hayashi A, Sakamoto A, Kiga D, Nakayama H, et al. 2002. Site-specific incorporation of an unnatural amino acid into proteins in mammalian cells. *Nucleic Acids Res.* 30:4692–99
133. Liu W, Brock A, Chen S, Chen S, Schultz PG. 2007. Genetic incorporation of unnatural amino acids into proteins in mammalian cells. *Nat. Methods* 4:239–44
134. Wang W, Takimoto JK, Louie GV, Baiga TJ, Noel JP, et al. 2007. Genetically encoding unnatural amino acids for cellular and neuronal studies. *Nat. Neurosci.* 10:1063–72
135. Kwon I, Tirrell DA. 2007. Site-specific incorporation of tryptophan analogues into recombinant proteins in bacterial cells. *J. Am. Chem. Soc.* 129:10431–37
136. Kwon I, Wang P, Tirrell DA. 2006. Design of a bacterial host for site-specific incorporation of *p*-bromophenylalanine into recombinant proteins. *J. Am. Chem. Soc.* 128:11778–83
137. Oki K, Sakamoto K, Kobayashi T, Sasaki HM, Yokoyama S. 2008. Transplantation of a tyrosine editing domain into a tyrosyl-tRNA synthetase variant enhances its specificity for a tyrosine analog. *Proc. Natl. Acad. Sci. USA* 105:13298–303
138. Liu W, Alfonta L, Mack AV, Schultz PG. 2007. Structural basis for the recognition of para-benzoyl-L-phenylalanine by evolved aminoacyl-tRNA synthetases. *Angew. Chem. Int. Ed. Engl.* 46:6073–75
139. Turner JM, Graziano J, Spraggon G, Schultz PG. 2006. Structural plasticity of an aminoacyl-tRNA synthetase active site. *Proc. Natl. Acad. Sci. USA* 103:6483–88
140. Turner JM, Graziano J, Spraggon G, Schultz PG. 2005. Structural characterization of a *p*-acetylphenylalanyl aminoacyl-tRNA synthetase. *J. Am. Chem. Soc.* 127:14976–77
141. Cusack S, Yaremchuk A, Tukalo M. 2000. The 2 A crystal structure of leucyl-tRNA synthetase and its complex with a leucyl-adenylate analogue. *EMBO J.* 19:2351–61
142. Ryu Y, Schultz PG. 2006. Efficient incorporation of unnatural amino acids into proteins in *Escherichia coli*. *Nat. Methods* 3:263–65
143. Bossi L. 1983. Context effects: translation of UAG codon by suppressor tRNA is affected by the sequence following UAG in the message. *J. Mol. Biol.* 164:73–87
144. Taira H, Hohsaka T, Sisido M. 2006. In vitro selection of tRNAs for efficient four-base decoding to incorporate non-natural amino acids into proteins in an *Escherichia coli* cell-free translation system. *Nucleic Acids Res.* 34:1653–62
145. Magliery TJ, Anderson JC, Schultz PG. 2001. Expanding the genetic code: selection of efficient suppressors of four-base codons and identification of "shifty" four-base codons with a library approach in *Escherichia coli*. *J. Mol. Biol.* 307:755–69
146. Kwon I, Kirshenbaum K, Tirrell DA. 2003. Breaking the degeneracy of the genetic code. *J. Am. Chem. Soc.* 125:7512–13
147. Kudla G, Murray AW, Tollervey D, Plotkin JB. 2009. Coding-sequence determinants of gene expression in *Escherichia coli*. *Science* 324:255–58
148. Coleman JR, Papamichail D, Skiena S, Futcher B, Wimmer E, Mueller S. 2008. Virus attenuation by genome-scale changes in codon pair bias. *Science* 320:1784–87
149. Bain JD, Switzer C, Chamberlin AR, Benner SA. 1992. Ribosome-mediated incorporation of a non-standard amino acid into a peptide through expansion of the genetic code. *Nature* 356:537–39
150. Hirao I, Ohtsuki T, Fujiwara T, Mitsui T, Yokogawa T, et al. 2002. An unnatural base pair for incorporating amino acid analogs into proteins. *Nat. Biotechnol.* 20:177–82
151. Hirao I, Harada Y, Kimoto M, Mitsui T, Fujiwara T, Yokoyama S. 2004. A two-unnatural-base-pair system toward the expansion of the genetic code. *J. Am. Chem. Soc.* 126:13298–305
152. Liu CC, Cellitti S, Geierstanger BH, Schultz PG. 2009. Efficient expression of tyrosine-sulfated proteins in *E. coli* using an expanded genetic code. *Nat. Protoc.* 4:1784–89
153. Chen S, Schultz PG, Brock A. 2007. An improved system for the generation and analysis of mutant proteins containing unnatural amino acids in *Saccharomyces cerevisiae*. *J. Mol. Biol.* 371:112–22
154. Wang Q, Wang L. 2008. New methods enabling efficient incorporation of unnatural amino acids in yeast. *J. Am. Chem. Soc.* 130:6066–67
155. Kobayashi T, Nureki O, Ishitani R, Yaremchuk A, Tukalo M, et al. 2003. Structural basis for orthogonal tRNA specificities of tyrosyl-tRNA synthetases for genetic code expansion. *Nat. Struct. Biol.* 10:425–32

156. Guo J, Melancon CE, Lee HS, Groff D, Schultz PG. 2009. Evolution of amber suppressor tRNAs for efficient bacterial production of proteins containing nonnatural amino acids. *Angew. Chem. Int. Ed. Engl.* 48:9148–51
157. Young TS, Ahmad I, Yin JA, Schultz PG. 2009. An enhanced system for unnatural amino acid mutagenesis in *E. coli*. *J. Mol. Biol.* 395:361–74
158. Rackham O, Chin JW. 2005. A network of orthogonal ribosome x mRNA pairs. *Nat. Chem. Biol.* 1:159–66
159. Wang K, Neumann H, Peak-Chew SY, Chin JW. 2007. Evolved orthogonal ribosomes enhance the efficiency of synthetic genetic code expansion. *Nat. Biotechnol.* 25:770–77
160. Doi Y, Ohtsuki T, Shimizu Y, Ueda T, Sisido M. 2007. Elongation factor Tu mutants expand amino acid tolerance of protein biosynthesis system. *J. Am. Chem. Soc.* 129:14458–62
161. Sella G, Ardell DH. 2006. The coevolution of genes and genetic codes: Crick's frozen accident revisited. *J. Mol. Evol.* 63:297–313
162. Vetsigian K, Woese C, Goldenfeld N. 2006. Collective evolution and the genetic code. *Proc. Natl. Acad. Sci. USA* 103:10696–701
163. Lu Y, Freeland S. 2006. On the evolution of the standard amino-acid alphabet. *Genome Biol.* 7:102.1–.6
164. Weber AL, Miller SL. 1981. Reasons for the occurrence of the twenty coded protein amino acids. *J. Mol. Evol.* 17:273–84
165. Ye S, Huber T, Vogel R, Sakmar TP. 2009. FTIR analysis of GPCR activation using azido probes. *Nat. Chem. Biol.* 5:397–99
166. Groff D, Thielges MC, Cellitti S, Schultz PG, Romesberg FE. 2009. Efforts toward the direct experimental characterization of enzyme microenvironments: tyrosine100 in dihydrofolate reductase. *Angew. Chem. Int. Ed. Engl.* 48:3478–81
167. Mills JH, Lee HS, Liu CC, Wang J, Schultz PG. 2009. A genetically encoded direct sensor of antibody-antigen interactions. *ChemBioChem* 10:2162–64
168. Lee HS, Dimla RD, Schultz PG. 2009. Protein-DNA photo-crosslinking with a genetically encoded benzophenone-containing amino acid. *Bioorg. Med. Chem. Lett.* 19:5222–24
169. Hino N, Okazaki Y, Kobayashi T, Hayashi A, Sakamoto K, Yokoyama S. 2005. Protein photo-cross-linking in mammalian cells by site-specific incorporation of a photoreactive amino acid. *Nat. Methods* 2:201–6
170. Grunewald J, Tsao ML, Perera R, Dong L, Niessen F, et al. 2008. Immunochemical termination of self-tolerance. *Proc. Natl. Acad. Sci. USA* 105:11276–80
171. Grunewald J, Hunt GS, Dong L, Niessen F, Wen BG, et al. 2009. Mechanistic studies of the immunochemical termination of self-tolerance with unnatural amino acids. *Proc. Natl. Acad. Sci. USA* 106:4337–42
172. Liu CC, Brustad E, Liu W, Schultz PG. 2007. Crystal structure of a biosynthetic sulfo-hirudin complexed to thrombin. *J. Am. Chem. Soc.* 129:10648–49
173. Liu H, Wang L, Brock A, Wong CH, Schultz PG. 2003. A method for the generation of glycoprotein mimetics. *J. Am. Chem. Soc.* 125:1702–3
174. Brustad EM, Lemke EA, Schultz PG, Deniz AA. 2008. A general and efficient method for the site-specific dual-labeling of proteins for single molecule fluorescence resonance energy transfer. *J. Am. Chem. Soc.* 130:17664–65
175. Ye S, Kohrer C, Huber T, Kazmi M, Sachdev P, et al. 2008. Site-specific incorporation of keto amino acids into functional G protein–coupled receptors using unnatural amino acid mutagenesis. *J. Biol. Chem.* 283:1525–33
176. Guo J, Wang J, Lee JS, Schultz PG. 2008. Site-specific incorporation of methyl- and acetyl-lysine analogues into recombinant proteins. *Angew. Chem. Int. Ed. Engl.* 47:6399–401
177. Deiters A, Cropp TA, Summerer D, Mukherji M, Schultz PG. 2004. Site-specific PEGylation of proteins containing unnatural amino acids. *Bioorg. Med. Chem. Lett.* 14:5743–45
178. Love KR, Swoboda JG, Noren CJ, Walker S. 2006. Enabling glycosyltransferase evolution: a facile substrate-attachment strategy for phage-display enzyme evolution. *ChemBioChem* 7:753–56
179. Liu CC, Mack AV, Tsao ML, Mills JH, Lee HS, et al. 2008. Protein evolution with an expanded genetic code. *Proc. Natl. Acad. Sci. USA* 105:17688–93
180. Liu CC, Mack AV, Brustad E, Mills JH, Groff D, et al. 2009. Evolution of proteins with genetically encoded "chemical warheads." *J. Am. Chem. Soc.* 131:9616–17

181. Liu CC, Choe H, Farzan M, Schultz PG. 2009. Mutagenesis and evolution of sulfated antibodies using an expanded genetic code. *Biochemistry* 48:8891–98

182. Donia MS, Hathaway BJ, Sudek S, Haygood MG, Rosovitz MJ, et al. 2006. Natural combinatorial peptide libraries in cyanobacterial symbionts of marine ascidians. *Nat. Chem. Biol.* 2:729–35

183. Lee HS, Schultz PG. 2008. Biosynthesis of a site-specific DNA cleaving protein. *J. Am. Chem. Soc.* 130:13194–95

Bacterial Nitric Oxide Synthases

Brian R. Crane, Jawahar Sudhamsu, and Bhumit A. Patel

Department of Chemistry and Chemical Biology, Cornell University, Ithaca, New York 14853; email: bc69@cornell.edu

Annu. Rev. Biochem. 2010. 79:445–70

First published online as a Review in Advance on April 6, 2010

The *Annual Review of Biochemistry* is online at biochem.annualreviews.org

This article's doi:
10.1146/annurev-biochem-062608-103436

0066-4154/10/0707-0445$20.00

Key Words

bioptein, catalysis, electron transfer, heme oxygenase, regulation, signaling

Abstract

Nitric oxide synthases (NOSs) are multidomain metalloproteins first identified in mammals as being responsible for the synthesis of the widespread signaling and protective agent nitric oxide (NO). Over the past 10 years, prokaryotic proteins that are homologous to animal NOSs have been identified and characterized, both in terms of enzymology and biological function. Despite some interesting differences in cofactor utilization and redox partners, the bacterial enzymes are in many ways similar to their mammalian NOS (mNOS) counterparts and, as such, have provided insight into the structural and catalytic properties of the NOS family. In particular, spectroscopic studies of thermostable bacterial NOSs have revealed key oxyheme intermediates involved in the oxidation of substrate L-arginine (Arg) to product NO. The biological functions of some bacterial NOSs have only more recently come to light. These studies disclose new roles for NO in biology, such as taking part in toxin biosynthesis, protection against oxidative stress, and regulation of recovery from radiation damage.

Contents

NITRIC OXIDE: A BRIEF OVERVIEW

Nitric oxide (NO) was first identified as a biological product in 1967, when it was shown to be an intermediate of denitrification in the marine bacterium *Pseudomonas perfectomarinus* (1). Denitrification is an important arm of the environmental nitrogen cycle in which bacteria reduce nitrate to N_2O and N_2 as either a means of acquiring energy or balancing the redox state during anaerobic respiration (2, 3). As part of this process, NO is produced by one-electron reduction of nitrite by heme or Cu-containing nitrite reductases and then consumed by further reduction to N_2O by the NO reductases (4, 5). NO is generally regarded as a reactive, highly diffusible entity, long considered a poison and pollutant (6–8). It is not surprising then that even in the early days of denitrification, acceptance of NO as a metabolic intermediate was not immediate (2). The discovery of NO as an essential regulator in mammalian biology came later in the mid-1980s from three convergent lines of research: the identification of NO as endothelium relaxing factor (ERF) in the vascular system, the identification of NO as a key cytotoxic agent of the immune system, and the finding that NO acts as a signaling molecule in the nervous system (9–17). Development in these three, now overlapping areas, roughly coincided with the discovery of three isozymes responsible for the regulated synthesis of NO in mammals, endothelial nitric oxide synthase (eNOS, or NOS-III), inducible NOS (iNOS or NOS-II), and neuronal NOS (nNOS or NOS-I) (9–16). Over the past two decades, studies of NOS have associated NO with a tremendous number of processes in higher organisms including control of vascular tone and blood pressure, protection against pathogens and cancer, hormone regulation, nerve cell transmission, and angiogenesis (9–17). It is only in the past few years that proteins homologous to mammalian NOS (mNOS) have been identified in prokaryotic sources and studied. The characterization of these enzymes has shed new light on mNOS function but also has revealed new roles for NO in microbial systems. We have previously published two smaller reviews of prokaryotic NOSs (18, 19); here, we cover the salient areas in more detail and report on new recent developments regarding the enzymology and biological functions of bacterial NOSs.

NOS: nitric oxide synthase

Arg: L-arginine

Cit: L-citrulline

NOHA: N^{ω}-hydroxy-L-arginine

NOS_{ox}: NOS oxygenase domain

NOS_{red}: NOS reductase domain

Mammalian Nitric Oxide Synthase

NOSs (EC 1.14.13.39) are highly regulated, multidomain metalloenzymes that catalyze the oxidation of L-arginine (Arg) to NO and L-citrulline (Cit) via the stable intermediate N^{ω}-hydroxy-L-arginine (NOHA) (14, 15, 20–24) (**Figure 1*a***). The three mNOS isozymes are homodimers that contain a NOS oxygenase (NOS_{ox}) domain and a C-terminal flavoprotein reductase (NOS_{red}) domain

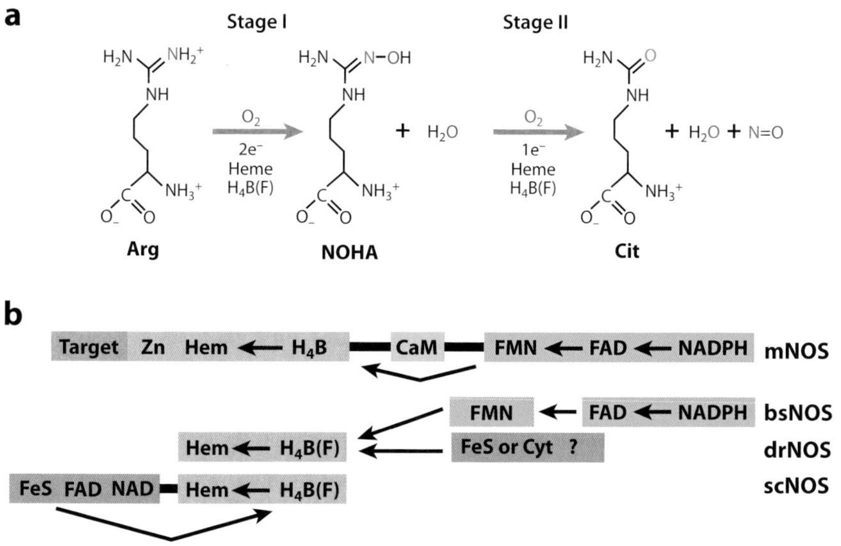

Figure 1

Nitric oxide synthase (NOS) chemistry and protein architechture. (*a*) NOS-catalyzed two-stage oxidation of L-arginine (Arg) to L-citrulline (Cit) and NO via the stable intermediate N^{ω}-hydroxy-L-arginine (NOHA). Three exogenously supplied electrons are required, two for stage I and one for stage II, as well as a heme and a reduced pterin cofactor, tetrahydrobiopterin (H_4B) or tetrahydrofolate (H_4F). (*b*) Domain organizations and redox partners for mammalian NOS (mNOS), and NOSs from *Bacillus subtilis* (bsNOS), *Deinococcus radiodurans* NOS (drNOS) and *Sorangium cellulosom* (scNOS); arrows denote electron flow. Abbreviations: FAD, flavine adenine dinucleotide; FeS, iron-sulfur cluster; FMN, flavine mononucleotide.

(**Figure 1*b***). NOS_{ox} binds Arg, heme, and the redox-active cofactor 6R-tetrahydrobiopterin (H_4B) and contains the catalytic center of the enzyme (**Figures 1** and **2**). NOS_{red} has binding sites for flavine adenine dinucleotide (FAD), flavine mononucleotide (FMN), and NADPH and acts as a source of reducing equivalents for oxygen activation at the heme center (**Figure 1**). A calmodulin interaction sequence connects NOS_{ox} to NOS_{red} and regulates the reduction of NOS_{ox} by NOS_{red} in response to calcium (14, 15, 21–23, 25). The catalytic mechanism of NO formation by mNOS involves two heme-based oxygenation reactions at the same active site, with conversion of Arg to NOHA in stage I, and conversion of NOHA to NO and Cit in stage II (**Figures 1** and **3**) (14, 15, 20–24). Considerable mechanistic work has been performed on the mNOSs, which has established that both stage I and stage II involve heme-based oxygenation reactions with parallels to those reactions carried out by cytochrome P450 (14, 15, 20, 23, 26). In both stage I and II, the NOS ferric heme (**1**) undergoes reduction to Fe^{2+} (**2**) in order to bind molecular oxygen (**3**; **Figure 3**). A second electron, critical for oxygen activation, is delivered to the active center by H_4B in both stages of the reaction (22; and see below), and this involvement of the pterin cofactor provides NOSs with a unique reactivity profile compared to other monoxygenases.

BACTERIAL NITRIC OXIDE SYNTHASE

Investigations of NOS in bacteria began with biochemical attempts to detect NOS activity from various culture isolates. This led to identifications that were less than definitive. Genome sequencing quickly focused efforts on organisms containing putative NOS homologs and resulted in the cloning of genes, followed by the recombinant expression and characterization of NOS-like proteins.

H_4B: tetrahydrobiopterin

FAD: flavine adenine dinucleotide

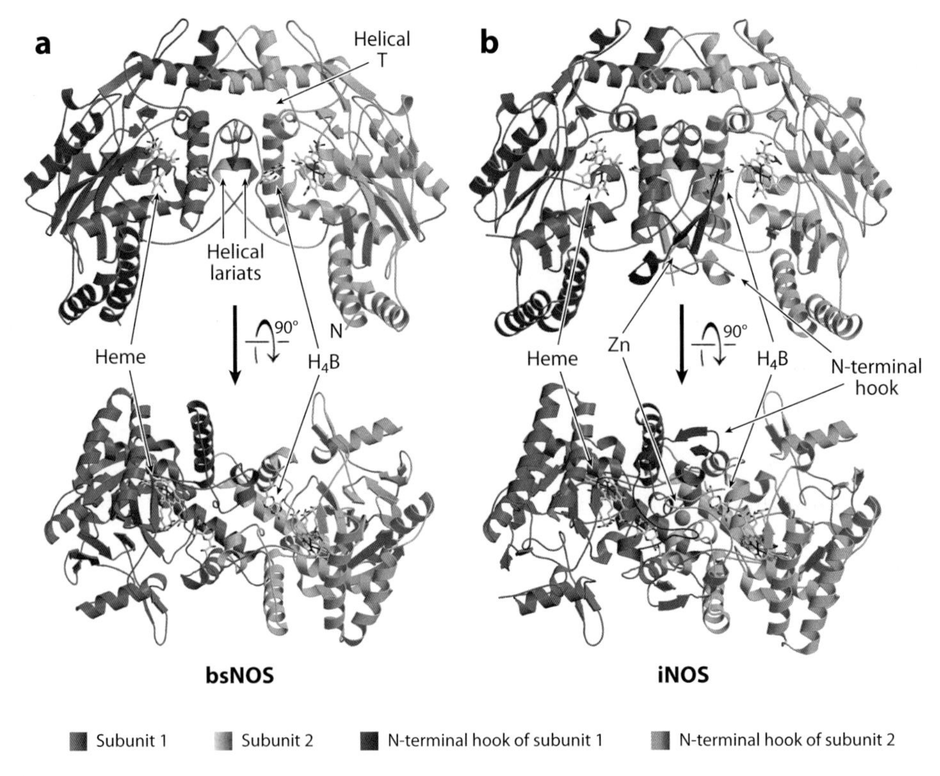

Figure 2

Structures of the N-terminal heme oxygenase (NOS_{ox}) domains of (*a*) *B. subtilis* nitric oxide synthase (bsNOS) [Protein Data Bank (PDB) code: 1M7V)] and (*b*) human inducible nitric oxide synthase (iNOS) with heme, Arg, and tetrahydrobiopterin (H_4B) bound (PDB code: 1NSI). The structures are highly similar with the exception of the missing N-terminal hook and zinc-binding region in bsNOS. Abbreviations: Helical T, secondary structure elements; Zn, zinc site.

Early Biochemistry

The first report of NOS-like activity in a bacterium came from studies on a *Nocardia* species (27). A 52-kDa enzyme was purified by following NADPH oxidation activity in cell lysate and was shown to form Cit from radiolabeled Arg (28). mNOS inhibitors inhibited product formation. However, none of the *Nocardia* species whose genomes have been sequenced to date contain a gene similar to that of the animal NOSs. There are a number of similar reports where several bacterial strains have been demonstrated to have NOS-like enzyme activity, but the associated genomes contain no obvious NOS homolog. For example, in *Lactobacillus fermentum* (29) NO production was

→

Figure 3

Nitric oxide synthase (NOS) catalytic mechanism. (*a*) Initial steps of reduction and oxygen activation common to stage I and stage II of NOS catalysis. (*b*) Stage I and II of NOS catalysis. In stage I, compound I (**6**) is the active species, whereas in stage II it is an Fe^{3+}-hydroperoxo (state **5**). In stage I, two protons are ultimately taken up from bulk solvent (H_3O^+). Conversion of state **5** to state **7**[1] could be highly concerted, and thus the proton transfers shown in these steps, although likely involving substrate, are hypothetical. Abbreviations: NOHA, N^{ω}-hydroxy-L-arginine; H_4B, tetrahydrobiopterin.

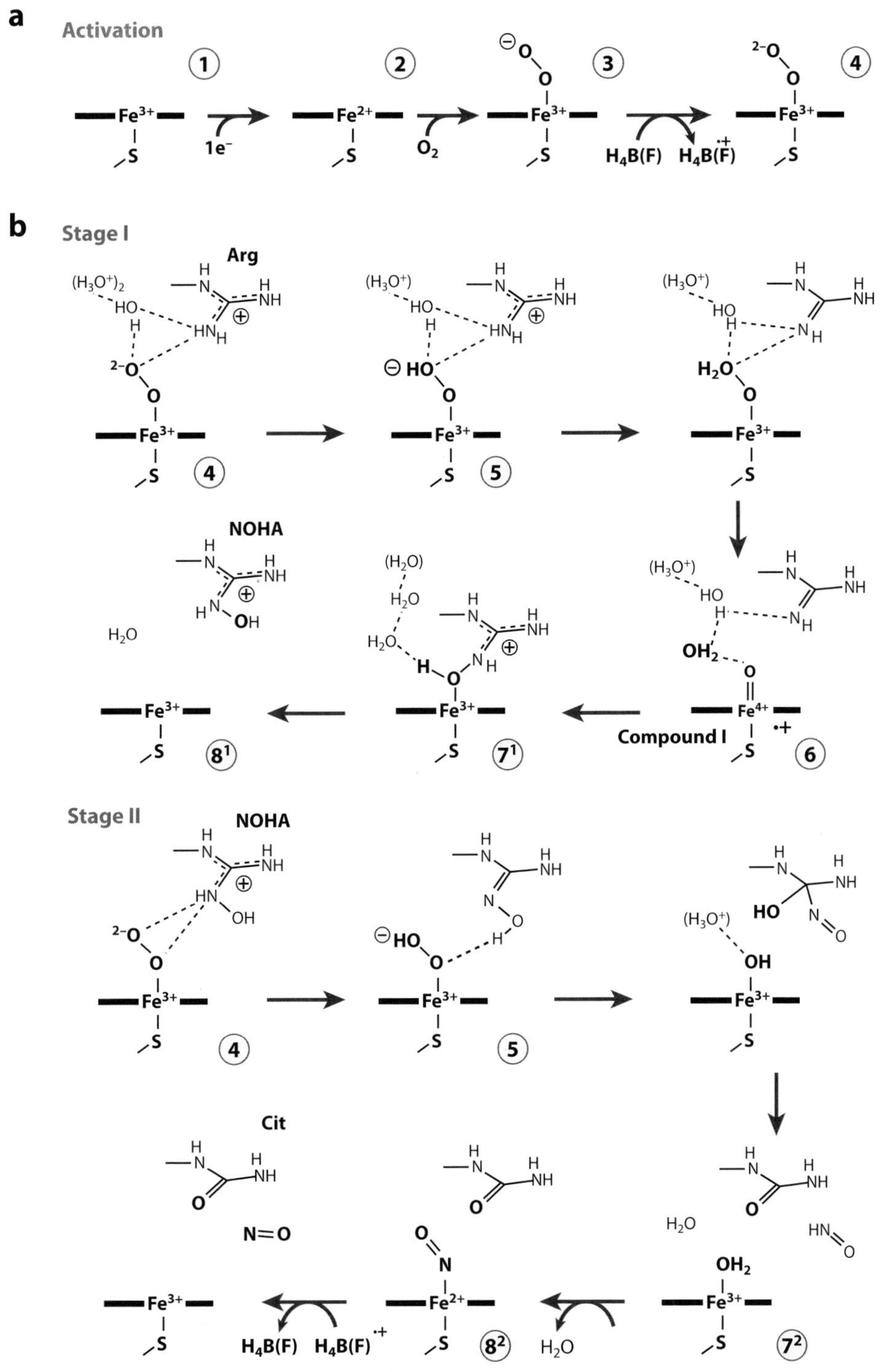

a Activation

① ② ③ ④

Fe^{3+} → (1e⁻) Fe^{2+} → (O_2) Fe^{3+}–O–O^{-} → ($H_4B(F)$ → $H_4B(F)^{\bullet+}$) Fe^{3+}–O–O^{2-}

b Stage I

Arg

$(H_3O^+)_2$ ... ^{2-}O–O–Fe^{3+} ④ → (H_3O^+) ... $\ominus$ HO–O–Fe^{3+} ⑤ → (H_3O^+) ... H_2O–O–Fe^{3+}

NOHA

H_2O ... Fe^{3+} 8^1 ← (H_2O), H_2O, H_2O ... H–O–Fe^{3+} 7^1 ← Compound I (H_3O^+) ... OH_2 ... O=Fe^{4+} $\bullet+$ ⑥

Stage II

NOHA

^{2-}O–O–Fe^{3+} ④ → $\ominus$ HO–O–Fe^{3+} ⑤ → (H_3O^+) HO ... OH–Fe^{3+}

Cit

N=O Fe^{3+} ← ($H_4B(F)$ ← $H_4B(F)^{\bullet+}$) O=N–Fe^{2+} 8^2 ← (H_2O) H_2O, HN=O, OH_2–Fe^{3+} 7^2

NO_x: nitrogen oxides of higher oxidation state than NO

detected by the conversion of metmyoglobin to nitrosylmyoglobin. Nitrite (an oxidized end product of NO formation) was also shown to build up in the culture media. The amount of nitrite formed was increased twofold on addition of the mNOS cofactors NADPH, FAD, FMN and H_4B (29), but ^{15}N nitrite levels only increased by 0.32% when the growth media was supplemented with ^{15}N-Arg. To date, no direct formation of NO by a NOS-like enzyme from this bacterium has been shown, and in addition, none of the *Lactobacillus* species sequenced contain an mNOS homolog. Similarly, formation of citrulline from both Arg and NOHA was detected in cell lysates of *Rhodococcus* sp. *R312* (30). This activity was again inhibited by mNOS inhibitors, and H_4B increased the activity. A *Rhodococcus* protein (~100 kDa) was recognized by a human iNOS antibody but not by antibodies raised against bovine eNOS and rat nNOS. Arg-dependent NO production was also shown in *Rhodococcus* sp. *APG1* using the fluorescein-based fluorophore, DAF (31). The genomes of these bacteria have not been sequenced, although the genome of *Rhodococcus* sp. *RHA1*, does contain a NOS-like sequence. A NOS-like enzyme (~93 kDa) was partially purified from *Salmonella typhimurium* and appeared to produce citrulline and NO (32). But, in this case too, the genome does not contain a NOS homolog. Similar studies have observed NOS-like activities in the protozoans *Entamoeba histolytica* (33) and *Toxoplasma gondii* (34), but again neither of these genomes contain a NOS homolog. A NOS-like enzyme (~64 kDa) was identified from *Staphylococcus aureus* using an mNOS antibody (35), and NOS activity was established. Addition of methylesters of Arg and *N*-nitro-L-arginine to growth media induced protein expression and increased NOS activity (36). This same enzyme, later shown to be a complex of two separate polypeptide chains of 64 kDa and 67 kDa, was purified from cell lysate and characterized to oxidize Arg when mNOS cofactors were supplied (37). It is now well established that *S. aureus* contains an mNOS homolog in its genome (38–40), but the subunit molecular weight is much less than those reported above. When considering these studies, one should keep in mind that bacteria can and do produce NO from a variety of pathways, many of which are not dependent on NOS. Furthermore, NOS-independent chemistry can also convert Arg to citrulline in the urea cycle, either in one step by arginine deaminase or in two steps by argininosuccinate synthetase and lyase (41–43). In summary, without complimentary biochemistry and genetics, it can be difficult to assign NOS-like activity to a true NOS homolog.

Genome Mining

The first definitive evidence for a NOS-like protein in prokaryotes came from genomic sequencing, which revealed bacterial open reading frames coding for proteins with high sequence similarity to $mNOS_{ox}$ (44–46). Conservation of all key catalytic residues among prokaryotic and mNOSs suggested that all prokaryotic NOSs produce nitrogen oxides (NO_x) from Arg and NOHA. These NOS_{ox}-like proteins are mostly found in gram-positive bacteria (e.g., including the orders Bacillales, Actinobacteria, and Deinococcus), although a gram-negative bacterium, a cyanobacterium, and an archeon also contain a NOS sequence (19). It is immediately apparent from the gene sequences that very few of the proteins have a covalently attached reductase domain (for a notable exception, see below). In addition, the proteins lack the N-terminal motif of $mNOS_{ox}$ that coordinates a tetrahedral zinc site on the dimer interface and participates in binding the H_4B side chain through a motif called the N-terminal hook (**Figure 2**) (24, 47). Interestingly, the NOS from *Streptomyces* also has a truncated N terminus compared to mNOS but may contain an N-terminal zinc site, in which one of the two conserved Cys is replaced by a His residue (48). Phylogenetic analysis of the sequences show that they are highly similar, with nearly all heme-binding and active-site residues conserved among the homologs. The *nos* sequences cluster in a manner quite different than that predicted by 16s rRNA analysis from the same organisms (19). This suggests that the NOS genes have been acquired in at least some species by horizontal gene transfer. It is also true that many of the organisms that contain NOS sequences live in soil environments and thus share locality and perhaps similar environmental pressures, which drive gene acquisition by lateral transfer.

Cloning and Enzymology

The first NOS-like protein to be cloned, expressed in *Escherichia coli* and purified, was that from the radiation-resistant bacteria *Deinococcus radiodurans* (drNOS) (49). The protein was shown to contain heme, form a stable dimer, and produce liganded states with ultraviolet-visable spectroscopy (UV/Vis) properties very similar to those of $nNOS_{ox}$ (49). drNOS produced nitrite/nitrate from Arg with a surrogate mammalian reductase domain (49) and was dependent on either H_4B or the closely related pterin, tetrahydrofolate (H_4F), which both bind in the 10–20 μM range (49, 50). Owing to its redox role in oxygen activation, H_4B accelerates the decay rate of the ferrous oxy species (**3**, **Figure 3**); in nNOS, the acceleration is 60-fold, and in drNOS, the acceleration is three-fold relative to the pterin-free enzyme (49, 50). However, with H_4F, the decay of **3** in drNOS increases ~4.5-fold compared to the pterin-free enzyme (49, 50) and increases 25-fold compared to drNOS loaded with the redox-inactive folate H_2F (51). Thus, occupying the pterin site intrinsically stabilizes state **3** (**Figure 3**) in drNOS, and H_4F is a more facile donor to the heme center than H_4B (51). The NOS from *Bacillus subtilis* (bsNOS) was first demonstrated to quantitatively produce NOHA and NO in a $H_4B(F)$-dependent fashion (50). NO has also subsequently been observed as a bona fide product of both in vitro and in vivo reactions of other bacterial NOSs (51–56). Once NO is formed, it binds the heme to form an inhibitory Fe^{3+}-NO complex; the dissociation rate of this species is one of the major factors that determines net NO production by the enzyme (57, 58). One significant difference between $mNOS_{ox}$ and bsNOS is a 15–25 times slower NO dissociation in the bacterial protein (50). This sets a slower steady-state flux of NO from the heme and provides more time for heme reduction—and eventual oxidation of the resulting Fe^{2+}-NO to higher NO_xs, such as peroxynitrate and nitrate (57, 58). Subsequent characterization of *S. aureus* NOS (saNOS) (38, 59, 60), *Bacillus anthrasis* NOS (baNOS) (60, 61), *Streptomyces turgidiscabies* NOS (stNOS) (48), and *Geobacillus stearothermophilis* NOS (gsNOS) (52) revealed properties similar to drNOS and bsNOS, but also some interesting differences, as discussed below.

H_4F: tetrahydrofolate

Spectroscopy

Like their mammalian counterparts, the heme centers of bacterial NOSs have been investigated by a number of spectroscopies, including UV/Vis, resonance Raman (RR), Fourier-transform infrared (FTIR), electron paramagnetic resonance (EPR), and electron-nuclear double resonance (ENDOR) spectroscopies (the latter is discussed below in the context of a catalytic mechanism). Initial UV/Vis spectral characterization of overexpressed drNOS (49), saNOS (59, 60), bsNOS (50, 62), and baNOS (60, 61) showed that the proteins have properties typical of $mNOS_{ox}$, such as a five-coordinate thiolate-ligated high-spin ferric heme iron (Fe^{3+}) in the presence of substrate Arg and a low-spin six-coordinate heme that is very similar to $mNOS_{ox}$ in the presence of heme ligands, such as imidazole, NO, CO, O_2, DTT. There are some differences among the bacterial NOSs in the optical properties of the substrate-free (sf) enzymes, which probably indicate altered active-site solvation. sf-saNOS has a six-coordinate low-spin or a high-spin/low-spin-mixed electronic configuration (59, 60), which suggests that a water molecule is coordinated to the heme iron on the side opposite of the thiolate ligand; whereas sf-bsNOS has a five-coordinate high-spin heme, which suggests that water does not interact as strongly in the distal ligand-binding site (62). Vibrational spectroscopy (RR and FTIR) of diatomic heme ligands bound to the open coordination site has been useful in probing structure and reactivity of the heme center. In particular Fe-CO and FeC-O stretching frequencies (ν) report on the electron-donating ability of the heme- and hydrogen-bonding environment in the distal pocket. Similar to mNOS (63, 64), RR studies of the CO adducts of both bsNOS (62) and saNOS (59) found two different Fe-CO stretching modes (νFe-CO) in each enzyme. Arg binding merges the two νFe-CO modes into a single mode, with a frequency similar to that observed in nNOS (59, 62). Thus, substrate interacts closely with heme ligands and with binding causes two conformers of the Fe-CO moiety to become one.

In thermostable gsNOS, two active-site conformations were also revealed by two distinct ν(Fe-CO)/ν(C-O) ratios (65). One set of resonances was characteristic of that found for mNOS, whereas the other resembled that found in bsNOS. Arg stabilizes the mNOS-like conformation, and NOHA stabilizes the bsNOS-like conformation. Furthermore, Arg decreases the ν(Fe-Cys) stretch, whereas NOHA increases ν(Fe-Cys). Thus, Arg and NOHA interact differently with heme ligands and also produce different perturbations to the proximal Fe-Cys bond. The latter likely results from substrate-induced modulation of a hydrogen bond between the proximal Cys and a conserved Trp indole nitrogen (**Figure 4**) that is known to influence electronics of the mNOS heme center (65). This ability of the proximal Trp to reduce the basicity of the axial Cys was confirmed in saNOS, where mutation of this residue to nonhydrogen-donating residues reduced the νFe-NO and νFe-CO frequencies (66).

Despite their overall similarities, there is a range of active-site properties among the bacterial NOSs, not unlike that observed in mNOS isozymes. In addition to sf-saNOS assuming a hexacoordinated low-spin ferric state (59, 60) and native bsNOS (62) and baNOS being mainly high spin, H_4B and related pterins show different degrees of cooperativity with substrate binding for the different bacterial NOSs. Pterins increase Arg affinity of saNOS but not of baNOS (60). Furthermore, saNOS can bind large ligands, such as nitrosoalkanes and tert-butylisocyanide, whereas baNOS cannot (60). Thus, saNOS exhibits properties very similar to those of the oxygenase domain of iNOS, where subunit association reflects heme center accessibility and is coupled with pterin- and ligand-binding properties (59, 60, 67). In contrast, bsNOS and baNOS have properties typical of a NOS loaded with pterin, even in their cofactor-free state (60, 62).

Structure

The first crystal structure of a bacterial NOS was determined for bsNOS (68) and then soon after for saNOS (38) and subsequently for the thermostable gsNOS (52). All proteins are similar in structure and strongly resemble $mNOS_{ox}$ (47, 69–71), with the notable absence of the N-terminal hook and tetrahedral zinc center, which participate in binding the dihydroxylpropyl side chain of H_4B. Otherwise, like mNOS, the heme is cradled in a "catcher's mitt" fold, with a winged β-sheet and a helix-turn motif that frame the active center deep within each subunit (**Figure 2**). The dimer interface is primarily composed of a hydrophobic "helical lariat," which forms the pterin-binding site, surrounded by two perpendicular long helices arranged in a "T" shape; the top two helices symmetrically associate across the two subunits (**Figure 2**). A conserved cysteine thiolate binds the heme on the proximal side, where a conserved Trp residue stacks with the heme and hydrogen bonds to the thiolate sulfur through its indole nitrogen (**Figure 4**).

Substrate Recognition

Like in mNOSs, substrates Arg and NOHA both bind with their guanindinium and hydroxyguanidinium groups projected into the back of a mostly hydrophobic pocket over the heme and the amino acid moieties held at the mouth of the cavity through interactions with a number of hydrophilic side chains (**Figure 4**). A highly conserved glutamate residue anchors the Arg guanidinium ring over the heme adjacent to the open coordination site of the heme iron. The

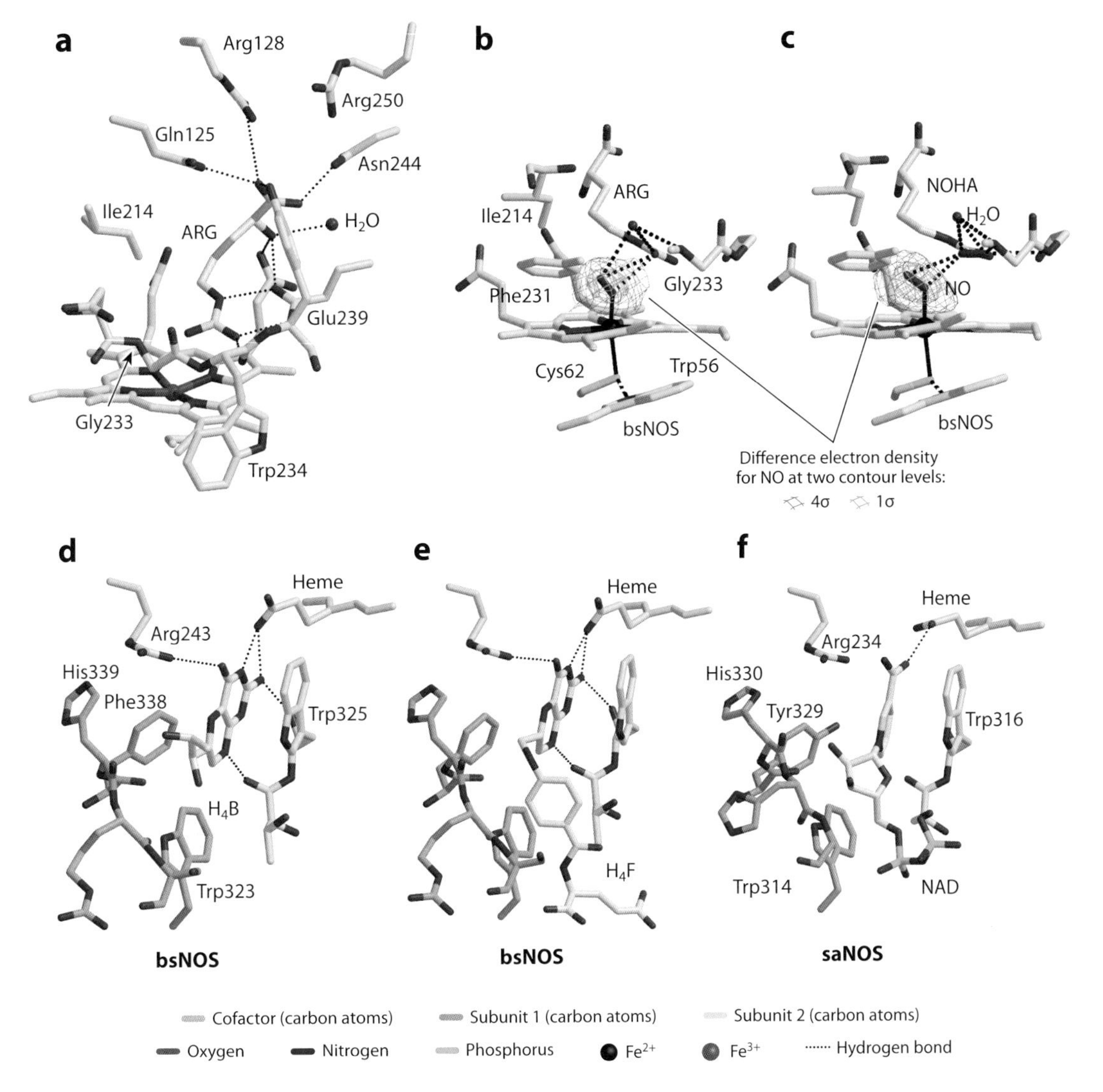

Figure 4

Substrate- and cofactor-binding sites of bacterial nitric oxide synthase (NOS). (*a*) Substrate L-arginine (Arg) bound to *Bacillus subtilis* NOS (bsNOS, PDB code 1M7V). Hydrogen bonds (*black dotted lines*) between protein and substrate are very similar for Arg and N^{ω}-hydroxy-L-arginine (NOHA), with key interactions of the guanidinium group involving a conserved Glu, the protein backbone, and a heme carboxylate. An Ile residue, normally a Val in mammalian NOS, juxtaposes the substrate and partially occludes the heme coordination site from the active center channel. (*b*, *c*) Fe^{2+}-nitrosyl complexes of (*b*) bsNOS with Arg (PDB code 2FC1); (*c*) bsNOS with NOHA (PDB code 2FBZ). Diatomic ligands, such as nitric oxide (NO) and O_2, coordinate the heme iron and interact directly with the substrate guanidinium group, as well as with the active center water. A conserved Trp hydrogen bonds to the proximal heme thiolate and thereby modulates the electronic properties of the heme center. Cofactor-binding sites of (*d*) bsNOS with tetrahydrobiopterin (H_4B, PDB code 1M7V), (*e*) bsNOS with tetrahydrofolate (H_4F, PDB code 1M7Z) and (*f*) *S. aureus* NOS (saNOS) with NAD (PDB code 1MJT). Cofactors (*gray*) are bound at the interface between two subunits (*orange* and *yellow*) and hydrogen bond with a heme carboxylate.

Arg guanidinium group also hydrogen bonds to one of the heme carboxylate groups and the backbone carbonyl of a conserved Trp, whose side chain forms the back of the heme pocket. In the case of NOHA, the hydroxyl group extends toward the back of the pocket and hydrogen bonds with the amino nitrogen of a conserved Gly residue immediately N-terminal to the Trp, mentioned above. The pyramidalization of the NOHA N^{ω}-nitrogen indicates its protonation in the active center. The final product citrulline also sits above the heme of gsNOS much like Arg and NOHA, with the two urea nitrogens hydrogen bonding to the conserved Glu and the urea oxygen projected over the heme in a position similar to the substrate nitrogen of Arg (72). Interestingly, low-temperature EPR/ENDOR studies of gsNOS suggest a different configuration for Cit, with the terminal nitrogen coordinated to the ferric heme iron (73).

Cofactor Binding

The pterin cofactor-binding site is also similar among bacterial and mNOSs, with both subunits of the dimer providing key interactions in the form of hydrophobic side chains that stack on both sides of the bound cofactor ring (**Figure 4**). BsNOS structures contain either H_4B or H_4F (68), whereas saNOS was crystallized with the nicotinamide ring of NAD bound in the cofactor site (38). In this latter case, NAD was an additive that facilitated crystallization and probably does not represent a biologically relevant cofactor. The cofactor-devoid structures of gsNOS (and bsNOS) are very similar to those with cofactors bound, indicating that these pockets are largely preformed. Again, the absence of the N-terminal hook found in mNOS removes interactions to the dihydroxypropyl side chain of H_4B and opens up the active-site pocket in the bacterial enzymes. Importantly, the 3,4 amide group of both H_4B and H_4F hydrogen bonds directly with the heme carboxylate, an interaction thought to be key that allows these cofactors to donate electrons to the heme. H_4F has a much larger glutamyl *p*-amino benzoic acid (pABA) side chain than the dihydroxylpropyl side chain of H_4B, and in the structure of H_4F-bound bsNOS, the pABA side chain resides at the surface of the cofactor pocket and is not well ordered. In mNOS, the pABA side chain would severely clash with the N-terminal hook, explaining why only the bacterial enzymes can utilize this cofactor (49). Although most of the prokaryotes that have a *nos* gene also contain some H_4B biosynthesis genes as a part of the folate biosynthesis pathway, only the *Bacillus* genus has an obvious homolog of sepiapterin reductase, the enzyme needed for the final step in H_4B biosynthesis. In contrast, all organisms, including *D. radiodurans*, produce H_4F. Furthermore, drNOS synthesizes NO more efficiently from H_4F than from H_4B (51). These considerations have led to the suggestion that H_4F is perhaps the natural cofactor drNOS (49, 51) but not necessarily for bsNOS, which does not appear to provide strong, specific interactions for the pABA side chain (68) and binds H_4B more tightly (50).

Reductase Partners

Early studies of drNOS demonstrated that NO synthesis could be supported by a surrogate mammalian reductase domain supplied in *trans* (49). This suggested that, although not covalently attached, the reductase partners of bacterial NOSs may be similar to $mNOS_{red}$. In support of this, the bsNOS structure conserves surface features surrounding an exposed edge of the heme, which has been implicated in electron transfer between NOS_{red} and NOS_{ox} (68). Flavin-containing reductases from *B. subtilis* were tested for their ability to reduce bsNOS in vitro. One such FMN containing flavodoxin, YkuN, was shown to donate electrons to bsNOS and support NO synthesis with activity and heme-reduction rates that rivaled mNOS (74). The multiflavin-containing reductases of *B. subtilis* (including sulfite reductase) were unable to act nearly as effectively as reductase partners, which goes against the supposition that $mNOS_{red}$-related sulfite reductases are redox partners for bacterial NOSs (75). However, when the *ykuN* gene is deleted, *B. subtilis* continues to show behavior associated with

NOS function (54). Furthermore, bacterial NOSs expressed in *E. coli* produce NO in vivo, presumably utilizing *E. coli* reductases (54). The *nos* gene in certain *Streptomyces* strains lies on a pathogenicity island that transfers among species, yet the *nos* gene contains no obvious reductase partner for NOS (48). Also, *D. radiodurans* contains no obvious flavodoxin-like homologs, despite NOS-dependent production of NO (56). Thus, it appears that many different types of reductase proteins can support NO synthesis and that, in the case of *B. subtilis*, a dedicated reductase may not be required at all (54). Some caution should be taken here as the biological roles of NOS have probably not been fully explored in any organism, and certain functions may in fact require dedicated reductase proteins.

There is one exception to the stand-alone bacterial NOS proteins, and that is a *nos* sequence found in the genome of a gram-negative bacterium, *Sorangium cellulosum* (76). The *S. cellulosum* NOS (scNOS) contains a covalently attached reductase module. However, the domain organization and cofactor complement in scNOS is unlike that found in other NOSs (**Figure 1*b***). The first ~450 amino acids of scNOS (1163 residues) compose a domain of unpredicted function, followed by an Fe_2S_2 cluster, a FAD-binding motif, an NAD-binding motif, and finally a C-terminal NOS_{ox}. Thus, compared to mNOS, the reductase domain of scNOS is N-terminal, rather than C-terminal, to NOS_{ox}, and the flavodoxin module in mNOS is replaced by an FeS cluster, which is also capable of one-electron chemistry. The FAD-binding ferredoxin reductase module appears to be conserved by scNOS, allowing NADPH to be the ultimate electron donor. A mixing and matching of redox partners is a common feature of P450 systems, and it appears that functional NOSs are assembled with a similar strategy (**Figure 2**) (23).

Dimerization Properties

The ability of bacterial NOSs to form stable dimers in the absence of the N-terminal hook can be explained by residue substitutions at the dimer interface (77). The mNOS isozymes have very different dimer stabilities and degrees of cooperativity among subunit association, substrate binding, and pterin binding (67). iNOS forms the weakest dimer, and in cells, an active dimeric iNOS depends on H_4B incorporation (14, 67). Recombinant expression of $iNOS_{ox}$ revealed a "loose" dimer state in the absence of pterin, with perturbed heme electronic properties and heightened sensitivity to proteolysis (78). In particular, the heme center is more solvent exposed, which implies a lower heme redox potential. This latter feature may have physiological significance as the lower potential prevents heme reduction and hence limits reaction with oxygen, which could lead to oxidative damage. Insight into the structure of the NOS loose dimer state has come from the crystallographic characterization of bsNOS in an unusual, alternative subunit association mode (77). This bsNOS loose dimer is expanded relative to the normal tight dimer, with the heme irons ~10 Å further apart. The subunits still associate via helical T structures that border the cofactor sites, but the helical lariat, which binds the pterin, is highly disordered, rendering the heme center and substrate-binding site much more exposed. Addition of substrates and cofactors converts the loose dimer to a tight dimer in both solution and the crystal, thereby demonstrating cooperativity among substrate recognition, pterin binding, and subunit association, which is also displayed by iNOS (67). It is interesting that the bsNOS subunits remain associated in the loose dimer, despite substantial disorder of the hydrophobic core at the center of the dimer interface. Under limiting cofactor and substrate conditions, the prokaryotic NOSs could have perturbed properties caused by these large-scale structural changes.

CATALYTIC MECHANISM

The catalytic mechanism of mNOS has attracted tremendous interest owing to its chemical complexity, cofactor requirements, distinguishing features from other heme-based

Cpd I: compound I

oxidation reactions, and high biological importance. As the mammalian and bacterial NOSs are very similar in their structure and chemical properties, this work provides a solid foundation for understanding bacterial NOS reactivity; the reader is referred to a number of excellent reviews on the subject (15, 20–24, 79–81).

By and large, the data indicate that bacterial NOS and mNOS have analogous reactivities. Despite subtle difference in heme electronic states and ligand stabilities, characterized bacterial NOSs produce NO from Arg via NOHA in a manner dependent on H_4B or H_4F and an external reductase. Bacterial NOSs will also produce nitrate/nitrite from Arg and peroxide with an activity that is roughly equivalent to H_4B-free mNOS (38, 82–84). Structurally, the catalytic centers are also nearly identical, with the exception perhaps of the Val → Ile substitution adjacent to the heme iron that is found in nearly all bacterial NOS sequences (**Figure 4**) (68, 85). Thus, what we know about the mNOS catalytic mechanism applies to that of bacterial NOS, and conversely, what we learn from bacterial NOSs regarding NO synthesis likely applies to the mammalian enzymes. This is fortunate as the bacterial NOSs offer some advantages for studying the reaction mechanism: Many are well expressed in *E. coli* and readily purified; they form stable, proteolysis-resistant dimers in the absence of pterin cofactors; they produce crystals that diffract to high resolution; and in at least one case, they can be derived from thermophilic organisms and hence provide hightened stability and slow reaction kinetics at ambient temperatures.

These qualities have been important for a series of recently reported EPR/ENDOR cryoreduction/annealing experiments on the NOS from gsNOS aimed at elucidating intermediates in the catalytic cycle. gsNOS forms an especially stable oxy complex (state **3**, **Figure 3**) compared to many eukaryotic NOSs (52). The stability of state **3** makes gsNOS a favorable subject for cryoreduction EPR/ENDOR studies, which have been instrumental in revealing catalytic intermediates important for substrate transformations in cytochrome P450 and heme oxygenase (86–88). In these experiments, oxy- Fe^{2+} NOS is prepared at low temperature, flash cooled to 77 K, and then further reduced through solvent photolysis by a γ-radiation source, such as ^{60}Co. Subsequent temperature step annealing allows for the accumulation of enzymatic intermediates and product states for EPR/ENDOR analysis, which together provide signature spectra that distinguish heme-oxy redox states and also give positional information of ^1H and ^{15}N nuclei relative to the spin centers. These experiments and others are discussed below in the context of our current understanding of the stage I and II mechanisms.

Stage I

The hydroxylation of Arg is thought to proceed via a P-450-type monoxygenation of Arg (14, 20, 26), where O_2 binds the reduced heme to form an Fe^{2+}-O_2 complex that is best represented as a ferric-iron superoxy state [Fe^{3+}-O_2^-, **Figure 3**, (**3**)] (89). The second electron for catalysis is delivered by H_4B to produce a ferric-iron peroxo state [Fe^{3+}-O_2^-, **Figure 3**, (**4**)] that was observed in low-temperature cryoreduction experiments of eNOS (90). The heme center then hydroxylates Arg, presumably through conversion to the iron-oxo, porphyrin radical state [Fe^{4+} = O $Por^{\cdot+}$, **Figure 3**, (**6**)] known as compound I (Cpd I). In mNOS, the H_4B^+ radical is then rereduced by the reductase domain. Support for a Cpd I intermediate in stage I draws largely from analogies to P450 chemistry. The assignment of Cpd I as the reactive species responsible for converting Arg to NOHA also is consistent with a recent single-turnover experiment of an $mNOS_{ox}$ variant, in which hydrogen bonding is altered to the proximal Cys by substituting the indole of Trp for the imidazole of histidine (91). This variant forms a heme-oxygen intermediate on the same timescale as H_4B oxidation that has some spectral features characteristic of state **6**. However, a few observations argue against Cpd I in stage I, which has promoted alternative mechanisms to be proposed (79).

Unlike P450s, shunting mNOS with peroxide, which should direct conversion to state **6**, produces little (82, 83) or no (92) NOHA, and reaction with iodosobenzene, an oxo donor, also does not produce a hydroxylated product, as it does with P450s (92). Work with alternative substrates demonstrates that the ability of substrate to provide a proton is key for effective turnover. For example, when amidine replaces the guanidino group, there is no longer a proton source on the substrate, and the enzyme carries out a heme-oxygenase-like reaction that hydroxylates the heme (93). Finally, quantum mechanics/molecular mechanics (QM/MM) calculations do support Cpd I as the active intermediate in stage I (94, 95).

Initial radiolytic reduction of oxy eNOS oxygenase domain both with bound Arg in the presence of 4-amino tetrahydropterin (an inactive H_4B analog) produced a peroxyferri NOS intermediate that then converted upon annealing to a product with EPR spectra characteristic of NOHA coordinated to heme (90). Unlike reactions with cytochrome P450 and heme oxygenase, one does not observe a hydroperoxo species (**5**), a key intermediate in these reactions. Because of its stability and slower catalytic profile, a number of additional states were observed with gsNOS (**Figure 3**) (73). Cryoreduction of state **3** at 77 K yields not state **4**, but rather the Fe^{3+}-hydroperoxo state **5**, which then converts to a product state at a very low temperature of 145 K in which NOHA, confirmed by product analysis, is bound with its guanidinium hydroxyl coordinated to the ferric iron ($\mathbf{7^1}$, **Figure 3**). NOHA does not coordinate the heme when complexed with the ferric enzyme, and thus, this observed nonequilibirum geometry must result from oxo insertion into the guanidine group from a Cpd I–type species (**6**). This step proceeds with a large solvent isotope effect and substantial pH dependence, which indicates protonation of state **5** upon its conversion to state **6**. Given the low temperature of the conversion of states **5** → **6**, the immediate proton source is likely local to the Arg-oxyheme complex and, given the pH dependence, probably involves either protein or ordered solvent. Product NOHA is protonated when bound to NOS, and calculations suggest that, although Arg may assist in formation of state **6**, it likely does not lose a net proton during hydroxylation (94).

CN-orn: L-cyanoornithine

Stage II

The oxidation of NOHA to Cit and NO shares features with another reaction catalyzed by cytochrome P450s: the oxidative deformylation of female steroid hormones by P450 aromatase (**Figure 3**) (96). A number of lines of evidence have implicated a nucleophilic ferrous-iron peroxo species (**4**) in this reaction, and such a species has been proposed to act as the active intermediate in stage II (13, 20, 77, 97). Oxygen activation proceeds as in stage I, with electron donation from H_4B generating the Fe^{2+}-$O_2{}^{2-}$ state. Rather than decomposing to Cpd I, it is thought that state **4** reacts directly with NOHA to produce Cit and NO. This reaction can be shunted with peroxide; however, peroxide favors formation of a side product in addition to Cit: L-cyanoornithine (CN-orn) (98, 99). The proportion of CN-orn to Cit correlates with perturbations that affect conversion of states **5** → **6** (**Figure 3**). In particular, residue substitutions that increase peroxidase activity (hence formation of Cpd I, state **6**) generate more CN-orn, whereas changes expected to stabilize the hydroperoxo (state **5**) (metal-substituted porphyrins and alternative substrates) favor Cit (99). A key observation associated with stage II is need for H_4B oxidation and thus formation of state **4** as an early intermediate, despite the fact that the active center is now one electron reduced compared to the final product state (100). At the end of the normal mNOS cycle, the H_4B^+ radical is rereduced by the heme center, perhaps by the Fe^{2+}-NO species, which rapidly forms after Cit oxidation (22). Direct evidence for the active oxygen species in stage II was lacking from mNOS, and again, work with thermostable bacterial NOSs has provided important insight in this regard.

In cryoreduction experiments with gsNOS, state **4**, not state **5**, forms at 77 K and then converts to state **5** at 160 K with a large solvent isotope effect but without pH dependence. This implies that a proton transfer is involved in the rate-determining step but that this proton comes, not from solvent, but directly from NOHA, which is protonated on N^{ω} when bound in the NOS active center. State **5** then converts to a product state with H_2O bound to low-spin Fe^{3+} heme, which then produces an Fe^{2+}-NO complex. Unlike in stage I, where product coordinates the heme, water is found coordinated to Fe^{3+} after the initial reaction (**7**^{2}, **Figure 3**), and this species then leads to formation of Fe^{2+}-NO (**8**^{2}). Water coordination indicates that state **5** is the reactive species for stage II and not state **6 (Figure 3**). In stage II, the active site requires only one net exogenous electron, taken up on formation of state **2** at the beginning of the reaction. Nonetheless, in the course of the catalytic process, the heme-oxy complex (**3**) must be reduced to the peroxo state by a second electron for products to form. This explains why, in single-turnover studies of gsNOS (and mNOSs) with NOHA, H_4B is required to generate NO (52). Overreduction of the active center ultimately causes the primary product of NOHA oxidation to be HNO/NO^-. However, under physiological conditions, the second electron comes from the reduced pterin, and the resulting pterin radical then oxidizes HNO/NO^- to produce NO as the final product.

These experiments demonstrate the ability of the substrate to control the reactive oxidizing species. The arrangement and delivery of protons is key to this process and depends upon which substrate (Arg or NOHA) is bound. Importantly, in stage I, NOS facilitates proton delivery to state **4**, favoring formation of Cpd I (**6**), but in stage II, NOS protects state **4** from initial solvent-mediated protonation, allowing the proton to be derived from NOHA itself. It is interesting that further reaction of state **5** with NOHA in the cryoreduction experiment of gsNOS is gated by proton transfer from NOHA to state **4**. This reaction has been thought to involve a nucleophilic attack of the heme-peroxo (**4**) on the Arg guanidinium (14, 15, 20, 79, 97, 98, 101), and indeed, reducing the electron density on NOHA by fluorination increases the rate of the reaction (99). But proton transfer from NOHA to state **4** reduces the nucleophilicity of state **4**. Thus, the reaction is perhaps better viewed as an electrophilic attack of hydroperoxide on the oxime-like moiety of the hydroxyguanidinium. Evidence for participation of electrophilic hydroperoxo is mounting in P450 chemistry in reactions involving sulfoxidation, alkene epoxidation, and hydroxylation of cyclopropyl derivatives (96). The reason fluorinated NOHA produces more Cit than CN-orn, with the peroxide shunt, could be related to the shifted pK_a of the fluorinated substrate (102) and its reduced tendency to protonate **5** and favor conversion to **6**. Interestingly, when Fe^{3+}-NO bsNOS was bound to NOHA, crystallographic studies of the complex indicated some nitrosation of NOHA, which may reflect the tendency of NOHA to react with heme-bound electrophiles in the NOS active center (72).

As an indication of the perturbations to oxy-NOS that could be incurred by switching Arg for NOHA, crystal structures of Fe^{2+}-NO bsNOS with Arg and NOHA show that hydrogen bonds among the two substrates, heme-bound NO, and associated water molecules are significantly different (**Figure 4**). Arg hydrogen bonds more strongly with the peripheral oxygen of bound NO, whereas protonated NOHA interacts more closely with the proximal nitrogen (72). If a similar effect is seen with the oxy complex, it may facilitate further protonation and breakdown of state **5** to state **6** with Arg as well as stabilization of state **5** with NOHA. RR studies on saNOS provide support for differential H bonding to the distal and proximal oxygen atoms of liganded O_2 in the Arg and NOHA complexes (40, 103). In comparing the oxy complexes of Arg and NOHA (**3**) to sf oxy, the O-O bond is weaker and the Fe-O bond is unaffected in the presence of Arg, whereas the Fe-O bond is stronger and the O-O bond unaffected in the presence of NOHA (40, 103).

Given the above discussion, the source and movement of protons in the NOS active center is an important unresolved issue in the NOS mechanism. Potential ionizable groups that could construct a proton transfer network with substrate and active-site water include the heme carboxylates and the converved Glu residue that binds the substrate guanidine group, all of which are predicted to be ~6.5 Å from the site of oxygen binding (**Figure 4**) (73). One observation suggests that proton delivery is different in bacterial compared to mNOSs. Unlike gsNOS and cytochrome P450, in cryoreduction and annealing experiments on eNOS, the Fe^{3+}-hydroperoxo species (**5**) does not form at 77 K and is never observed upon annealing (90); only the Fe^{3+}-peroxo (**4**) is observed (90). This suggests that the protein and solvent network that delivers a proton to state **4** is different in gsNOS compared to eNOS. A notable structural difference in the active center between the mNOSs and the bacterial NOSs could influence solvent-mediated proton transfer to oxy-heme complexes. A Val that juxtaposes the oxygen coordination site in mNOS is replaced by an Ile in bacterial NOSs (Ile224 in **Figure 4**), and this change correlates with the binding properties of heme ligands (85).

The Need for Pterin

Given the analogies of the NOS reactions to those carried out by cytochrome P450, the requirement of an additional redox-active cofactor is a curious feature of NOS catalysis. The pterin requirement may stem from complications associated with producing NO. NO has a high affinity for ferrous hemes and is thus a very effective inhibitor of NOS, especially when the heme is reduced to Fe^{2+}. Thus, two electron donors are needed: (*a*) a slow donor (reductase) to initially reduce Fe^{3+}, but not before the product from the previous cycle, Fe^{3+}-NO, can dissociate; and (*b*) a fast donor for oxygen activation, to out compete autoxidation of state **3** to Fe^{3+} and O_2^- (57, 58). Interestingly, the slow NO release rates of the bacterial enzymes (50, 85) would then favor oxidation of NO to other more reactive species, raising the possibility that the desired product might not be NO per se but some further oxidized NO_X species. It should also be noted that in solution, H_4B exclusively undergoes two-electron concerted redox chemistry, and thus to achieve the radical state important for NOS catalysis, the enzyme must substantially alter the reactivity by this cofactor, presumably by controlling its electrostatic environment, and most importantly its protonation state (104).

The timing of electron and proton delivery to the heme center can influence the product ultimately formed by NOS (e.g., NO versus HNO/NO^- and Cit versus CN-orth); this should be kept in mind when considering that the natural cofactors and redox partners of bacterial NOSs are largely unresolved. Formation of nitroxyl may be preferred for some functionalities, especially if the goal of the reaction is to produce a nitrating agent (15, 57). It is not inconceivable that some prokaryotic NOSs do not function to produce NO, but rather HNO/NO-, and, as such, may react with peroxide directly or incorporate a redox-inactive cofactor. Nonetheless, the current data suggest that NO is produced in vivo by at least three bacterial NOSs (53, 55, 56), although many more remain to be characterized.

BIOLOGICAL FUNCTION

Perhaps the most intriguing questions surrounding prokaryotic NOS proteins concern their uses. Emerging data from a number of organisms suggest that, although production of NO may be a commonality, the purpose for NO production may vary considerably.

Streptomyces turgidiscabies

The first insight into bacterial NOS biological function came from the discovery that an open reading frame for a NOS was contained on a pathogenicity island common to certain *Streptomyces* strains that cause potato scab disease (**Figure 5**) (48). The pathogenicity

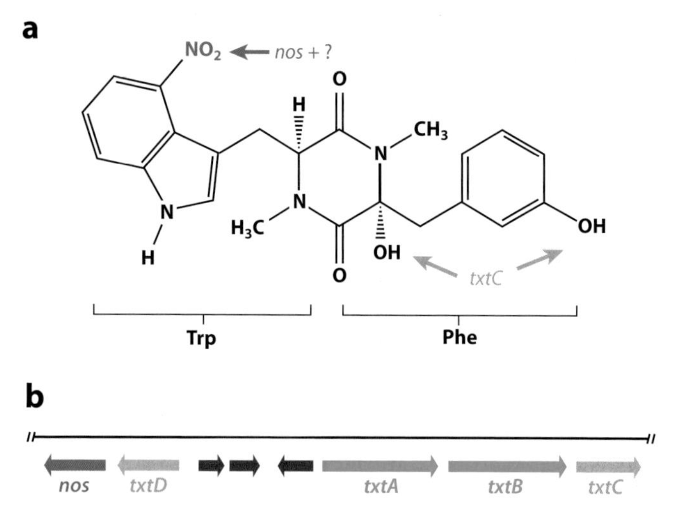

Figure 5

Nitric oxide synthase (NOS) participates in thaxtomin A synthesis. (*a*) Structure of thaxtomin A, a nonribosomal peptide, that is produced by two nonribosomal peptide synthases (txtA and txtB) and the decorating enzymes NOS, TxtC, and TxtD. (*b*) The thaxtomin-producing pathogenicity island in *Streptomyces*. The gene for *nos* (*blue*) along with the *txtA*, *txtB*, and *txtC* genes cluster together on the ~400-kb island.

island can transfer among *Streptomyces* strains and, in doing so, confers the ability to synthesize a class of plant toxins, called thaxtomins, that interfere with plant cell wall synthesis (105). Thaxtomins are dipeptides, derivatives of cyclo-[L-tryptophanyl-L-phenylananyl], produced by nonribosomal peptide synthesis (106). Most interestingly, the tryptophanyl moiety is nitrated at the four position (**Figure 5**) (106). The location of *nos* on the island in close proximity to the two nonribosomal peptide synthases of the thaxtomin biosynthetic pathway strongly suggested that *nos* might be involved in nitration of thaxtomins. Biosynthetic nitration reactions are not common in natural product synthesis and, where known, involve oxidation of an amine (107); thus participation of NO in such a process was unprecedented. Disruption of the *nos* gene nearly abrogated all thaxtomin production, which was substantially restored by *nos* complementation (48). Inhibitors of mNOSs, added to the growth media of *S. turgidiscabies*, resulted in a decrease in thaxtomin production without having detrimental effects on the cell growth (48, 108). It was further demonstrated by a feeding study with ^{15}N-Arg (labeled on the terminal guanidinium nitrogens) that the thaxtomin nitro group nitrogen originates from the terminal guanidinium nitrogen of Arg. NOS is the only known enzyme that oxidizes a terminal guanidinium nitrogen of Arg to NO. Thus, it was concluded that the position of *nos* on the thaxtomin pathogenicity island was because of its involvement in the nitration of the Trp moiety (48). It is notable that, in the *nos* knockout strain, a small amount of thaxtomin is still produced, which suggests that NOS is not the sole source of NO within *Streptomyces*. Furthermore, the *nos* gene is closely associated with an open reading frame for a predicted P450 (txtD), whose function in thaxtomin biosynthesis is not yet assigned. This may be significant because NO will not react directly to nitrate indole, and thus, some further degree of oxidation is required either of the indole moiety or NO itself. Oxidized forms of NO, such as nitrosonium (NO^+), peroxynitrite ($ONOO^-$), nitronium (NO_2^+), and nitrogen dioxide (NO_2), are known to actively nitrate aromatic groups (7, 8). Thus, much remains to be learned regarding the biosynthesis of this phytotoxin and the importance of *nos* in this process.

Production of NOS-derived NO was detected from *Streptomyces* spp. cultures by chemiluminescence gas-phase detection and in situ spin-trapping techniques (53). Bacterial hyphae produce NO in response to the plant cell wall component cellobiose at contact sites with root tips, root hairs, and elongation zones, i.e., the host pathogen interface of the growing plant. *Streptomyces* also produce NO in excess of that needed for toxin biosynthesis and at a time that precedes the onset of toxin synthesis (53). In plants, NO is known to act as a signal for the growth and extension of root tips (109), raising the interesting possibility that *Streptomyces* synthesizes NO not only for toxin biosynthesis, but also to promote the growth of tissue bacteria can infect.

Deinococcus radiodurans

In addition to the role of stNOS in thaxtomin biosynthesis, another link between bacterial NOS and Trp nitration was suggested by interactions between drNOS and a tryptophanyl tRNA synthetase protein (TrpRS II) (110). Pull-down experiments with recombinantly expressed, affinity-tagged drNOS showed that drNOS binds TrpRS II, one of two tryptophanyl tRNA synthetases found in *D. radiodurans*. The other *D. radiodurans* TrpRS, TrpRS I, has higher sequence homology to typical bacterial TrpRSs and higher activity for producing Trp-$tRNA^{Trp}$; but both TrpRS I and II are capable of adenylating Trp and charging $tRNA^{Trp}$ (110). TrpRS II increases the affinity of drNOS for Arg, and in turn, drNOS perturbs fluorescent ATP analogs bound to TrpRS II. The structure of TrpRS II revealed alternations to the substrate-binding pocket, which suggested the enzyme may be able to accommodate modified forms of Trp (111, 112). Indeed the enzyme was shown to charge tRNA equally well with Trp 4-nitro-Trp and 5-hydroxy-Trp (111). When tested for the ability to produce a modified form of Trp that could be a substrate for TrpRS II, drNOS was demonstrated to catalyze production of small quantities of 4-nitro-Trp when coupled to a mammalian reductase domain or when shunted by hydrogen peroxide (84). The levels of nitro-Trp produced increased in the presence of TrpRS II and were diminished by H_4B. This combined biochemical data suggested that drNOS and TrpRS II may work together to produce 4-nitro-Trp-$tRNA^{Trp}$, which could be used as a substrate for protein or secondary metabolite biosynthesis (84, 110). However, to date, no such products have been identified in *D. radiodurans*. Furthermore, a recent report was unable to reproduce nitro-Trp production from drNOS with peroxide (51); the reactivity of drNOS and TrpRS II remains under investigation.

Genetic experiments have provided greater insight into the biological role of drNOS. A strain of *D. radiodurans* in which the *nos* gene has been deleted (Δnos) is severely compromised in its ability to recover from exposure to UV radiation (56). *D. radiodurans* can withstand stress conditions, such as desiccation, oxidative damage, and radiation exposure, to a far greater extent than most organisms (113, 114). The Δnos strain has a small growth defect under rich media, nonstressed conditions when compared to wild type; however, the defect is more dramatic after cells have been exposed to UV light. The growth defect can be complemented by recombinant expression of NOS from an exogenous plasmid and can be rescued by the addition of exogenous NO. The fact that cells recover whether NO is added 10 min prior, during, or even hours after UV exposure suggests that NO does not act to protect or prevent radiation-induced damage but rather sends a signal to resume cell proliferation. Consistent with this, the *nos* gene is strongly induced by UV radiation, reaching a peak of expression a few hours after exposure (56). NOS expression coincides with intracellular NO production. Transcriptional profiling experiments verified by targeted mRNA analysis revealed that some genes were differentially regulated in Δnos after irradiation. In particular, a general growth regulator, Obg, was upregulated by UV exposure in the wild type but not in the Δnos strain (56). Sustained overexpression of Obg alleviated much of the growth defect in Δnos. Obg GTPases are involved in bacterial growth proliferation and stress response (115, 116). For example, in *B. subtilis*, Obg participates in the activation of stress response transcription factor σ_B, regulation of DNA replication, monitoring the intracellular GTP levels, and maintenance of ribosome function (115). Thus, it appears that in *D. radiodurans*, UV radiation exposure induces NO by increasing *nos* expression, and NO then acts to upregulate transcription of an important growth factor involved in cell proliferation. Unlike the case in *B. subtilis* (see below), *nos* deletion in *D. radiodurans* does not increase sensitivity to oxidative damage, and hydrogen peroxide does not induce *nos* expression. Many issues remain to be resolved, such as how is *nos* induced by UV? How is NO affecting gene transcription?

Are there are other roles for NO in this stress response?

Bacilli

bsNOS and baNOS protect *Bacilli* against oxidative stress (55, 117). As such, exposure of cells to mM concentrations of H_2O_2 produces DNA damage and cell death. However, if *B. subtilis* cells are pretreated with ~30 μM NO 5 s prior to peroxide exposure, ~100-fold more cells survive (117). Addition of NO either simultaneously with H_2O_2 or after H_2O_2 does not increase survival (117). Although NO is known to activate certain genes in *B. subtilis* and *E. coli* to protect the cells from oxidative and nitrosative stress (117), the very short time delay between NO treatment and protection suggests that NO acts on factors or cell states already in place at the time of exposure. NOS was implicated as the source of protective NO in *B. subtilis* through the properties of a *nos* deletion mutant that showed susceptibility to oxidative damage under conditions where reduced thiols were upregulated (117). Reduced thiols recycle ferrous iron and thereby promote Fenton chemistry, which generates damaging hydroxyl radicals (118). NO was proposed to block Fe^{2+} recycling by *S*-nitrosation of free thiols and, in particular, to interrupt the thioredoxin system. In addition, NO directly activated a specific *B. subtilis* catalase, whose enzymatic activity breaks down peroxide. The pathogenic bacterium *S. aureus* showed similar NO-mediated protection phenotypes in similar experiments (117). Related to these findings, NO appears to have antioxidant properties in *B. subtilis* and *S. aureus* in that it protects against irreversible thiol oxidation induced by peroxide (119). However, the mechanism underlying this latter antioxidant effect is currently unclear.

An NO-mediated oxidative protection mechanism could be important during a host:pathogen interaction because hosts often protect themselves from infection by oxidatively damaging their assailants. For example, mammalian phagocytes produce reactive oxygen and nitrogen species in response to infection. Survival of pathogenic *B. anthracis* depends on its own NOS activity in the presence of phagocytes (55). Spores of the *B. anthracis* Δ*nos* mutant lose their virulence in a mouse model of systemic infection and have compromised survival when germinating in macrophages (55). Furthermore, NO production is induced in the pathogen in response to the oxidative burst of the host defense mechanism (55). It is somewhat paradoxical that pathogen-derived NO is important for pathogen survival in the early stages of infection, yet host-derived NO is an important cytotoxic agent for host defense. Similar to the cases of *Streptomyces* and plants, NO plays a complex role in the interaction between *B. anthracis* and phagocytes. The ability of NOS to protect *Bacilli* and *Staphyloccuos* strains against oxidative damage could have important implications for antibiotic treatment, as many antibiotics produce some degree of oxidative damage in their targets (120).

TARGETS FOR NITRIC OXIDE SYNTHASE-DERIVED NITRIC OXIDE

Potential mechanisms through which NO could affect bacterial regulators include direct reaction with metallocofactors or through *S*-nitrosation of cysteine residues. In mammals, NO is known to regulate phosphatases, kinases, and transcription factors, such as Hif-1, NF-κB, and Nrf-Keap1 by *S*-nitrosation (121, 122). In bacteria, some NO-sensitive transcription factors have been identified: NorR is a dedicated NO sensor that activates three different enzymes used to metabolize NO (NO reductase, flavorubredoxin, flavohemoglobin); NsrR is likely an NO and/or a nitrite sensor that also activates NO protection mechanisms; and NnrR activates transcription of denitrification genes in the presence of NO (123). In addition, several other NO-responsive transcriptional regulators have been found, such as SoxR, OxyR, FNR, MetR, and Fur, although the primary function of each regulator is to sense another signal (superoxide, hydrogen peroxide, oxygen, homocysteine, and iron, respectively) (123). Several bacterial sensor kinase systems have been characterized, such as the DosS/DosT systems of *Mycobacterium tuberculosis* (124) and the H-NOX-histidine kinase pair, which serves as an environment sensor by binding NO (125). The H-NOX domain is a heme-binding module found in many bacteria, but also represented in the animal soluble guanylate cyclase, an important and specific target of NO in mammals (126). Given the number of regulatory systems sensitive to NO, it is not surprising that bacteria mount complex responses to this reactive agent. For the NOS-containing *B. subtilis*, NO generates a broad transcriptional response, which includes NO induction of *hmp*, the gene for NO-detoxifying flavohemoglobin, genes regulated by the Fe^{2+}-containing transcription factors Fur and PerR, and the σ^B general stress regulon (127, 128). Similarly, the NO donor compound MAHMA NONOnate upregulated proteins controlled by the Fur, PerR, OhrR, Spx, and σ^B regulons (119). Given the large number of intrinsic NO-sensitive pathways in bacteria, introduction of a *nos* gene into a genome would provide the potential to control a large number of stress responses—provided *nos* expression could come under the influence of the correct stimuli. For example, PerR controls antioxidant responses in *B. subtilis*, including the upregulation of the catalase gene *katA* (129, 130); NO-mediated induction of PerR may relate to the protective effect of NOS against oxidative damage in *Bacilli*. As described above, in *D. radiodurans*, NOS-derived NO is a signal for regulation of genes important to recovery from radiation damage and cell proliferation, i.e., *obg* and potentially others (56). There are also analogies to a radioprotective role for NO in mammalians, where UV radiation induces iNOS and thereby leads to *S*-nitrosation and activation of the transcription factor Hif-1α (131).

SUMMARY POINTS

1. Bacterial nitric oxide synthases (NOSs) are homologs of mammalian NOS (mNOS) oxygenase domains that are found in a subset of mostly gram-positive bacteria. They have structural and spectroscopic properties similar to those of mNOS and catayze the conversion of Arg to nitric oxide (NO) via the intermediate N^{ω}-hydroxy-L-arginine (NOHA) in reconstituted and in vivo systems.
2. Not all bacteria produce the NOS-essential pterin cofactor H_4B; however, bacterial NOSs can employ the related, ubiquitous cofactor, H_4F.
3. Mechanistic studies on a thermostable bacterial NOS have revealed that compound I is the reactive heme-oxygen intermediate in the conversion of Arg to NOHA but that a ferric-heme hydroperoxide is responsible for conversion of NOHA to NO and Cit.
4. Bacterial NOSs employ different classes of reductases in different organisms and may operate without dedicated reductase proteins.
5. Bacterial NOSs produce NO in vivo, but the functions of this NO are varied and different than in mammals. They include toxin biosynthesis and protection against oxidative damage, and act as a signal to regulate growth responses. Given their importance in stress and growth responses, bacterial NOSs could be good targets for antibiotics.

FUTURE ISSUES

The emergence of prokaryotic NOSs has expanded our view of the NOS enzyme family by setting a new range for reactivity and function, while also establishing the essential requirements for NO synthesis from Arg. Although the key parameters surrounding their reactivity have been determined, there is still much to be learned about the prokaryotic proteins. Ultimately, these questions must feed from the central issue of their biological functions, and if mammals can serve as a precedent, the roles for NOS-derived NO could be extensive and varied in microorganisms. In these many contexts, an important question is how do cofactor requirements and redox partners figure in function and regulation? Clearly, some variation must be present in the electron sources and cofactor relays among the prokaryotic NOSs, but how extensive is this variation, and does it reflect redundancy or true deviations in biological activity? The interplay between pathogen- and host-derived NO is particularly engaging as the same molecule appears to serve different masters to cross-purposes. Linked to these questions is the need to identify the immediate targets of NOS-derived NO, be they transcription factors, biosynthetic enzymes, metalloproteins, or kinases. Also, questions of dispersion and descent inevitably come to mind. Why are NOSs found only in subsets of bacteria? Why are they found in some species within a genus, and not others? And finally, what is the evolutionary relationship between the animal and prokaryotic proteins? The coming years will surely advance our knowledge in all of these areas, and we hope this review inspires some new participants to enter the fray.

DISCLOSURE STATEMENT

The authors are not aware of any affiliations, memberships, funding, or financial holdings that might be perceived as affecting the objectivity of this review.

NOTE ADDED IN PROOF

After this review was submitted, two key papers on bacterial NOS were published. The Sorangium FeS-containing NOS has been biochemically characterized by Marletta and colleagues and shown to indeed contain a reductase module (132), and Nudler and colleagues have demonstrated that NOS activity helps protect *Bacilli* against certain types of antibiotics (133).

LITERATURE CITED

1. Barbaree JM, Payne WJ. 1967. Products of denitrification by a marine bacterium as revealed by gas chromatography. *Marine Biol.* 1:136–39
2. Payne WJ. 1983. Bacterial denitrification: asset or defect. *Bioscience* 33:319–25
3. Richardson D, Felgate H, Watmough N, Thomson A, Baggs E. 2009. Mitigating release of the potent greenhouse gas N_2O from the nitrogen cycle—could enzymic regulation hold the key? *Trends Biotechnol.* 27:388–97
4. Cabello P, Roldán MD, Moreno-Vivián C. 2004. Nitrate reduction and the nitrogen cycle in Archaea. *Microbiology* 150:3527–46
5. Tavares P, Pereira AS, Moura JJG, Moura I. 2006. Metalloenzymes of the denitrification pathway. *J. Inorg. Biochem.* 100:2087–100
6. Patel RP, McAndrew J, Sellak H, White CR, Jo HJ, et al. 1999. Biological aspects of reactive nitrogen species. *Biochim. Biophys. Acta Bioenerg.* 1411:385–400

7. Koppenol WH. 1998. The basic chemistry of nitrogen monoxide and peroxynitrite. *Free Radic. Biol. Med.* 25:385–91
8. Hughes MN. 2008. Chemistry of nitric oxide and related species. *Methods Enzymol.* 436:3–19
9. Barbato JE, Tzeng E. 2004. Nitric oxide and arterial disease. *J. Vasc. Surg.* 40:187–93
10. Nott A, Riccio A. 2009. Nitric oxide-mediated epigenetic mechanisms in developing neurons. *Cell Cycle* 8:725–30
11. Yun HY, Dawson VL, Dawson TM. 1996. Neurobiology of nitric oxide. *Crit. Rev. Neurobiol.* 10:291–316
12. Yun HY, Dawson VL, Dawson TM. 1997. Nitric oxide in health and disease of the nervous system. *Mol. Psychiatry* 2:300–10
13. Feldman PL, Griffith OW, Stuehr DJ. 1993. The surprising life of nitric oxide. *Chem. Eng. News* 71:26–38
14. Griffith OW, Stuehr DJ. 1995. Nitric oxide synthases: properties and catalytic mechanism. *Annu. Rev. Physiol.* 57:707–36
15. Alderton WK, Cooper CE, Knowles RG. 2001. Nitric oxide synthases: structure, function and inhibition. *Biochem. J.* 357:593–615
16. Pfeiffer S, Mayer B, Hemmens B. 1999. Nitric oxide: chemical puzzles posed by a biological messenger. *Angew. Chem. Int. Ed.* 38:1715–31
17. Lipton SA. 2001. Physiology: nitric oxide and respiration. *Nature* 413:118–21
18. Crane BR. 2008. The enzymology of nitric oxide in bacterial pathogenesis and resistance. *Biochem. Soc. Trans.* 36:1149–54
19. Sudhamsu J, Crane BR. 2009. Bacterial nitric oxide synthases: What are they good for? *Trends Microbiol.* 17:212–18
20. Marletta MA. 1993. Nitric-oxide synthase structure and mechanism. *J. Biol. Chem.* 268:12231–34
21. Masters BSS. 1994. Nitric oxide synthases: Why so complex? *Annu. Rev. Nutr.* 14:131–45
22. Stuehr DJ, Santolini J, Wang ZQ, Wei CC, Adak S. 2004. Update on mechanism and catalytic regulation in the NO synthases. *J. Biol. Chem.* 279:36167–70
23. Gorren ACF, Mayer B. 2007. Nitric-oxide synthase: a cytochrome P450 family foster child. *Biochim. Biophys. Acta* 1770:432–45
24. Li HY, Poulos TL. 2005. Structure-function studies on nitric oxide synthases. *J. Inorg. Biochem.* 99:293–305
25. Stevens-Truss R, Marletta MA, Brownlow KC. 2005. Calcium-binding sites of calmodulin and electron transfer by inducible nitric oxide synthase. *Biochemistry.* 44:7593–601
26. Sono M, Roach MP, Coulter ED, Dawson JH. 1996. Heme-containing oxygenases. *Chem. Rev.* 96:2841–88
27. Chen YJ. 1994. A bacterial, nitric-oxide synthase from a *Nocardia* species. *Biochem. Biophys. Res. Commun.* 203:1251–58
28. Chen YJ, Rosazza JP. 1995. Purification and characterization of nitric-oxide synthase (NOSNoc) from a *Nocardia* species. *J. Bacteriol.* 177:5122–28
29. Morita H. 1997. Synthesis of nitric oxide from the two equivalent guanidino nitrogens of L-arginine by *Lactobacillus fermentum*. *J. Bacteriol.* 179:7812–15
30. Sari MA. 1998. Detection of a nitric oxide synthase possibly involved in the regulation of the *Rhodococcus* sp. R312 nitrile hydratase. *Biochem. Biophys. Res. Commun.* 250:364–68
31. Cohen MF, Yamasaki H. 2003. Involvement of nitric oxide synthase in sucrose-enhanced hydrogen peroxide tolerance of *Rhodococcus* sp. strain APG1, a plant-colonizing bacterium. *Nitric Oxide* 9:1–9
32. Choi DW, Oh HY, Hong SY, Han JW, Lee HW. 2000. Identification and characterization of nitric oxide synthase in *Salmonella typhimurium*. *Arch. Pharmacol. Res.* 23:407–12
33. Hernández-Campos ME, Campos-Rodríguez R, Tsutsumi V, Shibayama M, Garcia-Latorre E, et al. 2003. Nitric oxide synthase in *Entamoeba histolytica*: its effect on rat aortic rings. *Exp. Parasitol.* 104:87–95
34. Gutierrez-Escobar AJ, Arenas AF, Villoria-Guerrero Y, Padilla-Londoño JM, Gómez-Marin JE. 2008. *Toxoplasma gondii*: Molecular cloning and characterization of a nitric oxide synthase-like protein. *Exp. Parasitol.* 119:358–63
35. Choi WS. 1997. Identification of nitric oxide synthase in *Staphylococcus aureus*. *Biochem. Biophys. Res. Commun.* 237:554–58

36. Choi WS. 1998. Methylesters of L-arginine and *N*-nitro-L-arginine induce nitric oxide synthase in *Staphylococcus aureus*. *Biochem. Biophys. Res. Commun.* 246:431–35
37. Hong IS. 2003. Purification and characterization of nitric oxide synthase from *Staphylococcus aureus*. *FEMS Microbiol. Lett.* 222:177–82
38. Bird LE, Ren J, Zhang J, Foxwell N, Hawkins AR, et al. 2002. Crystal structure of SANOS, a bacterial nitric oxide synthase oxygenase protein from *Staphylococcus aureus*. *Structure* 10:1687–96
39. Chartier FJM, Couture M. 2004. Stability of the heme environment of the nitric oxide synthase from *Staphylococcus aureus* in the absence of pterin cofactor. *Biophys. J.* 87:1930–50
40. Chartier FJM, Blais SP, Couture M. 2006. A weak Fe-O bond in the oxygenated complex of the nitric-oxide synthase of *Staphylococcus aureus*. *J. Biol. Chem.* 281:9953–62
41. Jansson E, Lindblad P. 1998. Cloning and molecular characterization of a presumptive *argF*, a structural gene encoding ornithine carbamoyl transferase (OCT), in the cyanobacterium *Nostoc* sp. PCC 73102. *Physiol. Plant.* 103:347–53
42. Wei YZ, Zhou H, Sun Y, He YB, Luo YZ. 2007. Insight into the catalytic mechanism of arginine deiminase: functional studies on the crucial sites. *Proteins* 66:740–50
43. Viator RJR, Rest RF, Hildebrandt E, McGee DJ. 2008. Characterization of *Bacillus anthracis* arginase: effects of pH, temperature, and cell viability on metal preference. *BMC Biochem.* 9:15
44. Raman CS, Martaskek P, Masters BSS. 2000. Structural themes determining functions in nitric oxide synthases. In *Biochemistry and Binding: Activation of Small Molecules*, Vol. 4. *Porphyrin Handbook*, ed. KM Kadish, KM Smith, R Guilard, pp. 293–341. New York: Academic
45. Kunst F, Ogasawara N, Moszer I, Albertini AM, Alloni G, et al. 1997. The complete genome sequence of the gram-positive bacterium *Bacillus subtilis*. *Nature* 390:249–56
46. White O, Eisen JA, Heidelberg JF, Hickey EK, Peterson JD, et al. 1999. Genome sequence of the radioresistant bacterium *Deinococcus radiodurans* R1. *Science* 286:1571–77
47. Crane BR, Arvai AS, Ghosh DK, Wu CQ, Getzoff ED, et al. 1998. Structure of nitric oxide synthase oxygenase dimer with pterin and substrate. *Science* 279:2121–26
48. Kers JA, Wach MJ, Krasnoff SB, Widom J, Cameron KD, et al. 2004. Nitration of a peptide phytotoxin by bacterial nitric oxide synthase. *Nature* 429:79–82
49. Adak S, Bilwes AM, Panda K, Hosfield D, Aulak KS, et al. 2002. Cloning, expression, and characterization of a nitric oxide synthase protein from *Deinococcus radiodurans*. *Proc. Natl. Acad. Sci. USA* 99:107–12
50. Adak S, Aulak KS, Stuehr DJ. 2002. Direct evidence for nitric oxide production by a nitric-oxide synthase-like protein from *Bacillus subtilis*. *J. Biol. Chem.* 277:16167–71
51. Reece SY, Woodward JJ, Marletta MA. 2009. Synthesis of nitric oxide by the NOS-like protein from *Deinococcus radiodurans*: a direct role for tetrahydrofolate. *Biochemistry* 48:5483–91
52. Sudhamsu J, Crane BR. 2006. Structure and reactivity of a thermostable prokaryotic nitric-oxide synthase that forms a long-lived oxy-heme complex. *J. Biol. Chem.* 281:9623–32
53. Johnson EG, Sparks JP, Dzikovski B, Crane BR, Gibson DM, Loria R. 2008. Plant-pathogenic *Streptomyces* species produce nitric oxide synthase-derived nitric oxide in response to host signals. *Chem. Biol.* 15:43–50
54. Gusarov I, Starodubtseva M, Wang ZQ, McQuade L, Lippard SJ, et al. 2008. Bacterial nitric-oxide synthases operate without a dedicated redox partner. *J. Biol. Chem.* 283:13140–47
55. Shatalin K, Gusarov I, Avetissova E, Shatalina Y, McQuade LE, et al. 2008. *Bacillus anthracis*–derived nitric oxide is essential for pathogen virulence and survival in macrophages. *Proc. Natl. Acad. Sci. USA* 105:1009–13
56. Patel BA, Moreau M, Widom J, Chen H, Yin L, et al. 2009. Endogenous nitric oxide regulates the recovery of the radiation-resistant bacterium *D. radiodurans* from exposure to UV light. *Proc. Natl. Acad. Sci. USA* 106:18183–88
57. Santolini J, Meade AL, Stuehr DJ. 2001. Differences in three kinetic parameters underpin the unique catalytic profiles of nitric-oxide synthases I, II, and III. *J. Biol. Chem.* 276:48887–98
58. Santolini J, Adak S, Curran CML, Stuehr DJ. 2001. A kinetic simulation model that describes catalysis and regulation in nitric-oxide synthase. *J. Biol. Chem.* 276:1233–43
59. Chartier FJM, Couture M. 2004. Stability of the heme environment of the nitric oxide synthase from *Staphylococcus aureus* in the absence of pterin cofactor. *Biophys. J.* 87:1939–50

60. Salard I, Mercey E, Rekka E, Boucher JL, Nioche P, et al. 2006. Analogies and surprising differences between recombinant nitric oxide synthase-like proteins from *Staphylococcus aureus* and *Bacillus anthracis* in their interactions with L-arginine analogs and iron ligands. *J. Inorg. Biochem.* 100:2024–33
61. Midha S, Mishra R, Aziz MA, Sharma M, Mishra A, et al. 2005. Cloning, expression, and characterization of recombinant nitric oxide synthase-like protein from *Bacillus anthracis*. *Biochem. Biophys. Res. Commun.* 336:346–56
62. Santolini J, Roman M, Stuehr DJ, Mattioli TA. 2006. Resonance Raman study of *Bacillus subtilis* NO synthase-like protein: similarities and differences with mammalian NO synthases. *Biochemistry* 45:1480–89
63. Li D, Kabir M, Stuehr DJ, Rousseau DL, Yeh SR. 2007. Substrate- and isoform-specific dioxygen complexes of nitric oxide synthase. *J. Am. Chem. Soc.* 129:6943–51
64. Rousseau DL, Li D, Couture M, Yeh SR. 2005. Ligand-protein interactions in nitric oxide synthase. *J. Inorg. Biochem.* 99:306–23
65. Kabir M, Sudhamsu J, Crane BR, Yeh SR, Rousseau DL. 2008. Substrate-ligand interactions in *Geobacillus stearothermophilus* nitric oxide synthase. *Biochemistry* 47:12389–97
66. Lang J, Driscoll D, Gelinas S, Rafferty SP, Couture M. 2009. Trp180 of endothelial NOS and Trp56 of bacterial saNOS modulate sigma bonding of the axial cysteine to the heme. *J. Inorg. Biochem.* 103:1102–12
67. Panda K, Rosenfeld RJ, Ghosh S, Meade AL, Getzoff ED, Stuehr DJ. 2002. Distinct dimer interaction and regulation in nitric-oxide synthase types I, II, and III. *J. Biol. Chem.* 277:31020–30
68. Pant K, Bilwes AM, Adak S, Stuehr DJ, Crane BR. 2002. Structure of a nitric oxide synthase heme protein from *Bacillus subtilis*. *Biochemistry* 41:11071–79
69. Crane BR, Arvai AS, Gachhui R, Wu C, Ghosh DK, et al. 1997. The structure of nitric oxide synthase oxygenase domain and inhibitor complexes. *Science* 278:425–31
70. Raman CS, Li H, Martasek P, Kral V, Masters BS, Poulos TL. 1998. Crystal structure of constitutive endothelial nitric oxide synthase: a paradigm for pterin function involving a novel metal center. *Cell* 95:939–50
71. Fischmann TO, Hruza A, Niu XD, Fossetta JD, Lunn CA, et al. 1999. Structural characterization of nitric oxide synthase isoforms reveals striking active-site conservation. *Nat. Struct. Biol.* 6:233–42
72. Pant K, Crane BR. 2006. Nitrosyl-heme structures of *Bacillus subtilis* nitric oxide synthase have implications for understanding substrate oxidation. *Biochemistry* 45:2537–44
73. Davydov R, Sudhamsu J, Crane BR, Hoffman BR. 2009. EPR and ENDOR characterization of the reactive intermediates in the generation of NO by cryoreduced oxy-nitric oxide synthase from *G. stearothermophilus*. *J. Am. Chem. Soc.* 131:14493–507
74. Wang ZQ, Lawson RJ, Buddha MR, Wei CC, Crane BR, et al. 2007. Bacterial flavodoxins support nitric oxide production by *Bacillus subtilis* nitric-oxide synthase. *J. Biol. Chem.* 282:2196–202
75. Zemojtel T, Wade RC, Dandekar T. 2003. In search of the prototype of nitric oxide synthase. *FEBS Lett.* 554:1–5
76. Schneiker S, Perlova O, Kaiser O, Gerth K, Alici A, et al. 2007. Complete genome sequence of the myxobacterium *Sorangium cellulosum*. *Nat. Biotechnol.* 25:1281–89
77. Pant K, Crane BR. 2005. Structure of a loose dimer: an intermediate in nitric oxide synthase assembly. *J. Mol. Biol.* 352:932–40
78. Li D, Hayden EY, Panda K, Stuehr DJ, Deng HT, et al. 2006. Regulation of the monomer-dimer equilibrium in inducible nitric-oxide synthase by nitric oxide. *J. Biol. Chem.* 281:8197–204
79. Zhu YQ, Silverman RB. 2008. Revisiting heme mechanisms. A perspective on the mechanisms of nitric oxide synthase (NOS), heme oxygenase (HO), and cytochrome p450s (CYP450s). *Biochemistry* 47:2231–43
80. Werner ER, Gorren ACF, Heller R, Werner-Felmayer G, Mayer B. 2003. Tetrahydrobiopterin and nitric oxide: mechanistic and pharmacological aspects. *Exp. Biol. Med.* 228:1291–302
81. Sessa WC. 1994. The nitric-oxide synthase family of proteins. *J. Vasc. Res.* 31:131–43
82. Adhikari S, Ray S, Gachhui R. 2000. Catalase activity of oxygenase domain of rat neuronal nitric oxide synthase. Evidence for product formation from L-arginine. *FEBS Lett.* 475:35–8
83. Adak S, Wang Q, Stuehr DJ. 2000. Arginine conversion to nitroxide by tetrahydrobiopterin-free neuronal nitric-oxide synthase: implications for mechanism. *J. Biol. Chem.* 275:33554–61

84. Buddha MR, Tao T, Parry RJ, Crane BR. 2004. Regioselective nitration of tryptophan by a complex between bacterial nitric-oxide synthase and tryptophanyl-tRNA synthetase. *J. Biol. Chem.* 279:49567–70
85. Wang ZQ, Wei CC, Sharma M, Pant K, Crane BR, Stuehr DJ. 2004. A conserved Val to Ile switch near the heme pocket of animal and bacterial nitric-oxide synthases helps determine their distinct catalytic profiles. *J. Biol. Chem.* 279:19018–25
86. Davydov R, Kofman V, Fujii H, Yoshida T, Ikeda-Saito M, Hoffman BM. 2002. Catalytic mechanism of heme oxygenase through EPR and ENDOR of cryoreduced oxy-heme oxygenase and its Asp 140 mutants. *J. Am. Chem. Soc.* 124:1798–808
87. Davydov R, Makris TM, Kofman V, Werst DE, Sligar SG, Hoffman BM. 2001. Hydroxylation of camphor by reduced oxy-cytochrome P450cam: mechanistic implications of EPR and ENDOR studies of catalytic intermediates in native and mutant enzymes. *J. Am. Chem. Soc.* 123:1403–15
88. Makris TM, Davydov R, Denisov IG, Hoffman BM, Sligar SG. 2002. Mechanistic enzymology of oxygen activation by the cytochromes P450. *Drug Metab. Rev.* 34:691–708
89. Couture M, Stuehr DJ, Rousseau DL. 2000. The ferrous dioxygen complex of the oxygenase domain of neuronal nitric-oxide synthase. *J. Biol. Chem.* 275:3201–5
90. Davydov R, Ledbetter-Rogers A, Martasek P, Larukhin M, Sono M, et al. 2002. EPR and ENDOR characterization of intermediates in the cryoreduced oxy-nitric oxide synthase heme domain with bound L-arginine or N^G-hydroxyarginine. *Biochemistry* 41:10375–81
91. Tejero JS, Biswas A, Wang ZQ, Page RC, Haque MM, et al. 2008. Stabilization and characterization of a heme-oxy reaction intermediate in inducible nitric-oxide synthase. *J. Biol. Chem.* 283:33498–507
92. Pufahl RA, Wishnok JS, Marletta MA. 1995. Hydrogen peroxide-supported oxidation of N^G-hydroxy-L-arginine by nitric-oxide synthase. *Biochemistry* 34:1930–41
93. Zhu Y, Nikolic D, Van Breemen RB, Silverman RB. 2005. Mechanism of inactivation of inducible nitric oxide synthase by amidines. Irreversible enzyme inactivation without inactivator modification. *J. Am. Chem. Soc.* 127:858–68
94. Cho KB, Carvajal MA, Shaik S. 2009. First half-reaction mechanism of nitric oxide synthase: the role of proton and oxygen coupled electron transfer in the reaction by quantum mechanics/molecular mechanics. *J. Phys. Chem. B* 113:336–46
95. Cho KB, Derat E, Shaik S. 2007. Compound I of nitric oxide synthase: the active site protonation state. *J. Am. Chem. Soc.* 129:3182–88
96. Jin SX, Bryson TA, Dawson JH. 2004. Hydroperoxoferric heme intermediate as a second electrophilic oxidant in cytochrome P450-catalyzed reactions. *J. Biol. Inorg. Chem.* 9:644–53
97. Korth HG, Sustmann R, Thater C, Butler AR, Ingold KU. 1994. On the mechanism of the nitric oxide synthase-catalyzed conversion of N^{ω}-hydroxy-L-arginine to citrulline and nitric oxide. *J. Biol. Chem.* 269:17776–79
98. Clague MJ, Wishnok JS, Marletta MA. 1997. Formation of N^{δ}-cyanoornithine from N^G-hydroxy-L-arginine and hydrogen peroxide by neuronal nitric oxide synthase: implications for mechanism. *Biochemistry* 36:14465–73
99. Woodward JJ, Chang MM, Martin NI, Marletta MA. 2009. The second step of the nitric oxide synthase reaction: evidence for ferric-peroxo as the active oxidant. *J. Am. Chem. Soc.* 131:297–305
100. Wei CC, Wang ZQ, Hemann C, Hille R, Stuehr DJ. 2003. A tetrahydrobiopterin radical forms and then becomes reduced during N^{ω}-hydroxyarginine oxidation by nitric-oxide synthase. *J. Biol. Chem.* 278:46668–73
101. Stuehr DJ, Kwon NS, Nathan CF, Griffith OW, Feldman PL, Wiseman J. 1991. N^{ω}-hydroxy-L-arginine is an intermediate in the biosynthesis of nitric oxide from L-arginine. *J. Biol. Chem.* 266:6259–63
102. Martin NI, Woodward JJ, Winter MB, Marletta MA. 2009. 4,4-Difluorinated analogues of L-arginine and N^G-hydroxy-L-arginine as mechanistic probes for nitric oxide synthase. *Bioorg. Med. Chem. Lett.* 19:1758–62
103. Chartier FJM, Couture M. 2007. Interactions between substrates and the haem-bound nitric oxide of ferric and ferrous bacterial nitric oxide synthases. *Biochem. J.* 401:235–45
104. Hoke KR, Crane BR. 2009. The solution electrochemistry of tetrahydrobiopterin revisited. *Nitric Oxide* 20:79–87

105. Hogenhout SA, Loria R. 2008. Virulence mechanisms of gram-positive plant pathogenic bacteria. *Curr. Opin. Plant Biol.* 11:449–56
106. Healy FG, Wach M, Krasnoff SB, Gibson DM, Loria R. 2000. The *txtAB* genes of the plant pathogen *Streptomyces acidiscabies* encode a peptide synthetase required for phytotoxin thaxtomin A production and pathogenicity. *Mol. Microbiol.* 38:794–804
107. Carter GT, Nietsche JA, Goodman JJ, Torrey MJ, Dunne TS, et al. 1989. Direct biochemical nitration in the biosynthesis of dioxapyrrolomycin. A unique mechanism for the introduction of nitro groups in microbial products. *J. Chem. Soc. Chem. Commun.*:1271–73
108. Wach MJ, Kers JA, Krasnoff SB, Loria R, Gibson DM. 2005. Nitric oxide synthase inhibitors and nitric oxide donors modulate the biosynthesis of thaxtomin A, a nitrated phytotoxin produced by *Streptomyces* spp. *Nitric Oxide* 12:46–53
109. Tewari RK, Hahn EJ, Paek KY. 2008. Function of nitric oxide and superoxide anion in the adventitious root development and antioxidant defence in *Panax ginseng*. *Plant Cell Rep.* 27:563–73
110. Buddha MR, Keery KM, Crane BR. 2004. An unusual tryptophanyl tRNA synthetase interacts with nitric oxide synthase in *Deinococcus radiodurans*. *Proc. Natl. Acad. Sci. USA* 101:15881–86
111. Buddha MR, Crane BR. 2005. Structure and activity of an aminoacyl-tRNA synthetase that charges tRNA with nitro-tryptophan. *Nat. Struct. Mol. Biol.* 12:274–75
112. Buddha MR, Crane BR. 2005. Structures of tryptophanyl-tRNA synthetase II from *Deinococcus radiodurans* bound to ATP and tryptophan: insight into subunit cooperativity and domain motions linked to catalysis. *J. Biol. Chem.* 280:31965–73
113. Cox MM, Battista JR. 2005. *Deinococcus radiodurans*—the consummate survivor. *Nat. Rev. Microbiol.* 3:882–92
114. Battista JR. 1997. Against all odds: the survival strategies of *Deinococcus radiodurans*. *Annu. Rev. Microbiol.* 51:203–24
115. Czyz A, Wegrzyn G. 2005. The Obg subfamily of bacterial GTP-binding proteins: essential proteins of largely unknown functions that are evolutionarily conserved from bacteria to humans. *Acta Biochim. Pol.* 52:35–43
116. Foti JJ, Schienda J, Sutera VA Jr, Lovett ST. 2005. A bacterial G protein–mediated response to replication arrest. *Mol. Cell* 17:549–60
117. Gusarov I, Nudler E. 2005. NO-mediated cytoprotection: instant adaptation to oxidative stress in bacteria. *Proc. Natl. Acad. Sci. USA* 102:13855–60
118. Aruoma OI, Halliwell B, Gajewski E, Dizdaroglu M. 1989. Damage to the bases in DNA induced by hydrogen peroxide and ferric ion chelates. *J. Biol. Chem.* 264:20509–12
119. Hochgrafe F, Wolf C, Fuchs S, Liebeke M, Lalk M, et al. 2008. Nitric oxide stress induces different responses but mediates comparable protein thiol protection in *Bacillus subtilis* and *Staphylococcus aureus*. *J. Bacteriol.* 190:4997–5008
120. Kohanski MA, Dwyer DJ, Hayete B, Lawrence CA, Collins JJ. 2007. A common mechanism of cellular death induced by bactericidal antibiotics. *Cell* 130:797–810
121. Marshall HE, Merchant K, Stamler JS. 2000. Nitrosation and oxidation in the regulation of gene expression. *FASEB J.* 14:1889–900
122. Li C, Kim MY, Godoy LC, Thiantanawat A, Trudel LJ, Wogan GN. 2009. Nitric oxide activation of Keap1/Nrf2 signaling in human colon carcinoma cells. *Proc. Natl. Acad. Sci. USA* 106:14547–51
123. Spiro S. 2007. Regulators of bacterial responses to nitric oxide. *FEMS Microbiol. Rev.* 31:193–211
124. Kumar A, Toldeo JC, Patel RP, Lancaster JR Jr, Steyn AJC. 2007. *Mycobacterium tuberculosis* DosS is a redox sensor and DosT is a hypoxia sensor. *Proc. Natl. Acad. Sci. USA* 104:11568–73
125. Price MS, Chao L, Marletta MA. 2007. *Shewanella oneidensis* MR-1 H-NOX regulation of a histidine kinase by nitric oxide. *Biochemistry* 46:13677–83
126. Boon EM, Marletta MA. 2005. Ligand discrimination in soluble guanylate cyclase and the H-NOX family of heme sensor proteins. *Curr. Opin. Chem. Biol.* 9:441–46
127. Rogstam A, Larsson JT, Kjelgaard P, von Wachenfeldt C. 2007. Mechanisms of adaptation to nitrosative stress in *Bacillus subtilis*. *J. Bacteriol.* 189:3063–71
128. Moore CM, Nakano MM, Wang T, Ye RW, Helmann JD. 2004. Response of *Bacillus subtilis* to nitric oxide and the nitrosating agent sodium nitroprusside. *J. Bacteriol.* 186:4655–64

129. Helmann JD, Wu MF, Gaballa A, Kobel PA, Morshedi MM, et al. 2003. The global transcriptional response of *Bacillus subtilis* to peroxide stress is coordinated by three transcription factors. *J. Bacteriol.* 185:243–53

130. Loewen PC, Switala J. 1987. Multiple catalases in *Bacillus subtilis*. *J. Bacteriol.* 169:3601–7

131. Li F, Sonveaux P, Rabbani ZN, Liu S, Yan B, et al. 2007. Regulation of HIF-1alpha stability through *S*-nitrosylation. *Mol. Cell* 26:63–74

132. Agapie T, Suseno S, Woodward JJ, Stoll S, Britt RD, Marletta MA. 2009. NO formation by a catalytically self-sufficient bacterial nitric oxide synthase from *Sorangium cellulosum*. *Proc. Natl. Acad. Sci. USA* 106:16221–26

133. Gusarov I, Shatalin K, Starodubtseva M, Nudler E. 2009. Endogenous nitric oxide protects bacteria against a wide spectrum of antibiotics. *Science* 325:1380–84

Enzyme Promiscuity: A Mechanistic and Evolutionary Perspective

Olga Khersonsky and Dan S. Tawfik

Department of Biological Chemistry, Weizmann Institute of Science, Rehovot 76100, Israel; email: tawfik@weizmann.ac.il

Annu. Rev. Biochem. 2010. 79:471–505

First published online as a Review in Advance on March 17, 2010

The *Annual Review of Biochemistry* is online at biochem.annualreviews.org

This article's doi: 10.1146/annurev-biochem-030409-143718

0066-4154/10/0707-0471$20.00

Key Words

enzyme mechanism, protein evolution, superfamilies

Abstract

Many, if not most, enzymes can promiscuously catalyze reactions, or act on substrates, other than those for which they evolved. Here, we discuss the structural, mechanistic, and evolutionary implications of this manifestation of infidelity of molecular recognition. We define promiscuity and related phenomena and also address their generality and physiological implications. We discuss the mechanistic enzymology of promiscuity—how enzymes, which generally exert exquisite specificity, catalyze other, and sometimes barely related, reactions. Finally, we address the hypothesis that promiscuous enzymatic activities serve as evolutionary starting points and highlight the unique evolutionary features of promiscuous enzyme functions.

Contents

INTRODUCTION

Enzymes are traditionally referred to as remarkably specific catalysts. Yet the notion that many enzymes are capable of catalyzing other reactions and/or transforming other substrates, in addition to the ones for which they are physiologically specialized, or evolved, is definitely not new. Early examples of enzyme promiscuity include pyruvate decarboxylase (1), carbonic anhydrase (2), pepsin (3), chymotrypsin (4), and L-asparaginase (5). Nonetheless, the notion of "one enzyme—one substrate—one reaction" dominated, and still dominates the textbooks, and until recently, the wider implications of the "darker" side of enzyme promiscuity were largely ignored.

The idea of nature as an opportunistic modifier of preexisting suboptimal functions is also relatively old and has been formulated by Jacob in his classical note "Evolution and Tinkering" (6). The first direct connection between promiscuity and protein evolution was made, to our knowledge, in 1976 by Jensen (7). Jensen boldly forwarded the hypothesis that, unlike modern enzymes that tend to specialize in one substrate and reaction, the primordial, ancient enzymes possessed very broad specificities. Thus, relatively few rudimentary enzymes acted on multiple substrates to afford a wider range of metabolic capabilites. Divergence of specialized enzymes, via duplication, mutation, and selection, led to the current diversity of enzymes and to increased metabolic efficiency.

During the past decade, protein, and especially enzyme, promiscuity received considerable attention. Reviews by O'Brien & Herschlag (8) and Copley (9) were the first to highlight the potential mechanistic and evolutionary implications of promiscuity from an enzymologist's point of view. More recent reviews focused on practical implications of promiscuity (1, 10–12), on promiscuity and divergence in specific enzyme families (13–16), on mechanistic aspects of promiscuity (1, 17), and on promiscuity in the context of protein evolution and design (17, 18).

Here, we focus on the structural and mechanistic aspects of promiscuity as well as its role in the evolution of new functions. New enzymes have constantly emerged throughout the natural history of this planet. Enzymes that degrade synthetic chemicals introduced to the biosystem during the last decades (19–24), enzymes associated with drug resistance (25–28), and enzymes in plant secondary metabolism (29–31) provide vivid examples of how fast and efficient the evolution of new enzymatic functions can be. Indeed, extensive research since Jensen's article provided ample evidence for the notion that promiscuity is a key factor in the evolution of new protein functions. Here, we attempt to summarize this accumulating knowledge and point out some open questions in this emerging field of research.

DEFINING AND QUANTIFYING PROMISCUITY

The term enzyme promiscuity (8) is loosely applied and is used to describe a wide range of fundamentally different phenomena. We, and several others (8, 9, 32), use promiscuity to only describe enzyme activities other than the activity for which an enzyme evolved and that are not part of the organism's physiology. Thus, enzymes, such as glutathione *S*-transferases (GSTs) and cytochrome P450s (33), which a priori evolved to transform a whole range of substrates, are not promiscuous; they are multispecific or broad-specificity enzymes.

Degree of promiscuity refers to the level of specificity breach, namely, how diverse are the promiscuous activities of a given enzyme (34), and how different are the native and promiscuous functions. The degree of promiscuity can be assessed by examining the type of bonds that are being formed or broken and by differences in the mechanism between the native and promiscuous reactions (10). An "index of promiscuity," which computes the degree of variability between different substrates, has also been proposed (35). However, this method assumes that the same chemical transformation occurs on all substrates. As such, it is more suitable for the analysis of multispecific enzymes such as GSTs (as originally demonstrated), rather than promiscuity. We proposed a simple, relatively objective, way of assessing the degree of promiscuity by comparing differences in the Enzyme Commission (EC) numbers (33). In enzymes exhibiting multispecificity, or substrate ambiguity, EC numbers for the various substrates should be the same, or differ only by the fourth digit (which generally distinguishes between enzymes of the same class). Catalytic promiscuity generally correlates with cases in which the EC numbers of the various substrates and reactions catalyzed by the same enzyme differ in the second, or the third, digits (which refer to different classes of substrates) or even by the first digit (which indicates a different reaction category). (For examples see **Supplemental Table 1**. Follow the **Supplemental Material link** from the Annual Reviews home page at **http://www.annualreviews.org**.)

Magnitude of promiscuity refers to the kinetic parameters for the promiscuous activity relative to the native one. Whereas most enzymes exhibit k_{cat}/K_M values in the order of 10^5–10^8 $M^{-1}s^{-1}$ for their native substrates, the magnitude of promiscuous activities varies over more orders of magnitude, in absolute terms and also relative to the native activity. Catalytic proficiency ($k_{cat}/K_M/k_{uncat}$) and rate acceleration (k_{cat}/k_{uncat}) can provide a measure of the magnitude of catalytic effects exerted on native versus promiscuous substrates. In many cases, although the k_{cat}/K_M values for the promiscuous substrates are very low, and hence might have little physiological relevance, the rate accelerations and catalytic proficiencies are impressively high (34, 36–38).

Promiscuity: coincidental catalysis of reactions other than the reaction(s) for which an enzyme evolved

Multispecific or broad-specificity enzymes: enzymes performing the same reaction on a whole range of substrates, usually with similar efficiency

Substrate ambiguity: the activity of enzymes with substrates whose structure resembles the native substrate

PROMISCUITY: RULE OR EXCEPTION?

Numerous examples for enzyme promiscuity are currently known, but these are anecdotal and hardly provide an indication for the scale of this phenomenon. Systematic, high-throughput screens for promiscuous enzymatic activities are not a feasible option at present; no single detection method is available that can detect the whole range of different substrates and reactions. In contrast, high-throughput screens for binding cross-reactivities are relatively straightforward. These reveal a clear

Native function(s): physiologically relevant chemical transformation(s) and substrate(s) for which an enzyme evolved and is maintained under selection

Secondary function: an additional function in secondary metabolism or signaling. Secondary functions are also defined as native

Primary function: a well-defined function, often in central metabolism, typically shared by all orthologs

trend whereby the number of identified cross-reactants (small molecules or proteins) increase exponentially with the number of tested ligands (39–41). Several theoretical models account for these observations (42–44; see also 45, 46). Future screens, using large diversities of substrates and reactions performed with enzymes, are likely to reveal that essentially every enzyme exhibits a range of promiscuous functions.

Despite the absence of systematic data, outlined below are several arguments in favor of the notion that promiscuity is a wide phenomenon and thus should be regarded as a rule, rather than an exception.

Specificity Is Context Dependent

High specificity can bear a high cost in substrate-binding energies, thereby resulting in higher activation energies and lower turnover rates (k_{cat}) for the cognate substrate (47) (for an alternative mode whereby noncognate substrates exhibit low k_{cat} values due to poor positioning relative to the active site's catalytic residues, see Kinetic Parameters for Native versus Promiscuous Functions, below). Even the most specific enzymes, e.g., enzymes involved in DNA or protein synthesis, exhibit measurable substrate infidelities, often at surprisingly high rates. High fidelity is often achieved via proofreading, or proofediting, mechanisms that reverse the process and redo it (47). For example, the selectivity of aminoacyl-tRNA synthetases is under tight selection—having the wrong amino acid loaded onto a given tRNA yields a mutated protein. Because of the close similarity of certain amino acids, proofediting mechanisms have evolved whereby formation of a noncognate aminoacyl-tRNA leads to its rapid hydrolysis and resynthesis of the aminoacyl-tRNA at the cost of ATP (48). Similarly, the proofreading domain of polymerases is an exonuclease that can digest parts of the extended strand.

Specificity is shaped by natural selection. Promiscuous activities that are harmful were selected against. The adenylation domain *TycA* is highly selective for its cognate amino acid L-phenylalanine, primarily with respect to naturally occurring amino acids (e.g., L-tyrosine exhibits ~800-fold lower k_{cat}/K_M). However, an artificial substrate D-phenylalanine, to which the enzyme has probably never been exposed, is accommodated by *TycA* with k_{cat}/K_M only twofold lower than that of L-phenylalanine (49).

Many enzymes perform secondary tasks (50, 51) that are likely to have stemmed from their promiscuity. Examples include enzymes that have been under intense selection for high specificity, such as aminoacyl-tRNA synthetases. Lysyl-tRNA-synthetase, for example, mediates the synthesis of the signaling molecule Ap4A (two adenosines linked via four phosphates) (52, 53), and so do most other tRNA synthetases (54). This side reaction occurs within the same active site. In the absence of tRNA, the aminoacyl-AMP intermediate reacts with a second ATP molecule to generate the free amino acid and Ap4A. Certain aminoacyl-tRNA synthetases bind DNA or mRNA and thus regulate transcription, splicing, and translation, or they act as cofactors in RNA trafficking (51). It is likely that these oft-called secondary functions were recruited well after the primary function had emerged. Once recruited, they remained under selection and therefore became an additional native function of the enzyme.

Regulation and Masking of Promiscuity

Few of the promiscuous activities found in vitro bear a physiological or evolutionary meaning. Even those that might are not manifested in vivo (this is, by definition, what promiscuous activities are). A primary factor to consider is regulation, which prevents many of the undesirable outcomes of promiscuity. Of the entire enzyme diversity available to organisms, only a small fraction is accessible and active at a given time and cellular location. Regulation at the level of expression prevents the spending of unnecessary resources (51, 55), but the fitness costs associated with unncessary transcription and translation are relatively low (56). However, regulation regimes are also the key in controlling enzyme activity, especially with enzymes whose specificity is broad. For example, *Escherichia coli* has 23 different haloacid dehalogenase (HAD)-like hydrolases. Most of these are phosphatases exhibiting very broad substrate specificity (57, 58), but these operate under different regulation, and specificity is achieved via regulation and not by restricting enzyme specificity (58).

Regulation occurs also at the protein level, such as allosteric regulation that prevents the wasteful conversion of costly metabolites. As expected, this regulation is mostly product controlled. But in some cases, the substrate is an allosteric regulator of its own enzyme. Why would such a regulatory mechanism evolve? In the absence of its substrate, an enzyme is supposed to remain silent. Preventing active sites from promiscuously reacting with undesirable substrates could be one of the driving forces for the evolution of substrate-dependent allosteric regulation.

Promiscuity within Living Cells

Despite the action of natural selection to increase enzyme selectivity by various means, ranging from shaping the active site itself to regulation of enzyme expression and activity, numerous cross-reactions and breaches of specificity occur, not just in vitro, but also within living cells. Such cross-reactivities are often unraveled by the analysis of auxotrophic knockout strains that lack a crucial enzyme. Such deficiencies are often complemented by other enzymes, or even other enzyme pathways, sometimes in an unexpected manner. For example, knockouts of the *phn* operon in *E. coli* that utilizes phosphite (HPO_3^{2-}) led to the identification of promiscuous phosphite-dependent hydrogenase activity in alkaline phosphatase (see Mechanistic Aspects of Promiscuity, below) (59). Glutamyl phosphate reductase (*ProA*) exhibits low promiscuous activity with *N*-acetylglutamyl phosphate, the substrate for *ArgC* (*N*-acetylglutamyl phosphate reductase). Following a single mutation in the enzyme's active site, and changes in regulation, *ProA* could complement the *ArgC* knockout from a single-copy plasmid (60).

The level of cross-reactivity between different metabolic pathways was also indicated by an in silico experiment that attempted to dock 125 common metabolites into the active sites of 120 key metabolic enzymes. Numerous potential cross-reactions were found among these 15,000 potential pairs. Although docking has obvious limitations, this study further highlights the potential for promiscuity (12, 61). Complementation of *E. coli* knockout strains by selection from a library of *E. coli*'s own genes under overexpression from a multiple-copy plasmid revealed a similar picture (32). The deleted gene and its suppressor were, in most cases, unrelated. Complementation was achieved through the promiscuous action of other enzymes, through increased transport (and not necessarily of the deficient metabolite), and, most often, by an alternative metabolic pathway. Thus, promiscuity is not necessarily limited to the single enzyme level, but often, whole pathways act promiscuously, namely, outside their ordinary functional role. Other examples of metabolic plasticity, or "underground metabolism," are reviewed in References 7, 13, and 62.

The "Flexible Metabolome"

The above observations led to new hypotheses that suggest that genetic and metabolic pathways are inherently ambiguous and stochastic. By these hypotheses, the well-defined linear pathways described in textbooks are cross wired in a variety of unexpected ways. Evolution may capitalize on these cross-wirings, as a way of adaptive plasticity (i.e., with no genetic changes to begin with), to generate new metabolic capabilities (63). Phenomena similar to underground metabolism and adaptive plasticity were also observed in genetic analyses, wherein altered phenotypes turned out to be correlated with changes in many different genes, including genes from unrelated pathways. As is the case with enzymes and metabolic pathways, genome flexibility is an inevitable outcome of limited specificity, or promiscuity, of gene action and of intergenic interactions (64, 65). Thus, it appears that, beyond the well-studied, linear pathways, there exist flexible genomes (64), as well as flexible proteomes and flexible metabolomes, whose contribution to evolutionary adaptation is still understudied.

MECHANISTIC ASPECTS OF PROMISCUITY

How does the very same active site and catalytic machinery show exquisite specificity with respect to the native substrate but still promiscuously catalyze other, often completely unrelated, reactions? The answer to this question is complex, and different scenarios, mechanisms, and other aspects of the specificity-promiscuity dichotomy are outlined below.

Specificity and Promiscuity Coincide within the Same Active Site

Conformational diversity. The role of structural plasticity in facilitating enzyme action, promiscuity, and evolution is discussed in several reviews (67–69). In many cases, promiscuity is linked to diverse conformations, whereby the native and the promiscuous functions are mediated by different active-site configurations (**Figure 1**). For example, isopropylmalate isomerase is an enzyme with dual-substrate specificity, where a loop structure depends on the substrate present (70). In sulfotransferase SULT1A1, conformational changes enable the same enzyme to accommodate a range of different substrates (66), as is the case with glutathione-*S*-transferases GSTA1-1 and GSTA4-4 (71) and with certain P450s (72). And, in β-lactamase, an expanded spectrum of antibiotic substrates is accommodated through increased flexibility and altered dynamics (73, 74).

Accommodating alternative substrates. In many cases, promiscuous activities share

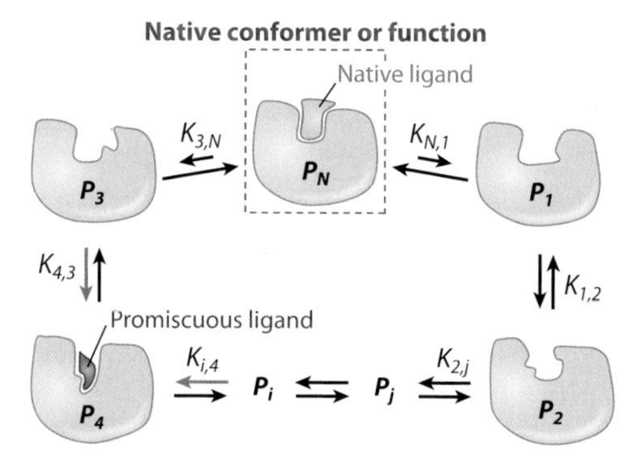

Figure 1

Protein promiscuity, evolvability, and conformational diversity. Proteins exist as an ensemble of different conformations (depicted as P_1, $P_2..P_j$) that exchange via the respective equilibrium constants (Kij). The primary conformation is the native state (P_N), which interacts with the native ligand. The alternative conformers relate to structural variations spanning from different side chain rotamers and active-site loop rearrangements to more profound fold transitions. Minor conformers (e.g., P_4) may mediate alternative functions, such as binding of a promiscuous ligand. Mutations can gradually alter this equilibrium such that scarcely populated conformers become more favorable with significant effects on the corresponding promiscuous function (e.g., an increase in occupancy of P_4 from 0.01 to 0.1 can yield a tenfold increase in the overall level of promiscuous function). The relative occupancy of the native conformer would be hardly affected (e.g., from 0.5 to ≥ 0.41, leading to $<20\%$ loss of the native function). This model also accounts for weak negative trade-offs between the existing and evolving functions as well as the evolutionary potential of neutral mutations. Adapted from Reference 69.

the same active-site configuration and main active-site features with the native activity. For example, guanidine-transferring enzymes utilize the same catalytic triad in their promiscuous action on various derivatives of arginine (75). In this case of substrate ambiguity, the active-site residues that bind the Cα-carboxyl and the guanidino-NH_2 of different substrates are different. Another case where the network of hydrogen bonds is the main feature that differentiates the native reaction from the promiscuous one is D-2-keto-3-deoxygluconate aldolase (**Figure 2**) (78).

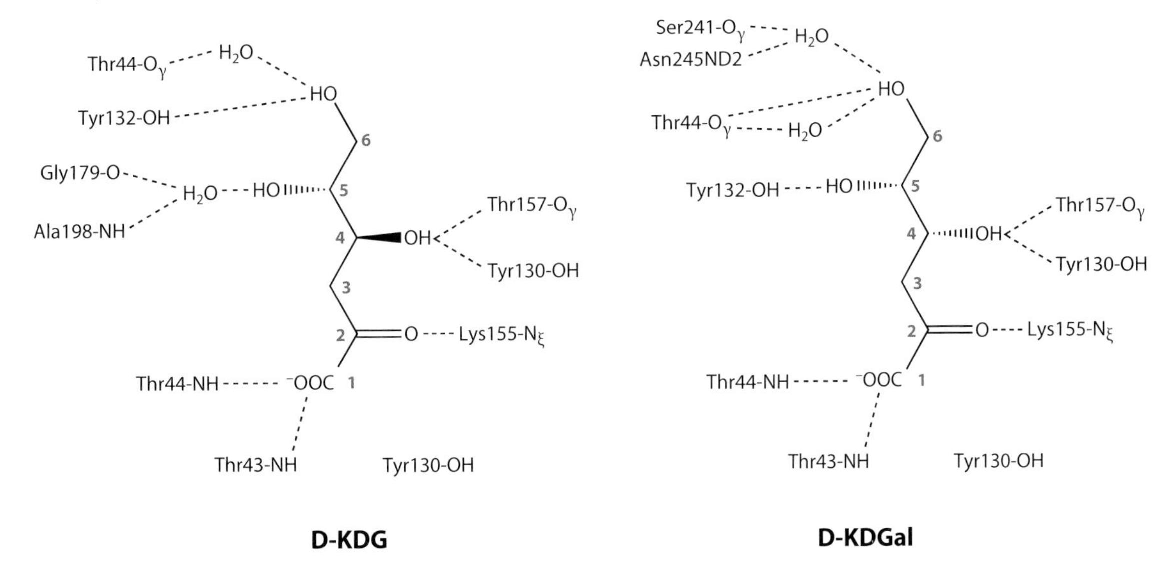

Figure 2

Schematic summary of the different interactions made in the active site of D-2-keto-3-deoxy-gluconate aldolase (KDGA) from *Sulfolobus solfataricus* (adapted from Reference 78). This enzyme transforms both D-2-keto-3-deoxy-gluconate (D-KDG) and D-2-keto-3-deoxy-galactonate (D-KDGal) with similar rates (176). The mechanism with both substrates involves Schiff base formation by Lys155 and subsequent hydration and cleavage. The differences between the gluconate and galactonate substrates are in the hydrogen bonds formed with KDGA's active site and, in particular, in the manner by which the 5′ and 6′ hydroxyl groups are bound.

Other examples include enzymes that apply nucleophilic catalysis, such as alkaline phosphatase, a highly proficient ($k_{cat}/K_M > 10^7$ M^{-1} s^{-1}) phosphate monoesterase that promiscuously hydrolyzes phosphodiesters, phosphoamides, and sulfate esters (36, 38, 79), as well as phosphite (while reducing water to release hydrogen) (**Figure 3*a***) (59). The catalytic mechanism is presumed to be similar for all these reactions and involves nucleophilic attack by Ser102 and stabilization of the negatively charged intermediate by the active site Zn^{2+} ions and Arg166 (**Figure 3*b***) (38, 59). Comparison between the activities revealed that, although these substrates all bind in a similar mode, the interactions with both Zn^{2+} ions and Arg166 are much more favorable for the native phosphate-monoester substrates than for other promiscuous substrates (36, 38). This difference accounts for the orders-of-magnitude higher rates and catalytic proficiencies of the native substrates versus the promiscuous ones.

The very same active site can therefore offer several different modes of interactions, and some of these might be utilized by promiscuous substrates. It should be noted, however, that most of the above describes cases analyzed by kinetics and site-directed mutagenesis. Because very few structures of the enzyme-substrate or enzyme-transition state analog complexes exist for both the native and promiscuous substrates, subtle changes of the active site's conformation cannot be excluded.

Different protonation states. The same catalytic residue can act in a different protonation state in the native compared to the promiscuous function. In the tautomerase superfamily, various enzymes share the catalytic Pro residue at the enzyme's amino terminus, but the mechanism of catalysis depends on its pK_a. In the 4-oxalocrotonate tautomerase (4-OT) the pK_a of Pro1 is ~6.4, and it acts as a general base. In *trans*-3-chloroacrylic acid

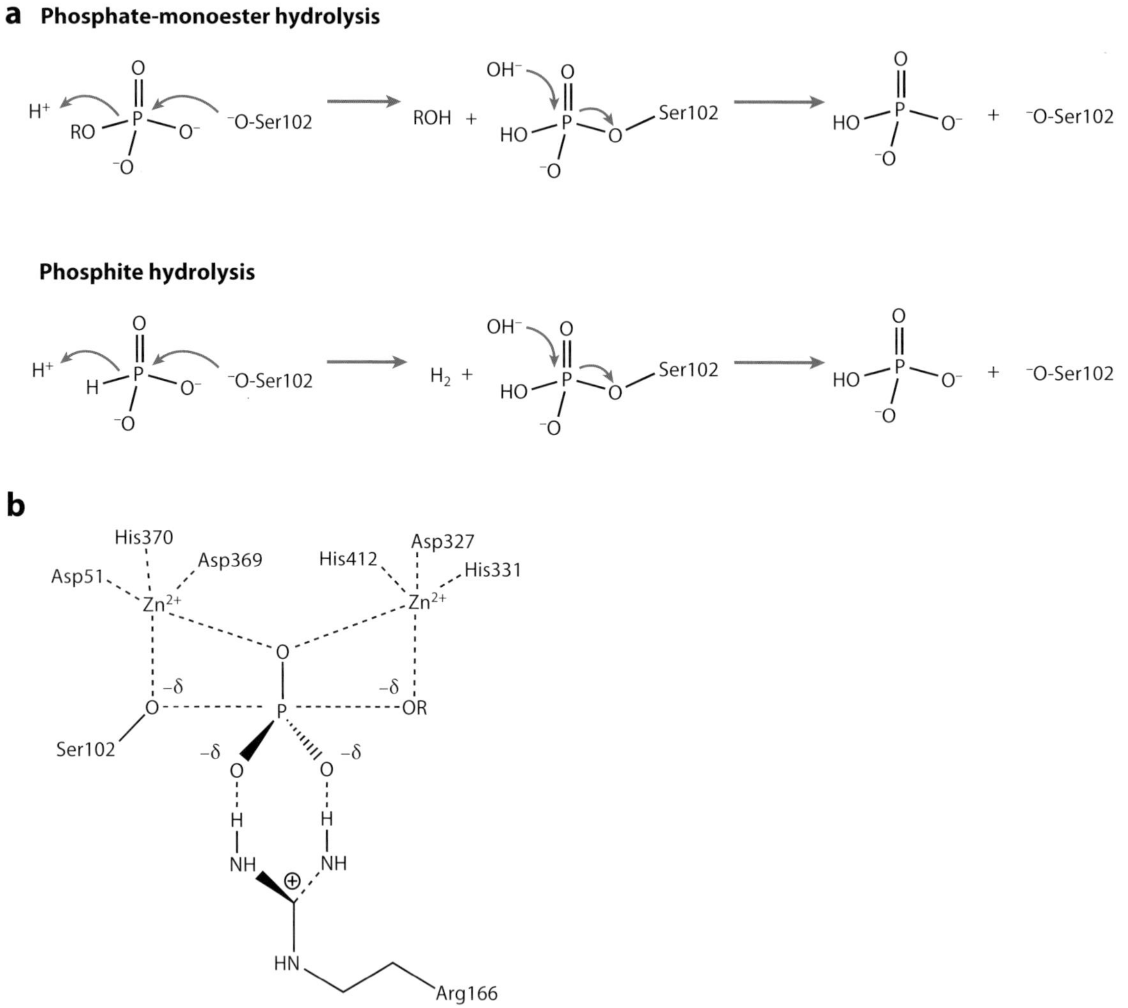

a Phosphate-monoester hydrolysis

Phosphite hydrolysis

b

Figure 3

(*a*) The native monoester phosphatase activity and the promiscuous phosphite hydrolysis reactions catalyzed by alkaline phosphatase (adapted from Reference 59). (*b*) The active-site arrangement of alkaline phosphatase with a bound transition state model (adapted from Reference 36).

dehalogenase (CaaD), which catalyzes the hydrolytic halogenation of haloacrylates, Pro1 is protonated (pKa ~9.2) and serves as a general acid (80, 81). Because in 4-OT only a small fraction of Pro1 is present in the protonated state, it exhibits very weak promiscuous general acid catalysis of the hydratase activity. However, another family member, malonate semialdehyde decarboxylase (MSAD), exhibits a substantial promiscuous hydratase activity, primarily because Pro1 is protonated and serves as a general acid in the mechanism of the enzyme's native activity (82, 83).

Different subsites within the same active site. In several cases, although both the original and promiscuous activities reside within the same active site and rely on its major feature (e.g., an oxyanion hole), other key parts of the catalytic machinery differ. One example is serum paraoxonase (PON1), a mammalian lactonase with promiscuous esterase and phosphotriesterase (PTE) activities (**Figure 4*a***) (84). The coordination of the phosphoryl/carbonyl oxygen to the active-site calcium is a feature shared by all the activities. However, whereas the hydrolysis of lactones and esters is mediated by a His115-His134 dyad, the promiscuous phosphotriesterase activity is mediated by another set of residues (84, 85), possibly via a nucleophilic attack of Asp269 (86).

An analogous example is *Candida antarctica* lipase B (CAL-B) whose native activity (lipids hydrolysis) is mediated by a Ser105-His224-Asp187 catalytic triad. Using its oxyanion hole, CAL-B also catalyzes various carbon-carbon bond formation reactions, such as Michael additions and aldol condensations (87–89). However, in these reactions, the nucleophilic serine takes no role, and acid-base transfer is presumably mediated by His224 in conjunction with Asp187 (**Figure 4*b***).

Promiscuity via alternative cofactors and amino acids. In a particular case of cofactor ambiguity, changes in enzyme specificity can also be induced by metal substitutions. Following work by Kaiser & Lawrence (90), the introduction of copper ions induced promiscuous oxidase activities in several hydrolytic enzymes (91, 92). In carbonic anhydrase, substitution of the native Zn^{2+} by Mn^{2+}

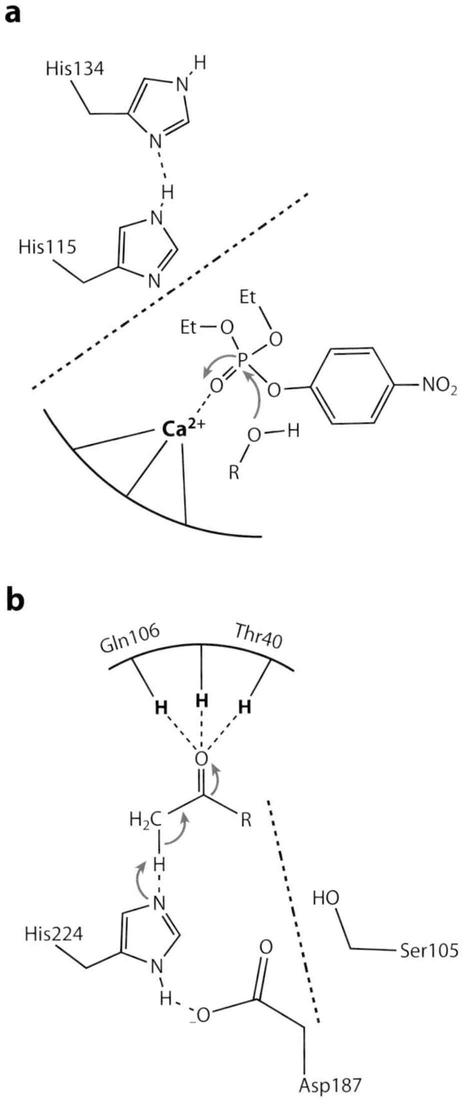

Figure 4

Different subsites within the same active site. (*a*) The main active-site feature of the serum paraoxonase (PON1) is the catalytic calcium ion, which lies at the bottom of a deep and hydrophobic active site and is thought to act as the "oxyanion hole" of PONs. The native function, hydrolysis of lactones, is mediated by a His115-His134 dyad, which deprotonates a water molecule to generate the attacking hydroxide. Although the same dyad appears to mediate the promiscuous arylesterase activity of PON1, the promiscuous phosphotriesterase activity (shown here for paraoxon as substrate) is independent and is mediated by other residues that act as base or nucleophile (84, 174). Indeed, mutations of both His residues diminish the lactonase activity but may increase the promiscuous phosphotrieterase activity by up to 300-fold with certain organophosphate substrates (85, 175). (*b*) A similar scenario has been described for *Candida antarctica* lipase B (CAL-B). Its native activity (lipid hydrolysis) is mediated by the Ser105-His224-Asp187 triad, and the acyl-enzyme intermediate is stabilized by its oxyanion hole. CAL-B also catalyzes promiscuous C-C bond formation reactions. In these promiscuous activities, the oxyanion hole is also utilized for negative charge stabilization (shown here). However, the catalytic serine takes no part, and acid-base transfer is thought to be mediated by His224 in conjunction with Asp187 (87–89).

enabled the catalysis of styrene epoxidation (93), and rhodium-substituted carbonic anhydrase acts as a hydrogen-utilizing reductase (94). Similarly, incorporating selenocysteine into the active sites of subtilisin (95), glyceraldehyde-3-phosphate dehydrogenase (96), and GST (97) endowed these enzymes with novel peroxidase activities.

Water-assisted promiscuity. Although the native substrate may interact directly with active-site residues, accidental hydrogen bonds mediated by water molecules may play a role in promiscuous interactions. Water molecules can buffer opposing dipoles or charges between the substrate and active-site residues, or they can act as acid, base, or nucleophile in the catalysis of promiscuous reactions. Indeed, spatially defined active-site water molecules have catalytic powers that are comparable to amino acid residues, and localized water molecules may have played a key role in primordial enzymatic active sites (47). A study of the molecular dynamics of the *Bacillus subtilis* esterase suggested that promiscuous amide hydrolysis is mediated by a network of water-mediated hydrogen bonds that are not involved in the esterase reaction (98). Further evidence for water-mediated promiscuity awaits more structures of enzymes complexed with promiscuous substrates.

Enzyme Mechanisms Analyzed by Studying Promiscuous Functions

Enzymologists have discovered that a systematic research of the hidden skills of enzymes can provide valuable insights regarding their catalytic mechanisms. For example, the promiscuous hydrolysis of phosphonate diesters by *Tetrahymena thermophila* ribozyme provided key insights regarding the relative importance of transition state geometry versus charge (99). In another study, the promiscuous chorismate mutase activity of PchB was used to derive mechanistic insights into its native activity (isochorismate pyruvate lyase) (100).

Kinetic Parameters for Native versus Promiscuous Functions

Differences between the efficiency of promiscuous and native activities can be manifested in differences in either k_{cat} or K_M. Although it is expected that promiscuous substrates that bind weakly will exhibit high K_M values, many promiscuous substrates are characterized by low k_{cat} values. Thus, specificity may result not only from substrate binding interactions per se, but also from appropriate positioning relative to the catalytic machinery. For example, analysis of substrates of PON1, the primary function of which is lipophilic lactonase, indicated that all promiscuous aryl esters and phosphotriester substrates exhibit K_M values in the mM range (0.8–5 mM), and the differences in reactivity are primarily due to k_{cat} values that vary by >1000-fold (101). For the promiscuous substrates, substrate binding is driven primarily by nonspecific hydrophobic forces within the deep and hydrophobic active site of PON1. However, promiscuous substrates are inadequately positioned relative to the catalytic machinery and therefore exhibit low k_{cat} values. Interestingly, for the lactones that comprise the native substrate of this enzyme, K_M values vary by $\sim$200-fold (from about 0.1 up to 20 mM), whereas the variations in k_{cat} values are an order of magnitude lower ($\sim$10–200 s^{-1}). Indeed, binding of the native substrate is typically mediated by enthalpy-driven interactions, such as hydrogen bonds, whereas for the promiscuous substrates, hydrophobic and other entropy-driven interactions play a key role (12, 102).

PROMISCUITY AND DIVERGENCE OF ENZYME SUPERFAMILIES

An enzyme superfamily combines dozens to thousands of enzymes that, although distant in sequence and catalyzing different chemical transformations of many different substrates, share the same fold and a common catalytic strategy (e.g., abstraction of a proton from a position alpha to a carboxylate, and stabilization of

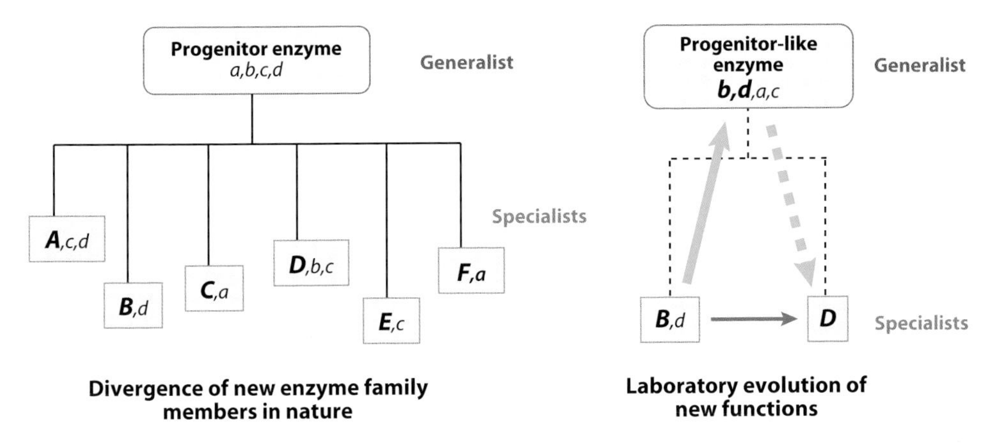

Figure 5

Divergence of a generalist progenitor enzyme to a family of specialist enzymes. (*left*) Jensen's hypothesis (7) surmises that, in nature, an ancestor protein displaying a low level of a range of activities (denoted as *a*, *b*, *c*, *d*) has been subjected to selection pressures for those activities, thus duplicating and diverging into a family of potent and highly specialized enzymes of the kind seen today (denoted *A*, *B*, etc.). In the course of divergence, new activities that were not present in the progenitor may also emerge (denoted as *E* and *F*). Today's specialists may still retain some of the functions of the common ancestor (denoted in lower case letters) as low levels of promiscuous activities. Indeed, several reports indicate a low level of shared activities within a family and, in particular, that the native activity of one member is the promiscuous activity of another, and vice versa (**Table 1**). (*right*) Additional support to the above model comes from the results of many directed evolution experiments. Direct switches of specificity, e.g., from B to D (*blue arrow*) are rare and are typically seen following a parallel selection for an increase in the target activity and elimination of the original one. Upon mutation and selection for an increase of a promiscuous activity (*green arrow*), the resulting variants usually show significant increases in the target activity and a smaller decrease in the original one, thus yielding, in effect, a generalist intermediate exhibiting both *d* and *b* at relatively high levels (the weak negative trade-off line in **Figure 6**). Such intermediates are often observed in the lab; some even gain other activities, for which they were never selected (denoted *a*, *c*), and may therefore resemble the progenitor of this enzyme family or node intermediates along past routes of its divergence. Adapted from Reference 17.

the resulting enolate intermediate, in the enolase superfamily) (14). Analysis of enzyme families and superfamilies provides ample evidence for the role of promiscuity in the evolution of new functions. Specifically, the identification of promiscuous activities, or cross-reactivities, between different members of the same enzyme family or superfamily and the ability to evolve these promiscuous activities in the laboratory provide important hints regarding evolutionary, structural, and mechanistic relationships within enzyme superfamilies (**Figure 5**). Examples of the promiscuous catalytic activities within enzyme families and superfamilies are listed in **Table 1**. Conclusions supported by these data are summarized below:

1. The primary, or native, function of one family member is often identified as a promiscuous activity in other family members (**Table 1**, entries 2 and 5–9). This overlap may reflect the common catalytic strategy that underlines these families and superfamilies as well as a common evolutionary origin (**Figure 5**). It can therefore guide the identification of the native function of new superfamily members by virtue of its similarity to the promiscuous activity of related family members. This principle was demonstrated in an attempt to trace the origins of a bacterial phosphotriesterase (PTE from *Pseudomonas diminuta*), an enzyme thought to have evolved for the degradation of paraoxon, an insecticide introduced in the twentieth century. PTE possesses a promiscuous lactonase

Table 1 Examples for promiscuous activities within enzyme families and superfamilies

Entry number	Family/superfamily	Enzymes	Native activity (substrate, k_{cat}/K_M in $M^{-1}s^{-1}$)	Promiscuous activity (substrate, k_{cat}/K_M in $M^{-1}s^{-1}$)	References
1	Mammalian paraoxonases (PONs)	PON1 (serum paraoxonase)	Lipo-lactonase (γ-dodecanoic lactone, 1.2×10^5)	PON1: aryl esterase (phenyl acetate, $\sim 6 \times 10^5$) Phosphotriesterase (paraoxon, 6×10^3)	101, 109
		PON2		PON2: barely detectable aryl esterase; no phosphotriesterase	
		PON3		PON3: low aryl esterase; barely detectable phosphotriesterase	
2	Tautomerase superfamily	Malonate semialdehyde decarboxylase (MSAD)	Decarboxylation (malonate semialdehyde, 2.2×10^7)	Hydration (2-oxo-3-pentynoate, 6×10^2)	80–83
		4-oxalocrotonate tautomerase (4-OT)	Isomerization (2-oxo-4E-hexenedioate, 2.0×10^7)	Hydrolytic dehalogenation (CaaD activity) (3E-chloroacrylate, 2.6×10^{-2})	
		YwhB tautomerase (4-OT analog)	Isomerization (2-oxo-4E-hexenedioate, 2.8×10^4)	Hydrolytic dehalogenation (CaaD activity) (3E-chloroacrylate, 4.4×10^{-2})	
		trans-3-chloroacrylic acid dehalogenase (CaaD)	Hydrolytic dehalogenation (3E-chloroacrylate, 1.2×10^5)	Hydration (2-oxo-3-pentynoate, 6.4×10^3)	
3	ROK family (repressor, open reading frame, kinase)	NanK	*N*-acetyl-D-mannosamine kinase (*N*-acetyl-D-mannosamine, 2.7×10^5)	Glucose kinase (glucose, 5.1×10^2)	123, 127
		YajF	Fructose kinase (fructose, 1.1×10^4)	Glucose kinase (glucose, 2×10^2)	
		YcfX	Unknown	Glucose kinase (glucose, 2.4×10^3)	
		AlsK	Allose kinase (allose, 6.5×10^4)	Glucose kinase (glucose, 15)	
4	Enolase superfamily: MLE (muconate lactonizing enzyme) subgroup	*o*-succinylbenzoate synthase (OSBS)	Dehydration [2-succinyl-6R-hydroxy-2,4-cyclohexadiene-1R-carboxylate (SHCHC), 2.5×10^5]	*N*-acylaminoacid racemase (NAAAR) (*N*-acetyl methionine isomers, $4.9–5.9 \times 10^2$)	16, 177

5	Amido hydrolase superfamily	Phosphotriesterase (PTE)	Phosphotriesterase (paraoxon, 4×10^7)	Aryl esterase (2-naphthyl acetate, 500); lactonase (dihydrocoumarin, 6.5×10^5)	103, 104
		Phosphotriesterase homology protein	Unknown	Aryl esterase (2-naphthyl acetate, 70)	
		Dihydroorotase	Dihydroorotic acid hydrolysis (dihydroorotic acid, 1.2×10^6)	Phosphotriesterase (paraoxon, 2.8)	
		AhlA; a member of the PLL family (PTE-like lactonases)	Lactonase (*N*-3-oxooctanoyl L-homoserine lactone, 0.7×10^6)	Phosphotriesterase (paraoxon, 0.5)	
		PPH; a member of the PLL family (PTE-like lactonases)	Lactonase (*N*-3-oxooctanoyl L-homoserine lactone, 0.55×10^5)	Phosphotriesterase (paraoxon, 8.6)	
		PTE-like lactonase SsoPox	Lactonase (*N*-3-oxooctanoyl L-homoserine lactone, $>10^6$)	Aryl esterase (naphthyl acetate, 400); phosphotriesterase (paraoxon, 4000)	
6	Orotidine 5′ monophosphate decarboxylase suprafamily (OMPDC)	3′ keto L-gluconate 6-phosphate decarboxylase (KGPDS)	Decarboxylation (3′ keto L-gluconate 6-phosphate, 7.7×10^4)	HPS activity, aldol condensation (D-ribulose 5-phosphate and formaldehyde, 8.2×10^{-2})	178
		D-*arabino*-hex-3-ulose 6-phosphate synthase (HPS)	Aldol condensation (D-ribulose 5-phosphate and formaldehyde, 1.6×10^4)	KGPDS activity, decarboxylation (3′ keto L-gluconate 6-phosphate, 2.3×10^3)	
7	*N*-acetyl-neuraminate lyase (NAL) family, pyruvate-dependent aldolases	*N*-acetyl-neuraminate lyase	Cleavage of *N*-acetyl-neuraminate (3.1×10^3)	DHDPS activity, aldol condensation (pyruvate and L-aspartate-β-semialdehyde, 20)	179
		Dihydrodipicolinate synthase (DHDPS)	Aldol condensation (pyruvate and L-aspartate-β-semialdehyde	—	
8	Alkaline phosphatase superfamily	Alkaline phosphatase	Phosphomonoesters hydrolysis (p-nitrophenyl phosphate, 3.3×10^7)	Phosphodiesters hydrolysis (bis-p-nitrophenyl phosphate, 5×10^{-2})	36–38, 79
				Sulfate ester hydrolysis (p-nitrophenyl sulfate, 1×10^{-2})	
		Arylsulfatase	Sulfate ester hydrolysis (p-nitrophenyl sulfate, 5×10^7)	Phosphomonoesters hydrolysis (p-nitrophenyl phosphate, 790)	
		Nucleotide pyrophosphatase-phosphodiesterase	Phosphodiesters hydrolysis (thymidine 5′-monophosphate 4-nitrophenyl ester, 1.6×10^6)	Phosphomonoesters hydrolysis (p-nitrophenyl phosphate, 1.1)	

(Continued)

Table 1 (*Continued*)

Entry number	Family/superfamily	Enzymes	Native activity (substrate, k_{cat}/K_M in $M^{-1}s^{-1}$)	Promiscuous activity (substrate, k_{cat}/K_M in $M^{-1}s^{-1}$)	References
9	Guanidino-modifying enzyme superfamily, hydrolase branch	Arginine deiminase (PaADI)	Arginine hydrolysis (arginine, 4.5×10^4)	PaDDAH activity, N^w,N^w-dimethylarginine hydrolysis (N^w,N^w-dimethylarginine, 1.8×10^3)	75
		Agmantine deiminase (PaAgDI)	Agmantine hydrolysis (agmantine, 7×10^3)	None	
		N^w,N^w-dimethylarginine dimethylaminohydrolase (PaDDAH)	N^w,N^w-dimethylarginine hydrolysis (N^w,N^w-dimethylarginine, 1.8×10^3)	PaADI activity, arginine hydrolysis (arginine, 1.8)	
10	Pyridoxal 5′-phosphate-dependent transferases superfamily	Dopa decarboxylase	Decarboxylation of L-aromatic amino acids into aromatic amines [3,4-dihydroxyphenylalanine (Dopa), 6.1×10^4]	Half-transamination of D-aromatic amino acids (5-hydroxytryptophan, 1.3) Oxidative deamination of aromatic amines (5-hydroxytryptamine, 35)	76, 77
		Phenylacetaldehyde synthase (PAAS)	Coupled decarboxylation and amine oxidation (phenylalanine, 667)		
11	C-C hydrolase family (branch of α/β hydrolase superfamily)	C-C hydrolase *MhpC*	C-C bond cleavage (2-hydroxy 6-keto-nona-2,4-dienoic acid, 28 units)	Esterase (monoethyl adipate, 0.0027 units)	180–182
				Thioesterase (thioethyl adipate, 0.46 units)	
				Hydroxamic acid formation (monoethyl adipate + NH_4OH, 0.013 units)	
		Haloperoxidase/esterase *ThcF*	Haloperoxidase (monochlorodimedon, V_{max} = 0.45 nmol/min)	Esterase (p-nitrophenyl acetate, V_{max} = 2.58 nmol/min)	
		Lactonase	Lactonase (3,4-dihydrocoumarin, V_{max} = 4760 units)	Haloperoxidase (monochlorodimedon, V_{max} = 199 units)	
12	Glutathione-*S*-transferase (GST) superfamily (zeta class)	TCHQ dehalogenase	Dehalogenation [tetrachlorohydroquinone (TCHQ), 3.6×10^4]	Isomerization of double bonds (maleylacetone, 410) Dehalogenation [dichloroacetic acid (DCA), 23]	114
		MAA isomerase	Isomerization of double bonds [maleylacetoacetate (MAA), 1670]	—	
		Zeta GST	Dehalogenation [dichloroacetic acid (DCA), 8500]	—	

activity (103) that could comprise a vestige of its progenitor. Indeed, three homologs from the same superfamily (amidohydrolase) turned out to be representatives of a new group of microbial lactonases, dubbed PTE-like lactonases (PLLs) (104). These three PLLs, and some newly identified ones (105–108), proficiently hydrolyze lactones, particularly *N*-acyl homoserine quorum-sensing lactones, and exhibit weaker promiscuous PTE activities. PLLs share key sequence and active-site features with PTE and differ primarily by an insertion in one active-site loop (104, 107, 108). Given their function and phylogeny, PLLs emerged dozens of millions of years ago. The latent promiscuous phosphotriesterase activity of a yet-to-be-identified PLL served as the essential starting point for the evolution of PTE (104).

2. The same promiscuous activity is often shared by more than one family member (**Table 1**; entries 2, 3, 5, and 11).
3. The magnitude of promiscuous functions varies dramatically between family members (**Table 1**, entries 1, 5, and 9). For example, in the mammalian paraoxonases family, the promiscuous PTE activity is high in one paralog (PON1; k_{cat}/K_M $\sim 10^4$ $M^{-1}s^{-1}$) and barely detectable or undetectable in the two other paralogs. Indeed, the consistency of the lactonase function in all PON paralogs and orthologs and the haphazardness of others' activities (phosphotriesterase and aryl esterase; **Table 1**, entry 1) prompted the identification of the lactonase as the native function of PONs (101, 109). This pattern is consistent with promiscuous activities not being under selection and also with the observation that promiscuous activities show large increases and decreases in response to one or a few mutations that are neutral with respect to the primary function (110, 111).
4. Laboratory evolution of one promiscuous activity often leads, indirectly, to the appearance of other promiscuous activities thus yielding "generalist" intermediates (see Evolutionary Aspects of Promiscuity, below) (112). Activities found in these generalist intermediates can be shared by other family members, as either their native or promiscuous function (103, 113).

EVOLUTIONARY ASPECTS OF PROMISCUITY

Studies of divergent evolution within enzyme families and superfamilies support the hypothesis that throughout evolution promiscuous activities served as the starting points for the divergence of new functions and that broad-specificity enzymes served as progenitors for today's specialized enzymes (7). Evidence of this, however, is largely circumstantial and provides little insight into the mechanisms and mutational paths that underlined these processes of divergence. Describing the mutational paths is a particular challenge, because in today's enzymes, even within the same superfamily, different functions imply sequence differences ranging from 30% up to 80%. In addition, most of these sequence changes relate to "drift" rather than change of function. Furthermore, paths leading from one function to another are most likely to be gradual (one mutation at a time) and smooth (via intermediates that are all folded and functional to some degree) (115). A detailed discussion of evolutionary mechanisms is beyond the scope of this review but we do outline several key points (for additional information see References 116 and 117).

That natural paths of divergence are most likely to be gradual is also supported by laboratory evolution, where it seems that one "should select what is already there" (118), i.e., evolving an existing weak, promiscuous function is the most feasible option. Indeed, to our knowledge, there exists only one example for the laboratory evolution of an enzymatic function in a noncatalytic fold (RNA ligase evolved from a zinc finger scaffold), and this evolution demanded the exploration of genetic diversity ($>10^{12}$ library variants) that exceeds natural diversities

(119). When no initial activity was present, incorporation of a new function demanded intensive sequence alterations, including deletion and insertion of active-site loops, even within an enzyme from the same superfamily (120). Generation of novel enzymes by computational design involved the simultaneous exchange of 8–20 amino acids (121, 122). Most notably, all the above noted cases involve starting points, and/or intermediates, that possess no activity, or even folding capability, whatsoever.

Evolvability of Promiscuous Functions: The Three Basic Postulates

Given the likelihood of gradual, smooth transitions, it is likely that natural evolution routinely takes advantage of promiscuous activities as starting points for the divergence of new enzymes. However, for promiscuity to lead the divergence of new enzyme functions, the following three basic prerequisites (discussed in detail in sections below) should be met.

1. The promiscuous activities provide an immediate physiological advantage and could thus become selected.
2. Once a promiscuous function becomes physiologically relevant, it can be improved through one, or just few, mutation(s), initially without abolishing the primary, native function of the enzyme.
3. The divergence path can be completed to give a newly specialized enzyme, for which the promiscuous activity became the native one.

Promiscuous Functions Can Provide an Immediate Advantage

Many reports indicate that weak promiscuous activities can provide an immediate selective advantage to an organism, typically following a deficiency created by a genetic manipulation in the laboratory. A systematic study conducted by Patrick et al. (32) is discussed in the section Promiscuity: Rule or Exception? In an *E. coli* strain deficient in glucokinase activity, several sugar kinases were found that promiscuously phosphorylate glucose (123). These promiscuous activities are notably weak (**Table 1**); the k_{cat}/K_M values of the promiscuous sugar kinase *YajF* are in the range of 10^2 $M^{-1}s^{-1}$ and are $\sim10^4$ lower than that of the primary *E. coli* glucokinase (*Glk*). Indeed, in these cases, overexpression of the promiscuous enzyme from a multiple-copy plasmid was necessary, as low catalytic efficiency can be clearly compensated by higher enzyme levels (60, 124).

Another notable example is alkaline phosphatase, whose promiscuous phosphite oxidation complemented the growth deficiency of *E. coli phn* knockout strains (**Figure 3*a***). The ability to grow on phosphite as the sole source of inorganic phosphorous occurred via the chromosomal gene of alkaline phosphatase, owing to the extremely high expression levels of the native alkaline phosphatase under phosphate starvation (59). In other cases, changes in regulation of chromosomal genes, leading to higher expression, were observed (60). In Hall's classical experiment (125, 126) of the emergence of an alternative β-galactosidase, mutations increased a weak promiscuous β-galactosidase activity in *egb* (a glycosylase whose native function remains unknown). The first mutation dramatically increased the expression of *egb* by disabling its repressor (125). A promoter mutation in a complementing plasmid also led to ~100-fold increase in expression level of the promiscuous glucokinase *YajF* (127). Gene duplication is another abundant event, leading to increased enzyme levels (117, 128–130). Thus, if and when a new function becomes necessary, the combination of a weak promiscuous activity with an increase in enzyme levels via regulatory mutations and/or gene duplication can provide the organism an immediate advantage.

Negative Trade-offs and the Evolvability of Promiscuous Functions

The second postulate regarding the evolvability of promiscuous functions is that promiscuous functions can be readily improved through one, or just a few, mutations and that mutations leading to improvements in promiscuous functions need not induce parallel decreases in the native function. Strong negative trade-offs between the evolving trait and existing traits are a dominant factor in evolution (131). Hence, gene duplication, and a split of the original and evolving functions between the two copies, is considered a prerequisite for adaptation. The weak trade-off hypothesis allows alternative modes for emergence of new genes carrying new functions.

Evolvability, or evolutionary adaptability, is the capacity of biological systems, whether they are organisms, cells, or proteins, to evolve. Evolvability comprises two elements: plasticity and robustness (51, 132). Plasticity is the induction of novel phenotypic traits by a relatively small number of mutations. This property of promiscuous enzyme functions has been demonstrated by numerous laboratory evolution experiments. Moreover, it seems that the more promiscuous and versatile is a metabolic pathway, the more evolvable are the enzymes within it (133). However, plasticity is in conflict with the fact that most mutations are deleterious (134–136). Organisms constantly endure mutations while maintaining fitness. They therefore maintain a certain level of resistance to the effects of mutations (robustness). These two features may appear to be conflicting: Can mutations simultaneously induce no phenotypic changes and significant changes? It appears that biological systems, including proteins, exhibit both traits, namely plasticity and robustness, and the two are not necessarily mutually exclusive (51, 137). The promiscuous, accidental functions of the protein are highly plastic and can be reshaped through a few mutations. However, these mutations need not have a large effect on the protein's native activity. Indeed, many directed evolution experiments indicate that, in contrast to the large shifts observed with the promiscuous substrates, native activities taking place in the same active site show comparatively small changes. This robustness of the native function was observed, although the only selection criterion applied in these experiments was an increase in a promiscuous activity of the target enzyme.

The weak trade-off trend was first described in three different enzymes subjected to a selection for an increase in six different promiscuous activities (138), yet it was also observed in many other laboratory experiments (**Table 2**). On average, mutations increased the promiscuous activity under selection by 10–10^6-fold, whereas the original activity decreased by 0.8–42-fold. In most cases, the ratio of increase in the evolving promiscuous function versus decrease in the original function is ≥ 10. Similar trends were seen in receptors, where mutations leading to the binding of a new ligand initially broadened the spectrum of bound ligands while retaining the original one (139, 140). In bacterial transcription factors, new effector specificities were acquired during natural or laboratory evolution based on existing promiscuous effectors, and with weak trade-offs with respect to the original effector (141–143).

Original, or existing, function: the native function in the evolutionary context of divergence of new functions

The different effects of mutations on the native versus the promiscuous functions are particularly striking in view of the fact that many of these mutations are found within the active site's wall and perimeter. Structural and thermodynamic insights into the effects of these generalist mutations are needed before any definite statements can be made. Yet, it seems likely that the plasticity of the mutated residues is related to the fact that they are not part of the protein's scaffold or of the catalytic machinery of the enzyme. They are typically located on surface loops that exhibit high conformational flexibility and comprise the substrate-binding part of the active site (66, 67, 138, 144, 145). Indeed, conformational plasticity provides a straightforward explanation for weak trade-offs at the early stages of divergence (**Figure 1**). For example, in α-lytic protease, structural flexibility of the substrate-binding loops (146) enabled a single amino acid substitution to increase the activity toward promiscuous substrates by a factor of 10^5, whereas the native activity was reduced by only twofold (147). In an evolved aminoacyl-tRNA synthetase, the disruption of an α-helix introduced structural plasticity to the enzyme's active site and thus enabled it to accept a range of unnatural amino acid substrates (148).

Table 2 Examples of directed evolution of promiscuous enzyme functions and their trade-offs with the native function[a]

Entry number	Enzyme	Native activity (catalytic efficiency of wild type)	Promiscuous activity (catalytic efficiency of wild type)	Mutations in selected variants for higher promiscuous activity	Changes in the evolved promiscuous activity ($k_{cat}/K_M^{variant}$/ k_{cat}/K_M^{wt})	Changes in native activity ($k_{cat}/K_M^{variant}$/ k_{cat}/K_M^{wt})	Comments	References
1	Aspartate amino-transferase (AATase)	Transamination of dicarboxylic substratrates (k_{cat}/K_M = $9.1 M^{-1}s^{-1}$)	Transamination of tyrosine (k_{cat}/K_M = 0.055 $M^{-1}s^{-1}$) and phenylalanine (k_{cat}/K_M = 0.012 $M^{-1}s^{-1}$)	*Pro13Thr* *Asn69Ser* *Gly72Asp* *Arg129Gly* *Thr167Ala* *Ala293Val* *Asn297Ser* *Asn339Ser* *Ala381Val* *Asn396Asp* *Ala398Val*	130- and 270-fold higher, respectively	1.2-fold higher	This is a clear example of a generalist intermediate. The in vitro–evolved enzyme exhibits wild-type-like AATase activity and TATase activity that is >10% of that of wild-type TATase	113
2	Muconate lactonizing enzyme (MLE II)	Cycloisomerization (k_{cat}/K_M = 2×10^4 $M^{-1}s^{-1}$)	β-elimination (o-succinylbenzoate synthase, OSBS activity). No detectable promiscuous activity (nondetectable) (k_{cat}/K_M < 1.5×10^{-3} $M^{-1}s^{-1}$)	*Glu323Gly*	>1.2 million-fold higher	15-fold lower	The corresponding mutation in a homologous enzyme decreased the native function far more significantly (see Reference 158)	183
3	Galactokinase (GalK)	Phosphorylation of D-galactose to produce α-D-galactose-1-phosphate (k_{cat}/K_M = 860 $M^{-1}s^{-1}$)	Phosphorylation of C5- or C6-substituted sugars (k_{cat}/K_M = 138 $M^{-1}s^{-1}$ for 2-deoxy D-galactose, and lower for the other substrates)	*Tyr371His*	Fivefold higher for 2-deoxy D-galactose, and higher improve-ments for the other target substrates	1.3-fold lower	This variant accommodates an expanded spectrum of substrates, including substrates that were not used in the screen	184

4	β-glucuronidase (GUS)	Hydrolysis of β-glucuronides (k_{cat}/K_M = 8.3×10^5 $M^{-1}s^{-1}$)	Hydrolysis of pNP-galactoside (k_{cat}/K_M = 2.3 $M^{-1}s^{-1}$)	*Ile12Val* *Phe365Ser* *Trp529Leu* *Ser557Pro* *Ile560Val*	16-fold higher	8.3-fold lower	Larger increases in the evolving promiscuous galactosidase function of *E. coli* GUS, with smaller changes of the native function, and acquisition of specificities, which were not selected for, were also described (112)	185
5	*SinI* DNA-methyltransferase	Methylation of the internal cytosine of the GG(A/T)CC sequence (k_{cat}/K_M = 2.9×10^5 $M^{-1}s^{-1}$)	Relaxation of sequence specificity toward GG(N)CC (k_{cat}/K_M = 2×10^3 $M^{-1}s^{-1}$)	*Leu214Ser* *Tyr229His*	18.5-fold higher for the GG(G/C)CC sequence	4.5-fold lower	Specificity broadening was also observed with *Hae*III methyltransferase, where the native activity increased together with the acquisition of higher promiscuous activities	186, 187
6	Phosphotriesterase (PTE)	Phosphotriesterase (paraoxon, k_{cat}/K_M = 4×10^7 $M^{-1}s^{-1}$)	Ester hydrolysis (2-naphthyl acetate, k_{cat}/K_M = 480 $M^{-1}s^{-1}$)	*His254Arg* *Phe306Cys* *Pro342Ala*	13-fold higher	Threefold lower	≤150-fold higher activities were observed with other esters, for which there was no selection	103, 138
7	Human carbonic anhydrase (hCAII)	Bicarbonate dehydration (k_{cat}/K_M = 3×10^7 $M^{-1}s^{-1}$)	Esterase (p-nitrophenyl acetate, k_{cat}/K_M = 2×10^3 $M^{-1}s^{-1}$)	*Ala65Val* *Asp110Asn* *Thr200Ala*	40-fold higher	Twofold lower	Mutations in conserved regions of the active site did not affect the highly proficient native activity	138, 157

(*Continued*)

Table 2 (*Continued*)

Entry number	Enzyme	Native activity (catalytic efficiency of wild type)	Promiscuous activity (catalytic efficiency of wild type)	Mutations in selected variants for higher promiscuous activity	Changes in the evolved promiscuous activity ($k_{cat}/K_M{}^{variant}/k_{cat}/K_M{}^{wt}$)	Changes in native activity ($k_{cat}/K_M{}^{variant}/k_{cat}/K_M{}^{wt}$)	Comments	References
8	Mammalian serum paraoxonase (PON1)	Lipo-lactonase[b] (δ-valerolactone, $k_{cat}/K_M = 1.3 \times 10^5$ $M^{-1}s^{-1}$; and γ-heptanolide, $k_{cat}/K_M = 2 \times 10^4$)	Thiolactonase (γ-butyryl thiolactone, $k_{cat}/K_M = 75$ $M^{-1}s^{-1}$)	*Ile291Leu* *Thr332Ala*	80-fold higher	No change[b]	The selected mutations are located on surface loops that contain the substrate-binding pocket	138, 188
			Esterase (2-naphthyl octanoate, $k_{cat}/K_M = 1.5 \times 10^3$ $M^{-1}s^{-1}$)	*Phe292Val* *Tyr293Asp*	31-fold higher	No change		
			Esterase (7-acetoxy coumarin, $k_{cat}/K_M = 1.2 \times 10^5$ $M^{-1}s^{-1}$)	*Phe292Ser* *Val346Met*	62-fold higher	~22-fold lower		
			Phosphotri-esterase (7-diethyl-phosphoro 4-cyano-7-hydroxycoumarin, $k_{cat}/K_M = 9 \times 10^3$ $M^{-1}s^{-1}$)	*Leu69Val* *Ser138Leu* *Ser193Pro* *Asn287Asp*	155-fold higher	2.6-fold lower		

9	Deacetoxy cephalosporin C synthase (DAOCS)	Ring expansion of penicillin N into deacetoxy-cephalosporin C ($k_{cat}/K_M = 2.2 \times 10^4\ M^{-1}s^{-1}$)	Ring expansion of penicillin G into phenylacetyl-7-aminodeacetoxy cephalosporanic acid ($k_{cat}/K_M = 18\ M^{-1}s^{-1}$)	*Val275Ile Ile305Met*	32-fold higher	1.1-fold higher		189
				Cys155Tyr Tyr184His Val275Ile Cys281Tyr	41-fold higher	42-fold lower		
10	β-lactamase TEM-1	Ampicillin hydrolysis ($k_{cat}/K_M = 4.18 \times 10^7\ M^{-1}s^{-1}$)	Cefotaxime hydrolysis ($k_{cat}/K_M = 2.07 \times 10^3\ M^{-1}s^{-1}$)	*Gly238Ser*	86-fold higher	6.2-fold lower		27
			Ceftazidime hydrolysis ($k_{cat}/K_M = 32.1\ M^{-1}s^{-1}$)	*Gly238Ser*	19-fold higher			
			Cefotaxime hydrolysis ($k_{cat}/K_M = 2.07 \times 10^3\ M^{-1}s^{-1}$)	*Gly238Ser Glu104Lys*	806-fold higher	29-fold lower		
			Ceftazidime hydrolysis ($k_{cat}/K_M = 32.1\ M^{-1}s^{-1}$)	*Gly238Ser Glu104Lys*	284-fold higher			
11	Extended-spectrum β-lactamase CTX-M	Hydrolysis of cephalothin and cefotaxime (4×10^6–2×10^7)	Hydrolysis of ceftazidime ($k_{cat}/K_M = 3.3 \times 10^3\ M^{-1}s^{-1}$)	*Gln87Leu His112Tyr Thr230Ile Ala231Val Asp240Gly Arg276His*	24-fold higher	1.5-fold higher, and 1.4-fold lower for cephalothin and cefotaxime, respectively		190

(*Continued*)

Table 2 (*Continued*)

Entry number	Enzyme	Native activity (catalytic efficiency of wild type)	Promiscuous activity (catalytic efficiency of wild type)	Mutations in selected variants for higher promiscuous activity	Changes in the evolved promiscuous activity ($k_{cat}/K_M^{variant}$/ k_{cat}/K_M^{wt})	Changes in native activity ($k_{cat}/K_M^{variant}$/ k_{cat}/K_M^{wt})	Comments	References
12	*Not*I endonuclease	Recognition and cleavage of GCGGCCGC DNA sequence (5×10^5 units/mg enzyme)	Recognition and cleavage of altered 8-bp sequence (no detectable star activity)	*Met91Val Glu156Gly*	>32-fold higher than the Glu156Gly intermediate with GCTGCCGC sequence	23-fold lower	Although a considerable reduction in the rate of cleavage of the original sequence is reported, the cleavage specificity of the *Met91Val/ Glu156Gly* mutant appears to be relaxed toward a whole set of 8-bp sequence targets, with a distinct preference for the original target	191
13	D-allose kinase (AlsK)	Phosphorylation of D-allose (k_{cat}/K_M = 2.5×10^5 $M^{-1}s^{-1}$)	Phosphorylation of D-glucose (k_{cat}/K_M = 3.4×10^2 $M^{-1}s^{-1}$)	*Ala73Gly*	62-fold higher	1.25-fold lower		145
				Phe145Leu	11.4-fold higher	1.28-fold higher		
	N-acetyl D-mannosamine kinase (NanK)	Phosphorylation of *N*-acetyl D-mannosamine (k_{cat}/K_M = 1.5×10^5 $M^{-1}s^{-1}$)	Phosphorylation of D-glucose (k_{cat}/K_M = 3.4×10^3 $M^{-1}s^{-1}$)	*Leu84Pro*	11.8-fold higher	Twofold lower		
				Val138Met	6.4-fold higher	1.25-fold lower		

14	ProFAR isomerase (HisA)	Isomerization of N'-[(5'-phosphoribosyl) formimino]-5-aminoimidazole-4-carboxamide ribonucleotide ($k_{cat}/K_M = 1.2 \times 10^6\ M^{-1}s^{-1}$)	Isomerization of phosphoribosylanthranilate: TrpF activity is below detection limits	*Asp127Val*	Wild-type activity is below detection limits	~10^4-fold lower	Almost all the original HisA activity was lost	192

[a]These are examples from the past few years for which kinetic parameters are available for both the promiscuous activity under selection and the original activity, which was not subjected to selection. For additional examples, see supplementary Table 8 in Reference 138. Because the above analysis aims at providing insights to the evolution of new enzyme functions in nature, the examples selected involve selection for only one parameter, an increase in a promiscuous activity, and make use of gene libraries prepared by mutagenesis in a completely random manner (point mutations or shuffling) and throughout the gene length.

[b]Since the publication of Reference 138, it has been established that serum paraoxonase (PON1) is a lipo-lactonase, and its preferred substrates are 5- and 6-membered ring lactones, typically with aliphatic side chains (101, 109, 193). In the original article (138), data for trade-offs with the native activity were presented with both the aromatic lactone dihydrocoumarin and aliphatic lactones. However, more recent works indicated that dihydrocoumarin is not binding PON1's active site in the same mode as aliphatic lactones (84, 101). Thus, the trade-offs presented here are the average values of two aliphatic lactones (δ-valerolactone and γ-heptanolide).

In addition, as discussed above, there seem to be fundamental differences between the mode of binding of the native substrate versus the promiscuous substrates, and it is therefore likely that the same mutation could differently affect the native and the promiscuous substrates. Better understanding of the effect of mutations awaits a sufficient number of structures of both the wild-type enzyme and its evolved mutants in complex with analogs of both the native and promiscuous substrates.

Something for Nothing: For How Long?

Ultimately, the acquisition of a proficient new activity must come at the expense of the old one. Yet, the relative rate by which a new function is gained, and the old one is lost, matters (**Figure 6**). In those cases, where the negative trade-off is initially weak (convex route), the divergence of new function proceeds via a generalist intermediate exhibiting broad specificity. This route suggests that, under selection for increasingly higher proficiency, specialists might evolve spontaneously (i.e., without an explicit selection against the original function) because at a certain point increases in the new function will be accompanied by large losses in the original one. At present, however, laboratory evolution experiments demonstrate that generalist intermediates re-specialize primarily upon dual selection for an increase in the newly evolving activity and a decrease in the original activity (149, 150–153). However, in a living cell, the toll of a generalist on fitness might be too high, and the driving force for specialization is likely to be stronger than observed in vitro (150).

Altogether, the above observations support the hypothesis of evolutionary progenitors and intermediates being of broad specificity or high promiscuity (7) and that a frequent (but not exclusive) evolutionary route leads from a specialist to a generalist and, in turn, to a new specialist (**Figures 5** and **6**). The reconstruction of evolutionary ancestors of both enzymes and receptors (154, 155), as well as laboratory evolution of protein-protein interfaces (149),

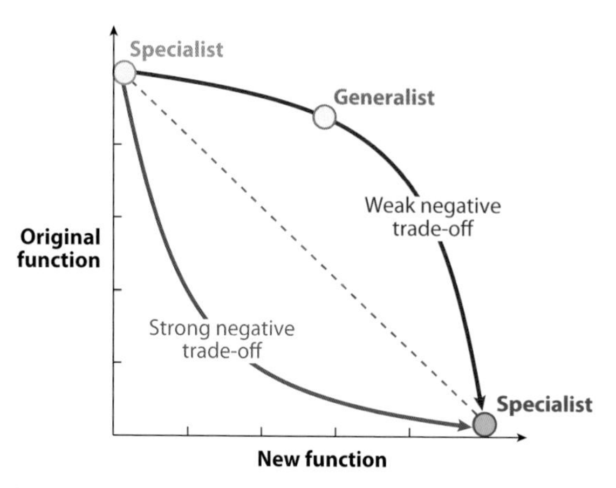

Figure 6

Possible routes to the divergence of a new function. Under selection, a weak, promiscuous activity of a protein with a given primary function (*blue circle*) gradually evolves. By the end of this process, which typically requires many generations of mutation and selection, a new protein emerges with a new function that replaced the original one (*green circle*). The dynamics of this divergence process may vary. The gain-loss of the new versus old function and the conversion of one specialist protein into another may trade off linearly (*dotted line*) or follow either the concave or the convex route. The convex route (weak negative trade-off) is supported by the observation that large increases in the promiscuous function under selection for a new function are often accompanied by significantly smaller decreases in the original function (**Table 2**). By virtue of gaining a new function without losing the original one (and often gaining other new functions that were not selected for), the intermediates of these routes are generalists, and their evolution can therefore proceed prior to gene duplication. In contrast, the concave route implies that gene duplication (or other means of significantly increasing enzyme levels) is a necessary prerequisite because acquisition of even low levels of the new function is accompanied by large losses of the original one. Adapted from Reference 17.

and transcription factors (142, 143) supports the idea of generalist progenitors.

Exceptions to Weak Negative Trade-offs

Although weak negative trade-offs are common, this generalization has notable exceptions. In few cases, a single amino acid exchange can completely switch the specificity of an enzyme (152). For example, the *His89Phe* mutation in the active site of tyrosine ammonia lyase switched its substrate selectivity from tyrosine to phenylalanine, with kinetic parameters and selectivity comparable to phenylalanine ammonia lyase (156). In another example, a mutant (*Glu383Ala*) of *proA* that exhibits higher promiscuous activity with *N*-acetylglutamyl phosphate (*argC* activity) traded off strongly with the original *proA* function (glutamyl phosphate as substrate). In this case, upregulation of the mutated *proA*'s facilitated growth despite the overall low rates with both acetylglutamyl phosphate (the new activity) as well as a significance decrease in the original one (60).

Size and charge considerations. The magnitude of trade-offs may depend on differences in size and charge between the native and promiscuous substrates (157, 158). Most reported studies involve promiscuous substrates that are larger than the native one and cases in which both the native and the promiscuous substrates are uncharged. In these cases, a mutation that makes the active site larger may increase activity toward the promiscuous substrate with no drastic effect on the native substrate. However, in cases where promiscuous substrates are smaller than the native ones, mutations that reshape the active site to increase contacts with the smaller substrate can reduce the activity with the larger native substrate. Still, examples for weak trade-offs at the early steps of directed evolution for smaller substrates exist (159). Other cases in which the native and promiscuous activities might trade off strongly involve differences in charge, e.g., a charged native substrate and a neutral, hydrophobic promiscuous substrate. Mutations that favor the charged substrate are likely to restrict binding of the hydrophobic one, and vice versa (60).

Targeted versus random mutagenesis. Mutations incorporated through rational design (and probably by computational design) show larger trade-offs relative to mutations obtained by selection from random repertoires, **Supplemental Table 2**. Follow the **Supplemental Material link** from the Annual Reviews home page at **http://www.annualreviews.org.** This difference can be largely ascribed to the location of the function-altering mutations. Evolutionary processes, in the laboratory or in nature, usually involve mutations in the active-site periphery (second and third shell mutations) with more subtle effects. However, rational design aims at the replacement of key active-site residues (first shell), and such exchanges yield more drastic changes of specificity. An interesting exception is the introduction of glutathione transferase activity in a glutathione-dependent peroxidase via a single point mutation, which may relate to the evolutionary history of these enzymes (160).

Stability Trade-offs

An important facet of the trade-off that is not reflected in the kinetic parameters per se is the effect of mutations on stability. Most mutations are destabilizing, and mutations that affect function often exhibit even higher destabilizing effects (136). Destabilization usually results in reduced cellular enzyme levels, owing to misfolding and aggregation, proteolytic digestion, or clearance. This phenomenon was first highlighted through the analysis of mutations found in clinical isolates of TEM-1 β-lactamase (27). Thus, although in terms of k_{cat} and K_M the trade-off between the native and the promiscuous functions of TEM-1 mutants is weak (**Table 2**, entry 10), the function-altering mutations are destabilizing, leading to much reduced enzyme levels and slow bacterial growth. For the evolutionary process to continue (in nature or in the laboratory), this loss of stability must be compensated. Indeed, many mutations that appear in directed evolution variants with no obvious role in the new function exert compensatory stabilizing effects (136, 161). A recent review addresses in detail the stability effects of mutations on protein evolution (162).

Promiscuity and Mechanisms for Divergence of New Gene Functions

The mechanisms governing the divergence of new gene-protein functions are a central part of evolutionary theory, a discipline that is unfamiliar to most biochemists. However, the notions of protein promiscuity, and the unique evolutionary features of promiscuous functions, have fundamental implications for this theory. These are briefly mentioned below.

The textbook paradigm, Ohno's model (163), assumes that duplication is a frequent event which is largely neutral, i.e., initially, duplication provides no fitness advantage, or disadvantage, and occurs under no selection (164). The duplicated copy is redundant and free from the burden of selection, and it can therefore accumulate mutations, including deleterious ones. If and when the need arises, variants carrying duplicated genes with mutations that endow a new function become under positive, adaptive selection, thus leading to the divergence of the new gene, protein, and function.

The prerequisites of duplication and relief from selection stem from the negative trade-off assumption: Selection for the existing function purges mutations with adaptive potential, and such mutations can only accumulate in a redundant copy. However, as described above, many promiscuous functions further evolve with little effect on the original function. Ohno's hypothesis—that gene duplication and the subsequent mutational drift occur under no selection—is being questioned for additional reasons. First, most duplicated genes found in existing genomes appear to be under functional selection that purges deleterious mutations (128, 165, 166). Second, expression of redundant mRNA and protein copies carries substantial energetic costs (56, 167), and there exists a selection pressure to inactivate their expression (55, 168). Third, gene duplication is often not a neutral event but is rather positively selected under demands for higher protein doses (129, 130). Last, over a third of the random mutations in a given protein are deleterious (134, 135, 169), whereas beneficial mutations promoting new functions are scarce (estimated frequency of $\sim 10^{-3}$). Thus, when drifting in the absence of any selection, loss of all functions (nonfunctionalization) because of mutations that undermine folding and stability (162, 170) is orders of magnitude more likely than neofunctionalization (171).

Moonlighting: utilization of protein parts outside the active site for other functions, mostly regulatory and structural

The above observations prompted a number of alternative scenarios, which include the following:

1. Gene sharing model—a gene with a given function is recruited for a different, moonlighting function without any changes in the coding region (50, 172).
2. Divergence prior to duplication model—this model (131, 138), the parallel innovation-amplification-divergence model (130), the escape from adaptive conflict model (173), and to a degree the Hughes' model (165) assume that the very first step toward divergence is the selection of a mutant protein with sufficiently high secondary, promiscuous activity, while retaining the original, primary function. Duplication follows and enables the complete re-specialization of the diverging function at the expense of the original function (**Figure 6**).
3. Duplication is a positively selected event, leading to increased variability. When divergence is capitalizing on a weak promiscuous activity in an existing protein, immediate selective advantage can be provided by increasing protein doses. Thus, duplication and the resulting higher protein levels have key roles in enabling promiscuous functions to become physiologically relevant and in enabling a wider variety of function altering mutations to accumulate. Despite the generally weak trade-offs, at the end of the day, mutations that endow new enzymatic functions often have a measurable effect on the existing enzymatic function and/or on the enzyme's stability and expression levels. The acquisition of potentially beneficial mutations can only continue as long as the existing function is reduced to an extent that does not severely compromise organismic fitness. By virtue of two genes carrying the same level of function, duplication offers a margin that allows a larger variety of potentially beneficial mutations to accumulate provided that, contrary to Ohno's model, the two genes remain under selection (117, 128, 171).

SUMMARY POINTS

1. Promiscuity regards reactions that an enzyme performs, although it never evolved to do so (as opposed to its original, native activity).
2. Promiscuous activities are not rare exceptions but are rather widely spread, inherent features of enzymes, and proteins in general.
3. Specificity and promiscuity can reside within the same active site. Promiscuous enzymatic functions may utilize different active-site conformers, and their mechanisms can overlap, partly overlap, or differ altogether from the mechanism by which an enzyme performs its native function.
4. Promiscuous enzyme functions provide immediately accessible starting points for the evolution of new functions via a gradual mutational path that eventually converts a weak, promiscuous function into the primary, native function.
5. A promiscuous function of an enzyme can be a vestige of the function of its ancestor. Promiscuous activities shared by members of same enzyme family and/or superfamily correlate with their divergence from a common ancestor.
6. Mutations that increase a promiscuous activity and have little effect on the primary, native function (weak trade-off) underlie the divergence of a new enzymatic function via a generalist intermediate.

7. The notion of promiscuity as the seed of new gene functions has significant implications for evolutionary theory. Although gene duplication is the key to divergence of new gene functions, when and how duplication occurs and how a new enzyme diverges from an existing one are still a matter of debate.

FUTURE ISSUES

1. Rigorous and quantitative measures of promiscuity are needed, including ways of systematically measuring the magnitude and degree of promiscuity in a wide range of proteins.
2. Are promiscuous functions executed in modes (structural, thermodynamic, kinetic, and/or mechanistic) that fundamentally differ from the modes of primary, native function?
3. Better physicochemical understanding of the effects of mutations on native versus promiscuous activities, and of the origins of the weak trade-offs between the evolving promiscuous activity and the original activity, is needed.
4. Clear cut cases of natural enzymes that diverged from other natural enzymes by virtue of a latent promiscuous activity might be identified from inferred ancestors, recently evolved bacterial enzymes that degrade anthropogenic or xenobiotic chemicals, or secondary metabolism of plants.
5. It remains unclear whether there are fundamental structural and mechanistic differences between generalists and specialists and whether the evolutionary history of an enzyme dictates its future. Are highly specialized enzymes of the central metabolism (enzymes that experienced little change) less promiscuous and less evolvable than secondary metabolism enzymes that have constantly evolved under changing environments?
6. Complete evolutionary trajectories from one specialist to another specialist, whereby the promiscuous activity becomes primary, and vice versa, need to be reproduced in the laboratory (**Figure 6**). Such experiments can unravel the molecular basis of conversion and possible reversion of traits such as robustness and evolvability.
7. The roles of promiscuity of individual enzymes (flexible proteomes) and of cross-wiring of metabolic pathways (flexible metabolomes), in both physiological and evolutionary adaptations, need to be examined.

DISCLOSURE STATEMENT

The authors are not aware of any affiliations, memberships, funding, or financial holdings that might be perceived as affecting the objectivity of this review.

ACKNOWLEDGMENTS

Financial support by the Sasson and Marjorie Peress Philanthropic Fund, the Israel Science Foundation, the European Union via the MiFEM network, and the Adams Fellowship to O.K. is gratefully acknowledged. We thank Shelley Copley for reviewing the manuscript and for her constructive criticism and thank John Gerlt, Patsy Babbitt, and our lab members for insightful discussions.

LITERATURE CITED

1. Hult K, Berglund P. 2007. Enzyme promiscuity: mechanism and applications. *Trends Biotechnol.* 25:231–38
2. Pocker Y, Stone JT. 1965. The catalytic versatility of erythrocyte carbonic anhydrase. The enzyme-catalyzed hydrolysis of rho-nitrophenyl acetate. *J. Am. Chem. Soc.* 87:5497–98
3. Reid TW, Fahrney D. 1967. The pepsin-catalyzed hydrolysis of sulfite esters. *J. Am. Chem. Soc.* 89:3941–43
4. Nakagawa Y, Bender ML. 1969. Modification of alpha-chymotrypsin by methyl p-nitrobenzenesulfonate. *J. Am. Chem. Soc.* 91:1566–67
5. Jackson RC, Handschumacher RE. 1970. *Escherichia coli* L-asparaginase. Catalytic activity and subunit nature. *Biochemistry* 9:3585–90
6. Jacob F. 1977. Evolution and tinkering. *Science* 196:1161–66
7. Jensen RA. 1976. Enzyme recruitment in evolution of new function. *Annu. Rev. Microbiol.* 30:409–25
8. O'Brien PJ, Herschlag D. 1999. Catalytic promiscuity and the evolution of new enzymatic activities. *Chem. Biol.* 6:R91–105
9. Copley SD. 2003. Enzymes with extra talents: moonlighting functions and catalytic promiscuity. *Curr. Opin. Chem. Biol.* 7:265–72
10. Bornscheuer UT, Kazlauskas RJ. 2004. Catalytic promiscuity in biocatalysis: using old enzymes to form new bonds and follow new pathways. *Angew. Chem. Int. Ed. Engl.* 43:6032–40
11. Kazlauskas RJ. 2005. Enhancing catalytic promiscuity for biocatalysis. *Curr. Opin. Chem. Biol.* 9:195–201
12. Nobeli I, Favia AD, Thornton JM. 2009. Protein promiscuity and its implications for biotechnology. *Nat. Biotechnol.* 27:157–67
13. Bonner CA, Disz T, Hwang K, Song J, Vonstein V, et al. 2008. Cohesion group approach for evolutionary analysis of TyrA, a protein family with wide-ranging substrate specificities. *Microbiol. Mol. Biol. Rev.* 72:13–53
14. Glasner ME, Gerlt JA, Babbitt PC. 2006. Evolution of enzyme superfamilies. *Curr. Opin. Chem. Biol.* 10:492–97
15. O'Brien PJ. 2006. Catalytic promiscuity and the divergent evolution of DNA repair enzymes. *Chem. Rev.* 106:720–52
16. Palmer DR, Garrett JB, Sharma V, Meganathan R, Babbitt PC, Gerlt JA. 1999. Unexpected divergence of enzyme function and sequence: "*N*-acylamino acid racemase" is *o*-succinylbenzoate synthase. *Biochemistry* 38:4252–58
17. Khersonsky O, Roodveldt C, Tawfik DS. 2006. Enzyme promiscuity: evolutionary and mechanistic aspects. *Curr. Opin. Chem. Biol.* 10:498–508
18. Toscano MD, Woycechowsky KJ, Hilvert D. 2007. Minimalist active-site redesign: teaching old enzymes new tricks. *Angew. Chem. Int. Ed. Engl.* 46:3212–36
19. Hartley CJ, Newcomb RD, Russell RJ, Yong CG, Stevens JR, et al. 2006. Amplification of DNA from preserved specimens shows blowflies were preadapted for the rapid evolution of insecticide resistance. *Proc. Natl. Acad. Sci. USA* 103:8757–62
20. Janssen DB, Dinkla IJ, Poelarends GJ, Terpstra P. 2005. Bacterial degradation of xenobiotic compounds: evolution and distribution of novel enzyme activities. *Environ. Microbiol.* 7:1868–82
21. Newcomb RD, Campbell PM, Ollis DL, Cheah E, Russell RJ, Oakeshott JG. 1997. A single amino acid substitution converts a carboxylesterase to an organophosphorus hydrolase and confers insecticide resistance on a blowfly. *Proc. Natl. Acad. Sci. USA* 94:7464–68
22. Raushel FM, Holden HM. 2000. Phosphotriesterase: an enzyme in search of its natural substrate. *Adv. Enzymol. Relat. Areas Mol. Biol.* 74:51–93
23. Wackett LP. 2009. Questioning our perceptions about evolution of biodegradative enzymes. *Curr. Opin. Microbiol.* 12:244–51

24. Copley SD. 2009. Evolution of efficient pathways for degradation of anthropogenic chemicals. *Nat. Chem. Biol.* 5:559–66
25. Barlow M, Hall BG. 2002. Phylogenetic analysis shows that the OXA beta-lactamase genes have been on plasmids for millions of years. *J. Mol. Evol.* 55:314–21
26. Hall BG. 2004. Predicting the evolution of antibiotic resistance genes. *Nat. Rev. Microbiol.* 2:430–35
27. Wang X, Minasov G, Shoichet BK. 2002. Evolution of an antibiotic resistance enzyme constrained by stability and activity trade-offs. *J. Mol. Biol.* 320:85–95
28. Weinreich DM, Delaney NF, Depristo MA, Hartl DL. 2006. Darwinian evolution can follow only very few mutational paths to fitter proteins. *Science* 312:111–14
29. O'Maille PE, Malone A, Dellas N, Andes Hess B Jr, Smentek L, et al. 2008. Quantitative exploration of the catalytic landscape separating divergent plant sesquiterpene synthases. *Nat. Chem. Biol.* 4:617–23
30. Austin MB, O'Maille PE, Noel JP. 2008. Evolving biosynthetic tangos negotiate mechanistic landscapes. *Nat. Chem. Biol.* 4:217–22
31. Field B, Osbourn AE. 2008. Metabolic diversification—independent assembly of operon-like gene clusters in different plants. *Science* 320:543–47
32. Patrick WM, Quandt EM, Swartzlander DB, Matsumura I. 2007. Multicopy suppression underpins metabolic evolvability. *Mol. Biol. Evol.* 24:2716–22
33. Khersonsky O, Tawfik DS. 2010. Enzyme promiscuity—evolutionary and mechanistic aspects. In *Comprehensive Natural Products Chemistry*, Vol. 8, ed. D. Barton, O. Meth-Cohn. Oxford: Elsevier. In press
34. van Loo B, Jonas S, Babtie AC, Benjdia A, Berteau O, et al. 2010. An efficient, multiply promiscuous hydrolase in the alkaline phosphatase superfamily. *Proc. Natl. Acad. Sci. USA* 107:2740–45
35. Nath A, Atkins WM. 2008. A quantitative index of substrate promiscuity. *Biochemistry* 47:157–66
36. Catrina I, O'Brien PJ, Purcell J, Nikolic-Hughes I, Zalatan JG, et al. 2007. Probing the origin of the compromised catalysis of *E. coli* alkaline phosphatase in its promiscuous sulfatase reaction. *J. Am. Chem. Soc.* 129:5760–65
37. Olguin LF, Askew SE, O'Donoghue AC, Hollfelder F. 2008. Efficient catalytic promiscuity in an enzyme superfamily: an arylsulfatase shows a rate acceleration of 10^{13} for phosphate monoester hydrolysis. *J. Am. Chem. Soc.* 130:16547–55
38. O'Brien PJ, Herschlag D. 2001. Functional interrelationships in the alkaline phosphatase superfamily: phosphodiesterase activity of *Escherichia coli* alkaline phosphatase. *Biochemistry* 40:5691–99
39. Michaud GA, Salcius M, Zhou F, Bangham R, Bonin J, et al. 2003. Analyzing antibody specificity with whole proteome microarrays. *Nat. Biotechnol.* 21:1509–12
40. Varga JM, Kalchschmid G, Klein GF, Fritsch P. 1991. Mechanism of allergic cross-reactions. II. Cross-stimulation, by chemically unrelated ligands, of rat basophilic leukemia cells sensitized with an anti-DNP IgE antibody. *Mol. Immunol.* 28:655–59
41. Varga JM, Kalchschmid G, Klein GF, Fritsch P. 1991. Mechanism of allergic cross-reactions. I. Multi-specific binding of ligands to a mouse monoclonal anti-DNP IgE antibody. *Mol. Immunol.* 28:641–54
42. Inman J. 1978. The antibody combining region: Speculation on the hypothesis of general multispecificity. In *Theoretical Immunology*, ed. GI Bell, AS Perelson, Pimbley GH Jr, pp. 243–78. New York: Dekker
43. Lancet D, Sadovsky E, Seidemann E. 1993. Probability model for molecular recognition in biological receptor repertoires: significance to the olfactory system. *Proc. Natl. Acad. Sci. USA* 90:3715–19
44. Perelson AS, Oster GF. 1979. Theoretical studies of clonal selection: minimal antibody repertoire size and reliability of self-non-self discrimination. *J. Theor. Biol.* 81:645–70
45. Griffiths AD, Tawfik DS. 2000. Man-made enzymes–from design to in vitro compartmentalisation. *Curr. Opin. Biotechnol.* 11:338–53
46. James LC, Tawfik DS. 2003. The specificity of cross-reactivity: Promiscuous antibody binding involves specific hydrogen bonds rather than nonspecific hydrophobic stickiness. *Protein Sci.* 12:2183–93
47. Fersht A. 1999. *Structure and Mechanism in Protein Science*. New York: Freeman
48. Kotik-Kogan O, Moor N, Tworowski D, Safro M. 2005. Structural basis for discrimination of L-phenylalanine from L-tyrosine by phenylalanyl-tRNA synthetase. *Structure* 13:1799–807
49. Villiers BR, Hollfelder F. 2009. Mapping the limits of substrate specificity of the adenylation domain of TycA. *ChemBioChem* 10.671–82

50. Piatigorsky J. 2007. *Gene Sharing and Evolution: The Diversity of Protein Functions*. Cambridge, MA/London: Harvard Univ. Press. 320 pp.
51. Wagner A. 2005. Robustness, evolvability, and neutrality. *FEBS Lett.* 579:1772–78
52. Brevet A, Plateau P, Cirakoglu B, Pailliez JP, Blanquet S. 1982. Zinc-dependent synthesis of 5′,5′-diadenosine tetraphosphate by sheep liver lysyl- and phenylalanyl-tRNA synthetases. *J. Biol. Chem.* 257:14613–15
53. Lee YN, Nechushtan H, Figov N, Razin E. 2004. The function of lysyl-tRNA synthetase and Ap4A as signaling regulators of MITF activity in FcεRI-activated mast cells. *Immunity* 20:145–51
54. Guo RT, Chong YE, Guo M, Yang XL. 2009. Crystal structures and biochemical analyses suggest a unique mechanism and role for human glycyl-tRNA synthetase in Ap4A homeostasis. *J. Biol. Chem.* 284:28968–76
55. Dekel E, Alon U. 2005. Optimality and evolutionary tuning of the expression level of a protein. *Nature* 436:588–92
56. Stoebel DM, Dean AM, Dykhuizen DE. 2008. The cost of expression of *Escherichia coli* lac operon proteins is in the process, not in the products. *Genetics* 178:1653–60
57. Kuznetsova E, Proudfoot M, Gonzalez CF, Brown G, Omelchenko MV, et al. 2006. Genome-wide analysis of substrate specificities of the *Escherichia coli* haloacid dehalogenase-like phosphatase family. *J. Biol. Chem.* 281:36149–61
58. Tremblay LW, Dunaway-Mariano D, Allen KN. 2006. Structure and activity analyses of *Escherichia coli* K-12 NagD provide insight into the evolution of biochemical function in the haloalkanoic acid dehalogenase superfamily. *Biochemistry* 45:1183–93
59. Yang K, Metcalf WW. 2004. A new activity for an old enzyme: *Escherichia coli* bacterial alkaline phosphatase is a phosphite-dependent hydrogenase. *Proc. Natl. Acad. Sci. USA* 101:7919–24
60. McLoughlin SY, Copley SD. 2008. A compromise required by gene sharing enables survival: implications for evolution of new enzyme activities. *Proc. Natl. Acad. Sci. USA* 105:13497–502
61. Macchiarulo A, Nobeli I, Thornton JM. 2004. Ligand selectivity and competition between enzymes in silico. *Nat. Biotechnol.* 22:1039–45
62. D'Ari R, Casadesus J. 1998. Underground metabolism. *BioEssays* 20:181–86
63. Kurakin A. 2007. Self-organization versus Watchmaker: ambiguity of molecular recognition and design charts of cellular circuitry. *J. Mol. Recognit.* 20:205–14
64. Greenspan RJ. 2001. The flexible genome. *Nat. Rev. Genet.* 2:383–87
65. van Swinderen B, Greenspan RJ. 2005. Flexibility in a gene network affecting a simple behavior in *Drosophila melanogaster*. *Genetics* 169:2151–63
66. Gamage NU, Tsvetanov S, Duggleby RG, McManus ME, Martin JL. 2005. The structure of human SULT1A1 crystallized with estradiol. An insight into active site plasticity and substrate inhibition with multi-ring substrates. *J. Biol. Chem.* 280:41482–86
67. James LC, Tawfik DS. 2003. Conformational diversity and protein evolution—a 60-year-old hypothesis revisited. *Trends Biochem. Sci.* 28:361–68
68. Meier S, Ozbek S. 2007. A biological cosmos of parallel universes: Does protein structural plasticity facilitate evolution? *BioEssays* 29:1095–104
69. Tokuriki N, Tawfik DS. 2009. Protein dynamism and evolvability. *Science* 324:203–7
70. Yasutake Y, Yao M, Sakai N, Kirita T, Tanaka I. 2004. Crystal structure of the *Pyrococcus horikoshii* isopropylmalate isomerase small subunit provides insight into the dual substrate specificity of the enzyme. *J. Mol. Biol.* 344:325–33
71. Hou L, Honaker MT, Shireman LM, Balogh LM, Roberts AG, et al. 2007. Functional promiscuity correlates with conformational heterogeneity in A-class glutathione *S*-transferases. *J. Biol. Chem.* 282:23264–74
72. Zhao Y, Sun L, Muralidhara BK, Kumar S, White MA, et al. 2007. Structural and thermodynamic consequences of 1-(4-chlorophenyl)imidazole binding to cytochrome P450 2B4. *Biochemistry* 46:11559–67
73. Tomatis PE, Fabiane SM, Simona F, Carloni P, Sutton BJ, Vila AJ. 2008. Adaptive protein evolution grants organismal fitness by improving catalysis and flexibility. *Proc. Natl. Acad. Sci. USA* 105:20605–10

74. Delmas J, Chen Y, Prati F, Robin F, Shoichet BK, Bonnet R. 2008. Structure and dynamics of CTX-M enzymes reveal insights into substrate accommodation by extended-spectrum beta-lactamases. *J. Mol. Biol.* 375:192–201
75. Lu X, Li L, Wu R, Feng X, Li Z, et al. 2006. Kinetic analysis of *Pseudomonas aeruginosa* arginine deiminase mutants and alternate substrates provides insight into structural determinants of function. *Biochemistry* 45:1162–72
76. Bertoldi M, Gonsalvi M, Contestabile R, Voltattorni CB. 2002. Mutation of tyrosine 332 to phenylalanine converts Dopa decarboxylase into a decarboxylation-dependent oxidative deaminase. *J. Biol. Chem.* 277:36357–62
77. Kaminaga Y, Schnepp J, Peel G, Kish CM, Ben-Nissan G, et al. 2006. Plant phenylacetaldehyde synthase is a bifunctional homotetrameric enzyme that catalyzes phenylalanine decarboxylation and oxidation. *J. Biol. Chem.* 281:23357–66
78. Theodossis A, Walden H, Westwick EJ, Connaris H, Lamble HJ, et al. 2004. The structural basis for substrate promiscuity in 2-keto-3-deoxygluconate aldolase from the Entner-Doudoroff pathway in *Sulfolobus solfataricus*. *J. Biol. Chem.* 279:43886–92
79. O'Brien PJ, Herschlag D. 1998. Sulfatase activity of *E. coli* alkaline phosphatase demonstrates a functional link to arylsulfatases, an evolutionarily related enzyme family. *J. Am. Chem. Soc.* 120:12369–70
80. Wang SC, Johnson WH Jr, Whitman CP. 2003. The 4-oxalocrotonate tautomerase- and YwhB-catalyzed hydration of 3E-haloacrylates: implications for the evolution of new enzymatic activities. *J. Am. Chem. Soc.* 125:14282–83
81. Wang SC, Person MD, Johnson WH Jr, Whitman CP. 2003. Reactions of trans-3-chloroacrylic acid dehalogenase with acetylene substrates: consequences of and evidence for a hydration reaction. *Biochemistry* 42:8762–73
82. Poelarends GJ, Serrano H, Johnson WH Jr, Hoffman DW, Whitman CP. 2004. The hydratase activity of malonate semialdehyde decarboxylase: mechanistic and evolutionary implications. *J. Am. Chem. Soc.* 126:15658–59
83. Poelarends GJ, Serrano H, Johnson WH Jr, Whitman CP. 2005. Inactivation of malonate semialdehyde decarboxylase by 3-halopropiolates: evidence for hydratase activity. *Biochemistry* 44:9375–81
84. Khersonsky O, Tawfik DS. 2006. The histidine 115-histidine 134 dyad mediates the lactonase activity of mammalian serum paraoxonases. *J. Biol. Chem.* 281:7649–56
85. Yeung DT, Lenz DE, Cerasoli DM. 2005. Analysis of active-site amino-acid residues of human serum paraoxonase using competitive substrates. *FEBS J.* 272:2225–30
86. Blum MM, Timperley CM, Williams GR, Thiermann H, Worek F. 2008. Inhibitory potency against human acetylcholinesterase and enzymatic hydrolysis of fluorogenic nerve agent mimics by human paraoxonase 1 and squid diisopropyl fluorophosphatase. *Biochemistry* 47:5216–24
87. Branneby C, Carlqvist P, Magnusson A, Hult K, Brinck T, Berglund P. 2003. Carbon-carbon bonds by hydrolytic enzymes. *J. Am. Chem. Soc.* 125:874–75
88. Carlqvist P, Svedendahl M, Branneby C, Hult K, Brinck T, Berglund P. 2005. Exploring the active-site of a rationally redesigned lipase for catalysis of Michael-type additions. *ChemBioChem* 6:331–36
89. Torre O, Alfonso I, Gotor V. 2004. Lipase catalysed Michael addition of secondary amines to acrylonitrile. *Chem. Commun.* 15:1724–25
90. Kaiser ET, Lawrence DS. 1984. Chemical mutation of enzyme active sites. *Science* 226:505–11
91. Bakker M, van Rantwijk F, Sheldon RA. 2002. Metal substitution in thermolysin: catalytic properties of tungstate thermolysin in sulfoxidation with H_2O_2. *Can. J. Chem. Rev. Can. Chim.* 80:622–25
92. da Silva GF, Ming LJ. 2005. Catechol oxidase activity of di-Cu^{2+}-substituted aminopeptidase from *Streptomyces griseus*. *J. Am. Chem. Soc.* 127:16380–81
93. Fernandez-Gacio A, Codina A, Fastrez J, Riant O, Soumillion P. 2006. Transforming carbonic anhydrase into epoxide synthase by metal exchange. *ChemBioChem* 7:1013–16
94. Jing Q, Okrasa K, Kazlauskas RJ. 2009. Stereoselective hydrogenation of olefins using rhodium-substituted carbonic anhydrase—a new reductase. *Chemistry* 15:1370–76
95. Wu ZP, Hilvert D. 1990. Selenosubtilisin as a glutathione-peroxidase mimic. *J. Am. Chem. Soc.* 112:5647–48

96. Boschi-Muller S, Muller S, Van Dorsselaer A, Bock A, Branlant G. 1998. Substituting selenocysteine for active site cysteine 149 of phosphorylating glyceraldehyde 3-phosphate dehydrogenase reveals a peroxidase activity. *FEBS Lett.* 439:241–45
97. Yu HJ, Liu JQ, Bock A, Li J, Luo GM, Shen JC. 2005. Engineering glutathione transferase to a novel glutathione peroxidase mimic with high catalytic efficiency. Incorporation of selenocysteine into a glutathione-binding scaffold using an auxotrophic expression system. *J. Biol. Chem.* 280:11930–35
98. Kourist R, Bartsch S, Fransson L, Hult K, Bornscheuer UT. 2008. Understanding promiscuous amidase activity of an esterase from *Bacillus subtilis*. *ChemBioChem* 9:67–69
99. Forconi M, Herschlag D. 2005. Promiscuous catalysis by the tetrahymena group I ribozyme. *J. Am. Chem. Soc.* 127:6160–61
100. Kunzler DE, Sasso S, Gamper M, Hilvert D, Kast P. 2005. Mechanistic insights into the isochorismate pyruvate lyase activity of the catalytically promiscuous PchB from combinatorial mutagenesis and selection. *J. Biol. Chem.* 280:32827–34
101. Khersonsky O, Tawfik DS. 2005. Structure-reactivity studies of serum paraoxonase PON1 suggest that its native activity is lactonase. *Biochemistry* 44:6371–82
102. James LC, Tawfik DS. 2005. Structure and kinetics of a transient antibody binding intermediate reveal a kinetic discrimination mechanism in antigen recognition. *Proc. Natl. Acad. Sci. USA* 102:12730–35
103. Roodveldt C, Tawfik DS. 2005. Shared promiscuous activities and evolutionary features in various members of the amidohydrolase superfamily. *Biochemistry* 44:12728–36
104. Afriat L, Roodveldt C, Manco G, Tawfik DS. 2006. The latent promiscuity of newly identified microbial lactonases is linked to a recently diverged phosphotriesterase. *Biochemistry* 45:13677–86
105. Chow JY, Wu L, Yew WS. 2009. Directed evolution of a quorum-quenching lactonase from *Mycobacterium avium* subsp. *paratuberculosis* K-10 in the amidohydrolase superfamily. *Biochemistry* 48:4344–53
106. Hawwa R, Aikens J, Turner RJ, Santarsiero BD, Mesecar AD. 2009. Structural basis for thermostability revealed through the identification and characterization of a highly thermostable phosphotriesterase-like lactonase from *Geobacillus stearothermophilus*. *Arch. Biochem. Biophys.* 488:109–20
107. Hawwa R, Larsen SD, Ratia K, Mesecar AD. 2009. Structure-based and random mutagenesis approaches increase the organophosphate-degrading activity of a phosphotriesterase homologue from *Deinococcus radiodurans*. *J. Mol. Biol.* 393:36–57
108. Xiang DF, Kolb P, Fedorov AA, Meier MM, Federov LV, et al. 2009. Functional annotation and three-dimensional structure of Dr0930 from *Deinococcus radiodurans*: a close relative of phosphotriesterase in the amidohydrolase superfamily. *Biochemistry* 48:2237–47
109. Draganov DI, Teiber JF, Speelman A, Osawa Y, Sunahara R, La Du BN. 2005. Human paraoxonases (PON1, PON2, and PON3) are lactonases with overlapping and distinct substrate specificities. *J. Lipid Res.* 46:1239–47
110. Amitai G, Devi-Gupta R, Tawfik DS. 2007. Latent evolutionary potentials under the neutral mutational drift of an enzyme. *HFSP J.* 1:67–78
111. Bloom JD, Romero PA, Lu Z, Arnold FH. 2007. Neutral genetic drift can alter promiscuous protein functions, potentially aiding functional evolution. *Biol. Direct.* 2:17
112. Matsumura I, Ellington AD. 2001. In vitro evolution of beta-glucuronidase into a beta-galactosidase proceeds through non-specific intermediates. *J. Mol. Biol.* 305:331–39
113. Rothman SC, Kirsch JF. 2003. How does an enzyme evolved in vitro compare to naturally occurring homologs possessing the targeted function? Tyrosine aminotransferase from aspartate aminotransferase. *J. Mol. Biol.* 327:593–608
114. Anandarajah K, Kiefer PM Jr, Donohoe BS, Copley SD. 2000. Recruitment of a double bond isomerase to serve as a reductive dehalogenase during biodegradation of pentachlorophenol. *Biochemistry* 39:5303–11
115. Smith JM. 1970. Natural selection and concept of a protein space. *Nature* 225:563–64
116. Copley SD. 2010. Evolution and the enzyme. In *Comprehensive Natural Products Chemistry*, Vol. 8, ed. D Barton, O Meth-Cohn. Oxford: Elsevier. In press
117. Conant GC, Wolfe KH. 2008. Turning a hobby into a job: How duplicated genes find new functions. *Nat. Rev. Genet.* 9:938–50
118. Peisajovich SG, Tawfik DS. 2007. Protein engineers turned evolutionists. *Nat. Methods* 4:991–94

119. Seelig B, Szostak JW. 2007. Selection and evolution of enzymes from a partially randomized non-catalytic scaffold. *Nature* 448:828–31
120. Park HS, Nam SH, Lee JK, Yoon CN, Mannervik B, et al. 2006. Design and evolution of new catalytic activity with an existing protein scaffold. *Science* 311:535–38
121. Jiang L, Althoff EA, Clemente FR, Doyle L, Röthlisberger D, et al. 2008. De novo computational design of retro-aldol enzymes. *Science* 319:1387–91
122. Röthlisberger D, Khersonsky O, Wollacott AM, Jiang L, DeChancie J, et al. 2008. Kemp elimination catalysts by computational enzyme design. *Nature* 453:190–95
123. Miller BG, Raines RT. 2004. Identifying latent enzyme activities: substrate ambiguity within modern bacterial sugar kinases. *Biochemistry* 43:6387–92
124. James LC, Tawfik DS. 2001. Catalytic and binding poly-reactivities shared by two unrelated proteins: the potential role of promiscuity in enzyme evolution. *Protein Sci.* 10:2600–7
125. Hall BG. 1999. Experimental evolution of Ebg enzyme provides clues about the evolution of catalysis and to evolutionary potential. *FEMS Microbiol. Lett.* 174:1–8
126. Hall BG. 1982. Evolution on a petri dish: the evolved B-galactosidase system as a model for studying acquisitive evolution in the laboratory. *Evol. Biol.* 15:85–150
127. Miller BG, Raines RT. 2005. Reconstitution of a defunct glycolytic pathway via recruitment of ambiguous sugar kinases. *Biochemistry* 44:10776–83
128. Scannell DR, Wolfe KH. 2008. A burst of protein sequence evolution and a prolonged period of asymmetric evolution follow gene duplication in yeast. *Genome Res.* 18:137–47
129. McLoughlin SY, Ollis DL. 2004. The role of inhibition in enzyme evolution. *Chem. Biol.* 11:735–37
130. Bergthorsson U, Andersson DI, Roth JR. 2007. Ohno's dilemma: evolution of new genes under continuous selection. *Proc. Natl. Acad. Sci. USA* 104:17004–9
131. Kondrashov FA. 2005. In search of the limits of evolution. *Nat. Genet.* 37:9–10
132. Kirschner M, Gerhart J. 1998. Evolvability. *Proc. Nat. Acad. Sci. USA* 95:8420–27
133. Umeno D, Tobias AV, Arnold FH. 2005. Diversifying carotenoid biosynthetic pathways by directed evolution. *Microbiol. Mol. Biol. Rev.* 69:51–78
134. Bershtein S, Segal M, Bekerman R, Tokuriki N, Tawfik DS. 2006. Robustness-epistasis link shapes the fitness landscape of a randomly drifting protein. *Nature* 444:929–32
135. Camps M, Herman A, Loh E, Loeb LA. 2007. Genetic constraints on protein evolution. *Crit. Rev. Biochem. Mol. Biol.* 42:313–26
136. Tokuriki N, Stritcher F, Serrano L, Tawfik DS. 2008. How protein stability and new functions trade off. *PloS Comput. Biol.* 4:e1000002
137. Wagner A. 2008. Robustness and evolvability: a paradox resolved. *Proc. Biol. Sci.* 275:91–100
138. Aharoni A, Gaidukov L, Khersonsky O, McQ GS, Roodveldt C, Tawfik DS. 2005. The 'evolvability' of promiscuous protein functions. *Nat. Genet.* 37:73–76
139. Adami C. 2006. Evolution: reducible complexity. *Science* 312:61–63
140. Bridgham JT, Carroll SM, Thornton JW. 2006. Evolution of hormone-receptor complexity by molecular exploitation. *Science* 312:97–101
141. Galvao TC, de Lorenzo V. 2006. Transcriptional regulators a la carte: engineering new effector specificities in bacterial regulatory proteins. *Curr. Opin. Biotechnol.* 17:34–42
142. Ju KS, Parales JV, Parales RE. 2009. Reconstructing the evolutionary history of nitrotoluene detection in the transcriptional regulator NtdR. *Mol. Microbiol.* 74:826–43
143. Kivisaar M. 2009. Degradation of nitroaromatic compounds: a model to study evolution of metabolic pathways. *Mol. Microbiol.* 74:777–81
144. Bertini I, Calderone V, Cosenza M, Fragai M, Lee YM, et al. 2005. Conformational variability of matrix metalloproteinases: beyond a single 3D structure. *Proc. Natl. Acad. Sci. USA* 102:5334–39
145. Larion M, Moore LB, Thompson SM, Miller BG. 2007. Divergent evolution of function in the ROK sugar kinase superfamily: role of enzyme loops in substrate specificity. *Biochemistry* 46:13564–72
146. Bone R, Frank D, Kettner CA, Agard DA. 1989. Structural analysis of specificity: alpha-lytic protease complexes with analogues of reaction intermediates. *Biochemistry* 28:7600–9
147. Bone R, Silen JL, Agard DA. 1989. Structural plasticity broadens the specificity of an engineered protease. *Nature* 339:191–95

148. Turner JM, Graziano J, Spraggon G, Schultz PG. 2006. Structural plasticity of an aminoacyl-tRNA synthetase active site. *Proc. Natl. Acad. Sci. USA* 103:6483–88
149. Bernath Levin K, Dym O, Albeck S, Magdassi S, Keeble AH, et al. 2009. Following evolutionary paths to protein-protein interactions with high affinity and selectivity. *Nat. Struct. Mol. Biol.* 16:1049–55
150. O'Loughlin TL, Greene DN, Matsumura I. 2006. Diversification and specialization of HIV protease function during in vitro evolution. *Mol. Biol. Evol.* 23:764–72
151. Ran N, Draths KM, Frost JW. 2004. Creation of a shikimate pathway variant. *J. Am. Chem. Soc.* 126:6856–57
152. Varadarajan N, Gam J, Olsen MJ, Georgiou G, Iverson BL. 2005. Engineering of protease variants exhibiting high catalytic activity and exquisite substrate selectivity. *Proc. Natl. Acad. Sci. USA* 102:6855–60
153. Varadarajan N, Rodriguez S, Hwang BY, Georgiou G, Iverson BL. 2008. Highly active and selective endopeptidases with programmed substrate specificities. *Nat. Chem. Biol.* 4:290–94
154. Thornton JW, Need E, Crews D. 2003. Resurrecting the ancestral steroid receptor: ancient origin of estrogen signaling. *Science* 301:1714–17
155. Wouters MA, Liu K, Riek P, Husain A. 2003. A despecialization step underlying evolution of a family of serine proteases. *Mol. Cell* 12:343–54
156. Watts KT, Mijts BN, Lee PC, Manning AJ, Schmidt-Dannert C. 2006. Discovery of a substrate selectivity switch in tyrosine ammonia-lyase, a member of the aromatic amino acid lyase family. *Chem. Biol.* 13:1317–26
157. Gould SM, Tawfik DS. 2005. Directed evolution of the promiscuous esterase activity of carbonic anhydrase II. *Biochemistry* 44:5444–52
158. Vick JE, Schmidt DM, Gerlt JA. 2005. Evolutionary potential of (beta/alpha)8-barrels: in vitro enhancement of a "new" reaction in the enolase superfamily. *Biochemistry* 44:11722–29
159. Fasan R, Meharenna YT, Snow CD, Poulos TL, Arnold FH. 2008. Evolutionary history of a specialized p450 propane monooxygenase. *J. Mol. Biol.* 383:1069–80
160. Zhang ZR, Perrett S. 2009. Novel glutaredoxin activity of the yeast prion protein Ure2 reveals a native-like dimer within fibrils. *J. Biol. Chem.* 284:14058–67
161. Berstein S, Goldin K, Tawfik DS. 2008. Intense neutral drifts yield robust and evolvable consensus proteins. *J. Mol. Biol.* 379:1029–44
162. Tokuriki N, Tawfik DS. 2009. Stability effects of mutations and protein evolvability. *Curr. Opin. Struct. Biol.* 19:596–604
163. Ohno S. 1970. *Evolution by Gene Duplication*. London-New York: Allen & Unwin/Springer-Verlag. 160 pp.
164. Kimura M, Ota T. 1974. On some principles governing molecular evolution. *Proc. Natl. Acad. Sci. USA* 71:2848–52
165. Hughes AL. 2002. Adaptive evolution after gene duplication. *Trends Genet.* 18:433–34
166. Lynch M, Katju V. 2004. The altered evolutionary trajectories of gene duplicates. *Trends Genet.* 20:544–49
167. Wagner A. 2005. Energy constraints on the evolution of gene expression. *Mol. Biol. Evol.* 22:1365–74
168. Cooper VS, Lenski RE. 2000. The population genetics of ecological specialization in evolving *Escherichia coli* populations. *Nature* 407:736–39
169. Tokuriki N, Stricher F, Schymkowitz J, Serrano L, Tawfik DS. 2007. The stability effects of protein mutations appear to be universally distributed. *J. Mol. Biol.* 369:1318–32
170. Yue P, Li Z, Moult J. 2005. Loss of protein structure stability as a major causative factor in monogenic disease. *J. Mol. Biol.* 353:459–73
171. Bershtein S, Tawfik DS. 2008. Ohno's model revisited: measuring the frequency of potentially adaptive mutations under various mutational drifts. *Mol. Biol. Evol.* 25:2311–18
172. Piatigorsky J, O'Brien WE, Norman BL, Kalumuck K, Wistow GJ, et al. 1988. Gene sharing by delta-crystallin and argininosuccinate lyase. *Proc. Natl. Acad. Sci. USA* 85:3479–83
173. Des Marais DL, Rausher MD. 2008. Escape from adaptive conflict after duplication in an anthocyanin pathway gene. *Nature* 454:762–65

174. Blum MM, Lohr F, Richardt A, Ruterjans H, Chen JC. 2006. Binding of a designed substrate analogue to diisopropyl fluorophosphatase: implications for the phosphotriesterase mechanism. *J. Am. Chem. Soc.* 128:12750–57
175. Amitai G, Gaidukov L, Adani R, Yishay S, Yacov G, et al. 2006. Enhanced stereoselective hydrolysis of toxic organophosphates by directly evolved variants of mammalian serum paraoxonase. *FEBS J.* 273:1906–19
176. Lamble HJ, Heyer NI, Bull SD, Hough DW, Danson MJ. 2003. Metabolic pathway promiscuity in the archaeon *Sulfolobus solfataricus* revealed by studies on glucose dehydrogenase and 2-keto-3-deoxygluconate aldolase. *J. Biol. Chem.* 278:34066–72
177. Taylor Ringia EA, Garrett JB, Thoden JB, Holden HM, Rayment I, Gerlt JA. 2004. Evolution of enzymatic activity in the enolase superfamily: functional studies of the promiscuous *o*-succinylbenzoate synthase from *Amycolatopsis*. *Biochemistry* 43:224–29
178. Yew WS, Akana J, Wise EL, Rayment I, Gerlt JA. 2005. Evolution of enzymatic activities in the orotidine 5′-monophosphate decarboxylase suprafamily: enhancing the promiscuous D-arabino-hex-3-ulose 6-phosphate synthase reaction catalyzed by 3-keto-L-gulonate 6-phosphate decarboxylase. *Biochemistry* 44:1807–15
179. Joerger AC, Mayer S, Fersht AR. 2003. Mimicking natural evolution in vitro: an *N*-acetylneuraminate lyase mutant with an increased dihydrodipicolinate synthase activity. *Proc. Natl. Acad. Sci. USA* 100:5694–99
180. De Mot R, De Schrijver A, Schoofs G, Parret AHA. 2003. The thiocarbamate-inducible *Rhodococcus* enzyme ThcF as a member of the family of alpha/beta hydrolases with haloperoxidative side activity. *FEMS Microbiol. Lett.* 224:197–203
181. Kataoka M, Honda K, Shimizu S. 2000. 3,4-Dihydrocoumarin hydrolase with haloperoxidase activity from *Acinetobacter calcoaceticus* F46. *Eur. J. Biochem.* 267:3–10
182. Li C, Hassler M, Bugg TDH. 2008. Catalytic promiscuity in the alpha/beta-hydrolase superfamily: hydroxamic acid formation, C-C bond formation, ester and thioester hydrolysis in the C-C hydrolase family. *ChemBioChem* 9:71–76
183. Schmidt DM, Mundorff EC, Dojka M, Bermudez E, Ness JE, et al. 2003. Evolutionary potential of (beta/alpha)8-barrels: functional promiscuity produced by single substitutions in the enolase superfamily. *Biochemistry* 42:8387–93
184. Hoffmeister D, Yang J, Liu L, Thorson JS. 2003. Creation of the first anomeric D/L-sugar kinase by means of directed evolution. *Proc. Natl. Acad. Sci. USA* 100:13184–89
185. Rowe LA, Geddie ML, Alexander OB, Matsumura I. 2003. A comparison of directed evolution approaches using the beta-glucuronidase model system. *J. Mol. Biol.* 332:851–60
186. Cohen HM, Tawfik DS, Griffiths AD. 2004. Altering the sequence specificity of *Hae*III methyltransferase by directed evolution using in vitro compartmentalization. *Protein Eng. Des. Sel.* 17:3–11
187. Timar E, Groma G, Kiss A, Venetianer P. 2004. Changing the recognition specificity of a DNA-methyltransferase by in vitro evolution. *Nucleic Acids Res.* 32:3898–903
188. Aharoni A, Gaidukov L, Yagur S, Toker L, Silman I, Tawfik DS. 2004. Directed evolution of mammalian paraoxonases PON1 and PON3 for bacterial expression and catalytic specialization. *Proc. Natl. Acad. Sci. USA* 101:482–87
189. Wei CL, Yang YB, Deng CH, Liu WC, Hsu JS, et al. 2005. Directed evolution of *Streptomyces clavuligerus* deacetoxycephalosporin C synthase for enhancement of penicillin G expansion. *Appl. Environ. Microbiol.* 71:8873–80
190. Delmas J, Robin F, Carvalho F, Mongaret C, Bonnet R. 2006. Prediction of the evolution of ceftazidime resistance in extended-spectrum beta-lactamase CTX-M-9. *Antimicrob. Agents Chemother.* 50:731–38
191. Samuelson JC, Morgan RD, Benner JS, Claus TE, Packard SL, Xu SY. 2006. Engineering a rare-cutting restriction enzyme: genetic screening and selection of NotI variants. *Nucleic Acids Res.* 34:796–805
192. Jurgens C, Strom A, Wegener D, Hettwer S, Wilmanns M, Sterner R. 2000. Directed evolution of a $(\beta\alpha)_8$-barrel enzyme to catalyze related reactions in two different metabolic pathways. *Proc. Natl. Acad. Sci. USA* 97:9925–30
193. Gaidukov L, Tawfik DS. 2005. The high affinity, stability and lactonase activity of serum paraoxonase (PON1) anchored on HDL with ApoA-I. *Biochemsitry* 44:11843–54

Hydrogenases from Methanogenic Archaea, Nickel, a Novel Cofactor, and H_2 Storage

Rudolf K. Thauer, Anne-Kristin Kaster, Meike Goenrich, Michael Schick, Takeshi Hiromoto, and Seigo Shima

Max Planck Institute for Terrestrial Microbiology, D-35043 Marburg, Germany; email: thauer@mpi-marburg.mpg.de

Annu. Rev. Biochem. 2010. 79:507–36

First published online as a Review in Advance on March 17, 2010

The *Annual Review of Biochemistry* is online at biochem.annualreviews.org

This article's doi:
10.1146/annurev.biochem.030508.152103

0066-4154/10/0707-0507$20.00

Key Words

H_2 activation, energy-converting hydrogenase, complex I of the respiratory chain, chemiosmotic coupling, electron bifurcation, reversed electron transfer

Abstract

Most methanogenic archaea reduce CO_2 with H_2 to CH_4. For the activation of H_2, they use different [NiFe]-hydrogenases, namely energy-converting [NiFe]-hydrogenases, heterodisulfide reductase-associated [NiFe]-hydrogenase or methanophenazine-reducing [NiFe]-hydrogenase, and F_{420}-reducing [NiFe]-hydrogenase. The energy-converting [NiFe]-hydrogenases are phylogenetically related to complex I of the respiratory chain. Under conditions of nickel limitation, some methanogens synthesize a nickel-independent [Fe]-hydrogenase (instead of F_{420}-reducing [NiFe]-hydrogenase) and by that reduce their nickel requirement. The [Fe]-hydrogenase harbors a unique iron-guanylylpyridinol cofactor (FeGP cofactor), in which a low-spin iron is ligated by two CO, one $C(O)CH_2$-, one $S\text{-}CH_2$-, and a sp^2-hybridized pyridinol nitrogen. Ligation of the iron is thus similar to that of the low-spin iron in the binuclear active-site metal center of [NiFe]- and [FeFe]-hydrogenases. Putative genes for the synthesis of the FeGP cofactor have been identified. The formation of methane from 4 H_2 and CO_2 catalyzed by methanogenic archaea is being discussed as an efficient means to store H_2.

Contents

INTRODUCTION

In 1933, Stephenson & Stickland (1) enriched from river sediments methane-forming microorganisms that grow on H_2 and CO_2 (Reaction 1) and concluded that these methanogens must contain hydrogenases that activate H_2 (Reaction 2).

$$4\,H_2 + CO_2 \rightarrow CH_4 + 2\,H_2O \quad \Delta G^{\circ\prime} = -131\,\text{kJ mol}^{-1}. \qquad 1.$$

$$H_2 \rightleftharpoons 2\,e^- + 2\,H^+ \quad E_o' = -414\,\text{mV}. \qquad 2.$$

Black and white smokers: chimney-like structures formed around hydrothermal vents, where superheated mineral rich water from below Earth's crust comes through the ocean floor

The name hydrogenase was coined in 1931 by Stephenson & Stickland (2, 3) for an activity in anaerobically grown *Escherichia coli* cells mediating the reversible reduction of dyes with H_2. Dye reduction was reversibly inhibited by CO, indicating the involvement of a transition metal in H_2 activation (4). The transition metal later turned out to be nickel in a binuclear nickel-iron center in the case of [NiFe]-hydrogenases (5–8), iron in a binuclear iron-iron center in the case of [FeFe]-hydrogenases (9–11), and iron in a mononuclear iron center in the case of [Fe]-hydrogenase (12–14), which are the three different types of hydrogenases known to date (**Figure 1**) (15, 16).

From the work of Stephenson, it became evident that methane formation from biomass in river sediments is at least in part the result of the syntrophic interaction of H_2-forming bacteria such as *E. coli* and H_2-consuming methanogens. And indeed in later studies, it turned out that interspecies hydrogen transfer is a quantitatively important process in the carbon cycle despite the fact that for thermodynamic and kinetic reasons the H_2 concentration in anaerobic habitats is generally very low ($pH_2 < 10$ Pa; $E'(H^+/H_2) = -300$ mV) (17, 18). H_2 (see the sidebar titled Properties of H_2) (19, 20), even at low concentrations, is an ideal electron carrier between organisms because it can freely diffuse through cytoplasmic membranes. Estimates are that approximately 150 million tons of H_2 are annually formed by microorganisms and used to fuel methanogens (17). The combustion of 150 million tons H_2 yields 18×10^{18} J, an energy amount that is 3.75% of the primary energy consumed in 2006 by the world population (455×10^{18} J).

Today on Earth, most of the H_2 used by methanogens is of biological origin. Only some of the H_2 that sustains the growth of methanogens is geochemically generated, e.g., in black and white smokers. However, in the Archaeozoic (4 to 2.5 billion years back), when the different lineages of microbes on Earth evolved and when the temperatures were much higher than today, geochemically formed H_2 probably predominated that of biological origin and

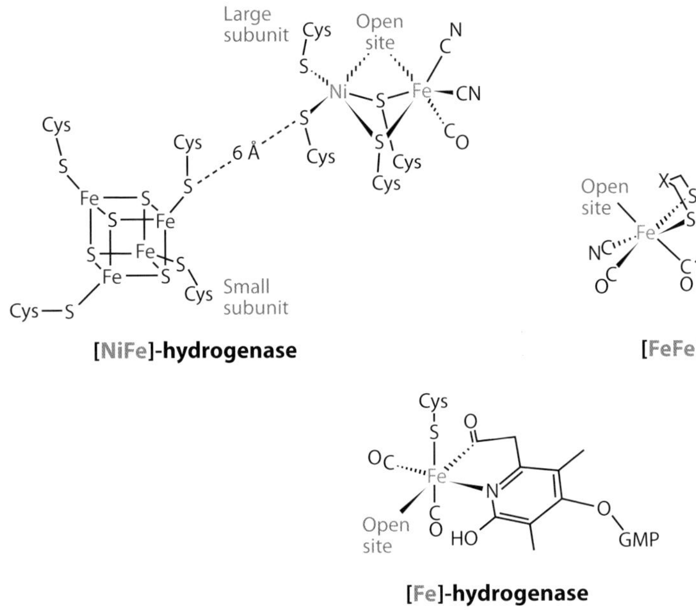

[NiFe]-hydrogenase

[FeFe]-hydrogenase

[Fe]-hydrogenase

Figure 1

The metal sites of the three types of hydrogenases involved in interspecies hydrogen transfer (see **Figure 2**) have unusual structural features in common, such as intrinsic CO ligands. Despite this fact, [NiFe]-hydrogenases (5–8), [FeFe]-hydrogenases (9–11), and [Fe]-hydrogenase (12–14) are not phylogenetically related at the level of their primary structure or at the level of the enzymes involved in their active-site biosynthesis (12). Abbreviation: GMP, guanylyl rest.

fueled the growth of methanogens. Consistently, among recent hydrogenotrophic methanogens, there are many hyperthermophiles, such as *Methanopyrus kandleri* (98°C optimum growth temperature) and *Methanocaldococcus jannaschii* (85°C optimum growth temperature), and these hyperthermophiles branch off the 16S phylogenetic tree relatively early.

This review highlights the properties of the five different hydrogenases found in methanogens within the context of their function in metabolism. Four of the enzymes are [NiFe]-hydrogenases with some properties similar and others dissimilar to those of related [NiFe]-hydrogenases in bacteria. It was in methanogens that nickel was first found to be required for hydrogenase activity (21, 22). The fifth enzyme is a [Fe]-hydrogenase (23) that is unique to methanogens and functional in these only under conditions of nickel limitation. [FeFe]-hydrogenases, which are present in Bacteria and lower Eukarya, have not yet been found in Archaea (15, 16).

PROPERTIES OF H_2

H_2 is a colorless gas with a boiling point at 22.28 K (−250.87°C). Its Bunsen coefficient α in water at 20°C is 0.018 (0.8 mM at 1 bar), and its diffusion coefficient D_w in water at 20°C is near 4×10^{-9} m^2 s^{-1}. The homolytic cleavage of H_2 in the gas phase is endergonic by +436 kJ mol^{-1}, and the heterolytic cleavage in water at 20°C is endergonic by about +200 kJ mol^{-1} (pK_a near 35) (19). The combustion energy of H_2 is 120 MJ kg^{-1}. The activation of H_2 is mechanistically challenging, and the catalytic mechanism is of considerable interest. The H_2 generated, e.g., by electrolysis or photolysis of water, is presently discussed as an environmentally clean energy carrier for use in fuel cell-powered electrical cars. Before H_2 can be used in fuel cells, cheap catalysts still have to be developed, and it is hoped that the active-site structure of hydrogenases will show how to proceed (20).

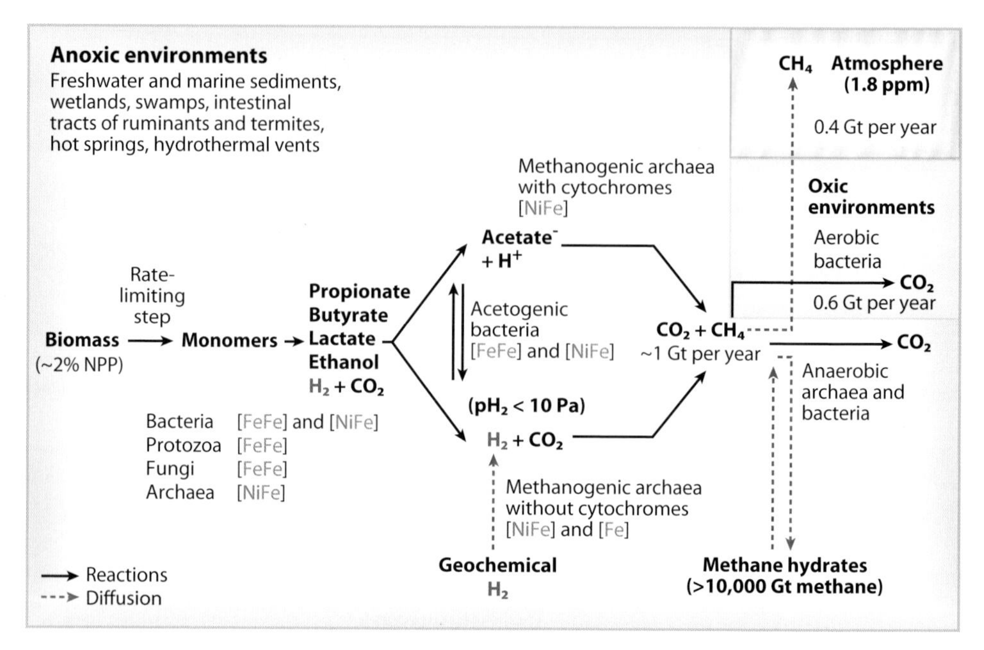

Figure 2

Approximately 2% of the net primary production (NPP) of plants, algae, and cyanobacteria are fermented in anoxic environments by a syntrophic association of anaerobic microorganisms with methane, in a process that involves interspecies H_2 transfer. The kinetics and thermodynamics of the process are such that the steady-state H_2 concentration remains below 0.1 μM (<10 Pa) (17, 18). At these low H_2 concentrations, the redox potential of the H^+/H_2 couple at pH 7 is near −300 mV. The three types of hydrogenases involved are abbreviated [NiFe], [FeFe], and [Fe], respectively (see **Figure 1**). In the intestinal tract of termites and ruminants, methanogens converting acetate to methane and CO_2 are lacking. Therefore, acetate, propionate, butyrate, lactate, and ethanol concentrations build up and can be used by the host as anabolic and catabolic substrates.

H_2 AS AN INTERMEDIATE IN CH_4 FORMATION AND THE ORGANISMS INVOLVED

Approximately 2% of the net primary production of plants, algae, and cyanobacteria (70 billion tons C per year) are remineralized via methane in anoxic environments such as freshwater and marine sediments, wetlands, swamps, sewage digesters, landfills, hot springs, and the intestinal tract of ruminants and termites (**Figure 2**). From the biomass, which consists of 60%–70% cellulose, approximately one billion tons of methane are generated per year; 60% is oxidized to CO_2 by microorganisms, and 40% escapes into the atmosphere, where its concentration almost doubled within the last hundred years (17). This is of concern since methane is an effective greenhouse gas.

Geochemically formed H_2: H_2 generated abiotically from H_2S or by reaction of H_2O with ultramafic rocks (serpentinization)

In a rate-limiting step, the biomass is degraded by extracellular hydrolytic enzymes excreted by anaerobic bacteria and protozoa to monomers, which after uptake by these microorganisms are primarily fermented to lactic acid, propionic acid, butyric acid, ethanol, and acetic acid with the concomitant formation of CO_2, formic acid, and some H_2. This process also involves anaerobic fungi in the rumen and anaerobic archaea in hot springs. Of these products, lactic, propionic, and butyric acids and ethanol serve syntrophic bacteria as substrates, which ferment them to acetic acid, CO_2, H_2, and formic acid.

From acetic acid, H_2, CO_2, and formic acid, methane is then formed by methanogenic archaea, of which there are two types, those with and those without cytochromes. Acetic acid is converted to CO_2 and methane only by the methanogens with cytochromes, and H_2, CO_2 and formate are converted to methane mainly by those without cytochromes. None of the methanogens can use lactic, propionic, or butyric acid as energy substrates. But by consuming H_2, acetic acid and formic acid, the methanogens keep the H_2 partial pressure between 1 Pa and 10 Pa and the acetic and formic acid concentrations well below 0.1 mM, enabling the syntrophic bacteria to convert lactic, propionic, and butyric acid and ethanol to acetic acid, H_2, and CO_2. Only at low concentrations of H_2 and acetic acid are the fermentations of the syntrophs exergonic enough to sustain their growth (18).

In the intestinal tract of ruminants and termites, methanogens with cytochromes are not present, and therefore methanogenesis from acetate does not occur. The reason for this is probably that the growth rate of acetoclastic methanogens is generally lower than the dilution rate in the intestinal tract, and therefore, the acetoclastic methanogens are continuously washed out. Because of the lack of methanogenesis from acetic acid the concentration of acetic acid builds up considerably (>10 mM) with the result that, for the thermodynamic reasons discussed above, lactic, propionic, and butyric acid, therefore also increase in their concentrations. The organic acids are resorbed by the ruminants and insects from their intestinal tracts and used for gluconeogenesis (lactic and propionic acid) and ATP synthesis (acetic acid and butyric acid).

In sediments of hot springs, in which cellulose is completely converted to methane and CO_2, surprisingly, at temperatures above 60°C, acetoclastic methanogens are absent for reasons not yet fully understood. Methanogens with cytochromes growing above 60°C have yet to be found. In hot sediments, acetic acid is converted to two CO_2 and four H_2 by bacteria related to acetogenic bacteria, and the H_2 and CO_2 thus formed are then converted to methane by methanogens without cytochromes, which have many thermophilic and hypothermophilic species. Thus, in hot springs, the conversion of glucose from cellulose to three CO_2 and three CH_4 involves 12 H_2 as intermediates, which underlines the quantitative importance of H_2 as electron carrier between fermenters and methanogens.

All methanogens are known to belong to the domain of Archaea and to the kingdom of Euryarchaeota. From the latter lineage, the Methanopyrales branch off first, followed by the orders Methanococcales and Methanobacteriales, and then by Methanomicrobiales and Methanosarcinales. Only the members of the Methanosarcinales contain cytochromes and can use acetic acid as methanogenic substrate. Methanogenesis from acetate is therefore believed to be a late invention. The ability to use acetate as methanogenic substrate was associated with a change in the mechanism of energy conservation, as electron transport now involves cytochromes. The altered mechanism allowed the methanogens with cytochromes to also use methanol, methylamines, and methylthiols as energy substrates. But it also had a price, namely the loss of the ability to use H_2 down to partial pressures below 10 Pa (for an explanation, see below), which is a characteristic of methanogens without cytochromes that are specialized on H_2 plus CO_2 and/or formate as energy sources. Members of the Methanosarcinales that can grow on H_2 and CO_2 do this only at significantly higher H_2 concentrations than the members of the other orders, which lack cytochromes. This is why the Methanosarcinales do not contribute to methane formation from H_2 and CO_2 in most anoxic environments (**Figure 2**) (17).

HYDROGENASES FOUND IN METHANOGENS AND THEIR FUNCTION

The genomes of several members of each of the five known orders of methanogens have been sequenced, and the genes putatively encoding hydrogenases have been identified. Biochemical studies of the hydrogenases have concentrated mainly on a few

EchA-F, EhaA-T, EhbA-Q, and MbhA-N: energy-converting [NiFe]-hydrogenases

MvhADG: heterodisulfide reductase-associated [NiFe]-hydrogenase

HdrABC and HdrDE: heterodisulfide reductases

VhtACG: methanophenazine-reducing [NiFe]-hydrogenase

FrhABG: an F_{420}-reducing [NiFe]-hydrogenase

species, namely *Methanothermobacter thermautotrophicus*, *Methanothermobacter marburgensis*, *Methanococcus maripaludis*, *Methanosarcina barkeri*, *and Methanosarcina mazei*. Genetic analyses have been restricted to *Methanococcus voltae*, *M. maripaludis*, *Methanosarcina acetivorans*, *M. mazei*, and *M. barkeri*. From these studies, a partially coherent picture has emerged.

Four different subtypes of [NiFe]-hydrogenases and one [Fe]-hydrogenase are found in methanogens. The four [NiFe]-subtypes are (*a*) the membrane-associated, energy-converting [NiFe]-hydrogenases (EchA-F, EhaA-T, EhbA-Q, and MbhA-N) for the reduction of ferredoxin with H_2; (*b*) the cytoplasmic [NiFe]-hydrogenase (MvhADG) associated with the heterodisulfide reductase (HdrABC) for the coupled reduction of ferredoxin and of the heterodisulfide CoM-S-S-CoB with H_2; (*c*) the membrane-associated, methanophenazine-reducing [NiFe]-hydrogenase (VhtACG); and (*d*) the cytoplasmic coenzyme F_{420}-reducing [NiFe]-hydrogenases (FrhABG). Not all five hydrogenases are found in all methanogens. Thus, the methanophenazine-reducing [NiFe]-hydrogenase is restricted to methanogens with cytochromes, and the cytoplasmic [Fe]-hydrogenase, which together with F_{420}-dependent methylenetetrahydromethanopterin dehydrogenase substitute for the F_{420}-reducing [NiFe]-hydrogenase under nickel-limiting growth conditions (see below), is only present in some methanogens without cytochromes. Genes for [Fe]-hydrogenase synthesis are lacking in methanogens with cytochromes and in most members of the Methanomicrobiales (15, 23).

The function of the different [NiFe]-hydrogenases in methanogenesis from H_2 and CO_2 in methanogens with cytochromes can be deduced from **Figure 3*a*** and in methanogens without cytochromes from **Figure 3*b***. The differences outlined in the two schemes are based, among many other observations, on the finding that the growth yield of cytochrome-containing methanogens on H_2 and CO_2 (maximally 6.4 g per mole CH_4) is more than twice as high as that of methanogens without cytochromes (maximally 3 g per mole CH_4), indicating that the ATP gain per mole methane is approximately 0.5 in methanogens with cytochromes and 1 to 1.5 in methanogens without cytochromes (17). The low ATP gain of 0.5 allows the methanogens without cytochromes to grow on H_2 and CO_2 at H_2 partial pressures of 5 Pa at which methanogenesis from CO_2 and H_2 is exergonic by -25 kJ per mole, which is just sufficient to drive the synthesis of 0.5 mole ATP ($\Delta G' = -50$ kJ per mole ATP). Conversely, an ATP gain of 1 to 1.5 is only thermodynamically possible if the H_2 concentration is >100 Pa ($\Delta G < -63$ kJ per mole CH_4). And indeed, methanogens with cytochromes are known to have a much higher H_2 threshold concentration (>100 Pa) than methanogens without cytochromes <10 Pa) (17). The relatively high threshold concentration for H_2 can explain why members of the Methanosarcinales are generally not involved in methanogenesis from H_2 and CO_2 in most methanogenic habitats (**Figure 2**) and why in some members, e.g., in *M. acetivorans*, transcription of the genes for the hydrogenases are permanently turned off (24).

For an understanding of the function of the different hydrogenases in the proposed two metabolic schemes (**Figure 3*a*,*b***) the energetics of ferredoxin reduction with H_2 are of special importance. Under physiological standard conditions ($pH_2 = 10^5$ Pa; pH 7; $Fd_{ox}/Fd_{red} = 1$), the reduction of ferredoxin ($E'_o = -420$ mV) with H_2 ($E'_o = -414$ mV) is neither endergonic nor exergonic. However, under in vivo conditions ($pH_2 = 10$ Pa; pH 7; $Fd_{ox}/Fd_{red} < 0.01$), the reduction of ferredoxin with H_2 is strongly endergonic with E' of the H^+/H_2 couple $= -300$ mV and that of the Fd_{ox}/Fd_{red} couple $= -500$ mV. In methanogens, fully reduced ferredoxin is required for the reduction of CO_2 to formylmethanofuran (CHO-MFR) ($E'_o = -500$ mV), which is the first step in methanogenesis from CO_2 (**Figure 3**), for the reduction of CO_2 to CO ($E'_o = -520$ mV), for the reduction of acetyl coenzyme A (acetyl-CoA) and CO_2 to pyruvate ($E'_o = -500$ mV), and—in most methanogens—for the reduction

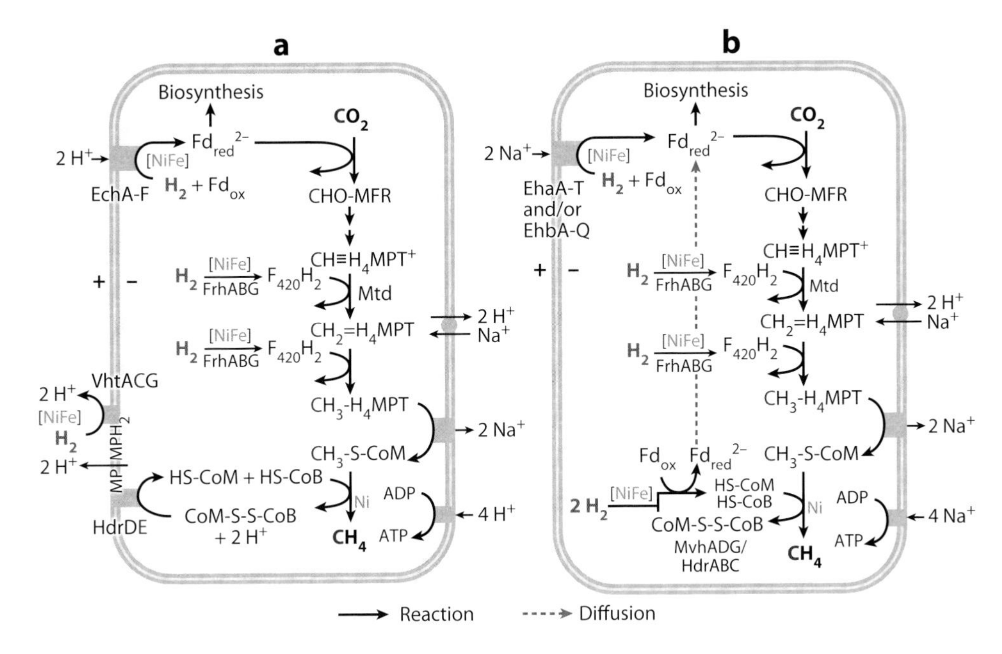

Figure 3

The proposed function and localization within the cell of the [NiFe]-hydrogenases involved in methanogenesis from H_2 and CO_2 are shown for methanogens (*a*) with cytochromes and (*b*) without cytochromes. The cations translocated and the exact stoichiometry of the translocation reactions are still a matter of dispute, and the stoichiometry of electron bifurcation in the MvhADG/HdrABC complex remains to be ascertained (17). In some members of the Methanomicrobiales, genes for MvhA and MvhG are not found, and energy-converting hydrogenases other than Eha or Ehb can be present (25). For interpretations of these novel findings, see the section Heterodisulfide Reductase-Associated [NiFe]-Hydrogenase MvhADG, below. Members of the Methanosarcinales and Methanococcales contain tetrahydrosarcinapterin rather than tetrahydromethanopterin (H_4MPT). The two pterins have identical functions in C_1-unit transformation. Abbreviations: CHO-MFR, formylmethanofuran; $CH{\equiv}H_4MPT^+$, methenyltetrahydromethanopterin; $CH_2{=}H_4MPT$, methylenetetrahydromethanopterin; CH_3-H_4MPT, methyltetrahydromethanopterin; CoB-SH, coenzyme B with its thiol group; CoM-SH, coenzyme M with its thiol group; CoM-S-S-CoB, heterodisulfide; EchA-F, EhaA-T, and EhbA-Q, energy-converting [NiFe]-hydrogenases; Fd, ferredoxin with two [4Fe4S]-clusters; FrhABG, F_{420}-reducing [NiFe]-hydrogenase; HdrABC and HdrDE, heterodisulfide reductases; MP, methanophenazine; MvhADG, heterodisulfide reductase-associated [NiFe]-hydrogenase; VhtACG, methanophenazine-reducing [NiFe]-hydrogenase; VhtC and HdrE, *b*-type cytochromes.

of succinyl-CoA and CO_2 to 2-oxoglutarate (E'_o = −500 mV). In methanogens growing on H_2 and CO_2, the latter three reduction reactions participate in autotrophic CO_2 fixation. All of these ferredoxin-dependent oxidoreductase reactions are catalyzed by cytoplasmic enzymes. Therefore, it is the reduction of ferredoxin with H_2 that must be energy driven and the site of energy coupling. As outlined below, in energy-converting [NiFe]-hydrogenases, the mechanism of energy coupling is chemiosmotic (**Figure 3*a,b***), and in the MvhADG/HdrABC complex, the mechanism of coupling is by electron bifurcation (**Figure 3*b***).

Whereas the reduction of ferredoxin with H_2 in methanogens is strongly endergonic, that of methanophenazine (E'_o = −170 mV) and of the heterodisulfide CoM-S-S-CoB (E'_o = −140 mV) with H_2 (E'_o = −414 mV) is a strongly exergonic reaction (26). Consistently, methanophenazine and CoM-S-S-CoB reduction with H_2 are coupled with energy conservation (**Figure 3*a,b***). Of the hydrogenase-catalyzed reactions, only the reduction of

Electron bifurcation: the disproportionation of two electrons at the same redox potential to one electron with a higher and one with a lower redox potential coenzyme F_{420} ($E'_o = -360$ mV) with H_2 is not coupled with energy conversion (27). Under in vivo conditions ($pH_2 = 10$ Pa; $F_{420}/F_{420}H_2$ <0.1), the free energy change associated with the reaction is essentially zero. The energetic differences of the hydrogenase-catalyzed reactions in methanogenesis from H_2 and CO_2 can therefore explain why there are at least three different hydrogenases in hydrogenotrophic methanogens.

THE FOUR SUBTYPES OF [NiFe]-HYDROGENASES IN METHANOGENS

The crystal structures of the [NiFe]-hydrogenases found in methanogens have not been determined. Currently, only structures of [NiFe]-hydrogenases from sulfate-reducing bacteria are available (5–8). However, on the basis of sequence comparisons, all [NiFe]-hydrogenases appear to be phylogenetically related, although the sequence similarity is sometimes restricted to the sequences around the N-terminal and C-terminal CxxC motifs, RxCGxCxxxH and DPCxxCxxH/R, respectively, involved in [NiFe]-center coordination. Nevertheless, it is generally assumed that the active-site structures of all [NiFe]-hydrogenases are very similar (**Figure 1*a***); but in one case (soluble [NiFe]-hydrogenase from *Ralstonia eutropha*), there is spectroscopic evidence that the ligand structure could be substantially different (28, 29).

[NiFe]-hydrogenases are minimally composed of two subunits, a large one (40–68 kDa) and a small one (16–30 kDa). The large subunit harbors the [NiFe]-binuclear active-site center. The small subunit generally contains three linearly arranged and evenly spaced iron-sulfur clusters, a proximal and a distal [4Fe4S]-cluster, and one central [3Fe4S]-cluster (8). In energy-converting [NiFe]-hydrogenases, the small subunit contains only the proximal [4Fe4S]-cluster, which appears to be necessary and sufficient for [NiFe]-hydrogenase function. In the heterodimer, the [NiFe]-center is buried and located close to the large interface between the two subunits and close to the proximal [4Fe4S]-cluster of the small subunit (**Figure 1*a***). A gas channel connects the surface with the active site (30).

The large subunit of most [NiFe]-hydrogenases is synthesized as a preprotein. The C-terminal extension after H/R of the DPCxxCxxH/R motif is clipped off proteolytically in the maturation process (31–33). The gene coding the large subunit of some of the energy-converting hydrogenases (34, 35) and some of the H_2-sensory [NiFe]-hydrogenases (36) ends with a stop codon directly after the nucleotide sequence for the DPCxxCxxH/R motif. Therefore, synthesis of these [NiFe]-hydrogenases appears to be independent of this proteolytic maturation step.

In the next paragraphs, we summarize what is known about the four different subtypes of [NiFe]-hydrogenases found in methanogens: (*a*) energy-converting [NiFe]-hydrogenases, (*b*) heterodisulfide reductase-associated [NiFe]-hydrogenase, (*c*) methanophenazine-reducing [NiFe]-hydrogenase, and (*d*) F_{420}-reducing [NiFe]-hydrogenase.

Energy-Converting [NiFe]-Hydrogenases

Energy-converting [NiFe]-hydrogenases from methanogens are membrane associated and catalyze the reversible reduction of ferredoxin ($E' \approx -500$ mV) with H_2 ($E' = -300$ mV), driven by a proton or sodium ion motive force (Reaction 3) (**Figure 4**) (34, 35).

$$Fd_{ox} + H_2 + \Delta\mu H^+/Na^+ \rightleftharpoons Fd_{red}{}^{2-} + 2\,H^+. \quad 3.$$

Related enzymes are found in some hydrogenotrophic bacteria and in some H_2-forming bacteria and archaea. In the H_2-forming microorganisms, the enzyme catalyzes the reverse of Reaction 3. Most convincing is the energy-converting function of the [NiFe]-hydrogenase in the gram-negative *Rhodospirillum rubrum* (37) and *Rubrivivax gelatinosus* (38) as well as in the gram-postive *Carboxidothermus hydrogenoformans* (39). These anaerobic bacteria can grow chemolithoautotrophically on

CO, with H_2 and CO_2 being the only catabolic end products formed ($CO + H_2O \rightleftharpoons CO_2 + H_2$ $\Delta G^{\circ\prime} = -20$ kJ mol^{-1}) (40). The fermentation, which involves only a cytoplasmic carbon monoxide dehydrogenase (CooS), a cytoplasmic polyferredoxin (electron transfer protein) (CooF), and a membrane-associated, energy-converting [NiFe]-hydrogenase (CooHKLMUX), is coupled with chemiosmotic energy conservation as evidenced by growth and uncoupling experiments (37–39).

The energy-converting hydrogenases (EchA-F, EhaA-T, EhbA-Q, and MbhA-N) from methanogens contain six conserved core subunits (**Figure 4**) and up to 14 additional subunits. The six core subunits show sequence similarity to the six subunits of the carbon monoxide dehydrogenase (CooS)-associated [NiFe]-hydrogenase (CooHKLMUX) from bacteria, to the five subunits of the formate dehydrogenase-associated [NiFe]-hydrogenase (HycCDEFG) from *E. coli*, and to six of the core subunits of the NADH:ubiquinone oxidoreductase (NuoA-N) (complex I of the respiratory chain) from *E. coli* (41). Of the conserved subunits, two are integral membrane proteins (the larger one most probably involved in cation translocation), and four are hydrophilic proteins (**Figure 4**). Of the hydrophilic proteins, one is the [NiFe]-hydrogenase large subunit, one the hydrogenase small subunit with only one [4Fe4S]-cluster (the proximal one), one an iron-sulfur protein with two [4Fe4S]-clusters, and one a subunit without a prosthetic group. In complex I, the subunit NuoD homologous to the [NiFe]-hydrogenase large subunit lacks the N- and C-terminal CxxC motifs for [NiFe]-center formation (35).

None of the genes encoding energy-converting [NiFe]-hydrogenases in Bacteria or Archaea show a twin arginine translocation (Tat) motif-encoding sequence. The lack of the Tat motif-encoding sequence (42) indicates that the large subunit and the small subunit of the energy-converting [NiFe]-hydrogenases are not translocated from the cytoplasm to the periplasm and are therefore oriented toward the cytoplasm.

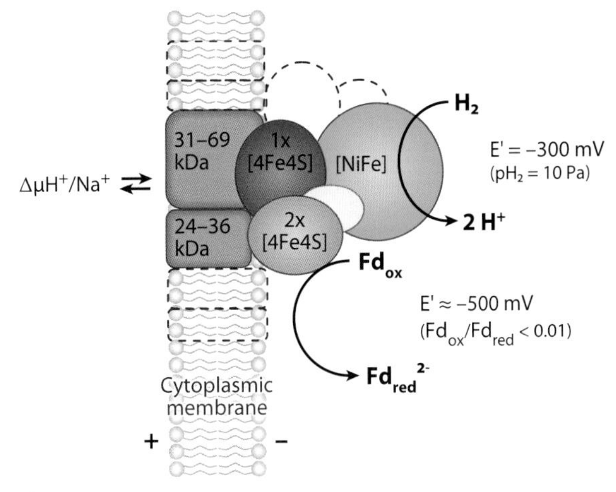

Figure 4

Schematic representation of the structure and function of the energy-converting [NiFe]-hydrogenases EchA-F, EhaA-T, EhbA-Q, and MbhA-N found in methanogenic archaea. The energy-converting hydrogenase EchA-F is composed only of the six conserved core subunits, which are highlighted in color. The energy-converting hydrogenases EhaA-T, EhbA-Q, and MbhA-N also contain several hydrophobic and hydrophilic subunits of unknown function. These subunits are symbolized by areas with dashed boundaries. Abbreviation: Fd, ferredoxin with two [4Fe4S]-clusters.

EchA-F. Genes for this type of energy-converting [NiFe]-hydrogenase are found in *M. barkeri* and *M. mazei* but not in *M. acetivorans*. They are also present in the genome of all members of the Methanomicrobiales. The enzyme in these organisms is most similar to the energy-converting hydrogenase CooHKLMUX from *R. rubrum* (37) and *C. hydrogenoformans* (39). EchA-F differs, however, in not forming a tight complex with its ferredoxin and functionally associated oxidoreductase. The lack of complex formation is because in methanogens the reduced ferredoxin, generated by the energy-converting hydrogenase, is used in electron transfer to more than one oxidoreductase. A 6 kDa 2[4Fe4S]-ferredoxin from *M. barkeri*, which is most probably the ferredoxin reduced by H_2 via the energy-converting hydrogenase EchA-F, has been characterized (43).

The enzyme complex EchA-F has been purified from *M. barkeri* and is composed of six different subunits (**Figure 4**) encoded by the *echA-F* operon (44, 45). EchA (69 kDa) and EchB (32 kDa) are the two integral

Tat: twin arginine translocation

membrane proteins. EchA is predicted to have 17 membrane-spanning α-helices and shows 30% sequence identity to a putative Na^+/H^+ translocator component in *Bacillus subtilis* (44). EchE is the large subunit, EchC is the small subunit, and these harbor the [NiFe]-center and a [4Fe4S]-cluster, respectively. EchF contains two [4Fe4S]-clusters. EchD is the soluble subunit without a prosthetic group. Chemical analyses of the purified complex have revealed the presence of nickel, nonheme iron, and acid-labile sulfur in a ratio of 1:12.5:12, substantiating the presence of three [4Fe4S]-clusters in addition to the [NiFe]-center. Like CooH, the large subunit EchE is synthesized without a C-terminal extension; the gene ends directly after the DPCxxCxxR motif with a stop codon. The evidence for reversed electron transfer in ferredoxin reduction with H_2 comes from biochemical and genetic studies. Cell suspensions of *M. barkeri* catalyze the reduction of CO_2 to CO ($E_o' = -520$ mV) with H_2 ($E_o' = -414$ mV), involving a cytoplasmic CO dehydrogenase, ferredoxin, and EchA-F. The reaction is driven by a proton motive force (46). The cells also catalyze the reverse reaction, the dehydrogenation of CO to CO_2 and H_2, which is coupled with the buildup of a proton motive force (47, 48). Δ*ech* mutants did not catalyze the forward or the backward reaction and also did not catalyze one of the other ferredoxin-dependent reductions mentioned above (49, 50). There are conflicting reports with respect to the coupling ion used by Ech in *M. barkeri*. The experiments investigating the reversible conversion of H_2 and CO_2 to CO and H_2O are more in favor of protons (46–48), whereas those addressing the reduction of CO_2 with H_2 to formylmethanofuran are more in favor of sodium ions (51, 52).

The iron-sulfur centers in the EchA-F complex have been characterized by electron paramagnetic resonance spectroscopy, revealing that two of the [4Fe4S]-clusters show pH-dependent redox potentials (53) and indicating that these clusters mediate electron and proton transfer. The [4Fe4S]-clusters were assigned to the individual subunits via site-directed mutants (54). Insights into the mechanism of ion translocation come also from inhibition experiments with dicyclohexylcarbodiimide (DCCD), which specifically modifies protonated carboxyl residues located in a hydrophobic environment. Labeling studies of Ech with [^{14}C]-DCCD showed that the inhibition of the enzyme was associated with a specific labeling of the two integral membrane subunits EchA and B, particularly of EchA (35). The inhibition of Ech by DCCD indicates that the electron transfer reaction in this enzyme is strictly coupled to cation translocation.

EhaA-T and EhbA-Q. The [NiFe]-hydrogenase EhaA-T and EhbA-Q differ from the energy-converting hydrogenase EchA-F by having up to 14 additional subunits; many of these subunits are integral membrane proteins, and some are iron-sulfur proteins (55). A function of the additional subunits is difficult to envisage. Interestingly, the number of subunits of complex I of the respiratory chain also varies significantly without an apparent change in properties; thus, complex I in *E. coli* is composed of 14 subunits and in mitochondria of more than 40 subunits (56, 57).

Genes encoding EhaA-T and/or EhbA-Q are found in *M. kandleri*, in all members of the Methanococcales and Methanobacteriales, and in some members of the Methanomicrobiales (but not in *Methanosphaerula palustris* and *Methanoregula boonei*), and *eha* and/or *ehb* genes are not found in members of the Methanosarcinales. In some methanogens, e.g., in *M. marburgensis*, the genes are clustered and form a transcription unit, and in others, e.g., *M. jannaschii*, some of the genes are in separate loci.

The *eha* operon (12.5 kb) in *M. marburgensis* is composed of 20 open reading frames that form a transcription unit (55). Sequence analysis indicates that four of the genes encode proteins with high sequence similarity to four of the six different subunits characteristic for energy-converting hydrogenases (**Figure 4**): *ehaO* encodes the large [NiFe]-center harboring subunit as a preprotein; *ehaN* encodes the small subunit with only one [4Fe4S]-cluster.

ehaH and *ehaJ* encode the two conserved integral membrane proteins (24 kDa and 31 kDa); EhaJ is probably involved in ion translocation. A gene encoding the subunit with two [4Fe4S]-clusters appears to be lacking. Instead, the gene cluster harbors a gene for a 6[4Fe4S]-polyferredoxin (EhaP) and one for a 10[4Fe4S]-polyferredoxin (EhaQ). In addition to these subunits, the *eha* operon encodes four nonconserved hydrophilic subunits and ten nonconserved integral membrane proteins.

The *ehb* operon (9.6 kb) in *M. marburgensis* is composed of 17 open reading frames (55). The gene *ehbN* is predicted to encode the large subunit as a preprotein and *ehbM* the small subunit; *ehbF* and *ehbO* encode the two integral membrane proteins (53 kDa and 36 kDa), with the larger one probably involved in ion translocation; *ehbL* encodes a 2[4Fe4S]-cluster-containing protein (**Figure 4**). The *ehb* operon additionally encodes a 14[4Fe4S]-polyferredoxin (EhbK), two nonconserved hydrophilic subunits, and nine nonconserved integral membrane proteins.

Deletion of the *ehb* genes in *M. maripaludis* revealed a function of Ehb in autotrophic CO_2 fixation. The mutant was an acetate auxotroph. Deletion of the *eha* genes was not possible (58).

In Methanomicrobiales, in contrast to Methanobacteriales, the *ehaO* gene encoding the large subunit lacks the 3′ extension; it ends with the sequence motif DPCxxCxxR. The subunit with the [NiFe]-center in Methanomicrobiales is therefore predicted not to be synthesized as a preprotein.

MbhA-N. Genes encoding MbhA-N are found in two members of the Methanomicrobiales (*Methanospirillum hungatei* and *Methanocorpusculum labreanum*) but not in members of the four other orders (25). The *mbhA-N* gene cluster found in *M. hungatei* and *M. labreanum* is composed of 14 open reading frames. Six of the deduced proteins correspond to subunits conserved in all energy-converting hydrogenases (**Figure 4**): MbhL (large subunit with [NiFe]-center), MbhJ (small subunit with one [4Fe4S]-cluster), MbhN (subunit with two [4Fe4S]-clusters), MbhK (hydrophilic subunit without a prosthetic group), and MbhM and MbhH (conserved integral membrane proteins of 36 kDa and 56 kDa, respectively). The other eight deduced proteins all appear to be nonconserved integral membrane proteins. The gene *mbhL* for the large subunit encodes a protein with a C-terminal extension.

An enzyme complex with a similar composition was partially purified from *Pyrococcus furiosus*, which ferments glucose at 100°C to two acetic acids, two CO_2, and four H_2 (59). The enzyme complex was shown to use ferredoxin as electron donor. Addition of reduced ferredoxin to inverted membrane vesicles resulted in the generation of a proton or sodium motive force (inside positive) that could drive ATP synthesis (60). The ATPase involved was shown to translocate sodium ions (61). *P. furiosus* is a member of the order Thermococcales, which—like the five orders of methanogenic archaea—belong to the kingdom of Euryarchaeota. As in the Methanomicrobiales, the large [NiFe]-center harboring subunit (MbhL) in *P. furiosus* is synthesized as a preprotein.

Heterodisulfide Reductase-Associated [NiFe]-Hydrogenase MvhADG

This cytoplasmic heterodisulfide reductase-associated [NiFe]-hydrogenase (MvhADG) catalyzes the reduction of dyes such as methyl viologen with H_2. The physiological electron acceptor is most probably the cytoplasmic heterodisulfide reductase HdrABC (Reaction 4) with which MvhADG forms a tight complex (**Figure 5*a***). In *Methanothermobacter* grown under nickel-limiting growth conditions, almost all of the MvhADG is found in complex with HdrABC (62). When the cells are grown in media with excess nickel, in addition to the MvhADG/HdrABC complex, free MvhADG and HdrABC are also present in varying amounts (62, 63). The MvhADG/HdrABC complex (62, 64) as well as MvhADG (62, 65) and HdrABC (66, 67) have also been purified and characterized.

$$H_2 + HdrABC_{ox} \rightleftharpoons HdrABC_{red}^{2-} + 2\,H^+. \quad 4.$$

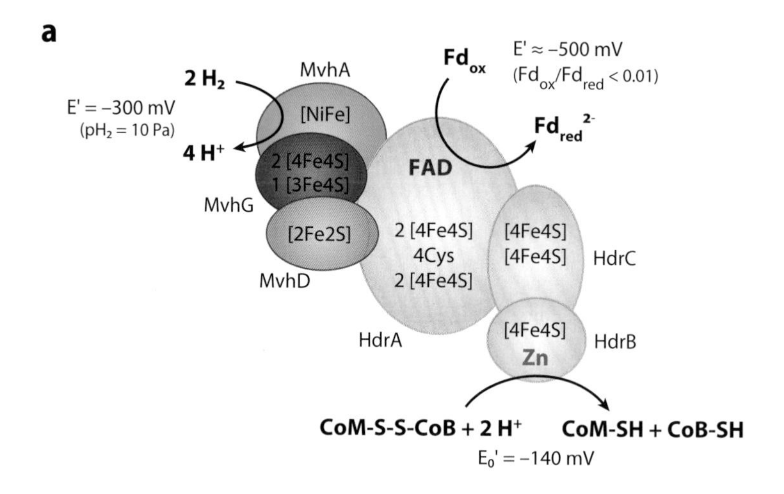

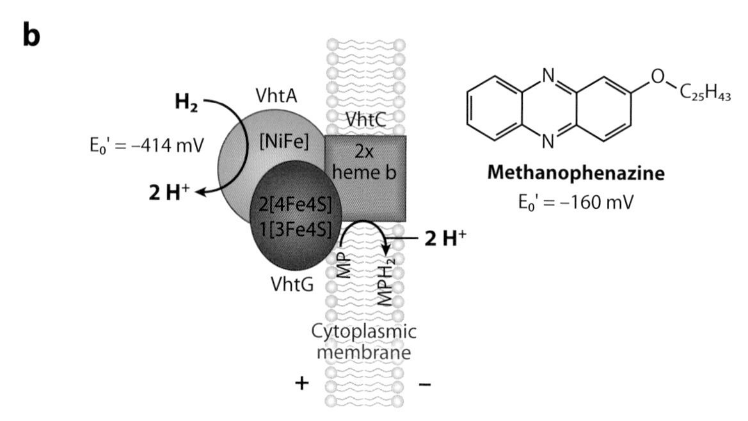

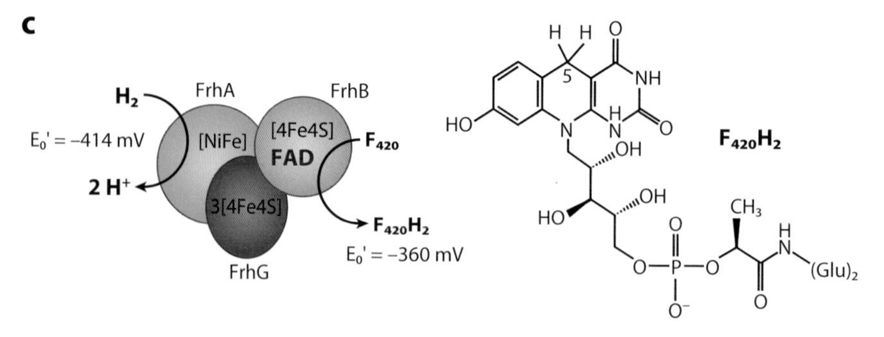

Figure 5

The structures and functions of (*a*) the MvhADG/HdrABC complex, (*b*) the VhtACG complex, and (*c*) the FrhABG complex involved in H_2 uptake in methanogenic archaea are schematically shown. The Mvh/Hdr complex is found mainly in methanogens without cytochromes, and the Vht complex is found only in methanogens with cytochromes. The stoichiometry of the MvhADG/HdrABC-catalyzed reaction has not yet been ascertained. Abbreviations: CoM-SH, coenzyme M with its thiol group; CoB-SH, coenzyme B with its thiol group; F_{420}, coenzyme F_{420}; Fd, ferredoxin with two [4Fe4S]-clusters; FrhA, FrhB, and FrhG, F_{420}-reducing [NiFe]-hydrogenase subunits; HdrA, HdrB, and HdrC, heterodisulfide reductase subunits; MP, methanophenazine; MvhA, MvhD, and MvhG, heterodisulfide reductase-associated [NiFe]-hydrogenase subunits; VhtA, VhtC, and VhtG, methanophenazine-reducing [NiFe]-hydrogenase subunits.

MvhA is the large subunit with the [NiFe]-center. It is synthesized as a preprotein. MvhG is the small subunit with one [3Fe4S]-cluster and two [4Fe4S]-clusters. MvhD is a subunit with one [2Fe2S]-cluster that mediates electron transfer from MvhG to HdrABC (**Figure 5*a***). None of the other [NiFe]-hydrogenases from methanogens contain a subunit with a [2Fe2S]-cluster. The presence of [2Fe2S]-clusters in iron-sulfur proteins of Archaea is the exception.

HdrB harbors the active site for CoM-S-S-CoB reduction. It contains two cysteine-rich sequence motifs $Cx_{31-39}CCx_{35-36}CxxC$ designated as CCG domains. The C-terminal CCG domain is involved in the binding of an unusual [4Fe4S]-cluster, and the N-terminal one is involved in zinc binding (68). HdrC harbors two [4Fe4S]-clusters, and HdrA contains four [4Fe4S]-clusters and a FAD that is only loosely bound but essential for activity. In addition, a conserved sequence motif with four cysteines is found. In HdrA from *Methanococcus* species, one of the four cysteines is a selenocysteine (**Figure 5*a***).

The purified MvhADG/HdrABC complex catalyzes the reduction of the heterodisulfide CoM-S-S-CoB with H_2 at only low specific activity (62, 64). The complex also catalyzes a CoM-S-S-CoB-dependent reduction of clostridial ferredoxin with H_2 at high specific activity. In the presence of ferredoxin, the specific rate of CoM-S-S-CoB reduction is increased (A.-K. Kaster, unpublished results). The complex thus appears to couple the endergonic reduction of ferredoxin ($E' \approx -500$ mV) with H_2 ($E'_0 = -414$ mV) to the exergonic reduction of CoM-S-S-CoB ($E'_0 = -140$ mV) (26) with H_2. The coupling probably involves the FAD in HdrA as the center of electron bifurcation (17, 69, 70). The exact stoichiometry of the reaction has not yet been ascertained but is, analogous to a ferredoxin-dependent crotonyl-CoA reduction with NADH (70), presently assumed to be two H_2 that reduce one clostridial ferredoxin (with two one-electron-accepting [4Fe4S]-clusters) and one CoM-S-S-CoB (**Figure 5*a***).

In *Methanothermobacter*, the ferredoxin reduced by the MvhADG/HdrABC complex is most probably the 12[4Fe4S] polyferredoxin encoded by the *mvhB* gene of the *mvhDGAB* operon (71). The polyferredoxin partially copurifies with the MvhADG/HdrABC complex (72, 73), which is why the MvhADG/HdrABC preparations always contain polyferredoxin—albeit in substoichiometric amounts (64, 65).

Genes encoding the MvhADG/HdrABC complex are also found in some cytochrome-containing methanogens, e.g., *M. barkeri*, *M. mazei*, and in Rice cluster I methanogens, and in some nonmethanogenic archaea, e.g., *Archaeoglobus fulgidus* (17). In *M. barkeri* and *A. fulgidus*, the homolog of *mvhD* is fused to the 3′ end of an *hdrA* homolog (62). In the Methanobacteriales, Methanopyrales, and Methanococcales, the genes are generally organized in three transcription units, *mvhDGAB*, *hdrBC*, and *hdrA*, which are not located adjacent to one another (67). In the Methanomicrobiales, the three *hdr* genes are juxtapositioned.

In the *Methanococcus* species and *M. kandleri*, there are two versions of MvhADG, one designated VhuADG in which the large [NiFe]-center harboring subunit A shows a C-terminal DPUxxCxxH motif (U for selenocysteine) and one abbreviated VhcADG in which the subunit A shows a C-terminal DPCxxCxxH motif (74, 75). When sufficient selenium is in the medium, only the [NiFeSe]-hydrogenase VhuADG is formed (76). In *Methanocaldococcus* species, there is only the selenoprotein version. The three other orders of methanogens do not contain selenoproteins (77).

Interestingly, in *Methanococcus* species, the gene for the large subunit of the [NiFeSe]-hydrogenase is split, and therefore the large subunit consists of two polypeptides, each contributing two ligands to the [NiFeSe]-center. A fusion of the two proteins was shown to be without effect on the kinetic and spectroscopic properties of the [NiFeSe]-hydrogenase VhuADG (78).

The genomes of most members of the cytochrome-less Methanomicrobiales—an exception being *Methanoculleus marisnigri*—lack the genes for MvhA and MvhG but contain the genes for MvhD and HdrABC, which are juxtapositioned. It has therefore been proposed that in these methanogens a MvhD/HdrABC complex is associated with one of the energy-converting hydrogenases EchA-F, EhaA-T, or MbhA-N (25). As a consequence, in most Methanomicrobiales, heterodisulfide reduction with H_2 would be energy consuming. This is the consequence of the finding that the different energy-converting hydrogenases all have the same topology and should therefore have the same function, namely to catalyze the oxidation of H_2 in a reaction requiring, rather than generating, energy. Consistent with this interpretation is that *M. hungatei* (one of the Methanomicrobiales without *mvhA* and *mvhG* genes) is known to grow on H_2 and CO_2 at very low H_2 partial pressures (79, 80), indicating a very low ATP gain (17).

An alternative hypothesis is that, in cytochrome-less methanogens lacking the genes for MvhA and MvhG, these two subunits are substituted by FrhA (large [NiFe]-hydrogenase subunit) and FrhG (small [NiFe]-hydrogenase subunit) of the F_{420}-reducing hydrogenase (FrhABG) (see below). All methanogens without cytochromes contain genes for this enzyme. FrhA and G would thus be present both in a putative FrhAG/MvhD/HdrABC complex and in the FrhABG complex. There are precedents for such subunit sharing. Thus, the molybdenum-containing formylmethanofuran dehydrogenase FwdA/FmdBC and the tungsten-containing formylmethanofuran dehydrogenase FwdABC from *M. marburgensis* share the subunit FwdA and the pyruvate dehydrogenase complex, and the 2-oxoglutarate dehydrogenase complex from *E. coli* shares the lipoamide dehydrogenase subunit (81). Interesting in this respect is that in the genome of *M. boonei* putative genes for a large subunit (NCBI Mboo_2023) and one for a small subunit (NCBI Mboo_1398) of the F_{420}-reducing [NiFe]-hydrogenase are found in addition to the *frhADGB* transcription unit.

Methanophenazine-Reducing [NiFe]-Hydrogenase VhtACG

This membrane-associated, cytochrome *b*-containing [NiFe]-hydrogenase catalyzes the reduction of methanophenazine with H_2 (Reaction 5) and couples this reaction with the buildup of an electrochemical proton potential (**Figure 5*b***). Methanophenazine is a 2-hydroxyphenazine derivative that is connected via an ether bridge to a pentaprenyl side chain (**Figure 5*b***) (82, 83). Like ubiquinone (E'_o = +110 mV) and menaquinone (E'_o = −80 mV), methanophenazine (E'_o = −170 mV) (26) is a lipid-soluble electron and proton carrier; the difference is that methanophenazine's redox potential is much lower. Methanophenazine (shown as MP in Reaction 5) is only found in the Methanosarcinales, i.e., in methanogens that contain cytochromes (84, 85).

$$H_2 + MP \rightarrow MPH_2 \quad \Delta G^{\circ\prime} = -50\,\text{kJ mol}^{-1}. \qquad 5.$$

The methanophenazine-reducing hydrogenase VhtACG from *M. barkeri* has been characterized (**Figure 5*b***) (86). VhtA is the [NiFe]-center harboring large subunit, which is synthesized as a preprotein. VhtG is the [4Fe4S]/[3Fe4S]/[4Fe4S]-cluster harboring small subunit. VhtC is a cytochrome *b* that is integrated into the membrane. The gene *vhtG* contains at its 5′ end a sequence encoding a Tat signal (DRRTFM/I). Genetic and biochemical studies indicate that in such cases the large subunit is cotranslocated with the small subunit across the cytoplasmic membrane (87, 88). The subunits with the [NiFe] active site thus face the periplasm. As a consequence, the protons generated upon H_2 oxidation are released outside. VhtACG thus has a topology similar to that described for [NiFe]-hydrogenase-1 and [NiFe]-hydrogenase-2 in *E. coli* and for the membrane-associated [NiFe]-hydrogenase in *Ralstonia* (32).

In some *Methanosarcina* species, e.g., *M. mazei*, the genome harbors two sets of genes, *vhtGACD* and *vhoGAC* (the latter encodes a Vht isoenzyme); each set is a transcription unit. In the *vho* operon, a *vhtD*-like gene is not present. *vhtD* is homologous to *hoxM* from *R. eutropha* and to *hyaD* from *E. coli*, which encode specific maturation endopeptidases. In *M. mazei*, the *vho* operon is transcribed constitutively, whereas the *vht* operon is transcribed only during growth on methanol and H_2/CO_2 rather than on acetate (89, 90).

The cytoplasmic membrane of all methanogens with cytochromes contains an associated methanophenazine-dependent heterodisulfide reductase, HdrDE (**Figure 3*a***). The subunit HdrE is a cytochrome *b* that is integrated into the membrane, and HdrD is the peripheral subunit that catalyzes CoM-S-S-CoB reduction. HdrD combines the sequences of HdrB and HdrC of the cytoplasmic heterodisulfide reductase HdrABC. The gene encoding HdrD lacks a Tat sequence, indicating that the HdrD subunit faces the cytoplasm, which is consistent with its function as a catalyst of the reduction of CoM-S-S-CoB, generated by methyl-coenzyme M reduction with coenzyme B in the cytoplasm (91, 92).

Thus, whereas the active-site-harboring subunit of the methanophenazine-reducing hydrogenase is located on the periplasmic side of the membrane, that of the methanophenazine-oxidizing heterodisulfide reductase is located on the cytoplasmic side (**Figure 3*a***). Both complexes are electrically connected via the lipid-soluble methanophenazine, which is reduced by the cytochrome *b* of the hydrogenase and is reoxidized by the cytochrome *b* of the heterodisulfide reductase. Experimental evidence has been provided that per heterodisulfide reduced with H_2 in this system four electrogenic protons are generated, which can be used to drive the synthesis of one ATP via a proton-translocating A_1A_0-ATPase (84, 85, 93, 94).

In Rice cluster I, which belongs to the Methanosarcinales, the gene for HdrE (cytochrome *b*) is lacking, indicating that in this methanogen the methanophenazine reduced by VhtACG cannot be reoxidized. In agreement with this prediction is the finding that in Rice cluster I the genes for a functional MvhADG/HdrABC complex are present (17).

Coenzyme F_{420}-Reducing [NiFe]-Hydrogenases FrhABG

This cytoplasmic [NiFe]-hydrogenase catalyzes the reversible reduction of coenzyme F_{420} with H_2 (Reaction 6). Coenzyme F_{420} is a 5-deazaflavin (**Figure 5*c***) found in high concentrations in methanogenic archaea and in *A. fulgidus* and in low concentrations also in other archaea and in some bacteria. Although structurally resembling a flavin, F_{420} is functionally more like the pyridine nucleotide NAD(P) in transferring two electrons plus a proton (a hydride) rather than single electrons. The functional difference from NAD(P) is, however, that the redox potential of the $F_{420}/F_{420}H_2$ couple is −360 mV and thus 40 mV more negative than that of the NAD(P)/NAD(P)H_2 couple (27).

$$H_2 + F_{420} \rightleftharpoons F_{420}H_2 \quad \Delta G^{\circ\prime} = -11\,\text{kJ mol}^{-1}. \qquad 6.$$

In methanogenic archaea, $F_{420}H_2$ is involved in two reduction steps of methanogenesis from CO_2 (**Figure 3**) and also in several anabolic reduction reactions, e.g., in the $F_{420}H_2$:NADP oxidoreductase reaction, the F_{420}-dependent glutamate synthase reaction, the F_{420}-dependent sulfite reductase reaction, and the $F_{420}H_2:O_2$ oxidoreductase reaction. Under conditions of H_2 limitation, transcription of the genes for the F_{420}-reducing hydrogenase are upregulated (95, 96) and under conditions of nickel limitation downregulated (97).

During growth of methanogenic archaea on formate, F_{420} reduction is catalyzed by a cytoplasmic F_{420}-dependent formate dehydrogenase FdhABC. Under these conditions, the F_{420}-reducing hydrogenase catalyzed the formation of H_2 (Reaction 6) with the H_2 used via intraspecies hydrogen transfer as electron donor for the coupled reduction of ferredoxin and heterodisulfide catalyzed by the cytoplasmic MvhAGD/HdrABC complex (**Figure 5*a***) (98, 99).

The FrhABG complex has been purified, and the encoding genes have been determined (100). The genes are organized in a transcription unit *frhADGB*, where *frhA* encodes the large subunit with the [NiFe]-center, *frhG* encodes the small subunit with three [4Fe4S]-clusters, and *frhB* encodes an iron-sulfur flavoprotein with one [4Fe4S]-cluster and one FAD, which functions as a one electron/two electron switch in F_{420} reduction (**Figure 5*c***). *frhD* encodes an endopeptidase (homologous to HycI from *E. coli*), which is required to clip off the C-terminal extension in the FrhA preprotein.

In the genome of *M. barkeri*, a *frhADGB* operon and a *freAEGB* operon are found, the latter encoding a Frh isoenzyme. The *freAEGB* operon lacks a gene homolog of *frhD* for the endopeptidase (101). Genetic evidence has recently been found that *freAEGB* is expressed functionally only if the *frhADGB* operon is simultaneously expressed, indicating that FrhD is also involved in FreA maturation (102). The function of the *freE* gene (123 bp) is not known.

The small subunit of most [NiFe]-hydrogenases harbors two [4Fe4S]-clusters and one central [3Fe4S]-cluster. However, in the small subunit of the F_{420}-reducing hydrogenase, the central cluster is always a [4Fe4S]-cluster. Mutational studies, in which the middle cluster was converted to a [3Fe4S]-cluster, revealed significant changes in electron transport rates (103).

In the cytoplasm of methanogenic archaea, FrhABG is aggregated to a complex with a molecular mass of >900 kDa (63, 104). Upon ultracentrifugation of cell extracts, the F_{420}-reducing hydrogenase is recovered in the membrane fraction, which is why it was long believed that this enzyme is membrane associated.

In most *Methanococcus* species and *M. kandleri*, there are two versions of F_{420}-reducing hydrogenases, FrcABG and FruABG. The large subunit FrcA has a C-terminal DPCxxCxxH motif, and the large subunit FruA has a C-terminal DPUxxCxxH motif (U for selenocysteine) (74, 75). When selenium is in the medium, only the [NiFeSe]-hydrogenase (FruABG) is formed (76). In *Methanococcus aeolicus* and in *Methanocaldococcus* species, there is only the selenoprotein version.

Genes Involved in [NiFe]-Hydrogenase Maturation

In *E. coli*, for the synthesis of the [NiFe]-center in the large subunit of hydrogenase-3 at least six proteins are required: HypA and HypB for nickel insertion, HypE and HypF for the synthesis of the cyanide ligand from carbamoyl phosphate, and HypC and HypD for the transfer of the cyanide to the active site (31, 32, 105–107). The *hyp* genes are also found in all methanogenic archaea, although not clustered as in *E. coli*, e.g., in *M. marburgensis* only the *hypAB* genes form a transcription unit. Despite this fact, it is very likely that in methanogens the synthesis of the [NiFe]-center proceeds in principle as has been described for hydrogenase-3 from *E. coli*.

It is not yet known how in *E. coli* the CO ligand of iron in the [NiFe]-hydrogenases is generated. Carbamoyl phosphate was excluded as a precursor, and free CO was shown to be incorporated (108, 109). Labeling experiments with acetate indicate that in *Allochromatium vinosum* the CO in the [NiFe]-center is derived from the carboxyl group of acetate (110). A hypothesis is that the iron, which at the end carries two cyanide ligands and one CO ligand (**Figure 1*a***), reacts with acetyl-CoA, yielding a acetyl-iron complex (CH_3CO-Fe), which after methyl group migration affords the CO iron complex and methanol. An acyl iron complex (-CH_2CO-Fe) is found in [Fe]-hydrogenase (**Figure 1*c***).

Some methanogens, examples include *Methanobrevibacter smithii* and *Methanosphaera stadtmanae*, growing on H_2 and CO_2 as energy sources require acetate as a carbon source. These methanogens lack genes for carbon monoxide dehydrogenase and acetyl-CoA synthase/decarbonylase but contain active [NiFe]-hydrogenases, indicating that the two nickel enzymes are not involved in the synthesis of CO for the [NiFe]-center.

The large subunits of most of the [NiFe]-hydrogenases in methanogens are synthesized as preproteins from which a C-terminal extension has to be clipped off after completion of

[NiFe]-center synthesis. The endopeptidase gene *vhtD*, required for the maturation of methanophenazine-reducing hydrogenase, was found in the *vhtGACD* operon, and the gene *frhD* for the maturation of the F_{420}-reducing hydrogenase was found in the *frhADGB* operon. Whereas VhtD shows the best hits to the endopeptidase HyaD from *E. coli* (involved in hydrogenase-1 maturation) and HoxM from *Ralstonia* (involved in membrane-bound hydrogenase maturation), FrhD is more similar to the endopeptidase HycI from *E. coli* (involved in hydrogenase-3 maturation) (111, 112). The endopeptidase genes for the other [NiFe]-hydrogenases in methanogens have not yet been found. The *mvhDGAB* operon lacks a gene for an endopeptidase, and none of the genes in the *eha* or *ehb* operons show homology to genes for known endopeptidases or proteases.

In the genomes of some methanogens, aside from the gene clusters for the various hydrogenases, an open reading frame predicted to encode for an endopeptidase is found. The putative endopeptidase has sequence similarity to HycI involved in hydrogenase-3 maturation in *E. coli*. This gene is not associated with any other gene cluster from which a function could be deduced. Whether the *hycI* homolog outside the hydrogenase gene clusters has a function in [NiFe]-hydrogenase maturation remains to be shown.

Nickel Regulation

Nickel is a relatively abundant metal, although its concentration in freshwater and marine environments can be very low (<10 nM). Because the requirement of microorganisms for nickel is generally also low and because most microorganisms including methanogens have active, high-affinity nickel-uptake transporters (113), it was long overlooked that nickel is an essential trace element for most prokaryotes. It was the finding in 1979 that growth of methanogens is dependent on nickel that changed the picture (114). In addition to the [NiFe]-hydrogenases, methanogens contain three other nickel enzymes for methanogenesis and autotrophic CO_2 fixation, namely methyl-coenzyme M reductase (**Figure 3**), carbon monoxide dehydrogenase, and acetyl-CoA synthase/decarbonylase. The nickel enzymes are required in such high concentrations that nickel has to be added to growth media in over 1 μM concentrations in order for nickel not to become growth limiting. Therefore, in their natural habitats, methanogens have to continuously cope with the problem of nickel famine, and they probably have had to do so for the past 2.4 billion years, since the time of the so-called Great Oxidation Event. Recent evidence indicates that this was when the concentration of nickel in the oceans dropped from 400 nM to below 200 nM within 100 million years and subsequently to the modern day value of 9 nM by 550 Mya (115). It is argued that, as the rate of methanogenesis became nickel limited, the high concentrations of methane in the Precambrian atmosphere decreased, allowing the atmospheric O_2 concentration to build up. Methane reacts in the atmosphere with O_2 in a photochemical reaction cycle to become CO_2 and H_2O.

Great Oxidation Event: Earth's atmospheric oxygen rose from $<10^{-5}$ PAL (present atmospheric level) to between 0.1 and 0.2 PAL

Methanogenic archaea respond to changing nickel concentrations in the growth medium. Under conditions of nickel limitation, for example, the transcription of the genes for [Fe]-hydrogenase and F_{420}-dependent methylene-tetrahydromethanopterin dehydrogenase are upregulated, and those for F_{420}-reducing hydrogenase (FrhABG) are downregulated. This has been shown for *M. marburgensis* (97, 116) as well as for *M. maripaludis* and *M. jannaschii* (A.-K. Kaster, unpublished results).

In *E. coli*, there are two nickel-responsive transcriptional regulators, NikR, which suppresses transcription only in the presence of nickel (117), and RcnR, which only allows transcription in the presence of nickel (118). NikR and RcnR bind nickel reversibly with high affinity. Genes for only one of the two transcriptional regulators, namely NikR, are found in the genomes of methanogens. In many methanogens, several copies for NikR are present. The hypothesis therefore is that NikR is involved in the transcriptional regulation of the synthesis of [Fe]-hydrogenase and of F_{420}-dependent methylenetetrahydromethanopterin dehydrogenase, which are upregulated under conditions of nickel limitation.

hmd: gene encoding [Fe]-hydrogenase

The nucleotide sequence in the promoter region, to which NikR binds, has been mapped in *E. coli*. In this γ-proteobacteria, the NikR box is a 28-bp palindromic operator sequence (GTATGA-N_{16}-TCATAC) (119). In other taxonomic groups, the palindrome sequences differ somewhat, the palindrome can be less complete, and the space between the dyad-symmetric consensus sequences can be 12 bp to 16 bp (120). In some cases, the genes regulated by NikR do not even contain identifiable symmetric recognition sites (121). With this caveat, a putative NikR box was identified in the promoter region of the *hmd* gene in *M. marburgensis*, *M. maripaludis* and *M. jannaschii* (A.-K. Kaster, unpublished results).

How the synthesis of the F_{420}-reducing hydrogenase in methanogens is downregulated under nickel-limiting growth conditions is not known to date. In the nickel-limited growth of *M. marburgensis*, neither the enzyme nor the transcript of the *frhADGB* operon was found (97).

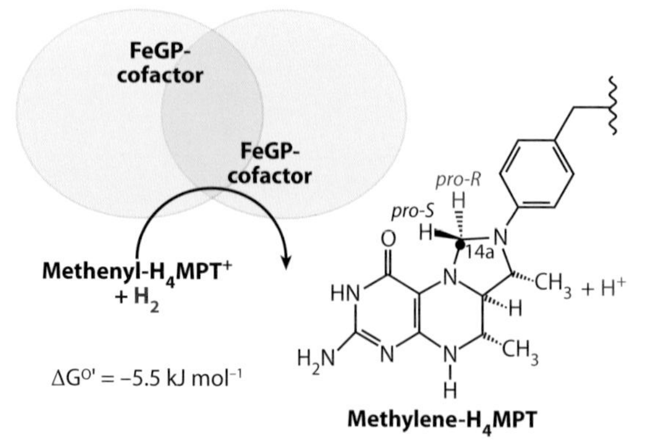

Figure 6

The structure and function of the homodimeric [Fe]-hydrogenase are schematically shown with its two active sites. For the structure of the iron-guanylylpyridinol cofactor (FeGP cofactor) see **Figure 1c**. [Fe]-hydrogenase catalyzes the reversible transfer of a hydride from H_2 into the *pro-R* side of methenyl-tetrahydromethanopterin (methenyl-H_4MPT^+) yielding methylene-H_4MPT (122).

[Fe]-HYDROGENASE IN METHANOGENS WITHOUT CYTOCHROMES

When methanogens without cytochromes grow under conditions of nickel limitation, some of them synthesize the nickel-free [Fe]-hydrogenase instead of the F_{420}-reducing [NiFe]-hydrogenase (97, 116). [Fe]-hydrogenase catalyzes the reversible transfer of a hydride from H_2 to methenyltetrahydromethanopterin (methenyl-H_4MPT^+), which is reduced to methylene-H_4MPT (Reaction 7) (**Figure 6**) (122, 123).

$$H_2 + \text{methenyl-}H_4MPT^+ \rightleftharpoons \text{methylene-}H_4MPT + H^+ \quad \Delta G^{\circ\prime} = -5.5\ \text{kJ mol}^{-1} \qquad 7.$$

Together with the F_{420}-dependent methylenetetrahydromethanopterin dehydrogenase (Reaction 8), [Fe]-hydrogenase catalyzes the reduction of F_{420} with H_2 (Reaction 6) (116). Consistent with this function are the findings that the synthesis of both [Fe]-hydrogenase and F_{420}-dependent methylenetetrahydromethanopterin dehydrogenase are upregulated under nickel-limiting growth conditions (116) and that in *M. maripaludis* it has been possible to knock out the genes for F_{420}-reducing hydrogenase or the gene for [Fe]-hydrogenase or that for F_{420}-dependent methylenetetrahydromethanopterin dehydrogenase with only minor effects on growth on H_2 and CO_2, but it has not been possible to knock out two of these genes (99, 124).

$$\text{Methylene-}H_4MPT + F_{420} + H^+ \rightleftharpoons \text{methenyl-}H_4MPT^+ + F_{420}H_2 \quad \Delta G^{\circ\prime} = -5.5\ \text{kJ mol}^{-1}. \qquad 8.$$

[Fe]-hydrogenase has a more than 20-fold higher K_m for H_2 (0.2 mM) than the F_{420}-reducing [NiFe]-hydrogenase (0.01 mM). As a compensation, cells grown with limited nickel have more than 40 times the specific [Fe]-hydrogenase activity (65 μmol min^{-1} mg $protein^{-1}$) than nickel-sufficient cells have F_{420}-reducing hydrogenase activity (1.6 μmol min^{-1} mg $protein^{-1}$) (values for *M. marburgensis*) (116). Thus the catalytic efficiency of H_2 uptake is maintained more or less constant (15).

Structural Properties

When discovered in 1990, [Fe]-hydrogenase was found to contain two moles iron per mole homodimer of 76 kDa but not to contain iron-sulfur clusters (23) and was therefore named iron-sulfur-cluster-free hydrogenase (125). The single iron per subunit is low spin and not redox active. A catalytic mechanism was proposed that did not require a redox active iron (126). Therefore, the iron was initially thought not to have a catalytic function, which is why the enzyme was dubbed "metal-free hydrogenase" (123, 126). However, it had been overlooked that the enzyme is inhibited by CO, albeit only at relatively high concentrations (K_i >0.5 mM), indicating an involvement of the iron in H_2 activation (127).

It is now known that [Fe]-hydrogenase harbors a novel iron-guanylylpyridinol (FeGP) cofactor covalently bound to the [Fe]-hydrogenase only via the thiol/thiolate group of a cysteine residue (**Figure 1*c***). In the cofactor, a low-spin iron (II) is ligated by two CO, one $C(O)CH_2$-, one $S-CH_2$-, and a sp^2-hybridized nitrogen of the pyridinol ring (12–14, 127–131). After protein unfolding, the cofactor can be released from the protein in the presence of thiol reagents under mild alkaline conditions or in the presence of acids in the absence of thiols (S. Shima, unpublished). When the FeGP cofactor is added to apoenzyme heterologously produced in *E. coli*, an active holoenzyme is formed (132), which has allowed the investigation of [Fe]-hydrogenase from methanogens that are difficult to grow and the performance of genetic analysis of the active-site amino acids involved in catalysis (13, 131). In [FeFe]-hydrogenases, the [FeFe]-center is also covalently attached to the protein only via a single cysteine residue (**Figure 1*b***). However, until now, it has not been possible to reversibly detach the center from this enzyme.

Catalytic Properties

As mentioned above, [Fe]-hydrogenase is reversibly inhibited by CO as are most [NiFe]- and [FeFe]-hydrogenases. [Fe]-hydrogenase is also reversibly inhibited by cyanide (K_i = 0.1 mM) and by isocyanides. With cyclohexylisocyanide, a specific, highly effective inhibitor with a K_i of <0.1 μM was recently found (S. Shima, unpublished result). Cyanide and isocyanides do not appear to inhibit [NiFe]-hydrogenases or [FeFe]-hydrogenases.

All three types of hydrogenases are rapidly inactivated by copper ions and by the superoxide anion radical $O_2^{\cdot-}$ (E'_o of the $O_2^{\cdot-}/H_2O_2$ couple = +890 mV). Whereas, in the presence of H_2, [FeFe]- and most [NiFe]-hydrogenases are rapidly inactivated by O_2, most probably in part owing to the reduction of O_2 with H_2 to $O_2^{\cdot-}$ (E'_o of the $O_2/O_2^{\cdot-}$ couple = −330 mV), purified [Fe]-hydrogenase remains active in the presence of O_2 both in the presence and absence of H_2 (15, 127).

[Fe]-hydrogenase shows a ternary complex catalytic mechanism (127). It does not catalyze the reduction of dyes with H_2, the exchange of protons of water into H_2, or the conversion of *para*-H_2 to *ortho*-H_2 (spin isotopomers of H_2), three reactions characteristically catalyzed by [FeFe]- and [NiFe]-hydrogenases. However, in the presence of its substrate methenyl-H_4MPT^+, [Fe]-hydrogenase catalyzes the exchange of H^+ from water into H_2 and the conversion of *para*-H_2 to *ortho*-H_2, with kinetics almost indistinguishable from those of the two other types of hydrogenases (133). The enzyme also catalyzes a stereospecific exchange of the *pro-R* hydrogen of methylene-H_4MPT (see **Figure 6**) with protons of water (134). A catalytic mechanism consistent with these results was recently deduced from the crystal structure of the [Fe]-hydrogenase-methylene-H_4MPT complex (14).

hcg: *hmd* co-occurring genes

SAM: *S*-adenosylmethionine

The *hmd* gene for [Fe]-hydrogenase is present in the genomes of *M. kandleri*, all members of the Methanococcales, most members of the Methanobacteriales, and only one member of the Methanomicrobiales (*M. labreanum*). The *hmd* gene has not yet been found in the genome of one of the members of the Methanosarcinales.

Genes Involved in FeGP Cofactor Biosynthesis

The genes involved in the biosynthesis of the FeGP cofactor (**Figure 1*c***) have not yet been determined experimentally. However, an *in silico* analysis indicates that there are seven genes present in all methanogens with a *hmd* gene, with one exception (see below). These genes are tentatively designated *hcg* genes (*hmd* co-occurring genes). In many of the methanogens, the seven *hcg* genes neighbor the *hmd* gene and are clustered (**Figure 7**). Despite being juxtapositioned to *hcgABCDEF* in *M. marburgensis*, the *hmd* gene is transcribed monocistronically (97). The exception is *M. hungatei*. This methanogen lacks a *hmd* gene but harbors a *hcgCDEFG* gene cluster without, however, having the genes *hcgA* and *hcgB*.

The gene *hcgA* is predicted to encode a protein with a sequence similar to the radical-SAM (*S*-adenosylmethionine) iron-sulfur protein BioB, which is involved in sulfur insertion in biotin biosynthesis. However, HcgA lacks the N-terminal signature CX_3CX_2C motif or CX_4CX_2C motif that is characteristic for the radical-SAM protein superfamily and that coordinates a [4Fe4S]-cluster essential for radical formation (135, 136). Instead, HcgA universally harbors a unique CX_5CX_2C motif (137). Some radical-SAM enzymes have a function in methylation reactions (138). One of the two methyl groups attached to the pyridinol ring in the FeGP cofactor is derived from the methyl group of methionine (see below). It is therefore likely that HcgA is involved in this methylation reaction. Interestingly, a BioB homolog is also involved in [FeFe]-hydrogenase maturation (139–141).

The crystal structure of HcgB from *M. thermautotrophicus* has been determined within a structural genomics project (142). The homodimeric protein contains three bound

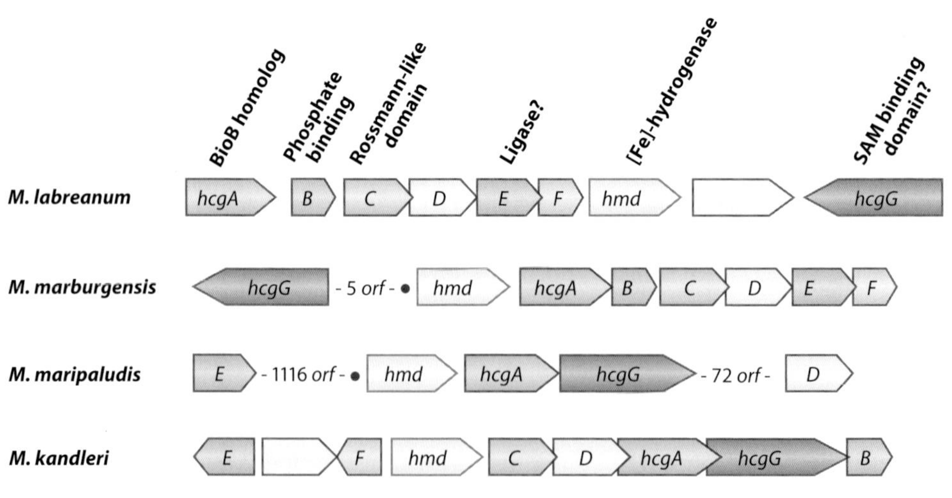

Figure 7

Genes neighboring the *hmd* gene for [Fe]-hydrogenase in *Methanocorpusculum labreanum*, *Methanothermobacter marburgensis*, *Methanococcus maripaludis*, and *Methanopyrus kandleri* are shown. The red dot indicates the presence of a putative NikR box to which the transcriptional regulator NikR of methanogens could bind and stop transcription when nickel is present. *hcg* is the abbreviation for *hmd* co-occurring genes.

phosphates and shares structural similarities with pyrophosphatases. The gene *hcgC* encodes a hypothetical protein with a putative NAD(P)-binding Rossmann-like domain. The presumed protein HcgD has a sequence similar to a protein that in yeast interacts with the transcriptional activator NGG1p. HcgE shows sequence similarity to proteins that catalyze ubiquitin activation with ATP. The gene *hcgF* is without a recognizable function. HcgG is annotated as a fibrillarin-like protein with a C-terminal domain that could bind SAM.

All the *hcg* genes and the *hmd* gene in *M. labreanum* (the only member of the Methanomicrobiales with these genes) show higher sequence similarity at the protein level to the respective genes in *M. marburgensis* than to the respective genes in *M. kandleri*, in the Methanococcales, and in *M. smithii*. These findings are interpreted to indicate that the *hcg* gene cluster in *M. labreanum* (**Figure 7**) has been acquired by this methanogen from a *Methanothermobacter* species via lateral gene transfer (M. Schick, unpublished results). On the same basis, this is also likely for the gene cluster *hcgCDEFG* in *M. hungatei*.

Labeling studies with [1-^{13}C]-acetate, [2-^{13}C]-acetate, [1-^{13}C]-pyruvate and L-[methyl-D_3]-methionine, performed mainly with the acetate auxotroph *M. smithii*, have revealed via mass spectrometry that of the nine carbons in the pyridinol moiety of the FeGP cofactor (**Figure 1*c***) three are derived from C1 of acetate, two from C2 of acetate, one from the methyl group of methionine, two from the carboxyl group of pyruvate, and one from CO_2 (M. Schick, unpublished results).

In the genomes of some methanogens that contain an *hmd* gene, one or two genes homologous to *hmd* are found (97). The two encoded proteins, designated HmdII and HmdIII, show only low sequence identity (<20%) to [Fe]-hydrogenase but share high sequence identity (80%) with each other. The homologs are not found in methanogens without an *hmd* gene. Structure predictions indicate that HmdII and HmdIII have an intact site for FeGP cofactor binding (14, 15). Consistently, HmdII was found to bind the FeGP cofactor. However, neither HmdII nor HmdIII catalyzed the reduction of methenyl-H_4MPT^+ with H_2. These results were interpreted to indicate that HmdII and HmdIII could be scaffold proteins involved in FeGP cofactor biosynthesis (15). In [NiFe]-hydrogenase maturation, there is a precedent for this. The synthesis of the [NiFe]-center of the membrane-associated hydrogenase from *R. eutropha* involves HoxV, which is a HoxA homolog (large subunit carrying the [NiFe]-center), as the scaffold (143).

M. labreanum and *M. smithii*, which both can synthesize active [Fe]-hydrogenase (M. Schick, unpublished data), do not contain *hmdII* or *hmdIII* genes, which does not support the scaffold hypothesis. However, in *Ralstronia*, the scaffold (HoxV) is required only for the maturation of the membrane-bound hydrogenase and not for the soluble one, and in *E. coli*, a scaffold protein homologous to the large subunit is not involved in the synthesis of any of the three [NiFe]-hydrogenases. It could therefore be that in some methanogens [Fe]-hydrogenase synthesis is independent of Hmd homologous putative scaffold proteins.

H_2 STORAGE VIA CH_4 FORMATION

The formation of methane from 4 H_2 and CO_2 (Reaction 1) catalyzed by methanogenic archaea is being discussed as an efficient means to store H_2 (144). The combustion of 4 H_2 with 2 O_2 to 4 H_2O yields 949 kJ mol^{-1} and that of CH_4 with 2 O_2 to CO_2 and 2 H_2O yields 818 kJ mol^{-1} free energy. Thus, most of the combustion energy of H_2 is conserved in methane. Compared to H_2, methane is relatively easy to store and to transport. From methane, H_2 can be regenerated in a reforming process ($CH_4 + H_2O \rightarrow 3\ H_2 + CO$), followed by the shift reaction ($CO + H_2O \rightarrow CO_2 + H_2$), which is standard technology.

The chemical reduction of CO_2 with H_2 to methane requires very high temperatures and pressures. By contrast, methanogenic archaea catalyze the process at room temperature and at H_2 pressures way below 1 bar. The rate of biological CO_2 reduction to methane in cell suspensions, e.g., of *M. marburgensis*, can be as high as 3 μmol per minute and milligram of cells (dry mass) (145). Thus, with 100 g cells of this methanogen per day, approximately 7 kg methane can be formed from 4 H_2 and CO_2; this is equivalent to 350 MJ energy, which is only somewhat lower than the amount of primary energy consumed by an average person each day in Germany (474 MJ per person per day) (144).

The idea is to use CO_2 from coal power plants and H_2 generated either via reforming of biomass or via photolysis or electrolysis of water (the electricity required for H_2O electrolysis could be provided by solar or wind energy). After storage, when the methane is burned, no more CO_2 is released into the atmosphere than was used in the formation of methane from H_2 and CO_2.

In contrast to the [Fe]-hydrogenase, all the [NiFe]-hydrogenases present in methanogens are rapidly inactivated by O_2 in the presence of H_2 and even the [Fe]-hydrogenase is inactivated by the O_2-reduction product $O_2^{\cdot -}$. Therefore, there is no potential to employ these enzymes in vitro in large-scale technical processes. However, within the cells, the hydrogenases are much more robust because methanogens contain enzymes that reduce O_2 and $O_2^{\cdot -}$ to H_2O (96, 146–148). There are even some methanogens that can thrive on H_2 and CO_2 in the presence of O_2 (146, 148). The use of methanogenic archaea in converting energy from H_2 to methane is thus not an illusion. The economic feasibility, however, remains to be shown.

SUMMARY POINTS

1. There are three types of enzymes that activate H_2, namely [NiFe]-hydrogenases, [FeFe]-hydrogenases, and [Fe]-hydrogenase, which have emerged by convergent evolution. Of the three types, only [NiFe]- and [Fe]-hydrogenases are found in methanogenic archaea.
2. In methanogenic archaea, there are four different [NiFe]-hydrogenases, of which the F_{420}-reducing hydrogenase and the heterodisulfide reductase-associated hydrogenase are cytoplasmic, and the energy-converting hydrogenases and methanophenazine-reducing hydrogenase are membrane-associated proteins. The [NiFe]-center harboring subunit of the energy-converting hydrogenases faces the cytoplasm, and the active-site harboring subunit of the methanophenazine-reducing hydrogenase is oriented toward the periplasm.
3. The energy-converting [NiFe]-hydrogenases are proton or sodium ion pumps from which complex I of the respiratory chain has evolved. The reduction of ferredoxin with H_2, catalyzed by the energy-converting hydrogenases, is energy consuming.
4. The F_{420}-reducing hydrogenase in methanogens is unique in that its small subunit contains three [4Fe4S]-clusters, the energy-converting hydrogenase is unique in that its small subunit contains only one [4Fe4S]-cluster, and the heterodisulfide reductase-associated hydrogenase is unique in that it contains a subunit, MvhD, which harbors a [2Fe2S]-cluster seldomly found in archaeal proteins.
5. The gene clusters encoding F_{420}-reducing hydrogenase and methanophenazine-reducing hydrogenase each harbor a gene for an endopeptidase involved in [NiFe]-hydrogenase maturation. The gene clusters encoding the energy-converting hydrogenases and the heterodisulfide reductase-associated hydrogenase lack such a gene. Maturation of some of the energy-converting [NiFe]-hydrogenases appears not to require a protease step.

6. In all methanogens investigated, homologs of the genes *hypA-F* are found, which in *E. coli* encode proteins necessary, and apparently sufficient, for [NiFe]-center biosynthesis.
7. The [Fe]-hydrogenase is of functional importance in methanogens under nickel-limiting growth conditions, which appear to prevail in many natural habitats. The enzyme harbors a novel iron-guanylylpyridinol cofactor covalently attached to the protein only via one cysteine sulfur ligand to iron. In the presence of thiol reagents or acids, the cofactor can be reversibly detached. Genes putatively involved in cofactor biosynthesis have been identified.
8. Methanogenic archaea can catalyze the formation of methane from CO_2 and H_2 at specific activities high enough to be considered as catalysts in industrial energy transformation.

FUTURE ISSUES

1. The crystal structure of the energy-converting hydrogenase of the EchA-F type from methanogens would be of great interest because EchA-F is phylogenetically related to complex I of the respiratory chain whose crystal structure for the complete enzyme is still unknown (57).
2. The crystal structure of the MvhADG/HdrABC complex is required to help understand how the electrons from H_2 are bifurcated such that both ferredoxin and heterodisulfide are reduced by H_2 in an energy-coupled reaction. The stoichiometry of ferredoxin and heterodisulfide reduction with H_2 remains to be ascertained.
3. A crystal structure of [Fe]-hydrogenase in complex with its substrate methylenetetrahydromethanopterin is required in a closed form. The recently published structure is in the open form, in which the bound methylenetetrahydromethanopterin does not interact with the active-site iron.
4. There is a need to isolate the hydrogenase/heterodisulfide reductase complex from *M. hungatei* and from other members of the Methanomicrobiales that lack the genes for MvhA and MvhG. Is there an FrhAG/MvhD/HdrABC complex as speculated in this review?
5. Seven genes co-occurring with the gene *hmd* encoding [Fe]-hydrogenase have been proposed to be involved in the FeGP cofactor biosynthesis. Gene knockout experiments in *M. maripaludis*, for which a genetic system has been developed, could help clarify this point. In parallel, attempts are needed to heterologously express the seven putative cofactor synthesis genes together with the *hmd* gene in *E. coli*.

DISCLOSURE STATEMENT

The authors are not aware of any affiliations, memberships, funding, or financial holdings that might be perceived as affecting the objectivity of this review.

ACKNOWLEDGMENTS

This work was supported by the Max Planck Society and by the Fonds der Chemischen Industrie.

LITERATURE CITED

1. Stephenson M, Stickland LH. 1933. Hydrogenase III. The bacterial formation of methane by the reduction of one-carbon compounds by molecular hydrogen. *Biochem. J.* 27:1517–27
2. Stephenson M, Stickland LH. 1931. Hydrogenase: a bacterial enzyme activating molecular hydrogen. I. The properties of the enzyme. *Biochem. J.* 25:205–14
3. Green DE, Stickland LH. 1934. Studies on reversible dehydrogenase systems. I. The reversibility of the hydrogenase system of *Bacterium coli*. *Biochem. J.* 28:898–900
4. Warburg O. 1946. Absorptionsspektrum eines wasserstoffentwickelnden Gärungsferments. In *Schwermetalle als Wirkungsgruppe von Fermenten*, pp. 157–62. Berlin: Verlag Saenger
5. Volbeda A, Charon MH, Piras C, Hatchikian EC, Frey M, Fontecilla-Camps JC. 1995. Crystal structure of the nickel-iron hydrogenase from *Desulfovibrio gigas*. *Nature* 373:580–87
6. Higuchi Y, Yagi T, Yasuoka N. 1997. Unusual ligand structure in [Ni-Fe] active center and an additional Mg site in hydrogenase revealed by high resolution X-ray structure analysis. *Structure* 5:1671–80
7. Frey M. 2002. Hydrogenases: hydrogen-activating enzymes. *ChemBioChem* 3:153–60
8. Fontecilla-Camps JC, Volbeda A, Cavazza C, Nicolet Y. 2007. Structure/function relationships of [NiFe]- and [FeFe]-hydrogenases. *Chem. Rev.* 107:4273–303
9. Peters JW, Lanzilotta WN, Lemon BJ, Seefeldt LC. 1998. X-ray crystal structure of the Fe-only hydrogenase (Cpl) from *Clostridium pasteurianum* to 1.8 angstrom resolution. *Science* 282:1853–58
10. Nicolet Y, Lemon BJ, Fontecilla-Camps JC, Peters JW. 2000. A novel FeS cluster in [Fe]-only hydrogenases. *Trends Biochem. Sci.* 25:138–43
11. Pandey AS, Harris TV, Giles LJ, Peters JW, Szilagyi RK. 2008. Dithiomethylether as a ligand in the hydrogenase H-cluster. *J. Am. Chem. Soc.* 130:4533–40
12. Shima S, Pilak O, Vogt S, Schick M, Stagni MS, et al. 2008. The crystal structure of [Fe]-hydrogenase reveals the geometry of the active site. *Science* 321:572–75
13. Hiromoto T, Ataka K, Pilak O, Vogt S, Stagni MS, et al. 2009. The crystal structure of C176A mutated [Fe]-hydrogenase suggests an acyl-iron ligation in the active site iron complex. *FEBS Lett.* 583:585–90
14. Hiromoto T, Warkentin E, Moll J, Ermler U, Shima S. 2009. The crystal structure of an [Fe]-hydrogenase substrate complex reveals the framework for H_2-activation. *Angew. Chem. Int. Ed. Engl.* 48:6457–60
15. Shima S, Thauer RK. 2007. A third type of hydrogenase catalyzing H_2 activation. *Chem. Rec.* 7:37–46
16. Vignais PM, Billoud B. 2007. Occurrence, classification, and biological function of hydrogenases: an overview. *Chem. Rev.* 107:4206–72
17. Thauer RK, Kaster A-K, Seedorf H, Buckel W, Hedderich R. 2008. Methanogenic archaea: ecologically relevant differences in energy conservation. *Nat. Rev. Microbiol.* 6:579–91
18. Stams AJ, Plugge CM. 2009. Electron transfer in syntrophic communities of anaerobic bacteria and archaea. *Nat. Rev. Microbiol.* 7:568–77
19. Kubas GJ. 1988. Molecular hydrogen complexes—coordination of a sigma bond to transition metals. *Acc. Chem. Res.* 21:120–28
20. Tard C, Pickett CJ. 2009. Structural and functional analogues of the active sites of the [Fe]-, [NiFe]-, and [FeFe]-hydrogenases. *Chem. Rev.* 109:2245–74
21. Graf EG, Thauer RK. 1981. Hydrogenase from *Methanobacterium thermoautotrophicum*, a nickel-containing enzyme. *FEBS Lett.* 136:165–69
22. Albracht SPJ, Graf EG, Thauer RK. 1982. The electron paramagnetic-resonance properties of nickel in hydrogenase from *Methanobacterium thermoautotrophicum*. *FEBS Lett.* 140:311–13
23. Zirngibl C, Hedderich R, Thauer RK. 1990. N^5,N^{10}-methylenetetrahydromethanopterin dehydrogenase from *Methanobacterium thermoautotrophicum* has hydrogenase activity. *FEBS Lett.* 261:112–16
24. Guss AM, Kulkarni G, Metcalf WW. 2009. Differences in hydrogenase gene expression between *Methanosarcina acetivorans* and *Methanosarcina barkeri*. *J. Bacteriol.* 191:2826–33

25. Anderson I, Ulrich LE, Lupa B, Susanti D, Porat I, et al. 2009. Genomic characterisation of Methanomicrobiales reveals three classes of methanogens. *PLoS ONE* 4:e5797
26. Tietze M, Beuchle A, Lamla I, Orth N, Dehler M, et al. 2003. Redox potentials of methanophenazine and CoB-S-S-CoM, factors involved in electron transport in methanogenic archaea. *ChemBioChem* 4:333–35
27. Thauer RK. 1998. Biochemistry of methanogenesis: a tribute to Marjory Stephenson. *Microbiology* 144:2377–406
28. van der Linden E, Burgdorf T, de Lacey AL, Buhrke T, Scholte M, et al. 2006. An improved purification procedure for the soluble [NiFe]-hydrogenase of *Ralstonia eutropha*: new insights into its (in)stability and spectroscopic properties. *J. Biol. Inorg. Chem.* 11:247–60
29. Saggu M, Zebger I, Ludwig M, Lenz O, Friedrich B, et al. 2009. Spectroscopic insights into the oxygen-tolerant membrane-associated [NiFe]-hydrogenase of *Ralstonia eutropha* H16. *J. Biol. Chem.* 284:16264–76
30. Fontecilla-Camps JC, Amara P, Cavazza C, Nicolet Y, Volbeda A. 2009. Structure-function relationships of anaerobic gas-metabolizing metalloenzymes. *Nature* 460:814–22
31. Böck, A, King PW, Blokesch M, Posewitz MC. 2006. Maturation of hydrogenases. *Adv. Microb. Physiol.* 51:1–71
32. Forzi L, Sawers RG. 2007. Maturation of [NiFe]-hydrogenases in *Escherichia coli*. *Biometals* 20:565–78
33. Leach MR, Zamble DB. 2007. Metallocenter assembly of the hydrogenase enzymes. *Curr. Opin. Chem. Biol.* 11:159–65
34. Hedderich R. 2004. Energy-converting [NiFe]-hydrogenases from Archaea and extremophiles: ancestors of complex I. *J. Bioenerg. Biomembr.* 36:65–75
35. Hedderich R, Forzi L. 2005. Energy-converting [NiFe] hydrogenases: more than just H_2 activation. *J. Mol. Microbiol. Biotechnol.* 10:92–104
36. Friedrich B, Buhrke T, Burgdorf T, Lenz O. 2005. A hydrogen-sensing multiprotein complex controls aerobic hydrogen metabolism in *Rastonia eutropha*. *Biochem. Soc. Trans.* 33:97–101
37. Singer SW, Hirst MB, Ludden PW. 2006. CO-dependent H_2 evolution by *Rhodospirillum rubrum*: role of CODH:CooF complex. *Biochim. Biophys. Acta* 1757:1582–91
38. Maness PC, Huang J, Smolinski S, Tek V, Vanzin G. 2005. Energy generation from the CO oxidation-hydrogen production pathway in *Rubrivivax gelatinosus*. *Appl. Environ. Microbiol.* 71:2870–74
39. Soboh B, Linder D, Hedderich R. 2002. Purification and catalytic properties of a CO-oxidizing:H_2-evolving enzyme complex from *Carboxydothermus hydrogenoformans*. *Eur. J. Biochem.* 269:5712–21
40. Oelgeschläger E, Rother M. 2008. Carbon monoxide-dependent energy metabolism in anaerobic bacteria and archaea. *Arch. Microbiol.* 190:257–69
41. Nakamaru-Ogiso E, Matsuno-Yagi A, Yoshikawa S, Yagi T, Ohnishi T. 2008. Iron-sulfur cluster N5 is coordinated by an HXXXCXXCXXXXXC motif in the NuoG subunit of *Escherichia coli* NADH:quinone oxidoreductase (complex I). *J. Biol. Chem.* 283:25979–87
42. Berks BC, Sargent F, Palmer T. 2000. The Tat protein export pathway. *Mol. Microbiol.* 35:260–74
43. Daas PJH, Hagen WR, Keltjens JT, Vogels GD. 1994. Characterization and determination of the redox properties of the 2[4Fe-4S] ferredoxin from *Methanosarcina barkeri* strain MS. *FEBS Lett.* 356:342–44
44. Künkel A, Vorholt JA, Thauer RK, Hedderich R. 1998. An *Escherichia coli* hydrogenase-3-type hydrogenase in methanogenic archaea. *Eur. J. Biochem.* 252:467–76
45. Meuer J, Bartoschek S, Koch J, Künkel A, Hedderich R. 1999. Purification and catalytic properties of Ech hydrogenase from *Methanosarcina barkeri*. *Eur. J. Biochem.* 265:325–35
46. Bott M, Thauer RK. 1987. Protonmotive force driven formation of CO from CO_2 and H_2 in methanogenic bacteria. *Eur. J. Biochem.* 168:407–12
47. Bott M, Eikmanns B, Thauer RK. 1986. Coupling of carbon monoxide oxidation to CO_2 and H_2 with the phosphorylation of ADP in acetate-grown *Methanosarcina barkeri*. *Eur. J. Biochem.* 159:393–98
48. Bott M, Thauer RK. 1989. Proton translocation coupled to the oxidation of carbon monoxide to CO_2 and H_2 in *Methanosarcina barkeri*. *Eur. J. Biochem.* 179:469–72
49. Meuer J, Kuettner HC, Zhang JK, Hedderich R, Metcalf WW. 2002. Genetic analysis of the archaeon *Methanosarcina barkeri* Fusaro reveals a central role for Ech hydrogenase and ferredoxin in methanogenesis and carbon fixation. *Proc. Natl. Acad. Sci. USA* 99:5632–37

50. Stojanowic A, Hedderich R. 2004. CO_2 reduction to the level of formylmethanofuran in *Methanosarcina barkeri* is non-energy driven when CO is the electron donor. *FEMS Microbiol. Lett.* 235:163–67
51. Kaesler B, Schönheit P. 1989. The role of sodium ions in methanogenesis. Formaldehyde oxidation to CO_2 and 2 H_2 in methanogenic bacteria is coupled with primary electrogenic Na^+ translocation at a stoichiometry of 2–3 Na^+/CO_2. *Eur. J. Biochem.* 184:223–32
52. Kaesler B, Schönheit P. 1989. The sodium cycle in methanogenesis. CO_2 reduction to the formaldehyde level in methanogenic bacteria is driven by a primary electrochemical potential of Na^+ generated by formaldehyde reduction to CH_4. *Eur. J. Biochem.* 186:309–16
53. Kurkin S, Meuer J, Koch J, Hedderich R, Albracht SPJ. 2002. The membrane-bound [NiFe]-hydrogenase (Ech) from *Methanosarcina barker*: unusual properties of the iron-sulphur clusters. *Eur. J. Biochem.* 269:6101–11
54. Forzi L, Koch J, Guss AM, Radosevich CG, Metcalf WW, Hedderich R. 2005. Assignment of the [4Fe-4S]-clusters of Ech hydrogenase from *Methanosarcina barkeri* to individual subunits via the characterization of site-directed mutants. *FEBS J.* 272:4741–53
55. Tersteegen A, Hedderich R. 1999. *Methanobacterium thermoautotrophicum* encodes two multisubunit membrane-bound [NiFe]-hydrogenases. Transcription of the operons and sequence analysis of the deduced proteins. *Eur. J. Biochem.* 264:930–43
56. Friedrich T, Böttcher B. 2004. The gross structure of the respiratory complex I: a Lego System. *Biochim. Biophys. Acta* 1608:1–9
57. Zickermann V, Dröse S, Tocilescu MA, Zwicker K, Kerscher S, Brandt U. 2008. Challenges in elucidating structure and mechanism of proton pumping NADH:ubiquinone oxidoreductase (complex I). *J. Bioenerg. Biomembr.* 40:475–83
58. Porat I, Kim W, Hendrickson EL, Xia QW, Zhang Y, et al. 2006. Disruption of the operon encoding Ehb hydrogenase limits anabolic CO_2 assimilation in the archaeon *Methanococcus maripaludis*. *J. Bacteriol.* 188:1373–80
59. Chou CJ, Jenney FE Jr, Adams MW, Kelly RM. 2008. Hydrogenesis in hyperthermophilic microorganisms: implications for biofuels. *Metab. Eng.* 10:394–404
60. Sapra R, Bagramyan K, Adams MWW. 2003. A simple energy-conserving system: proton reduction coupled to proton translocation. *Proc. Natl. Acad. Sci. USA* 100:7545–50
61. Pisa KY, Huber H, Thomm M, Müller V. 2007. A sodium ion-dependent A_1A_0-ATP synthase from the hyperthermophilic archaeon *Pyrococcus furiosus*. *FEBS J.* 274:3928–38
62. Stojanowic A, Mander GJ, Duin EC, Hedderich R. 2003. Physiological role of the F_{420}-non-reducing hydrogenase (Mvh) from *Methanothermobacter marburgensis*. *Arch. Microbiol.* 180:194–203
63. Farhoud MH, Wessels HJCT, Steenbakkers PJM, Mattijssen S, Wevers RA, et al. 2005. Protein complexes in the archaeon *Methanothermobacter thermautotrophicus* analyzed by blue native/SDS-PAGE and mass spectrometry. *Mol. Cell. Proteomics* 4:1653–63
64. Setzke E, Hedderich R, Heiden S, Thauer RK. 1994. H_2:heterodisulfide oxidoreductase complex from *Methanobacterium thermoautotrophicum*: composition and properties. *Eur. J. Biochem.* 220:139–48
65. Woo GJ, Wasserfallen A, Wolfe RS. 1993. Methyl viologen hydrogenase II, a new member of the hydrogenase family from *Methanobacterium thermoautotrophicum* delta H. *J. Bacteriol.* 175:5970–77
66. Hedderich R, Berkessel A, Thauer RK. 1990. Purification and properties of heterodisulfide reductase from *Methanobacterium thermoautotrophicum* (strain Marburg). *Eur. J. Biochem.* 193:255–61
67. Hedderich R, Koch J, Linder D, Thauer RK. 1994. The heterodisulfide reductase from *Methanobacterium thermoautotrophicum* contains sequence motifs characteristic of pyridine nucleotide-dependent thioredoxin reductases. *Eur. J. Biochem.* 225:253–61
68. Hamann N, Mander GJ, Shokes JE, Scott RA, Bennati M, Hedderich R. 2007. A cysteine-rich CCG domain contains a novel [4Fe-4S] cluster binding motif as deduced from studies with subunit B of heterodisulfide reductase from *Methanothermobacter marburgensis*. *Biochemistry* 46:12875–85
69. Herrmann G, Jayamani E, Mai G, Buckel W. 2008. Energy conservation via electron-transferring flavoprotein in anaerobic bacteria. *J. Bacteriol.* 190:784–91
70. Li F, Hinderberger J, Seedorf H, Zhang J, Buckel W, Thauer RK. 2008. Coupled ferredoxin and crotonyl coenzyme A (CoA) reduction with NADH catalyzed by the butyryl-CoA dehydrogenase/Etf complex from *Clostridium kluyveri*. *J. Bacteriol.* 190:843–50

71. Reeve JN, Beckler GS, Cram DS, Hamilton PT, Brown JW, et al. 1989. A hydrogenase-linked gene in *Methanobacterium thermoautotrophicum* strain delta H encodes a polyferredoxin. *Proc. Natl. Acad. Sci. USA* 86:3031–35
72. Hedderich R, Albracht SPJ, Linder D, Koch J, Thauer RK. 1992. Isolation and characterization of polyferredoxin from *Methanobacterium thermoautotrophicum*. The *mvhB* gene product of the methylviologen-reducing hydrogenase operon. *FEBS Lett.* 298:65–68
73. Steigerwald VJ, Pihl TD, Reeve JN. 1992. Identification and isolation of the polyferredoxin from *Methanobacterium thermoautotrophicum* strain delta H. *Proc. Natl. Acad. Sci. USA* 89:6929–33
74. Halboth S, Klein A. 1992. *Methanococcus voltae* harbors four gene clusters potentially encoding 2[NiFe] and 2[NiFeSe] hydrogenases, each of the cofactor F_{420}-reducing or F_{420}-non-reducing types. *Mol. Gen. Genet.* 233:217–24
75. Slesarev AI, Mezhevaya KV, Makarova KS, Polushin NN, Shcherbinina OV, et al. 2002. The complete genome of hyperthermophile *Methanopyrus kandleri* AV19 and monophyly of archaeal methanogens. *Proc. Natl. Acad. Sci. USA* 99:4644–49
76. Sun J, Klein A. 2004. A lysR-type regulator is involved in the negative regulation of genes encoding selenium-free hydrogenases in the archaeon *Methanococcus voltae*. *Mol. Microbiol.* 52:563–71
77. Stock T, Rother M. 2009. Selenoproteins in Archaea and gram-positive bacteria. *Biochim. Biophys. Acta* 1790:1520–32
78. Bingemann R, Pierik AJ, Klein A. 2000. Influence of the fusion of two subunits of the F_{420}-non-reducing hydrogenase of *Methanococcus voltae* on its biochemical properties. *Arch. Microbiol.* 174:375–78
79. Scholten JCM, Conrad R. 2000. Energetics of syntrophic propionate oxidation in defined batch and chemostat cocultures. *Appl. Environ. Microbiol.* 66:2934–42
80. Plugge CM, Jiang B, de Bok FAM, Tsai C, Stams AJM. 2009. Effect of tungsten and molybdenum on growth of a syntrophic coculture of *Syntrophobacter fumaroxidans* and *Methanospirillum hungatei*. *Arch. Microbiol.* 191:55–61
81. Hochheimer A, Linder D, Thauer RK, Hedderich R. 1996. The molybdenum formylmethanofuran dehydrogenase operon and the tungsten formylmethanofuran dehydrogenase operon from *Methanobacterium thermoautotrophicum*. Structures and transcriptional regulation. *Eur. J. Biochem.* 242:156–62
82. Abken HJ, Tietze M, Brodersen J, Bäumer S, Beifuss U, Deppenmeier U. 1998. Isolation and characterization of methanophenazine and function of phenazines in membrane-bound electron transport of *Methanosarcina mazei* Gö1. *J. Bacteriol.* 180:2027–32
83. Beifuss U, Tietze M, Bäumer S, Deppenmeier U. 2000. Methanophenazine: structure, total synthesis, and function of a new cofactor from methanogenic archaea. *Angew. Chem. Int. Ed. Engl.* 39:2470–72
84. Deppenmeier U. 2004. The membrane-bound electron transport system of *Methanosarcina* species. *J. Bioenerg. Biomembr.* 36:55–64
85. Deppenmeier U, Müller V. 2008. Life close to the thermodynamic limit: how methanogenic archaea conserve energy. In *Results and Problems in Cell Differentiation*, eds. G Schafer, HS Penefsky, 45:123–52. Berlin: Springer Verlag
86. Deppenmeier U, Blaut M, Schmidt B, Gottschalk G. 1992. Purification and properties of a F_{420}-nonreactive, membrane-bound hydrogenase from *Methanosarcina* strain Gö1. *Arch. Microbiol.* 157:505–11
87. Wu LF, Chanal A, Rodrigue A. 2000. Membrane targeting and translocation of bacterial hydrogenases. *Arch. Microbiol.* 173:319–24
88. Schubert T, Lenz O, Krause E, Volkmer R, Friedrich B. 2007. Chaperones specific for the membrane-bound [NiFe]-hydrogenase interact with the Tat signal peptide of the small subunit precursor in *Ralstonia eutropha* H16. *Mol. Microbiol.* 66:453–67
89. Deppenmeier U. 1995. Different structure and expression of the operons encoding the membrane-bound hydrogenases from *Methanosarcina mazei* Gö1. *Arch. Microbiol.* 164:370–76
90. Deppenmeier U, Blaut M, Lentes S, Herzberg C, Gottschalk G. 1995. Analysis of the VhoGAC and VhtGAC operons from *Methanosarcina mazei* strain Gö1, both encoding a membrane-bound hydrogenase and a cytochrome *b*. *Eur. J. Biochem.* 227:261–69
91. Heiden S, Hedderich R, Setzke E, Thauer RK. 1994. Purification of a two-subunit cytochrome *b*–containing heterodisulfide reductase from methanol grown *Methanosarcina barkeri*. *Eur. J. Biochem.* 221:855–61

92. Künkel A, Vaupel M, Heim S, Thauer RK, Hedderich R. 1997. Heterodisulfide reductase from methanol-grown cells of *Methanosarcina barkeri* is not a flavoenzyme. *Eur. J. Biochem.* 244:226–34
93. Deppenmeier U, Blaut M, Gottschalk G. 1991. H_2-heterodisulfide oxidoreductase, a second energy-conserving system in the methanogenic strain Gö1. *Arch. Microbiol.* 155:272–77
94. Ide T, Bäumer S, Deppenmeier U. 1999. Energy conservation by the H_2:heterodisulfide oxidoreductase from *Methanosarcina mazei* Gö1: identification of two proton-translocating segments. *J. Bacteriol.* 181:4076–80
95. Hendrickson EL, Haydock AK, Moore BC, Whitman WB, Leigh JA. 2007. Functionally distinct genes regulated by hydrogen limitation and growth rate in methanogenic archaea. *Proc. Natl. Acad. Sci. USA* 104:8930–34
96. Kato S, Kosaka T, Watanabe K. 2008. Comparative transcriptome analysis of responses of *Methanothermobacter thermautotrophicus* to different environmental stimuli. *Environ. Microbiol.* 10:893–905
97. Afting C, Kremmer E, Brucker C, Hochheimer A, Thauer RK. 2000. Regulation of the synthesis of H_2-forming methylenetetrahydromethanopterin dehydrogenase (Hmd) and of HmdII and HmdIII in *Methanothermobacter marburgensis*. *Arch. Microbiol.* 174:225–32
98. Wood GE, Haydock AK, Leigh JA. 2003. Function and regulation of the formate dehydrogenase genes of the methanogenic archaeon *Methanococcus maripaludis*. *J. Bacteriol.* 185:2548–54
99. Lupa B, Hendrickson EL, Leigh JA, Whitman WB. 2008. Formate-dependent H_2 production by the mesophilic methanogen *Methanococcus maripaludis*. *Appl. Environ. Microbiol.* 74:6584–90
100. Alex LA, Reeve JN, Orme-Johnson WH, Walsh CT. 1990. Cloning, sequence determination and expression of the genes encoding the subunits of the nickel-containing 8-hydroxy-5-deazaflavin reducing hydrogenase from *Methanobacterium thermoautotrophicum* delta H. *Biochemistry* 29:7237–44
101. Vaupel M, Thauer RK. 1998. Two F_{420}-reducing hydrogenases in *Methanosarcina barkeri*. *Arch. Microbiol.* 169:201–5
102. Kulkarni G, Kridelbaugh DM, Guss AM, Metcalf WW. 2009. Hydrogen is a preferred intermediate in the energy conserving electron transport chain of *Methanosarcina barkeri*. *Proc. Natl. Acad. Sci. USA* 106:15915–20
103. Bingemann R, Klein A. 2000. Conversion of the central [4Fe-4S] cluster into a [3Fe-4S] cluster leads to reduced hydrogen-uptake activity of the F_{420}-reducing hydrogenase of *Methanococcus voltae*. *Eur. J. Biochem.* 267:6612–18
104. Fox JA, Livingston DJ, Orme-Johnson WH, Walsh CT. 1987. 8-Hydroxy-5-deazaflavin-reducing hydrogenase from *Methanobacterium thermoautotrophicum*.1. Purification and Characterization. *Biochemistry* 26:4219–27
105. Watanabe S, Matsumi R, Arai T, Atomi H, Imanaka T, Miki K. 2007. Crystal structures of [NiFe]-hydrogenase maturation proteins HypC, HypD, and HypE: insights into cyanation reaction by thiol redox signaling. *Mol. Cell* 27:29–40
106. Dias AV, Mulvihill CM, Leach MR, Pickering IJ, George GN, Zamble DB. 2008. Structural and biological analysis of the metal sites of *Escherichia coli* hydrogenase accessory protein HypB. *Biochemistry* 47:11981–91
107. Rangarajan ES, Asinas A, Proteau A, Munger C, Baardsnes J, et al. 2008. Structure of [NiFe]-hydrogenase maturation protein HypE from *Escherichia coli* and its interaction with HypF. *J. Bacteriol.* 190:1447–58
108. Forzi L, Hellwig P, Thauer RK, Sawers RG. 2007. The CO and CN^- ligands to the active site Fe in [NiFe]-hydrogenase of *Escherichia coli* have different metabolic origins. *FEBS Lett.* 581:3317–21
109. Lenz O, Zebger I, Hamann J, Hildebrandt P, Friedrich B. 2007. Carbamoylphosphate serves as the source of CN^-, but not of the intrinsic CO in the active site of the regulatory [NiFe]-hydrogenase from *Ralstonia eutropha*. *FEBS Lett.* 581:3322–26
110. Roseboom W, Blokesch M, Böck A, Albracht SPJ. 2005. The biosynthetic routes for carbon monoxide and cyanide in the Ni-Fe active site of hydrogenases are different. *FEBS Lett.* 579:469–72
111. Theodoratou E, Huber R, Böck A. 2005. [NiFe]-hydrogenase maturation endopeptidase: structure and function. *Biochem. Soc. Trans.* 33:108–11
112. Devine E, Holmqvist M, Stensjo K, Lindblad P. 2009. Diversity and transcription of proteases involved in the maturation of hydrogenases in *Nostoc punctiforme* ATCC 29133 and *Nostoc* sp. strain PCC 7120. *BMC Microbiol.* 9:53

113. Zhang Y, Rodionov DA, Gelfand MS, Gladyshev VN. 2009. Comparative genomic analyses of nickel, cobalt and vitamin B_{12} utilization. *BMC Genomics* 10:78
114. Schönheit P, Moll J, Thauer RK. 1979. Nickel, cobalt, and molybdenum requirement for growth of *Methanobacterium thermoautotrophicum*. *Arch. Microbiol.* 123:105–7
115. Konhauser KO, Pecoits E, Lalonde SV, Papineau D, Nisbet EG, et al. 2009. Oceanic nickel depletion and a methanogen famine before the Great Oxidation Event. *Nature* 458:750–53
116. Afting C, Hochheimer A, Thauer RK. 1998. Function of H_2-forming methylenetetrahydromethanopterin dehydrogenase from *Methanobacterium thermoautotrophicum* in coenzyme F_{420} reduction with H_2. *Arch. Microbiol.* 169:206–10
117. Wang SC, Dias AV, Zamble DB. 2009. The "metallo-specific" response of proteins: a perspective based on the *Escherichia coli* transcriptional regulator NikR. *Dalton Trans.* 14:2459–66
118. Iwig JS, Leitch S, Herbst RW, Maroney MJ, Chivers PT. 2008. Ni(II) and Co(II) sensing by *Escherichia coli* RcnR. *J. Am. Chem. Soc.* 130:7592–606
119. Schreiter ER, Wang SC, Zamble DB, Drennan CL. 2006. NikR-operator complex structure and the mechanism of repressor activation by metal ions. *Proc. Natl. Acad. Sci. USA* 103:13676–81
120. Rodionov DA, Hebbeln P, Gelfand MS, Eitinger T. 2006. Comparative and functional genomic analysis of prokaryotic nickel and cobalt uptake transporters: evidence for a novel group of ATP-binding cassette transporters. *J. Bacteriol.* 188:317–27
121. Dosanjh NS, West AL, Michel SLJ. 2009. *Helicobacter pylori* NikR's interaction with DNA: a two-tiered mode of recognition. *Biochemistry* 48:527–36
122. Schleucher J, Griesinger C, Schworer B, Thauer RK. 1994. H_2-forming N^5, N^{10}-methylenetetrahydromethanopterin dehydrogenase from *Methanobacterium thermoautotrophicum* catalyzes a stereoselective hydride transfer as determined by two-dimensional NMR spectroscopy. *Biochemistry* 33:3986–93
123. Thauer RK, Klein AR, Hartmann GC. 1996. Reactions with molecular hydrogen in microorganisms: evidence for a purely organic hydrogenation catalyst. *Chem. Rev.* 96:3031–42
124. Hendrickson EL, Leigh JA. 2008. Roles of coenzyme F_{420}-reducing hydrogenases and hydrogen- and F_{420}-dependent methylenetetrahydromethanopterin dehydrogenases in reduction of F_{420} and production of hydrogen during methanogenesis. *J. Bacteriol.* 190:4818–21
125. Zirngibl C, Vandongen W, Schwörer B, Vonbunau R, Richter M, et al. 1992. H_2-forming methylenetetrahydromethanopterin dehydrogenase, a novel type of hydrogenase without iron-sulfur clusters in methanogenic archaea. *Eur. J. Biochem.* 208:511–20
126. Berkessel A, Thauer RK. 1995. On the mechanism of catalysis by a metal-free hydrogenase from methanogenic archaea: enzymatic transformation of H_2 without a metal and its analogy to the chemistry of alkanes in superacidic solution. *Angew. Chem. Int. Ed. Engl.* 34:2247–50
127. Lyon EJ, Shima S, Buurman G, Chowdhuri S, Batschauer A, et al. 2004. UV-A/blue-light inactivation of the 'metal-free' hydrogenase (Hmd) from methanogenic archaea: The enzyme contains functional iron after all. *Eur. J. Biochem.* 271:195–204
128. Lyon EJ, Shima S, Boecher R, Thauer RK, Grevels FW, et al. 2004. Carbon monoxide as an intrinsic ligand to iron in the active site of the iron-sulfur-cluster-free hydrogenase H_2-forming methylenetetrahydromethanopterin dehydrogenase as revealed by infrared spectroscopy. *J. Am. Chem. Soc.* 126:14239–48
129. Shima S, Lyon EJ, Sordel-Klippert MS, Kauss M, Kahnt J, et al. 2004. The cofactor of the iron-sulfur cluster free hydrogenase Hmd: structure of the light-inactivation product. *Angew. Chem. Int. Ed. Engl.* 43:2547–51
130. Shima S, Lyon EJ, Thauer RK, Mienert B, Bill E. 2005. Mössbauer studies of the iron-sulfur cluster-free hydrogenase: the electronic state of the mononuclear Fe active site. *J. Am. Chem. Soc.* 127:10430–35
131. Korbas M, Vogt S, Meyer-Klaucke W, Bill E, Lyon EJ, et al. 2006. The iron-sulfur cluster-free hydrogenase (Hmd) is a metalloenzyme with a novel iron binding motif. *J. Biol. Chem.* 281:30804–13
132. Buurman G, Shima S, Thauer RK. 2000. The metal-free hydrogenase from methanogenic archaea: evidence for a bound cofactor. *FEBS Lett.* 485:200–4
133. Vogt S, Lyon EJ, Shima S, Thauer RK. 2008. The exchange activities of [Fe]-hydrogenase (iron-sulfur-cluster-free hydrogenase) from methanogenic archaea in comparison with the exchange activities of [FeFe] and [NiFe] hydrogenases. *J. Biol. Inorg. Chem.* 13:97–106

134. Schleucher J, Schwörer B, Thauer RK, Griesinger C. 1995. Elucidation of the stereochemical course of chemical reactions by magnetic labeling. *J. Am. Chem. Soc.* 117:2941–42
135. Frey PA, Hegeman AD, Ruzicka FJ. 2008. The radical SAM superfamily. *Crit. Rev. Biochem. Mol. Biol.* 43:63–88
136. Chatterjee A, Li Y, Zhang Y, Grove TL, Lee M, et al. 2008. Reconstitution of ThiC in thiamine pyrimidine biosynthesis expands the radical SAM superfamily. *Nat. Chem. Biol.* 4:758–65
137. McGlynn SE, Boyd ES, Shepard EM, Lange RK, Gerlach R, et al. 2010. Identification and characterization of a novel member of the radical AdoMet enzyme superfamily and implication for the biosynthesis of the Hmd hydrogenase active site cofactor. *J. Bacteriol.* 192:595–98
138. Chew AGM, Frigaard NU, Bryant DA. 2007. Bacteriochlorophyllide *c* C-8^2 and C-12^1 methyltransferases are essential for adaptation to low light in *Chlorobaculum tepidum*. *J. Bacteriol.* 189:6176–84
139. King PW, Posewitz MC, Ghirardi ML, Seibert M. 2006. Functional studies of [FeFe] hydrogenase maturation in an *Escherichia coli* biosynthetic system. *J. Bacteriol.* 188:2163–72
140. Nicolet Y, Rubach JK, Posewitz MC, Amara P, Mathevon C, et al. 2008. X-ray structure of the [FeFe]-hydrogenase maturase HydE from *Thermotoga maritima*. *J. Biol. Chem.* 283:18861–72
141. Pilet E, Nicolet Y, Mathevon C, Douki T, Fontecilla-Camps JC, Fontecave M. 2009. The role of the maturase HydG in [FeFe]-hydrogenase active site synthesis and assembly. *FEBS Lett.* 583:506–11
142. Zhang R, Evdokimova E, Kudritska M, Savchenko A, Edwards AM, Joachimiak A. 2008. Crystal structure of a conserved protein of unknown function from *Methanobacterium thermoautotrophicum*. RCSB Protein Data Bank. DOI:10.2210/pdb3brc/pdb. **http://www.rcsb.org/pdb/explore.do?structureId=3BRC**
143. Ludwig M, Schubert T, Zebger I, Wisitruangsakul N, Saggu M, et al. 2009. Concerted action of two novel auxiliary proteins in assembly of the active site in a membrane-bound [NiFe]-hydrogenase. *J. Biol. Chem.* 284:2159–68
144. Thauer RK. 2008. Biologische Methanbildung: Eine erneuerbare Energiequelle von Bedeutung? In *Die Zukunft der Energie*. eds. F Schüth, P Gruss, pp. 119–37. München, Ger.: Verlag Beck
145. Perski HJ, Schönheit P, Thauer RK. 1982. Sodium dependence of methane formation in methanogenic bacteria. *FEBS Lett.* 143:323–26
146. Shima S, Sordel-Klippert M, Brioukhanov A, Netrusov A, Linder D, Thauer RK. 2001. Characterization of a heme-dependent catalase from *Methanobrevibacter arboriphilus*. *Appl. Environ. Microbiol.* 67:3041–45
147. Seedorf H, Dreisbach A, Hedderich R, Shima S, Thauer RK. 2004. $F_{420}H_2$ oxidase (FprA) from *Methanobrevibacter arboriphilus*, a coenzyme F_{420}-dependent enzyme involved in O_2 detoxification. *Arch. Microbiol.* 182:126–37
148. Tholen A, Pester M, Brune A. 2007. Simultaneous methanogenesis and oxygen reduction by *Methanobrevibacter cuticularis* at low oxygen fluxes. *FEMS Microbiol. Ecol.* 62:303–12

RELATED RESOURCES

1. Sigel A, Sigel H, Sigel R, eds. 2009. *Metal Ions in Life Sciences*. Vol. 6. Cambridge, UK: RCS Publ.
2. Fontecilla-Camps JC. 2009. Structure and function of [NiFe]-hydrogenases. See Related Resources Ref. 1, 6:151–78
3. Peters J. 2009. Carbon monoxide and cyanide ligands in the active site of [FeFe]-hydrogenase. See Related Resources Ref. 1, 6:179–218
4. Shima S, Thauer RK, Ermler U. 2009. Carbon monoxide as intrinsic ligand to iron in the active site of [Fe]-hydrogenase. See Related Resources Ref. 1, 6:219–40
5. Goldman AD, Leigh JA, Samudrala R. 2009. Comprehensive computational analysis of Hmd enzymes and paralogs in methanogenic archaea. *BMC Evol. Biol.* 9:99–111

Copper Metallochaperones

Nigel J. Robinson[1] and Dennis R. Winge[2]

[1]Institute for Cell and Molecular Biosciences, Medical School, Newcastle University, NE2 4HH, United Kingdom; email: N.J.Robinson@ncl.ac.uk

[2]University of Utah Health Sciences Center, Salt Lake City, Utah 84132; email: dennis.winge@hsc.utah.edu

Annu. Rev. Biochem. 2010. 79:537–62

First published online as a Review in Advance on March 5, 2010

The *Annual Review of Biochemistry* is online at biochem.annualreviews.org

This article's doi: 10.1146/annurev-biochem-030409-143539

0066-4154/10/0707-0537$20.00

Key Words

P_1-type ATPase, cytochrome oxidase, Cox17, CopZ, Atx1, Ccs1

Abstract

The current state of knowledge on how copper metallochaperones support the maturation of cuproproteins is reviewed. Copper is needed within mitochondria to supply the Cu_A and intramembrane Cu_B sites of cytochrome oxidase, within the *trans*-Golgi network to supply secreted cuproproteins and within the cytosol to supply superoxide dismutase 1 (Sod1). Subpopulations of copper-zinc superoxide dismutase also localize to mitochondria, the secretory system, the nucleus and, in plants, the chloroplast, which also requires copper for plastocyanin. Prokaryotic cuproproteins are found in the cell membrane and in the periplasm of gram-negative bacteria. Cu(I) and Cu(II) form tight complexes with organic molecules and drive redox chemistry, which unrestrained would be destructive. Copper metallochaperones assist copper in reaching vital destinations without inflicting damage or becoming trapped in adventitious binding sites. Copper ions are specifically released from copper metallochaperones upon contact with their cognate cuproproteins and metal transfer is thought to proceed by ligand substitution.

Contents

INTRODUCTION

As cofactors for enzymes, copper ions are required for processes ranging from oxidative phosphorylation, mobilization of iron, connective tissue cross-linking, pigment formation, neuropeptide amidation, catecholamine synthesis, and antioxidant defense. Copper has additional biological roles that may be distinct from serving as a catalytic moiety in cuproenzymes, for example, in the innate immune response, in the modulation of synaptic transmission, and in angiogenesis. A consequence of these vital functions is that copper deficiency has profound clinical outcomes often associated with neurodegeneration. Dietary intake of copper salts generally exceeds tissue demands, so homeostatic mechanisms exist to modulate uptake and facilitate export through the bile. Impairment in biliary copper excretion results in liver and brain copper overload in patients with Wilson's disease. Copper's ability to accept and donate single electrons make it an ideal redox cofactor, but copper ions are also complicit in the Fenton reaction and hence capable of driving the generation of deleterious hydroxyl radicals (**Figure 1*a***). Cu(II) is located at the top of the Irving-Williams series, and Cu(I) is also highly competitive, and hence, all copper ions have the potential to displace less competitive metals from metalloproteins (**Figure 1*b***). For all of these reasons, cellular copper pools must be tightly balanced to sustain a sufficent supply while minimizing toxic effects.

In the 1990s, a class of protein, designated copper metallochaperones, was discovered that functions as intracellular Cu(I) ion shuttles distributing Cu(I) ions to specific partner proteins, thereby overcoming a high copper chelation capacity of the cytoplasm. In many cells, the abundance of Cu(I)-buffering metallothionein and glutathione (GSH) molecules results in a cytoplasm depleted of free copper ions. The discoveries of the Atx1 and Ccs1 metallochaperones and the elucidation of their structures were reviewed in the 2001 edition of the *Annual Review of Biochemistry* (1). This review focuses on subsequent advances, including the mechanisms of copper transfer, copper metallochaperone-dependent and -independent maturation of superoxide dismutase and the assembly of the copper sites of cytochrome oxidase.

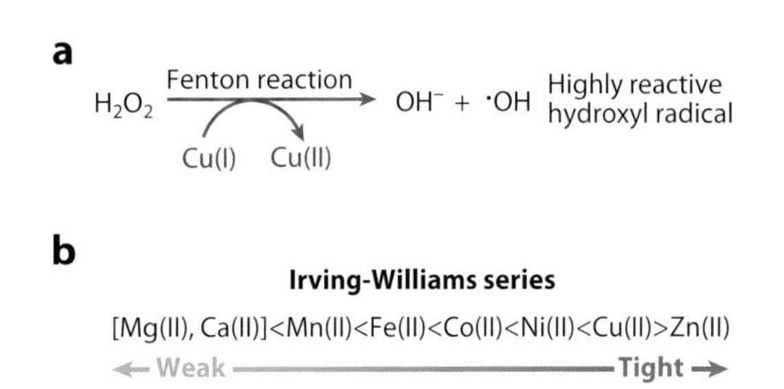

Figure 1

(*a*) Copper catalyzes production of hydroxyl radicals via the Fenton reaction. (*b*) Copper has a tendency to form stable complexes relative to other essential divalent metals, and hence it is at the top of the Irving-Williams series. (*b*) Monovalent copper is also competitive, forming tight protein complexes.

The Copper Proteome

Attempts have been made to define the copper proteome (cuproproteome) using genomic sequence data (2, 3). Ten distinct Cu-containing proteins have been identified in prokaryotes, but wide fluctuations exist in the distribution of these proteins within species (2). Analyses of prokaryotic and archean genomes indicated that nearly 72% of bacteria and 31% of archean species utilize copper (2). The cuproproteome distribution may actually be wider if novel copper proteins exist. Bioinformative searches using 39 Cu-binding motif signatures revealed additional putative Cu-binding proteins in Bacteria and Archaea (3). Over 70% of the putative cuproproteins identified in prokaryotes have homologs in Archaea and Eukaryotes. The most frequently utilized cuproprotein is cytochrome *c* oxidase. The second most common cuproprotein is NADH dehydrogenase-2 (2). Secretory cuproenzymes include dopamine-β-hydroxylase, tyrosinase, ceruloplasmin, lysyl oxidase, and amine oxidases. Certain bacterial phyla (Thermotogae, Chlorobi, Lactobacillales, and Mollicutes) lack known copper enzymes or copper metallochaperones. However, most of these species contain copper effluxers likely as a defense against the deleterious effect of copper ions in the cytoplasm.

Compartmentalization of the eukaryotic cuproproteome is well established (**Figure 2**). Cuproenzymes are localized within the plasma membrane and the periplasm of gram-negative bacteria. Within plants, copper is required in six compartments, including the cytoplasm, endoplasmic reticulum, mitochondria, chloroplast stroma, thylakoid lumen, and apoplast (4). Copper redistribution between compartments can occur upon transitions to copper deficiency (5).

Cuproprotein Metallation Reactions

In general, metallation reactions are expected to occur during protein biosynthesis, or shortly thereafter, because many metal cofactors influence protein stability. On cytoplasmic ribosomes, protein biosynthesis and chain elongation are coupled to chaperone-mediated protein folding for many proteins. Nascent polypeptides are protected within the ribosome and can only fold as they emerge from the exit tunnel. Subsequently, molecular chaperones bind nascent polypeptides and actively guide the polypeptides in folding through cycles of binding and release. Whereas cytosolic metalloproteins are likely metallated at this stage, because many cuproenzymes are either compartmentalized or secreted their folding and metallation must occur after translation and translocation. Secretory cuproenzymes pass through the endoplasmic reticulum lumen, where folding reactions commence, and maturation continues during transit through

Atx1: the yeast copper metallochaperone for P_1-type ATPases

Ccs1: a copper metallochaperone for Cu-Zn superoxide dismutase

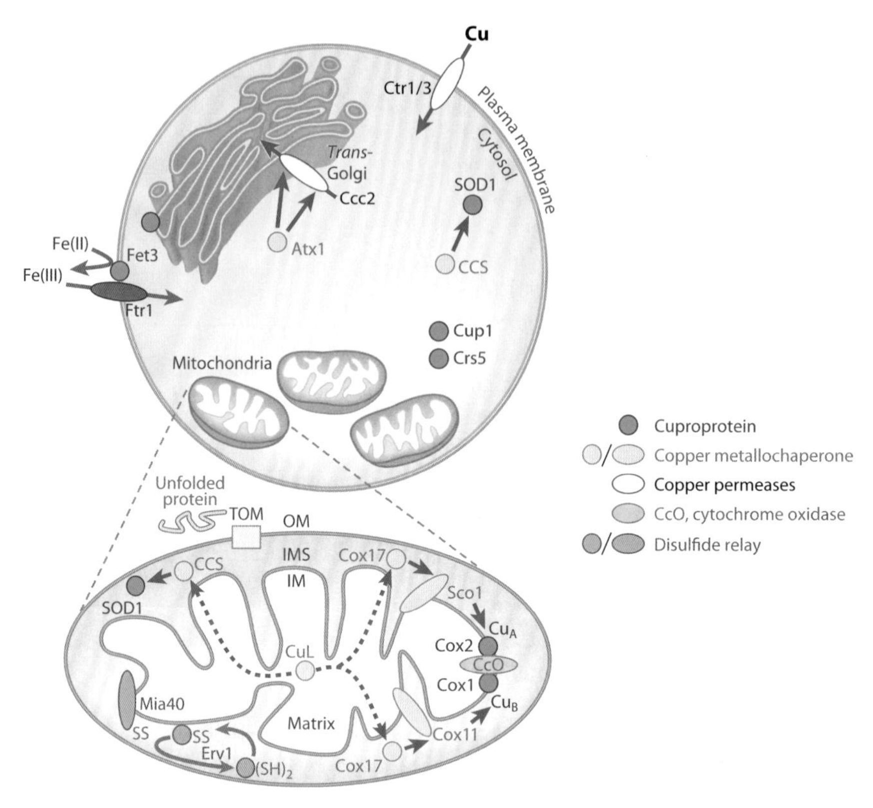

Figure 2

Copper is passed (*green arrows*) to cuproproteins (*dark green circles*), including metallothioneins Crs5 and Cup1, in for example *Saccharomyces cerevisiae*, by copper metallochaperones (*pale green circles*, or *ovals* if membrane associated). Copper permeases (*white ovals*) import copper into the cell or deliver Cu(I) to compartments, such as the *trans*-Golgi network where secreted cuproproteins (such as Fet3) acquire copper. Fet3 oxidizes ferrous ions (*curved brown arrow*) to provide substrate for high-affinity iron import by Ftr1. P_I-type ATPases (such as Ccc2) have one or more soluble N-terminal domains, which engage in regulatory interactions with copper metallochaperones (such as Atx1), but additional Cu(I) donation (hence *two arrows*) may supply the *trans*-membrane regions for Cu(I) transport. Metallation of the cytochrome *c* oxidase complex (CcO) in mitochondria (OM, outer membrane of mitochondria) involves a pathway via Cox11 for the Cu_B site of the Cox1 subunit, embedded in the inner membrane (IM), and via Sco1 for the Cu_A site of Cox2, protruding into the intermembrane space (IMS). Cox17 supplies both pathways. Nuclear encoded mitochondrial proteins, such as Cox17 and Ccs1, are imported across the OM unfolded (*pale green line*) via the TOM translocase then captured in the IMS, following introduction of disulfide bonds (SS) through the actions of Mia40. A sulfhydryl oxidase Erv1 generates a reactive disulfide on Mia40. A small copper ligand (CuL) supplies Cu(I) (*dashed arrows*) to the IMS copper metallochaperones.

the Golgi apparatus. Copper metallation occurs within Golgi vesicles.

Proteins discriminate between metal ions imperfectly, at least in vitro, yet biology seems to achieve precision in metallation reactions. Metallation is influenced by preferences in coordination geometry and in polarizability of ligand donor atoms. Computational chemistry studies infer that the specificity of a metal for a set of ligands in a protein cavity depends mainly on the metal's natural abundance in the biological locality, tailored by the properties of the protein and metal (6). For example, two structurally related cyanobacterial proteins, MncA and CucA, are the predominant manganese- and copper-containing proteins in the periplasm but have similar metal-binding sites (7). The mangano-protein prefers to bind Cu(II), or Cu(I), or Zn(II) rather than Mn(II) in vitro (7). However, precise metallation is achieved by differential compartmentalization during the metallation reactions. The mangano-protein, MncA, is dependent on incorporation of Mn(II) within the cytoplasm, where copper and zinc are highly restricted. Metallated MncA is then secreted into the periplasm as a holoenzyme via the Tat pathway. In contrast, metallation of the copper protein, CucA, occurs after export of the nascent chain via the Sec system. Another strategy for precision of metallation is through the actions of metallochaperones.

COPPER METALLOCHAPERONES FOR COPPER TRANSPORTERS

Bacterial *copZ* was discovered in the nucleotide sequences adjacent to a gene encoding a copper-transporting P_1-type ATPase (8), whereas yeast Atx1 was cloned as a suppressor of oxygen sensitivity in strains lacking superoxide dismutase (9). Studies of these two small proteins (69 residues for CopZ and 73 for Atx1) established the concept of copper metallochaperones. Atx1 and CopZ are similar in amino acid sequence to the N-terminal regions of metal-transporting P_1-type ATPases, including the one (CopA) from the same operon as CopZ, human Menkes and Wilson ATPases and the P-type ATPase from yeast, Ccc2. All of these proteins have a metal-binding motif MXCXXC (where X is any amino acid) associated with the first loop of a $\beta\alpha\beta\beta\alpha\beta$ (ferredoxin) structural fold. The single $\beta\alpha\beta\beta\alpha\beta$-fold of Atx1 or CopZ is replicated once (CopA), or twice (Ccc2), or six (Menkes and Wilson) times in the cytosolic regions of the ATPases (1). Copper metallochaperones that are variants of this structural theme occur in plants and Archeae. CopZ or Atx1-like copper metallochaperones are not obligatory partners for copper-transporting P_1-type ATPases since some genomes encode the ATPases but no homologs of these metallochaperones (10).

Yeast mutants deficient in Atx1 are deficient in iron uptake owing to impaired copper supply to the multicopper oxidase Fet3 (**Figure 2**). Extracellular Fet3 activity is needed to generate ferric substrate for the high-affinity iron importer Ftr1. As noted above, copper metallation of secreted proteins occurs within Golgi vesicles, and this is true of Fet3. The Ccc2 ATPase transports copper into this compartment, and thus Atx1 is phenotypically linked to the activity of the ATPase. Evidence that Atx1 forms associations with the N-terminal region of Ccc2 and that the human homolog Atox1 (sometimes called Hah1) forms associations with the Menkes or Wilson ATPases was obtained from coimmunoprecipitations and Cu-dependent two-hybrid interactions, whereas copper transfer between pairs of recombinant proteins was observed in vitro. A model was proposed in which these copper metallochaperones donate metal ions to the N-terminal regions of the ATPases via a sequence of ligand-exchange reactions. At the time of the 2001 *Annual Review of Biochemistry* article (1), solution structures had been obtained for apo-Atx1, Cu(I)-Atx1, as well as for the apo and Cu(I) forms of one domain of Ccc2 (Ccc2a), plus an X-ray crystal structure was known for the homodimer of Cu(I)-Hah1, among others. The homodimer structure was used to generate models of docked heterodimers (11). Solution structural studies have now examined these anticipated heterodimers.

CopZ: a bacterial protein similar to Atx1 and Atox1

$\beta\alpha\beta\beta\alpha\beta$-fold: ferredoxin-fold metal-binding domains of P_1-type ATPases and some copper metallochaperones

Atox1: the human copper metallochaperone for P_1-type ATPases

Visualization of Adducts with Cytosolic Domains of P_1-Type ATPases

A solution structure of an Atx1-Cu(I)-Ccc2a heterodimer has been generated (**Figure 3**) (12). The interaction is dependent upon copper, with the first cysteine of each protein (Cys15 of Atx1 and Cys13 of Ccc2a) being essential for complex formation (12). This seems reasonable because Cys15 of Atx1 is the more solvent accessible of its cysteine pair. A collection of lysine residues (residues 24, 28, 62, and 65) confer complementary surface charge to a negative surface of Ccc2a. Conversion of these lysines to glutamate impairs Atx1 activity as

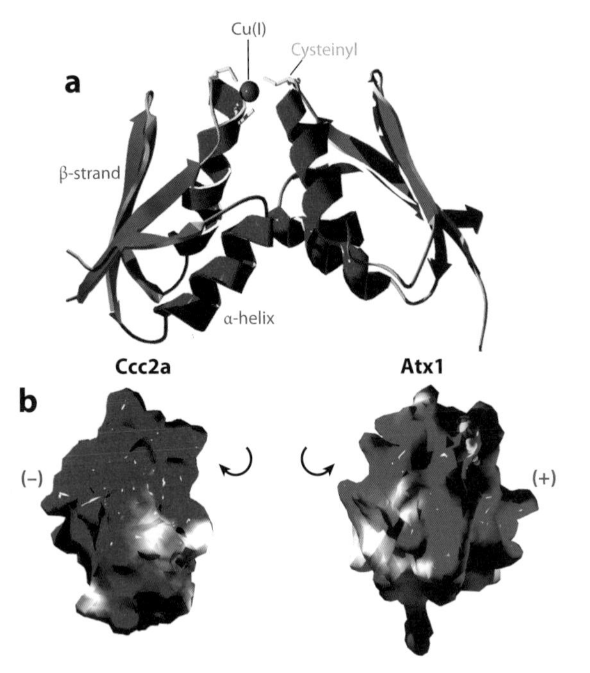

Figure 3

(*a*) NMR solution structural model [Protein Data Bank (PDB) code 2GGP] of a heterodimer of the *S. cerevisiae* copper metallochaperone Atx1 and one of two N-terminal cytosolic domains of its cognate P_1-type copper-transporting ATPase Ccc2. Cysteinyl (*yellow*) thiol coordinated Cu(I) (*sphere*) stabilizes a transient complex between similar $\beta\alpha\beta\beta\alpha\beta$ (ferredoxin)-structural folds (*red*, α-helix; *blue*, β-strand). (*b*) The interacting faces of Ccc2a and Atx1 have complementary (*red*, negative; *blue*, positive) electrostatic potentials.

assessed by loss of yeast two-hybrid interactions with Ccc2a and poor ferrous ^{55}Fe uptake by yeast strains carrying these Atx1 variants (13). Thus, electrostatics enables encounter complexes to form between the copper metallochaperone and the P_1-type ATPase (**Figure 3**). However, the encounter complex must be stabilized by coincident coordination of copper through cysteine residues from both partners if the transient complex is to persist for a sufficient length of time to enable detection by NMR. The complex is also predicted to be transient in vivo as Atx1 is largely found in the cytoplasm rather than associated with Golgi membrane preparations. A possible sequence for the exchange of metal-binding ligands between the two proteins was proposed with Cys13 of Ccc2a first invading a digonally coordinated Cu(I)-Atx1 site after which the second Cys of Ccc2a replaces the second Cys of Atx1. An expectation is that the heterodimer dissociates upon formation of a bisthiolate Cu(I) complex with Ccc2a (**Figure 4**), although there is only a small change in free energy associated with the transfer of Cu(I) from Atx1 to Ccc2a (14).

The Transfer of Copper

Copper transfer from Atx1 to Ccc2a in vitro was unaffected by the presence of a 50-fold molar excess of GSH, which was added to chelate any released Cu(I) (14). The off rate from Cu(I)-Atx1 to solvent is thought to be negligible, such that transfer to Ccc2a can only occur by direct protein contact and not via free solution. Presumably, the off rate for copper is enhanced in the heterodimer owing to conformation changes induced in Atx1 by its partner. This predicted switch in the Cu(I)-Atx1 off rate has not been measured, although structural data have suggested what might trigger a change. Loop 5 of Atx1 contains a lysine residue (Lys65), which is conspicuously conserved in many homologs from other species. Lys65 is thought to serve as a counter ion, balancing the negative charge of a metal site in which Cu(I) is bound to two thiolates and shielding the site. In adducts with Ccc2a, this residue is displaced away from the metal-binding site (12). Thus, movement of Lys65 in adducts opens the Atx1 metal site for ligand invasion by Ccc2a. The metal-binding site of apo-Atx1 is disordered, whereas that of apo-Ccc2a is structured (12), perhaps also contributing a small free-energy change to encourage metal release from the former to the latter.

The bacterial copper metallochaperones, CopZ from *Enterococcus hirae* (8, 15) and from *Bacillus subtilis* (16, 17), plus ScAtx1 from *Synechocystis* PCC 6803 (18), have a tendency to multimerize when bound to copper. These copper metallochaperone homodimers are presumed to dissociate to allow heterodimers to form with the ATPases. In the bacterial systems, the electrostatics are reversed relative to yeast Atx1-Cu(I)-Ccc2a, with CopZ and ScAtx1 presenting negative faces to positive ones on the respective P-type ATPase domains. Bioinformatics initially suggested that the CopA P-type ATPase of *B. subtilis* may import copper; however, ^{15}N chemical shift differences attributed to CopZ-Cu(I)-CopA showed that the metal-binding region of CopZ adopts an apo-like form and that CopA adopts a Cu(I)-like form in the heterodimer, more consistent with a vector for copper transfer from the copper metallochaperone to the ATPase (17). In support of this prediction, Δ*copA B. subtilis* was subsequently shown to be hypersensitive to elevated copper and deficient in copper export not import (19). Indeed, there is now uncertainty about whether any P_1-type ATPases import copper, although this was first proposed for CopA from *E. hirae* and remains the simplest model for cyanobacterial CtaA on the basis of the phenotypes of Δ*ctaA* strains.

Copper metallochaperones can direct Cu(I) to bona fide destinations but also keep it away from aberrant ones. Zinc-transporting P_1-type ATPases, such as ZntA and ZiaA, also possess $\beta\alpha\beta\beta\alpha\beta$-folds plus CXXC metal-binding sites, akin to those of copper transporters (20, 21). In ZiaA, from the cyanobacterium *Synechocystis* PCC 6803, the metal site has been shown to have a preference for Cu(I) over Zn(II). However, the N-terminal domain of ZiaA does not form heterodimers with the copper metallochaperone Cu(I)-ScAtx1 (22). Electrostatics repulses encounter complexes between ZiaA and ScAtx1, while complexes between ScAtx1 and related copper transporters CtaA and PacS are not repulsed. This mechanism provides a kinetic barrier to inhibit aberrant association of Cu(I) with ZiaA (21). Whereas most bacteria have no cytoplasmic membranous compartments and no known copper-requiring enzymes in the cytoplasm, cyanobacteria, such as *Synechocystis* PCC 6803, are peculiar in possessing thylakoid membranes housing plastocyanin and a *caa*$_3$-type cytochrome oxidase. CtaA, PacS, and ScAtx1 act positively with respect to thylakoid cuproproteins, and crucially, the ATPases are nonredundant (22, 23). PacS has been localized to thylakoid membranes (24), suggesting a model in which CtaA imports copper at plasma membranes while PacS loads

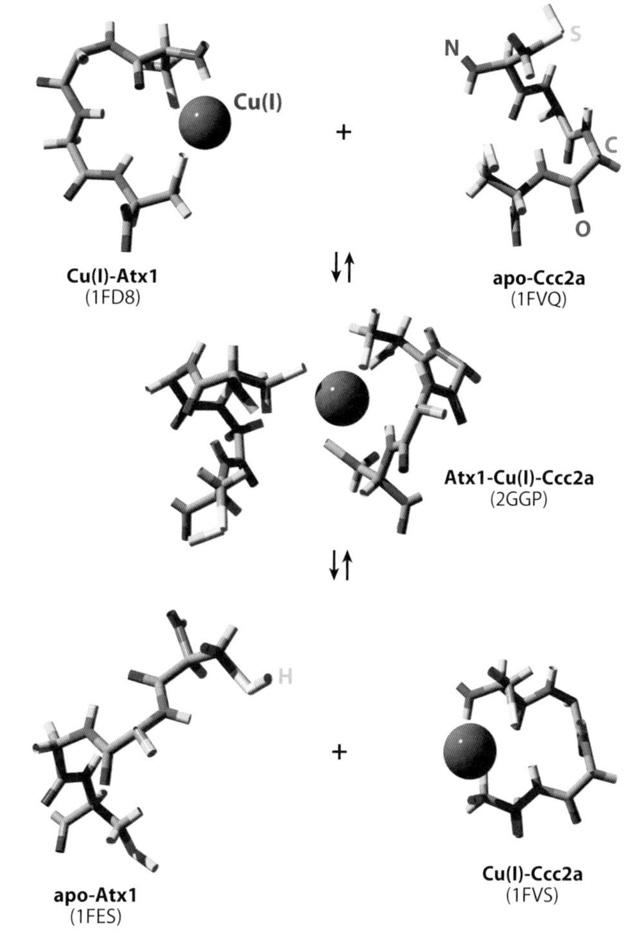

Figure 4

Visualization of the Cys-Xaa-Xaa-Cys motifs taken from NMR solution structures (Protein Data Bank codes indicated) of the apo-, Cu(I)-, and heterodimeric forms of Atx1 and Ccc2a to represent the transfer of Cu(I) from Atx1 to Ccc2a via ligand-exchange reactions.

it into thylakoids, analogous to the supply of copper to cuproproteins within the Golgi in eukaryotes. This model, however, relies upon CtaA acting to import copper, which has not been demonstrated for a P_1-type ATPase. A credible alternative model is that CtaA transports copper into thylakoids but at a different stage in their biogenesis than when PacS is active. For example, copper must be supplied twice to tyrosinase in mammals, transiently by the Menkes ATPase in Golgi and again by the Menkes ATPase in specialized melanosome compartments derived from the

Golgi (25). Copper binding is necessary to stabilize tyrosinase in the Golgi, but its folding state, coupled with the ionic conditions in the earlier Golgi network, is thought to lead to release of the initial copper atoms.

Copper Donation to Membrane Sites

Because there is only a small free-energy change associated with metal transfer from copper metallochaperones to the N-terminal regions of P_1-type ATPase, it has been widely assumed that irreversible membrane translocation driven by the hydrolysis of ATP is what shifts the equilibrium by removing the Cu(I)-ATPase product. However, there is no direct evidence that copper acquired by an N-terminal metal-binding domain of a P_1-type ATPase is transferred to *trans*-membrane regions for transport. CopZ from *Archaeoglobus fulgidus* interacts with, and can donate metal to, soluble $\beta\alpha\beta\beta\alpha\beta$-domains of CopA from the same organism (26). Importantly, Cu(I)-CopZ can also activate ATPase activity of isolated CopA, even when the ATPase is missing its own soluble $\beta\alpha\beta\beta\alpha\beta$-domain or when Cu-binding Cys residues have been converted to Ala. In contrast, the purified Cu(I)-bound $\beta\alpha\beta\beta\alpha\beta$-domain of CopA, expressed without the rest of the ATPase, does not activate the variant CopA missing the same domain. Under nonturnover conditions (in the absence of Mg^{2+}-ATP), irreversible copper transfer can be detected between Cu(I)-CopZ and the membrane metal-binding sites of CopA (26). It is possible that the transfer of metal from copper metallochaperones to the N-terminal regions of P_1-type ATPases serves regulatory roles and that separate donation events to the metal sites located in the membranous regions might be a common requirement.

The metal-binding domain of CopA interacts with the nucleotide-binding domain of the ATPase, and this interaction is impaired upon Cu(I) binding or nucleotide binding (27, 28). Thus, Cu(I) donation to the N-terminal region could serve to regulate Cu(I) transport by modulating this internal interaction within the ATPase. It is notable that CopZ from *A. fulgidus* is atypical in structure, having an extra 130-residue N-terminal domain that contains both zinc and a 2Fe2S cluster (29).

The cytosolic domain of the Wilson ATPase containing its ATP-binding site also interacts with the N-terminal $\beta\alpha\beta\beta\alpha\beta$-regions in vitro, and these interactions are also inhibited by Cu(I) binding (30). Donation of copper to the N-terminal regions may thus expose the ATP-binding site to assist catalytic turnover in response to increased abundance of Cu-bound metallochaperone. Both the Menkes and Wilson ATPases also undergo metal-dependent shifts in their cellular localization when copper becomes elevated, and the N-terminal $\beta\alpha\beta\beta\alpha\beta$-regions of the Menkes ATPase are known to influence trafficking (31–33).

The V_{max} of CopA from *A. fulgidus* was greater with Cu(I)-CopZ rather than Cu(I) alone (26). The dynamics of copper release from human Atox1 have been investigated by using the Cu(I) chelator bicinchonic acid (BCA), which forms a Cu(I) BCA_2 complex (34). Transfer progresses via formation of an Atox1-Cu(I)-BCA intermediate before ligand invasion by a second molecule of BCA. Lack of release to GSH presumably implies that the thiol of GSH is unable to invade the Atox1 Cu(I) site in the same manner as BCA. BCA acquires copper more rapidly from Cu(I)-Atox1 than it does from buffer alone because Cu(I)-dithiothreitol complexes dissociate more slowly than the copper metallochaperone. The rate of release to BCA is further enhanced using Atox1 mutant proteins in which alanine residues are substituted for either the shielding lysine or the methionine of the MXCXXC motif. In both cases, rapid release from mutant *Atox1* occurs because the second molecule of BCA more readily displaces Atox1 from the intermediate adduct relative to wild-type Atox1 (34). Copper metallochaperones can thus make copper swiftly accessible to the correct substrates capable of ligand invasion, presumably including membranous metal-binding sites of P_1-type ATPases.

Copper Acquisition by Atx1-Like Copper Metallochaperones

Yeast mutants missing any of the diverse types of copper permeases still retain functional copper metallochaperones, implying that there is no single preferred donor (35). Furthermore, because the permeases are so varied, it seems improbable that they can all support specific interactions with copper metallochaperones. The electron crystallographic structure of the Ctr1 copper pore may reveal a cytosolic domain with a candidate docking site for Atx1 (36). Copper transfer between the cytosolic regions of Ctr1 and Atx1 does occur in vitro, although the estimated affinities of the two proteins are similar at about 10^{-19} M (37, 38). Constitutive overexpression of competing Cu(I)-binding proteins, such as the metallothioneins Crs5 and Cup1, in the cytosol does not interfere with Cu(I) acquisition by copper metallochaperones. One of the burning questions is how do cells prioritize copper supply? An intriguing notion is that this could be achieved if some, but not all, copper metallochaperones gained access to metal ions released at cuproprotein turnover. Atx1 targets would be subordinate if Atx1 solely obtained copper from de novo import, and the primacy of cytochrome oxidase could be sustained provided its copper metallochaperones had preferential access to recycled copper.

Handling Copper in a More Oxidizing Periplasm

A five-stranded β-barrel protein, CusF, involved in the efflux of surplus copper from the periplasm of gram-negative bacteria, such as *Escherichia coli*, has been added to the catalog of copper metallochaperones. CusF has an unusual copper site, involving a histidine imidazole, two methionine thioethers, plus a π-interaction from the aromatic ring of tryptophan (39–41). The methionine thioether group appears as a Cu ligand in sites located within more oxidizing environments, for example, in copper permeases such as Ctr1 (42). The *E. coli* CusBCA proteins are thought to form a complex that straddles the inner and outer membranes to couple the inner-membrane (IM) proton motive force to copper efflux across the outer membrane. It is predicted that the substrate is acquired in the periplasm rather than the cytoplasm. By using selenomethionine derivatives to distinguish the metal sites, it was possible to follow movement of copper from Cu(I)-SeMet-CusF to apo-CusB by X-ray absorption spectroscopy (43). The proteins have closely matched affinities, leading to a suggestion that metal transfer might be driven by removal of the Cu(I)-CusB product through CusBCA-mediated efflux, analogous to the arguments proposed for the driving force for copper transfer from Atx1 to Ccc2.

Cox17: a eukaryotic copper metallochaperone for Sco1 and Cox11

Cox11: a copper metallochaperone for the Cu_B site of cytochrome oxidase

Sco1: a copper metallochaperone for the Cu_A site of cytochrome oxidase

COPPER METALLOCHAPERONES FOR CYTOCHROME OXIDASE

Unlike Atx1 and Ccs1 that donate Cu(I) directly to the target molecule, Cox17-mediated Cu(I) donation to cytochrome *c* oxidase (CcO) subunits involves two accessory factors Cox11 and Sco1, which function in the metallation of the Cu_B site in Cox1 and Cu_A site in Cox2, respectively. Both Cox11 and Sco1 are IM-associated proteins with Cu(I)-binding globular domains protruding into the intermembrane space (IMS). Cox17-mediated Cu(I) transfer to Cox11 and Sco1 is the initial event, followed by the subsequent transfer to Cox1 and Cox2, respectively. Cu(I) donation to Cox11 and Sco1 is believed to impart specificity to the Cu(I) transfer to CcO subunits during CcO biogenesis. One difference between the Cu(I) metallation of Cox1 and Cox2 is that the Cu_A site in Cox2 is localized 8 Å above the IM, whereas the Cu_B site in Cox1 is buried 13 Å below the membrane surface (44).

Cox17

Cox17 was cloned from *Saccharomyces cerevisiae* from a CcO deficiency mutant. Cells lacking Cox17 were respiratory deficient and unable to propagate on nonfermentable carbon sources (45). Mitochondria isolated from the mutant cells contained both mitochondrial and nuclear CcO subunits, precluding a role for Cox17 in the expression or import of the subunit polypeptides. The mutant cells lacked the CcO-specific heme *a*, but not heme *b* found in other respiratory complexes or cytochrome *c* (45). Insight into the role of Cox17 in CcO biogenesis was gleaned by the observation that *cox17*Δ cells regained an ability to propagate on glycerol and ethanol medium with the addition of 0.4% copper salts. Because the mutant cells contained a functional Sod1 superoxide dismutase, the role of Cox17 is specific to CcO and/or mitochondria. Mice lacking Cox17 die during early embryogenesis at a stage similar to mice lacking the copper transporter Ctr1 (46, 47). Prior to embryonic death, CcO activity was severely impaired in *COX17 -/-* mice. Depletion of *COX17* mRNA, using siRNA, in HeLa cells resulted in an induced CcO deficiency (48). Steady-state levels of Cox1 and Cox2 were diminished following the knockdown, but the other respiratory complexes retained the same magnitude of abundance as in wild-type (WT) cells.

Cox17 is a small hydrophilic protein of 63 residues in humans. It contains a conserved twin CX_9C motif that forms a helical hairpin configuration (**Figure 5**). This helical hairpin is also referred as a CHCH (coiled-coil$_1$ helix$_1$ coiled-coil$_2$ helix$_2$) motif (49). The twin CX_9C structural motif is found in numerous IMS proteins; several of these, including Cox19, Cox23, and Pet191, are important in CcO biogenesis. The Cox12 structural subunit of CcO projecting into the IMS also contains the twin CX_9C motif. The cysteines within the twin CX_9C motif are present in two disulfides, stabilizing the helical hairpin of Cox17 (**Figure 5*a***) (50, 51). The redox couple of the double disulfide configuration of Cox17 to the fully reduced state has a midpoint potential of −294 mV, consistent with the dual disulfide molecule being the likely species in vivo, considering that the IMS redox potential is approximately −255 mV unlike the more-reducing potential of the cytoplasm (−296 mV) (52, 53). In support of the disulfide configuration of Cox17 within the IMS is the observation that other Cys-rich

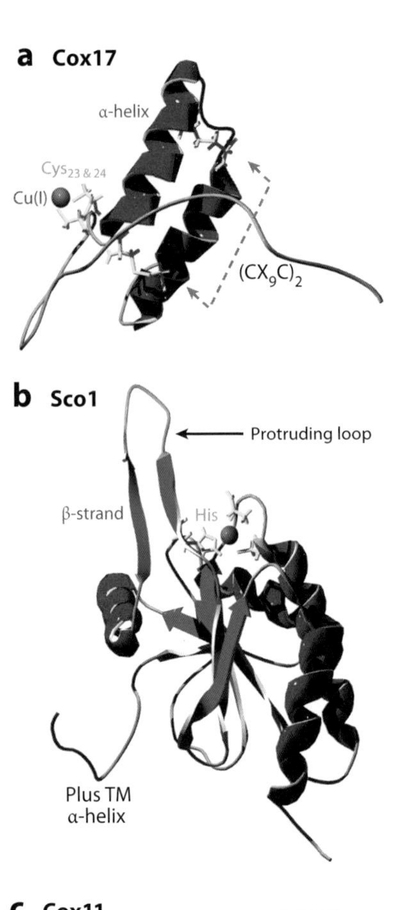

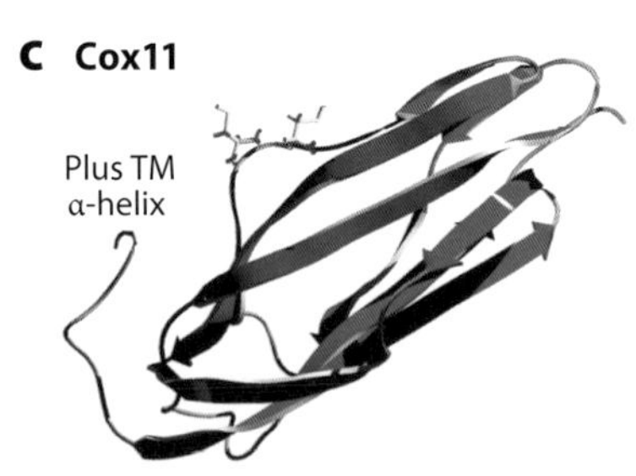

Figure 5

Copper metallochaperones for CcO. (*a*) Disulfide bonds between two pairs of cysteine residues (*yellow*) on antiparallel α-helices (*red*) create a helical hairpin of Cox17 [Protein Data Bank (PDB) code 2RNB]. Cu(I) is coordinated to a pair of cysteine residues. (*b*) Sco1 is tethered to the inner membrane (IM) (hydrophobic α-helix not shown) with a thioredoxin fold (β-strand, *dark blue*; histidine, *pale blue*) in the intermembrane space (PDB 2GQM). (*c*) Cox11 is tethered to the IM (hydrophobic α-helix not shown) with an immunoglobulin-like β-barrel in the IMS (PDB 1SPO).

helical hairpin proteins within the IMS with a twin CX_3C motif form heterohexameric complexes in which disulfides are essential to stabilize both the structure and function of the complex (55). Although Cox17 likely exists with twin disulfides, Cox17 is functional without either of the two disulfides (54).

Cu(I) binding to Cox17 occurs through two vicinal cysteine residues in a digonal Cu(I)-thiolate complex (**Figures 5*a*** and **6**) (50, 51, 55). The mononuclear Cu(I)-Cox17 has been characterized only after in vitro Cu(I) reconstitution. Recombinant expression of Cox17 in *E. coli* leads to the accumulation of a Cu-Cox17 oligomeric, tetranuclear Cu(I) complex in cells cultured in Cu-supplemented medium (56). Recombinant expression in bacteria without copper supplementation results in the isolation of metal-free protein. This is presumably due to the absence of available cytosolic copper ions in *E. coli* (57). The polycopper protein is in a dimer/tetramer equilibrium, with the Cu(I)-thiolate cluster likely existing at the dimer interface (56). The tetracopper cluster conformer requires multiple cysteine residues to be in the reduced thiolate state. Cox17 purified from the IMS is largely devoid of bound Cu(I) and is monomeric (58). Cu-reconstituted human Cox17 shows a S-Cu-S angle of 130° (55). The bent coordination may permit an exogenous thiolate to provide a third ligand. The Cu(I) thiolate center is charge stabilized by two adjacent Lys residues. Lys20 may mimic the stabilizing Lys65 in the Cu-Atx1 complex. In addition, a hydrophobic patch, formed by ends of the C-terminal helix and the N-terminal segment, orients the Cys thiolates for Cu(I) binding. The Cu(I) center is solvent accessible yet avidly bound with a K_d value of 6.4×10^{-15} M (59). The Cu(I)-binding affinity must be poised for transfer to target proteins Sco1 and Cox11. Another well-defined digonal Cu(I)-thiolate complex exists in the *E. coli* CueR transcriptional regulator (57). This MerR-based regulator exhibits an essentially linear S-Cu-S bond angle of 176° with short Cu-S bond distances of 2.13 Å. Eight residues separate the two coordinating cysteines. The linearity of the Cu(I)

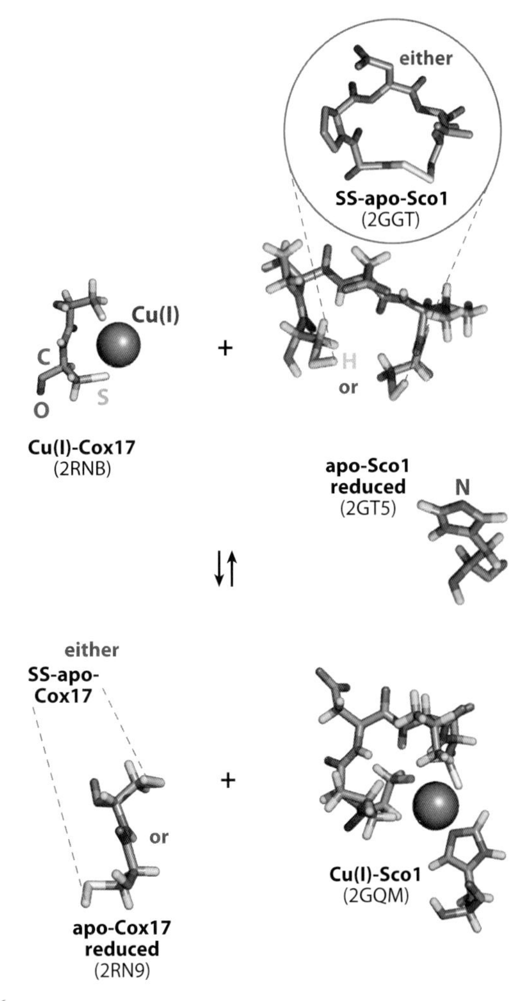

Figure 6

Visualization of metal-binding motifs taken from structures (Protein Data Bank codes indicated) of the apo and Cu(I) forms of Cox17 and Sco1 to represent the transfer of Cu(I) from Cox17 to Sco1 and, potentially, to represent the coincident transfer of two electrons when apo-Sco1 is oxidized.

coordination results in a Cu(I) K_d near 10^{-21} M. In contrast, the bent coordination of Cu(I) by the vicinal Cys residues in Cox17 reduces the binding affinity perhaps to permit transfer.

The disulfides in Cox17 are introduced during import into the IMS. The IMS import of small cysteine-rich polypeptides with either twin CX_9C or twin CX_3C structural motifs is mediated by a disulfide relay system involving the mitochondrial intermembrane space assembly (MIA) machinery consisting of at least

Mia40 and Erv1 (**Figure 2**) (60–64). Transit of these proteins, as unfolded polypeptides through the TOM translocase in the OM, results in transient capture of the imported molecules by Mia40 through an intermolecular disulfide. Mia40 contains two structural disulfide bonds formed by its twin CX_9C motif as well as a N-terminal labile disulfide. The reactive disulfide lies within a CPC sequence motif and has a reduction potential of −200 mV (65). The nonoptimal stereochemistry of the disulfide with a single residue spacer likely contributes to its instability. The sulfhydryl oxidase Erv1 catalyzes formation of a reactive disulfide in Mia40 that is responsible for transient capture of imported protein with reduced thiolates (60, 61, 66, 67). The second Cys of the CPC motif is the active Cys forming the transient disulfide with Cox17 and other target proteins (65). Protein release from the Mia40 complex is facilitated by disulfide exchange reactions resulting in disulfides in the imported proteins. The oxidative folding of Cox17 and other Mia40 targets traps the proteins within the IMS. It remains unclear whether the Cu(I)-binding Cys23 and Cys24 residues of Cox17 are also initially oxidized. If a disulfide forms with Cys23 and Cys24, reduction would be required for Cu(I) binding. Information is lacking on what redox couples exist within the IMS. GSH reductase and thioredoxin reductase are not known to be present in this compartment.

Cox17 is also present in the yeast cytoplasm, initially encouraging a hypothesis that Cox17 may shuttle Cu(I) to mitochondria (58). However, restricting Cox17 to mitochondria through tethering to the IM retained normal CcO biogenesis (68). Import of Cox17 through the TOM translocase occurs as an unfolded and, presumably, apopolypeptide. The source of copper for metallation of Cox1 and Cox2 during CcO biogenesis arises from a mitochondrial matrix copper pool. A significant fraction of mitochondrial copper is localized within the matrix as a soluble, low-molecular-weight ligand complex (CuL), which is conserved from yeast to humans. Information is lacking on whether copper within the CuL complex is imported as ionic Cu(I) or as a CuL complex. The ligand appears to be present in the apo state within the cytoplasm, so an attractive postulate is that formation of the CuL complex occurs within the cytoplasm and the complex is imported into the mitochondrial matrix through an unknown carrier. Heterologous expression of two Cu(I)-binding proteins in the mitochondrial matrix was shown to titrate out the matrix Cu(I)-ligand complex and induce a respiratory deficiency (69, 70). These effects are reversed by supplementation of cell cultures with $CuSO_4$. Attenuation of mitochondrial copper in *Podospora anserina* by a heterologous Cu(I)-binding protein also induces a CcO deficiency (71). In addition, the presence of the matrix-targeted Crs5 in yeast results in diminished Sod1 protein levels within the IMS and impaired activity of an IM-tethered hSod1.

Cu-Cox17- and Sco1-Mediated Cu(I) Transfer to Cox2

The recipient of Cu(I) from Cu-Cox17 is Sco1 in yeast, and a combination of Sco1 and Sco2 in mammalian cells. Overexpression of Sco1 restored respiratory function in yeast lacking Cox17, thereby linking Cox17 and Sco1 (72). Sco1 participates in the formation of the mixed valent Cu_A site in Cox2. In the absence of Sco1, yeast are stalled in CcO biogenesis, and the complex is degraded (73, 74). Yeast also possess Sco2, but cells lacking Sco2 are not respiratory deficient. Both human Sco1 and Sco2 proteins are required for CcO biogenesis and have nonoverlapping functions (75). Mutations in either gene results in a fatal encephalomyopathy, although the clinical presentation differs between h*SCO1* and h*SCO2* patients (76, 77). In addition to the CcO deficiency, patient tissues show a tissue-dependent copper deficiency (78).

Sco1 and Sco2 are tethered to the IM by a single *trans*-membrane helix with a globular domain protruding into the IMS containing a single Cu(I) site (79, 80). Cu(I) is trigonally coordinated by two cysteinyl residues within a CPDVC sequence motif plus a histidine imidazole (**Figures 5b** and **6**) (80). Mutation of the Cys or His residues abrogates Cu(I) binding, resulting in a nonfunctional CcO (79, 80). The Cu(I) ion is partially shielded by the first Cys and the adjacent two residues (81). The structures of the metal-free human Sco1 and Cu_I Sco1 complex are similar with the exception of one protruding loop that adopts two different conformations (81) (see the loop marked by an arrow in **Figure 5b**). In the apo conformation, the Cu(I)-binding region is more disordered, and the conformer, frozen in the crystal structure of hSco1, contained the His rotated out of position for planar Cu(I) coordination (82–84). The movement of the protruding loop orients the Cu(I) binding His residue for metal binding. The dynamic properties of this loop suggest it could influence interactions involved in Cu(I) transfer (81).

Human Sco2 shows greater conformational dynamics than hSco1 (85). These proteins have thioredoxin folds, raising questions about whether Sco proteins may function in redox reactions. Recent evidence suggests that hSco2 acts as a thioldisulfide oxidoreductase in CcO biogenesis, whereas hSco1 may be the dominant Cu(I) donor to Cox2 (86). Thioredoxins typically contain a *cis*-Pro in juxtaposition to the redox active cysteinyl residues. The *cis*-Pro precludes metal ion binding (87). The corresponding positions in Sco1 and Sco2 are the Cu(I) ligand His, consistent with their Cu(I)-binding function. If Sco2 is primarily an oxidoreductase, its function remains dependent on the Cys and His residues. Substrate recognition by protein disulfide isomerases, which share the thioredoxin fold (88), may also be dissimilar to partner recognition by copper metallochaperones with the former being dominated by hydrophobic interactions. The reduction potential of the Cys residues in hSco1 and hSco2 are −280 mV and −310 mV, respectively. Considering the reduction potential of the IMS is approximately −255 mV (52), it is predicted that Sco proteins may exist with at least some Cys residues oxidized and therefore not poised for Cu(I) binding. Sco1 and Sco2 in fibroblasts do have a mixture of oxidized and reduced Cys residues (86).

Direct Cu(I) transfer from Cox17 to Sco1 has been shown in vitro and in vivo. This likely occurs through a highly transient interaction of the two molecules (59). The interaction presumably imparts specificity to the Cu(I) transfer step. The specificity of the transfer reaction was demonstrated by using mutant proteins (89). Cys57 forms a disulfide with Cys26 in the twin CX_9C motif of Cox17. Whereas a $C^{57}S$ substitution maintains function, a $C^{57}Y$ substitution abrogates Cox17 function (45, 54). The nonfunctional $C^{57}Y$ Cox17 mutant failed to mediate Cu(I) transfer to Sco1 in both in vivo and in vitro assays despite binding Cu(I) normally (89). However, the mutant retained the ability to transfer Cu(I) to its second target Cox11. Sco1 $P^{174}L$ is an allele of patients with CcO deficiency (90–92). Pro174 is adjacent to the second Cys in the Cu-binding CXXXC motif in Sco1, and this mutant has impaired Cu(I) acquisition from Cox17 (59, 92). A transient interaction persists between $P^{174}L$ Sco1 and Cox17, but K_{Cu} is weaker, perhaps too weak to allow Cu(I) acquisition from a, now relatively tighter, Cu(I) site in Cox17 (59).

Recently, Cu-Cox17 was shown to effectively transfer both Cu(I) and reducing equivalents to oxidized apo-Sco1, leading to Cu-Sco1 formation and fully oxidized Cox17 (93). The Cu-free conformer of Cox17 failed to reduce oxidized Sco1, suggesting that Cu(I) is necessary for the disulfide exchange mechanism. The ability of Cu-Cox17 to mediate Cu(I) transfer to oxidized Sco1 abrogates a necessary mechanism to maintain reduced thiolates in Sco1. Although Cu-Cox17 can also transfer Cu(I) to Sco2 with reduced thiolates (94), it fails to transfer to oxidized Sco2 (93).

The Cu(I) ions transferred from Cox17 to Sco1 appear to be donated subsequently to Cox2 as an intermediate step in CcO assembly (**Figure 2**). The Sco1-mediated metallation of Cox2 likely occurs after extrusion of the Cox2-soluble domain into the IMS by the Cox18 translocase. Information on the Sco1-mediated metallation of Cox2 in eukaryotes is limited, although a group of conserved residues on the dynamic protruding loop in Sco1 was shown to be important for function with Cox2 (95). Sco proteins are widely distributed phylogenically. In prokaryotes, Sco1 is often codistributed with CcO and linked to the Cu_A center in Cox2 (96). However, Sco1 in *Rhodobacter capulatus* is associated with a *cbb*$_3$ CcO lacking a Cu_A site (97). The Cu-binding function of this Sco protein, designated SenC, is required for maturation of the *cbb*$_3$ CcO. One scenario is that SenC is the Cu(I) donor to the Cu_B center. The *B. subtilis* Sco1 is implicated in formation of the Cu_A site in CcO. The intriguing aspect of the *Bacillus* Sco1 is its dramatic preference for Cu(II) over Cu(I) (98). The marked stability of the Cu(II)-Sco1 complex may necessitate a conversion step to an intermediate state of Cu(I)-Sco1 compatible with Cu transfer (99).

Cox17 is restricted to eukaryotes. A bioinformatic search for genes associated with Sco1 in bacteria revealed a protein of unknown function with a conserved $H(M)X_{10}MX_{21}HXM$ sequence motif present in both gram-negative and gram-positive organisms (100). The protein, designated DR1885, has a single-candidate *trans*-membrane helix followed by a globular domain folded into a Greek key β-barrel conformation capable of binding a single Cu(I) ion with a Met_3His ligand set. The protein is analogous to other bacterial periplasmic Cu-binding proteins that use Met-rich motifs to coordinate Cu(I). The surface-exposed Cu(I) site makes it accessible for Cu(I) transfer. A homologous protein in *Thermus*, designated PCu_AC for the periplasmic Cu_A chaperone, was shown to bind Cu(I) with a subpicomolar affinity (101). Cu-PCu_AC was able to transfer Cu(I) in vitro to the reduced, soluble Cu_A domain of CcO subunit II of *Thermus* without the assistance of Sco1 (101). Successive Cu(I) transfer steps assembled the binuclear Cu_A center in subunit II. No transfer occurred when the Cys residues in subunit II were oxidized, unless reduced apo-Sco1 was present, leading to the oxidation of Sco1. Thus, *Thermus* Sco1 functions as an oxidoreductase, rather than a Cu(I) donor, to form the Cu_A center in CcO. The direct Cu(I) transfer from Cu-PCu_AC to subunit II may involve ligand-exchange steps with the two Cys residues in the Cu_A center. Cu(I) coordination by a Met_3His protein is preferred over a Cys-based donor like Cox17 in the more oxidizing periplasm. The DR1885/PCu_AC family of proteins may functionally replace Cox17 in prokaryotes as Cu(I) donors to Cu_A centers that face the periplasmic space. A clear prediction is that bacterial cells lacking DR1885/PCu_AC may be impaired in CcO biogenesis.

The implications of Sco1 having a redox role rather than a Cu(I) transfer function in *Thermus* are unclear. Maintaining Sco1 in the reduced state competent for a redox function would require reducing equivalents in the periplasm. In yeast, Sco1 has a Cu(I) transfer function, and it remains to be demonstrated whether it possesses a redox function. In humans, Sco2 is implicated in a redox function (86).

Cox11-Mediated Cu_B Site Formation

Copper metallation of the Cu_B site in Cox1 requires Cox11 (**Figure 2**). *S. cerevisiae* lacking Cox11 have impaired CcO activity and less Cox1 (102). CcO isolated from *Rhodobacter sphaeroides cox11Δ* cells lacked Cu_B but contained both hemes, and the environment of the heme in the Cu_B-heme *a*$_3$ site was altered (103). The Cu_A site in *R. sphaeroides* CcO was unaffected in cells lacking Cox11. Thus, the absence of Cox11 appears to specifically preclude Cu_B site formation.

Cox11 is tethered to the IM by a single *trans*-membrane helix with a C-terminal domain protruding into the IMS (104, 105). The Cu(I)-binding cysteinyl residues lie within this C-terminal domain. The structure of the globular domain of the Cox11 homolog from *Sinorhizobium meliloti* adopts an immunoglobulin-like β-barrel fold (**Figure 5*c***) (106). Removal of the *trans*-membrane domain of yeast Cox11 yields a soluble protein that dimerizes upon Cu(I) binding (104). The Cu(I) sites in each monomer are closely juxtaposed as the dimeric complexes can form a binuclear Cu(I) thiolate cluster at the dimer interface. Mutations of two Cys residues within a CXC sequence motif abrogate Cu(I) binding and formation of active CcO. A third functionally important Cys is spatially removed from the CXC motif, yet when mutated, this Cys results in substoichiometric Cu(I) binding.

Cu_B site formation occurs in Cox1 prior to addition of the Cox2 or Cox3 subunits. The Cu_B-heme a_3 binuclear site forms in an early assembly intermediate of Cox1 that is stabilized by the Shy1 assembly factor. Cox11 forms a transient interaction with Shy1, which may be important to coordinate Cu_B site formation. This proposition is supported by the observation that accumulation of the nonfunctional CcO complexes in *Rhodobacter* and *Paracoccus surf1*Δ (Shy1 homolog) cells is compromised in both Cu_B and heme a_3 (107, 108).

The membrane-embedded Cu_B site in Cox1 consists of three His ligands lying within a 12-helical bundle. In the absence of Cox2, a partially occluded channel forms from the IMS side of Cox1, where the heme *a* and Cu(I) moieties enter. The putative physical transfer from Cu-Cox11 to the buried Cu_B site in Cox1 may involve the third Cys residue that was removed from the CXC motif. This Cys is predicted to be juxtaposed to the IM and may mediate the Cu(I) transfer. However, it is not clear whether it would mediate Cu(I) transfer through the channel or laterally through an accessible port of the helical bundle. Cox11 appears to transiently occlude the Cox1 channel, as mutant alleles lacking the Cu-binding CXC residues confer resistance to the pro-oxidant heme a_3:Cox1 assembly intermediate (109).

Summary of the Key Steps in Copper Metallation of CcO

Copper-transfer reactions are initiated within the IMS by Cu-Cox17, which acquires Cu(I) from a small molecule (**Figure 2**). Cox17 transfers Cu(I) ions to IM-associated Sco1 and Cox11. Cu-Cox11 is the donor to the Cu_B site in Cox1, and this site is formed in a Cox1 assembly intermediate that may contain only the Cox5a and Cox6 peripheral CcO subunits. The assembly intermediate is fully inserted within the IM and is associated with CcO assembly factors, including Shy1. Upon hemylation of the two heme *a* centers and formation of Cu_B, Cox1 is poised for the addition of Cox2. It is not clear whether the binuclear Cu_A site in Cox2 is formed in an isolated Cox2 or as Cox2 is associated with metallated Cox1. Sco1 mediates Cu_A site formation, and this reaction may occur as a ternary complex of Cu-Cox17, Sco1, and Cox2. Once the redox copper and heme cofactors are added to Cox1 and Cox2, the remaining subunits are added for final CcO maturation.

COPPER METALLOCHAPERONE FOR SUPEROXIDE DISMUTASE

The copper metallochaperone for Cu,Zn-superoxide dismutase is Ccs1, a 27-kDa three-domain polypeptide (**Figure 2**) (110, 111). Two of the domains, domains 1 and 3, bind Cu(I), whereas the central domain is a key domain for interaction with Sod1 (112, 113). The physiological significance of Ccs1 has been defined in multiple species. Yeast lacking Ccs1 is devoid of Sod1 enzymatic activity unless cells are cultured with high levels of copper (114, 115). *Drosophila melanogaster* deficient in Ccs1 lacks Sod1 activity and shows exquisite sensitivity to redox cycling of paraquat (116). Mice with a targeted disruption of *CCS* have a markedly attenuated level of Sod1 activity and exhibit no apparent radio-copper incorporation into Sod1 (117). Thus, a strong correlation exists that Ccs1 is required for Sod1 activation.

Role of Ccs1 in Superoxide Dismutase 1 Activation

Cu-Ccs1 induces the activation of monomeric reduced Sod1, facilitating both Cu(I) transfer and disulfide bond formation (118). Active Sod1 requires binding of both Zn(II) and Cu and formation of an intrasubunit disulfide bond between Cys57 and Cys146 (118, 119). Sod1 is unusual in possessing such an essential disulfide bond in the reducing cytoplasm. The reduction potential of the Sod1 disulfide is

−230 mV, whereas the potential in the cytoplasm is approximately −290 mV. The stability of the Sod1 disulfide may arise from the low solvent accessibility of Cys146 in the disulfide pair as well as from the stabilization imparted by Sod1 dimerization. The copper center in Sod1 is sunken within an active site channel that superoxide anions enter for dismutation (120, 121). In the absence of the metal cofactors and upon reduction of the disulfide, the Sod1 dimer is destabilized and exists as an inactive monomer (118, 122). Metal binding and/or disulfide bond formation stabilizes the β-barrel and loop conformations within a dimeric enzyme.

Domain Structure of Ccs1

The N-terminal domain 1 of Ccs1 is an Atx1-like $\beta\alpha\beta\beta\alpha\beta$-fold capable of binding a single Cu(I) ion (123). Domain 1 of yeast Ccs1 is required for Sod1 activation in Cu-limited cells, but this domain is not essential under normal growth conditions (114). In the absence of domain 1, no radiocopper incorporation into mouse Sod1 was observed in cultured mouse fibroblasts, but copper salts restored limited active Sod1 (124). *Drosophila* and *Anopheles gambiae* Ccs1 proteins lack domain 1, yet these proteins activate Sod1 efficiently (116). Human mutant Ccs1 with Cys substitutions in the MXCXXC motif and Cu(I) bound to domain 3 retain the in vitro activation function of Sod1 (123).

Domain 2 adopts an eight-stranded β-barrel structure analogous to Sod1 (125, 126). The dimeric domain 2 in human Ccs1 contains a Zn(II) site similar to Sod1 but lacks the ligand set for Cu coordination. In contrast, yeast Ccs1 domain 2 lacks both the Zn and Cu centers. Domain 2 is important for docking with Sod1 during the activation reaction. Mutations that compromise Ccs1 dimerization also abrogate both Ccs1 and Sod1 interaction and Ccs1-mediated Sod1 activation (111, 124).

Domain 3 is a short C-terminal segment containing a critical CXC motif that is unstructured in the yeast apo-Ccs1 protein (125), but domain 3 is ordered in the co-structure of yeast Ccs1 and a mutant Sod1 that stabilizes the intermolecular complex (113). Ccs1 lacking the domain 3 CXC motif fails to transfer Cu(I) or to induce disulfide formation in Sod1 (114, 123, 124). The binding of Ccs1 to Sod1 involves both domains 2 and 3, and binding is retained if the Cys residues are mutated (111). Whereas the domain 3 CXC motif may suggest digonal Cu(I)-thiolate coordination, EXAFS analyses of the Cu(I) complex clearly showed a polycopper cluster. The cluster formed with only domain 3 Cys residues is consistent with a binuclear Cu(I)-thiolate center bridging two Ccs1 molecules (123). Polycopper cluster formation is lost when the domain 3 Cys residues are mutated.

The Sequence of Events in Ccs1-Mediated Activation of Superoxide Dismutase 1

The Ccs1-Sod1 complex is transient and becomes detectable with reduced Sod1 lacking a Cu site or in mutant Sod1 lacking one Cys of the disulfide pair (112, 118, 127). In the absence of bound Cu in Sod1, complex formation between Ccs1 and Sod1 is dependent on the disulfide being reduced. A loop in Sod1 containing Cys57, one of the disulfide-forming residues, is more flexible when the disulfide bond is reduced (128). Ccs1 is postulated to recognize a partially folded nascent conformation of Sod1 through mobile loop elements (127). Ccs1 interaction with Sod1 is abrogated when the loops become ordered in the copper-bound, disulfide-bonded fold. The Ccs1 interaction with Sod1 is independent of Zn(II) binding to Sod1. Because Zn binding to Sod1 stabilizes a folded conformation (129), Ccs1 must be competent to interact with a partially folded Sod1 conformer, but this structure has reduced thiolates and an empty copper site. The activation process involves copper insertion and disulfide bond formation.

Cu site formation appears to be concurrent with disulfide bond formation. Sod1 mutants blocked in Cu site formation remain largely in the reduced state in yeast containing Ccs1

(127). Likewise, in vitro studies revealed that reduced Zn-Sod1 protein is inefficiently converted to the oxidized state aerobically in the absence of Ccs1 (118). Sod1 oxidation requires the intact CXC motif in Ccs1 (130). The addition of Cu-Ccs1 to Zn-Sod1 fails to induce disulfide bond formation anaerobically, but this process is facile with the addition of air (118). A putative intermediate in the oxidation reaction is an intermolecular disulfide between Cys57 of Sod1 and Cys229 of Ccs1 (113). Formation of this intermolecular disulfide may resolve to the C57-C146 disulfide by a disulfide exchange reaction.

Ccs1-Independent Activation of Superoxide Dismutase 1

Although Ccs1 has a key role in Sod1 activation, it is now clear that a secondary mechanism exists for Sod1 activation independent of Ccs1 in multicellular enkaryotes. The *Caenorhabditis elegans* genome completely lacks Ccs1 (131). Mice lacking Ccs1 have 10% to 30% WT Sod1 activity in various tissues (132). Likewise, Ccs1-null flies show residual Sod1 activity such that the phenotype of the null flies is less severe compared to flies lacking Sod1 (116). Human Sod1 is capable of being partially activated independently of Ccs1 when expressed in yeast or flies (116, 133). This Ccs1-independent pathway of Sod1 activation does not require Atx1 or Cox17 (133).

Yeast Sod1 differs from human and fly Sod1 in depending on Ccs1 for activation (115, 131). Two C-terminal Pro residues in yeast Sod1 preclude Ccs1-independent activation. Replacement of the yeast Pro residues restores Ccs1-independent activation, and insertion of Pro residues at the corresponding positions in human Sod1 abrogates Ccs1-independent activation (133). The Ccs1-independent pathway requires GSH, although the role of GSH in Cu insertion and/or disulfide bond formation is unclear. Because Sod1 proteins lacking the C-terminal Pro residues are Cu activated in *ccs1*Δ yeast cells (116, 131), there must be some bioavailable Cu(I) complexes in the yeast cytosol. The identity of the bioavailable Cu(I) complex is a major unresolved question. Cu metallation of Sod1 by such complexes is less efficient than by Cu-Ccs1, but once it occurs, disulfide bond formation also proceeds. The facile oxidation of worm Sod1 in the absence of Ccs1 implies a secondary pathway (131). Disulfide bond formation is typically a facilitated process, but sulfhydryl oxidases are not known in the cytoplasm.

Activation of Superoxide Dismutase 1 in the Mitochondrial Intermembrane Space by Ccs1

Ccs1 is key to activation of Sod1 within the mitochondrial IMS, where 1% to 5% of Sod1 resides (**Figure 2**). The presence of Sod1 within the IMS is largely dependent on Ccs1 (134). In the absence of Ccs1, only minimal levels of Sod1 are apparent in the IMS (131). The limiting factor for the import of Ccs1 into the IMS is Mia40 (135–137). Overexpression of Mia40 in yeast or mammalian cells enhances IMS levels of Ccs1, and depletion of Mia40 has the opposite effect (136, 137). Import and capture of Ccs1 in the IMS are likely achieved through disulfide bonding to Mia40, followed by transfer of the disulfide to Ccs1 either in domain 1 or 3 (137). Two factors question whether oxidative folding of Ccs1 is the major mechanism of its retention within the IMS. First, it is unclear whether disulfide bonding within the CXXC in domain 1 or CXC in domain 3 would be sufficient to trap Ccs1 within the IMS as the cross-links would not stabilize disperse segments of the Ccs1 polypeptide to sterically block diffusion back out through the TOM translocase. Second, the stability of those candidate disulfides may be low because of unsatisfactory spacing between Cys residues (138). This also supports a transient disulfide introduced in domain 1 of Ccs1 upon IMS import. However, retention of Ccs1 in the IMS must involve general folding of domains 1 and 2. Because the Zn site in domain 2 is not conserved in all Ccs1 orthologs, metal binding is not an obvious trigger for IMS folding.

If transient disulfides form in domain 1 of Ccs1, it is unlikely this disulfide is transferred initially to imported Sod1 as Cu insertion must precede disulfide bond formation in Sod1. It is conceivable that Cu(I) insertion into Sod1 is mediated by domain 3 of Ccs1 and that the disulfide is subsequently transferred from the Ccs1 domain 1. Sod1 is not an apparent substrate of the Mia40/Erv1 system (137), so disulfide bond formation in Sod1 must occur through a Ccs1-dependent pathway as in the cytoplasm. Human Sod1 contains two additional Cys residues (C6 and C111), and neither of these is involved in a disulfide bond, yet retention of human Sod1 in the mitochondria of cultured cells is dependent on both (136). Cys111 is near the dimer interface, and the distance between the two Cys side chains is about 10 Å. Cys111 was reported to be part of a nonnative Cu(II)-binding site, leading to speculation that it may mediate Cu insertion into the active site (139).

Key Features of Ccs1-Mediated Superoxide Dismutase 1 Activation

Ccs1 binds Cu(I) in either domains 1 or 3. Domain 1 binding must necessitate transfer to the domain 3 CXC motif for subsequent transfer to Sod1. Cu-charged Ccs1 interacts with newly synthesized Sod1 polypeptides prior to completion of its native fold. The transient interaction is mediated by domains 2 and 3 of Ccs1. Zn(II) binding to Sod1 can occur either prior to Ccs1 complex formation or concomitantly. Cu(I) transfer occurs from Cu(I) associated in the domain 3 CXC motif. The only characterized Cu(I)-bound state in domain 3 is the binuclear Cu(I) thiolate cluster bridging a Ccs1 dimer. A second candidate Cu(I) donor is a mononuclear digonal Cu(I) complex, but this state has not been reported to date. Cu(I) transfer to Sod1 is accompanied by disulfide bond formation in an oxygen- or superoxide-dependent manner. Major unresolved questions are the mechanism by which the disulfide bond is formed and the sequence of Cu(I)-ligand exchange. The major questions in the Ccs1-independent pathway of Sod1 activation include the identity of the bioavailable copper pool(s) used in Sod1 metallation and the mechanism of disulfide bond formation.

ADDITIONAL TARGETS FOR COPPER METALLOCHAPERONES

Atx1 is implicated in copper supply to amine oxidase 1 (Cao1) in *Schizosaccharomyces pombe*, which has been localized (as a green fluorescent protein fusion) to the cytosol rather than the Golgi or some other internal membranous compartment (140). Of the copper metallochaperones, only disruption of Atx1 caused a loss of enzymatic activity, and a yeast two-hybrid interaction has been detected between Atx1 and Cao1. *Arabidopsis thaliana* has a second Atx1, CCH, with an extra C-terminal domain that is inhibitory to its interactions with the P_1-type ATPases HMA5 (4, 141–143). CCH is expressed in the vicinity of xylem and phloem elements, and one hypothesis is that the C-terminal region enables movement through plasmodesmata, suggesting a role for CCH in systemic Cu trafficking.

Mutants of *E. hirae* missing CopZ show low expression of the *cop* operon, including *copA* and *copB*, plus hypersensitivity to copper resulting from insufficient copper-exporting P_1-type ATPase activity (8). If CopZ solely donated Cu(I) ions to a copper exporter, then intracellular copper levels would be elevated in Δ*copZ*, and copper-responsive expression of the *cop* operon would be predicted to become higher. An alternate target for CopZ has been suggested. Transcriptional regulation of the *cop* operon is mediated by the DNA-binding repressor CopY. Purified CopY contains one Zn(II) ion, which can be displaced by two copper ions to form a Cu(I)-thiolate cluster (144, 145). Binding of CopY to nucleotide sequences from the *cop* operator promoter region is impaired by the replacement of zinc with copper. The second $\beta\alpha\beta\beta\alpha\beta$-domain of the Menkes ATPases expressed in isolation is incapable of donating Cu(I) to CopY, although a variant in which four residues are replaced with lysines does donate (145). This supports the idea of direct protein contact mediated by complementary electrostatic surfaces on CopZ and CopY. As with analogous Cu(I) transfer experiments, the metal might not attain a similar equilibrium by exchanging via free solution owing to its kinetic properties, notably the copper metallochaperone off rate for Cu(I), that is unique to the heterodimer interface.

Atox1 can affect the transcriptional control of cyclin D1 and mouse embryonic fibroblast proliferation (146). A yeast two-hybrid interaction also occurs between Ccs1 and the β-secretase (BACE1) that cleaves the amyloid precursor protein and generates the Aβ peptides that aggregate in Alzheimer's disease senile plaques. Ccs1 and BACE1 can be coimmunoprecipitated in extracts from rat brains (147). A short cytosolic tail of BACE1 contains three Cys residues that bind Cu(I), and this same short region associates with domain 1 of Ccs1. Fluorescent fusions of the two proteins track together along axons of neuronal cells, heading toward the synapse, implying that Ccs1 forms a stable association with secretory vesicles containing BACE1. Perhaps the intermolecular interaction between domain 1 of Ccs1 and BACE1's tail is structurally similar to transient intramolecular interactions with domain 3 of Ccs1. Some detected copper metallochaperone interactions might not be physiological, but aberrations linked to grave pathologies demand consideration.

SUMMARY POINTS

1. The switch between cuprous and cupric ions is exploited in oxidoreductases but can also catalyze deleterious redox chemistry, for example via the Fenton reaction.
2. Cu(II) is at the top of the Irving-Williams series, and Cu(I) also has a tendency to form tight thiol complexes; hence copper is liable to outcompete other metals in metalloproteins.
3. The cytosol contains numerous potential copper-binding sites, and the estimated available copper concentration is negligible, probably subfemtomolar.
4. Copper is passed to cuproproteins by ligand-exchange reactions from copper metallochaperones to avoid copper becoming kinetically trapped, or engaging in deleterious interactions while en route with docking between a metallochaperone and its partner, presumed to encourage the release of copper from the metallochaperone.
5. Solution structural studies have visualized transient complexes between N-terminal domains of P_1-type ATPases and Atx1-type copper metallochaperones to reveal changes predicted to aid Cu(I) release and to reveal interacting surfaces with complementary electrostatics.
6. Interactions between Atx1-type copper metallochaperones and N termini of P_1-type ATPase are regulatory, and Cu(I) can be separately donated to ATPase membrane sites.
7. Cox17, Sco1, and Cox11 in the mitochondrial IMS form two copper supply pathways for Cu_A and Cu_B, sites of cytochrome oxidase.
8. Ccs1 introduces an essential disulfide bond into Sod1 and supplies Cu(I) to Sod1 both in the cytosol and in the mitochondrial IMS, although several Sod1 species can acquire Cu(I) and the critical disulfide without Ccs1 provided GSH is present.

FUTURE ISSUES

1. Studies over the coming years should aim to further visualize heterodimer complexes between Ccs1 and Sod1, the mitochondrial copper metallochaperones for cytochrome oxidase and their partners, along with CopZ and membranous regions of P_1-type ATPases.
2. The effect that docking between a metallochaperone and its partner has on K_{OFF} for Cu(I) from the metallochaperone needs to quantified, perhaps by stopped-flow methods.
3. In addition to determining the sequence of Cu(I)-ligand exchange, proton and electron migration that prepares recipient Cys residues for Cu(I) awaits a full description, as does the coupling of Cu(I) transfer from Ccs1 to Sod1 with disulfide bond formation in Sod1.
4. Additional twin CX_9C proteins within the IMS contribute to CcO biogenesis, including Cox19, Cox23, and Pet191, but their actions remain to be defined.
5. The sources of copper for each metallochaperone remain to be unequivocally defined, the low-molecular-weight mitochondrial matrix copper ligand needs to be structurally characterized, and mechanisms that prioritize different cuproprotein destinations are yet to be discovered.

DISCLOSURE STATEMENT

The authors are not aware of any affiliations, memberships, funding, or financial holdings that might be perceived as affecting the objectivity of this review.

ACKNOWLEDGMENTS

The authors are supported by the Biotechnology and Biological Sciences Research Council (BB/E001688/1) to N.J.R. and grant ES 03817 from the National Institutes of Environmental Health Sciences, NIH to D.R.W.

LITERATURE CITED

1. Huffman DL, O'Halloran TV. 2001. Function, structure, and mechanism of intracellular copper trafficking proteins. *Annu. Rev. Biochem.* 70:677–701
2. Ridge PG, Zhang Y, Gladyshev VN. 2008. Comparative genomic analyses of copper transporters and cuproproteomes reveal evolutionary dynamics of copper utilization and its link to oxygen. *PLoS ONE* 3:e1378
3. Andreini C, Banci L, Bertini I, Rosato A. 2008. Occurrence of copper proteins through the three domains of life: a bioinformatic approach. *J. Proteome Res.* 7:209–16
4. Burkhead JL, Reynolds KA, Abdel-Ghany SE, Cohu CM, Pilon M. 2009. Copper homeostasis. *New. Phytol.* 182:799–816
5. Merchant SS, Allen MD, Kropat J, Moseley JL, Long JC, et al. 2006. Between a rock and a hard place: trace element nutrition in Chlamydomonas. *Biochim. Biophys Acta.* 1763:578–94
6. Dudev T, Lim C. 2008. Metal binding affinity and selectivity in metalloproteins: insights from computational studies. *Annu. Rev. Biophys.* 37:97–116
7. Tottey S, Waldron KJ, Firbank SJ, Reale B, Bessant C, et al. 2008. Protein-folding location can regulate manganese-binding versus copper- or zinc-binding. *Nature* 455:1138–42
8. Odermatt A, Solioz M. 1995. Two *trans*-acting metalloregulatory proteins controlling expression of the copper-ATPases of *Enterococcus hirae*. *J. Biol. Chem.* 270:4349–54

9. Lin SJ, Culotta VC. 1995. The ATX1 gene of *Saccharomyces cerevisiae* encodes a small metal homeostasis factor that protects cells against reactive oxygen toxicity. *Proc. Natl. Acad. Sci. USA* 92:3784–88
10. Zhang Y, Gladyshev VN. 2009. Comparative genomics of trace elements: emerging dynamic view of trace element utilization and function. *Chem. Rev.* 109:4828–61
11. Wernimont AK, Huffman DL, Lamb AL, O'Halloran TV, Rosenzweig AC. 2000. Structural basis for copper transfer by the metallochaperone for the Menkes/Wilson disease proteins. *Nat. Struct. Biol.* 7:766–71
12. Banci L, Bertini I, Cantini F, Felli IC, Gonnelli L, et al. 2006. The Atx1-Ccc2 complex is a metal-mediated protein-protein interaction. *Nat. Chem. Biol.* 2:367–68
13. Portnoy ME, Rosenzweig AC, Rae T, Huffman DL, O'Halloran TV, Culotta VC. 1999. Structure-function analyses of the ATX1 metallochaperone. *J. Biol. Chem.* 274:15041–45
14. Huffman DL, O'Halloran TV. 2000. Energetics of copper trafficking between the Atx1 metallochaperone and the intracellular copper transporter, Ccc2. *J. Biol. Chem.* 275:18611–14
15. Wimmer R, Herrmann T, Solioz M, Wuthrich K. 1999. NMR structure and metal interactions of the CopZ copper chaperone. *J. Biol. Chem.* 274:2597–603
16. Zhou L, Singleton C, Le Brun NE. 2008. High Cu(I) and low proton affinities of the CXXC motif of *Bacillus subtilis* CopZ. *Biochem. J.* 413:459–65
17. Banci L, Bertini I, Ciofi-Baffoni S, Del Conte R, Gonnelli L. 2003. Understanding copper trafficking in bacteria: interaction between the copper transport protein CopZ and the N-terminal domain of the copper ATPase CopA from *Bacillus subtilis*. *Biochemistry* 42:1939–49
18. Banci L, Bertini I, Ciofi-Baffoni S, Kandias NG, Robinson NJ, et al. 2006. The delivery of copper for thylakoid import observed by NMR. *Proc. Natl. Acad. Sci. USA* 103:8320–25
19. Radford DS, Kihlken MA, Borrelly GP, Harwood CR, Le Brun NE, Cavet JS. 2003. CopZ from *Bacillus subtilis* interacts in vivo with a copper exporting CPx-type ATPase CopA. *FEMS Microbiol. Lett.* 220:105–12
20. Banci L, Bertini I, Ciofi-Baffoni S, Finney LA, Outten CE, O'Halloran TV. 2002. A new zinc-protein coordination site in intracellular metal trafficking: solution structure of the apo and Zn(II) forms of ZntA (46–118). *J. Mol. Biol.* 323:883–97
21. Banci L, Bertini I, Ciofi-Baffoni S, Poggi L, Vanarotti M, et al. 2010. NMR structural analysis of the soluble domain of ZiaA-ATPase and the basis of selective interactions with copper metallochaperone Atx1. *J. Biol. Inorg. Chem.* 15:87–98
22. Tottey S, Rondet SA, Borrelly GP, Robinson PJ, Rich PR, Robinson NJ. 2002. A copper metallochaperone for photosynthesis and respiration reveals metal-specific targets, interaction with an importer, and alternative sites for copper acquisition. *J. Biol. Chem.* 277:5490–97
23. Tottey S, Rich PR, Rondet SAM, Robinson NJ. 2001. Two Menkes-type ATPases supply copper for photosynthesis in *Synechocystis* PCC 6803. *J. Biol. Chem.* 276:19999–20004
24. Kanamaru K, Kashiwagi S, Mizuno T. 1994. A copper-transporting P-type ATPase found in the thylakoid membrane of the cyanobacterium *Synechococcus* species PCC7942. *Mol. Microbiol.* 13:369–77
25. Setty SR, Tenza D, Sviderskaya EV, Bennett DC, Raposo G, Marks MS. 2008. Cell-specific ATP7A transport sustains copper-dependent tyrosinase activity in melanosomes. *Nature* 454:1142–46
26. González-Guerrero M, Arguello JM. 2008. Mechanism of Cu^+-transporting ATPases: soluble Cu^+ chaperones directly transfer Cu^+ to transmembrane transport sites. *Proc. Natl. Acad. Sci. USA* 105:5992–97
27. Wu CC, Rice WJ, Stokes DL. 2008. Structure of a copper pump suggests a regulatory role for its metal-binding domain. *Structure* 16:976–85
28. Gonzalez-Guerrero M, Hong D, Arguello JM. 2009. Chaperone-mediated Cu^+ delivery to Cu^+ transport ATPases: requirement of nucleotide binding. *J. Biol. Chem.* 284:20804–11
29. Sazinsky MH, LeMoine B, Orofino M, Davydov R, Bencze KZ, et al. 2007. Characterization and structure of a Zn^{2+} and [2Fe-2S]-containing copper chaperone from *Archaeoglobus fulgidus*. *J. Biol. Chem.* 282:25950–59
30. Tsivkovskii R, MacArthur BC, Lutsenko S. 2001. The Lys1010-Lys1325 fragment of the Wilson's disease protein binds nucleotides and interacts with the N-terminal domain of this protein in a copper-dependent manner. *J. Biol. Chem.* 276:2234–42

31. Petris MJ, Mercer JFB, Culvenor JG, Lockhart P, Gleeson PA, Camakaris J. 1996. Ligand-regulated transport of the Menkes copper P-type ATPase efflux pump from the Golgi apparatus to the plasma membrane: a novel mechanism of regulated trafficking. *EMBO J.* 15:6084–95
32. Voskoboinik I, Strausak D, Greenough M, Brooks H, Petris M, et al. 1999. Functional analysis of the N-terminal CXXC metal-binding motifs in the human Menkes copper-transporting P-type ATPase expressed in cultured mammalian cells. *J. Biol. Chem.* 274:22008–12
33. Guo Y, Nyasae L, Braiterman LT, Hubbard AL. 2005. NH_2-terminal signals in ATP7B Cu-ATPase mediate its Cu-dependent anterograde traffic in polarized hepatic cells. *Am. J. Physiol. Gastrointest. Liver Physiol.* 289:G904–16
34. Hussain F, Olson JS, Wittung-Stafshede P. 2008. Conserved residues modulate copper release in human copper chaperone Atox1. *Proc. Natl. Acad. Sci. USA* 105:11158–63
35. Portnoy ME, Schmidt PJ, Rogers RS, Culotta VC. 2001. Metal transporters that contribute copper to metallochaperones in *Saccharomyces cerevisiae*. *Mol. Genet. Genomics* 265:873–82
36. De Feo CJ, Aller SG, Siluvai GS, Blackburn NJ, Unger VM. 2009. Three-dimensional structure of the human copper transporter hCTR1. *Proc. Natl. Acad. Sci. USA* 106:4237–42
37. Xiao Z, Wedd AG. 2002. A C-terminal domain of the membrane copper pump Ctr1 exchanges copper(I) with the copper chaperone Atx1. *Chem. Commun.* 6:588–89
38. Xiao Z, Loughlin F, George GN, Howlett GJ, Wedd AG. 2004. C-terminal domain of the membrane copper transporter Ctr1 from *Saccharomyces cerevisiae* binds four Cu(I) ions as a cuprous-thiolate polynuclear cluster: sub-femtomolar Cu(I) affinity of three proteins involved in copper trafficking. *J. Am. Chem. Soc.* 126:3081–90
39. Loftin IR, Franke S, Blackburn NJ, McEvoy MM. 2007. Unusual Cu(I)/Ag(I) coordination of *Escherichia coli* CusF as revealed by atomic resolution crystallography and X-ray absorption spectroscopy. *Protein Sci.* 16:2287–93
40. Loftin IR, Franke S, Roberts SA, Weichsel A, Heroux A, et al. 2005. A novel copper-binding fold for the periplasmic copper resistance protein CusF. *Biochemistry* 44:10533–40
41. Xue Y, Davis AV, Balakrishnan G, Stasser JP, Staehlin BM, et al. 2008. Cu(I) recognition via cation-pi and methionine interactions in CusF. *Nat. Chem. Biol.* 4:107–9
42. Davis AV, O'Halloran TV. 2008. A place for thioether chemistry in cellular copper ion recognition and trafficking. *Nat. Chem. Biol.* 4:148–51
43. Bagai I, Rensing C, Blackburn NJ, McEvoy MM. 2008. Direct metal transfer between periplasmic proteins identifies a bacterial copper chaperone. *Biochemistry* 47:11408–14
44. Tsukihara T, Aoyama H, Yamashita E, Tomizaki T, Yamaguchi H, et al. 1995. Structures of metal sites of oxidized bovine heart cytochrome c oxidase at 2.8 A. *Science* 269:1069–74
45. Glerum DM, Shtanko A, Tzagoloff A. 1996. Characterization of *COX17*, a yeast gene involved in copper metabolism and assembly of cytochrome oxidase. *J. Biol. Chem.* 271:14504–9
46. Takahashi Y, Kako K, Kashiwabara S-I, Takehara A, Inada Y, et al. 2002. Mammalian copper chaperone Cox17 has an essential role in activation of cytochrome *c* oxidase and embryonic development. *Mol. Cell. Biol.* 22:7614–21
47. Lee J, Prohaska JR, Thiele DJ. 2001. Essential role for mammalian copper transporter Ctr1 in copper homeostasis and embryonic development. *Proc. Natl. Acad. Sci. USA* 98:6842–47
48. Oswald C, Krause-Buchholz U, Rödel G. 2009. Knockdown of human COX17 affects assembly and supramolecular organization of cytochrome c oxidase. *J. Mol. Biol.* 389:470–79
49. Banci L, Bertini I, Ciofi-Baffoni S, Tokatlidis K. 2009. The coiled coil-helix-coiled coil-helix proteins may be redox proteins. *FEBS Lett.* 583:1699–702
50. Abajian C, Yatsunyk LA, Ramirez BE, Rosenzweig AC. 2004. Yeast Cox17 solution structure and copper(I) binding. *J. Biol. Chem.* 279:53584–92
51. Arnesano F, Balatri E, Banci L, Bertini I, Winge DR. 2005. Folding studies of Cox17 reveal an important interplay of cysteine oxidase and copper binding. *Structure* 13:713–22
52. Hu J, Dong L, Outten CE. 2008. The redox environment in the mitochondrial intermembrane space is maintained separately from the cytosol and matrix. *J. Biol. Chem.* 283:29126–34
53. Palumaa P, Zovo K. 2008. Modulation of redox switches of copper chaperone Cox17 by Zn(II) ions, determined by new ESI MS based approach. *Antioxid. Redox Signal.* 11:985–95

54. Heaton D, Nittis T, Srinivasan C, Winge DR. 2000. Mutational analysis of the mitochondrial copper metallochaperone Cox17. *J. Biol. Chem.* 275:37582–87
55. Banci L, Bertini I, Ciofi-Baffoni S, Janicka A, Martinelli M, et al. 2008. A structural-dynamical characterization of human Cox17. *J. Biol. Chem.* 283:7912–20
56. Heaton DN, George GN, Garrison G, Winge DR. 2001. The mitochondrial copper metallochaperone Cox17 exists as an oligomeric polycopper complex. *Biochem.* 40:743–51
57. Changela A, Chen K, Xue Y, Holschen J, Outten CE, et al. 2003. Molecular basis of metal-ion selectivity and zeptomolar sensitivity by CueR. *Science* 301:1383–87
58. Beers J, Glerum DM, Tzagoloff A. 1997. Purification, characterization, and localization of yeast Cox17p, a mitochondrial copper shuttle. *J. Biol. Chem.* 272:33191–96
59. Banci L, Bertini I, Ciofi-Baffoni S, Leontari I, Martinelli M, et al. 2007. Human Sco1 functional studies and pathological implications of the P174L mutant. *Proc. Natl. Acad. Sci. USA* 104:15–20
60. Rissler M, Wiedemann N, Pfannschmidt S, Gabriel K, Guiard B, et al. 2005. The essential mitochondrial protein Erv1 cooperates with Mia40 in biogenesis of intermembrane proteins. *J. Mol. Biol.* 353:485–92
61. Chacinska A, Pfannschmidt S, Wiedemann N, Kozjak V, Sanjuan Szklarz LK, et al. 2004. Essential role of Mia40 in import and assembly of mitochondrial intermembrane space proteins. *EMBO J.* 23:3735–46
62. Allen S, Balabanidou V, Sideris DP, Lisowsky T, Tokatidis K. 2005. Erv1 mediates the Mia40-dependent protein import pathway and provides a functional link to the respiratory chain by shuttling electrons to cytochrome *c*. *J. Mol. Biol.* 353:937–44
63. Tokatlidis K. 2005. A disulfide relay system in mitochondria. *Cell* 121:965–67
64. Terziyska N, Grumbt B, Bien M, Neupert W, Herrmann JM, Hell K. 2007. The sulfhydryl oxidase Erv1 is a substrate of the Mia40-dependent protein translocation pathway. *FEBS Lett.* 581:1098–102
65. Banci L, Bertini I, Cefaro C, Ciofi-Baffoni S, Gallo A, et al. 2009. MIA40 is an oxidoreductase that catalyzes oxidative protein folding in mitochondria. *Nat. Struct. Mol. Biol.* 16:198–206
66. Mesecke N, Terziyska N, Kozany C, Baumann F, Neupert W, et al. 2005. A disulfide relay system in the intermembrane space of mitochondria that mediates protein import. *Cell* 121:1059–69
67. Terziyska N, Lutz T, Kozany C, Mokranjac D, Mesecke N, et al. 2005. Mia40, a novel factor for protein import into the intermembrane space of mitochondria is able to bind metal ions. *FEBS Lett.* 579:179–84
68. Maxfield AB, Heaton DN, Winge DR. 2004. Cox17 is functional when tethered to the mitochondrial inner membrane. *J. Biol. Chem.* 279:5072–80
69. Cobine PA, Ojeda LD, Rigby KM, Winge DR. 2004. Yeast contain a non-proteinaceous pool of copper in the mitochondrial matrix. *J. Biol. Chem.* 279:14447–55
70. Cobine PA, Pierrel F, Bestwick ML, Winge DR. 2006. Mitochondrial matrix copper complex used in metallation of cytochrome oxidase and superoxide dismutase. *J. Biol. Chem.* 281:36552–59
71. Scheckhuber CQ, Grief J, Boilan E, Luce K, Debacq-Chainiaux F, et al. 2009. Age-related cellular copper dynamics in the fungal ageing model *Podospora anserina* and in ageing human fibroblasts. *PLoS ONE* 4:e4919
72. Glerum DM, Shtanko A, Tzagoloff A. 1996. *SCO1* and *SCO2* act as high copy suppressors of a mitochondrial copper recruitment defect in *Saccharomyces cerevisiae*. *J. Biol. Chem.* 271:20531–35
73. Schulze M, Rödel G. 1988. *SCO1*, a yeast nuclear gene essential for accumulation of mitochondrrial cytochrome *c* oxidase subunit II. *Mol. Gen. Genet.* 211:492–98
74. Krummeck G, Rödel G. 1990. Yeast SCO1 protein is required for a post-translational step in the accumulation of mitochondrial cytochrome *c* oxidase subunits I and II. *Curr. Genet.* 18:13–15
75. Leary SC, Kaufman BA, Pellechia G, Gguercin G-H, Mattman A, et al. 2004. Human SCO1 and SCO2 have independent, cooperative functions in copper delivery to cytochrome *c* oxidase. *Hum. Mol. Genet.* 13:1839–48
76. Shoubridge EA. 2001. Cytochrome *c* oxidase deficiency. *Am. J. Med. Genet.* 106:46–52
77. Stiburek L, Vesela K, Hansikova H, Hulkova H, Zeman J. 2009. Loss of function of Sco1 and its interaction with cytochrome *c* oxidase. *Am. J. Physiol. Cell Physiol.* 296:C1218–26
78. Leary SC, Cobine PA, Kaufman BA, Guercin GH, Mattman A, et al. 2007. The human cytochrome c oxidase assembly factors SCO1 and SCO2 have regulatory roles in the maintenance of cellular copper homeostasis. *Cell Metab.* 5:9–20

79. Rentzsch N, Krummeck-Weiß G, Hofer A, Bartuschka A, Ostermann K, Rödel G. 1999. Mitochondrial copper metabolism in yeast: mutational analysis of Sco1p invovled in the biogenesis of cytochrome *c* oxidase. *Curr. Genet.* 35:103–8
80. Nittis T, George GN, Winge DR. 2001. Yeast Sco1, a protein essential for cytochrome *c* oxidase function is a Cu(I)-binding protein. *J. Biol. Chem.* 276:42520–26
81. Banci L, Bertini I, Calderone V, Ciofi-Baffoni S, Mangani S, et al. 2006. A hint for the function of human Sco1 from different structures. *Proc. Natl. Acad. Sci. USA* 103:8595–600
82. Williams JC, Sue C, Banting GS, Yang H, Glerum DM, et al. 2005. Crystal structure of human SCO1: implications for redox signaling by a mitochondrial cytochrome *c* oxidase "assembly" protein. *J. Biol. Chem.* 280:15202–11
83. Abajian C, Rosenzweig AC. 2006. Cystal structure of yeast Sco1. *J. Biol. Inorg. Chem.* 11:459–66
84. Banci L, Bertini I, Calderone V, Ciofi-Baffoni S, Mangani S, et al. 2006. A hint for the function of human Sco1 from different strucures. *Proc. Natl. Acad. Sci. USA* 103:8595–600
85. Banci L, Bertini I, Ciofi-Baffoni S, Gerothanassis IP, Leontari I, et al. 2007. A structural characterization of human SCO2. *Structure* 15:1132–40
86. Leary SC, Sasarman F, Nishimura T, Shoubridge EA. 2009. Human SCO2 is required for the synthesis of CO II and as a thiol-disulphide oxidoreductase for SCO1. *Hum. Mol. Genet.* 18:2230–40
87. Su D, Berndt C, Fomenko DE, Holmgren A, Gladyshev VN. 2007. Conserved cis-proline precludes metal binding by the active site thiolates in members of the thioredoxin family of proteins. *Biochemistry* 46:6903–10
88. Riemer J, Bulleid N, Herrmann JM. 2009. Disulfide formation in the ER and mitochondria: two solutions to a common process. *Science* 324:1284–87
89. Horng YC, Cobine PA, Maxfield AB, Carr HS, Winge DR. 2004. Specific copper transfer from the Cox17 metallochaperone to both Sco1 and Cox11 in the assembly of yeast cytochrome *c* oxidase. *J. Biol. Chem.* 279:35334–40
90. Valnot I, Osmond S, Gigarel N, Mehaye B, Amiel J, et al. 2000. Mutations of the *SCO1* gene in mitochondrial cytochrome *c* oxidase deficiency with neonatal-onset hepatic failure and encephalopathy. *Am. J. Hum. Genet.* 67:1104–9
91. Paret C, Lode A, Krause-Buchholz U, Rödel G. 2000. The P(174)L mutation in the human *hSCO1* gene affects the assembly of cytochrome *c* oxidase. *Biochem. Biophys. Res. Commun.* 279:341–47
92. Cobine PA, Pierrel F, Leary SC, Sasarman F, Horng YC, et al. 2006. The P174L mutation in human Sco1 severely compromises Cox17-dependent metallation but does not impair copper binding. *J. Biol. Chem.* 281:12270–76
93. Banci L, Bertini I, Ciofi-Baffoni S, Hadjiloi T, Martinelli M, Palumaa P. 2008. Mitochondrial copper(I) transfer from Cox17 to Sco1 is coupled to electron transfer. *Proc. Natl. Acad. Sci. USA* 105:6803–8
94. Horng YC, Leary SC, Cobine PA, Young FB, George GN, et al. 2005. Human Sco1 and Sco2 function as copper-binding proteins. *J. Biol. Chem.* 280:34113–22
95. Rigby K, Cobine PA, Khalimonchuk O, Winge DR. 2008. Mapping the functional interaction of Sco1 and Cox2 in cytochrome oxidase biogenesis. *J. Biol. Chem.* 283:15015–22
96. Banci L, Bertini I, Cavallaro G, Rosato A. 2007. The functions of Sco proteins from genome-based analysis. *J. Proteome Res.* 6:1568–79
97. Swem DL, Swem LR, Setterdahl A, Bauer CE. 2005. Involvement of SenC in assembly of cytochrome *c* oxidase in *Rhodobacter capsulatus*. *J. Bacteriol.* 187:8081–87
98. Davidson DE, Hill BC. 2009. Stability of oxidized, reduced and copper bound forms of *Bacillus subtilis* Sco. *Biochim. Biophys. Acta.* 1794:275–81
99. Cawthorn TR, Poulsen BE, Davidson DE, Andrews D, Hill BC. 2009. Probing the kinetics and thermodynamics of copper(II) binding to *Bacillus subtilis* Sco, a protein involved in the assembly of the Cu(A) center of cytochrome *c* oxidase. *Biochemistry* 48:4448–54
100. Banci L, Bertini I, Ciofi-Baffoni S, Katsari E, Katsaros N, et al. 2005. A copper(I) protein possibly involved in the assembly of Cu_A center of bacterial cytochrome *c* oxidase. *Proc. Natl. Acad. Sci. USA* 102:3994–99
101. Abriata LA, Banci L, Bertini I, Ciofi-Baffoni S, Gkazonis P, et al. 2008. Mechanism of Cu(A) assembly. *Nat. Chem. Biol.* 4:599–601

102. Tzagoloff A, Capitanio N, Nobrega MP, Gatti D. 1990. Cytochrome oxidase assembly in yeast requires the product of COX11, a homolog of the *P. denitrificans* protein encoded by ORF3. *EMBO J.* 9:2759–64
103. Hiser L, Di Valentin M, Hamer AG, Hosler JP. 2000. Cox11 is required for stable formation of the Cu_B and magnesium centers of cytochrome *c* oxidase. *J. Biol. Chem.* 275:619–23
104. Carr HS, George GN, Winge DR. 2002. Yeast Cox11, a protein essential for cytochrome *c* oxidase assembly, is a Cu(I) binding protein. *J. Biol. Chem.* 277:31237–42
105. Carr HS, Maxfield AB, Horng Y-C, Winge DR. 2005. Functional analysis of the domains of Cox11. *J. Biol. Chem.* 280:22664–69
106. Banci L, Bertini I, Cantini F, Ciofi-Baffoni S, Gonnelli L, Mangani S. 2004. Solution structure of Cox11: a novel type of β-immunoglobulin-like fold involved in Cu_B site formation of cytochrome *c* oxidase. *J. Biol. Chem.* 279:34833–39
107. Smith D, Gray J, Mitchell L, Antholine WE, Hosler JP. 2005. Assembly of cytochrome *c* oxidase in the absence of the assembly protein Surf1p leads to loss of the active site heme. *J. Biol. Chem.* 280:17652–56
108. Bundschuh FA, Hoffmeier K, Ludwig B. 2008. Two variants of the assembly factor Surf1 target specific terminal oxidases in *Paracoccus denitrificans*. *Biochim. Biophys. Acta* 1777:1336–43
109. Khalimonchuk O, Bird A, Winge DR. 2007. Evidence for a pro-oxidant intermediate in the assembly of cytochrome oxidase. *J. Biol. Chem.* 282:17442–49
110. Culotta VC, Klomp LWJ, Strain J, Casareno RLB, Krems B, Gitlin JD. 1997. The copper chaperone for superoxide dismutase. *J. Biol. Chem.* 272:23469–72
111. Schmidt PJ, Kunst C, Culotta VC. 2000. Copper activation of superoxide dismutase 1 (SOD1) in vivo. *J. Biol. Chem.* 275:33771–76
112. Lamb AL, Torres AS, O'Halloran TV, Rosenzweig AC. 2000. Heterodimer formation between superoxide dismutase and its copper chaperone. *Biochemistry* 39:14720–27
113. Lamb AL, Torres AS, O'Halloran TV, Rosenzweig AC. 2001. Heterodimeric structure of superoxide dismutase in complex with its metallochaperone. *Nat. Struct. Biol.* 8:751–55
114. Schmidt PJ, Rae TD, Pufahl RA, Hamma T, Strain J, et al. 1999. Multiple protein domains contribute to the action of the copper chaperone for superoxide dismutase. *J. Biol. Chem.* 274:23719–25
115. Rae RD, Schmidt PJ, Pufahl RA, Culotta VC, O'Halloran TV. 1999. Undetectable intracellular free copper: the requirement of a copper chaperone for superoxide dismutase. *Science* 284:805–7
116. Kirby K, Jensen LT, Binnington J, Hilliker AJ, Ulloa J, et al. 2008. Instability of superoxide dismutase 1 of *Drosophila* in mutants deficient for its cognate copper chaperone. *J. Biol. Chem.* 283:35393–401
117. Wong PC, Waggoner D, Subramaniam JR, Tessarollo L, Bartnikas TB, et al. 2000. Copper chaperone for superoxide dismutase is essential to activate mammalian Cu/Zn superoxide dismutase. *Proc. Natl. Acad. Sci. USA* 97:2886–91
118. Furukawa Y, Torres AS, O'Halloran TV. 2004. Oxygen-induced maturation of SOD1: a key role for disulfide formation by the copper chaperone CCS. *EMBO J.* 23:2872–81
119. Fridovich I. 1997. Superoxide anion radical (O_2^-), superoxide dismutases and related matters. *J. Biol. Chem.* 272:18515–17
120. Tainer JA, Getzoff ED, Beem KM, Richardson JS, Richardson DC. 1982. Determination and analysis of the 2 A-structure of copper, zinc superoxide dismutase. *J. Mol. Biol.* 160:181–217
121. Tainer JA, Getzoff ED, Richardson JS, Richardson DC. 1983. Structure and mechanism of copper, zinc superoxide dismutase. *Nature* 306:284–87
122. Forman HJ, Fridovich I. 1973. On the stability of bovine superoxide dismutase. The effects of metals. *J. Biol. Chem.* 248:2645–49
123. Stasser JP, Siluvai GS, Barry AN, Blackburn NJ. 2007. A multinuclear copper(I) cluster forms the dimerization interface in copper-loaded human copper chaperone for superoxide dismutase. *Biochemistry* 46:11845–56
124. Caruano-Yzermans AL, Bartnikas TB, Gitlin JD. 2006. Mechanisms of the copper-dependent turnover of the copper chaperone for superoxide dismutase. *J. Biol. Chem.* 281:13581–87
125. Lamb AL, Wernimont AK, Pufahl RA, O'Halloran TV, Rosenzweig AC. 1999. Crystal structure of the copper chaperone for superoxide dismutase. *Nat. Struct. Biol.* 6:724–29
126. Lamb AL, Wernimont AK, Pufahl RA, O'Halloran TV, Rosenzweig AC. 2000. Crystal structure of the second domain of the human copper chaperone for superoxide dismutase. *Biochemistry* 39:1589–95

127. Winkler DD, Schuermann JP, Cao X, Holloway SP, Borchelt DR, et al. 2009. Structural and biophysical properties of the pathogenic SOD1 variant H46R/H48Q. *Biochemistry* 48:3436–47
128. Banci L, Bertini I, Cantini F, D'Onofrio M, Viezzoli MS. 2002. Structure and dynamics of copper-free SOD: the protein before binding copper. *Protein Sci.* 11:2479–92
129. Potter SZ, Zhu H, Shaw BF, Rodriguez JA, Doucette PA, et al. 2007. Binding of a single zinc ion to one subunit of copper-zinc superoxide dismutase apoprotein substantially influences the structure and stability of the entire homodimeric protein. *J. Am. Chem. Soc.* 129:4575–83
130. Proescher JB, Son M, Elliott JL, Culotta VC. 2008. Biological effects of CCS in the absence of SOD1 enzyme activation: implications for disease in a mouse model for ALS. *Hum. Mol. Genet.* 17:1728–37
131. Jensen LT, Culotta VC. 2005. Activation of CuZn superoxide dismutases from *Caenorhabditis elegans* does not require the copper chaperone CCS. *J. Biol. Chem.* 280:41373–79
132. Subramaniam JR, Lyons WE, Liu J, Bartnikas TB, Rothstein J, et al. 2002. Mutant SOD1 causes motor neuron disease independent of copper chaperone-mediated copper loading. *Nat. Neurosci.* 5:301–7
133. Carroll MC, Girouard JB, Ulloa JL, Subramaniam JR, Wong PC, et al. 2004. Mechanisms for activating Cu- and Zn-containing superoxide dismutase in the absence of the CCS Cu chaperone. *Proc. Natl. Acad. Sci. USA* 101:5964–69
134. Field LS, Furukawa Y, O'Halloran TV, Culotta VC. 2003. Factors controlling the uptake of yeast copper/zinc superoxide dismutase into mitochondria. *J. Biol. Chem.* 278:28052–59
135. Son M, Puttaparthi K, Kawamata H, Rajendran B, Boyer PJ, et al. 2007. Overexpression of CCS in G93 A-SOD1 mice leads to accelerated neurological deficits with severe mitochondrial pathology. *Proc. Natl. Acad. Sci. USA* 104:6072–77
136. Kawamata H, Manfredi G. 2008. Different regulation of wild-type and mutant Cu Zn superoxide dismutase localization in mammalian mitochondria. *Hum. Mol. Genet.* 17:3303–17
137. Reddehase S, Grumbt B, Neupert W, Hell K. 2009. The disulfide relay system of mitochondria is required for the biogenesis of mitochondrial Ccs1 and Sod1. *J. Mol. Biol.* 385:331–38
138. Zhang R, Snyder GH. 1991. Factors governing selective formation of specific disulfides in synthetic variants of α-conotoxin. *Biochemistry* 30:11343–48
139. Liu H, Zhu H, Eggers DK, Nersissian AM, Faull KF, et al. 2000. Copper^{2+} binding to the surface residue cysteine 111 of His46Arg human copper-zinc superoxide dismutase, a familial amyotrophic lateral sclerosis mutant. *Biochemistry* 39:8125–32
140. Peter C, Laliberte J, Beaudoin J, Labbe S. 2008. Copper distributed by Atx1 is available to copper amine oxidase 1 in *Schizosaccharomyces pombe*. *Eukaryot. Cell* 7:1781–94
141. Puig S, Penarrubia L. 2009. Placing metal micronutrients in context: transport and distribution in plants. *Curr. Opin. Plant Biol.* 12:299–306
142. Puig S, Mira H, Dorcey E, Sancenon V, Andres-Colas N, et al. 2007. Higher plants possess two different types of ATX1-like copper chaperones. *Biochem. Biophys. Res. Commun.* 354:385–90
143. Andres-Colas N, Sancenon V, Rodriguez-Navarro S, Mayo S, Thiele DJ, et al. 2006. The *Arabidopsis* heavy metal P-type ATPase HMA5 interacts with metallochaperones and functions in copper detoxification of roots. *Plant J.* 45:225–36
144. Cobine P, Wickramasinghe WA, Harrison MD, Weber T, Solioz M, Dameron CT. 1999. The *Enterococcus hirae* copper chaperone CopZ delivers copper(I) to the CopY repressor. *FEBS Lett.* 445:27–30
145. Cobine PA, George GN, Jones CE, Wickramasinghe WA, Solioz M, Dameron CT. 2002. Copper transfer from the Cu(I) chaperone, CopZ, to the repressor, Zn(II)CopY: metal coordination environments and protein interactions. *Biochemistry* 41:5822–29
146. Itoh S, Kim HW, Nakagawa O, Ozumi K, Lessner SM, et al. 2008. Novel role of antioxidant-1 (Atox1) as a copper dependent transcription factor involved in cell proliferation. *J. Biol. Chem.* 283:9157–67
147. Angeletti B, Waldron KJ, Freeman KB, Bawagan H, Hussain I, et al. 2005. BACE1 cytoplasmic domain interacts with the copper chaperone for superoxide dismutase-1 and binds copper. *J. Biol. Chem.* 280:17930–37

High-Throughput Metabolic Engineering: Advances in Small-Molecule Screening and Selection

Jeffrey A. Dietrich,[1,2,3] Adrienne E. McKee,[2,3] and Jay D. Keasling[1,2,3,4,5]

[1]UCSF-UCB Joint Graduate Group in Bioengineering, Berkeley, California 94720; email: jadietrich@berkeley.edu

[2]Synthetic Biology Department, Physical Biosciences Division, Lawrence Berkeley National Laboratory, Berkeley, California 94710; email: AEMcKee@lbl.gov

[3]DOE Joint BioEnergy Institute, Emeryville, California 94208

[4]Department of Chemical Engineering and [5]California Institute for Quantitative Biomedical Research, University of California, Berkeley, California 94720; email: keasling@berkeley.edu

Annu. Rev. Biochem. 2010. 79:563–90

First published online as a Review in Advance on April 2, 2010

The *Annual Review of Biochemistry* is online at biochem.annualreviews.org

This article's doi: 10.1146/annurev-biochem-062608-095938

0066-4154/10/0707-0563$20.00

Key Words

biosensors, directed evolution, FACS, synthetic biology, transcription factors

Abstract

Metabolic engineering for the overproduction of high-value small molecules is dependent upon techniques in directed evolution to improve production titers. The majority of small molecules targeted for overproduction are inconspicuous and cannot be readily obtained by screening. We provide a review on the development of high-throughput colorimetric, fluorescent, and growth-coupled screening techniques, enabling inconspicuous small-molecule detection. We first outline constraints on throughput imposed during the standard directed evolution workflow (library construction, transformation, and screening) and establish a screening and selection ladder on the basis of small-molecule assay throughput and sensitivity. An in-depth analysis of demonstrated screening and selection approaches for small-molecule detection is provided. Particular focus is placed on in vivo biosensor-based detection methods that reduce or eliminate in vitro assay manipulations and increase throughput. We conclude by providing our prospectus for the future, focusing on transcription factor-based detection systems as a natural microbial mode of small-molecule detection.

Contents

1. INTRODUCTION: THE CASE FOR SMALL-MOLECULE SCREENS

Natural selection, the force behind the amazing breadth of phenotypic variation in the living world, has long been a source of motivation for the engineering of synthetic biological systems. By mimicking the processes of mutation, recombination, and selection found in nature, directed evolution is used to impart industrial microbes with user-defined phenotypes. Frequently regarded as the First Law of Directed Evolution (1), the central tenet "you get what you screen for" draws attention to the paramount importance in finding an appropriate screen or selection assay to sift through vast libraries of variant hosts. A great body of work has been devoted to development of highly tailored assays specific for detection of single proteins or functions (2).

Throughput: total number of library variants screened per unit time (experiment or day)

As applied to metabolic engineering, directed evolution is focused on improving small-molecule biosynthesis. Improving product yields or pathway efficiencies, however, can be a daunting task. Only a small subset of targeted compounds are natural chromophores or fluorophores that can be readily screened for using standard assay techniques. The majority of small-molecule targets for overproduction today do not illicit a conspicuous phenotype. For inconspicuous targets, chromatography-mass spectrometry methods have been the primary mode of detection; although they are nearly ubiquitously applied in small-molecule detection and quantification, these assays are inherently low throughput (3). Screenable library sizes are generally limited to less than 10^3 variants. At this level of throughput, only a paltry number of rational modifications can be introduced into the panoply of host biosynthetic machinery, leaving the majority of sequence space untouched and unexplored.

Today, a discussion of improvements in small-molecule detection assays is set against the backdrop of increased focus in the field on microbial biosynthesis of commodity chemicals and fuels. Microbially produced alcohols, fatty acids, and alkanes are targeted for use as petroleum-derived fuel substitutes (4, 5), aromatics (6), and diols (7); and polyhydroxyalkanoates (8) are being targeted for use as bioplastics. Spurred by an increased demand for renewable, green alternatives to petroleum-derived production routes, microbial production routes are competing economically against entrenched industry stakeholders (9). Product yields and pathway efficiencies demanded from biosynthetic routes are pushing the limits of what can be achieved using existing metabolic engineering technologies. Efforts in the field are currently defined by the implementation of a series of rational design-based strategies to modify the host genome and heterologous pathway enzymes to achieve moderate product titers. However, metabolic engineering is a highly complex process, and product yields are dictated by a host of parameters. Biosynthetic pathways comprise multiple native and heterologous catalytic steps; each step is a potential bottleneck when directing carbon flux toward target small-molecule production. Furthermore, the host organism's native genetic network, regulation, and their interaction with the target pathway can all impact product yields. To achieve higher productivities, metabolic engineering must follow the decades-long trail of successes in protein engineering and develop more elegant approaches to high-throughput screening. Directed evolution through random and targeted mutagenesis of the host genome, overexpressed operon(s), and enzyme-encoded genes followed by high-throughput screening or selection is a requisite step in strain development.

In this review, we focus on advances in small-molecule screens using *Escherichia coli* and *Saccharomyces cerevisiae* as model prokaryotic and eukaryotic hosts. These organisms were selected both for their demonstrated application in industrial fermentation processes and because the vast majority of novel screening and selection processes are first demonstrated in these hosts. Given the immense number of mutable genetic elements to be targeted in metabolic pathway evolution, special consideration must be given to library design and sequence coverage. Screening efficacy has been demonstrated to increase in libraries that are maximally diverse, providing evidence for increased library size and mutation rate as methods for improving the diversity in the sample population (10). Technical limitations imposed during library generation, transformation, and screening technologies are addressed below, and we focus in particular on screening and selection as rate-limiting steps in directed evolution efforts for small-molecule overproduction. Throughout our discussion, we highlight novel biosensor-based assays that enable inconspicuous small-molecule detection.

2. STATISTICAL LIMITATIONS ON LIBRARY SIZE AND SEQUENCE SPACE COVERAGE

Any directed evolution experiment, regardless of target, requires a thoughtful analysis of the library size required to gain significant coverage of the targeted sequence space. For good reason, most directed evolution efforts use focused libraries, mutating a relatively small number of preselected positions for saturation mutagenesis. However, even the most straightforward efforts to introduce a small number of substitutions are subject to harsh statistical realities. Simultaneous alteration of n selected positions in a given sequence necessitates the creation and screening of a library of size L, according to Equation 1:

$$L = X^{n}, \text{ where } X = \begin{Bmatrix} 4 \text{ for, nucleotide substitution} \\ 20 \text{ for, amino acid substitution} \\ 2 \text{ for, genome (binary output)} \end{Bmatrix}. \qquad 1.$$

Here, X corresponds to the number of possible genetic elements or states that may be present. The four naturally occurring nucleotides and the 20 naturally occurring amino acids provide the set of states for standard DNA and protein mutagenesis, respectively. The number of states for genome modifications, two, is modeled on deletion and insertion libraries. Although our discussion here has focused predominantly on randomization of existing genetic elements, a similar analysis can be applied to other, equally as important diversity-generating techniques, including insertions (11, 12), deletions (11–13), and recombination (14, 15), among others.

Exhaustive sequence coverage when randomization is not targeted to specific positions, but is instead applied uniformly over the full length of a target sequence, is a difficult prospect. When there exists little-to-no basis for rationalized substitutions using structure elucidation, or otherwise, the researcher may opt to introduce multiple, random mutations across a sequence of length K. The binomial coefficient describes the number of possible variants, L, given N randomizations with X genotypic states. Library sizes now scale according to Equation 2:

$$L = X^n \frac{K!}{N!(K-N)!}. \qquad 2.$$

Exhaustive library coverage for an indicated sequence space differs tremendously between the focused and random mutagenesis approaches. For example, mutation of 2 positions in a 100-amino acid protein (i.e., $X = 20$) yields a library of 400 unique members when the amino acid positions have been preselected and 1.98×10^6 unique members when mutations are randomly incorporated over the length of the entire protein-coding sequence. The four order-of-magnitude difference in library size is of note, but underlying this analysis is the finding that most random mutations have neutral or deleterious effects on protein function (16). Presupposing that some knowledge of protein structure or function can be used to guide a focused mutagenesis strategy, screening efficiency can be dramatically improved. These arguments exemplify why the vast majority of protein engineering efforts possess either a strong screening/selection assay or introduce focused mutations on a small subset of the total sequence space.

Given its small number of genotypic states, statistically speaking, targeted mutagenesis of the genome appears to provide the most straightforward approach. However, a number of caveats must be considered. First, the simplifying assumption was made that genomic elements have only on and off states, a route that ignores the importance of intermediate behaviors. For example, in the case of promoter insertions, induction profiles can range from all-or-none to graded responses (17). Although library size is dependent only on the presence or absence of a promoter at a given position, assay size scales with the number of induction conditions tested. An additional caveat is that relatively little is known about the function of a large fraction of the genomes in experimental and industrial-use host microbes, making prediction of the effects of a targeted mutagenesis strategy difficult. For example, the genome of *E. coli* K-12 MG1655, the most well-studied microbe, contains approximately 4288 annotated protein-coding genes; of these, 19.5% remain of unknown function (18, 19). Without substantial a priori knowledge, the host genome, pathway operons, and its constituent enzymes all remain tenable targets for mutagenesis.

3. TECHNICAL LIMITATIONS IN LIBRARY CONSTRUCTION AND SCREENING

Even though statistical limitations establish a glass ceiling, hindering the exhaustive search of large sequence spaces, technical limitations in library generation and screening are the significant bottlenecks in practice. The workflow for a standard directed evolution assay can be divided into discrete segments: (*a*) in vitro diversity generation, (*b*) transformation and in vivo small-molecule production, and (*c*) screening and selection. Each step can impose significant technical limitations on the efficient exploration of sequence space.

3.1. Diversity Generation

Diversity generation stands as perhaps the most robust step in mutant library screening. Methodologies employed to generate in vitro or in vivo genetic diversity are described in the box Methodologies to Generate Sequence or Genetic Diversity. To a first approximation, technologies for diversity generation are agnostic to a genetic element's downstream, in vivo end function. For example, an experimentalist's success in introducing genotypic variability into a protein-coding sequence using error-prone polymerase chain reaction (PCR) is independent of the downstream function of the protein. Disconnect between diversity generation and end function is due, in part, to segregation of in vitro diversity generation from downstream in vivo expression. Creating genetic diversity in vitro allows for an extremely large number of variants to be created; plasmid libraries on the order of approximately 10^{14} molecules can be tractably prepared, amounting to 1 mg of plasmid DNA (20). For everyday benchtop experiments, however, an upper limit of 10^{12} molecules is more commonly observed.

Until recently, a lack of computational estimates of library sequence diversity, using various methods, left experimentalists to wander through sequence space. Although still seemingly underutilized, in silico approaches to model diversity generation in error-prone PCR, gene shuffling, and oligonucleotide-directed randomization have been exhaustively reviewed (21–23). When coupled with a readily accessible user interface, computational methods can be valuable guides to choose mutagenesis methods satisfying an experimentalist's library size, diversity, and coverage objectives. To this end, a suite of user-friendly diversity analysis software tools have been made readily available online (24). The programs provide a useful statistical analysis of library diversity and sequence space coverage, which can guide library construction and screening when using error-prone PCR, site-directed mutagenesis, and in vitro recombination (10, 25, 26). There still remains an upper limit on the sequence space that can be analyzed by computationally predictive methods; one recently published method for modeling random point mutagenesis establishes an upper limit of analyzing 2000 amino acids or 16,000 nucleotides, and 10^9–10^{10} individual sequences (27). This level of computation power continues to meet or exceed the reasonable number of DNA variants that can be constructed using existing DNA synthesis technologies.

METHODOLOGIES TO GENERATE SEQUENCE OR GENETIC DIVERSITY

Error-prone polymerase chain reaction (PCR) is performed with a low-fidelity polymerase (e.g., with $MnCl_2$, low amounts of template, or with unequal concentrations of nucleotides) to introduce copying errors.

Oligonucleotide-mediated mutagenesis, through PCR, incorporates user-defined nucleotide substitutions.

DNA shuffling involves randomly digested gene libraries, which are rejoined to combine mutations.

Mutator strains exploit deficient DNA repair pathways to generate replication errors in plasmids or genomes (30). Deletions, substitutions, and frameshifts are possible. Inducible mutator strains may also be used that express a dominant-negative mutator under select conditions.

Error-prone DNA polymerase, i.e., a highly inaccurate DNA polymerase I, is used to initiate low-fidelity replication of the target sequence in vivo (31).

Somatic hypermutation is an in vivo approach that makes use of the properties that generate immunoglobulin genes to create sequence diversity in targeted sequences (32).

Multiplex automated genome engineering (MAGE) permits numerous and simultaneous chromosomal changes across a population of cells (34). The bacteriophage λ-Red ssDNA-binding protein β is used to mediate oligonucleotide-based allelic replacement. The strength of MAGE is in the rapid and continuous generation of genetic variants.

3.2. Transformation Efficiency

The efficiency of introducing variability into the host to gain in vivo functionality provides the next significant bottleneck. The method most commonly employed for both *S. cerevisiae* and *E. coli* library incorporation is nucleotide transformation, our focus here. For *E. coli*, the maximum library size is estimated to be on the order of 10^{12} molecules (28), but libraries on the order of 10^9 transformants are more readily realized at the benchtop scale of everyday experiments. Library sizes in *S. cerevisiae* expression systems are subject to decreased transformation efficiencies, and maximum library sizes are approximately an order of magnitude less than those witnessed for *E. coli*.

Transformation of an in vitro library into the in vivo screening context can entail significant losses in library size, and steps can be taken to circumvent transformation inefficiencies. One option is to generate and screen a library in vitro; the in vitro compartmentalization, mRNA display, and ribosome display

Sensitivity: the slope of the assay's response curve, describing the minimal differentiable change in input product concentration between two samples

Dynamic range: the difference in biosensor output signal between the maximum and background states

Linear range of detection: the range of input small-molecule productivities that can be linearly correlated with the output reporter signal

Transfer function: the mathematical relationship between biosensor input and output

methods have been reviewed elsewhere (29). Commonly used for directed evolution of single proteins, completely in vitro assays are not generally applicable to metabolic engineering for small-molecule production and ignore in vivo biological context and regulation. The second, more commonly employed approach to avoiding library transformation inefficiencies is to develop the library diversity in vivo. Published methods include employing engineered strains of *E. coli* with increased mutation rates (30), an error-prone *E. coli* polymerase I (Pol I) for more targeted diversity generation in plasmids containing a ColE1 origin (31), and somatic hypermutation in mammalian expression hosts (32, 33). The general drawback to in vivo diversity generation is the introduction of untargeted, sometimes genome-wide, mutations. Circumventing this issue, multiplex automated genome engineering (MAGE) (34) uses transformed ss-DNA to target individual genomic regions for mutagenesis at high efficiency (>30%).

3.3. Small-Molecule Screening

Last in the directed evolution flowchart stand screening and selection. Technologies for small-molecule screening have long lagged behind those for diversity generation, predominantly owing to a need to independently tailor assay methods for application toward different target small molecules. For this reason, screening and selection processes are the most significant bottleneck in directed evolution efforts for small-molecule detection and quantification.

Although throughput can be generalized for specific screening technologies (**Figure 1**), small-molecule assays are burdened by additional, equally important parameters. Screen and selection strategies are only as good as their sensitivity and selectivity: A desired phenotype that remains undetected will not be captured, regardless of the number of variants analyzed. For this reason, assay sensitivity, dynamic range, and linear range of detection—parameters commonly used to describe transfer functions in genetic circuit design (17, 35–38)—are also apt for the characterization of small-molecule screens and selections (**Figure 2**). While virtually unreported in phenotypic assays described to date, this quantitative framework provides for a minimal set of parameters that enable more accurate comparisons between different assay methodologies. The caveat being that, given the aforementioned molecule-tailored nature of most screens, codified rules regarding screen sensitivity and dynamic range can be ineffectual generalizations.

Foundational work in directed evolution for small-molecule production has been focused on conspicuous targets, those that can be optically detected. Without the need for additional synthetic chemistry or biotransformation, conspicuous products can be accurately measured using standard high-throughput colorimetric and fluorometric assays. In contrast, inconspicuous targets, a class to which the majority of small molecules belong, emit no spectral signature suitable for existing high-throughput screening technologies. Inconspicuous small molecules can be converted into detectable outputs through the activity of biosensors. Biosensors take form as single- or multistep enzymatic pathways, inducible expression systems, and entire host organisms; an accurate description of the correlation between a biosensor's inputs and outputs is provided by its transfer function (**Figure 2**) (38). In the remainder of this review, we provide a discussion of small-molecule screens and selections and present methods for moving up the screening and selection ladder (**Figure 1**). This qualitative metric seeks to incorporate throughput, sensitivity, and dynamic range into a generalized rank of assay strength and molecule applicability. Small-molecule screens exhibiting high-throughput and sensitive analyte detection for a broad spectrum of compounds rank higher than less-specific, lower-throughput assays. In our analysis of the various small-molecule screening methods commonly employed, we highlight the role of biosensor-driven assays that enable inconspicuous small-molecule targets to leapfrog rungs in the ladder.

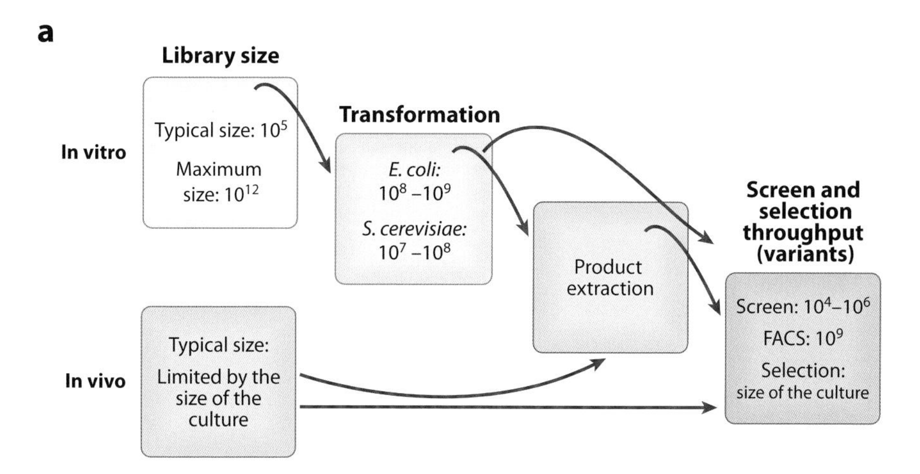

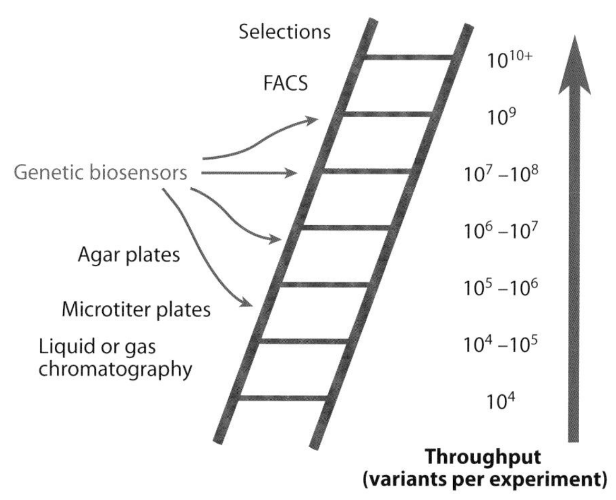

Figure 1

Inefficiencies in library screening. (*a*) The in vitro and in vivo steps associated with standard directed evolution dictate the scope of sequence space that can be explored. Moving between in vitro and in vivo compartments imposes significant losses in library size and diversity. In vivo methods of library generation avoid inefficiencies of transformation and product extraction; however, mutations are randomly inserted across a target sequence and cannot be focused to a few key positions. (*b*) The choice of screening assay will ultimately have the biggest impact on throughput; genetic biosensors converting small-molecule concentration into a readily detectable reporter molecule serve as one method to move away from low-throughput chromatography techniques. Abbreviation: FACS, fluorescence-activated cell sorting.

4. THE SCREENING AND SELECTION LADDER

We present here colorimetric, fluorometric, and growth-complementation assays as the foundation for small-molecule assays in metabolic engineering. Excluded from this review are gas and liquid chromatography and H^1- and C^{13}-based NMR techniques for small-molecule identification and quantification. Although ubiquitous in the field and offering unparalleled accuracy and precision, their extremely low throughput limits their

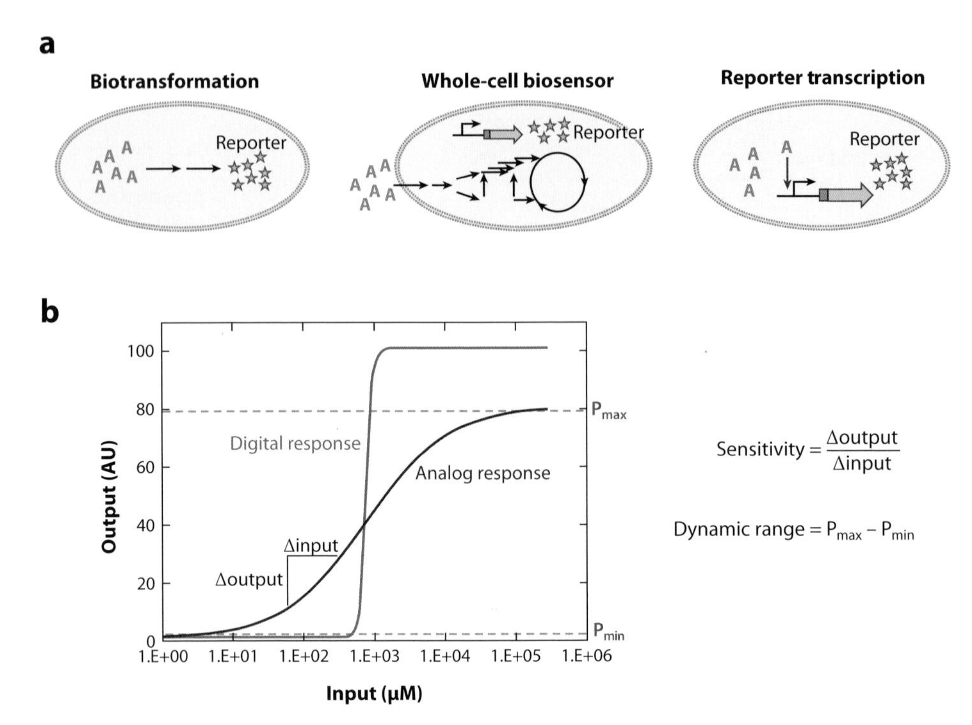

Figure 2

Transfer curves enable a quantitative characterization of performance features. (*a*) Biosynthesis of an inconspicuous small-molecule, A, results in a detectable signal by transformation using a series of biosensor constructs. Small-molecule A can be transformed into a detectable reporter molecule through the action of single- and multistep enzymatic pathways. Whole-cell biosensors couple growth (and a constitutively expressed reporter) with a host microbe's productivity. Lastly, reporter transcription can be achieved using classically regulated promoter systems induced by small-molecule A. (*b*) The correlation between input product concentration and output biosensor response (AU, arbitrary reporter units) provides the biosensor transfer function. Transfer functions provide information on product sensitivity, the linear range of detection, and the detection threshold, which can guide assay design.

application to assays with small sets ($<10^3$) of modifications.

In engineering terms, we define our system as an individual microbe, with the target product being an output. As such, we focus only on whole-cell assays for target small-molecule production. End-product screens detect the desired output (i.e., improved molecule production) and are applicable with directed evolution of any upstream biosynthetic machinery. In this sense, small-molecule screens offer greater versatility and application than screens for intermediate functions. We leave untouched assays that analyze single or intermediate steps in the conversion of a carbon source to product.

For each assay technology, examples of conspicuous small-molecule detection are used to provide a backdrop for development of next-generation, biosensor-driven approaches. Screen throughput, sensitivity, and dynamic range, where available, are highlighted from exemplary studies in the field.

4.1. Colorimetric and Fluorometric Plate-Based Screens

The relative ease of conducting colorimetric and fluorometric assays has established their position as the a posteriori techniques of choice for proof-of-principle experiments in novel mutagenesis and directed evolution (39).

Assay sensitivity, linear range of detection, and, to some extent, throughput will vary depending on the method used and the compound being interrogated. Individual variants are monitored as liquid cultures on microtiter plates or as colonies on solid, agarose media. Photometric assays using microtiter plates are highly robust and provide the distinct advantage over other techniques in being able to broaden the linear range of detection by diluting or concentrating the sample. Assay throughput on microtiter plates is moderate (approximately 10^5 variants per experiment) and is strongly affected by an oft requisite in vitro product extraction step. In comparison, screening colonies on solid media provides increased throughput, and with modern robotics upward of 10^6 variants are screened (40). The increase in throughput comes at the expense of a greatly diminished sensitivity, however, and small differences in intercolony productivities can be overlooked. Thus, the decision between liquid and solid media assays is largely dictated by available equipment and a compromise between throughput and sensitivity.

4.1.1. The upper limit of plate-based screening: carotenoid biosynthesis.

Detection of microbially produced photophores provides upper bounds with regard to throughput and sensitivity as no additional chemical transformations are required for product detection. From the expansive body of metabolic engineering work on production of natural photophores, we focus here on lessons learned during directed evolution for improved carotenoid biosynthesis. The carotenoids, including lycopene, β-carotene, and astaxanthin, were recognized early on as high-valued neutraceuticals with remarkable antioxidant properties (41). This feature, coupled with their relative ease in detection, has driven extensive research into establishing microbial production routes and, in large part, has laid the foundation for many aspects of metabolic engineering. To date, lycopene has been the major carotenoid of focus for production in microbial hosts. The majority of titer-oriented research for lycopene has been conducted in *E. coli* and has highlighted a number of the performance features of colorimetric screens. Lycopene production can be monitored colorimetrically by measuring the absorbance at 470 nm following extraction into an organic solvent (42). The assay is highly sensitive, differentiating between submilligram per liter differences in lycopene yields (43) and achieving a level of sensitivity more than adequate for directed evolution. Detracting from the lycopene screen, however, is a requisite organic solvent extraction that drives up assay price while decreasing throughput.

The success of plate-based photometric screening is exemplified by use of the carotenoids in various proof-of-principle experiments in organism and protein engineering. The broad spectrum of colored pigments found in the carotenoid family was utilized to demonstrate strategies in combinatorial biosynthesis (44, 45), a powerful approach wherein promiscuous enzymes serve as molecular scaffolds on which nonnative small molecules are synthesized. Additionally, advances in mRNA-based regulation of pathway flux (46), model and combinatorial-driven methods in genome engineering (47), engineered metabolic control (48), and multiplex genome engineering (34) have been enabled.

4.1.2. In vitro synthetic chemistry and enzyme-coupled assay design.

In contrast to the carotenoids, the vast majority of primary and secondary metabolites targeted for overproduction in the laboratory are not natural chromo- or fluorophores. For these molecules, in vitro synthetic and enzyme-coupled catalyses may offer an alternative approach to direct product detection. Using synthetic chemistry, a target molecule-specific chemical moiety reacts with an exogenously added reagent to yield a detectable product. Fluorescent and colorimetric detection of specific chemical moieties and compound classes is an area of great interest in organic synthesis. A wealth of both demonstrated and potential high-throughput assays can be borrowed from the field, including thiols (49), cyclic and linear amines

(50, 51), carboxylic acids (52, 53), alcohols (54), aldehydes (55), and glycosaminoglycans (56), among others (57, 58). Alternatively, indirect methods for product detection can also be employed, such as product-associated changes in pH (59, 60) or coproduction of H_2O_2 (61). The application of synthetic chemistry-based detection to metabolic engineering strategies, however, is not without limitation. Accurate product quantification in synthetic chemistry-based detection schemes may be impeded by a low signal-to-noise ratio, as the majority of moieties being targeted are naturally found at high abundance in the microbial intracellular environment. This issue may be minimized by increasing production titers. Lastly, reagent cost when incorporating in vitro synthetic chemistry may become a significant factor, a problem exacerbated with increasing library size.

When synthetic catalyses are either unavailable or cost prohibitive, or when higher substrate specificity is required to alleviate background noise, enzyme-coupled assays provide another approach. Coupling enzymes are added to the analyte solution to form single- or multistep pathways that yield detectable photophores or utilize traceable cofactors in their catalysis. The distinct advantage in cofactor-dependent assays is the broad range of enzymes and reaction types for which they are necessary; thus, cofactor monitoring can be viewed as a strategy more universally applicable than direct detection methods. Photometric assays have been described for quantification of many enzyme cofactors, ATP and ADP (62–64), reduced and oxidized states of NAD and NADP (65–67), and free coenzyme A (68). When background noise from native, endogenous small-molecule production is not an issue, using coupling enzymes with broad substrate acceptance is an interesting option. For example, substrate promiscuity is a hallmark characteristic of the P450 superfamily (69), and NADH or NADPH consumption in P450-catalyzed reactions is frequently used as an indirect measure for substrate oxidation (70, 71). Similarly, *S*-adenosyl-L-methionine is a cofactor for small-molecule methyltransferases and can be monitored following multienzyme biotransformations to homocysteine or hypoxanthine (49, 72, 73). Although inherently indirect measures (and as such push the boundaries of the rule "you get what you screen for"), the cofactor assays nonetheless are often highly robust, accurate proxies for direct product detection (62, 72, 74).

Design of in vitro small-molecule assays using either synthetic or biocatalyzed transformations requires significant thought with regard to the reagents and enzymes used. Although not a prerequisite, driving a reaction to completion is highly desirable and will facilitate a more accurate back calculation of target small-molecule production titers. For this reason, use of irreversible enzymes without product inhibition is the preferred enzymatic route. The choice of catalyst also determines the reaction conditions. Under the best-case scenario, assays are performed directly in the growth medium without cell removal, product extraction, or pH adjustment; however, optimal reaction conditions always need to be determined experimentally. These considerations become increasingly important as library sizes grow, as each additional in vitro manipulation adds both significant consumable and throughput costs. Lastly, the synthetic reagents or coupling enzymes must also be economically synthesized, purified, or purchased to screen a complete set of library variants.

4.1.3. In vivo biosensor-driven assay design. By shifting from an in vitro, enzyme-coupled product detection regime to an in vivo context, significant advantages can be garnered in terms of both throughput and cost. Extractions, enzyme purifications, and in vitro manipulations are minimized or eliminated. Catalytic biosensors, single and multistep in vivo biosynthetic pathways with inconspicuous substrate inputs and detectable product outputs, have just begun to be explored. Examples to date include assays for intermediates in an engineered *Pseudomonas putida* paraoxon catabolic pathway (75), strictosidine glucosidase-catalyzed transformation of tryptamine analogs (76), and a tyrosinase-catalyzed transformation of the amino acid L-tyrosine into melanin (77).

Directed evolution of *E. coli* for improved L-tyrosine production exemplifies the role enzyme-based biosensors can assume during inconspicuous small-molecule detection. Tyrosine is a colorless, essential metabolite of great import in the synthesis of a wide range of value-added pharmaceuticals and commodity chemicals, including the morphine alkaloids and *p*-hydroxystyrene (78). Rational engineering strategies, targeting both tyrosine and its intermediates, have been extensively explored and reviewed (79, 80). Concomitant work in the Stephanopoulos lab (77, 81) in both rational engineering and directed evolution of *E. coli* tyrosine biosynthesis enables a thorough comparison of the two approaches. Using strictly rational metabolic engineering, two tyrosine-overproducing strains of *E. coli* were reported (82). The first, T1, incorporated feedback-inhibition-resistant derivatives of pathway enzymes and eliminated native pathway regulation. Building on T1, strain T2 increased the availability of the central metabolite precursors D-erythrose-4-phosphate and phosphoenolpyruvate necessary for tyrosine biosynthesis. Tyrosine production titers in minimal media culture flasks reached 346 ± 26 and 621 ± 26 $mg{\cdot}L^{-1}$ for strains T1 and T2, respectively.

Before using random mutagenesis to build on their rationally designed strains, the authors developed a pair of high-throughput screening assays. The first, although not biosensor based, is an interesting application of an in vitro synthetic chemistry approach. It has long been known that 1-nitroso-2-naphthol reacts with parasubstituted phenols, such as tyrosine with high specificity, yielding a red-orange product (83). By modifying the published assay methods, the authors developed a high-throughput, microtiter-based fluorometric screen for L-tyrosine (81). This assay was then used to screen a small, combinatorial library of amino acid biosynthetic genes overexpressed in *E. coli* (84). The authors report that their screen accurately differentiates between product concentrations as low as 50 $mg{\cdot}L^{-1}$ over a linear range of 0.05–0.5 $g{\cdot}L^{-1}$ tyrosine (81). As with lycopene, throughput is limited to the order of 10^5 variants per experiment owing to requisite use of microtiter plates and in vitro synthetic chemistry.

More recently, Santos & Stephanopoulos (77) describe a catalytic biosensor approach using an in vivo expressed tyrosinase to use melanin as a reporter of tyrosine productivity (**Figure 3**). A black, insoluble pigment, melanin, can be readily assayed in colonies grown on solid media without need for additional in vitro synthetic chemistry or solvent extractions. Starting with a base *E. coli* strain producing 347 $mg{\cdot}L^{-1}$ tyrosine (similar to the above-mentioned rationally engineered strain T1), the authors constructed a transposon-mediated knockout library of the *E. coli* genome and screened 21,000 colonies on agar plates. Over two five-day rounds of screening, 30 variants were selected for a more rigorous characterization; two mutants were discovered that individually produced 57% and 71% higher L-tyrosine titers than the background strain. The associated chromosomal knockouts occurred in *dnaQ* and *ygdT*, whose gene products are part of the epsilon subunit of PolIII and a hypothetical protein, respectively. Under similar culture conditions, the *yadT* knockout strain exhibited product yields comparable to those measured in the highest producing rationally engineered strains (77, 82). Neither of these genomic knockouts could have been predicted using rational design or flux analysis approaches, a feature that draws attention to the potential value of strong-screening assays. The performance features for this assay were not provided and thus inhibit comparison to their synthetic chemistry-based route; however, visual assessment of colonies is in general a much less sensitive detection method.

4.2. Growth Complementation

Strains auxotrophic for essential small molecules are natural biosensors and have long been used to build strong selection assays (85). These strains provide what is perhaps

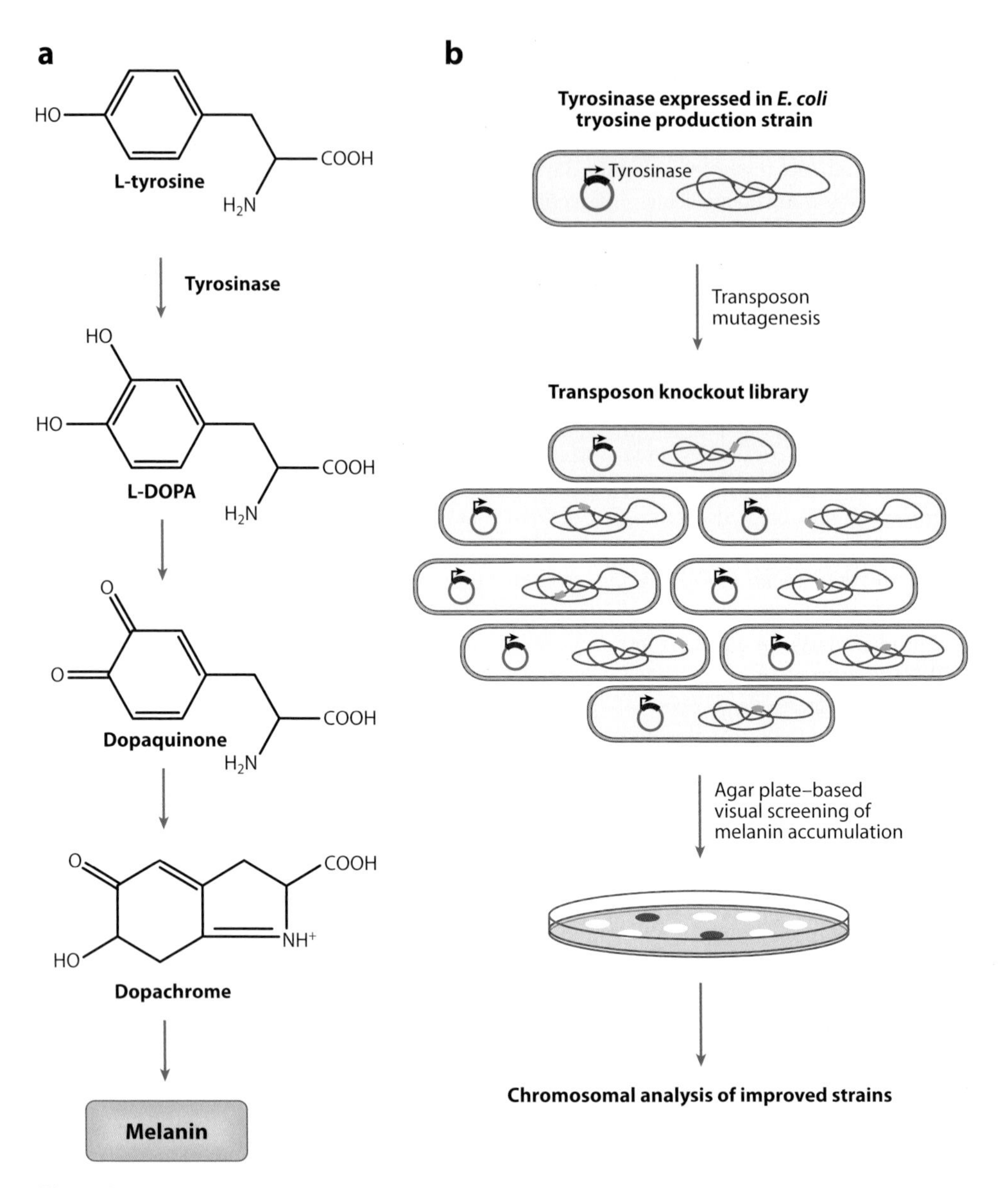

Figure 3

Development of a catalytic biosensor. Santos & Stephanopoulos (77) coupled production of the colorless amino acid L-tryosine to synthesis of the black, diffusible pigment melanin through heterologous expression of a tryosinase in *E. coli*. (*a*) Tyrosinases use molecular oxygen to catalyze the *ortho*-hydroxylation of L-tyrosine to L-DOPA, followed by its oxidation to dopachrome. The generated reactive quinones then polymerize to form melanin. (*b*) Expression of a bacterial tryosinase (*blue plasmid*) enabled a high-throughput screen, as melanin was used to report on the production of L-tyrosine. Agar plate-based screening of a transposon-mediated knockout library identified strains with increased tyrosine production. Analysis of the chromosomal mutations in the improved strains revealed interruptions in two genes, *dnaQ* and *yadT*. Neither gene would have been predicted to impact aromatic amino acid production through rational design or flux analysis strategies. This work illustrates the importance of strong screening assays in uncovering untapped genotypic improvements.

the most readily discernable phenotype, growth, if the auxotrophy is relieved by complementation of lost enzymatic function. Growth-complementation assays have been used extensively to engineer hosts for catabolism of nonnative or nonideal carbon, nitrogen, and phosphate sources (86, 87). Assay development is relatively straightforward: By supplementing the growth medium with a single molecular source of an essential element, only those host variants engineered or evolved for its catabolism survive.

Although highly successful in engineering novel catabolic activities, coupling anabolism to growth has proven to be a more difficult task. Oftentimes, an overproduction phenotype is deleterious to the host strain's reproductive fitness (88, 89), a characteristic easily inferred by depressed growth curves in overproducing strains. To screen for anabolic activity, small-molecule production must complement the auxotrophy and be strongly correlated with the specific growth rate to enrich cultures for high producers.

Auxotrophies of components of amino acid biosynthesis pathways have been successfully utilized in metabolic engineering applications because the pathways are strongly coupled to growth. An early study using growth-complementation knocked out the gene encoding for branched chain amino acid aminotransferase *ilvE*; this strain was then used to evolve an aspartate aminotransferase to recognize branched amino acids (90). Following five rounds of DNA shuffling, the catalytic efficiency for transamination between 2-oxovaline and aspartate was improved by five orders of magnitude. Further selection increased the catalytic efficiency of the final mutant variant by an additional order of magnitude (91). Other examples from amino acid biosynthesis, which used similar strategies, include aminotransferases (92), an alanine racemase (93), and evolution of chorismate mutase (94, 95).

Intermediates in amino acid biosynthesis, keto acids, have been more recently explored in a metabolic engineering context. Atsumi et al. (96) rationally engineered 1-propanol and 1-butanol production in *E. coli* using 2-ketobutyrate derived from threonine degradation. A more direct route to 2-ketobutyrate, however, is the citramalate pathway, identified in *Leptospira interrogans* and *Methanocaldococcus jannaschii* but was not present in *E. coli*. An *E. coli* strain auxotrophic for isoleucine would require flux to be rerouted through a transformed citramalate pathway to restore growth. This strategy was employed in an *E. coli* host for functional expression and directed evolution of a heterologous citramalate biosynthetic pathway leading to 2-ketobutyrate (97). The evolved strain exhibited 9- and 22-fold improvements in 1-propanol and 1-butanol productivities, respectively, over the wild-type citramalate base strain. Furthermore, when similar media conditions and genetic backgrounds were employed, the evolved citramalate pathway provided greater than 35-fold improvement in 1-propanol yield and an eightfold improvement in 1-butanol yield. Given the large number of industrially important compounds that can be derived from amino acid biosynthetic pathways and the success of auxotrophic selection assays for these compounds, there is likely to be continued focus on these assays in the future.

4.2.1. Auxotrophic reporter strains. A more generalized growth-complementation strategy is the use of an auxotrophic reporter strain. Here, a producer strain is engineered for over-expression of the target pathway, and a second reporter strain is constructed that constitutively expresses a detectable market (i.e., green fluorescent protein, GFP) and is auxotrophic for an essential metabolite found in the target pathway (**Figure 2**). When cocultured, the reporter strain's growth—and thus reporter output—is coupled to the metabolite yield achieved by the producer strain. This strategy was successfully employed in the development of a whole-cell biosensor for mevalonate (**Figure 4**) (98). A 10% difference in mevalonate production could be discerned between liquid cell cultures with 95% confidence. When utilized in a spray-on technique for solid media screens, however, the assay was only effective in distinguishing

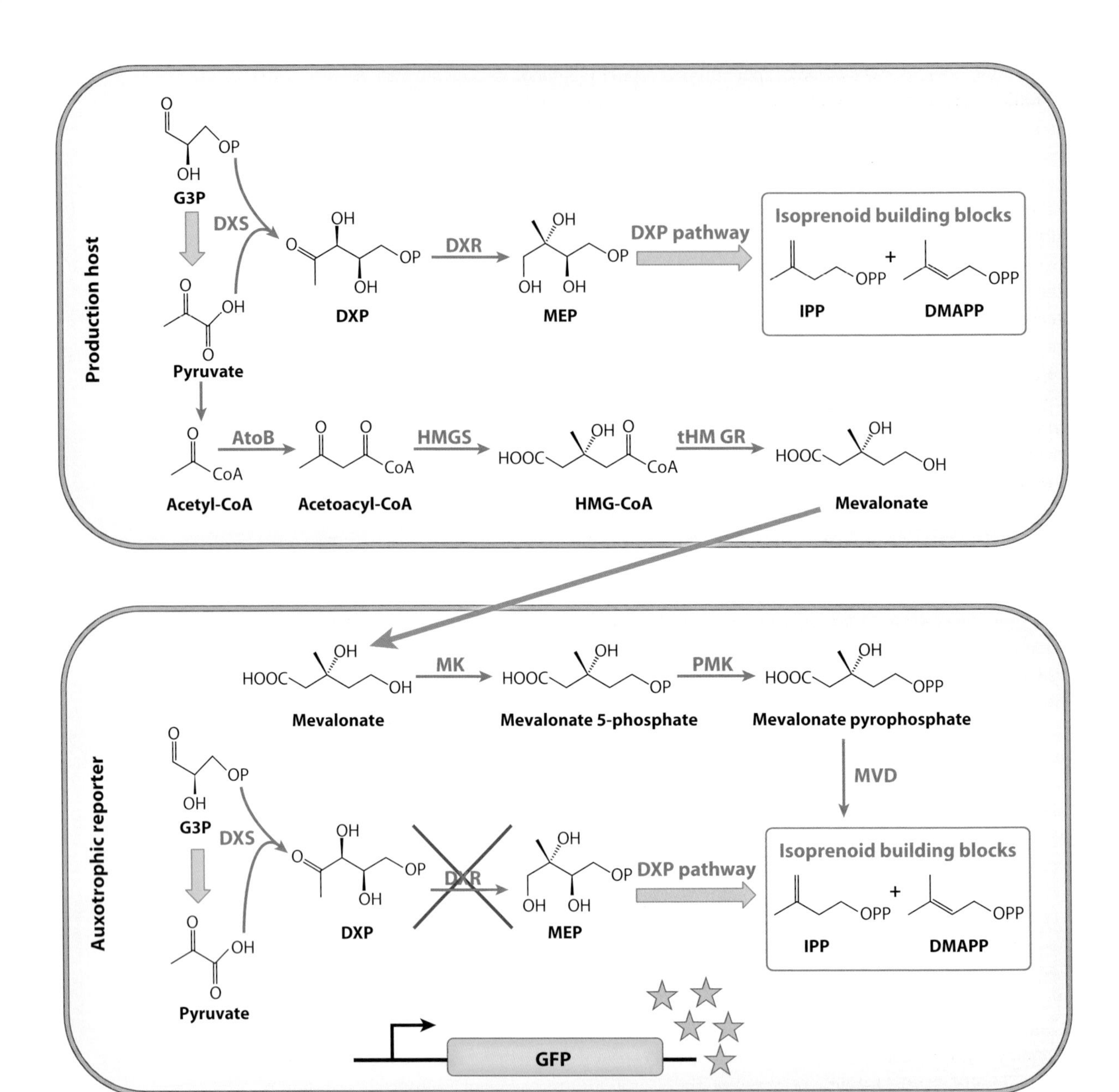

Figure 4

Development of a mevalonate biosensor. The isoprene subunits isopentenyl diphosphate (IPP) and dimethylallyl diphosphate (DMAPP) are essential for *E. coli* growth. Natively, IPP and DMAPP are produced in *E. coli* through the 1-deoxy-D-xylulose-5-phosphate (DXP) pathway; however, expression of a heterologous mevalonate pathway provides an alternative route. By concomitantly knocking out the DXP pathway while overexpressing the enzymes necessary for mevalonate utilization, a mevalonate biosensor was constructed (98). The mevalonate biosensor also contained a constitutively expressed gene for green fluorescent protein (GFP) to provide a measurable readout of cell growth on solid media plates. Mevalonate was supplied by a production host containing genes necessary for transformation of acetyl-CoA to mevalonate. Abbreviations: AtoB, acetoacetyl-CoA thiolase; DMAPP, dimethylallyl diphosphate; DXR, 1-deoxy-D-xylulose-5-phosphate reductoisomerase; DXS, 1-deoxyxylulose-5-phosphate synthase; G3P, glyceraldehyde-3-phosphate; HMGS, HMG-CoA synthase; IPP, isopentenyl diphosphate; MEP, 2-C-methyl-D-erythritol-4-phosphate; MK, mevalonate kinase; MVD, mevalonate pyrophosphate decarboxylase; PMK, phosphomevalonate kinase; tHMGR, truncated HMG-CoA reductase.

mevalonate producers from nonproducers (98); a finding that again highlights the higher sensitivity observed in liquid versus solid media plate-based assays.

In general, auxotrophic strains provide a unique, high-throughput approach for the screening or selection of small-molecule overproducing strains. However, there remain some significant drawbacks when considering their broad-scale adoption. First, the essential small molecule must be native to the host or reporter microbe; thus, growth complementation is more suitable to building platform production strains for metabolite precursors than exotic secondary metabolites. Second, the dynamic range in growth-complementation assays can be limited. From a protein engineering perspective, even slightly functional biocatalysts can be sufficient to restore cell growth; in practice, this places a glass ceiling on the small-molecule product yields that can be accurately screened or selected. Addressing this challenge, user-defined transcription and enzyme degradation tags have been explored as a means of increasing selective pressures and assay dynamic range (95). Other readily accessible approaches for fine-tuning protein expression levels include decreasing plasmid copy number, modifying RNA stability, and modifying translation initiation efficiency (99).

4.3. Fluorescence-Activated Cell Sorting-Based Screening

Near the top of the phenotypic screening and selection ladder stands fluorescence-activated cell sorting (FACS). Through rapid analysis of size and fluorescence measurements of single cells, this technique possesses nearly ideal high-throughput screening characteristics (100). Libraries are commonly first passed through a primary FACS screen, thereby enriching the population between 80- to 5000-fold for the 0.5% to 1.0% cells exhibiting the highest measured fluorescence (101–105). Library sizes between 10^8 and 10^9 variants have been readily screened in assays for modified protein specificity (101, 106), achieving the realistic 10^9 variant cutoff incurred owing to low microbial transformation efficiencies (28). Those cells surviving the primary screen are subjected to a secondary screen to provide phenotypic confirmation. In practice, between 10 and 10^3 clones enriched by the primary FACS screen are analyzed by gas chromatography-mass spectrometry (GC-MS) for identification or quantification (107). Although FACS is ultra high throughput, a drawback is overlapping fluorescence profiles, and aberrant fluorescence can lead to a high rate of false positives during screening. This arises from the distribution of fluorescence values found in a population when single cells are analyzed, a problem not witnessed in plate-based screens because of population averaging (**Figure 5**). For this reason, FACS-based screens are used to enrich for a target population and are then followed by secondary, orthogonal screens to provide a more accurate validation.

FACS: fluorescence-activated cell sorting

4.3.1. Direct and coupled-fluorescence detection. Direct detection of fluorescent proteins and metabolites using FACS is a straightforward application and has been well explored in both protein and metabolic engineering. In metabolic engineering, naturally fluorescent metabolites can be directly screened so long as they are intracellular and so long as their emission spectra are not masked by that of the native host. Again, the carotenoids have served as model small molecules for isolation of hyperproducing yeast strains. The inherent fluorescent property of astaxanthin was used to recover yeast strains at high sensitivity, capturing productivity improvements as low as 260 $\mu g \cdot g^{-1}$ dry cell weight (108, 109). A recent effort to improve astaxanthin production from *Xanthophyllomyces dendrorhous* achieved a 3.8-fold production improvement and reported fivefold improvement in screening efficiency as compared to plate-based methods (110). Beyond the carotenoids, FACS screening has been restricted to a small number of natural and synthetically coupled fluorophores, including gramicidin S (111), polyhydroxyalkanoates

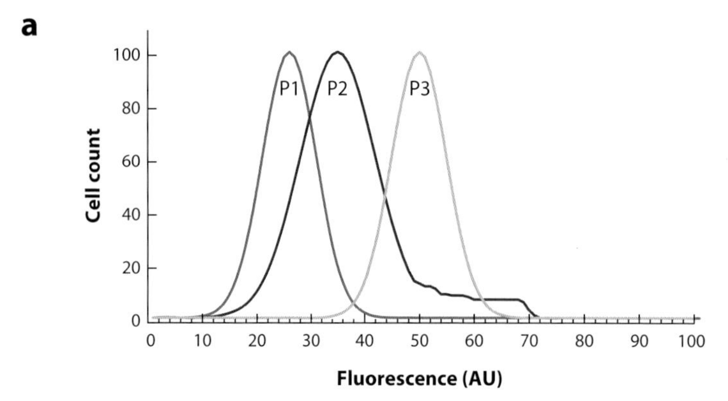

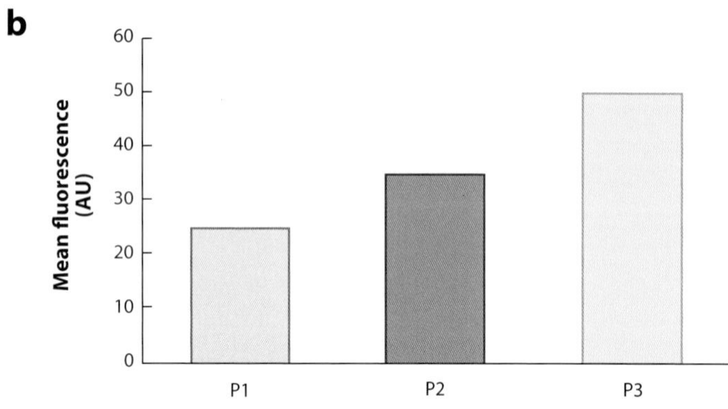

Figure 5

Population distributions can inhibit efficient cell sorting. (*a*) Representative data collected from fluorescence-activated cell sorting (FACS) and (*b*) fluorometer analyses of output reporter fluorescence highlight the difficulties in efficient screening. We examine the hypothetical strains P1, P2, and P3, displaying different levels of target small-molecule production. Although the mean fluorescence for all cell populations does not change between fluorometer and FACS instruments, population distributions inhibit efficient cell sorting by FACS. Overlapping population profiles (separating P2 from P1 and P3) and aberrant fluorescence (the right-hand tail on P2) can decrease enrichment efficiencies. Lower-throughput, orthogonal secondary screens must be used to further purify the target population. Abbreviations: P1, -2, -3, hypothetical populations.

(112), poly(β-hydroxybutyrate) (113), and lipids (114, 115).

4.3.2. Physiology-associated fluorescence detection. Another route for FACS-based screening is to assess those changes to cellular physiology induced by small-molecule production or cell stress. In this indirect method, fluorescent dyes act as indicators for changes in intracellular pH or viability (116). Following this strategy, strains of *S. cerevisiae* capable of improved lactic acid tolerance and production have been isolated from a mutant library, using both ethidium bromide staining and a pH-dependent fluorescent probe (117). The fluorescent probe used, cSNARF-4F, exhibits a shift in the spectral emission depending on intracellular pH; the ratio of the two local maxima indicate the pH (118). Because the assay is physiology dependent and not product dependent, it can be extrapolated toward detection of a range of microbially synthesized weak organic acids and bases.

The approach of screening for a downstream effect of molecule overproduction, and not for the small molecule itself, may be a versatile strategy given that entire classes of molecules can produce similar changes in physiology. For instance, FACS-based screens could prove effective in cases of product-induced oxidative stress, a common side effect of P450 overexpression (119), using in vivo fluorescent probes for reactive oxygen species (120, 121). Similarly, membrane fluidity (122), intracellular magnesium (123), and intracellular calcium (124) can also be quantified using fluorescent dyes. Despite the straightforwardness of this screening approach, it is hampered by the dearth of available in vivo reporter probes. In addition, it remains unknown whether a strong, reproducible correlation can be drawn between small-molecule biosynthesis and a microbe's physiological state over a large set of molecule classes.

4.3.3. Biosensor-driven fluorophore production. Genetic biosensors can take inconspicuous small-molecule inputs and output one of several available fluorescent proteins as reporters. In general, methods described to date have focused on transcriptional read through from product-encoding operons as a proxy for end function. One interesting approach has been construction of genetic traps, screens used to isolate genetic elements from metagenomic libraries that encode for user-desired phenotypes. SIGREX, or substrate-induced gene-expression screening, uses a fluorescent reporter housed behind a multiple cloning site into which a metagenomic library is inserted (125, 126). When a small-molecule substrate is exogenously added, only clones harboring metagenomic inserts whose transcription is activated by the substrate small molecule will express the downstream fluorescent reporter. A second, negative screen under uninduced conditions can remove variants with constitutive fluorescence. For the insert to produce a fluorescent signal, it must encode for a responsive transcription factor-promoter system. The promoter must also be oriented in the correct direction to transcribe the downstream reporter gene. The success of this method hinges on the overarching assumption that the requisite transcription factor is housed proximal to its cognate promoter controlling GFP expression (126), a guilt-by-association strategy that is by no means a genotypic rule (127).

User-selected biosensors can also be applied for isolation of anabolic operons from metagenomic libraries encoding for biosynthesis of biosensor-responsive small molecules (128). This was demonstrated using the well-characterized acyl-homoserine lactone-responsive LuxR- P_{luxI} genetic actuator isolated from *Vibrio fischeri* (129). By placing GFP under control of P_{luxI}, a metagenomic library was transformed into *E. coli* and screened for novel inhibitors of the quorum-sensing response. Approximately 10,000 colonies survived the preliminary FACS assay by inhibiting GFP production in the presence of exogenously added acyl-homoserine lactone. Only a meager 0.2%, or two colonies, produced confirmable inhibition of transcription from P_{luxI}, a finding that draws light toward the high rate of false positives when using FACS for phenotypic characterization. A similar approach was recently used to screen for alkane biosynthesis and transport in engineered strains of *E. coli* using the AlkR transcription factor isolated from *Acinetobacter* (130). In cases where a target small molecule lacks no known cognate transcription factor, FACS-coupled directed evolution of the transcription factor's effector recognition domain can realize minor perturbations in ligand specificity (131). Nearly all genetic biosensors described to date have been used to obtain an all-or-none fluorescent response profile, which is well suited for gene discovery but poorly suited for mechanistic studies where intermediate behaviors provide information equally as important as the final product. As discussed in detail below, however, genetic biosensors providing a linear response profile can have also been characterized and can be used to elucidate intermediate behaviors.

5. FUTURE DIRECTIONS: TRANSCRIPTION FACTOR-BASED SMALL-MOLECULE SCREENING

The screening strategies described above have demonstrated success in identifying proteins or strains capable of increased small-molecule production titers. Many small molecules, however, are not natively chromophoric, fluorescent, or essential for growth. Furthermore, only a limited few can be readily transformed into compounds possessing these properties. Thus, a universal high-throughput screening platform for small molecules remains out of reach. In spite of that, to expand our existing repertoire of detection techniques, we can turn to nature, the apt experimentalist, and borrow from the plethora of molecular devices evolved for small-molecule recognition.

Nature has evolved RNA, DNA, and protein devices for the binding of small molecules and for activation of a downstream transcriptional response. Some molecular devices, including the aptamer (132) and FRET biosensors (133), can be engineered for small-molecule detection and deserve further exploration. One strategy deserving special attention is the use of transcription factors, proteins that regulate a promoter's transcriptional output in response to a small-molecule ligand, to report on in vivo small-molecule production. Bioengineers have long used transcription factors to construct whole-cell biosensors for the detection of environmental small-molecule pollutants (134). However, this same approach has remained largely untranslated toward library screening and directed evolution purposes.

5.1. Biosensor Response Characterization

Screening methods utilizing transcription factors possess many characteristics ideal for developing small-molecule screens and selections. The reporter gene housed downstream of an inducible promoter can be user selected to encode for colorimetric, fluorescent, or growth-coupled responses. Additionally, because the transcription factor-promoter pair is responsible for direct, in vivo detection of the target analyte, the need for downstream synthetic chemistry or in vitro manipulation is eliminated.

The first task when designing a screen is to understand the relationship between small-molecule input and reporter protein outputs. This process is greatly facilitated in transcription factor–based detection technologies by continued work in genetic circuit design. In part, this is due to a trend in the synthetic biology community toward better characterization of standardized hosts and biological parts (17, 135, 136). For example, a thorough characterization of the frequently used LuxR-P_{LuxR} expression system for *E. coli* provides a solid framework for biosensor design and characterization (37). The mathematical description of an expression system's transfer function can elucidate the conditions under which a biosensor can provide robust, reproducible function. As with optical detection techniques, biosensor transfer functions are characterized by their dynamic range, sensitivity, small-molecule specificity, and range of linear analyte detection (**Figure 2**) (38). All of the relevant parameters can be readily calculated using existing models that describe a range of regulatory schemes (137–139).

Even in *E. coli*, unfortunately, standardized promoter characterization remains scant to date and remains virtually nonexistent for yeast expression systems. We provide here a retrospective analysis of published response curves and provide estimates for a minimum set of governing parameters for many of the most commonly used biosensor-ligand induction systems (**Table 1**). Shown are the sensitivity, dynamic range, and linear detection range for a number of promoter-induction systems when fit using the Hill equation. Few of the small-molecule ligands for which these systems were designed to detect are industrially relevant; thus, their cognate biosensors are unlikely to be used for small-molecule assay design. However, we can still glean useful information on generalized trends that will facilitate biosensor-based assay design.

From this data there emerge a number of key trends that can help guide more efficient biosensor design. Perhaps the most important performance feature describing a transfer function is the apparent Hill coefficient, n_{app}. By definition n_{app} describes the assay sensitivity, but information on the linear range of detection, the detection threshold, and the dynamic range can be inferred. For example, a large n_{app} indicates a high slope within the linear portion of the transfer curve and a sensitive assay, but it is also associated with a small-linear induction window, a large dynamic range, and a low detection threshold. In electric and genetic circuit design, a large n_{app} describes digital-like behavior. The tight regulatory control displayed by digital logic gates has made them a staple for in vivo genetic circuit design; the P_{Tet}-TetR and P_{LuxR}-LuxR systems, in particular, are archetypal model systems providing robust, user-defined behaviors (140–143). Analog-like behavior is also witnessed in our selection of expression systems. For example, the P_{BAD} and P_{PRO} systems surveyed here have low sensitivity and a broad window for which the output response is linearly correlated to the input concentration. In general, analog-type responses are more in line with established screening assays for small-molecule detection. Along a similar vein, analog logic gates are suitable for use in directed evolution assays, where phenotypic improvements are typically small and incremental.

Although broad-reaching conclusions remain elusive, evolutionary arguments support some general trends seen among the transfer

Table 1 Transfer function–derived performance parameters for common induction systems

Induction system	Ligand	Sensitivity (n_{app})	Dynamic range (fold increase)	Linear detection range	Reference
$P_{LtetO-1}$-TetR	Anhydrotetracycline	10.43	2535	≈0–4.5 nM	149
P_{TETO7}-rtTA[a]	Doxycycline	1.12	28	0.01–2.5 μM	144
P_{TETO2}-rtTA[a]	Doxycycline	0.96	15	0.05–11 μM	144
P_{TETO1}-rtTA[a]	Doxycycline	0.70	15	0.25–35 μM	144
$P_{LlacO-1}$-LacR	Isopropyl β-D-1-thiogalacto-pyranoside (IPTG)	1.14	620	0.01–1 mM	149
P_{BAD}-AraC	Arabinose	0.43	5.5	0.08–20 mM	150
P_{prpB}-PrpR	Propionate	2.25	6.5	0.2–12.6 mM	150
P_{Trc}-lacI	Isopropyl β-D-1-thiogalacto-pyranoside (IPTG)	0.52	5	4–6.3 μM	150
P_{LuxR}-LuxR	3-Oxohexanoyl-L-homoserine lactone ($3OC_6HSL$)	1.6	500	0.5–10 nM	37
P_{pu}-XylR	Toluene	2.3	16	50–300 μM	151
P_{pu}-XylR	Benzene	3.2	9.8	—	151
P_{pu}-XylR	4-Xylene	2.0	26	—	151
P_{pu}-XylR	o-xylene	0.79	20.3 ± 1.6	0.05–5 mM	146
P_{po},-XylR	o-xylene	[b]	3151 ± 156	0.1–10 mM	146
P_{ps}-XylR	o-xylene	1.41	32.5 ± 6.1	0.1–5 mM	146

[a]The O# in P_{TETO} indicates the number of operator sites included in the promoter design.

[b]Curve could not be accurately fit using the provided data points, however an $n_{app} > 10$ is estimated.

functions between various expression systems. Evolution has tailored the regulation of each promoter to function in a specific environmental context. Digital behavior is more often witnessed in genetic systems evolved for antibiotic, quorum sensing, and sugar utilization. In contrast, analog behavior is more typical of nonideal carbon source utilization. Of course, when the system is removed from its original genetic context (i.e., removing a transcription factor from control by its native promoter), these observations may not hold. Transcriptional logic gates have proven to be readily engineered genetic parts, and transfer functions can be altered for more digital- or analog-like profiles, as needed for the application at hand. A shift toward a more digital response was observed by increasing the number of operator sites on a doxycycline-induced P_{Tet} promoter (144), which manifested as an increase in n_{app}, a lower detection threshold, and an increase in dynamic range (**Table 1**). Similarly, for transcription factors exhibiting a poor binding profile to the target ligand, directed protein evolution has been demonstrated to increase ligand specificity and biosensor sensitivity (145). Perhaps more often than realized, response profiles can be quickly adjusted without the need for laborious protein or promoter engineering. Poor ligand-transcription factor binding is often a characteristic of an analog response, suggesting transcription factors with promiscuous ligand binding could serve as ideal analog devices using weak inducers. Lastly, in the case of the transcription factor XylR, just moving between one of its three native promoter systems dramatically changed the response profiles in an o-xylene-induced system (146). Whether analog or digital, both types of behaviors can be either pilfered from nature's abundant stock or engineered from preexisting, standardized biological parts.

5.2. Applications Using Digital and Analog Biosensor Response Profiles

Digital and analog biosensor response modes are both equally important for in vivo small-molecule assays; digital behavior has demonstrated a value for construction of genetic traps, whereas analog behavior is well poised for applications in directed evolution. Mohn et al. (147) describe both digital and analog transcription factor–based biosensors in their detection of the biotransformation of linade into 1,2,4-trichlorobenzene. The authors first construct their biosensor using an evolved mutant variant of the σ^{54}-dependent transcription factor XylR, which is sensitive to 1,2,4-trichlorobenzene but not to lindane. Using this system, three reporter strains were created: two *P. putida* strains harboring luciferase or LacZ reporters and an *E. coli* strain harboring a LacZ fusion protein reporter. To achieve a strong digital response, the biosensor's dynamic range was increased by eliminating sources of background noise caused by leaky transcription from their promoter. In the luciferase reporter strain, the authors couple their evolved XylR protein with the *Po* promoter, which exhibits a substantially lower level of leaky expression as compared to the more commonly used *Pu* promoter (146, 148). In the *E. coli* reporter strain, LacZ was fused to XylU, the first native gene product downstream of the *Pu* promoter. Strains harboring the fused reporter construct were reported to exhibit background expression levels below the detectable limit of a β-galactosidase assay (147). When cotransformed with a lindane dehydrochlorinase from *Sphingomonas paucimobilis* UT26, the strains provided proof-of-concept demonstration of digital biosensor response using both colorimetric and growth-complementation assays. Although not employed for purposes of directed evolution, the *E. coli* biosensor system yielded a near-linear relationship between output β-galactosidase activity and concentration of exogenously added lindane. Linearity was held over a two orders-of-magnitude concentration range, providing strong evidence this strain could have a use as an in vivo screen for improved lindane or 1,2,4-trichlorobenzene biosynthesis.

6. CONCLUSIONS

Metabolic engineers require high-throughput screens to facilitate the directed evolution of microbial strains for small-molecule overproduction. The sheer number of variables that can be modified within either the genome and heterologous pathway are astounding; furthermore, biosynthetic pathways comprise multiple enzymes subject to regulation, which is too often poorly understood. Without detailed experimental evidence, there is little hope for rational mutagenesis strategies to substantially improve small-molecule yields and pathway efficiencies.

When coupled with a high-throughput screen, random mutagenesis can be a powerful method to explore sequence space with little-to-no a priori knowledge of the subject's structure or function. The walk through sequence space, however, is limited by the underlying statistics behind library generation and screening assay throughput. For the vast majority of small-molecule targets, screening throughput stands as the rate-limiting step, with low-throughput liquid and gas chromatography being the modus operandi in the field today. Recent innovations are enabling experimentalists to "climb the screening and selection ladder," the underlying principle being the design of biosensors to transform inconspicuous small molecules into more readily detectable reporters.

SUMMARY POINTS

1. A continued focus in metabolic engineering on directed evolution for small-molecule-producing microbes demands improved high-throughput screening technologies; whole-cell small-molecule screens and selections address this issue.

2. An exhaustive search of sequence space is limited by choice of mutagenesis target, library construction method, and screening technology. When sufficient a priori knowledge is available, a targeted mutagenesis strategy can substantially reduce the required library size.
3. In vitro manipulations (transformation and screening) are the most significant bottlenecks restricting assay throughput. When possible, assays should be designed to minimize the number of in vitro manipulations.
4. Transfer functions serve as important guides in the design of biosensor-based screening techniques. The mathematical relationship between biosensor input and output describes key performance features: dynamic range, sensitivity, and linear range of detection.
5. Colorimetric and fluorescence plate-based screens are the most well-explored screening assays; throughput is moderate (10^5–10^{6+} variants), and sensitivity can vary from low (colony analysis on solid media plates) to high (product extraction and concentration). These methods are supported by a wealth of experience in small-molecule labeling from synthetic chemistry literature.
6. Growth-coupled assays can be constructed by utilizing strains auxotrophic for essential metabolites in pathways required for target small-molecule biosynthesis. Throughput can be extremely high—limited only be the size of the culture and mutagenesis technique—when selections are employed. Assay sensitivity varies but can be low if only a moderate level of activity is required to restore growth.
7. FACS-based screens provide the highest throughput and sensitivity. Up to 10^9 cells can be screened per experiment with single-cell resolution. FACS screening is limited by the requirement for in vivo localization of a fluorophore.
8. Transcription factor-based screens can be used to create whole-cell biosensors coupled to colorimetric, fluorescent, and growth-coupled detection assays. Progress in this area is ongoing and builds on biological parts characterization and circuit design in the synthetic biology community.

FUTURE ISSUES

1. Can in vitro compartmentalization and phage display technologies be adapted to be more readily applicable to small-molecule screening?
2. Can in silico enzyme design be coupled to existing high-throughput screening approaches to breach the current limits on the maximum screenable library size?
3. In our directed evolution flow diagram, throughput losses are incurred when in vitro steps are added. Are there in vivo methods for the simultaneous selection and mutagenesis of targeted genetic elements (i.e., accelerated evolution)?
4. Currently, mutagenesis of genomic targets remains more difficult than mutagenesis of plasmid-borne targets. Are there approaches for the simultaneous modification (both up- and downregulation) of multiple genomic targets?

5. A more rigorous standard in small-molecule assay characterization needs to be designed and applied. What additional performance features need to be included? Can elements from the high-throughput screening of small-molecule targets for pharmaceutical development be incorporated?
6. As production titers are improved, product-induced toxicity is becoming a reoccurring theme. In addition to standard selection assays for improved tolerance, can existing themes in small-molecule detection be adapted to transporter engineering?
7. Implementation of transcription factor–derived biosensors will require a greater understanding of our ability to modify transfer curves. To what extent can analog and digital behaviors be engineered? Is there modularity within transcription factor families that can be used to screen for undiscovered ligand receptor domains?

DISCLOSURE STATEMENT

J.D.K. has financial interests in Amyris and LS9. J.A.D. and A.E.M are not aware of any affiliations, memberships, funding, or financial holdings that might be perceived as affecting the objectivity of this review.

ACKNOWLEDGMENTS

This work was part of the DOE Joint BioEnergy Institute supported by the U. S. Department of Energy, Office of Science, Office of Biological and Environmental Research through the contract DE-AC02-05CH11231 between Lawrence Berkeley National Laboratory and the U. S. Department of Energy.

LITERATURE CITED

1. Schmidt-Dannert C, Arnold FH. 1999. Directed evolution of industrial enzymes. *Trends Biotechnol.* 17:135–36
2. Arnold FH, Georgiou G, eds. 2003. *Directed Enzyme Evolution: Screening and Selection Methods*. Totowa, NJ: Humana
3. Niessen WMA. 2006. *Liquid Chromatography-Mass Spectrometry*. Boca Raton, FL: CRC Press. 608 pp.
4. Fortman JL, Chhabra S, Mukhopadhyay A, Chou H, Lee TS, et al. 2008. Biofuel alternatives to ethanol: pumping the microbial well. *Trends Biotechnol.* 26:375–81
5. Atsumi S, Liao JC. 2008. Metabolic engineering for advanced biofuels production from *Escherichia coli*. *Curr. Opin. Biotechnol.* 19:414–19
6. Barker JL, Frost JW. 2001. Microbial synthesis of *p*-hydroxybenzoic acid from glucose. *Biotechnol. Bioeng.* 76:376–90
7. Nakamura CE, Whited GM. 2003. Metabolic engineering for the microbial production of 1,3-propanediol. *Curr. Opin. Biotechnol.* 14:454–59
8. Aldor IS, Keasling JD. 2003. Process design for microbial plastic factories: metabolic engineering of polyhydroxyalkanoates. *Curr. Opin. Biotechnol.* 14:475–83
9. Dale BE. 2003. 'Greening' the chemical industry: research and development priorities for biobased industrial products. *J. Chem. Technol. Biotechnol.* 78:1093–103
10. Patrick WM, Firth AE, Blackburn JM. 2003. User-friendly algorithms for estimating completeness and diversity in randomized protein-encoding libraries. *Protein Eng.* 16:451–57
11. Murakami H, Hohsaka T, Sisido M. 2002. Random insertion and deletion of arbitrary number of bases for codon-based random mutation of DNAs. *Nat. Biotechnol.* 20:76–81

12. Pikkemaat MG, Janssen DB. 2002. Generating segmental mutations in haloalkane dehalogenase: a novel part in the directed evolution toolbox. *Nucleic Acids Res.* 30:e35
13. Osuna J, Yáñez J, Soberón X, Gaytán P. 2004. Protein evolution by codon-based random deletions. *Nucleic Acids Res.* 32:e136
14. Stemmer WPC. 1994. Rapid evolution of a protein in vitro by DNA shuffling. *Nature* 370:389–91
15. Zhao H, Giver L, Shao Z, Affholter JA, Arnold FH. 1998. Molecular evolution by staggered extension process (StEP) in vitro recombination. *Nat. Biotechnol.* 16:258–61
16. Voigt C, Kauffman S, Wang Z. 2000. Rational evolutionary design: the theory of in vitro protein evolution. *Adv. Protein Chem.* 55:79–160
17. Voigt CA. 2006. Genetic parts to program bacteria. *Curr. Opin. Biotechnol.* 17:548–57
18. Blattner FR, Plunkett G, Bloch CA, Perna NT, Burland V, et al. 1997. The complete genome sequence of *Escherichia coli* K-12. *Science* 277:1453–62
19. Serres MH, Gopal S, Nahum LA, Liang P, Gaasterland T, Riley M. 2001. A functional update of the *Escherichia coli* K-12 genome. *Genome Biol.* 2:research0035.1–.7
20. Lin H, Cornish VW. 2002. Screening and selection methods for large-scale analysis of protein function. *Angew. Chem. Int. Ed. Engl.* 41:4402–25
21. Wong TS, Roccatano D, Schwaneberg U. 2007. Steering directed protein evolution: strategies to manage combinatorial complexity of mutant libraries. *Environ. Microbiol.* 9:2645–59
22. Shivange AV, Marienhagen J, Mundhada H, Schenk A, Schwaneberg U. 2009. Advances in generating functional diversity for directed protein evolution. *Curr. Opin. Chem. Biol.* 13:19–25
23. Patrick WM, Firth AE. 2005. Strategies and computational tools for improving randomized protein libraries. *Biomol. Eng.* 22:105–12
24. Firth AE. 2005. *Statistics of randomized library construction.* **http://guinevere.otago.ac.nz/aef/STATS/index.html**
25. Firth AE, Patrick WM. 2008. GLUE-IT and PEDEL-AA: new programmes for analyzing protein diversity in randomized libraries. *Nucleic Acids Res.* 36:W281–85
26. Firth AE, Patrick WM. 2005. Statistics of protein library construction. *Bioinformatics* 21:3314–15
27. Volles MJ, Lansbury PT. 2005. A computer program for the estimation of protein and nucleic acid sequence diversity in random point mutagenesis libraries. *Nucleic Acids Res.* 33:3667–77
28. Sidhu SS, Lowman HB, Cunningham BC, Wells JA. 2000. Phage display for selection of novel binding peptides. *Methods Enzymol.* 328:333–58
29. Leemhuis H, Stein V, Griffiths AD, Hollfelder F. 2005. New genotype-phenotype linkages for directed evolution of functional proteins. *Curr. Opin. Struct. Biol.* 15:472–78
30. Greener A, Callahan M, Jerpseth B. 1997. An efficient random mutagenesis technique using an *E. coli* mutator strain. *Mol. Biotechnol.* 7:189–95
31. Camps M, Naukkarinen J, Johnson BP, Loeb LA. 2003. Targeted gene evolution in *Escherichia coli* using a highly error-prone DNA polymerase I. *Proc. Natl. Acad. Sci. USA* 100:9727–32
32. Wang L, Jackson WC, Steinbach PA, Tsien RY. 2004. Evolution of new nonantibody proteins via iterative somatic hypermutation. *Proc. Natl. Acad. Sci. USA* 101:16745–49
33. Wang L, Tsien RY. 2006. Evolving proteins in mammalian cells using somatic hypermutation. *Nat. Protoc.* 1:1346–50
34. Wang HH, Isaacs FJ, Carr PA, Sun ZZ, Xu G, et al. 2009. Programming cells by multiplex genome engineering and accelerated evolution. *Nature* 460:894–98
35. Yokobayashi Y, Weiss R, Arnold FH. 2002. Directed evolution of a genetic circuit. *Proc. Natl. Acad. Sci. USA* 99:16587–91
36. Weiss R, Homsy GE, Knight TF. 1999. Toward in vivo digital circuits. *DIMACS Workshop Evol. Comput.* 1:1–18
37. Canton B, Labno A, Endy D. 2008. Refinement and standardization of synthetic biological parts and devices. *Nat. Biotechnol.* 26:787–93
38. Tabor JJ, Groban ES, Voigt CA. 2009. Performance characteristics for sensors and circuits used to program *E. coli*. In *Systems Biology and Biotechnology of Escherichia coli*, ed. SY Lee, pp. 402–33. New York: Springer

39. Sylvestre J, Chautard H, Cedrone F, Delcourt M. 2006. Directed evolution of biocatalysts. *Org. Process Res. Dev.* 10:562–71
40. Turner NJ. 2006. Agar plate-based assays. In *Enzyme Assays: High-Throughput Screening, Genetic Selection and Fingerprinting*, ed. J-L Reymond, pp. 139–61. Weinheim, Ger.: WILEY-VCH Verlag GmbH KGaA
41. Mascio PD, Kaiser S, Sies H. 1989. Lycopene as the most efficient biological carotenoid singlet oxygen quencher. *Arch. Biochem. Biophys.* 274:532–38
42. Kim S-W, Keasling JD. 2000. Metabolic engineering of the nonmevalonate isopentenyl diphosphate synthesis pathway in *Escherichia coli* enhances lycopene production. *Biotechnol. Bioeng.* 72:408–15
43. Harker M, Bramley PM. 1999. Expression of prokaryotic 1-deoxy-D-xylulose-5-phosphatases in *Escherichia coli* increases carotenoid and ubiquinone biosynthesis. *FEBS Lett.* 448:115–19
44. Albrecht M, Takaichi S, Steiger S, Wang Z-Y, Sandmann G. 2000. Novel hydroxycarotenoids with improved antioxidative properties produced by gene combination in *Escherichia coli. Nat. Biotechnol.* 18:843–46
45. Sandmann G. 2002. Combinatorial biosynthesis of carotenoids in a heterologous host: a powerful approach for the biosynthesis of novel structures. *ChemBioChem* 3:629–35
46. Smolke CD, Martin VJJ, Keasling JD. 2001. Controlling the metabolic flux through the carotenoid pathway using directed mRNA processing and stabilization. *Metab. Eng.* 3:313–21
47. Alper H, Miyaoku K, Stephanopoulos G. 2005. Construction of lycopene-overproducing *E. coli* strains by combining systematic and combinatorial gene knockout targets. *Nat. Biotechnol.* 23:612–16
48. Farmer WR, Liao JC. 2000. Improving lycopene production in *Escherichi coli* by engineering metabolic control. *Nat. Biotechnol.* 18:533–37
49. Wang C, Leffler S, Thompson DH, Hrycyna CA. 2005. A general fluorescence-based coupled assay for *S*-adenosylmethionine-dependent methyltransferases. *Biochem. Biophys. Res. Commun.* 331:351–56
50. Zatar NA, Abu-Zuhri AZ, Abu-Shaweesh AA. 1998. Spectrophotometric determination of some aromatic amines. *Talanta* 47:883–90
51. Benson JR, Hare PE. 1975. *o*-phthalaldehyde: fluorogenic detection of primary amines in the picomole range. Comparison with fluorescamine and ninhydrin. *Proc. Natl. Acad. Sci. USA* 72:619–22
52. Mukherjee PS, Karnes HT. 1998. Ultraviolet and fluorescence derivatization reagents for carboxylic acids suitable for high performance liquid chromatography: a review. *Biomed. Chromatogr.* 10:193–204
53. Toyo'oka T. 2002. Fluorescent tagging of physiologically important carboxylic acids, including fatty acids, for their detection in liquid chromatography. *Anal. Chim. Acta* 465:111–30
54. Takadate A, Rikura M, Suehiro T, Fujino H, Goya S. 1985. New labeling reagents for alcohols in fluorescence high-performance liquid chromatography. *Chem. Pharm. Bull.* 33:1164–69
55. Vogel M, Büldt A, Karst U. 2000. Hydrazine reagents as derivatizing agents in environmental analysis—a critical review. *Fresenius J. Anal. Chem.* 366:781–91
56. Yu H, Tyo K, Alper H, Klein-Marcuschamer D, Stephanopoulos G. 2008. A high-throughput screen for hyaluronic acid accumulation in recombinant *Escherichia coli* transformed by libraries of engineered sigma factors. *Biotechnol. Bioeng.* 101:788–96
57. Danielson ND, Targove MA, Miller BE. 1988. Pre- and postcolumn derivatization chemistry in conjunction with HPLC for pharmaceutical analysis. *J. Chromatogr. Sci.* 26:362–71
58. Alcalde M, Bulter T, Arnold FH. 2002. Colorimetric assays for biodegradation of polycyclic aromatic hydrocarbons by fungal laccases. *J. Biomol. Screen.* 7:547–53
59. Bornscheuer UT, Altenbuchner J, Meyer HH. 1999. Directed evolution of an esterase: screening of enzyme libraries based on pH-indicators and a growth assay. *Bioorgan. Med. Chem.* 7:2169–73
60. Miesenböck G, Angelis DAd, Rothman JE. 1998. Visualizing secretion and synaptic transmission with pH-sensitive green fluorescent proteins. *Nature* 394:192–95
61. Zhou M, Diwu Z, Panchuk-Voloshina N, Haugland RP. 1997. A stable nonfluorescent derivative of resorufin for the fluorometric determination of trace hydrogen peroxide: applications in detecting the activity of phagocyte NADPH oxidase and other oxidases. *Anal. Biochem.* 253:162–68
62. Charter NW, Kauffman L, Singh R, Eglen RM. 2006. A generic, homogenous method for measuring kinase and inhibitor activity via adenosine 5'-diphosphate accumulation. *J. Biomol. Screen.* 11:390–99
63. Koresawa M, Okabe T. 2004. High-throughput screening with quantitation of ATP consumption: a universal non-radioisotope, homogeneous assay for protein kinase. *Assay Drug Dev. Technol.* 2:153–60

64. Kleman-Leyer KM, Klink TA, Kopp AL, Westermeyer TA, Koeff MD, et al. 2009. Characterization and optimization of a red-shifted fluorescence polarization ADP detection assay. *Assay Drug Dev. Technol.* 7:56–67
65. Klingenberg M. 1974. Nicotinamide-adenine-dinucleotides (NAD, NADP, NADH, NADPH): spectrophotometric and fluorimetric methods. In *Methods of Enzymatic Analysis*. ed. H Bergmeyer, 4:2045–59. New York: Academic
66. Molnos J, Gardiner R, Dale GE, Lange R. 2003. A continuous coupled enzyme assay for bacterial malonyl–CoA:acyl carrier protein transacylase (FabD). *Anal. Biochem.* 319:171–76
67. Smith BC, Hallows WC, Denu JM. 2009. A continuous microplate assay for sirtuins and nicotinamide-producing enzymes. *Anal. Biochem.* 394:101–9
68. Hulcher FH, Oleson WH. 1973. Simplified spectrophotometric assay for microsomal 3-hydroxy-3-methylglutaryl CoA reductase by measurement of coenzyme A. *J. Lipid Res.* 14:625–31
69. Nath A, Atkins WM. 2008. A quantitative index of substrate promiscuity. *Biochemistry* 47:157–66
70. Glieder A, Farinas ET, Arnold FH. 2002. Laboratory evolution of a soluble, self-sufficient, highly active alkane hydroxylase. *Nat. Biotechnol.* 20:1135–39
71. Dietrich J, Yoshikuni Y, Fisher K, Woolard F, Ockey D, et al. 2009. A novel semi-biosynthetic route for artemisinin production using engineered substrate-promiscuous $P450_{BM3}$. *ACS Chem. Biol.* 4:261–67
72. Hendricks CL, Ross JR, Pichersky E, Noel JP, Zhou ZS. 2004. An enzyme-coupled colorimetric assay for *S*-adenosylmethionine-dependent methyltransferases. *Anal. Biochem.* 326:100–5
73. Dorgan KM, Wooderchak WL, Wynn DP, Karschner EL, Alfaro JF, et al. 2006. An enzyme-coupled continuous spectrophotometric assay for *S*-adenosylmethionine-dependent methyltransferases. *Anal. Biochem.* 350:249–55
74. Wagschal K, Franqui-Espiet D, Lee CC, Robertson GH, Wong DWS. 2005. Enzyme-coupled assay for β-xylosidase hydrolysis of natural substrates. *Appl. Environ. Microbiol.* 71:5318–23
75. Mattozzi M, Tehara SK, Hong T, Keasling JD. 2006. Mineralization of paraoxon and its use as a sole C and P source by a rationally designed catabolic pathway in *Pseudomonas putida*. *Appl. Environ. Microbiol.* 72:6699–706
76. Bernhardt P, McCoy E, O'Connor SE. 2007. Rapid identification of enzyme variants for reengineered alkaloid biosynthesis in periwinkle. *Chem. Biol.* 14:888–97
77. Santos CNS, Stephanopoulos G. 2008. Melanin-based high-throughput screen for L-tyrosine production in *Escherichia coli*. *Appl. Environ. Microbiol.* 74:1190–97
78. Sariaslani FS. 2007. Development of a combined biological and chemical process for production of industrial aromatics from renewable resources. *Annu. Rev. Microbiol.* 61:51–69
79. Lütke-Eversloh T, Nicole C, Santos S, Stephanopoulos G. 2007. Perspectives of biotechnological production of L-tyrosine and its applications. *Appl. Microbiol. Biotechnol.* 77:751–62
80. Sprenger GA. 2007. From scratch to value: engineering *Escherichia coli* wild type cells to the production of L-phenylalanine and other fine chemicals derived from chorismate. *Appl. Microbiol. Biotechnol.* 75:739–49
81. Lütke-Eversloh T, Stephanopoulos G. 2007. A semi-quantitative high-throughput screening method for microbial L-tyrosine production in microtiter plates. *J. Ind. Microbiol. Biotechnol.* 34:807–11
82. Lütke-Eversloh T, Stephanopoulos G. 2007. L-tyrosine production by deregulated strains of *Escherichia coli*. *Appl. Microbiol. Biotechnol.* 75:103–10
83. Knight JA, Robertson G, Wu JT. 1983. The chemical basis and specificity of the nitrosonaphthol reaction. *Clin. Chem.* 29:1969–71
84. Lütke-Eversloh T, Stephanopoulos G. 2008. Combinatorial pathway analysis for improved L-tyrosine production in *Escherichia coli*: identification of enzymatic bottlenecks by systematic gene overexpression. *Metab. Eng.* 10:69–77
85. Hall BG. 1981. Changes in the substrate specificities of an enzyme during directed evolution of new functions. *Biochemistry* 20:4042–49
86. Singh S, Kang SH, Mulchandani A, Chen W. 2008. Bioremediation: environmental clean-up through pathway engineering. *Curr. Opin. Biotechnol.* 19:437–44
87. Parales RE, Bruce NC, Schmid A, Wackett LP. 2002. Biodegradation, biotransformation, and biocatalysis (B3). *Appl. Environ. Microbiol.* 68:4699–709

88. Martin VJJ, Pitera DJ, Withers ST, Newman JD, Keasling JD. 2003. Engineering a mevalonate pathway in *Escherichia coli* for production of terpenoids. *Nat. Biotechnol.* 21:796–802
89. Pitera DJ, Paddon CJ, Newman JD, Keasling JD. 2007. Balancing a heterologous mevalonate pathway for improved isoprenoid production in *Escherichia coli*. *Metab. Eng.* 9:193–207
90. Yano T, Oue S, Kagamiyama H. 1998. Directed evolution of an aspartate aminotransferase with new substrate specificities. *Proc. Natl. Acad. Sci. USA* 95:5511–15
91. Oue S, Okamoto A, Yano T, Kagamiyama H. 1999. Redesigning the substrate specificity of an enzyme by cumulative effects of the mutations of non-active site residues. *J. Biol. Chem.* 274:2344–49
92. Rothman SC, Kirsch JF. 2003. How does an enzyme evolved in vitro compare to naturally occurring homologs possessing the targeted function? Tyrosine aminotransferase from aspartate aminotransferase. *J. Mol. Biol.* 327:593–608
93. Jua J, Misonob H, Ohnishic K. 2005. Directed evolution of bacterial alanine racemases with higher expression level. *J. Biosci. Bioeng.* 100:246–54
94. MacBeath G, Kast P, Hilvert D. 1998. Exploring sequence constraints on an interhelical turn using in vivo selection for catalytic activity. *Protein Sci.* 7:325–35
95. Neuenschwander M, Butz M, Heintz C, Kast P, Hilvert D. 2007. A simple selection strategy for evolving highly efficient enzymes. *Nat. Biotechnol.* 25:1145–47
96. Atsumi S, Hanai T, Liao JC. 2008. Non-fermentative pathways for synthesis of branched-chain higher alcohols as biofuels. *Nature* 451:86–90
97. Atsumi S, Liao JC. 2008. Directed evolution of *Methanococcus jannaschii* citramalate synthase for biosynthesis of 1-propanol and 1-butanol by *Escherichia coli*. *Appl. Environ. Microbiol.* 2008:7802–8
98. Pfleger BF, Pitera DJ, Newmana JD, Martin VJJ, Keasling JD. 2007. Microbial sensors for small molecules: development of a mevalonate biosensor. *Metab. Eng.* 9:30–38
99. Kudla G, Murray AW, Tollervey D, Plotkin JB. 2009. Coding-sequence determinants of gene expression in *Escherichia coli*. *Science* 324:255–58
100. Becker S, Schmoldt H-U, Adams TM, Wilhelm S, Kolmar H. 2004. Ultra-high-throughput screening based on cell-surface display and fluorescence-activated cell sorting for the identification of novel biocatalysts. *Curr. Opin. Biotechnol.* 15:323–29
101. Santoro SW, Schultz PG. 2002. Directed evolution of the site specificity of Cre recombinase. *Proc. Natl. Acad. Sci. USA* 99:4185–90
102. Olsen MJ, Stephens D, Griffiths D, Daugherty P, Georgiou G, Iverson BL. 2000. Function-based isolation of novel enzymes from a large library. *Nat. Biotechnol.* 18:1071–74
103. Aharoni A, Thieme K, Chiu CPC, Buchini S, Lairson LL, et al. 2006. High-throughput screening methodology for the directed evolution of glycosyltransferases. *Nat. Methods* 3:609–14
104. van den Berg S, Löfdahl PÅ, Härd T, Berglund H. 2006. Improved solubility of TEV protease by directed evolution. *J. Biotechnol.* 121:291–98
105. Cormack BP, Valdivia RH, Falkow S. 1996. FACS-optimized mutants of the green fluorescent protein (GFP). *Gene* 173:33–38
106. Santoro SW, Wang L, Herberich B, King DS, Schultz PG. 2002. An efficient system for the evolution of aminoacyl-tRNA synthetase specificity. *Nat. Biotechnol.* 20:1044–48
107. Olsen MJ, Gam J, Iverson BL, Georgiou G. 2003. High-throughput FACS method for directed evolution of substrate specificity. See Ref. 2, pp. 329–42
108. Nonomura AM, Coder DM. 1988. Improved phycocatalysis of carotene production by flow cytometry and cell sorting. *Biocatal. Biotransform.* 1:333–38
109. An G-H, Bielich J, Auerbach R, Johnson EA. 1991. Isolation and characterization of carotenoid hyper-producing mutants of yeast by flow cytometry and cell sorting. *Nat. Biotechnol.* 9:70–73
110. Ukibe K, Katsuragi T, Tani Y, Takagi H. 2008. Efficient screening for astaxanthin-overproducing mutants of the yeast *Xanthophyllomyces dendrorhous* by flow cytometry. *FEMS Microbiol. Lett.* 286:241–48
111. Azuma T, Harrison G, Demain A. 1992. Isolation of a gramicidin S hyperproducing strain of *Bacillus brevis* by use of a flourescence activated cell sorting system. *Appl. Microbiol. Biotechnol.* 38:173–78
112. Vidal-Mas J, Resina-Pelfort O, Haba E, Comas J, Manresa A, Vives-Rego J. 2001. Rapid flow cytometry—Nile red assessment of PHA cellular content and heterogeneity in cultures of *Pseudomonas aeruginosa* 47T2 (NCIB 40044) grown in waste frying oil. *Antonie Van Leeuwenhoek* 80:57–63

113. Fouchet P, Jan S, Courtois J, Courtois B, Frelat G, Barbotin JN. 2006. Quantitative single-cell detection of poly(β-hydroxybutyrate) accumulation in *Rhizobium meliloti* by flow cytometry. *FEMS Microbiol. Lett.* 126:31–35
114. Silva TL, Reis A, Medeiros R, Oliveira AC, Gouveia L. 2008. Oil production towards biofuel from autotrophic microalgae semicontinuous cultivations monitorized by flow cytometry. *Appl. Biochem. Biotechnol.* 159:568–78
115. Gouveia L, Marques AE, Silva TL, Reis A. 2009. *Neochloris oleabundans* UTEX #1185: a suitable renewable lipid source for biofuel production. *J. Ind. Microbiol. Biotechnol.* 36:821–26
116. Nebe-von-Caron G, Stephens P, Hewitt C, Powell J, Badley R. 2000. Analysis of bacterial function by multi-colour fluorescence flow cytometry and single cell sorting. *J. Microbiol. Methods* 42:97–114
117. Valli M, Sauer M, Branduardi P, Borth N, Porro D, Mattanovich D. 2006. Improvement of lactic acid production in *Saccharomyces cerevisiae* by cell sorting for high intracellular pH. *Appl. Environ. Microbiol.* 72:5492–99
118. Valli M, Sauer M, Branduardi P, Borth N, Porro D, Mattanovich D. 2005. Intracellular pH distribution in *Saccharomyces cerevisiae* cell populations, analyzed by flow cytometry. *Appl. Environ. Microbiol.* 71:1515–21
119. Lewis DFV. 2002. Oxidative stress: the role of cytochromes P450 in oxygen activation. *J. Chem. Technol. Biotechnol.* 77:1095–100
120. Halliwell B, Whiteman M. 2009. Measuring reactive species and oxidative damage in vivo and in cell culture: How should you do it and what do the results mean? *Br. J. Pharmacol.* 142:231–55
121. Soh N. 2006. Recent advances in fluorescent probes for the detection of reactive oxygen species. *Anal. Bioanal. Chem.* 386:532–43
122. Laroche C, Beney L, Marechal PA, Gervais P. 2001. The effect of osmotic pressure on the membrane fluidity of *Saccharomyces cerevisiae* at different physiological temperatures. *Appl. Microbiol. Biotechnol.* 56:249–54
123. Kolisek M, Zsurka G, Samaj J, Weghuber J, Schweyen RJ, Schweigel M. 2003. Mrs2p is an essential component of the major electrophoretic Mg^{2+} influx system in mitochondria. *EMBO J.* 22:1235–44
124. Gangola P, Rosen B. 1987. Maintenance of intracellular calcium in *Escherichia coli*. *J. Biol. Chem.* 262:12570–74
125. Uchiyama T, Abe T, Ikemura T, Watanabe K. 2005. Substrate-induced gene-expression screening of environmental metagenome libraries for isolation of catabolic genes. *Nat. Biotechnol.* 23:88–93
126. Uchiyama T, Watanabe K. 2008. Substrate-induced gene expression (SIGEX) screening of metagenome libraries. *Nat. Protoc.* 3:1202–12
127. Galvão TC, Mohn WW, Lorenzo Vd. 2005. Exploring the microbial biodegradation and biotransformation gene pool. *Trends Biotechnol.* 23:497–506
128. Williamson LL, Borlee BR, Schloss PD, Guan C, Allen HK, Handelsman J. 2005. Intracellular screen to identify metagenomic clones that induce or inhibit a quorum-sensing biosensor. *Appl. Environ. Microbiol.* 71:6335–44
129. Sitnikov DM, Schineller JB, Baldwin TO. 1995. Transcriptional regulation of bioluminesence genes from *Vibrio fischeri*. *Mol. Microbiol.* 17:801–12
130. Schirmer A, Hu Z, Da Costa B. 2008. *U.S. Patent No. 20,080,293,060*
131. Beggah S, Vogne C, Zenaro E, van der Meer JR. 2008. Mutant HbpR transcription activator isolation for 2-chlorobiphenyl via green fluorescent protein-based flow cytometry and cell sorting. *Microb. Biotechnol.* 1:68–78
132. Breaker RR. 2004. Natural and engineered nucleic acids as tools to explore biology. *Nature* 432:838–45
133. Lalonde S, Ehrhardt DW, Frommer WB. 2005. Shining light on signaling and metabolic networks by genetically encoded biosensors. *Curr. Opin. Plant Biol.* 8:574–81
134. Gu MB, Mitchell RJ, Kim BC. 2004. Whole-cell-based biosensors for environmental biomonitoring and application. In *Biomanufacturing*, ed. J-J Zhong, pp. 269–305. Berlin: Springer-Verlag
135. Endy D. 2005. Foundations for engineering biology. *Nature* 438:449–53
136. Ellis T, Wang X, Collins JJ. 2009. Diversity-based, model-guided construction of synthetic gene networks with predicted functions. *Nat. Biotechnol.* 27:465–71
137. Bintu L, Buchler N, Garcia H, Gerland U, Hwa T, et al. 2005. Transcriptional regulation by the numbers: models. *Curr. Opin. Genet. Dev.* 15:116–24

138. Bintu L, Buchler NE, Garcia HG, Gerland U, Hwa T, et al. 2005. Transcriptional regulation by the numbers: applications. *Curr. Opin. Genet. Dev.* 15:125–35
139. Levitzki A, Schlessinger J. 1974. Cooperativity in associating proteins. Monomer-dimer equilibrium coupled to ligand binding. *Biochemistry* 13:5214–19
140. Gardner TS, Cantor CR, Collins JJ. 2000. Construction of a genetic toggle switch in *Escherichia coli. Nature* 403:339–42
141. Guet C, Elowitz M, Hsing W, Leibler S. 2002. Combinatorial synthesis of genetic networks. *Science* 296:1466–70
142. Basu S, Mehreja R, Thiberge S, Chen M-T, Weiss R. 2004. Spatiotemporal control of gene expression with pulse-generating networks. *Proc. Natl. Acad. Sci. USA* 101:6355–60
143. Basu S, Gerchman Y, Collins CH, Arnold FH, Weiss R. 2005. A synthetic multicellular system for programmed pattern formation. *Nature* 434:1130–34
144. Becskei A, Kaufmann BB, Oudenaarden Av. 2005. Contributions of low molecule number and chromosomal positioning to stochastic gene expression. *Nat. Genet.* 37:937–44
145. Collins CH, Leadbetter JR, Arnold FH. 2006. Dual selection enhances the signaling specificity of a variant of the quorum-sensing transcriptional activator LuxR. *Nat. Biotechnol.* 24:708–12
146. Kim MN, Park HH, Lim WK, Shin HJ. 2005. Construction and comparison of *Escherichia coli* whole-cell biosensors capable of detecting aromatic compounds. *J. Microbiol. Methods* 60:235–45
147. Mohn WW, Garmendia J, Galvao TC, de Lorenzo V. 2008. Surveying biotransformations with à la carte genetic traps: translating dehydrochlorination of lindane (gamma-hexachlorocyclohexane) into *lacZ*-based phenotypes. *Environ. Microbiol.* 8:546–55
148. Fernández S, Shingler V, de Lorenzo V. 1994. Cross-regulation by XylR and DmpR activators of *Pseudomonas putida* suggests that transcriptional control of biodegradative operons evolves independently of catabolic genes. *J. Bacteriol.* 176:5052–58
149. Lutz R, Bujard H. 1997. Independent and tight regulation of transcriptional units in *Escherichia coli* via the LacR/O, the TetR/O and AraC/I1-I2 regulatory elements. *Nucleic Acids Res.* 25:1203–10
150. Lee S, Keasling JD. 2006. Propionate-regulated high-yield protein production in *Escherichia coli. Biotechnol. Bioeng.* 93:912–18
151. Willardson BM, Wilkins JF, Rand TA, Schupp JM, Hill KK, et al. 1998. Development and testing of a bacterial biosensor for toluene-based environmental contaminants. *Appl. Environ. Microbiol.* 64:1006–12

Botulinum Neurotoxin: A Marvel of Protein Design

Mauricio Montal

Section of Neurobiology, Division of Biological Sciences, University of California San Diego, La Jolla, California 92093-0366; email: mmontal@ucsd.edu

Annu. Rev. Biochem. 2010. 79:591–617

First published online as a Review in Advance on March 16, 2010

The *Annual Review of Biochemistry* is online at biochem.annualreviews.org

This article's doi: 10.1146/annurev.biochem.051908.125345

0066-4154/10/0707-0591$20.00

Key Words

channel, chaperone, complexity, membrane protein, modules, unfolding/refolding

Abstract

Botulinum neurotoxin (BoNT), the causative agent of botulism, is acknowledged to be the most poisonous protein known. BoNT proteases disable synaptic vesicle exocytosis by cleaving their cytosolic SNARE (soluble NSF attachment protein receptor) substrates. BoNT is a modular nanomachine: an N-terminal Zn^{2+}-metalloprotease, which cleaves the SNAREs; a central helical protein-conducting channel, which chaperones the protease across endosomes; and a C-terminal receptor-binding module, consisting of two subdomains that determine target specificity by binding to a ganglioside and a protein receptor on the cell surface and triggering endocytosis. For BoNT, functional complexity emerges from its modular design and the tight interplay between its component modules—a partnership with consequences that surpass the simple sum of the individual component's action. BoNTs exploit this design at each step of the intoxication process, thereby achieving an exquisite toxicity. This review summarizes current knowledge on the structure of individual modules and presents mechanistic insights into how this protein machine evolved to this level of sophistication. Understanding the design principles underpinning the function of such a dynamic modular protein remains a challenging task.

Contents

Botulinum neurotoxin (BoNT) serotypes: BoNT occurs in seven antigenic types denoted as serotypes A to G (e.g., BoNT/A)

Botulism: an acute neuroparalytic disease characterized by symmetric descending flaccid paralysis that typically involves the cranial nerve musculature

Light chain (LC): a Zn^{2+} endopeptidase that selectively cleaves the SNARE substrates

H_N: translocation domain

TRI-MODULAR ARCHITECTURE OF THE MATURE TOXIN

Botulinum neurotoxins (BoNTs), a family of bacterial proteins produced by the anaerobic bacterium *Clostridium botulinum*, are potent blockers of synaptic transmission in peripheral cholinergic nervous system synapses (1). There are seven serologically distinct BoNT isoforms (denoted A-G), which exhibit strong amino acid sequence similarity (2). Numerous subtypes have been identified for at least six of the seven serotypes with significant differences at the amino acid level (3, 4). Four of the BoNT serotypes (A, B, E, and F) cause human botulism, a neuroparalytic disease (5). Each BoNT isoform is synthesized as a single polypeptide chain with a molecular mass of ~150 kDa. The inactive precursor protein is cleaved either by clostridial or tissue proteases into a 50-kDa light chain (LC) and a 100 kDa heavy chain (HC) linked by an essential interchain disulfide bridge and by the belt, a loop from the HC that wraps around the LC (1, 6). Structurally, the activated mature toxin consists of three modules (2, 7–9): an N-terminal LC Zn^{2+}-metalloprotease and the HC that encompasses the N-terminal ~50-kDa translocation domain (H_N), and the C-terminal ~50-kDa receptor-binding domain (H_C). The H_C comprises two subdomains—a β-sheet jelly roll fold, denoted H_{CN}, and a β-tree foil fold carboxy subdomain, known as H_{CC}.

The crystal structures of BoNT/A [Protein Data Bank number (PDB) 3BTA] (2, 7), BoNT/B (PDB 1EPW) (8), and BoNT/E (PDB 3FFZ) (9) holotoxins validated a tri-modular architecture of the neurotoxin protein. **Figure 1*a*** shows the side view of the structure of BoNT/A, the archetype (7). The global folds of holotoxins A (7) and B (8) are practically superimposable; the root-mean-square deviation (rmsd) between the LC, H_N, and H_C are 1.43, 1.56, and 1.43 Å, respectively (8). The central H_N, highlighted by the presence of two long coiled-coil helices, demarcates the confinement of the protease and binding domains on either side of the tripartite molecule. The active site of the LC is partially occluded by the belt in the unreduced holotoxin. Although the architecture of the individual, functionally conserved modules of BoNT/E is similar to those of BoNT/A and BoNT/B, their spatial arrangement within the global fold of the holotoxin is unique (**Figure 1*b*,*c***) (9, 10). For BoNT/E, the protease and binding domains appear closely apposed to each other and to H_N, with which they both share a common boundary. The H_N belt in BoNT/E surrounds the LC as it does in BoNT/A and BoNT/B; by contrast, the N-terminal segment, including approximately Ser460 to Leu483, is confined between the LC and the H_C, generating an interface (9) not present in BoNT/A (7) or BoNT/B (8). These structural features of BoNT/E imply an intricate association of the three modules that constrains the global fold. The functional significance of the singular architecture of BoNT/E is unknown.

MULTISTEP MECHANISM OF CELLULAR INTOXICATION

The BoNT tri-modular architecture has a physiological representation. BoNTs exert their neurotoxic effect by a multistep mechanism (1, 11): binding, internalization,

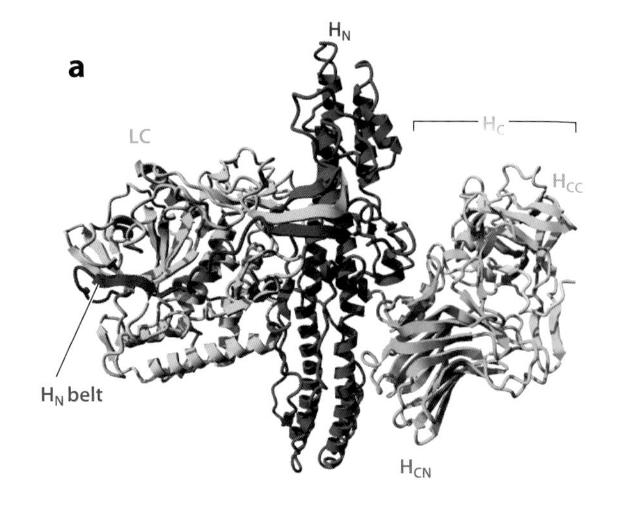

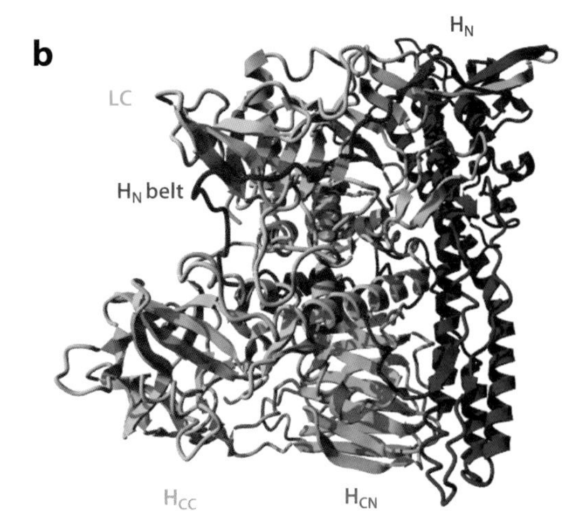

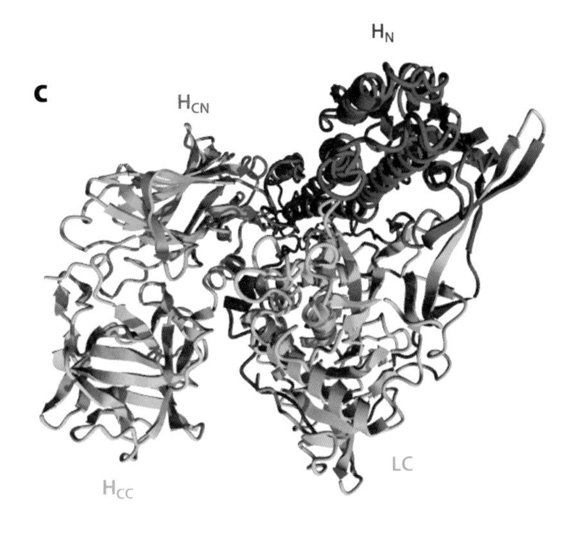

Figure 1

(*a*) Structure of BoNT/A (PDB 3BTA) (7). (*b*) Side view and (*c*) end view of the structure of BoNT/E (PDB 3FFZ) (9). The Cα backbone is represented as ribbons with the light chain (LC) in cyan (residues 1–439 numbered for the archetypal BoNT/A), H_N in dark blue (residues 449–870), and H_C in a green-to-yellow gradient that highlights the two subdomains—H_{CN} and H_{CC} (residues 871–1296). The H_N belt is depicted in red. All images were rendered on YASARA (**http://www.yasara.org**).

H_C: receptor-binding domain

H_{CN}: N-terminal subdomain of the receptor-binding domain

H_{CC}: C-terminal subdomain of the receptor-binding domain

rmsd: root-mean-square deviation

Gangliosides: oligosaccharide-rich sphingolipids that contain sialic acid and are coreceptors for BoNT entry into neurons

Syt: synaptotagmin

SV: synaptic vesicle

Soluble N-ethylmaleimide-sensitive factor attachment protein receptor (SNARE) complex: the synaptic vesicle docking/fusion complex that mediates membrane fusion and is essential for neurotransmitter release

SNAP-25: synaptosome-associated protein of 25 kDa

membrane translocation, intracellular traffic, and proteolytic degradation of target. H_C determines the cellular specificity mediated by the high-affinity interaction with a surface protein receptor, and a ganglioside (GD1b or GT1b) coreceptor (12, 13, 14, 15). The protein receptors are SV2 for BoNT/A (12, 13), BoNT/E (16), and BoNT/F (17), and synaptotagmin (Syt) I and II for BoNT/B (18–20) and BoNT/G (14). Then, BoNTs enter sensitive cells via receptor-mediated endocytosis (1, 11, 21). Exposure of the BoNT-receptor complex to the acidic milieu of endosomes induces a conformational change, leading to the insertion of the HC into the endosomal membrane (11, 21–24), thereby forming a transmembrane protein-conducting channel that translocates the LC to the cytosol where it acts (1, 25). The LCs are sequence-specific endoproteases that cleave unique components of the synaptic vesicle (SV) docking-fusion complex known as the SNARE (soluble *N*-ethylmaleimide-sensitive factor attachment protein receptor) complex. The SNARE complex is widely considered to be the catalyst of membrane fusion and is essential for neurotransmitter release (26–28). BoNT/A and BoNT/E cleave the plasma membrane-associated protein SNAP-25 (synaptosome-associated protein of 25 kDa), whereas BoNT types B, D, F, and G proteolyze synaptobrevin, a vesicle-associated membrane protein, also known as VAMP (1), the most abundant SV entity (29). Unique among the BoNTs is BoNT/C, which cleaves both SNAP-25 and syntaxin, another plasma membrane-anchored SNARE. Cleavage of the SNAREs abrogates vesicle fusion and synaptic transmission, which thereby causes the severe paralysis pathognomonic of botulism.

EMERGENCE OF BoNT FUNCTIONAL COMPLEXITY FROM ITS MODULAR DESIGN

The question arises as to how this modular arrangement determines the diverse activities expressed by the holotoxin in intoxicated cells. Each module has been generated from the holotoxin or produced as a recombinant protein, and the isolated entities have been shown nontoxic to cells. At the neutral pH of the extracellular environment, the holotoxin is a water-soluble protein; by contrast, during its intracellular journey, BoNT becomes a membrane-embedded protein. This drastic change in solvent environment is a requirement for cellular toxicity. The isolated protease is impermeable across membranes and for this reason cannot reach its cytosolic substrates. In contrast, in the context of the holotoxin, the LC is delivered by H_N to the cytosol after receptor-mediated endocytosis. Neither the isolated H_N nor the H_C exhibit SNARE proteolytic activity and, therefore, are nontoxic.

Initial studies indicated that collapsing the transmembrane pH gradient (ΔpH) prevalent across endosomes prevented intoxication (11, 21). This was accomplished by pretreating the neuromuscular junction with agents, such as chloroquine or methylamine, that equilibrate the endosomal ΔpH (21) or by using bafilomycin to inhibit the vacuolar proton pump responsible for endosomal acidification (30). The implication is that the ΔpH across endosomes triggers a conformational change of the endocytosed toxin that promotes its insertion into the membrane and eventual translocation of the protease into the cytosol. There is evidence that the BoNT aggregates and in due turn precipitates in aqueous solutions below pH ~5.5 (24) and that the isolated H_N precipitates in the pH range of 5.5–3.5 with a concomitant increase in its propensity to insert into membranes (31). For the isolated LC of BoNT/A (LC/A) (32), there is a decrease in helicity at pH 5.0 and an increase at pH 4.5, the pH at which its protease activity is abrogated (33). The inference is that a ΔpH-induced concerted insertion of H_N into the membrane with a partial unfolding of the LC are critical events in the translocation of active protease across endosomes. A strong body of evidence has accumulated supporting the notion that H_N is a protein-conducting channel that chaperones the protease across endosomes (**Figure 2**) acidic conditions existent inside lysosomes. H_C dictates the target cell specificity and, during cell binding and intracellular traffic, serves to chaperone the LC and H_N, which ensures that partial unfolding of the LC is concomitant with H_N channel formation, thereby promoting productive LC translocation (35–37). We turn now to examine the structure of individual modules and to draw inferences about plausible functional consequences. In doing so, we follow the path of a toxin after its encounter with sensitive neurons.

PE: phosphatidylethanolamine

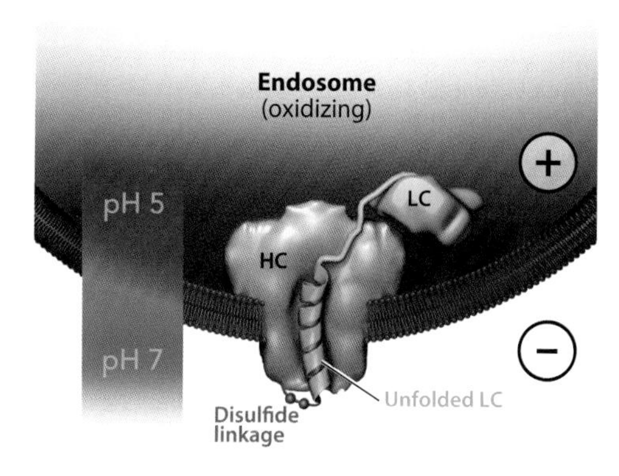

Figure 2

Representation of a molecular chaperone driven by ΔpH across endosomes. The heavy chain (HC) channel (depicted in *orange*) prevents the aggregation of the light chain (LC) cargo (depicted in *cyan*) in the acidic endosomal lumen, maintaining the unfolded or partially folded conformation (*cyan ribbon*) during translocation and releasing the LC after it refolds at the neutral cytosolic pH. In the cycle, the H_N channel is occluded by the LC during transit and open after completion of translocation and release of cargo. The LC is illustrated as both a ribbon and a space-filling shape to highlight the unfolding/refolding dynamics of cargo during transit and the associated transient interactions between cargo and channel. The disulfide linkage between the HC and the LC is shown in green. The + and – signs denote the polarity of the membrane potential across endosomes. Craig Foster of Foster Medical Communications prepared the artwork.

(33–38). This constitutes a fascinating example of molecular partnership: A protein-conducting channel driven by ΔpH mediates cargo unfolding, maintains an unfolded conformation during translocation, and releases cargo after it refolds in the cytosol. It is remarkable that the endosomal pathway is ideally suited for BoNT processing: The ΔpH of early endosomes is finely tuned to elicit drastic conformational changes, leading to the insertion of BoNT into the membrane, while it is auspiciously set to interrupt further processing in the harsh

THE RECEPTOR-BINDING DOMAIN: TOXIN ENTRY INTO SENSITIVE CELLS

BoNTs gain entry into neurons by exploiting the SV recycling pathway. What are the determinants for the distinctive neuronal specificity of BoNTs?

Ganglioside Coreceptor

Binding of BoNTs to the peripheral neuromuscular junction is the initial step of intoxication (1). This event involves the tight association between BoNTs with complex polysialogangliosides known to be enriched in neurons (39). Among these, disialo (GD1b)- and trisialogangliosides (GT1bs) exhibit BoNT-binding affinities in the nM range and thus establish an initial anchorage to the neuronal membrane. Indeed, a monoclonal antibody to GT1b hinders the action of BoNT/A on cervical ganglionic neurons (40). And, inhibition of ganglioside biosynthesis in the neuroblastoma cell line Neuro2a renders them insensitive to BoNT/A (41). Overall, BoNT types A, B, C, and F bind GT1b, GD1b, and GD1a. BoNT/E binds GT1b and GD1a, whereas BoNT/G recognizes all gangliosides with approximately similar affinity. By contrast, BoNT/D binds phosphatidylethanolamine (PE) and by inference lacks the ganglioside-binding pocket characteristic of other BoNTs.

The ganglioside-binding pocket is confined to H_{CC} (42–44). The structure of BoNT/A

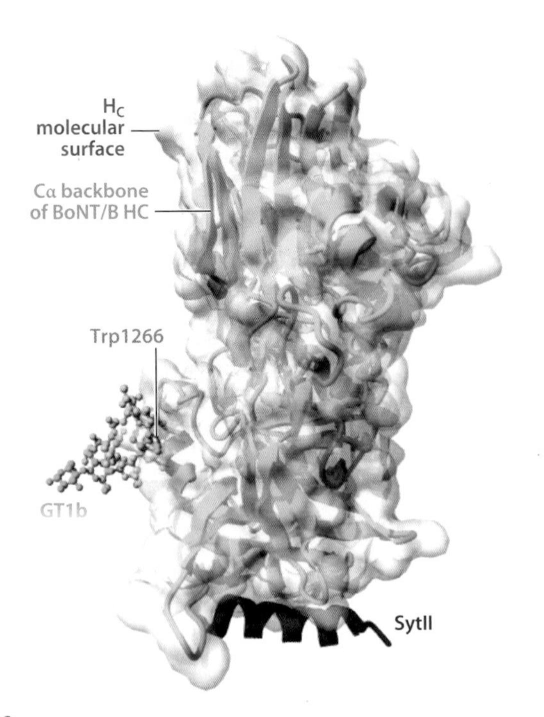

Figure 3

The two-receptor model for BoNT recognition at the neuronal membrane: the protein receptor SytII and the coreceptor GT1b bound onto H_C. Superposition of the structures of BoNT/A H_C (residues 873–1297) in complex with GT1b (PDB 2VU9) (45) and of the BoNT/B H_C (residues 858–1291) in complex with the SytII peptide (PDB 2NM1) (47). The Cα backbone of the BoNT/B H_C is represented as cyan ribbons, and its molecular surface in transparent gray. GT1b is displayed in a green-yellow rainbow using the ball-and-stick representation, and Trp1266 is shown in magenta. For the composite display, the Cα backbone of BoNT/A H_C was removed. For the superposition, the backbone atoms were used for the best fit between the structures.

H_C alone (PDB 2VUA) and in complex with GT1b (PDB 2VU9) was determined at 1.6-Å resolution (**Figure 3**) (45). The overall structures of the isolated, recombinant H_C/A (PDB 3FUO) (17) and H_C/B (PDB 1Z0H) (46) are similar to those in the context of their respective holotoxins, BoNT/A (PDB 3BTA) (7) and BoNT/B (PDB 1EPW) (8). The structure of H_C from BoNT/B (PDB 2NM1) (47), BoNT/E (PDB 3FFZ) (9), and BoNT/F (PDB 3FUQ) (17) supports the view that the H_C fold is highly conserved. H_{CC} exhibits a β-tree foil fold. GT1b binds on a crevice formed by Trp1266 and Tyr1267 on one face and Glu1203, His1253, and Phe1252 on the other for BoNT/A (45). Trp1266 (magenta on **Figure 3**) and Tyr1267, conserved among all BoNT serotypes, constitute key residues of a lactose-binding motif (H...SXWY...G) (48) that contribute to the hydrophobic character of the binding cavity via extensive interactions with GT1b (45). The H_C structures in the presence and absence of GT1b reveal only minor differences, with an overall rmsd of 0.3 Å (45). This indicates that the fold is robust and undergoes minimal structural fluctuations upon GT1b binding. The structure of H_C from BoNT/B in complex with the trisaccharide sialyllactose, a mimic of GT1b (PDB 1F31), displays a similar binding cavity with corresponding residues Trp1261, Tyr1262, and His1240 (8). Mutational analysis had previously identified these residues as critical for GT1b binding for BoNT/A and BoNT/B (48). Such analysis, conducted on BoNT/C, indicates that the corresponding residues Trp1257 and Tyr1258 are deterministic for GT1b binding (49, 50). In addition, the crystal structure of H_C from BoNT/F displays a ganglioside-binding pocket with corresponding residues Trp1250, Tyr1251, and His1241 (17). The implication is that the GT1b-binding pocket for all these BoNTs is similar (45). In contrast, BoNT/D, which interacts with PE and not with GT1b, lacks these key residues and shows a dependency on Lys1117 and Lys1135 for PE binding (50).

Ganglioside binding, however, does not account entirely for the neurotropism of BoNT or that of the related *Clostridium tetani* tetanus neurotoxin (TeNT) (1). And, given the trypsin-sensitivity of BoNT (51) and TeNT (51, 52), binding to neuronal membranes led to the explicit formulation of the two-receptor hypothesis: a common ganglioside coreceptor and a high-affinity protein receptor that confer serotype specificity (53).

Surface Protein Receptor

Conditions that increase SV recycling, such as neuronal activity, lead to enhanced uptake of BoNTs into neurons (11, 54). This is consistent with the notion that protein components of the SV membrane are candidates for the elusive BoNT protein receptor given that they are transiently integrated into the plasma membrane upon vesicle fusion. Indeed, Syt, widely considered the Ca^{2+} sensor that triggers SV fusion (26, 55), was initially identified as the protein receptor for BoNT/B (15, 19, 20). Another family of SV membrane proteins, SV2, was later identified as the protein receptor for BoNT/A (SV2A) (12), BoNT/E (glycosylated SV2A and SV2B) (16), and recently BoNT/F (glycosylated SV2) (17). Interestingly, the BoNTs' serotypes that exhibit highest sequence similarity share the same protein receptor, i.e., BoNT types A, E, and F bind SV2, whereas BoNT types B and G bind SytI and II (56). The protein receptor(s) for BoNT/C and BoNT/D remain unknown. The BoNT-binding segment of both SytII and SV2 proteins is confined to the domain that is exposed to the SV luminal space. The BoNT/B-binding segment of SytII is specifically delimited to the 20 residues proximal to the membrane (residues 40–60) with the sequence GESQEDMFAKLKEKFFNEINK (18), and is further restricted to residues 47–60 (57). The identical segment of SytI and II was revealed as the specific determinant for BoNT/G (14, 58). A turning point was achieved with the determination of the crystal structure of BoNT/B in complex with a peptide corresponding to SytII residues 40–60 (PDB 2NP0) (57) and of a binary complex between BoNT/B H_C and SytII residues 8–61 (PDB 2NM1) (47). For the latter, residues 8–43 were not discerned, yet the electron density map allowed the unambiguous assignment of residues 44–60 (47). Significantly, both studies showed that the isolated SytII peptide is unstructured in solution; nevertheless, it adopts an α-helical conformation upon binding to the toxin (59): an α-helix encompassing Asp45 and Asn59 in the first structure (57) and Phe47 and Ile58 in the second (**Figure 3**) (47). The SytII helix binds to an adjacent but different site than GT1b on a hydrophobic groove formed by two β-strands at the C end of the H_{CC} β-tree foil.

The H_{CC} structures in presence and absence of the SytII peptide are extremely similar (overall rmsd of 0.73 Å) (47, 57). This finding, analogous to that with GT1b, implies not only that the H_{CC} fold is rigid but that the binding sites for GT1b and SytII are nonoverlapping and noninteracting, which thereby precludes potential allosteric effects between protein receptor and coreceptor. Such inference affords the opportunity to model both entities onto H_{CC}. Unfortunately, the information is incomplete: There is no structure of a BoNT/A–SV2 complex nor of a BoNT/B–GT1b complex. What we have are the structures of the BoNT/A H_{CC}–GT1b complex (PDB 2VU9) (45) and of the BoNT/B H_{CC}–SytII peptide complex (PDB 2NM1) (47), allowing us to build a heuristic model of SytII and GT1b onto H_{CC} shown in **Figure 3**. This rendering embodies the two-receptor model at atomic resolution. The two independent determinants of BoNT neurotropism are a ganglioside anchorage site, which presumably hinders the lateral diffusion of the bound BoNT in the plane of the neuronal membrane and concurrently enriches its surface density, and a protein receptor that dictates serotype specificity. The distance between SytII and GT1b is 22 Å (45); this delimits the attachment area of the holotoxin bound with its two receptors onto the presynaptic membrane. Intriguingly, an intervening hydrophobic loop present between the SytII- and GT1b-binding sites (loop 1250 on BoNT/B) may insert into the membrane and thereby bring the H_N long helices into proximity to the membrane (45). Alternatively, the hydrophobic faces of the H_N helices oriented parallel to the membrane plane may juxtapose to it, promoting their insertion (47). A global protein dipole with the positive charge on H_{CC} may orient BoNTs for enhanced binding to the presynaptic membrane (60). Given the helix-inducing effect of H_{CC} on its partner receptor, it is conceivable that the solvent-accessible surface of each BoNT serotype is unique, e.g., the surface of the BoNT/A-SV2-GT1b ternary complex would differ from BoNT/E-glycosylated SV2A-GT1b. Each complex may expose distinct charged residues that by interacting with phospholipid headgroups may further facilitate membrane insertion (60). Together, the ternary complex acquires new surface features that confer a propensity for novel interactions and that endow it with orientational information suitable for membrane insertion (45, 47, 53, 57, 60).

Heavy chain (HC) channel/chaperone: a protein-conducting channel that preserves partially unfolded and native-like conformers of cargo during translocation across membranes

What is the role then of H_{CN}? H_{CN} adopts a jelly roll fold, a recurrent motif found in numerous proteins involved in recognition of diverse ligands. A phosphatidylinositol phosphate (PIP)-binding site was identified on BoNT/A H_{CN} on the basis of its specific binding to sphingomyelin-enriched membrane microdomains (61). This result is significant in a number of ways. First, phosphoinositides are recognized as key intracellular signals relevant to vesicular traffic (62). Second, a PIP-binding motif similar to that present in the structure of the PIP-binding protein ING2 (63) was discerned on BoNT/A H_{CN} (61), revealing an array of positive residues suitable for interactions with the PIP phosphate (Arg892, Lys896, Lys902, and Lys910) (61). This cluster is exposed on the face of H_{CN} that is opposite to the one on which the GT1b-binding pocket of H_{CC} is located. If this motif is involved in anchoring H_N to the membrane, then a significant rotation of H_{CN} with respect to H_{CC} would be required to accommodate the two subdomains on an equivalent orientation with respect to the membrane. Is this flexibility compatible with the robustness of the fold? Third, PIP optimizes channel formation by diphtheria toxin (DT) (64): PIP is required on the opposite side of the membrane to that in which DT is inserted, implying that specific lipid-protein interactions have a role in protein translocation (64). These considerations suggest a role for a PIP-binding motif on BoNT/A H_{CN}, presumably by promoting the proximity of H_C to the membrane via a dual lipid anchorage mediated through PIP (H_{CN}) and GT1b (H_{CC}) (61). Accordingly, a preinsertion conformation of H_N may be primed for insertion once the conditions present across endosomes are satisfied to reach a translocation-competent conformation (**Figure 2**). Further analysis is required to explore this notion.

THE TRANSLOCATION DOMAIN: INTRACELLULAR PROCESSING OF THE TOXIN

The pathway that BoNTs follow after receptor-mediated endocytosis embodies a steadfastly decoded, multistep process. A crucial intracellular event for BoNT neurotoxicity is the translocation of the LC protease across endosomes. The collective evidence, although still incomplete, suggests a hypothetical minimum scenario considered next.

The structure of H_N (**Figure 4**) is available from crystals of holotoxin types A (7), B (8), and E (9) and of a truncated version of BoNT/A consisting only of the LC and H_N (LC-H_N) (65). No structure is available for the isolated HC or H_N of any of the BoNT serotypes. H_N exhibits high sequence conservation among all the serotypes with the exception of the belt that diverges.

LC translocation by the HC is stringently dependent on conditions that mimic those prevalent across endosomes, which include the following: (*a*) a pH gradient, acidic on the inside and neutral on the cytosol; (*b*) a redox gradient, oxidizing on the inside and reducing on the cytosol; and (*c*) a transmembrane potential, positive inside (**Figure 2**) (33–36). Under these conditions, retrieval of a folded, catalytically active LC endopeptidase was demonstrated after completion of cargo translocation across lipid bilayers and release from the channel (33). Implementation of a single-molecule translocation assay afforded real-time measurements of LC translocation of both BoNT types A and E by their respective HC in membrane patches of neuroblastoma cells as a progressive increase of channel conductance (10, 33–36). A model of the sequence of events was formulated (**Figure 5**) (36). Step 1 illustrates BoNT/A prior to membrane insertion, followed by an entry event (step 2) with the LC (cyan) trapped within the HC channel (orange), a series of transfer steps (steps 3 and 4), and an exit event (step 5). The disulfide bridge (green) between the LC and the HC is intact in the low pH oxidizing environment of the endosomal lumen (*cis*-compartment). The presence of reductant and neutral pH in the cytosol (*trans*-compartment) promotes release of the LC from the HC after completion of translocation. The channel recordings displayed under each step indicate the progressive increase in single-channel conductance. This observable is interpreted as the progress of LC translocation, during which the HC channel is transiently occluded by the LC during transit, then unoccluded after completion of translocation and release of cargo. During translocation, the HC channel conducts gradually more Na^+ around the unfolded LC (illustrated as a helix in steps 2, 3, and 4) before entering an exclusively ion-conductive state (step 5). Given that unreduced BoNT/A does not form channels, that prereduced BoNT/A forms channels with properties equivalent to those of the isolated HC (33–36), and that the HC is a channel irrespective of the redox state (33, 34, 36) implies that in unreduced holotoxin the anchored LC cargo occludes the HC channel (step 2) (10, 33–36).

Information about unfolding and refolding pathways is limited. There is evidence that only the unfolded conformation of the LC correlates with both channel and protease activities of BoNT/A (33). This condition necessarily constrains the cargo to be either extended or α-helical segments to fit into a channel of ~15 Å in diameter (10, 33, 35, 36). The importance of unfolding for efficient translocation was also documented for BoNT/D (66). Similarly, a requirement for acid-induced unfolding was reported for the translocation of the catalytic moieties of anthrax [the lethal factor (LF)] (67, 68), the *C. botulinum* C2 toxin (69, 70), and DT (71) through their respective translocation pores.

Figure 4

Structure of BoNT/A H_N. The Cα backbone is depicted in dark blue for residues Asn450–Glu491, in red for the belt residues Asn492–Asp545, and in cyan for Lys546–Lys870. (*a*) Side view of BoNT/A H_N and (*b*) is the end view. Overall, the H_N fold consists of two long (105-Å) coiled-coil helices, two short helices on the N (residues 602–617) and C (residues 834–844) boundaries, and the belt—a ~50-residue-long unstructured loop or helix that encircles the protease. Note the cavity created by the belt that is occupied by the light chain (LC) in the holotoxin structure (PDB 3BTA) (7) and the structure of the LC-H_N (PDB 2W2D) (65). The long helices are kinked in the center, hinting to a potentially susceptible site to fold over into a four-helix bundle. Intriguingly, this is roughly the location at which the belt starts and ends (shown in *red*).

What is the role of the interchain disulfide linkage in translocation? For BoNT/A, the disulfide cross-link must be on the cytosolic compartment to achieve productive cargo translocation (step 5) (33, 35, 36). Disulfide reduction prior to translocation dissociates the LC from the HC, thereby generating a channel devoid of translocation activity (33, 34). Disulfide disruption within the bilayer during translocation aborts it (35). The implication is that completion of translocation occurs as the disulfide bridge, the C terminus of the LC, enters the cytosol and is the last portion to be translocated and to exit the channel (step 5). Accordingly, LC refolding in the cytosol may be interpreted as a trap that precludes retrotranslocation and dictates the unidirectional nature of the process. The disulfide linkage is, therefore, a crucial aspect of the BoNT toxicity (72–74) and is required for chaperone function, acting as a principal determinant for cargo translocation and release. BoNT/A is cleaved to the mature dichain toxin within the *Clostridium* bacteria; in contrast, BoNT/E is not (6). For BoNT/E, completion of LC translocation occurs only after proteolytic cleavage of the LC from the HC and disulfide reduction in the cytosol, implying that release of cargo from chaperone is necessary for productive translocation (step 5) (**Figure 5**) (35, 36).

These findings pose a new set of questions. How is cargo conformation protected by the channel during translocation to ensure proper refolding after translocation is complete? What is the significance of the intermediate states identified during translocation (10, 34–36)? The lifetime of each intermediate may reflect the conformational changes of cargo within the chaperone pore that determine the outcome of translocation. The occurrence of intermediate states reflects a propensity to preserve partially unfolded

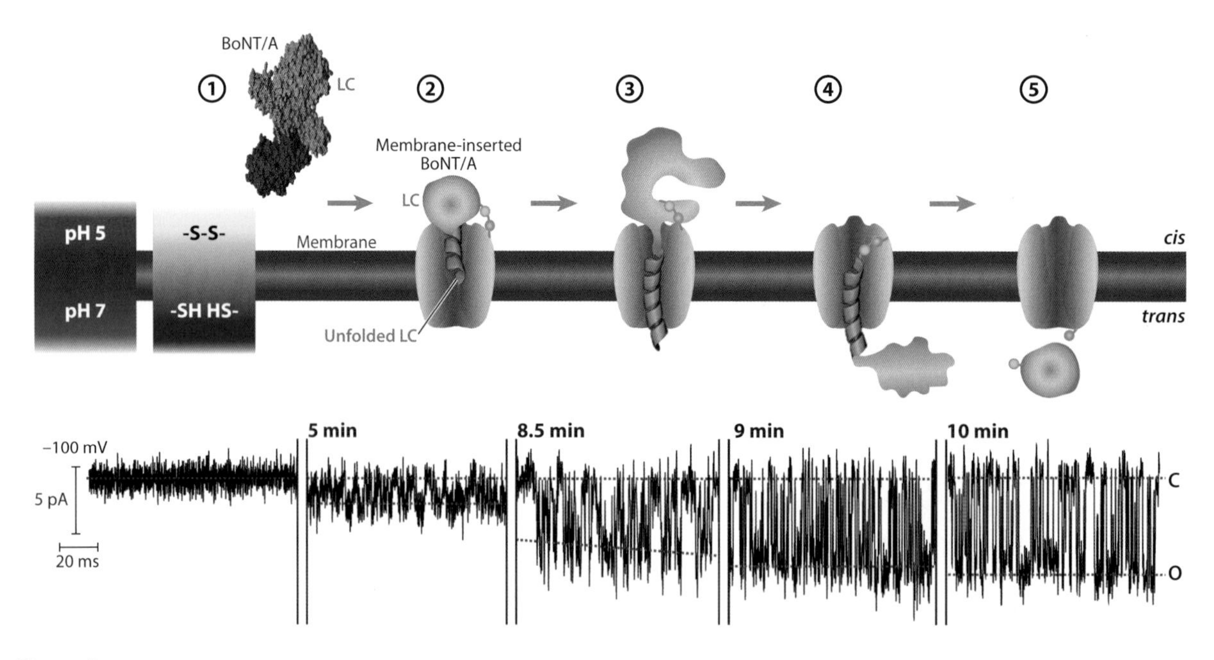

Figure 5

Sequence of events underlying BoNT LC translocation through the HC protein-conducting channel. Step ① BoNT/A holotoxin prior to insertion in the membrane; BoNT/A is represented by the crystal structure (PDB 3BTA) (7). Then, there is a schematic representation of the membrane-inserted BoNT/A during an entry event step ②, a series of transfer steps ③ and ④, and an exit event in step ⑤, under conditions that recapitulate those across endosomes. Segments of typical channel currents, recorded at –100 mV, are displayed under the corresponding interpretation for each step. Downward deflections reflect channel openings; C and O denote closed and open states. Modified and reproduced with permission from Reference 36. Copyright © 2007, National Academy of Sciences, U.S.A.

conformers, (indicated by the occluded intermediates) (**Figure 5**, steps 2–4) (36) and native-like conformers (demonstrated by the recovery of LC protease activity at the end of translocation) (**Figure 5**, step 5) (33). The evidence supports the notion of an interdependent, intimate interaction between the HC chaperone and the cargo preventing LC aggregation and dictating the outcome of translocation, i.e., the productive passage of cargo or abortive channel occlusion by cargo.

The modular design of BoNTs implies that channel formation may be dissociated from cargo translocation. Indeed, the channel-forming entity is confined to H_N (**Figure 4**), which, at variance with the holotoxin and the dimodular HC (H_N–H_C) (**Figure 1**), displays channel activity irrespective of a transmembrane ΔpH (35–37). For the HC, the pH threshold for membrane insertion and channel formation is modulated by the interactions between the H_N and H_C modules. H_C is dispensable for both channel activity and LC translocation under mild acidic conditions (pH ~6) (37). In contrast, H_C restricts insertion of H_N into the membrane until it is localized to acidic endosomes, where a pH ~5 induces channel insertion concurrent with partial cargo unfolding, thereby triggering productive LC translocation. In addition to their individual functions, each of the three BoNT modules acts as a chaperone for the others, working in concert to achieve productive intoxication (35–37); H_C insures that H_N channel formation occurs concurrently with LC unfolding to a translocation-competent conformation. At the positive membrane potential prevailing across endosomes, the H_N channel would be closed until it is gated by the LC to initiate its translocation into the cytosol (34). H_N protects the LC within the acidic lumen of endosomes, chaperones it to the cytosol, and delivers it in an enzymatically active conformation to access its substrates (33). The H_N belt may act as a surrogate pseudosubstrate for the LC until localization within the cytosol (75) (see below).

What is the fate and function of the H_N channel in the endosome after completion of translocation? The HC and H_N form transmembrane channels implying that they remain embedded in the membrane until degraded; there is no known driving force that would promote membrane ejection. Accordingly, while resident in endosomes, the BoNT channel is either open by the LC to initiate translocation or closed after completion of LC translocation, thereby precluding disruption of cellular homeostasis (34, 36, 37).

CYTOSOLIC EVENTS AFTER LIGHT CHAIN PROTEASE TRANSLOCATION

It is clear that the LC released after completion of translocation refolds in the cytosol given that it is competent in SNARE cleavage (33, 35–37). However, information about cytosolic events proceeding after translocation that lead to inhibition of neurotransmitter release is scarce. Studies using lipid bilayers devoid of additional cellular components demonstrated that productive translocation of active protease is an inherent property of the holotoxin and proceeds autonomously in the absence of auxiliary proteins (33). LC refolding inside neurons is presumably a highly regulated process facilitated by proteins enriched at the surface of endosomes or resident in the highly concentrated environment of the cytosol. Key unsolved questions include the following: What determines cargo refolding after translocation? When is refolding initiated? What is the role of cytosolic chaperones in LC refolding, trafficking, transport to sites of substrate enrichment, and encounter of specific substrates? Does substrate cleavage occur near the membrane surface?

The involvement of cytosolic chaperones has been documented for DT (76) and for *C. botulinum* C2 toxin, a binary actin-ADP-ribosylating toxin in which the activated heptameric C2IIa-binding component translocates the catalytic moiety C2I into the cytosol (77). The heat shock protein Hsp90 is required for productive translocation of the catalytic domains of both the DT (DTA) (76) and the C2 toxin (77) from the lumen of acidified endosomes to the cytosol. A 10-amino acid motif in the DT transmembrane helix 1, conserved in anthrax toxin LF and edema factor, and also in BoNT types A, C, and D, was identified as a mediator of the toxin-chaperone interactions. This led to the subsequent identification of COPI coatomer complex proteins as catalysts for the delivery of anthrax LF to the cytosol in a manner analogous to that exhibited by DTA (78). No specific cytosolic chaperones for any of the BoNTs are known, though the density of chaperones attached to the SV surface is significant (29).

The selective encounter of the LC with its substrates prior to SNARE complex assembly presumably is guided by unidentified proteins that mediate trafficking and specific delivery. The location of the target substrate is distinct for different BoNT serotypes, namely the cytosolic face of SVs for synaptobrevin and of the plasma membrane for SNAP-25 and syntaxin. Synaptobrevin is the most abundant SV component (about 70 copies per vesicle) (29). Its exposed cytosolic domain, the target for BoNT types B, D, F, and G, may be shielded by other SV resident proteins that are also present in high copy numbers (e.g., synaptophysin and Syt, ~31 and ~15 copies per SV, respectively) (29). SNAP-25 and syntaxin have a plethora of interacting partners, both cytosolic and membrane bound. Prominent among these are regulators of the SNARE complex's assembly—the Sec1- and Munc18-like proteins, complexin, Syt (26, 79, 80), adaptor proteins, and small Rab GTPases (26, 29). Munc18-1 is a deterministic factor for the plasma membrane localization of syntaxin-1 in neuroendocrine cells (81). A myriad of potential interactive proteins (82) may occupy the substrate unfolded domains required to fold and bind to the exosites of the LC upon its encounter (see below).

How is LC degradation avoided? This is an area of great interest and intense research because the lifetime of active protease inside neurons dictates the ultimate duration and severity of its action, a crucial determinant in the ever-growing number of clinical applications of BoNT as a drug (83). The mechanism underlying the disparity of longevity of different BoNT serotypes is currently an enigma. Intriguingly, cleavage of the same substrate by BoNT/A and BoNT/E produces the longest (up to 12 months) and the shortest (few weeks) durations of neurotransmitter blockade (84). Several factors contribute to this discrepancy. The two serotypes cleave SNAP-25 at different sites (85), producing SNAP-25 fragments of different sizes that inhibit exocytosis by competing for SNARE complex assembly (86–88). Syntaxin and SNAP-25 form a stable dimer prior to the assembly of the ternary SNARE complex and colocalize in extensive clusters at the plasma membrane (89). BoNT/E dissociates the dimer and disrupts the clusters (89), implying that the 26 C-terminal residues of SNAP-25 are essential for SNARE assembly. The subcellular distribution of the two proteases is different: The BoNT/A LC colocalizes with the truncated SNAP-25 product at the plasma membrane, whereas BoNT/E appears to be diffusely distributed in the cytosol (90). For BoNT/E, entry into neurons and intracellular action are faster than for BoNT/A (54, 91, 92). Chimeric constructs, consisting of the protease and H_N of BoNT/E fused to the H_C of BoNT/A, displayed equivalent characteristics to holotoxin E, entering neurons faster than BoNT/A; in contrast, a chimera containing the LC and H_N of BoNT/A joined to H_C of BoNT/E entered neurons more slowly, yet its paralytic effect on mice was longer—akin to that produced by holotoxin A (92). These findings are consistent with the view that H_N dictates the speed of intoxication (type E faster than type A) (92), whereas LC-H_N determines the differential pH sensing for translocation (37, 92). Furthermore, BoNT/E requires host cell proteases to cleave its LC from the HC after translocation, potentially yielding a population of LCs with unique C termini and resulting sensitivity to modification or degradation.

Is the longevity of active protease or its degradation modulated by covalent protein modifications? Presumably one or more of the known protein modification pathways are usurped by BoNTs to prolong their residence inside neurons. Occurrence of a tyrosine-phosphorylated LC inside neurons was demonstrated for types A and E (93). Such modification increases both their catalytic activity and thermal stability, hinting that the biologically significant form of BoNT inside neurons is phosphorylated (93). Ubiquitylation is more profuse for type E than for type A (94), which may determine the extent of proteasomal degradation (G. Oyler, unpublished data). Palmitoylation of Cys residues of protease types A and E may contribute to trafficking to the plasma membrane (G. Oyler, unpublished data), yet it is not clear if this determines the disparity of subcellular localization (90). S-nitrosylation of BoNTs is unknown, yet it is anticipated on the basis of its ubiquitous nature and possibly as a means for redox-based regulation (95). A number of channels and proteases are regulated by S-nitrosylation, e.g., the ryanodine receptor/Ca^{2+} release channel (96). The occurrence of prenylation, SUMOylation, sulfation, arginine methylation (97), or lysine acetylation (98) is unknown. Presence of BoNT HC in endosomes or the LC in the cytosol during intoxication may disrupt the regulatory networks involved in neuronal protein homeostasis (99) and may trigger a stress response comparable to the endoplasmic reticulum "unfolded protein response" (100). The functional and structural differences between BoNT/A and BoNT/E (**Figure 1**) highlight the elegance of BoNT molecular design, which appears to subvert the host cellular systems to its own function at each step of the intoxication process.

THE PROTEASE MODULE

The ultimate consequence of BoNT action on nerve terminals is abrogation of neurotransmitter release. This is attained by the LC acting on the SNARE substrates (26, 27, 101–103).

The SNARE Proteins: The Intracellular Substrates of the Light Chain Protease

The SNARE complex folds into a parallel four-helix bundle (27): two helices are contributed by a molecule of SNAP-25, denoted sn1 (residues 7–83) and sn2 (residues 141–204) for the N- and C-terminal helices; the other two by synaptobrevin-2 (residues 30–85) and syntaxin-1a (residues 183–256). These helical regions are referred to as the SNARE motifs. Of the four helices, three have a central Gln (Q) residue, and one contains a central Arg (R) residue in the hydrophobic, completely buried ionic "0" layer thought to be essential for SNARE complex stability (27) and function (104). SNAP-25 (Q53 and Q174) and syntaxin-1a (Q226) donate the Q-containing coils (Q-SNAREs), whereas synaptobrevin-2 provides the R-containing helix (R56) (R-SNARE) (105). Syntaxin-1a and synaptobrevin-2 are each anchored to their respective cell compartments by a single transmembrane segment, whereas SNAP-25 associates less intimately with the cell membrane, either through four palmitoylated Cys residues in the nonhelical region linking sn1 and sn2 or through its association with syntaxin-1a. The four helices are thought to intertwine with a zippering action in the N- to C- terminal direction, bringing into juxtaposition the surfaces of the apposed SV and neuronal lipid bilayers via the force exerted on the transmembrane anchors of syntaxin-1a and synaptobrevin-2 (26–28, 105–107). The detailed mechanism of bilayer fusion is still elusive (26, 79, 80, 107, 108). Amazingly, the scissile bonds hydrolyzed by all BoNTs are confined in the region delimited by the C-terminal membrane anchors and the ionic 0 layer. BoNTs cleave only the unpaired SNAREs before the assembly of the SNARE complex; once assembled, the target sites are inaccessible to the BoNTs, which renders the complex resistant to proteolysis (109).

The Light Chain Module Is a Zn^{2+}-Metalloprotease with a Thermolysin-Like Fold

The structures of the seven BoNT LCs have been solved at atomic resolution (7, 8, 110–117). They share a similar three-dimensional structure, irrespective of the distinct SNARE substrate specificity, and display structural

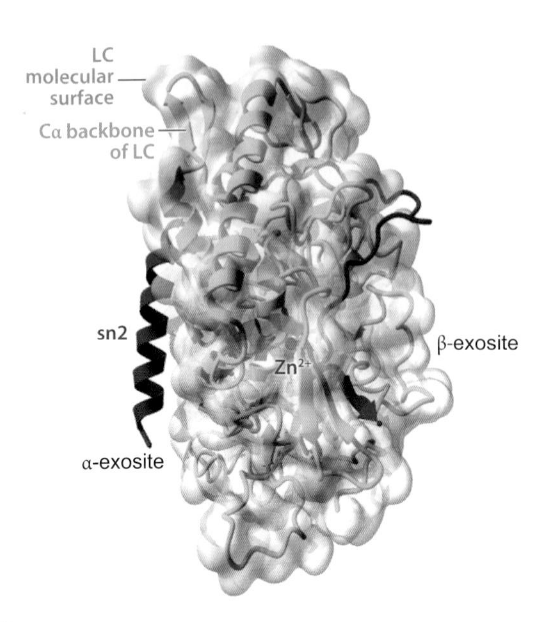

Figure 6

Structure of the BoNT/A-light chain (LC) in complex with the sn2 segment of SNAP-25 (PDB 1XTG) (115). The Cα backbone of the LC is represented as cyan ribbons, and its molecular surface is in transparent gray. sn2 is depicted in red, and the catalytic Zn^{2+} at the active site as a purple sphere. Modified and reproduced from Reference 75. Copyright © 2007, *PLoS Pathogens*.

similarities with the Zn^{2+}-metalloprotease thermolysin. The LC is a compact globular structure composed of a combination of β-strands and α-helices. The fold of the isolated LCs of type A (**Figure 6**) (PDB 1XTG) (115, 118), type B (PDB 1F82) (119, 120), and type E (PDB 1T3A) (111) is similar to the corresponding structures in the context of the respective holotoxin (7–9). Limited yet significant changes are discerned on the active-site conformation of these isolated LCs with respect to their cognates in the holotoxins, primarily in the proximity of the substrate access site. The Zn^{2+} protease motif HEXXH present in other Zn^{2+} endopeptidases is conserved among all BoNTs (2, 112). For BoNT/A, the catalytic Zn^{2+} is bound to the active site by interactions with the imidazole side chains of His222 and His226 and with the carboxyl side chain of Glu261. Completing the tetrahedral Zn^{2+} coordination is a water molecule bound to Glu223 of the motif, which plays a crucial role in proteolysis (7). The architecture of the active site is similar to that of thermolysin, suggesting that the proteolytic blueprint emerged from an ancestral thermolysin-like peptidase (112), which evolved to express an astonishingly high cleavage site specificity. For example, BoNT/A and BoNT/C1 cleave adjacent peptide bonds on SNAP-25 (the P1 and P1′ residues are Gln197–Arg198 for type A and Arg198–Ala199 for type C1); a similar feature prevails for BoNT/D and BoNT/F, which cleave synaptobrevin-2 (the P1 and P1′ residues are Lys59–Leu60 for type D and Gln58–Lys59 for type F). The substrate and scissile bonds for the other BoNTs are type E, SNAP-25 Arg180–Ile181; type C1, syntaxin Lys253–Ala254; type B, synaptobrevin-2 Gln76–Phe77; and type G, synaptobrevin-2 Ala81–Ala82.

Structure of the Enzyme-Substrate Complex: The Exosites

At variance with other metalloproteases, BoNTs require long stretches of the substrate for optimal cleavage (85, 121–123), and their catalytic efficiency is sharply attenuated by mutations in SNARE regions distant from the scissile bond (115, 122, 124). The X-ray structure of a binary complex between a mutated LC/A and sn2 revealed an extensive array of exosites, which are substrate-binding sites distant from the active site that orient the substrate into the vicinity of the active site and determine the target specificity (115). Both SNARE motifs of SNAP-25, sn1 and sn2, are unstructured in the uncomplexed SNAP-25 (125–127), whereas they are helical in the context of the SNARE complex four-helix bundle (27, 107). By contrast, in the structure of the LC/A-sn2 binary complex (115), sn2 adopts a different conformation (**Figure 6**): A short α-helix at its N terminus (residues 147–167) is bound to the α-exosite, a site remote from the active site; it is followed by an extended segment encompassing residues 168–200 that loops around a crevice on the protease surface to reach the active site and is ultimately anchored by a β-sheet at the LC β-exosite near its C terminus (residues 201–204).

The α-helix and the β-sheet validate the occurrence of the α- and β-exosites inferred from biochemical evidence (124).

The structure of the enzyme-substrate complex highlights the extensive interface shared by the partners, which ultimately determines catalytic specificity and efficiency. The sequence of events leading to complex formation is unknown; however, the acquisition of a secondary structure by SNAP-25 upon binding to the LC is a crucial event that presumably orients it and anchors it in the vicinity of the active site (115, 128). A plausible scenario may involve the initial alignment of the P5 residue of SNAP-25 (Asp193) with the S5 pocket residue (Arg177) via a salt bridge, which orients the P4′-residue (Lys201) and the S4′ residue at the active site of LC/A (Asp257). These interactions may open the active-site crevice rendering the P1′-residue (Arg198) accessible to the S1′ pocket residue Asp370 (128). Hydrophobic interactions between the P3 residue (Ala195) and the S3 pocket optimize alignment of the scissile bond for cleavage (128). Similar considerations appear valid for BoNT/E (129). Proteolysis is likely to proceed by a general base-catalyzed mechanism (110, 130, 131). The LC/A residues Arg362 and Tyr365 interact with the carbonyl oxygens of the P1 (Gln197) and P1′ (Arg198) residues of SNAP-25 and stabilize the oxyanion in the transition state; corresponding residues for the BoNT/E LC (LC/E) are Arg347 and Tyr350 (110). Hydrolysis is initiated by the water molecule bound to Glu143 of the Zn^{2+} protease motif and by Zn^{2+}.

Mutagenesis and kinetics studies demonstrated a similar interaction pattern between BoNT/B and synaptobrevin-2 (132). The structure of a binary complex between the BoNT/B LC and a synaptobrevin peptide (PDB 3G94) has been retracted (120). The cocrystal structure of the BoNT/F LC and two substrate-based peptide inhibitors, active in the nM range, was determined at ~2-Å resolution (PDB 3FIE and 3FII) (133). The substrates are synaptobrevin residues 22–58 and 27–58 in which the native Gln58 was substituted by D-Cys (134). The structure reveals the occurrence of three exosites involved in positioning and anchoring the substrate for optimal catalysis, a pattern analogous to that identified for sn2 on the LC/A (115). Importantly, the exosites are different from those on LC/A and underlie the selective exclusion of SNAP-25 by the BoNT/F LC. The orientation of substrate binding is conserved in the BoNT/D LC, exhibiting the substrate in an extended β-strand conformation and antiparallel to the active-site β-strand (112). Exosites' diversity appears to determine serotype substrate-specific binding, whereas sequence variations in the vicinity of the active site dictate scissile bond specificity (115, 133). Thus, the structural basis for substrate selectivity appears generally valid for all BoNTs, yet it is finely tuned by unique, serotype-specific residues for both recognition and cleavage.

The Translocation Domain Belt: An Intramolecular Chaperone for the Light Chain Protease

The belt, an enigmatic integral constituent of H_N, may be viewed as a crucial link between the protease and the H_N channel. The belt in the crystal structures of BoNT/A (7), BoNT/B (8), and BoNT/E (9) is a largely unstructured segment (consisting of residues 492–545 for BoNT/A, 481–532 for BoNT/B, and 460–507 for BoNT/E) that embraces the protease and occupies the enzyme surface allocated for substrate binding (**Figure 1**). An outcome of determining the LC/A-sn2 enzyme-substrate complex structure (115) is the realization that sn2 occupies a long cleft along the same trajectory where the belt resides in holotoxin/A (**Figures 1** and **6**) (75, 128). And, correspondingly, the belt occupies the exosites, yet it does not contain the scissile bond, thus potentially acting as a protease inhibitor (7–9, 75, 115, 135). This relationship between partners is conserved in the enzyme-substrate complex structure of LC/F-synaptobrevin (133) and, by inference, in holotoxin/B, which also cleaves synaptobrevin (8). Significantly, both sn2 and a peptide corresponding to the sequence of the

Intrinsically unstructured proteins (IUPs): flexible protein segments that are unstructured in isolation yet adopt secondary structure features upon interaction with partners

H_N/A belt (residues 492–545) are unstructured in isolation, yet adopt a secondary structure upon forming a complex with their LC partner (75). This extraordinary feature shared by a number of proteins and conceptually denoted as "intrinsically unstructured proteins" (IUPs) (59), implies that function emerges from disorder (entropy); upon a proximity encounter with suitable partners, they transiently bind and fold into lower-energy structures that express diverse functions (e.g., substrate for sn2 and pseudosubstrate/chaperone for the belt).

The belt embodies a convergence of chaperone and surrogate substrate mechanisms, acting as a surrogate pseudosubstrate inhibitor of the protease (75, 135) and as its chaperone during translocation across endosomes (33, 36, 75). The belt qualifies as a pseudosubstrate inhibitor given that it presumably restricts access of substrate to the protease both by occupying the binding interfaces of the exosites and by lacking the scissile bond (compare **Figure 1** and **Figure 6**) (8, 75, 118, 128, 133, 135). Protein-assisted unfolding and pseudosubstrate-assisted refolding of the protease could be added attributes of its chaperone action. There is precedence for protease inhibitors acting as intramolecular chaperones (136–138); these peptides are effectively IUPs (59). This notion was recently extended to the calpains, the Ca^{2+}-dependent proteases involved in a wealth of important Ca^{2+}-triggered intracellular processes, and to their inhibition by calpastatin. Two crystal structures of the complex (139, 140) hint a mechanism that fits well with that formulated for the BoNT belt (33, 36, 75). Calpastatin (an IUP), upon binding to Ca^{2+}-bound calpain, induces formation of three helices: Two terminal helices anchor the inhibitory domain to two different subunits of calpain while the intervening helix wraps around the other subunits blocking access to the active site and avoiding cleavage by looping away from it (139, 140).

This brings us to the other mainspring of the chaperone, its trigger by the conditions across endosomes (33, 35, 36). Plausibly, a ΔpH-induced transition of the belt from coil to helix triggers the insertion of H_N into the membrane (31, 141). Because the belt embraces the LC, it may facilitate the concerted unfolding of the LC at the endosomal acidic pH and direct the beginning of its translocation by the H_N channel through the membrane. Significantly, the belt does not appear to confer LC serotype specificity given that the TeNT protease, which exhibits a similar protein fold to the BoNT proteases (114, 116) and whose substrate is synaptobrevin (103), is effectively translocated and delivered by the BoNT/A HC into peripheral motor neurons (142).

The occurrence of the belt integrated into the modular organization of the BoNT highlights the functional role of flexible, disordered segments (59) in the tight yet dynamic association between modules and the emergence of functional complexity following the interactions between partners. Such complexity is expressed as a pseudosubstrate of the protease concomitant to a trigger of folding transitions of the protease and of its interactions as an unfolded cargo within the protein-conducting channel (33, 36, 37, 75). Such molecular sophistication inherent to the design cannot simply be imagined from the structure and the function of individual modules; it arises from their interactions. The design adds functional value to the linked, integral holotoxin in terms of efficiency, which is paramount to its biological action. BoNT is an astonishing nanomachine that unites recognition, trafficking, unfolding, translocation, refolding, and catalysis in a single entity.

MODULAR DESIGN OF BoNT AS A TOOL FOR BIOMOLECULE DELIVERY TO PREDETERMINED TARGET CELLS

Modularity is a structural principle that embodies rules governing the emergence of complexity from simple combinations of individual modules (143). An inherent feature of the modular design of BoNT is that by clever combinations of individual modules, a number of new entities with desired properties may be engineered. Indeed, biologically active chimeric constructs consisting of equivalent modules (LCs) from similar modular toxins (BoNT and TeNT) are a practical reality (142). A different itinerary for engineering target cell specificity is exploiting the H_C folds. Hierarchical representation was validated by using chimeras of BoNT/A and BoNT/E in which the H_C of one serotype was recombinantly transplanted to the dimodular LC-H_N of the other serotype (143); the chimeras faithfully express the translocation features of the primary LC-H_N (92). This version was developed as a retargeting strategy by which the sensitivity of nociceptive neurons to inflammatory or pain stimuli was mitigated by LC/E-H_N/E-H_C/A but not by BoNT/A (144). A conjugate of the dimodular LC-H_N of type A with *Erythrina cristagalli* lectin expressed in vitro selectivity for nociceptive afferents and attenuated the release of substance P and glutamate from cultured dorsal root ganglion neurons (145). These designs hint to new paths in pain therapy. A conjugate of wheat germ agglutinin with the type A LC-H_N conferred selectivity for a variety of secretory cells. Noteworthy is the conversion of a pancreatic β-cell line from BoNT/A insensitive to responsive to the type A LC-H_N in terms of mitigation of stimulated insulin release (146). Another approach involves expanding the protease substrate repertoire: a mutated LC/E (Lys224Asp) was engineered to exhibit substrate specificity for both SNAP-25 and SNAP-23, a nonneuronal SNARE protein involved in secretion of airway mucus. Delivery of the mutated LC/E to cultured epithelial cells inhibited secretion of mucin and IL-8 (147), thereby opening a path for the application of this type of engineered protease in the treatment of hypersecretion diseases, such as asthma.

For holotoxin type D, translocation of different cargos attached to the protease module led to the delivery of enzymatically active proteins to neurons (66). This modality dictated a constraint for flexible, unfolded cargo attached to the protease for productive translocation (66), implying that reversible unfolding/refolding is deterministic (33, 36). Conceivably, the H_N module could be engineered into a widespread delivery system for a diverse array of cargo to the target tissue of choice, provided cargo proteins reversibly unfold and refold at the beginning and the end of translocation, thereby retaining a tight association with the channel throughout the process (33, 36). The isolated H_N forms channels at neutral pH and in the absence of a ΔpH (37). It is plausible that a proper linkage to the H_N of a suitable cargo and a ligand for selective cell targeting may yield an array of novel molecules with interesting applications for cytosolic delivery of cargo that circumvent the endocytic pathway. The implication is that the modular organization of BoNTs is robust and intrinsically adaptable; the emergence of new designs is not only imaginable but imminent.

INHIBITORS AND MODULATORS OF BoNTs: TOWARD THE DESIGN OF ANTIDOTES

BoNT, one of the most feared biological weapons of the twenty-first century (5), is also a wonderful drug (83). This is a conundrum: Mass vaccination remains unlikely given the paucity of the disease and the preclusion of its subsequent medical use. The established therapy against botulism is the antitoxin (5), and currently, there is no specific small-molecule inhibitor of the BoNT action for prophylactic or therapeutic use. Knowledge on the structure and function of individual modules provides templates for drug design and discovery.

Receptor-Binding Module

Antibodies (3, 148) and lectins (11) have been intensely explored as inhibitors of the recognition step between BoNT and its cellular receptors. This constitutes a robust research field aimed toward the development of therapeutic antibodies and ligands. The two-receptor model (**Figure 3**) outlines a blueprint to design small-molecule inhibitors of BoNT entry. The information at hand underscores that, despite the fold robustness, the binding of competitive entities to the protein receptor site may be more difficult than imagined given the structure-inducing action of interacting partners. Nevertheless, this remains an attractive target and one for which candidate inhibitors are eagerly anticipated. In this regard, the structure of a complex of BoNT/B and doxorubicin, a DNA intercalator (149), showed that indeed doxorubicin interacts in the GT1b-binding cavity with Trp1261 and His1240 (8).

Translocation Domain Module

Given that the translocation process is essential for neurotoxicity, the protein-conducting channel emerges as a potential target. A semisynthetic strategy (38) identified inhibitors using toosendanin, a traditional Chinese medicine reported to protect monkeys from BoNT intoxication (150). Toosendanin and a more potent tetrahydrofuran analog selectively arrest translocation of the LC/A and the LC/E with subnanomolar potency (38). By contrast, after completion of LC translocation, toosendanin stabilized a channel-opened state apparent at higher concentrations (~2000-fold) than required to arrest translocation. Such bimodal modulation of the protein-conducting channel, namely the transformation from a cargo-dependent inhibitor of translocation to a cargo-free channel activator, is determined by the conformation of cargo within the chaperone (38). The dynamic interplay between modules is, therefore, not only necessary for function, yet it is an important consideration in inhibitor design; activity detected on assays of the isolated modules may not translate to efficacy when in the context of the holotoxin (151). "Smart screening" of natural products (152) for this class of channel blockers may evolve into a platform for antidote discovery.

Protease Module

The fact that Zn^{2+} in the active site is purely catalytic (131, 153) has attracted attention to Zn^{2+}-coordinating compounds, with the caveat that these may inhibit host cell proteases and be inadequate candidates for drug development. The design of small molecules that include the hydroxamate Zn^{2+}-binding functionality linked to a scaffold aimed to confer specificity led to L-arginine hydroxamic acid, an LC/A inhibitor in the high μM range (154). A cocrystal structure with LC/A showed that the inhibitor carbonyl and N-hydroxyl groups bind to the Zn^{2+} in a bidentate manner, and the Arg moiety is liganded to Asp369, suggesting that the inhibitor-bound structure mimics a catalytic intermediate with Arg binding at the P1′ site (131). In agreement, a cocrystal structure of LC/A with 4-chlorocinnamic hydroxamate, a derivative displaying modest in vivo efficacy in mice, identified the P1′-binding pocket of the protease (155).

An extensive *in silico* screening of the National Cancer Institute database into the active site of LC/A led to the identification of five quinolinol-based protease inhibitors (156). These inhibitors also hindered the BoNT/A toxicity in two in vivo assays in the sub-μM range. The study highlights the power of vast arrays of candidate small molecules, suspected to inhibit the target, that provide large structure-activity relationship databases.

A peptidomimetic, denoted I1 and patterned after the sequence of a heptapeptide from SNAP-25, which includes the scissile bond ($Q^{197}RATKML^{203}$), was designed (157). I1 selectively inhibits LC/A in the nM range and is inactive against the LCs of BoNT types B, D, E, and F. The cocrystal structure with LC/A disclosed a 3_{10} helix for I1 bound to the active site near the scissile bond. The conformation of the corresponding sn2 in the LC/A-sn2 complex is, by contrast, extended (115). The implication is that I1 induces binding pockets in the protease, and these pockets are absent both in the isolated LC/A and the LC/A-sn2 complexes (115). Inhibition is proposed to arise by displacement of the catalytic water, which interacts with the side chain of Glu224 at the active site (157). Cocrystal structures of LC/A with weak inhibitory heptapeptide (N-Ac-CRATKML) (153) and hexapeptides ($Q^{197}RATKM^{202}$ and RRATKM) (158) are instructive. In the LC/A-N-Ac-CRATKML complex, the peptide is bound with the Cys Sγ atom coordinating the Zn^{2+} (153). For the LC/A-$Q^{197}RATKM^{202}$ complex, the amino nitrogen and carbonyl oxygen of P1 (Gln197) chelate the Zn^{2+} and replace the nucleophilic water (158). For BoNT/A (115, 128), BoNT/F (133), and by inference for BoNT/B (8), the exosites may be superior targets. However, the specific serotypes have different substrates, requiring the design of seven unique inhibitors.

Together, these advances represent new paradigms emerging from the function and structure of individual modules. This is an exciting time to exploit the new knowledge, which in concert with structure-based design and other innovative technologies may uncover compounds that embody safe and realistic antidotes for all BoNT serotypes.

CLOSING REMARK

The analysis described here represents one particular direction of search for a fundamental principle underlying the action of BoNT. The focus has been on how the complexity of its biological activity emerges from the simplicity of its modular design (143). To paraphrase Wolfgang Pauli, understanding phenomena requires adequate concepts; modularity may represent such a concept for BoNT.

SUMMARY POINTS

1. BoNTs abrogate neurotransmitter release at peripheral nerve terminals by a multistep mechanism involving binding, internalization, translocation across endosomes, cytosolic traffic, and proteolytic degradation of SNARE substrates.
2. Mature BoNT consists of three modules: an N-terminal LC Zn^{2+}-metalloprotease, the HC that encompasses the N-terminal ~50-kDa translocation domain (H_N), and the C-terminal ~50-kDa receptor-binding domain (H_C).
3. H_{CC}, the C-terminal subdomain of H_C, contains two independent determinants of BoNT neurotropism: a ganglioside anchorage site, conjectured to hinder the lateral diffusion of the bound BoNT in the plane of the neuronal membrane and concurrently enrich its surface density, combined with a protein receptor that dictates serotype specificity.
4. H_N is a helical bundle that chaperones the LC protease across endosomes. It embodies a fascinating example of molecular partnership: a protein-conducting channel driven by a transmembrane proton gradient that mediates cargo unfolding, maintains an unfolded LC conformation during translocation, and releases cargo after it refolds in cytosol.
5. The LC–HC interchain disulfide linkage is a crucial component of the BoNT toxicity and is required for chaperone function, acting as a principal determinant for cargo translocation and release.
6. The H_N belt embodies a combination of surrogate pseudosubstrate inhibitor of the protease and its chaperone during translocation across endosomes.
7. The LC executes the ultimate action of BoNTs on nerve terminals by cleaving the SNAREs. Diversity of the protease exosites determines serotype substrate-specific binding, whereas sequence variations in the vicinity of the active site dictate scissile bond specificity.
8. BoNT is an astonishing modular nanomachine that unites recognition, trafficking, unfolding, translocation, refolding, and catalysis in a single entity.

FUTURE ISSUES

1. Is the H_N channel a transmembrane helical bundle? Is the H_N channel a monomeric entity or does it operate as an oligomer during translocation? The crystal structure of the H_N channel in a membrane-embedded environment is not available yet clearly vital.
2. Is the H_N belt cleaved by cytosolic proteases after completion of translocation and released as a binary complex with the cargo to accompany it through its cytosolic journey? If the belt were to remain linked to the membrane-embedded H_N, then there may be another protein that occupies its location on the protease during trafficking to be eventually displaced by the specific substrate upon its encounter.
3. The isolated H_N does not translocate cargo, yet the LC-H_N does; it is not known if a beltless holotoxin or a beltless LC-H_N translocates cargo—just as it is not established if a beltless holotoxin is neurotoxic.
4. Is the lifetime of an active protease or its degradation inside neurons modulated by covalent protein modifications? Is this a pathway for intervention to attenuate intoxication, or conversely, to prolong the persistence of active BoNT inside neurons?
5. Are BoNTs active on the central nervous system? Is it plausible that retrograde axonal transport and transcytosis of catalytically active BoNTs to afferent neurons proceed after their uptake at the peripheral neuromuscular junction? This is reminiscent of the pathway used by TeNT to reach central synapses. The intricacies of intraneuronal trafficking for BoNTs and TeNT are unknown.

DISCLOSURE STATEMENT

The author is not aware of any affiliations, memberships, funding, or financial holdings that might be perceived as affecting the objectivity of this review.

ACKNOWLEDGMENTS

I thank Audrey Fischer, Lilia Koriazova, and Myrta Oblatt-Montal for their devoted participation in the BoNT project. My work was supported by the National Institutes of Health Pacific Southwest Regional Center of Excellence (AI065359). I offer sincere apologies to authors whose contribution was not cited because of space limits.

LITERATURE CITED

1. Rossetto O, Montecucco C. 2008. Presynaptic neurotoxins with enzymatic activities. *Handb. Exp. Pharmacol.* 184:129–70
2. Lacy DB, Stevens RC. 1999. Sequence homology and structural analysis of the clostridial neurotoxins. *J. Mol. Biol.* 291:1091–104
3. Arndt JW, Jacobson MJ, Abola EE, Forsyth CM, Tepp WH, et al. 2006. A structural perspective of the sequence variability within botulinum neurotoxin subtypes A1–A4. *J. Mol. Biol.* 362:733–42
4. Hill KK, Smith TJ, Helma CH, Ticknor LO, Foley BT, et al. 2007. Genetic diversity among botulinum neurotoxin-producing clostridial strains. *J. Bacteriol.* 189:818–32
5. Arnon SS, Schechter R, Inglesby TV, Henderson DA, Bartlett JG, et al. 2001. Botulinum toxin as a biological weapon: medical and public health management. *JAMA* 285:1059–70

6. Sathyamoorthy V, DasGupta BR. 1985. Separation, purification, partial characterization and comparison of the heavy and light chains of botulinum neurotoxin types A, B, and E. *J. Biol. Chem.* 260:10461–66
7. Lacy DB, Tepp W, Cohen AC, DasGupta BR, Stevens RC. 1998. Crystal structure of botulinum neurotoxin type A and implications for toxicity. *Nat. Struct. Biol.* 5:898–902
8. Swaminathan S, Eswaramoorthy S. 2000. Structural analysis of the catalytic and binding sites of *Clostridium botulinum* neurotoxin B. *Nat. Struct. Biol.* 7:693–99
9. Kumaran D, Eswaramoorthy S, Furey W, Navaza J, Sax M, Swaminathan S. 2009. Domain organization in *Clostridium botulinum* neurotoxin type E is unique: its implication in faster translocation. *J. Mol. Biol.* 386:233–45
10. Fischer A, Garcia-Rodriguez C, Geren I, Lou J, Marks JD, et al. 2008. Molecular architecture of botulinum neurotoxin E revealed by single particle electron microscopy. *J. Biol. Chem.* 283:3997–4003
11. Simpson LL. 2004. Identification of the major steps in botulinum toxin action. *Annu. Rev. Pharmacol. Toxicol.* 44:167–93
12. Dong M, Yeh F, Tepp WH, Dean C, Johnson EA, et al. 2006. SV2 is the protein receptor for botulinum neurotoxin A. *Science* 312:592–96
13. Mahrhold S, Rummel A, Bigalke H, Davletov B, Binz T. 2006. The synaptic vesicle protein 2C mediates the uptake of botulinum neurotoxin A into phrenic nerves. *FEBS Lett.* 580:2011–14
14. Rummel A, Karnath T, Henke T, Bigalke H, Binz T. 2004. Synaptotagmins I and II act as nerve cell receptors for botulinum neurotoxin G. *J. Biol. Chem.* 279:30865–70
15. Nishiki T, Tokuyama Y, Kamata Y, Nemoto Y, Yoshida A, et al. 1996. Binding of botulinum type B neurotoxin to Chinese hamster ovary cells transfected with rat synaptotagmin II cDNA. *Neurosci. Lett.* 208:105–8
16. Dong M, Liu H, Tepp WH, Johnson EA, Janz R, Chapman ER. 2008. Glycosylated SV2A and SV2B mediate the entry of botulinum neurotoxin E into neurons. *Mol. Biol. Cell* 19:5226–37
17. Fu Z, Chen C, Barbieri JT, Kim JJ, Baldwin MR. 2009. Glycosylated SV2 and gangliosides as dual receptors for botulinum neurotoxin serotype F. *Biochemistry* 48:5631–41
18. Dong M, Richards DA, Goodnough MC, Tepp WH, Johnson EA, Chapman ER. 2003. Synaptotagmins I and II mediate entry of botulinum neurotoxin B into cells. *J. Cell Biol.* 162:1293–303
19. Nishiki T, Kamata Y, Nemoto Y, Omori A, Ito T, et al. 1994. Identification of protein receptor for *Clostridium botulinum* type B neurotoxin in rat brain synaptosomes. *J. Biol. Chem.* 269:10498–503
20. Nishiki T, Tokuyama Y, Kamata Y, Nemoto Y, Yoshida A, et al. 1996. The high-affinity binding of *Clostridium botulinum* type B neurotoxin to synaptotagmin II associated with gangliosides G_{T1b}/G_{D1a}. *FEBS Lett.* 378:253–57
21. Simpson LL. 1983. Ammonium chloride and methylamine hydrochloride antagonize clostridial neurotoxins. *J. Pharmacol. Exp. Ther.* 225:546–52
22. Lawrence G, Wang J, Chion CK, Aoki KR, Dolly JO. 2007. Two protein trafficking processes at motor nerve endings unveiled by botulinum neurotoxin E. *J. Pharmacol. Exp. Ther.* 320:410–18
23. Montecucco C, Schiavo G, Dasgupta BR. 1989. Effect of pH on the interaction of botulinum neurotoxins A, B and E with liposomes. *Biochem. J.* 259:47–53
24. Puhar A, Johnson EA, Rossetto O, Montecucco C. 2004. Comparison of the pH-induced conformational change of different clostridial neurotoxins. *Biochem. Biophys. Res. Commun.* 319:66–71
25. Finkelstein A. 1990. Channels formed in phospholipid bilayer membranes by diphtheria, tetanus, botulinum and anthrax toxin. *J. Physiol.* 84:188–90
26. Sudhof TC, Rothman JE. 2009. Membrane fusion: grappling with SNARE and SM proteins. *Science* 323:474–77
27. Sutton RB, Fasshauer D, Jahn R, Brunger AT. 1998. Crystal structure of a SNARE complex involved in synaptic exocytosis at 2.4 Å resolution. *Nature* 395:347–53
28. Weber T, Zemelman BV, McNew JA, Westermann B, Gmachl M, et al. 1998. SNAREpins: minimal machinery for membrane fusion. *Cell* 92:759–72
29. Takamori S, Holt M, Stenius K, Lemke EA, Gronborg M, et al. 2006. Molecular anatomy of a trafficking organelle. *Cell* 127:831–46

30. Simpson LL, Coffield JA, Bakry N. 1994. Inhibition of vacuolar adenosine triphosphatase antagonizes the effects of clostridial neurotoxins but not phospholipase A2 neurotoxins. *J. Pharmacol. Exp. Ther.* 269:256–62
31. Galloux M, Vitrac H, Montagner C, Raffestin S, Popoff MR, et al. 2008. Membrane interaction of botulinum neurotoxin A translocation (T) domain. The belt region is a regulatory loop for membrane interaction. *J. Biol. Chem.* 283:27668–76
32. Li L, Singh BR. 2000. Spectroscopic analysis of pH-induced changes in the molecular features of type A botulinum neurotoxin light chain. *Biochemistry* 39:6466–74
33. Koriazova LK, Montal M. 2003. Translocation of botulinum neurotoxin light chain protease through the heavy chain channel. *Nat. Struct. Biol.* 10:13–18
34. Fischer A, Montal M. 2006. Characterization of Clostridial botulinum neurotoxin channels in neuroblastoma cells. *Neurotox. Res.* 9:93–100
35. Fischer A, Montal M. 2007. Crucial role of the disulfide bridge between botulinum neurotoxin light and heavy chains in protease translocation across membranes. *J. Biol. Chem.* 282:29604–11
36. Fischer A, Montal M. 2007. Single molecule detection of intermediates during botulinum neurotoxin translocation across membranes. *Proc. Natl. Acad. Sci. USA* 104:10447–52
37. Fischer A, Mushrush DJ, Lacy DB, Montal M. 2008. Botulinum neurotoxin devoid of receptor binding domain translocates active protease. *PLoS Pathog.* 4:e1000245
38. Fischer A, Nakai Y, Eubanks LM, Clancy CM, Tepp WH, et al. 2009. Bimodal modulation of the botulinum neurotoxin protein-conducting channel. *Proc. Natl. Acad. Sci. USA* 106:1330–35
39. Simpson LL, Rapport MM. 1971. The binding of botulinum toxin to membrane lipids: sphingolipids, steroids and fatty acids. *J. Neurochem.* 18:1751–59
40. Kozaki S, Kamata Y, Watarai S, Nishiki T, Mochida S. 1998. Ganglioside GT1b as a complementary receptor component for *Clostridium botulinum* neurotoxins. *Microb. Pathog.* 25:91–99
41. Yowler BC, Kensinger RD, Schengrund CL. 2002. Botulinum neurotoxin A activity is dependent upon the presence of specific gangliosides in neuroblastoma cells expressing synaptotagmin I. *J. Biol. Chem.* 277:32815–19
42. Simpson LL. 1984. Botulinum toxin and tetanus toxin recognize similar membrane determinants. *Brain Res.* 305:177–80
43. Lalli G, Herreros J, Osborne SL, Montecucco C, Rossetto O, Schiavo G. 1999. Functional characterisation of tetanus and botulinum neurotoxins binding domains. *J. Cell Sci.* 112(Part 16):2715–24
44. Kohda T, Ihara H, Seto Y, Tsutsuki H, Mukamoto M, Kozaki S. 2007. Differential contribution of the residues in C-terminal half of the heavy chain of botulinum neurotoxin type B to its binding to the ganglioside GT1b and the synaptotagmin 2/GT1b complex. *Microb. Pathog.* 42:72–79
45. Stenmark P, Dupuy J, Imamura A, Kiso M, Stevens RC. 2008. Crystal structure of botulinum neurotoxin type A in complex with the cell surface co-receptor GT1b—insight into the toxin-neuron interaction. *PLoS Pathog.* 4:e1000129
46. Jayaraman S, Eswaramoorthy S, Ahmed SA, Smith LA, Swaminathan S. 2005. N-terminal helix reorients in recombinant C-fragment of *Clostridium botulinum* type B. *Biochem. Biophys. Res. Commun.* 330:97–103
47. Jin R, Rummel A, Binz T, Brunger AT. 2006. Botulinum neurotoxin B recognizes its protein receptor with high affinity and specificity. *Nature* 444:1092–95
48. Rummel A, Mahrhold S, Bigalke H, Binz T. 2004. The HCC-domain of botulinum neurotoxins A and B exhibits a singular ganglioside binding site displaying serotype specific carbohydrate interaction. *Mol. Microbiol.* 51:631–43
49. Tsukamoto K, Kohda T, Mukamoto M, Takeuchi K, Ihara H, et al. 2005. Binding of *Clostridium botulinum* type C and D neurotoxins to ganglioside and phospholipid. Novel insights into the receptor for clostridial neurotoxins. *J. Biol. Chem.* 280:35164–71
50. Tsukamoto K, Kozai Y, Ihara H, Kohda T, Mukamoto M, et al. 2008. Identification of the receptor-binding sites in the carboxyl-terminal half of the heavy chain of botulinum neurotoxin types C and D. *Microb. Pathog.* 44:484–93
51. Evans DM, Williams RS, Shone CC, Hambleton P, Melling J, Dolly JO. 1986. Botulinum neurotoxin type B. Its purification, radioiodination and interaction with rat-brain synaptosomal membranes. *Eur. J. Biochem.* 154:409–16

52. Yavin E, Nathan A. 1986. Tetanus toxin receptors on nerve cells contain a trypsin-sensitive component. *Eur. J. Biochem.* 154:403–7
53. Montecucco C. 1986. How do tetanus and botulinum toxins bind to neuronal membranes? *Trends Biochem. Sci.* 11:314–17
54. Simpson LL. 1980. Kinetic studies on the interaction between botulinum toxin type A and the cholinergic neuromuscular junction. *J. Pharmacol. Exp. Ther.* 212:16–21
55. Chapman ER. 2008. How does synaptotagmin trigger neurotransmitter release? *Annu. Rev. Biochem.* 77:615–41
56. Binz T, Rummel A. 2009. Cell entry strategy of clostridial neurotoxins. *J. Neurochem.* 109:1584–95
57. Chai Q, Arndt JW, Dong M, Tepp WH, Johnson EA, et al. 2006. Structural basis of cell surface receptor recognition by botulinum neurotoxin B. *Nature* 444:1096–100
58. Rummel A, Eichner T, Weil T, Karnath T, Gutcaits A, et al. 2007. Identification of the protein receptor binding site of botulinum neurotoxins B and G proves the double-receptor concept. *Proc. Natl. Acad. Sci. USA* 104:359–64
59. Dyson HJ, Wright PE. 2005. Intrinsically unstructured proteins and their functions. *Nat. Rev. Mol. Cell Biol.* 6:197–208
60. Fogolari F, Tosatto SC, Muraro L, Montecucco C. 2009. Electric dipole reorientation in the interaction of botulinum neurotoxins with neuronal membranes. *FEBS Lett.* 583:2321–25
61. Muraro L, Tosatto S, Motterlini L, Rossetto O, Montecucco C. 2009. The N-terminal half of the receptor domain of botulinum neurotoxin A binds to microdomains of the plasma membrane. *Biochem. Biophys. Res. Commun.* 380:76–80
62. Zoncu R, Perera RM, Balkin DM, Pirruccello M, Toomre D, De Camilli P. 2009. A phosphoinositide switch controls the maturation and signaling properties of APPL endosomes. *Cell* 136:1110–21
63. Gozani O, Karuman P, Jones DR, Ivanov D, Cha J, et al. 2003. The PHD finger of the chromatin-associated protein ING2 functions as a nuclear phosphoinositide receptor. *Cell* 114:99–111
64. Donovan JJ, Simon MI, Montal M. 1982. Insertion of diphtheria toxin into and across membranes: role of phosphoinositide asymmetry. *Nature* 298:669–72
65. Masuyer G, Thiyagarajan N, James PL, Marks PM, Chaddock JA, Acharya KR. 2009. Crystal structure of a catalytically active, non-toxic endopeptidase derivative of *Clostridium botulinum* toxin A. *Biochem. Biophys. Res. Commun.* 381:50–53
66. Bade S, Rummel A, Reisinger C, Karnath T, Ahnert-Hilger G, et al. 2004. Botulinum neurotoxin type D enables cytosolic delivery of enzymatically active cargo proteins to neurones via unfolded translocation intermediates. *J. Neurochem.* 91:1461–72
67. Krantz BA, Finkelstein A, Collier RJ. 2006. Protein translocation through the anthrax toxin transmembrane pore is driven by a proton gradient. *J. Mol. Biol.* 355:968–79
68. Basilio D, Juris SJ, Collier RJ, Finkelstein A. 2009. Evidence for a proton-protein symport mechanism in the anthrax toxin channel. *J. Gen. Physiol.* 133:307–14
69. Blocker D, Pohlmann K, Haug G, Bachmeyer C, Benz R, et al. 2003. *Clostridium botulinum* C2 toxin: Low pH-induced pore formation is required for translocation of the enzyme component C2I into the cytosol of host cells. *J. Biol. Chem.* 278:37360–67
70. Haug G, Wilde C, Leemhuis J, Meyer DK, Aktories K, Barth H. 2003. Cellular uptake of *Clostridium botulinum* C2 toxin: membrane translocation of a fusion toxin requires unfolding of its dihydrofolate reductase domain. *Biochemistry* 42:15284–91
71. Ren J, Kachel K, Kim H, Malenbaum SE, Collier RJ, London E. 1999. Interaction of diphtheria toxin T domain with molten globule-like proteins and its implications for translocation. *Science* 284:955–57
72. de Paiva A, Poulain B, Lawrence GW, Shone CC, Tauc L, Dolly JO. 1993. A role for the interchain disulfide or its participating thiols in the internalization of botulinum neurotoxin A revealed by a toxin derivative that binds to ecto-acceptors and inhibits transmitter release intracellularly. *J. Biol. Chem.* 268:20838–44
73. Antharavally B, Tepp W, DasGupta BR. 1998. Status of Cys residues in the covalent structure of botulinum neurotoxin types A, B, and E. *J. Protein Chem.* 17:187–96
74. Shi X, Garcia GE, Neill RJ, Gordon RK. 2009. TCEP treatment reduces proteolytic activity of BoNT/B in human neuronal SHSY-5Y cells. *J. Cell Biochem.* 107:1021–30

75. Brunger AT, Breidenbach MA, Jin R, Fischer A, Santos JS, Montal M. 2007. Botulinum neurotoxin heavy chain belt as an intramolecular chaperone for the light chain. *PLoS Pathog.* 3:1191–94
76. Ratts R, Zeng H, Berg EA, Blue C, McComb ME, et al. 2003. The cytosolic entry of diphtheria toxin catalytic domain requires a host cell cytosolic translocation factor complex. *J. Cell Biol.* 160:1139–50
77. Haug G, Leemhuis J, Tiemann D, Meyer DK, Aktories K, Barth H. 2003. The host cell chaperone Hsp90 is essential for translocation of the binary *Clostridium botulinum* C2 toxin into the cytosol. *J. Biol. Chem.* 278:32266–74
78. Tamayo AG, Bharti A, Trujillo C, Harrison R, Murphy JR. 2008. COPI coatomer complex proteins facilitate the translocation of anthrax lethal factor across vesicular membranes in vitro. *Proc. Natl. Acad. Sci. USA* 105:5254–59
79. Giraudo CG, Garcia-Diaz A, Eng WS, Chen Y, Hendrickson WA, et al. 2009. Alternative zippering as an on-off switch for SNARE-mediated fusion. *Science* 323:512–16
80. Maximov A, Tang J, Yang X, Pang ZP, Sudhof TC. 2009. Complexin controls the force transfer from SNARE complexes to membranes in fusion. *Science* 323:516–21
81. Arunachalam L, Han L, Tassew NG, He Y, Wang L, et al. 2008. Munc18-1 is critical for plasma membrane localization of syntaxin1 but not of SNAP-25 in PC12 cells. *Mol. Biol. Cell* 19:722–34
82. Jonikas MC, Collins SR, Denic V, Oh E, Quan EM, et al. 2009. Comprehensive characterization of genes required for protein folding in the endoplasmic reticulum. *Science* 323:1693–97
83. Jankovic J, Albanese A, Atassi MZ, Dolly JO, Hallett M, Mayer NH, eds. 2009. *Botulinum toxin: Therapeutic Clinical Practice and Science*. Philadelphia, PA: Saunders, Elsevier. 493 pp.
84. Foran PG, Mohammed N, Lisk GO, Nagwaney S, Lawrence GW, et al. 2003. Evaluation of the therapeutic usefulness of botulinum neurotoxin B, C1, E, and F compared with the long lasting type A. Basis for distinct durations of inhibition of exocytosis in central neurons. *J. Biol. Chem.* 278:1363–71
85. Vaidyanathan VV, Yoshino K, Jahnz M, Dorries C, Bade S, et al. 1999. Proteolysis of SNAP-25 isoforms by botulinum neurotoxin types A, C, and E: domains and amino acid residues controlling the formation of enzyme-substrate complexes and cleavage. *J. Neurochem.* 72:327–37
86. Ferrer-Montiel AV, Gutiérrez LM, Apland JP, Canaves JM, Gil A, et al. 1998. The 26-mer peptide released from SNAP-25 cleavage by botulinum neurotoxin E inhibits vesicle docking. *FEBS Lett.* 435:84–88
87. Bajohrs M, Rickman C, Binz T, Davletov B. 2004. A molecular basis underlying differences in the toxicity of botulinum serotypes A and E. *EMBO Rep.* 5:1090–95
88. Keller JE, Neale EA. 2001. The role of the synaptic protein SNAP-25 in the potency of botulinum neurotoxin type A. *J. Biol. Chem.* 276:13476–82
89. Rickman C, Meunier FA, Binz T, Davletov B. 2004. High affinity interaction of syntaxin and SNAP-25 on the plasma membrane is abolished by botulinum toxin E. *J. Biol. Chem.* 279:644–51
90. Fernandez-Salas E, Steward LE, Ho H, Garay PE, Sun SW, et al. 2004. Plasma membrane localization signals in the light chain of botulinum neurotoxin. *Proc. Natl. Acad. Sci. USA* 101:3208–13
91. Keller JE, Cai F, Neale EA. 2004. Uptake of botulinum neurotoxin into cultured neurons. *Biochemistry* 43:526–32
92. Wang J, Meng J, Lawrence GW, Zurawski TH, Sasse A, et al. 2008. Novel chimeras of botulinum neurotoxins A and E unveil contributions from the binding, translocation, and protease domains to their functional characteristics. *J. Biol. Chem.* 283:16993–7002
93. Ferrer-Montiel AV, Canaves JM, DasGupta BR, Wilson MC, Montal M. 1996. Tyrosine phosphorylation modulates the activity of clostridial neurotoxins. *J. Biol. Chem.* 271:18322–25
94. Raiborg C, Stenmark H. 2009. The ESCRT machinery in endosomal sorting of ubiquitylated membrane proteins. *Nature* 458:445–52
95. Hess DT, Matsumoto A, Kim SO, Marshall HE, Stamler JS. 2005. Protein S-nitrosylation: purview and parameters. *Nat. Rev. Mol. Cell Biol.* 6:150–66
96. Durham WJ, Aracena-Parks P, Long C, Rossi AE, Goonasekera SA, et al. 2008. RyR1 S-nitrosylation underlies environmental heat stroke and sudden death in Y522S RyR1 knockin mice. *Cell* 133:53–65
97. Bedford MT, Clarke SG. 2009. Protein arginine methylation in mammals: who, what, and why. *Mol. Cell* 33:1–13

98. Yang XJ, Seto E. 2008. Lysine acetylation: codified crosstalk with other posttranslational modifications. *Mol. Cell* 31:449–61
99. Balch WE, Morimoto RI, Dillin A, Kelly JW. 2008. Adapting proteostasis for disease intervention. *Science* 319:916–19
100. Korennykh AV, Egea PF, Korostelev AA, Finer-Moore J, Zhang C, et al. 2009. The unfolded protein response signals through high-order assembly of Ire1. *Nature* 457:687–93
101. Blasi J, Chapman ER, Link E, Binz T, Yamasaki S, et al. 1993. Botulinum neurotoxin A selectively cleaves the synaptic protein SNAP-25. *Nature* 365:160–63
102. Brunger AT. 2005. Structure and function of SNARE and SNARE-interacting proteins. *Q. Rev. Biophys.* 38:1–47
103. Schiavo G, Benfenati F, Poulain B, Rossetto O, Polverino de Laureto P, et al. 1992. Tetanus and botulinum-B neurotoxins block neurotransmitter release by proteolytic cleavage of synaptobrevin. *Nature* 359:832–35
104. Wei S, Xu T, Ashery U, Kollewe A, Matti U, et al. 2000. Exocytotic mechanism studied by truncated and zero layer mutants of the C-terminus of SNAP-25. *EMBO J.* 19:1279–89
105. Fasshauer D, Sutton RB, Brunger AT, Jahn R. 1998. Conserved structural features of the synaptic fusion complex: SNARE proteins reclassified as Q- and R-SNAREs. *Proc. Natl. Acad. Sci. USA* 95:15781–86
106. Fasshauer D, Eliason WK, Brünger AT, Jahn R. 1998. Identification of a minimal core of the synaptic SNARE complex sufficient for reversible assembly and disassembly. *Biochemistry* 37:10354–62
107. Stein A, Weber G, Wahl MC, Jahn R. 2009. Helical extension of the neuronal SNARE complex into the membrane. *Nature* 460:525–28
108. Brunger AT, Weninger K, Bowen M, Chu S. 2009. Single-molecule studies of the neuronal SNARE fusion machinery. *Annu. Rev. Biochem.* 78:903–28
109. Hayashi T, McMahon H, Yamasaki S, Binz T, Hata Y, et al. 1994. Synaptic vesicle membrane fusion complex: action of clostridial neurotoxins on assembly. *EMBO J.* 13:5051–61
110. Agarwal R, Binz T, Swaminathan S. 2005. Structural analysis of botulinum neurotoxin serotype F light chain: implications on substrate binding and inhibitor design. *Biochemistry* 44:11758–65
111. Agarwal R, Eswaramoorthy S, Kumaran D, Binz T, Swaminathan S. 2004. Structural analysis of botulinum neurotoxin type E catalytic domain and its mutant Glu212→Gln reveals the pivotal role of the Glu212 carboxylate in the catalytic pathway. *Biochemistry* 43:6637–44
112. Arndt JW, Chai Q, Christian T, Stevens RC. 2006. Structure of botulinum neurotoxin type D light chain at 1.65 Å resolution: repercussions for VAMP-2 substrate specificity. *Biochemistry* 45:3255–62
113. Arndt JW, Yu W, Bi F, Stevens RC. 2005. Crystal structure of botulinum neurotoxin type G light chain: serotype divergence in substrate recognition. *Biochemistry* 44:9574–80
114. Breidenbach MA, Brunger AT. 2005. 2.3 Å crystal structure of tetanus neurotoxin light chain. *Biochemistry* 44:7450–57
115. Breidenbach MA, Brunger AT. 2004. Substrate recognition strategy for botulinum neurotoxin serotype A. *Nature* 432:925–29
116. Rao KN, Kumaran D, Binz T, Swaminathan S. 2005. Structural analysis of the catalytic domain of tetanus neurotoxin. *Toxicon* 45:929–39
117. Jin R, Sikorra S, Stegmann CM, Pich A, Binz T, Brunger AT. 2007. Structural and biochemical studies of botulinum neurotoxin serotype C1 light chain protease: implications for dual substrate specificity. *Biochemistry* 46:10685–93
118. Segelke B, Knapp M, Kadkhodayan S, Balhorn R, Rupp B. 2004. Crystal structure of *Clostridium botulinum* neurotoxin protease in a product-bound state: evidence for noncanonical zinc protease activity. *Proc. Natl. Acad. Sci. USA* 101:6888–93
119. Hanson MA, Stevens RC. 2000. Cocrystal structure of synaptobrevin-II bound to botulinum neurotoxin type B at 2.0 Å resolution. *Nat. Struct. Biol.* 7:687–92
120. Hanson MA, Stevens RC. 2009. Retraction: cocrystal structure of synaptobrevin-II bound to botulinum neurotoxin type B at 2.0 Å resolution. *Nat. Struct. Mol. Biol.* 16:795
121. Foran P, Shone CC, Dolly JO. 1994. Differences in the protease activities of tetanus and botulinum B toxins revealed by the cleavage of vesicle-associated membrane protein and various sized fragments. *Biochemistry* 33:15365–74

122. Schmidt JJ, Bostian KA. 1995. Proteolysis of synthetic peptides by type A botulinum neurotoxin. *J. Protein Chem.* 14:703–8
123. Schmidt JJ, Stafford RG, Bostian KA. 1998. Type A botulinum neurotoxin proteolytic activity: development of competitive inhibitors and implications for substrate specificity at the S1′ binding subsite. *FEBS Lett.* 435:61–64
124. Rossetto O, Schiavo G, Montecucco C, Poulain B, Deloye F, et al. 1994. SNARE motif and neurotoxins. *Nature* 372:415–16
125. Fasshauer D, Otto H, Eliason WK, Jahn R, Brunger AT. 1997. Structural changes are associated with soluble N-ethylmaleimide-sensitive fusion protein attachment protein receptor complex formation. *J. Biol. Chem.* 272:28036–41
126. Fasshauer D, Bruns D, Shen B, Jahn R, Brunger AT. 1997. A structural change occurs upon binding of syntaxin to SNAP-25. *J. Biol. Chem.* 272:4582–90
127. Cánaves JM, Montal M. 1998. Assembly of a ternary complex by the predicted minimal coiled-coil-forming domains of syntaxin, SNAP-25, and synaptobrevin. A circular dichroism study. *J. Biol. Chem.* 273:34214–21
128. Chen S, Kim JJ, Barbieri JT. 2007. Mechanism of substrate recognition by botulinum neurotoxin serotype A. *J. Biol. Chem.* 282:9621–27
129. Chen S, Barbieri JT. 2007. Multiple pocket recognition of SNAP25 by botulinum neurotoxin serotype E. *J. Biol. Chem.* 282:25540–47
130. Binz T, Bade S, Rummel A, Kollewe A, Alves J. 2002. Arg(362) and Tyr(365) of the botulinum neurotoxin type A light chain are involved in transition state stabilization. *Biochemistry* 41:1717–23
131. Fu Z, Chen S, Baldwin MR, Boldt GE, Crawford A, et al. 2006. Light chain of botulinum neurotoxin serotype A: structural resolution of a catalytic intermediate. *Biochemistry* 45:8903–11
132. Chen S, Hall C, Barbieri JT. 2008. Substrate recognition of VAMP-2 by botulinum neurotoxin B and tetanus neurotoxin. *J. Biol. Chem.* 283:21153–59
133. Agarwal R, Schmidt JJ, Stafford RG, Swaminathan S. 2009. Mode of VAMP substrate recognition and inhibition of *Clostridium botulinum* neurotoxin F. *Nat. Struct. Mol. Biol.* 16:789–94
134. Schmidt JJ, Stafford RG. 2005. Botulinum neurotoxin serotype F: identification of substrate recognition requirements and development of inhibitors with low nanomolar affinity. *Biochemistry* 44:4067–73
135. Ahmed SA, Byrne MP, Jensen M, Hines HB, Brueggemann E, Smith LA. 2001. Enzymatic autocatalysis of botulinum A neurotoxin light chain. *J. Protein Chem.* 20:221–31
136. Bryan P, Wang L, Hoskins J, Ruvinov S, Strausberg S, et al. 1995. Catalysis of a protein folding reaction: mechanistic implications of the 2.0 Å structure of the subtilisin-prodomain complex. *Biochemistry* 34:10310–18
137. Kojima S, Iwahara A, Yanai H. 2005. Inhibitor-assisted refolding of protease: a protease inhibitor as an intramolecular chaperone. *FEBS Lett.* 579:4430–36
138. Shinde U, Fu X, Inouye M. 1999. A pathway for conformational diversity in proteins mediated by intramolecular chaperones. *J. Biol. Chem.* 274:15615–21
139. Hanna RA, Campbell RL, Davies PL. 2008. Calcium-bound structure of calpain and its mechanism of inhibition by calpastatin. *Nature* 456:409–12
140. Moldoveanu T, Gehring K, Green DR. 2008. Concerted multi-pronged attack by calpastatin to occlude the catalytic cleft of heterodimeric calpains. *Nature* 456:404–8
141. Mueller M, Grauschopf U, Maier T, Glockshuber R, Ban N. 2009. The structure of a cytolytic α-helical toxin pore reveals its assembly mechanism. *Nature* 459:726–30
142. Weller U, Dauzenroth ME, Gansel M, Dreyer F. 1991. Cooperative action of the light chain of tetanus toxin and the heavy chain of botulinum toxin type A on the transmitter release of mammalian motor endplates. *Neurosci. Lett.* 122:132–34
143. Simon HA. 1962. The architecture of complexity. *Proc. Am. Philos. Soc.* 106:467–82
144. Meng J, Ovsepian SV, Wang J, Pickering M, Sasse A, et al. 2009. Activation of TRPV1 mediates calcitonin gene-related peptide release, which excites trigeminal sensory neurons and is attenuated by a retargeted botulinum toxin with anti-nociceptive potential. *J. Neurosci.* 29:4981–92

145. Chaddock JA, Purkiss JR, Alexander FC, Doward S, Fooks SJ, et al. 2004. Retargeted clostridial endopeptidases: inhibition of nociceptive neurotransmitter release in vitro, and antinociceptive activity in in vivo models of pain. *Mov. Disord.* 19(Suppl 8):S42–47
146. Chaddock JA, Purkiss JR, Friis LM, Broadbridge JD, Duggan MJ, et al. 2000. Inhibition of vesicular secretion in both neuronal and nonneuronal cells by a retargeted endopeptidase derivative of *Clostridium botulinum* neurotoxin type A. *Infect. Immun.* 68:2587–93
147. Chen S, Barbieri JT. 2009. Engineering botulinum neurotoxin to extend therapeutic intervention. *Proc. Natl. Acad. Sci. USA* 106:9180–84
148. Garcia-Rodriguez C, Levy R, Arndt JW, Forsyth CM, Razai A, et al. 2007. Molecular evolution of antibody cross-reactivity for two subtypes of type A botulinum neurotoxin. *Nat. Biotechnol.* 25:107–16
149. Eswaramoorthy S, Kumaran D, Swaminathan S. 2001. Crystallographic evidence for doxorubicin binding to the receptor-binding site in *Clostridium botulinum* neurotoxin B. *Acta Crystallogr. D Biol. Crystallogr.* 57:1743–46
150. Shi YL, Wang ZF. 2004. Cure of experimental botulism and antibotulismic effect of toosendanin. *Acta Pharmacol. Sin.* 25:839–48
151. Eubanks LM, Hixon MS, Jin W, Hong S, Clancy CM, et al. 2007. An in vitro and in vivo disconnect uncovered through high-throughput identification of botulinum neurotoxin A antagonists. *Proc. Natl. Acad. Sci. USA* 104:2602–7
152. Li JW, Vederas JC. 2009. Drug discovery and natural products: end of an era or an endless frontier? *Science* 325:161–65
153. Silvaggi NR, Wilson D, Tzipori S, Allen KN. 2008. Catalytic features of the botulinum neurotoxin A light chain revealed by high resolution structure of an inhibitory peptide complex. *Biochemistry* 47:5736–45
154. Dickerson TJ, Janda KD. 2006. The use of small molecules to investigate molecular mechanisms and therapeutic targets for treatment of botulinum neurotoxin A intoxication. *ACS Chem. Biol.* 1:359–69
155. Silvaggi NR, Boldt GE, Hixon MS, Kennedy JP, Tzipori S, et al. 2007. Structures of *Clostridium botulinum* neurotoxin serotype A light chain complexed with small-molecule inhibitors highlight active-site flexibility. *Chem. Biol.* 14:533–42
156. Roxas-Duncan V, Enyedy I, Montgomery VA, Eccard VS, Carrington MA, et al. 2009. Identification and biochemical characterization of small-molecule inhibitors of *Clostridium botulinum* neurotoxin serotype A. *Antimicrob. Agents Chemother.* 53:3478–86
157. Zuniga JE, Schmidt JJ, Fenn T, Burnett JC, Arac D, et al. 2008. A potent peptidomimetic inhibitor of botulinum neurotoxin serotype A has a very different conformation than SNAP-25 substrate. *Structure* 16:1588–97
158. Kumaran D, Rawat R, Ahmed SA, Swaminathan S. 2008. Substrate binding mode and its implication on drug design for botulinum neurotoxin A. *PLoS Pathog.* 4:e1000165

Chemical Approaches to Glycobiology

Laura L. Kiessling[1,2] and Rebecca A. Splain[1]

[1]Department of Chemistry, [2]Department of Biochemistry, University of Wisconsin–Madison, Wisconsin 53706; email: kiessling@chem.wisc.edu

Annu. Rev. Biochem. 2010. 79:619–53

First published online as a Review in Advance on April 8, 2010

The *Annual Review of Biochemistry* is online at biochem.annualreviews.org

This article's doi:
10.1146/annurev.biochem.77.070606.100917

0066-4154/10/0707-0619$20.00

Key Words

array, glycan, glycomimetic, glycosylation, lectin, multivalency

Abstract

Glycans are ubiquitous components of all organisms. Efforts to elucidate glycan function and to understand how they are assembled and disassembled can reap benefits in fields ranging from bioenergy to human medicine. Significant advances in our knowledge of glycan biosynthesis and function are emerging, and chemical biology approaches are accelerating the pace of discovery. Novel strategies for assembling oligosaccharides, glycoproteins, and other glycoconjugates are providing access to critical materials for interrogating glycan function. Chemoselective reactions that facilitate the synthesis of glycan-substituted imaging agents, arrays, and materials are yielding compounds to interrogate and perturb glycan function and dysfunction. To complement these advances, small molecules are being generated that inhibit key glycan-binding proteins or biosynthetic enzymes. These examples illustrate how chemical glycobiology is providing new insight into the functional roles of glycans and new opportunities to interfere with or exploit these roles.

Contents

Glycan: a generic term referring to a monosaccharide, oligosaccharide, polysaccharide, or its conjugate (e.g., glycolipid, glycoprotein, or other glycoconjugate)

Glycoconjugate: one or more saccharide units (glycone) covalently linked to a noncarbohydrate moiety (aglycone)

INTRODUCTION

Glycans, which are compounds that include monosaccharides, oligosaccharides, polysaccharides, and their conjugates, are critical constituents of all organisms. Members of a glycan subset, the polysaccharides, are the most abundant organic compounds on Earth. Glycoconjugates (e.g., peptidoglycan, glycolipids, glycoproteins) also are prevalent. In humans, for example, half of all proteins are glycosylated (1). Consistent with glycan abundance in nature, data from genomic sequencing indicate that approximately 1% of each genome, from eubacteria to archea and eukaryotes, is dedicated to sugar-processing enzymes (2). Moreover, these genes can be highly conserved, as the components of few other biochemical pathways are so invariant as those responsible for glycan biosynthesis (3). The importance of this conservation is underscored by data indicating that defects in the glycan biosynthetic machinery in humans, known as congenital disorders of glycosylation, are rare and generally have severe deleterious consequences (4).

Genomic analysis is a powerful means to identify enzymes that generate or degrade glycans and the proteins that recognize the glycan products. Still, it does not reveal what glycans are present in a cell or organism because the synthesis of glycans is not template directed. As a result, elucidating the molecular mechanisms that underlie glycan function has been a challenge. Nevertheless, researchers have uncovered numerous roles for glycans, including those in fertilization and development, hormone function, cell proliferation and organization, host-pathogen interactions, and the inflammatory and immune responses (3). These findings are providing additional impetus to devise new approaches that meet the challenges of elucidating and manipulating glycan function.

The increased appreciation for the ubiquity of glycans and their importance to human health has spawned the field of chemical glycobiology. Because of the complexities of glycans, their study has compelled researchers to pursue interdisciplinary approaches. Since the pioneering contributions of 1902 Nobel Laureate Emil Fischer, it has been apparent that our understanding of glycan function can be advanced using approaches that span biology and chemistry. The nucleation of the discipline of chemical biology is yielding new and innovative strategies to probe glycan function (5). Indeed, there has been an explosion of research in this area. As a result, this review cannot provide comprehensive coverage of the field but rather offers an overview of select advances that illustrate the unique contributions and exciting opportunities within the field of chemical glycobiology.

The state of the art of chemical glycobiology is focused on key questions: How are glycans made and degraded, what are their biological roles once in place, and how can these roles be exploited? To address these questions, researchers have employed the complementary strategies of interrogation and perturbation (**Figure 1**). The interrogation strategy strives to understand endogenous interactions between natural glycans and their cognate enzymes or binding partners. Access to naturally occurring and novel glycans provides the means to examine protein-glycan or enzyme-glycan interactions. Arrays composed of glycoconjugates (**Figure 1*a***) or lectins (**Figure 1*b***) are valuable tools for interrogating protein-binding specificity or cellular glycosylation patterns. With the complementary perturbation approach, inhibitors, analogs, or other nonnatural substrates can serve as probes of both the biosynthesis and the biological roles of glycans. Indeed, novel nonnatural oligosaccharide mimics or synthetic glycoconjugates can inhibit or encourage specific biomolecular interactions within cells and organisms (**Figure 1*c*,*d***). Moreover, compounds have been identified that can block key steps within glycan biosynthetic pathways (**Figure 1*e***). Finally, carbohydrate analogs can be incorporated into glycans using the cellular biosynthetic machinery (**Figure 1*f***). Such agents can be used for purposes ranging from imaging glycans to cross-linking them to their binding partners. Together, these chemical strategies are illuminating the molecular mechanisms that underlie glycan function.

Glycobiology: the study of sugars in biological systems, including their structures, biosynthesis, and physiological roles

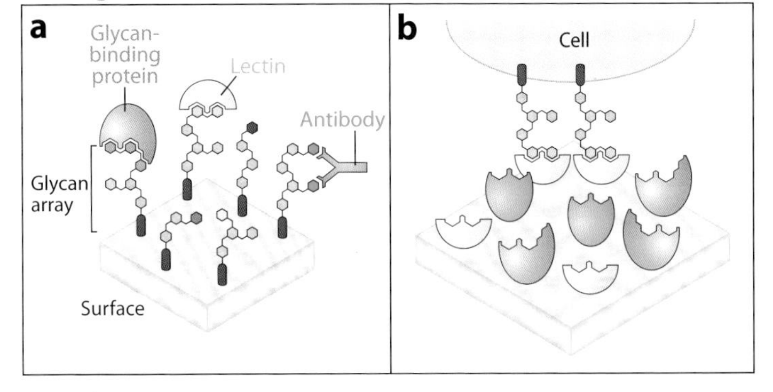

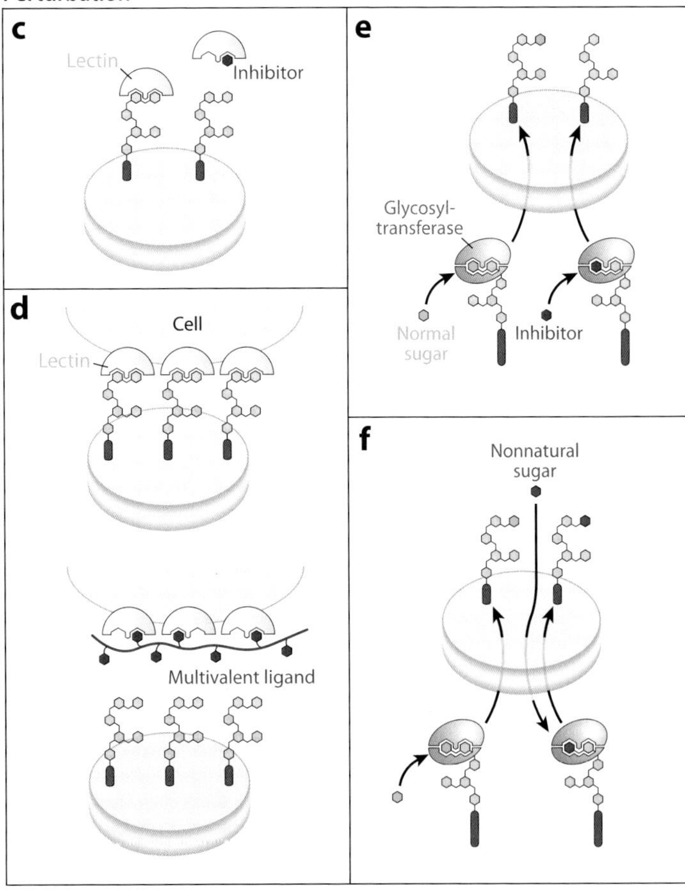

Figure 1

Interrogation and perturbation in chemical glycobiology. (*a*) Glycan arrays have been developed to interrogate the binding specificities of lectins (*blue*), antibodies (*green*), and other glycan-binding proteins (*orange*). (*b*) Lectin arrays can be used to fingerprint cell-surface or pathogen glycosylation patterns. Monovalent (*c*) and multivalent (*d*) ligands for glycan-binding proteins can perturb protein-glycan interactions. (*e*) Inhibitors can prevent key steps in glycan biosynthesis, thereby reducing the production of specific glycan structures. (*f*) Nonnatural monosaccharides can serve as substrates for biosynthetic enzymes and thereby be incorporated into glycans.

Lectin: a glycan-binding protein of nonimmune origin

GLYCAN SYNTHESIS

Defined oligosaccharides and glycoconjugates are critical for unraveling the function of glycans. Obtaining these entities from natural sources is difficult because their production generally involves the participation of multiple transporters and enzymes (6). This complexity is illustrated by the pathway for eukaryotic glycoprotein synthesis (**Figure 2**). The saccharide building blocks (typically nucleotide sugars) must be generated and then transported to the appropriate cellular location, where they can be used by glycosyltransferases. The efficiency of producing any particular glycan depends on the concentration of building blocks, what glycosyltransferases and other biosynthetic enzymes are present, and the K_m values of those building blocks for the glycosyltransferases that use them. Pathways for the production of *N*-glycoproteins, *O*-glycoproteins, glycolipids, glycosylphosphatidylinositol anchors, proteoglycans, and polysaccharides are influenced by accessibility of the nucleotide donors, but the mechanisms governing the regulation of these pathways are still being elucidated. Thus, it is difficult to obtain sufficient quantities of glycans for study from biological sources.

Chemical strategies are addressing this deficiency by providing the means to generate an ever-increasing diversity of glycans. Naturally occurring glycans can be synthesized, as can derivatives. In this way, critical structure-activity relationships can be elucidated. There are two general approaches for the synthesis of oligosaccharides: chemical and enzymatic. Here, we outline some of the major advances that have occurred on both fronts. More detailed information can be found in several excellent reviews (7–10).

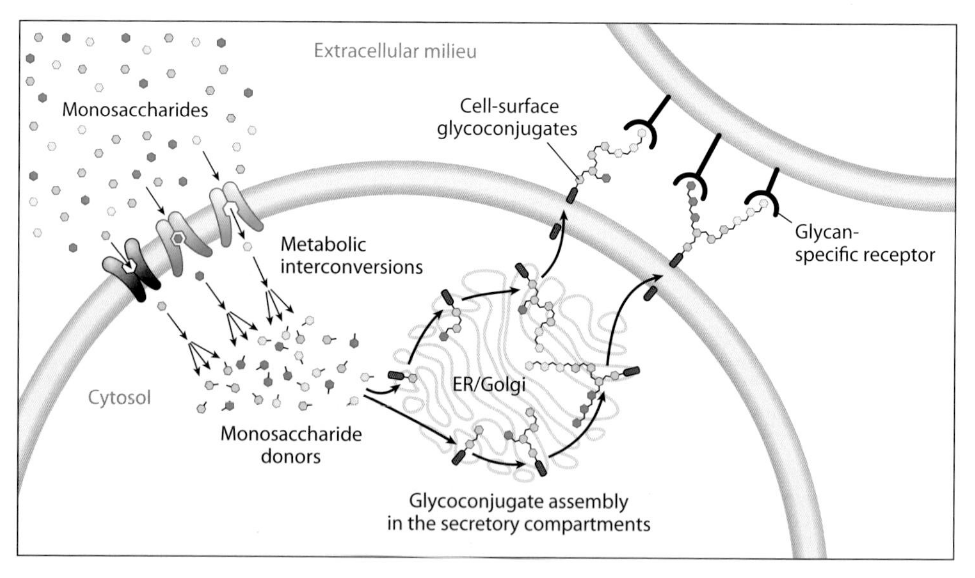

Figure 2

Schematic depiction of glycoconjugate biosynthesis and cell-surface recognition of glycans. Most exogenously supplied monosaccharides are taken up by cells and converted to monosaccharide donors in the cytosol. The donors are imported into the endoplasmic reticulum (ER) and Golgi compartments, where they are used by glycosyltransferases to assemble glycoconjugates. In the case of N-linked glycoproteins, a core oligosaccharide is assembled in the cytosol, transported into the ER where it is processed by glycosidases, and further elaborated by glycosyltransferases. Once displayed in fully mature forms on the cell surface, the glycoconjugates can serve as ligands for soluble lectins, cell-surface glycan-binding proteins, or glycan-binding proteins on other cells or pathogens. In principle, chemical glycobiology can yield molecules that can be used to inhibit or promote any stage of this process.

Chemical Synthesis of Oligosaccharides

The chemical synthesis of oligosaccharides offers tremendous flexibility. It can give rise to diverse glycans, including those available in minute quantities from biological sources or those for which the biosynthetic enzymes are unknown. Moreover, chemical synthesis provides the means to test the importance of different functional groups because nonnatural sugars can be introduced. The power of chemical synthesis is illustrated by the development of potent and effective sulfated saccharides as anticoagulants (11). These defined compounds were inspired by analysis of the properties of heparin, an anionic glycosaminoglycan (GAG) that has long been used as an anticoagulant. Heparin can bind antithrombin III, thereby producing a complex that blocks blood clotting. Pharmaceutical-grade heparin is typically isolated from porcine intestinal mucosa as a mixture of sulfated polysaccharides. Chemical synthesis was important in verifying that heparin's activity resides in a critical pentasaccharide recognition sequence. These findings indicate that proteins like antithrombin III can recognize specific sulfated oligosaccharide sequences within GAGs. Additionally, they led to development of the defined anticoagulant drug Arixtra®. The recent contamination of heparin isolated from biological sources, in which the presence of other sulfated polysaccharides led to over 100 deaths, highlights the utility of therapeutics that are based on defined, synthetic glycans (12).

The chemical synthesis of oligosaccharides appears deceptively simple. It involves the formation of glycosyl bonds, a reaction first described in 1893 (13). The approach that was employed then is similar to that used by nature and remains the preferred strategy for chemical synthesis: A donor monosaccharide, equipped with a leaving group at the anomeric position, undergoes reaction with a nucleophilic group on an acceptor (**Figure 3*a***). In chemical synthesis, a promoter is added to the donor monosaccharide to facilitate the departure of the leaving group.

The simplicity of this approach belies its complexity. Glycosylation reactions are regulated by the reactivity-selectivity principle of organic chemistry. The anomeric group must be sufficiently prone to leaving, such that a relatively poor nucleophile like a hydroxyl group can engage in bond formation; however, the donor must not be so reactive that bond formation occurs without stereocontrol. Thus, whether the reaction occurs via an S_N2-like (inversion) or S_N1-like (oxocarbenium intermediate) pathway has a critical influence on the stereochemical outcome. For this reason, the dual problems of regiochemical control and stereoselectivity of glycosylation reactions are intertwined. Fraser-Reid and coworkers' classification of donors as "armed" (fast reacting) or "disarmed" (slow reacting) has led to major advances in the field (14) because it offers insight into how the electronics of the glycosyl donor can be manipulated to control glycosylation reaction outcomes. The reactivity of the donor can be tuned by modifying several factors, including the electron-withdrawing ability of the protecting groups, the lability of the anomeric leaving group, the method of leaving group activation, the conformation of the donor or acceptor, and the nature of the solvent. These changes in reactivity impact glycosylation reaction stereochemistry because they influence the reaction mechanism (S_N2-like versus S_N1-like) (**Figure 3*a***). It is possible to tune a series of glycosylation reactions to gain outstanding stereoselectivity.

A longstanding yet clever means to achieving stereocontrol is to exploit protecting groups that can influence glycosylation stereochemistry via neighboring group participation (**Figure 3*b***). In the paradigmatic example, a 2-acyl group forms an acetoxonium ion by attack onto the anomeric carbon. With a protected glucose derivative, the 1,2-acetoxonium ion blocks nucleophilic attack of the acceptor from the α-face and results in the formation of β-glucosides; for mannose, the β-face is blocked, and α-mannosides are produced. This strategy has been used extensively to generate β-glucosides, β-galactosides, α-mannosides, and

GAG: glycosaminoglycan

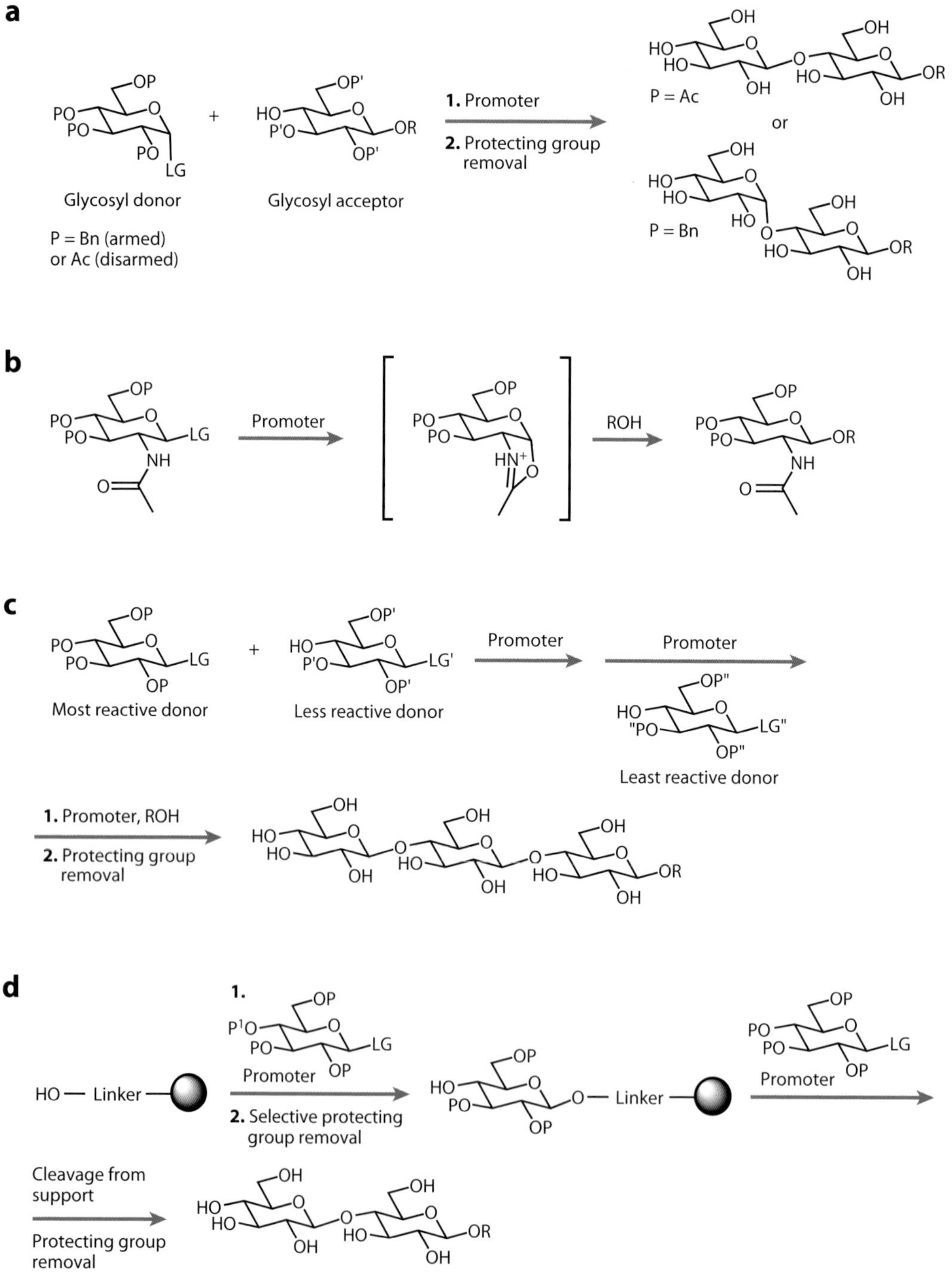

α-rhamnosides. More recently, novel variations on the neighboring group participation strategy have been employed to access a wider array of glycosidic linkages, including α-glucosides and α-galactosides (7). Still, many key glycosidic linkages cannot be formed via neighboring group participation strategies. Important examples include β-mannosides, β-rhamnosides, and sialic acid derivatives. New insight into the interplay of stereoelectronic and conformational effects is providing strategies to assemble these kinds of linkages (15). The synthesis of sialic acid derivatives has been especially challenging because the anomeric position is more hindered and possesses an electron-withdrawing group, but the use of protecting groups to alter thiosialoside conformation in combination with novel conditions for donor activation has led to dramatic improvements in glycosylation yields (16–18). The value of these new approaches is illustrated by the synthesis of an α2,9-trisialic acid oligomer in a single reaction vessel (one-pot) (19).

Efforts to streamline the chemical synthesis of oligosaccharides have focused on minimizing purification steps. One approach is to conduct the kind of one-pot glycosylation reactions described above, in which multiple glycosidic bonds are made without isolation or purification of intermediates (**Figure 3*c***). There are three general strategies to achieve this end. First, the relative reactivities of the glycosyl donors can be varied by protecting group selection, such that the addition of a promoter triggers the most armed glycosyl donor first, and the most disarmed donor eventually engages in the final glycosylation reaction. Second, glycosyl donors can be preactivated before exposure to a glycosyl acceptor, and the order and timing of their addition determine the reaction outcome. Third, orthogonal anomeric leaving groups can be selected that are activated by different promoters. Impressive one-pot syntheses of biologically relevant oligosaccharides are becoming more common. Two notable examples include generation of the branched hexasaccharides GM1 and the tumor-associated carbohydrate antigen Globo-H (20). In the former, three components were joined in a single reaction; while in the latter, four building blocks were linked (21).

Another strategy for oligosaccharide synthesis that circumvents the need for multiple purification steps is solid-supported synthesis (22). In this manifold, reactions can be driven to completion by the addition of an excess of one partner. In the typical configuration, a nucleophilic acceptor substrate is appended onto a solid support and exposed to an excess of activated donor in solution (**Figure 3*d***). Subsequent steps involve hydroxyl protecting group removal followed by glycosylation. The multiple sites of reactivity and branching found in oligosaccharides require monomers that possess orthogonal protecting groups, which can be masked and unmasked at appropriate stages of glycan construction. Thus, the solid-supported synthesis of oligosaccharides is complicated by the need for diverse building blocks. Nevertheless, the potential of solid-supported synthesis has continued to spark advances, including methods to automate the process. Automated synthesis now can be used to prepare even complex oligosaccharides (22), such as a branched β-glucan dodecasaccharide; blood group oligosaccharides Lewis x, Lewis y, and the Lewis x-Lewis y nonasaccharide; and tumor-associated carbohydrate antigens Gb-3 and Globo-H. These latter examples are compounds with multiple types of glycoside linkages, and their successful synthesis demonstrates that glycosylation

Figure 3

In its most versatile form, the chemical synthesis of oligosaccharides depends upon glycosylation reactions that involve activation of a glycosyl donor. (*a*) Treatment with a promoter facilitates the departure of the anomeric leaving group (LG) such that a glycosidic bond can be formed in a substitution reaction with a nucleophilic glycosyl acceptor. Depending on the electron-withdrawing potential of the protecting groups (P) on the glycosyl donor, the reaction can proceed through an S_N2-like mechanism (P = Ac, acetate) or an S_N1-like mechanism in which the stereochemical outcome is controlled by the anomeric effect (P = Bn, benzyl). (*b*) Neighboring group participation can dictate product stereochemistry. One-pot methods (*c*) and solid-phase synthesis (*d*) are approaches that eliminate purification steps and thereby facilitate glycan assembly. Abbreviation: ROH, an alcohol, including a hydroxyl group of a protected sugar derivative.

reactions in solid-supported synthesis can occur with excellent stereoselectivity.

There has yet to emerge a universal strategy to form glycosidic bonds with the requisite regio- and stereocontrol. Suitable reaction conditions must be optimized for each glycosidic variation. Thus, current methods are focused on developing sets of standard building blocks that can be used to generate key bioactive oligosaccharides. To this end, many of the targets assembled to date possess linkages that can be formed reliably, such as α-mannosides, β-galactosides, and β-glucosides. It is estimated that approximately 500 orthogonally protected monosaccharides would be needed to synthesize the bioactive oligosaccharides found in mammals (23), although a recent analysis suggests that 36 building blocks could generate 75% of the known mammalian oligosaccharides (24). The need for glycans that reflect the diversity of physiological systems is driving efforts to develop methods to prepare all the relevant glycosidic bonds, including those that have been challenging (e.g., β-mannosides, sialic acid derivatives). Progress on this front has made accessible biologically important glycans, such as sulfated GAG sequences and protein glycosylphosphatidylinositol anchors. Despite the rapid development in generating mammalian oligosaccharides, there has been less emphasis on assembling glycans found in microbes. Many of these contain noncanonical sugars (e.g., deoxysugars and furanoses) that pose unique challenges. Methods to assemble these glycans are needed to elucidate their roles in microbes, to probe host-pathogen interactions, and to investigate novel antimicrobial strategies.

Engineering Enzymes for Glycan Synthesis

Enzymatic and chemoenzymatic methods for glycan assembly complement those from chemical synthesis. These approaches harness the components used by physiological systems to generate glycans. Specifically, glycosyltransferases transfer nucleotide-sugar donors onto glycone or aglycone acceptors. The use of these enzymes can facilitate the chemoenzymatic synthesis of glycans, and recent advances have increased the utility of this approach. Historically, nucleotide-sugar donors can only be generated with naturally occurring sugars, although efforts in enzyme engineering indicate that this problem can be overcome (25). Indeed, the utility of enzymes for generating oligosaccharides has increased in the last decade (26, 27), owing, in part, to the ability to identify glycosyltransferases from sequence data.

Bacterial glycosyltransferases and related biosynthetic enzymes have proved especially useful for glycan assembly. These enzymes and their variants can be produced and purified more readily than their eukaryotic counterparts, and they often act on a broad array of substrates. In a powerful example, Chen and coworkers (28) used three classes of bacterial sugar-processing enzymes (a sialic acid aldolase, a cytidine 5′-monophosphate-sialic acid synthetase, and a sialyltransferase) to produce a library of 72 biotinylated sialosides in an array format. By screening this array, detailed information about the binding preferences of a key immunomodulatory protein, human CD22, could be gleaned. Another example in which the broad substrate specificity of bacterial glycosyltransferases was exploited is in the production of 70 glycoforms of the natural products calicheamicin and vancomycin (29). Because glycosylation can influence natural product biological activity, specificity, and pharmacology, the ability to introduce different saccharide substituents is valuable. The enzymes can be engineered to increase their substrate tolerance even further (30).

Another chemoenzymatic approach to oligosaccharides is based upon glycosidases. Nature's antipode to the glycosyltransferase is the glycosidase, an enzyme that catalyzes the hydrolytic cleavage of glycosidic bonds. Replacement of the active-site water nucleophile with a glycosyl acceptor can result in transglycosylation (26, 27). Because glycosidases can readily be produced, often are highly soluble and stable, and tend to be more promiscuous than glycosyltransferases, they are attractive as catalysts. Unfortunately, glycosidase-catalyzed transglycosylation reactions suffer from low yields and product hydrolysis because the products are themselves substrates. A major breakthrough in the field occurred with the invention of nonhydrolyzing glycosidases, or glycosynthases.

Glycosynthases were developed by exploiting key features of glycosidase mechanisms. Glycosidases come in two varieties, retaining and inverting (26, 27). In general, both possess active sites in which catalytic carboxylic acid residues are proximal (**Figure 4**). Most glycosynthases are based on retaining glycosidases, which use a two-step, double-substitution mechanism that proceeds through a covalent carbohydrate-protein adduct (**Figure 4*b***). Substitution of the nucleophilic active site aspartate or glutamate with a small hydrophobic residue

Glycosynthase: engineered glycosidase capable of catalyzing transglycosylation reactions

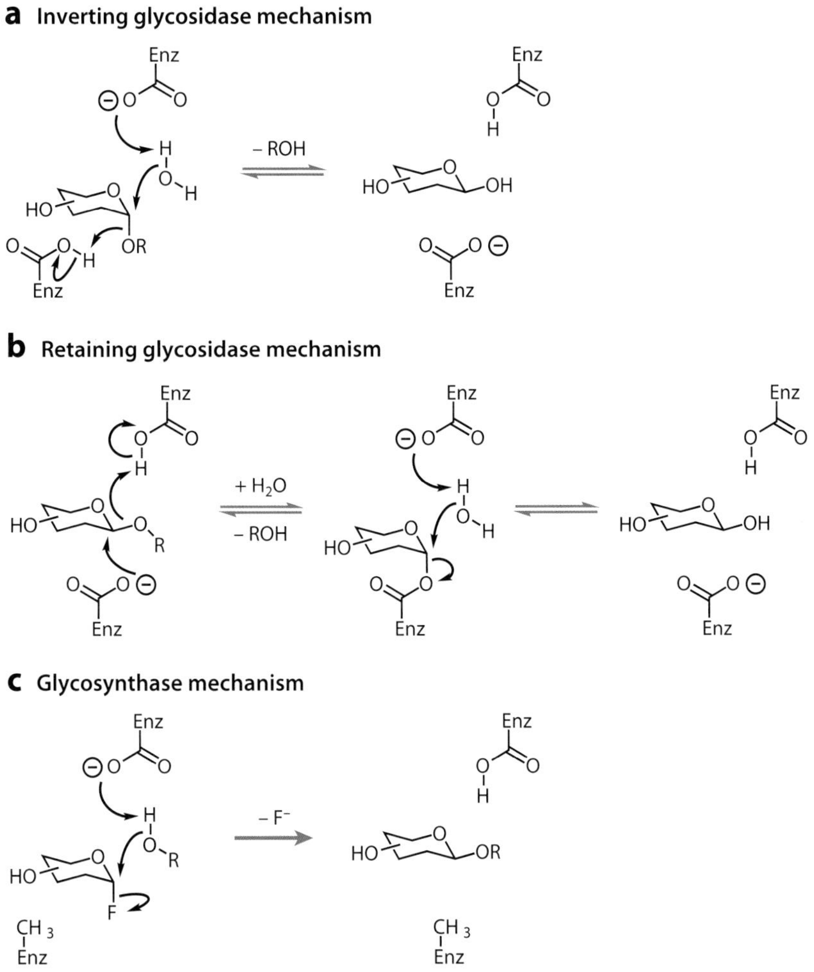

Figure 4

Catalytic mechanisms for glycosidases and glycosynthases. (*a*) Inverting glycosidases use two catalytic carboxylate residues positioned proximally. (*b*) Retaining glycosidases use a two-step, double-substitution mechanism, with a covalent carbohydrate-enzyme adduct. (*c*) Substitution of the nucleophilic active-site carboxylate of retaining glycosidases with a nonpolar side chain affords glycosynthases capable of transglycosylation. Abbreviations: Enz, enzyme; ROH, an alcohol, including a hydroxyl group of a protected sugar derivative.

(i.e., alanine) (**Figure 4*c***) renders the enzyme incapable of hydrolysis. These variants can catalyze the formation of glycosidic bonds between an acceptor and an α-glycosyl fluoride donor. In the engineered enzyme, the product is no longer susceptible to hydrolysis; therefore, transglycosylation reactions can occur in high yields and with high levels of regio- and stereoselectivity.

Glycosynthases engineered to process alternative substrates can be generated by either rational mutagenesis or directed evolution. Because libraries of enzyme variants can be readily prepared, the major roadblock in the discovery of novel glycosynthase enzymes has been the development of high-throughput screens. The enzyme-catalyzed glycosylation reactions are not accompanied by the release of a chromophore, so novel screens were needed. The approaches that have been devised fall into three categories: a yeast three-hybrid chemical complementation assay (31), an assay based on pH changes (32), and a fluorescence-activating cell sorting (FACS) assay (33). In all three approaches, glycosynthase activity is evaluated in whole cells; therefore, protein isolation is not required.

Glycosynthases have been identified (26, 27) that act on a range of nucleophiles, including glycone and aglycone acceptors. A notable feature of glycosynthases is that they can produce oligosaccharides that are difficult to obtain using chemical synthesis (e.g., β-mannosides) (34). In addition, disaccharide fluoride donors and acceptors can be used in transglycosylation reactions, thereby enabling the rapid assembly of complex oligosaccharides (26). Glycosynthases also can be used to generate oligosaccharides that contain β-glucuronic acid or β-galacturonic acid residues, suggesting they can be used for GAG assembly (26). Notably, retaining endoglycosidases have also been devised that catalyze the convergent assembly of *N*-glycans. Specifically, these modified enzymes promote the reaction of oxazolines derived from 2-deoxy *N*-acetyl glucosamine-containing substrates with asparagine-containing peptides (35). To complement the linkages that can be formed by engineered retaining glycosidases, inverting glycosidases have been generated that afford α-linkages (36, 37). Thus, advances in the identification and engineering of both glycosyltransferases and glycosynthases are extending the range of accessible glycans.

Glycoprotein and Glycopeptide Synthesis

In parallel with the development of new methods for oligosaccharide assembly, there have emerged new approaches for glycoconjugate synthesis. The prevalence of glycosylated proteins and the benefits of access to single-protein glycoforms have inspired the development of methods to generate N- and O-glycosylated peptides and proteins. Protein glycosylation can influence the pharmacological properties of therapeutic proteins, including their serum half-lives, their ability to target specific cells or organs, and their modes of clearance. Glycosylation can also exert an influence by playing a direct role in recognition, such that whole glycosylated protein is more than the sum of the parts. A notable example of the latter involves P-selectin, a protein involved in the inflammatory response, which binds to a highly O-glycosylated protein (a mucin) bearing the tetrasaccharide sialyl Lewis x. The tightest complexes between P-selectin and its ligand are formed when specific tyrosine residues adjacent to a sialyl Lewis x motif are sulfated (38). Thus, the identity of the glycoconjugate as a whole is important for recognition (39). Together, the mode of P-selectin recognition and the requirement for therapeutic glycoproteins with optimized properties underscore the need to obtain defined glycoconjugates.

Several strategies have emerged that yield defined glycoproteins and glycopeptides. One approach is to employ engineered cell lines or recombinant enzymes to obtain glycoproteins. For instance, complete heterogeneous glycoproteins can be expressed, isolated, and trimmed via glycosidases to bear individual monosaccharide moieties (40). These monosaccharides serve as starting points for embellishment by recombinant glycosyltransferases or by transglycosylation using endoglycosidases. The latter strategy is especially useful for the rapid assembly of complex glycoproteins because a large glycan motif can be added in a single step (as described above).

Synthetic chemists also have taken on the challenge of glycopeptide and glycoprotein preparation. In focusing on various complex glycopeptides and glycoproteins as targets for synthesis, the Danishefsky research group (41) has pushed the limits of existing synthetic methods. In pursuing their complex targets, including prostate-specific antigen, gp120 fragments, and erythropoietin, they have developed new strategies for the assembly of multiple peptide precursors.

The chemical synthesis of glycoproteins is fueled both by methods to construct complex glycopeptide fragments and the advent of chemoselective ligation reactions to join them. Solid-phase peptide synthesis is generally limited to glycopeptides <50 residues. Chemical ligation reactions, such as native chemical ligation (NCL), alleviate this limitation because they can be used to link peptide fragments together (**Figure 5*a***) (42). The NCL process involves the transthioesterification reaction of a C-terminal thioester with an N-terminal cysteine residue of a second peptide. The resulting thioester intermediate subsequently undergoes an intramolecular transacylation reaction to produce a stable peptide bond. One valuable variation on NCL is expressed protein ligation (EPL), in which the peptide component bearing the C-terminal thioester is produced using recombinant DNA methods (43). Although both NCL and EPL increase the scope of glycopeptide synthesis, they require a cysteine residue, a relatively rare amino acid, at the ligation junction.

The Staudinger ligation of peptide thioesters circumvents the need for cysteine at the junction, as the two peptides couple when a C-terminal phosphinothioester undergoes reaction with an N-terminal azide. The utility of the Staudinger ligation for glycopeptide synthesis is under investigation (44). Other approaches have been described that capitalize on removable or transient auxiliaries (45–47). One of these, sugar-assisted ligation is particularly useful in the construction of N-linked glycans (**Figure 5*b***) (48). Thus, the means to construct larger glycopeptides and glycoproteins are available and can be used to examine the influence of glycosylation on protein function.

Chemical Glycobiology of Glycolipids

Glycolipids have been implicated in many critical processes, but identifying their precise physiological roles has been difficult. Recent discoveries have revealed that glycolipids can serve as critical immunomodulators. Natural killer T (NKT) cells are a class of T cells that play a central role regulating the immune response, and NKT cells can recognize glycolipids displayed by CD1d-positive antigen-presenting cells. Both endogenous and exogenous glycolipids can serve as CD1d ligands and thereby activate NKT cells. An endogenous glycolipid is presumably necessary for positive selection of NKT cells in the thymus, and NKT cells can recognize exogenous lipopolysaccharides from bacterial pathogens (49). The synthetic glycolipid antigen KRN7000 (**Figure 6**) and related compounds are illuminating the critical features of the glycolipid that result in NKT cell activation. This understanding can lead to the development of new immunomodulators. Additionally, because glycolipid trafficking and degradation are involved in several diseases, glycolipid analogs serve as probes and therapeutic leads (50).

Chemoselective Reactions to Modify Glycans

Glycoconjugates are critical tools in the examination of glycan function. They can be immobilized for affinity isolation of glycan-binding proteins, used to generate glycan arrays, or converted to natural or nonnatural probes. Such probes can be generated from

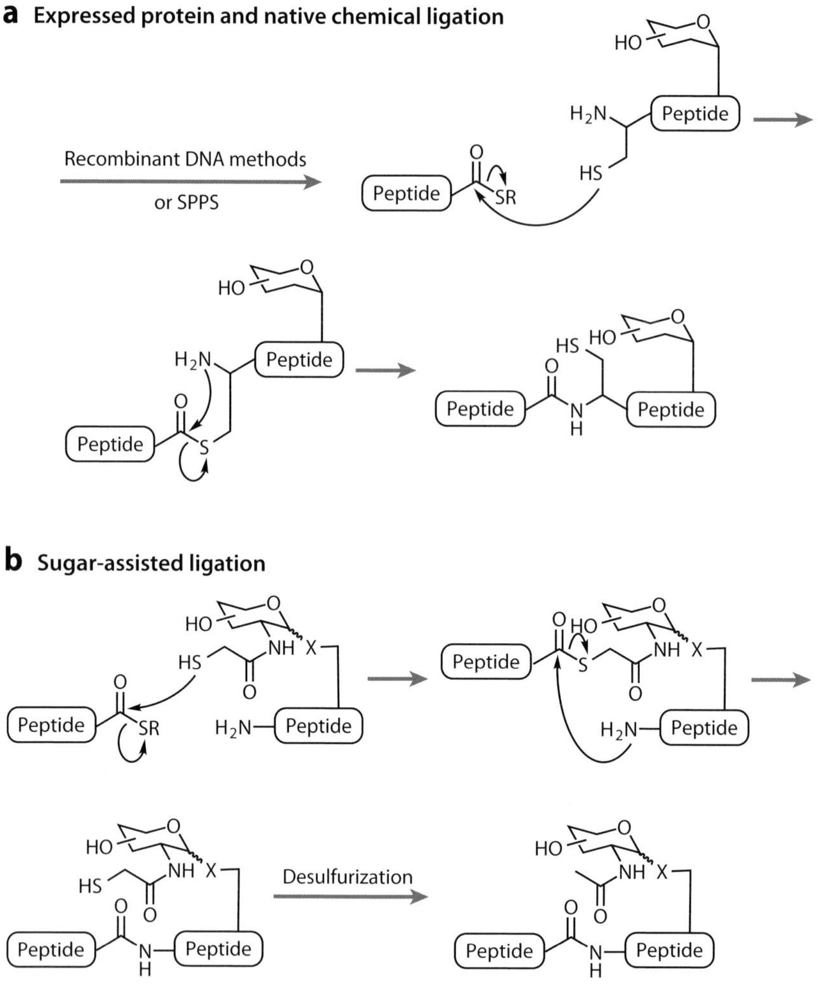

Figure 5

Ligation strategies for glycopeptide and glycoprotein synthesis. (*a*) In native chemical ligation (NCL), the peptide components are obtained by solid-phase peptide synthesis (SPPS). For expressed protein ligation (EPL), recombinant DNA methods are used to produce a peptide or protein fragment with a C-terminal thioester. For both EPL and NCL, the thioester is captured by an N-terminal cysteine residue, and the incipient thioester conjugate rearranges to the amide. (*b*) In sugar-assisted ligation, a carbohydrate bearing a thiol substituent serves in the same capacity as the Cys side chain.

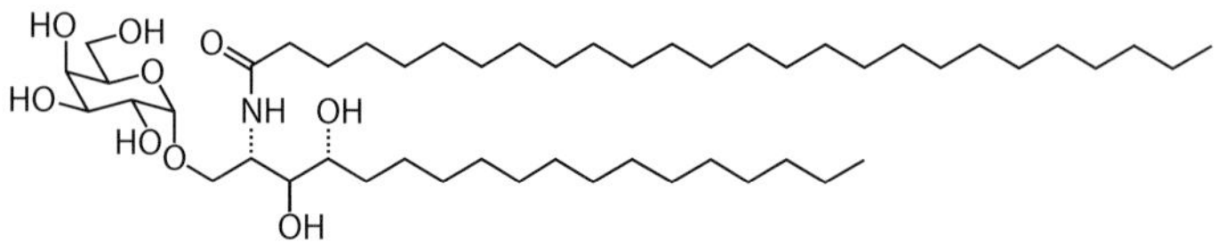

Figure 6

The structure of glycolipid antigen KRN7000, which functions as an immunomodulator that leads to natural killer T cell activation.

chemoselective reactions (i.e., reactions that occur among select functional groups in the presence of others) of natural and synthetic oligosaccharides (51). One of the most common strategies is to use the intrinsic reactivity of oligosaccharides, which contain an electrophile at the reducing end, most commonly, an aldehyde. This masked carbonyl group is susceptible to nucleophilic addition, which has been exploited for conjugate production. Nucleophiles, such as alkoxylamine, hydrazine, or acylhydrazine derivatives, can be employed to afford glycoconjugates containing oxime, hydrazone, or hydrazide linkages, respectively (**Figure 7*a***). These functional groups vary in their stability and in whether the reducing-end sugar exists in the open or closed form. Thus, the mode of conjugation can be chosen for a specific purpose.

An alternative strategy is to generate a glycan that bears a linker that possess a functional group that can undergo reaction with a coupling partner in the presence of hydroxyl, acetamide, carboxylate, and other common carbohydrate functional groups. Perhaps the most commonly used linker, and that used by the Consortium for Functional Glycomics (CFG, **http://www.functionalglycomics.org**), is an aminopropyl group appended to the anomeric position. The resulting amine-bearing oligosaccharides can undergo reaction with several types of partners, including those bearing *N*-hydroxysuccinimidyl esters, aldehydes (followed by reductive amination), or dimethyl squarate (**Figure 7*b***) (52–55). Other chemoselective linkage strategies rely on the unique reactivity of thiol-containing saccharides, which can undergo conjugate addition

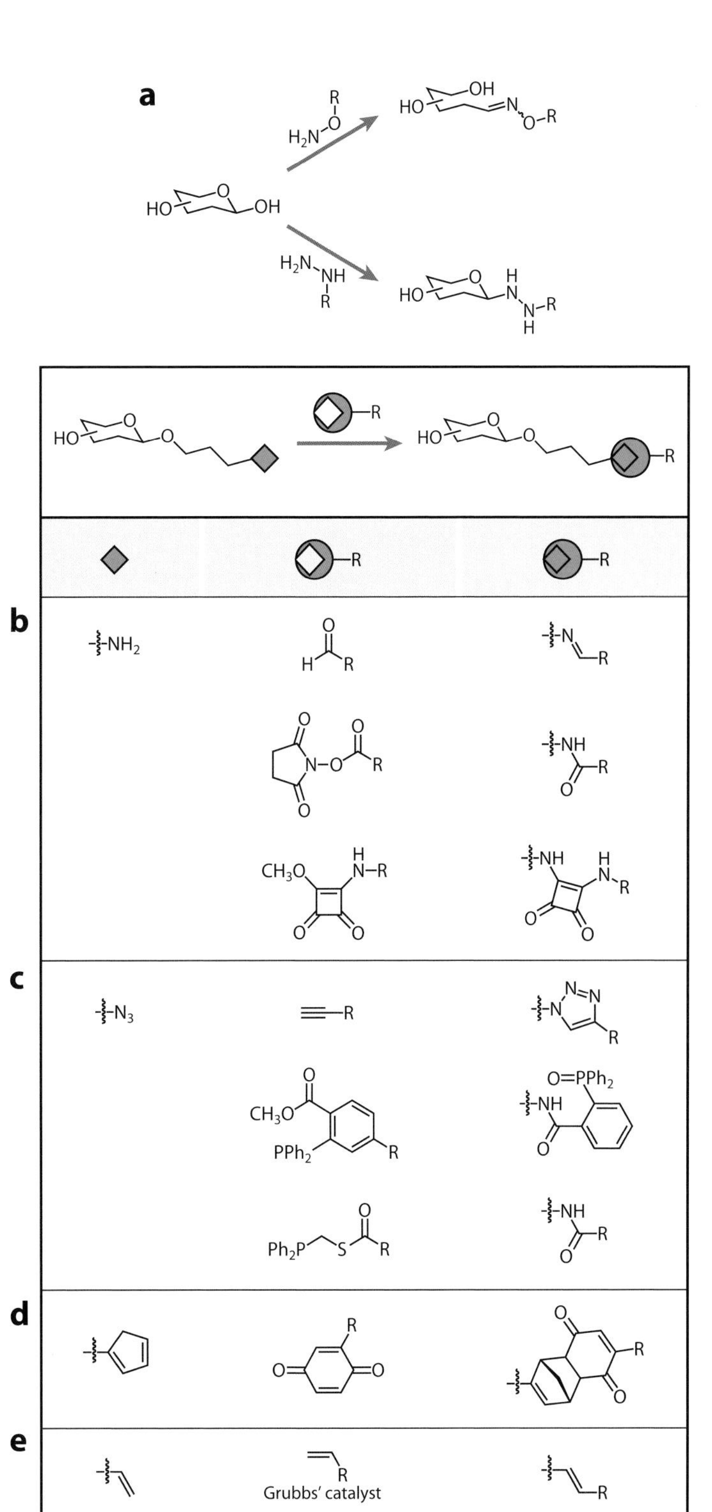

Figure 7

Chemoselective reactions used to generate glycoconjugates. (*a*) Glycans with free reducing ends can undergo reaction with aminooxy- and hydrazine-bearing linkers to form oxime and hydrazone linkages. (*b-e*) Different reactant sets for the general reaction shown at the top of the table. (*b*) Glycans bearing amino groups can attack *N*-hydroxysuccinimidyl esters, aldehydes, or dimethyl squarate to generate adducts. (*c*) Glycans that possess azide functional groups can engage in azide-alkyne cycloaddition [Cu(I)-catalyzed or strain-promoted reactions] and Staudinger ligation reactions. Diels-Alder cycloadditions (*d*) and olefin metathesis reactions (*e*) are other examples of chemoselective methods for glycan attachment.

CFG: Consortium for Functional Glycomics

to maleimide-bearing glycans (56, 57) or form disulfide-linked conjugates (58). Cycloaddition reactions, including Cu(I)-catalyzed azide-alkyne (**Figure 7*c***) and Diels-Alder reactions (**Figure 7*d***), also have been used (59, 60). Although olefin metathesis depends upon a metal carbene catalyst, remarkably, it has been shown to be compatible with carbohydrate functionality (**Figure 7*e***), including sulfate groups (22, 61). The Staudinger ligation reactions of azides with phosphinoesters or phosphinothioesters also are useful chemoselective reactions, and the phosphinoester version has been used in diverse contexts (**Figure 7*c***) (62, 63). Some examples of how the different aforementioned chemoselective reaction processes have been exploited to investigate glycan function are described in subsequent sections.

INTERROGATION OF GLYCAN RECOGNITION

Glycans are present both inside and outside of cells. Within cells, glycosylation is critical for protein trafficking, and more recently, it has been found to influence gene expression (64). Glycans on the surface of pathogens can serve both as a protective shield and as a means for recognizing and entering target cells. Similarly, protein-glycoconjugate interactions are a major line of communication between cells and their environment. Lectins, proteins of nonimmune origin that bind to specific glycan structural motifs, typically use solvent-exposed binding sites to interact with their target oligosaccharide ligands (65). As a result, they bind weakly to single carbohydrate residues and even oligosaccharides. Indeed, monovalent protein-glycan binding dissociation constants are often in the range of 10^{-4} to 10^{-3} M (66). These low affinities might suggest that protein-glycan complexes are not important, yet weak binding is ideal for mediating cell adhesion. When cells interact, glycans on one cell surface can bind to multiple copies of a lectin on another, thereby increasing the apparent binding constant (functional affinity). The advantage of using low-affinity interactions for cell-cell recognition is that, when each individual receptor-ligand interaction is weak, binding will be kinetically labile. In this way, only cells with the correct combination of receptor-ligand pairs will interact stably.

The involvement of multivalent binding in many protein-glycan interactions complicates the identification of the relevant endogenous ligands. Standard receptor-ligand assays lack the necessary sensitivity to monitor low-affinity binding. Accordingly, many methods to assess protein-glycan interactions depend upon multivalent display of one or both binding partners. Such assays have higher sensitivity and can have even higher specificity (67, 68). Moreover, they can minimize the amount of material required, which is especially important considering the challenges associated with the acquisition of glycans. Still, they are best used to compare compounds because determining the true equilibrium constant for a multivalent interaction is complicated. Indeed, many distinct types of binding modes can contribute to the strength of a multivalent interaction (69, 70). Glycan arrays are a technology that invokes multivalent binding and allows many different samples to be compared simultaneously. New tools for array fabrication and analysis have been advanced that depend on a combination of analytical, biochemical, and synthetic methods. The topic of glycan arrays has been reviewed extensively (71–75), and our goal is to highlight relevant contributions of chemistry to their development.

Glycan Arrays

Glycan arrays have been widely embraced as platforms suited to rapid screening of protein binding to carbohydrates. On the basis of principles developed for DNA and protein microarrays, glycan arrays have emerged as tools to assess the specificities of lectins, antibodies, and other glycan-binding proteins. There are many methods for fabricating glycan arrays, yet all have the same overall features. Specifically, natural or synthetic glycans are immobilized onto a surface through either covalent or noncovalent attachment. The resulting glycan surfaces are treated with whole cells, complex biological samples such as sera, or isolated glycan-binding proteins. Binding can be assessed using fluorescence or another type of reporter.

The first challenge to creating glycan arrays was to develop a method to spatially pattern various oligosaccharides or glycoconjugates on a surface. To date, the methods implemented for array fabrication fall into three general categories: (*a*) immobilization by physical adsorption, (*b*) immobilization via high-affinity, specific noncovalent interactions, or (*c*) covalent capture, in which complementary reactants are displayed on the glycan and the surface. Physical adsorption, which exploits the ability of glycosylated proteins or glycolipids to adhere to the surface, helped to establish the utility of glycan arrays as a multivalent assay platform (76–78). Nonnatural glycolipids can be generated from oligosaccharides that possess a free reducing end (i.e., a masked aldehyde) using lipids that bear nucleophiles, such as amines (79) or alkoxylamines (80). In an innovative variation of the adsorption approach, fluorous lipid tags, which can be used both for synthesis and immobilization, have been employed (81). An alternative approach is to immobilize a glycoconjugate through specific noncovalent complexes, such as biotin-streptavidin binding or DNA hybridization (82). In general, however, in most arrays, the glycan is linked to the surface by covalent bond formation.

A common means for glycan array construction is to exploit the unique reactivity of the anomeric position. For example, oligosaccharides can be appended to a surface that presents nucleophiles via reaction with the reducing end (83). Similarly, the reducing end can undergo reductive amination with 2-aminobenzamine derivatives in solution (84, 85), and these fluorescent saccharides can be subsequently attached to the surface. In the latter approach, the fluorescent tag serves multiple purposes: It provides a means to purify heterogeneous pools of natural glycans, it can react with an electrophilic (e.g., succinimidyl ester- or epoxide-functionalized) surface, and it provides a means of quantifying the immobilization efficiency. Reductive amination reactions of oligosaccharides with the lysine side chains of proteins also have been exploited to yield multivalent glycoconjugates that were subsequently immobilized; the presentation of these conjugates on the surface can mimic that of glycosylated proteins (86–88).

When glycans are generated by chemical synthesis, tailored functional groups for immobilization can be introduced. As mentioned previously, the mostly widely used strategy is to introduce a linker bearing an amine group and exploit its nucleophilicity with *N*-hydroxysuccinimidyl ester-coated slides. A complementary approach is to build oligosaccharides directly onto the array surface (56). This general strategy provides not only a means to construct known glycan structures, but also the opportunity to interrogate the selectivities of glycosyltransferases (89).

Many relevant protein-glycan interactions can be uncovered using glycan arrays as experiments focused on Tn antigen illustrate. The Tn antigen is rarely expressed in normal tissues but is associated with several cancers. Glycan arrays were used to reveal that only a subset of prostate tumors display the Tn antigen, a finding that may have therapeutic implications (90). Another instructive feature of the Tn antigen studies is that variations in the specificities determined by individual research groups were different. This unexpected outcome highlights the variability that can arise from glycan array data and the need for standardization.

Despite the challenges, glycan arrays are providing new insight into the protein-binding specificities in complex systems, including those involving highly anionic sulfated GAG sequences. GAGs such as heparin, heparan sulfate, and chondroitin sulfate are involved in processes ranging from development, angiogenesis, cancer metastasis, wound healing, and viral invasion (91). GAGs can be composed of heterogeneous sequences, but the idea that specific sequences are recognized selectively has been controversial owing to a paucity of supporting data. To investigate this hypothesis,

GAG arrays have been assembled from the attachment of di- to hexasaccharides bearing amine- and aminooxy-terminated linkers to electrophilic surfaces (52, 53, 92). The resulting arrays are bolstering the hypothesis that specific GAG sequences have defined physiological functions (93–95).

With all the advances in array fabrication, the most significant barrier to the widespread adoption of glycan arrays is the limited availability of oligosaccharide structures. The CFG provides arrays for the nonspecialist that focus on human and mammalian glycans. The version currently available to researchers (Version 4.0) displays 442 mammalian glycans, whereas the pathogen array presents 96 glycans derived from seven pathogen species. Researchers need to continue to identify and generate a broad array of glycans to extend further the utility of the array platform. Additionally, although the array surface is suited to multivalent interactions, it is unclear how the mode of glycan display influences protein recognition. New technologies to address the role of presentation include the immobilization of multivalent glycosylated scaffolds, such as proteins and peptides (87, 88) or polymer ligands (86). Additionally, fluidic arrays have been introduced that are designed to mimic the mobility of glycoconjugates imbedded in a lipid bilayer (96). As glycan array technology continues to evolve, standard methods, from fabrication to interpretation, will undoubtedly emerge.

Lectin Arrays

Lectin arrays provide a means to assess lectin binding to individual glycoconjugates, patterns of cell-surface glycosylation, and pathogen-lectin interactions (73). They can provide important structural information about uncharacterized glycans and serve as multivalent and sensitive monitors of protein-glycan interactions. Lectin arrays are typically fabricated using commercially available carbohydrate-binding proteins of defined specificity. Mahal and coworkers (97) have pioneered a two-color technique for the analysis of glycans from mixtures, similar to that employed for DNA arrays. The ratiometric data obtained from a pair of dye-swapped arrays afford reproducible data. This method was used to support the hypothesis that the human immunodeficiency virus (HIV) co-opts the microvesicular exocytic mechanism to exit T cells. Lectin array technology also has been used to elucidate differences in sialic acid expression between nontumorigenic and adenocarcinoma cells (98).

Although lectin array technology is more nascent than that of glycan arrays, some of the challenges are shared. Notably, just as glycan arrays are limited by the availability of oligosaccharides, lectin arrays are restricted by the recognition specificities of known lectins. Most commercially available lectins are isolated from plants, but pathogenic organisms often contain unique glycan structures that are not recognized by the available and characterized lectins. As more carbohydrate-binding proteins become available, the value of these arrays and their ability to distinguish between different cell types will increase.

Generally, glycan and lectin arrays use fluorescence to detect protein-glycan interactions. Methods of introducing probes include conjugation of fluorescent tags to cells or lectins, cell staining, and incubation with labeled antibodies. Modification-free techniques for array analysis are also under development, and these include evanescent-field fluorescence detection (99, 100), surface plasmon resonance (101), and fluorescence interference contrast microscopy (102).

PERTURBATION OF GLYCAN FUNCTION

A traditional approach to perturb protein function is to delete a protein or proteins of interest within a cell or organism. The application of RNA interference or gene knockouts can provide insight into the importance of a lectin or an enzyme involved in glycan biosynthesis (103). Drawing conclusions from single gene deletions of enzymes involved in the biosynthesis or processing of glycans, however, can be difficult, as studies of CD22 illustrate. Mice that lack CD22 are hypersensitive to B cell antigen receptor stimulation, but those that lack the glycosyltransferase (ST6Gal-I) that generates the CD22 ligand have compromised B cell responses to antigen (104). To determine that CD22-ligand interactions suppress B cell activation required further experimentation. Another complication of genetic experiments is the masking of phenotypic changes because of compensation by other enzymes, which can camouflage the function of the protein of interest. Alternatively, a single protein may have multiple roles, but a null mutant lacks all of them. Perhaps most significantly, genetic methods were devised to examine protein function; even though they can be applied to a particular protein that binds or generates a glycan, they do not report on the function of the glycan itself. Thus, although genetic methods are powerful, complementary strategies are needed. One alternative is to perturb glycan function with compounds that disrupt or alter specific protein-glycan interactions or the production of specific glycans. Such perturbations can shed light on the physiological processes mediated by a glycan and also provide therapeutic leads. In the following sections, we outline recent advances in the use of synthetic molecules to probe glycan function.

Perturbation of Protein-Glycan Recognition with Monovalent Ligands

The recognition that protein-glycan complexes are critical in physiological and therapeutically important processes, which include inflammation, immune system function, cancer, and host-pathogen interactions, has fueled efforts to generate inhibitors. The aforementioned features of the proteins that bind glycans—their low-affinity and solvent-exposed binding sites—render the generation of effective ligands a formidable challenge. Still, inhibitors with the requisite attributes are emerging from an enhanced understanding of glycan-lectin interactions coupled with advances in high-throughput methods for identifying them.

Oligosaccharides based on endogenous glycans are an obvious starting point (105), but converting these polar molecules into potent inhibitors has been difficult. The aim is to devise molecules with improved affinity and selectivity, reduced polarity, and greater stability than naturally occurring glycans. One strategy to address these issues is to apply molecular design principles. Although the rationale used to optimize a glycomimetic generally is tailored to the specific lectin target, analysis of the successful design efforts to date reveals some common strategies (105). First, either structure-function relationship data or the structure of the complex is used to identify glycan functional groups that are critical for binding. Second, nonessential polar functional groups (e.g., hydroxyl and acetamido groups) are removed to increase lipophilicity. Third, conformational control elements are introduced to preorganize the oligosaccharide to adopt the active, bound conformation. Fourth, the observation that many glycan-binding sites are lined with aromatic residues can be exploited by introducing aromatic substituents at key positions to enhance binding affinity. Some examples that illustrate successful implementation of these design elements follow.

Glycomimetic: a small molecule designed to mimic the function of a carbohydrate with improved pharmacological properties

The development of galectin inhibitors highlights the value of the aforementioned strategies. Galectins are a class of glycan-binding proteins found in multicellular organisms; humans possess 12 genes encoding galectin family members (106). Their name comes from their propensity to bind β-galactose-containing oligosaccharides, although individual galectins can exhibit distinct selectivities. These proteins are involved in a range of physiological processes, including regulation of cell growth, differentiation and apoptosis, cell adhesion, chemoattraction, and cell migration (107, 108). They also are implicated in the inflammatory response and tumor progression. Unlike most mammalian lectins, galectins are not membrane bound but rather are produced in the cytosol and then secreted. Consistent with their ability to occupy two different cellular locations, galectins appear to have important intracellular and extracellular functions. Still, their functional roles have been difficult to discern, and the relevance of glycan binding is not always clear. Thus, inhibitors that could interact selectively with different members of the galectin family could serve as valuable biological probes.

Several of the galectins, including galectin-3, form complexes in which the 4- and 6-hydroxyl groups of galactose form hydrogen bonds to the protein, whereas the remaining hydroxyl groups do not make direct contact (109). These nonessential ligand hydroxyl groups serve as points for modification, and it was postulated that aromatic substituents at the 3-position would enhance binding. Inhibitors of this type possess dissociation constants that are 1000-fold more potent (K_d < 50 nM) than *N*-acetyllactosamine (110). They also show selectivity (~100-fold) for galectin-3 over other galectins. A galectin-3 ligand of this type revealed that glycan binding by this lectin plays a role in alternative macrophage activation (111). Alternative macrophage activation is linked to processes ranging from asthma to wound repair and fibrosis; therefore, these studies suggest that galectin inhibitors could have beneficial therapeutic effects.

Many efforts to devise glycomimetic inhibitors have focused on the C-type lectin family, whose members require Ca^{2+} for binding. Three lectins from this group, E-, L-, and P-selectin, have served as a major testing ground for glycomimetic design. The selectins have been targets because of their participation in recruiting leukocytes to inflamed tissue and their putative roles in tumor cell migration. Each selectin can bind to the structurally related tetrasaccharides sialyl Lewis x and sialyl Lewis a, and extensive structure-activity studies with oligosaccharide derivatives established the key features that contribute to binding. For E-selectin complexation, critical attributes include the carboxylic acid group, the three hydroxyl groups of fucose, and the 4- and 6-hydroxyl groups of galactose. This motif was used to guide the design of a glycomimetic, in which the relevant groups were presented on a scaffold that is preorganized for binding (**Figure 8*b***) (105). Ligands with even less resemblance to the saccharide residues they were designed to mimic also were effective. For example, peptide motifs appended to fucose bind to P-selectin. These also exhibit selectivity for P- over E-selectin, which is consistent with the ability of the former to recognize an epitope encompassing a carbohydrate and a peptide backbone (**Figure 8*c***) (112). As other examples of

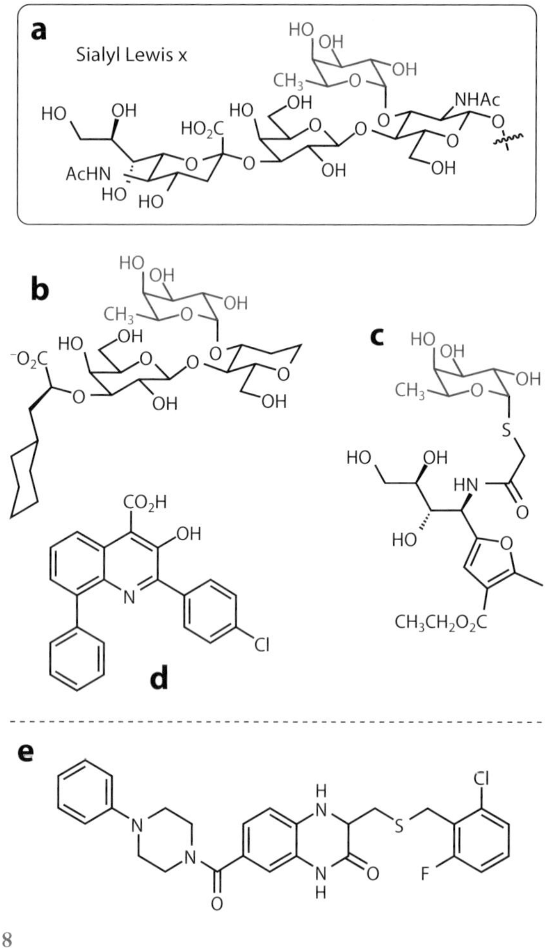

Figure 8

Monovalent ligands for perturbation of protein-glycan recognition. Glycomimetics that present key functional groups in specific orientations have been designed. The tetrasaccharide sialyl Lewis x (*a*) binds to the selectins, and compounds *b–d* have been designed to mimic critical attributes of the oligosaccharide. The naturally occurring oligosaccharide ligands are boxed, and important functional groups that have been incorporated into the glycomimetic are highlighted in red. Compound *e* binds to another member of the C-type (Ca^{2+}-dependent) lectin family, DC-SIGN. Abbreviation: Ac, acetate.

noncarbohydrate ligands, two small-molecule inhibitors of P-selectin were generated from a quinoline salicylic acid scaffold (**Figure 8*d***). These compounds have progressed into clinical trials for rheumatoid arthritis and atherothrombotic vascular events (113).

To date, most glycomimetic design strategies have focused on individual protein-saccharide complexes. In contrast, peptidomimetics often model structural elements (e.g., a β-turn) known to be critical for protein-protein contacts. There are common features of the C-type lectin complexes that might be exploited in inhibitor design. Many C-type lectin complexes use adjacent hydroxyl groups on fucose to coordinate the protein-bound calcium ion, which suggests that scaffolds that possess key Ca^{2+}-coordinating groups can be modified to enhance affinity or specificity. One such strategy has been described that employs focused libraries of glycomimetics using shikimic acid as a building block (114). These have yielded inhibitors of the prototypical lectin mannose-binding protein A. The identification of other approaches that can be applied broadly to other lectin classes could accelerate the pace of glycomimetic generation.

An alternative to the design approach is to identify small-molecule inhibitors of lectins through screening. This strategy could be valuable if cell-permeable ligands could be found; to date, however, a limited number of such compounds have been described. Studies of the selectins have yielded some positive results (105), as have investigations focused on the C-type lectin DC-SIGN. DC-SIGN facilitates several host-pathogen interactions, including dissemination of HIV (115). Inhibitors of DC-SIGN were identified from a 35,000-compound small-molecule library using a high-throughput fluorescence competition assay (116). Seven compounds were identified that are ≥100-fold more potent than *N*-acetylmannosamine for DC-SIGN (**Figure 8*e***). None of the small molecules that bind the selectins or DC-SIGN resembles carbohydrates, an observation that provides impetus to use high-throughput screens to search for effective inhibitors of other carbohydrate-binding proteins.

Perturbation of Protein-Glycan Recognition with Multivalent Ligands

An alternative strategy to overcome low-affinity protein-glycan interactions is to employ multivalent ligands. This approach can be especially effective for blocking protein-glycan engagement at the cell surface. Naturally occurring, multivalent glycan displays are widespread; representatives include glycosylated proteins, the glycan coats of bacteria, viruses, other pathogens, and the surfaces of mammalian cells. Many carbohydrate-binding proteins are oligomeric and therefore are present in multiple copies on the cell surface. In this way, both cell-surface glycans and lectins are poised to engage in multivalent binding.

Multivalent carbohydrate derivatives can exploit unique modes of recognition not available to their monovalent counterparts. Many lectins contain more than one saccharide-binding site or can oligomerize to form larger structures with multiple binding sites. Multivalent ligands that can span the distance between binding sites have an advantage over their monovalent counterparts. This chelation mechanism is advantageous because the translational entropy cost is paid with the first receptor-ligand contact (70, 117). Nevertheless, the apparent affinity of a multivalent interaction often is less than might be expected, presumably because of the conformational entropy restrictions incurred by multipoint binding. Multivalent ligands also can exhibit functional affinity enhancements by occupying secondary binding sites. Alternatively, glycan-binding proteins may cluster in a membrane microdomain either in response to a multivalent ligand or in response to cellular signals. Although multivalent ligands can be potent inhibitors, their ability to cluster glycan-binding receptors allows them also to serve as activators of signaling pathways (69). Thus, depending on their binding modes, multivalent ligands can exhibit a wide range of different activities.

Chemical synthesis can provide architecturally diverse multivalent ligands, including low-molecular-weight displays, dendrimers, polymers, liposomes, and proteins (69, 118). This diversity can be used to optimize a synthetic ligand for a given application. For example, unlike naturally occurring, multivalent glycan ligands, the valency of a synthetic ligand can be altered systematically by varying the length or size of the scaffold. Polymers of defined lengths or dendrimers of different generations will possess different valencies and therefore differing activities. Evaluating the impact of these changes on the biological response can illuminate the mechanisms underlying the function of natural protein-glycan interactions and lead to highly efficacious inhibitors.

Potent multivalent inhibitors for several medically relevant protein-glycan interactions have been identified. One target that has been explored is influenza virus hemagglutinin, and a number of multivalent sialic acid derivatives have been generated that block the interaction of the virus with cells (119). Other host-pathogen interactions also can be inhibited with multivalent ligands, and representative examples include compounds that prevent *Pseudomonas aeruginosa* adhesion or the binding of uropathogenic *Escherichia coli* (120). Although the influence of scaffold structure is just beginning to be explored (121), the inhibitors of the AB_5 bacterial toxin family highlight the benefits of multivalent ligand design. This toxin family is characterized by one active component (A) and a pentamer of subunits (B) that bind carbohydrates displayed on cellular surfaces. The AB_5 toxins are responsible for diseases ranging from travelers' diarrhea (heat-labile enterotoxin), to acute kidney failure in children (*Shiga*-like toxins), to fatal cholera (Cholera toxin). Efforts from several groups have highlighted the importance of distance between binding motifs (118), a parameter that appears to be at least as important as the identity of the ligands themselves. For example, a pentacyclen core was used to display five galactose residues that were tethered using a range of linker lengths (122). The activity of the multivalent ligands against heat-labile enterotoxin depended on the linker. The most potent ligand possessed the longest linker and was 10^5-fold more active than the corresponding monovalent galactose derivative. Dynamic light scattering experiments indicated a 1:1 protein-ligand complex, suggesting the efficacy of the ligand is the result of the chelate effect. In another example, Bundle and coworkers (123) used glucose as a core structure to display two trisaccharides per glucose oxygen on long spacer arms (**Figure 9*a***). This STARFISH ligand was designed to occupy a *Shiga*-like toxin through both the primary binding site and a subsite. It was a highly effective inhibitor (IC_{50} 0.24 nM). Unexpectedly, however, X-ray crystallographic analysis revealed that the designed multivalent ligand did not bind a single pentamer but rather it dimerized two copies by occupying all five B subunits. Thus, for both AB_5 toxin ligands, the spacing between the binding elements was critical for activity, even though their modes of multivalent binding differ.

Most applications of multivalency in glycobiology involve the use of multivalent ligands as inhibitors, but multivalency also can be used to activate particular cellular processes. An example involves blocking the action of L-selectin, which mediates leukocyte migration and recruitment from the blood to lymphatic tissues and sites of inflammation (124). The natural ligands for L-selectin are mucins that present a multivalent display of sialyl Lewis x derivatives. Experiments using mucin mimics highlight the importance of multivalency for L-selectin recognition (**Figure 9*b***). In this study, synthetic multivalent ligands were generated using the ring-opening metathesis polymerization (ROMP). This polymerization is especially valuable for multivalent ligand synthesis because the length, and therefore valency of the ligands, can be controlled. Interestingly, the ROMP-derived polymers not only bind L-selectin, but also promote its proteolytic release, or shedding, from the cell surface (125). These results suggest that clustering L-selectin may signal for its cleavage.

Multivalent displays of oligosaccharides also are being tested as vaccines against bacteria, parasites, and cancers (7, 9, 126, 127). Although oligosaccharides typically do not elicit robust immune responses, they are effective when appended to an immunogenic carrier protein. To boost immunity further, novel multivalent conjugates are being developed. An innovative example is a conjugate that simultaneously displays multiple groups: a B cell epitope, a T helper epitope, and a toll receptor ligand (128). These groups can synergistically augment immune responses by recruiting different aspects of the innate and adaptive immune responses.

Multivalent ligands also can be used to recruit lectins to signaling complexes. Such assemblies are useful in controlling cellular responses because some glycan-binding proteins enhance signaling and others diminish it. Although vaccine designs are necessarily focused on augmenting the immune response, compounds that suppress autoimmune responses also are needed. The aforementioned CD22, which dampens immune activation, is a sialic acid–binding lectin from the Siglec family. A CD22 ligand [i.e., *N*-acetylneuraminic acid-α(2,6)-galactose-β(1,4)-glucose] was attached to a multivalent antigen, such that the resulting polymer engaged both CD22 and the B cell antigen receptor (**Figure 9*b***) (129). This sialylated antigen inhibited B cell activation. These results identify a mechanism by which antigen glycosylation can suppress immune activation. Moreover, they highlight how multivalent ligands can be used to perturb the assembly of glycan-binding proteins on the cell surface.

A new but related aspect of multivalent protein-glycan interactions involves exploiting noncovalent interactions to create functional supramolecular protein-glycan assemblies. Two different general strategies have emerged; both

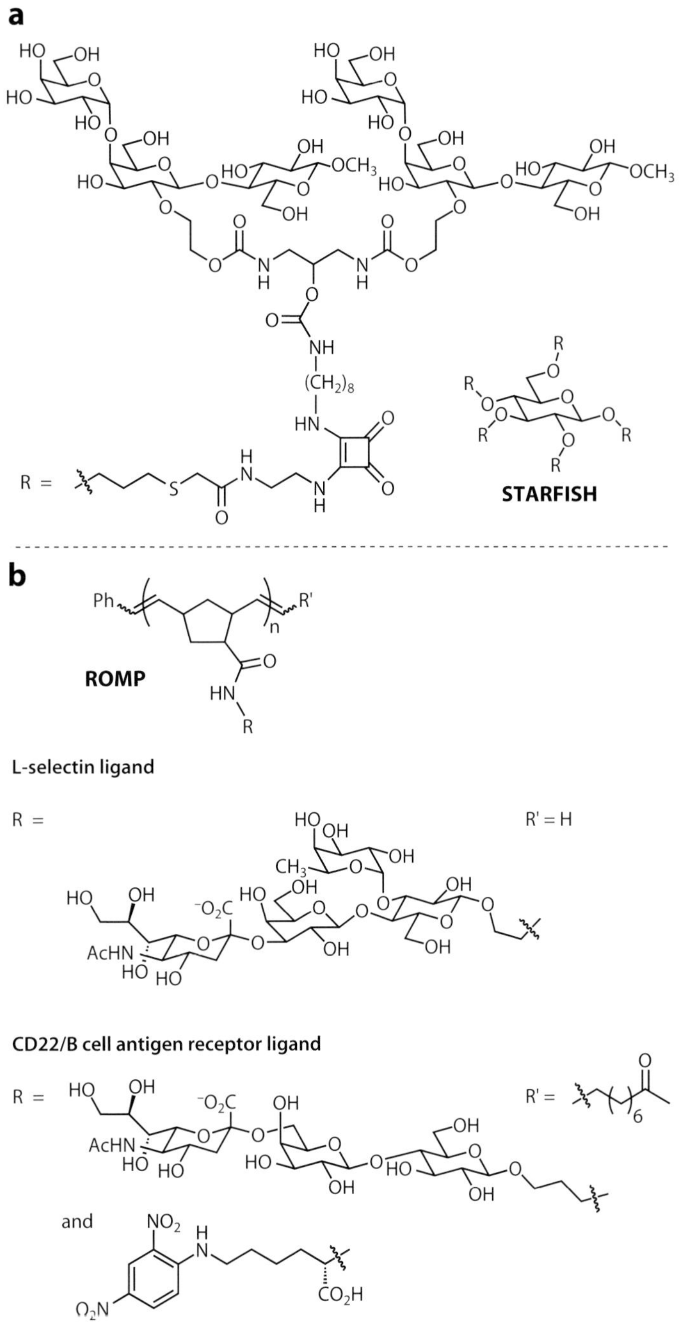

a

R =

STARFISH

b

ROMP

L-selectin ligand

R =

R' = H

CD22/B cell antigen receptor ligand

R =

R' =

and

Figure 9

Selected examples of multivalent ligands that can be used to perturb protein-glycan recognition. (*a*) The pentameric STARFISH core has been used to devise potent inhibitors of the AB_5 toxin family.
(*b*) Polymers are useful scaffolds for producing multivalent ligands. The structures depicted were generated using the ring-opening metathesis polymerization (ROMP), which differs from most polymerization reactions in that it can afford defined ligands. Abbreviation: Ac, acetate.

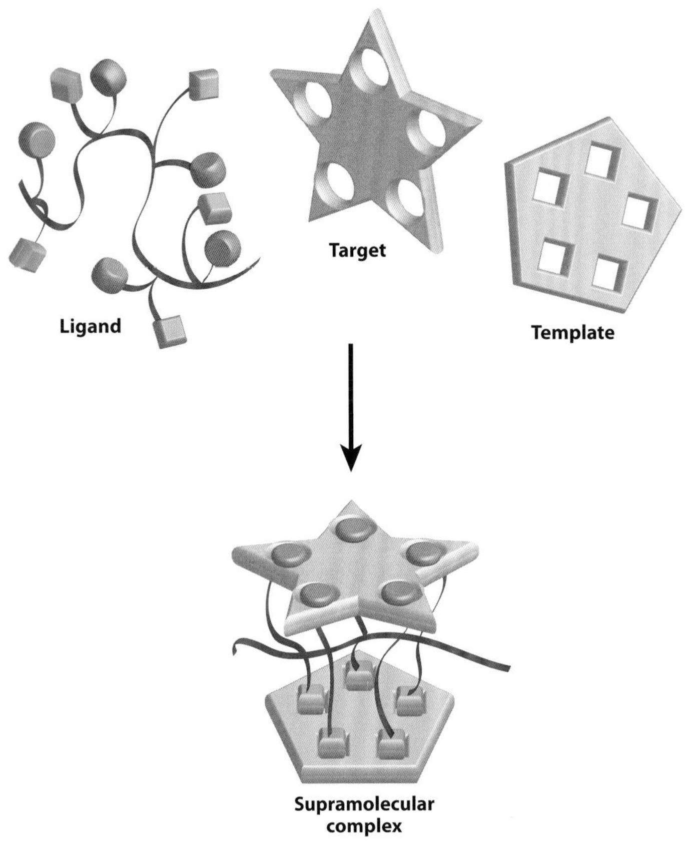

Figure 10

Polymeric ligands possessing a preorganized structure can form supramolecular complexes that display increased functional binding affinity or increased serum half-life.

rely on bifunctional ligands. In the first, polymeric ligands that display multiple copies of two distinct recognition elements, such as the sialylated antigens described above, promote macromolecular protein assemblies. Multivalent assemblies of this type were used to address the disappointing in vivo activity of the aforementioned pentavalent inhibitors of *Shiga* toxin. Polymeric ligands that possess a preorganized architecture promote the formation of complexes of *Shiga* toxin 1 and an endogenous circulating pentavalent receptor human serum amyloid P (**Figure 10**) (130). When tested in mice, a stable macromolecular complex was generated with a prolonged half-life in circulation. Most importantly, the animals were protected from the toxin.

A second approach is to employ compounds that are not polymeric yet possess two different epitopes (54, 131). A bifunctional ligand was used to selectively kill tumor cells via recruitment of a glycan-binding antibody, which subsequently led to complement activation. The bifunctional ligand was designed such that, upon binding to a cell-surface receptor associated with cancer, it could present the antigenic epitope galactosyl-α(1,3)-galactose (α-Gal), which would allow cells to be recognized as foreign by the naturally occurring anti-α-Gal antibody (54). The α-Gal epitope is not found in humans (132), but human serum contains a significant level of anti-α-Gal antibody (about 3% of circulating antibody). Cells displaying α-Gal epitopes are subject to complement-mediated cell killing. Importantly, the recruitment of anti-α-Gal to the cell surface requires a multivalent presentation of glycan residues. The final conjugate consists of an α-Gal epitope linked to an integrin $\alpha_v\beta_3$-binding ligand, as this integrin is upregulated on tumor cells. When cells are treated with the bifunctional ligand, complement-mediated lysis occurs. Cells with low levels of the integrin receptor were spared, whereas tumor cells displaying high levels of integrin were killed selectively (54), and this selectivity underscores the advantage of using multivalent interactions for cell targeting. Thus, ligands that promote macromolecular assemblies of carbohydrate binding-proteins can co-opt the function of glycans and their receptors for new purposes.

Perturbation of Glycan Assembly

An emerging strategy to understand glycan function is to block selected steps in glycoconjugate assembly or disassembly. To this end, efforts have been launched to develop inhibitors of specific glycosyltransferases, glycosidases, and other sugar-processing enzymes. Inhibitors can illuminate both the biological roles of glycans and the mechanisms behind their turnover and assembly. Indeed, compounds that selectively target major pathways (including the production of *N*-glycoproteins and O-glycosylated mucins, as well as the attachment of *O*-GlcNAc residues) could be used to probe the roles of different glycan classes. Moreover, inhibitors of the production of specific glycans in tumors or pathogens could serve as therapeutic leads.

Nature has provided some design strategies for generating compounds that inhibit N-glycosylation in the form of natural products. For example, tunicamycin (**Figure 11*a***) blocks a crucial transphosphorylation reaction between UDP-GlcNAc and dolichol-phosphate that generates dolichol-PP-GlcNAc. This step is required for the synthesis of *N*-glycoproteins, and tunicamycin has therefore been used to illuminate the consequences of deficiencies in *N*-glycan production (133). Another strategy to inhibit N-glycosylation is to block the activity of the oligosaccharyltransferase complex, and this mode of inhibition is exhibited by a cyclic peptide that adopts the conformation of the peptide acceptor (134). Cell-permeable compounds that block N-glycosylation selectively in either bacteria or eukaryotes would be useful probes (135).

Inhibitors of glycosyltransferases involved in *O*-glycan biosynthesis also have been sought (136). Given that *O*-glycans are involved in processes from pathogen binding to cancer, inhibitors will be useful for investigating the physiological and pathophysiological roles of this important class of glycans. For these targets, there are no obvious starting points because natural product inhibitors have not been identified, and the peptide sequence requirements for glycosylation have not been fully delineated. Thus, high-throughput screens were employed. One innovative assay exploits the observation that the rate of proteolysis of a glycosylated peptide is diminished relative to that of its unmodified counterpart (136). By appending N- and C-terminal fluorophores that can engage in Förster resonance energy transfer, a peptide's susceptibility to proteolytic cleavage can be assessed. When a glycosyltransferase is present, cleavage is decreased, whereas the presence of a glycosyltransferase inhibitor restores rapid proteolysis. This assay format has been employed to identify compounds that target *O*-GlcNAc transferase (OGT) (**Figure 12**). OGT is an essential enzyme that regulates signaling through its ability to mediate intracellular O-glycosylation. Compounds that block this enzyme can serve as valuable probes that complement those known for the corresponding *O*-GlcNAc residue hydrolyzing glycosidase. These studies illustrate the value of high-throughput screens to identify inhibitors, and a number of strategies have been implemented including those that rely on glycan arrays (72, 137), activity assays (138), and binding assays (139).

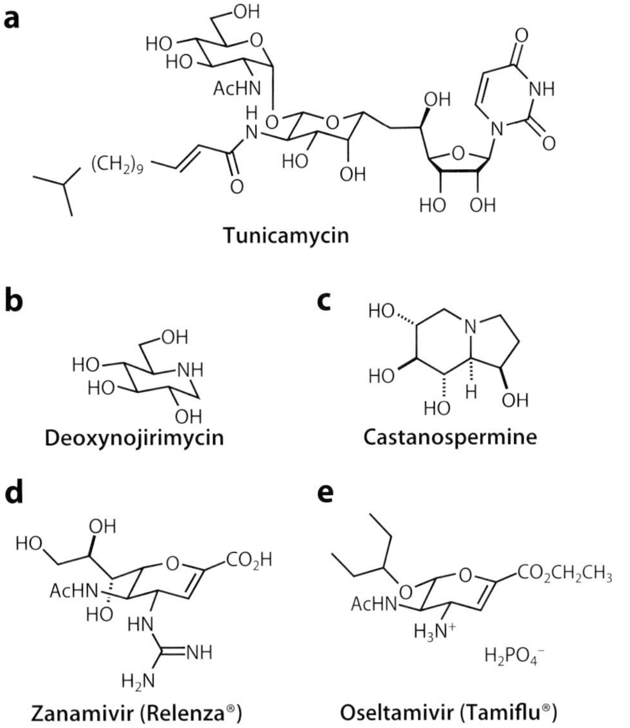

The enzymes responsible for glycan assembly or modification can be critical for microbes and therefore represent attractive targets. One such target is the influenza viral coat protein neuraminidase, whose function is to catalyze

Figure 11

Natural products (compounds 11*a*–11*c*) have inspired the generation of drugs (compounds *d* and *e*) that act as transition-state analogs and thereby inhibit influenza virus neuraminidase. Inhibitors of this type can be used to perturb glycan assembly.

UDP-GlcNAc: uridine 5′-diphosphate-*N*-acetylglucosamine

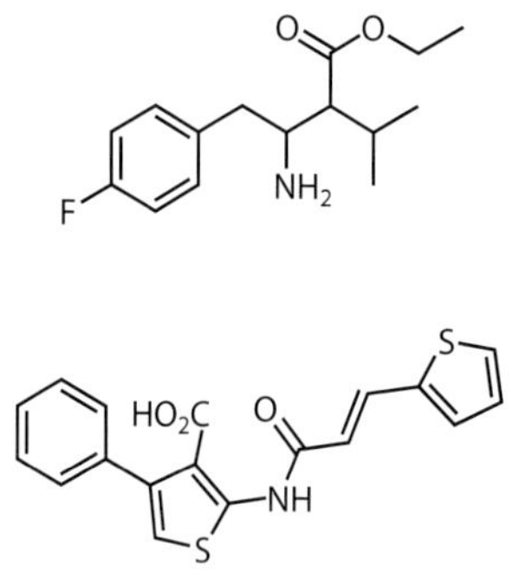

Figure 12

Compounds identified in a high-throughput screen as inhibitors of the essential glycosyltransferase *O*-GlcNAc transferase (OGT).

the cleavage of sialic acids on the host cell surface to enable viral infection. The design of these compounds was inspired by natural product glycosidase inhibitors, such as deoxynojirimycin (**Figure 11*b***) and castanospermine (**Figure 11*c***). These compounds are amines that are protonated at neutral pH, and they are thought to mimic an oxocarbenium transition state (140). Design strategies inspired by transition state models have led to valuable drugs; these include zanamivir (marketed as Relenza®) (**Figure 11*d***) and oseltamivir (marketed as Tamiflu®) (**Figure 11*e***) for the treatment of influenza (141). These compounds, which block viral replication and infection of new host cells, underscore the value of targeting enzymes that operate on glycans. Moreover, this strategy has been applied to afford inhibitors of other glycosyltransferases or glycosidases (142).

Carbohydrate-modifying enzymes critical for cell wall biosynthesis in microbes also have been the objects of inhibition studies. Several investigations in this area have employed ligand displacement assays that rely upon fluorescence polarization (FP). FP assays serve as a general method to study enzymes that use nucleotide sugar substrates, which are prevalent yet poorly understood. FP assays have been used to identify inhibitors of two different enzymes essential for cell wall biosynthesis: UDP-galactopyranose mutase (UGM), which is found in mycobacteria and catalyzes the interconversion of the isomeric compounds UDP-galactopyranose and UDP-galactofuranose, and the glycosyltransferase MurG, which is a bacterial glucosaminotransferase involved in the biosynthesis of the crucial bacterial cell wall component peptidoglycan.

In the case of MurG, a fluorescent probe was designed based on the structure of the donor, UDP-GlcNAc, complexed to the enzyme (**Figure 13**) (139). A library of 64,000 compounds was screened, and subsequent analysis revealed many of the most potent binders possessed a common scaffold: a 1,3-disubstituted heterocyclic core. Because glycosaminotransferases are present in all organisms, probes or lead compounds must bind the bacterial enzyme selectively. Importantly, the inhibitors are selective over other nucleotide-sugar-processing enzymes (143).

In another example, a UDP-fluorescein probe was used in a 16,000-compound, high-throughput inhibitor screen of the UGM from *Mycobacterium tuberculosis* (144). The active compounds identified share structural features with those found to block MurG, including a five-membered thiazolidinone heterocycle, substituents at the 1- and 3-positions, and at least one aromatic group that might mimic the uracil (**Figure 13**). This class of thiazolidinone derivatives is subject to reaction with nucleophiles in a biological milieu (144), which limits their utility as biological probes or therapeutic leads. Still, their identification in two independent screens suggests that the shape of these compounds is well suited for targeting nucleotide-sugar-utilizing enzymes. Further design and optimization based on this hypothesis resulted in the identification of compounds that block mycobacterial growth (**Figure 13**) (145). These studies underscore that high-throughput screens can be used as tools to aid in identifying new therapeutic targets and, more generally, as probes of glycan biosynthesis.

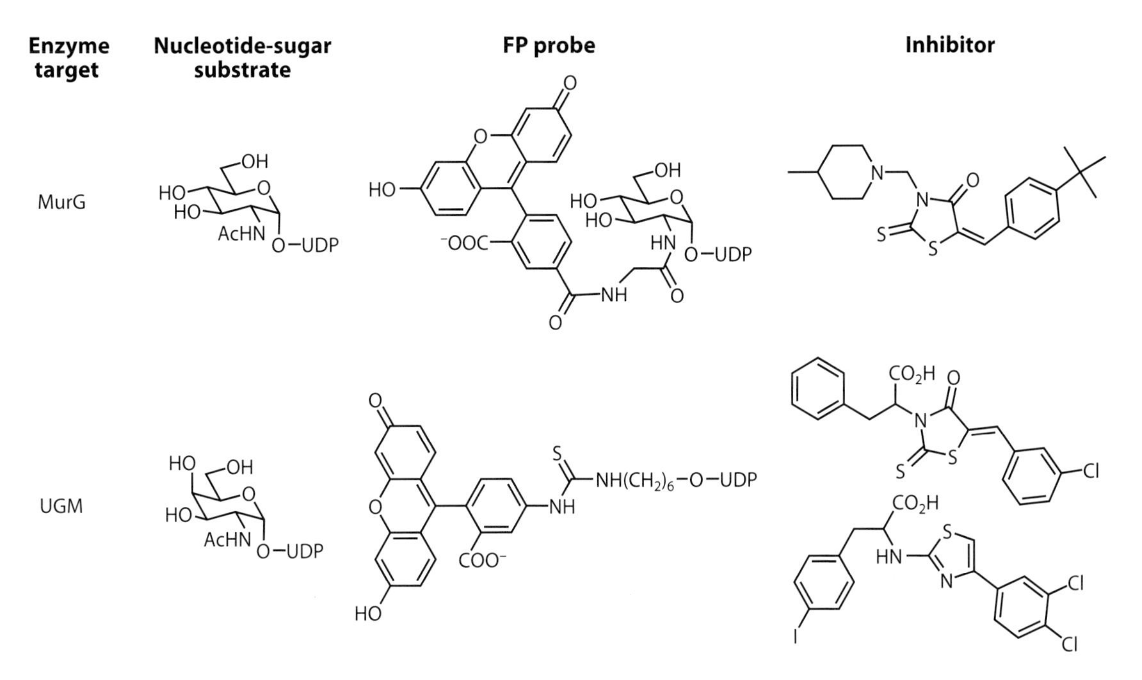

Figure 13

Fluorescence polarization (FP) has been used to identify inhibitors for sugar-processing enzymes with essential roles in bacterial cell wall biosynthesis. These enzymes include the glycosyltransferase MurG and the isomerase UDP-galactopyranose mutase (UGM).

Exploiting Alternative Substrates in Glycan Biosynthesis

Modified glycans can be used to investigate glycan recognition, biosynthesis, cellular or organismal localization, and turnover. For example, primer saccharides can be introduced that compete with endogenous substrates for the glycosyltransferases (146). These compounds serve as decoys by preventing the production of physiological glycoconjugates. This approach can block the generation of specific cell-surface glycoconjugates, but the consequences of harboring the resulting glycan chains within cells have not been elucidated fully. Glycan biosynthetic pathways also can be co-opted to incorporate monosaccharide analogs, thereby generating modified glycoconjugates in cells or organisms. The feasibility of this strategy was demonstrated by exposing cells to *N*-propanoylmannosamine, an analog of *N*-acetylmannosamine, a key intermediate in the biosynthesis of sialylated glycans. With this treatment, the cell-surface glycans generated bear sialic acid residues with *N*-propionamide groups (147). This general strategy has subsequently been exploited for diverse applications.

The incorporation of a modified building block can interfere with subsequent glycosylation reactions, thereby blocking the production of specific glycan structures. In this way, glycan functional roles can be interrogated. For example, when cells are treated with *N*-butanoylmannosamine, they display truncated polysialic acid chains and provide a means to assess the influence of the length of polysialic acid on neuronal plasticity (148). Another example of altering glycan production employs 2-deoxygalactose, which precludes the generation of the fucosyl-α(1,2)-galactose epitope. Investigations using this deoxysugar revealed that fucosylation prevents the proteolytic degradation of synapsin, a critical regulator of neuronal function (149). The incorporation of modified saccharide residues also has been used to investigate the requirements for glycan recognition. For example, the use of metabolic engineering to generate *N*-glycosyl-substituted sialylglycoconjugates disrupts the interaction of neural glycoproteins with myelin-associated glycoprotein (150). *N*-Acetylmannosamine derivatives also have been used to decorate tumor cell surfaces with immunogenic glycans (151). Thus, altering glycan biosynthetic pathways can provide insight into the functional roles of specific carbohydrate epitopes.

Oligosaccharide biosynthesis also can be co-opted to introduce unique functional groups into cellular glycans. The Bertozzi group (152) has investigated different aspects of glycobiology using metabolic labeling combined with chemoselective functionalization reactions. In one example, they showed that *N*-azidoacetylglucosamine (GlcNAz) could be incorporated at sites subject to *O*-GlcNAc modification. Azides are valuable handles because they are absent from biological systems and are relatively unreactive toward most biological function groups. As stated above, they can undergo transformations in the presence of other nonphysiological function groups. Specifically, an azide-bearing compound can be coupled using a Staudinger ligation reaction with a phosphinoester bearing a reporter (e.g., biotin) to generate a conjugate. This reaction scheme was utilized to profile proteins modified with *O*-GlcNAc. Cells were treated with GlcNAz to afford azide-substituted glycoproteins that were subsequently modified for isolation and characterization (153, 154). A similar strategy has been used with *N*-azidoacetylgalactosamine (GalNAz) to detect mucins. Metabolic labeling also can be used to introduce photoaffinity labels, such that cross-linking to glycan-binding proteins can be carried out. For example, aryl-azide (155) or diazirine (156) moieties can be incorporated into sialic acid-bearing glycans at the C9 or C5 positions of sialic acid, respectively. When these photoactivable groups are displayed on cell-surface glycans, they can be used to identify glycan-binding proteins by covalent trapping. The ability to modify glycans by metabolic incorporation also can be used to introduce groups to alter cell adhesion. The display of *N*-thioglycolylneuraminic acid in cell-surface glycans causes self-assembly into large clustered cell aggregates, which may prove valuable for tissue engineering purposes (157).

Illuminating Glycan Biosynthesis

Our understanding of glycans would be advanced by strategies to visualize and track their cellular and organismal distribution in development or disease. The biosynthesis of glycans containing specific residues can be visualized by the incorporation of functional groups that can be used to append reporter groups. Both the Staudinger ligation and the Huisgen 1,3-dipolar cycloaddition reaction of azides with alkynes (often referred to as click chemistry) have been employed in biological environments (**Figure 7*c***) (51). A major issue in these transformations is the rate of reaction, which determines the labeling time and sensitivity. Attempts to increase reaction efficiency have focused on the azide-alkyne 1,3-dipolar cycloaddition. Although this reaction can be catalyzed by copper(I), these conditions are deleterious to cells. Using a more reactive, strained difluorinated cyclooctyne partner circumvents the requirement for copper catalysis (**Figure 14*a***). A variety of sugar building blocks bearing azido groups can be incorporated, including derivatives corresponding to sialic acid, *N*-acetylgalactosamine (GalNAc), GlcNAc, and fucose (152). The azide moiety can then undergo reaction with an alkyne bearing a fluorophore or other reporter group. The advantages of this labeling strategy are illustrated by its use in visualizing glycans in whole animals. Specifically, zebrafish embryos were incubated with GalNAz to generate glycans that bear azide groups. The GalNAz residues could be visualized by their ability to undergo a cycloaddition reaction with fluorophore-labeled cyclooctynes. By using two different fluorescent tags, changes in O-glycosylation during the course of zebrafish development could be observed. An alternative approach for glycan tagging in cells takes advantage of mild oxidation, which occurs at endogenous sialic acid residues, to afford an aldehyde at the C7

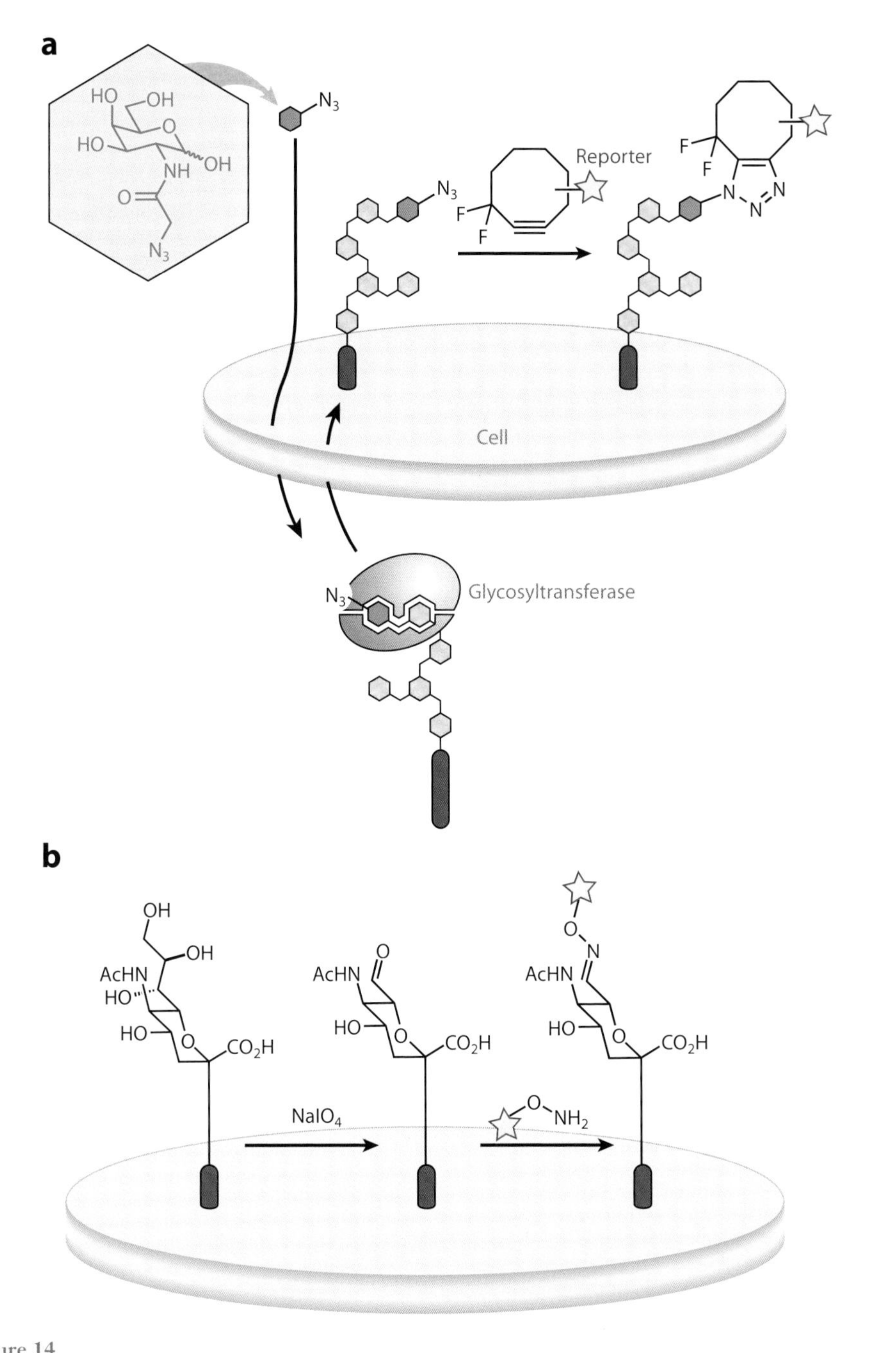

Figure 14

Strategies for visualizing glycans can be achieved by (*a*) metabolic incorporation of a sugar bearing a latent reactive group (e.g., azide) or (*b*) chemical modification of endogenous glycans to introduce a chemical handle that can be functionalized with fluorophores or other reporter molecules. Abbreviation: Ac, acetate.

position (158). Aniline-catalyzed oxime ligation with an aminooxy-modified molecular tag provides a chemical handle that can be used for visualization or tracking of glycans (**Figure 14*b***). This method is complementary to metabolic labeling; not all systems will tolerate the mild oxidation step, but the labeling can be done on a faster timescale. Together, these results emphasize the advances that have been made in strategies to observe glycan production and turnover.

CONCLUSION

Emil Fischer, a pioneer in recognizing the critical role of chemistry in understanding carbohydrates, stated in his Nobel Lecture (159), "... the chemical enigma of Life will not be solved until organic chemistry has mastered another, even more difficult subject, the proteins, in the same way as it has mastered the carbohydrates." Though one can only admire Fischer, today his statement appears ironic, especially given the formidable challenges of elucidating glycan function that remain. These challenges are being met by an assemblage of interdisciplinary approaches. Research in chemical glycobiology is driving discovery by offering new approaches and tools to explore and exploit the functions of glycans.

SUMMARY POINTS

1. Glycans are involved in myriad specific molecular and cellular events from development to disease.
2. The field of glycobiology demands the development of new approaches to elucidate glycan function. Chemical biology approaches provide critical tools and strategies that can be used to probe or perturb glycan function.
3. The synthesis of glycans and the development of chemoselective reactions have greatly expanded the repertoire of oligosaccharides and glycoconjugates available.
4. Glycan and lectin arrays have emerged as valuable platforms to interrogate protein-glycan interactions.
5. The biosynthesis of glycans can be exploited to introduce new functionality that can be used for analysis or imaging.
6. Inhibitors, both rationally designed glycomimetics and small molecules, are promising new tools with which to investigate and interfere with the biological roles of glycans.
7. Synthetic multivalent ligands are valuable tools for investigating the low-affinity interactions characteristic of glycan-binding events.

FUTURE ISSUES

1. The continued development of new strategies and approaches in both chemical and chemoenzymatic syntheses will provide increased access to complex glycans. Methods are needed to assemble less explored glycans, including sulfated oligosaccharides and the unique glycans found in microbes and other pathogens.
2. Glycans lack functional group handles that can be used to install fluorophores or other reporters to directly assess binding or enzyme-catalyzed modification reactions. Thus, new strategies for devising high throughput are needed to identify and engineer enzymes for the synthesis of glycans and to promote the discovery of probes of glycan function.

3. Methods for monitoring glycan degradation are needed, as this process is critical for human health and has ramifications for harvesting energy from plant cell walls.
4. Although great strides have been made in the development of small-molecule ligands for glycan-binding proteins, inhibitors exist for only a handful of lectins. Selective inhibitors for a wide range of lectins are needed.
5. Small molecules that interfere with glycan processing are valuable, as the drugs that block influenza virus neuraminidase illustrate. Recent studies indicate that small-molecule inhibitors of key enzymes in glycan biosynthesis can be found, but more probes could expedite the elucidation of glycan function.
6. Great strides are being made in devising glycoconjugates that can elicit or inhibit immune responses. Promising glycolipids and synthetic glycoconjugates are being designed as vaccines, adjuvants, and immunomodulators.
7. With an increased understanding of how glycans function, chemistry can provide new tools to co-opt these functions for new purposes. From the metabolic incorporation of unique functional groups to the use of ligands that recruit anticarbohydrate antibodies to kill tumor cells, glycan function can be used to elicit new and valuable biological responses in cells and organisms.

DISCLOSURE STATEMENT

The authors are not aware of any affiliations, memberships, funding, or financial holdings that might be perceived as affecting the objectivity of this review.

ACKNOWLEDGMENTS

We apologize to authors whose contributions were omitted from this review owing to limitations of space and limitations in the number of references. This research was supported by the National Institutes of Health (GM49974, GM55984, AI063596, and AI055258) to L.L.K. R.A.S. acknowledges the American Chemical Society Division of Medicinal Chemistry for a fellowship. We thank C.D. Brown, S.L. Mangold, L. Li, and M.R. Levengood for their comments on the manuscript.

LITERATURE CITED

1. Apweiler R, Hermjakob H, Sharon N. 1999. On the frequency of protein glycosylation, as deduced from analysis of the SWISS-PROT database. *Biochim. Biophys. Acta* 1473:4–8
2. Coutinho PM, Deleury E, Davies GJ, Henrissat B. 2003. An evolving hierarchical family classification for glycosyltransferases. *J. Mol. Biol.* 328:307–17
3. Varki A. 1993. Biological roles of oligosaccharides: All of the theories are correct. *Glycobiology* 3:97–130
4. Jaeken J, Matthijs G. 2007. Congenital disorders of glycosylation: a rapidly expanding disease family. *Annu. Rev. Genomics Hum. Genet.* 8:261–78
5. Bertozzi CR, Kiessling LL. 2001. Chemical glycobiology. *Science* 291:2357–64
6. Baum LG. 2002. Developing a taste for sweets. *Immunity* 16:5–8
7. Boltje TJ, Buskas T, Boons G-J. 2009. Opportunities and challenges in synthetic oligosaccharide and glycoconjugate research. *Nat. Chem.* 1:611–22
8. Bernardes GJ, Castagner B, Seeberger PH. 2009. Combined approaches to the synthesis and study of glycoproteins. *ACS Chem. Biol.* 4:703–13

9. Galonic DP, Gin DY. 2007. Chemical glycosylation in the synthesis of glycoconjugate antitumour vaccines. *Nature* 446:1000–7
10. Zhu X, Schmidt RR. 2009. New principles for glycoside-bond formation. *Angew. Chem. Int. Ed.* 48:1900–34
11. Petitou M, Duchaussoy P, Driguez P-A, Jaurand G, Herault J-P, et al. 1998. First synthetic carbohydrates with the full anticoagulant properties of heparin. *Angew. Chem. Int. Ed.* 37:3009–14
12. Guerrini M, Zhang Z, Shriver Z, Naggi A, Masuko S, et al. 2009. Orthogonal analytical approaches to detect potential contaminants in heparin. *Proc. Natl. Acad. Sci. USA* 106:16956–61
13. Fischer E. 1893. Ueber die Gluooside der Alkohole. *Ber. Dtsch. Chem. Ges.* 26:2400–12
14. Mootoo DR, Konradsson P, Udodong U, Fraser-Reid B. 1988. Armed and disarmed *n*-pentenyl glycosides in saccharide couplings leading to oligosaccharides. *J. Am. Chem. Soc.* 110:5583–84
15. Crich D, Smith M. 2001. 1-Benzenesulfinyl piperidine/trifluoromethanesulfonic anhydride: a potent combination of shelf-stable reagents for the low-temperature conversion of thioglycosides to glycosyl triflates and for the formation of diverse glycosidic linkages. *J. Am. Chem. Soc.* 123:9015–20
16. Ikeda K, Aizawa M, Sato K, Sato M. 2006. Novel glycosylation reactions using glycosyl thioimidates of *N*-acetylneuraminic acid as sialyl donors. *Bioorg. Med. Chem. Lett.* 16:2618–20
17. Crich D, Li W. 2006. Efficient glycosidation of a phenyl thiosialoside donor with diphenyl sulfoxide and triflic anhydride in dichloromethane. *Org. Lett.* 8:959–62
18. Meijer A, Ellervik U. 2004. Interhalogens (ICl/IBr) and AgOTf in thioglycoside activation; synthesis of bislactam analogues of ganglioside GD3. *J. Org. Chem.* 69:6249–56
19. Tanaka H, Tateno Y, Nishiura Y, Takahashi T. 2008. Efficient synthesis of an α(2,9) trisialic acid by one-pot glycosylation and polymer-assisted deprotection. *Org. Lett.* 10:5597–600
20. Mong TKK, Lee HK, Duron SG, Wong CH. 2003. Reactivity-based one-pot total synthesis of fucose GM_1 oligosaccharide. *Proc. Natl. Acad. Sci. USA* 100:797–802
21. Wang Z, Zhou LY, El-Boubbou K, Ye XS, Huang XF. 2007. Multi-component one-pot synthesis of the tumor-associated carbohydrate antigen Globo-H based on preactivation of thioglycosyl donors. *J. Org. Chem.* 72:6409–20
22. Seeberger PH. 2008. Automated oligosaccharide synthesis. *Chem. Soc. Rev.* 37:19–28
23. Sears P, Wong CH. 2001. Toward automated synthesis of oligosaccharides and glycoproteins. *Science* 291:2344–50
24. Werz DB, Ranzinger R, Herget S, Adibekian A, von der Lieth C-W, Seeberger PH. 2007. Exploring the structural diversity of mammalian carbohydrates ("glycospace") by statistical databank analysis. *ACS Chem. Biol.* 2:685–91
25. Koeller KM, Wong CH. 2000. Synthesis of complex carbohydrates and glycoconjugates: enzyme-based and programmable one-pot strategies. *Chem. Rev.* 100:4465–94
26. Shaikh FA, Withers SG. 2008. Teaching old enzymes new tricks: engineering and evolution of glycosidases and glycosyl transferases for improved glycoside synthesis. *Biochem. Cell Biol.* 86:169–77
27. Wang LX, Huang W. 2009. Enzymatic transglycosylation for glycoconjugate synthesis. *Curr. Opin. Chem. Biol.* 13:592–600
28. Chokhawala HA, Huang S, Lau K, Yu H, Cheng J, et al. 2008. Combinatorial chemoenzymatic synthesis and high-throughput screening of sialosides. *ACS Chem. Biol.* 3:567–76
29. Zhang C, Griffith BR, Fu Q, Albermann C, Fu X, et al. 2006. Exploiting the reversibility of natural product glycosyltransferase-catalyzed reactions. *Science* 313:1291–94
30. Williams GJ, Gantt RW, Thorson JS. 2008. The impact of enzyme engineering upon natural product glycodiversification. *Curr. Opin. Chem. Biol.* 12:556–64
31. Tao H, Peralta-Yahya P, Decatur J, Cornish VW. 2008. Characterization of a new glycosynthase cloned by using chemical complementation. *ChemBioChem* 9:681–84
32. Ben-David A, Shoham G, Shoham Y. 2008. A universal screening assay for glycosynthases: directed evolution of glycosynthase XynB2(E335G) suggests a general path to enhance activity. *Chem. Biol.* 15:546–51
33. Aharoni A, Thieme K, Chiu CP, Buchini S, Lairson LL, et al. 2006. High-throughput screening methodology for the directed evolution of glycosyltransferases. *Nat. Methods* 3:609–14
34. Sasaki A, Ishimizu T, Geyer R, Hase S. 2005. Synthesis of β-mannosides using the transglycosylation activity of endo-β-mannosidase from *Lilium longiflorum*. *FEBS J.* 272:1660–68

35. Huang W, Li C, Li B, Umekawa M, Yamamoto K, et al. 2009. Glycosynthases enable a highly efficient chemoenzymatic synthesis of *N*-glycoproteins carrying intact natural *N*-glycans. *J. Am. Chem. Soc.* 131:2214–23
36. Wada J, Honda Y, Nagae M, Kato R, Wakatsuki S, et al. 2008. 1,2-α-L-Fucosynthase: a glycosynthase derived from an inverting α-glycosidase with an unusual reaction mechanism. *FEBS Lett.* 582:3739–43
37. Honda Y, Fushinobu S, Hidaka M, Wakagi T, Shoun H, et al. 2008. Alternative strategy for converting an inverting glycoside hydrolase into a glycosynthase. *Glycobiology* 18:325–30
38. Leppanen A, White SP, Helin J, McEver RP, Cummings RD. 2000. Binding of glycosulfopeptides to P-selectin requires stereospecific contributions of individual tyrosine sulfate and sugar residues. *J. Biol. Chem.* 275:39569–78
39. Herzner H, Reipen T, Schultz M, Kunz H. 2000. Synthesis of glycopeptides containing carbohydrate and peptide recognition motifs. *Chem. Rev.* 100:4495–538
40. Gamblin DP, Scanlan EM, Davis BG. 2009. Glycoprotein synthesis: an update. *Chem. Rev.* 109:131–63
41. Kan C, Danishefsky SJ. 2009. Recent departures in the synthesis of peptides and glycopeptides. *Tetrahedron* 65:9047–65
42. Dawson PE, Kent SBH. 2000. Synthesis of native proteins by chemical ligation. *Annu. Rev. Biochem.* 69:923–60
43. Muir TW. 2003. Semisynthesis of proteins by expressed protein ligation. *Annu. Rev. Biochem.* 72:249–89
44. Liu L, Hong ZY, Wong CH. 2006. Convergent glycopeptide synthesis by traceless Staudinger ligation and enzymatic coupling. *ChemBioChem* 7:429–32
45. Macmillan D, Anderson DW. 2004. Rapid synthesis of acyl transfer auxiliaries for cysteine-free native glycopeptide ligation. *Org. Lett.* 6:4659–62
46. Crich D, Banerjee A. 2007. Native chemical ligation at phenylalanine. *J. Am. Chem. Soc.* 129:10064–65
47. Chen J, Wan Q, Yuan Y, Zhu J, Danishefsky SJ. 2008. Native chemical ligation at valine: a contribution to peptide and glycopeptide synthesis. *Angew. Chem. Int. Ed.* 47:8521–24
48. Bennett CS, Dean SM, Payne RJ, Ficht S, Brik A, Wong CH. 2008. Sugar-assisted glycopeptide ligation with complex oligosaccharides: scope and limitations. *J. Am. Chem. Soc.* 130:11945–52
49. Savage PB, Teyton L, Bendelac A. 2006. Glycolipids for natural killer T cells. *Chem. Soc. Rev.* 35:771–79
50. Lahiri S, Futerman AH. 2007. The metabolism and function of sphingolipids and glycosphingolipids. *Cell. Mol. Life Sci.* 64:2270–84
51. Sletten EM, Bertozzi CR. 2009. Bioorthogonal chemistry: fishing for selectivity in a sea of functionality. *Angew. Chem. Int. Ed.* 48:6974–98
52. Gama CI, Tully SE, Sotogaku N, Clark PM, Rawat M, et al. 2006. Sulfation patterns of glycosaminoglycans encode molecular recognition and activity. *Nat. Chem. Biol.* 2:467–73
53. Noti C, de Paz JL, Polito L, Seeberger PH. 2006. Preparation and use of microarrays containing synthetic heparin oligosaccharides for the rapid analysis of heparin-protein interactions. *Chem. Eur. J.* 12:8664–86
54. Carlson CB, Mowery P, Owen RM, Dykhuizen EC, Kiessling LL. 2007. Selective tumor cell targeting using low-affinity, multivalent interactions. *ACS Chem. Biol.* 2:119–27
55. Kitov PI, Shimizu H, Homans SW, Bundle DR. 2003. Optimization of tether length in nonglycosidically linked bivalent ligands that target sites 2 and 1 of a *Shiga*-like toxin. *J. Am. Chem. Soc.* 125:3284–94
56. Park S, Lee MR, Pyo SJ, Shin I. 2004. Carbohydrate chips for studying high-throughput carbohydrate-protein interactions. *J. Am. Chem. Soc.* 126:4812–19
57. Verez-Bencomo V, Fernandez-Santana V, Hardy E, Toledo M, Rodriguez M, et al. 2004. A synthetic conjugate polysaccharide vaccine against *Haemophilus influenzae* type b. *Science* 305:522–25
58. Gamblin DP, Garnier P, van Kasteren S, Oldham NJ, Fairbanks AJ, Davis BG. 2004. Glyco-SeS: selenenylsulfide-mediated protein glycoconjugation—a new strategy in post-translational modification. *Angew. Chem. Int. Ed.* 43:828–33
59. Bryan MC, Fazio F, Lee HK, Huang CY, Chang A, et al. 2004. Covalent display of oligosaccharide arrays in microtiter plates. *J. Am. Chem. Soc.* 126:8640–41
60. Fazio F, Bryan MC, Blixt O, Paulson JC, Wong CH. 2002. Synthesis of sugar arrays in microtiter plate. *J. Am. Chem. Soc.* 124:14397–402
61. Mortell KH, Gingras M, Kiessling LL. 1994. Synthesis of cell agglutination inhibitors by ring-opening metathesis polymerization. *J. Am. Chem. Soc.* 116:10253–54

62. Nilsson BL, Kiessling LL, Raines RT. 2001. High-yielding Staudinger ligation of a phosphinothioester and azide to form a peptide. *Org. Lett.* 3:9–12
63. Saxon E, Armstrong JI, Bertozzi CR. 2000. A "traceless" Staudinger ligation for the chemoselective synthesis of amide bonds. *Org. Lett.* 2:2141–43
64. Hart GW, Housley MP, Slawson C. 2007. Cycling of O-linked β-*N*-acetylglucosamine on nucleocytoplasmic proteins. *Nature* 446:1017–22
65. Weis WI, Drickamer K. 1996. Structural basis of lectin-carbohydrate recognition. *Annu. Rev. Biochem.* 65:441–73
66. Lundquist JJ, Toone EJ. 2002. The cluster glycoside effect. *Chem. Rev.* 102:555–78
67. Mortell KH, Weatherman RV, Kiessling LL. 1996. Recognition specificity of neoglycopolymers prepared by ring-opening metathesis polymerization. *J. Am. Chem. Soc.* 118:2297–98
68. Lee YC, Lee RT. 1995. Carbohydrate-protein interactions: basis of glycobiology. *Acc. Chem. Res.* 28:321–27
69. Kiessling LL, Gestwicki JE, Strong LE. 2006. Synthetic multivalent ligands as probes of signal transduction. *Angew. Chem. Int. Ed.* 45:2348–68
70. Mammen M, Choi S-K, Whitesides GM. 1998. Polyvalent interactions in biological systems: Implications for design and use of multivalent ligands and inhibitors. *Angew. Chem. Int. Ed.* 37:2754–94
71. Horlacher T, Seeberger PH. 2008. Carbohydrate arrays as tools for research and diagnostics. *Chem. Soc. Rev.* 37:1414–22
72. Liang PH, Wu CY, Greenberg WA, Wong CH. 2008. Glycan arrays: biological and medical applications. *Curr. Opin. Chem. Biol.* 12:86–92
73. Krishnamoorthy L, Mahal LK. 2009. Glycomic analysis: an array of technologies. *ACS Chem. Biol.* 4:715–32
74. Park S, Sung JW, Shin I. 2009. Fluorescent glycan derivatives: their use for natural glycan microarrays. *ACS Chem. Biol.* 4:699–701
75. Oyelaran O, Gildersleeve JC. 2009. Glycan arrays: recent advances and future challenges. *Curr. Opin. Chem. Biol.* 13:406–13
76. Willats WG, Rasmussen SE, Kristensen T, Mikkelsen JD, Knox JP. 2002. Sugar-coated microarrays: a novel slide surface for the high-throughput analysis of glycans. *Proteomics* 2:1666–71
77. Bryan MC, Plettenburg O, Sears P, Rabuka D, Wacowich-Sgarbi S, Wong CH. 2002. Saccharide display on microtiter plates. *Chem. Biol.* 9:713–20
78. Wang DN, Liu SY, Trummer BJ, Deng C, Wang AL. 2002. Carbohydrate microarrays for the recognition of cross-reactive molecular markers of microbes and host cells. *Nat. Biotechnol.* 20:275–81
79. Fukui S, Feizi T, Galustian C, Lawson AM, Chai W. 2002. Oligosaccharide microarrays for high-throughput detection and specificity assignments of carbohydrate-protein interactions. *Nat. Biotechnol.* 20:1011–17
80. Liu Y, Feizi T, Carnpanero-Rhodes MA, Childs RA, Zhang Y, et al. 2007. Neoglycolipid probes prepared via oxime ligation for microarray analysis of oligosaccharide-protein interactions. *Chem. Biol.* 14:847–59
81. Mamidyala SK, Ko KS, Jaipuri FA, Park G, Pohl NL. 2006. Noncovalent fluorous interactions for the synthesis of carbohydrate microarrays. *J. Fluor. Chem.* 127:571–79
82. Zhang J, Pourceau G, Meyer A, Vidal S, Praly J, et al. 2009. DNA-directed immobilisation of glycomimetics for glycoarrays application: comparison with covalent immobilisation, and development of an on-chip IC_{50} measurement assay. *Biosens. Bioelectron.* 24:2515–21
83. Zhi ZL, Powell AK, Turnbull JE. 2006. Fabrication of carbohydrate microarrays on gold surfaces: direct attachment of nonderivatized oligosaccharides to hydrazide-modified self-assembled monolayers. *Anal. Chem.* 78:4786–93
84. Song X, Lasanajak Y, Xia B, Smith DF, Cummings RD. 2009. Fluorescent glycosylamides produced by microscale derivatization of free glycans for natural glycan microarrays. *ACS Chem. Biol.* 4:741–50
85. de Boer AR, Hokke CH, Deelder AM, Wuhrer M. 2007. General microarray technique for immobilization and screening of natural glycans. *Anal. Chem.* 79:8107–13
86. Gestwicki JE, Cairo CW, Mann DA, Owen RM, Kiessling LL. 2002. Selective immobilization of multivalent ligands for surface plasmon resonance and fluorescence microscopy. *Anal. Biochem.* 305:149–55

87. Oyelaran O, Li Q, Farnsworth D, Gildersleeve JC. 2009. Microarrays with varying carbohydrate density reveal distinct subpopulations of serum antibodies. *J. Proteome Res.* 8:3529–38
88. Godula K, Rabuka D, Nam KT, Bertozzi CR. 2009. Synthesis and microcontact printing of dual end-functionalized mucin-like glycopolymers for microarray applications. *Angew. Chem. Int. Ed.* 48:4973–76
89. Park S, Shin I. 2007. Carbohydrate microarrays for assaying galactosyltransferase activity. *Org. Lett.* 9:1675–78
90. Manimala JC, Li Z, Jain A, VedBrat S, Gildersleeve JC. 2005. Carbohydrate array analysis of anti-Tn antibodies and lectins reveals unexpected specificities: implications for diagnostic and vaccine development. *ChemBioChem* 6:2229–41
91. Fuster MM, Esko JD. 2005. The sweet and sour of cancer: glycans as novel therapeutic targets. *Nat. Rev. Cancer* 5:526–42
92. de Paz JL, Noti C, Seeberger PH. 2006. Microarrays of synthetic heparin oligosaccharides. *J. Am. Chem. Soc.* 128:2766–67
93. Tully SE, Rawat M, Hsieh-Wilson LC. 2006. Discovery of a TNF-α antagonist using chondroitin sulfate microarrays. *J. Am. Chem. Soc.* 128:7740–41
94. Shipp EL, Hsieh-Wilson LC. 2007. Profiling the sulfation specificities of glycosaminoglycan interactions with growth factors and chemotactic proteins using microarrays. *Chem. Biol.* 14:195–208
95. de Paz JL, Moseman EA, Noti C, Polito L, von Andrian UH, Seeberger PH. 2007. Profiling heparin-chemokine interactions using synthetic tools. *ACS Chem. Biol.* 2:735–44
96. Zhu XY, Holtz B, Wang Y, Wang L, Orndorff PE, Guo A. 2009. Quantitative glycomics from fluidic glycan microarrays. *J. Am. Chem. Soc.* 131:13646–50
97. Pilobello KT, Slawek DE, Mahal LK. 2007. A ratiometric lectin microarray approach to analysis of the dynamic mammalian glycome. *Proc. Natl. Acad. Sci. USA* 104:11534–39
98. Chen S, Zheng T, Shortreed MR, Alexander C, Smith LM. 2007. Analysis of cell surface carbohydrate expression patterns in normal and tumorigenic human breast cell lines using lectin arrays. *Anal. Chem.* 79:5698–702
99. Ebe Y, Kuno A, Uchiyama N, Koseki-Kuno S, Yamada M, et al. 2006. Application of lectin microarray to crude samples: differential glycan profiling of Lec mutants. *J. Biochem.* 139:323–27
100. Kuno A, Uchiyama N, Koseki-Kuno S, Ebe Y, Takashima S, et al. 2005. Evanescent-field fluorescence-assisted lectin microarray: a new strategy for glycan profiling. *Nat. Methods* 2:851–56
101. Smith EA, Thomas WD, Kiessling LL, Corn RM. 2003. Surface plasmon resonance imaging studies of protein-carbohydrate interactions. *J. Am. Chem. Soc.* 125:6140–48
102. Godula K, Umbel ML, Rabuka D, Botyanszki Z, Bertozzi CR, Parthasarathy R. 2009. Control of the molecular orientation of membrane-anchored biomimetic glycopolymers. *J. Am. Chem. Soc.* 131:10263–68
103. Ohtsubo K, Marth JD. 2006. Glycosylation in cellular mechanisms of health and disease. *Cell* 126:855–67
104. Collins BE, Blixt O, Bovin NV, Danzer CP, Chui D, et al. 2002. Constitutively unmasked CD22 on B cells of ST6Gal I knockout mice: novel sialoside probe for murine CD22. *Glycobiology* 12:563–71
105. Ernst B, Magnani JL. 2009. From carbohydrate leads to glycomimetic drugs. *Nat. Rev. Drug Discov.* 8:661–77
106. Paulson JC, Blixt O, Collins BE. 2006. Sweet spots in functional glycomics. *Nat. Chem. Biol.* 2:238–48
107. Yang R-Y, Liu F-T. 2003. Galectins in cell growth and apoptosis. *Cell. Mol. Life Sci.* 60:267–76
108. Liu F-T, Rabinovich GA. 2005. Galectins as modulators of tumour progression. *Nat. Rev. Cancer* 5:29–41
109. Kiessling LL, Carlson EE. 2007. The search for chemical probes to illuminate carbohydrate function. In *Chemical Biology. From Small Molecules to System Biology and Drug Design*, ed. SL Schreiber, T Kapoor, G Wess, 2:635–67. Weinheim: Wiley-VCH
110. Cumpstey I, Salomonsson E, Sundin A, Leffler H, Nilsson UJ. 2008. Double affinity amplification of galectin-ligand interactions through arginine-arene interactions: synthetic, thermodynamic, and computational studies with aromatic diamido thiodigalactosides. *Chem. Eur. J.* 14:4233–45
111. MacKinnon AC, Farnworth SL, Hodkinson PS, Henderson NC, Atkinson KM, et al. 2008. Regulation of alternative macrophage activation by galectin-3. *J. Immunol.* 180:2650–58
112. Moreno-Vargas AJ, Molina L, Carmona AT, Ferrali A, Lambelet M, et al. 2008. Synthesis and biological evaluation of S neofucopeptides as E- and P-selectin inhibitors. *Eur. J. Org. Chem.* 2008:2973–82

113. Kaila N, Janz K, Huang A, Moretto A, DeBernardo S, et al. 2007. 2-(4-Chlorobenzyl)-3-hydroxy-7,8,9,10-tetrahydrobenzo[*H*]quinoline-4-carboxylic acid (PSI-697): identification of a clinical candidate from the quinoline salicylic acid series of P-selectin antagonists. *J. Med. Chem.* 50:40–64
114. Schuster MC, Mann DA, Buchholz TJ, Johnson KM, Thomas WD, Kiessling LL. 2003. Parallel synthesis of glycomimetic libraries: targeting a C-type lectin. *Org. Lett.* 5:1407–10
115. van Kyook Y, Geijtenbeek TBH. 2003. DC-SIGN: escape mechanisms for pathogens. *Nat. Rev. Immunol.* 3:697–709
116. Borrok MJ, Kiessling LL. 2007. Non-carbohydrate inhibitors of the lectin DC-SIGN. *J. Am. Chem. Soc.* 129:12780–85
117. Page MI, Jencks WP. 1971. Entropic contributions to rate accelerations in enzymic and intramolecular reactions and the chelate effect. *Proc. Natl. Acad. Sci. USA* 68:1678–83
118. Pieters RJ. 2009. Maximising multivalency effects in protein-carbohydrate interactions. *Org. Biomol. Chem.* 7:2013–25
119. Kiessling LL, Gestwicki JE, Strong LE. 2000. Synthetic multivalent ligands in the exploration of cell-surface interactions. *Curr. Opin. Chem. Biol.* 4:696–703
120. Imberty A, Chabre YM, Roy R. 2008. Glycomimetics and glycodendrimers as high affinity microbial anti-adhesins. *Chem. Eur. J.* 14:7490–99
121. Gestwicki JE, Cairo CW, Strong LE, Oetjen KA, Kiessling LL. 2002. Influencing receptor-ligand binding mechanisms with multivalent ligand architecture. *J. Am. Chem. Soc.* 124:14922–33
122. Fan EK, Zhang ZS, Minke WE, Hou Z, Verlinde C, Hol WGJ. 2000. High-affinity pentavalent ligands of *Escherichia coli* heat-labile enterotoxin by modular structure-based design. *J. Am. Chem. Soc.* 122:2663–64
123. Kitov PI, Sadowska JM, Mulvey G, Armstrong GD, Ling H, et al. 2000. *Shiga*-like toxins are neutralized by tailored multivalent carbohydrate ligands. *Nature* 403:669–72
124. McEver RP. 2002. Selectins: lectins that initiate cell adhesion under flow. *Curr. Opin. Cell Biol.* 14:581–86
125. Mowery P, Yang ZQ, Gordon EJ, Dwir O, Spencer AG, et al. 2004. Synthetic glycoprotein mimics inhibit L-selectin-mediated rolling and promote L-selectin shedding. *Chem. Biol.* 11:725–32
126. Ouerfelli O, Warren JD, Wilson R, Danishefsky SJ. 2005. Synthetic carbohydrate-based antitumor vaccines: challenges and opportunities. *Expert Rev. Vaccines* 4:677–85
127. Seeberger PH, Werz DB. 2007. Synthesis and medical applications of oligosaccharides. *Nature* 446:1046–51
128. Ingale S, Wolfert MA, Gaekwad J, Buskas T, Boons G-J. 2007. Robust immune responses elicited by a fully synthetic three-component vaccine. *Nat. Chem. Biol.* 3:663–67
129. Courtney AH, Puffer EB, Pontrello JK, Yang Z-Q, Kiessling LL. 2009. Sialylated multivalent antigens engage CD22 in *trans* and inhibit B cell activation. *Proc. Natl. Acad. Sci. USA* 106:2500–5
130. Kitov PI, Mulvey GL, Griener TP, Lipinski T, Solomon D, et al. 2008. In vivo supramolecular templating enhances the activity of multivalent ligands: a potential therapeutic against the *Escherichia coli* O157 AB_5 toxins. *Proc. Natl. Acad. Sci. USA* 105:16837–42
131. O'Reilly MK, Collins BE, Han S, Liao L, Rillahan C, et al. 2008. Bifunctional CD22 ligands use multimeric immunoglobulins as protein scaffolds in assembly of immune complexes on B cells. *J. Am. Chem. Soc.* 130:7736–45
132. Galili U, Shohet SB, Kobrin E, Stults CLM, Macher BA. 1988. Man, apes, and Old World monkeys differ from other mammals in the expression of α-galactosyl epitopes on nucleated cells. *J. Biol. Chem.* 263:17755–62
133. Tkacz JS, Lampen O. 1975. Tunicamycin inhibition of polyisoprenyl *N*-acetylglucosaminyl pyrophosphate formation in calf-liver microsomes. *Biochem. Biophys. Res. Commun.* 65:248–57
134. Eason PD, Imperiali B. 1999. A potent oligosaccharyl transferase inhibitor that crosses the intracellular endoplasmic reticulum membrane. *Biochemistry* 38:5430–37
135. Elbein AD. 1987. Inhibitors of the biosynthesis and processing of N-linked oligosaccharide chains. *Annu. Rev. Biochem.* 56:497–534
136. Gross BJ, Swoboda JG, Walker S. 2008. A strategy to discover inhibitors of O-linked glycosylation. *J. Am. Chem. Soc.* 130:440–41
137. Laurent N, Voglmeir J, Flitsch SL. 2008. Glycoarrays—tools for determining protein-carbohydrate interactions and glycoenzyme specificity. *Chem. Commun.* 37:4400–12

138. Rose NL, Zheng RB, Pearcey J, Zhou R, Completo GC, Lowary TL. 2008. Development of a coupled spectrophotometric assay for GlfT2, a bifunctional mycobacterial galactofuranosyltransferase. *Carbohydr. Res.* 343:2130–39
139. Helm JS, Hu Y, Chen L, Gross B, Walker S. 2003. Identification of active-site inhibitors of MurG using a generalizable, high-throughput glycosyltransferase screen. *J. Am. Chem. Soc.* 125:11168–69
140. Lillelund VH, Jensen HH, Liang X, Bols M. 2002. Recent developments of transition-state analogue glycosidase inhibitors of non-natural product origin. *Chem. Rev.* 102:515–53
141. Asano N. 2003. Glycosidase inhibitors: update and perspectives on practical use. *Glycobiology* 13:R93–104
142. Compain P, Martin OR. 2003. Design, synthesis and biological evaluation of iminosugar-based glycosyltransferase inhibitors. *Curr. Top. Med. Chem.* 3:541–60
143. Hu Y, Helm JS, Chen L, Ginsberg C, Gross B, et al. 2004. Identification of selective inhibitors for the glycosyltransferase MurG via high-throughput screening. *Chem. Biol.* 11:703–11
144. Carlson EE, May JF, Kiessling LL. 2006. Chemical probes of UDP-galactopyranose mutase. *Chem. Biol.* 13:825–37
145. Dykhuizen EC, May JF, Tongpenyai A, Kiessling LL. 2008. Inhibitors of UDP-galactopyranose mutase thwart mycobacterial growth. *J. Am. Chem. Soc.* 130:6706–7
146. Brown JR, Crawford BE, Esko JD. 2007. Glycan antagonists and inhibitors: a fount for drug discovery. *Crit. Rev. Biochem. Mol. Biol.* 42:481–515
147. Kayser H, Zeitler R, Kannicht C, Grunow D, Nuck R, Reutter W. 1992. Biosynthesis of a nonphysiological sialic acid in different rat organs, using *N*-propanoyl-D-hexosamines as precursors. *J. Biol. Chem.* 267:16934–38
148. Mahal LK, Charter NW, Angata K, Fukuda M, Koshland DE, Bertozzi CR. 2001. A small-molecule modulator of poly-α2,8-sialic acid expression on cultured neurons and tumor cells. *Science* 294:380–82
149. Murrey H, Gama C, Kalovidouris S, Luo W, Driggers E, et al. 2006. Protein fucosylation regulates synapsin Ia/Ib expression and neuronal morphology in primary hippocampal neurons. *Proc. Natl. Acad. Sci. USA* 103:21–26
150. Collins B, Yang L, Mukhopadhyay G, Filbin M, Kiso M, et al. 1997. Sialic acid specificity of myelin-associated glycoprotein binding. *J. Biol. Chem.* 272:1248–55
151. Liu T, Guo Z, Yang Q, Sad S, Jennings H. 2000. Biochemical engineering of surface α2–8 polysialic acid for immunotargeting tumor cells. *J. Biol. Chem.* 275:32832–36
152. Agard NJ, Bertozzi CR. 2009. Chemical approaches to perturb, profile, and perceive glycans. *Acc. Chem. Res.* 42:788–97
153. Vocadlo DJ, Hang HC, Kim E-J, Hanover JA, Bertozzi CR. 2003. A chemical approach for identifying *O*-GlcNAc-modified proteins in cells. *Proc. Natl. Acad. Sci. USA* 100:9116–21
154. Hang H, Yu C, Kato D, Bertozzi C. 2003. A metabolic labeling approach toward proteomic analysis of mucin-type O-linked glycosylation. *Proc. Natl. Acad. Sci. USA* 100:14846–51
155. Han S, Collins BE, Bengtson P, Paulson JC. 2005. Homomultimeric complexes of CD22 in B cells revealed by protein-glycan cross-linking. *Nat. Chem. Biol.* 1:93–97
156. Tanaka Y, Kohler JJ. 2008. Photoactivatable crosslinking sugars for capturing glycoprotein interactions. *J. Am. Chem. Soc.* 130:3278–79
157. Sampathkumar S-G, Li AV, Jones MB, Sun Z, Yarema KJ. 2006. Metabolic installation of thiols into sialic acid modulates adhesion and stem cell biology. *Nat. Chem. Biol.* 2:149–52
158. Zeng Y, Ramya TNC, Dirksen A, Dawson PE, Paulson JC. 2009. High-efficiency labeling of sialylated glycoproteins on living cells. *Nat. Methods* 6:207–9
159. Fisher E. 1902. Syntheses in the purine and sugar group. *Nobel lect.* **http://nobelprize.org/nobel_prizes/chemistry/laureates/1902/fischer-lecture.pdf**

Cellulosomes: Highly Efficient Nanomachines Designed to Deconstruct Plant Cell Wall Complex Carbohydrates

Carlos M.G.A. Fontes[1] and Harry J. Gilbert[2]

[1]CIISA, Faculdade de Medicina Veterinária, Universidade Técnica de Lisboa, 1300-477 Lisboa, Portugal; email: cafontes@fmv.utl.pt

[2]Complex Carbohydrate Research Center, University of Georgia, Athens, Georgia 30602-4712; email: hgilbert@ccrc.uga.edu

Annu. Rev. Biochem. 2010. 79:655–81

First published online as a Review in Advance on April 7, 2010

The *Annual Review of Biochemistry* is online at biochem.annualreviews.org

This article's doi:
10.1146/annurev-biochem-091208-085603

0066-4154/10/0707-0655$20.00

Key Words

protein:protein interactions, cohesin, dockerin, bioenergy, multienzyme complexes, nanomachines

Abstract

Cellulosomes can be described as one of nature's most elaborate and highly efficient nanomachines. These cell bound multienzyme complexes orchestrate the deconstruction of cellulose and hemicellulose, two of the most abundant polymers on Earth, and thus play a major role in carbon turnover. Integration of cellulosomal components occurs via highly ordered protein:protein interactions between cohesins and dockerins, whose specificity allows the incorporation of cellulases and hemicellulases onto a molecular scaffold. Cellulosome assembly promotes the exploitation of enzyme synergism because of spatial proximity and enzyme-substrate targeting. Recent structural and functional studies have revealed how cohesin-dockerin interactions mediate both cellulosome assembly and cell-surface attachment, while retaining the spatial flexibility required to optimize the catalytic synergy within the enzyme complex. These emerging advances in our knowledge of cellulosome function are reviewed here.

Contents

INTRODUCTION

The synthesis of organic carbon is a major biological process and the primary source of energy for life. Through photosynthesis plants convert solar energy into organic carbon that can be further utilized by heterotrophic organisms. In plants, cell wall polysaccharides, primarily cellulose and hemicellulose, are a major reservoir of carbon and energy. However, only a restricted number of microorganisms have acquired the capacity to deconstruct these structural carbohydrates. The chemical and physical complexity of plant cell walls restricts their accessibility to enzyme attack, and thus, the recycling of photosynthetically fixed carbon is a relatively inefficient biological process (1).

A common feature of all plant cell wall–degrading organisms is that they harness extensive consortia of extracellular enzymes that act in synergy to degrade the recalcitrant amorphous and crystalline substrates present in these composite structures (see Reference 2 for a review). Cellulases and hemicellulases are remarkably elaborate enzymes, displaying complex molecular architectures in which catalytic modules are appended to noncatalytic modules that participate in pivotal protein:carbohydrate [carbohydrate-binding modules (CBMs)] or protein:protein (dockerins) interactions (2). Significantly, the plant cell wall–degrading apparatus of aerobic and anaerobic microorganisms differ considerably in their macromolecular organization. Thus, cellulases and hemicellulases synthesized by anaerobes, particularly clostridia and rumen microorganisms, frequently assemble into a large multienzyme complex (molecular weight >3 MDa) termed the cellulosome [see previous reviews (3–7)]. It is likely that anaerobic environments impose selective pressures that have led to the formation of cellulosomes; however, the nature of the evolutionary drivers that have resulted in the formation of these enzyme complexes is currently unclear.

Historical Perspective

In the early 1980s, Bayer & Lamed and their colleagues (8–10) identified and characterized the first cellulosome, on the basis of studies of the cellulolytic system of the anaerobic thermophilic bacterium *Clostridium thermocellum* (**Figure 1**). Indeed, the bacterium's multienzyme complex remains the paradigm for understanding the mechanism of cellulosome assembly and function. In these pioneering and inspiring experiments, the authors defined the cellulosome as a "discrete, cellulose-binding, multienzyme complex that mediates the degradation of cellulosic substrates," pointing to the molecular ordering of the cellulosome components (10). Initially, the cellulosome was believed to exclusively degrade cellulose, but soon it was recognized that the complex contains not only cellulases but also a large array of hemicellulases (11, 12) and even pectinases (13), with enzyme activities that include polysaccharide lyases, carbohydrate esterases, and glycoside hydrolases. Throughout the 1980s and 1990s, there was a sustained effort to clone cellulosomal genes and characterize the encoded proteins. Such studies rapidly identified the molecular mechanisms by which the cellulosome assembles and how the enzyme complex is presented on the surface of the host bacterium (**Figure 2**). Thus, it was soon evident that the cellulosomal catalytic components contain noncatalytic modules, called dockerins, which bind to the cohesin modules, located in a large noncatalytic protein that acts as a scaffoldin (**Figure 2**). The tight protein:protein interaction established between dockerins and cohesins allows the integration of the hydrolytic enzymes into the complex (14, 15), and a similar interaction, mediated by membrane-associated proteins and an atypical cohesin in the scaffoldin, tethers the cellulosome to the surface of the bacterium (5). In addition, scaffoldins usually contain a noncatalytic CBM that anchors the entire complex onto crystalline cellulose (16). More recently, a range of anaerobic bacteria and fungi were shown to produce cellulosome systems similar to those of *C. thermocellum*, particularly the bacteria *Clostridium cellulovorans*, *Clostridium cellulolyticum*, *Clostridium acetobutylicum*, *Clostridium josui*, *Clostridium papyrosolvens*, *Acetivibrio cellulolyticus*, *Bacteroides cellulosolvens*, *Ruminococcus albus*, *Ruminococcus flavefaciens*, and the anaerobic fungi of the genera *Neocalimastix*, *Piromyces*, and *Orpinomyces* (reviewed in Reference 3). The genome sequences of *C. thermocellum*, *C. acetobutylicum*, *R. flavefaciens*, and *C. cellulolyticum* are already known, and others will follow soon; these sequences will provide a complete view of the molecular components of the cellulosome of each organism.

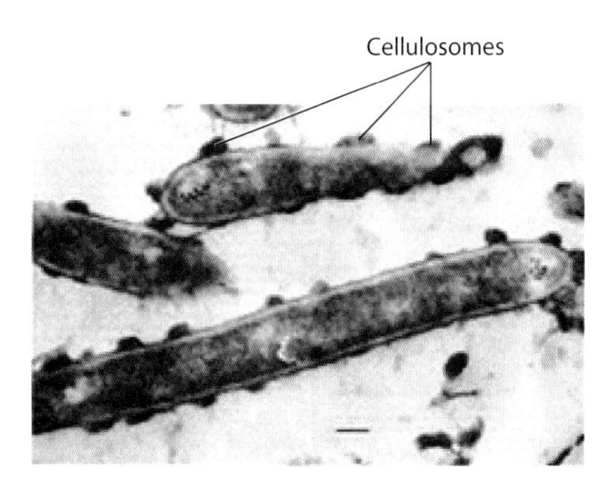

Figure 1

Cellulosomes at the surface of *Clostridium thermocellum*.

Biological Functions of Cellulosomes

It has been argued that cellulosomes are more efficient at deconstructing plant structural polysaccharides than the corresponding "free" enzyme systems produced by aerobic bacteria and fungi. For example, *C. thermocellum* exhibits one of the highest rates of cellulose utilization known, and the cellulosome of the bacterium is reported to display a specific activity against crystalline cellulose that is 50-fold higher than the corresponding *Tricoderma* system (17). It is possible that the energetic constraints imposed by the anaerobic environment have led to the evolution of highly efficient plant cell wall–degrading

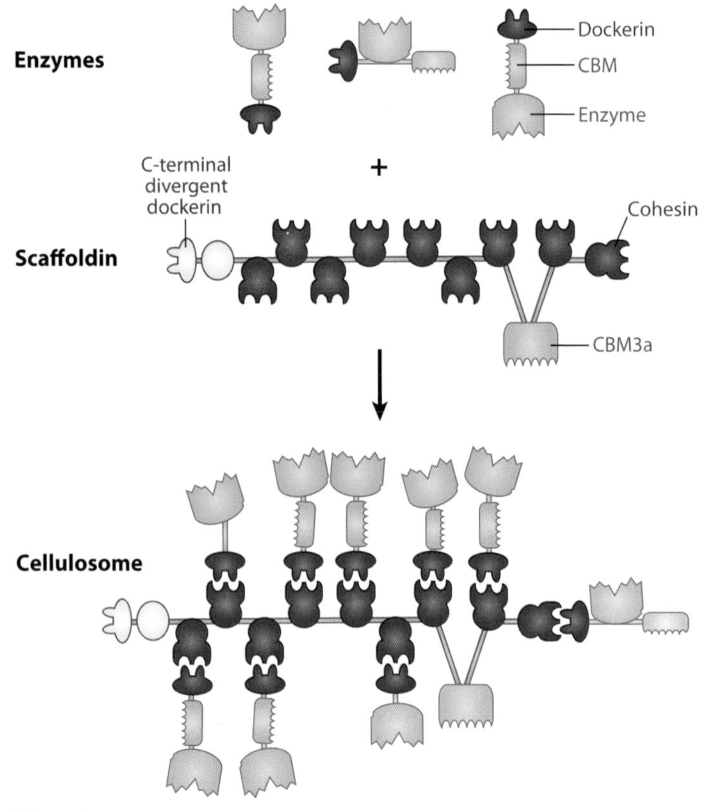

Figure 2

Mechanism of cellulosome assembly. Modular cellulases and hemicellulases produced by anaerobic microbes contain a dockerin appended to catalytic (enzyme) and noncatalytic carbohydrate-binding modules (CBMs). Dockerins bind the cohesins (*red*) of a noncatalytic scaffoldin, providing a mechanism for cellulosome assembly. In general, scaffoldins also contain a cellulose-specific family 3 CBM (CBM3a) and a C-terminal divergent dockerin that target the cellulosome to the plant cell wall and the bacterial cell envelope, respectively. The linkers joining the modules in the scaffoldin and catalytic subunits are shown as orange and blue lines, respectively.

cellulosomes. Indeed, it has been proposed that the grafting of plant cell wall–degrading enzymes onto a macromolecular complex leads to a spatial enzyme proximity that potentiates the synergistic interactions between the cellulosomal catalytic units, which are further augmented by enzyme-substrate targeting through the scaffoldin-borne CBM (3). It remains unclear, however, whether such hypotheses are, indeed, tenable. The rate of cellulose hydrolysis is dependent on numerous highly variable factors, such as the source of the polysaccharide, quantification of the degradative process, and the period of the assay. Thus, cellulose hydrolysis rates are likely to vary between laboratories. Indeed, others have reported very similar crystalline cellulase activities for *C. thermocellum* and *Trichoderma* (18). Furthermore, the importance of the assembly of the enzymes into a complex is questioned by recent studies showing that *Anaerocellum thermophilum*, a highly efficient cellulose-degrading anaerobic bacterium, synthesizes enzymes similar to *C. thermocellum* that do not physically associate into cellulosomes (19, 20). It is evident, however, that the bacterium does synthesize an unusually large number of enzymes with multiple catalytic modules, similar to its close relative *Caldicellulosiruptor saccharolyticus* (21), and thus, synergistic interactions between the catalytic components may be enhanced through their physical linkage. Compelling evidence supporting the importance of recruiting cellulases into the cellulosome to mediate efficient hydrolysis of the crystalline substrate was provided by Schwarz and colleagues (22). They showed that transposon insertions in the *CipA* gene, although not an influence on the activity of the bacterium against soluble β-glucan, reduced the capacity of the organism to hydrolyze crystalline cellulose 15-fold (22).

An alternative possible function of the cellulosome, at least in *C. thermocellum*, is the presentation of the enzyme system on the surface of the bacterium, which may enhance the capacity of the host bacterium to preferentially utilize the mono- and oligosaccharides released from the cell wall. Indeed, Lynd and colleagues (23) showed that when the cellulosome is tethered to the outer membrane of *C. thermocellum* the rate of cellulose hydrolysis was about twofold higher than when the complex was released into the culture medium. It was suggested that this increase in rate may reflect enhanced released of glucose and cellobiose, which are known to inhibit cellulose action. It is now apparent, however, that the surface of *C. thermocellum* contains several noncellulosomal receptors for plant cell wall–degrading enzymes (3). Thus, it is possible that the enhanced hydrolysis of cellulose reflects the interplay between the multienzyme complex and other glycoside hydrolases on the surface of the bacterium. It should also be noted that the cellulosome in *C. thermocellum* is released from the bacterium in the latter part of the growth phase, which may reflect the requirement for a more mobile enzyme complex to locate more recalcitrant and scarce forms of cellulose. Indeed, the importance of cell association across the cellulosome landscape remains unclear as in some organisms, such as *C. cellulolyticum*, there is little evidence for membrane attachment of the cellulosome.

Cellulosomes and Bioenergy

Society today faces the challenging problem of finding alternative and renewable energy sources to the conventional and still widely used fossil fuels. The current energy crisis requires the development of a combination of processes that are based on renewable substrates. Annually, $\sim 10^{11}$ tons of plant biomass, comprising mainly plant cell walls, are hydrolyzed by microbes, and the energy released by this process corresponds to 640 billion barrels of crude oil (24, 25). Hence, conversion of lignocellulosic biomass to fermentable sugars may represent a viable alternative for the production of renewable fuels such as ethanol. Nonetheless, hydrolysis of structural polysaccharides is still the rate-limiting step in the conversion of lignocellulose into fuels, which will require the development of more efficient enzyme systems. Cellulosomes can be described as one of nature's most elegant and elaborate nanomachines. Since they were first discovered, the intricacy of these multicomponent systems has highlighted the requirement for extensive enzyme consortia to overcome the recalcitrant, and chemically complex, nature of plant cell wall hydrolysis, which is a key biological and biotechnological macromolecule.

CELLULOSOME ASSEMBLY IN DIFFERENT BACTERIA

The mechanism of cellulosome assembly is an area of considerable biological interest, and such knowledge can also be exploited in the construction of industrially significant nanomachines. To date, studies on the mechanism by which the cellulosome assembles have focused on *Clostridia*, primarily *C. thermocellum*; however, it is evident that structural information on the components that mediate the assembly of non-*Clostridial* cellulosomes is rapidly emerging.

The Cohesin-Dockerin Interaction

Cellulosome assembly is mediated by a high-affinity protein:protein interaction ($>10^9$ M^{-1}) between dockerin modules in the cellulosomal catalytic components and cohesin modules on the scaffoldins (26, 27). Dockerins, which were shown to play a noncatalytic function in a variety of cellulosomal enzymes (28, 29), consist of $\sim$70 amino acids that contain two duplicated segments, each of about 22 residues (15, 30). Dockerins are usually present in a single copy at the C terminus of cellulosomal enzymes. The first 12 residues of each duplicated segment resemble the calcium-binding loop of EF-hand motifs in which the calcium-binding residues, aspartate or asparagine, are highly conserved. More recently, structural data suggested that the homology to the EF-hand motif is restricted to the calcium-binding loop and the F helix, thereby suggesting an F-hand motif (31). Calcium was shown to be pivotal for dockerin stability and function; in the presence of EDTA, which chelates calcium, dockerins are unable to interact with cohesins (32, 33).

Cohesins are 150-residue modules that are usually present as tandem repeats in scaffoldins. Seminal work by Béguin and colleagues (14, 15) demonstrated that the internal repeats of scaffoldins (the cohesins) specifically bound to the noncatalytic dockerin modules identified in cellulosomal cellulases and hemicellulases. Thus, scaffoldins, which can be defined as cohesin-containing proteins, fulfill a major role in cellulosome assembly and have now been identified in most cellulosome-producing bacteria. The number of cohesin modules in scaffoldins may vary from 1 to 11 but is usually higher than 4 (see

Reference 3 for a review). Significantly, in an individual scaffoldin, cohesin domains are unable to discriminate between the dockerins present in the various cellulosomal enzymes (34, 35). Both cohesins and dockerins are highly homologous within the same species, and the residues directly involved in protein:protein recognition are highly conserved within a species (see below). For example, the CelS (Cel48A) dockerin, the major cellulosomal catalytic component, can bind any of the cohesins of CipA, the *C. thermocellum* scaffoldin. Lack of specificity in cohesin-dockerin recognition suggests that cellulosome complexes may comprise a different ensemble of catalytic subunits that is influenced by the induction of specific genes by plant cell wall polymers. Indeed, proteomic analysis of individual *C. thermocellum*, *C. cellulolyticum*, and *C. cellulovorans* cellulosomes revealed that different cellulosomes contain a distinct repertoire of enzymes (36–38). Although differences in cellulosome composition are also apparent when organisms are grown in different carbon sources, the composition of individual cellulosome molecules remains heterogenous (34).

Noncellulosomal Cohesins and Dockerins

Until recently, the presence of cohesins and dockerins in a bacterial proteome was believed to be a definitive signature of a cellulosome-producing microbe. Surprisingly, screening microbial genomes for the presence of dockerin- and cohesin-encoding genes revealed that cohesins and dockerins are quite widespread in Archaea, Bacteria, and Eukarya (39). However, although a large number of proteins containing cohesin and dockerin was observed in a variety of organisms, most of them are components of enzymes that are not involved in carbohydrate metabolism. In addition, there is little evidence for the widespread presence of scaffoldins, proteins containing tandem cohesins, in most of the organisms; in the majority of cases, a few proteins contain a limited number of cohesin and dockerin domains. Therefore, cohesin-dockerin interaction is quite ubiquitous in nature, although there are relatively few examples of dockerin and cohesin modules directly involved in the assembly of cellulosomes. Clearly, cohesins and dockerins seem to serve a variety of, mostly unknown, functions that are distinct from the cellulosome paradigm. Recently, the cohesin-dockerin interactions were shown to be pivotal for toxin production in *Clostridium perfringens* (40), while other functions for the cohesin-dockerin complexes in cellular metabolism remain to be identified.

Cellulosome Cell-Surface Attachment

When the gene encoding *C. thermocellum* scaffoldin protein CipA was initially cloned, it was found to be part of an operon containing several other genes. Gene walking allowed the cloning and the subsequent sequencing of all the genes located in CipA operon (41, 42). Significantly, all the proteins encoded by the CipA-containing operon also had cohesin domains. More recently, it became apparent that several cellulosome-producing microbes express more than one type of scaffoldin. Thus, scaffoldins, such as CipA, that bind cellulases and hemicellulases, the cellulosomal catalytic components, were termed primary scaffoldins (3). In *C. thermocellum*, cellulosomes were located at the cell surface in the early stages of growth. The primary scaffoldin, CipA, contains, in addition to its nine type I cohesins (that recognize the type I dockerins in the catalytic subunits), a C-terminal type II dockerin that does not recognize CipA cohesins. Instead, the CipA dockerin recognizes type II cohesins located in cell-surface proteins (43); in *C. thermocellum*, most of these proteins are encoded by the CipA-containing operon (41, 42). Thus, proteins that bind primary scaffoldins, although they do not interact with the cellulosome catalytic components, are termed anchoring scaffoldins. Significantly, type I and type II cohesin/dockerin partners do not interact, ensuring a clear distinction between the mechanism for cellulosome assembly and cell-surface attachment (**Figure 3**) (43).

Bacterial cellulosomes may be classified in two major types: those that present multiple types of scaffoldins such as *C. thermocellum* and those that contain a single scaffoldin, which are characteristic of most mesophilic *Clostridia* (*C. cellulolyticum*, *C. cellulovorans*, *C. josui*, and *C. acetobutylicum*). Cellulosomes that assemble via a single primary scaffoldin are the simplest and may contain six to nine catalytic components, reflecting the number of cohesins in the primary scaffoldin. Apart from *C. thermocellum*, there are no obvious molecular mechanisms by which the scaffoldin of clostridial cellulosomes is anchored to the bacterial cell wall, which may explain why these enzyme complexes are found in the culture media, particularly in the late exponential and stationary growth phases. Bacteria producing these simple cellulosomes only contain type I cohesins and lack the type II dockerins that contribute to cellulosome assembly on the bacterial surface in *C. thermocellum*. In *C. cellulovorans*, however, a cellulosomal cellulase was found to contain three repeated domains with homology to the anchor domains found in bacterial surface (S)-layer proteins (SLH, from S-layer homology), although it remains to be established whether this enzyme contributes to the anchoring of the complete cellulosome onto the cell surface of the prokaryote (44, 45).

Bacteria expressing cell-surface cellulosomes, such as *C. thermocellum*, *R. flavefaciens*, *A. cellulolyticus*, and *B. cellulosolvens*, contain a single primary scaffoldin and multiple anchoring scaffoldins. The majority of the anchoring scaffoldins contain a threefold reiterated segment, the SLH modules or SLH domains, which are usually present in the majority of S-layer proteins (46, 47). SLH domains mediate the attachment of the structural proteins to the bacterial cell wall and may bind the peptidoglycan layer or secondary cell wall polysaccharides. In *C. thermocellum*, there are four anchoring scaffoldins containing one, two, or seven (two of them contain seven) type II cohesin domains (**Figure 3**). One of the scaffoldins containing seven type II cohesins (Cthe_0736), which was shown to be quite prominent in the

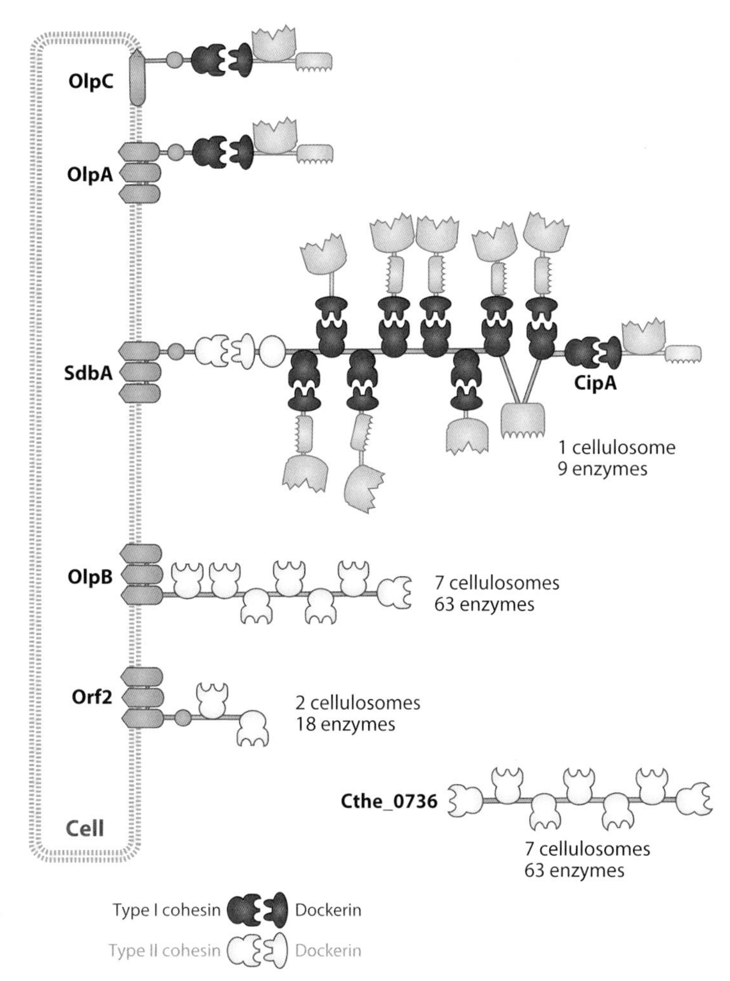

Figure 3

Organization of *Clostridium thermocellum* cellulases and hemicellulases in cellulosomes. The *C. thermocellum* scaffoldin (CipA) contains nine type I cohesins and thus organizes a multiprotein complex with nine enzymes. The C-terminal type II dockerin domain of CipA binds specifically type II cohesin domains found in cell-surface proteins (SdbA, OlpB, and Orf2) or in the extracellular Cthe_0736. Because the anchoring scaffoldins OlpB, Orf2, and Cthe_0736 contain more than one type II cohesin domain, they appear to contribute to the assembly of polycellulosomes that may contain up to 63 catalytic subunits. Nevertheless, cellulosomal enzymes may adhere directly to the bacterium cell surface by binding the single type I cohesin domains found in OlpA and OlpC. The linkers joining the modules in the scaffoldin and catalytic subunits are shown as orange and blue lines, respectively.

C. thermocellum proteome (38), does not appear to be cell associated and may allow the formation of an extracellular cell-free complex. The presence of tandem-repeated type II cohesins in anchoring scaffoldins allows for the formation of polycellulosomes. In the case of *C. thermocellum* if seven cellulosomes assembled with the seven cohesin-anchoring scaffoldin, a polycellulosome of 63 catalytic units may be produced (**Figure 3**). Such a molecular organization may explain why the *C. thermocellum* cellulosomes range in size from 1.5 to 6 MDa (48), where a single CipA molecule is attached to a monovalent anchoring scaffoldin, up to 100 MDa in size, which signifies thc binding of seven primary scaffoldins to a heptavalent secondary scaffoldin.

The Diversity of Bacterial Cellulosomes

In recent years, the complexity and diversity of bacterial cellulosomes have become apparent through the cloning and sequencing of the multiple scaffoldins from *A. cellulolyticus* and *B. cellulosolvens* and the genome sequencing of *R. flavefaciens*. *B. cellulosolvens* and *C. thermocellum* cellulosomes are remarkably similar, although in *B. cellulosolvens* the primary scaffoldin contains type II cohesins and a type I dockerin, and the anchoring scaffoldin contains type I cohesins (49). The biological significance, if any, of this unique and reverse disposition of type I and type II cohesins remains opaque. Indeed, the *B. cellulosolvens* polycellulosomes may contain more than 100 enzymes because the bacterium anchoring protein (ScaB) contains 10 cohesins and the primary scaffoldin (ScaA) contains 11 cohesins (49, 50). In *A. cellulolyticus*, three different cohesin-dockerin specificities have been identified, which are assembled into two distinct cellulosome systems. In the first system, the cell-surface scaffoldin, ScaD, contains two type II cohesins, which interact directly with the primary scaffoldin (ScaA), and a third type I cohesin, which specifically recognizes the type I dockerin of a cellulosomal catalytic subunit. Thus, ScaD is a bifunctional anchoring scaffoldin as it contains cohesins that display different specificities (51). The primary scaffoldin, which contains a GH9 cellulase, is bound to seven additional enzymes through its type I cohesins. The ScaD cellulosome therefore contains 17 enzymes. The other cell-surface scaffoldin, ScaC, binds to ScaA indirectly through an "adaptor" scaffoldin (ScaB) that contains four type II cohesins and an atypical dockerin that recognizes the three cohesins on SacC, which, similarly, display structural deviation from the canonical type I and type II cohesins present in SacA and ScaD, respectively. ScaB contains four type II cohesins and is therefore able to recruit a total of 12 ScaA primary scaffoldin molecules (three ScaBs bind to each ScaD surface protein), each containing eight enzymes, onto ScaD. Thus, the ScaD-centered cellulosome contains a total of 96 enzymes (52).

The most intricate, and potentially versatile, cellulosomal complex described to date is that of *R. flavefaciens* (**Figure 4**). Initial characterization of *R. flavefaciens* cohesins suggested clear sequence and structural differences to

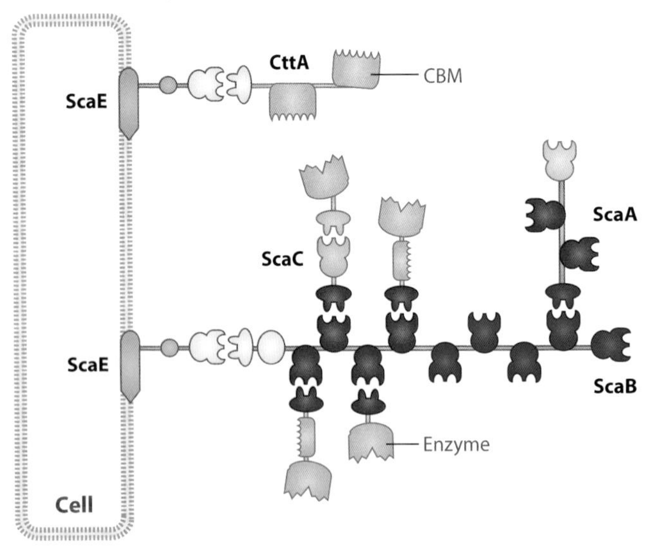

Figure 4

The complexity of *Ruminococcus flavefaciens* strain FD-1 cellulosome. The single cell-surface scaffoldin, ScaE, may bind CttA, which carries the noncatalytic carbohydrate-binding modules (CBMs) that mediate the primary anchorage to the plant cell wall or to ScaB. ScaB contains cohesins with two different specificities; one (*dark blue*) exclusively interacts with the adaptor scaffoldin ScaA. ScaA contains two cohesins that present a similar specificity to the second set of cohesins of ScaB (*red*). These cohesins bind cellulosomal enzymes or ScaC. As ScaA, ScaC is an adaptor scaffoldin that recognizes a different set of dockerin-containing proteins. Other adaptor scaffoldins, presenting a similar structure to ScaC but displaying a yet unknown specificity, exist in *R. flavefaciens*.

the previously described type I and type II cohesin-dockerin complexes, and thus cohesin-dockerin pairs of *R. flavefaciens* were termed type III modules (53). The *R. flavefaciens* FD-1 genome encodes over 200 dockerin-containing proteins, most of them of unknown function. In addition, a diversity of potential substrate specificities is displayed by the enzymes bearing homology with glycoside hydrolases, carbohydrate esterases, and polysaccharide lyases of known function. Elegant studies by White, Flint, Lamed, Bayer, and their colleagues (54–56) unraveled the mechanism of cellulosome assembly in *R. flavefaciens*. In strain FD-1, the anchoring scaffoldin ScaE contains a single cohesin domain. This protein is covalently attached to the cell surface of the bacterium through a sortase-mediated transpeptidation reaction (56), whereas in *Clostridia*, cellulosome anchoring is via the noncovalent binding of SLH domains to the S-layer of the host cell (41). The ScaE cohesin binds to the C-terminal dockerin of the primary scaffoldin, ScaB, which contains nine cohesins. Surprisingly, in strain FD-1, ScaB cohesins present two different specificities; four recognize the dockerins of the catalytic subunits, and five bind to ScaA, which functions as an adaptor and primary scaffoldin. ScaA contains two cohesins that bind the cellulosomal catalytic subunits, and thus, ScaA (similar to ScaB in *A. cellulolyticus*) amplifies the number of enzymes in the *R. flavefaciens* cellulosome (54–56). A variety of other adaptor scaffoldins, such as ScaC, are also produced in *R. flavefaciens* strain FD-1, and these scaffoldins contain a single dockerin connected to a single cohesin. The ScaC dockerin displays specificity similar to those present on the catalytic subunits and thus binds to the primary scaffoldin ScaB or ScaA. ScaC, however, contains a divergent cohesin that binds a different and, as yet, unknown group of dockerins (54, 55). The functional significance for this large array of adaptor scaffoldins, which lead to highly complex cellulosome structures, is currently unclear. However, it is apparent that changes to the expression of ScaC-like adaptors will have a significant influence on cellulosome composition, which may enable the bacterium to tune the structure of its cellulosomes to reflect the nature of the plant cell wall presented to the organism. Significantly, cellulosome structural organization varies between strains of *R. flavefaciens*. Thus, in strain 17, the ScaB ortholog contains cohesins that bind exclusively to ScaA, which in turn recognizes the dockerins presented on catalytic subunits and additional adaptor scaffoldins such as ScaC (54, 55). *R. flavefaciens* cellulosome strain heterogeneity may reflect the complexity and diversity of the lignocellulosic substrate found in the rumen.

How Cellulosomes Bind to the Plant Cell Wall

CBMs play a key role in the deconstruction of complex insoluble composites exemplified by the plant cell wall. CBMs have been grouped into 59 sequence-based families (57). Biochemical and structural studies, reviewed in Reference 58, revealed that these modules display three distinct specificities, and thus, CBMs have also been categorized into three types. The type A CBMs interact with crystalline polysaccharides, primarily cellulose, type B modules bind to internal regions of single glycan chains, and type C CBMs recognize small saccharides or, in the context of complex polysaccharides, the termini of these polymers.

Initial studies by Bayer & Lamed and their colleagues (8–10) showed that bacterial cellulosomes bound tightly to cellulose, and thus, these enzyme complexes were referred to as the *C. thermocellum* "cellulose-binding factor." It was later recognized that the attachment of the cellulosome to the plant cell wall is primarily mediated by a family 3 CBM (CBM3) located in scaffoldins (16). CBM3s are generally classified as type A modules (except CBM3c, as discussed below) (59, 60) and bind tightly to the surface of crystalline cellulose. Indeed, CBM3s bind more extensively to cellulosic structures in plant cell walls than other type A CBMs (61). The crystal structure of CBM3 was first elucidated from the *C. thermocellum* scaffoldin, CipA (62).

The protein module displays a nine-stranded β-sandwich fold with one of the β-sheets presenting a planar topology reflecting the topology of crystalline cellulose. The side chains of three aromatic residues form a planar strip, which is predicted to stack against the glucose rings of the cellulose chains. In addition, several polar amino acids on the binding surface are likely to make polar contacts with the hydroxyl groups and, possibly, with the endocyclic oxygen of the glucose residues. The topology of the binding interface of CBM3 modules precludes their interaction with single β1,4 glucan chains, which adopt a more helical conformation (63).

It is recognized that plant cell walls are highly heterogeneous macromolecules comprising a large variety of interacting polysaccharides. Thus, upon binding to crystalline cellulose, cellulosomes require additional supramolecular targeting to enable the catalytic components of the enzyme complex to be brought into proximity with their specific substrate. Fine-tuning polysaccharide recognition is accomplished by a variety of type B and type C CBMs located in the cellulosomal enzymes (see below). For example, in *C. thermocellum*, several cellulosomal enzymes were shown to contain cellulose (single chain), xyloglucan, xylan, and pectin-specific CBMs that target their appended catalytic modules to their target substrates. In contrast to *Clostridial* scaffoldins, *R. flavefaciens* scaffoldins do not generally contain CBMs, and this bacterium seems to have developed a different mechanism for crystalline cellulose recognition. A protein termed CttA, which contains two putative CBMs, also has a dockerin that specifically recognizes the cohesin of ScaE, which is attached to the surface of the bacterium (**Figure 4**) (64). The CBMs may recruit the plant cell wall to the cellulosomes presented on the cell envelope of the bacterium. Indeed, in the related bacterium, *R. albus*, CBM37s, which are present in several cellulosomal proteins and display broad ligand specificity (65), appear to mediate attachment of the cellulosome to the bacterial surface (66).

The Specificity of the Type I Cohesin-Dockerin Interaction

Supramolecular cellulosome architecture is orchestrated by the specificity of the various cohesin-dockerin interactions involved in complex assembly. As described above, type I dockerins do not interact with type II cohesins, and vice versa, allowing for the correct assembly and cell-surface attachment of bacterial cellulosomes (26, 27). In addition, type III modules do not interact with both type I and type II domains (47). However, ligand specificities in type I cohesin-dockerin interactions were shown to vary between different species (67). This is in clear contrast with the type II interactions, which demonstrated relatively extensive cross-species plasticity (68). For example, a type II dockerin of *A. cellulolyticus* binds both *A. cellulolyticus* and *C. thermocellum* type II cohesins (68). In addition, the type II cohesin of the *C. thermocellum* anchoring scaffoldin SdbA binds not only to the *C. thermocellum* CipA type II dockerin but also to both *B. cellulosolvens* and *A. cellulolyticus* type II cohesins (68). The biological relevance of the promiscuous type II cohesin-dockerin interaction remains unknown.

It has been extensively established that *C. thermocellum* type I dockerins do not bind *C. cellulolyticum* (and also *C. josui*) cohesins, and *C. cellulolyticum* dockerins do not cross-react with the thermophilic cohesins (31, 67). Hence, *C. thermocellum* enzymes are not assembled into *C. cellulolyticum* cellulosomes and vice versa, which suggests that cellulosomal enzyme sharing is not an evolutionary driver. Analysis of the primary sequences of type I dockerins displaying distinct specificities informed various site-directed mutagenesis studies designed to identify the amino acids that interacted with the type I cohesin partner. Thus, residues at positions 11 and 12 of *C. cellulolyticum* were changed to the equivalent amino acids in *C. thermocellum* dockerins and vice versa, and the resultant variants recognized cohesins from both clostridia. A refinement of this residue-swapping strategy identified a single amino acid substitution,

T12L, in a type I *C. thermocellum* dockerin that conferred high affinity for a *C. cellulolyticum* cohesin (67). It was apparent, however, that, although the mutated dockerins were able to recognize the noncognate cohesins, they still bound the cohesins of their own species. Later studies and structural data (see below) suggested that residues not only at positions 11 and 12 but also amino acids at positions 18, 19, and 23 of the dockerin are involved in species-specific ligand recognition. With respect to cohesin engineering, a gene-swapping experiment allowed the identification of a three-residue substitution in the *C. cellulolyticum* module that conferred recognition of *C. thermocellum* dockerins with an affinity that was only fivefold less than *C. thermocellum* cohesins (67). The biological significance of this lack of cross-species specificity in cohesin-dockerin complexes is not entirely clear. It is likely that the grafting of enzymes with different origins in the same complex may adversely affect the catalytic efficiency of the whole cellulosome. In contrast, it could be argued that microbes inhabiting the same ecological niche might benefit from the sharing of cellulosomal enzymes. This is supported by the observation that *C. cellulolyticum* and *C. josui* dockerins display considerable sequence homology, and there is some evidence for cross specificity (69), while the organisms share the same ecological niche.

Fungal Cellulosomes

Anaerobic fungi effectively hydrolyze a large range of polysaccharides, including cellulose and hemicellulose, and are usually present in the gastrointestinal tract, especially in the rumen and cecum of herbivorous animals (70). It is believed that, in anaerobic ecosystems, anaerobic fungi are the initial colonizers of lignocellulose and play a key role in fiber digestion together with bacteria and other microorganisms. Similar to many anaerobic bacteria, anaerobic fungal plant cell wall–degrading enzymes are organized into cellulosomes that have estimated sizes ranging from 3 to 80 MDa (71, 72). However, in contrast to the bacterial cellulosomes, fungal cellulosomes are much less-well characterized. A number of putative dockerin sequences have been identified in plant cell wall–degrading enzymes from fungi of the *Neocallimastix*, *Orpinomyces*, and *Piromyces* genera (73–75). These noncatalytic modules were identified as dockerins through their capacity to bind an ~100-kDa protein in Western blot experiments (76). Significantly, the amino acid sequences of fungal dockerins are completely unrelated to their bacterial counterparts. This is rather surprising if we consider that the majority of fungal cellulases and hemicellulases were acquired by horizontal gene transfer. Indeed, the evolutionary selection for the grafting of a convergent set of enzymes on divergent dockerins is currently unclear. It is possible that the dockerins and their receptor(s) mediated protein-protein interactions in the fungi prior to acquisition of the plant cell wall–degrading apparatus, and these modules were grafted onto the acquired enzymes to facilitate the formation of complexes. The much-awaited genome sequence of these organisms will provide insight into the extent to which these dockerin sequences are present in proteins that are not associated with plant cell wall degradation. Another feature of fungal dockerins that distinguishes them from the corresponding bacterial modules is that enzymes generally contain two copies of these modules, each consisting of ~40 highly conserved residues joined together by short linker sequences (70). Exceptions to the two dockerin modules are evident, and there is a relationship between the affinity of a cellulosomal protein for the apparent scaffoldin and the number of dockerins present in the polypeptide.

The structure of dockerin domains from *Piromyces equi* endoglucanase Cel45A were solved by NMR, which shed light onto the mechanism of fungal cellulosome assembly (77, 78). Fungal dockerins consist of a three-stranded β-sheet and a short helix, held together by two disulfides. These modules were therefore identified as a small disulfide-rich protein domain (**Figure 5**) (78). Significantly, these putative protein docking sequences were classified in the same structural family as a

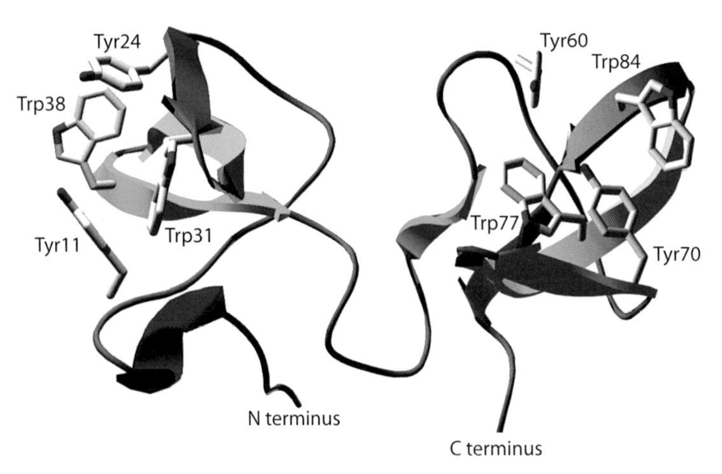

Figure 5

The structure of the double dockerin from cellulase Cel45A of *Piromyces equi*. The structure is shown with color graduating from blue to red. Residues at the two dockerins that make contacts with the ligand are shown in stick format. The figure demonstrates that the two ligand-binding interfaces are highly similar and are positioned on adjacent surfaces.

family 10 CBM from the bacterium *Cellvibrio japonicus*, which is involved in protein:carbohydrate recognition (79). Reduction of the disulfides results in protein unfolding. The ligand-binding site contains the side chains of four aromatic residues, Tyr11, Tyr24, Trp31, and Trp38, which are highly conserved in all fungal dockerins and, through site-directed mutagenesis, were shown to play a key role in ligand recognition (78). The solution structure of a double-dockerin construct from Cel45A allowed visualization of a flexible linker between the two domains that is short enough to keep the two binding sites on adjacent surfaces (**Figure 5**). Surprisingly, the double Cel45A dockerin was shown to bind to the native form of a GH3 cellulosomal β-glucosidase, which was thereby identified as a potential scaffoldin (77). The observed carbohydrate recognition is consistent with the Rosetta-like topology of the dockerin ligand-binding site. Thus, cellulosome assembly in fungi may not involve cohesin modules. Rather, it has been proposed that the dockerin domains bind the carbohydrate decorations on GH3, although it is difficult to decipher how this type of apparently nonspecific molecular interaction will lead to the precise ordering of cellulosomal subunits seen in bacterial cellulosomes. Clearly, more work is required to address several unresolved issues concerning fungal cellulosomes, and the genome sequence of cellulolytic anaerobic fungi may help in the development of a unifying hypothesis for cellulosome assembly in these microbial eukaryotes.

CELLULOSOME CATALYTIC COMPONENTS: DIVERSITY AND ABUNDANCE

Cellulosomal plant cell wall hydrolases are modular enzymes containing, in addition to dockerin domains, one or, in some instances, two catalytic modules and one or more CBMs. Gene cloning of cellulosome-containing microbes has led to the identification of numerous proteins that are components of these enzyme complexes. However, the complete repertoire of cellulosomal components has only been revealed by genome sequencing, which was recently achieved for *C. thermocellum*, *C. cellulolyticum*, and *R. flavefaciens*. The number of cellulosome-associated components produced by a given bacterium is surprisingly high, reflecting diversity in the complement of enzymes. For example, *C. thermocellum* produces 72 cellulosomal enzymes (polypeptides with type I dockerins), whereas in *R. flavefaciens*, cellulosomes may comprise more than 200 dockerin-containing subunits. This extraordinary diversity of enzymes, which are also inherently complex because of their multimodular organization, may simply reflect the chemical and structural intricacy of the cellulosome substrate, the plant cell wall. Thus, the array of polysaccharides in the plant cell wall is matched by the complexity and diversity of the cellulosomal catalytic machinery, which, for example, includes cellulases, β-glucanases, xylanases, mannanases, galactanases, xyloglucanases, arabinofuranosidases, arabinanases, xylan and pectic esterases, pectate lyases, rhamnogalacturan lyases, and proteinases (3, 17, 70).

Cellulose, the most abundant plant cell wall polymer, is chemically relatively simple, but its generally crystalline structure makes it highly resistant to biological degradation. Amorphous portions of the cellulose fibrils are relatively susceptible to hydrolysis, especially via endoglucanases. However, the predominant crystalline regions of cellulose impose significant steric constraints to enzyme attack (2, 3, 70). Thus, the synergistic interaction of a defined repertoire of complementary enzymes [extensively reviewed previously (2, 3, 70, 80–82)], usually displaying exo and endo activities, is required for the efficient hydrolysis of crystalline cellulose. It is generally assumed that endo-acting cellulases produce new chain ends in the internal portions of the polysaccharides, which act as substrate for the exo-acting biocatalysts, which because of their processive mode of action are able to move from amorphous regions into the crystalline structures of cellulose. This view, however, is likely to be a simplification of the interactions between these enzyme species as GH6 cellobiohydrolases have been shown to display endo activity (83, 84), a view supported by structural data revealing an opening of the loops that form the tunnel over the active site (85). Furthermore, the synergy observed between exocellulases, acting at the reducing (GH7 and GH48) and nonreducing (GH6) ends of cellulose chains, respectively (86), may be explained by the partial endo activity displayed by GH6 cellobiohydrolases (84). Microorganisms generally synthesize a range of different endoglucanases from distinct CAZy families. Although the biological significance for the plethora of cellulases is not entirely clear, it is becoming increasingly apparent that specific GH9 cellulases that display an apparent endo-processive mode of action, so named as these enzymes are able to cleave internal regions of the cellulose chain, but are then able to act in a processive manner, are particularly important components of cellulase systems (87). Cellulosomes appear to be efficient at degrading crystalline cellulose, which likely reflects the composition of the enzyme consortium and the efficient targeting of substrate, which, in *C. thermocellum* and other clostridia, is mediated through the scaffoldin CBM3 and enzyme-borne cellulose-specific CBMs. The products released by the cellulosomal cellulases are hydrolyzed to glucose by β-glucosidases, which usually are cell associated. *C. thermocellum*, for example, expresses at least four exoglucanases, which act in synergy with more than 10 different endoglucanases.

Various proteomic studies indicate that Cel48A, formerly known as CelS, and Cel8A, previously designated CelA, are the major exo- and endocellulases, respectively, found in *C. thermocellum* cellulosomes (37). The structure of Cel48A revealed an $(\alpha/\alpha)_6$ barrel fold with a tunnel-shaped substrate-binding region (88–90). The enzyme attacks the reducing end of cellulose chains. The polymer threads through the tunnel, and the glycosidic bond between the second and third glucose residue is cleaved, releasing cellobiose. The catalytic machinery is located inside the tunnel, preventing the enzyme from attacking cellulose chains internally. Cel48A is consequently a processive cellulase. In contrast, the structure of Cel8A, which displays a similar fold, contains a groove-shaped substrate-binding region with an open cleft, explaining why the enzyme can cleave the internal regions of the polysaccharide (91). In addition to Cel8A, the *C. thermocellum* cellulosome contains nine GH5 endoglucanases and twelve GH9 enzymes that may function in an endo or processive mode of action. The majority of these GH9 enzymes are fused to a CBM3c, which does not bind directly to crystalline cellulose. Instead, CBM3c acts in concert with the adjacent catalytic domain by transiently binding the incoming cellulose chain that is progressively fed into the GH9 active site prior to hydrolysis. Deletion of CBM3c modules from GH9 enzymes renders these enzymes less active and is associated with a switch to an endo mode of attack (59, 60). Thus, through the modulator effect of CBM3c, GH9 enzymes function as processive cellulases acting in an exo-mode fashion. However, products resulting from GH9-CBM3c hydrolysis may differ. For example, CelR, a major cellulosomal GH9 endoglucanase, produces cellotetraose as its primary hydrolysis product (92). It remains unclear, however, why *C. thermocellum* requires such a large array of GH5 and GH9 enzymes to mediate plant cell wall hydrolysis. The elegant quantitative proteomic studies, developed by Mielenz and colleagues (38), suggest that four GH5 and six GH9 enzymes were among the top 20 catalytic components of cellulosomes isolated from *C. thermocellum*, irrespective of the carbon source. GH5 abundance increases when the bacterium is grown in the presence of cellulose, and GH9 enzymes are particularly abundant when the carbon source is crystalline cellulose. Thus, it was suggested that GH9 enzymes play a major role in the hydrolysis of crystalline cellulose. It is unclear whether the GH9 and GH5 present subtle differences in substrate specificity, possibly reflecting the multiple conformations adopted by cellulose chains in amorphous and paracrystalline regions.

The functional significance of individual cellulases has been explored through the in vitro construction of minicellulosomes. The data showed that combining the GH48 cellobiohydrolase with an endo-processive GH9 resulted in the highest degree of synergy, although the cellulase activity against crystalline substrates was considerably lower than the native cellulosome (87, 93, 94). These series of elegant experiments also showed that integrating these enzymes onto the cellulosome scaffold enhanced activity both by bringing these enzymes into close proximity and through the targeting of the scaffoldin CBM (87). It should be emphasized, however, that these experiments only analyzed a fraction of the "cellulosome space" (nine enzyme receptors on the scaffoldin and 72 potential components of the cellulosomes equates to 72^9 combinations), and thus, it remains possible that cellulosome assembly may mediate synergistic interactions between protein combinations that were not explored by the in vitro experiments. Further evidence for the importance of GH9 and GH48 cellulases in the hydrolysis of crystalline cellulose is derived from genetic and proteomic studies on *R. albus*. Mutants of the bacterium, which displays a significant reduction in activity against crystalline cellulose, do not express two major cellulases, the endo-processive GH9 and a GH48 exocellulase (95). Neither enzyme is a component of the cellulosome, but they both contain a CBM37 that, as stated above, contributes to the attachment of the bacterium to crystalline cellulose (60, 96). Thus, the loss in crystalline cellulase activity likely reflects both the carbohydrate binding and catalytic function of these two enzymes.

It is interesting that, although unable to grow on xylan or their monosaccharide constituents (primarily xylose and arabinose), *C. thermocellum* expresses a large number of hemicellulases, and many of these target xylan. Indeed, when cultured on cellobiose, 22% of the cellulosomal enzymes are xylanases, and this percentage is reduced to 12% when the substrate is Avicel (38). It appears, therefore, that the bacterium removes the hemicellulosic polysaccharides to expose the cellulose microfibrils, which it uses as its primary carbon and energy source. This is consistent with the observation that the incorporation of a xylanase into a minicellulosome, containing a pair of cellulases, increased the release of soluble sugars fourfold (94). It was suggested that the xylanase increased the activity of the cellulase pair on this composite substrate. Such a conclusion, however, may be a little premature as the identity of the soluble sugars was not determined, and thus, it is possible that a component of the saccharides released is derived from xylans.

The majority of *C. thermocellum* cellulosomal enzymes display rather complex multimodular architectures containing a variety of CBMs. These CBMs fine-tune the interaction of the multifunctional complex with the diversity of polysaccharides in the plant cell wall by directing the appended catalytic domains to their target substrates. For example, CBM3c feeds the active site of GH9 catalytic domains with isolated cellulose chains, whereas xylanases contain xylan-binding CBMs that are backbone or side chain specific, depending on the activity of the associated catalytic domain (97), and CBM11, CBM30, and CBM44, present in endoglucanases, bind amorphous cellulose or β-1,3–1,4-glucan chains (98, 99). Indeed, recent studies have shown that a CBM35 targets a cellulosomal pectin acetylesterase to regions of pectin that have been cleaved by a pectate lyase (100), and the CE2 esterase, appended to a GH5 cellulase, displays a catalytic function, hydrolyzing acetyl groups from xylan and gluomannan, and a cellulose-targeting CBM function, which enhances the activity of the cellulase (101, 102). Interestingly, cellulosomal GH9 endo-processive cellulases contain CBM3c modules but lack other cellulose-specific CBMs. In contrast, the equivalent non-cellulosomal cellulases contain, in addition to the CBM3c, a CBM3a module that binds to crystalline cellulose (103), suggesting that all the cellulosomal cellulases utilize the scaffoldin CBM3a to target crystalline cellulose. Thus, numerous type B and type C CBMs play a subtle role in cellulosome function by directing the various catalytic subunits of the enzyme complex, which is bound to crystalline cellulose, to their target substrates.

STRUCTURAL BASIS FOR CELLULOSOME ASSEMBLY AND CELL-SURFACE ATTACHMENT

Since the early 1990s, when the importance of cohesin-dockerin interactions was established in cellulosome assembly and attachment to the bacterial surface, the mechanism by which these heterodimers form has been an area of intensive study. It was soon recognized, using site-directed mutagenesis, that two adjacent amino acids in the duplicated dockerin sequence played a key role in cohesin specificity. The mutagenesis studies also showed that the duplicated sequence in type I dockerins introduced an element of functional redundancy. The crystal structure of cohesin-dockerin partners, described below, has now provided a more complete mechanistic understanding of how these proteins associate and display species specificity.

The Structure of the Type I Cohesin-Dockerin Complex

The structure of cohesin and dockerin modules, individually or in heterodimeric complexes, has revealed the molecular mechanisms by which cellulosome assembly occurs (**Figure 6**). The presence of a duplicated 22-residue segment in the dockerins suggested that the structure of these modules displays a twofold symmetry. This prediction is consistent with the crystal structure of the type I dockerin from *C. thermocellum* Xyn10B in complex with the second cohesin of the scaffoldin protein CipA (103). The dockerin folds into three α-helices, with helices 1 and 3 comprising the first and second duplicated segment, respectively (**Figure 6**). Each duplicated segment displays remarkable structural conservation and also contributes an F-hand calcium-binding motif. In the absence of its protein partner, dockerins display a great deal a structural flexibility (104) and are largely disordered in the absence of calcium (33). The 147-residue type I cohesin displays a flattened β-barrel fold, which is defined by two β-sheets, one of which presents a planar dockerin-binding surface. The crystal structure of the cohesin-dockerin complex (103) revealed that, although the dockerin presents a remarkable internal symmetry, cohesin recognition is mediated mainly by helix 3; only the C-terminal region of helix 1 contributes to ligand binding (**Figure 6**). Hydrophobic forces dominate the cohesin-dockerin interaction, which is consistent with the negative heat capacity associated with the binding event. The hydrogen-bonding network is dominated by the highly conserved Ser-Thr pair located in helix 3 of the dockerin, which confers species specificity among the type I dockerins. Analysis of *C. thermocellum* cohesin and dockerin modules revealed a high degree of conservation in the residues involved in cellulosome assembly. This observation is consistent with the inability of CipA cohesins to discriminate between the different dockerins appended to the cellulosomal enzymes (34).

Plasticity in the Type I Cohesin-Dockerin Interaction

The visualization of the first structure of a type I cohesin-dockerin complex revealed that, indeed, the two duplicate regions of the

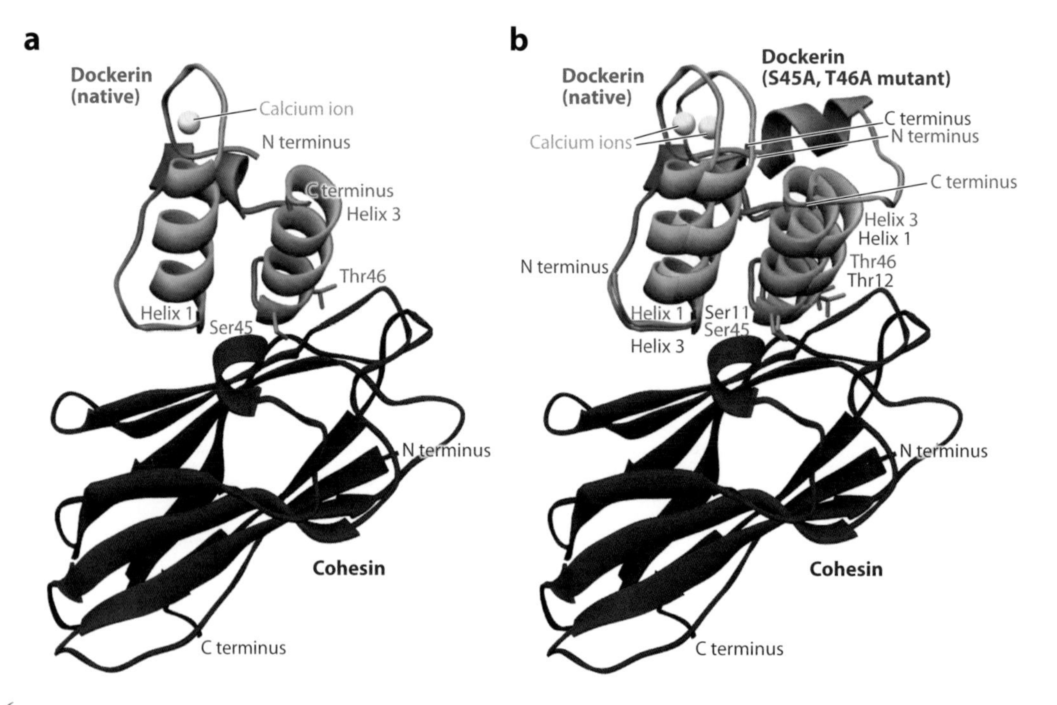

Figure 6

The cohesin-dockerin complex and the two cohesin-binding interfaces of type I dockerins. (*a*) This image represents the structure of *C. thermocellum* Xyn10B dockerin in complex with the second cohesin of CipA [Protein Data Bank (PDB), 1ohz]. The dockerin recognizes the cohesin predominantly through contacts at helix 3. The hydrogen bond network is dominated by Ser45 and Thr46 dockerin residues. (*b*) This blue image represents the structure of the S45A/T46A Xyn10B dockerin mutant in complex with the same cohesin that allowed the visualization of the second dockerin-binding interface (PDB, 2ccl). The structure is overlayed with the native dockerin from panel *a*. The internal symmetry of the dockerin leads to a dual ligand interacting interface manifested by a 180° rotation of the dockerin. Labels are colored according to the represented modules.

dockerin display significant structural conservation. Hence, the dockerin possesses a near-perfect internal twofold symmetry such that residues 1–22 (helix 1) overlay residues 35–56 (helix 3) and vice versa. Thus, cohesin-interacting residues, particularly the Ser-Thr pair, and residues involved in calcium coordination are absolutely conserved when the structures of the two duplicated segments are overlaid. Because the dockerin interacts with the cohesin predominantly through helix 3, it was suggested that the rotation of the dockerin by 180° on the top of the cohesin would allow cohesin recognition to switch to helix 1 rather than helix 3 (105). The second binding mode of the dockerin to the cohesin was visualized by solving the structure of a cohesin-dockerin complex where the Ser-Thr pair of helix 3 was substituted by alanine (**Figure 6**). According to the prediction, the mutated dockerin was rotated by 180° with helix 1 in the position of helix 3, and the Ser-Thr pair of the first duplicated segment dominated the hydrogen bond network (105). Thus, although the mutated dockerin retains its internal symmetry, residues in helix 1, instead of helix 3, now dominate cohesin recognition (**Figure 6**). The dockerin dual binding mode is entirely consistent with the elegant site-directed mutagenesis and thermodynamic studies developed by Béguin and colleagues (26, 27). Their data showed that substitution of residues at positions 11 and 12 at one of the helices, with the equivalent amino acids in type II dockerins, had no major impact on the cohesin-dockerin interaction. Only when substitutions occurred in both Ser-Thr pairs was there a significant reduction in cohesin affinity (27). The existence of two cohesin-binding sites in dockerins suggests that these modules might be able to bind two cohesins simultaneously, providing an additional mechanism for polycellulosome assembly. However, the stoichiometry of the binding of a variety of type I cohesin-dockerin complexes is consistently 1:1, suggesting that the two binding sites are not able to bind ligand simultaneously (105, 106).

Biological Function of Dockerin Dual Binding Mode

The dual binding mode displayed by *C. thermocellum* type I dockerins seems to be a common feature of, at least, clostridial dockerins. The dockerin sequences of cellulosomal enzymes from a variety of bacteria reveal a remarkable conservation in the sequence duplication, first observed in *C. thermocellum* (reviewed previously in Reference 3). The structure of *C. cellulolyticum* type I cohesin-dockerin complexes confirmed that the *cellulolyticum* dockerin also displays an internal structural symmetry, which is compatible with the presence of two essentially identical cohesin-binding surfaces (107). Through the introduction of mutations at each of the dockerin putative-binding interfaces, it was possible to visualize the two binding modes. Binding through the C-terminal helix, observed in the A16S-A17T dockerin mutant, is predominantly hydrophobic and mediated by Leu50. In the other binding mode, visualized through the A47S-A48T dockerin variant, the dockerin is rotated 180°, and the dockerin interacts with the cohesin via the N-terminal helix, with Phe-19 dominating the cohesin contacts. The two dockerin apolar residues occupy a hydrophobic pocket on the surface of the cohesin formed by Leu87 and Leu89. Again, there is a strong conservation on the dockerin (at the two binding sites) and cohesin-interacting residues in *C. cellulolyticum*, indicating that there is little, if any, specificity between type I cohesin-dockerin partners in the cellulosome of this bacterium. Although the dockerins from nonclostridial species also display significant sequence duplication, the variation in the primary structure of the two segments is more evident than in the clostridial dockerins. Thus, it is currently unclear whether the dual binding mode, evident in *C. thermocellum* and *C. cellulotyicum* dockerins, is an invariant feature of cellulosomes.

From the above discussion, it is clear that there is a strong evolutionary driver for the retention of two identical cohesin-binding sites in clostridial dockerins. The functional significance of this plasticity in cohesin-dockerin recognition is, however, unknown. It is likely, however, that the dockerin dual binding mechanism confers flexibility in cellulosome function and assembly. Small-angle X-ray scattering studies with cellulosomal components suggested that linker sequences separating the catalytic and dockerin domains when incorporated into a cellulosome adopt a more rigid conformation, and thus, the motions of catalytic modules with respect to the scaffoldin interface are highly restricted (108). The dual binding mode may therefore confer significant conformational flexibility to the assembly of the catalytic subunits into cellulosome. The steric constraints imposed by the appended catalytic and noncatalytic domains may restrict the incorporation of certain enzyme combinations into a single cellulosome molecule. The dual binding mode could overcome these spatial limitations and may contribute to the plasticity required for the incorporation of specific enzyme combinations. Finally, the efficiency of cellulosome function may require, temporally, the switching of the enzymatic subunits from one cellulosome position to another to optimize the synergy between specific enzymes, notably cellulases. Because the cohesin-dockerin interaction is extremely tight, the existence of a second ligand-binding surface in dockerins may facilitate the switching of the appended enzymes onto a different cellulosomal cohesin. This mechanism may contribute to the continuous modulation of the cellulosome such that the spatial locations of its catalytic components are optimized for the substrate, whose surface is undergoing continuous modification during the degradative process. These hypotheses remain, however, to be tested, and there is an urgent need to investigate the biological significance of this observation.

Cohesin-Dockerin Interaction and Cellulosome Cell-Surface Attachment

Cell-surface attachment of cellulosomes is required for the efficient degradation of plant cell wall polysaccharides, as discussed above. Type II cohesin-dockerin interactions are typically associated with cellulosome cell-surface attachment (43). Smith and colleagues (109) elucidated the crystal structure of a type II complex, which provides insights into the mechanistic basis for the exquisite specificity displayed by type I and type II proteins. Type II dockerins are usually present at the C terminus of a module of unknown function, termed the X module; the only exceptions are type II dockerins of *B. cellulosolvens* present in the bacterium's primary scaffoldin. The importance of the X module in the type II cohesin-dockerin interactions was demonstrated by Smith and colleagues (109) through the resolution of the structure of a trimodular complex. The type II dockerin, which displays a fold similar to its type I counterpart, establishes an extensive range of interactions with the X module that adopts an immunoglobulin-like fold (**Figure 7**). However, the cohesin-binding region of the type II dockerin is more extensive than in type I dockerins, and both the N- and C-terminal helices contact the cohesin surface. The type II cohesin also presents an extended binding surface because the C-terminal region of the β-strand 8,

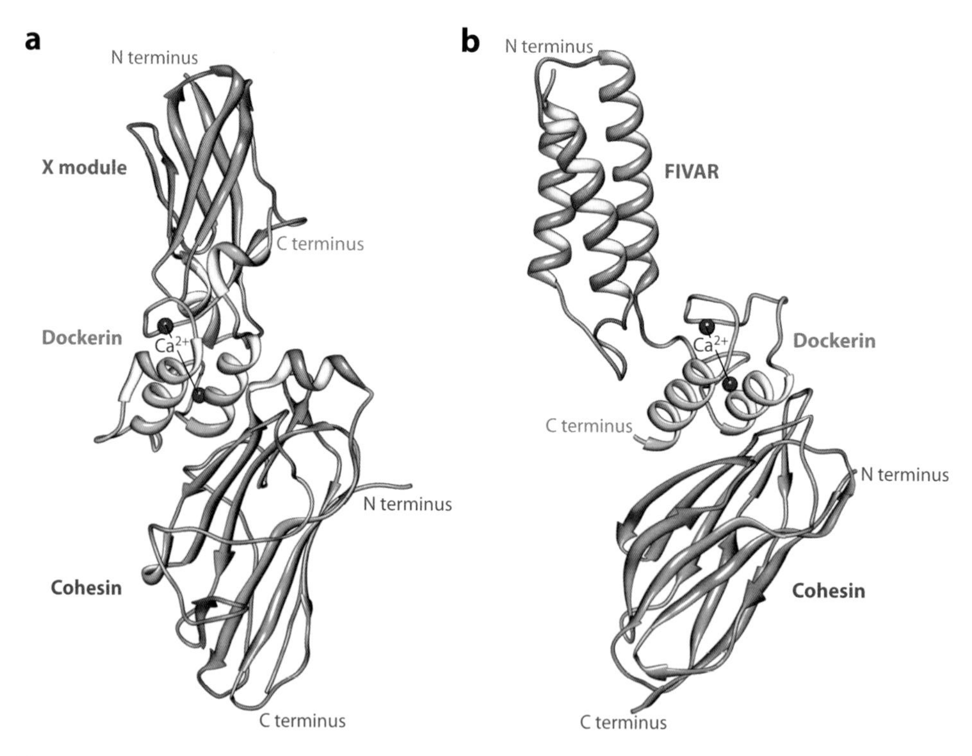

Figure 7

Structure of the type II cohesin-dockerin complexes of *C. thermocellum* and *C. perfringens*. (*a*) Structure of the type II cohesin-dockerin complex of *C. thermocellum* (PDB, 2B59). Ribbon representation depicting the cohesin, the dockerin, and the X module. (*b*) Structure of a *C. perfringens* cohesin-dockerin complex that is involved in toxin formation (PDB, 2OZN). Ribbon representation depicting the cohesin, the dockerin, and the FIVAR module. The N and C termini of both complexes are labeled.

which is absent in type I cohesins, makes several contacts with its protein partner (**Figure 7**). The type II cohesin-dockerin interaction is also much more hydrophobic and tighter than the corresponding type I heterodimer. In part, the increase in affinity displayed by the type II complexes derives from the presence of the X module as the affinity of the dockerin alone for the cohesin is considerably lower than that of the X module-dockerin bimodular protein. Thus, the X module contributes to the structural stability of the adjacent dockerin, which is inherently flexible in solution, thus restricting the entropic penalty incurred by conformational restriction upon cohesin binding. The type II cohesin-dockerin interaction is not only tighter but seems to be considerably less flexible when compared with the type I complexes. Hence, structural analysis revealed that, although a considerable degree of symmetry exists in the type II dockerin molecule, there is no evidence for a dual binding mode in type II complexes. The apparent lack of plasticity in the type II cohesin-dockerin interaction may indicate that flexibility is required in the binding of the cellulosomal catalytic subunits to the scaffoldins but is actually selected against in the anchorage of cellulosomes to the cell surface.

Structure of Noncellulosomal Cohesin-Dockerin Complexes

Cohesin and dockerin modules displaying distant similarity with the type II *C. thermocellum* cellulosomal protein partners were identified in *C. perfringens*. Both cohesin (previously termed X82 domains) and dockerin-like modules were found in a range of glycoside hydrolases that belong to a repertoire of carbohydrate-active toxins, whose catalytic properties are consistent with the degradation of the mucosal layer of the human gut and eukaryotic glycans (39). Dockerins and cohesin modules were shown to form an ultratight interaction that is responsible for the association of large noncovalent-associated, multitoxin complexes that enhance the effectiveness of the individual toxins by combining complementary toxin specificities (40). The structure of one of those cohesin-dockerin complexes was solved (**Figure 7**), revealing an extensive hydrophobic interface between the protein partners with the dockerin rotated by 180° relative to the position of the type II dockerin in the cellulosomal complex described above (40). Again the *C. perfringens* dockerins do not display the internal symmetry, characteristic of type I dockerins, and both dockerin helices are directly involved in the cohesin recognition fold (**Figure 7**). Thus, the dual binding mode does not seem to be a feature of cohesin recognition by *C. perfringens* dockerins but, rather, appears to be restricted to cellulosomal cohesin-dockerin pairs. These elegant studies show the importance of cohesin-dockerin interactions in the organization of noncellulosomal, noncellulolytic systems involved in *C. perfringens* infection, demonstrating that this high-affinity protein interaction, which is quite widespread in nature, might fulfill a variety of relevant cellular functions in a diversity of biological processes.

FUTURE PERSPECTIVES: APPLICATIONS OF CELLULOSOMES

The construction of multiprotein complexes is one of the key emerging fields in nanotechnology and modern chemistry. Bacterial cellulosomes are stunning examples of highly elaborate naturally evolved nanomachines that could be used as a blueprint for the design, construction, and exploitation of tailor-made catalytic multiprotein complexes with precise functions. Initial studies toward this goal have led to the production of cellulosome chimeras, or mini-cellulosomes, composed of a miniscaffoldin containing two or more cohesins of different specificities that anchor different dockerin-containing enzymes in precise locations. Cellulosome chimeras were shown to exhibit enhanced activities on recalcitrant cellulose owing to enzyme proximity and enzyme substrate targeting through a scaffoldin-borne CBM (87). However, the efficiency of the most active cellulosome chimeras is still much lower than natural occurring cellulosomes (87). Nevertheless, it is clear that the same concept could be used for the production of cellulosome chimeras for particular biotechnological applications, such as paper pulp production using a repertoire of hemicellulases. In addition, designer cellulosomes may provide an attractive alternative for the emerging bioenergy field, in particular, in the production of highly efficient enzyme systems to generate bioethanol from lignocellulosic biomass. Cellulosomes integrating fungal and bacterial enzymes from nonaggregating systems, which display particular promise in biomass saccharification, and other enzymes from noncellulosomal systems could be generated to increase hydrolytic activities (110, 111). In an attractive alternative strategy, the genetic capacity to synthesize cellulosomes could be introduced into microorganisms that, although capable of fermenting simple sugars into industrially useful molecules such as butanol, do not have a functional endogenous plant cell wall–degrading apparatus (112, 113). Alternatively, cellulosome-containing bacteria, such as *C. thermocellum*, could be engineered to improve their capacity to synthesize ethanol from lignocelluloses.

In the 25 years since cellulosomes were first identified, in vitro molecular biology, structural biology, and, latterly, genomics, in harness with continued detailed biochemical analysis, have resulted in a sophisticated understanding of the mechanism by which these protein complexes assemble and the activities displayed by the catalytic components. It is apparent, however, that numerous questions remain unresolved concerning the duplication of the ligand-binding site in the dockerins, the evident redundancy in the enzyme systems, and the significance of the retention of these structures on the surface of the host bacterium. To address these functional questions, there is an urgent need to develop genetic tools so that gene knock-ins and null mutations can be introduced into these organisms. Recent studies by Minton and colleagues indicate that sophisticated clostridial genetic systems will be available in the near future (114, 115). Notwithstanding the potential of these emerging applications, cellulosomes provide an attractive framework for the development of nanotechnology. The cohesin-dockerin interaction allows the self-assembly of different types of molecules capable of forming higher-order complexes that can be used in a variety of nanotechnological applications (116), particularly in systems where substrate and product channeling can lead to improvements in tightly linked catalytic processes.

SUMMARY POINTS

1. Cellulosomes are remarkably efficient nanomachines produced by anaerobic microbes to deconstruct plant cell wall polysaccharides.
2. Molecular integration of cellulases and hemicellulases into cellulosomal multienzyme complexes results from the high-affinity interaction established between type I dockerin domains of the modular enzymes and type I cohesin modules of a noncatalytic scaffoldin.
3. Cellulosomes bind the plant cell wall primarily through a scaffoldin-borne carbohydrate-binding module (CBM). In addition, cellulosomes may be anchored to the microbial cell surface through the interaction of a type II dockerin located in the scaffoldin and type II cohesin domains located at the cell envelope.
4. Genome sequencing revealed that cellulosomes may be composed of a large consortium of enzymes that, in *R. flavefaciens*, consists of more than 200 different polypeptides.
5. Integration of the microbial biocatalysts into cellulosomes potentiates catalysis through the maximization of enzyme synergism afforded by enzyme proximity and efficient substrate targeting.

6. Cellulosomal dockerins display a dual cohesin-binding interface that may introduce enhanced flexibility in the quaternary organization of the multienzyme complex.
7. Recent evidence suggests that the cohesin-dockerin interaction is quite widespread in nature and may fulfill a plethora of, mostly currently unknown, functions in addition to the assembly of highly complex nanomachines such as the cellulosome.

FUTURE ISSUES

Future issues include answering the following questions:

1. How do cellulosomal enzymes displaying such a diversity of substrate specificities cooperate during substrate hydrolysis?
2. What is the molecular basis for the organization of fungal cellulosomes?
3. What is the biological significance of the dockerin dual binding mode? Are dockerins displaying two cohesion-binding interfaces restricted to cellulosomes?
4. What is the three-dimensional structure of the *Ruminococcus* type III cohesin-dockerin complexes? What is the mechanism of cellulosome assembly in *Ruminococcus*?
5. How can we extrapolate our current knowledge about the structure and function of cellulosomes to engineer nanomachines for applied processes, such as the use of lignocellulosic residues for bioethanol production?

DISCLOSURE STATEMENT

The authors are not aware of any affiliations, memberships, funding, or financial holdings that might be perceived as affecting the objectivity of this review.

ACKNOWLEDGMENTS

We wish to thank Fundação para a Ciência ea Tecnologi and the Biotechnological and Biological Sciences Research Council for continued support of the cellulosome programs in Portugal and the United Kingdom, respectively.

LITERATURE CITED

1. Brett CT, Waldren K. 1996. *Physiology and Biochemistry of Plant Cell Walls: Topics in Plant Functional Biology*. London: Chapman & Hall. 2nd ed.
2. Warren RA. 1996. Microbial hydrolysis of polysaccharides. *Annu. Rev. Microbiol.* 50:183–212
3. Bayer EA, Belaich JP, Shoham Y, Lamed R. 2004. The cellulosomes: multienzyme machines for degradation of plant cell wall polysaccharides. *Annu. Rev. Microbiol.* 58:521–54
4. Bayer EA, Chanzy H, Lamed R, Shoham Y. 1998. Cellulose, cellulases and cellulosomes. *Curr. Opin. Struct. Biol.* 8:548–57
5. Béguin P, Lemaire M. 1996. The cellulosome: an exocellular, multiprotein complex specialized in cellulose degradation. *Crit. Rev. Biochem. Mol. Biol.* 31:201–36
6. Gilbert HJ. 2007. Cellulosomes: microbial nanomachines that display plasticity in quaternary structure. *Mol. Microbiol.* 63:1568–76
7. Shoham Y, Lamed R, Bayer EA. 1999. The cellulosome concept as an efficient microbial strategy for the degradation of insoluble polysaccharides. *Trends Microbiol.* 7:275–81

8. Bayer EA, Kenig R, Lamed R. 1983. Adherence of *Clostridium thermocellum* to cellulose. *J. Bacteriol.* 156:818–27
9. Lamed R, Setter E, Bayer EA. 1983. Characterization of a cellulose-binding, cellulase-containing complex in *Clostridium thermocellum*. *J. Bacteriol.* 156:828–36
10. Lamed R, Setter E, Kenig R, Bayer EA. 1983. The cellulosome: a discrete cell surface organelle of *Clostridium thermocellum* which exhibits separate antigenic, cellulose-binding and various cellulolytic activities. *Biotechnol. Bioeng. Symp.* 13:163–81
11. Kosugi A, Murashima K, Doi RH. 2002. Xylanase and acetyl xylan esterase activities of XynA, a key subunit of the *Clostridium cellulovorans* cellulosome for xylan degradation. *Appl. Environ. Microbiol.* 68:6399–402
12. Morag E, Bayer EA, Lamed R. 1990. Relationship of cellulosomal and noncellulosomal xylanases of *Clostridium thermocellum* to cellulose-degrading enzymes. *J. Bacteriol.* 172:6098–105
13. Tamaru Y, Doi RH. 2001. Pectate lyase A, an enzymatic subunit of the *Clostridium cellulovorans* cellulosome. *Proc. Natl. Acad. Sci. USA* 98:4125–29
14. Salamitou S, Raynaud O, Lemaire M, Coughlan M, Béguin P, Aubert JP. 1994. Recognition specificity of the duplicated segments present in *Clostridium thermocellum* endoglucanase CelD and in the cellulosome-integrating protein CipA. *J. Bacteriol.* 176:2822–27
15. Tokatlidis K, Salamitou S, Béguin P, Dhurjati P, Aubert JP. 1991. Interaction of the duplicated segment carried by *Clostridium thermocellum* cellulases with cellulosome components. *FEBS Lett.* 291:185–88
16. Poole DM, Morag E, Lamed R, Bayer EA, Hazlewood GP, Gilbert HJ. 1992. Identification of the cellulose-binding domain of the cellulosome subunit S1 from *Clostridium thermocellum* YS. *FEMS Microbiol. Lett.* 78:181–86
17. Demain AL, Newcomb M, Wu JH. 2005. Cellulase, clostridia, and ethanol. *Microbiol. Mol. Biol. Rev.* 69:124–54
18. Ng TK, Zeikus JG. 1981. Comparison of extracellular cellulase activities of *Clostridium thermocellum* LQRI and *Trichoderma reesei* QM9414. *Appl. Environ. Microbiol.* 42:231–40
19. Kataeva IA, Yang SJ, Dam P, Poole FL II, Yin Y, et al. 2009. Genome sequence of the anaerobic, thermophilic, and cellulolytic bacterium "*Anaerocellum thermophilum*" DSM 6725. *J. Bacteriol.* 191:3760–61
20. Yang SJ, Kataeva I, Hamilton-Brehm SD, Engle NL, Tschaplinski TJ, et al. 2009. Efficient degradation of lignocellulosic plant biomass, without pretreatment, by the thermophilic anaerobe "*Anaerocellum thermophilum*" DSM 6725. *Appl. Environ. Microbiol.* 75:4762–69
21. van de Werken HJ, Verhaart MR, VanFossen AL, Willquist K, Lewis DL, et al. 2008. Hydrogenomics of the extremely thermophilic bacterium *Caldicellulosiruptor saccharolyticus*. *Appl. Environ. Microbiol.* 74:6720–29
22. Zverlov VV, Klupp M, Krauss J, Schwarz WH. 2008. Mutations in the scaffoldin gene, *cipA*, of *Clostridium thermocellum* with impaired cellulosome formation and cellulose hydrolysis: insertions of a new transposable element, IS*1447*, and implications for cellulase synergism on crystalline cellulose. *J. Bacteriol.* 190:4321–27
23. Lu Y, Zhang YH, Lynd LR. 2006. Enzyme-microbe synergy during cellulose hydrolysis by *Clostridium thermocellum*. *Proc. Natl. Acad. Sci. USA* 103:16165–69
24. Boudet AM, Kajita S, Grima-Pettenati J, Goffner D. 2003. Lignins and lignocellulosics: a better control of synthesis for new and improved uses. *Trends Plant Sci.* 8:576–81
25. Ragauskas AJ, Williams CK, Davison BH, Britovsek G, Cairney J, et al. 2006. The path forward for biofuels and biomaterials. *Science* 311:484–89
26. Miras I, Schaeffer F, Béguin P, Alzari PM. 2002. Mapping by site-directed mutagenesis of the region responsible for cohesin-dockerin interaction on the surface of the seventh cohesin domain of *Clostridium thermocellum* CipA. *Biochemistry* 41:2115–19
27. Schaeffer F, Matuschek M, Guglielmi G, Miras I, Alzari PM, Béguin P. 2002. Duplicated dockerin subdomains of *Clostridium thermocellum* endoglucanase CelD bind to a cohesin domain of the scaffolding protein CipA with distinct thermodynamic parameters and a negative cooperativity. *Biochemistry* 41:2106–14

28. Grepinet O, Chebrou MC, Béguin P. 1988. Nucleotide sequence and deletion analysis of the xylanase gene (*xynZ*) of *Clostridium thermocellum*. *J. Bacteriol.* 170:4582–88
29. Hall J, Hazlewood GP, Barker PJ, Gilbert HJ. 1988. Conserved reiterated domains in *Clostridium thermocellum* endoglucanases are not essential for catalytic activity. *Gene* 69:29–38
30. Salamitou S, Tokatlidis K, Béguin P, Aubert JP. 1992. Involvement of separate domains of the cellulosomal protein S1 of *Clostridium thermocellum* in binding to cellulose and in anchoring of catalytic subunits to the cellulosome. *FEBS Lett.* 304:89–92
31. Pages S, Belaich A, Belaich JP, Morag E, Lamed R, et al. 1997. Species-specificity of the cohesin-dockerin interaction between *Clostridium thermocellum* and *Clostridium cellulolyticum*: prediction of specificity determinants of the dockerin domain. *Proteins* 29:517–27
32. Choi SK, Ljungdahl LG. 1996. Structural role of calcium for the organization of the cellulosome of *Clostridium thermocellum*. *Biochemistry* 35:4906–10
33. Lytle BL, Volkman BF, Westler WM, Wu JH. 2000. Secondary structure and calcium-induced folding of the *Clostridium thermocellum* dockerin domain determined by NMR spectroscopy. *Arch. Biochem. Biophys.* 379:237–44
34. Ciruela A, Gilbert HJ, Ali BR, Hazlewood GP. 1998. Synergistic interaction of the cellulosome integrating protein (CipA) from *Clostridium thermocellum* with a cellulosomal endoglucanase. *FEBS Lett.* 422:221–24
35. Yaron S, Morag E, Bayer EA, Lamed R, Shoham Y. 1995. Expression, purification and subunit-binding properties of cohesins 2 and 3 of the *Clostridium thermocellum* cellulosome. *FEBS Lett.* 360:121–24
36. Fendri I, Tardif C, Fierobe HP, Lignon S, Valette O, et al. 2009. The cellulosomes from *Clostridium cellulolyticum*: identification of new components and synergies between complexes. *FEBS J.* 276:3076–86
37. Gold ND, Martin VJ. 2007. Global view of the *Clostridium thermocellum* cellulosome revealed by quantitative proteomic analysis. *J. Bacteriol.* 189:6787–95
38. Raman B, Pan C, Hurst GB, Rodriguez M Jr, McKeown CK, et al. 2009. Impact of pretreated switchgrass and biomass carbohydrates on *Clostridium thermocellum* ATCC 27405 cellulosome composition: a quantitative proteomic analysis. *PLoS One* 4:e5271
39. Peer A, Smith SP, Bayer EA, Lamed R, Borovok I. 2009. Noncellulosomal cohesin- and dockerin-like modules in the three domains of life. *FEMS Microbiol. Lett.* 291:1–16
40. Adams JJ, Gregg K, Bayer EA, Boraston AB, Smith SP. 2008. Structural basis of *Clostridium perfringens* toxin complex formation. *Proc. Natl. Acad. Sci. USA* 105:12194–99
41. Fujino T, Béguin P, Aubert JP. 1993. Organization of a *Clostridium thermocellum* gene cluster encoding the cellulosomal scaffolding protein CipA and a protein possibly involved in attachment of the cellulosome to the cell surface. *J. Bacteriol.* 175:1891–99
42. Lemaire M, Ohayon H, Gounon P, Fujino T, Béguin P. 1995. OlpB, a new outer layer protein of *Clostridium thermocellum*, and binding of its S-layer-like domains to components of the cell envelope. *J. Bacteriol.* 177:2451–59
43. Leibovitz E, Béguin P. 1996. A new type of cohesin domain that specifically binds the dockerin domain of the *Clostridium thermocellum* cellulosome-integrating protein CipA. *J. Bacteriol.* 178:3077–84
44. Kosugi A, Amano Y, Murashima K, Doi RH. 2004. Hydrophilic domains of scaffolding protein CbpA promote glycosyl hydrolase activity and localization of cellulosomes to the cell surface of *Clostridium cellulovorans*. *J. Bacteriol.* 186:6351–59
45. Kosugi A, Murashima K, Tamaru Y, Doi RH. 2002. Cell-surface-anchoring role of N-terminal surface layer homology domains of *Clostridium cellulovorans* EngE. *J. Bacteriol.* 184:884–88
46. Rincon MT, Ding SY, McCrae SI, Martin JC, Aurilia V, et al. 2003. Novel organization and divergent dockerin specificities in the cellulosome system of *Ruminococcus flavefaciens*. *J. Bacteriol.* 185:703–13
47. Rincon MT, Martin JC, Aurilia V, McCrae SI, Rucklidge GJ, et al. 2004. ScaC, an adaptor protein carrying a novel cohesin that expands the dockerin-binding repertoire of the *Ruminococcus flavefaciens* 17 cellulosome. *J. Bacteriol.* 186:2576–85
48. Felix CR, Ljungdahl LG. 1993. The cellulosome: the exocellular organelle of *Clostridium*. *Annu. Rev. Microbiol.* 47:791–819

49. Xu Q, Bayer EA, Goldman M, Kenig R, Shoham Y, Lamed R. 2004. Architecture of the *Bacteroides cellulosolvens* cellulosome: description of a cell surface-anchoring scaffoldin and a family 48 cellulase. *J. Bacteriol.* 186:968–77
50. Ding SY, Bayer EA, Steiner D, Shoham Y, Lamed R. 2000. A scaffoldin of the *Bacteroides cellulosolvens* cellulosome that contains 11 type II cohesins. *J. Bacteriol.* 182:4915–25
51. Xu Q, Barak Y, Kenig R, Shoham Y, Bayer EA, Lamed R. 2004. A novel *Acetivibrio cellulolyticus* anchoring scaffoldin that bears divergent cohesins. *J. Bacteriol.* 186:5782–89
52. Xu Q, Gao W, Ding SY, Kenig R, Shoham Y, et al. 2003. The cellulosome system of *Acetivibrio cellulolyticus* includes a novel type of adaptor protein and a cell surface anchoring protein. *J. Bacteriol.* 185:4548–57
53. Ding SY, Rincon MT, Lamed R, Martin JC, McCrae SI, et al. 2001. Cellulosomal scaffoldin-like proteins from *Ruminococcus flavefaciens*. *J. Bacteriol.* 183:1945–53
54. Jindou S, Borovok I, Rincon MT, Flint HJ, Antonopoulos DA, et al. 2006. Conservation and divergence in cellulosome architecture between two strains of *Ruminococcus flavefaciens*. *J. Bacteriol.* 188:7971–76
55. Jindou S, Brulc JM, Levy-Assaraf M, Rincon MT, Flint HJ, et al. 2008. Cellulosome gene cluster analysis for gauging the diversity of the ruminal cellulolytic bacterium *Ruminococcus flavefaciens*. *FEMS Microbiol. Lett.* 285:188–94
56. Rincon MT, Cepeljnik T, Martin JC, Lamed R, Barak Y, et al. 2005. Unconventional mode of attachment of the *Ruminococcus flavefaciens* cellulosome to the cell surface. *J. Bacteriol.* 187:7569–78
57. Cantarel BL, Coutinho PM, Rancurel C, Bernard T, Lombard V, Henrissat B. 2009. The Carbohydrate-Active EnZymes database (CAZy): an expert resource for glycogenomics. *Nucleic Acids Res.* 37:D233–38
58. Boraston AB, Bolam DN, Gilbert HJ, Davies GJ. 2004. Carbohydrate-binding modules: fine-tuning polysaccharide recognition. *Biochem. J.* 382:769–81
59. Sakon J, Irwin D, Wilson DB, Karplus PA. 1997. Structure and mechanism of endo/exocellulase E4 from *Thermomonospora fusca*. *Nat. Struct. Biol.* 4:810–18
60. Burstein T, Shulman M, Jindou S, Petkun S, Frolow F, et al. 2009. Physical association of the catalytic and helper modules of a family-9 glycoside hydrolase is essential for activity. *FEBS Lett.* 583:879–84
61. Blake AW, McCartney L, Flint JE, Bolam DN, Boraston AB, et al. 2006. Understanding the biological rationale for the diversity of cellulose-directed carbohydrate-binding modules in prokaryotic enzymes. *J. Biol. Chem.* 281:29321–29
62. Tormo J, Lamed R, Chirino AJ, Morag E, Bayer EA, et al. 1996. Crystal structure of a bacterial family-III cellulose-binding domain: a general mechanism for attachment to cellulose. *EMBO J.* 15:5739–51
63. Charnock SJ, Bolam DN, Nurizzo D, Szabo L, McKie VA, et al. 2002. Promiscuity in ligand-binding: the three-dimensional structure of a *Piromyces* carbohydrate-binding module, CBM29-2, in complex with cello- and mannohexaose. *Proc. Natl. Acad. Sci. USA* 99:14077–82
64. Rincon MT, Cepeljnik T, Martin JC, Barak Y, Lamed R, et al. 2007. A novel cell surface-anchored cellulose-binding protein encoded by the *sca* gene cluster of *Ruminococcus flavefaciens*. *J. Bacteriol.* 189:4774–83
65. Xu Q, Morrison M, Nelson KE, Bayer EA, Atamna N, Lamed R. 2004. A novel family of carbohydrate-binding modules identified with *Ruminococcus albus* proteins. *FEBS Lett.* 566:11–16
66. Ezer A, Matalon E, Jindou S, Borovok I, Atamna N, et al. 2008. Cell surface enzyme attachment is mediated by family 37 carbohydrate-binding modules, unique to *Ruminococcus albus*. *J. Bacteriol.* 190:8220–22
67. Mechaly A, Fierobe HP, Belaich A, Belaich JP, Lamed R, et al. 2001. Cohesin-dockerin interaction in cellulosome assembly: a single hydroxyl group of a dockerin domain distinguishes between nonrecognition and high affinity recognition. *J. Biol. Chem.* 276:9883–88
68. Haimovitz R, Barak Y, Morag E, Voronov-Goldman M, Shoham Y, et al. 2008. Cohesin-dockerin microarray: diverse specificities between two complementary families of interacting protein modules. *Proteomics* 8:968–79
69. Jindou S, Soda A, Karita S, Kajino T, Béguin P, et al. 2004. Cohesin-dockerin interactions within and between *Clostridium josui* and *Clostridium thermocellum*: binding selectivity between cognate dockerin and cohesin domains and species specificity. *J. Biol. Chem.* 279:9867–74
70. Ljungdahl LG. 2008. The cellulase/hemicellulase system of the anaerobic fungus *Orpinomyces* PC-2 and aspects of its applied use. *Ann. N.Y. Acad. Sci.* 1125:308–21

71. Ali BR, Zhou L, Graves FM, Freedman RB, Black GW, et al. 1995. Cellulases and hemicellulases of the anaerobic fungus *Piromyces* constitute a multiprotein cellulose-binding complex and are encoded by multigene families. *FEMS Microbiol. Lett.* 125:15–21
72. Wood CA, Wood TM. 1992. Studies on the cellulase of the rumen anaerobic fungus *Neocallimastix frontalis*, with special reference to the capacity of the enzyme to degrade crystalline cellulose. *Enzyme Microb. Technol.* 14:258–64
73. Black GW, Hazlewood GP, Xue GP, Orpin CG, Gilbert HJ. 1994. Xylanase B from *Neocallimastix patriciarum* contains a non-catalytic 455-residue linker sequence comprised of 57 repeats of an octapeptide. *Biochem. J.* 299:381–87
74. Millward-Sadler SJ, Hall J, Black GW, Hazlewood GP, Gilbert HJ. 1996. Evidence that the *Piromyces* gene family encoding endo-1,4-mannanases arose through gene duplication. *FEMS Microbiol. Lett.* 141:183–88
75. Zhou L, Xue GP, Orpin CG, Black GW, Gilbert HJ, Hazlewood GP. 1994. Intronless celB from the anaerobic fungus *Neocallimastix patriciarum* encodes a modular family A endoglucanase. *Biochem. J.* 297:359–64
76. Fanutti C, Ponyi T, Black GW, Hazlewood GP, Gilbert HJ. 1995. The conserved noncatalytic 40-residue sequence in cellulases and hemicellulases from anaerobic fungi functions as a protein docking domain. *J. Biol. Chem.* 270:29314–22
77. Nagy T, Tunnicliffe RB, Higgins LD, Walters C, Gilbert HJ, Williamson MP. 2007. Characterization of a double dockerin from the cellulosome of the anaerobic fungus *Piromyces equi*. *J. Mol. Biol.* 373:612–22
78. Raghothama S, Eberhardt RY, Simpson P, Wigelsworth D, White P, et al. 2001. Characterization of a cellulosome dockerin domain from the anaerobic fungus *Piromyces equi*. *Nat. Struct. Biol.* 8:775–78
79. Raghothama S, Simpson PJ, Szabo L, Nagy T, Gilbert HJ, Williamson MP. 2000. Solution structure of the CBM10 cellulose binding module from *Pseudomonas* xylanase A. *Biochemistry* 39:978–84
80. Davies G, Henrissat B. 1995. Structures and mechanisms of glycosyl hydrolases. *Structure* 3:853–59
81. Tomme P, Warren RA, Gilkes NR. 1995. Cellulose hydrolysis by bacteria and fungi. *Adv. Microbiol. Physiol.* 37:1–81
82. Wood TM. 1992. Fungal cellulases. *Biochem. Soc. Trans.* 20:46–53
83. Armand S, Drouillard S, Schulein M, Henrissat B, Driguez H. 1997. A bifunctionalized fluorogenic tetrasaccharide as a substrate to study cellulases. *J. Biol. Chem.* 272:2709–13
84. Boisset C, Fraschini C, Schulein M, Henrissat B, Chanzy H. 2000. Imaging the enzymatic digestion of bacterial cellulose ribbons reveals the endo character of the cellobiohydrolase Cel6A from *Humicola insolens* and its mode of synergy with cellobiohydrolase Cel7A. *Appl. Environ. Microbiol.* 66:1444–52
85. Varrot A, Schulein M, Davies GJ. 1999. Structural changes of the active site tunnel of *Humicola insolens* cellobiohydrolase, Cel6A, upon oligosaccharide binding. *Biochemistry* 38:8884–91
86. Wood TM, McCrae SI. 1986. The cellulase of *Penicillium pinophilum*. Synergism between enzyme components in solubilizing cellulose with special reference to the involvement of two immunologically distinct cellobiohydrolases. *Biochem. J.* 234:93–99
87. Fierobe HP, Bayer EA, Tardif C, Czjzek M, Mechaly A, et al. 2002. Degradation of cellulose substrates by cellulosome chimeras. Substrate targeting versus proximity of enzyme components. *J. Biol. Chem.* 277:49621–30
88. Guimarães BG, Souchon H, Lytle BL, David Wu JH, Alzari PM. 2002. The crystal structure and catalytic mechanism of cellobiohydrolase CelS, the major enzymatic component of the *Clostridium thermocellum* cellulosome. *J. Mol. Biol.* 320:587–96
89. Parsiegla G, Juy M, Reverbel-Leroy C, Tardif C, Belaich JP, et al. 1998. The crystal structure of the processive endocellulase CelF of *Clostridium cellulolyticum* in complex with a thiooligosaccharide inhibitor at 2.0 Å resolution. *EMBO J.* 17:5551–62
90. Parsiegla G, Reverbel C, Tardif C, Driguez H, Haser R. 2008. Structures of mutants of cellulase Cel48F of *Clostridium cellulolyticum* in complex with long hemithiocellooligosaccharides give rise to a new view of the substrate pathway during processive action. *J. Mol. Biol.* 375:499–510
91. Guérin DMA, Lascombe M-B, Costabel M, Souchon H, Lamzin V, et al. 2002. Atomic (0.94 Å) resolution structure of an inverting glycosidase in complex with substrate. *J. Mol. Biol.* 316:1061–69

92. Zverlov VV, Schantz N, Schwarz WH. 2005. A major new component in the cellulosome of *Clostridium thermocellum* is a processive endo-beta-1,4-glucanase producing cellotetraose. *FEMS Microbiol. Lett.* 249:353–58
93. Fierobe HP, Mechaly A, Tardif C, Belaich A, Lamed R, et al. 2001. Design and production of active cellulosome chimeras. Selective incorporation of dockerin-containing enzymes into defined functional complexes. *J. Biol. Chem.* 276:21257–61
94. Fierobe HP, Mingardon F, Mechaly A, Belaich A, Rincon MT, et al. 2005. Action of designer cellulosomes on homogeneous versus complex substrates: controlled incorporation of three distinct enzymes into a defined trifunctional scaffoldin. *J. Biol. Chem.* 280:16325–34
95. Devillard E, Goodheart DB, Karnati SK, Bayer EA, Lamed R, et al. 2004. *Ruminococcus albus* 8 mutants defective in cellulose degradation are deficient in two processive endocellulases, Cel48A and Cel9B, both of which possess a novel modular architecture. *J. Bacteriol.* 186:136–45
96. Xu Q, Morrison M, Nelson KE, Bayer EA, Atamna N, Lamed R. 2004. A novel family of carbohydrate-binding modules identified with *Ruminococcus albus* proteins. *FEBS Lett.* 566:11–16
97. Charnock SJ, Bolam DN, Turkenburg JP, Gilbert HJ, Ferreira LM, et al. 2000. The X6 "thermostabilizing" domains of xylanases are carbohydrate-binding modules: structure and biochemistry of the *Clostridium thermocellum* X6b domain. *Biochemistry* 39:5013–21
98. Carvalho AL, Goyal A, Prates JA, Bolam DN, Gilbert HJ, et al. 2004. The family 11 carbohydrate-binding module of *Clostridium thermocellum* Lic26A-Cel5E accommodates beta-1,4- and beta-1,3-1,4-mixed linked glucans at a single binding site. *J. Biol. Chem.* 279:34785–93
99. Najmudin S, Guerreiro CI, Carvalho AL, Prates JA, Correia MA, et al. 2006. Xyloglucan is recognized by carbohydrate-binding modules that interact with beta-glucan chains. *J. Biol. Chem.* 281:8815–28
100. Montanier C, van Bueren AL, Dumon C, Flint JE, Correia MA, et al. 2009. Evidence that family 35 carbohydrate binding modules display conserved specificity but divergent function. *Proc. Natl. Acad. Sci. USA* 106:3065–70
101. Durrant AJ, Hall J, Hazlewood GP, Gilbert HJ. 1991. The non-catalytic C-terminal region of endoglucanase E from *Clostridium thermocellum* contains a cellulose-binding domain. *Biochem. J.* 273:289–93
102. Montanier C, Money VA, Pires VM, Flint JE, Pinheiro BA, et al. 2009. The active site of a carbohydrate esterase displays divergent catalytic and noncatalytic binding functions. *PLoS Biol.* 7:e71
103. Gilad R, Rabinovich L, Yaron S, Bayer EA, Lamed R, et al. 2003. CelI, a noncellulosomal family 9 enzyme from *Clostridium thermocellum*, is a processive endoglucanase that degrades crystalline cellulose. *J. Bacteriol.* 185:391–98
104. Lytle BL, Volkman BF, Westler WM, Heckman MP, Wu JH. 2001. Solution structure of a type I dockerin domain, a novel prokaryotic, extracellular calcium-binding domain. *J. Mol. Biol.* 307:745–53
105. Carvalho AL, Dias FM, Prates JA, Nagy T, Gilbert HJ, et al. 2003. Cellulosome assembly revealed by the crystal structure of the cohesin-dockerin complex. *Proc. Natl. Acad. Sci. USA* 100:13809–14
106. Carvalho AL, Dias FM, Nagy T, Prates JA, Proctor MR, et al. 2007. Evidence for a dual binding mode of dockerin modules to cohesins. *Proc. Natl. Acad. Sci. USA* 104:3089–94
107. Pinheiro BA, Proctor MR, Martinez-Fleites C, Prates JA, Money VA, et al. 2008. The *Clostridium cellulolyticum* dockerin displays a dual binding mode for its cohesin partner. *J. Biol. Chem.* 283:18422–30
108. Hammel M, Fierobe HP, Czjzek M, Finet S, Receveur-Brechot V. 2004. Structural insights into the mechanism of formation of cellulosomes probed by small angle X-ray scattering. *J. Biol. Chem.* 279:55985–94
109. Adams JJ, Pal G, Jia Z, Smith SP. 2006. Mechanism of bacterial cell-surface attachment revealed by the structure of cellulosomal type II cohesin-dockerin complex. *Proc. Natl. Acad. Sci. USA* 103:305–10
110. Caspi J, Irwin D, Lamed R, Li Y, Fierobe HP, et al. 2008. Conversion of *Thermobifida fusca* free exoglucanases into cellulosomal components: comparative impact on cellulose-degrading activity. *J. Biotechnol.* 135:351–57
111. Mingardon F, Chanal A, Lopez-Contreras AM, Dray C, Bayer EA, Fierobe HP. 2007. Incorporation of fungal cellulases in bacterial minicellulosomes yields viable, synergistically acting cellulolytic complexes. *Appl. Environ. Microbiol.* 73:3822–32

112. Mingardon F, Perret S, Belaich A, Tardif C, Belaich JP, Fierobe HP. 2005. Heterologous production, assembly, and secretion of a minicellulosome by *Clostridium acetobutylicum* ATCC 824. *Appl. Environ. Microbiol.* 71:1215–22

113. Perret S, Casalot L, Fierobe HP, Tardif C, Sabathe F, et al. 2004. Production of heterologous and chimeric scaffoldins by *Clostridium acetobutylicum* ATCC 824. *J. Bacteriol.* 186:253–57

114. Heap JT, Pennington OJ, Cartman ST, Carter GP, Minton NP. 2007. The ClosTron: a universal gene knock-out system for the genus *Clostridium*. *J. Microbiol. Methods* 70:452–64

115. Heap JT, Pennington OJ, Cartman ST, Minton NP. 2009. A modular system for *Clostridium* shuttle plasmids. *J. Microbiol. Methods* 78:79–85

116. Heyman A, Barak Y, Caspi J, Wilson DB, Altman A, et al. 2007. Multiple display of catalytic modules on a protein scaffold: nano-fabrication of enzyme particles. *J. Biotechnol.* 131:433–39

Somatic Mitochondrial DNA Mutations in Mammalian Aging

Nils-Göran Larsson

Max Planck Institute for Biology of Ageing, Cologne D-50931, Germany; email: larsson@age.mpg.de

Annu. Rev. Biochem. 2010. 79:683–706

First published online as a Review in Advance on March 29, 2010

The *Annual Review of Biochemistry* is online at biochem.annualreviews.org

This article's doi: 10.1146/annurev-biochem-060408-093701

0066-4154/10/0707-0683$20.00

Key Words

mitochondria, oxidative phosphorylation, respiratory chain

Abstract

Mitochondrial dysfunction is heavily implicated in the multifactorial aging process. Aging humans have increased levels of somatic mtDNA mutations that tend to undergo clonal expansion to cause mosaic respiratory chain deficiency in various tissues, such as heart, brain, skeletal muscle, and gut. Genetic mouse models have shown that somatic mtDNA mutations and cell type-specific respiratory chain dysfunction can cause a variety of phenotypes associated with aging and age-related disease. There is thus strong observational and experimental evidence to implicate somatic mtDNA mutations and mosaic respiratory chain dysfunction in the mammalian aging process. The hypothesis that somatic mtDNA mutations are generated by oxidative damage has not been conclusively proven. Emerging data instead suggest that the inherent error rate of mitochondrial DNA (mtDNA) polymerase γ (Pol γ) may be responsible for the majority of somatic mtDNA mutations. The roles for mtDNA damage and replication errors in aging need to be further experimentally addressed.

Contents

INTRODUCTION

The cause of aging has attracted significant interest for thousands of years. One of the earliest documented examples is the first Chinese emperor Qin Shi Huang (259–210 BCE), who desperately searched for the elixir of life and tried to prolong his life by ingestion of mercury. Needless to say, his attempt had the opposite effect, and subsequent treatments to achieve human immortality have also failed. There have been many theories put forward to explain the phenomenon of aging, as exemplified by the now largely refuted rate of living theory (1). There is an emerging consensus that aging is a multifactorial process, which is not caused by a genetic program, although genes certainly influence the aging process (2).

It was reported in the 1930s that dietary restriction prolongs life in rats, and this intervention has since been reported to have a similar effect in a large number of animals (3). Dietary restriction does not only lead to life span extension but also reduces the severity of various age-associated pathologies (3). In 1956, Harman (4) proposed that reactive oxygen species (ROS) drive aging by creating cumulative damage to vital cellular processes. The theory was later amended to include mitochondria as the main source of ROS production, and this has remained popular for decades (5). However, recent experimental results have suggested that reduced oxidative damage may not invariably prolong life span, and the free radical theory of aging has therefore started to lose support (5).

It came as a surprise when screens in the worm *Caenorhabditis elegans* identified specific mutations that could prolong the life span drastically (6). Today, we know of many such mutations, and a well-studied example is reduced insulin signaling that prolongs life span in worms, flies, and mice and therefore likely represents an evolutionarily conserved longevity pathway (7). Another example is that moderately reduced oxidative phosphorylation capacity prolongs life span in worms (8). There is some evidence that this mechanism may also be conserved in mammals because mice with moderately reduced oxidative phosphorylation have improved glucose homeostasis (9–11) and live longer (12). These findings must be reconciled with other findings suggesting that mitochondrial dysfunction may promote aging.

Human aging is strongly associated with the occurrence of acquired damage to mitochondrial DNA (mtDNA) and clonal expansion of these somatic mutations, leading to focal respiratory chain deficiency (5, 13, 14). Age-associated damage to mtDNA and decline of respiratory chain function are also found in a variety of other mammalian species, such as mouse (15), rat (16), and rhesus monkey (17). Furthermore, experimental studies in the mouse have shown that genetic defects that increase levels of somatic mtDNA mutations cause a premature aging syndrome (18). There are thus both observational and experimental support for the hypothesis that mitochondrial dysfunction may be an important mechanism in mammalian aging. This review summarizes the current knowledge of regulation of maintenance and expression of mtDNA in mammals and then discusses how somatic mtDNA mutations may promote mammalian aging.

BIOGENESIS OF THE OXIDATIVE PHOSPHORYLATION SYSTEM

Mitochondria form a dynamic network within the cell, and specialized transport machineries make sure that mitochondria will reach appropriate subcellular localizations, such as nerve endings (19, 20). The mitochondrial fusion and fission machineries have predominantly been studied in yeast (19) but are to a large extent conserved in mammals (20). The mitochondrial network in a somatic mammalian cell contains thousand of copies of mtDNA and harbors the oxidative phosphorylation system in the inner mitochondrial membrane (**Figure 1**). The mtDNA only encodes 13 proteins that all are components of the oxidative phosphorylation system. Nuclear genes encode the remaining $\sim 10^2$ proteins of the oxidative phosphorylation system and an additional $\sim 10^3$ proteins that have various functions in mitochondria, such as maintenance and expression of mtDNA (21), mitochondrial protein synthesis (22), mitochondrial proteolysis (23), mitochondrial protein import (24), mitochondrial morphology (19, 20), iron-sulfur cluster synthesis (25), citric acid cycle metabolism, fatty acid oxidation, and additional metabolic pathways.

The oxidative phosphorylation system is situated in the inner mitochondrial membrane and produces the energy currency ATP (**Figure 2**). The respiratory chain consists of complexes I–IV, coenzyme Q, and cytochrome *c*. The respiratory chain receives electrons harvested from the citric acid cycle and fatty acid oxidation and transfers these electrons through a series of coupled reduction-oxidation reactions until finally molecular oxygen is reduced to water. This electron transport results in translocation of protons across the inner mitochondrial membrane, and the resulting proton gradient is used by the ATP synthase (complex V) to drive ATP production. In case of uncoupling, the protons re-enter the mitochondrial matrix, e.g., through a specialized uncoupling protein, and the electron transfer is no longer coupled to ATP synthesis (**Figure 2**).

Dietary restriction: a diet regimen that restricts food intake without causing malnutrition

ROS: reactive oxygen species

Oxidative phosphorylation: the process whereby the respiratory chain couples electron transport to ATP synthesis

mtDNA: mitochondrial DNA

Clonal expansion: preferential amplification of a mutated mtDNA species in a tissue cell

H: heavy strand of mtDNA

L: light strand of mtDNA

Maintenance and Expression of Mammalian mtDNA

The two strands of mtDNA (**Figure 3**) are for historical reasons denoted the heavy (H) and light (L) strands owing to their different base compositions, which lead to different densities in alkaline cesium chloride gradients (21). The mtDNA is densely packed with genes and contains only one longer noncoding region, denoted the displacement loop (D loop). The D loop contains the promoters for transcription of the L and H strands (LSP and HSP) and the origin of replication of the H strand (O_H). Transcription from LSP provides primers for initiation of mtDNA replication at O_H. However, this DNA replication is often abortive and stops at the end of the D loop, thus producing a triple-stranded D-loop structure containing a nascent H strand. Transcription of the circular mtDNA molecule produces 13 mRNAs, which all encode oxidative phosphorylation components, and 2 rRNAs and 22 tRNAs needed for mitochondrial translation of these mRNAs (21).

All of the proteins involved in replication and transcription of mtDNA and mitochondrial

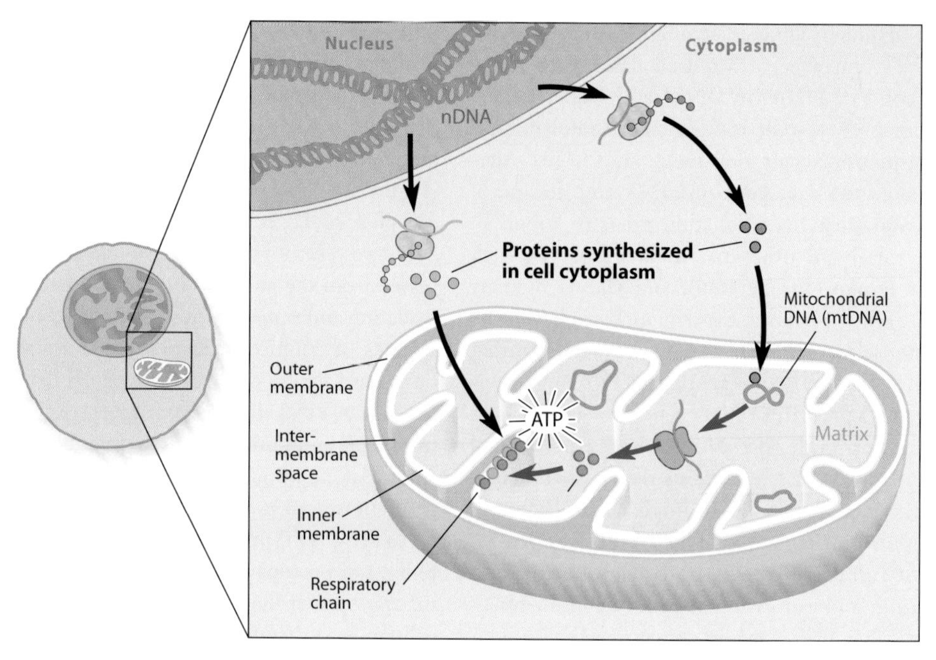

Figure 1

Biogenesis of the oxidative phosphorylation system. The respiratory chain is situated in the inner mitochondrial membrane, and its protein components are encoded by either mtDNA or nuclear DNA. Transcription of mtDNA produces 13 mRNAs, which all encode oxidative phosphorylation system subunits as well as all RNA components (22 tRNAs and 2 rRNAs) needed for translation of these mRNAs on the mitochondrial ribosomes. The mRNAs produced by transcription of nuclear genes are exported to the cytoplasm and translated on cytoplasmic ribosomes, which are distinct from the mitochondrial ribosomes. The nucleus-encoded mitochondrial proteins are imported into mitochondria and constitute the vast majority of the ~1000 proteins present in mammalian mitochondria. Nucleus-encoded mitochondrial proteins include the majority of the oxidative phosphorylation system subunits, all proteins needed for expression and maintenance of mtDNA, and all proteins of the mitochondrial ribosomes. The mtDNA-encoded respiratory chain and ATP synthase subunits are assembled together with the nuclear subunits to form a functional oxidative phosphorylation system. Illustration by Annika Röhl.

translation are encoded by nuclear genes and imported into mitochondria (21). The mitochondrial machineries for transcription and DNA replication are not similar to the nuclear ones (21); instead, several of the components are related to proteins replicating and transcribing DNA in bacteriophages (26). Mitochondria also contain their own ribosomes, which are distinct from the cytoplasmic ones, as well as a variety of translation factors (22). The mitochondrial ribosomes have low RNA content and some differences in structure in comparison with cytoplasmic and bacterial ribosomes (22, 27). The mitochondrial transcription and translation machineries are not compartmentalized as both processes occur within the confines of the mitochondrial matrix. This prokaryote-like organization makes it likely that there is a coupling between both processes, similar to the situation in bacteria (28).

POLRMT: mitochondrial RNA polymerase

Mitochondrial Transcription

Transcription of mtDNA is necessary for mtDNA gene expression and also produces the RNA primers necessary for initiation of mtDNA replication at O_H (21). The basal mitochondrial transcription machinery consists of the mitochondrial RNA polymerase (POLRMT), mitochondrial transcription

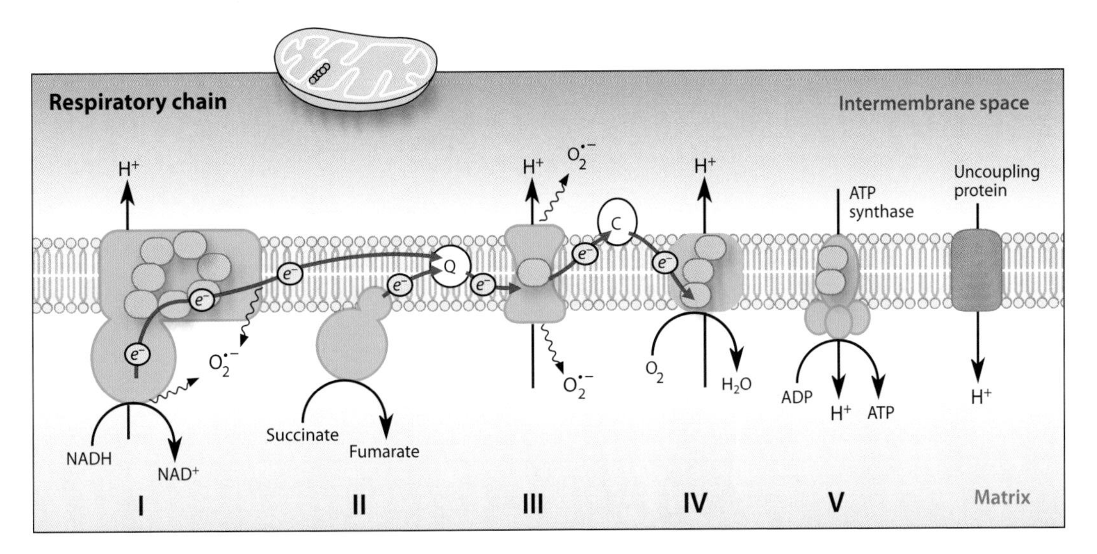

Figure 2

The mammalian oxidative phosphorylation system consists of five enzyme complexes. Complexes I–IV constitute the respiratory chain, whereas complex V is the ATP synthase. Complex I (NADH dehydrogenase) and complex II (succinate dehydrogenase) receive electrons (e^-) from intermediary metabolism and translocate these electrons to coenzyme Q, which, in turn, delivers the electrons to complex III (cytochrome *c* reductase). The electron shuttling protein cytochrome *c* then transfers the electrons to complex IV (cytochrome *c* oxidase), which constitutes the final step in the electron transport chain that reduces oxygen (O_2) to water (H_2O). The electron transport is coupled to proton (H^+) pumping across the inner mitochondrial membrane by complexes I, III, and IV. The resulting proton gradient drives ATP synthesis, and there is thus a coupling between electron transport and ATP synthesis. Protons can be translocated from the intermembrane space to the mitochondrial matrix through the activated uncoupling protein or by other uncoupling mechanisms, which leads to a dissociation (uncoupling) of electron transport and ATP synthesis. The mtDNA encodes critical subunits of complexes I, III, IV, and V (*orange dots*). Electrons may also exit the respiratory chain at the level of complex I or III to form the reactive oxygen species (ROS) superoxide, $O_2^{\cdot -}$. Illustration by Annika Röhl.

factor B2 (TFB2M), and mitochondrial transcription factor A (TFAM) (29, 30). These three factors are sufficient and necessary for promoter-specific initiation of mtDNA transcription in a pure recombinant in vitro system. POLRMT and TFB2M interact in vitro and form a heterodimer (29). TFB1M is a paralog of TFB2M, but has no role in mtDNA transcription (29, 30). Instead, TFB1M functions as a 12S rRNA methyltransferase essential for the integrity of the small subunit of the mammalian mitochondrial ribosome (30). The TFAM protein is absolutely essential for transcription initiation, and there is not even abortive transcription in its absence (21). In addition, TFAM also has a direct role in packaging mtDNA (29). The activity of the basal mitochondrial transcription machinery is likely modulated by additional factors that affect both transcription initiation and termination. The mitochondrial transcription termination factor 1 (MTERF1) binds downstream of the rRNA genes and has been proposed to have roles in termination of both H and L strand transcription (31). There is a family of four MTERF proteins (MTERF1–4) in mammals (32). The MTERF3 protein has been shown to be a potent repressor of mitochondrial transcription (33). The MTERF2 protein has also been implicated in transcription regulation (34), whereas the function of MTERF4 remains to be established. Studies are ongoing in our laboratory and elsewhere to identify additional factors important in coordinating mitochondrial transcription.

Replication and Repair of mtDNA

The mammalian mtDNA polymerase γ (Pol γ) is heterotrimeric and consists of a catalytic subunit (Pol γA) and a dimeric accessory subunit

TFAM: mitochondrial transcription factor A

Pol γ: mitochondrial DNA polymerase γ

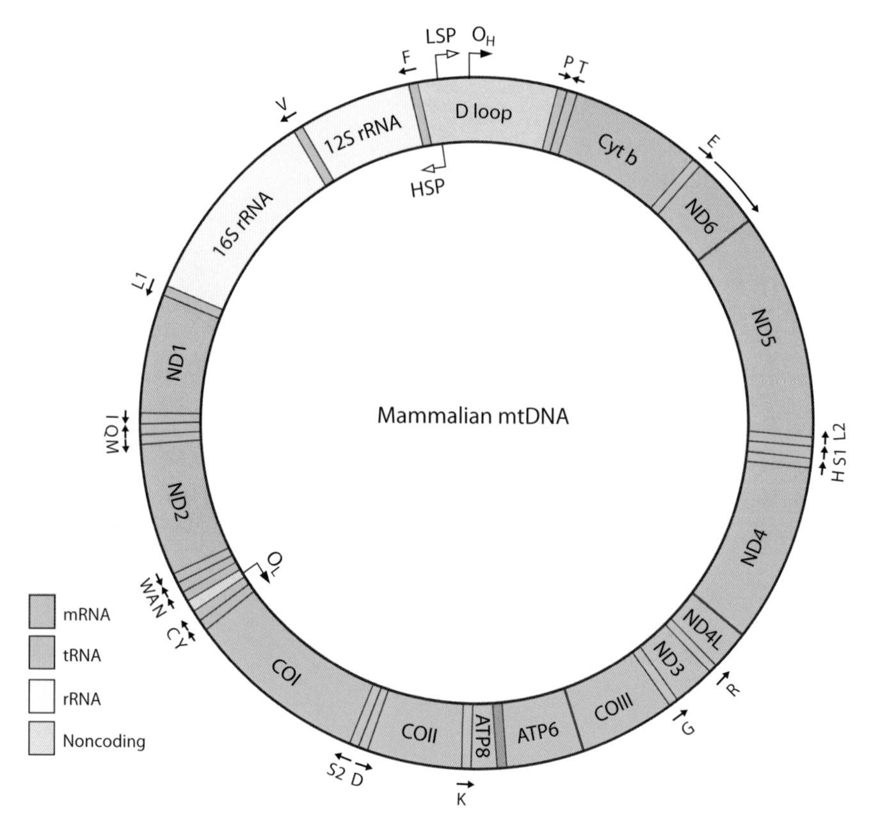

Figure 3

Mammalian mtDNA is a double-stranded circular molecule. The two strands are denoted the heavy (H) and light (L) strand due to different buoyant densities. The only longer noncoding region, the displacement loop (D loop), contains a triplex structure with a nascent H strand. The D loop contains the promoters for transcription of the H and L strand (HSP and LSP) as well as the origin of replication of the leading strand of mtDNA (O_H). The origin of replication of the lagging strand (O_L) is located in a cluster of tRNA genes. Transcription from HSP produces 2 rRNAs (12S and 16S rRNA), 12 mRNAs (ND1–5, ND4L, Cyt b, COI–III, ATP6, and ATP8), and 14 tRNAs (F, V, L1, I, M, W, D, K, G, R, H, S1, L2, T). Transcription from LSP has a dual function. First, it produces RNA primers needed for initiation of replication at O_H. Second, it is needed to produce one mRNA (ND6) and eight tRNAs (P, E, S2, Y, C, N, A, Q). Illustration by Annika Röhl.

(Pol γB) (35–38). The twinkle helicase forms a hexamer that unwinds mtDNA in the 5′–3′ direction (39). The Pol γA subunit harbors a 3′–5′ exonucleolytic proofreading activity in three domains containing conserved aspartate residues. Knockin mice with severe reduction of the Pol γ proofreading activity develop substantial levels of mtDNA mutations and a premature aging syndrome, thus demonstrating the in vivo importance of active proofreading during mtDNA replication (18). A minimal replisome consisting of Pol γA, Pol γB, twinkle, and the mitochondrial single-stranded binding protein (mtSSB) has been reconstituted in vitro and is capable of synthesizing leading-strand DNA products with a length of ~16 kb at a velocity of ~180 bp per min (35). Recent data show that POLRMT can function as a primase to provide primers for initiation of lagging-strand DNA synthesis (40). In addition, POLRMT specifically interacts with O_L to synthesize RNA primers (41), consistent with previous observations that O_L is a preferred site for initiation of lagging-strand DNA synthesis (42, 43).

The first model described for mammalian mtDNA replication is often referred to as the asymmetric replication model. It is based on extensive studies of replication intermediates by electron microscopy, biochemical characterization of nucleic acids, and pulse-chase labeling experiments (42, 44). According to this model, initiation of leading-strand replication occurs at O_H and is dependent on an RNA primer formed by transcription from LSP. Initiation of lagging-strand replication occurs when the leading-strand replication has reached two-thirds around the mtDNA circle and O_L is activated. The strand-asymmetric replication model predicts that the lagging strand also undergoes continuous DNA synthesis without the need for repeated priming and thus without formation of Okazaki fragments. More recently, a second model has emerged on the basis of results from neutral two-dimensional agarose gel electrophoresis of replication intermediates. This latter model argues that mtDNA replication occurs by coupled leading- and lagging-strand synthesis (45). In addition, transient ribonucleotide incorporation on the lagging strand has been suggested to play an important role (46). It should be noted that studies of human patients with clonal expansion of single large deletions of mtDNA have shown that LSP, O_H, and O_L are always retained in the deleted molecules (HSP is dispensable) and therefore must have important functions in replication (47, 48). The D-loop structure with its nascent H strand represents an abortive DNA replication event, and its presence is consistent with initiation of leading-strand mtDNA replication at O_H, as predicted by the asymmetric replication model. The role for the D loop in the coupled leading- and lagging-strand replication model needs to be defined. The mode of mtDNA replication has been subject to an intense debate (49–51), and it is beyond the scope of this review to elaborate further on this topic. It is, however, important to point out that both models predict the involvement of a limited number of enzymes, and further work with in vitro biochemical systems and mouse genetics should therefore help clarify the situation.

It was for a long time assumed that mammalian mitochondria were essentially devoid of DNA repair machineries. However, this is not true, and the existence of short-patch base excision repair systems removing damaged bases in mammalian mtDNA is now well documented (52, 53). Recently, the existence of mitochondrial long-patch base excision repair was reported (54). However, several other systems, such as nucleotide excision repair, mismatch repair, homologous recombination, and nonhomologous end joining, are either absent or remain to be molecularly defined in mammalian mitochondria (52, 53).

Packaging of mtDNA into Nucleoids

The mtDNA in mammalian cell is not naked; instead it is packaged in DNA-protein aggregates referred to as nucleoids (55). TFAM is not only essential for transcription, as discussed above, but also has a direct role in mtDNA maintenance. The TFAM protein and its budding yeast homolog ABF2 belong to the high-mobility-group domain protein family and have the capacity to bend, unwind, and package DNA (56, 57). ABF2 is dispensable for transcription but is essential for mtDNA maintenance (58), whereas the *Xenopus laevis* homolog (xl-mtTFA) activates transcription (59) and also packages mtDNA (60). Similar to the situation in yeast and frogs, mammalian mtDNA is likely coated with TFAM as there is one TFAM molecule per 10–20 bp of mtDNA in mouse tissues (61, 62). Disruption of the TFAM gene is embryonically lethal and leads to complete loss of mtDNA expression and mtDNA (63). This result does not distinguish between the roles for TFAM in providing primers for initiation of replication and its direct role in packaging mtDNA. However, additional genetic experiments do support the idea that TFAM is a dual-function protein with distinct roles in gene expression and mtDNA copy number regulation. Bacterial artificial clone (BAC) transgenic mice expressing a TFAM variant, with poor capacity to activate transcription, develop an increased mtDNA copy number in

Nucleoids: DNA-protein aggregates that organize and package mtDNA

Heteroplasmy: a mixture of normal and mutated mtDNA in a cell, tissue, or individual

proportion to the amount of expressed TFAM protein (61), but this increase of mtDNA copy number occurs without increased oxidative phosphorylation capacity or mitochondrial mass (61). Thus, the biophysical properties of the TFAM protein (56, 57) and the results from genetic studies, wherein TFAM expression has been manipulated in the mouse (61, 63), make it likely that TFAM has a major role in packaging and compaction of mammalian mtDNA. The presence of TFAM in nucleoids has been confirmed by immunofluorescence microscopy and biochemical cross-linking approaches (55). A large number of other proteins have also been suggested to be constituents of nucleoids (64, 65), but functional roles for most of these other putative components need to be established in mammalian tissues. Proteins that can be chemically cross-linked to mtDNA have traditionally been considered as part of the nucleoid. However, this definition is not necessarily functionally meaningful as many such proteins may have no role in packaging and organizing mtDNA. Future studies must therefore rely on a more stringent definition of the nucleoid and also take into account the biochemical and biophysical properties of isolated proteins.

SOMATIC mtDNA MUTATIONS IN AGING

In 1988, two different groups reported that high levels of deletions (66) and point mutations (67) of mtDNA cause human mitochondrial disease. The same year, a much less recognized paper by Piko and coworkers (15) reported electron microscopy studies of denatured-reannealed duplexes of mtDNA from livers of young and senescent mice. The results revealed an increased, but still rather low, frequency of deletions of mtDNA in senescent mice (15). When the polymerase chain reaction (PCR) method was invented in the late 1980s, it was rapidly applied to studies of mtDNA, and it was soon reported that low levels of mtDNA deletions could be detected in heart and brain of aging humans (68). Additional reports showed that the levels of deleted mtDNA varied in different regions of the adult human brain, with the highest levels in regions such as the putamen, cerebral cortex, and substantia nigra (69, 70). Meanwhile, the understanding of the role of pathogenic mtDNA mutations in human mitochondrial disease had become a bit clearer. It was recognized that mutated mtDNA often coexists with normal mtDNA, a condition referred to as heteroplasmy, and that the levels of mutated mtDNA could vary dramatically between tissues (71, 72). It was also discovered that pathogenic mtDNA mutations only cause respiratory chain dysfunction if they are present above a certain threshold level, which is >60% for single large mtDNA deletions (73) and >90% for certain point mutations in tRNA genes (74). Single large mtDNA deletions, which always remove one or several tRNA genes, as well as point mutations of tRNA genes lead to impaired mitochondrial translation and respiratory chain deficiency (73, 75). Skeletal muscle fiber segments containing an accumulation of abnormal mitochondria, denoted ragged-red fibers (RRFs), are considered a morphological hallmark of mitochondrial disease. The formation of RRFs was explained by the finding that they contain very high proportions of mutated mtDNA in comparison with adjacent normal-appearing fibers (75). It is assumed that the respiratory chain deficiency in RRFs activates mitochondrial biogenesis as a compensatory but futile response, which explains the massive amount of abnormal mitochondria in RRFs. Consistent with this assumption, biochemical measurement of respiratory chain function in mice with mitochondrial myopathy has demonstrated that the increased amount of mitochondria at least partially compensates for a decrease of oxidative phosphorylation capacity in skeletal muscle (76, 77).

Mitotic Segregation of Mutated mtDNA

The replication of mtDNA is not linked to the cell cycle (44), and a particular mtDNA molecule may be replicated many times or not at all as a cell divides (**Figure 4**). The daughter

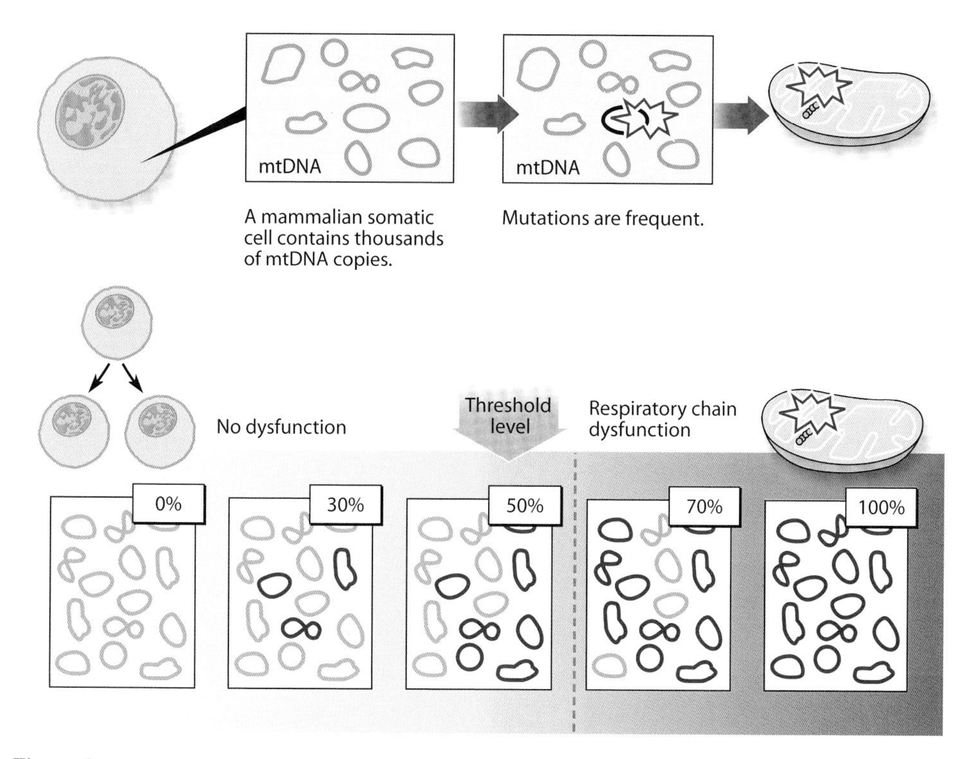

Figure 4

Clonal expansion of mutated mtDNA molecules. Mutations of mtDNA are frequent, but a single mutated mtDNA molecule in a cell likely has no effect owing to the very high mtDNA copy number. The replication of mtDNA is not correlated with the cell cycle, and there is no counting mechanism to ensure that all mtDNA molecules in a cell are replicated once during the cell cycle. In addition, mtDNA replication also occurs in postmitotic cells. The poorly understood process of mitotic segregation of mtDNA can lead to loss of a somatic mtDNA mutation in some cells and clonal expansion in other cells. A heteroplasmic mtDNA mutation only causes respiratory chain dysfunction if present above a certain minimal threshold level. Illustration by Annika Röhl.

cells therefore obtain very different levels of mutated mtDNA after repeated cell divisions (**Figure 4**). The turnover rates for mtDNA in differentiated tissues are not well defined, but it is important to recognize that mtDNA is also replicated in postmitotic cells and thus can undergo similar types of segregation in the absence of cell division. Mathematical modeling has suggested that random intracellular drift can fully account for the occurrence of clonally expanded somatic mtDNA mutations (78). According to this model, the expansions are predicted to predominantly result from mutations acquired in childhood or early adult life (78). However, there are also reports that certain apparently neutral mtDNA sequence variants can be selected in a tissue-specific manner depending on the nuclear background in the mouse (79). As discussed above, the mutated mtDNA has to be present at a certain minimal threshold level to cause respiratory chain deficiency (**Figure 4**). The mechanisms for mitotic segregation of mtDNA mutations are incompletely understood and need to be studied further.

Mosaic Respiratory Chain Deficiency in Aging

Studies in human patients with mitochondrial disease were thus able to establish clear cause-and-effect relationships between mtDNA mutations and respiratory chain dysfunction, and

Mosaic respiratory chain deficiency: a mixture of normal and respiratory chain–deficient cells in a tissue

it was also discovered that levels of mutated mtDNA often vary substantially between different cells of the same tissue in affected patients. This progress in understanding the pathophysiology of mitochondrial disease led to a widespread belief that the low levels of mutated mtDNA found in aging humans would have no impact on oxidative phosphorylation. However, other types of correlative data, such as reports of a moderate decline of respiratory chain function with age (80) and a mosaic occurrence of respiratory chain-deficient cells in aging tissues (81, 82), started to appear in support of the idea that mitochondrial dysfunction nevertheless is important in human aging. Further studies also showed that skeletal muscle fiber segments with focal respiratory chain deficiency in aged humans contain clonal accumulations of deleted mtDNA (83). The realization that age-associated somatic mtDNA mutations tend to undergo clonal expansion and thereby cause focal respiratory chain deficiency in skeletal muscle stimulated many additional studies. Mosaic respiratory chain deficiency has now been reported in many different types of aging tissues in humans (**Figure 5**), e.g., heart (81), skeletal muscle (83), hippocampal neurons (84), choroid plexus (84), midbrain dopaminergic neurons (85), and colon (86). In most published cases, the clonally expanded mutations are single large mtDNA deletions. The deletions differ in various respiratory chain-deficient cells of

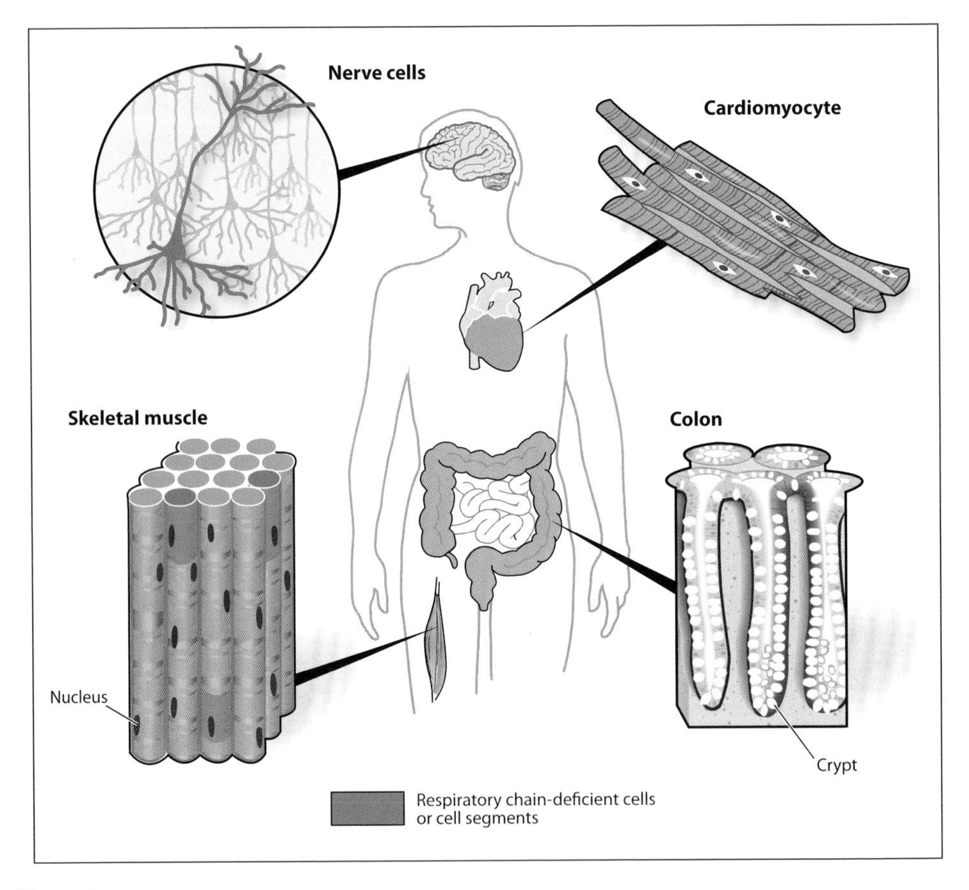

Figure 5

Mosaic respiratory chain deficiency in human aging. Clonal expansion of somatic mtDNA mutations leads to the formation of mosaic respiratory chain deficiency in aging humans. Examples of affected tissues include brain, heart, colon, and skeletal muscle. Illustration by Annika Röhl.

the same tissue, which is in agreement with the somatic nature of the mutations (**Figure 4**). By contrast, colonic crypts contain various clonally expanded point mutations that likely originate in stem cells at the base of the crypts (86). This type of focal respiratory chain deficiency is found in >15% of all colonic crypts of humans at ages >80 years (86). The frequency of respiratory chain-deficient dopamine neurons is ~3% in Parkinson's disease patients and ~1% in age-matched controls (85). Point mutations are rare in respiratory chain-deficient dopamine neurons (87), whereas mtDNA deletions are common (88). The reasons why some aging cell types preferentially accumulate point mutations whereas others accumulate deletions are not understood but could well be related to the mitotic activity of the tissue. Studies of rats, rhesus monkeys, and humans have shown that accumulation of deleted mtDNA colocalizes not only with respiratory chain deficiency (16, 89) but also with areas of fiber atrophy and splitting, thus suggesting a role in age-associated sarcopenia (90). Around 30% and 60% of all skeletal muscle fibers in old humans and rhesus monkeys, respectively, contain at least one respiratory chain-deficient segment (89, 90). It can thus be concluded that focal respiratory chain deficiency is a ubiquitous phenomenon in aged humans (5). In addition, the frequency of respiratory chain-deficient cells is substantial and could well contribute to age-associated organ dysfunction (5).

There are many reports describing somatic mtDNA deletions and point mutations in the aging human skin (91, 92), but it is not yet clear if these mutations are clonally expanded in certain cell types. Interestingly, the levels of somatic mtDNA mutations in skin are increased in regions exposed to UV radiation (91–93). A specific point mutation (T414G) in the control region of mtDNA in fibroblasts was reported to correlate strongly with aging (94). Later studies of old individuals have shown that the T414G mutation was absent in brain (95) but present in muscle (95, 96) and skin (96), particularly if sun exposed (93, 97). The role, if any, for this and other mutations in the control region is unclear (96). Possibilities include interference with regulatory processes such as mtDNA transcription and/or replication.

MOUSE MODELS

The studies of the connection between mtDNA mutations and human aging, as described above, are correlative, and causality is unclear. Validation of cause-and-effect relationships therefore necessitates further studies by using experimental genetics in model organisms. Such studies have, for example, shown that somatic mtDNA mutations can directly induce a premature aging syndrome in the mouse (18), thus supporting a role for somatic mtDNA mutations in mammalian aging. Mouse genetics is a wonderful tool that is increasingly used to study molecular mechanisms in mitochondrial biology and aging. However, interpreting the available literature in this area requires some basic knowledge of potential pitfalls associated with the use of the mouse as an experimental organism.

Genetic Background

The genetic background is important as phenotypes can vary quite dramatically depending on the mouse strain used (98). The issue of genetic background is less of a problem if the phenotype differences are drastic but becomes a big problem if the phenotypes are subtle. Effects on body weight are especially difficult to assess in mixed genetic backgrounds owing to the many genes involved (99). Backcrossing is time consuming but can be facilitated by high-speed congenic strategies (100). It should also be pointed out that backcrossed strains, such as C57/BL6, often vary considerably between different commercial sources.

Pronuclear Injection Strategies

It is important to critically scrutinize the transgenic strategy used. This is of course a complicated area, and I only give some information here. The most problematic approach

Cre-loxP system: a system for conditional deletion of DNA sequences; Cre recombinase can specifically excise DNA flanked by loxP sequences

is pronuclear injection of small DNA constructs, typically consisting of a short promoter sequence and a cDNA. It is important to recognize that such a strategy creates a forced expression that largely bypasses physiological control mechanisms regulating gene expression. In addition, the random integration of the construct invariably leads to positional effects that affect the expression of the transgene and in some cases the integration also disrupts an endogenous gene. Forced expression from strong promoters can indeed create toxic effects unrelated to the expected protein function, as exemplified by the finding that overexpression of green fluorescent protein can cause dilated cardiomyopathy (101). Forced expression of the mtDNA maintenance factor ABF2 in the yeast leads to loss of mtDNA (102), likely because other important maintenance proteins will not have access to the mtDNA. Another example is that overexpression of wild-type Pol γA causes mtDNA depletion in flies, perhaps owing to an imbalance between the amounts of the catalytic and accessory subunits (36). The knockin strategy to target the Pol γA gene in the mouse, as discussed below, has the advantage that the gene maintains a normally regulated expression from its endogenous promoter, thus avoiding overexpression artifacts. Another problematic area is to manipulate the expression of mitochondrial inner membrane proteins, e.g., inner membrane transporters. It is known that abundantly expressed forms of inner membrane transporters significantly contribute to the basal proton leak (103), and it is therefore difficult to use forced expression approaches to establish putative roles for proposed uncoupling proteins (104). Some of the problems listed above can be circumvented by the use of BAC transgenic strategies. Pronuclear injection of BAC clones has the advantage that the positional effects are less problematic, and in addition, the introduced gene is regulated by its own promoter (105). Only a few copies of the BAC are integrated, and a gene dosage effect with moderate overexpression is the net result (61, 106). The BAC clones also have the advantage that they can be engineered in bacteria prior to pronuclear injection (107). The drawbacks are that the BAC may contain other unwanted genes and that the insertion of the BAC into the mouse genome may disrupt endogenous genes.

Constitutive and Conditional Knockout Strategies

The classical mouse knockout approach, whereby a gene is disrupted by the insertion of an antibiotic resistance marker that is not subsequently removed, has three main problems. First, there may be effects on neighboring genes that create phenotypes unrelated to the manipulated gene, e.g., loss of the complex IV assembly factor SURF1 was initially reported to be predominantly embryonically lethal, but this effect was later attributed to effects on neighboring genes from an introduced DNA sequence (12, 108). Second, the use of embryonic stem cells from the 129 mouse strain creates a knockout allele flanked by 129 genes, even after backcrossing (98). This problem can be circumvented by the use of C57BL6 embryonic stem cells. Third, the introduced DNA sequences may introduce novel splice sites, making it difficult to predict the gene expression pattern from the targeted allele (109). The conditional knockout strategy has the advantages that the antibiotic resistance gene can be removed and that the control and knockout animals have the same genes flanking the targeted gene. The antibiotic resistance gene is typically flanked by frt sites and is removed by flp-mediated recombination in vivo (18, 30, 33). Failure to remove the antibiotic resistance gene may result in aberrant splicing and a hypomorphic allele (109). Drawbacks with the conditonal knockout system are that the breeding is tedious and complicated and that the Cre-loxP system sometimes only gives a partial (mosaic) knockout (110).

The mtDNA Mutator Mouse

The mtDNA mutator mouse was developed to critically test whether somatic mtDNA mutations can elicit aging phenotypes and has a homozygous knockin mutation that changes a highly conserved amino acid (D257A) necessary for the proofreading activity of Pol γ (18). These mice appear normal until the age of ~25 weeks, and thereafter, they develop a progressive premature aging syndrome with weight loss, reduced subcutaneous fat, alopecia, kyphosis, osteporosis, anemia, reduced fertility, heart enlargement, graying of the hair, and hearing loss (18). Sequence analysis of mtDNA with the PCR and cloning method (see below) demonstrated an elevated amount of random point mutations with 15–25 mutations per mtDNA molecule (18). The point mutations are prevalent already in midgestation and increase further in adult life (111). A similar lag phase between onset of the genetic defect and the occurrence of symptoms was previously observed in conditional knockout mouse strains that have disruptions of nuclear genes controlling mtDNA expression and is likely the result of compensatory responses (30, 33, 76, 112). A second, independently generated mtDNA mutator strain has essentially the same phenotype as the first strain and thus provides important confirmation that mtDNA mutations can directly cause a premature aging syndrome (113).

In addition to point mutations, the mtDNA mutator mice also contain multiple linear deleted mtDNA species of ~11 kb extending from O_L to O_H (18). These deleted mtDNA molecules represent ~30% of the total mtDNA in all investigated tissues at different ages (18) and likely represent a continuously formed replication intermediate (18, 114). One recent report has claimed that circular mtDNA molecules with deletions are driving the phenotype of mtDNA mutator mice (115). These circular mtDNA molecules with large deletions are proposed to represent a third form of mutated mtDNA, besides two forms (full-length mtDNA with abundant point mutations and linear deleted mtDNA) described in the original characterization of the mtDNA mutator mouse (18). However, this report (115) did not measure absolute levels of circular mtDNA molecules with deletions, and the findings have been refuted by two recent reports, which show no or extremely low amounts of such deleted molecules (116, 117). Another strong argument against a causative role for circular mtDNA molecules with deletions is that two mouse models with high levels of such mutations display no premature aging phenotypes (118, 119). A recent extended characterization of the respiratory chain phenotype in mtDNA mutator mice has given novel mechanistic insights (116). As discussed above, large deletions of mtDNA are known to cause problems with mitochondrial translation owing to the lack of certain tRNAs encompassed by the deletion (73, 75). However, mitochondrial translation is normal in mtDNA mutator mice, arguing against a role for the linear or circular deleted mtDNA molecules in creating the observed respiratory chain deficiency (116). The respiratory chain subunits are in fact synthesized at normal rates in mtDNA mutator mice, but the resulting respiratory chain complexes are not stable (116). The likely explanation is that the point mutations of mtDNA lead to synthesis of respiratory chain subunits with amino acid substitutions that impair the assembly and/or stability of the respiratory chain complexes (116). This can be exemplified by complex IV, where the mtDNA-encoded subunits are highly conserved and provide the catalytic function. The formation of complex IV starts with assembly of the mtDNA-encoded subunits to provide a platform for the subsequent assembly of the nuclear subunits (12). It is thus not difficult to envision that amino acid substutions of the mtDNA-encoded subunits of complex IV can be deleterious for the function and stability of this complex. It is also interesting to note that amino acid substitutions of respiratory chain subunits are strongly selected against in the germ line, whereas mutations in tRNAs and rRNA genes seem to be better tolerated (120, 121). The occurrence of somatic mtDNA mutations causing amino acid substitutions in normal aging should be further studied and may provide a mechanism explaining the age-associated decline of respiratory chain function.

Mice heterozygous for the mtDNA mutator allele have increased levels of somatic mtDNA point mutations and a normal life span (122). These findings have been interpreted as evidence that somatic mtDNA mutations do not cause aging (122). However, these results should be interpreted with great caution given the many inherent problems with the random mutation capture (RMC) method used, as discussed below.

Mice with Tissue-Specific Respiratory Chain Deficiency

Experimental strategies in the mouse have been used to study phenotypes caused by organ-specific mitochondrial dysfunction, and the results have implications for our understanding of the age-associated disease processes. Mice with disruption of TFAM in pancreatic β-cells develop an impaired stimulus-secretion coupling followed by β-cell death, similar to the insulin secretion phenotype found in patients with age-associated diabetes (112). Mice with disruption of TFAM in dopamine neurons of the midbrain develop a very slowly progressive movement disorder with many similarities to Parkinson's disease in humans (123). However, decline of respiratory chain function does not always induce aging phenotypes as exemplified by the finding that mice with respiratory chain deficiency in muscle have improved glucose homeostasis (9–11). These findings also provide a strong argument against the idea that age-associated decline of respiratory chain function causes insulin resistance (9–11).

Studies of chimeric mice with mosaic respiratory chain deficiency in the neocortex have shown that a low proportion of respiratory chain-deficient cortical neurons (>20%) can cause symptoms (124). Importantly, respiratory chain-deficient neurons also cause an increased death of neurons with normal respiratory chain function, thus eliciting a transneuronal degeneration effect that may also occur in normal aging (124). A similar experimental approach would also be valuable in studying the pathophysiology of mosaic respiratory chain dysfunction in other organs, such as heart, skeletal muscle, and colon.

Reactive Oxygen Species, Apoptosis, and Respiratory Chain Dysfunction

The downstream effects of respiratory chain dysfunction are complex and not limited to deficient ATP production (125). Mice with abolished mtDNA expression, caused by conditional knockout of TFAM, have increased apoptosis but very little evidence of oxidative stress (126). Similarly, the mtDNA mutator mice display a massive increase of apoptosis without much evidence of oxidative stress (111, 113, 127). The complex I–deficient NDUFS4 (128) and AIF (10) knockout mouse strains also do not have substantially increased ROS production or oxidative damage. It can thus be concluded that apoptotic cell loss is a common feature in respiratory chain deficiency, whereas increased ROS production is usually not present (5), although there are rare exceptions (129).

HOW ARE MUTATIONS OF mtDNA GENERATED?

The point mutation rate of mtDNA has been reported to be two orders of magnitude greater than the corresponding rate for nuclear DNA in humans (130). The reason for this difference is unclear and has been attributed to differences in damage mechanisms, endogenous repair capacity, and enzymatic processes in DNA replication. It has been known for a time that specific types of mutagens cause different mutation signatures and that spontaneous mutations also have predominant patterns (130, 131). Different types of mutational processes can therefore be predicted to leave specific DNA signatures, and this circumstance has been used to predict causes, although the interpretation is complicated by repair processes (130). There is a surprisingly large number of oxidative lesions to DNA, and a recent review catalogs 24 major products of oxidative damage to DNA bases and 13 major products of oxidative damage to the sugar moiety (132). Most of these products have been reported to occur in mammalian cell lines or tissues (132), but the biological significance for the majority of the lesions remains unknown. The focus has mostly been on the guanine adduct 7,8-dihydro-8-oxo-deoxyguanosine (8-oxodG) (132). The in vivo levels of 8-oxodG were initially thought to be very high (133), but this result has been questioned and was attributed to methodological problems that overestimated levels by at least two orders of magnitude (134). The 8-oxodG lesion has the propensity to cause G:C to T:A transversion mutations (135). The glycosylase OGG1 is involved in repair of 8-oxodG lesions in mitochondria, and mice lacking this enzyme have increased levels of 8-oxodG in mtDNA (136) but, surprisingly, appear healthy and have normal respiratory chain function (137). An analysis of the mutational spectra generated by wild-type Pol γ in vitro showed a good concordance with the observed human in vivo spectra, including a paucity of G:C to T:A transversions. This result argues that most of the human mtDNA mutations are generated by replication errors and not by damage (138). Consistent with this, biochemical experiments have shown that Pol γ synthesis past 8-oxodG in the template is stalled in the vast majority of cases (139). In addition, most of the mutations generated in mtDNA mutator mice are transitions (18), and the mutation pattern after germ line transmission of these mutations is reminiscent of the mutation spectra observed in natural populations of mice and humans (120, 121). There are thus several lines of evidence to support the hypothesis that most mtDNA mutations are produced by replication errors and that oxidative damage in the form of 8-oxodG is not a major contributor. A mouse expressing a mutated version of Pol γA with increased proofreading activity should provide a critical test of the replication error hypothesis.

Slipped mispairing during mtDNA replication has been suggested to be an important mechanism for formation of mtDNA deletions (140). Most human mtDNA deletions are indeed flanked by repeated sequences, consistent with such a mechanism (48). Experimental studies in the mouse have demonstrated that mitochondrially targeted restriction enzymes can induce mtDNA deletion formation, although the mechanism is not understood (141). Other scenarios have been proposed (142), but slipped mispairing during mtDNA replication still remains as the simplest explanation for the formation of deletions.

WHEN ARE mtDNA MUTATIONS GENERATED?

It is often assumed that the age-associated somatic mtDNA mutations are generated due to damage and therefore start to accumulate in aging adults. This idea has been supported by a wealth of correlative data showing higher levels of somatic mtDNA mutations in older than in younger humans and other mammals. However, an alternative possibility is that somatic mtDNA mutations are generated by replication errors and that many of the somatic mtDNA mutations in adults can be traced back to embryonic development or early postnatal life. The mtDNA mutator mouse gives some support for this suggestion as it contains abundant mtDNA mutations in midgestation but develops symptoms and focal respiratory chain deficiency in adult life (18). This hypothesis is also supported by mathematical modeling (78) but needs further experimental testing. Studies in this area are complicated by the fact that the true turnover rate of mtDNA in mammalian tissues is largely unknown.

HOW CAN LEVELS OF SOMATIC mtDNA MUTATIONS BE MEASURED?

Measurement of somatic mtDNA mutation levels is technically complicated, and available approaches are very much debated. There is yet no standard method to sequence individual nonamplified mtDNA molecules, and all available protocols therefore involve PCR and/or cloning procedures. There are no available standards with precisely defined amounts of low levels of mtDNA mutations. This precludes definition of detection limits (sensitivity) and performance of the methods across a range of defined mtDNA mutation levels (accuracy). Instead of using standards, intellectual arguments are often put forward to argue that for theoretical reasons one method is superior to another one. However, theories cannot be used to validate methods, and future progress in this area will instead depend on the development of standards shared between laboratories and analyzed in a blind fashion. There is currently very rapid development of different high-throughput sequencing techniques, but these technologies also have inherent problems (e.g., intrinsic error rates). Therefore, the development of new and elegant technology cannot on its own solve the problem of accurate quantification of levels of somatic mtDNA mutations, but analysis results must always be validated by comparison with defined standards (143). In the absence of defined standards, different laboratories should share samples to validate the reproducibility of various methods.

Quantifying Levels of mtDNA Deletions

There are many mitochondrial disease patients with single large or multiple mtDNA deletions. These mutations have been extensively characterized by various methods and can be detected by Southern blot analysis if present at a fraction of more that 5% of total mtDNA. Southern blot analysis is thus a robust method for detecting mtDNA deletions, but it is often too insensitive to detect the low levels occurring in aging. Instead, quantitative PCR approaches have been used. There are many inherent problems with the long-extension PCR technique (144) and qPCR approaches (145). Also, in this case, standards are useful, e.g., generated by diluting mtDNA with known proportions of deleted mtDNA with wild-type samples (116, 146).

Quantifying Levels of mtDNA Point Mutations

Standard sequencing protocols, such as the dideoxy chain terminator method (developed by Sanger), applied to isolated (not cloned or PCR-amplified) mtDNA are not applicable as mutated mtDNA present below 20% of total mtDNA will escape detection (143). To circumvent this problem, the PCR and cloning method has been developed. DNA preparations from whole cells or mitochondria are subjected to PCR amplification, and the resulting mtDNA fragments are cloned and sequenced. This method has been criticized as damage to the templates or the PCR reaction itself may introduce mutations (147). However, the method seems to give reproducible results because two different laboratories have reported rather similar mutation levels in two different strains of mice hetero- and homozygous for the mtDNA mutator allele (18, 113). Control experiments whereby cloned mtDNA fragments have been analyzed with the PCR and cloning method (clone of a clone) showed no mutations (147) or one mutation (148) in $\sim 4 \times 10^4$ bases of sequence, indicating that this method is rather robust. However, this method, like all of the other methods, still needs to be validated by analysis of defined standards.

The single-molecule PCR (sm-PCR) technique is based on serial dilution of undigested mtDNA samples to levels where amplification is believed to occur from a single template (149). The sm-PCR technique typically does not involve restriction or topoisomerase enzyme treatments of templates, which complicate the interpretation of results as a substantial portion of mtDNA is present as catenated (interlocked) circles (150) or other types of aggregates (151) in mammalian mtDNA preparations.

The RMC method is based on gene capture and has been reported to detect random mtDNA mutations at a frequency of $1{:}10^9$ bp (122, 152). Despite its high sensitivity, a drawback of the method is that it only surveys a target sequence of 4 bp (a Taq I restriction enzyme site), and it may therefore not predict mutation events elsewhere in the genome (152). In addition, the RMC assay requires knowledge of the exact mtDNA copy number and is therefore dependent on the accuracy of DNA quantification methods (147). A recent study showed that the RMC method detects on average 100-fold less mutations than the PCR and cloning method (147). In addition, experiments with colon crypts with known levels of clonally expanded mtDNA mutations showed that the RMC method severely underestimated the mutant fraction (147).

Another problem that must be taken into account when detecting mtDNA mutations is that the nuclear genome contains abundant mtDNA pseudogenes that may cause false-positive results (153). Bioinformatics approaches may be of value to exclude nuclear pseudogenes, but the situation is sometimes complicated as long stretches of mtDNA may be integrated in the nucleus.

SUMMARY POINTS

1. A substantial amount of observations and experimental data support a role for somatic mtDNA mutations in mammalian aging.
2. Somatic mtDNA mutations expand clonally to cause mosaic respiratory chain deficiency in different aging organs.
3. Experimental genetic studies in the mouse have demonstrated that respiratory chain dysfunction leads to increased apoptosis, whereas no clear role for ROS has been established.

FUTURE ISSUES

1. The performance of methods used to measure levels of somatic mtDNA mutations is much discussed. There is a clear need to develop accurate standards to evaluate these methods concerning sensitivity, accuracy, and other parameters.
2. The hypothesis that somatic mtDNA mutations are generated by oxidative damage is not conclusively proven. Instead, several lines of evidence suggest that the inherent error rate of Pol γ may be responsible for creating the majority of all somatic mtDNA mutations. The hypothesis that replication errors create most of the somatic mtDNA mutations should be experimentally tested by developing mice expressing Pol γ with increased proofreading capacity (antimutator Pol γ).
3. The role for mosaic respiratory chain dysfunction in aging is poorly understood, and it is unclear how this phenomenon impacts organ function. Additional experimental studies in the mouse (e.g., mouse chimeras, tissue-specific mutator alleles, and conditional kockouts) should help reveal to what extent focal respiratory chain deficiency can cause organ dysfunction and age-associated degenerative disease.
4. Respiratory chain dysfunction impairs organ function, but the role for ROS is highly circumstantial. Most previous experiments have focused on manipulating ROS-scavenging mechanisms. However, the future development of mutant mice with increased ROS production at defined sites of the respiratory chain or elsewhere would be a valuable tool to study the possible role of ROS in aging.

DISCLOSURE STATEMENT

Nils-Göran Larsson holds stock in Kampavata AB.

ACKNOWLEDGMENTS

I dedicate this article to the loving memory of my parents Nils and Märta Larsson, whose support throughout the years has been a great encouragement for me. I thank Dan Bogenhagen and Jim Stewart for critically reading the manuscript. I apologize to all fellow scientists whose work has not been cited because of space constraints.

LITERATURE CITED

1. Austad SN. 1999. *Why We Age: What Science Is Discovering about the Body's Journey Through Life*. New York/Chichester: Wiley
2. Kirkwood TB. 2005. Understanding the odd science of aging. *Cell* 120:437–47
3. Weindruch R, Walford RL. 1988. *The Retardation of Aging and Disease by Dietary Restriction*. Springfield IL: Thomas
4. Harman D. 1956. Aging: a theory based on free radical and radiation chemistry. *J. Gerontol.* 11:298–300
5. Trifunovic A, Larsson NG. 2008. Mitochondrial dysfunction as a cause of ageing. *J. Int. Med.* 263:167–78
6. Klass MR. 1983. A method for the isolation of longevity mutants in the nematode *Caenorhabditis elegans* and initial results. *Mech. Ageing Dev.* 22:279–86
7. Piper MD, Selman C, McElwee JJ, Partridge L. 2008. Separating cause from effect: How does insulin/IGF signalling control lifespan in worms, flies and mice? *J. Intern. Med.* 263:179–91
8. Dillin A, Hsu AL, Arantes-Oliveira N, Lehrer-Graiwer J, Hsin H, et al. 2002. Rates of behavior and aging specified by mitochondrial function during development. *Science* 298:2398–401
9. Wredenberg A, Freyer C, Sandstrom ME, Katz A, Wibom R, et al. 2006. Respiratory chain dysfunction in skeletal muscle does not cause insulin resistance. *Biochem. Biophys. Res. Commun.* 350:202–7
10. Pospisilik JA, Knauf C, Joza N, Benit P, Orthofer M, et al. 2007. Targeted deletion of AIF decreases mitochondrial oxidative phosphorylation and protects from obesity and diabetes. *Cell* 131:476–91
11. Freyer C, Larsson NG. 2007. Is energy deficiency good in moderation? *Cell* 131:448–50
12. Dell'Agnello C, Leo S, Agostino A, Szabadkai G, Tiveron C, et al. 2007. Increased longevity and refractoriness to Ca^{2+}-dependent neurodegeneration in Surf1 knockout mice. *Hum. Mol. Genet.* 16:431–44
13. Terzioglu M, Larsson NG. 2007. Mitochondrial dysfunction in mammalian ageing. *Novartis Found. Symp.* 287:197–208; discuss. 208–13
14. Krishnan KJ, Greaves LC, Reeve AK, Turnbull D. 2007. The ageing mitochondrial genome. *Nucleic Acids Res.* 35:7399–405
15. Piko L, Hougham AJ, Bulpitt KJ. 1988. Studies of sequence heterogeneity of mitochondrial DNA from rat and mouse tissues: evidence for an increased frequency of deletions/additions with aging. *Mech. Ageing Dev.* 43:279–93
16. Wanagat J, Cao Z, Pathare P, Aiken JM. 2001. Mitochondrial DNA deletion mutations colocalize with segmental electron transport system abnormalities, muscle fiber atrophy, fiber splitting, and oxidative damage in sarcopenia. *FASEB J.* 15:322–32
17. Schwarze SR, Lee CM, Chung SS, Roecker EB, Weindruch R, Aiken JM. 1995. High levels of mitochondrial DNA deletions in skeletal muscle of old rhesus monkeys. *Mech. Ageing Dev.* 83:91–101
18. Trifunovic A, Wredenberg A, Falkenberg M, Spelbrink JN, Rovio AT, et al. 2004. Premature ageing in mice expressing defective mitochondrial DNA polymerase. *Nature* 429:417–23
19. Hoppins S, Lackner L, Nunnari J. 2007. The machines that divide and fuse mitochondria. *Annu. Rev. Biochem.* 76:751–80
20. Detmer SA, Chan DC. 2007. Functions and dysfunctions of mitochondrial dynamics. *Nat. Rev. Mol. Cell Biol.* 8:870–79
21. Falkenberg M, Larsson NG, Gustafsson CM. 2007. DNA replication and transcription in mammalian mitochondria. *Annu. Rev. Biochem.* 76:679–99
22. Rorbach J, Soleimanpour-Lichaei R, Lightowlers RN, Chrzanowska-Lightowlers ZM. 2007. How do mammalian mitochondria synthesize proteins? *Biochem. Soc. Trans.* 35:1290–91
23. Tatsuta T, Langer T. 2008. Quality control of mitochondria: protection against neurodegeneration and ageing. *EMBO J.* 27:306–14

24. Bolender N, Sickmann A, Wagner R, Meisinger C, Pfanner N. 2008. Multiple pathways for sorting mitochondrial precursor proteins. *EMBO Rep.* 9:42–49
25. Lill R, Muhlenhoff U. 2008. Maturation of iron-sulfur proteins in eukaryotes: mechanisms, connected processes, and diseases. *Annu. Rev. Biochem.* 77:669–700
26. Cermakian N, Ikeda TM, Cedergren R, Gray MW. 1996. Sequences homologous to yeast mitochondrial and bacteriophage T3 and T7 RNA polymerases are widespread throughout the eukaryotic lineage. *Nucleic Acids Res.* 24:648–54
27. Sharma MR, Koc EC, Datta PP, Booth TM, Spremulli LL, Agrawal RK. 2003. Structure of the mammalian mitochondrial ribosome reveals an expanded functional role for its component proteins. *Cell* 115:97–108
28. Gowrishankar J, Harinarayanan R. 2004. Why is transcription coupled to translation in bacteria? *Mol. Microbiol.* 54:598–603
29. Falkenberg M, Gaspari M, Rantanen A, Trifunovic A, Larsson NG, Gustafsson CM. 2002. Mitochondrial transcription factors B1 and B2 activate transcription of human mtDNA. *Nat. Genet.* 31:289–94
30. Metodiev MD, Lesko N, Park CB, Cámara Y, Shi Y, et al. 2009. Methylation of 12S rRNA is necessary for in vivo stability of the small subunit of the mammalian mitochondrial ribosome. *Cell Metab.* 9:386–97
31. Fernandez-Silva P, Martinez-Azorin F, Micol V, Attardi G. 1997. The human mitochondrial transcription termination factor (mTERF) is a multizipper protein but binds to DNA as a monomer, with evidence pointing to intramolecular leucine zipper interactions. *EMBO J.* 16:1066–79
32. Linder T, Park CB, Asin-Cayuela J, Pellegrini M, Larsson NG, et al. 2005. A family of putative transcription termination factors shared amongst metazoans and plants. *Curr. Genet.* 48:265–69
33. Park CB, Asin-Cayuela J, Cámara Y, Shi Y, Pellegrini M, et al. 2007. MTERF3 is a negative regulator of mammalian mtDNA transcription. *Cell* 130:273–85
34. Wenz T, Luca C, Torraco A, Moraes CT. 2009. mTERF2 regulates oxidative phosphorylation by modulating mtDNA transcription. *Cell Metab.* 9:499–511
35. Korhonen JA, Pham XH, Pellegrini M, Falkenberg M. 2004. Reconstitution of a minimal mtDNA replisome in vitro. *EMBO J.* 23:2423–29
36. Kaguni LS. 2004. DNA polymerase γ, the mitochondrial replicase. *Annu. Rev. Biochem.* 73:293–320
37. Carrodeguas JA, Theis K, Bogenhagen DF, Kisker C. 2001. Crystal structure and deletion analysis show that the accessory subunit of mammalian DNA polymerase gamma, Pol gamma B, functions as a homodimer. *Mol. Cell* 7:43–54
38. Copeland WC. 2008. Inherited mitochondrial diseases of DNA replication. *Annu. Rev. Med.* 59:131–46
39. Korhonen JA, Gaspari M, Falkenberg M. 2003. TWINKLE has $5' \rightarrow 3'$ DNA helicase activity and is specifically stimulated by mitochondrial single-stranded DNA-binding protein. *J. Biol. Chem.* 278:48627–32
40. Wanrooij S, Fuste JM, Farge G, Shi Y, Gustafsson CM, Falkenberg M. 2008. Human mitochondrial RNA polymerase primes lagging-strand DNA synthesis in vitro. *Proc. Natl. Acad. Sci. USA* 105:11122–27
41. Fusté JM, Wanrooij S, Jemt E, Granycome CE, Cluett TJ, et al. 2010. Mitochondrial RNA polymerase is needed for activation of the origin of light-strand DNA replication. *Mol. Cell* 37:67–78
42. Clayton DA. 1991. Replication and transcription of vertebrate mitochondrial DNA. *Annu. Rev. Cell Biol.* 7:453–78
43. Brown TA, Cecconi C, Tkachuk AN, Bustamante C, Clayton DA. 2005. Replication of mitochondrial DNA occurs by strand displacement with alternative light-strand origins, not via a strand-coupled mechanism. *Genes Dev.* 19:2466–76
44. Clayton DA. 1982. Replication of animal mitochondrial DNA. *Cell* 28:693–705
45. Yang MY, Bowmaker M, Reyes A, Vergani L, Angeli P, et al. 2002. Biased incorporation of ribonucleotides on the mitochondrial L-strand accounts for apparent strand-asymmetric DNA replication. *Cell* 111:495–505
46. Yasukawa T, Reyes A, Cluett TJ, Yang MY, Bowmaker M, et al. 2006. Replication of vertebrate mitochondrial DNA entails transient ribonucleotide incorporation throughout the lagging strand. *EMBO J.* 25:5358–71
47. Moraes CT, Andreetta F, Bonilla E, Shanske S, DiMauro S, Schon EA. 1991. Replication-competent human mitochondrial DNA lacking the heavy-strand promoter region. *Mol. Cell. Biol.* 11:1631–37

48. Mita S, Rizzuto R, Moraes CT, Shanske S, Arnaudo E, et al. 1990. Recombination via flanking direct repeats is a major cause of large-scale deletions of human mitochondrial DNA. *Nucleic Acids Res.* 18:561–67
49. Bogenhagen DF, Clayton DA. 2003. The mitochondrial DNA replication bubble has not burst. *Trends Biochem. Sci.* 28:357–60
50. Holt IJ, Jacobs HT. 2003. Response: The mitochondrial DNA replication bubble has not burst. *Trends Biochem. Sci.* 28:355–56
51. Bogenhagen DF, Clayton DA. 2003. Concluding remarks: The mitochondrial DNA replication bubble has not burst. *Trends Biochem. Sci.* 28:404–5
52. Bogenhagen DF. 1999. Repair of mtDNA in vertebrates. *Am. J. Hum. Genet.* 64:1276–81
53. de Souza-Pinto NC, Wilson DM 3rd, Stevnsner TV, Bohr VA. 2008. Mitochondrial DNA, base excision repair and neurodegeneration. *DNA Repair* 7:1098–109
54. Zheng L, Zhou M, Guo Z, Lu H, Qian L, et al. 2008. Human DNA2 is a mitochondrial nuclease/helicase for efficient processing of DNA replication and repair intermediates. *Mol. Cell* 32:325–36
55. Garrido N, Griparic L, Jokitalo E, Wartiovaara J, van der Bliek AM, Spelbrink JN. 2003. Composition and dynamics of human mitochondrial nucleoids. *Mol. Biol. Cell* 14:1583–96
56. Fisher RP, Lisowsky T, Parisi MA, Clayton DA. 1992. DNA wrapping and bending by a mitochondrial high mobility group-like transcriptional activator protein. *J. Biol. Chem.* 267:3358–67
57. Kaufman BA, Durisic N, Mativetsky JM, Costantino S, Hancock MA, et al. 2007. The mitochondrial transcription factor TFAM coordinates the assembly of multiple DNA molecules into nucleoid-like structures. *Mol. Biol. Cell* 18:3225–36
58. MacAlpine DM, Kolesar J, Okamoto K, Butow RA, Perlman PS. 2001. Replication and preferential inheritance of hypersuppressive petite mitochondrial DNA. *EMBO J.* 20:1807–17
59. Antoshechkin I, Bogenhagen DF, Mastrangelo IA. 1997. The HMG-box mitochondrial transcription factor xl-mtTFA binds DNA as a tetramer to activate bidirectional transcription. *EMBO J.* 16:3198–206
60. Shen EL, Bogenhagen DF. 2001. Developmentally-regulated packaging of mitochondrial DNA by the HMG-box protein mtTFA during *Xenopus* oogenesis. *Nucleic Acids Res.* 29:2822–28
61. Ekstrand MI, Falkenberg M, Rantanen A, Park CB, Gaspari M, et al. 2004. Mitochondrial transcription factor A regulates mtDNA copy number in mammals. *Hum. Mol. Genet.* 13:935–44
62. Pellegrini M, Asin-Cayuela J, Erdjument-Bromage H, Tempst P, Larsson NG, Gustafsson CM. 2009. MTERF2 is a nucleoid component in mammalian mitochondria. *Biochim. Biophys. Acta* 1787:296–302
63. Larsson NG, Wang J, Wilhelmsson H, Oldfors A, Rustin P, et al. 1998. Mitochondrial transcription factor A is necessary for mtDNA maintenance and embryogenesis in mice. *Nat. Genet.* 18:231–36
64. Wang Y, Bogenhagen DF. 2006. Human mitochondrial DNA nucleoids are linked to protein folding machinery and metabolic enzymes at the mitochondrial inner membrane. *J. Biol. Chem.* 281:25791–802
65. Bogenhagen DF, Rousseau D, Burke S. 2008. The layered structure of human mitochondrial DNA nucleoids. *J. Biol. Chem.* 283:3665–75
66. Holt IJ, Harding AE, Morgan-Hughes JA. 1988. Deletions of muscle mitochondrial DNA in patients with mitochondrial myopathies. *Nature* 331:717–19
67. Wallace DC, Singh G, Lott MT, Hodge JA, Schurr TG, et al. 1988. Mitochondrial DNA mutation associated with Leber's hereditary optic neuropathy. *Science* 242:1427–30
68. Cortopassi GA, Arnheim N. 1990. Detection of a specific mitochondrial DNA deletion in tissues of older humans. *Nucleic Acids Res.* 18:6927–33
69. Soong NW, Hinton DR, Cortopassi G, Arnheim N. 1992. Mosaicism for a specific somatic mitochondrial DNA mutation in adult human brain. *Nat. Genet.* 2:318–23
70. Corral-Debrinski M, Horton T, Lott MT, Shoffner JM, Beal MF, Wallace DC. 1992. Mitochondrial DNA deletions in human brain: regional variability and increase with advanced age. *Nat. Genet.* 2:324–29
71. Shoffner JM, Lott MT, Lezza AMS, Seibel P, Ballinger SW, Wallace DC. 1990. Myoclonic epilepsy and ragged-red fiber disease (MERRF) is associated with a mitochondrial DNA tRNALys mutation. *Cell* 61:931–37
72. Goto YI, Nonaka I, Horai S. 1990. A mutation in the $tRNA^{Leu(UUR)}$ gene associated with the MELAS subgroup of mitochondrial encephalomyopathies. *Nature* 348:651–53

73. Hayashi J-I, Ohta S, Kikuchi A, Takemitsu M, Goto Y-I, Nonaka I. 1991. Introduction of disease-related mitochondrial DNA deletions into HeLa cells lacking mitochondrial DNA results in mitochondrial dysfunction. *Proc. Natl. Acad. Sci. USA* 88:10614–18
74. Chomyn A, Martinuzzi A, Yoneda M, Daga A, Hurko O, et al. 1992. MELAS mutation in mtDNA binding site for transcription termination factor causes defects in protein synthesis and in respiration but no change in levels of upstream and downstream mature transcripts. *Proc. Natl. Acad. Sci. USA* 89:4221–25
75. Moraes CT, Ricci E, Petruzzella V, Shanske S, DiMauro S, et al. 1992. Molecular analysis of the muscle pathology associated with mitochondrial DNA deletions. *Nat. Genet.* 1:359–67
76. Wredenberg A, Wibom R, Wilhelmsson H, Graff C, Wiener HH, et al. 2002. Increased mitochondrial mass in mitochondrial myopathy mice. *Proc. Natl. Acad. Sci. USA* 99:15066–71
77. Wenz T, Diaz, Speigelman BM, Moraes CT. 2008. Activation of the PPAR/PGC-1alpha pathway prevents a bioenergetic deficit and effectively improves a mitochondrial myopathy phenotype. *Cell Metab.* 8:249–56
78. Elson JL, Samuels DC, Turnbull DM, Chinnery PF. 2001. Random intracellular drift explains the clonal expansion of mitochondrial DNA mutations with age. *Am. J. Hum. Genet.* 68:802–6
79. Jenuth JP, Peterson AC, Shoubridge EA. 1997. Tissue-specific selection for different mtDNA genotypes in heteroplasmic mice. *Nat. Genet.* 16:93–95
80. Trounce I, Byrne E, Marzuki S. 1989. Decline in skeletal muscle mitochondrial respiratory chain function: possible factor in ageing. *Lancet* 1:637–39
81. Müller-Höcker J. 1989. Cytochrome-c-oxidase deficient cardiomyocytes in the human heart—an age-related phenomenon. A histochemical ultracytochemical study. *Am. J. Pathol.* 134:1167–73
82. Müller-Höcker J. 1990. Cytochrome *c* oxidase deficient fibres in the limb muscle and diaphragm of man without muscular disease: an age related alteration. *J. Neurol. Sci.* 100:14–21
83. Fayet G, Jansson M, Sternberg D, Moslemi AR, Blondy P, et al. 2002. Ageing muscle: Clonal expansions of mitochondrial DNA point mutations and deletions cause focal impairment of mitochondrial function. *Neuromuscul. Disord.* 12:484–93
84. Cottrell DA, Blakely EL, Johnson MA, Ince PG, Turnbull DM. 2001. Mitochondrial enzyme-deficient hippocampal neurons and choroidal cells in AD. *Neurology* 57:260–64
85. Bender A, Krishnan KJ, Morris CM, Taylor GA, Reeve AK, et al. 2006. High levels of mitochondrial DNA deletions in substantia nigra neurons in aging and Parkinson disease. *Nat. Genet.* 38:515–17
86. Taylor RW, Barron MJ, Borthwick GM, Gospel A, Chinnery PF, et al. 2003. Mitochondrial DNA mutations in human colonic crypt stem cells. *J. Clin. Investig.* 112:1351–60
87. Reeve AK, Krishnan KJ, Taylor G, Elson JL, Bender A, et al. 2009. The low abundance of clonally expanded mitochondrial DNA point mutations in aged substantia nigra neurons. *Aging Cell* 8:496–98
88. Bender A, Schwarzkopf RM, McMillan A, Krishnan KJ, Rieder G, et al. 2008. Dopaminergic midbrain neurons are the prime target for mitochondrial DNA deletions. *J. Neurol.* 255:1231–35
89. Bua E, Johnson J, Herbst A, Delong B, McKenzie D, et al. 2006. Mitochondrial DNA-deletion mutations accumulate intracellularly to detrimental levels in aged human skeletal muscle fibers. *Am. J. Hum. Genet.* 79:469–80
90. Pak JW, Herbst A, Bua E, Gokey N, McKenzie D, Aiken JM. 2003. Mitochondrial DNA mutations as a fundamental mechanism in physiological declines associated with aging. *Aging Cell* 2:1–7
91. Berneburg M, Plettenberg H, Medve-König K, Pfahlberg A, Gers-Barlag H, et al. 2004. Induction of the photoaging-associated mitochondrial common deletion in vivo in normal human skin. *J. Investig. Dermatol.* 122:1277–83
92. Krishnan KJ, Birch-Machin MA. 2006. The incidence of both tandem duplications and the common deletion in mtDNA from three distinct categories of sun-exposed human skin and in prolonged culture of fibroblasts. *J. Investig. Dermatol.* 126:408–15
93. Birket MJ, Birch-Machin MA. 2007. Ultraviolet radiation exposure accelerates the accumulation of the aging-dependent T414G mitochondrial DNA mutation in human skin. *Aging Cell* 6:557–64
94. Michikawa Y, Mazzucchelli F, Bresolin N, Scarlato G, Attardi G. 1999. Aging-dependent large accumulation of point mutations in the human mtDNA control region for replication. *Science* 286:774–79

95. Murdock DG, Christacos NC, Wallace DC. 2000. The age-related accumulation of a mitochondrial DNA control region mutation in muscle, but not brain, detected by a sensitive PNA-directed PCR clamping based method. *Nucleic Acids Res.* 28:4350–55
96. Wang Y, Michikawa Y, Mallidis C, Bai Y, Woodhouse L, et al. 2001. Muscle-specific mutations accumulate with aging in critical human mtDNA control sites for replication. *Proc. Natl. Acad. Sci. USA* 98:4022–27
97. Birket MJ, Passos JF, von Zglinicki T, Birch-Machin MA. 2009. The relationship between the aging- and photo-dependent T414G mitochondrial DNA mutation with cellular senescence and reactive oxygen species production in cultured skin fibroblasts. *J. Investig. Dermatol.* 129:1361–66
98. Rivera J, Tessarollo L. 2008. Genetic background and the dilemma of translating mouse studies to humans. *Immunity* 28:1–4
99. Brockmann GA, Bevova MR. 2002. Using mouse models to dissect the genetics of obesity. *Trends Genet.* 18:367–76
100. Wong GT. 2002. Speed congenics: applications for transgenic and knock-out mouse strains. *Neuropeptides* 36:230–36
101. Huang WY, Aramburu J, Douglas PS, Izumo S. 2000. Transgenic expression of green fluorescence protein can cause dilated cardiomyopathy. *Nat. Med.* 6:482–83
102. Zelenaya-Troitskaya O, Newman SM, Okamoto K, Perlman PS, Butow RA. 1998. Functions of the high mobility group protein, Abf2p, in mitochondrial DNA segregation, recombination and copy number in *Saccharomyces cerevisiae*. *Genetics* 148:1763–76
103. Brand MD, Pakay JL, Ocloo A, Kokoszka J, Wallace DC, et al. 2005. The basal proton conductance of mitochondria depends on adenine nucleotide translocase content. *Biochem. J.* 392:353–62
104. Cannon B, Shabalina IG, Kramarova TV, Petrovic N, Nedergaard J. 2006. Uncoupling proteins: a role in protection against reactive oxygen species–or not? *Biochim. Biophys. Acta* 1757:449–58
105. Van Keuren ML, Gavrilina GB, Filipiak WE, Zeidler MG, Saunders TL. 2009. Generating transgenic mice from bacterial artificial chromosomes: transgenesis efficiency, integration and expression outcomes. *Transgenic Res.* 18:769–85
106. Silva JP, Shabalina IG, Dufour E, Petrovic N, Backlund EC, et al. 2005. SOD2 overexpression: enhanced mitochondrial tolerance but absence of effect on UCP activity. *EMBO J.* 24:4061–70
107. Lee EC, Yu D, Martinez de Velasco J, Tessarollo L, Swing DA, et al. 2001. A highly efficient *Escherichia coli*–based chromosome engineering system adapted for recombinogenic targeting and subcloning of BAC DNA. *Genomics* 73:56–65
108. Agostino A, Invernizzi F, Tiveron C, Fagiolari G, Prelle A, et al. 2003. Constitutive knockout of *Surf1* is associated with high embryonic lethality, mitochondrial disease and cytochrome *c* oxidase deficiency in mice. *Hum. Mol. Genet.* 12:399–413
109. Meyers EN, Lewandoski M, Martin GR. 1998. An Fgf8 mutant allelic series generated by Cre- and Flp-mediated recombination. *Nat. Genet.* 18:136–41
110. Ekstrand M, Larsson NG. 2002. Breeding and genotyping of Tfam conditional knockout mice. *Methods Mol. Biol.* 197:391–400
111. Trifunovic A, Hansson A, Wredenberg A, Rovio AT, Dufour E, et al. 2005. Somatic mtDNA mutations cause aging phenotypes without affecting reactive oxygen species production. *Proc. Natl. Acad. Sci. USA* 102:17993–98
112. Silva JP, Kohler M, Graff C, Oldfors A, Magnuson MA, et al. 2000. Impaired insulin secretion and beta-cell loss in tissue-specific knockout mice with mitochondrial diabetes. *Nat. Genet.* 26:336–40
113. Kujoth GC, Hiona A, Pugh TD, Someya S, Panzer K, et al. 2005. Mitochondrial DNA mutations, oxidative stress, and apoptosis in mammalian aging. *Science* 309:481–84
114. Bailey LJ, Cluett TJ, Reyes A, Prolla TA, Poulton J, et al. 2009. Mice expressing an error-prone DNA polymerase in mitochondria display elevated replication pausing and chromosomal breakage at fragile sites of mitochondrial DNA. *Nucleic Acids Res.* 37:2327–35
115. Vermulst M, Wanagat J, Kujoth GC, Bielas JH, Rabinovitch PS, et al. 2008. DNA deletions and clonal mutations drive premature aging in mitochondrial mutator mice. *Nat. Genet.* 40:392–94

116. Edgar D, Shabalina I, Camara Y, Wredenberg A, Calvaruso MA, et al. 2009. Random point mutations with major effects on protein-coding genes are the driving force behind premature aging in mtDNA mutator mice. *Cell Metab.* 10:131–38
117. Kraytsberg Y, Simon DK, Turnbull DM, Khrapko K. 2009. Do mtDNA deletions drive premature aging in mtDNA mutator mice? *Aging Cell* 8:502–6
118. Inoue K, Nakada K, Ogura A, Isobe K, Goto Y, et al. 2000. Generation of mice with mitochondrial dysfunction by introducing mouse mtDNA carrying a deletion into zygotes. *Nat. Genet.* 26:176–81
119. Tyynismaa H, Mjosund KP, Wanrooij S, Lappalainen I, Ylikallio E, et al. 2005. Mutant mitochondrial helicase Twinkle causes multiple mtDNA deletions and a late-onset mitochondrial disease in mice. *Proc. Natl. Acad. Sci. USA* 102:17687–92
120. Stewart JB, Freyer C, Elson JL, Larsson NG. 2008. Purifying selection of mtDNA and its implications for understanding evolution and mitochondrial disease. *Nat. Rev. Genet.* 9:657–62
121. Stewart JB, Freyer C, Elson JL, Wredenberg A, Cansu Z, et al. 2008. Strong purifying selection in transmission of mammalian mitochondrial DNA. *PLoS Biol.* 6:e10
122. Vermulst M, Bielas JH, Kujoth GC, Ladiges WC, Rabinovitch PS, et al. 2007. Mitochondrial point mutations do not limit the natural lifespan of mice. *Nat. Genet.* 39:540–43
123. Ekstrand MI, Terzioglu M, Galter D, Zhu S, Hofstetter C, et al. 2007. Progressive parkinsonism in mice with respiratory-chain-deficient dopamine neurons. *Proc. Natl. Acad. Sci. USA* 104:1325–30
124. Dufour E, Terzioglu M, Sterky FH, Sorensen L, Galter D, et al. 2008. Age-associated mosaic respiratory chain deficiency causes trans-neuronal degeneration. *Hum. Mol. Genet.* 17:1418–26
125. Smeitink JA, Zeviani M, Turnbull DM, Jacobs HT. 2006. Mitochondrial medicine: a metabolic perspective on the pathology of oxidative phosphorylation disorders. *Cell Metab.* 3:9–13
126. Wang J, Silva JP, Gustafsson CM, Rustin P, Larsson NG. 2001. Increased in vivo apoptosis in cells lacking mitochondrial DNA gene expression. *Proc. Natl. Acad. Sci. USA* 98:4038–43
127. Niu X, Trifunovic A, Larsson NG, Canlon B. 2007. Somatic mtDNA mutations cause progressive hearing loss in the mouse. *Exp. Cell Res.* 313:3924–34
128. Kruse SE, Watt WC, Marcinek DJ, Kapur RP, Schenkman KA, Palmiter RD. 2008. Mice with mitochondrial complex I deficiency develop a fatal encephalomyopathy. *Cell Metab.* 7:312–20
129. Geromel V, Kadhom N, Cebalos-Picot I, Ouari O, Polidori A, et al. 2001. Superoxide-induced massive apoptosis in cultured skin fibroblasts harboring the neurogenic ataxia retinitis pigmentosa (NARP) mutation in the ATPase-6 gene of the mitochondrial DNA. *Hum. Mol. Genet.* 10:1221–28
130. Khrapko K, Coller HA, Andre PC, Li XC, Hanekamp JS, Thilly WG. 1997. Mitochondrial mutational spectra in human cells and tissues. *Proc. Natl. Acad. Sci. USA* 94:13798–803
131. Benzer S, Freese E. 1958. Induction of specific mutations with 5-bromouracil. *Proc. Natl. Acad. Sci. USA* 44:112–19
132. Evans MD, Dizdaroglu M, Cooke MS. 2004. Oxidative DNA damage and disease: induction, repair and significance. *Mutat. Res.* 567:1–61
133. Richter C, Park J-W, Ames BN. 1988. Normal oxidative damage to mitochondrial and nuclear DNA is extensive. *Proc. Natl. Acad. Sci. USA* 85:6465–67
134. Hamilton ML, Guo Z, Fuller CD, van Remmen H, Ward WF, et al. 2001. A reliable assessment of 8-oxo-2-deoxyguanosine levels in nuclear and mitochondrial DNA using the sodium iodide method to isolate DNA. *Nucleic Acids Res.* 29:2117–26
135. Pinz KG, Shibutani S, Bogenhagen DF. 1995. Action of mitochondrial DNA polymerase gamma at sites of base loss or oxidative damage. *J. Biol. Chem.* 270:9202–6
136. de Souza-Pinto NC, Eide L, Hogue BA, Thybo T, Stevnsner T, et al. 2001. Repair of 8-oxodeoxyguanosine lesions in mitochondrial DNA depends on the oxoguanine DNA glycosylase (*OGG1*) gene and 8-oxoguanine accumulates in the mitochondrial DNA of OGG1-defective mice. *Cancer Res.* 61:5378–81
137. Stuart JA, Bourque BM, de Souza-Pinto NC, Bohr VA. 2005. No evidence of mitochondrial respiratory dysfunction in OGG1-null mice deficient in removal of 8-oxodeoxyguanine from mitochondrial DNA. *Free Radic. Biol. Med.* 38:737–45
138. Zheng W, Khrapko K, Coller HA, Thilly WG, Copeland WC. 2006. Origins of human mitochondrial point mutations as DNA polymerase gamma-mediated errors. *Mutat. Res.* 599:11–20

139. Graziewicz MA, Bienstock RJ, Copeland WC. 2007. The DNA polymerase gamma Y955C disease variant associated with PEO and parkinsonism mediates the incorporation and translesion synthesis opposite 7,8-dihydro-8-oxo-2′-deoxyguanosine. *Hum. Mol. Genet.* 16:2729–39
140. Madsen CS, Ghivizzani SC, Hauswirth WW. 1993. In vivo and in vitro evidence for slipped mispairing in mammalian mitochondria. *Proc. Natl. Acad. Sci. USA* 90:7671–75
141. Fukui H, Moraes CT. 2009. Mechanisms of formation and accumulation of mitochondrial DNA deletions in aging neurons. *Hum. Mol. Genet.* 18:1028–36
142. Krishnan KJ, Reeve AK, Samuels DC, Chinnery PF, Blackwood JK, et al. 2008. What causes mitochondrial DNA deletions in human cells? *Nat. Genet.* 40:275–79
143. Hancock DK, Tully LA, Levin BC. 2005. A standard reference material to determine the sensitivity of techniques for detecting low-frequency mutations, SNPs, and heteroplasmies in mitochondrial DNA. *Genomics* 86:446–61
144. Kajander OA, Poulton J, Spelbrink JN, Rovio A, Karhunen PJ, Jacobs HT. 1999. The dangers of extended PCR in the clinic. *Nat. Med.* 5:965–66
145. Chan SW, Chen JZ. 2009. Measuring mtDNA damage using a supercoiling-sensitive qPCR approach. *Methods Mol. Biol.* 554:183–97
146. Larsson N-G, Eiken HG, Boman H, Holme E, Oldfors A, Tulinius MH. 1992. Lack of transmission of deleted mtDNA from a woman with Kearns-Sayre syndrome to her child. *Am. J. Hum. Genet.* 50:360–63
147. Greaves LC, Beadle NE, Taylor GA, Commane D, Mathers JC, et al. 2009. Quantification of mitochondrial DNA mutation load. *Aging Cell* 8:566–72
148. Wilding CS, Cadwell K, Tawn EJ, Relton CL, Taylor GA, et al. 2006. Mitochondrial DNA mutations in individuals occupationally exposed to ionizing radiation. *Radiat. Res.* 165:202–7
149. Kraytsberg Y, Bodyak N, Myerow S, Nicholas A, Ebralidze K, Khrapko K. 2009. Quantitative analysis of somatic mitochondrial DNA mutations by single-cell single-molecule PCR. *Methods Mol. Biol.* 554:329–69
150. Clayton DA, Smith CA. 1975. Complex mitochondrial DNA. *Int. Rev. Exp. Pathol.* 14:1–67
151. Pohjoismaki JL, Goffart S, Tyynismaa H, Willcox S, Ide T, et al. 2009. Human heart mitochondrial DNA is organized in complex catenated networks containing abundant four-way junctions and replication forks. *J. Biol. Chem.* 284:21446–57
152. Vermulst M, Bielas JH, Loeb LA. 2008. Quantification of random mutations in the mitochondrial genome. *Methods* 46:263–68
153. Hirano M, Shtilbans A, Mayeux R, Davidson MM, DiMauro S, et al. 1997. Apparent mtDNA heteroplasmy in Alzheimer's disease patients and in normals due to PCR amplification of nucleus-embedded mtDNA pseudogenes. *Proc. Natl. Acad. Sci. USA* 94:14894–99

Physical Mechanisms of Signal Integration by WASP Family Proteins

Shae B. Padrick and Michael K. Rosen

Howard Hughes Medical Institute and Department of Biochemistry, University of Texas Southwestern Medical Center, Dallas, Texas 75390;
email: Shae.Padrick@UTSouthwestern.edu, Michael.Rosen@UTSouthwestern.edu

Annu. Rev. Biochem. 2010. 79:707–35

First published online as a Review in Advance on April 5, 2010

The *Annual Review of Biochemistry* is online at biochem.annualreviews.org

This article's doi:
10.1146/annurev.biochem.77.060407.135452

0066-4154/10/0707-0707$20.00

Key Words

actin regulation, allostery, Arp2/3 complex, Rho GTPase, signal transduction

Abstract

The proteins of the Wiskott-Aldrich syndrome protein (WASP) family are activators of the ubiquitous actin nucleation factor, the Arp2/3 complex. WASP family proteins contain a C-terminal VCA domain that binds and activates the Arp2/3 complex in response to numerous inputs, including Rho family GTPases, phosphoinositide lipids, SH3 domain–containing proteins, kinases, and phosphatases. In the archetypal members of the family, WASP and N-WASP, these signals are integrated through two levels of regulation, an allosteric autoinhibitory interaction, in which the VCA is sequestered from the Arp2/3 complex, and dimerization/oligomerization, in which multi-VCA complexes are better activators of the Arp2/3 complex than monomers. Here, we review the structural, biochemical, and biophysical details of these mechanisms and illustrate how they work together to control WASP activity in response to multiple inputs. These regulatory principles, derived from studies of WASP and N-WASP, are likely to apply broadly across the family.

Contents

WASP: Wiskott-Aldrich syndrome protein

Arp2/3 complex: the actin-related protein 2/actin-related protein 3 complex

INTRODUCTION

Cells are constantly bathed in a multitude of signals from the environment. Proper cellular function requires integration of these signals to yield complex processes and behaviors such as differentiation, division, and movement. Often signal integration is the specific response of individual molecules to multiple simultaneous physical interactions. The mechanisms of this processing are a fascinating and important area of research in biochemistry, biophysics, and structural biology.

Dynamic rearrangements of the actin cytoskeleton underlie many cellular processes, including division, endocytosis, and movement. Control over actin dynamics is achieved through the integration of numerous signals, which together guide the assembly, disassembly, architecture, and movement of the actin filament network in a spatially and temporally defined manner.

Members of the Wiskott-Aldrich syndrome protein (WASP) family are central hubs in the signaling networks that control actin. These proteins integrate a huge range of inputs. In response, they activate the ubiquitous actin-nucleating machine, the Arp2/3 complex (1, 2). By controlling the degree, rate, and location of filament nucleation by the Arp2/3 complex, WASP proteins shape the structure and dynamics of filament networks throughout biology.

Several excellent reviews have recently discussed the cellular functions of WASP proteins, highlighting genetic and cell biological findings (2–6). Here, we focus on complementary structural, biophysical, and biochemical findings, with an eye toward illustrating how these data inform biology. The central concept that pervades our discussion is that the activity of WASP proteins toward the Arp2/3 complex is controlled at two levels. First, an allosteric process involving transitions between inactive and active conformations controls the accessibility of the Arp2/3-stimulatory element of WASP proteins. Second, oligomerization modulates that ability of an active WASP to stimulate the Arp2/3 complex, with dimers or higher-order oligomers having much greater potency than monomers. These two mechanisms function synergistically to allow WASP family members to integrate diverse signals and control Arp2/3 activity with precision. This view provides a unifying lens through which WASP biology can be understood in quantitative terms.

DOMAIN ORGANIZATION OF WASP AND NEURAL WASP

The WASP family is defined by a conserved C-terminal VCA domain (**Figure 1*a***) named for its three sub-elements, the verprolin (V) homology (also called WASP homology 2 or WH2), central (C) hydrophobic, and acidic (A) regions. The VCA domain strongly stimulates actin nucleation by the Arp2/3 complex (1, 7, 8). The three elements have distinct functions during this process. The C-terminal A region contributes substantial binding affinity toward the Arp2/3 complex (9–12). There is also a connection between the A region and activity; different WASP family VCAs have quite different maximal activities, which correlate with the A region sequence (11). The C region also contributes both to binding affinity and to an affinity-independent aspect of activation (10, 13), as well as to autoinhibitory regulation (see below). The C region appears to contact the Arp2/3 complex using the hydrophobic face of an amphipathic helix (13).

The V region binds to G-actin and delivers it to the Arp2 and/or Arp3 subunits of the Arp2/3 complex (both are homologous to actin), creating a pseudotrimer/tetramer nucleus for filament growth (7, 9, 10, 14–18). A recent small-angle X-ray scattering study suggests that a high-affinity site for the VCA:actin complex is

VCA domain: the verprolin homology, central hydrophobic, and acidic C-terminal regions of WASP family proteins

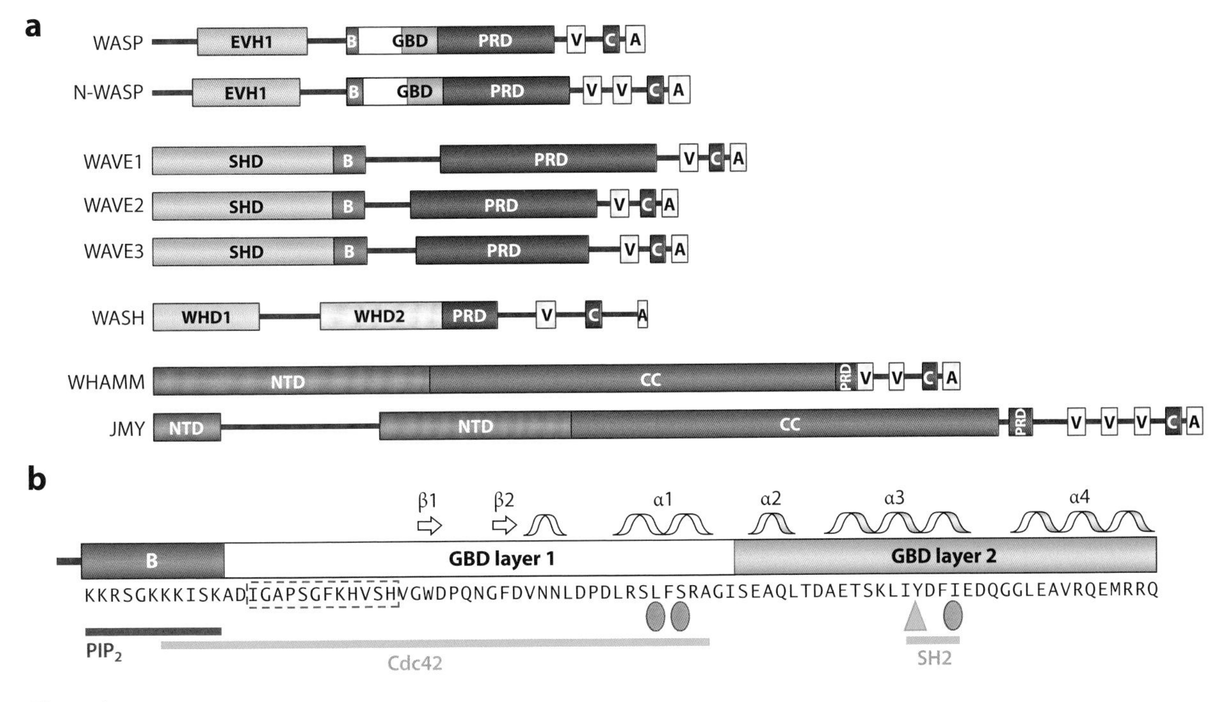

Figure 1

The Wiskott-Aldrich syndrome protein (WASP) family of proteins. (*a*) Domain structure of human WASP family proteins. Different domains are abbreviated as follows: A, acidic region; B, basic region; C, central hydrophobic region; CC, coiled-coil domain; EVH1, Ena/VASP homology domain 1; GBD, GTPase-binding domain; NTD, N-terminal homology domain; PRD, proline-rich domain; SHD, Scar homology domain; V (also WH2), verprolin homology domain; WHD1, WASH homology domain 1; WHD2, WASH homology domain 2. In JMY, the NTD has an insertion of >100 amino acids when compared to WHAMM. There are multiple closely related WASH proteins in humans; shown here is WASH1 (National Center for Biotechnology Information reference number NP_878908). (*b*) WASP has distinct binding sites for different ligands. Human WASP B-GBD sequence is shown with domains colored as in panel *a*. The secondary structure from the WASP GBD-C-autoinhibited structure is shown (see **Figure 2*c–e***). Bars indicate the binding sites for phosphatidylinositol 4,5-diphosphate (PIP2), Cdc42, and SH2 domains. Other annotations: dashed box, Cdc42-Rac-interactive-binding or CRIB; orange ovals, X-linked neutropenia WASP mutations; green triangle, tyrosine 291.

WIP: WASP-interacting protein

positioned to deliver actin to the Arp2 subunit (18); it is also possible that a second VCA binding to a different site (see below) could deliver actin to Arp3. V region peptides form an amphipathic helix that inserts into the hydrophobic groove between actin subunits 1 and 3, followed by an extended segment that traverses up the actin face toward the nucleotide-binding cleft (**Figure 2*a***) (16, 19; see also the Research Collaboratory for Structural Bioinformatics, **http://www.rcsb.org**, database entry for Protein Data Bank number 2vcp). As detailed in recent reviews, V/WH2 repeats are also found in many other proteins, and can act to either sequester an actin monomer or, when in tandem arrays, nucleate filaments directly (20, 21).

An important caveat to this straightforward characterization of the V, C, and A functions is

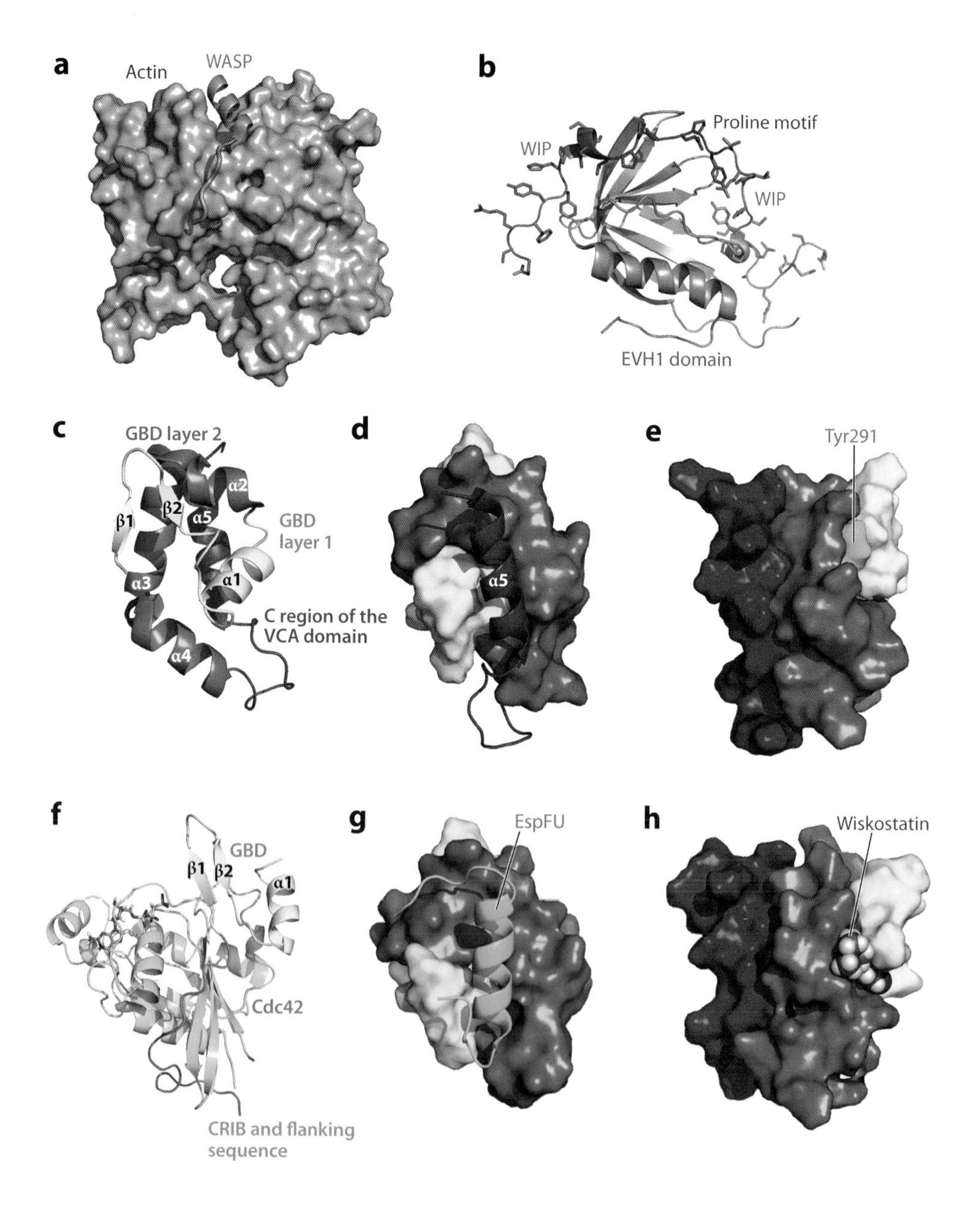

that nearly all experiments to date do not take into account the recent discovery of a second VCA-binding site on the Arp2/3 complex (22). This finding requires new experiments (and reconsideration of previous data) to analyze these functions in the context of the individual sites. In general, the VCAs could have quite different actions at the two sites, perhaps even promoting temporally distinct steps in the nucleation process. Understanding both the thermodynamic and kinetic aspects of Arp2/3 activation through the two VCA-binding sites represents an important avenue for future research.

N-terminal elements of WASP family proteins control localization, association with ligands, and biochemical activity of the VCA. These elements are divergent among different family members, affording the proteins distinct molecular details of regulation. However, as argued below, the physical regulatory principles are likely similar across the family. These principles have emerged from studies of the archetypal members, WASP and neural Wiskott-Aldrich syndrome protein (N-WASP).

The N-terminal Ena/VASP homology 1 (EVH1) domain of WASP and N-WASP (**Figure 1**) binds to a conserved proline motif in members of the WIP family (23). This interaction stabilizes and localizes WASP and N-WASP in cells (23–25). Structural studies show that the WIP proline motif contacts the canonical EVH1 ligand-binding site through a type II polyproline helix (**Figure 2*b***) (26). The flanking sequences extend linearly around the EVH1 domain in opposite directions, wrapping over halfway around the domain. Mutations that disrupt this interface decrease WIP binding, leading in vivo to proteolytic degradation of WASP (25) and many forms of WAS (27). WIP binding also negatively regulates the N-WASP VCA (28, 29).

PIP_2: phosphatidylinositol 4,5-diphosphate

The EVH1 domain is followed by a hydrophilic, low-complexity sequence, without an ascribed function for either WASP or N-WASP. Following this is a sequence of basic residues known as the basic (B) region (**Figure 1**). In both proteins, the B region stabilizes autoinhibitory interactions (see below) (30–32). In N-WASP, it is clear that this element also binds PIP_2 and mediates many PIP_2 effects on activity (31–34) (see below). The element probably plays a similar role in WASP, but additional regions appear necessary for this functional interaction (35). As elaborated below, the B region also plays an important role in selective binding of the WASP activator, Cdc42 (36).

Figure 2

Structures of Wiskott-Aldrich syndrome protein (WASP). (*a*) The structure of WASP verprolin homology/WASP homology 2 (V/WH2) region in complex with actin. Actin is shown as a gray surface, and WASP is shown as an orange ribbon. (*b*) Structure of the neural Wiskott-Aldrich syndrome protein Ena/VASP homology 1 domain (EVH1) in complex with WASP-interacting protein (WIP) residues 451–485 (only the ordered WIP residues 454–481 are shown). The EVH1 domain is gold, and the proline motif (461DLPPPEPY468) and flanking regions of WIP are magenta and blue, respectively. (*c-e*) Structure of autoinhibited WASP GTPase-binding domain fused to the C-region of the VCA (GBD-C). (*c*) Structure of the GBD-C protein. The three layers of this structure are colored yellow (GBD layer 1), blue (GBD layer 2) and red (C region of the VCA domain), respectively. Secondary structure elements are labeled. (*d*) Structure of the GBD-C protein, rotated 180° from panel *c*. GBD and C elements are shown in surface and ribbon representations, respectively. (*e*) Surface representation of GBD-C protein, rotated ~90° from panel *c*, showing the Y291 phosphorylation site in green. (*f,g*) Structures of active WASP complexes. (*f*) Structure of the Cdc42:WASP GBD complex. Cdc42 is a green ribbon, with GMPPNP in sticks. Ordered GBD residues (231–277) are shown in orange [Cdc42-Rac-interactive-binding (CRIB) motif and flanking sequence] and yellow (layer one as in panel *c*) ribbon. (*g*) Structure of the WASP GBD:EspFU 1R complex. GBD is colored as in panel *c* and shown in surface representation. EspFU is shown as a green ribbon. (*h*) Structure of the WASP GBD-C:wiskostatin complex, orientation and color as in panel *e*, with wiskostatin shown as van der Waals spheres.

GBD: GTPase-binding domain

PRD: proline-rich domain

After the B region, WASP and N-WASP contain an ~85-residue element termed the GTPase-binding domain (GBD) (**Figure 1**). The first 20 residues of this element contain a Cdc42-Rac-interactive-binding (CRIB) motif (37), which mediates nucleotide-dependent binding of numerous effectors of the Cdc42 and Rac GTPases. The GBD in WASP and N-WASP binds the VCA intramolecularly and plays a central role in controlling activity toward the Arp2/3 complex (38, 39).

Between the GBD and the VCA, WASP and N-WASP contain a 100–125-residue proline-rich domain (PRD) (**Figure 1**). The PRD contains approximately six canonical binding sites for SH3 domains and three profilin-binding sites (40). This segment mediates interactions with SH3 domains of a large and diverse range of ligands. As WIP also contains a PRD, WASP:WIP complexes will have even more SH3-binding sites.

The five functional elements of WASP allow binding to multiple ligands simultaneously. A central concept developed in the following sections is that the ligands can act through two distinct regulatory mechanisms to control the biochemical activity of WASP toward the Arp2/3 complex: allostery, which controls accessibility of the VCA to the Arp2/3 complex, and oligomerization, which controls how strongly the exposed VCA can stimulate the Arp2/3 complex. When multiple ligands engage one or both of these mechanisms simultaneously, they act cooperatively to control WASP activity, thus integrating multiple signals.

ALLOSTERIC REGULATION OF WASP AND N-WASP

The idea of autoinhibition in WASP, mediated by intramolecular binding of the GBD to the VCA, has guided thinking in the field since its discovery in 1998 (38). A large body of work has revealed structural and thermodynamic mechanisms of autoinhibition and its relief by upstream signals. We discuss these data in the context of single activators here and show how autoinhibition contributes to integrative behavior, below.

The WASP Autoinhibitory Mechanism

Physical studies have shown that isolated GBD and VCA peptides are largely unfolded in solution (39, 41). Upon binding, the C region of the VCA and most of the GBD, except for the CRIB motif, fold together to make a small domain (39) (**Figures 2*c*,*d***). From a functional standpoint, the structure can be considered in three layers. The first consists of a short β-hairpin and an α-helix (α1). This layer represents the minimal high-affinity Cdc42-binding element. The second layer is formed by three additional helices organized into a planar C shape. Layer three consists of the C helix region of the VCA. This element lies behind layer two with its hydrophobic face contacting predominantly helices α2–α4 and also the C terminus of helix α1 through the open end of the layer two C. A number of C region side chains, which are buried in the layer two-layer three interface (**Figures 2*c*,*d***), also contribute substantially to the activity and affinity of the VCA for the Arp2/3 complex (10, 13). Thus, occlusion of these residues in the GBD:VCA complex is an important mechanistic component of autoinhibition in WASP and N-WASP. Mutations of the GBD, including *L270P*, *I294T*, and *S272P*, disrupt autoinhibition and cause the immunodeficiency disease X-linked neutropenia (**Figure 1**) (42, 43).

A variety of data suggest that additional elements in WASP contribute to binding of the GBD to the VCA, and consequently to autoinhibition. Partial deletion of the A region increases N-WASP activity in vitro (12), and complete removal (including several residues typically classified as being in the C region) greatly decreases the affinity of GBD-containing constructs for the VCA in both WASP and N-WASP (31, 39). Similarly, partial truncation of the B region in WASP and N-WASP decreased autoinhibition (30–32), concomitant with decreased unfolding stability of the BGBD-VCA WASP (30). The potential for favorable electrostatic contacts between the A and B regions has led to suggestions that these elements may interact to stabilize the BGBD-VCA. Finally, the EVH1 domain can bind weakly to the VCA, and GBD constructs containing the EVH1 domain bind the VCA more strongly than those that do not (31). Truncation of the EVH1 has also been reported to cause a small increase in the basal activity of N-WASP (31), although this has not been universally observed (12). The structural basis for the interaction of the EVH1 domain with the VCA and the contribution of the EVH1 domain to autoinhibition remain unknown.

The thermodynamics of autoinhibition in WASP have been analyzed through a quantitative two-state allosteric model, which is based on classical Monod-Wyman-Changeux (MWC) formalism (**Figure 3*a***) (30, 44, 45). In this model, WASP exists in equilibrium between an inhibited state, where the GBD

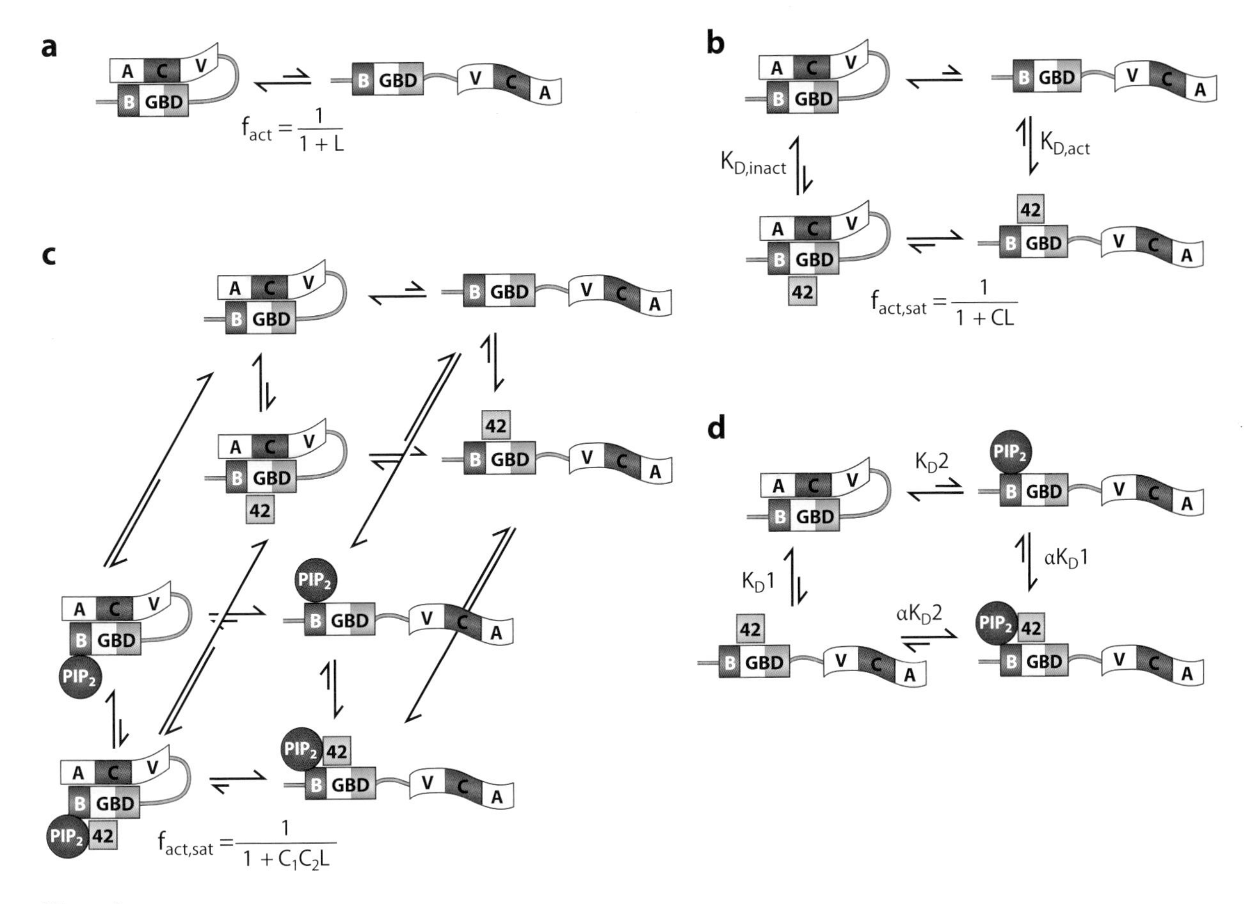

Figure 3

Thermodynamic models of allostery in a Wiskott-Aldrich syndrome protein (WASP). (*a*) The two-state Monod-Wyman-Changeux (MWC)-based allosteric model for WASP regulation by controlling access to the C-helix through association with the GTPase-binding domain (GBD). (*b*) Two-state allosteric model with one activator. (*c*) Two-state allosteric model with two activators. Equations describing the fraction of WASP in the active state, either free (f_{act}) or in the presence of saturating ligand ($f_{act,sat}$) are given below the images. L is defined as $[WASP_{inactive}]/[WASP_{active}]$; C is the ratio of dissociation constants of a ligand for the active and inactive states, $K_{D,act}/K_{D,inact}$. (*d*) Simplified allosteric model for response to two activators. K_D1 and K_D2 are the dissociation constants of the two ligands for free WASP; α is the cooperativity constant between the ligands. Abbreviations: A, acidic region; B, basic region; V, verprolin homology/WH2 region; 42, Cdc42.

is bound to the VCA, and an active state, where the elements are dissociated. Because the GBD and the VCA are both unfolded in isolation, the allosteric equilibrium can be described as the folding equilibrium. Thus, for any given WASP, the free energy of the allosteric transition is equal to the stability of the GBD-VCA domain against unfolding. As the latter can be directly measured by standard denaturation assays, the model provides a ready means of quantifying the allosteric equilibrium in WASP. This model quantitatively predicted the activity of many WASP proteins simply on the basis of their free energies of unfolding (30). These proteins contained various linkers between the GBD and the VCA, different portions of the B region, and different destabilizing mutations in the GBD. As described below, this thermodynamic view of WASP as existing in equilibrium between a folded, inhibited state and an unfolded, active state is also an effective means of understanding activation by effectors.

The combined structural, biochemical, and biophysical data have led to a straightforward mechanism of autoinhibition in WASP. The structure of the GBD-C construct shows that residues of the C region needed for activation of and affinity toward the Arp2/3 complex are buried in the interface (10, 13, 39). Biochemical data show that BGBD fragments compete with the Arp2/3 complex for VCA binding in *cis* (30) and *trans* (31, 35). Finally, the activity of a large range of GBD-VCA and BGBD-VCA proteins toward the Arp2/3 complex correlates quantitatively with the allosteric equilibrium of the free proteins (30). Thus, autoinhibition is due to sequestration of the C region by the GBD. Additional functional regions outside of the GBD enhance autoinhibition by further stabilizing the GBD-C fold (e.g., interactions between the B and A regions in BGBD-VCA proteins).

A distinct inhibitory mechanism has been proposed in which the closed protein (BGBD bound to the VCA) binds to and actively inhibits the Arp2/3 complex through contacts of the B and A regions (32). It is difficult to reconcile this second model with the data described above, and we favor the first model for its consistency with the more extensive and quantitative body of evidence. Importantly, both models lead to essentially identical thermodynamic descriptions of the system, because both invoke inhibited states that are stabilized by multiple interactions, and an active state involving the free VCA. Thus, they lead to nearly identical descriptions of signal integration, and we will not distinguish them in our discussions of this issue below.

Mechanism of Allosteric Activation and Inhibition

Physical studies have revealed diverse structural mechanisms by which the allosteric equilibrium in WASP/N-WASP can be controlled by ligands. The unfolded nature of the GBD in active WASP plays an important role in enabling this diversity by allowing a wide range of conformations to be recognized. This property also affords novel means of WASP inhibition through conformational stabilization. These ideas can be incorporated into thermodynamic descriptions of WASP regulation.

Structural basis of Cdc42 activation. Cdc42 is the best-characterized activator of WASP and N-WASP. GTP-bound Cdc42 binds the WASP GBD, destabilizing its interactions with the VCA and leading to activation.

In the solution structure of the Cdc42:GBD complex (**Figure 2*f***), the CRIB motif extends in a linear fashion along the β2 strand and contiguous switch I region of the GTPase, forming an irregular β-strand (41). This interaction is very similar to that seen in complexes of Cdc42 with the CRIB domains from PAK1 (46, 47), PAK6 (X. Yang, E. Ugochukwu, J. Elkins, M. Soundararajan, J. Eswaran, A.C.W. Pike, N. Burgess, J.E. Debreczeni, O. Gileadi, S. Knapp, D. Doyle, unpublished, RCSB accession number 2odb) and PAR6 (48), and the interaction is more distantly related to the complex with the longer CRIB motif of ACK (49). Contacts to switch I enable CRIB effectors to distinguish the GTP and GDP nucleotide states of Cdc42 and Rac (41, 49). Hydrogen bonds between the side chains of the conserved HXXH element in the CRIB motif with the side chain of D38 in Cdc42 (and Rac) appear to select against Rho, which contains a Glu at this position (41). WASP sequences between the CRIB motif and B region contribute to selectivity against Rac (50). Contacts of the B region to the β2-β3 turn and α5 also make significant contributions to the interactions between WASP and Cdc42 (36). Neutralizing mutations to the B region of WASP can decrease the association rate and affinity for Cdc42 by over two orders of magnitude. This electrostatic steering effect also contributes to the specificity of WASP for Cdc42 over the closely related GTPase, TC10.

The CRIB motif is unstructured in the autoinhibited GBD-VCA protein (39). Thus, its interactions with Cdc42 do not drive WASP activation directly. In the Cdc42:WASP complex, WASP residues following the CRIB motif form a β-hairpin and α-helix, which make hydrophobic contacts to the switch I and switch II regions of Cdc42 (41). These interactions also likely contribute to nucleotide switch sensitivity. The secondary structure elements are nearly identical to those observed in the autoinhibited structure (compare panels *c* and *f* in **Figure 2**). However, packing of these conserved elements against Cdc42 is incompatible with the autoinhibited domain. Extraction of the β-hairpin and α1 from the autoinhibited domain by Cdc42 thus greatly destabilizes the domain, releasing the VCA. This structural model is consistent with a 400-fold increase in hydrogen exchange rates in the core of the GBD upon binding to Cdc42 (45). A highly analogous regulatory mechanism is also found for Pak, where Cdc42 binding disrupts a similar autoinhibitory domain, leading to activation of the adjacent kinase domain (51).

Structural basis of activation by Src family SH2 domains. In response to various stimuli, the GBD of WASP/N-WASP becomes phosphorylated by Src family kinases on a conserved tyrosine side chain, Y291/Y256 (52–56). This modification is needed for processes that include T cell activation (52), neuronal differentiation (53), and intracellular movement of the pathogen *Shigella flexneri* (54), and provides a mechanism for WASP activation by SH2 domains (57, 58).

In the autoinhibited GBD-C structure, Y291 (on α3) is >90% buried in the interface between layers one and two, with only the phenolic hydroxyl group exposed (**Figure 2*e***). The sequence surrounding Y291 comprises a consensus binding motif for the SH2 domains of Src family kinases (59). Structural studies have shown that Src family SH2 domains bind their ligands in extended conformation (60), which is incompatible with the helical conformation of α3 in the autoinhibited structure. Thus, SH2 domain binding destabilizes the α3 helix of phosphorylated WASP/N-WASP, disrupting the GBD-C domain, leading to release of the VCA (57, 58).

Structural basis of activation by $EspF_U$. Enterohemorrhagic *E. coli* (EHEC) is a food-borne human pathogen that adheres to the surface of intestinal epithelial cells and injects a protein named $EspF_U$ into the host cytoplasm (61, 62). $EspF_U$ binds and locally activates N-WASP, leading to formation of actin-rich "pedestals" beneath the bound bacterium (63).

$EspF_U$ contains 2–6 nearly exact repeats of a 27-residue hydrophobic segment and a 20-residue proline-rich segment (63, 64). The hydrophobic segment binds the GBD of WASP and N-WASP with nanomolar affinity, displacing the VCA (65, 66). In the solution structure of a complex of the WASP GBD bound to a single $EspF_U$ repeat, the GBD adopts a conformation that is essentially identical to that observed in the autoinhibited GBD-VCA structure (**Figure 2*g***) (65). The hydrophobic $EspF_U$ segment forms an amphipathic helix, which closely mimics the GBD layer two interactions with the VCA. Mutagenesis of the contact sites showed that side chains of the helix contribute appreciably to affinity between the two molecules (65, 66). $EspF_U$ residues C-terminal to the helix form an extended arm, which lies across the top of the GBD and contributes an additional $\sim$4 kcal·mol^{-1} to the stability of the complex. $EspF_U$ has thus developed a high-affinity ligand for WASP/N-WASP that acts by mimicking the VCA and by competitively displacing it from its binding site on the GBD.

Structural basis of inhibition by wiskostatin. Chemical screens for small-molecule inhibitors of actin assembly by the PIP_2/Cdc42/N-WASP/Arp2/3 system have yielded a cyclic peptide, termed 187-1 (67), and a modified carbazole compound, named wiskostatin (68), which both target N-WASP. Both compounds appear to act by stabilizing interactions between the GBD and the VCA.

In the solution structure of a complex between wiskostatin and WASP GBD-C, the small molecule binds in a shallow hydrophobic pocket between layers one and two, at the interface of the β-hairpin and α3 helix, with the carbazole ring directed inward and the amino alcohol side chain directed toward solvent (**Figure 2*b***) (68). Wiskostatin makes no direct contacts to the VCA or the GBD-VCA interface. However, NMR studies showed that wiskostatin can stabilize the autoinhibited conformation of the isolated GBD (68). Such stabilization of the VCA-bound conformation should increase the affinity of the GBD for the VCA, accounting for increased autoinhibition.

One important caveat regarding wiskostatin is that in both published (69) and unpublished (C. Wülfing, personal communication) accounts, the effects of the compound in cells do not seem to be specific to inhibition of WASP/N-WASP. Thus, wiskostatin may not be a useful reagent to probe the function of these proteins in cells. Nevertheless, its discovery and effects in vitro demonstrate the basic principle that allosteric equilibriums can be targeted by small molecules for inhibitory effect (70).

Thermodynamics of activation and inhibition. In MWC formalism, activity of an allosteric protein in the presence of a given concentration of regulatory ligand can be described by three parameters. These are the allosteric equilibrium constant in the free protein (L), the affinity of the ligand for the inhibited state of the protein, and the affinity of the ligand for the active state of the protein (44). The ratio of the affinities ($K_{D,active}/K_{D,inhibited}$), defined as C, provides the driving force to change the allosteric equilibrium (**Figure 3*b***). Activators have higher affinity for the active state ($C < 1$), and inhibitors have higher affinity for the inactive state ($C > 1$). As described above, in the allosteric model for WASP, L is equal to the folding stability of the GBD-VCA domain and can be measured directly (45). For WASP activation, C can also be measured directly, on the basis of the affinity of Cdc42:GTP for the GBD (as a mimic of the active state) and GBD-C (as a hyperstabilized mimic of the fully inhibited state) (30, 45). These measurements have provided $C \sim 2.5 \times 10^{-3}$, indicating that at saturation Cdc42:GTP will shift the allosteric equilibrium by $1/C \sim 400$-fold. Using this parameter, the model was able to quantitatively predict the effect of Cdc42 on hydrogen exchange rates in cores of the GBD-C and the GBD-VCA, the affinities of Cdc42 for WASP proteins with a large range of stabilities, and the effect of Cdc42 on the activity of WASP toward the Arp2/3 complex (30, 45).

One surprising finding of this work was that the nucleotide state of Cdc42 controls not only affinity for WASP, as had been previously shown for the GBD, but also the degree to which the GTPase can drive the allosteric equilibrium (30). That is, Cdc42:GTP not only has a higher affinity for WASP than does Cdc42:GDP, but also a fivefold smaller value of C. Thus, for a given degree of saturation, Cdc42:GTP causes a larger shift in the allosteric equilibrium of WASP upon binding than does Cdc42:GDP. This effect was postulated to increase the fidelity of WASP signaling to the nucleotide switch in Cdc42.

ACTIVATION BY OLIGOMERIZATION

Autoinhibition and its relief by upstream signals can explain many aspects of WASP/N-WASP function. However, several observations could not be explained by this mechanism. This conundrum was resolved through discovery of an additional regulatory mechanism in which dimerization of active WASP proteins greatly increases their potency toward the Arp2/3 complex (22). Below, we explain this effect and show how it clarifies the actions of several WASP activators.

The VCA Dimerization Effect

In 2000, Higgs & Pollard (35) reported that glutathione *S*-transferase (GST)-VCA dimers are roughly 100-fold more potent activators of the Arp2/3 complex than VCA monomers. This observation, which suggested a potentially important aspect of regulation independent of autoinhibitory control, was recently pursued systematically (22). These efforts led to the realization that enhanced potency results generally from VCA dimerization. Thus, several VCAs dimerized through GST, or covalently or using the FKBP:mTOR:rapamycin system, gave similarly high activity. Surprisingly, light-scattering and sedimentation velocity ultracentrifugation experiments showed that a single Arp2/3 complex binds to a VCA dimer, suggesting the presence of a previously unrecognized VCA-binding site on the assembly. Competition experiments suggested that this site is the same as that occupied by a distinct class of Arp2/3 activators, the cortactin proteins, which are known to bind the complex simultaneously with a VCA monomer. Importantly, fluorescence binding data showed that VCA dimers bind the Arp2/3 complex with ~100-fold higher affinity than do monomers. Higher affinity provides a potential explanation for the enhanced actin polymerization kinetics produced by VCA dimers, and current research in several laboratories aims to understand this effect in detail.

The increased affinity of VCA dimers for the Arp2/3 complex provides several predictions regarding the biochemical behavior of dimerizing WASP ligands (22). First, at low WASP concentrations, ligands that create WASP dimers (**Figure 4*a***) can produce higher Arp2/3 activity than can a VCA monomer (**Figure 4*b***). Such ligand-mediated stimulation of WASP beyond the level of the free VCA is termed hyperactivation. Next, at high concentrations, a dimerizing ligand competes for its own binding site and breaks up the WASP dimer (**Figure 4*a***), reducing observed activity (**Figure 4*b***). Thus, titrations of WASP with dimerizing ligands do not produce a monotonic increase in Arp2/3 activity but rather show a peaked response as $WASP_2$ species are populated and then decay (**Figure 4*b***). Finally, monovalent ligands can compete with their divalent counterparts for a binding site on WASP (**Figure 4*a***). Thus, addition of a monovalent ligand to a hyperactivated system disrupts WASP dimers and returns the activity to that of a monomeric VCA or less (**Figure 4*c***), a process termed back titration. These effects have been explicitly researched in three disparate systems, $EspF_U$, SH3 ligands, and membrane phosphoinositides.

Hyperactivation: activity of a WASP protein above that of an equal concentration of isolated VCA, usually owing to oligomerization or clustering

Back titration: the reduced activity of a WASP protein that has been hyperactivated by a dimerizing ligand upon addition of an analogous monomeric ligand

Oligomerization by $EspF_U$

The strongest evidence that dimerization plays a role in WASP/N-WASP regulation derives from studies of $EspF_U$ and its function during pathogenesis of EHEC. $EspF_U$ contains multiple repeat elements, each of which is capable of driving allosteric activation (65). However, constructs with two repeats (2R) or more are better activators of both WASP and N-WASP (22, 66, 71, 72). Consistent with the dimerization model, 2R produces a peaked activity profile when added to N-WASP in Arp2/3-mediated actin assembly assays, with strong hyperactivation at the maximum (22). 1R can block hyperactivation by 2R, resulting in back-titration behavior. 2R:$(N\text{-}WASP)_2$ and 2R:$(N\text{-}WASP)_2$:(Arp2/3) complexes have been directly observed by analytical ultracentrifugation and gel filtration chromatography, respectively (66,

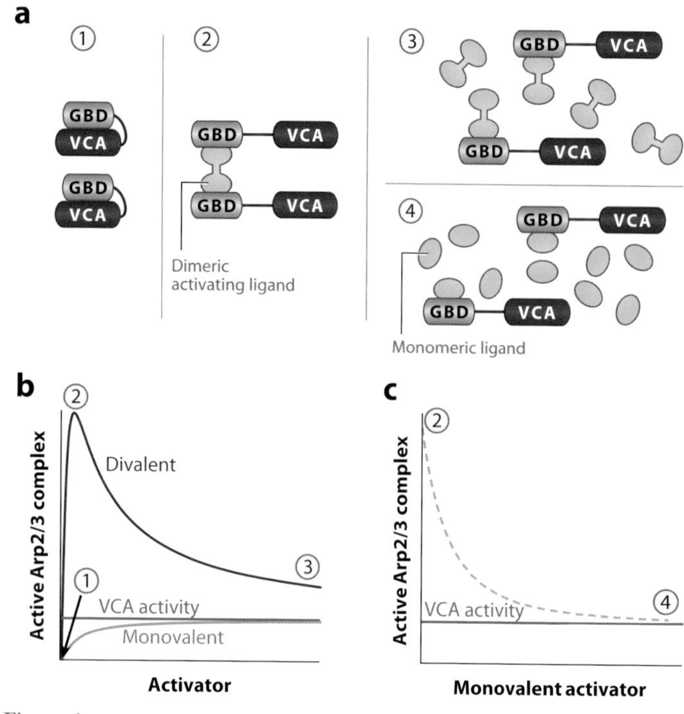

Figure 4

Simulated titrations of mono- and divalent Wiskott-Aldrich syndrome protein (WASP) activators. (*a*) Cartoons of the WASP:activator mixtures. WASP is shown as a blue GTPase-binding domain (GBD) fused to a red VCA. Inactive WASP (1) is bound and activated by either a dimeric activating ligand [*green dumbbells* in (2) and (3)] or a monomeric ligand [*green ovals* in (4)]. (*b*) Computationally modeled titration of a divalent activator into autoinhibited WASP (*purple line*). Allosteric activation of WASP proteins occurs concomitantly with activator binding. Activity is modeled by the concentration of WASP bound to the Arp2/3 complex. In the absence of activator, only inactive WASP is present, and activity is low (1). As divalent activator is added, active dimeric WASP species are formed (2), and activation exceeds that of free VCA (*blue line*). When high concentrations of divalent activator are added, competition for WASP proteins results in disassembly of the dimers, and activity approaches that of free VCA (3). Monovalent activator titration (*orange line*) monotonically increases WASP activity to that of free VCA. (*c*) Computationally modeled titration of monovalent activator into a maximal activity mixture of WASP and divalent activator (*dashed gray line*). The initial conditions are identical to those at (2) in panel *b*. As monovalent activator is added, dimers are disrupted, and activity drops to that of free VCA (*blue line*). Mixtures indicated by circled numbers in panels *b* and *c* are those in the subpanels of *a* with the same number. Figure adapted with permission from Cell Press, *Molecular Cell* (22), copyright © 2008.

72). Finally, increasing $EspF_U$ repeat number correlates with increased actin assembly in cells (22, 66, 71, 72). These data indicate that $EspF_U$ activates WASP/N-WASP by engaging both the allosteric and dimerization mechanisms.

BAR domain: Bin, amphiphysin, and Rvs161/167 domain

Oligomerization by SH3 Ligands

WASP family proteins bind dozens of SH3 domain–containing ligands through the numerous proline motifs in the PRD and in the distinct proline-rich domain of the constitutively associated WIP (23). These ligands can largely be grouped into three categories as recently detailed (6). Most contain multiple SH3 domains, either by the nature of their primary sequence or through homo-oligomerization. In the first category are the multi-SH3 signaling proteins, including Nck and Grb2. The second category, oligomerized SH3 ligands, includes many proteins that contain the dimeric BAR domain, such as PACSIN, Toca-1, and SNX9 proteins. A final major category of WASP family SH3 ligands are the nonreceptor tyrosine kinases, which typically contain only a single SH3 domain. Activation of WASP proteins has been studied for about half of the known SH3 ligands. Of these, Nck, Grb2, WISH, Abi, Cortactin, Abp1, and a number of BAR proteins have been shown to directly enhance WASP activity in vitro (22, 29, 73–84).

Members of the dimeric BAR domain family coordinate membrane dynamics and actin assembly in a variety of processes (6, 85). They act by simultaneously binding membrane through the BAR domain and WASP proteins through an adjacent SH3 domain, potentially dimerizing the WASP proteins. Several BAR proteins directly activate WASP proteins: FBP17 (29, 83), SNX9 (84), PACSIN2/syndaptin (22), EndophilinA (86), and Nwk (79). When examined with an eye toward dimerization, this activation shows the hallmarks of the dimerization mechanism. PACSIN2 and CIP4 can hyperactivate an N-WASP mutant in which autoinhibition is disrupted (22; S.B. Padrick, unpublished). In addition, hyperactivation by PACSIN is lost at high concentrations or when the free PACSIN SH3 domain is added, fulfilling the peaked titration and back-titration predictions of the dimerization mechanism, respectively. Analogous behavior is seen for GST-Nck (22), suggesting that dimerization-mediated N-WASP activation is likely to be widespread among SH3 proteins.

Interestingly, cortactin, which also appears to have two binding sites on the Arp2/3 complex (S.B. Padrick, unpublished), shows similar peaked activation through SH3-proline interactions by GST fusions with faciogenital dysplasia protein (Fgd1) (87).

Activation by High-Density PIP_2

N-WASP is both allosterically activated and oligomerized by the phosphoinositide PIP_2 (31, 32). The N-WASP basic region is a disordered cluster of arginine and lysine residues that likely engages multiple headgroups simultaneously. Thus, the affinity of N-WASP for liposomes depends on their fractional PIP_2 content (i.e., the PIP_2 surface density), as shown in a detailed analysis (34). However, in that study, although strong activation was observed at high PIP_2 density, intermediate densities that were still capable of binding N-WASP did not activate well (34), suggesting that density itself is a parameter of activation. Interpreted in light of the dimerization model, at high density, VCA domains from two different N-WASP molecules could have access to the same Arp2/3 complex. This will result in high-level activation through effective dimerization.

In more recent work, high-density PIP_2 liposomes were found to hyperactivate an N-WASP mutant lacking GBD-VCA contacts (22). Activation declined at the highest liposome concentrations. Similar effects were observed previously in PIP_2-mediated activation of WASP purified from tissue (35). This behavior is similar to the peaked titrations seen with dimerizing ligands, where N-WASP dimers were disrupted by excess ligand. Here, effective N-WASP dimers were disrupted by increased vesicle surface area and consequent decreased N-WASP density. Hyperactivation, but not vesicle binding of N-WASP, was lost when PIP_2 density on individual vesicles was decreased, holding total PIP_2 in solution constant (22). This is analogous to the back-titration phenomenon with dimeric ligands. Thus, activation of WASP proteins by phosphoinositides, and likely membrane recruitment in general, shows the hallmarks of dimerization-mediated activation. We note that a similar effect was also suggested in recent studies of actin assembly by the intracellular pathogen, *Listeria monocytogenes* (88). During infection, *Listeria* displays its own Arp2/3 activator, the ActA protein, on its surface to usurp the actin cytoskeleton. ActA is arrayed at a density equivalent to a GST-mediated dimer, suggesting that the bacterium uses the density-mediated dimerization effect to more potently activate the Arp2/3 complex.

Peaked titration: the activity of a WASP protein that rises and then falls as a dimerizing ligand is added

INTEGRATION OF MULTIPLE SIGNALS BY WASP PROTEINS

The combination of allostery and oligomerization leads to a hierarchical mechanism of WASP regulation (**Figure 5**) (22). Within this mechanism, allostery controls availability of the VCA to the Arp2/3 complex, and dimerization controls affinity for the Arp2/3 complex.

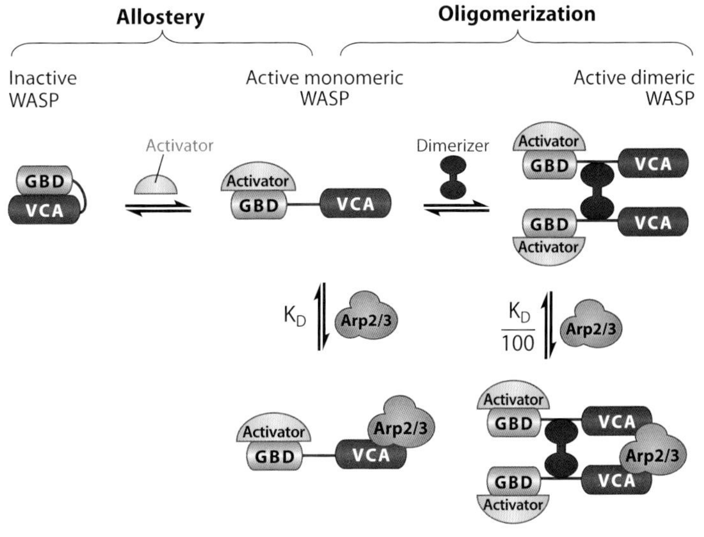

Figure 5

The hierarchical model of Wiskott-Aldrich syndrome protein (WASP) regulation. WASP proteins are regulated hierarchically. An inner layer of allostery controls access to the VCA (verprolin homology, central hydrophobic, and acidic regions), and an outer layer of dimerization/oligomerization controls the affinity of the VCA for the Arp2/3 complex. The two layers are thermodynamically coupled, so that binding of the Arp2/3 complex to a VCA dimer will also shift the allosteric equilibrium toward the active state. Figure adapted with permission from Cell Press, *Molecular Cell* (22), copyright 2008. Abbreviations: Arp2/3, the actin-related protein 2/actin-related protein 3 complex; K_D, dissociation constant; GBD, GTPase-binding domain.

Maximal activation of the WASP-Arp2/3 pathway will occur when both branches are simultaneously engaged. We note that the two branches are mechanistically distinct, but thermodynamically coupled, such that binding of the Arp2/3 complex to a VCA dimer will also shift the allosteric equilibrium toward the active state. Here, we show how cooperativity between multiple inputs arises in each branch, and how the branches function cooperatively together. These two layers of regulation thus allow WASP proteins to act as powerful signal integration devices in the cell.

L parameter: in the MWC-based model of two-state allosteric equilibrium, the population ratio of inactive to active states of the free protein

C parameter: in the MWC-based model of two-state allosteric equilibrium, the ratio of dissociation constants of a ligand for active and inactive states ($K_{D,act}/K_{D,inact}$)

Allosteric Activation of WASP by Two Ligands

The autoinhibited state of WASP is stabilized by multiple inhibitory interactions (e.g., GBD:C and B:A contacts, see above) that act together to repress the activity of the VCA. Activators such as Cdc42, PIP_2, and SH2 domains target distinct sequences in the autoinhibited fold. However, because the ligands all act on the same allosteric equilibrium, their binding and effects on that equilibrium are thermodynamically coupled. Thus, two ligands acting together will provide greater engagement and activation of WASP than either acting alone.

Two related formalisms have been developed to describe this underlying concept quantitatively (**Figures 3*c*,*d***). The first posits that WASP acts in a binary fashion, where it is completely inactive in free form and completely active when bound to any single ligand (32). Because activating ligands bind more tightly to the active form, when one ligand activates WASP, the second can bind with higher affinity. The degree of thermodynamic coupling between the two ligands can be expressed as a cooperativity factor whereby the affinity of a second activator is enhanced by a first (**Figure 3*d***) (32). As described below, this formalism nicely explains cooperative binding to and synergistic activation of WASP/N-WASP by the Cdc42-PIP_2 and Cdc42-SH2 activator pairs (32, 58).

The second formalism extends the first by including the WASP allosteric equilibrium quantitatively (30, 45), giving a multiligand version of the MWC allosteric model described in previous sections (**Figure 3*c***). In this model, the autoinhibitory equilibrium in the presence of multiple saturating ligands is given by $L\prod C_i$, where L is the equilibrium constant in free WASP ([inactive]/[active]), and C_i is the ratio of affinity for the active and inactive states of WASP for each ligand, respectively. Thus, the cooperative effects of multiple ligands are reflected in their multiplicative action on the regulatory equilibrium in WASP. This model allows inhibitory ligands such as WIP to be accounted for, in addition to activators; activation of a WASP:WIP complex (where $C_{WIP} > 1$, favoring the inhibited state of WASP) can be described as requiring an activator with a strong binding preference for the active state (smaller $C \ll 1$) or cooperation between multiple activators (each with their own $C < 1$). In general, this model explains how the balance of L and C parameters tunes the system to respond to one or more activators.

Synergy between Cdc42 and PIP_2. The minimal autoinhibited GBD-VCA WASP construct is further stabilized by additional inhibitory contacts such as those of the B region. Although B region inhibition may derive from B-A contacts or from B-Arp2/3-A contacts (see above), the effects on binding and activation are the same. Activation by Cdc42 is understood structurally (39). PIP_2 vesicles engage the allosteric mechanism by interacting with the B region (31), thus destabilizing its contacts with A or the Arp2/3 complex. They also engage the oligomerizing mechanism (see above). Individually, Cdc42 and PIP_2 can each activate weakly inhibited systems (e.g., the *trans* complex of BGBD + the VCA). But combinations of the two are required to activate more strongly inhibited constructs (32, 35, 89). Importantly, Cdc42 and PIP_2 activate WASP and N-WASP by binding to distinct regions of the BGBD element (**Figure 1**). Thus, their binding is coupled through the autoinhibitory equilibrium. Systematic measurement of N-WASP activation by the two ligands found a binding cooperativity factor of 350 (**Figure 3*d***) (32). In the MWC formalism, the B region shifts the autoinhibitory equilibrium toward the inhibited state such that $L \gg 1$ (30). Thus $C_{Cdc42}{}^{*}L$ remains greater than one, and the GTPase alone cannot strongly activate. PIP_2 has a similarly small effect. However, the combination of the two together can activate N-WASP (29, 32) (at saturation $C_{Cdc42}C_{PIP2}L < 1$). Even though this analysis only considers the allosteric part of the interaction, it explains the requirement for multiple allosteric effectors in activation of the endogenous WASP:WIP complex, which has additional autoinhibitory contacts.

Synergy between Cdc42 and WASP phosphorylation. Phosphorylation of WASP/N-WASP on the conserved Y291/Y256 (see above) exemplifies signal integration through the allosteric equilibrium in several respects. Tyrosine 291 is >90% buried by layer two-layer three contacts in the autoinhibited GBD-C structure and thus is a poor substrate for kinases and phosphatases (**Figure 2*e***) (57). Destabilization of the GBD-C domain by Cdc42 accelerates phosphorylation and dephosphorylation by 40-fold and 30-fold, respectively (58). The fact that Y291/pY291 are only accessible in the open state couples allosteric effectors to covalent modifications. This results in WASP having a covalently encoded memory of recent activation.

Phosphorylation of Y291/Y256 has two biochemical effects (57). First, phosphorylated GBD-VCA is destabilized, enhancing the basal activity of WASP (i.e., decreasing L in the MWC formalism). Second, phosphorylation creates a binding site for SH2 domains, which serve as allosteric activators of WASP (discussed above). Because SH2 and Cdc42 bind different regions of the GBD (**Figure 1**), they can bind simultaneously and thus act cooperatively on the allosteric equilibrium. A cooperativity factor of 50 was measured for the two ligands acting on phosphorylated WASP (58). Thus, the location of Y291/256 in the autoinhibited fold enables WASP/N-WASP to integrate signals in both the phosphorylation event itself and in the functional consequences of phosphorylation.

Engineering WASP. WASP has proven to be an excellent platform for illustrating the principles of allostery and its evolution through protein design (34, 90, 91). By appending additional interaction modules to the GBD-VCA core, Lim and colleagues (91) showed that a wide variety of switch-like behaviors can be readily created. These include "OR gates," where only a single ligand can activate; "AND gates," where multiple ligands are required for activation; and even antagonistic functionality, where multiple ligands oppose each other's actions, depending on the balance between autoinhibitory stability and cooperativity in effector binding. When multiple identical inhibitory modules are appended, ultrasensitivity akin to that observed in hemoglobin can be created (90). These data suggest that evolutionary shuffling of modular elements in signaling proteins provides a mechanism to readily create allosteric systems with complex behaviors.

Integration of Multiple Oligomerizing Signals

When multiple dimeric/oligomeric ligands contact distinct regions of WASP simultaneously, their binding will occur cooperatively. That is, when one ligand dimerizes WASP, a second will bind tightly owing to an avidity effect. Thus, ligands acting in combination will produce WASP oligomers at lower concentrations than any individual ligand acting alone. This will provide an additional mechanism of integrating signals to enhance WASP activity that is distinct from the allosteric mechanisms discussed above.

SH3 interactions are a likely source of such cooperativity, but similar principles will hold for any combination of oligomerizing factors. WASP proteins and WASP:WIP complexes (23) have numerous proline motifs that bind SH3 domains. The specificity of an SH3 dimer for the different motifs will determine its propensity to engage a single WASP molecule in bivalent fashion or to dimerize two WASP molecules. Only the latter binding mode will strongly affect WASP activity (see above). Thus, the activation strength of a given SH3 dimer will be dictated significantly by its binding specificity. An analogous but more complex argument can be made for SH3 ligands with multiple, different SH3 domains (e.g., Nck, which has three SH3 domains). This reasoning is consistent with the wide variability in the effects of SH3 dimers on WASP activity (e.g., Reference 74). SH3 specificity will also affect cooperativity between different ligands. For example, two dimeric SH3 ligands, each highly specific for a single proline motif in WASP, will strongly reinforce each other's binding and produce an avidity-enhanced multi-WASP complex. But ligands with overlapping specificities will compete for the same proline motifs and will not act cooperatively. SH3 ligand specificity for the different proline motifs in the mammalian WASP proteins has not been extensively explored (although see Reference 92 for a recent example). However, such an analysis was done for the proline motifs in the *Saccharomyces cerevisiae* WASP protein, Las17p/Bee1 (93). There, the five proline motifs in the PRD were systematically mapped against all SH3 proteins in the yeast genome. Some SH3 domains interacted strongly with only a single motif (e.g., Bbc1, Sho1, and Myo3), and others (e.g., Ygr136, Rvs167, and Bzz1) bound multiple motifs. Three of the motifs bound only a single SH3 domain, whereas two bound many. Clearly, there is potential for cooperative, neutral, and competitive interactions in this system. This type of analysis will be instrumental in predicting the effects of SH3 proteins acting individually and in combination to control WASP protein assembly and activity.

Localization of WASP proteins to membranes can similarly cooperate with dimerizing ligands. Oligomerization can allow WASP proteins to bind membranes in multivalent fashion, increasing affinity. Thus, assembly and membrane binding of complexes can be synergistic. For example, association of N-WASP with PIP_2 vesicles is relatively weak (34). However, N-WASP dimers created by GST-Nck will have two basic regions and thus a greater affinity for PIP_2 vesicles, consistent with strongly cooperative activation of N-WASP by GST-Nck and PIP_2 (80).

Combining Allosteric Activation and Oligomerization

Few WASP activators are purely allosteric or purely oligomerizing factors. At high density, membrane-associated Cdc42 will be both an allosteric activator and a clustering effector. This is consistent with its high activity relative to soluble Cdc42 (35). Likewise, PIP_2 both destabilizes BGBD-VCA interactions (31) and can be an oligomerizer at high density (22, 34, 80). Thus, the actions of ligands functioning individually and in combination should be considered in terms of both aspects of regulation.

IQGAP1 is a well-studied signaling protein that can activate N-WASP. IQGAP1 functions through allostery, as it can activate a BGBD:VCA complex assembled in *trans* (94). However, full-length IQGAP1 is oligomerized through an N-terminal domain, and its activation of N-WASP showed a peaked titration similar to that of PACSIN2 (22, 95). Thus, IQGAP1 likely acts through both altering the allosteric equilibrium in N-WASP and creating high-potency N-WASP oligomers.

The Toca proteins (Toca-1, CIP4, and FBP17) provide a more complex example (**Figure 6*a***). Toca proteins have BAR, SH3, and Cdc42-binding HR1 domains (29, 96). When active Cdc42 and PIP_2 are both present in membranes, N-WASP, Toca, Cdc42, and PIP_2 will assemble a multiply reinforced complex. Toca will bind membranes through cooperative BAR-lipid and HR1-Cdc42 interactions. N-WASP can then simultaneously associate in a high-affinity, trivalent fashion with the Toca SH3 domain, PIP_2, and Cdc42, through its PRD, B region, and GBD, respectively. This will provide high-level engagement of N-WASP by Cdc42 and consequently potent allosteric activation. Finally, as the Toca protein is a dimer, two N-WASP molecules

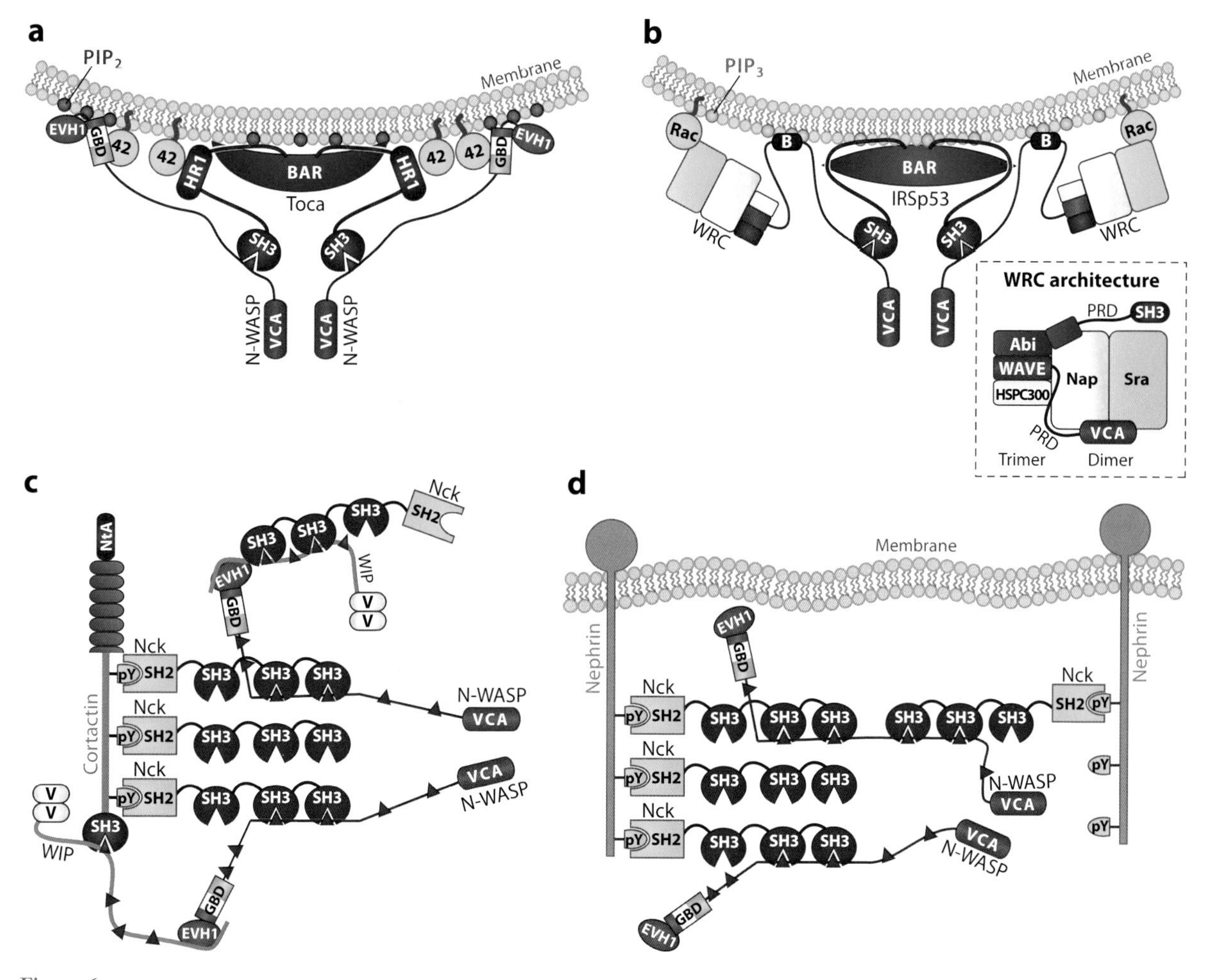

Figure 6

Cooperative activation of the N-WASP/WAVE regulatory complex (WRC) by higher-order complex formation. Multiprotein complexes bring two or more VCAs (verprolin homology, central hydrophobic, and acidic regions) together. (*a*) Schematic illustration of the complex formed by N-WASP, Toca (*dark blue*, with the BAR, SH3 and HR1 domains labeled), Cdc42 (42), and PIP_2 (*red circles*) at membranes. (*b*) Schematic illustration of the complex formed by the WRC, IRSp53 (*blue*, with the BAR and SH3 domains labeled), Rac, and PIP_3 (*orange circles*) at membranes. Inset shows organization of the WRC. (*c*) Schematic illustration of the complex formed by phosphorylated Cortactin, Nck, N-WASP, and WIP. (*d*) Schematic illustration of the complex formed by phosphorylated nephrin, Nck (*purple ovals* and *green square*) and N-WASP (domain colors as in **Figure 1**). Abbreviations: Abi, Abl interactor; B, basic region; BAR, Bin/amphiphysin/Rvs homology domain; EVH1, Ena/VASP homology 1; GBD, GTPase-binding domain; HR1, protein kinase C-related kinase homology region 1; IRSp53, insulin receptor tyrosine kinase substrate p53; Nap1, Nck-associated protein; N-WASP, neural Wiskott-Aldrich syndrome protein; PIP_3, phosphatidylinositol 3,4,5-triphosphate; PRD, proline-rich domain; pY, phosphotyrosine; Rac, GTP bound Rac GTPase; SH3, Src-homology domain 3; Sra1, specifically Rac-associated protein; Toca, transducer of Cdc42-dependent actin assembly family protein; V, verprolin homology region; WAVE, WASP family verprolin-homologous protein; WIP, WASP interactin protein.

can be brought together to allow potent recruitment and activation of the Arp2/3 complex. As these interactions all contribute thermodynamically to a single assembly, the individual contacts will reinforce one another. Furthermore, the multivalency of N-WASP (PRD) and Toca (SH3) could also provide mechanisms of higher-order oligomerization, leading to spatial clustering of all components and sustaining transiently high local activation levels. As discussed below, such assemblies are abundant in WASP biology and likely play an important role in governing the specificity and dynamics of actin regulatory signaling.

HIGHER-ORDER OLIGOMERIZATION FURTHER STIMULATES WASP ACTIVATION

Allostery and dimerization are the core elements of the hierarchical model of WASP regulation. However, a probablistic argument suggests that higher-order oligomerization may further enhance WASP activity by increasing the likelihood that two active WASP molecules will be spatially proximal (22). This effect may explain the frequent appearance of WASP proteins in large assemblies in vivo. Below we explain two related versions of higher-order oligomerization that lead to WASP activation and illustrate how this effect could play a role in a number of signaling systems.

Clustering of $EspF_U$

Two-repeat fragments of $EspF_U$ hyperactivate WASP by engaging both allostery and dimerization. However, $EspF_U$ proteins with >2 repeats produce even greater activation (22, 62, 66, 72). Statistical considerations provide a likely explanation for this effect. When the fractional saturation of each $EspF_U$ repeat by WASP is small ($P \ll 1$), the fraction of $EspF_U$ 2R molecules with two bound WASP molecules is equal to P^2. However, as the number of repeats (n) in $EspF_U$ increases, there are more ways that any two repeats can bind WASP. So the probability of creating a WASP dimer increases. For a fixed total repeat concentration, the fraction of complexes containing two or more WASP molecules scales with (n-1) (22). Thus, higher-order multimerization of WASP further enhances actin polymerization by $EspF_U$.

$EspF_U$ and its relative EspF from enteropathogenic *E. coli* have mechanisms to undergo oligomerization in vivo. Each repeat of these proteins contains both a WASP-binding element and a proline-rich element. The latter has been shown to bind the SH3 domain of dimeric BAR proteins, EspF to Snx9 (97) and $EspF_U$ to IRSp53 (92) and IRTKS (98). These interactions, which can provide additional clustering of EspF/$EspF_U$-bound WASP molecules (98), are necessary for bacterially induced actin assembly in cells (92, 97, 98). Moreover, EspF and $EspF_U$ are both recruited to the transmembrane protein Tir, which is itself clustered through extracellular interactions with the intimin protein expressed on the bacterial surface (63, 92, 98). Thus, actin is assembled by this system through N-WASP molecules that are clustered at multiple levels, consistent with the idea that creation of large WASP arrays can enhance activation of the Arp2/3 complex.

Clustering by Endogenous Effectors

WASP proteins oligomerized by natural ligands can be viewed analogously (22). In a WASP cluster of size n, if the probability of any molecule being in the active state is P (dictated by the allosteric equilibrium), then the probability of any two molecules being simultaneously active is approximately $(n-1) \times P^2$ (for low values of P). For example, a cluster containing ten WASP proteins is nine times more likely to have an active VCA dimer than a cluster containing only two WASP proteins. Thus, oligomerization, or arraying WASP proteins in close proximity on a membrane, lowers the degree of WASP allosteric activation needed for high-level stimulation of the Arp2/3 complex. Geometric considerations should limit this effect in very large clusters. But the long linkers connecting the VCA to other elements, and the ability of some clusters to rearrange, suggest that it should remain significant out to large sizes (see below).

This effect is likely significant to the actions of BAR proteins, which bind membranes in close-packed, high-density arrays, causing formation of tubular structures (83, 99, 100). Many BAR proteins contain SH3 domains that recruit WASP/N-WASP and the Arp2/3 complex to these arrays, causing actin polymerization (85, 101). Recent studies have shown a high degree of cooperativity between membrane binding of BAR proteins and actin polymerization through the N-WASP-Arp2/3 complex (29, 86). The surface of BAR-induced membrane tubules is coated with SH3 domains at high density. These structures are known to recruit N-WASPs (83, 102). Given the geometry of the BAR domain array on membranes (100), and the spacing between VCA elements in known hyperactive N-WASP dimers, clusters of more than ten N-WASP molecules should be able to access a single Arp2/3 complex (22). Thus, the dimerization effect likely contributes to actin assembly by BAR domain arrays.

The idea that SH3 clusters will result in larger, better Arp2/3-activating WASP complexes is consistent with in vitro biochemical data. Dimeric SNX9 was shown to activate N-WASP in solution, but SNX9 aggregates were more potent (84). Addition of PIP_2 lipids produced even more potent SNX9-dependent activation. Similarly, Toca-1 and FBP17 activated the N-WASP:WIP complex in the presence of added membranes (29) (discussed above). The less-autoinhibited N-WASP (without WIP) was activated by FBP17 alone, but FBP17 activation was enhanced by a phospholipid surface. Finally, endophilin A showed a similar enhancement of N-WASP stimulatory activity in the presence of acidic lipids (86). Thus, activation of N-WASP by many BAR proteins appears to be generally enhanced by assembly onto liposomes.

A distinct assembly containing cortactin, Nck, WIP, and N-WASP has been elegantly dissected biochemically (**Figure 6*c***) (81). Cortactin is a weak activator of the Arp2/3 complex that can also act as a molecular scaffold. Cortactin is phosphorylated on three tyrosine sites near its C terminus upon cell stimulation, and these sites bind the Nck SH2 domain (81). The SH3 domains in the p-cortactin:Nck complex (one from cortactin and three from each Nck) can bind the proline motifs in WIP (103) and N-WASP (80) to form a large assembly. The high multivalency in this system (three pTyr on p-cortactin; numerous proline motifs in N-WASP and WIP; and three SH3 domains in Nck) suggests a very cooperative assembly of the components, consistent with surface plasmon resonance data (81). Moreover, the size of the complex, although not measured in this work, is expected to be large. A key finding in this study was that phosphorylation of cortactin greatly increased the Arp2/3-mediated actin assembly of mixtures of cortactin, Nck, and N-WASP. Thus, assembling Nck:N-WASP complexes on p-cortactin increases activity toward the Arp2/3 complex. The further enhancement seen when WIP was also added suggests an advantage to creating even higher-order assemblies. In the future, it will be interesting to examine correlations between complex stoichiometry, size, and activity in this system.

A conceptually related system controls the assembly of actin-rich podocytes in the kidney (**Figure 6*d***) (104, 105). These cellular structures require phosphorylation of multiple tyrosine residues on the intracellular tail of the cell surface receptor, nephrin (104, 105). Analogous to p-cortactin, these phosphotyrosines each recruit the SH2 domain of Nck, allowing activated nephrin to display an array of Nck SH3 domains. This results in actin assembly, likely through recruitment of N-WASP:WIP complexes. Interestingly, strong actin assembly by this system in cells requires multiplicity, either in the number of pTyr sites on nephrin or the number of SH3 domains in Nck (104), consistent with the idea that larger clusters of N-WASP have higher activity and are functionally important. This effect is similar to the antibody-dependent clustering to Nck, leading to N-WASP-dependent actin polymerization (106).

Finally, there are additional multicomponent complexes that probably contain multiple WASP proteins but have not been characterized in detail. These include the WASP:WIP:Nck:SLAP:SLP76:VASP assembly formed in T cells upon T cell receptor ligation (24, 107), and the Las17p:Vrp:Myo3:Myo5:Bzz1:Bbc1p:Hsp70p complex that can be isolated from budding yeast (108). Physical characterization of these complexes to understand their energetics, stoichiometry, structural organization, and dynamics as well as the relation of these properties to activity represents an exciting and challenging area for future investigation.

WAVE: WASP family verprolin homologous protein (also known as Scar)

WAVE regulatory complex (WRC): a complex comprising WAVE/Scar, Abi, HSPC300, Nap1/Hem2, and Sra1/Pir121/CYFIP

PIP_3: phosphatidylinositol 3,4,5-triphosphate

OTHER WASP FAMILY MEMBERS

In addition to WASP and N-WASP, the WAVE, WASH, WHAMM, and JMY proteins have VCA domains capable of activating the Arp2/3 complex. Many of the regulatory principles discussed above for WASP and N-WASP appear to hold for these proteins as well.

Regulation of WAVE

Shortly after the discovery of WASP and N-WASP, the WAVE/Scar (WASP family verprolin-homologous protein/suppressor of cyclic AMP receptor) proteins were identified in *Dictyostelium* (109) and vertebrates (**Figure 1**) (15, 109, 110). A variety of data indicate the WAVEs function downstream of the Rac GTPase is to stimulate actin assembly through the Arp2/3 complex (3, 4). WAVE proteins contain a C-terminal VCA domain, but unlike WASP, isolated WAVEs are not autoinhibited (8). Instead, WAVEs are regulated by constitutive incorporation into a heteropentameric complex (**Figure 6*b***, inset), consisting of WAVE, Abi (Abl interactor), HSPC300 (hematopoietic stem progenitor cell 300), Nap/Hem (Nck-associated protein/hematopoietic protein) and Sra1/Pir121/CYFIP (specifically Rac-associated protein/p53-inducible mRNA 121/cytoplasmic FMRP interacting protein), which we term the WAVE regulatory complex (WRC) (111). The complex appears to be organized as a trimer held together by contacts between HSPC300 and N-terminal elements of WAVE and Abi, and an Sra:Nap dimer (22, 112–116). Contacts between Nap and Abi link the dimer and trimer (112). Homologs of WRC components all incorporate into similar complexes (111, 112, 116, 117). Since its initial biochemical purification (111), genetic studies have shown that the WRC is conserved in plants, slime mold, flies, worms, and vertebrates (3, 4). As reviewed elsewhere, cell biological and genetic data have established the functional importance of the WRC in cellular processes, including spreading, motility, and cell-cell adhesion (3, 4).

Recent studies have shown that, although the molecular details are different, the basic principles of WAVE and WASP regulation are highly analogous. Although the basal activity of the WRC toward the Arp2/3 complex has been controversial (111, 115, 118, 119), recent, definitive studies have shown that both the human and fly complexes are strongly inhibited (114, 120). Inhibition involves binding of the Sra:Nap dimer to a constitutively active trimer (114). This behavior suggests an analogy to GBD-VCA contacts in WASP, although the molecular mechanism of inhibition awaits further studies. Highly purified inactive WRC can be activated in a nucleotide-dependent fashion by Rac, which is known to bind Sra1 (121), similar to WASP activation by Cdc42 (111, 114). Rac may function with other factors, as activation requires relatively high concentrations of the GTPase (5–10 μM). For example, PIP_3 binds a conserved basic region near the center of WAVE (122). This interaction is needed for proper WAVE localization and also contributes to in vitro activation by Rac, although the mechanism of this effect is not known (122, 123). WAVE is phosphorylated at numerous tyrosine, serine, and threonine sites in its N-terminal PRD and VCA elements (reviewed in Reference 3). VCA phosphorylation increases affinity for the Arp2/3 complex but paradoxically decreases activity (124, 125). WAVE2 can be phosphorylated on the conserved Tyr150 by Abl upon cell stimulation, but its effects on WRC activity have not been examined (113). It is not yet known whether any of the phosphorylation events occur cooperatively with Rac binding, as in the Cdc42-WASP system (57). Finally, the large PRD in WAVE and an analogous region in Abi endow the WRC with numerous SH3-binding sites. Many SH3 proteins bind these elements, including multi-SH3 adaptors [e.g., Nck (111) and Grb2 (126)] and dimeric BAR proteins [e.g., IRSp53 (123), Wrp (127), and Tuba (128)]. Thus, the WRC appears to be regulated by the same hierarchical process as WASP, with collections of allosteric and dimerizing/oligomerizing ligands acting in a cooperative fashion to control activity toward the Arp2/3 complex. Both WASP and WAVE have similar responses to the same classes of signals, suggesting that there is an evolutionary driving force for WASP proteins to stimulate actin assembly in response to phosphoinositide, SH3, and Rho GTPase cues.

Recent reports illustrate how many of these interactions could work together to cluster and activate the WRC at membranes (**Figure 6*b***). Takenawa and colleagues (123) have shown that the combination of Rac, IRSp53, and PIP_3 activates the WRC better than any of the individual effectors alone. This system is highly analogous to the activation of N-WASP by Cdc42, Toca-1, and PIP_2 (see above). In both cases, the GTPase, BAR protein, and phospholipid recognize different elements of the WRC/N-WASP and likely can bind it simultaneously. When the ligands are present together at a membrane, they should act in a multivalent fashion to strongly recruit the WRC/N-WASP. The ligands become coupled through WRC/N-WASP, strengthening membrane association of the BAR protein and causing a clustering of all components. Clustering should be further enhanced by the dimeric nature of the BAR protein. In the case of the WRC, these cooperative interactions should effectively overcome the modest affinity of Rac, enabling the WRC to be engaged and allosterically activated to a high degree. Thus, through the combination of all three ligands, both the allosteric and oligomerization mechanisms can be engaged, leading to strong, localized activation of the Arp2/3 complex.

Other Family Members

Several vertebrate WASP family members have been discovered only recently: WASH (129), WHAMM (130), and JMY (**Figure 1**) (131). All of these proteins have a C-terminal VCA domain, which in isolation can activate the Arp2/3 complex, and an adjacent PRD that can bind SH3 domains. Their more N-terminal elements are distinct, and their molecular mechanisms of regulation are not yet known. Current work shows that WASH is incorporated into a multiprotein complex of a size and complexity similar to the WRC (D. Jia, T. Gomez, D. Billadeau, unpublished), as is WHAMM (K. Campellone and M. Welch, personal communication). However, it is not known whether these assemblies are active toward the Arp2/3 complex and thus how functionally analogous they are to the WRC. WASH activity has also been linked to the GTPase Rho (132), suggesting analogies to WASP and WAVE, although it is unclear if Rho acts directly or indirectly. Characterization, reconstitution, and assay of these complexes is still needed to understand their mechanisms of regulation and their relationships to WASP and/or WAVE.

ACTIN FEEDBACK

This review focuses on upstream signals that act on WASP proteins. However, there is increasing evidence that the downstream actin network also feeds back into the WASP proteins and their assemblies. Taunton and coworkers (17) have shown that the V region of membrane-associated N-WASP, after delivering actin to the Arp2/3 branch, can remain attached to the filament-barbed end. This tethers the actin network to membranes and may enhance membrane recruitment and/or clustering of WASP proteins. This idea is consistent with recent data showing Arp2/3 complex-dependent polarization of N-WASP on rocketing vesicles (133). Interactions with the actin network should further stabilize the multiprotein assemblies that recruit and activate WASP at membranes. Paradoxically, the interaction of the VCA with the Arp2/3 complex is also necessary for rapid exchange of N-WASP into and out of membrane complexes created by vaccinia viruses (134). N-WASP exchange is necessary for efficient actin-based movement of the virus. Such negative feedback between the actin-Arp2/3 complex and the VCA may also account for the rapid membrane dissociation of the WRC, observed in mobile waves of actin assembly during cell spreading and motility (135). As more complete biochemical descriptions of WASP activation emerge, the feedback from the actin network needs to be included to advance beyond bulk biochemical descriptions into descriptions of cellular processes.

CONCLUDING REMARKS

The hierarchical model provides a unifying regulatory framework for the WASP family. The allosteric arm quantitatively explains autoinhibition and its relief by individual or combined ligands that act on the folding equilibrium of the GBD-VCA element. The dimerization/oligomerization arm explains how formation of assemblies of WASP proteins by multivalent ligands and membrane association enhances stimulation of the Arp2/3 complex. Together, the two arms explain how WASP proteins are endowed with the ability to integrate a large number of diverse signals through interactions of multiple domains to precisely control actin assembly in cells. The model also raises many new questions. How do two WASP molecules engage and activate the Arp2/3 complex? What are the physical properties of the large, multivalent assemblies that contain WASP proteins; how do these affect and respond through feedback to downstream actin dynamics? Can other more recently discovered WASP family members also be described by the hierarchical model, and if so, what signals control the two arms and allow them to work together? These questions and others that extend from them will drive research in this exciting area for years to come.

SUMMARY POINTS

1. WASP proteins are regulated by two processes, allostery and dimerization.
2. Allostery controls access of the VCA to the Arp2/3 complex and is understood structurally and energetically.
3. Dimerization controls affinity of WASP protein for the Arp2/3 complex and explains aspects of WASP activation by a variety of ligands.
4. Allostery and dimerization work together to integrate multiple, simultaneous, diverse signaling inputs.
5. Probablistic considerations suggest that higher-order assemblies will further enhance WASP activity and sensitize it toward allosteric activators.
6. The five-component WAVE assembly is regulated analogously to WASP.

FUTURE ISSUES

1. Where are the two VCA-binding sites on the Arp2/3 complex, what are the functions of each VCA, and which elements of the VCA are needed at each site? What is the kinetic pathway of nucleation?
2. What are the quantitative principles that underlie coupling between membrane binding, oligomerization, and allosteric activation of WASP proteins?
3. What are the physical properties (stoichiometries, stabilities, dynamics) of multi-WASP assemblies, and how do these lead to enhanced actin assembly activity?

4. How does interaction with the actin network modulate the kinetic and thermodynamic properties of WASP assemblies?
5. How is the activity of the WAVE regulatory complex controlled structurally and energetically?
6. How are the recently discovered WASP family members WASH, WHAMM, and JMY regulated? What are the components of their regulatory complexes, and how do these control VCA activity and respond to signals? What signals control them?

DISCLOSURE STATEMENT

The authors are not aware of any affiliations, memberships, funding, or financial holdings that might be perceived as affecting the objectivity of this review.

ACKNOWLEDGMENTS

M.K.R. thanks all past and present members of the Rosen lab for their efforts, insights, and enthusiasm in understanding regulation of actin assembly by proteins in the WASP family. Work in the Rosen lab is supported by the Howard Hughes Medical Institute and grants from the National Institutes of Health (NIH) (R01-GM56322) and Welch Foundation (I-1544). S.B.P. was supported by a fellowship from the NIH (1F32-GM06917902).

LITERATURE CITED

1. Pollard TD. 2007. Regulation of actin filament assembly by Arp2/3 complex and formins. *Annu. Rev. Biophys. Biomol. Struct.* 36:451–77
2. Chhabra ES, Higgs HN. 2007. The many faces of actin: matching assembly factors with cellular structures. *Nat. Cell Biol.* 9:1110–21
3. Pollitt AY, Insall RH. 2009. WASP and SCAR/WAVE proteins: the drivers of actin assembly. *J. Cell Sci.* 122:2575–78
4. Kurisu S, Takenawa T. 2009. The WASP and WAVE family proteins. *Genome Biol.* 10:226
5. Stradal TE, Scita G. 2006. Protein complexes regulating Arp2/3-mediated actin assembly. *Curr. Opin. Cell Biol.* 18:4–10
6. Takenawa T, Suetsugu S. 2007. The WASP-WAVE protein network: connecting the membrane to the cytoskeleton. *Nat. Rev. Mol. Cell Biol.* 8:37–48
7. Machesky LM, Insall RH. 1998. Scar1 and the related Wiskott-Aldrich syndrome protein, WASP, regulate the actin cytoskeleton through the Arp2/3 complex. *Curr. Biol.* 8:1347–56
8. Machesky LM, Mullins RD, Higgs HN, Kaiser DA, Blanchoin L, et al. 1999. Scar, a WASp-related protein, activates nucleation of actin filaments by the Arp2/3 complex. *Proc. Natl. Acad. Sci. USA* 96:3739–44
9. Hufner K, Higgs HN, Pollard TD, Jacobi C, Aepfelbacher M, Linder S. 2001. The verprolin-like central (VC) region of Wiskott-Aldrich syndrome protein induces Arp2/3 complex-dependent actin nucleation. *J. Biol. Chem.* 276:35761–67
10. Marchand JB, Kaiser DA, Pollard TD, Higgs HN. 2000. Interaction of WASP/Scar proteins with actin and vertebrate Arp2/3 complex. *Nat. Cell Biol.* 3:76–82
11. Zalevsky J, Lempert L, Kranitz H, Mullins RD. 2001. Different WASP family proteins stimulate different Arp2/3 complex-dependent actin-nucleating activities. *Curr. Biol.* 11:1903–13
12. Suetsugu S, Miki H, Takenawa T. 2001. Identification of another actin-related protein (Arp) 2/3 complex binding site in neural Wiskott-Aldrich syndrome protein (N-WASP) that complements actin polymerization induced by the Arp2/3 complex activating (VCA) domain of N-WASP. *J. Biol. Chem.* 276:33175–80

13. Panchal SC, Kaiser DA, Torres E, Pollard TD, Rosen MK. 2003. A conserved amphipathic helix in WASP/Scar proteins is essential for activation of Arp2/3 complex. *Nat. Struct. Biol.* 10:591–98
14. Miki H, Miura K, Takenawa T. 1996. N-WASP, a novel actin-depolymerizing protein, regulates the cortical cytoskeletal rearrangement in a PIP_2-dependent manner downstream of tyrosine kinases. *EMBO J.* 15:5326–35
15. Miki H, Suetsugu S, Takenawa T. 1998. WAVE, a novel WASP-family protein involved in actin reorganization induced by Rac. *EMBO J.* 17:6932–41
16. Chereau D, Kerff F, Graceffa P, Grabarek Z, Langsetmo K, Dominguez R. 2005. Actin-bound structures of Wiskott-Aldrich syndrome protein (WASP)-homology domain 2 and the implications for filament assembly. *Proc. Natl. Acad. Sci. USA* 102:16644–49
17. Co C, Wong DT, Gierke S, Chang V, Taunton J. 2007. Mechanism of actin network attachment to moving membranes: barbed end capture by N-WASP WH2 domains. *Cell* 128:901–13
18. Boczkowska M, Rebowski G, Petoukhov MV, Hayes DB, Svergun DI, Dominguez R. 2008. X-ray scattering study of activated Arp2/3 complex with bound actin-WCA. *Structure* 16:695–704
19. Hertzog M, van Heijenoort C, Didry D, Gaudier M, Coutant J, et al. 2004. The beta-thymosin/WH2 domain; structural basis for the switch from inhibition to promotion of actin assembly. *Cell* 117:611–23
20. Qualmann B, Kessels MM. 2009. New players in actin polymerization—WH2-domain-containing actin nucleators. *Trends Cell Biol.* 19:276–85
21. Chesarone MA, Goode BL. 2009. Actin nucleation and elongation factors: mechanisms and interplay. *Curr. Opin. Cell Biol.* 21:28–37
22. Padrick SB, Cheng HC, Ismail AM, Panchal SC, Doolittle LK, et al. 2008. Hierarchical regulation of WASP/WAVE proteins. *Mol. Cell* 32:426–38
23. Ramesh N, Geha R. 2009. Recent advances in the biology of WASP and WIP. *Immunol. Res.* 44:99–111
24. Sasahara Y, Rachid R, Byrne MJ, de la Fuente MA, Abraham RT, et al. 2002. Mechanism of recruitment of WASP to the immunological synapse and of its activation following TCR ligation. *Mol. Cell* 10:1269–81
25. Konno A, Kirby M, Anderson SA, Schwartzberg PL, Candotti F. 2007. The expression of Wiskott-Aldrich syndrome protein (WASP) is dependent on WASP-interacting protein (WIP). *Int. Immunol.* 19:185–92
26. Peterson FC, Deng Q, Zettl M, Prehoda KE, Lim WA, et al. 2007. Multiple WASP-interacting protein recognition motifs are required for a functional interaction with N-WASP. *J. Biol. Chem.* 282:8446–53
27. Jin Y, Mazza C, Christie JR, Giliani S, Fiorini M, et al. 2004. Mutations of the Wiskott-Aldrich syndrome protein (WASP): hotspots, effect on transcription, and translation and phenotype/genotype correlation. *Blood* 104:4010–19
28. Martinez-Quiles N, Rohatgi R, Anton IM, Medina M, Saville SP, et al. 2001. WIP regulates N-WASP-mediated actin polymerization and filopodium formation. *Nat. Cell Biol.* 3:484–91
29. Takano K, Toyooka K, Suetsugu S. 2008. EFC/F-BAR proteins and the N-WASP-WIP complex induce membrane curvature-dependent actin polymerization. *EMBO J.* 27:2817–28
30. Leung DW, Rosen MK. 2005. The nucleotide switch in Cdc42 modulates coupling between the GTPase-binding and allosteric equilibria of Wiskott-Aldrich syndrome protein. *Proc. Natl. Acad. Sci. USA* 102:5685–90
31. Rohatgi R, Ho HY, Kirschner MW. 2000. Mechanism of N-WASP activation by CDC42 and phosphatidylinositol 4,5-bisphosphate. *J. Cell Biol.* 150:1299–310
32. Prehoda KE, Scott JA, Mullins RD, Lim WA. 2000. Integration of multiple signals through cooperative regulation of the N-WASP-Arp2/3 complex. *Science* 290:801–6
33. Suetsugu S, Miki H, Yamaguchi H, Takenawa T. 2001. Requirement of the basic region of N-WASP/WAVE2 for actin-based motility. *Biochem. Biophys. Res. Commun.* 282:739–44
34. Papayannopoulos V, Co C, Prehoda KE, Snapper S, Taunton J, Lim WA. 2005. A polybasic motif allows N-WASP to act as a sensor of PIP_2 density. *Mol. Cell* 17:181–91
35. Higgs HN, Pollard TD. 2000. Activation by Cdc42 and PIP_2 of Wiskott-Aldrich syndrome protein (WASp) stimulates actin nucleation by Arp2/3 complex. *J. Cell Biol.* 150:1311–20
36. Hemsath L, Dvorsky R, Fiegen D, Carlier MF, Ahmadian MR. 2005. An electrostatic steering mechanism of Cdc42 recognition by Wiskott-Aldrich syndrome proteins. *Mol. Cell* 20:313–24

37. Burbelo PD, Drechsel D, Hall A. 1995. A conserved binding motif defines numerous candidate target proteins for both Cdc42 and Rac GTPases. *J. Biol. Chem.* 270:29071–74
38. Miki H, Sasaki T, Takai Y, Takenawa T. 1998. Induction of filopodium formation by a WASP-related actin-depolymerizing protein N-WASP. *Nature* 391:93–96
39. Kim AS, Kakalis LT, Abdul-Manan N, Liu GA, Rosen MK. 2000. Autoinhibition and activation mechanisms of the Wiskott-Aldrich syndrome protein. *Nature* 404:151–58
40. Li SS. 2005. Specificity and versatility of SH3 and other proline-recognition domains: structural basis and implications for cellular signal transduction. *Biochem. J.* 390:641–53
41. Abdul-Manan N, Aghazadeh B, Liu GA, Majumdar A, Ouerfelli O, et al. 1999. Structure of Cdc42 in complex with the GTPase-binding domain of the 'Wiskott-Aldrich syndrome' protein. *Nature* 399:379–83
42. Devriendt K, Kim AS, Mathijs G, Frints SG, Schwartz M, et al. 2001. Constitutively activating mutation in WASP causes X-linked severe congenital neutropenia. *Nat. Genet.* 27:313–17
43. Ancliff PJ, Blundell MP, Cory GO, Calle Y, Worth A, et al. 2006. Two novel activating mutations in the Wiskott-Aldrich syndrome protein result in congenital neutropenia. *Blood* 108:2182–89
44. Monod J, Wyman J, Changeux JP. 1965. On the nature of allosteric transitions: a plausible model. *J. Mol. Biol.* 12:88–118
45. Buck M, Xu W, Rosen MK. 2004. A two-state allosteric model for autoinhibition rationalizes WASP signal integration and targeting. *J. Mol. Biol.* 338:271–85
46. Morreale A, Venkatesan M, Mott HR, Owen D, Nietlispach D, et al. 2000. Structure of Cdc42 bound to the GTPase binding domain of PAK. *Nat. Struct. Biol.* 7:384–88
47. Gizachew D, Guo W, Chohan KK, Sutcliffe MJ, Oswald RE. 2000. Structure of the complex of Cdc42Hs with a peptide derived from P-21 activated kinase. *Biochemistry* 39:3963–71
48. Garrard SM, Capaldo CT, Gao L, Rosen MK, Macara IG, Tomchick DR. 2003. Structure of Cdc42 in a complex with the GTPase-binding domain of the cell polarity protein, Par6. *EMBO J.* 22:1125–33
49. Mott HR, Owen D, Nietlispach D, Lowe PN, Manser E, et al. 1999. Structure of the small G protein Cdc42 bound to the GTPase-binding domain of ACK. *Nature* 399:384–88
50. Owen D, Mott HR, Laue ED, Lowe PN. 2000. Residues in Cdc42 that specify binding to individual CRIB effector proteins. *Biochemistry* 39:1243–50
51. Lei M, Lu W, Meng W, Parrini MC, Eck MJ, et al. 2000. Structure of PAK1 in an autoinhibited conformation reveals a multistage activation switch. *Cell* 102:387–97
52. Badour K, Zhang J, Shi F, Leng Y, Collins M, Siminovitch KA. 2004. Fyn and PTP-PEST-mediated regulation of Wiskott-Aldrich syndrome protein (WASp) tyrosine phosphorylation is required for coupling T cell antigen receptor engagement to WASp effector function and T cell activation. *J. Exp. Med.* 199:99–112
53. Suetsugu S, Hattori M, Miki H, Tezuka T, Yamamoto T, et al. 2002. Sustained activation of N-WASP through phosphorylation is essential for neurite extension. *Dev. Cell* 3:645–58
54. Burton EA, Oliver TN, Pendergast AM. Abl kinases regulated actin comet tail elongation via an N-Wasp-dependent pathway. *Mol. Cell. Biol.* 25:8834–43
55. Wu X, Suetsugu S, Cooper LA, Takenawa T, Guan JL. 2004. Focal adhesion kinase regulation of N-WASP subcellular localization and function. *J. Biol. Chem.* 279:9565–76
56. Yokoyama N, Lougheed J, Miller WT. 2005. Phosphorylation of WASP by the Cdc42-associated kinase ACK1: dual hydroxyamino acid specificity in a tyrosine kinase. *J. Biol. Chem.* 280:42219–26
57. Torres E, Rosen MK. 2003. Contingent phosphorylation/dephosphorylation provides a mechanism of molecular memory in WASP. *Mol. Cell* 11:1215–27
58. Torres E, Rosen MK. 2006. Protein-tyrosine kinase and GTPase signals cooperate to phosphorylate and activate Wiskott-Aldrich syndrome protein (WASP)/neuronal WASP. *J. Biol. Chem.* 281:3513–20
59. Yaffe MB, Leparc GG, Lai J, Obata T, Volinia S, Cantley LC. 2001. A motif-based profile scanning approach for genome-wide prediction of signaling pathways. *Nat. Biotechnol.* 19:348–53
60. Waksman G, Shoelson SE, Pant N, Cowburn D, Kuriyan J. 1993. Binding of a high affinity phosphotyrosyl peptide to the Src SH2 domain: crystal structures of the complexed and peptide-free forms. *Cell* 72:779–90

61. Campellone KG, Robbins D, Leong JM. 2004. $EspF_U$ is a translocated EHEC effector that interacts with Tir and N-WASP and promotes Nck-independent actin assembly. *Dev. Cell* 7:217–28
62. Garmendia J, Phillips AD, Carlier MF, Chong Y, Schuller S, et al. 2004. TccP is an enterohaemorrhagic *Escherichia coli* O157:H7 type III effector protein that couples Tir to the actin-cytoskeleton. *Cell Microbiol.* 6:1167–83
63. Hayward RD, Leong JM, Koronakis V, Campellone KG. 2006. Exploiting pathogenic *Escherichia coli* to model transmembrane receptor signalling. *Nat. Rev. Microbiol.* 4:358–70
64. Garmendia J, Ren Z, Tennant S, Midolli Viera MA, Chong Y, et al. 2005. Distribution of *tccP* in clinical enterohemorrhagic and enteropathogenic *Escherichia coli* isolates. *J. Clin. Microbiol.* 43:5715–20
65. Cheng HC, Skehan BM, Campellone KG, Leong JM, Rosen MK. 2008. Structural mechanism of WASP activation by the enterohaemorrhagic *E. coli* effector $EspF_U$. *Nature* 454:1009–13
66. Sallee NA, Rivera GM, Dueber JE, Vasilescu D, Mullins RD, et al. 2008. The pathogen protein $EspF_U$ hijacks actin polymerization using mimicry and multivalency. *Nature* 454:1005–8
67. Peterson JR, Lokey RS, Mitchison TJ, Kirschner MW. 2001. A chemical inhibitor of N-WASP reveals a new mechanism for targeting protein interactions. *Proc. Natl. Acad. Sci. USA* 98:10624–29
68. Peterson JR, Bickford LC, Morgan D, Kim AS, Ouerfelli O, et al. 2004. Chemical inhibition of N-WASP by stabilization of a native autoinhibited conformation. *Nat. Struct. Mol. Biol.* 11:747–55
69. Bompard G, Rabeharivelo G, Morin N. 2008. Inhibition of cytokinesis by wiskostatin does not rely on N-WASP/Arp2/3 complex pathway. *BMC Cell Biol.* 9:42
70. Hardy JA, Wells JA. 2004. Searching for new allosteric sites in enzymes. *Curr. Opin. Struct. Biol.* 14:706–15
71. Garmendia J, Carlier MF, Egile C, Didry D, Frankel G. 2006. Characterization of TccP-mediated N-WASP activation during enterohaemorrhagic *Escherichia coli* infection. *Cell Microbiol.* 8:1444–55
72. Campellone KG, Cheng HC, Robbins D, Siripala AD, McGhie EJ, et al. 2008. Repetitive N-WASP-binding elements of the enterohemorrhagic *Escherichia coli* effector $EspF_U$ synergistically activate actin assembly. *PLoS Pathog.* 4:e1000191
73. Carlier MF, Nioche P, Broutin-L'Hermite I, Boujemaa R, Le Clainche C, et al. 2000. GRB2 links signaling to actin assembly by enhancing interaction of neural Wiskott-Aldrich syndrome protein (N-WASp) with actin-related protein (ARP2/3) complex. *J. Biol. Chem.* 275:21946–52
74. Fukuoka M, Suetsugu S, Miki H, Fukami K, Endo T, Takenawa T. 2001. A novel neural Wiskott-Aldrich syndrome protein (N-WASP) binding protein, WISH, induces Arp2/3 complex activation independent of Cdc42. *J. Cell Biol.* 152:471–82
75. Innocenti M, Gerboth S, Rottner K, Lai FP, Hertzog M, et al. 2005. Abi1 regulates the activity of N-WASP and WAVE in distinct actin-based processes. *Nat. Cell Biol.* 7:969–76
76. Kowalski JR, Egile C, Gil S, Snapper SB, Li R, Thomas SM. 2005. Cortactin regulates cell migration through activation of N-WASP. *J. Cell Sci.* 118:79–87
77. Martinez-Quiles N, Ho HY, Kirschner MW, Ramesh N, Geha RS. 2004. Erk/Src phosphorylation of cortactin acts as a switch on-switch off mechanism that controls its ability to activate N-WASP. *Mol. Cell. Biol.* 24:5269–80
78. Pinyol R, Haeckel A, Ritter A, Qualmann B, Kessels MM. 2007. Regulation of N-WASP and the Arp2/3 complex by Abp1 controls neuronal morphology. *PLoS ONE* 2:e400
79. Rodal AA, Motola-Barnes RN, Littleton JT. 2008. Nervous wreck and Cdc42 cooperate to regulate endocytic actin assembly during synaptic growth. *J. Neurosci.* 28:8316–25
80. Rohatgi R, Nollau P, Ho HY, Kirschner MW, Mayer BJ. 2001. Nck and phosphatidylinositol 4,5-bisphosphate synergistically activate actin polymerization through the N-WASP-Arp2/3 pathway. *J. Biol. Chem.* 276:26448–52
81. Tehrani S, Tomasevic N, Weed S, Sakowicz R, Cooper JA. 2007. Src phosphorylation of cortactin enhances actin assembly. *Proc. Natl. Acad. Sci. USA* 104:11933–38
82. Tomasevic N, Jia Z, Russell A, Fujii T, Hartman JJ, et al. 2007. Differential regulation of WASP and N-WASP by Cdc42, Rac1, Nck, and PI(4,5)P2. *Biochemistry* 46:3494–502
83. Tsujita K, Suetsugu S, Sasaki N, Furutani M, Oikawa T, Takenawa T. 2006. Coordination between the actin cytoskeleton and membrane deformation by a novel membrane tubulation domain of PCH proteins is involved in endocytosis. *J. Cell Biol.* 172:269–79

84. Yarar D, Waterman-Storer CM, Schmid SL. 2007. SNX9 couples actin assembly to phosphoinositide signals and is required for membrane remodeling during endocytosis. *Dev. Cell* 13:43–56
85. Frost A, Unger VM, De Camilli P. 2009. The BAR domain superfamily: membrane-molding macromolecules. *Cell* 137:191–96
86. Otsuki M, Itoh T, Takenawa T. 2003. Neural Wiskott-Aldrich syndrome protein is recruited to rafts and associates with endophilin A in response to epidermal growth factor. *J. Biol. Chem.* 278:6461–69
87. Kim K, Hou P, Gorski JL, Cooper JA. 2004. Effect of Fgd1 on cortactin in Arp2/3 complex-mediated actin assembly. *Biochemistry* 43:2422–27
88. Footer MJ, Lyo JK, Theriot JA. 2008. Close packing of *Listeria monocytogenes* ActA, a natively unfolded protein, enhances F-actin assembly without dimerization. *J. Biol. Chem.* 283:23852–62
89. Rohatgi R, Ma L, Miki H, Lopez M, Kirchhausen T, et al. 1999. The interaction between N-WASP and the Arp2/3 complex links Cdc42-dependent signals to actin assembly. *Cell* 97:221–31
90. Dueber JE, Mirsky EA, Lim WA. 2007. Engineering synthetic signaling proteins with ultrasensitive input/output control. *Nat. Biotechnol.* 25:660–62
91. Dueber JE, Yeh BJ, Chak K, Lim WA. 2003. Reprogramming control of an allosteric signaling switch through modular recombination. *Science* 301:1904–8
92. Weiss SM, Ladwein M, Schmidt D, Ehinger J, Lommel S, et al. 2009. IRSp53 links the enterohemorrhagic *E. coli* effectors Tir and $EspF_U$ for actin pedestal formation. *Cell Host Microbe* 5:244–58
93. Tong AH, Drees B, Nardelli G, Bader GD, Brannetti B, et al. 2002. A combined experimental and computational strategy to define protein interaction networks for peptide recognition modules. *Science* 295:321–24
94. Le Clainche C, Schlaepfer D, Ferrari A, Klingauf M, Grohmanova K, et al. 2007. IQGAP1 stimulates actin assembly through the N-WASP-Arp2/3 pathway. *J. Biol. Chem.* 282:426–35
95. Bensenor LB, Kan HM, Wang N, Wallrabe H, Davidson LA, et al. 2007. IQGAP1 regulates cell motility by linking growth factor signaling to actin assembly. *J. Cell Sci.* 120:658–69
96. Ho HY, Rohatgi R, Lebensohn AM, Le M, Li J, et al. 2004. Toca-1 mediates Cdc42-dependent actin nucleation by activating the N-WASP-WIP complex. *Cell* 118:203–16
97. Alto NM, Weflen AW, Rardin MJ, Yarar D, Lazar CS, et al. 2007. The type III effector EspF coordinates membrane trafficking by the spatiotemporal activation of two eukaryotic signaling pathways. *J. Cell Biol.* 178:1265–78
98. Vingadassalom D, Kazlauskas A, Skehan B, Cheng HC, Magoun L, et al. 2009. Insulin receptor tyrosine kinase substrate links the *E. coli* O157:H7 actin assembly effectors Tir and $EspF_U$ during pedestal formation. *Proc. Natl. Acad. Sci. USA* 106:6754–59
99. Itoh T, Erdmann KS, Roux A, Habermann B, Werner H, De Camilli P. 2005. Dynamin and the actin cytoskeleton cooperatively regulate plasma membrane invagination by BAR and F-BAR proteins. *Dev. Cell* 9:791–804
100. Frost A, Perera R, Roux A, Spasov K, Destaing O, et al. 2008. Structural basis of membrane invagination by F-BAR domains. *Cell* 132:807–17
101. Dawson JC, Legg JA, Machesky LM. 2006. Bar domain proteins: a role in tubulation, scission and actin assembly in clathrin-mediated endocytosis. *Trends Cell Biol.* 16:493–98
102. Shin N, Ahn N, Chang-Ileto B, Park J, Takei K, et al. 2008. SNX9 regulates tubular invagination of the plasma membrane through interaction with actin cytoskeleton and dynamin 2. *J. Cell Sci.* 121:1252–63
103. Kinley AW, Weed SA, Weaver AM, Karginov AV, Bissonette E, et al. 2003. Cortactin interacts with WIP in regulating Arp2/3 activation and membrane protrusion. *Curr. Biol.* 13:384–93
104. Blasutig IM, New LA, Thanabalasuriar A, Dayarathna TK, Goudreault M, et al. 2008. Phosphorylated YDXV motifs and Nck SH2/SH3 adaptors act cooperatively to induce actin reorganization. *Mol. Cell. Biol.* 28:2035–46
105. Jones N, Blasutig IM, Eremina V, Ruston JM, Bladt F, et al. 2006. Nck adaptor proteins link nephrin to the actin cytoskeleton of kidney podocytes. *Nature* 440:818–23
106. Rivera GM, Briceno CA, Takeshima F, Snapper SB, Mayer BJ. 2004. Inducible clustering of membrane-targeted SH3 domains of the adaptor protein Nck triggers localized actin polymerization. *Curr. Biol.* 14:11–22

107. Krause M, Sechi AS, Konradt M, Monner D, Gertler FB, Wehland J. 2000. Fyn-binding protein (Fyb)/SLP-76-associated protein (SLAP), Ena/vasodilator-stimulated phosphoprotein (VASP) proteins and the Arp2/3 complex link T cell receptor (TCR) signaling to the actin cytoskeleton. *J. Cell Biol.* 149:181–94
108. Soulard A, Friant S, Fitterer C, Orange C, Kaneva G, et al. 2005. The WASP/Las17p-interacting protein Bzz1p functions with Myo5p in an early stage of endocytosis. *Protoplasma* 226:89–101
109. Bear JE, Rawls JF, Saxe CL III. 1998. SCAR, a WASP-related protein, isolated as a suppressor of receptor defects in late *Dictyostelium* development. *J. Cell Biol.* 142:1325–35
110. Suetsugu S, Miki H, Takenawa T. 1999. Identification of two human WAVE/SCAR homologues as general actin regulatory molecules which associate with the Arp2/3 complex. *Biochem. Biophys. Res. Commun.* 260:296–302
111. Eden S, Rohatgi R, Podtelejnikov AV, Mann M, Kirschner MW. 2002. Mechanism of regulation of WAVE1-induced actin nucleation by Rac1 and Nck. *Nature* 418:790–93
112. Gautreau A, Ho HY, Li J, Steen H, Gygi SP, Kirschner MW. 2004. Purification and architecture of the ubiquitous Wave complex. *Proc. Natl. Acad. Sci. USA* 101:4379–83
113. Leng Y, Zhang J, Badour K, Arpaia E, Freeman S, et al. 2005. Abelson-interactor-1 promotes WAVE2 membrane translocation and Abelson-mediated tyrosine phosphorylation required for WAVE2 activation. *Proc. Natl. Acad. Sci. USA* 102:1098–103
114. Ismail AM, Padrick SB, Chen B, Umetani J, Rosen MK. 2009. The WAVE regulatory complex is inhibited. *Nat. Struct. Mol. Biol.* 16:561–63
115. Innocenti M, Zucconi A, Disanza A, Frittoli E, Areces LB, et al. 2004. Abi1 is essential for the formation and activation of a WAVE2 signalling complex. *Nat. Cell Biol.* 6:319–27
116. Stovold CF, Millard TH, Machesky LM. 2005. Inclusion of Scar/WAVE3 in a similar complex to Scar/WAVE1 and 2. *BMC Cell Biol.* 6:11
117. Weiner OD, Rentel MC, Ott A, Brown GE, Jedrychowski M, et al. 2006. Hem-1 complexes are essential for Rac activation, actin polymerization, and myosin regulation during neutrophil chemotaxis. *PLoS Biol.* 4:e38
118. Kim Y, Sung JY, Ceglia I, Lee KW, Ahn JH, et al. 2006. Phosphorylation of WAVE1 regulates actin polymerization and dendritic spine morphology. *Nature* 442:814–17
119. Steffen A, Rottner K, Ehinger J, Innocenti M, Scita G, et al. 2004. Sra-1 and Nap1 link Rac to actin assembly driving lamellipodia formation. *EMBO J.* 23:749–59
120. Derivery E, Lombard B, Loew D, Gautreau A. 2009. The Wave complex is intrinsically inactive. *Cell. Motil. Cytoskelet.* 66:777–90
121. Kobayashi K, Kuroda S, Fukata M, Nakamura T, Nagase T, et al. 1998. p140Sra-1 (specifically Rac1-associated protein) is a novel specific target for Rac1 small GTPase. *J. Biol. Chem.* 273:291–95
122. Oikawa T, Yamaguchi H, Itoh T, Kato M, Ijuin T, et al. 2004. PtdIns(3,4,5)P3 binding is necessary for WAVE2-induced formation of lamellipodia. *Nat. Cell Biol.* 6:420–26
123. Suetsugu S, Kurisu S, Oikawa T, Yamazaki D, Oda A, Takenawa T. 2006. Optimization of WAVE2 complex-induced actin polymerization by membrane-bound IRSp53, PIP(3), and Rac. *J. Cell Biol.* 173:571–85
124. Nakanishi O, Suetsugu S, Yamazaki D, Takenawa T. 2007. Effect of WAVE2 phosphorylation on activation of the Arp2/3 complex. *J. Biochem.* 141:319–25
125. Pocha SM, Cory GO. 2009. WAVE2 is regulated by multiple phosphorylation events within its VCA domain. *Cell. Motil. Cytoskelet.* 66:36–47
126. Miki H, Fukuda M, Nishida E, Takenawa T. 1999. Phosphorylation of WAVE downstream of mitogen-activated protein kinase signaling. *J. Biol. Chem.* 274:27605–9
127. Soderling SH, Binns KL, Wayman GA, Davee SM, Ong SH, et al. 2002. The WRP component of the WAVE-1 complex attenuates Rac-mediated signalling. *Nat. Cell Biol.* 4:970–75
128. Salazar MA, Kwiatkowski AV, Pellegrini L, Cestra G, Butler MH, et al. 2003. Tuba, a novel protein containing Bin/amphiphysin/Rvs and Dbl homology domains, links dynamin to regulation of the actin cytoskeleton. *J. Biol. Chem.* 278:49031–43
129. Linardopoulou EV, Parghi SS, Friedman C, Osborn GE, Parkhurst SM, Trask BJ. 2007. Human subtelomeric WASH genes encode a new subclass of the WASP family. *PLoS Genet.* 3:e237

130. Campellone KG, Webb NJ, Znameroski EA, Welch MD. 2008. WHAMM is an Arp2/3 complex activator that binds microtubules and functions in ER to Golgi transport. *Cell* 134:148–61
131. Zuchero JB, Coutts AS, Quinlan ME, Thangue NB, Mullins RD. 2009. p53-cofactor JMY is a multifunctional actin nucleation factor. *Nat. Cell Biol.* 11:451–59
132. Liu R, Abreu-Blanco MT, Barry KC, Linardopoulou EV, Osborn GE, Parkhurst SM. 2009. Wash functions downstream of Rho and links linear and branched actin nucleation factors. *Development* 136:2849–60
133. Delatour V, Helfer E, Didry D, Le KH, Gaucher JF, et al. 2008. Arp2/3 controls the motile behavior of N-WASP-functionalized GUVs and modulates N-WASP surface distribution by mediating transient links with actin filaments. *Biophys. J.* 94:4890–905
134. Weisswange I, Newsome TP, Schleich S, Way M. 2009. The rate of N-WASP exchange limits the extent of ARP2/3-complex-dependent actin-based motility. *Nature* 458:87–91
135. Weiner OD, Marganski WA, Wu LF, Altschuler SJ, Kirschner MW. 2007. An actin-based wave generator organizes cell motility. *PLoS Biol.* 5:e221

Amphipols, Nanodiscs, and Fluorinated Surfactants: Three Nonconventional Approaches to Studying Membrane Proteins in Aqueous Solutions

Jean-Luc Popot

Laboratoire de Physico-Chimie Moléculaire des Protéines Membranaires, Unité Mixte de Recherche 7099, Centre National de la Recherche Scientifique and Université Paris-7 Denis Diderot, Institut de Biologie Physico-Chimique, F-75005 Paris, France;
e-mail: jean-luc.popot@ibpc.fr

Annu. Rev. Biochem. 2010. 79:737–75

First published online as a Review in Advance on March 18, 2010

The *Annual Review of Biochemistry* is online at biochem.annualreviews.org

This article's doi:
10.1146/annurev.biochem.052208.114057

0066-4154/10/0707-0737$20.00

Key Words

nanolipoprotein particles, nanoscale apolipoprotein-bound bilayers, reconstituted high-density lipoprotein particles, hemifluorinated surfactants

Abstract

Membrane proteins (MPs) are usually handled in aqueous solutions as protein/detergent complexes. Detergents, however, tend to be inactivating. This situation has prompted the design of alternative surfactants that can be substituted for detergents once target proteins have been extracted from biological membranes and that keep them soluble in aqueous buffers while stabilizing them. The present review focuses on three such systems: *Amphipols* (APols) are amphipathic polymers that adsorb onto the hydrophobic transmembrane surface of MPs; *nanodiscs* (NDs) are small patches of lipid bilayer whose rim is stabilized by amphipathic proteins; *fluorinated surfactants* (FSs) resemble detergents but interfere less than detergents do with stabilizing protein/protein and protein/lipid interactions. The structure and properties of each of these three systems are described, as well as those of the complexes they form with MPs. Their respective usefulness, constraints, and prospects for functional and structural studies of MPs are discussed.

Contents

Detergent: a surfactant with the ability to solubilize fats

MP: membrane protein

1. INTRODUCTION: STRANGER IN A STRANGE LAND

Thirty-five years have elapsed since the art of using detergents to handle membrane proteins (MPs) emerged from the "cooking recipe" age and entered that of physical chemistry (1, 2). Yet, most biochemists will confess to a feeling of nervousness when compelled to deal with membrane-associated proteins. MPs indeed have earned a well-deserved reputation for being hard to handle once extracted from their natural environment and made water soluble, and the search for the detergent and conditions that will confer upon them a modicum of stability is known to be time-consuming and, more often than not, frustrating.

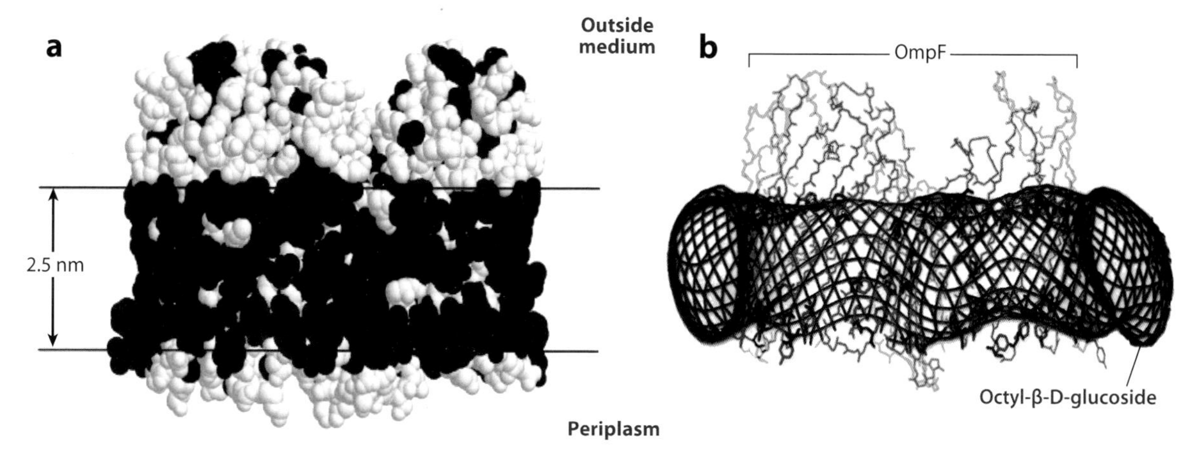

Figure 1

How surfactants associate with a membrane protein. (*a*) Space-filling model of *Escherichia coli*'s trimeric outer membrane protein OmpF [Protein Data Bank (PDB) accession code 2OMF; see Reference 154]. Hydrophobic and aromatic amino acid side chains, shown in black, form an ~2.5-nm wide belt, which, in situ, faces the hydrophobic interior of the membrane (approximated by horizontal lines). Reprinted from Reference 11. (*b*) The distribution of the detergent octyl-β-D-glucoside (*purple cage*) around OmpF (in skeleton representation), as determined by neutron crystallography, closely follows the belt of hydrophobic residues. Adapted from Reference 155.

A glimpse at **Figure 1*a*** makes it immediately obvious why integral MPs—the only ones we are concerned with in this review—cannot be handled in pure water: The part of their surface that, in situ, is in contact with lipid acyl chains and/or the transmembrane surface of other proteins is, as a rule, highly hydrophobic. A solution of MPs in an aqueous buffer does not stay monomeric because the hydrophobic effect, which tends to minimize the number of water molecules in contact with apolar surfaces (3), will drive MP transmembrane surfaces to interact with one another. This results in aggregation and, most often, in precipitation. Adding surfactants to the solution prevents this phenomenon. Surfactants are molecules that comprise at least one polar and at least one apolar moiety. In aqueous solutions, the polar groups are readily solvated, whereas the apolar ones are pushed toward the air-water interface from which they displace water molecules. This reduces the free-energy cost of creating this interface, which lowers the surface tension, hence the name surfactant.[1] Surfactants also displace water from the hydrophobic surface of MPs onto which they adsorb, making it more polar. If their concentration is high enough, they may form a continuous belt that covers what, in situ, was the transmembrane surface (**Figure 1*b***). The resulting complex can be readily soluble. For the protein to remain monomeric, surfactant/protein interactions must overcome protein/protein ones, a process that depends both on the properties of the surfactant and on its concentration.

Before being handled in aqueous solutions, MPs must be extracted from membranes. That is, the protein-lipid and protein-protein interactions that anchor them in the membrane must be replaced with protein-surfactant ones. This is usually achieved using a special class of surfactants that biologists call detergents (physical chemists hate the term, which they reserve for laundry). Detergents are surfactants whose solution properties allow them to disperse fats and other hydrophobic molecules. In contrast to the lipids that comprise biological membranes, which, as a rule, form bilayers, detergents self-organize in water in the form of small aggregates, comprising typically 40–100 molecules, called micelles. Their apolar

Integral membrane protein: a protein that is in contact with the hydrophobic interior of a biological membrane

Surfactant: a compound that lowers the surface tension of water

[1]Most proteins are surfactants, but they are not included under this term in the present text.

Critical micellar concentration (cmc): the concentration of a surfactant above which its molecules start to organize into micelles

moieties are grouped in the core of the micelle, away from water, and the polar groups face the solution. The reason why biological lipids form extended planar structures and detergents form small closed ones is geometric: In projection on the plane of the membrane, the polar and apolar moieties of lipids occupy comparable areas—that is, on average, lipid molecules look more or less cylindrical—so that the juxtaposition of many molecules forms a flat monolayer, two of which appose to form the membrane. In detergents, the apolar part is less bulky than the polar one, which bends the interface, generating spheres, ellipsoids, or cylinders, called micelles (3, 4). The concentration above which micelles form—the critical micellar concentration (cmc)—depends on a balance between the hydrophobic effect and translational entropy, the first effect driving detergent molecules to assemble, whereas the second one favors their dispersion. Micelles are in rapid equilibrium with monomers (see, e.g., Reference 5). For straightforward thermodynamic (entropic) reasons, the formation of micelles essentially buffers the concentration of monomers at the cmc, so that every new addition of detergent to the solution results in the formation of more micelles. Above the cmc, the chemical potential of the detergent remains almost constant (3). Detergents dissolve hydrophobic or amphipathic molecules that partition into micelles.

Because they partition efficiently into membranes while having a molecular shape different from that of lipids, detergents stress the membrane/water interface, which they tend to bend. Ultimately, if the detergent is "strong" enough, the planar structure becomes energetically unfavorable, holes form, and membrane constituents become dispersed into mixed micelles comprising detergent, lipids, and MPs. For a detergent to be solubilizing, the membrane has to break up before micelles of pure detergent appear in the solution. If such is not the case, membranes into which some detergent has partitioned will coexist with micelles. This happens with some "weak" detergents and with non-detergent surfactants. For detergents to efficiently resolve a membrane into its constituents, protein/detergent and lipid/detergent interactions have to overcome the protein/protein, protein/lipid, and lipid/lipid interactions that keep the membrane together. Detergents used to solubilize biological membranes are, therefore, out of necessity dissociating.

It has long been observed that, once exposed to detergents, most MPs rapidly lose their functionality (for two examples among hundreds of studies, see, e.g., References 6 and 7). Why this is so is seldom studied in detail, is probably variable from one protein to the next, and is not the object of a consensus among membrane biochemists (for reviews, see, e.g., References 8–11). An extensive discussion of this matter is beyond the scope of the present review, and I present only my own views about it. Much of the data that my coworkers and I have collected in the course of more than 30 years of work with half-a-dozen MPs suggests that a major contribution to the destabilization of MPs by detergents is the dissociating properties of the latter, i.e., their ability to disrupt protein/protein and protein/lipid interactions, a property that is the very reason why they are used in the first place. Protein/protein and protein/lipid interactions, however, are essential to MP stability: Oligomeric MPs usually feature subunit/subunit interactions in the transmembrane region; the three-dimensional (3D) structure of monomeric MPs, as well as that of subunits, depends on protein/protein interactions between their transmembrane segments; and most MPs are extracted from membranes along with bound lipids, which stabilize them. Detergents compete with all of these interactions, and micelles act as a "hydrophobic sink" for molecules that, initially, were associated with the MP under study. Delipidation is one of the most, if not the most, common causes of MP inactivation. The destabilizing effect of diluting lipids, subunits, and/or hydrophobic or amphipathic cofactors among detergent micelles explains the fact that, for a given detergent, working close to the cmc, i.e., in the presence of few micelles, will often improve the stability of MPs.

The obvious underlying mechanism is that reducing the concentration of micelles limits the hydrophobic volume into which lipids, cofactors, and subunits can diffuse or, stated another way, increases their concentration in the mixed micelles. This favors their sticking together rather than coming apart. An equally frequent observation is that supplementing an MP in detergent solution with lipids very often will stabilize it. This can be due to the preservation of interactions with the lipids that help the protein to keep its 3D structure and/or to the lipids preventing access of the detergent to vulnerable spots at the transmembrane surface. Other causes of destabilization can also be considered. For example, it can be argued that the dynamics of transmembrane α-helices within helix bundles is restricted by the geometry of the bilayer more than it is within a detergent micelle or that the lateral pressure gradient within the bilayer is important for their stability (for a discusssion of these effects, see Reference 12). Our own work has been based on the belief that finding ways to limit the disruption of protein/protein and protein/lipid interactions would be a decisive step toward improving MP stability.

2. KEEPING MEMBRANE PROTEINS WATER SOLUBLE IN THE ABSENCE OF DETERGENTS

Because dealing with solubilized MPs is necessary to understanding their structure and function and because most MPs become unstable once solubilized, many attempts have been made to develop countermeasures. One approach is to make the protein itself more stable, either by selecting it from appropriate organisms, e.g., hyperthermophiles, or by engineering it. The latter strategy has led to remarkable success (see, e.g., References 13–16), and it certainly holds great promise for the future.

An alternative or complementary approach is to make the environment less destabilizing. After being extracted from their original membrane environment, MPs traditionally are kept water soluble with a detergent. Investigators may resort to the same detergent that was used for extraction, which they generally use at a lower concentration, so as to limit MP destabilization. Alternatively, MPs, once solubilized, are often transferred to another less aggressive detergent, a detergent mixture, or a lipid/detergent mixture. There is indeed no fundamental necessity to use a strongly dissociating detergent once a protein has been extracted. Weak detergents, such as digitonin or surfactants of the Tween series, have long been used for this purpose, despite their chemical heterogeneity and less than satisfying micellar properties (very low cmc, large micelles). Once the protein has been extracted, however, what the biochemist needs is to keep it soluble and to prevent it from aggregating, conditions that can be provided by surfactants that are not necessarily able to solubilize membranes. This has led to the development of such molecules as tripod amphiphiles (17–19), surfactants with rigid hydrophobic tails containing cycles (20), peptitergents (21), lipopeptides (22, 23), peptergents (24), fluorinated surfactants (FSs) (25, 26), or amphipathic polymers called amphipols (APols) (27), most of which are not good membrane solubilizers (for brief overviews, see References 11 and 28). A major part of this review is devoted to the latter two approaches. Earlier reviews on APols and FSs are found in References 11 and 28–30.

Because replacing biological lipids with other surfactants is generally detrimental to MP stability, an obvious alternative is to reinsert them, after extraction, into a lipid environment. This may seem self-defeating as the lipids an MP is normally in contact with generally organize into bilayers, which can be dispersed as vesicle suspensions but do not form the small entities most suitable for biochemical and biophysical approaches. Bilayers can, however, be fragmented into small patches by mixing lipids with certain surfactants, e.g., bile salts or short-chain lipids such as dihexanoylphosphatidylcholine (DHPC), which tend to segregate from the surfactant-saturated lipid bilayers and form the rim of small lamellar discs called bicelles. MPs inserted into these discs find themselves

Fluorinated surfactant (FS): a surfactant whose hydrophobic chain is fluorinated

Amphipol (APol): an amphipathic polymer that can keep membrane proteins water soluble in detergent-free solutions as small individual entities by adsorbing onto their transmembrane surface

Bicelle: a patch of bilayer whose rim is stabilized by small surfactants such as DHPC

Nanodisc (ND): a small patch of bilayer stabilized by amphipathic rim proteins

Nonconventional surfactant (NCS): as used in this text, either a nanodisc, an amphipol, or a (hemi)fluorinated surfactant

Membrane scaffolding proteins (MSPs): nanodisc rim proteins derived from apoA-1

surrounded by a bilayer-like environment. Bicelles are a highly interesting medium that has been used, among other things, for MP NMR studies (31–36), as well as for MP crystallization (37, 38). For want of space, bicelles are not discussed here.

Over the past decade, a variant of bicelles has been actively investigated, the rim of which is formed by amphipathic proteins. These structures, which can integrate MPs up to a certain size determined by the structure of the rim proteins, are variously called nanodiscs,[2] nanolipoprotein particles, nanoscale apolipoprotein-bound bilayers, or reconstituted high-density lipoprotein particles, hereafter lumped under the collective name of nanodiscs (NDs). NDs have been the object of two recent reviews (39, 40). This article includes a brief description of what NDs are and which MP studies have been done thanks to them and also compares their prospects with those of APols and FSs.

Throughout this text, APols, NDs, and FSs are collectively referred to as nonconventional surfactants (NCSs).

3. PRINCIPLE AND MOLECULAR ORGANIZATION OF NONCONVENTIONAL SURFACTANTS

NDs, FSs, and APols have quite different chemical compositions and molecular structures, which are reflected in the very distinct ways they assemble into particles when they are dispersed in aqueous solutions. This, in turn, has profound effects on the nature and properties of the complexes they form with MPs.

3.1. Nanodiscs

NDs were introduced by S.G. Sligar (University of Illinois) and his coworkers as the spin-off of a very large body of work on high-density lipoproteins (for a review, see Reference 39). Small NDs consist of a patch of, typically, 130–160 lipids, organized as a bilayer and surrounded by stabilizing proteins. The latter are often the so-called membrane scaffolding proteins (MSPs), which are derived from human high-density lipoprotein apoA-1 by modifications such as pruning away some undesired domains (41) and duplicating other domains so as to increase the protein's length and, thereby, the perimeter of the ND (42). MSPs can also be endowed with a polyhistidine tag (42). Other proteins can be used as well (see References 43–45). The structure of NDs, whether empty or containing MPs, has been intensely studied by such approaches as small-angle X-ray scattering (SAXS), atomic force microscopy (AFM), size exclusion chromatography (SEC), native gel electrophoresis, electron microscopy (EM), solid-state NMR (ssNMR), Fourier-transform infrared spectroscopy (FTIR), and various other types of optical spectroscopy (reviewed in References 39, 40). The thickness of an empty ND is that of a bilayer (42, 46, 47). The size of the disc depends on the rim proteins, and for a given protein the number of encapsulated lipids depends on the surface area of the lipids (46). Overall diameters reported to date vary between ~10 and ~20 nm (reviewed in References 39, 40). How variable it is from disc to disc in a given population is under investigation (see Reference 47, and the references therein). The molecular mass of the smallest NDs is ~150 kDa. ND lipids undergo phase transitions similar to those of extended bilayers, although, for entropic reasons, the transitions are broader, and the behavior of the first two rows of lipids in contact with the rim proteins is altered (48, 49).

Original versions of NDs contained two copies of MSP per disc, but these can also be fused one to another, resulting in objects containing a single protein per disc (42). Various models have been proposed for the arrangement of the MSPs in NDs, and extensive molecular dynamics (MD) simulations have been carried out. Most data, particularly ssNMR determination of the conformation of specific classes of MSP amino acid residues (50) and MD simulations (reviewed in

[2]The name Nanodisc™ is a trademark of Nanodisc Inc.

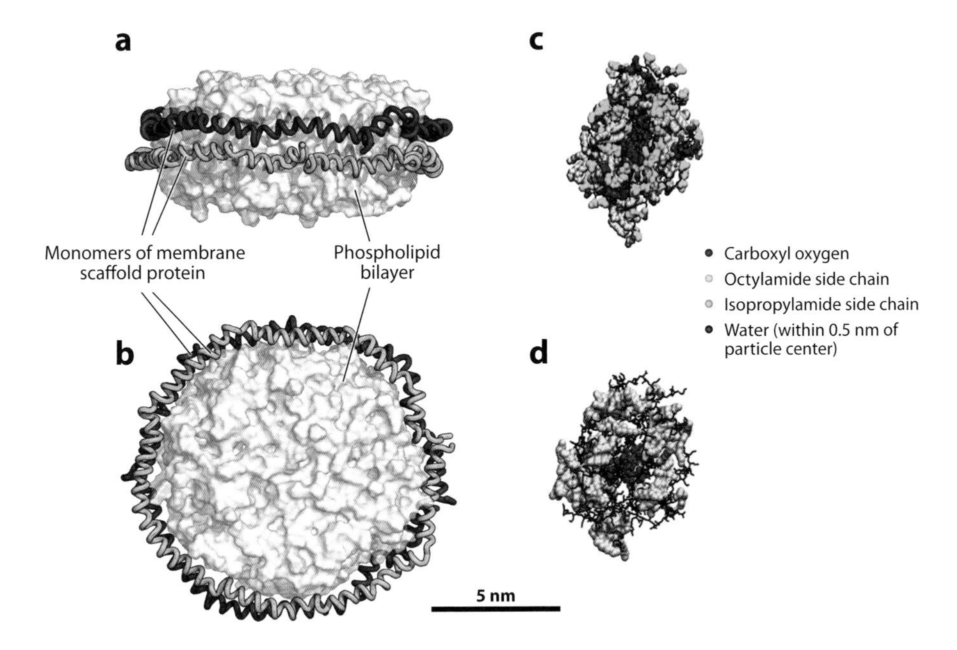

Figure 2

Molecular dynamics (MD) structures of nonconventional surfactants. (*a*, *b*) MD structure of a nanodisc. Model viewed (*a*) perpendicular to the bilayer and (*b*) in the plane of the bilayer, based on the molecular belt model of discoidal high-density lipoprotein (156). Two monomers of the membrane scaffold protein form an amphipathic helical belt around a segment of phospholipid bilayer. Model by S.C. Harvey from Reference 39. (*c*, *d*) MD structure of an amphipol (APol) particle. A snapshot from the end of a 50-ns MD simulation of a fully hydrated A8-35 particle. The particle ($R_g \approx 2.4$ nm) consists of four identical 10-kDa A8-35 molecules, whose side chains were distributed randomly along the backbone. In panel *c*, groups are colored as described in the key, with black lines indicating the polyacrylate backbone. Surrounding waters, hydrogens, and ions have been excluded for clarity. In panel *d*, only the octylamide side chains, backbone (*black lines*), and water within 0.5 nm of the particle center are shown in order to highlight the formation of submicellar domains and a hydrated particle core. The scale bar represents ~5 nm (J.N. Sachs, personal communication). Nanodiscs and APols are represented approximately to scale.

Reference 51), favor the "molecular belt" model in which the protein (or proteins) wraps around the rim of the disc (**Figure 2*a*,*b***) (reviewed in References 39, 40, 51).

NDs have two unique properties compared to the two other systems discussed in this review. First, the MPs they harbor find themselves surrounded by a medium that is extremely similar to a normal bilayer. MP functions that depend, in one way or another, on interactions with lipids therefore stand a better chance to be faithfully reproduced than in APols or FSs. Second, the defined size of the rim proteins sets that of the disc and, therefore, the dimensions of the MP or MP complexes that can be encapsulated. In contrast to detergents, APols, or FSs, MSPs cannot accommodate MPs beyond a certain size, and the probability that discs can fuse, allowing MPs trapped in different discs to temporarily occupy the same disc, seems remote. NDs can be used, therefore, to investigate issues where the degree of association of MPs with one another is of primary importance (Section 7.3). The flip side, of course, is that, as of now, MPs with large transmembrane domains cannot be handled in NDs (Section 5.2).

3.2. Fluorinated Surfactants

The chemical structure of FSs resembles those of classical detergents, but their hydrophobic tails contains fluorine atoms (**Figure 3**). The rationale behind the use of such compounds is based on the poor miscibility of alkanes and perfluorinated alkanes (52–55). FSs indeed are not cytolytic (they are not detergents) because they do not partition well into lipid membranes (25, 55–61). Because they are not detergents, FSs originally attracted little attention from membrane biochemists (see, e.g., References 25, 62, 63), with the exception of perfluorooctanoate, which has been used for MP electrophoresis (64, 65). It could nevertheless be hoped that FSs might be less prone than detergents to destabilize MPs for the following two reasons: (*a*) Lipids, subunits, and hydrophobic cofactors should partition less favorably into FS micelles; and (*b*) because fluorinated alkyl chains are more bulky and more rigid than alkanes and have little affinity for hydrogenated transmembrane protein segments, they ought to be less efficient than detergents at disrupting protein/protein interactions. By the same token, however, FSs bearing perfluorinated chains could be expected to be ineffective in preventing MPs from aggregating, which early data seemed to bear out (25, 63). To try to improve interactions with the methyl group-covered transmembrane surfaces of MPs while preserving the overall lyophobic ("lipid-hating") character of FS micelles, a hydrogenated tip was grafted onto the fluorinated tail, yielding "hemifluorinated surfactants" (HFSs) (**Figure 3**). Hereafter, FSs and HFSs are collectively referred to as (H)FSs.

The development of (H)FSs, which has gone through three main phases, has been carried out through a long-term collaboration between our laboratory and that of

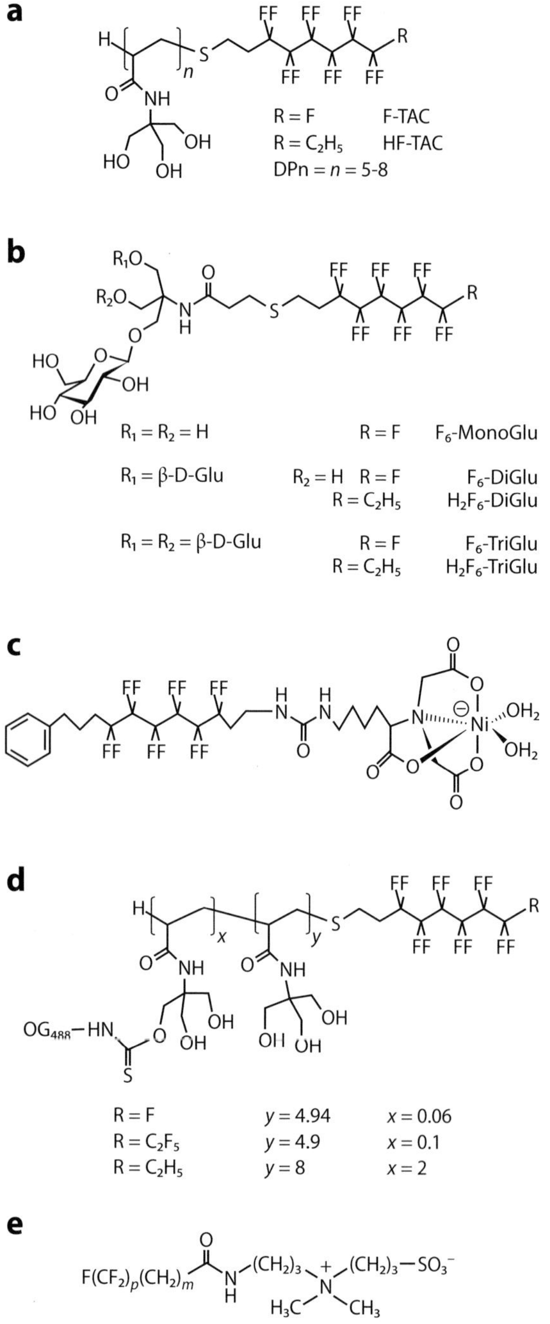

a

$R = F$ F-TAC
$R = C_2H_5$ HF-TAC
DPn = n = 5-8

b

$R_1 = R_2 = H$ — $R = F$ — F_6-MonoGlu
R_1 = β-D-Glu — $R_2 = H$ — $R = F$ — F_6-DiGlu
$R = C_2H_5$ — H_2F_6-DiGlu
$R_1 = R_2$ = β-D-Glu — $R = F$ — F_6-TriGlu
$R = C_2H_5$ — H_2F_6-TriGlu

c

d

$R = F$ — $y = 4.94$ — $x = 0.06$
$R = C_2F_5$ — $y = 4.9$ — $x = 0.1$
$R = C_2H_5$ — $y = 8$ — $x = 2$

e

$F(CF_2)_p(CH_2)_m$–CO–NH–$(CH_2)_3$–$N^+(CH_3)_2$–$(CH_2)_3$–SO_3^-

Figure 3

Chemical structure of some fluorinated surfactants. (*a*) F-TAC (C_8F_{17}-C_2H_4-*S*-poly-*tris*-(hydroxymethyl)aminomethane) (see References 25, 56, and 57) and HF-TAC [C_2H_5-C_6F_{12}-C_2H_4-*S*-poly-*tris*-(hydroxymethyl)aminomethane] (see References 26 and 55). (*b*) (H)F-Mono-, Di- and TriGlu; F_6- and H_2F_6-, as used in the text and in **Figure 7**, refer to hydrophobic moieties, where R = F and R = C_2H_5, respectively. See References 71 and 72. (*c*) Phenyl-HF-NTANi. See Reference 138. (*d*) (H)F-TACs labeled with Oregon Green (OG_{448}). See Reference 61. (*e*) A fluorinated amidosulfobetaine, FASB-p,m. See Reference 74.

B. Pucci (University of Avignon). The first molecules to be tested had as a polar head group a short hydrophilic oligomer derived from *tris*-(hydroxymethyl)aminomethane (THAM) and a perfluorinated hydrophobic moiety (**Figure 3*a***) (25, 56, 57). In a second stage, the tail was hemifluorinated (**Figure 3*a***) (26, 57, 66). These molecules yielded very promising results, and most of the applications of (H)FSs that have been explored to date have been developed using them (Section 7). The oligomeric polar head of (H)F-TAC—where "(H)F-" refers indifferently to the fluorinated or the hemifluorinated form—however, has been a concern from the start because it is chemically polydisperse. From the biochemist's point of view, this means that (H)F-TAC batches consist of a mixture of molecules with slightly different properties and that this mixture will never be exactly the same from one batch to the next.

Replacing the poly-THAM oligomer with a monodisperse headgroup initially seemed a straightforward proposition, which was explored as a third stage of development. This step turned out to be highly frustrating. Indeed, the surfactants obtained upon grafting (hemi)fluorinated chains onto monodisperse polar heads that, when associated to alkyl chains, yield efficient detergents, such as an aminoxyde (67), a monodisperse polyethyleneglycol group (see Reference 28), or saccharidic groups derived from galactose (25, 68), lactose (69), or maltose (70) featured unsatisfactory properties. A recurrent problem was that most of the molecules thus obtained tended to form huge polydisperse micelles by themselves and highly polydisperse MP/(H)FS complexes (for a discussion, see Reference 71). This behavior suggested that the bulky hydrophobic moiety of (H)FSs requires a bulkier hydrophilic head than classical detergents do to create the overall molecular asymmetry that leads to the formation of small globular micelles (4). A systematic investigation was, therefore, undertaken in which polar heads carrying one, two, or three glucose moieties were grafted onto perfluorinated, hemifluorinated, or hydrogenated hydrophobic chains (72). This study led to the identification of two chemically defined (H)FSs, F_6-DiGlu and H_2F_6-DiGlu (**Figure 3*b***), which, as described below (Sections 5.3 and 6.2), form with MPs small, well-defined complexes in which MPs are stabilized as compared to detergent solutions (71).

A new kind of hemifluorinated surfactant has recently been introduced in which the tip of the hydrophobic chain is perfluorinated, but a more or less extended hydrocarbon region is inserted between it and an amidosulfobetaine polar head (**Figure 3*e***) (73, 74). To avoid confusion, these compounds are designated here by the name used by their developers, FASBs (fluorinated amidosulfobetaines). As (H)FSs, FASBs do not by themselves extract MPs (74).

Lyophobic: having little affinity for lipids; perfluoroalkanes are both hydrophobic and lyophobic

Hemifluorinated surfactant (HFS): as used in this text, a surfactant with a fluorinated hydrophobic chain that ends with a hydrogenated tip

(Hemi)fluorinated surfactant, abbreviated (H)FS: is used whenever the matter under discussion applies indifferently to FSs and HFSs

THAM: *tris*-(hydroxymethyl)aminomethan

(H)F-TAC: an (H)FS with an oligomeric polar head derived from THAM

3.3. Amphipathic Polymers (Amphipols)

Detergents and FSs are in a constant equilibrium between monomers, micelles, and the surfactant layer that covers the transmembrane region of the protein and makes it hydrophilic. MPs will aggregate if the surfactant concentration drops below its cmc, meaning that MP/detergent complexes must be handled in the presence of free micelles. The initial idea behind the concept of APols was to design molecules that would have such a high affinity for the surface of the protein that traces of free surfactant in the solution would suffice to keep the protein soluble. An MP transferred to such a medium would face no difficulty retaining its associated lipids, cofactors, and/or subunits and, therefore, should be strongly stabilized. This concept led my colleagues C. Tribet and R. Audebert (ESPCI, Paris) and myself to devise a family of short amphipathic polymers that carry a large number of hydrophobic chains and thus can associate with the transmembrane surface of MPs by multiple contact points. This was expected to result in a vanishing low rate of spontaneous desorption and a very high affinity for MP transmembrane surfaces. The new molecules were dubbed amphipols (APols) to distinguish them from other

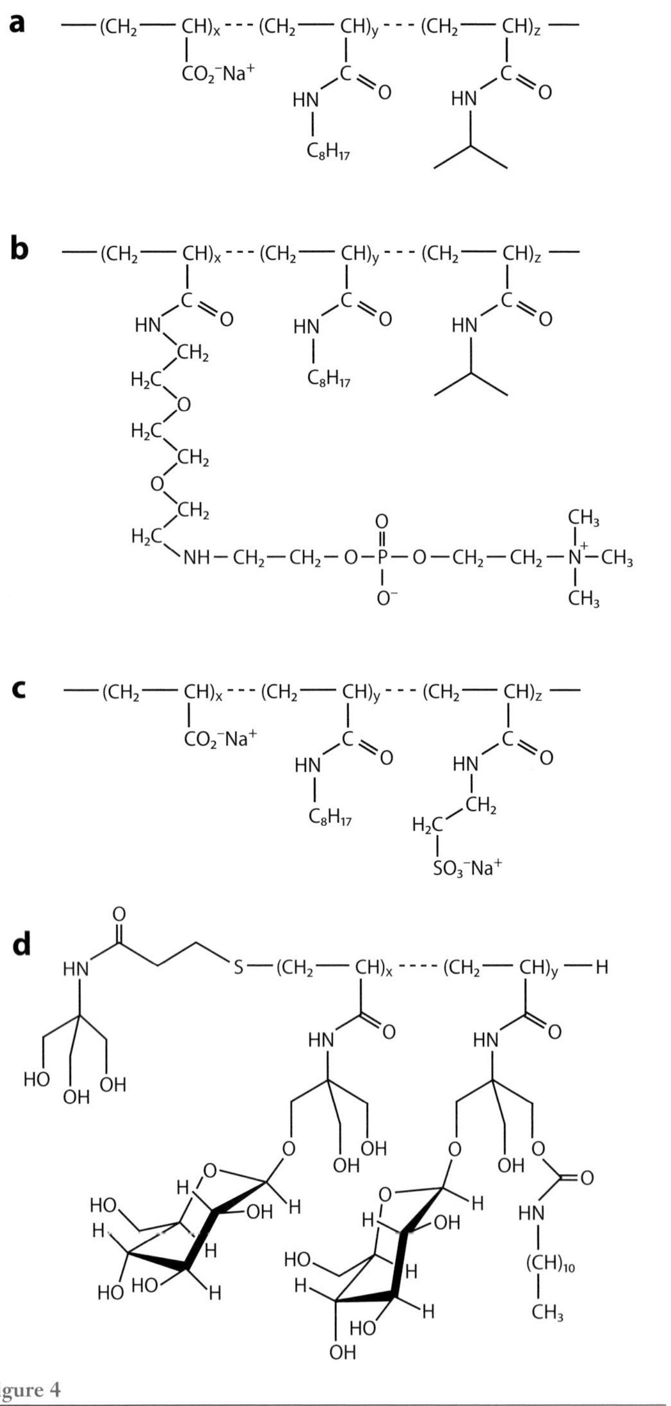

Figure 4

Chemical structures of amphipols (APols). (*a*) A8-35, a polyacrylate-based APol. See References 27 and 76. (*b*) C22-43, a phosphorylcholine-based APol. See References 82 and 85. (*c*) A sulfonated APol. See Reference 84. (*d*) A glucosylated, nonionic APol (NAPol). See References 87–89.

amphipathic polymers, such as those used in the industry to stabilize dispersions of mineral particles or droplets or to control the rheology of solutions.

For an amphipathic polymer to form compact and stable complexes with an MP, the distribution of its hydrophobic chains must be dense, its solubility in water must be high, and it must be highly flexible. It is also desirable that the synthesis of tens of grams be reasonably simple and reproducible. A first series of APols was designed in 1994. One of its members, A8-35 (27), has since become by far the most extensively studied APol (75, 76). A8-35 (**Figure 4*a***) is composed of a relatively short polyacrylate chain (~70 residues), in which some (~17) of the carboxylates have been grafted at random with octylamine and some (~28) with isopropylamine. The ~25 free acid groups are charged in aqueous solutions (75), which makes the polymer highly water soluble, whereas the octylamide moieties, which are spaced along the chain about every nanometer (statistically), render it highly amphipathic. The incorporation of isopropylamide groups is not essential: Indeed, the sister structure, A8-75, which does not contain any such groups, is just as good as A8-35 when it comes to keeping MPs water soluble (27, 77). However, A8-35 features a lower charge density along the chain than A8-75, which seems to have a favorable effect on the stability of MPs (J.-L. Popot and coworkers, unpublished observations). The average molecular weight of a molecule of A8-35 is 9–10 kDa. Batches of A8-35 are a complex mixture of molecules with a variable overall length and a variable distribution of lateral chains. Whether this heterogeneity represents a favorable or an unfavorable feature from the biochemist's and biophysicist's points of view remains an open question because it is not known to which extent the protein may select among the variety of chains offered to it. It is worth noting, however, that A8-35 batches with a much narrower length distribution did not appear to behave differently from more polydisperse batches (C. Tribet & F. Giusti, unpublished observations).

A8-35 is highly water soluble (>200 g.L^{-1}). In a process that is rather unusual for amphipathic polymers, it forms well-defined, small globular particles in aqueous solutions (76). Each A8-35 particle has a mass of ~40 kDa. It comprises, therefore, slightly more than four average A8-35 molecules and a total of ~80 octyl chains. The latter number is close to that of hydrophobic chains in a typical detergent micelle. Förster resonance energy transfer (FRET) studies of mixtures of A8-35 molecules labeled with pairs of complementary fluorophores have shown that their critical aggregation concentration (that above which individual molecules start self-assembling) is <2 mg.L^{-1} (F. Giusti, J.L. Popot & C. Tribet, unpublished observations), meaning that, under most usual conditions, nearly all of the free polymer is assembled into particles. An MD model of an A8-35 particle is shown in **Figure 2*c*,*d*** (J. N. Sachs, personal communication). Its calculated radius of gyration, $R_g \approx$ 2.4 nm, is identical to that measured experimentally (76). An unexpected feature of the MD model is the tendency of water molecules to occupy the center of the particle (**Figure 2*d***). There is also a marked tendency for octyl chains to form submicellar clusters in which octyl chains belonging to distinct APol molecules clump together (**Figure 2*d***).

Variations around the chemical structure of A8-35 have been experimented with. The original study included molecules that were longer than A8-35 and/or carried a higher charge density (27), some of which were used in subsequent works (77, 78, 79). Related structures have been proposed by others (80, 81). One of the constraints imposed by the chemical structure of A8-35 is that its solubility is provided by carboxylate groups. For this reason, it cannot be used below pH 7 (75, 76, 82, 83) nor in the presence of mM concentrations of Ca^{2+} (82, 84). These limitations have prompted the design of APols that are zwitterionic (**Figure 4*b***) (81, 82, 85), sulfonated (**Figure 4*c***) (84), or nonionic (**Figure 4*d***)—the latter also the fruit of a long-term collaboration between our laboratory and that of B. Pucci (86–89). All of these structures have proven able to trap MPs and, generally, have been found to be pH and calcium insensitive. Charged, amphipathic derivatives of pullulane (90), by contrast, turned out to be very inefficient at keeping MPs soluble (91). Most of the background studies and developments to date have been carried out using A8-35, but as discussed below (Section 7), there are some applications for which A8-35 is not the most suitable APol or cannot be used at all, making it highly desirable to develop and validate alternative APols.

NAPol: nonionic amphipol

Because APols are relatively large molecules, grafting a small functional group onto them will generally not affect their solution properties. A8-35 has thus been derivatized with biotin (92) or with various fluorophores (93; F. Giusti, unpublished data), opening the way to many interesting experiments (Sections 5.1 and 7.9). The polymerization process used for the synthesis of nonionic APols (NAPols), known as telomerization (87, 88), makes them easily amenable to stoichiometric functionalization with a single group per chain. A8-35 has also been labeled isotopically with ^{14}C (77), ^{3}H (83), or ^{2}H (75). Deuteration has been particularly useful for NMR (94, 95) and neutron scattering (75, 76, 83) studies.

4. TRANSFERRING MEMBRANE PROTEINS TO NONCONVENTIONAL SURFACTANTS

Although direct extraction of MPs by APols has been observed occasionally (reviewed in Reference 29), APols, NDs, and (H)FSs are not normally used to extract MPs. The usual procedure is to transfer the MP to the NCS following solubilization with a nondenaturing detergent. In the case of NDs and APols, scaffolding proteins and lipids or the polymer, respectively, are usually supplied to the protein in detergent solution. Upon detergent removal, the systems self-assemble (see, e.g., References 43, 44, 46, 83, 93, 94, 96). The ratio of NDs to MP can be adjusted so as to facilitate the trapping of oligomers or, on the contrary, to favor that of monomers (Section 7.3). With APols, an excess of polymer with respect to what the MP actually binds is necessary to obtain homogeneous MP/APol complexes, so that, after trapping, part—typically about half—of the APol is protein bound and part is present as free APol particles. Lipids, if retained by the protein or if present in the detergent solution, will be trapped along with the protein, forming ternary MP/lipid/APol complexes, which may be important for maintaining the stability and function of the protein (see, e.g., References 83 and 97–99 and Section 6.1). It is of great interest that complex mixtures of detergent-solubilized MPs, such as the whole supernatant obtained after solubilizing a biological membrane preparation, can be trapped, whether with NDs or with APols, in the form of discrete MPs or MP complexes, which can then be separated, e.g., by centrifugation in sucrose gradients, SEC, or isoelectrofocusing (IEF) (cf. Section 7.4) (89, 100).

Similar transfer procedures can be used for (H)FSs, but what has often been done in the course of developing these compounds has been to layer a ternary MP/detergent/(H)FS mixture on top of a sucrose gradient containing the (H)FS to be tested (25, 26, 71). Upon ultracentrifugation, MP/(H)FS complexes enter the gradient, leaving the detergent behind them (see below, Section 6.2). This procedure has the advantage of consuming relatively little (H)FS, as compared, for instance, to dialysis or molecular sieving, which is an important criterion when working with compounds whose synthesis is difficult and whose availability is generally limited.

As described below, MPs are not always transferred to NDs, APols, or (H)FSs fully folded from a solution in nondenaturing detergent: They can be directly folded in NCSs, either starting from the full-length, denatured protein in sodium dodecyl sulfate (SDS) or urea (Section 7.10) or in the course of in vitro cell-free protein synthesis (Section 7.11).

5. STRUCTURE OF MEMBRANE PROTEIN AND NONCONVENTIONAL SURFACTANT COMPLEXES

The composition, size, homogeneity, structure, and dynamics of MP/NCS complexes have been closely scrutinized because they largely determine the suitability of the complexes for various studies.

5.1. Membrane Protein/A8-35 Complexes

A8-35 has been shown to form a complex with, and maintain solubility of, a very broad range of MPs. Those include all of the 30-odd integral MPs that have been tested to date. Their molecular masses range from <5 kDa (a single transmembrane α-helix) to >1.1 MDa [mitochondrial complex I, which probably harbors ~70 transmembrane α-helices (101)], and they represent any conceivable diversity of origins, functions, and structures (for a review, see Reference 29). A8-35 has recently been used to trap and purify mitochondrial supercomplex B (T. Althoff & W. Kühlbrandt, personal communication; see Section 7.7), which comprises ~120 transmembrane helices and whose overall mass is ~1.7 MDa (102). The interactions of APols with other types of objects are beyond the scope of the present review. However, one may mention that APols have been used to stabilize in solution a signal sequence peptide (103), human apolipoprotein B-100 (I. Waldner, A. Kriško, A. Johs, & R. Prassl, unpublished data), plant lipid storage proteins (Y. Gohon, personal communication), and semiconductor quantum dots (104, 105). Several studies have been carried out of their interactions with lipid vesicles or cells (78, 79, 104, 106). Under most circumstances (for some exceptions, see Reference 78), APols will not solubilize preformed biological or lipid membranes (see Reference 29). They can, therefore, be applied to lipid vesicles, black films, or living cells without lysing them, which opens up some extremely interesting applications (Section 7.8).

The composition, structure, dynamics, and solution properties of MP/A8-35 complexes have been studied by SEC, analytical ultracentrifugation (AUC), small-angle neutron scattering (SANS), SAXS, FRET, and solution NMR, yielding a rather detailed picture of what these complexes look like. In NMR studies using as models the transmembrane domains of outer MP A from *Escherichia coli* (tOmpA) or *Klebsiella pneumoniae* (KpOmpA) and *E. coli*'s outer MP X (OmpX), each of which consists of an eight-strand β-barrel, the only detectable contacts between A8-35 and the protein were observed at the barrel's transmembrane surface (Section 7.5). AUC and SANS studies of bacteriorhodopsin (BR)/A8-35 complexes indicate that the polymer layer is compact and 1.5–2 nm thick (83), which is only slightly thicker than a detergent layer (107). Measuring precisely the amount of protein-bound APol is technically difficult, particularly for small proteins (83, 93). Estimates vary from ~2 g per g protein for a small, mainly transmembrane protein like BR (**Figure 5*c***) (83) to ~0.13 g per g for a complex with large extramembrane regions such as cytochrome bc_1 (**Figure 5*a***) (D. Charvolin, unpublished data). For large MPs, this corresponds to significantly less (by 2–3 times) octyl chains than the number of detergent molecules bound by the same protein in detergent solutions (29). For small MPs, the difference is much less (83, 93). In most experiments, such as those by SAXS, SANS, AUC, or solution NMR, MP/A8-35 complexes behave like compact, globular particles (83, 94, 108). Upon SEC, tOmpA/A8-35 and BR/A8-35 complexes migrate as though they are somewhat bigger than they actually are (83, 94).

When properly prepared, MP/A8-35 complexes are essentially homogeneous (83, 93), although they do not appear as narrowly distributed in SEC as MP/detergent complexes (93). It is, however, difficult to totally avoid the presence of minor fractions of small oligomers, which can seriously complicate, in particular, radiation scattering experiments (83). Among the factors that can lead to polydispersity, if not aggregation, are (*a*) the use of too little APol at the trapping step (93), (*b*) working at pH ≤ 7 (82, 83, 94), (*c*) the presence of Ca^{2+} ions (82, 84, 108), (*d*) drifts from the nominal composition of the polymer that result in increasing its hydrophobicity (83, 84), and (*e*) separating the complexes from the free polymer that coexists with MP/APol complexes at the end of a trapping experiment (83, 93). The oligomerization that follows the removal of free APol is reversible. It has been interpreted as resulting from the poor dispersive power of APols: In the competition between protein/protein and protein/APol interactions, the presence of an excess of polymer favors the formation of MP monomers, whereas its removal shifts the equilibrium toward aggregation (93). Increasing the ionic strength of the solutions progressively turns the interactions between cytochrome bc_1/A8-35 particles from a repulsive to an attractive mode (29). In a solution that contains both protein-bound and free APols, free and bound polymers exchange at the surface of the protein at a rate that, at least for A8-35, strongly depends on the ionic strength of the solution (minutes in 100 mM NaCl, hours at low ionic strength) (93). The underlying mechanism has not been studied in detail, but it probably involves collisions between MP/APol complexes and free APol particles, followed by fusion, mixing, and fission. Such a mechanism is consistent with the fact that APol particles can deliver retinal, a highly hydrophobic cofactor, to refolded bacterio-opsin (99). By contrast, in the absence of exchange with another surfactant, APols do not desorb from MPs even under extreme dilution or extensive washing (29, 77, 93, 109). A thermodynamic analysis of MP/detergent versus MP/APol interactions has led to the conclusion that the stability of MP/APol complexes is essentially of entropic origin: When a solution of MP in detergent is diluted below the cmc of the detergent, the release of tens of detergent molecules that accompanies the association with one another of two MPs creates a strong entropic drive toward aggregation; when two APol-trapped MPs aggregate, the desorption of one or two APol molecules, which

tOmpA: the transmembrane domain of outer membrane protein A from *Escherichia coli*

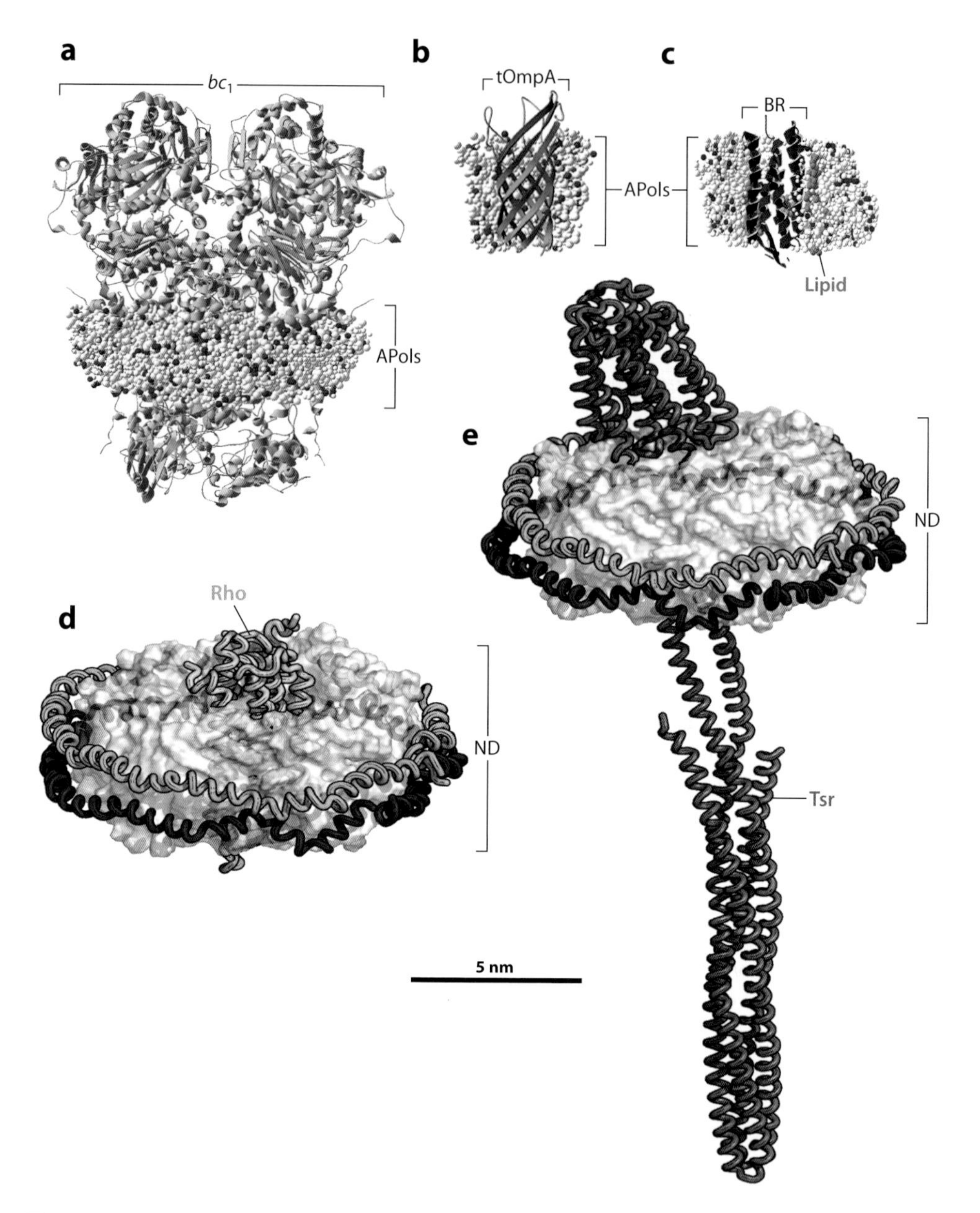

Figure 5

Models of complexes between membrane proteins (MPs) and amphipols (APols) or nanodiscs (NDs). (*a-c*) Models of MP/APol complexes. (*a*) Cytochrome bc_1/A8-35 complex (from Reference 29). (*b*) An "open" model of a complex between the transmembrane domain of *E. coli*'s OmpA (tOmpA) and A8-35 (from Reference 29). (*c*) Cross section through a bacteriorhodopsin (BR)/A8-35 complex. Lipids from the purple membrane, which were retained throughout the solubilization and trapping procedures, are shown in olive green (from Reference 83). The volume and distribution given to the APol belts in the three complexes are based on an ensemble of data including binding, SEC, AUC, SANS, and NMR measurements, obtained on each individual complex as well as on other MPs (see References 29 and 83 and Section 5.1) Models by D. Charvolin. (*d, e*) Models of rhodopsin (Rho) (*d*) and the serine chemotaxis receptor (Tsr) (*e*) embedded into NDs. From Reference 39. The five models are represented approximately to scale.

associate into particles, is entropically neutral (109).

Although few proteins have been tested yet, it seems that freezing MP/APol complexes is not detrimental, whereas lyophilizing them can be (83; Y. Gohon & L. Catoire, unpublished data).

In ternary mixtures containing an MP, a detergent, and an APol, the two surfactants mix at the transmembrane surface of the protein, forming ternary complexes (77, 93, 109). FRET and isothermal calorimetry studies show that mixing is almost ideal and that the exchange of one surfactant for the other is isoenthalpic: Detergents and APols mix freely, and the composition of the mixed belt of surfactant around the MP reflects that of the mixed, protein-free APol-detergent particles (93, 109). The exchange of APol for detergent at the surface of an MP is extremely rapid (<1 s) (93). Because APols have no strong preference for the MP-bound belt versus free APol/detergent particles, they are easily and rapidly washed away by an excess of detergent (77, 93, 97, 109). Interestingly, MPs can be stabilized biochemically even by mixtures of detergent and APols (97) (see Section 6.1 below). Upon SEC, MP/APol/detergent particles appear more homogeneous in size than pure MP/APol complexes (93). These two observations may have interesting implications for the choice of crystallization conditions (Section 7.6).

5.2. Membrane Protein/Nanodisc Complexes

MPs that have been reconstituted into NDs include a number of P450 cytochromes as well as the NADPH-cytochrome P450 reductase (reviewed in References 39 and 40), BR (110, 111), the bacterial Tar chemoreceptor (112), the SecYEG translocon complex (113), and several G protein–coupled receptors (GPCRs), including the β_2-adrenergic receptor (β_2-AR) (114–116), rhodopsin (44, 117), and a μ-opioid receptor (118). The size of the largest transmembrane domain that can be accommodated into a ND depends on the rim protein employed. In addition, recent single-particle data suggest that NDs can have a multimodal distribution of diameters and that MP encapsulation can shift this distribution toward larger diameters (45, 47). The largest system whose encapsulation has been reported to date is the BR trimer, which comprises 21 transmembrane helices (111). It is to be expected, however, that this constraint will be relaxed as protein engineering will produce MSPs and other proteins able to stabilize larger and larger discs (see, e.g., References 42, 43, and 45).

MP/ND complexes have been studied experimentally by SEC, SAXS, AFM, EM, circular dichroism (CD), fluorescence spectroscopy, and ssNMR (see, e.g., References 44, 47, 110–112, 115, 117, 118), as well as simulated by MD (119) (for reviews, see References 39 and 40). Essential to the use of NDs for functional studies is the knowledge of the number of copies of MPs they have captured. Thus, trapping conditions have been designed so as to trap either monomers or trimers of BR (110, 111), monomers of the β_2-AR (115) or the μ-opioid receptor (118), and monomers or dimers of rhodopsin (44, 117). The Tar chemoreceptor has been trapped either as dimers or trimers of dimers (112). Functional studies of biological systems where the number of copies of MP involved is a critical factor are one of the most exciting applications of NDs (see Section 7.3 below).

5.3. Membrane Protein/(Hemi)Fluorinated Surfactant Complexes

MP/(H)FS complexes have not yet been studied in great detail, particularly for those (H)FSs that appear to be most satisfying both from the point of view of the chemistry (defined chemical structure) (72) and biochemistry (small, well-defined MP/(H)FS complexes) (71). A striking observation is that MP/(H)FS complexes migrate much faster upon centrifugation in sucrose gradients than MP/detergent complexes

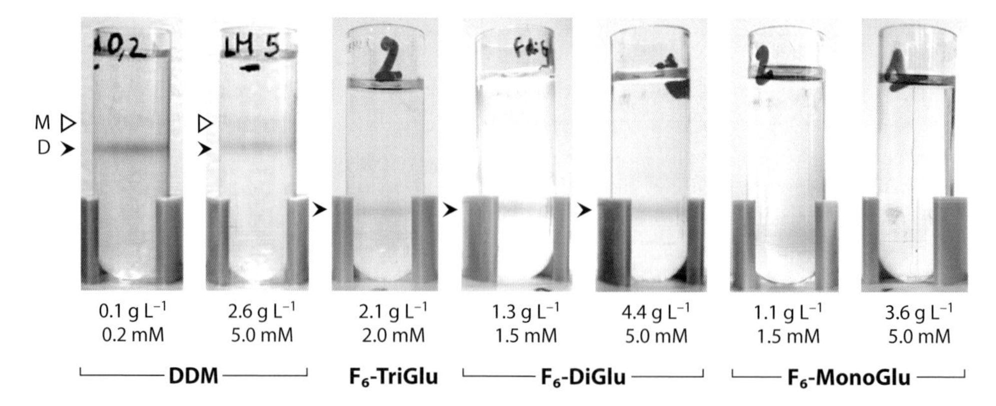

Figure 6

Behavior of cytochrome b_6f in sucrose gradients containing either a detergent, dodecylmaltoside (DDM), or glucosylated fluorinated surfactants at various concentrations. The chemical formulas of F_6-Mono-, -Di- and -TriGlu are shown in **Figure 3**. In DDM, the complex remains primarily a superdimer (D) at a concentration just above the critical micellar concentration and starts to fragment into monomers (M) at higher concentrations (cf. Reference 7). In F_6-TriGlu and F_6-DiGlu, it forms well-defined bands of dimer whatever the surfactant concentration. In F_6-MonoGlu, the bands are fuzzy, owing to the polydispersity of the complexes. From Reference 71.

(**Figure 6**) (25, 26, 69, 71). A priori, this could be due to an aggregation of the protein and/or to the binding of very large amounts of surfactant. However, a detailed study of the complexes formed between cytochrome b_6f and a lactose-derived HFS, HF-Lac, has shown that neither is true: The b_6f is dimeric in HF-Lac solutions, as it is in detergent ones as long as it retains its native structure (7), and it binds about the same number of HF-Lac molecules (~260) as of dodecylmaltoside ones (69). Instead, it is the much higher density of the surfactant [$\bar{v} \approx 0.6$ mL.g^{-1} for HF-Lac, F_6- and H_2F_6-DiGlu (69, 71)], owing to the presence of the fluorine atoms, which accounts for the increased sedimentation coefficient of the complexes.

In the course of comparing the behavior of (H)FSs carrying a variable number of glucose moieties, it was noted that at least two sugars are needed for (H)FSs to form small globular micelles and that only in these cases was it possible to obtain well-defined MP/(H)FS complexes. By contrast, molecules bearing a single glucose form long, cylindrical micelles and generate polydisperse MP/(H)FS complexes (71). This is illustrated in **Figure 6** in the case of cytochrome b_6f.

6. FUNCTIONALITY AND STABILITY OF MEMBRANE PROTEINS IN NONCONVENTIONAL SURFACTANTS

The stability and functionality of an MP in aqueous solution is extremely variable, depending both on the protein and on the surfactant that keeps it soluble, as well as on the concentration of the latter. Existing data indicate that, used at comparable ratios of "micellar" (assembled) surfactant to MP, the three types of NCSs considered here tend to be less inactivating than detergents—that is, they keep MPs from irreversible denaturation for a longer time. Data about the functionality of MPs associated to NCSs, which is a distinct question, are still relatively scarce but, on the whole, very encouraging.

6.1. Membrane Protein/Amphipol Complexes

Complexation by APols, in most cases, biochemically stabilizes MPs as compared to detergent solutions (see, e.g., References 27, 29, 83, 84, 88, 89, 97, and 120 and the references therein). This is illustrated in **Figure 7** for the sarcoplasmic reticulum calcium ATPase (**Figure 7*a***), BR (**Figure 7*b***), and a GPCR, the BLT1 receptor of leukotriene LTB_4 (**Figure 7*c***). As illustrated by the intermediate curve in **Figure 7*a***, even adding APols to an MP in detergent solution, without removing the detergent or diluting it, may have a stabilizing effect. A detailed discussion of the mechanisms of MP stabilization by APols is beyond the scope of the present review. Existing data indicate that the following effects may contribute to stabilization: (*a*) Retention of MP-associated lipids, cofactors, and subunits (a consequence of reducing the hydrophobic sink, but also of the poorly dissociating character of APols); (*b*) less efficient competition with protein/protein interactions, and (*c*) damping the dynamics of conformational excursions of transmembrane α-helix bundles, which limits opportunities for unfolding and/or aggregation (for discussions, see References 29 and 84). Stability can be further improved by forming ternary MP/lipid/APol complexes (**Figure 7*c***) (97, 120). Although data are still limited, a sprinkling of observations suggests that, as is the case for detergents, NAPols may be even less destabilizing than ionic ones are (86; Y. Pierre, unpublished observations).

The functionality of APol-trapped MPs is generally preserved (see, e.g., References 29, 81, 83, and 98). However, MPs whose functional cycle involves large rearrangements of the surface of their transmembrane region, as is the case for the sarcoplasmic calcium ATPase, may see their activity reversibly slowed or blocked, presumably because the adsorbed polymer damps such transconformations (for discussions, see References 29 and 84). For BR (83, 99) and, perhaps, the nicotinic acetylcholine receptor (nAChR) (98), indirect arguments suggest that transferring the protein from a detergent to an APol environment favors the rebinding of lipids at the surface of the protein, which probably contributes to restoring membrane-like functionality (Section 7.3). Ligand binding is, very generally, unaffected by APol trapping (see Sections 7.2, 7.9, and 7.10 below).

6.2. Membrane Protein/Hemifluorinated Surfactant Complexes

Following their transfer from detergent solutions to (H)FS ones, BR, cytochrome b_6f, the human GPCR Smoothened and the mitochondrial ATP synthase exhibit improved biochemical stability (25, 26, 69–71, 121, 122). Such is also the case of dimers of BLT1 (J.-L. Banères, personal communication). The improvement is usually limited when the surfactant concentration is close to the cmc, but it becomes obvious at higher concentrations (**Figures 6** and **7*d***), suggesting that it is mostly a consequence of the lyophobicity of (H)FS micelles, which makes them a poor sink for lipids. BR and b_6f are markedly stabilized in mono- and diglucosylated (H)FSs, but destabilized in triglucosylated ones (71). Although other hypotheses could be considered (see Reference 71), my favorite interpretation of the latter phenomenon is that too large a repulsion between polar heads, which favors the formation of particles with a small radius of curvature, may tend to either fragment and/or unfold MPs. The same phenomenon may well contribute to explaining the well-known destabilizing character of charged detergents. Because (H)FSs with a single glucose moiety form cylindrical micelles (Section 3.2) and polydisperse MP/(H)FS complexes (Section 5.3), and those with three glucose moieties are destabilizing, F- and HF-DiGlu (**Figure 3*b***) were identified as optimal among chemically well-defined (H)FSs (71).

The relative benefits of using either per- or hemifluorinated compounds should be further investigated. Among the factors to be taken into consideration are the following: (*a*) HFSs are much harder to synthesize than FSs; as a consequence, they are more costly and available in more limited quantities; this limitation is of particular importance if (H)FSs are to be used for MP purification, where large volumes of solution above the cmc of the surfactant are required. (*b*) Initial observations indicating that FSs are much less efficient at preventing MPs from aggregating (25) than HFSs are (26) have not been supported, or not strongly so, by more recent experiments, where the difference appears marginal (71), perhaps because MP preparations used in more recent studies contained less lipids. (*c*) A difference has been noted between the spectra of monomeric BR trapped in either HFSs or FSs: In HFSs, the spectrum of the protein resembles that of the native protein;

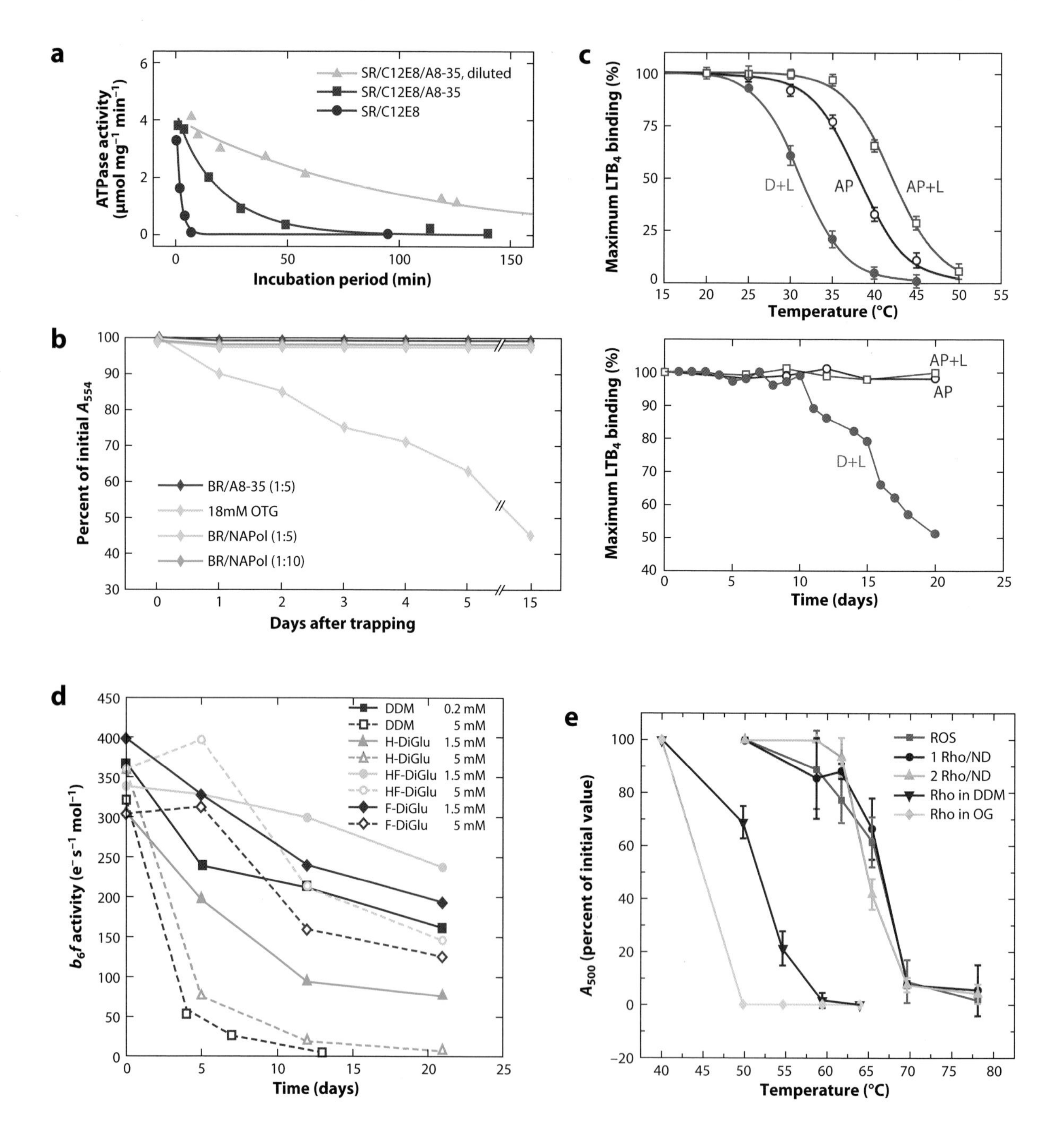

in FSs, it is shifted to the red ("blue BR") (71). The spectral change results from the protonation, owing to a pK shift, of Asp85, a residue that interacts electrostatically with the protonated Schiff base that associates the retinal to the protein (C. Breyton, unpublished results). This indicates that, directly or indirectly (e.g., via the lipids), FSs can affect the 3D structure of the MPs with which they are complexed. The fact that BR oligomers are not affected by this effect (71) suggests that perhaps this will not be a general concern but may simply reflect events occurring at one specific site at the surface of one particular protein.

6.3. Membrane Protein/Nanodisc Complexes

Because MPs trapped within NDs are not exposed to detergents, and because they are surrounded by an environment very similar to the natural one, they can be expected to exhibit improved stability. Few studies of this sort seem to have been reported. Rhodopsin, however, was indeed found to be much more resistant to thermal denaturation in NDs than in detergent solutions (**Figure 7*e***) (44).

Functional studies are one of the most interesting applications of NDs, and they have been pushed much farther than with any of the other NCSs considered here. They will be discussed below along with other applications (Section 7.3).

7. APPLICATIONS

Owing to space limitation, it is not possible to review here at length every application that has been explored using each of the three systems under discussion. My purpose is, rather, to illustrate the kinds of studies that may be facilitated by using one or the other NCSs instead of detergents and to provide directions for further reading. I will also try, even though I am fully conscious of the risks inherent in such an exercise, to present my feelings about the prospects and constraints of each of them.

A first benefit of transferring MPs to NCSs is the stabilization they afford, which may permit the study, using otherwise classical biochemical or biophysical approaches, of MPs that would be intractable in detergent solution. What one has to consider here is whether using NCSs rather than detergents comes at a price and, if so, what price? In addition, NCSs may permit studies for which detergents cannot be used (e.g., mediating MP immobilization, defining the maximal size of the objects trapped, or inserting MPs into membranes under equilibrium conditions) or are poorly efficient (e.g., MP folding).

7.1. Constraints for Optical Spectroscopy

A8-35 (93), NAPols (89), and (H)FSs (B. Pucci, unpublished data) do not absorb light significantly at wavelengths above ~245 nm. None

Figure 7

Stabilization of membrane proteins by nonconventional surfactants. (*a*) Stabilization of the sarcoplasmic reticulum calcium ATPase by A8-35. The destabilization of the ATPase was initiated by diluting solubilized sarcoplasmic reticulum (SR) into an ethyleneglycol-*O*, *O′*-*bis*(2-aminoethyl)-*N*, *N*, *N′*, *N′*-tetraacetic acid-containing solution, thereby leaving the ATPase in a Ca^{2+}-deprived, solubilized state, an environment known to lead to very rapid, irreversible inactivation. (*Lower curve*) The dilution medium contained 5 $g.L^{-1}$ (9.3 mM) $C_{12}E_8$; (*middle curve*) same medium but with the addition of 5 $g.L^{-1}$ A8-35; (*upper curve*) same medium as the latter sample, but incubation took place after a 250× dilution with surfactant-free buffer. From Reference 97. (*b*) Stabilization of BR by A8-35 and by nonionic amphipols (NAPols), as compared to 18 mM octylthioglucoside (OTG). From Reference 89. (*c*) Stabilization of the LTB_4 BLT1 receptor by A8-35 and by A8-35/lipid mixtures. (*top*) Stability upon incubation at increasing temperature; (*bottom*) stability upon extended storage at 4°C. From Reference 120. Abbreviations: D+L, detergent + lipids (fos-choline-16/asolectin, 2:1 w/w); AP, A8-35; AP+L, A8-35 + asolectin (5:1 w/w). (*d*) Stabilization of cytochrome $b_6 f$ by (hemi)fluorinated surfactants as compared to detergents [dodecylmaltoside (DDM) and H-DiGlu; H-DiGlu has the same chemical structure as F- and HF-DiGlu, except for a fully hydrogenated alkyl chain]. From Reference 71. (*e*) Stabilization of rhodopsin (Rho) upon integration into nanodiscs (NDs) (in this case, nanoscale apolipoprotein bound bilayers) containing either one or two copies of the protein per disc, versus in rod outer segments (ROS; rhodopsin's native membrane environment), DDM (29 mM) or octylglucoside (OG; 51 mM). From Reference 44.

of them fluoresces upon excitation at 280 nm (93; B. Pucci, unpublished data). A8-35 does not interfere with CD studies of proteins down to at least 180 nm (99, 103, 123, 124; T. Dahmane & F. Wien, unpublished data). (H)F-TACs are not expected to interfere significantly at these wavelengths. Glucosylated NAPols and (H)FSs, however, may be expected to contribute to CD spectra below ~195 nm, owing to the presence of the sugar residues (125). FTIR studies of MP/A8-35 complexes in the amide region are intractable because of the amide bonds of the polymer (Y. Gohon & E. Goormaghtigh, unpublished observations). The same difficulty will arise with sulfonated APols, NAPols, phosphorylcholine-based APols, and current (H)FSs.

NDs pose a special problem because they include scaffold proteins. It should be possible to reduce the contributions of MSPs by eliminating the tryptophan residues they contain using site-directed mutagenesis. The contribution of MSPs to IR and CD spectra will remain, but it can conceivably be subtracted.

7.2. Solution Studies of Membrane Protein Mass, Shape, and Interactions

All three NCS systems have been studied by SEC, AUC, SAXS, and/or SANS (see, e.g., References 71, 83, and 126). A priori, each of these techniques can be used to characterize NCS-complexed MPs. APols and (H)FSs have the advantages of adding less bulk and complexity to the system and of being more easily contrast matched. However, obtaining perfectly monodisperse preparations of MP/APol complexes is challenging, which makes molecular weight and R_g determination of the trapped MP by radiation scattering quite delicate (see Reference 83). Factors that have to be kept in mind when monodispersity is essential have been listed in Section 5.1.

APols do not interfere with most of the MP/ligand interactions that have been studied to date (92, 97, 98, 120) (see also Sections 7.9 & 7.10 below). This does not mean that one should not be aware of the possibility of such interferences (for discussions, see References 29 and 92). Thus, preliminary experiments with rhodopsin/A8-35 complexes suggested that A8-35 hinders the binding of transducin and arrestin (see Reference 29). This point has been reinvestigated recently with the leukotriene receptor BLT1. It was observed that receptor-catalyzed G protein activation was significantly slower with A8-35-trapped BLT1 than in detergent/lipid mixed micelles. In contrast, when folded in NAPols, BLT1 catalyzed GDP→GTP exchange on the $G_{\alpha i}$ subunit with kinetic features similar to those in lipid/detergent mixtures (J.-L. Banères, personal communication).

Particularly illustrative of the perspectives opened by the ND system are studies exploiting NDs to trap monomers or defined oligomers of various MPs. Some of the applications of this remarkable feature are described in Section 7.3.

Few detailed ligand-binding studies have been carried out in (H)FSs yet. The human Sonic Hedgehog receptor Patched was shown by surface plasmon resonance (SPR) to be able to interact with Sonic Hedgehog after complexation by FSs (127). The sensitivity of the mitochondrial ATP synthase to the inhibitors dicyclohexyl carbodiimide and oligomycin was found to be higher and to remain more stable over time in (H)F-TACs than it is in detergent solutions (122).

7.3. Functional Studies

As already mentioned, NDs provide a unique approach to examining the role of oligomerization in the biological function of MPs because they make it more straightforward and safer than any other system to prepare and handle monomers and various types of oligomers while limiting the risk that transient oligomerization complicates the interpretation of the experiments. In the case of GPCRs, NDs have been used to study the activation of G proteins by preparations of monomeric β_2-AR receptor, rhodopsin, and μ-opioid receptor (44, 114, 115, 117, 118, 128). Similarly, trapping of single copies of translocon in NDs made it possible to establish that SecYEG monomers are able to bring about the dissociation of dimeric SecA into monomers and (pre)-activate the SecA ATPase (113). Trapping the Tar chemoreceptor as either dimers or trimers of dimers led to the demonstration that the formation of the dimer suffices for transmembrane signal transduction (methylation and deamidation), but the super-trimer is required for downstream signaling (activation of histidine kinase) (112). A vast body of work has been carried out on P450 cytochromes and their reductase (reviewed in References 39 and 40). They concern, in particular, the origin of cooperativity in ligand binding curves and the role of lipids and the transmembrane anchor of the cytochromes in P450/reductase interactions. NDs have also proven useful in studying the activation of enzymes involved in blood clotting and the role played in it by specific lipids (129, 130). There seems to be little doubt that NDs will henceforth provide a privileged tool whenever such questions will have to be dissected.

To date, applications of APols to investigations of MP function have mostly focused on sorting out whether perturbations observed upon MP solubilization are the result of removing the membrane environment or of direct interference by the detergent. In the case of the nAChR, it was shown that allosteric equilibria, which are perturbed upon solubilization, are similar in the postsynaptic membrane and after trapping with A8-35, indicating that it is the presence of the detergent, and not the disappearance of physical constraints imposed by the membrane, that is responsible for the perturbation (98). It was hypothesized that detergents might act by displacing lipids from critical sites at the surface of the nAChR's transmembrane domain, where A8-35 would let them rebind. BR, a light-driven proton pump, undergoes severe perturbations of its photocycle upon solubilization by detergents. One such perturbation, the acceleration of the retinal's Schiff base deprotonation upon capture of a photon, largely persists after transfer to APols, suggesting that the conformational change that underlies this phenomenon results from the extraction of BR from the purple membrane rather than from detergent binding (83). The end of the cycle, when BR returns to its ground state, goes back to close to membrane-like features upon transfer from detergent solution to APols (83). Comparative studies of the photocycle of native BR, trapped in APols along with purple membrane lipids, and of BR refolded in APols in the presence or absence of lipids have led to the conclusion that it is probably the rebinding of lipids at critical sites at the surface of the protein upon transfer from detergent to APols that accounts for the recovery of normal kinetics in the last part of the photocycle (99).

7.4. Proteomics: Isoelectrofocusing, Two-Dimensional Gels, and Mass Spectrometry

Analysis of protein mixtures on two-dimensional (2D) gels usually starts with a separation by IEF followed by electrophoresis in polyacrylamide gels in the presence of sodium dodecylsulfate (SDS-PAGE) in a perpendicular direction. IEF of MPs using detergents is notoriously difficult (131). Most APols, which bear net charges, are not compatible with IEF, but NAPols are. The resolution of IEF observed with NAPols is similar to that in neutral detergents (89, 132). Whether the use of NAPols can improve on the yields obtained with detergents remains to be ascertained. However, they are certainly of potential interest as a tool to separate and analyze fragile MP complexes that do not resist purification in detergent solutions. (H)FSs could conceivably be used to the same end but are probably at more risk of raising aggregation problems. FASBs (see Section 3.2) have been used in 2D electrophoresis as a means to detach nonmembrane proteins from erythrocyte membranes prior to solubilizing MPs with a classical detergent (74). Separating MP/ND complexes by IEF may seem a priori an odd proposition because MSPs carry charges. However, provided a constant stoichiometry with the target MPs is maintained and the lipids are neutral, this should result in a simple shift of the isoelectric point of the complex as compared to that of the MP alone, without necessarily compromising the resolution.

In keeping with their displacement by non-ionic detergents (77, 93, 109), APols are washed away by an excess of SDS. This is reflected in the fact that, upon SDS-PAGE, MPs migrate at the same position whether they were initially complexed by A8-35 or not (123) and by the absence of a FRET signal and lack of any CD change when fluorescent A8-35 is added to an MP in SDS solution as a prelude to renaturation (89). In the second dimension of 2D gels, MPs, therefore, separate as they do in the absence of APols, and they can be analyzed by mass spectrometry under the same conditions (132).

Recent data indicate that native MPs complexed by either A8-35 or NAPols can be analyzed by matrix-assisted laser desorption/ionization time-of-flight (MALDI-TOF) mass spectrometry, with the APols remaining undetected (95; C. Béchara, G. Bolbach, & S. Sagan, unpublished data).

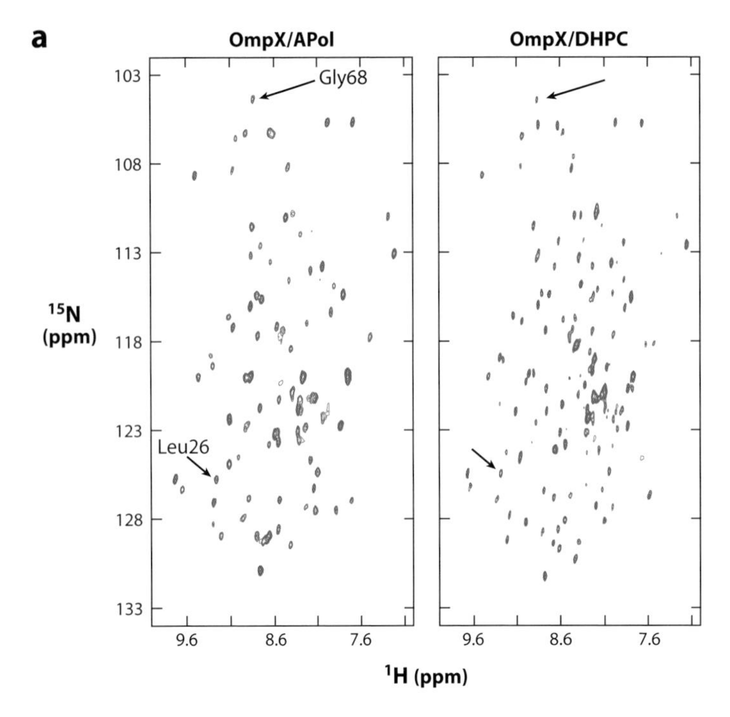

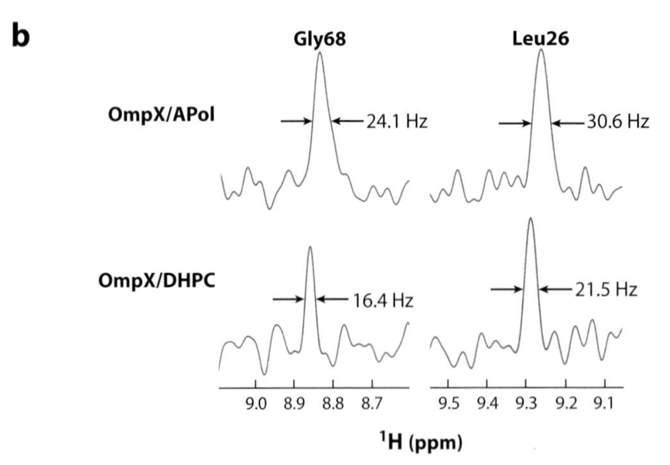

Figure 8

Solution NMR spectra of APol- versus detergent-complexed OmpX. (*a*) Two-dimensional [^{15}N,^{1}H]-TROSY spectra of [u-^{2}H,^{13}C,^{15}N]OmpX complexed by APol A8-35 (*left*; pH 8.0) or by DHPC (*right*; pH 6.8) collected at 30°C at 700 MHz. (*b*) A comparison of peak width in the direct dimension. From Reference 108.

7.5. Nuclear Magnetic Resonance Studies

APol-trapped MPs have been studied by solution NMR (94, 95, 108; M. Renault & A. Milon, personal communication). All published studies to date have been carried out using A8-35, and they all bear on β-barrel proteins. At this point, the objectives have been mainly to work out the technology and to obtain information about MP/APol complexes (size, folding state of the protein, protein/APol interactions, and protein dynamics). Not surprisingly, these studies have shown that the three proteins studied to date, *E. coli*'s tOmpA (94) and OmpX (95, 108) and *K. pneumoniae*'s KpOmpA (M. Renault & A. Milon, unpublished data), are correctly folded in A8-35 (**Figure 8*a***). Early studies with *E. coli*'s tOmpA were carried out in the absence of (ethylenedinitrilo)tetraacetic acid (EDTA) (94). In the presence of EDTA, the resolution of TROSY-HSQC spectra improves, even though the complexes still tumble more slowly than in DHPC solutions: For OmpX at 30°C, τ_c = 31 ns versus 23 ns (108); for KpOmpA at 40°C, τ_c = 30 ns versus 26 ns (M. Renault & A. Milon, unpublished data). For OmpX, the resulting broadening of the peaks in the

^1H dimension is ~35% as compared to their width in DHPC solutions (**Figure 8*b***) (108). Solution NMR has been used to map the contacts between the protein and the lateral chains of the polymer, based on amide ^1H peak broadening in hydrogenated versus deuteriated A8-35 (94), or on heteronuclear ^{13}C-^1H (95), or ^{15}N-edited ^1H-^1H (M. Renault & A. Milon, unpublished data) nuclear Overhauser effects (NOEs). In all three cases, A8-35 was found to interact exclusively with the transmembrane surface of the protein. No contacts are observed between KpOmpA and the APol's main chain (M. Renault & A. Milon, unpublished data). Whether contacts with the APol are homogeneously distributed over the protein's transmembrane surface and whether the polymer's octyl chains interact with it in the same way as detergent chains do are still relatively open questions (95; M. Renault & A. Milon, unpublished data). Studies of H/D exchange at amide bonds show that the dynamics and accessibility of OmpX's (108) and KpOmpA's (M. Renault & A. Milon, unpublished data) transmembrane β-strands are similar in DHPC and after trapping with A8-35. It is possible to determine, by transferred NOE studies, the structure of ligands bound to large APol-stabilized MPs (L. Catoire, unpublished observations).

In addition to the slower tumbling rate, an inconvenience that is likely to become less and less relevant as larger MPs are tackled and the technology improves, the major drawback of A8-35 in solution NMR studies is that it does not allow one to work at the slightly acidic pH at which the exchange of amide protons with the solution is slowest. This is of little import for transmembrane amide protons, which exchange slowly anyway, but it becomes critical for water-exposed protons in the extramembrane loops. This has prompted the development of alternative, pH-insensitive APols (Section 3.3). Recent data indicate that TROSY-HSQC spectra can be recorded of tOmpA trapped with sulfonated APols. At pH 6.8, it becomes possible to resolve loop amide protons that are not seen at pH 8.0. Encouraging data have been obtained recently with NAPols (L. Catoire, personal communication). It is to be expected that further progress in the chemistry of APols and their implementation will soon make it possible to establish by solution NMR the structure of MPs that are too unstable to be studied in detergent solutions.

Solution NMR studies in (H)FS solutions have not been attempted yet. A potential pitfall may be aggregation of the protein at the high concentrations required by NMR, but the attempt is certainly worthwhile. ^{19}F NMR potentially offers some original angles, e.g., to follow the distribution of (H)FSs whether in vitro or in vivo.

Solution NMR of MP/ND complexes is handicapped by the large size of NDs and possible anisotropy problems. Theoretical calculations indicate that the tumbling rate of an empty ND should be comparable to that of an ~200-kDa protein (133). In practice, the resolution obtained for a 13-residue peptide adsorbed onto NDs, although not as good as in bicelles, was somewhat better than expected, probably owing to the peptide moving with respect to the surface of the ND. Signal attenuation by a spin label dissolved in the lipid phase yielded information about the secondary structure of the peptide and its arrangement with respect to the ND surface (133). Similar results were obtained in the study of a transmembrane fragment of human CD4, in which the signals observed were thought to originate from flexible, solvent-exposed regions (134). Much more complete data were obtained recently using perdeuterated VDAC-1 (135). The rotational correlation time, τ_c, of VDAC-1 in NDs was estimated to be ~93 ns, only slightly longer that the value of ~70 ns observed in LDAO micelles. As larger and larger MPs are being studied, the extra bulk added by the MSPs and lipids ought to become less and less of a handicap. One may also note that, whereas the slow tumbling rate of MP/ND complexes complicates structural studies of the protein itself, it would not prevent, for instance, studying the structure of bound ligands by transferred NOE measurements.

ssNMR is well adapted to the study of large objects. It has been used, as mentioned above (Section 3.1), to study the structure of MSPs in NDs that had been precipitated using polyethylene glycol (50). Upon precipitation of NDs containing the human P450 cytochrome 3A4 (CYP3A4), the protein retained its ability to bind one of its ligands, bromocryptine. ^{13}C chemical shifts were consistent with the known X-ray structure of CYP3A4, illustrating the feasibility of studying by ssNMR the structure of ND-embedded MPs (136).

ssNMR of MP/APol complexes has not been attempted yet. It should be feasible, given that the complexes can be precipitated or frozen and, at least in some cases, lyophilized without denaturing the protein (Section 5.1). However, a great advantage of ssNMR is that it gives access to the protein in its natural environment, which is maintained in lipid bilayers and, to a large extent, in NDs and in bicelles, but not in APols. APols could, however, be useful, e.g., to study the structure of bound ligands or of small isotopically labeled subunits integrated into large unlabeled complexes, which current NDs would not be able to trap.

7.6. Crystallization

A8-35, the best-characterized APol, is not a priori an excellent candidate for forming 3D crystals of MP/APol complexes because of the high density of charges it carries (no structure of MP has ever been solved using a charged detergent). Indeed, attempts at crystallizing cytochrome bc_1/A8-35 complexes did not yield any crystals (D. Charvolin, unpublished data). Some poorly diffracting crystals were obtained in preliminary studies of ternary bc_1/A8-35/detergent complexes (D. Charvolin, A.-N. Galatanu, & M. Picard, unpublished data). The color and space group of these crystals, as well as their fluorescence when they were prepared using a fluorescent APol, indicated that they contained the protein, the polymer, and, not shown directly but necessarily (see Section 5.1), the detergent. It has not yet been possible, unfortunately, to determine whether the poor quality of the diffraction patterns observed (~20-Å resolution) is inherent to this type of crystals or results from their handling, from the freezing step, or, simply, from too few crystals having been examined. Why crystals were obtained in ternary mixtures and not with pure APols can have several origins. One is that diluting A8-35 with detergent will diminish the charge density of the surfactant belt surrounding the protein as well as facilitate charge redistribution in the crystal. Another may be a more homogeneous size of the complexes, as suggested by observations on tOmpA/A8-35 versus tOmpA/A8-35/C_8E_4 complexes (93) (Section 5.1). Crystallization attempts with NAPols may be expected to have greater chances of success. For the time being, APols cannot be advocated as a favorable medium for MP crystallization. However, because even mixtures of APols and detergent can have a stabilizing effect on MPs (Section 6.1), supplementing an MP in detergent solution with APols, used as an additive, may perhaps be attempted when protein instability is suspected to be the reason for the failure to grow good-quality crystals in pure detergent solution.

Attempts to grow 3D crystals from solutions of MPs in (H)FSs should preferably be carried out using chemically defined molecules, which have become available only very recently (71, 72). One may speculate that, because their affinity for MP transmembrane surfaces is likely to be relatively low, (H)FSs may favor the growth of type I crystals, which are made of stacks of 2D crystals.

Detergents do not displace or dissolve monolayers of (H)FSs, which are stronger surfactants and do not mix well with them. Nickel-bearing (H)FSs have been synthesized (137, 138). They can be used, as nickel-bearing fluorinated lipids have been in the past (139), to adsorb detergent-solubilized polyhistidine-tagged MPs at the air-water interface. Thus, His-tagged tOmpA/C_8E_4 complexes can diffuse within the aqueous space separating the two F-TAC monolayers of a Newton black film, where they form, depending on the conditions, one- or two-molecule thick layers, whose structure has been studied by X-ray reflectivity measurements (137).

HFS monolayers containing Phenyl-HF-NTANi (**Figure 3*c***) may hold interesting promises for forming 2D MP crystals (138). To date, HFS monolayers have been shown to allow adsorption and orientation of the sulfonylurea receptor, whose projection structure was then solved by single-particle EM imaging (C. Vénien-Bryan, unpublished data).

Solving MP structures by X-ray analysis of 3D crystals of MP/ND complexes would depend, in most cases, on the target MP locking itself in perfect register with the MSPs, which, although it could conceivably be engineered, may be a rare circumstance.

7.7. Single-Particle Electron Microscopy and Atomic Force Microscopy

As described in Section 3.1, empty NDs have been studied by EM and AFM. NDs can be used for single-particle analyses of embedded MPs, with the limitation that they will not be able to trap MPs with very large transmembrane domains. This is a handicap for EM single-particle studies, which work best on large objects. Upon exposure to a mica surface, NDs, whether or not they harbor an MP, will tend to lie flat on the surface, which greatly facilitates AFM studies of their structure and that of the MP. AFM, for instance, has been used to measure the extension of CYP2B4 and the cytochrome P450 reductase in a direction normal to the plane of the disc (Reference 41 and references therein). Single-particle EM imaging, coupled with the use of antibody Fab fragments, has shown that, in NDs containing two rhodopsin molecules, the two proteins adopt indifferently parallel or antiparallel orientations with respect to each other (44).

APols and (H)FSs have the potential to trap and stabilize very large MP complexes, which may facilitate their purification, the identification of their components, and the study of their structure by single particle EM and other approaches. This application, which seems both promising and relatively straightforward, has been underexploited. In an early study, scanning transmission EM, which is mostly useful to map the distribution of masses rather than for imaging, had been used to show that trapping with A8-35 captured the distribution of cytochrome b_6f dimers and monomers that preexisted in the original detergent solution (140). To date, four single-particle EM studies have been published, two studies of the A8-75-trapped proton ATP synthase (141, 142), a negative-stain study of A8-35-trapped BR and of the curious filaments these complexes form upon elimination of free APol (83), and a low-resolution cryo-EM study of A8-35-trapped mitochondrial complex I (143). As already mentioned, A8–35 has been used to trap and purify under a functional form mitochondrial supercomplex B, which comprises complexes I, III, and IV in a 1:2:1 molar ratio. The complexes were imaged both after negative staining and by cryo-EM (**Figure 9**); image reconstruction is in progress (T. Althoff & W. Kühlbrandt, personal communication).

No EM studies of MPs stabilized by (H)FSs have been reported yet.

7.8. Delivering Membrane Proteins to Preformed Lipid Bilayers

As mentioned above, APols and (H)FSs, under most conditions, will not dissolve lipid membranes, and they are not cytolytic (Sections 5.1 and 5.3). They can, therefore, be used to deliver MPs to lipid vesicles, planar bilayers, or cells without destroying them. This useful property has permitted integration of APol-trapped diacylglycerol kinase into lipid vesicles (80), of A8-35-trapped porins into lipid black films (123), of the mechanosensitive channel MscL, kept soluble by (H)F-TAC, into lipid vesicles (59), and of APol-trapped mimics of transmembrane peptides into living cells (G. Crémel, personal communication). In such studies, however, two caveats have to be kept in mind: (*a*) fragile MPs risk being denatured in the process (BR cannot

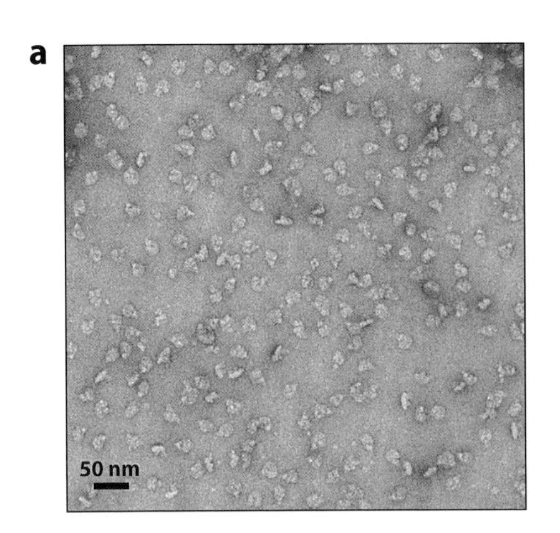

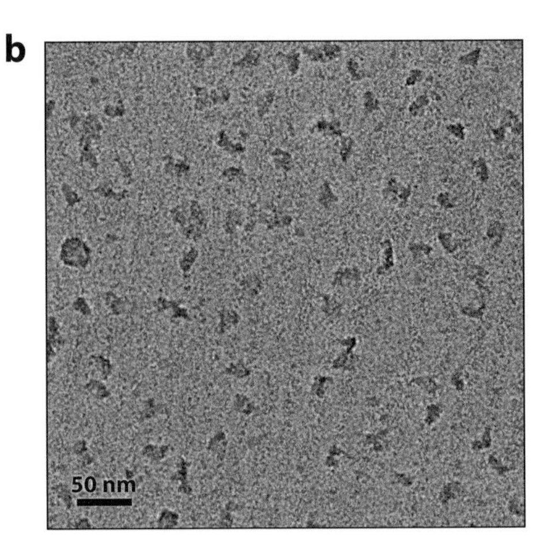

Figure 9

Electron microscopy of amphipol-trapped mitochondrial supercomplex B. The supercomplex (~1.7 MDa), which comprises one copy each of complexes I and IV and two copies of complex III, was trapped by APol A8-35, purified, and imaged either by negative staining (*a*) or in vitrous ice (*b*) (T. Althoff & W. Kühlbrandt, in preparation).

be directly transferred from (H)FSs to lipid vesicles without denaturation) (C. Breyton, unpublished observation); and (*b*) whereas (H)FSs can be washed away, as they will equilibrate between the solution and the membrane, APols will remain associated with the target membrane, which may have undesirable consequences (see Reference 123).

The unusual situation of being able to keep a hydrophobic protein soluble, thanks to a surfactant, without lysing the target membrane has been exploited in a series of studies of the thermodynamics of insertion of the T domain of diphteria toxin into membranes following a pH drop (**Figure 10**) (58, 61, 144). Protein/surfactant interactions were examined by FRET thanks to Oregon Green–labeled (H)F-TACs (**Figure 3*d***) (61).

Transfer of an MP from an MP/ND complex to a preformed membrane, an intriguing possibility, has not been reported yet. The absence of transfer, however, would make it feasible to study the interaction of a membrane-embedded MP with an ND-trapped one, or of two ND-trapped MPs with one another, while the two proteins are kept in distinct bilayers. This would provide a novel way to study interactions between MPs that in vivo belong to distinct membranes, as occurs, for instance, in processes involving membrane adhesion or fusion.

7.9. Immobilizing Membrane Proteins onto Solid Supports

Surface-based in vitro assays of ligand binding to proteins present multiple advantages, especially their adaptability to the use of minimal amounts of proteins and reagents and to multiplexing and high-throughput screening. Immobilizing an MP without modifying it or subjecting it to nonspecific, potentially denaturing interactions is, however, a challenge. Because the association of APols with MPs resists extensive washing with surfactant-free buffer (see Section 5.1), trapping an MP with a functionalized APol results in the permanent association with the protein of any functional group carried by the APol (93). Tagged APols thus can be used to immobilize MPs onto solid supports (92). The ligand-binding properties of the immobilized proteins can then be studied in detergent-free solutions, e.g., by SPR or fluorescence

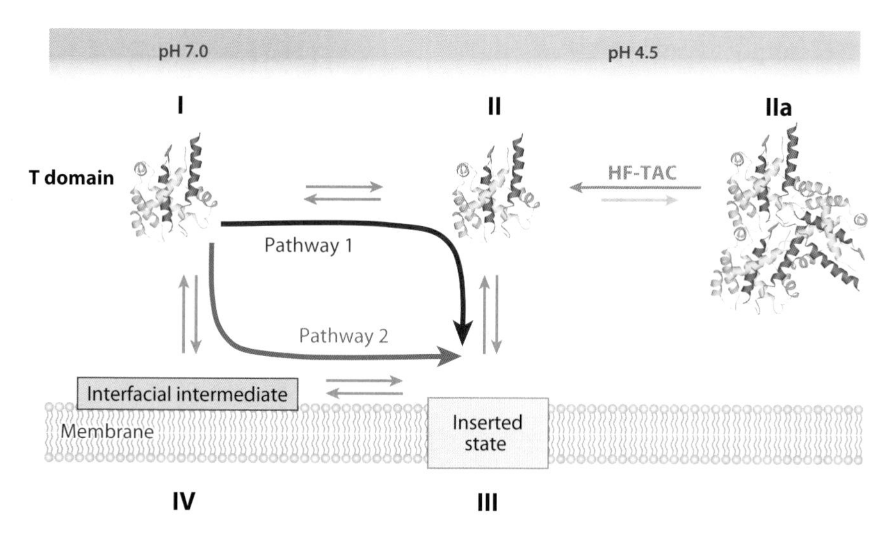

Figure 10

Using (H)FSs to study the thermodynamics of protein insertion into membranes. The pore-forming T domain of diphteria toxin, which is water soluble at pH 7.0 (I), changes conformation as the consequence of a pH drop, which renders it hydrophobic (II). In the absence of surfactants, a competition engages between membrane insertion (III, pathway 1) and nonproductive aggregation (IIa). HF-TAC keeps the T domain soluble and monomeric without solubilizing the target membrane, at variance with what a conventional detergent would do. This makes it possible to study the thermodynamics of insertion under equilibrium conditions (see Section 7.8). From Reference 58.

techniques, opening the way to applications in diagnostics, drug discovery, or the search for natural biological partners (**Figure 11**). The procedure is universal in the sense that any MP is a priori amenable to it. It is also mild, because the protein itself is not involved in any interaction with the support, and highly versatile.

NDs can adsorb onto solid supports, either spontaneously or via tagged lipids, which can be used to create patterned bilayer surfaces (145, 146). His-tagged scaffolding proteins were used to form NDs incorporating rhodopsin, which were immobilized onto a nickel-bearing support, and SPR and a modified version of MALDI-TOF mass spectrometry were used to demonstrate transducin binding upon exposure to light (147). Similarly, His-tagged NDs were used to capture glycolipid G_{M1}, the membrane receptor for cholera toxin. After immobilization of the complexes onto Ni-NTA chips, the kinetics of the interaction of the soluble toxin with the immobilized lipid was studied by SPR (148). Recent studies have described the screening of ligands of immobilized ND-trapped MPs by SPR (149) or NMR (150).

7.10. Folding Membrane Proteins from a Denatured State

Folding MPs to their native state in vitro provides precious insights into the respective roles of the amino acid sequence, the insertion machinery, and the membrane environment in determining the 3D fold adopted by the polypeptide. From a practical point of view, overexpressing MPs as inclusion bodies and folding them in vitro may allow production of large amounts of naturally rare MPs, which is often difficult to achieve when trying to express them directly in a functional state. Mild surfactants, such as APols and (H)FSs, may a priori provide favorable media for MPs to fold because they are expected to interfere less than detergents

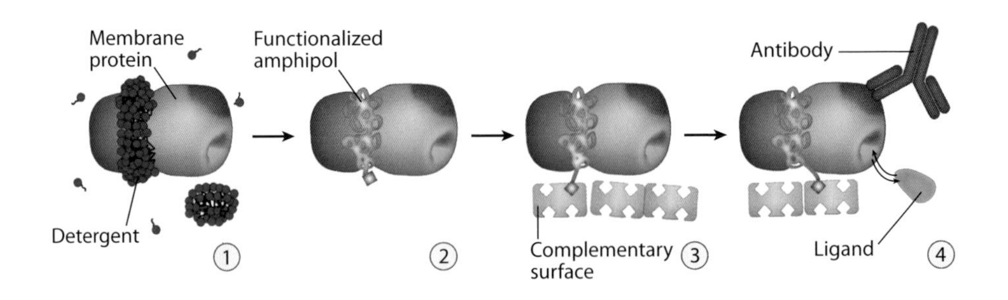

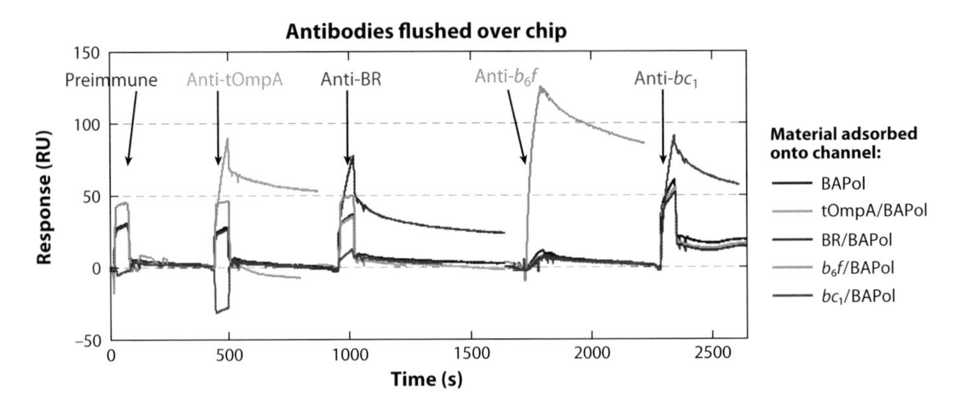

Figure 11

Using functionalized amphipols to attach membrane proteins onto solid supports and detect ligand binding to them. (*top*) Principle. A membrane protein (MP) in detergent solution ① is trapped with a biotinylated amphipol (BAPol) ②. It is then attached via the biotin groups to a streptavidin-coated chip ③. Ligands are flushed over it in plain, surfactant-free buffers, and their binding is detected by any of various methods ④. (*bottom*) Validation. Channels of a streptavidin-coated chip were exposed either to pure BAPol or to complexes of BAPol with any of four different MPs: tOmpA, BR, and cytochromes b_6f and bc_1. The five channels were then flushed with buffer containing either preimmune antibodies or antibodies raised against each of the four MPs. Antibody binding, measured in response units (RUs) by surface plasmon resonance, is protein specific. Adapted from Reference 92.

with the protein/protein and protein/lipid interactions that determine the native 3D structure of MPs and that stabilize them. APol A8-35 indeed has proven to be a highly efficient medium for folding BR, two porins, and several GPCRs (**Figure 12**) (120, 123). Generalization of this strategy could constitute a major breakthrough for structural and functional studies of many MPs. From a fundamental point of view, these experiments show that all of the chemical information needed for these MPs to correctly fold is stored in their sequences and that the highly specific and anisotropic environment provided by a lipid bilayer is not required to decode this information.

The use of (H)FSs to fold MPs has not been extensively investigated yet. Preliminary tests show that BR can be refolded with more than 50% yield in the presence of either F- or HF-TAC (60). tOmpA could be refolded into (H)FSs as well, albeit to a lesser extent than when using conventional detergents; HF-TAC was more efficient than F-TAC (60). The usefulness of (H)FSs for MP folding clearly deserves further investigation.

Refolding MPs into NDs is a more complex proposition because it will generally call for simultaneous refolding of the target MP and the scaffolding proteins as well as their reassembly with the lipids. Nevertheless, it may be worth

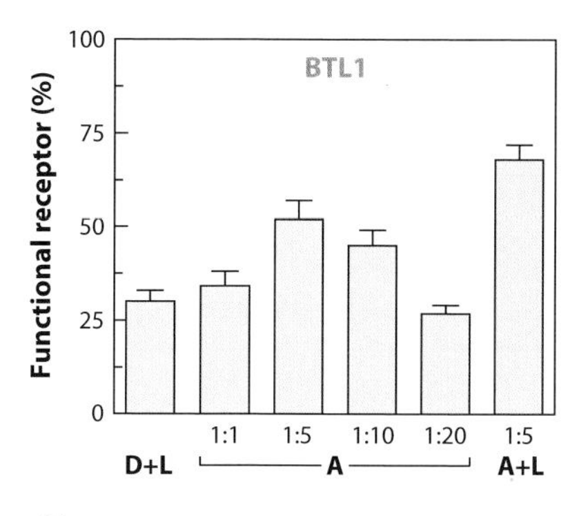

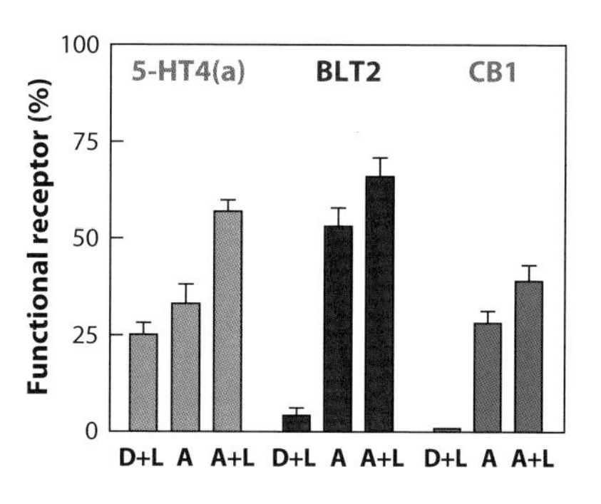

Figure 12

Amphipol-assisted folding of four G protein–coupled receptors (GPCRs) expressed in an inactive form in inclusion bodies. The BLT1 leukotriene receptor (*left*) and the 5-$HT_{4(a)}$ serotonin receptor, the BLT2 leukotriene receptor, and the CB1 cannabinoid receptor (*right*), all of them class A GPCRs, were expressed in inclusion bodies and purified in an inactive form in sodium dodecyl sulfate (SDS) solution. They were folded by substituting SDS either with a lipid/detergent mixture (D+L), with pure A8-35 (A) at different protein/APol mass ratios (BLT1), with A8-35 (A) at a 1:5 mass ratio (BLT2, 5-$HT_{4(a)}$ and CB1), or with A8-35 supplemented with asolectin in a 1:5:1 protein/APol/asolectin mass ratio (A+L). The extent of correct folding is expressed as the percentage of total receptor (on the basis of the protein concentration in the SDS solution) that is able to bind a specific ligand. Adapted from Reference 120.

examining whether NDs would not reform, for instance, from a mixture of their constituents in SDS solution, in which case MP folding experiments might be attempted.

7.11. Expressing Membrane Proteins in Cell-Free Systems

Cell-free expression of MPs in a lysate containing the transcription and translation machineries offers highly interesting opportunities, e.g., for expressing MPs that are toxic to cells as well as for specific labeling. Traditionally, MPs are expressed in vitro in the presence of a detergent, of lipid vesicles, or in the absence of any surfactant at all. In the latter case, MPs precipitate and are later solubilized using a detergent. Detergents have the drawback of being inactivating, and lipids have the disadvantage of offering limited yields. There is probably an advantage in the use of NCSs as less-aggressive environments, in which neosynthesized MPs should stand a better chance to correctly fold. All three systems can be used to this end. (H)F-TACs have been used to synthesize the mechanosensitive channel MscL (59). After synthesis, MscL was purified in solutions of the same surfactants and directly applied to liposomes, into which it spontaneously integrated. Its activity was then characterized by patch-clamp measurements (59). MscL synthesis has also been achieved using the new, chemically defined (H)FSs (71). NDs have been employed to synthesize under their functional form BR and the multidrug-resistance protein EmrE, as well as to express a large panel of other MPs (151–153). APols initially yielded disappointing results. Indeed, both A8-35 and sulfonated APols were found to inhibit the transcription-translation machinery (E. Billon-Denis & F. Zito, unpublished observations). More recently, however, it was observed that NAPols do not block protein synthesis and provide excellent yields of expression for several MPs (89; E. Billon-Denis, P. Bazzacco, & F. Zito, unpublished observations). BR expressed in the presence of NAPols is both functional and soluble.

These observations are potentially of great interest. Given that most MPs are more or less unstable in detergent solutions and that folding yields in detergent or detergent/lipid mixtures are generally low, it is to be expected that many, if not most, MPs, if exposed to detergents in

Table 1 **Opportunities and constraints associated with the use of amphipols, nanodiscs, and fluorinated surfactants**[a,b]

Technology	Amphipols	Nanodiscs[c]	Fluorinated surfactants	Sections[d]
Membrane protein (MP) stabilization	+	+	+	6
Functional studies	+	++	+	7.3
Mediating MP immobilization for ligand binding measurements	+	+	–	7.9
Optical spectroscopy (visible absorption spectrum)	+	+	+	7.1
Optical spectroscopy (UV, intrinsic MP fluorescence, circular dichroism)	+	±	+	7.1
Fluorescence spectroscopy using probes	+	+	+	7.1
Infrared spectroscopy	–	±	–	7.1
MP solution studies by AUC, SEC, SAXS, SANS	+	±	+	7.2
Solution NMR	+	±?	?	7.5
Solid-state NMR	+?	+	–	7.5
Three-dimensional crystallization	±	–?	+?	7.6
Two-dimensional crystallization	–	–?	+?	7.6
Trapping MP supercomplexes	+	±	+	5, 7.7
EM, AFM (single particles)	+	+	+	7.7
Transferring MPs to preformed membranes	+	?	+	7.8
Folding full-length MPs to native state	+	?	+	7.10
MP cell-free translation	+	+	+	7.11
Isoelectrofocusing with two-dimensional gels	+	+?	?	7.4
MP mass spectrometry	+	?	?	7.4

[a]+, ±, and – signs refer to how promising or problematical each application to MP studies looks: +, promising; ±, promising, but with limitations or difficulties; –, not promising if not plainly impossible.

[b]For those applications that have actually been tested, cells have been colored as follows: green, tested with success; yellow, shown to work, but with some caveats or limitations; pink, problematical if not impossible. A white background indicates that the +/- signs represent what appears to be reasonable expectations, but that these assessments are not currently backed up by actual data.

[c]In the case of nanodiscs, the table refers to the study of nanodisc-embedded MPs, not to that of rim proteins.

[d]Section of this article where this is discussed.

Abbreviations: AFM, atomic force microscopy; AUC, analytical ultracentrifugation; EM, electron microscopy; SANS, small-angle neutron scattering; SAXS, small-angle X-ray scattering; SEC, size exclusion chromatography; UV, ultraviolet.

the course of their synthesis, will not efficiently achieve their native 3D fold. NCSs may therefore offer an extremely attractive alternative for cell-free MP synthesis.

8. CONCLUSION: OPPORTUNITIES VERSUS CONSTRAINTS

It is my hope that the present article, which, unfortunately, had to limit itself to three among many more innovative systems, will convince membrane protein biochemists and biophysicists that the time is over when MPs could be handled in vitro only in a detergent solution or in a membrane-bound form. New tools have been developed and validated—painstakingly! These tools can at least partially circumvent the instability and other problems encountered with detergents. They can also provide totally new experimental opportunities. In **Table 1**, I have summarized my personal view of the usefulness of the three NCS systems discussed here as well as the constraints and difficulties with which each is associated. Needless to say, these prospects will evolve as improved molecules are developed and applications are more thoroughly explored.

As should be evident from the present review, each type of NCS has its own privileged applications. Although some uses, such as MP folding or in vitro synthesis, can probably be advantageously developed in parallel with the three of them, specialization will occur. One may speculate that, ten years from now, membrane biochemists and biophysicists may be in the habit of resorting to one or the other NCS depending on which particular problem they have at hand.

SUMMARY POINTS

1. Membrane proteins (MPs) can be handled in detergent-free aqueous solutions as individual, well-defined complexes after being transferred to nondetergent surfactants, including amphipols (APols), nanodiscs (NDs), or fluorinated surfactants (FSs).
2. Transfer is usually achieved by first solubilizing the target protein with a detergent and then replacing the detergent with any of these three media.
3. Transferring MPs to any of these three environments generally stabilizes them as compared to detergent solutions.
4. Some applications, such as cell-free synthesis of MPs, the study of fragile MP complexes, or single-particle imaging, have been shown or appear likely to be common to all three systems. Others are best performed using one or the other of them. The development of many applications, however, is still in its infancy.
5. NDs have the unique characteristic of featuring a well-defined area into which only MPs whose transmembrane domain does not exceed a defined size can be trapped in a membrane-like environment. This makes them exceptional tools to sort out functional issues related to MP oligomerization and interactions with lipids. Other applications whose development has started include electron microscopy, atomic force microscopy, solution and solid-state NMR, and MP immobilization.
6. Among applications for which APols appear particularly promising are MP folding, solution NMR, the study of large MP complexes, and immobilization of MPs onto solid supports.
7. Applications of FSs are at a less developed stage, but among those that may be particularly useful are the study of fragile complexes and the delivery of MPs to preexisting membranes.
8. All three systems have by now been studied extensively enough to be used for the exploration of biological systems whenever detergents are unsuitable.

FUTURE ISSUES

1. One can expect that future developments will take, for each of the three systems that have been discussed here, two main paths: improving the tools and exploring applications.
2. Regarding the development of applications, several suggestions have been evoked in Section 7 of the review. Exploring higher and less-permanent levels of organization than that of the single macromolecule is one of the frontiers of today's biology. Means to preserve fragile or transient interactions between MPs better than is possible using detergents, which nonconventional surfactants provide, are, therefore, particularly timely.

3. Regarding the development of the tools, the resources of genetic engineering will improve and diversify the proteins used to stabilize NDs, which will permit them to encapsulate larger MPs or to more precisely fix the size of the discs, to tailor their spectroscopic properties to the needs of biophysicists, or to functionalize them.
4. Functionalization is also a key to developing original applications of APols. The new APol generations, and in particular nonionic APols, will have to be as thoroughly validated as first-generation molecules have been, and optimized and functionalized in view of their specific uses.
5. Whether developing chemically well-defined APols—a difficult challenge to chemists—is worth the long effort required by such an attempt remains to date an open question. Developing and validating applications is probably more urgent.
6. Fluorinated surfactants have been brought to the stage where chemically defined molecules are, at long last, available. Much effort is still necessary, however, to more fully explore their specific applications.
7. Finally, one should not forget that the three systems analyzed here do not cover the whole range of approaches that have been explored to date and that novel ones will no doubt emerge.

DISCLOSURE STATEMENT

The author is coinventor on several granted or pending patents on amphipols and their uses.

ACKNOWLEDGMENTS

Particular thanks are due to T. Althoff, J.-L. Banères, S. Banerjee, P. Bazzacco, C. Béchara, E. Billon-Denis, G. Bolbach, J. Borch, C. Breyton, L. Catoire, P. Champeil, D. Charvolin, G. Crémel, T. Dahmane, F. Giusti, Y. Gohon, E. Goormaghtigh, J.-C. Guillemot, T. Hamann, A. Kriško, W. Kühlbrandt, A. Ladokhin, C. Le Bon, M. le Maire, K.L. Martinez, A. Milon, B.L. Møller, E. Pebay-Peyroula, M. Picard, Y. Pierre, R. Prassl, B. Pucci, M. Renault, C.M. Rienstra, J.N. Sachs, S. Sagan, T.P. Sakmar, C.R. Sanders, S.G. Sligar, R. Sunahara, C.G. Tate, C. Tribet, C. Vénien-Bryan, F. Wien, F. Winnik, F. Zito, and M. Zoonens for communication and permission to quote unpublished information and/or to reproduce published or unpublished figures, for help with the figures or the bibliography, and for comments on the manuscript. I am deeply indebted to J. Barra for her help with preparing the figures and with collecting the bibliography and to M.E. Dumont for his heroic attempts at improving the English wording. My own work has been mainly supported by the CNRS, the Human Frontier Science Program Organization (RG00223/2000-M), and the European Community (BIO4-CT98-0269 and STREP LSHG-CT-2005-513770 *Innovative Tools for Membrane Protein Structural Proteomics*).

This review is dedicated to the memory of my father, whose courage and thoroughness have been an inspiration for me.

LITERATURE CITED

1. Helenius A, Simons K. 1975. Solubilization of membranes by detergents. *Biochim. Biophys. Acta* 415:29–79
2. Tanford C, Reynolds JA. 1976. Characterization of membrane proteins in detergent solutions. *Biochim. Biophys. Acta* 457:133–70

3. Tanford C. 1980. *The Hydrophobic Effect: Formation of Micelles and Biological Membranes*. New York: Wiley. 233 pp.
4. Israelachvili JN, Mitchell DJ, Ninham BW. 1977. Theory of self-assembly of lipid bilayers and vesicles. *Biochim. Biophys. Acta* 470:185–201
5. Frindi M, Michels B, Zana R. 1992. Ultrasonic absorption studies of surfactant exchange between micelles and bulk phase in aqueous micellar solutions of nonionic surfactants with a short alkyl chain. 2. C_6E_3, C_8E_4 and C_8E_8. *J. Phys. Chem.* 96:6095–102
6. Brotherus JR, Jost PC, Griffith OH, Hokin LE. 1979. Detergent inactivation of sodium- and potassium-activated adenosinetriphosphatase of the electric eel. *Biochemistry* 18:5043–50
7. Breyton C, Tribet C, Olive J, Dubacq J-P, Popot J-L. 1997. Dimer to monomer transition of the cytochrome b_6f complex: causes and consequences. *J. Biol. Chem.* 272:21892–900
8. Bowie JU. 2001. Stabilizing membrane proteins. *Curr. Opin. Struct. Biol.* 11:397–402
9. Garavito RM, Ferguson-Miller S. 2001. Detergents as tools in membrane biochemistry. *J. Biol. Chem.* 276:32403–6
10. Rosenbusch JP. 2001. Stability of membrane proteins: relevance for the selection of appropriate methods for high-resolution structure determinations. *J. Struct. Biol.* 136:144–57
11. Gohon Y, Popot J-L. 2003. Membrane protein-surfactant complexes. *Curr. Opin. Colloid Interface Sci.* 8:15–22
12. Lee AG. 2010. How to understand lipid-protein interactions in biological membranes. In *Structure of Biological Membranes*, ed. P Yeagle. Boca Raton, FL: Taylor & Francis. In press.
13. Zhou Y, Bowie JU. 2000. Building a thermostable membrane protein. *J. Biol. Chem.* 275:6975–79
14. Serrano-Vega MJ, Magnani F, Shibata Y, Tate CG. 2008. Conformational thermostabilization of the β_1-adrenergic receptor in a detergent-resistant form. *Proc. Natl. Acad. Sci. USA* 105:877–82
15. Magnani F, Shibata Y, Serrano-Vega MJ, Tate CG. 2008. Co-evolving stability and conformational homogeneity of the human adenosine A2a receptor. *Proc. Natl. Acad. Sci. USA* 105:10744–49
16. Shibata Y, White JF, Serrano-Vega MJ, Magnani F, Aloia AL, et al. 2009. Thermostabilization of the neurotensin receptor NTS1. *J. Mol. Biol.* 390:262–77
17. Yu SM, McQuade DT, Quinn MA, Hackenberger CP, Krebs MP, et al. 2000. An improved tripod amphiphile for membrane protein solubilization. *Protein Sci.* 9:2518–27
18. Theisen MJ, Potocky TB, McQuade DT, Gellman SH, Chiu ML. 2005. Crystallization of bacteriorhodopsin solubilized by a tripod amphiphile. *Biochim. Biophys. Acta* 1751:213–16
19. Chae PS, Wander MJ, Bowling AP, Laible PD, Gellman SH. 2008. Glycotripod amphiphiles for solubilization and stabilization of a membrane-protein superassembly: importance of branching in the hydrophilic portion. *ChemBioChem* 9:1706–9
20. Schleicher A, Hofmann KP, Finkelmann H, Welte W. 1987. Deoxylysolecithin and a new biphenyl detergent as solubilizing agents for bovine rhodopsin. Functional test by formation of metarhodopsin II and binding of G-protein. *Biochemistry* 26:5908–16
21. Schafmeister CE, Miercke LJW, Stroud RA. 1993. Structure at 2.5 Å of a designed peptide that maintains solubility of membrane proteins. *Science* 262:734–38
22. McGregor C-L, Chen L, Pomroy NC, Hwang P, Go S, et al. 2003. Lipopeptide detergents designed for the structural study of membrane proteins. *Nat. Biotechnol.* 21:171–76
23. Privé G. 2009. Lipopeptide detergents for membrane protein studies. *Curr. Opin. Struct. Biol.* 19:1–7
24. Yeh JI, Du S, Tortajada A, Paulo J, Zhang S. 2005. Peptergents: peptide detergents that improve stability and functionality of a membrane protein, glycerol-3-phosphate dehydrogenase. *Biochemistry* 44:16912–19
25. Chabaud E, Barthélémy P, Mora N, Popot J-L, Pucci B. 1998. Stabilization of integral membrane proteins in aqueous solution using fluorinated surfactants. *Biochimie* 80:515–30
26. Breyton C, Chabaud E, Chaudier Y, Pucci B, Popot J-L. 2004. Hemifluorinated surfactants: a non-dissociating environment for handling membrane proteins in aqueous solutions? *FEBS Lett.* 564:312–18
27. Tribet C, Audebert R, Popot J-L. 1996. Amphipols: polymers that keep membrane proteins soluble in aqueous solutions. *Proc. Natl. Acad. Sci. USA* 93:15047–50

28. Breyton C, Pucci B, Popot J-L. 2010. Amphipols and fluorinated surfactants: two alternatives to detergents for studying membrane proteins in vitro. In *Membrane Protein Expression*, ed. I Mus-Veteau, pp. 219–45. Totowa, NJ: Humana.
29. Popot J-L, Berry EA, Charvolin D, Creuzenet C, Ebel C, et al. 2003. Amphipols: polymeric surfactants for membrane biology research. *Cell. Mol. Life Sci.* 60:1559–74
30. Sanders CR, Hoffmann AK, Gray DN, Keyes MH, Ellis CD. 2004. French swimwear for membrane proteins. *ChemBioChem* 5:423–26
31. Sanders CR, Prosser RS. 1998. Bicelles: a model membrane system for all seasons? *Structure* 6:1227–34
32. Sanders CR, Sönnichsen F. 2006. Solution NMR of membrane proteins: practice and challenges. *Magn. Reson. Chem.* 44:S24–40
33. Prosser RS, Evanics F, Kitevski JL, Al-Abdul-Wahid MS. 2006. Current applications of bicelles in NMR studies of membrane-associated amphiphiles and proteins. *Biochemistry* 45:8453–65
34. Poget SF, Girvin ME. 2007. Solution NMR of membrane proteins in bilayer mimics: small is beautiful, but sometimes bigger is better. *Biochim. Biophys. Acta* 1768:3098–106
35. De Angelis AA, Opella SJ. 2007. Bicelle samples for solid-state NMR of membrane proteins. *Nat. Protoc.* 2:2332–38
36. Kim HM, Howell SC, Van Horn WD, Jeon YH, Sanders CR. 2009. Recent advances in the application of solution NMR spectroscopy to multi-span integral membrane proteins. *Progr. Nucl. Magn. Reson. Spectrosc.* 55:335–60
37. Faham S, Bowie JU. 2002. Bicelle crystallization: a new method for crystallizing membrane proteins yields a monomeric bacteriorhodopsin structure. *J. Mol. Biol.* 316:1–6
38. Johansson LC, Wöhri AB, Katona G, Engström S, Neutze R. 2009. Membrane protein crystallization from lipidic phases. *Curr. Opin. Struct. Biol.* 19:372–78
39. Nath A, Atkins WM, Sligar SG. 2007. Applications of phospholipid bilayer nanodiscs in the study of membranes and membrane proteins. *Biochemistry* 46:2059–69
40. Borch J, Hamann T. 2009. The nanodisc: a novel tool for membrane protein studies. *Biol. Chem.* 390:805–14
41. Bayburt TH, Sligar SG. 2002. Single-molecule height measurements on microsomal cytochrome P450 in nanometer-scale phospholipid bilayer disks. *Proc. Natl. Acad. Sci. USA* 99:6725–30
42. Denisov IG, Grinkova YV, Lazarides AA, Sligar SG. 2004. Directed self-assembly of monodisperse phospholipid bilayer nanodiscs with controlled size. *J. Am. Chem. Soc.* 126:3477–87
43. Chromy BA, Arroyo E, Blanchette CD, Bench G, Benner H, et al. 2007. Different apolipoproteins impact nanolipoprotein particle formation. *J. Am. Chem. Soc.* 129:14348–54
44. Banerjee S, Huber T, Sakmar TP. 2008. Rapid incorporation of functional rhodopsin into nanoscale apolipoprotein-bound bilayer (NABB) particles. *J. Mol. Biol.* 377:1067–81
45. Blanchette CD, Law R, Benner WH, Pesavento JB, Cappuccio JA, et al. 2008. Quantifying size distributions of nanolipoprotein particles with single-particle analysis and molecular dynamic simulations. *J. Lipid Res.* 49:1420–30
46. Bayburt TH, Grinkova YV, Sligar SG. 2002. Self-assembly of discoidal phospholipid bilayer nanoparticles with membrane scaffold proteins. *Nano Lett.* 2:853–56
47. Blanchette CD, Cappuccio JA, Kuhn EA, Segelke BW, Benner WH, et al. 2009. Atomic force microscopy differentiates discrete size distributions between membrane protein-containing and empty nanolipoprotein particles. *Biochim. Biophys. Acta* 1788:724–31
48. Shaw AW, McLean MA, Sligar SG. 2004. Phospholipid phase transitions in homogeneous nanometer scale bilayers discs. *FEBS Lett.* 556:260–64
49. Denisov IG, McLean MA, Shaw AW, Grinkova YV, Sligar SG. 2005. Thermotropic phase transitions in soluble nanoscale lipid bilayers. *J. Phys. Chem. B.* 109:15580–88
50. Li Y, Kijac AZ, Sligar SG, Rienstra CM. 2006. Structural analysis of nanoscale self-assembled discoidal lipid bilayers by solid-state NMR spectroscopy. *Biophys. J.* 91:3819–28
51. Shih AY, Freddolino PL, Arkhipov A, Sligar SG, Schulten K. 2008. Molecular modeling of the structural properties and formation of high-density lipoprotein particles. In *Computational Modeling of Membrane Bilayers*, ed. SE Feller, pp. 313-42. London: Academic

52. Kissa E. 1994. Structure of micelles and mesophases. In *Fluorinated Surfactants: Synthesis, Properties, Applications*, pp. 264–82. New York: Dekker
53. Mukerjee P. 1994. Fluorocarbon-hydrocarbon interactions in micelles and other lipid assemblies, at interfaces, and in solutions. *Colloids Surf. A: Physicochem. Eng. Asp.* 84:1–10
54. Nakano TY, Sugihara G, Nakashima T, Yu SC. 2002. Thermodynamic study of mixed hydrocarbon/fluorocarbon surfactant system by conductometric and fluorimetric techniques. *Langmuir* 18:8777–85
55. Barthélémy P, Tomao V, Selb J, Chaudier Y, Pucci B. 2002. Fluorocarbon-hydrocarbon non-ionic surfactant mixtures: a study of their miscibility. *Langmuir* 18:2557–63
56. Pucci B, Maurizis J-C, Pavia AA. 1991. Télomères et co-télomères d'intérêt biologique et biomédical. IV. Les télomères du *tris*(hydroxyméthyl)-acrylamidométhane, nouveaux agents amphiphiles non-ioniques. *Eur. J. Polym.* 27:1101–6
57. Barthélémy P, Améduri B, Chabaud E, Popot J-L, Pucci B. 1999. Synthesis and preliminary assessment of ethyl-terminated perfluoroalkyl non-ionic surfactants derived from *tris*(hydroxymethyl)acrylamidomethane. *Org. Lett.* 1:1689–92
58. Palchevskyy SS, Posokhov YO, Olivier B, Popot J-L, Pucci B, Ladokhin AS. 2006. Chaperoning of membrane protein insertion into lipid bilayers by hemifluorinated surfactants: application to diphtheria toxin. *Biochemistry* 45:2629–35
59. Park K-H, Berrier C, Lebaupain F, Pucci B, Popot J-L, et al. 2007. Fluorinated and hemifluorinated surfactants as alternatives to detergents for membrane protein cell-free synthesis. *Biochem. J.* 403:183–87
60. Lebaupain F. 2007. *Développement de l'utilisation des tensioactifs fluorés pour la biochimie des protéines membranaires*. PhD thesis. Univ. Paris-7, Paris. 254 pp.
61. Rodnin MV, Posokhov YO, Contino-Pépin C, Brettmann J, Kyrychenko A, et al. 2008. Interactions of fluorinated surfactants with diphtheria toxin T-domain: testing new media for studies of membrane proteins. *Biophys. J.* 94:4348–57
62. Der Mardirossian C, Lederer F. 1996. Purification of flavocytochrome b_{558}: comparison between octylglucoside and a perfluoroalkyl detergent. *Biochem. Soc. Trans.* 24:32S
63. Der Mardirossian C, Krafft M-P, Gulik-Krzywicki T, le Maire M, Lederer F. 1998. On the lack of protein-solubilizing properties of two perfluoroalkylated detergents, as tested with neutrophil plasma membranes. *Biochimie* 80:531–41
64. Shepherd FH, Holzenburg A. 1995. The potential of fluorinated surfactants in membrane biochemistry. *Anal. Biochem.* 224:21–27
65. Ramjeesingh M, Huan LJ, Garami E, Bear CE. 1999. Novel method for evaluation of the oligomeric structure of membrane proteins. *Biochem. J.* 342(Part 1):119–23
66. Chaudier Y, Barthélémy P, Pucci B. 2001. Synthesis and preliminary assessment of hybrid hydrocarbon-fluorocarbon anionic and non-ionic surfactants. *Tetrahedron Lett.* 42:3583–85
67. Chaudier Y, Zito F, Barthélémy P, Stroebel D, Améduri B, et al. 2002. Synthesis and preliminary biochemical assessment of ethyl-terminated perfluoroalkylamine oxide surfactants. *Bioorg. Med. Chem. Lett.* 12:1587–90
68. Chabaud E. 2001. *Contribution à l'étude biochimique du complexe cytochrome* b_6f *de* Chlamydomonas reinhardtii*: manipulation in vitro et stabilisation, organisation des hèmes et spectroscopie de la chlorophylle a*. PhD thesis. Univ. Paris-VI. 180 pp.
69. Lebaupain F, Salvay AG, Olivier B, Durand G, Fabiano A-S, et al. 2006. Lactobionamide surfactants with hydrogenated, hemifluorinated or perfluorinated tails: physical-chemical and biochemical characterization. *Langmuir* 22:8881–90
70. Polidori A, Presset M, Lebaupain F, Améduri B, Popot J-L, et al. 2006. Fluorinated and hemifluorinated surfactants derived from maltose: synthesis and application to handling membrane proteins in aqueous solution. *Bioorg. Med. Chem. Lett.* 16:5827–31
71. Breyton C, Gabel F, Abla M, Pierre Y, Lebaupain F, et al. 2009. Micellar and biochemical properties of (hemi)fluorinated surfactants are controlled by the size of the polar head. *Biophys. J.* 97:1077–86
72. Abla M, Durand G, Pucci B. 2008. Glucose-based surfactants with hydrogenated, fluorinated, or hemifluorinated tails: synthesis and comparative physical-chemical characterization. *J. Org. Chem.* 73:8142–53

73. Thebault P, Taffin de Givenchy E, Starita-Geribaldi M, Guittard F, Geribaldi S. 2007. Synthesis and surface properties of new semifluorinated sulfobetaines potentially usable for 2D-electrophoresis. *J. Fluor. Chem.* 128:211–18
74. Starita-Geribaldi M, Thebault P, Taffin de Givenchy E, Guittard F, Geribaldi S. 2007. 2-DE using hemi-fluorinated surfactants. *Electrophoresis* 28:2489–97
75. Gohon Y, Pavlov G, Timmins P, Tribet C, Popot J-L, Ebel C. 2004. Partial specific volume and solvent interactions of amphipol A8–35. *Anal. Biochem.* 334:318–34
76. Gohon Y, Giusti F, Prata C, Charvolin D, Timmins P, et al. 2006. Well-defined nanoparticles formed by hydrophobic assembly of a short and polydisperse random terpolymer, amphipol A8–35. *Langmuir* 22:1281–90
77. Tribet C, Audebert R, Popot J-L. 1997. Stabilisation of hydrophobic colloidal dispersions in water with amphiphilic polymers: application to integral membrane proteins. *Langmuir* 13:5570–76
78. Ladavière C, Toustou M, Gulik-Krzywicki T, Tribet C. 2001. Slow reorganization of small phosphatidylcholine vesicles upon adsorption of amphiphilic polymers. *J. Colloid Interface Sci.* 241:178–87
79. Ladavière C, Tribet C, Cribier S. 2002. Lateral organization of lipid membranes induced by amphiphilic polymer inclusions. *Langmuir* 18:7320–27
80. Nagy JK, Kuhn Hoffmann A, Keyes MH, Gray DN, Oxenoid K, Sanders CR. 2001. Use of amphipathic polymers to deliver a membrane protein to lipid bilayers. *FEBS Lett.* 501:115–20
81. Gorzelle BM, Hoffman AK, Keyes MH, Gray DN, Ray DG, Sanders CR II. 2002. Amphipols can support the activity of a membrane enzyme. *J. Am. Chem. Soc.* 124:11594–95
82. Diab C, Tribet C, Gohon Y, Popot J-L, Winnik FM. 2007. Complexation of integral membrane proteins by phosphorylcholine-based amphipols. *Biochim. Biophys. Acta* 1768:2737–47
83. Gohon Y, Dahmane T, Ruigrok R, Schuck P, Charvolin D, et al. 2008. Bacteriorhodopsin/amphipol complexes: structural and functional properties. *Biophys. J.* 94:3523–37
84. Picard M, Dahmane T, Garrigos M, Gauron C, Giusti F, et al. 2006. Protective and inhibitory effects of various types of amphipols on the Ca^{2+}-ATPase from sarcoplasmic reticulum: a comparative study. *Biochemistry* 45:1861–69
85. Diab C, Winnik FM, Tribet C. 2007. Enthalpy of interaction and binding isotherms of non-ionic surfactants onto micellar amphiphilic polymers (amphipols). *Langmuir* 23:3025–35
86. Prata C, Giusti F, Gohon Y, Pucci B, Popot J-L, Tribet C. 2001. Non-ionic amphiphilic polymers derived from *tris*(hydroxymethyl)-acrylamidomethane keep membrane proteins soluble and native in the absence of detergent. *Biopolymers* 56:77–84
87. Sharma KS, Durand G, Giusti F, Olivier B, Fabiano A-S, et al. 2008. Glucose-based amphiphilic telomers designed to keep membrane proteins soluble in aqueous solutions: synthesis and physicochemical characterization. *Langmuir* 24:13581–90
88. Bazzacco P, Sharma KS, Durand G, Giusti F, Ebel C, et al. 2009. Trapping and stabilization of integral membrane proteins by hydrophobically grafted glucose-based telomers. *Biomacromolecules* 10:3317–26
89. Bazzacco P. 2009. *Non-ionic amphipols: new tools for in vitro studies of membrane proteins. Validation and development of biochemical and biophysical applications.* PhD thesis. Univ. Paris-7, Paris. 176 pp.
90. Duval-Terrié C, Cosette P, Molle G, Muller G, Dé E. 2003. Amphiphilic biopolymers (amphibiopols) as new surfactants for membrane protein solubilization. *Protein Sci.* 12:681–89
91. Picard M, Duval-Terrié C, Dé E, Champeil P. 2004. Stabilization of membranes upon interaction of amphipathic polymers with membrane proteins. *Protein Sci.* 13:3056–58
92. Charvolin D, Perez J-B, Rouvière F, Giusti F, Bazzacco P, et al. 2009. The use of amphipols as universal molecular adapters to immobilize membrane proteins onto solid supports. *Proc. Natl. Acad. Sci. USA* 106:405–10
93. Zoonens M, Giusti F, Zito F, Popot J-L. 2007. Dynamics of membrane protein/amphipol association studied by Förster resonance energy transfer. Implications for in vitro studies of amphipol-stabilized membrane proteins. *Biochemistry* 46:10392–404
94. Zoonens M, Catoire LJ, Giusti F, Popot J-L. 2005. NMR study of a membrane protein in detergent-free aqueous solution. *Proc. Natl. Acad. Sci. USA* 102:8893–98

95. Catoire LJ, Zoonens M, van Heijenoort C, Giusti F, Popot J-L, Guittet E. 2009. Inter- and intramolecular contacts in a membrane protein/surfactant complex observed by heteronuclear dipole-to-dipole cross-relaxation. *J. Magn. Res.* 197:91–95
96. Jonas A. 1986. Reconstitution of high-density lipoproteins. *Methods Enzymol.* 128:553–82
97. Champeil P, Menguy T, Tribet C, Popot J-L, le Maire M. 2000. Interaction of amphipols with the sarcoplasmic reticulum Ca^{2+}-ATPase. *J. Biol. Chem.* 275:18623–37
98. Martinez KL, Gohon Y, Corringer P-J, Tribet C, Mérola F, et al. 2002. Allosteric transitions of *Torpedo* acetylcholine receptor in lipids, detergent and amphipols: molecular interactions vs. physical constraints. *FEBS Lett.* 528:251–56
99. Dahmane T. 2007. *Protéines membranaires et amphipols: stabilisation, fonction, renaturation, et développement d'amphipols sulfonatés pour la RMN des solutions*. PhD thesis. Univ. Paris-7, Paris. 229 pp.
100. Civjan NR, Bayburt TH, Schuler MA, Sligar SG. 2003. Direct solubilization of heterologously expressed membrane proteins by incorporation into nanoscale lipid bilayers. *BioTechniques* 35:556–60, 62–63
101. Zickermann V, Kerscher S, Zwicker K, Tocilescua MA, Radermacher M, Brand U. 2009. Architecture of complex I and its implications for electron transfer and proton pumping. *Biochim. Biophys. Acta* 1787:574–83
102. Vonck J, Schäfer E. 2009. Supramolecular organization of protein complexes in the mitochondrial inner membrane. *Biochim. Biophys. Acta* 1793:117–24
103. Wolff N, Delepierre M. 1997. Conformation of the C-terminal secretion signal of the *Serratia marcescens* haem acquisition protein (HasA) in amphipols solution, a new class of surfactant. *J. Chim. Phys.* 95:437–42
104. Luccardini C, Tribet C, Vial F, Marchi-Artzner V, Dahan M. 2006. Size, charge, and interactions with giant lipid vesicles of quantum dots coated with an amphiphilic macromolecule. *Langmuir* 22:2304–10
105. Qi L, Gao X. 2008. Quantum dot-amphipol nanocomplex for intracellular delivery and real-time imaging of siRNA. *ACS Nano.* 2:1403–10
106. Vial F, Rabhi S, Tribet C. 2005. Association of octyl-modified poly(acrylic acid) onto unilamellar vesicles of lipids and kinetics of vesicle disruption. *Langmuir* 21:853–62
107. le Maire M, Champeil P, Møller JV. 2000. Interaction of membrane proteins and lipids with solubilizing detergents. *Biochim. Biophys. Acta* 1508:86–111
108. Catoire LJ, Zoonens M, van Heijenoort C, Giusti F, Guittet E, Popot J-L. 2009. Solution NMR mapping of water-accessible residues in the transmembrane β-barrel of OmpX. *Eur. Biophys. J.* PMID:19639312. In press
109. Tribet C, Diab C, Dahmane T, Zoonens M, Popot J-L, Winnik FM. 2009. Thermodynamic characterization of the exchange of detergents and amphipols at the surfaces of integral membrane proteins. *Langmuir* 25:12623–34
110. Bayburt TH, Sligar SG. 2003. Self-assembly of single integral membrane proteins into soluble nanoscale phospholipid bilayers. *Protein Sci.* 12:2476–81
111. Bayburt TH, Grinkova YV, Sligar SG. 2006. Assembly of single bacteriorhodopsin trimers in bilayer nanodiscs. *Arch. Biochem. Biophys.* 450:215–22
112. Boldog T, Grimme S, Li M, Sligar SG, Hazelbauer GL. 2006. Nanodiscs separate chemoreceptor oligomeric states and reveal their signaling properties. *Proc. Natl. Acad. Sci. USA* 103:11509–14
113. Alami M, Dalal K, Lelj-Garolla B, Sligar SG, Duong F. 2007. Nanodiscs unravel the interaction between the SecYEG channel and its cytosolic partner SecA. *EMBO J.* 26:1995–2004
114. Leitz AJ, Bayburt TH, Barnakov AN, Springer BA, Sligar SG. 2006. Functional reconstitution of β_2-adrenergic receptors utilizing self-assembling nanodisc technology. *Biotechniques* 40:601–12
115. Whorton MR, Bokoch MP, Rasmussen SG, Huang B, Zare RN, et al. 2007. A monomeric G protein-coupled receptor isolated in a high-density lipoprotein particle efficiently activates its G protein. *Proc. Natl. Acad. Sci. USA* 104:7682–87
116. Yao XJ, Vélez Ruiz G, Whorton MR, Rasmussen SG, DeVree BT, et al. 2009. The effect of ligand efficacy on the formation and stability of a GPCR-G protein complex. *Proc. Natl. Acad. Sci. USA* 106:9501–6
117. Bayburt TH, Leitz AJ, Xie G, Oprian DD, Sligar SG. 2007. Transducin activation by nanoscale lipid bilayers containing one and two rhodopsins. *J. Biol. Chem.* 282:14875–81

118. Kuszak AJ, Pitchiaya S, Anand JP, Mosberg HI, Walter NG, Sunahara RK. 2009. Purification and functional reconstitution of monomeric μ-opioid receptors: allosteric modulation of agonist binding by G_{i2}. *J. Biol. Chem.* 284:26732–41
119. Shih AY, Denisov IG, Phillips JC, Sligar SG, Schulten K. 2005. Molecular dynamics simulations of discoidal bilayers assembled from truncated human lipoproteins. *Biophys. J.* 88:548–56
120. Dahmane T, Damian M, Mary S, Popot J-L, Banères J-L. 2009. Amphipol-assisted in vitro folding of G protein-coupled receptors. *Biochemistry* 48:6516–21
121. Nehmé R, Joubert O, Bidet M, Lacombe B, Polidori A, et al. 2010. Stability study of the human G protein–coupled receptor, Smoothened. *Biochim. Biophys. Acta.* In press
122. Talbot J-C, Dautant A, Polidori A, Pucci B, Cohen-Bouhacina T, et al. 2009. Hydrogenated and fluorinated surfactants derived from *tris*(hydroxymethyl)-acrylamidomethane allow the purification of a highly active yeast F_1F_0 ATP synthase with an enhanced stability. *J. Bioenerg. Biomembr.* 41:349–60
123. Pocanschi CL, Dahmane T, Gohon Y, Rappaport F, Apell H-J, et al. 2006. Amphipathic polymers: tools to fold integral membrane proteins to their active form. *Biochemistry* 45:13954–61
124. Duarte AMS, Wolfs CJAM, Koehorsta RBM, Popot J-L, Hemminga MA. 2008. Solubilization of V-ATPase transmembrane peptides by amphipol A8-35. *J. Peptide Chem.* 14:389–93
125. Nelson RG, Johnson WC Jr. 1972. Optical properties of sugars. I. Circular dichroism of monomers at equilibrium. *J. Am. Chem. Soc.* 94:3343–45
126. Baas BJ, Denisov IG, Sligar SG. 2004. Homotropic cooperativity of monomeric cytochrome P450 3A4 in a nanoscale native bilayer environment. *Arch. Biochem. Biophys.* 430:218–28
127. Joubert O, Nehmé R, Fleury D, de Rivoyre M, Bidet M, et al. 2009. Functional studies of membrane-bound and purified human Hedgehog receptor Patched expressed in yeast. *Biochim. Biophys. Acta* 1788:1813–21
128. Whorton MR, Jastrzebska B, Park PS, Fotiadis D, Engel A, et al. 2008. Efficient coupling of transducin to monomeric rhodopsin in a phospholipid bilayer. *J. Biol. Chem.* 283:4387–94
129. Shaw AW, Pureza VS, Sligar SG, Morrissey JH. 2007. The local phospholipid environment modulates the activation of blood clotting. *J. Biol. Chem.* 282:6556–63
130. Morrissey JH, Pureza V, Davis-Harrison RL, Sligar SG, Ohkubo YZ, Tajkhorshid E. 2008. Blood clotting reactions on nanoscale phospholipid bilayers. *Thromb. Res.* 122(Suppl. 1):S23–66
131. Rabilloud T. 2009. Membrane proteins and proteomics: love is possible, but so difficult. *Electrophoresis* 30:S174–80
132. Gohon Y. 2002. *Etude structurale et fonctionnelle de deux protéines membranaires, la bactériorhodopsine et le récepteur nicotinique de l'acétylcholine, maintenues en solution aqueuse non détergente par des polymères amphiphiles*. PhD thesis. Univ. Paris-VI, Paris. 467 pp.
133. Lyukmanova EN, Shenkarev ZO, Paramonov AS, Sobol AG, Ovchinnikova TV, et al. 2008. Lipid-protein nanoscale bilayers: a versatile medium for NMR investigations of membrane proteins and membrane-active peptides. *J. Am. Chem. Soc.* 130:2140–41
134. Glück JM, Wittlich M, Feuerstein S, Hoffmann S, Willbold D, Koenig BW. 2009. Integral membrane proteins in nanodiscs can be studied by solution NMR spectroscopy. *J. Am. Chem. Soc.* 131:12060–61
135. Raschle T, Hiller S, Yu TY, Rice AJ, Walz T, Wagner G. 2009. Structural and functional characterization of the integral membrane protein VDAC-1 in lipid bilayer nanodiscs. *J. Am. Chem. Soc.* 131:17777–79
136. Kijac AZ, Li Y, Sligar SG, Rienstra CM. 2007. Magic-angle spinning solid-state NMR spectroscopy of nanodisc-embedded human CYP3A4. *Biochemistry* 46:13696–703
137. Petkova V, Benattar J-J, Zoonens M, Zito F, Popot J-L, et al. 2007. Free-standing films of fluorinated surfactants as 2D matrices for organizing detergent-solubilized membrane proteins. *Langmuir* 23:4303–9
138. Dauvergne J, Polidori A, Vénien-Bryan C, Pucci B. 2008. Synthesis of a hemifluorinated amphiphile designed for self-assembly and two-dimensional crystallization of membrane proteins. *Tetrahedron Lett.* 49:2247–50
139. Lebeau L, Lach F, Vénien-Bryan C, Renault A, Dietrich J, et al. 2001. Two-dimensional crystallization of a membrane protein on a detergent-resistant lipid monolayer. *J. Mol. Biol.* 308:639–47
140. Tribet C, Mills D, Haider M, Popot J-L. 1998. Scanning transmission electron microscopy study of the molecular mass of amphipol/cytochrome b_6f complexes. *Biochimie* 80:475–82

141. Wilkens S. 2000. F_1F_0-ATP synthase-stalking mind and imagination. *J. Bioenerg. Biomembr.* 32:333–39
142. Wilkens S, Zhou J, Nakayama R, Dunn SD, Capaldi RA. 2000. Localization of the δ subunit in the *Escherichia coli* F_1F_0-ATPsynthase by immuno-electron microscopy: The δ subunit binds on top of the F_1. *J. Mol. Biol.* 295:387–91
143. Flötenmeyer M, Weiss H, Tribet C, Popot J-L, Leonard K. 2007. The use of amphipathic polymers for cryo-electron microscopy of NADH:ubiquinone oxidoreductase (complex I). *J. Microsc.* 227:229–35
144. Posokhov YO, Rodnin MV, Das SK, Pucci B, Ladokhin AS. 2008. FCS study of the thermodynamics of membrane protein insertion into the lipid bilayer chaperoned by fluorinated surfactants. *Biophys. J.* 95:L54–56
145. Goluch ED, Shaw AW, Sligar SG, Liu C. 2008. Microfluidic patterning of nanodisc lipid bilayers and multiplexed analysis of protein interaction. *Lab Chip* 8:1723–28
146. Vinchurkar MS, Bricarello DA, Lagerstedt JO, Buban JP, Stahlberg H, et al. 2008. Bridging across length scales: multi-scale ordering of supported lipid bilayers via lipoprotein self-assembly and surface patterning. *J. Am. Chem. Soc.* 130:11164–69
147. Marin VL, Bayburt TH, Sligar SG, Mrksich M. 2007. Functional assays of membrane-bound proteins with SAMDI-TOF mass spectrometry. *Angew. Chem. Int. Ed. Engl.* 46:8796–98
148. Borch J, Torta F, Sligar SG, Roepstorff P. 2008. Nanodiscs for immobilization of lipid bilayers and membrane receptors: kinetic analysis of cholera toxin binding to a glycolipid receptor. *Anal. Chem.* 80:6245–52
149. Das A, Zhao J, Schatz GC, Sligar SG, Van Duyne RP. 2009. Screening of type I and II drug binding to human cytochrome P450-3A4 in nanodiscs by localized surface plasmon resonance spectroscopy. *Anal. Chem.* 81:3754–59
150. Früh V. 2009. *Application of fragment-based drug discovery to membrane proteins*. PhD thesis. Univ. Leiden, Leiden. 224 pp.
151. Katzen F, Fletcher JE, Yang JP, Kang D, Peterson TC, et al. 2008. Insertion of membrane proteins into discoidal membranes using a cell-free protein expression approach. *J. Proteome Res.* 7:3535–42
152. Cappuccio JA, Blanchette CD, Sulchek TA, Arroyo ES, Kralj JM, et al. 2008. Cell-free co-expression of functional membrane proteins and apolipoprotein, forming soluble nanolipoprotein particles. *Mol. Cell. Proteomics* 7:2246–53
153. Cappuccio JA, Hinz AK, Kuhn EA, Fletcher JE, Arroyo ES, et al. 2009. Cell-free expression for nano-lipoprotein particles: building a high-throughput membrane protein solubility platform. *Methods Mol. Biol.* 498:273–96
154. Cowan SW, Schirmer T, Rummel G, Steiert M, Ghosh R, et al. 1992. Crystal structures explain functional properties of two *E. coli* porins. *Nature* 358:727–33
155. Pebay-Peyroula E, Garavito RM, Rosenbusch JP, Zulauf M, Timmins P. 1995. Detergent structure in tetragonal crystals of OmpF porin. *Structure* 3:1051–59
156. Segrest JP, Jones MK, Klon AE, Sheldahl CJ, Hellinger M, et al. 1999. A detailed molecular belt model for apolipoprotein A-I in discoidal high density lipoprotein. *J. Biol. Chem.* 274:31755–58

Protein Sorting Receptors in the Early Secretory Pathway

Julia Dancourt and Charles Barlowe

Department of Biochemistry, Dartmouth Medical School, Hanover, New Hampshire 03755; email: barlowe@dartmouth.edu

Annu. Rev. Biochem. 2010. 79:777–802

First published online as a Review in Advance on April 5, 2010

The *Annual Review of Biochemistry* is online at biochem.annualreviews.org

This article's doi:
10.1146/annurev-biochem-061608-091319

0066-4154/10/0707-0777$20.00

Key Words

coat proteins, COPI, COPII, intracellular trafficking, vesicle transport

Abstract

Estimates based on proteomic analyses indicate that a third of translated proteins in eukaryotic genomes enter the secretory pathway. After folding and assembly of nascent secretory proteins in the endoplasmic reticulum (ER), the coat protein complex II (COPII) selects folded cargo for export in membrane-bound vesicles. To accommodate the great diversity in secretory cargo, protein sorting receptors are required in a number of instances for efficient ER export. These transmembrane sorting receptors couple specific secretory cargo to COPII through interactions with both cargo and coat subunits. After incorporation into COPII transport vesicles, protein sorting receptors release bound cargo in pre-Golgi or Golgi compartments, and receptors are then recycled back to the ER for additional rounds of cargo export. Distinct types of protein sorting receptors that recognize carbohydrate and/or polypeptide signals in secretory cargo have been characterized. Our current understanding of the molecular mechanisms underlying cargo receptor function are described.

Contents

INTRODUCTION

The eukaryotic secretory pathway is responsible for delivery of a tremendous variety of proteins to their proper cellular location and is essential for cellular function and multicellular development. Several lines of evidence indicate that secretory proteins contain sorting elements that are deciphered by the intracellular transport machinery at multiple stages to route proteins to their proper location. A central goal of cell biology has been to understand the cellular machinery that assembles and organizes distinct cellular structures. For membrane-bound compartments, coat protein complexes are major components of this organizational machinery that decode protein sorting signals presented on the surface of distinct membrane compartments. Three well-characterized coat complexes, clathrin and coat protein complexes I and II (COPI and COPII), have been described that are multisubunit assemblies that recognize specific protein sorting signals and can selectively sort proteins into dissociative carrier vesicles. Direct binding interactions between coat subunits and specific cargo often determine inclusion into the forming carrier vesicle. However, in certain instances, adaptor proteins or transmembrane receptors are needed for efficient linkage of cargo to a coat protein complex. For example, in the case of soluble proteins that are lumenally localized, transmembrane sorting receptors become critical for efficient linkage to cytoplasmic coat complexes. This review focuses on the COPII-dependent export process from the endoplasmic reticulum (ER) and on the mechanisms by which a diverse set of transmembrane sorting receptors link specific classes of cargo molecules to coat protein complexes. The cell biological, biochemical, and genetic approaches that have provided insight into receptor-mediated export from the ER and Golgi are discussed. Although molecular models have advanced in recent years, key questions remain regarding cargo recognition by protein sorting receptors and how interactions between sorting receptors and their cargo ligands are regulated.

COPI and COPII: coat protein complexes I and II

OVERVIEW OF THE EARLY SECRETORY PATHWAY

Nascent secretory proteins are translated and folded at the ER and then packaged into COPII vesicles for transport to pre-Golgi and Golgi compartments. A robust quality control system operates in the ER to ensure that nascent cargo

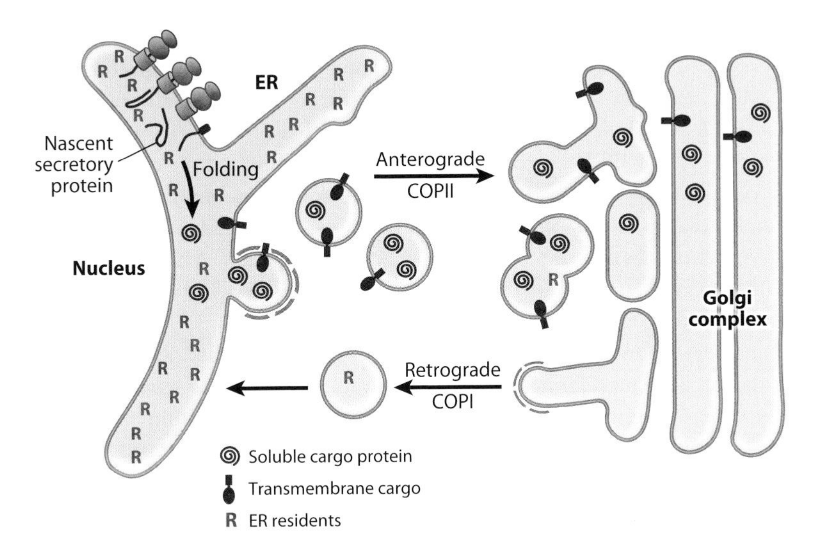

Figure 1

Model depicting bidirectional transport between the endoplasmic reticulum (ER) and Golgi compartments. After translation and folding of nascent secretory proteins, fully folded integral membrane cargo and soluble secretory cargo are exported from the ER in coat protein complex II (COPII)-formed transport vesicles. ER-derived vesicles traffic in an anterograde direction to fuse with or form pre-Golgi compartments. COPI coats bud retrograde-directed vesicles from pre-Golgi and Golgi compartments to recycle vesicle components and retrieve resident proteins (R) that have escaped the ER. This rapid cycling process allows net anterograde transport of secretory cargo, whereas resident proteins remain dynamically localized to early secretory compartments.

is retained and/or not recognized by the export machinery until the cargo is fully folded and assembled (1). Forward transport of folded secretory proteins in COPII vesicles is also balanced by a retrograde transport pathway that relies on COPI to recycle vesicle components and retrieve escaped ER resident proteins. There are several excellent reviews that describe the mechanisms underlying vesicle budding, targeting, and fusion in both anterograde and retrograde transport stages between the ER and Golgi compartments (2, 3). A simplified view of bidirectional transport in the early secretory pathway is provided in **Figure 1**. For the purposes of this review, it should be appreciated that transport between these compartments is extremely dynamic (4–6) but that biosynthetic secretory cargo efficiently advances through transient intermediates, whereas resident proteins of the early secretory pathway remain dynamically localized to their appropriate compartments. The COPII machinery plays a central role in anterograde protein sorting events at the ER, and the known molecular mechanisms by which this coat complex forms membrane vesicles and selects specific cargo for incorporation is considered next.

Coat Protein Complex II Vesicle Formation and Cargo Sorting Activities

The COPII budding machinery is organized at specialized regions of the ER known as transitional ER elements or ER exit sites (7–9). To initiate coat assembly at ER exit sites, the Sar1p GTPase is activated by the nucleotide exchange factor Sec12p. Activated Sar1p-GTP inserts a hydrophobic N-terminal amphipathic α-helix into the bilayer to initiate membrane curvature and also recruits the Sec23-Sec24 complex, a bifunctional cargo adaptor, and the Sar1-specific GTPase activating protein (GAP) complex. The outer layer Sec13-Sec31 complex

Retrograde transport: transport in the secretory pathway away from the cell surface and/or Golgi compartments and toward the ER assembles around Sar1-Sec23-Sec24 prebudding complexes, forming a caged structure that deforms the ER membrane bilayer and buds COPII vesicles (2, 10). Indeed, formation of the COPII cage (11) or COPII-coated vesicles (12) can be recapitulated using the five core-purified proteins (Sar1p, Sec23-Sec24, Sec13-Sec31) and synthetic liposomes.

Beyond the intrinsic properties that form coated membrane vesicles, COPII efficiently recognizes and segregates vesicle cargo away from ER resident proteins for incorporation into budding vesicles (13, 14). Several lines of investigation have identified sorting signals displayed on cytosolic surfaces of transmembrane cargo that direct these proteins into COPII vesicles (3). Subsequent biochemical studies demonstrated that the Sec23-Sec24 complex physically associates with transmembrane cargo proteins in a sorting signal-dependent manner (15, 16). Formation of ternary cargo complexes containing Sec23-Sec24 and Sar1p-GTP bound to cargo are stabilized by nonhydrolyzable forms of GTP. Finally, regulated hydrolysis of bound GTP by Sar1p allows for associations to be reversed such that cargo can be released from sorting subunits and COPII proteins released from budded vesicles for recycling at ER exit sites. A model depicting COPII vesicle formation and cargo sorting is shown in **Figure 2**.

Structural studies on COPII are now quite advanced and have revealed multiple cargo recognition sites, within the Sec24p subunit, that bind to defined sorting signals (17, 18). Moreover, cells are endowed with multiple Sec24p isoforms to greatly expand the diversity of export signals that can be recognized by the COPII sorting machinery (19, 20). However, not all secretory proteins have identifiable COPII sorting signals, and a large number of soluble secretory cargo proteins do not span ER

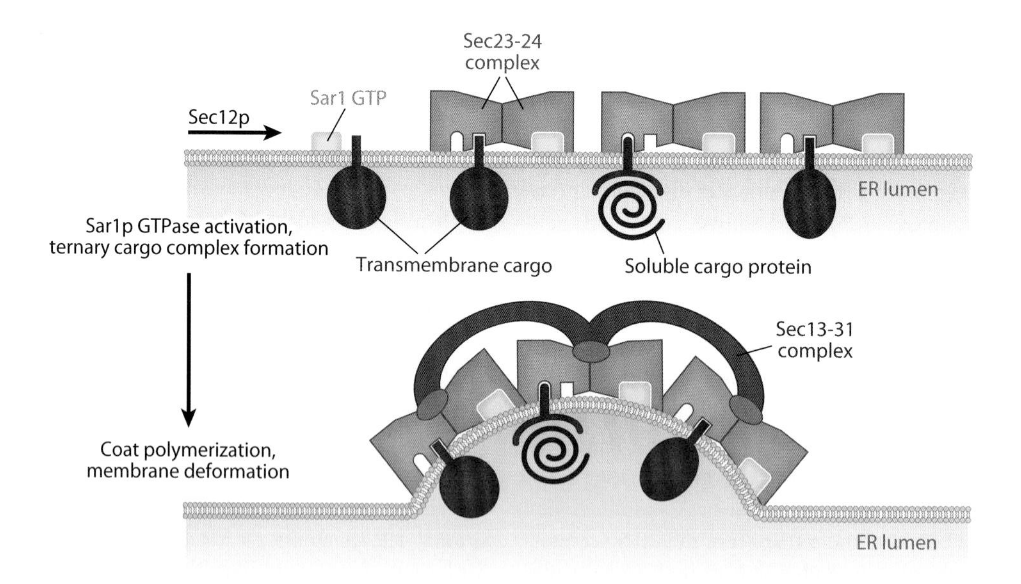

Figure 2

Model for coat protein complex II (COPII)-dependent cargo selection and vesicle formation. Sar1p is activated to Sar1p-GTP by the ER-localized nucleotide exchange factor Sec12p. Membrane-bound Sar1p-GTP assembles into prebudding cargo complexes with Sec23-Sec24 in which the Sec24p subunit binds to specific sorting signals displayed by export cargo. As shown, linkage of vesicle cargo to the Sec24p subunit may be direct or through transmembrane sorting receptors. Prebudding cargo complexes recruit the outer layer Sec13-Sec31 complex, leading to coat polymerization and vesicle formation. Abbreviation: ER, endoplasmic reticulum.

membranes and cannot be directly recognized by COPII subunits. Therefore, efficient ER export of several types of secretory cargo may depend on transmembrane adaptor proteins for linkage to the COPII budding machinery. We next consider distinct classes of secretory cargo that are likely to rely on COPII-dependent export from the ER.

Diversity of Cargo Packaged into Coat Protein Complex II Vesicles

The COPII export pathway must handle two general classes of secretory proteins that are translocated into the ER and folded: integral membrane proteins and soluble lumenal proteins. Within the integral membrane-spanning class, type I or type II topologies are established during translocation and orient the N terminus in or out of the ER lumen, respectively. Integral membrane proteins may possess single- as well as multipass membrane-spanning segments. For soluble secretory proteins, an N-terminal signal sequence targets polypeptides to the ER and is cleaved upon translocation into the ER lumen. Depending on additional properties and motifs in a newly synthesized secretory protein, posttranslational modifications that include attachment of N- and O-linked oligosaccharide, glycophosphatidylinositol (GPI) anchoring and disulfide bond formation may be introduced prior to export from the ER. Computational analysis of the human proteome indicates that out of 24,701 predicted full-length open reading frames, 11.2% encode soluble secretory proteins, 9.4% encode single pass transmembrane proteins, and 11.5% encode multispanning membrane proteins that are targeted to the ER and are distributed among various endomembrane compartments (21). This transmembrane prediction method includes GPI-anchored proteins in with soluble secretory proteins, which are predicted to represent 0.7% of the proteome (22). Analyses of genomes in other species indicate similar percentages of soluble and integral membrane secretory proteins. Although a fraction of these ER-targeted proteins ultimately reside in the ER, most are exported through a COPII-dependent packaging event. Even if one estimates that one-quarter of translocated polypeptides ultimately reside in the ER, this would indicate that ~6000 secretory proteins are exported from the ER in human cells. And, although some evidence suggests that there may be COPII-independent routes of secretion (23, 24), for a large majority of secretory proteins examined in the literature, ER export relies on COPII function. Thus, an important issue to appreciate at this juncture is that the COPII machinery must recognize and export several thousand distinctly folded secretory proteins of sizes ranging from ~2 nm to over 300 nm (25).

GPI: glycophosphatidylinositol

Concentrative transport: the process whereby cargo is packaged into vesicle intermediates at a concentration that is higher than the donor membrane compartment

Bulk-flow transport: the process whereby cargo is transported in vesicle intermediates at the prevailing concentration of the donor membrane compartment

Coat Protein Complex II–Dependent Concentration of Secretory Cargo

Initial stereological and biochemical studies indicated that transmembrane spike proteins of the Semliki Forest virus were concentrated during transport from the ER to the Golgi complex (26). Quantitative immunoelectron microscopy also documented that the vesicular stomatitis virus glycoprotein was concentrated tenfold during vesicle budding from the ER (27). These results coupled with in vitro assays of COPII activity (13, 15, 16, 28) and molecular genetic experiments (29, 30) provided strong evidence in support of COPII-dependent concentrative transport of integral membrane cargo during ER export. Indeed, as mentioned above, the mechanism for selection of specific signals displayed by transmembrane secretory cargo has been elucidated in molecular detail for the Sec24p subunit of COPII (17, 18). However, studies on export of soluble secretory proteins from the ER have indicated that both concentrative as well as passive or bulk-flow mechanisms may operate depending on the secretory cargo monitored (31). Assessment by quantitative immunoelectron microscopy showed that soluble serum albumin was concentrated into transport carriers that depart the ER in Hep-G2 cells (32). In contrast, export of amylase and chymotrypsinogen from the ER of pancreatic

ERGIC: ER-Golgi intermediate compartment

Erv protein: ER vesicle protein

exocrine cells did not indicate concentration into COPII-budded vesicles, but instead concentration occurred in vesicular tubular clusters en route to the Golgi complex (33). In vitro studies have revealed that a bulk-flow marker, such as ss-GFP, could be packaged into COPII vesicles but at relatively low efficiency of ~1%, whereas glyco-proα-factor (gpαf), a secreted pheromone in yeast, was much more efficiently packaged at ~28% (34). Thus bulk-flow and receptor-mediated transport pathways are not mutually exclusive mechanisms, and each may operate concurrently depending on the secretory protein and cell type. However, for certain soluble secretory cargo, including gpαf, compelling evidence supports receptor-dependent concentration during COPII budding from the ER (13, 15). The molecular identity of ER sorting receptors had not emerged through genetic screens, and receptors were not reported until after characterization of the COPII machinery. We next review the initial biochemical approaches that led to identification and characterization of ER sorting receptors.

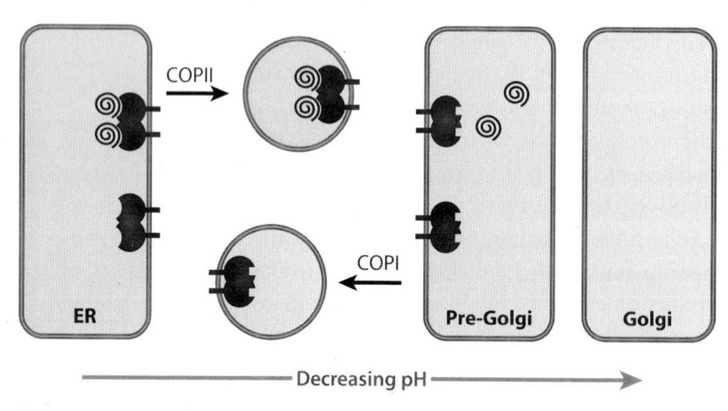

Figure 3

Basic model for regulation of binding interactions between sorting receptors and their cargo. For anterograde cargo receptors, the near neutral pH in the endoplasmic reticulum (ER) lumen (72) promotes binding of cargo ligands to their receptor. In pre-Golgi and Golgi compartments, a decreasing pH induces conformational changes in the sorting receptor to lower affinity for bound ligand. Abbreviation: COPI and –II, coat protein complexes I and II.

TRANSMEMBRANE SORTING RECEPTORS

Although experiments had indicated concentrative ER export of specific soluble secretory proteins (13, 32) and evidence for an ER export receptor bound to gpαf was reported (15), the molecular identity of transmembrane cargo receptors that link lumenal cargo to COPIIs was a later development. Biochemical approaches had identified several abundant membrane proteins that localized to early secretory compartments and to transport intermediates, which were potential receptor candidates, including the ERGIC-53 protein (35), the p24 proteins (36–38), and a set of ER vesicle (Erv) proteins (28, 39, 40). Different lines of investigation ultimately revealed that these proteins acted in cargo sorting in transport between the ER and Golgi. A key observation was that cells lacking a specific cargo receptor generally displayed selective sorting defects in which a subset of secretory proteins were inefficiently transported from the ER, whereas other secretory proteins trafficked at normal rates. In addition, cell biological studies indicated that sorting receptors cycle dynamically between the early secretory compartments owing to cytoplasmically exposed coat recognition signals. On the basis of these observations, ER sorting receptors were hypothesized to cycle between the ER and Golgi compartments in COPII- and COPI-derived vesicles, where the receptors would shuttle specific secretory cargo forward through cargo binding in the ER—followed by dissociation in the ERGIC or *cis*-Golgi compartments (**Figure 3**). Experimental evidence regarding the identification and function of ERGIC-53, the p24 proteins, and different Erv proteins is described in turn below.

ERGIC-53

ERGIC-53, named for the localization of this 53-kDa protein to ER-Golgi intermediate compartments, was initially identified through generation of monoclonal antibodies against purified Golgi membrane fractions (35, 41). ERGIC-53 is an oligomeric single-pass transmembrane protein with a larger N-terminal lumenal domain and a short, cytoplasmically exposed 12-amino acid C-terminal tail sequence (42). The cytoplasmic tail sequence terminates in a diphenylalanine motif, which is required for forward transport in COPII vesicles (30), as well as a dilysine signal, which acts in COPI-dependent retrograde movement (30, 42). The ERGIC-53 lumenal domain shares sequence and structural homology to a family of leguminous or L-type lectins that bind to high-mannose oligosaccharides in a Ca^{2+}-dependent manner (43–46). On the basis of initial biochemical studies showing mannose-selective binding, ERGIC-53 was proposed to act in ER export of fully folded glycoproteins (43). Subsequent studies demonstrated that overexpression of a dominant-negative ERGIC-53 mutation in cultured cells specifically blocked transport of the lysosomal glycoprotein cathepsin C (47). Pulse-labeling experiments showed that nascent cathepsin Z-related protein bound to ERGIC-53 in the ER and was dissociated in post-ER compartments (48) in accord with a cargo-receptor function. Additional studies demonstrated that cargo binding is pH dependent, that a conserved histidine residue at the carbohydrate-binding site of ERGIC-53 may serve as a pH sensor for cargo binding, and that elevation of lumenal pH in vivo specifically retards dissociation of cathepsin Z from ERGIC-53 (49). Finally, as observed for other lumenal lectins in the early secretory pathway, ERGIC-53 recognition of cargo relies on interactions with both N-linked carbohydrate as well as protein motifs. For procathepsin Z, a folded surface-exposed β-hairpin loop, which is next to the critical N-linked glycan site, was required for efficient binding (50).

Remarkably, independent lines of evidence for a cargo receptor function came from human genetic studies showing that mutations in ERGIC-53 caused a bleeding disorder owing to a combined deficiency of coagulation factor V and VIII in plasma (51). Over 30 distinct mutations that introduce nonsense and frameshift mutations have now been mapped in the gene (*LMAN1*) that encodes ERGIC 53, and some result in a complete absence of the protein. Genotype-phenotype comparisons indicate that the corresponding mutations reduce coagulation factor levels by 50% to 95% of normal (51, 52). Additional evidence indicated that ERGIC-53 loss of function caused inefficient ER export of the heavily glycosylated coagulation factors V and VIII (53) and presumably reduces their secretion from hepatocytes into the plasma (51). Indeed, chemical cross-linking studies have been used to detect the transient interactions between ERGIC-53 and factor V or factor VIII in early secretory compartments (54). These findings provide independent lines of evidence for ERGIC-53 function as a cargo receptor in ER export.

Further genetic analyses of the combined factor V and VIII deficiency in humans revealed that approximately one-third of the mutations mapped to a second novel gene that was termed *MCFD2*, for multiple coagulation factor deficiency gene 2 (55, 56). The symptoms displayed by mutations in this second gene are clinically indistinguishable from ERGIC-53 deficiencies. *MCFD2* encodes a soluble 16-kDa protein with an N-terminal signal sequence for ER targeting and tandem Ca^{2+}-binding EF-hand motifs. Interestingly, MCFD2 assembles into a Ca^{2+}-dependent complex with ERGIC-53 and normally colocalizes to the ERGIC. Furthermore, reduced ERGIC-53 levels cause mislocalization of MCFD2 and secretion from cells (54, 57). Structural studies have shown that MCFD2 binds Ca^{2+} in both EF-hand sites and that disease-causing point mutations disrupt Ca^{2+} binding and prevent proper folding, consistent with a Ca^{2+}-dependent requirement for interaction with ERGIC-53 (58, 59). Moreover, recent development of quantitative binding assays indicate that the high-mannose-type lectin activity of ERGIC-53 is enhanced by interaction with MCFD2 and that Ca^{2+} concentrations as well as pH can further modulate the associations between ERGIC-53 and high-mannose glycans (60). The collective biochemical and genetic experiments indicate a model in which ERGIC-53-MCFD2 forms a functional complex that recruits factor V and

L-type lectin: a carbohydrate-binding protein in the seeds of leguminous plants that contains a structural motif found in a variety eukaryotic cell lectins

***MCFD2*:** multiple coagulation factor deficiency gene 2

High-mannose glycan: a type of N-linked oligosaccharide that consists primarily of polymerized mannose residues with terminal α1–2 linkages factor VIII into forming COPII vesicles at the ER, where pH and Ca^{2+} concentrations are sufficiently high. In post-ER compartments, decreased pH and/or Ca^{2+} levels may trigger dissociation of bound cargo—followed by return of empty ERGIC-53 receptors to the ER via COPI-dependent retrograde transport (52, 61).

An important issue that remains to be resolved is how lumenal Ca^{2+} concentrations in the ER, ERGIC, and *cis*-Golgi compartments are controlled to regulate cargo receptor interactions among other calcium-dependent processes. Several lines of evidence indicate that Ca^{2+} transporters maintain Ca^{2+} stores of 0.4 mM to 0.3 mM in the ER and Golgi compartments, respectively (62–64). Moreover, distinct Ca^{2+} transporters localize to early secretory compartments to independently maintain lumenal Ca^{2+} stores even when vesicular transport is blocked (63). Interestingly, high-resolution calcium mapping suggests that Ca^{2+} concentrations are markedly lower in ERGIC structures than in the ER and Golgi compartments (64). These observations generally support the proposal that Ca^{2+} concentrations in the ER promote interactions between ERGIC-53 and specific glycoprotein cargo that are reversed in the ERGIC and then not favored in *cis*-Golgi compartments where pH and Ca^{2+} concentrations are lower (60).

In addition to factor V, factor VIII, cathepsin C, and cathespsin Z, other soluble glycosylated secretory proteins, including α1-antitrypsin, depend on ERGIC-53 for efficient export from the ER (65). Although MCFD2 plays a clear role in factor V and VIII transport, it is less clear if this subunit is required for transport of other glycoprotein cargo. Indeed, RNA-based knockdown studies in cultured human cells indicated that binding of cathespin C and cathepsin Z to ERGIC-53 was independent of MCFD2, suggesting that MCFD2 may play a specialized role in transport of certain glycoslyated secretory proteins (57). Key questions remain on the molecular mechanisms by which ERGIC-53 recognizes specific cargo and how this binding is regulated to efficiently shuttle cargo forward in the early secretory pathway. Structural studies on ERGIC-53 with a glycoprotein cargo molecule bound to the carbohydrate recognition domain should be informative and guide these efforts.

ERGIC-53 Family Members

ERGIC-53 belongs to a family of related Ca^{2+}-dependent L-type lectins that are widespread in nature and include ERGL, VIP36, and VIPL proteins in animal cells (45). ERGL (ERGIC-53-like) has not been extensively characterized but is reported to be differentially expressed, depending on cell type and tissue (66). VIP36 dynamically localizes to the ERGIC and *cis*-Golgi compartments, although modification of its N-linked carbohydrate and other evidence suggest a role in trafficking through later Golgi compartments (67–69). The VIP36-like protein, VIPL, is widely expressed in animal cells and localizes primarily to the ER (70). Quantitative comparisons of the carbohydrate recognition properties displayed by ERGIC-53, VIP36, and VIPL revealed that each of these L-type lectin family members has distinct binding activity toward high-mannose-type oligosaccharides (46). ERGIC-53 has low affinity and broad specificity for high-mannose oligosaccharide and does not distinguish between monoglucosylated and deglucosylated forms. In contrast, VIPL and VIP36 display a preference for deglucosylated forms of high-mannose oligosaccharide, a form generally thought to be released from the calreticulin and calnexin ER folding cycle (71). Interestingly, VIPL and VIP36 exhibit distinct pH profiles in binding assays with deglucosylated high-mannose carbohydrate such that VIPL binds optimally at pH ~7.5, whereas VIP36 binds optimally at pH ~ 6.5. Because the lumenal ER compartment is reported to maintain a pH of ~7.2 and Golgi compartments are in the range of pH ~6.4 (72), VIPL should have a binding preference for deglucosylated high-mannose carbohydrate in the ER, and VIP36 should prefer binding this carbohydrate structure in Golgi compartments. On the basis of intracellular localization, high-mannose-binding specificity, and pH optima, the following working model has been proposed (46). VIPL localizes to the ER and is thought to hand off fully folded and deglucosylated glycoproteins to ERGIC-53 for incorporation into COPII vesicles. In post-ER compartments, a reduction in pH releases glycoprotein ligands from ERGIC-53. Finally, VIP36 is proposed to act in Golgi compartments through recognition of misfolded glycoproteins for their retrieval back to the ER. Although the proposed model is appealing and consistent with current findings, experiments to directly test these ideas, using in vitro packaging into specific carrier vesicles and loss-of-function alleles for in vivo analyses, remain to be performed.

The p24 Proteins

Members of the p24 family of proteins were initially identified as abundant constituents of early secretory compartment membranes (36, 37) and as enriched membrane proteins on COPI vesicles (38) and COPII vesicles (39, 40). These proteins are in the 24-kDa molecular weight range and share a similar overall membrane topology with ERGIC-53 that includes a larger N-terminal lumenal domain, a single-pass transmembrane segment, and a cytoplasmically exposed C-terminal tail sequence of 12 to 20 amino acids. The C-terminal tail sequences contain specific sorting signals required for continual COPII- and COPI-dependent cycling between the ER and Golgi compartments (reviewed in Reference 73). On the basis of phylogenetic clustering, the p24 proteins have been divided into four subfamilies known as p24α, p24β, p24γ, and p24δ (74), which are present in most eukaryotic species with sequenced genomes. The number of representatives from each subfamily appears to vary across species, but there are approximately eight to ten family members per genome analyzed. Native p24 proteins assemble into heteromeric complexes (40, 74, 75) that are typically heterodimers (76), and the assembly of heteromeric complexes appears to be required for stable expression and proper localization.

A role for p24 proteins in cargo sorting was initially proposed based on genetic and biochemical analyses of yeast strains lacking Emp24p/p24β or Erv25p/p24δ, which were observed to display selective trafficking defects (39, 40). Specifically, the GPI-anchored protein Gas1p and secretory invertase were three- to fourfold delayed in transport from the ER to the Golgi complex when deletion strains were compared to a wild type. Interestingly, deletion of all eight p24 proteins in yeast produced viable strains with phenotypes that were no more severe than strains lacking Emp24p or Erv25p (77), suggesting that the two abundant p24 family members identified on COPI and COPII vesicles perform major roles. In vitro experiments with yeast components to further assess the role of the Emp24p-Erv25p heterodimer in ER export demonstrated that Emp24p was directly required for packaging of Gas1p into COPII-derived vesicles and that Gas1p was detected in complex with Emp24p and Erv25p in these vesicles (78). This biochemical evidence, combined with the defective sorting phenotypes displayed in yeast strains lacking Emp24p and Erv25p, provides the strongest indication for a cargo receptor function. The cargo receptor activity appears conserved across species as knockdown of p24δ in animal cells also produced selective sorting defects on transport of the GPI-anchored protein DAF from the ER (79). However, as with the ERGIC-53 model, key questions remain regarding how specific cargo are recognized by p24 proteins and how binding to this distinct class of receptors is regulated.

Although the evidence indicates that p24 proteins function as cargo receptors, additional studies document that loss of p24 protein function influences other cellular processes, and there is some debate as to whether these consequences arise because of primary sorting defects or whether the p24 proteins perform multiple roles (73). Notably, loss-of-function mutations in specific p24 proteins can influence structural organization of ER and Golgi compartments (80, 81). Studies have also shown that p24 proteins facilitate COPI vesicle formation, probably through recruitment of coatomer to Golgi membranes (82). Finally, evidence indicates that loss-of-function p24 mutations influence ER quality control mechanisms and permit increased transport of misfolded proteins from the ER (83, 84). The observation that yeast strains lacking all p24 proteins are viable suggests that p24 proteins do not play an essential role in COPI-dependent retrograde transport, although it is known that yeast activate the unfolded protein response pathway to cope with loss of p24 function (82, 85). It seems clear that p24 proteins are critical for intracellular trafficking and normal cellular function, as highlighted by the observation that knockout mice display embryonic lethality (86). However, further studies are required to understand if the influences on cell structure and quality control are caused by a primary function for p24 proteins in receptor-mediated export of cargo from the ER or if this ubiquitous family performs multiple cellular functions.

MULTISPANNING MEMBRANE SORTING RECEPTORS

Biochemical approaches were useful in identifying the p24 proteins that had been missed in classic forward genetic screens. Other abundant polypeptides identified on isolated COPII vesicles (28, 87) provided candidate proteins that could act as cargo receptors. Characterization of Erv proteins in yeast, coupled with reverse genetic methods, allowed for tests of function in transport and cargo sorting. This approach succeeded in identifying specific sorting defects connected to Erv29p, Erv26p, and Erv14p function. Functional roles of several other Erv proteins remain to be determined. For the set of yeast Erv proteins implicated in cargo sorting, similar general properties are shared, as described above, for the ERGIC-53 and p24 cargo sorting receptors. First, deletion of either Erv29p, Erv26p, or Erv14p causes distinct sorting defects such that specific secretory cargo accumulate in the ER, whereas other cargo are transported at wild-type or near wild-type rates. Second, these proteins are highly conserved in nature, suggesting a conserved function. Third, Erv proteins dynamically cycle between the ER and Golgi compartments through cytosolic sorting signals that are recognized and packaged into both COPII and COPI vesicles. And fourth, the multispanning membrane Erv proteins under native condition are detected as oligomers, a property that is shared among the ERGIC-53 and p24 protein families. The specific activities for Erv29p, Erv26p, and Erv14p are reviewed below.

Erv29p

Assays of gpαf transport from the ER to the Golgi complex in yeast have provided significant insights into the molecular mechanisms underlying protein transport through the early secretory pathway (2). Efficient in vitro packaging of gpαf into COPII vesicles initially provided strong evidence for a receptor-mediated export mechanism (13, 14, 28). Additional experimentation revealed that gpαf was captured in prebudding cargo complexes containing the Sec23-Sec24 complex and Sar1p when nonhydrolyzable analogs of GTP were used. Because prebudding cargo complexes are assembled on the cytosolic surface of ER membranes, a transmembrane receptor protein was hypothesized to link lumenal gpαf to COPII subunits. Initial characterization of the cargo receptor through chemical cross-linking indicated a transmembrane protein of approximately 30 kDa (15). Surprisingly, when yeast membranes lacking Erv29p were assessed for gpαf transport in vitro, COPII packaging of gpαf was strongly reduced, whereas other COPII vesicle proteins were packaged at normal levels (34, 88). Pulse-chase studies to monitor gpαf transport in vivo also revealed a tenfold reduction in transport rates when an *erv29Δ* strain was compared to wild type. Other soluble secretory proteins, including carboxypeptidase Y (CPY) and proteinase A, displayed reduced transport rates in the *erv29Δ* strain, whereas the GPI-anchored protein Gas1p and secretory invertase were transported at normal rates. Finally, chemical cross-linking experiments indicated that Erv29p physically associated with gpαf in ER membranes and produced a cross-linked adduct of the expected size (88). These findings indicate that Erv29p acts as a cargo receptor for gpαf in addition to other soluble secretory cargo.

Erv29, also known as Surf4 in mammals (81), is a tetramembrane spanning protein with both N and C termini exposed to the cytosol (89). In some species, the C-terminal tail of Erv29 terminates with a COPI-specific dilysine sorting signal, suggesting that this segment directs transport between the ER and Golgi compartments. The lumenal loop domains in Erv29p are proposed to recognize sorting elements in soluble secretory proteins, such as gpαf and CPY, for incorporation into COPII vesicles (88). Experiments to refine an ER export signal in gpαf revealed that the N-terminal proregion was necessary and sufficient for Erv29-dependent export. Moreover, specific hydrophobic residues (I39, L42, and V52) within a 25–amino acid segment of the gpαf proregion were required for Erv29p-dependent recognition (90). Secondary structure prediction programs indicate this segment of the proregion folds into an α-helical structure; therefore, the hydrophobic I-L-V residues may line one face of an α-helix that interacts with Erv29p. However, structural studies on the proregion are required to establish this feature. Erv29p-dependent transport of soluble cargo does not appear to involve carbohydrate recognition, because treatment with tunicamycin to inhibit N-linked glycosylation has only modest effects on transport of gpαf in vivo (91). In addition, the primary sequence of Erv29p does not contain apparent homology to any known carbohydrate-binding domains, as observed for the ERGIC-53 family of L-type lectins. Indeed, the Erv29p sequence does not share identity with any known structural motifs, providing few clues into binding mechanisms. Structural studies on multispanning membrane proteins are challenging, and yet this approach appears critical for elucidation of Erv29p-binding mechanisms.

Erv14p

Erv14p is a small hydrophobic protein with three transmembrane segments and a lumenally oriented C terminus. Erv14p belongs to a highly conserved family of proteins with *Drosophila* Cornichon as the founding member (92). In yeast, *erv14Δ* mutants display a defect in bud site selection because Ax12p, a transmembrane secretory protein, accumulates in the ER and is inefficiently delivered to its site of function on the cell surface (93). In *Drosophila*, *cornichon* mutants also display polarization defects because of a failure in ER export of transmembrane Gurken, a TGFα family member that is processed and delivered to the oocyte surface (92, 94, 95). Experiments in yeast demonstrated that Erv14p interacts with both the Axl2p cargo protein and a prebudding Sec23-Sec24-Sar1-GTP complex for incorporation into COPII vesicles. Putative COPII sorting signals are thought to reside in the cytosolically exposed second loop domain of Erv14p (96). Molecular dissection of the *Drosophila* Gurken protein indicated that the first 30 membrane proximal residues in the lumenal domain of this TGFα family member are required for interaction with the N-terminal half of Cornichon. On the basis of the collective findings, the Cornichon family of proteins is thought to recognize lumenal sorting signals in transmembrane cargo for efficient coupling to the ER export machinery. This mechanism appears to be highly conserved because studies of two of the four mammalian Cornichon paralogs show these proteins colocalize with early secretory pathway markers and selectively influence secretion of mammalian TGFα proteins (97). A more recent study in mammalian brain shows that the Cornichon proteins are required to increase the surface expression of glutamate receptors of the AMPA subtype (98). Moreover, these findings suggest that certain Cornichons may traffic with specific glutamate receptors to the cell surface. Other studies have indicated that Erv14p or Cornichon deficiencies produce more general trafficking defects (96, 99, 100), although further work is required to determine if these are secondary effects owing to impaired cargo sorting or if Cornichon proteins perform additional cellular functions. Interestingly, all of the known cargo that interact with Cornichon family members are transmembrane proteins and presumably could be recognized directly by COPII subunits. We speculate that certain classes of transmembrane cargo evolved through gene fusion events that placed lumenal sorting signals on transmembrane proteins. Alternatively, cargo-adaptor mechanisms may provide greater opportunities for regulation of ER export of cargo proteins. As discussed in a following section on the role of cargo receptors on ER quality control, these receptors appear to influence ER quality control events, and therefore, assembly of transmembrane receptor-cargo complexes may be an important aspect of ER quality control.

Erv26p

Erv26p is a polytopic membrane protein that was identified in yeast as an abundant polypeptide constituent on Sed5p-containing vesicles (101) and on COPII vesicles (102). The *ERV26* gene encodes a 26-kDa tetraspanning membrane protein and is required for efficient ER export of alkaline phosphatase (ALP) and the Ktr3p mannosyltransferase (102, 103). The Erv26p protein specifically associates with pro-ALP and Ktr3p in detergent-solubilized extracts (101, 102). Interestingly, both known Erv26p-dependent cargo proteins are type II membrane proteins (N terminus in the cytoplasm), which suggests that the Erv26p receptor may function more generally in export of this class of cargo. In the case of ALP, the short N-terminal cytoplasmic tail sequence has well-characterized sorting signals for clathrin-dependent transport from the Golgi to the yeast vacuole (104). Moreover, the short N-terminal tail of Ktr3p contains a signal that localizes glycosyltransferases to the Golgi complex in a Vps74p-dependent manner (105, 106). Therefore, this class of type II membrane proteins may rely on the Erv26p cargo receptor for linkage to COPII subunits because inclusion of ER export signals in such a short tail sequence is not compatible with other sorting signals needed to specify localization in later branches of the secretory pathway.

Molecular dissection of Erv26p has provided insights into how this tetraspanning receptor links cargo to COPII. The C-terminal tail sequence contains sorting signals that direct Erv26p into both COPI and COPII vesicles, whereas mutations in the third lumenal loop domain interfere with pro-ALP binding and incorporation into COPII vesicles (103). On the cargo side of the Erv26 sorting mechanism, studies on yeast ALP indicated that the cytosolic tail and transmembrane domain of pro-ALP are dispensable for COPII export. Instead, the ER export signal resides in the lumenal domain of pro-ALP, where Erv26p-dependent recognition relies on placement of the pro-ALP lumenal domain near the inner leaflet of the ER membrane (107). Although pro-ALP contains N-linked core oligosaccharides, mutations that remove these sites or inhibition of glycosylation with tunicamycin has mild influences on transport rates and indicates that carbohydrate recognition is unlikely to contribute to Erv26p-dependent sorting. Finally, it was observed that only assembled pro-ALP dimers bind to Erv26p and are exported from the ER, suggesting that Erv26p sorting may contribute to ER quality control (107). On the basis of the collective findings, lumenal domain interactions between Erv26p and fully assembled type II membrane cargo proteins are thought to link these cargoes to COPII. Most eukaryotic species with sequenced genomes contain a single high-sequence identity homolog of Erv26p, suggesting conservation of function. However, there are no studies that directly address cargo receptor activity of Erv26p in other species.

RETROGRADE SORTING RECEPTORS

The cargo receptors described at this stage of the review have generally functioned in efficient forward transport of nascent secretory cargo through the early secretory pathway. However, a robust COPI-dependent retrograde pathway operates to recycle vesicle proteins and lipids that are required for multiple rounds of transport between the ER and Golgi. COPI, or coatomer, was the first molecularly defined coat complex identified in the early secretory pathway (108). Assembly of COPI depends on the small GTPase Arf1 (109), and membrane-bound Arf1-GTP recruits the seven coatomer subunits *en bloc* to Golgi membranes to form COPI vesicles (110). Several Arf1-specific guanine nucleotide exchange factors and GTPase-activating proteins localize to early secretory compartments and regulate the COPI budding cycle to incorporate specific cargo into COPI vesicles (111, 112). COPI subunits also recognize defined retrograde sorting signals on transmembrane proteins including the dilysine and diphenylalanine motifs. Moreover, abundant cycling proteins that display these signals, such as the p24 family members, are thought to trigger COPI polymerization (112). The COPI retrograde pathway recycles anterograde machinery but is also critical to retrieve escaped ER-resident proteins through the action of retrograde sorting receptors.

Genetic approaches were most productive in identifying receptors that act in retrograde retrieval (113–115). Subsequent biochemical approaches also encountered retrograde cargo receptors on COPII-derived vesicles (85, 87). Soluble ER-resident proteins that terminate in the amino acid sequence KDEL, or related sequences, are recognized in the ERGIC and Golgi complex by the KDEL receptor and recycled back to the ER in COPI vesicles (116). In addition, the tetraspanning membrane protein Rer1p recognizes a subset of ER membrane proteins for COPI-dependent retrieval and localization to the ER (117). The KDEL and Rer1p receptors represent well-studied examples of retrograde sorting receptors and provide interesting contrasts to the anterograde receptors. Experimental evidence suggests several similarities between these classes of receptors, including mechanisms to link specific cargo to coat complexes and dynamic recycling routes between early secretory compartments. However, a primary distinction between the two classes of receptors appears to be the reciprocal affinities displayed for their bound ligands in that retrograde receptors preferentially bind ligands in post-ER compartments, whereas anterograde receptors favor dissociation of ligands in these same post-ER compartments. Our current understanding of mechanisms underlying the KDEL receptor and Rer1 in this context are considered below.

KDEL Receptor

The human KDEL receptor is predicted to contain seven transmembrane-spanning segments and terminates in a cytoplasmically exposed C-terminal tail that includes a probable COPI-binding dilysine motif (118). The KDEL family of proteins is highly conserved across species, and in fact, the first family member, Erd2p, was identified in yeast through a genetic screen for ER retention-defective (*erd*) mutants (113). In animal cells, the steady-state distribution of the KDEL receptor is predominantly to early Golgi compartments, but interestingly, the receptor is redistributed to the ER upon overexpression of KDEL-bearing secretory proteins (119). This ligand-mediated redistribution of the KDEL receptor is thought to arise through a ligand-induced clustering of the receptor (120, 121), which in turn recruits COPI budding components and stimulates the packaging of KDEL receptors into COPI-derived retrograde vesicles (122). Precisely how receptor clustering triggers COPI packaging and whether a protein kinase A site in the cytoplasic tail sequence of the receptor (123) plays a role in this event remain to be determined. However, there are some intriguing parallels in comparing ligand-dependent COPI packing of the multimembrane-spanning

Anterograde transport: transport in the secretory pathway away from the ER and toward the Golgi complex and/or cell surface

KDEL receptor and clathrin-dependent endocytosis of activated serpentine G protein–coupled receptors (124). Presumably, ligand-induced conformational changes in the receptor are transduced to cytoplasmic surfaces and initiate uptake into coated vesicles. This mechanism appears to be distinct when compared to the current information available on anterograde receptors, which are thought to traffic in a constitutive manner (125). However, no studies have specifically addressed the influence of overexpressed soluble cargo on distribution of anterograde cargo receptors or on packaging rates into COPII vesicles. There is evidence suggesting that increases in secretory cargo load in general stimulate the anterograde transport pathway (126, 127).

Interactions between the KDEL receptor and specific KDEL-bearing ligands have been extensively studied through molecular genetic and biochemical approaches (114, 118, 128, 129). Reconstitution of the purified human KDEL receptor into proteoliposomes demonstrated sufficiency for binding to KDEL ligands and permitted measurement of binding specificity and affinity. The receptor displayed dissociation constants in the range of 80 nM with specificity for KDEL and KDEL-like sequences. Receptor-binding studies also showed a strong dependency on pH with optimal binding affinity between pH 5.0 to 5.5 (129). Indeed, a pH gradient of increasing acidity from the ER through Golgi compartments has been proposed to regulate KDEL-receptor binding such that KDEL-bearing proteins bind receptors in Golgi compartments and KDEL sequences are released in the ER (128). It has been known for some time that different species use variations on the C-terminal KDEL motif to retrieve ER-resident proteins (114). Additional studies in animal cells have shown that distinct isoforms of the KDEL receptor within a species display preferences for different variants of the KDEL motif (130). Most human cells are now known to express three homologous KDEL receptors (known as ERD21, ERD22, and ERD23) with distinct but overlapping substrate specificities and affinities. Adding to the complexity of sorting in the early secretory pathway, recent studies indicate that over 50 variations of the KDEL motif are recognized and retrieved by these three human KDEL receptors (131). Molecular models depicting how these KDEL and KDEL-related sequences bind receptors and how binding is regulated to properly distribute proteins in the early secretory pathway are not well developed.

Rer1p

The *RER1* gene was initially identified in yeast through genetic screens for mutants that failed to retain the ER membrane protein Sec12p (115, 132). Rer1p is a highly conserved tetraspanning membrane protein with cytoplasmically exposed N and C termini (**Figure 4**). Under steady-state conditions, Rer1p localizes primarily to Golgi membranes, although the protein rapidly cycles between Golgi compartments and the ER through a COPI-dependent dilysine motif in its C-terminal tail (133, 134). In addition to Sec12p retention, subsequent studies have shown that Rer1p is required for ER localization of numerous transmembrane proteins, including Sec63p, Sec71p, and Mns1p (135, 136). It is also known that overexpression of specific Rer1p-dependent cargo saturates this retrograde receptor (134). However, to date, no studies have addressed whether retrograde movement of Rer1p is induced by ligand binding as observed for the KDEL receptor. Constitutive cycling of Rer1p could be adequate to retrieve escaped ER membrane proteins. Analysis of various genome databases indicates that most eukaryotic species contain a single Rer1p homolog of high sequence identity. Studies of human Rer1 demonstrate conserved function and subcellular distribution in early secretory compartments (137, 138). Importantly, human Rer1 has recently been implicated in regulation of γ-secretase activity. Experimental results indicate that human Rer1 binds directly to unassembled subunits of the γ-secretase complex, retrieves these subunits to ER, and thereby modulates cellular γ-secretase activity (138, 139). In addition to a role in retrieval of fully folded proteins, these findings suggest Rer1 may also assist in ER quality control (138).

Rer1p-dependent retrieval depends on direct binding to sorting motifs in the transmembrane domain of specific cargo (134). This binding property is distinct from all other cargo receptors described above, which appear to recognize lumenal sorting signals. What is the molecular basis for recognition of transmembrane sorting motifs? Mutational analysis of the transmembrane domains of Sec12p and Sec71p demonstrated that both depend on pairs of polar amino acid residues that must flank a hydrophobic cluster of 6–7 residues. However, the topologies of the Sec12p and Sec71p sorting motifs are opposite, suggesting different modes of binding to Rer1p. Indeed, mutational analysis of Rer1p indicated that a specific mutation (Y52L) in the fourth transmembrane segment interfered with Sec12p recognition but not Sec71p, supporting a model in which Rer1p binds to cargo through at least two distinct mechanisms (117). Studies in human cell lines indicate that the Rer1-dependent transmembrane sorting signals are conserved (138). Because Rer1p binds cargo through transmembrane domain interactions, less is known on how these associations are regulated to promote binding in post-ER compartments and then release retrieved cargo upon delivery to the ER. Although lumenal environments or coat binding could induce Rer1 conformational changes, it is also interesting to speculate that distinct lipid compositions between organelles could regulate the associations between Rer1p and bound cargo.

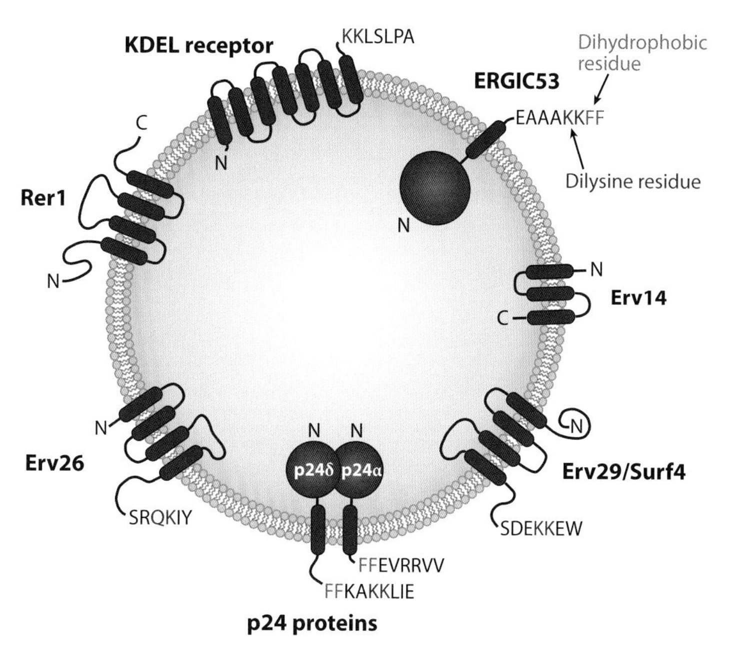

Figure 4

Model showing structures and membrane topologies of sorting receptors in the early secretory pathway. The N terminus and C terminus are indicated. ERGIC-53, Erv14p, Erv29p, Erv26p, and the p24 proteins act in anterograde transport of secretory proteins from the endoplasmic reticulum (ER). The KDEL receptor and Rer1p act in retrograde transport of proteins from Golgi compartments to the ER. However, these sorting receptors cycle between ER and Golgi compartments in vesicle intermediates formed by both coat protein complexes I and II (COPI and COPII). The C-terminal amino acid sequences from human proteins are shown for cargo receptors where sorting signals have been identified. Dilysine residues are implicated in COPI-dependent retrograde transport, whereas dihydrophobic residues are implicated in COPII-dependent export from the ER. Surf4 is the mammalian homolog of yeast Erv29p.

ADDITIONAL ENDOPLASMIC RETICULUM SORTING FACTORS

There are quite likely additional cargo sorting receptors in the early secretory pathway, beyond those listed in **Table 1**, that remain to be identified and/or characterized. Indeed, several candidate membrane proteins reported in the literature may function as cargo receptors in ER export. However, the analysis of candidates can be complicated because depletion of cargo-specific chaperones, which function in protein folding in the ER, produces phenotypic consequences very similar to depletion of export receptors. For example, the multispanning membrane protein Shr3p is required for ER export of specific amino acid permeases (e.g., Gap1p), whereas most other secretory proteins are transported at normal rates in an *shr3Δ* strain. But further experimentation revealed that Shr3p localizes to the ER and is not efficiently packaged into COPII vesicles and also that Shr3p depletion causes retention of misfolded Gap1p in the ER (140). Moreover, the misfolded amino acid permeases that accumulate in *shr3Δ* strains are

Table 1 Protein sorting receptors in the early secretory pathway[a]

Cargo receptor	Species	Cargo proteins transported	Signals recognized on cargo	Documented localization	Oligomerization state	Alternative names (species)	Regulation
Type I transmembrane, L-type lectins							
ERGIC-53	Mammals	Coagulation factors V and VIII, cathepsins C and Z, α1-antitrypsin	High-mannose *N*-glycans (mono- and deglucosylated) plus peptide β-hairpin	ERGIC, Golgi, COPI and COPII vesicles	Homodimers and hexamers	Emp46p/ Emp47p (sc)	pH (neutral pH optimum), Ca^{2+}, oligomerization, MCFD2
VIP36	Mammals	α-Amylase	Deglucosylated high-mannose *N*-glycans	ERGIC, Golgi	Unknown	–	pH (slightly acidic pH optimum)
VIPL	Mammals	Unidentified secreted glycoproteins	Deglucosylated high-mannose *N*-glycans	ER	Unknown	–	pH (neutral pH optimum)
p24 proteins							
Erv25p	Yeast	Gas1p (sc), GPI-anchored proteins (hs)	Unknown	ER, ERGIC, Golgi, COPI and COPII vesicles	Hetero-oligomer with Emp24p	p24δ (hs)	Unknown
Emp24p	Yeast	Gas1p (sc), GPI-anchored proteins (hs)	Unknown	ER, ERGIC, Golgi, COPI and COPII vesicles	Hetero-oligomer with Erv25p	p24β (hs)	Unknown
Erp1p	Yeast	Gas1p (sc), GPI-anchored proteins (hs)	Unknown	ERGIC, Golgi	Hetero-oligomer	p24α (hs)	Unknown
Erp2p	Yeast	Gas1p (sc), GPI-anchored proteins (hs)	Unknown	ERGIC, Golgi	Hetero-oligomer	p24γ (hs)	Unknown
Multispanning transmembrane cargo receptors							
Erv29p	Yeast	Soluble lumenal proteins, gpα-factor, CPY, proteinase A	I-L-V (α-factor)/α-helix	ER, Golgi, COPII vesicles	Oligomer	Surf4 (hs)	Unknown
Erv26p	Yeast	Type II transmembrane proteins, ALP, Ktr3p	Lumenal determinant	ER, Golgi, COPII vesicles	Oligomer	Svp26 (sc)	Unknown

Erv14p	Yeast	Transmembrane proteins, Axl2p, Sma2p	Lumenal determinant	ER, Golgi, COPII vesicles	Oligomer	Cornichon (dm)	Unknown
Erd2p	Yeast	Lumenal proteins with C-terminal HDEL	HDEL (yeast), KDEL (mammals)	ER, ERGIC, Golgi, COPI and COPII vesicles	Oligomer	KDEL receptor (hs)	pH, ligand-dependent oligomerization, phosphorylation
Rer1p	Yeast	Transmembrane proteins, Sec12p, Sec63p, Sec72p, Mns1p	Polar residues in transmembrane domain	ERGIC, Golgi, COPII vesicles	Unknown	Rer1 (hs)	Oligomerization of cargo

[a]Abbreviations: HDEL, a C-terminal amino acid sequence; dm, *Drosophila melanogaster*; ER, endoplasmic reticulum; ERGIC, ER-Golgi intermediate compartment; GPI, glycophosphatidylinositol; I-L-V, an amino acid sequence motif; hs, *Homo sapiens*; KDEL, an amino acid sequence; MCFD2, multiple coagulation factor deficiency 2; sc, *Saccharomyces cerevisiae*.

subjected to the ER-associated degradation pathway (141). Therefore, Shr3p is thought to function as an ER chaperone that assists in folding of specific membrane cargo. These properties are notably distinct from ER sorting receptors because receptors actively cycle between early secretory compartments, and the consequences of receptor depletion produce an accumulation of folded cargo in the ER that is slowly exported at a bulk-flow rate and apparently not subjected to ER-associated degradation (88, 102). With these distinctions in mind, other candidate membrane proteins that display selective ER export defects have been described in yeast. Gsf2p localizes to the ER and is required for export of hexose transporters from the ER (142). Chs7p is required for transport of chitin synthase III from the ER (143), whereas Pho86p acts in ER export of the Pho84p phosphate transporter (144). However, deletion of these export factors in *gsf2Δ*, *chs7Δ*, or *pho86Δ* cells causes the corresponding transmembrane cargo to aggregate in the ER, as observed for Gap1p in the *shr3Δ* strain (140). Further studies are required to fully understand the function of these ER export factors, although current findings suggest a role in folding and/or delivery of nascent secretory cargo to ER export sites instead of cargo receptor activity.

The ER membrane proteins BAP31 and BAP29 were initially identified as a part of the B cell receptor complex (145) and are reported to promote ER export of secretory proteins, including those in the major histocompatibility complex class I and cellubrevin (146, 147). Other studies suggest a role for BAP31 in ER retention of unassembled IgD molecules (148) and in degradation of mutant CFTR proteins when associated with the translocation apparatus (149). The BAP31 and BAP29 proteins may perform multiple roles to maintain ER homeostasis, although further work seems necessary to confirm their direct function as ER export receptors. A recent report identified TANGO1, an integral membrane protein that localizes to transitional ER sites, in a screen for components required for intracellular transport and Golgi organization. Depletion of TANGO1 also produces a selective secretory defect in transport of collagen from the ER. Interestingly, TANGO1 interacts with subunits of COPII but apparently is not packaged into COPII vesicles. Current proposals suggest that TANGO1 somehow guides a subset of secretory cargo into forming COPII vesicles (150). Additional studies are needed to fully understand the molecular mechanism by which TANGO1 operates in ER export.

THE ROLE OF CARGO RECEPTORS IN ENDOPLASMIC RETICULUM QUALITY CONTROL

Although cargo sorting receptors that cycle between ER and Golgi compartments do not seem to be required for folding of specific secretory cargo, there are obvious connections between sorting receptors and the ER quality control machinery. For example, *erv14Δ*, *erv25Δ*, *erv26Δ*, and *erv29Δ* strains all display an activated, unfolded protein response pathway (85, 102, 151), indicating quality control issues in the ER. In addition, the turnover rate of CPY*, a terminally misfolded ERAD substrate, is reduced in strains lacking Erv29p. Wild-type CPY normally depends on Erv29p for efficient export from the ER, and this stabilizing effect appears specific for CPY* as other ERAD substrates are not influenced by *erv29Δ* (152). Interestingly, a fraction of CPY* appears to bind to Erv29p and traffic to the Golgi complex in wild-type strains. These observations have led to the proposal that ER exit and retrieval from post-ER compartments makes CPY* more susceptible to ERAD (153). A similar relationship has been reported for the cargo receptor Erv26p and its misfolded cargo ligand ALP* (107). However, in this instance, binding of ALP* to Erv26p is greatly diminished compared to wild-type ALP. As an alternative explanation, weaker binding of misfolded cargo to the receptor could act to transfer cargo away from ER folding chaperones and into a position that is more susceptible to ERAD. The molecular understanding of the relationships between cargo receptors and ERAD are clearly at an early stage.

Several lines of evidence also indicate that cargo receptors assist quality control through efficient recognition of fully folded and assembled secretory cargo. In the case of ERGIC-53 cargo recognition, the ER export motif in cathepsin Z consists of both carbohydrate and a folded peptide determinant. An 18-amino acid β-hairpin loop sequence next to the N-linked glycosylation site at residue 184 of cathepsin Z was required for efficient binding of this ligand to ERGIC-53. Only fully folded cathepsin Z would present this conformation-based motif, which is proposed to act in concert with the ER quality control machinery to insure export of folded cargo (50). Similarly, a 25-amino acid α-helical motif in the proregion of gpαf is necessary and sufficient for Erv29p-dependent recognition and export from the ER (90). For Erv26p cargo recognition, a mutation in ALP that prevents dimerization results in inefficient recognition of unassembled monomeric ALP* by Erv26p (107). Therefore, in each case examined, cargo receptors display a clear preference for fully folded and assembled cargo proteins compared to their misfolded counterparts. Perhaps, reduced binding of misfolded cargo by their receptors prevents a needless tug-of-war between the folding and export machinery. Indeed, fusion of a cytoplasmically exposed COPII export signal on a misfolded lumenal cargo protein resulted in ER export and suggests that a potent export signal can override ER retention mechanisms (153). Although these collective observations indicate coordination between the ER folding machinery and cargo sorting receptors, these relationships are not clearly understood and represent an exciting area for further study.

CONCLUSIONS

With the great diversity in secretory cargo, there is a corresponding diversity in mechanisms by which cargo can be coupled to the COPI and COPII export machinery. However, some general themes seem important to consider. First, in cases where binding assays have been established between receptor-ligand pairs, pH and possibly Ca^{2+} gradients appear to influence binding in a manner that would favor net transfer of cargo ligands to the Golgi complex for anterograde receptors and to the ER for retrograde receptors (**Figure 3**). Second, all of the characterized sorting receptors are oligomeric protein complexes that may allow receptors to undergo significant conformational shifts, a common biological mechanism to regulate binding affinity. Indeed, changes in lumenal environments could trigger such conformational shifts. The very modest differences in pH between compartments of the early secretory pathway may only have modest influences on amino acid residues at specific ligand-binding sites, although slight pH differences could be amplified by inducing conformational changes. Third, the sorting signals in anterograde cargo seem to comprise folded epitopes and not the simple liner peptide sorting signals as observed for the signals recognized directly by cytoplasmic coat complexes. Although advances in the past decade have provided insight on sorting receptors and some themes are beginning to emerge, many exciting and challenging questions clearly lie ahead.

SUMMARY POINTS

1. Coat protein complexes play a central role in sorting proteins in the early secretory pathway, but diversity in cargo molecules necessitates sorting receptors for efficient linkage of diverse cargo to coat complexes.
2. Transmembrane sorting receptors cycle between early compartments of the secretory pathway and physically link specific cargo to COPII for export from the ER and to COPI for retrograde transport from post-ER compartments.
3. Sorting receptors recognize and bind to specific carbohydrate and/or protein motifs, primarily through lumenal domain interactions. Anterograde sorting receptors appear to recognize folded protein motifs and not short linear peptide signals.
4. Differences in lumenal pH and/or Ca^{2+} concentration between ER and Golgi compartments are thought to regulate ligand binding and favor net transport from a donor membrane compartment to an acceptor membrane compartment.
5. The oligomeric nature of sorting receptors is proposed to allow for conformational changes necessary to bind and release cargo ligands. Conformational changes may be induced by changes in the lumenal environment between compartments.

FUTURE ISSUES

1. How widespread is receptor-mediated export of secretory cargo from the ER relative to bulk-flow transport?
2. What are the minimal recognition motifs that are necessary and sufficient for receptor-mediated export?
3. Do anterograde sorting receptors cycle constitutively or does cargo binding stimulate packaging of receptors into carrier vesicles?
4. Are pH and/or Ca^{2+} gradients sufficient to regulate receptor-ligand interactions or are there other factors that regulate cargo binding?

5. How do sorting receptors interface with ER and post-ER quality control machinery?
6. What are the molecular mechanisms by which single and multispanning sorting receptors recognize and bind cargo ligands. High-resolution structural studies on membrane receptor proteins in complex with bound cargo represent major challenges.

DISCLOSURE STATEMENT

The authors are not aware of any affiliations, memberships, funding, or financial holdings that might be perceived as affecting the objectivity of this review.

LITERATURE CITED

1. Vembar SS, Brodsky JL. 2008. One step at a time: endoplasmic reticulum-associated degradation. *Nat. Rev. Mol. Cell Biol.* 9:944–57
2. Lee MC, Miller EA, Goldberg J, Orci L, Schekman R. 2004. Bi-directional protein transport between the ER and Golgi. *Annu. Rev. Cell Dev. Biol.* 20:87–123
3. Bonifacino JS, Glick BS. 2004. The mechanisms of vesicle budding and fusion. *Cell* 116:153–66
4. Sciaky N, Presley J, Smith C, Zaal KJ, Cole N, et al. 1997. Golgi tubule traffic and the effects of brefeldin A visualized in living cells. *J. Cell Biol.* 139:1137–55
5. Ward TH, Polishchuk RS, Caplan S, Hirschberg K, Lippincott-Schwartz J. 2001. Maintenance of Golgi structure and function depends on the integrity of ER export. *J. Cell Biol.* 155:557–70
6. Losev E, Reinke CA, Jellen J, Strongin DE, Bevis BJ, Glick BS. 2006. Golgi maturation visualized in living yeast. *Nature* 441:1002–6
7. Orci L, Ravazzola M, Meda P, Holcomb C, Moore HP, et al. 1991. Mammalian Sec23p homologue is restricted to the endoplasmic reticulum transitional cytoplasm. *Proc. Natl. Acad. Sci. USA* 88:8611–15
8. Bannykh SI, Rowe T, Balch WE. 1996. The organization of endoplasmic reticulum export complexes. *J. Cell Biol.* 135:19–35
9. Rossanese OW, Soderholm J, Bevis BJ, Sears IB, O'Connor J, et al. 1999. Golgi structure correlates with transitional endoplasmic reticulum organization in *Pichia pastoris* and *Saccharomyces cerevisiae*. *J. Cell Biol.* 145:69–81
10. Lee MC, Orci L, Hamamoto S, Futai E, Ravazzola M, Schekman R. 2005. Sar1p N-terminal helix initiates membrane curvature and completes the fission of a COPII vesicle. *Cell* 122:605–17
11. Stagg SM, Gürkan C, Fowler DM, LaPointe P, Foss TR, et al. 2006. Structure of the Sec13/31 COPII coat cage. *Nature* 439:234–38
12. Matsuoka K, Orci L, Amherdt M, Bednarek SY, Hamamoto S, et al. 1998. COPII-coated vesicle formation reconstituted with purified coat proteins and chemically defined liposomes. *Cell* 93:263–75
13. Salama NR, Yeung T, Schekman RW. 1993. The Sec13p complex and reconstitution of vesicle budding from the ER with purified cytosolic proteins. *EMBO J.* 12:4073–82
14. Barlowe C, Orci L, Yeung T, Hosobuchi M, Hamamoto S, et al. 1994. COPII: a membrane coat formed by Sec proteins that drive vesicle budding from the endoplasmic reticulum. *Cell* 77:895–907
15. Kuehn MJ, Herrmann JM, Schekman R. 1998. COPII-cargo interactions direct protein sorting into ER-derived transport vesicles. *Nature* 391:187–90
16. Aridor M, Weissman J, Bannykh S, Nuoffer C, Balch WE. 1998. Cargo selection by the COPII budding machinery during export from the ER. *J. Cell Biol.* 141:61–70
17. Mossessova E, Bickford LC, Goldberg J. 2003. SNARE selectivity of the COPII coat. *Cell* 114:483–95
18. Miller EA, Beilharz TH, Malkus PN, Lee MC, Hamamoto S, et al. 2003. Multiple cargo binding sites on the COPII subunit Sec24p ensure capture of diverse membrane proteins into transport vesicles. *Cell* 114:497–509

19. Miller E, Antonny B, Hamamoto S, Schekman R. 2002. Cargo selection into COPII vesicles is driven by the Sec24p subunit. *EMBO J.* 21:6105–13
20. Wendeler MW, Paccaud JP, Hauri HP. 2007. Role of Sec24 isoforms in selective export of membrane proteins from the endoplasmic reticulum. *EMBO Rep.* 8:258–64
21. Kanapin A, Batalov S, Davis MJ, Gough J, Grimmond S, et al. 2003. Mouse proteome analysis. *Genome Res.* 13:1335–44
22. Poisson G, Chauve C, Chen X, Bergeron A. 2007. FragAnchor: a large-scale predictor of glycosylphosphatidylinositol anchors in eukaryote protein sequences by qualitative scoring. *Genomics Proteomics Bioinformatics* 5:121–30
23. Reiterer V, Maier S, Sitte HH, Kriz A, Ruegg MA, et al. 2008. Sec24- and ARFGAP1-dependent trafficking of GABA transporter-1 is a prerequisite for correct axonal targeting. *J. Neurosci.* 28:12453–64
24. Nickel W, Rabouille C. 2009. Mechanisms of regulated unconventional protein secretion. *Nat. Rev. Mol. Cell Biol.* 10:148–55
25. Fromme JC, Schekman R. 2005. COPII-coated vesicles: flexible enough for large cargo? *Curr. Opin. Cell Biol.* 17:345–52
26. Quinn P, Griffiths G, Warren G. 1984. Density of newly synthesized plasma membrane proteins in intracellular membranes II. Biochemical studies. *J. Cell Biol.* 98:2142–47
27. Balch WE, McCaffery JM, Plutner H, Farquhar MG. 1994. Vesicular stomatitis virus glycoprotein is sorted and concentrated during export from the endoplasmic reticulum. *Cell* 76:841–52
28. Rexach MF, Latterich M, Schekman RW. 1994. Characteristics of endoplasmic reticulum-derived transport vesicles. *J. Cell Biol.* 126:1133–48
29. Nishimura N, Balch WE. 1997. A di-acidic signal required for selective export from the endoplasmic reticulum. *Science* 277:556–58
30. Kappeler F, Klopfenstein DRC, Foguet M, Paccaud JP, Hauri HP. 1997. The recycling of ERGIC-53 in the early secretory pathway. ERGIC-53 carries a cytosolic endoplasmic reticulum-exit determinant interacting with COPII. *J. Biol. Chem.* 272:31801–808
31. Barlowe C. 2003. Signals for COPII-dependent export from the ER: What's the ticket out? *Trends Cell Biol.* 13:295–300
32. Mizuno M, Singer SJ. 1993. A soluble secretory protein is first concentrated in the endoplasmic reticulum before transfer to the Golgi apparatus. *Proc. Natl. Acad. Sci. USA* 90:5732–36
33. Martínez-Menárguez JA, Geuze HJ, Slot JW, Klumperman J. 1999. Vesicular tubular clusters between the ER and Golgi mediate concentration of soluble secretory proteins by exclusion from COPI-coated vesicles. *Cell* 98:81–90
34. Malkus P, Jiang F, Schekman R. 2002. Concentrative sorting of secretory cargo proteins into COPII-coated vesicles. *J. Cell Biol.* 159:915–21
35. Schweizer A, Fransen JA, Bächi T, Ginsel L, Hauri HP. 1988. Identification, by a monoclonal antibody, of a 53-kD protein associated with a tubulo-vesicular compartment at the *cis*-side of the Golgi apparatus. *J. Cell Biol.* 107:1643–53
36. Wada I, Rindress D, Cameron PH, Ou WJ, Doherty JJ, et al. 1991. SSR alpha and associated calnexin are major calcium binding proteins of the endoplasmic reticulum membrane. *J. Biol. Chem.* 266:19599–610
37. Singer-Krüger B, Frank R, Crausaz F, Riezman H. 1993. Partial purification and characterization of early and late endosomes from yeast. Identification of four novel proteins. *J. Biol. Chem.* 268:14376–86
38. Stamnes MA, Craighead MW, Hoe MH, Lampen N, Geromanos S, et al. 1995. An integral membrane component of coatomer-coated transport vesicles defines a family of proteins involved in budding. *Proc. Natl. Acad. Sci. USA* 92:8011–15
39. Schimmöller F, Singer-Krüger B, Schröder S, Krüger U, Barlowe C, Riezman H. 1995. The absence of Emp24p, a component of ER-derived COPII-coated vesicles, causes a defect in transport of selected proteins to the Golgi. *EMBO J.* 14:1329–39
40. Belden WJ, Barlowe C. 1996. Erv25p, a component of COPII-coated vesicles, forms a complex with Emp24p that is required for efficient endoplasmic reticulum to Golgi transport. *J. Biol. Chem.* 271:26939–46
41. Schweizer A, Ericsson M, Bächi T, Griffiths G, Hauri HP. 1993. Characterization of a novel 63 kDa membrane protein. Implications for the organization of the ER-to-Golgi pathway. *J. Cell Sci.* 104:671–83

42. Schindler R, Itin C, Zerial M, Lottspeich F, Hauri HP. 1993. ERGIC-53, a membrane protein of the ER-Golgi intermediate compartment, carries an ER retention motif. *Eur. J. Cell Biol.* 61:1–9
43. Itin C, Roche AC, Monsigny M, Hauri HP. 1996. ERGIC-53 is a functional mannose-selective and calcium-dependent human homologue of leguminous lectins. *Mol. Biol. Cell* 7:483–93
44. Velloso LM, Svensson K, Schneider G, Pettersson RF, Lindqvist Y. 2002. Crystal structure of the carbohydrate recognition domain of p58/ERGIC-53, a protein involved in glycoprotein export from the endoplasmic reticulum. *J. Biol. Chem.* 277:15979–84
45. Nufer O, Mitrovic S, Hauri HP. 2003. Profile-based data base scanning for animal L-type lectins and characterization of VIPL, a novel VIP36-like endoplasmic reticulum protein. *J. Biol. Chem.* 278:15886–96
46. Kamiya Y, Kamiya D, Yamamoto K, Nyfeler B, Hauri HP, Kato K. 2008. Molecular basis of sugar recognition by the human L-type lectins ERGIC-53, VIPL, and VIP36. *J. Biol. Chem.* 283:1857–61
47. Vollenweider F, Kappeler F, Itin C, Hauri HP. 1998. Mistargeting of the lectin ERGIC-53 to the endoplasmic reticulum of HeLa cells impairs the secretion of a lysosomal enzyme. *J. Cell Biol.* 142:377–89
48. Appenzeller C, Andersson H, Kappeler F, Hauri HP. 1999. The lectin ERGIC-53 is a cargo transport receptor for glycoproteins. *Nat. Cell Biol.* 1:330–34
49. Appenzeller-Herzog C, Roche AC, Nufer O, Hauri HP. 2004. pH-induced conversion of the transport lectin ERGIC-53 triggers glycoprotein release. *J. Biol. Chem.* 279:12943–50
50. Appenzeller-Herzog C, Nyfeler B, Burkhard P, Santamaria I, Lopez-Otin C, Hauri HP. 2005. Carbohydrate- and conformation-dependent cargo capture for ER-exit. *Mol. Biol. Cell* 16:1258–67
51. Nichols WC, Seligsohn U, Zivelin A, Terry VH, Hertel CE, et al. 1998. Mutations in the ER-Golgi intermediate compartment protein ERGIC-53 cause combined deficiency of coagulation factors V and VIII. *Cell* 93:61–70
52. Zhang B. 2009. Recent developments in the understanding of the combined deficiency of FV and FVIII. *Br. J. Haematol.* 145:15–23
53. Moussalli M, Pipe SW, Hauri HP, Nichols WC, Ginsburg D, Kaufman RJ. 1999. Mannose-dependent endoplasmic reticulum (ER)-Golgi intermediate compartment-53-mediated ER to Golgi trafficking of coagulation factors V and VIII. *J. Biol. Chem.* 274:32539–42
54. Zhang B, Kaufman RJ, Ginsburg D. 2005. LMAN1 and MCFD2 form a cargo receptor complex and interact with coagulation factor VIII in the early secretory pathway. *J. Biol. Chem.* 280:25881–86
55. Zhang B, Cunningham MA, Nichols WC, Bernat JA, Seligsohn U, et al. 2003. Bleeding due to disruption of a cargo-specific ER-to-Golgi transport complex. *Nat. Genet.* 34:220–25
56. Seligsohn U, Ginsburg D. 2006. Deciphering the mystery of combined factor V and factor VIII deficiency. *J. Thromb. Haemost.* 4:927–31
57. Nyfeler B, Zhang B, Ginsburg D, Kaufman RJ, Hauri HP. 2006. Cargo selectivity of the ERGIC-53/MCFD2 transport receptor complex. *Traffic* 7:1473–81
58. Guy JE, Wigren E, Svärd M, Härd T, Lindqvist Y. 2008. New insights into multiple coagulation factor deficiency from the solution structure of human MCFD2. *J. Mol. Biol.* 381:941–55
59. Zhang B, McGee B, Yamaoka JS, Guglielmone H, Downes KA, et al. 2006. Combined deficiency of factor V and factor VIII is due to mutations in either LMAN1 or MCFD2. *Blood* 107:1903–7
60. Kawasaki N, Ichikawa Y, Matsuo I, Totani K, Matsumoto N, et al. 2008. The sugar-binding ability of ERGIC-53 is enhanced by its interaction with MCFD2. *Blood* 111:1972–79
61. Hauri HP, Kappeler F, Andersson H, Appenzeller C. 2000. ERGIC-53 and traffic in the secretory pathway. *J. Cell Sci.* 113:587–96
62. Osibow K, Malli R, Kostner GM, Graier WF. 2006. A new type of non-Ca^{2+}-buffering Apo(a)-based fluorescent indicator for intraluminal Ca^{2+} in the endoplasmic reticulum. *J Biol. Chem.* 281:5017–25
63. Pinton P, Pozzan T, Rizzuto R. 1998. The Golgi apparatus is an inositol 1,4,5-trisphosphate-sensitive Ca^{2+} store, with functional properties distinct from those of the endoplasmic reticulum. *EMBO J.* 17:5298–308
64. Pezzati R, Bossi M, Podini P, Meldolesi J, Grohovaz F. 1997. High-resolution calcium mapping of the endoplasmic reticulum-Golgi-exocytic membrane system. Electron energy loss imaging analysis of quick frozen-freeze dried PC12 cells. *Mol. Biol. Cell* 8:1501–12

65. Nyfeler B, Reiterer V, Wendeler MW, Stefan E, Zhang B, et al. 2008. Identification of ERGIC-53 as an intracellular transport receptor of alpha1-antitrypsin. *J. Cell Biol.* 180:705–12
66. Yerushalmi N, Keppler-Hafkemeyer A, Vasmatzis G, Liu XF, Olsson P, et al. 2001. ERGL, a novel gene related to ERGIC-53 that is highly expressed in normal and neoplastic prostate and several other tissues. *Gene* 265:55–60
67. Füllekrug J, Scheiffele P, Simons K. 1999. VIP36 localisation to the early secretory pathway. *J. Cell Sci.* 112:2813–21
68. Hara-Kuge S, Ohkura T, Ideo H, Shimada O, Atsumi S, Yamashita K. 2002. Involvement of VIP36 in intracellular transport and secretion of glycoproteins in polarized Madin-Darby canine kidney (MDCK) cells. *J. Biol. Chem.* 277:16332–39
69. Shimada O, Hara-Kuge S, Yamashita K, Tosaka-Shimada H, Yanchao L, et al. 2003. Localization of VIP36 in the post-Golgi secretory pathway also of rat parotid acinar cells. *J. Histochem. Cytochem.* 51:1057–63
70. Neve EP, Svensson K, Fuxe J, Pettersson RF. 2003. VIPL, a VIP36-like membrane protein with a putative function in the export of glycoproteins from the endoplasmic reticulum. *Exp. Cell Res.* 288:70–83
71. Helenius A, Aebi M. 2004. Roles of N-linked glycans in the endoplasmic reticulum. *Annu. Rev. Biochem.* 73:1019–49
72. Paroutis P, Touret N, Grinstein S. 2004. The pH of the secretory pathway: measurement, determinants, and regulation. *Physiology* 19:207–15
73. Strating JR, Martens GJ. 2009. The p24 family and selective transport processes at the ER-Golgi interface. *Biol. Cell.* 101:495–509
74. Dominguez M, Dejgaard K, Füllekrug J, Dahan S, Fazel A, et al. 1998. gp25L/emp24/p24 protein family members of the *cis*-Golgi network bind both COP I and II coatomer. *J. Cell Biol.* 140:751–65
75. Marzioch M, Henthorn DC, Herrmann JM, Wilson R, Thomas DY, et al. 1999. Erp1p and Erp2p, partners for Emp24p and Erv25p in a yeast p24 complex. *Mol. Biol. Cell* 10:1923–38
76. Jenne N, Frey K, Brugger B, Wieland FT. 2002. Oligomeric state and stoichiometry of p24 proteins in the early secretory pathway. *J. Biol. Chem.* 277:46504–11
77. Springer S, Chen E, Duden R, Marzioch M, Rowley A, et al. 2000. The p24 proteins are not essential for vesicular transport in *Saccharomyces cerevisiae*. *Proc. Natl. Acad. Sci. USA* 97:4034–39
78. Muniz M, Nuoffer C, Hauri HP, Riezman H. 2000. The Emp24 complex recruits a specific cargo molecule into endoplasmic reticulum–derived vesicles. *J. Cell Biol.* 148:925–30
79. Takida S, Maeda Y, Kinoshita T. 2008. Mammalian GPI-anchored proteins require p24 proteins for their efficient transport from the ER to the plasma membrane. *Biochem. J.* 409:555–62
80. Rojo M, Emery G, Marjomäki V, McDowall AW, Parton RG, Gruenberg J. 2000. The transmembrane protein p23 contributes to the organization of the Golgi apparatus. *J. Cell Sci.* 113:1043–57
81. Mitrovic S, Ben-Tekaya H, Koegler E, Gruenberg J, Hauri HP. 2008. The cargo receptors Surf4, endoplasmic reticulum-Golgi intermediate compartment (ERGIC)-53, and p25 are required to maintain the architecture of ERGIC and Golgi. *Mol. Biol. Cell* 19:1976–90
82. Aguilera-Romero A, Kaminska J, Spang A, Riezman H, Muñiz M. 2008. The yeast p24 complex is required for the formation of COPI retrograde transport vesicles from the Golgi apparatus. *J. Cell Biol.* 180:713–20
83. Elrod-Erickson MJ, Kaiser CA. 1996. Genes that control the fidelity of endoplasmic reticulum to Golgi transport identified as suppressors of vesicle budding mutations. *Mol. Biol. Cell* 7:1043–58
84. Wen C, Greenwald I. 1999. p24 proteins and quality control of LIN-12 and GLP-1 trafficking in *Caenorhabditis elegans*. *J. Cell Biol.* 145:1165–75
85. Belden WJ, Barlowe C. 2001. Deletion of p24 genes activates the unfolded protein response. *Mol. Biol. Cell* 12:957–969
86. Denzel A, Otto F, Girod A, Pepperkok R, Watson R, et al. 2000. The p24 family member p23 is required for early embryonic development. *Curr. Biol.* 10:55–58
87. Otte S, Belden WJ, Heidtman M, Liu J, Jensen ON, Barlowe C. 2001. Erv41p and Erv46p: new components of COPII vesicles involved in transport between the ER and Golgi complex. *J. Cell Biol.* 152:503–18
88. Belden WJ, Barlowe C. 2001. Role of Erv29p in collecting soluble secretory proteins into ER-derived transport vesicles. *Science* 294:1528–31

89. Foley DA, Sharpe HJ, Otte S. 2007. Membrane topology of the endoplasmic reticulum to Golgi transport factor Erv29p. *Mol. Membr. Biol.* 24:259–68
90. Otte S, Barlowe C. 2004. Sorting signals can direct receptor-mediated export of soluble proteins into COPII vesicles. *Nat. Cell Biol.* 6:1189–94
91. Julius D, Schekman R, Thorner J. 1984. Glycosylation and processing of prepro-alpha-factor through the yeast secretory pathway. *Cell* 36:309–18
92. Roth S, Neuman-Silberberg FS, Barcelo G, Schüpbach T. 1995. *cornichon* and the EGF receptor signaling process are necessary for both anterior-posterior and dorsal-ventral pattern formation in *Drosophila*. *Cell* 81:967–78
93. Powers J, Barlowe C. 1998. Transport of Axl2p depends on Erv14p, an ER-vesicle protein related to the *Drosophila cornichon* gene product. *J. Cell Biol.* 142:1209–22
94. Herpers B, Rabouille C. 2004. mRNA localization and ER-based protein sorting mechanisms dictate the use of transitional endoplasmic reticulum-Golgi units involved in Gurken transport in *Drosophila* oocytes. *Mol. Biol. Cell* 15:5306–17
95. Bökel C, Dass S, Wilsch-Bräuninger M, Roth S. 2006. *Drosophila* Cornichon acts as cargo receptor for ER export of the TGFα-like growth factor Gurken. *Development* 133:459–70
96. Powers J, Barlowe C. 2002. Erv14p directs a transmembrane secretory protein into COPII-coated transport vesicles. *Mol. Biol. Cell* 13:880–91
97. Castro CP, Piscopo D, Nakagawa T, Derynck R. 2007. Cornichon regulates transport and secretion of TGFα-related proteins in metazoan cells. *J. Cell Sci.* 120:2454–66
98. Schwenk J, Harmel N, Zolles G, Bildl W, Kulik A, et al. 2009. Functional proteomics identify cornichon proteins as auxiliary subunits of AMPA receptors. *Science* 323:1313–19
99. Gillingham AK, Tong AH, Boone C, Munro S. 2004. The GTPase Arf1p and the ER to Golgi cargo receptor Erv14p cooperate to recruit the golgin Rud3p to the *cis*-Golgi. *J. Cell Biol.* 167:281–92
100. Nakanishi H, Suda Y, Neiman AM. 2007. Erv14 family cargo receptors are necessary for ER exit during sporulation in *Saccharomyces cerevisiae*. *J. Cell Sci.* 120:908–16
101. Inadome H, Noda Y, Adachi H, Yoda K. 2005. Immunoisolaton of the yeast Golgi subcompartments and characterization of a novel membrane protein, Svp26, discovered in the Sed5-containing compartments. *Mol. Cell Biol.* 25:7696–710
102. Bue CA, Bentivoglio CM, Barlowe C. 2006. Erv26p directs pro-alkaline phosphatase into endoplasmic reticulum–derived coat protein complex II transport vesicles. *Mol. Biol. Cell* 17:4780–89
103. Bue CA, Barlowe C. 2009. Molecular dissection of Erv26p identifies separable cargo binding and coat protein sorting activities. *J. Biol. Chem.* 284:24049–60
104. Piper RC, Bryant NJ, Stevens TH. 1997. The membrane protein alkaline phosphatase is delivered to the vacuole by a route that is distinct from the VPS-dependent pathway. *J. Cell Biol.* 138:531–45
105. Schmitz KR, Liu J, Li S, Setty TG, Wood CS, et al. 2008. Golgi localization of glycosyltransferases requires a Vps74p oligomer. *Dev. Cell* 14:523–34
106. Tu L, Tai WC, Chen L, Banfield DK. 2008. Signal-mediated dynamic retention of glycosyltransferases in the Golgi. *Science* 321:404–7
107. Dancourt J, Barlowe C. 2009. Erv26p-dependent export of alkaline phosphatase from the ER requires lumenal domain recognition. *Traffic* 10:1006–18
108. Waters MG, Serafini T, Rothman JE. 1991. 'Coatomer': a cytosolic protein complex containing subunits of non-clathrin-coated Golgi transport vesicles. *Nature* 349:248–51
109. Serafini T, Orci L, Amherdt M, Brunner M, Kahn RA, Rothman JE. 1991. ADP-ribosylation factor is a subunit of the coat of Golgi-derived COP-coated vesicles: a novel role for a GTP-binding protein. *Cell* 67:239–53
110. Ostermann J, Orci L, Tani K, Amherdt M, Ravazzola M, et al. 1993. Stepwise assembly of functionally active transport vesicles. *Cell* 75:1015–25
111. Béthune J, Wieland F, Moelleken J. 2006. COPI-mediated transport. *J. Membr. Biol.* 211:65–79
112. Hsu VW, Yang JS. 2009. Mechanisms of COPI vesicle formation. *FEBS Lett.* 583:3758–63
113. Semenza JC, Hardwick KG, Dean N, Pelham HRB. 1990. *ERD2*, a yeast gene required for the receptor-mediated retrieval of luminal ER proteins from the secretory pathway. *Cell* 61:1349–57

114. Lewis MJ, Sweet DJ, Pelham HRB. 1990. The *ERD2* gene determines the specificity of the luminal ER protein retention system. *Cell* 61:1359–63
115. Nishikawa S, Nakano A. 1993. Identification of a gene required for membrane protein retention in the early secretory pathway. *Proc. Natl. Acad. Sci. USA* 90:8179–83
116. Pelham HR. 1996. The dynamic organisation of the secretory pathway. *Cell Struct. Funct.* 21:413–19
117. Sato K, Sato M, Nakano A. 2003. Rer1p, a retrieval receptor for ER membrane proteins, recognizes transmembrane domains in multiple modes. *Mol. Biol. Cell* 14:3605–16
118. Townsley FM, Wilson DW, Pelham HR. 1993. Mutational analysis of the human KDEL receptor: distinct structural requirements for Golgi retention, ligand binding and retrograde transport. *EMBO J.* 12:2821–29
119. Lewis MJ, Pelham HR. 1992. Ligand-induced redistribution of a human KDEL receptor from the Golgi complex to the endoplasmic reticulum. *Cell* 68:353–64
120. Aoe T, Cukierman E, Lee A, Cassel D, Peters PJ, Hsu VW. 1997. The KDEL receptor, ERD2, regulates intracellular traffic by recruiting a GTPase-activating protein for ARF1. *EMBO J.* 16:7305–16
121. Majoul I, Sohn K, Wieland FT, Pepperkok R, Pizza M, et al. 1998. KDEL receptor (Erd2p)-mediated retrograde transport of the cholera toxin A subunit from the Golgi involves COPI, p23, and the COOH terminus of Erd2p. *J. Cell Biol.* 143:601–12
122. Aoe T, Lee AJ, van Donselaar E, Peters PJ, Hsu VW. 1998. Modulation of intracellular transport by transported proteins: insight from regulation of COPI-mediated transport. *Proc. Natl. Acad. Sci. USA* 95:1624–29
123. Cabrera M, Muñiz M, Hidalgo J, Vega L, Martín ME, Velasco A. 2003. The retrieval function of the KDEL receptor requires PKA phosphorylation of its C-terminus. *Mol. Biol. Cell* 14:4114–25
124. Hanyaloglu AC, von Zastrow M. 2008. Regulation of GPCRs by endocytic membrane trafficking and its potential implications. *Annu. Rev. Pharmacol. Toxicol.* 48:537–68
125. Yeung T, Barlowe C, Schekman R. 1995. Uncoupled packaging of targeting and cargo molecules during transport vesicle budding from the endoplasmic reticulum. *J. Biol. Chem.* 270:30567–70
126. Forster R, Weiss M, Zimmermann T, Reynaud EG, Verissimo F, et al. 2006. Secretory cargo regulates the turnover of COPII subunits at single ER exit sites. *Curr. Biol.* 16:173–79
127. Farhan H, Weiss M, Tani K, Kaufman RJ, Hauri HP. 2008. Adaptation of endoplasmic reticulum exit sites to acute and chronic increases in cargo load. *EMBO J.* 27:2043–54
128. Wilson DW, Lewis MJ, Pelham HR. 1993. pH-dependent binding of KDEL to its receptor in vitro. *J. Biol. Chem.* 268:7465–68
129. Scheel AA, Pelham HR. 1996. Purification and characterization of the human KDEL receptor. *Biochemistry* 35:10203–9
130. Lewis MJ, Pelham HR. 1992. Sequence of a second human KDEL receptor. *J. Mol. Biol.* 226:913–16
131. Raykhel I, Alanen H, Salo K, Jurvansuu J, Nguyen VD, et al. 2007. A molecular specificity code for the three mammalian KDEL receptors. *J. Cell Biol.* 179:1193–204
132. Boehm J, Ulrich HD, Ossig R, Schmitt HD. 1994. Kex2-dependent invertase secretion as a tool to study the targeting of transmembrane proteins which are involved in ER→Golgi transport in yeast. *EMBO J.* 13:3696–710
133. Boehm J, Letourneur F, Ballensiefen W, Ossipov D, Démollière C, Schmitt HD. 1997. Sec12p requires Rer1p for sorting to coatomer (COPI)-coated vesicles and retrieval to the ER. *J. Cell Sci.* 110:991–1003
134. Sato K, Sato M, Nakano A. 2001. Rer1p, a retrieval receptor for endoplasmic reticulum membrane proteins, is dynamically localized to the Golgi apparatus by coatomer. *J. Cell Biol.* 152:935–44
135. Sato M, Sato K, Nakano A. 1996. Endoplasmic reticulum localization of Sec12p is achieved by two mechanisms: Rer1p-dependent retrieval that requires the transmembrane domain and Rer1p-independent retention that involves the cytoplasmic domain. *J. Cell Biol.* 134:279–93
136. Massaad MJ, Franzusoff A, Herscovics A. 1999. The processing alpha1,2-mannosidase of *Saccharomyces cerevisiae* depends on Rer1p for its localization in the endoplasmic reticulum. *Eur. J. Cell Biol.* 78:435–40
137. Füllekrug J, Boehm J, Röttger S, Nilsson T, Mieskes G, Schmitt HD. 1997. Human Rer1 is localized to the Golgi apparatus and complements the deletion of the homologous Rer1 protein of *Saccharomyces cerevisiae*. *Eur. J. Cell Biol.* 74:31–40

138. Spasic D, Raemaekers T, Dillen K, Declerck I, Baert V, et al. 2007. Rer1p competes with APH-1 for binding to nicastrin and regulates gamma-secretase complex assembly in the early secretory pathway. *J. Cell Biol.* 176:629–40
139. Kaether C, Scheuermann J, Fassler M, Zilow S, Shirotani K, et al. 2007. Endoplasmic reticulum retention of the gamma-secretase complex component Pen2 by Rer1. *EMBO Rep.* 8:743–48
140. Kota J, Ljungdahl PO. 2005. Specialized membrane-localized chaperones prevent aggregation of polytopic proteins in the ER. *J. Cell Biol.* 168:79–88
141. Kota J, Gilstring CF, Ljungdahl PO. 2007. Membrane chaperone Shr3 assists in folding amino acid permeases preventing precocious ERAD. *J. Cell Biol.* 176:617–28
142. Sherwood PW, Carlson M. 1999. Efficient export of the glucose transporter Hxt1p from the endoplasmic reticulum requires Gsf2p. *Proc. Natl. Acad. Sci. USA* 96:7415–20
143. Trilla JA, Duran A, Roncero C. 1999. Chs7p, a new protein involved in the control of protein export from the endoplasmic reticulum that is specifically engaged in the regulation of chitin synthesis in *Saccharomyces cerevisiae*. *J. Cell Biol.* 145:1153–63
144. Lau WT, Howson RW, Malkus P, Schekman R, O'Shea EK. 2000. Pho86p, an endoplasmic reticulum (ER) resident protein in *Saccharomyces cerevisiae*, is required for ER exit of the high-affinity phosphate transporter Pho84p. *Proc. Natl. Acad. Sci. USA* 97:1107–12
145. Kim KM, Adachi T, Nielsen PJ, Terashima M, Lamers MC, et al. 1994. Two new proteins preferentially associated with membrane immunoglobulin D. *EMBO J.* 13:3793–800
146. Paquet ME, Cohen-Doyle M, Shore GC, Williams DB. 2004. Bap29/31 influences the intracellular traffic of MHC class I molecules. *J. Immunol.* 172:7548–55
147. Annaert WG, Becker B, Kistner U, Reth M, Jahn R. 1997. Export of cellubrevin from the endoplasmic reticulum is controlled by BAP31. *J. Cell Biol.* 139:1397–410
148. Schamel WW, Kuppig S, Becker B, Gimborn K, Hauri HP, Reth M. 2003. A high-molecular-weight complex of membrane proteins BAP29/BAP31 is involved in the retention of membrane-bound IgD in the endoplasmic reticulum. *Proc. Natl. Acad. Sci. USA* 100:9861–66
149. Wang B, Heath-Engel H, Zhang D, Nguyen N, Thomas DY, et al. 2008. BAP31 interacts with Sec61 translocons and promotes retrotranslocation of CFTRDeltaF508 via the derlin-1 complex. *Cell* 133:1080–92
150. Saito K, Chen M, Bard F, Chen S, Zhou H, et al. 2009. TANGO1 facilitates cargo loading at endoplasmic reticulum exit sites. *Cell* 136:891–902
151. Jonikas MC, Collins SR, Denic V, Oh E, Quan EM, et al. 2009. Comprehensive characterization of genes required for protein folding in the endoplasmic reticulum. *Science* 323:1693–97
152. Caldwell SR, Hill KJ, Cooper AA. 2001. Degradation of endoplasmic reticulum (ER) quality control substrates requires transport between the ER and Golgi. *J. Biol. Chem.* 276:23296–303
153. Kincaid MM, Cooper AA. 2007. Misfolded proteins traffic from the endoplasmic reticulum (ER) due to ER export signals. *Mol. Biol. Cell* 18:455–63

Virus Entry by Endocytosis

Jason Mercer,[1] Mario Schelhaas,[2] and Ari Helenius[1]

[1]ETH Zurich, Institute of Biochemistry, CH-8093 Zurich, Switzerland;
email: ari.helenius@bc.biol.ethz.ch, jason.mercer@bc.biol.ethz.ch

[2]University of Münster, Institutes for Medical Biochemistry and Molecular Virology, Centre for Molecular Biology of Inflammation (ZMBE), D-48149 Münster, Germany;
email: schelhaas@uni-muenster.de

Annu. Rev. Biochem. 2010. 79:803–33

First published online as a Review in Advance on March 2, 2010

The *Annual Review of Biochemistry* is online at biochem.annualreviews.org

This article's doi:
10.1146/annurev-biochem-060208-104626

0066-4154/10/0707-0803$20.00

Key Words

caveolar/raft-dependent endocytosis, clathrin/caveolin-independent endocytosis, clathrin-mediated endocytosis, endosome network, macropinocytosis

Abstract

Although viruses are simple in structure and composition, their interactions with host cells are complex. Merely to gain entry, animal viruses make use of a repertoire of cellular processes that involve hundreds of cellular proteins. Although some viruses have the capacity to penetrate into the cytosol directly through the plasma membrane, most depend on endocytic uptake, vesicular transport through the cytoplasm, and delivery to endosomes and other intracellular organelles. The internalization may involve clathrin-mediated endocytosis (CME), macropinocytosis, caveolar/lipid raft-mediated endocytosis, or a variety of other still poorly characterized mechanisms. This review focuses on the cell biology of virus entry and the different strategies and endocytic mechanisms used by animal viruses.

Contents

INTRODUCTION

Viruses first bind to cell surface proteins, carbohydrates, and lipids. Interactions with virus receptors are often specific and multivalent, and these interactions lead to the activation of cellular signaling pathways. Cells respond by internalizing the viruses using one of several endocytic mechanisms. After arrival in the lumen of endosomes or the endoplasmic reticulum (ER), viruses receive cues in the form of exposure to low pH, proteolytic cleavage and activation of viral proteins, and/or association with cell proteins. These trigger changes in the virus particle, and the activated viruses penetrate the vacuolar membrane, delivering the viral genome, the capsid, or the intact viral particle into the cytosol. After penetration, most RNA viruses replicate in different locations within the cytosol, whereas most DNA viruses continue their journey to the nucleus. Parallel with the movement of the virus and viral capsids deeper into the cell, a process of stepwise disassembly and uncoating takes place, culminating in the controlled release of the genome and accessory proteins in a replication-competent form.

Detailed understanding of virus-host cell interactions is important for several reasons. First, the increase in world population, the explosion in international trade and travel, global warming, and other factors have led to an increased threat from infectious pathogens, including viruses (1). Any information that can help in the battle against existing and emerging viruses has high priority. Virus-cell interactions provide an area that is still incompletely explored and underexploited for antiviral strategies.

Second, viruses are increasingly used as tools in molecular medicine. They have evolved to master the art of entering cells and introducing "foreign" genes and macromolecules. Therefore, they are useful devices in gene therapy and in the delivery of macromolecules and drugs into cells. They also have the potential for targeted elimination of cancer cells. Finally, viruses continue to serve as important tools and model systems in the discovery of new concepts in molecular, structural, and cell biology.

ADVANTAGES OF USING ENDOCYTOSIS FOR ENTRY

Elegant morphological, genetic, and biochemical studies of bacteriophages in the 1960s uncovered mechanisms of infection of great complexity and ingenuity. They showed that coliphages T4 and T2 are constructed like hypodermic syringes with a contractile tail forming a DNA delivery apparatus capable of piercing through the two membranes of a gram-negative bacteria and injecting the DNA into the cytosol (2). Attachment to appropriate host cell receptors was found to serve as the cue that triggers the injection process. Animal viruses do not need such elaborate instruments for entry because their host cells lack the outer membrane and cell wall that, in bacteria, prevent direct access to the plasma membrane. Also, animal cells provide endocytic mechanisms that give incoming viruses advantages that bacteriophages do not have. Endocytic vesicles ferry incoming viruses from the periphery to the perinuclear area of the host cell, where conditions for infection are favorable and distance to the nucleus minimal. This allows viruses to bypass obstacles associated with cytoplasmic crowding and the meshwork of microfilaments in the cortex (3, 4). Transport in endocytic vesicles is particularly important for viruses that infect neurons, where long distances separate axons from the cell body. The maturation of endosomes with gradually changing conditions, such as a decreasing pH and a change in redox environment, allows viruses, moreover, to sense their location within the cell and the pathway and to use this information to set the time of penetration and uncoating (5). The presence of specific proteases, such as furin and cathepsins, provides necessary proteolytic activation of certain viruses (6, 7). Finally, by being endocytosed, animal viruses can avoid leaving evidence of their presence exposed on the plasma membrane, thus likely causing a delay in detection by immunosurveillance. Taken together, endocytosis is so advantageous that viruses, such as herpes simplex virus 1 and human immunodeficiency virus 1 (HIV-1), that are capable of entering directly often prefer to use endocytic pathways for productive entry (8–10).

The topic of viral endocytosis and related topics have been previously reviewed and discussed (5, 11–17), together with recent reviews on endocytic pathways (18–27). These reviews are highly recommended for more detail and for different points of view regarding this broad topic.

VIRUSES AS ENDOCYTIC CARGO

The size of animal viruses varies from about 30 nm for parvoviruses to 400 nm for poxviruses, with most viruses in the 60–150-nm range. Although usually roughly spherical in shape, viruses, such as *Filovirus*, *Paramyxovirus*, and influenza viruses, are (or can be) fibrous and highly elongated. As they bind to cells, viruses are not deformed; it is rather the plasma membrane that shows a tendency to invaginate to accommodate the shape of the virus. In some cases, such invagination is essential for endocytosis (see below).

ILV: Intralumenal vesicle

The surface of viruses is typically covered by receptor-binding proteins either in the form of capsid proteins in an icosahedral grid or as spike glycoproteins covering a viral envelope. Individual interactions with receptors are generally weak, but contact with multiple receptors make the avidity high and binding to cells virtually irreversible. Multivalent binding leads to receptor clustering, which in turn may result in association with lipid domains and activation of signaling pathways.

Once delivered to endosomes, most virus particles are similar in size to the intralumenal vesicles (ILVs). Too big to enter the narrow tubular extensions, they are generally localized to the bulbous, vacuolar domains of endosomes and are thus sorted to the degradative pathway.

Viruses have often been used as model cargo in endocytosis and membrane trafficking studies. They are easily recognized by electron microscopy (EM) and can be tagged with fluorescent groups or proteins, allowing single-particle detection and tracking in live cells. By providing a single spot-like source of light, the center of mass of fluorescent viruses can be precisely defined through the point-spread function (28). Owing to the amplification caused by infection, successful entry can be easily quantified, even with minute amounts of virus. In addition, the virology community has developed a large number of tools, such as virus mutants, fluorescent viruses, antibodies, expression systems, and modified host cells.

ATTACHMENT FACTORS AND RECEPTORS

Although some viruses take advantage of receptors with known endocytic receptor functions,

EE: early endosome

LE: late endosome

Clathrin-mediated endocytosis (CME): an endocytic process driven by the formation of a clathrin coat on the cytoplasmic leaflet of the plasma membrane

Macropinocytosis: actin-mediated endocytic process involving plasma membrane ruffles and internalization of fluid and particles in large, uncoated, endocytic vesicles

such as transferrin and low-density lipoprotein (LDL) receptors, most of the molecules that viruses bind to are involved in other functions such as cell-cell recognition, ion transport, and binding to the extracellular matrix (16). Often they are glycoconjugates (glycoproteins, glycolipids, proteoglycans), and the carbohydrate moieties can play a central role in virus binding.

From a conceptual point of view, it is useful to differentiate between attachment factors that merely bind viruses and thus help to concentrate the viruses on the cell surface, and virus receptors, which in addition serve to trigger changes in the virus, induce cellular signaling, or trigger penetration. One function that many receptors share is that they promote endocytosis and accompany the virus into the cell. Often entry starts with binding to attachment factors, followed by associations with one or more receptors. In practice, it is often difficult to differentiate between attachment factors and receptors because both contribute to the effectiveness of infection.

The most commonly encountered attachment factors are glycosaminoglycan (GAG) chains—especially heparan sulphate—in proteoglycans. Binding to these negatively charged polysaccharides is usually electrostatic and relatively nonspecific. It has been recognized in several cases that viruses evolve to use GAGs when adapting to growth in tissue culture (29, 30). Sialic acids constitute another common group of carbohydrates to which many viruses bind. As in the case of influenza and polyomaviruses, this binding is often highly specific and involves defined lectin domains or lectin sites (16).

The specificity of binding is a major factor in determining tropism and species specificity, and thus the nature of viral diseases. Another aspect of receptor specificity is that it determines the choice of endocytic pathway and intracellular routing that the incoming virus will take. For example, parvoviruses that bind to the transferrin receptor use a clathrin-mediated uptake pathway and are able to recycle to the cell surface with their receptor (31), whereas minor group rhinoviruses that bind to the LDL receptor dissociate from the receptor in early endosomes (EEs) and are transported to late endosomes (LEs) (32).

Some receptors are responsible for inducing changes in the virus that allow binding of the virus to a coreceptor, induction of endocytic uptake, or conversion to a membrane fusion-active conformation. The best-characterized case is HIV-1, where two receptors are required to induce conformational changes that trigger fusion (33). Adenoviruses 2 and 5 have two receptors that induce conformational changes and promote endocytosis (34). In the case of avian leukosis virus, the cues required for penetration are receptor binding combined with low pH (35).

The use of different receptors often correlates with the need for a virus to overcome barriers existing in the cell type or tissue that they infect. One well-studied example is the binding of Coxsackievirus B to decay-accelerating factor (DAF) in the apical surface of epithelial cells, and subsequently to the Coxsackievirus and adenovirus receptor (CAR), which is localized in the tight junction region. DAF helps to bring the virus to the tight junctions, and CAR induces a conformational change and promotes endocytosis (36). For hepatitis C virus (HCV), several different receptors are thought to have a similar "shuttle" function during infection of hepatocytes (37).

MECHANISMS OF ENDOCYTOSIS

Viruses typically make use of various pinocytic mechanisms of endocytosis that serve the cell by promoting the uptake of fluid, solutes, and small particles (**Figure 1**). Best studied are clathrin-mediated endocytosis (CME), macropinocytosis, and caveolar/raft-dependent endocytosis. There are, in addition, several clathrin- and caveolin/raft-independent mechanisms that are less well characterized but under active investigation. The formation of primary endocytic vesicles and vacuoles involves a large number of cellular factors; some are listed in **Table 1**. The factors are different for the various pathways and only partially known. The involvement of

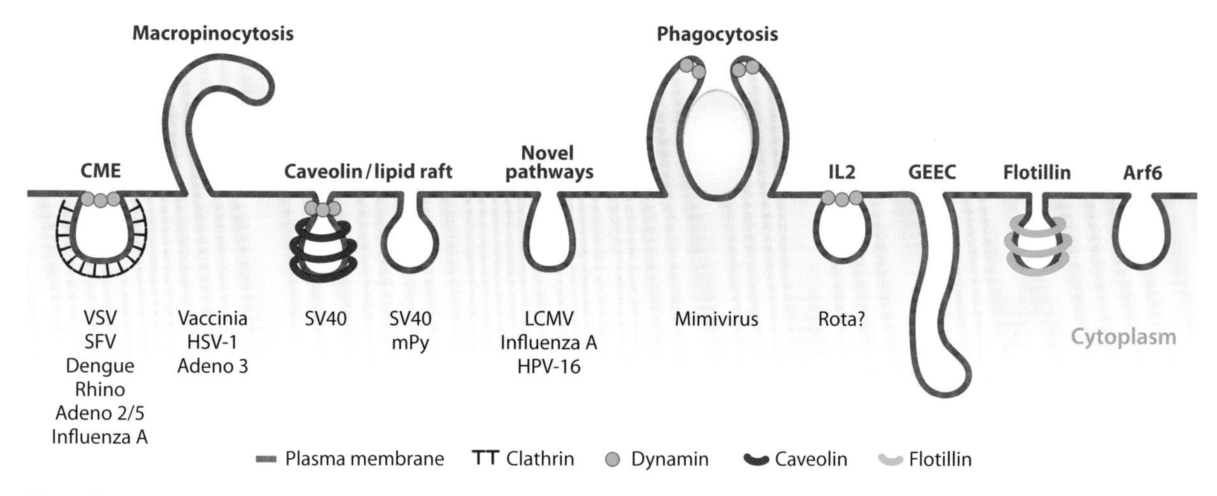

Figure 1

Endocytic mechanisms. Endocytosis in animal cells can occur via several different mechanisms. Multiple mechanisms are defined as pinocytic, i.e., they involve the uptake of fluid, solutes, and small particles. These include clathrin-mediated, macropinocytosis, caveolar/raft-mediated mechanisms, as well as several novel mechanisms. Some of these pathways involve dynamin-2 as indicated by the beads around the neck of the endocytic indentations. Large particles are taken up by phagocytosis, a process restricted to a few cell types. In addition, there are pathways such as IL-2, the so-called GEEC pathway, and the flotillin- and ADP-ribosylation factor 6 (Arf6)-dependent pathways that carry specific cellular cargo but are not yet used by viruses. Abbreviations: Adeno 2/5, adenovirus 2/5; Adeno 3, adenovirus 3; CME, clathrin-mediated endocytosis; HPV-16, human papillomavirus 16; HSV-1, herpes simplex virus 1; LCMV, lymphocytic choriomeningitis virus; mPy, mouse polyomavirus; SFV, Semliki Forest virus; SV40, simian virus 40; VSV, vesicular stomatis virus.

the various pathways in virus entry is discussed below.

It is generally assumed that viruses take advantage of existing mechanisms. Whereas, some are internalized by ongoing endocytic activities, many actually induce their own uptake by activating cellular signal transduction pathways. Caveolar/raft-dependent endocytosis, for example, is preceded by the activation of several tyrosine and other kinases (26, 38). Macropinocytosis of vaccinia virus requires activation of kinases and GTPases that regulate changes in actin dynamics (39). CME can also be virus triggered (14, 40, 41). For some of the novel virus entry pathways, the physiological functions are not yet known.

LOGISTICS OF THE ENDOSOME NETWORK

Once internalized within primary endocytic vesicles, the intracellular pathways followed by incoming viruses are the same as those used by physiological ligands and membrane components, such as nutrients and their carriers, hormones, growth factors, extracellular matrix components, plasma membrane factors, and lipids. The endosomal system in question is responsible for molecular sorting, recycling, degradation, storage, processing, and transcytosis of incoming substances, collectively called cargo.

The main organelle classes are EEs and LEs (the latter often taking the form of multivesicular bodies), recycling endosomes (REs), and lysosomes (**Figure 2**) (25, 42–47). Also included in the figure is a class of intermediate organelles between EEs and LEs, which have been called maturing endosomes (MEs) because they contain both Rab5 and Rab7 and serve as precursors for LEs (42, 48, 49). They are likely to play a role in the entry of several viruses. The endosome system is tightly connected with the secretory pathway through vesicles shuttling between endosomes and the *trans*-Golgi network (TGN), and also via the plasma membrane.

RE: recycling endosome

ME: maturing endosome

Table 1 Cellular endocytic pathways and their cellular factors

Endocytic pathway	Cellular factors
Clathrin-mediated	
Coat proteins	Clathrin heavy chain, clathrin light chain
Adaptors	AP2,[a] eps15,[a] epsin1[a]
Scission factors	Dynamin-2
Regulatory factors	PI(3,4)P, PI(4,5)P2, cholesterol,[a] cortactin,[a] Arp2/3[a]
Cytoskeleton	Actin,[a] microtubules[a]
Trafficking	Rab5, Rab7,[a] Rab4,[a] Rab11,[a] Rab22[a]
Macropinocytosis	
Coat proteins	None
Scission factors	Unknown
Regulatory factors	Tyrosine kinases, PAK1, PI(3)K, PKC, Ras, Rac1, Cdc42,[a] Rab34,[a] CtBP1, Na^+/H^+ exchange, cholesterol
Cytoskeleton	Actin, microtubules,[a] myosins[a]
Trafficking	Rab5,[a] Rab7,[a] Arf6[a]
CAV1 pathway	
Coat proteins	Caveolin-1
Scission factors	Dynamin-2
Regulatory factors	Tyrosine kinases, phosphatases, PKC, RhoA, cholesterol
Cytoskeleton	Actin, microtubules
Trafficking	Rab5
Lipid raft	
Coat proteins	None (clathrin and caveolin independent)
Scission factors	Unknown (dynamin-2 independent)
Regulatory factors	Tyrosine kinases, Rho A, cholesterol
Cytoskeleton	Actin
Trafficking	Unknown
IL-2 pathway	
Coat proteins	None
Scission factors	Dynamin-2
Regulatory factors	RhoA, lipid rafts, ubiquitination
Cytoskeleton	Actin
Trafficking	Rab5,[a] Rab 7
GEEC pathway	
Coat proteins	Unknown, GRAF1[a]
Scission factors	Unknown
Regulatory factors	Arf1, ARHGAP10, Cdc42, lipid rafts
Cytoskeleton	Actin
Trafficking	Rab5, PI(3)K
Flotillin pathway	
Coat proteins	Flotillin-1
Scission factors	Unknown
Regulatory factors	Fyn kinase, lipid rafts
Cytoskeleton	Unknown
Trafficking	Rab5, PI(3)K

(Continued)

Table 1 (*Continued*)

Endocytic pathway	Cellular factors
Arf6 pathway	
Coat proteins	None
Scission factors	Unknown (dynamin independent)
Regulatory factors	PIP5K, Arf6, Arf1, PI(4,5)P, PI(3)K, cholesterol, actin
Cytoskeleton	Actin
Trafficking	Rab5, Rab7 (degradation) Rab11, Rab22 (recycling only)
Phagocytosis	
Coat proteins	None (particle driven)
Adaptors	AP2[a]
Scission factors	Dynamin-2
Regulatory factors	Tyrosine kinases, PI(3)K, PKC, Ras, RhoA, RhoG, Rac1,[a] Cdc42,[a] Arf6,[a] cholesterol
Cytoskeleton	Actin, microtubules, myosins
Trafficking	Rab5, Rab7

[a]Factors that are required in a cell type or system-dependent fashion.

[b]Abbreviations: Arf1, -6, ADP-ribosylation factor-1 and 6; ARHGAP10, Rho GTPase-activating protein 1; Arp2/3, actin-related proteins 2/3; Cdc42, cell division control protein 42; CtBP1, C-terminal-binding protein 1; PI(3)K, phosphoinositide 3-kinase; PI(3,4)P, phosphoinositol-3,4 phosphate; PI(4,5)P2, phosphatidylinositol 4,5-bisphosphate; PIP5K, phosphatidylinositol-4-phosphate 5-kinase; PKC, protein kinase C; Rab4, -5, -7, -11, -22, Ras-related in brain; Rac1, Ras-related C3 botulinum toxin substrate 1; Ras, Rat sarcoma viral oncogene homolog; RhoA, Ras homolog gene family, member A; RhoG, Ras homolog gene family, member G.

Cargo delivered from the surface by CME typically reaches EEs in less than 2 min after internalization, MEs and LEs in the perinuclear region after 10–12 min, and the lysosomes within 30–60 min (50, 51). The different classes of endosomes are heterogeneous in composition, and their function and transit through the pathway are highly asynchronous (52). Recent studies in live cells suggest that there are at least two populations of EEs: highly motile and rapidly maturing as well as more static and slowly maturing (50).

To manage their various molecular sorting and trafficking functions, EEs have a complex structure with vacuolar elements and many long, narrow, often-branched tubes (53, 54). The membrane is composed of a patchwork of functionally different domains (20, 23, 44). These differ in composition, location, and structure (tubular or vacuolar), and EEs are often defined by different Rabs and their effectors. The domains are responsible for selective vesicular transport to distinct targets: the plasma membrane (Rab4), LEs (Rab7), REs (Rab22), and the TGN (Rab9 and the retromer complex). Most of the domains are located in the tubular part of the endosomes.

It is important to recognize that each endosome class corresponds to a heterogeneous collection of organelles and that they go through a program of changes with time. The conversion of EEs to LEs is a particularly complex process and the subject of a long-standing discussion: Does it represent maturation or vesicle transport (45, 48, 55)? The issue is partly semantic with current experimental data suggesting that both models apply (42, 48, 56). The process involves the formation of an ME, a hybrid endosome with Rab5 and Rab7 domains (**Figure 2**). The ME mainly contains the vacuolar component of EEs and the ILVs. It undergoes a maturation program of considerable significance for viruses such as influenza and simian virus 40 (SV40), which use MEs and LEs for entry. The repertoire of changes include the following:

1. There is a loss of recycling receptors, such as those for transferrin and LDL, and other membrane proteins and lipids targeted for recycling to the plasma membrane. These leave the EEs either by

direct vesicle transport or via the REs, often located in the perinuclear region. It has been found that lipids with short and unsaturated alkyl chains tend to sort to recycling endosomes, whereas lipids that have long and saturated alkyl chains, such as gangliosides, preferentially enter LEs (57).

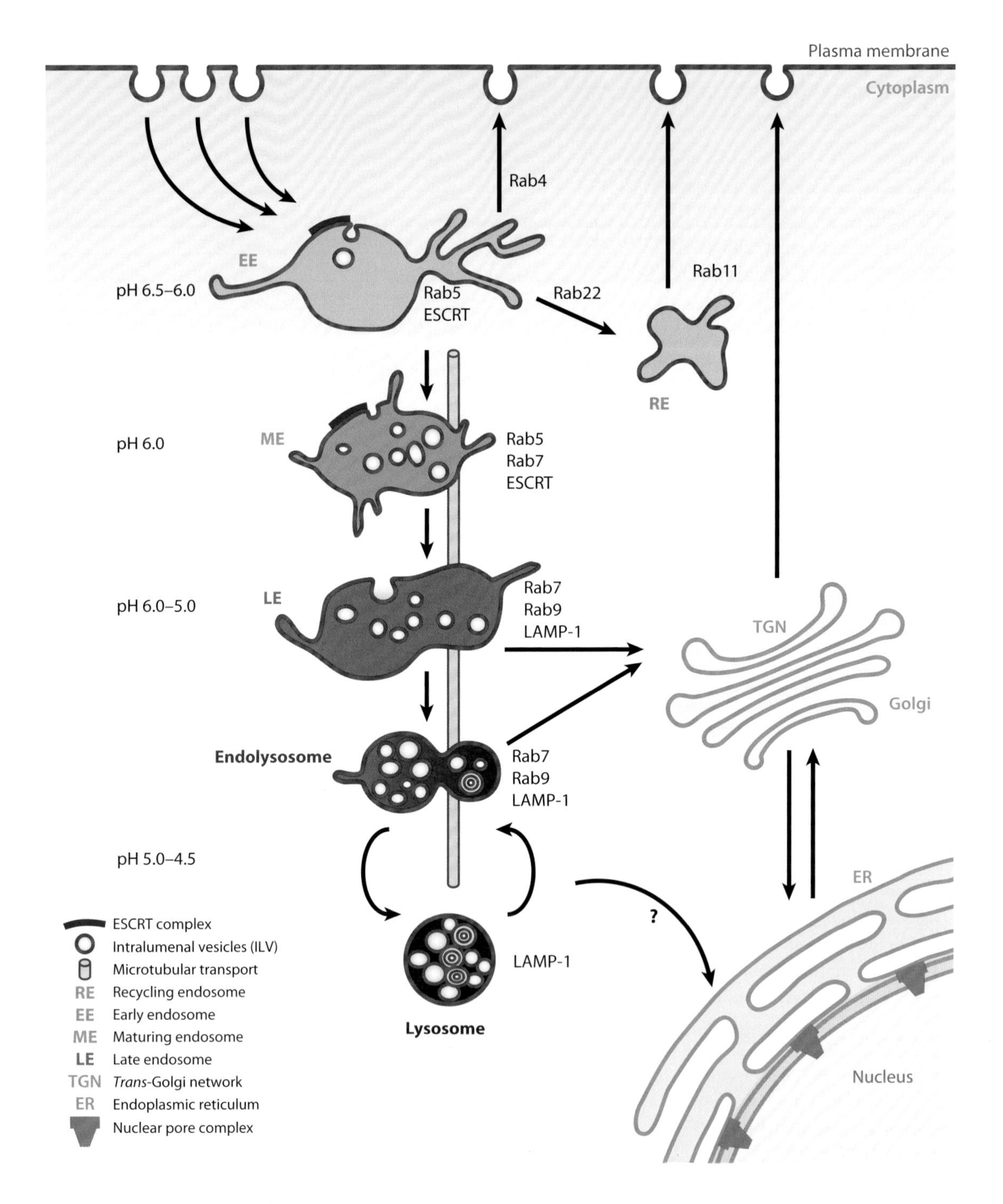

2. A gradual drop in internal pH occurs from mild acidity (pH 6.5–6.0) in EEs to values below 5. This change is most likely caused by variation in subunit composition and possibly concentration of the v-ATPase, as well as different isoforms of chloride channels (58, 59).
3. The formation of ILVs occurs. This is a function carried out by the endosomal sorting complex required for transport (ESCRT) complexes associated with the membrane of EEs and MEs (60). Monoubiquitin-tagged membrane proteins are selectively included in the ILVs. The result is the formation of LEs filled with vesicles (multivesicular bodies) destined for lysosomal degradation.
4. There is a switch of Rab subsets (i.e., Rab5 and Rab4 to Rab7 and Rab9) and their effectors. This also results in a change in predominant phosphatidylinositides from PI(3)P to PI(3,5)P2 (61, 62). Together, these changes have many consequences in defining associated peripheral proteins.
5. A change in the interaction with cytoskeletal elements (mainly microtubules and their motors) leads to endosome migration from the peripheral to the perinuclear area of the cell and, eventually, to fusion with lysosomes.

The elements in the maturation program are interconnected and interdependent in complex ways. This is seen when one of the functions is perturbed. The effect is often a disruption or retardation of the whole program. Disruptions can be induced by overexpressing different Rabs and their mutants, by preventing the formation of ILVs, by inhibiting endosome movement to the perinuclear region using inhibitors of microtubule assembly, by inhibiting PI(3)-kinases, by preventing acidification, by blocking maturation with a drop in temperature, and by other perturbations. For example, in HeLa cells, inhibition of the v-ATPase with bafilomycin A1 blocks formation of MEs (63–65), and the addition of nocodazole blocks LE formation and transport of cargo to REs (64). Wortmannin [a PI(3)-kinase inhibitor] causes a delay in EE to LE transport, and incubation at 20°C blocks LE fusion with lysosomes (64, 66). How useful such perturbations can be in the analysis of virus entry is exemplified by recent work on major and minor group human rhinoviruses (67).

It is important to recognize that the endosome network is a continuum with heterogeneous classes of organelles continuously undergoing changes through retrieval, recycling, dissociation, entrapment, processing, fusion, fission, and recruitment of molecular components (68). To a considerable degree, it is a self-organizing system able to adjust to

Figure 2

The endosome network. The main organelles of the endocytic pathway are the early endosomes (EEs), maturing endosomes (MEs), late endosomes (LEs), recycling endosomes (REs), and lysosomes. EEs are usually located in the periphery of the cytoplasm. They are complex organelles with several different domains (tubular and vacuolar). The tubular domains 50–90 nm in diameter and up to four microns in length contain most of the endosomal membrane and give rise to recycling vesicles and vesicles for transport to the REs and to the *trans*-Golgi network (TGN). The vacuolar domains 200–1000 nm in diameter contain most of the volume, the intralumenal vesicles (ILVs), and larger endocytosed particles. The vacuolar domains dissociate (with their contents) and undergo microtubule-mediated, dynein-dependent movement to the perinuclear region. These MEs contain markers of both early and late endosomes, such as Rab5 and Rab7. They also undergo further acidification and conversion to mature LEs, which can fuse with each other and eventually with lysosomes, generating endolysosomes in which active degradation takes place. The dense core lysosomes correspond to the end points of such degradation processes; they serve as a depository for lysosomal enzymes and membrane proteins awaiting fusion with incoming LEs. The majority of incoming membrane components undergo recycling to the plasma membrane. However, membrane proteins destined for degradation are first tagged with monoubiquitins, then recognized by ubiquitin-binding components of the endosomal sorting complex required for transport (ESCRT) machinery, and finally sequestered into ILVs. ILVs are formed by inward budding of EEs and LEs. They fill the lumen of the vacuolar domains, forming multivesicular bodies, and are eventually degraded in lysosomes. Abbreviation: LAMP-1, lysosomal-associated membrane protein.

changes in physiological conditions, the nature and quantity of incoming cargo, perturbations, and cell function. It is regulated by complex feed-back and feed-forward regulatory cycles involving numerous kinases, phosphatases, GTPases, and other factors (69, 70).

APPROACHES TO STUDY VIRUS ENTRY

Early studies of virus entry relied extensively on EM, a technology of continued importance because of the new techniques of great promise, such as cryo-EM, tomography, and focused ion beam-scanning electron microscopy (FIB-SEM) (71, 72). The newer techniques allow better organelle preservation, three-dimensional imaging, and correlative analysis with light microscopy. Light microscopy is also increasingly important because of its enhanced resolution and the possibility to follow incoming fluorescent viruses in live cells (56, 73, 74). With host components tagged with green fluorescent protein (GFP) and other fluorescent proteins, it is possible to track the fate of individual particles and determine which cell proteins are involved. Studies using single-particle tracking have greatly expanded our understanding of virus entry (73, 75–78). By taking advantage of the point-spread function, one can, under favorable conditions, define the center of a virus particle with a resolution of less than 5 nm (79).

Perturbations using inhibitors, dominant negative (D/N) or constitutively active (C/A) constructs of cellular proteins, siRNA silencing, and cell mutants provide another general approach commonly used. When adequately controlled and combined with specific, quantitative assays for infection, endocytosis, and penetration, these perturbations are quite powerful. The challenges and potential pitfalls include cell toxicity, cell type variability, off-target effects, unwanted side effects of protein tagging, and poor transfection or silencing efficiency. The compensatory activation of alternative pathways when one pathway is closed down is another problem, especially in long-lasting experiments (19).

Chemical inhibitors used today in virus entry span a wide array of effects, including signaling, regulation, endocytic machinery, cytoskeleton function, lipid composition of membranes, and others. Here, we will only mention a couple of inhibitor classes commonly employed because their use is not as straightforward as often thought. Lysosomotropic weak bases (such as ammoniun chloride, chloroquine, methylamine), carboxylic ionophores (monensin, nigericin), and v-ATPase inhibitors (bafilomycin A1 and concanamycin) are often employed to determine whether viruses require low pH for infection. The inhibition of virus infection by one or more of these pH perturbants does not automatically mean that the virus undergoes a pH-induced conformational change. Inhibition can also be caused by secondary effects, such as defective receptor recycling, inhibition of endosome maturation, inhibition of enzymes with a low pH optimum, inhibition of Ca^{2+} efflux from endosomes, and pH neutralization of nonendocytic compartments, such as the TGN.

Weak bases raise the pH almost instantly after addition, which makes them useful agents for determining the time course of the acid activation step (80). It is, however, important to note that they raise only endosomal and lysosomal pH reliably and inhibit virus penetration if the pH in the medium remains above 7.0 (81). Some of the confusion regarding the acid dependency of certain viruses is likely explained by the use of poorly buffered media in experiments involving lysosomotropic weak bases.

Inhibitors that cause cholesterol depletion are often used to test whether viruses enter by caveolar/raft-mediated pathways. The results depend on which agents and concentrations are used (19). Methyl-β-cyclodextran extracts the cholesterol from the plasma membrane rapidly and efficiently and inhibits not only caveolar/raft-mediated pathways but often clathrin-mediated pathways as well (82, 83). The effects of nystatin, an agent that binds cholesterol (often used in combination with progesterone, a cholesterol synthesis inhibitor), are generally milder.

Transfection with D/N and C/A constructs and siRNA allows targeted analysis of individual host cell proteins and pathways. The use of controls is critical, and poor cell viability can be a problem. The results obtained by silencing of individual genes using siRNA should be validated using immunoblotting or other techniques to quantify the amount of the target protein remaining in the cells. Multiple siRNA sequences are recommended against each gene to avoid off-target effects. Because the knockdown of proteins is never complete, negative data are only interpretable with proper controls.

Finally, it is important that viruses be properly stored. For acid-activated viruses, it is essential that they not be frozen in phosphate-buffered media such as MEM without added Hepes or Tris buffer, because the drop in pH during freezing may preacidify the virus, leading to losses in infectivity.

VIRUSES THAT USE THE CLATHRIN PATHWAY

Early EM studies revealed that certain viruses, such as adenovirus 2 and 5 and vesicular stomatitis virus (VSV), were present in thickened regions of the plasma membrane (84, 85), i.e., in regions that we now know as clathrin-coated pits (CCPs). That CME can be part of the productive infectious pathway was first shown for Semliki Forest virus (SFV), an enveloped animal virus (86), and this was followed by similar observations for several other viruses. The pathway is today the most commonly observed in virus entry. **Table 2** contains a partial list of viruses that have been reported to use this pathway. In some cases, CME serves as one of several pathways used by a virus.

The molecular information about the various steps in clathrin coat assemblies, induction of membrane curvature, cargo recruitment, vesicle fission, and coat disassembly is extensive, and there are several recent reviews (21, 24, 148). It is increasingly apparent that the composition of CCPs in the plasma membrane is not always the same. Once thought to be an integral part of all plasma membrane-coated pits, the adaptor complex AP2, for example, has turned out not to be an absolute requirement for the formation of coated pits and for the internalization of certain cargo (24, 149–151). This is also true for viruses; some depend on AP2, and others do not (**Table 1**).

The formation of clathrin-coated vesicles was initially thought to be a constitutive process that occurred whether cargo was present or not. It is now evident that cargo can actively initiate the formation of CCPs. This is particularly clear in the case of some viruses (reovirus, VSV, Influenza A) that induce de novo formation of a CCP at their site of binding (40, 41, 74, 88). Of clathrin-internalized influenza virus particles on BS-C-1 cells, 94% induce de novo assembly of a CCP centered under the virus (74). The adaptor protein required in this case is epsin-1 (111).

Similarly, clathrin, dynamin-2, and AP2 can be seen to accumulate transiently under or in close proximity to surface-bound VSV particles (41, 88). The assembly phase lasts about 110 s, which is somewhat longer than for normal coated pits (50 s). If the VSV particle cannot be internalized because it is fixed to the substratum, several cycles of "frustrated" CCP formation can be observed in the same spot. What causes coat assembly is not clear, but receptor clustering induced by the virus may give rise to a microdomain with properties different from that of the surrounding membrane.

CME of viruses is generally a rapid process, with surface-bound viruses entering within minutes after attachment, or after cell warming if binding was performed in the cold. Delivery to EEs follows within 1–2 min. For VSV and SFV, which have a high fusion pH threshold adjusted to the pH of EEs, the acid-activated step occurs within 1–5 min after internalization. Viruses that have a pH of fusion in the LE range, typically fuse 10–20 min after internalization.

A recent single-particle tracking study with dengue virus (serotype 2, strain 1) has allowed visualization of the various steps in the clathrin-mediated entry process (56). After binding,

Dynamin-2: a large GTPase required for various types of endocytosis and intracellular scission events. Dynamin-2 is indispensable for clathrin-mediated endocytosis

Table 2 Endocytic internalization of mammalian viruses. Listed are the primary endocytic mechanisms used by a variety of virus families

Primary pathway	Virus family (DNA/RNA)	Replication nucleus (Nuc) or cytoplasm (Cyto)	Example virus envelope (±)	Low pH ± dependency	Receptor molecules	Comments	References
CME	Rhabdo (ssRNA−)	Cyto	Vesicular stomatitis (+)	+ (<6.4)	?	Induces CCPs de novo; role of AP2 controversial; can fuse at EE; suggested fusion with ILVs	41, 87–89
CME	Alpha (ssRNA+)	Cyto	Semliki Forest (+)	+	GAGs ?	Induces CCPs de novo; transported to EEA1/Rab5 peripheral endosomes; fuses 5–8 min after internalization	86, 90–92; A. Vonderheit & A. Helenius, unpublished
CME	Alpha (ssRNA+)	Cyto	Sindbis (+)	+ (<6.2)	GAGs	Data on CME and acid activation contradicts previously published direct penetration model	93–96
CME	Flavi (ssRNA+)	Cyto	Dengue (+)	+	GAGs, DC-SIGN	Uses preformed CCPs; fuses at LE; entry similar to West Nile virus; in Vero cells clathrin, CAV1, and actin independent; dynamin dependent	56, 97, 98
CME	Flavi (ssRNA+)	Cyto	Hepatitis C (+)	+	GAGs, LDLR, SRB1, CD81	Fusion requires proteolytic activation	37, 99
CME	Adeno (dsDNA)	Nuc	Adenovirus 2 and 5 (−)	+ (<6.0)	CAR, αV integrins	Uptake into CCVs in parallel with induction of macropinocytosis (required for infection); penetration in EE	92, 100, 101
CME	Picorna (ssRNA+) Aphtho	Cyto	Hand, foot, and mouth disease (−)	+	αV integrins	Can enter EEs and REs; infection requires Rab 5 but not Rab 7, 4, or 11	102, 103

CME	Picorna (ssRNA+)	Cyto	Human rhino 2 minor group (−)	+ (<5.6)	LDL receptor	Virus and receptor separate in EEs; viral RNA deposited into cytosol from MEs; empty capsids degraded in LE/LY	104, 105
CME	Reo (dsRNA)	Cyto	Reovirus(+)	+	JAM1	Penetrates from EEA1 positive endosomes	40, 106
CME	Parvo (ssDNA)	Nuc	Canine parvo (−)	+	Transferrin receptor	CME is main pathway for parvo. Virus found in EEs and REs; penetration from LEs. Exception AAV5 (goes to Golgi)	31, 107–109
CME	Retro (ssRNA)	Nuc	Avian leukosis (+)	+	Transferrin receptor	Colocalizes with CCPs; chlorpromazine sensitive	35, 110
CME/Novel pathway	Myxo (ssRNA−)	Nuc	Influenza A (+)	+ (<5.6)	Sialoglyco-proteins	About 50% of particles or less use CME; induces CCPs de novo; epsin dependent, fuses at LEs. The rest enter clathrin independently	74, 111–113
Macro	Pox (dsDNA)	Cyto	Vaccinia MV (+)	± (<5.0)	GAGs?	Entry linked to membrane blebbing; apoptotic mimicry strategy; cellular requirements vary with strain	39, 114, 115
Macro	Adeno (dsDNA)	Nuc	Adenovirus 3 (−)	+ (<6.0)	CD46, αV integrins	Entry requires PAK1, CtBP1; cellular factors consistent with macropinocytosis	116–118
Macro	Picorna (ssRNA+)	Cyto	Echovirus 1 (−)	+	$\alpha 2\beta 1$ integrins	Entry requires PAK1, CtBP1; cellular factors consistent with macropinocytosis	119, 120
Macro	Picorna (ssRNA+)	Cyto	Coxsackievirus B (+)	±	CAR	Enters at tight junctions; does not require Rac1; requires Rab34	121, 122

(Continued)

Table 2 (*Continued*)

Primary pathway	Virus family (DNA/RNA)	Replication nucleus (Nuc) or cytoplasm (Cyto)	Example virus envelope (±)	Low pH ± dependency	Receptor molecules	Comments	References
Macro	Herpes (dsDNA)	Nuc	Herpes simplex virus 1 (+)	+ (<6.0)	GAGs?	Cell type specific; RTK and PI3K dependent; virions colocalize with fluid markers; needs further experimentation	8, 123, 124
Macro	Retro (ssRNA+)	Nuc	HIV-1 (+)	+	CD4, CCR5, CXCR4, CCR2	Cell type specific; Na^{+}/H^{+} dependent; virions colocalize with fluid markers; needs further experimentation	125–128
Macro	Herpes (dsDNA)	Nuc	Human herpes virus 8/(KSHV) (+)	+	GAGs?	Cell type specific; Rab34; cellular factors consistent with macropinocytosis	129
CAV1/Lipid raft	Polyoma (dsDNA)	Nuc	Simian virus 40 (–)	+	GM1	Parallel CAV1-dependent and –independent entry; transport via EE and LE to ER where penetration occurs	130–132; S. Engel, T. Heger, R. Mancini, J. Kartenbeck, F. Herzog, et al., in preparation
CAV1/Lipid raft	Papilloma (dsDNA)	Nuc	Human papillomavirus type 31 (–)	+	GAGs?	Exceptional pathway for HPVs; depends on CAV1, lipid rafts, dynamin; routed to EE	133
Lipid raft	Polyoma (dsDNA)	Nuc	Mouse polyomavirus (–)	+	GD1a	Transport via EE and RE to ER, where penetration occurs	134–136
IL-2 tentative	Reo (dsRNA)	Cyto	Rhesus rotavirus (–)	–	Sialic acid, integrins, Hsc70	Clathrin and CAV1 independent; cholesterol and dynamin dependent	137, 138

IL-2 tentative	Corona (ssRNA+)	Cyto	Feline infectious peritonitis virus, SARS (+)	?	DC-SIGN	Clathrin, CAV1, Rho GTPase independent; cholesterol and dynamin dependent. IL-2 pathway?	139, 140
Novel Pathway	Arena (ssRNA−)	Cyto	Lymphocytic choriomeningitis virus (+)	+	Alpha-dystroglycan	Independent of clathrin, CAV1, flotillin, lipid rafts, actin; routed directly to LE; minor fraction by CME	141, 142
Novel pathway	Papilloma (dsDNA)	Nuc	Human papillomavirus type 16 (−)	+	GAGs?	Noncoated vesicles; clathrin, CAV1, flotillin, lipid-raft, and dynamin independent; escape likely in LE/lysosomes	143; M. Schelhaas, M. Holzer, P. Blattmann, P.M. Day, J.T. Schiller, A. Helenius, submitted
Novel pathway	Circo (ssDNA−)	Nuc	Porcine circovirus 2 (−)	+	?	Cell-type dependent; in porcine monocytes by CME; in HeLa dependent on actin and Rho GTPase	144, 145
Phago	Herpes (dsDNA)	Nuc	Herpes simplex virus 1 (+)	+	HS, HVEM, Nectin1	Cell type specific; dynamin-2; cellular factors consistent with phagocytosis	146
Phago	Mimi (dsDNA)	Cyto	Mimivirus (+)	+	?	Amoebal pathogen; can only enter professional phagocytes; entry requires dynamin-2 and PI3K	147

Abbreviations: +, low pH dependent; −, low pH independent; AAV5, adeno-associated virus 5; AP2, adapter protein 2; CAR, coxsackievirus and adenovirus receptor; CAV1, caveolin-1; CCR2, -5, chemokine (CC motif) receptor 2 and 5; CCV, clathrin-coated vesicles; CD4, -46, cluster of differentiation 4 and 46; CCP, clathrin-coated pit; CD81, cluster of differentiation 81; CME, clathrin-mediated endocytosis; CtBP1, C-terminal-binding protein 1; CXCR4, CXC chemokine receptor 4; DC-SIGN, dendritic cell-specific intercellular adhesion molecule-3-grabbing non-integrin; EE, early endosome; EEA1, early endosome antigen 1; ER, endoplasmic reticulum; GAGs, glycosaminoglycans; GD1α, alpha series ganglioside; HHV8/KSHV, human herpesvirus 8/Kaposi's sarcoma-associated herpesvirus; HIV-1, human immunodeficiency virus 1; HPV, human papillomavirus; HS, heparan sulfate; Hsc70, human heat shock protein 70; HVEM, herpesvirus entry mediator; IL-2, interleukin-2; ILV, intralumenal vesicle; JAM1, junctional adhesion molecule-1; LDLR, low-density lipoprotein receptor; LY, lysosome; LE, late endosome; Macro, macropinocytosis; Mimi, Mimivirus; MV, mature virion; PI3K, phosphoinositide 3-kinase; PAK1, p21-activating kinase 1; Phago, phagocytosis; Rab5, -34, Ras-related in brain; RE, recycling endosome; RTK, receptor tyrosine kinase; SARS, severe acute respiratory syndrome; SRB1, scavenger receptor class B -1; SV40, simian virus 40.

Endosome network: a complex collection of membrane organelles responsible for sorting, recycling, degradation, storage, and processing of endocytic membranes and cargo

the virus particles move randomly along the plasma membrane for an average time of 110 s before associating with preexisting, clathrin-containing domains, where their movement is constrained. In the presence of chlorpromazine, an inhibitor of clathrin assembly, the movement continues without stopping. The viruses are next delivered to endosomes; 86% to Rab5-positive EEs and 14% to Rab5- and Rab7-positive MEs. The majority (80%) fuse within an average of 12.5 min after virus addition and 5.5 min after entry into Rab7-positive LEs. Of the fusion events, the majority occurred in peripherally located endosomes.

In **Table 2**, we summarize some of the information on a number of viruses that use CME. Some seem to depend exclusively on the clathrin pathway. Others, such as Influenza A virus (a myxovirus) and lymphocytic choriomeningitis virus (LCMV, an arena virus) can also exploit other pathways.

MACROPINOCYTOSIS AND VIRUS ENTRY

Under normal conditions, macropinocytosis can be defined as a transient, growth factor-induced, actin-dependent endocytic process that leads to internalization of fluid and membrane in large vacuoles. It is unique among the pinocytic processes in involving dramatic, cell-wide plasma membrane ruffling (152). Macropinocytic vacuoles (macropinosomes) are formed when membrane ruffles fold back onto the plasma membrane to form fluid-filled cavities that close by membrane fusion (153, 154). The ruffles can take the form of lamellipodia, filopodia, or blebs.

Macropinosome formation is not guided by a particle or a cytoplasmic coat, and this gives rise to their irregular size and shape. With a diameter up to 10 μm, they are large compared to other pinocytic vacuoles and vesicles. Their formation is associated with a transient 5- to 10-fold increase in cellular fluid uptake. After formation, macropinosomes move deeper into the cytoplasm, where they can undergo acidification as well as homo- and heterotypic fusion events (155). Depending on the cell type, they either recycle back to the cell surface or feed into the endosome network and mature with the gain and loss of classic EE and LE markers before fusing with lysosomes (155, 156). Although typically associated with growth factor–induced fluid uptake, a variety of particles, including apoptotic bodies, necrotic cells, bacteria, and viruses, can induce ruffling and thus promote their macropinocytic internalization together with fluid (157–162).

Like the factors involved in other endocytic mechanisms, the cellular components required for macropinocytosis are numerous, complex, interconnected, and somewhat variable depending on the cell type, internalized cargo, mode of stimulation, and cellular conditions. Activation involves cellular lipids, kinases, GTPases, Na^+/H^+ exchangers, adaptor molecules, actin, actin modulatory factors, myosins, as well as fusion and fission factors (see **Table 1**). For the numerous functions and molecular details, we recommend several recent reviews that focus on different aspects of the process (22, 26, 27, 163–167). The role of macropinocytosis in virus entry has been recently reviewed (17, 168).

Viruses from several different families make use of macropinocytosis either as a direct means of internalization or as an indirect mechanism to assist penetration of particles that have entered by some other form of endocytosis. Viruses reported to use macropinocytosis as a direct route of entry include vaccinia virus mature virions, species B human adenovirus serotype 3, echovirus 1, group B Coxsackieviruses, herpes simplex virus 1, Kaposi's sarcoma-associated herpesvirus, and HIV-1 (17, 129, 168).

The use of macropinocytosis by these viruses has been assessed using EM for the presence of particles in macropinosomes, light microscopy for membrane ruffling and colocalization of virus with macropinocytic markers, induction of fluid uptake, as well as perturbation analyses focusing on actin, microtubules, Na^+/H^+ exchange, Rho GTPases (Rac1 or Cdc42), and various families of cellular kinases, including tyrosine, serine, threonine, and PI(3) kinases. Although all of these viruses depend on actin and Na^+/H^+ exchange for entry, each has been analyzed to varying degrees of completeness. Many require further investigation before conclusive assignment to macropinocytic internalization can be made.

The best-characterized example of virus entry by macropinocytosis is the vaccinia virus (39). Although the cellular requirements of the canonical macropinocytosis pathway are conserved, vaccinia-induced macropinocytosis differs in several ways. Notably, the association of mature virions with cells triggers the formation of large transient plasma membrane blebs rather than classical lamellipodial ruffles. FIB-SEM suggests that uptake of vaccinia and fluid occurs as part of the bleb retraction process (72). Additionally, the membrane of vaccinia is enriched in phosphatidylserine, a phospholipid required for the macropinocytic clearance of apoptotic debris (160, 169). The requirement of phosphatidylserine for vaccinia virus entry suggests that vaccinia virions elicit the macropinocytic response in host cells by mimicking apoptotic bodies (39).

There are several reasons why viruses may have evolved to use macropinocytosis for internalization and entry. In the case of vaccinia and herpes simplex virus 1, the most obvious reason is particle size. These viruses are likely too large for uptake by most other forms of endocytosis. For other viruses, macropinocytic entry may serve as a mechanism to broaden their host range or tissue specificity. The macropinosomes undergo acidification, and many of the viruses are acid activated.

Some viruses that induce macropinocytosis require it for infectivity but do not use macropinosomes as a direct internalization pathway. They include species C human adenoviruses 2 and 5 and rubella virus. These are internalized by CME, do not enter macropinosomes, and do not colocalize with macropinosome markers. Yet, the cellular requirements for infection indicate that they need macropinocytosis to be infectious. For adenoviruses 2 and 5, it has been demonstrated that induction of macropinocytosis and rupture of newly formed macropinosomes are required for escape of the virus from EEs (92, 170). Thus, the induction of macropinocytosis, formation of macropinosomes, or the release of some macropinosomal component upon rupture is somehow required for productive penetration and infection.

Lipid raft: specialized cholesterol- and sphingolipid-enriched membrane microdomains that can influence membrane fluidity, receptor clustering, and assembly of signaling molecules

VIRUSES THAT USE CAVEOLAR/RAFT-DEPENDENT ENTRY

A characteristic of the caveolar/raft-dependent pathways is that formation of primary endocytic vesicles depends on cholesterol, lipid rafts, and a complex signaling pathway involving tyrosine kinases and phosphatases. The process is ligand triggered. Uptake may involve caveolae but can also occur without these lipid raft-containing microdomains. After internalization, the cargo passes through EEs and LEs, often followed by vesicle-mediated transport to the ER. Many of the viruses that use these pathways make use of glycosphingolipids as their receptors. Once in the ER, some exploit specific ER thiol oxidoreductases and related proteins to initiate uncoating and use components of the ER-associated protein degradation pathway for penetration into the cytosol (171–174).

Uptake and forward movement in this pathway are slow and asynchronous. Depending on the virus and the cell type, the penetration event in the ER can occur as late as 6–12 h after internalization (171). Nonviral ligands include bacterial toxins, such as cholera toxin and shiga toxin, which also bind to glycolipids, and some glycosylphosphatidylinositol (GPI)-anchored proteins and their ligands, including the autocrine motility factor and cytokines (175).

The best-studied viruses that make use of caveolar/raft-dependent pathways belong to the polyomavirus family (176). They include SV40, mouse polyomavirus (mPy), and two human pathogens, BK and JC viruses. All have different gangliosides as receptors (177–179). The association with the glycan moieties is highly specific but of relatively low affinity (180). With

CAV1: caveolin-1

multiple receptors, the avidity of virus binding is high. Viruses from other families have also been reported to use caveolar/raft-dependent pathways, but information about them is incomplete (181–183).

Polyomaviruses are nonenveloped DNA viruses that replicate in the nucleus. The main structural components are 72 homopentamers of the major capsid protein, VP1, icosahedrally arranged (184). Because each VP1 has a binding site for the carbohydrate moiety of a ganglioside (180, 185), the virus offers 320 receptor-binding sites about 9 nm apart in a fixed pattern over the curved surface of the 50-nm-diameter particle. The use of gangliosides as receptors is of great significance for the polyomaviruses as it determines their behavior on the cell surface, their mechanism of endocytosis, and the intracellular pathway from endosomes to the ER. Whereas SV40 needs the ganglioside GM1 for binding, internalization, and infection (186), recent evidence for mPy indicates that the receptor ganglioside GD1α is not required for binding and endocytosis but only for guiding the virus from LEs to the ER (134).

The viruses associate with detergent-resistant microdomains in the plasma membrane of cells and with liquid-ordered phases in giant unilamellar vesicles (186). Immediately after binding to cells, mPy particles undergo rapid, random, lateral movement for 5–10 s before stopping or changing to a slow drift (77, 134). Electron micrographs show that cell-associated viruses enter uncoated, tight-fitting indentations and pits of variable depth (187–190). The tight wrapping of the membrane around the particle gives the impression that the virus actually "buds" into the cell. Recent studies show that SV40, in fact, induces membrane curvature and thus promotes the formation of what later becomes the primary endocytic vesicle (186). For the final fission reaction generating a tight-fitting small vesicle, the virus depends, however, on cellular sources of energy and cellular factors recruited in response to activation of signaling pathways.

Several observations have led to a model in which SV40 entry involves caveolae and caveolar vesicles (38, 191–196). The association with caveolae is probably explained by the higher concentration of GM1 in caveolae (197) and possibly by a tendency of virus particles to take advantage of the already existing curvature, particularly because the dimensions of caveolae are close to those of the virus. Upon careful analysis, it has, however, become apparent that in many cell types the majority of surface-bound SV40 and mPy, although present in indentations of the plasma membrane, do not associate with caveolae (134, 136, 198). If colocalization with caveolin-1 (CAV1) occurs, it is usually at a later time when most of the viruses have already been internalized. Evidence for a second, closely related mechanism of SV40 uptake has emerged from studies with CAV1-deficient cell lines and primary fibroblasts from CAV1 knockout mice. These cells are efficiently infected by SV40 and mPy, and the viruses entered via virus-containing invaginations morphologically indistinguishable from those in CAV1-containing cells (136, 198). Although CAV1 and dynamin independent, the alternate mechanism shares many of the features described for caveolar endocytosis. It is, for example, cholesterol and tyrosine kinase dependent (198). The fraction of virus using caveolar entry may vary with cell type and virus because when CAV1 was expressed in caveolin-deficient Jurkat cells, infection by mPy was increased (136).

Taken together, the results indicate that only a fraction of SV40, mPy, and probably other polyomaviruses enter via caveolae. The rest use a related, CAV1-independent mechanism. Both mechanisms are inhibited—or considerably reduced—by genistein (a general tyrosine kinase inhibitor), by nystatin and progesterone (for depletion of cholesterol), by brefeldin A (an inhibitor of Arf1, a guanine nucleotide-binding protein), by bafilomycin A (an inhibitor of the vacuolar ATPase), by lactrunculin A or jaspakinolide (inhibitors of actin dynamics), and by reduction of temperature below 20°C (26, 192, 194, 199). Internalization is accelerated by okadaic acid and orthovanadate (inhibitors of Ser and Thr, or Tyr phosphatases, respectively) (196, 198, 200). All these agents and conditions also block infectivity. Internalization through caveolae is generally slower and dependent on dynamin-2 and also more dependent on actin dynamics than the noncaveolar pathway (194).

Kirkham & Parton (26) have suggested that the underlying mechanisms for the lipid raft-dependent, caveolar, and CAV1-independent endocytic pathways are fundamentally similar. They suggest that the basic pathway is a lipid raft-dependent, cargo-activated process to which CAV1 and dynamin provide an additional level of regulation. The regulatory effect is most likely inhibitory, as caveolae are normally highly stationary (19, 196, 200), and as expression of CAV1 generally tends to suppress lipid raft-mediated endocytosis (175, 201). For more detailed discussion of the field of caveolae and their endocytosis, the reader is referred to recent reviews (13, 19, 26, 202).

For these viruses, the site of penetration is the ER, which they reach after three hours or longer depending on cell type. Viruses accumulate in tubular, smooth membrane networks with complex geometry enriched for the smooth ER markers syntaxin 17 and BiP (188, 190, 198). What happens during the intervening hours when the virus is en route to the ER is not entirely clear. Recent studies with mPy and SV40 indicate that most of the time is spent in the organelles of the classical endocytic pathway, especially in LEs (134, 136; S. Engel, T. Heger, R. Mancini, F. Herzog, A. Hayer, & A. Helenius, in preparation).

That they can enter endosomes has been documented using EM, immunofluorescence microscopy, and live cell video microscopy (134–136, 188, 203). Quantitation indicates that 15% to 20% of the viruses colocalize with EE markers 1–2 h after cell warming. Starting at ~60 min, the virus begins to colocalize with LE markers, and after 2–4 h, SV40 enters tubular, vesicular extensions that detach from the LEs and move along microtubules (132, 134; S. Engel, T. Heger, R. Mancini, F. Herzog, A. Hayer, & A. Helenius, in preparation). The viruses then accumulate in the smooth ER. How the viruses move from LEs to the ER remains unclear. The formation of tubular, SV40-containing vesicles has been observed in live cells, followed by microtubule-mediated plus and minus end-directed movement (132, 198). Recent studies with mPy indicate that this step requires the presence of the GD1α receptor (134).

Our recent experiments indicate that in most cell types, the so called caveosomes that we described as an intermediate station in the pathway of SV40 entry correspond to modified LEs or endolysosomes (132; A. Hayer, D. Ritz, S. Engel, & A. Helenius, submitted; S. Engel, T. Heger, R. Mancini, F. Herzog, A. Hayer, & A. Helenius, in preparation). Although organelles with the characteristics of caveosomes are seen in cells overexpressing CAV1-GFP or other CAV1 constructs, the corresponding CAV1-containing organelles are not present in control cells with endogenous levels of CAV1. Instead, the distribution of CAV1 in the cytoplasm corresponds to much smaller structures, often associated with EEs (203). The generation of caveosome-like structures is caused by saturation of the caveolar assembly machinery in the Golgi complex and by release of unassembled CAV1 complexes to the cell surface from which they move into LEs (A. Hayer, D. Ritz, S. Engel, & A. Helenius, submitted).

VIRUSES THAT USE UNUSUAL ENDOCYTIC PATHWAYS

In addition to the established endocytic mechanisms described above, there is a growing body of data indicating involvement of additional mechanisms in the internalization of some viruses. These have in common the lack of detectable coats by EM, the lack of dependency on clathrin and caveolin, and the transfer of viruses to the endosomal network. However, there are also clear differences, implying the existence of variations of a common theme or multiple independent mechanisms. So far, the viruses that use these pathways include rotavirus, LCMV, human papillomavirus 16 (HPV-16), and Influenza A virus. In the case of influenza virus, the process operates in parallel with CME and serves as a pathway of productive entry (74, 111, 113), but there is little detailed information available.

Rotavirus is a nonenveloped RNA virus of the reovirus family. It is the leading cause of severe diarrhea among infants and young children. The particles interact sequentially on the cell surface with sialic acid-containing molecules, integrin $\alpha2\beta1$, Hsc70, and finally integrins $\alpha5\beta3$ and $\alpha X\beta2$ (204). These interactions trigger conformational changes in the capsid preparing the virus for uncoating and membrane penetration. The endocytic mechanism is dynamin-2 dependent and somewhat sensitive to cholesterol depletion, suggesting a role for lipid rafts. These features suggest a possible relationship with the uptake of interleukin 2, which is internalized by clathrin- and caveolin-independent pathways (205, 206).

Membrane penetration of rotavirus occurs by lysis or pore formation in endosomes, mediated by the trypsin-activated VP4 protein. Infection is not blocked by lysosomotropic, weak bases, suggesting that it is independent of acid activation (138). However, bafilomycin A1, an inhibitor of the endosomal proton pump, does block infection, possibly through a decrease in Ca^{2+} concentration (137).

LCMV is an enveloped RNA virus of the arenavirus family (207, 208). For cell entry, it binds to the carbohydrate moiety of α-dystroglycan (209). Endocytosis occurs in smooth, noncoated pits (142, 210). Cholesterol is required, but clathrin, caveolin, flotillin, Arf6, dynamin-2, and actin are not (141, 142). The available evidence suggests that a large fraction of the viruses are directly transferred to the LEs, where membrane fusion is activated by the low pH (142). Bypassing conventional EEs may provide cell surface components a fast route to degradation without recycling.

HPV-16 is a nonenveloped virus of the papillomavirus family that infects mucosal epithelia and causes warts as well as cervical and anogenital tumors (211). Virus particles contain two structural proteins (L1 and L2) that form an icosahedral ($T = 7$) particle of ~55 nm in diameter. Cell entry involves an initial interaction with heparan sulfate proteoglycans (HSPGs), followed by a sequence of structural changes caused by interaction with the sugar moiety of HSPGs, by cyclophilin B, and by activation by furin, a proprotein convertase (212–214).

Although HPV-16 was originally proposed to enter by CME, recent studies indicate endocytosis occurs by a clathrin-, caveolin-, flotillin-, lipid raft-, dynamin-independent mechanism distinct from macropinocytosis and phagocytosis (143; M. Schelhaas, M. Holcer, P. Blattmann, P.M. Day, J.T. Schiller, & A. Helenius, submitted). Uptake depends on actin polymerization but is independent of Rho-like GTPases. Endocytosis via small tubular pits requires PI3-kinase and protein kinase C activities, as well as a sodium-proton exchanger (M. Schelhaas, M. Holcer, P. Blattmann, P.M. Day, J.T. Schiller, & A. Helenius, submitted). Intracellular trafficking occurs through the endosomal network, where HPV-16 seems to follow the canonical transport from EEs and LEs to lysosomes. The virus is acid activated. Taken together, the requirements for infectious endocytosis and the morphology of vesicular carriers suggest an endocytic mechanism that combines features of macropinocytosis and the mechanism used by LCMV.

Although the information is still incomplete, it is likely that viruses have evolved to use mechanisms of endocytosis that have not yet been described in the cell biology literature. It will be interesting to learn more about these pathways and to identify their physiological ligands and functions. In the cell biological literature, there are, however, pathways with which no virus has yet been associated. These include the so called GEEC pathway involved in the endocytosis of GPI-anchored proteins, the flotillin pathway for GPI-anchored proteins and proteoglycans, the Arf6 pathway for the internalization of major histocompatibility antigens, and as mentioned above, the IL-2 pathway for the internalization of some cytokine receptors (**Figure 1** and **Table 1**) (summarized in References 21, 26, and 215).

VIRUS ENTRY BY PHAGOCYTOSIS

Phagocytosis is an endocytic mechanism classically considered distinct from the pinocytic mechanisms discussed above. It is used by specialized cell types, such as macrophages and amoeba, for the uptake of large particles, such as bacteria. It shares several features with macropinocytosis, including large vacuole size, transient activation, actin dependency, cellular factors, and regulatory components (**Table 1**). However, the molecular mechanism is fundamentally different because the attachment of the particle surface to the plasma membrane not only triggers the process but guides the actin-dependent formation of a tight-fitting endocytic vacuole around the particle (27, 216, 217). The process involves a major transient reorganization of the plasma membrane, which is localized only to the region in contact with the cargo particle.

To distinguish between phagocytic and macropinocytic entry of viruses, multiple experimental approaches can be used. These include EM and light microscopy, colocalization and internalization of phagocytic and fluid phase tracers, and inhibitor profiling. Using these techniques, the entry mechanism of herpes simplex virus 1 into nonprofessional phagocytes and of the amoebal pathogen mimivirus into macrophages were found to be consistent with phagocytosis (146, 147). As suggested for macropinocytic internalization, the huge size of mimiviruses may explain why they have evolved to make use of this unusual mechanism.

PERSPECTIVES

The viruses themselves may be simple, but the complexity of the pathways involved in endocytic membrane trafficking and their regulation pose major challenges for understanding virus entry and infection. Because the plasma membrane and its components provide the boundary of the cell to the outside world, any functions that it may have, including endocytosis, are highly regulated and carefully controlled. Powerful new technologies, such as siRNA silencing screens, provide unprecedented access to the underlying network of cellular factors involved. When applied to virus infection, validated, and followed up, the information obtained using such screening approaches will lay a new and much more complete foundation for understanding the cellular basis for infectious disease (218, 219).

To move forward, however, it is important to take advantage of novel technologies, including single-virus tracking, quantitative live cell imaging, and advanced EM techniques, combined with improved assays for following a virus step-by-step through its replication cycle. What is needed is the ambitious use of a full spectrum of cell biological approaches. The effort will be important and worthwhile because it will lead to new antiviral approaches and new defenses against existing and emerging viral diseases. The outcome will be new drugs that target critical functions in the host cell. Host-directed inhibitors have advantages in being less likely to generate resistance and in allowing inhibition of multiple viruses. However, to be realistic, a detailed understanding of virus-cell interactions is required.

Most of the studies on virus entry and endocytosis are currently performed in tissue culture cells with purified virus added to the medium. However, infection in situ in the tissues and live animal hosts raises many additional questions. How do viruses penetrate tissue barriers, and how do they target specific cells for infection? How do they avoid immune surveillance and other host defenses? To what extent do they take advantage of cells and cell mobility for transmission?

In this context, it is increasingly evident that for local infection virus particles do not need to be released from the surface of infected cells. The extracellular particles of poxviruses can, for example, induce the formation of actin-containing mobile protrusions that are thought to push the newly formed, surface-associated viruses into neighboring cells (220, 221). Other viruses can take advantage of filopodia and nanotubes as bridges for moving between cells

(222, 223). There is also evidence that motile cells, such as dendritic cells, carry viruses with them and deliver them in a targeted fashion to new host cells through formation of specialized cell contact regions, sometimes called virological synapses (222, 223). The cell biology of these types of processes is of great interest for future studies of viral pathogenesis.

SUMMARY POINTS

1. Endocytic internalization provides several advantages to incoming viruses. Most importantly, endocytosis leaves no evidence of virus entry, and primary endocytic vesicles and endosomes serve as intracellular transporters for the virus.
2. To define and compare the cell factors required for different endocytic pathways, viruses offer many advantages over traditional endogenous ligands. Individual virus particles can be monitored by different microscopy techniques, and infection provides a clear, highly sensitive, quantitative end point assay.
3. Eukaryotic cells possess numerous tightly regulated endocytic mechanisms that vary in regard to the cargo internalized, mechanisms of initiation, molecular factors utilized, intracellular trafficking, and cargo destination.
4. It is becoming increasingly clear that many viruses can activate cellular signaling processes to trigger their own endocytosis and prepare the cell for invasion.
5. After internalization, viruses are channeled to the endosome network, where they take advantage of the various sorting and trafficking functions of the system.
6. Recent evidence suggests that lipid raft-dependent endocytosis can be either CAV1 dependent or independent.
7. Several novel clathrin-independent mechanisms have been described, and some of them serve as important entry pathways for viruses. They are still poorly characterized.

FUTURE ISSUES

1. What is the molecular basis and advantage for viruses that make use of multiple endocytic mechanisms? Will this cause a problem in future antiviral efforts based on inhibiting virus entry?
2. siRNA silencing screens and other systems-based approaches used to identify cell proteins for the development of cell-based antiviral agents will require further development.
3. The influence of ligand type, ligand availability, and regulatory factors on the heterogeneity and interrelationship of different endocytic pathways and endosome populations must be addressed.
4. The cellular factors and roles of the novel endocytic pathways in normal cell life need to be further defined.
5. Virus endocytosis studies need to be moved into tissue explants and model organisms in order to provide detailed information about in vivo cell type specificity, virus cell-cell transfer, and the effectiveness of potential antiviral agents.

DISCLOSURE STATEMENT

The authors are not aware of any affiliations, memberships, funding, or financial holdings that might be perceived as affecting the objectivity of this review.

ACKNOWLEDGMENTS

We thank members of the Helenius laboratory for thoughtful discussions. A.H. is supported by LipidX, the Swiss National Science Foundation, and the European Research Council, M.S. by the German Science Foundation, and J.M. by the European Molecular Biology Organization.

LITERATURE CITED

1. Morens DM, Folkers GK, Fauci AS. 2008. Emerging infections: a perpetual challenge. *Lancet Infect. Dis.* 8:710–19
2. Simon LD, Anderson TF. 1967. The infection of *Escherichia coli* by T2 and T4 bacteriophages as seen in the electron microscope. II. Structure and function of the baseplate. *Virology* 32:298–305
3. Marsh M, Bron R. 1997. SFV infection in CHO cells: cell-type specific restrictions to productive virus entry at the cell surface. *J. Cell Sci.* 110(Part 1):95–103
4. Sodeik B. 2000. Mechanisms of viral transport in the cytoplasm. *Trends Microbiol.* 8:465–72
5. Marsh M, Helenius A. 1989. Virus entry into animal cells. *Adv. Virus Res.* 36:107–51
6. Chandran K, Sullivan NJ, Felbor U, Whelan SP, Cunningham JM. 2005. Endosomal proteolysis of the Ebola virus glycoprotein is necessary for infection. *Science* 308:1643–45
7. Ebert DH, Deussing J, Peters C, Dermody TS. 2002. Cathepsin L and cathepsin B mediate reovirus disassembly in murine fibroblast cells. *J. Biol. Chem.* 277:24609–17
8. Nicola AV, McEvoy AM, Straus SE. 2003. Roles for endocytosis and low pH in herpes simplex virus entry into HeLa and Chinese hamster ovary cells. *J. Virol.* 77:5324–32
9. Daecke J, Fackler OT, Dittmar MT, Krausslich HG. 2005. Involvement of clathrin-mediated endocytosis in human immunodeficiency virus type 1 entry. *J. Virol.* 79:1581–94
10. Miyauchi K, Kim Y, Latinovic O, Morozov V, Melikyan GB. 2009. HIV enters cells via endocytosis and dynamin-dependent fusion with endosomes. *Cell* 137:433–44
11. Sieczkarski SB, Whittaker GR. 2002. Dissecting virus entry via endocytosis. *J. Gen. Virol.* 83:1535–45
12. Marsh M, Helenius A. 2006. Virus entry: open sesame. *Cell* 124:729–40
13. Damm EM, Pelkmans L. 2006. Systems biology of virus entry in mammalian cells. *Cell Microbiol.* 8:1219–27
14. Greber UF. 2002. Signalling in viral entry. *Cell. Mol. Life Sci.* 59:608–26
15. Gruenberg J. 2009. Viruses and endosome membrane dynamics. *Curr. Opin. Cell Biol.* 21:582–88
16. Helenius A. 2007. Virus entry and uncoating. In *Fields Virology*, ed. DM Knipe, 1:99–118. Philadelphia: Wolters Kluwer/Lippincott Williams & Wilkins. 5th ed.
17. Mercer J, Helenius A. 2009. Virus entry by macropinocytosis. *Nat. Cell Biol.* 11:510–20
18. Gruenberg J. 2001. The endocytic pathway: a mosaic of domains. *Nat. Rev. Mol. Cell Biol.* 2:721–30
19. Sandvig K, Torgersen ML, Raa HA, van Deurs B. 2008. Clathrin-independent endocytosis: from nonexisting to an extreme degree of complexity. *Histochem. Cell Biol.* 129:267–76
20. van Meel E, Klumperman J. 2008. Imaging and imagination: understanding the endo-lysosomal system. *Histochem. Cell Biol.* 129:253–66
21. Doherty GJ, McMahon HT. 2009. Mechanisms of endocytosis. *Annu. Rev. Biochem.* 78:857–902
22. Mayor S, Pagano RE. 2007. Pathways of clathrin-independent endocytosis. *Nat. Rev. Mol. Cell Biol.* 8:603–12
23. Bonifacino JS, Rojas R. 2006. Retrograde transport from endosomes to the *trans*-Golgi network. *Nat. Rev. Mol. Cell Biol.* 7:568–79
24. Conner SD, Schmid SL. 2003. Regulated portals of entry into the cell. *Nature* 422:37–44
25. Mellman I. 1996. Endocytosis and molecular sorting. *Annu. Rev. Cell Dev. Biol.* 12:575–625

26. Kirkham M, Parton RG. 2005. Clathrin-independent endocytosis: new insights into caveolae and non-caveolar lipid raft carriers. *Biochim. Biophys. Acta* 1746:349–63
27. Swanson JA. 2008. Shaping cups into phagosomes and macropinosomes. *Nat. Rev. Mol. Cell Biol.* 9:639–49
28. Schnapp BJ, Gelles J, Sheetz MP. 1988. Nanometer-scale measurements using video light microscopy. *Cell Motil. Cytoskelet.* 10:47–53
29. Klimstra WB, Ryman KD, Johnston RE. 1998. Adaptation of Sindbis virus to BHK cells selects for use of heparan sulfate as an attachment receptor. *J. Virol.* 72:7357–66
30. Byrnes AP, Griffin DE. 1998. Binding of Sindbis virus to cell surface heparan sulfate. *J. Virol.* 72:7349–56
31. Cotmore SF, Tattersall P. 2007. Parvoviral host range and cell entry mechanisms. *Adv. Virus Res.* 70:183–232
32. Fuchs R, Blaas D. 2008. Human rhinovirus cell entry and uncoating. In *Structure-Based Study of Viral Replication*, ed. RH Cheng, T Miyamura, pp. 1-41. Singapore: World Sci.
33. Pierson TC, Doms RW. 2003. HIV-1 entry and its inhibition. *Curr. Top. Microbiol. Immunol.* 281:1–27
34. Nemerow GR. 2000. Cell receptors involved in adenovirus entry. *Virology* 274:1–4
35. Mothes W, Boerger AL, Narayan S, Cunningham JM, Young JA. 2000. Retroviral entry mediated by receptor priming and low pH triggering of an envelope glycoprotein. *Cell* 103:679–89
36. Coyne CB, Bergelson JM. 2006. Virus-induced Abl and Fyn kinase signals permit coxsakievirus entry through epithelial tight junctions. *Cell* 124:119–31
37. Helle F, Dubuisson J. 2008. Hepatitis C virus entry into host cells. *Cell. Mol. Life Sci.* 65:100–12
38. Pelkmans L, Fava E, Grabner H, Hannus M, Habermann B, et al. 2005. Genome-wide analysis of human kinases in clathrin- and caveolae/raft-mediated endocytosis. *Nature* 436:78–86
39. Mercer J, Helenius A. 2008. Vaccinia virus uses macropinocytosis and apoptotic mimicry to enter host cells. *Science* 320:531–35
40. Ehrlich M, Boll W, Van Oijen A, Hariharan R, Chandran K, et al. 2004. Endocytosis by random initiation and stabilization of clathrin-coated pits. *Cell* 118:591–605
41. Johannsdottir HK, Mancini R, Kartenbeck J, Amato L, Helenius A. 2009. Host cell factors and functions involved in vesicular stomatitis virus entry. *J. Virol.* 83:440–53
42. Vonderheit A, Helenius A. 2005. Rab7 associates with early endosomes to mediate sorting and transport of Semliki Forest virus to late endosomes. *PLoS Biol.* 3:e233
43. Gagescu R, Demaurex N, Parton RG, Hunziker W, Huber LA, Gruenberg J. 2000. The recycling endosome of Madin-Darby canine kidney cells is a mildly acidic compartment rich in raft components. *Mol. Biol. Cell* 11:2775–91
44. Zerial M, McBride H. 2001. Rab proteins as membrane organizers. *Nat. Rev. Mol. Cell Biol.* 2:107–17
45. Helenius A, Mellman I, Wall D, Hubbard A. 1983. Endosomes. *Trends Biochem. Sci.* 8:245–50
46. Luzio JP, Pryor PR, Bright NA. 2007. Lysosomes: fusion and function. *Nat. Rev. Mol. Cell Biol.* 8:622–32
47. Mukherjee S, Maxfield FR. 2004. Membrane domains. *Annu. Rev. Cell Dev. Biol.* 20:839–66
48. Rink J, Ghigo E, Kalaidzidis Y, Zerial M. 2005. Rab conversion as a mechanism of progression from early to late endosomes. *Cell* 122:735–49
49. Braulke T, Bonifacino JS. 2009. Sorting of lysosomal proteins. *Biochim. Biophys. Acta* 1793:605–14
50. Lakadamyali M, Rust MJ, Zhuang X. 2006. Ligands for clathrin-mediated endocytosis are differentially sorted into distinct populations of early endosomes. *Cell* 124:997–1009
51. Mukherjee S, Maxfield FR. 2004. Lipid and cholesterol trafficking in NPC. *Biochim. Biophys. Acta* 1685:28–37
52. Kielian M, Marsh M, Helenius A. 1986. Kinetics of endosome acidification detected by mutant and wild-type Semliki Forest virus. *EMBO J.* 5:3103–9
53. Marsh M, Griffiths G, Dean GE, Mellman I, Helenius A. 1986. Three-dimensional structure of endosomes in BHK-21 cells. *Proc. Natl. Acad. Sci. USA* 83:2899–903
54. Wall DA, Wilson G, Hubbard AL. 1980. The galactose-specific recognition system of mammalian liver: the route of ligand internalization in rat hepatocytes. *Cell* 21:79–93
55. Gu F, Gruenberg J. 1999. Biogenesis of transport intermediates in the endocytic pathway. *FEBS Lett.* 452:61–66
56. van der Schaar HM, Rust MJ, Chen C, van der Ende-Metselaar H, Wilschut J, et al. 2008. Dissecting the cell entry pathway of dengue virus by single-particle tracking in living cells. *PLoS Pathog.* 4:e1000244

57. Mukherjee S, Soe TT, Maxfield FR. 1999. Endocytic sorting of lipid analogues differing solely in the chemistry of their hydrophobic tails. *J. Cell Biol.* 144:1271–84
58. Jentsch TJ. 2008. CLC chloride channels and transporters: from genes to protein structure, pathology and physiology. *Crit. Rev. Biochem. Mol. Biol.* 43:3–36
59. Lafourcade C, Sobo K, Kieffer-Jaquinod S, Garin J, van der Goot FG. 2008. Regulation of the V-ATPase along the endocytic pathway occurs through reversible subunit association and membrane localization. *PLoS ONE* 3:e2758
60. Piper RC, Katzmann DJ. 2007. Biogenesis and function of multivesicular bodies. *Annu. Rev. Cell Dev. Biol.* 23:519–47
61. Gillooly DJ, Morrow IC, Lindsay M, Gould R, Bryant NJ, et al. 2000. Localization of phosphatidylinositol 3-phosphate in yeast and mammalian cells. *EMBO J.* 19:4577–88
62. Odorizzi G, Babst M, Emr SD. 1998. Fab1p PtdIns(3)P 5-kinase function essential for protein sorting in the multivesicular body. *Cell* 95:847–58
63. Clague MJ, Urbe S, Aniento F, Gruenberg J. 1994. Vacuolar ATPase activity is required for endosomal carrier vesicle formation. *J. Biol. Chem.* 269:21–24
64. Baravalle G, Schober D, Huber M, Bayer N, Murphy RF, Fuchs R. 2005. Transferrin recycling and dextran transport to lysosomes is differentially affected by bafilomycin, nocodazole, and low temperature. *Cell Tissue Res.* 320:99–113
65. Bayer N, Schober D, Prchla E, Murphy RF, Blaas D, Fuchs R. 1998. Effect of bafilomycin A1 and nocodazole on endocytic transport in HeLa cells: implications for viral uncoating and infection. *J. Virol.* 72:9645–55
66. Brabec M, Blaas D, Fuchs R. 2006. Wortmannin delays transfer of human rhinovirus serotype 2 to late endocytic compartments. *Biochem. Biophys. Res. Commun.* 348:741–49
67. Berka U, Khan A, Blaas D, Fuchs R. 2009. Human rhinovirus type 2 uncoating at the plasma membrane is not affected by a pH gradient but is affected by the membrane potential. *J. Virol.* 83:3778–87
68. Maxfield FR, McGraw TE. 2004. Endocytic recycling. *Nat. Rev. Mol. Cell Biol.* 5:121–32
69. Liberali P, Ramo P, Pelkmans L. 2008. Protein kinases: starting a molecular systems view of endocytosis. *Annu. Rev. Cell Dev. Biol.* 24:501–23
70. Spang A. 2009. On the fate of early endosomes. *Biol. Chem.* 390:753–59
71. Plitzko JM, Rigort A, Leis A. 2009. Correlative cryo-light microscopy and cryo-electron tomography: from cellular territories to molecular landscapes. *Curr. Opin. Biotechnol.* 20:83–89
72. Lucas M, Mercer J, Gasser P, Schertel A, Gunthert M, et al. 2008. Correlative 3D microscopy: CLSM and FIB/SEM tomography. A study of cellular entry of vaccinia virus. *Imaging Microsc.* 10:30–31
73. Lakadamyali M, Rust MJ, Babcock HP, Zhuang X. 2003. Visualizing infection of individual influenza viruses. *Proc. Natl. Acad. Sci. USA* 100:9280–85
74. Rust MJ, Lakadamyali M, Zhang F, Zhuang X. 2004. Assembly of endocytic machinery around individual influenza viruses during viral entry. *Nat. Struct. Mol. Biol.* 11:567–73
75. Suomalainen M, Nakano MY, Keller S, Boucke K, Stidwill RP, Greber UF. 1999. Microtubule-dependent plus- and minus end-directed motilities are competing processes for nuclear targeting of adenovirus. *J. Cell Biol.* 144:657–72
76. Brauchle C, Seisenberger G, Endress T, Ried MU, Buning H, Hallek M. 2002. Single virus tracing: visualization of the infection pathway of a virus into a living cell. *ChemPhysChem* 3:299–303
77. Ewers H, Smith AE, Sbalzarini IF, Lilie H, Koumoutsakos P, Helenius A. 2005. Single-particle tracking of murine polyoma virus-like particles on live cells and artificial membranes. *Proc. Natl. Acad. Sci. USA* 102:15110–15
78. Lehmann MJ, Sherer NM, Marks CB, Pypaert M, Mothes W. 2005. Actin- and myosin-driven movement of viruses along filopodia precedes their entry into cells. *J. Cell Biol.* 170:317–25
79. Kukura P, Ewers H, Müller C, Renn A, Helenius A, Sandoghdar V. 2009. High-speed nanoscopic tracking of the position and orientation of a single virus. *Nat. Methods* 6:923–27
80. Ohkuma S, Poole B. 1978. Fluorescence probe measurement of the intralysosomal pH in living cells and the perturbation of pH by various agents. *Proc. Natl. Acad. Sci. USA* 75:3327–31
81. Kielian M, Marsh M, Helenius A. 1986. Entry of alphaviruses. In *The Togaviridae and Flaviviridae*, ed. SS Schlesinger, MJ Schlesinger, pp. 91–119. New York: Plenum

82. Subtil A, Gaidarov I, Kobylarz K, Lampson MA, Keen JH, McGraw TE. 1999. Acute cholesterol depletion inhibits clathrin-coated pit budding. *Proc. Natl. Acad. Sci. USA* 96:6775–80
83. Rodal SK, Skretting G, Garred O, Vilhardt F, van Deurs B, Sandvig K. 1999. Extraction of cholesterol with methyl-beta-cyclodextrin perturbs formation of clathrin-coated endocytic vesicles. *Mol. Biol. Cell* 10:961–74
84. Dales S. 1973. Early events in cell-animal virus interactions. *Bacteriol. Rev.* 37:103–35
85. Dahlberg JE. 1974. Quantitative electron microscopic analysis of the penetration of VSV into L cells. *Virology* 58:250–62
86. Helenius A, Kartenbeck J, Simons K, Fries E. 1980. On the entry of Semliki Forest virus into BHK-21 cells. *J. Cell Biol.* 84:404–20
87. Matlin KS, Reggio H, Helenius A, Simons K. 1982. The pathway of vesicular stomatitis virus entry leading to infection. *J. Mol. Biol.* 156:609–31
88. Cureton DK, Massol RH, Saffarian S, Kirchhausen TL, Whelan SP. 2009. Vesicular stomatitis virus enters cells through vesicles incompletely coated with clathrin that depend upon actin for internalization. *PLoS Pathog.* 5:e1000394
89. Le Blanc I, Luyet PP, Pons V, Ferguson C, Emans N, et al. 2005. Endosome-to-cytosol transport of viral nucleocapsids. *Nat. Cell Biol.* 7:653–64
90. Doxsey JS, Brodsky FM, Blank GS, Helenius A. 1987. Inhibition of endocytosis by anti-clathrin antibodies. *Cell* 50:453–63
91. Marsh M, Bolzau E, Helenius A. 1983. Penetration of Semliki Forest virus from acidic prelysosomal vacuoles. *Cell* 32:931–40
92. Gastaldelli M, Imelli N, Boucke K, Amstutz B, Meier O, Greber UF. 2008. Infectious adenovirus type 2 transport through early but not late endosomes. *Traffic* 9:2265–78
93. Glomb-Reinmund S, Kielian M. 1998. The role of low pH and disulfide shuffling in the entry and fusion of Semliki Forest virus and Sindbis virus. *Virology* 248:372–81
94. DeTulleo L, Kirchhausen T. 1998. The clathrin endocytic pathway in viral infection. *EMBO J.* 17:4585–93
95. Abell BA, Brown DT. 1993. Sindbis virus membrane fudion is mediated by reduction of glycoprotein disulfide bridges at the cell surface. *J. Virol.* 67:5496–501
96. Smit JM, Bittman R, Wilschut J. 1999. Low-pH-dependent fusion of Sindbis virus with receptor-free cholesterol- and sphingolipid-containing liposomes. *J. Virol.* 73:8476–84
97. Chu JJ, Ng ML. 2004. Infectious entry of West Nile virus occurs through a clathrin-mediated endocytic pathway. *J. Virol.* 78:10543–55
98. Acosta EG, Castilla V, Damonte EB. 2009. Alternative infectious entry pathways for dengue virus serotypes into mammalian cells. *Cell Microbiol.* 11:1533–49
99. Meertens L, Bertaux C, Dragic T. 2006. Hepatitis C virus entry requires a critical postinternalization step and delivery to early endosomes via clathrin-coated vesicles. *J. Virol.* 80:11571–78
100. Stewart PL, Dermody TS, Nemerow GR. 2003. Structural basis of nonenveloped virus cell entry. *Adv. Protein Chem.* 64:455–91
101. Medina-Kauwe LK. 2003. Endocytosis of adenovirus and adenovirus capsid proteins. *Adv. Drug Deliv. Rev.* 55:1485–96
102. Johns HL, Berryman S, Monaghan P, Belsham GJ, Jackson T. 2009. A dominant-negative mutant of rab5 inhibits infection of cells by foot-and-mouth disease virus: implications for virus entry. *J. Virol.* 83:6247–56
103. O'Donnell V, LaRocco M, Duque H, Baxt B. 2005. Analysis of foot-and-mouth disease virus internalization events in cultured cells. *J. Virol.* 79:8506–18
104. Brabec-Zaruba M, Pfanzagl B, Blaas D, Fuchs R. 2009. Site of human rhinovirus RNA uncoating revealed by fluorescent in situ hybridization. *J. Virol.* 83:3770–77
105. Snyers L, Zwickl H, Blaas D. 2003. Human rhinovirus type 2 is internalized by clathrin-mediated endocytosis. *J. Virol.* 77:5360–69
106. Forzan M, Marsh M, Roy P. 2007. Bluetongue virus entry into cells. *J. Virol.* 81:4819–27
107. Suikkanen S, Antila M, Jaatinen A, Vihinen-Ranta M, Vuento M. 2003. Release of canine parvovirus from endocytic vesicles. *Virology* 316:267–80

108. Suikkanen S, Saajarvi K, Hirsimaki J, Valilehto O, Reunanen H, et al. 2002. Role of recycling endosomes and lysosomes in dynein-dependent entry of canine parvovirus. *J. Virol.* 76:4401–11
109. Bantel-Schaal U, Hub B, Kartenbeck J. 2002. Endocytosis of adeno-associated virus type 5 leads to accumulation of virus particles in the Golgi compartment. *J. Virol.* 76:2340–49
110. Diaz-Griffero F, Jackson AP, Brojatsch J. 2005. Cellular uptake of avian leukosis virus subgroup B is mediated by clathrin. *Virology* 337:45–54
111. Chen C, Zhuang X. 2008. Epsin 1 is a cargo-specific adaptor for the clathrin-mediated endocytosis of the influenza virus. *Proc. Natl. Acad. Sci. USA* 105:11790–95
112. Sieczkarski SB, Whittaker GR. 2002. Influenza virus can enter and infect cells in the absence of clathrin-mediated endocytosis. *J. Virol.* 76:10455–64
113. Matlin KS, Reggio H, Helenius A, Simons K. 1982. Infectious entry pathway of influenza virus in a canine kidney cell line. *J. Cell Biol.* 91:601–13
114. Huang CY, Lu TY, Bair CH, Chang YS, Jwo JK, Chang W. 2008. A novel cellular protein, VPEF, facilitates vaccinia virus penetration into HeLa cells through fluid phase endocytosis. *J. Virol.* 82:7988–99
115. Locker JK, Kuehn A, Schleich S, Rutter G, Hohenberg H, et al. 2000. Entry of the two infectious forms of vaccinia virus at the plasma membane is signaling-dependent for the IMV but not the EEV. *Mol. Biol. Cell* 11:2497–511
116. Amstutz B, Gastaldelli M, Kalin S, Imelli N, Boucke K, et al. 2008. Subversion of CtBP1-controlled macropinocytosis by human adenovirus serotype 3. *EMBO J.* 27:956–69
117. Wickham TJ, Mathias P, Cheresh DA, Nemerow GR. 1993. Integrins alpha v beta 3 and alpha v beta 5 promote adenovirus internalization but not virus attachment. *Cell* 73:309–19
118. Sirena D, Lilienfeld B, Eisenhut M, Kalin S, Boucke K, et al. 2004. The human membrane cofactor CD46 is a receptor for species B adenovirus serotype 3. *J. Virol.* 78:4454–62
119. Liberali P, Kakkonen E, Turacchio G, Valente C, Spaar A, et al. 2008. The closure of Pak1-dependent macropinosomes requires the phosphorylation of CtBP1/BARS. *EMBO J.* 27:970–81
120. Karjalainen M, Kakkonen E, Upla P, Paloranta H, Kankaanpaa P, et al. 2008. A Raft-derived, Pak1-regulated entry participates in $\alpha 2$ $\beta 1$ integrin-dependent sorting to caveosomes. *Mol. Biol. Cell* 19:2857–69
121. Coyne CB, Bergelson JM. 2006. Virus-induced Abl and Fyn kinase signals permit Coxsackievirus entry through epithelial tight junctions. *Cell* 124:119–31
122. Coyne CB, Shen L, Turner JR, Bergelson JM. 2007. Coxsackievirus entry across epithelial tight junctions requires occludin and the small GTPases Rab34 and Rab5. *Cell Host Microbe* 2:181–92
123. Nicola AV, Hou J, Major EO, Straus SE. 2005. Herpes simplex virus type 1 enters human epidermal keratinocytes, but not neurons, via a pH-dependent endocytic pathway. *J. Virol.* 79:7609–16
124. Butcher M, Raviprakash K, Ghosh HP. 1990. Acid pH-induced fusion of cells by herpes simplex virus glycoproteins gB an gD. *J. Biol. Chem.* 265:5862–68
125. Marechal V, Prevost MC, Petit C, Perret E, Heard JM, Schwartz O. 2001. Human immunodeficiency virus type 1 entry into macrophages mediated by macropinocytosis. *J. Virol.* 75:11166–77
126. Liu NQ, Lossinsky AS, Popik W, Li X, Gujuluva C, et al. 2002. Human immunodeficiency virus type 1 enters brain microvascular endothelia by macropinocytosis dependent on lipid rafts and the mitogen-activated protein kinase signaling pathway. *J. Virol.* 76:6689–700
127. Nguyen DG, Wolff KC, Yin H, Caldwell JS, Kuhen KL. 2006. "UnPAKing" human immunodeficiency virus (HIV) replication: using small interfering RNA screening to identify novel cofactors and elucidate the role of group I PAKs in HIV infection. *J. Virol.* 80:130–37
128. Fontenot DR, den Hollander P, Vela EM, Newman R, Sastry JK, Kumar R. 2007. Dynein light chain 1 peptide inhibits human immunodeficiency virus infection in eukaryotic cells. *Biochem. Biophys. Res. Commun.* 363:901–7
129. Raghu H, Sharma-Walia N, Veettil MV, Sadagopan S, Chandran B. 2009. Kaposi's sarcoma-associated herpesvirus utilizes an actin polymerization-dependent macropinocytic pathway to enter human dermal microvascular endothelial and human umbilical vein endothelial cells. *J. Virol.* 83:4895–911

130. Anderson HA, Chen Y, Norkin LC. 1996. Bound simian virus 40 translocates to caveolin-enriched membrane domains, and its entry is inhibited by drugs that selectively disrupt caveolae. *Mol. Biol. Cell* 7:1825–34
131. Stang E, Kartenbeck J, Parton RG. 1997. Major histocompatibility complex class I molecules mediate association of SV40 with caveolae. *Mol. Biol. Cell* 8:47–57
132. Pelkmans L, Kartenbeck J, Helenius A. 2001. Caveolar endocytosis of simian virus 40 reveals a new two-step vesicular-transport pathway to the ER. *Nat. Cell Biol.* 3:473–83
133. Smith JL, Campos SK, Ozbun MA. 2007. Human papillomavirus type 31 uses a caveolin 1- and dynamin 2-mediated entry pathway for infection of human keratinocytes. *J. Virol.* 81:9922–31
134. Qian M, Cai D, Verhey KJ, Tsai B. 2009. A lipid receptor sorts polyomavirus from the endolysosome to the endoplasmic reticulum to cause infection. *PLoS Pathog.* 5:e1000465
135. Mannova P, Forstova J. 2003. Mouse polyomavirus utilizes recycling endosomes for a traffic pathway independent of COPI vesicle transport. *J. Virol.* 77:1672–81
136. Liebl D, Difato F, Hornikova L, Mannova P, Stokrova J, Forstova J. 2006. Mouse polyomavirus enters early endosomes, requires their acidic pH for productive infection, and meets transferrin cargo in Rab11-positive endosomes. *J. Virol.* 80:4610–22
137. Chemello ME, Aristimuno OC, Michelangeli F, Ruiz MC. 2002. Requirement for vacuolar H^+-ATPase activity and Ca^{2+} gradient during entry of rotavirus into MA104 cells. *J. Virol.* 76:13083–87
138. Sánchez-San Martin C, López T, Arias CF, López S. 2004. Characterization of rotavirus cell entry. *J. Virol.* 78:2310–18
139. Van Hamme E, Dewerchin HL, Cornelissen E, Verhasselt B, Nauwynck HJ. 2008. Clathrin- and caveolae-independent entry of feline infectious peritonitis virus in monocytes depends on dynamin. *J. Gen. Virol.* 89:2147–56
140. Wang H, Yang P, Liu K, Guo F, Zhang Y, et al. 2008. SARS coronavirus entry into host cells through a novel clathrin- and caveolae-independent endocytic pathway. *Cell Res.* 18:290–301
141. Rojek JM, Perez M, Kunz S. 2008. Cellular entry of lymphocytic choriomeningitis virus. *J. Virol.* 82:1505–17
142. Quirin K, Eschli B, Scheu I, Poort L, Kartenbeck J, Helenius A. 2008. Lymphocytic choriomeningitis virus uses a novel endocytic pathway for infectious entry via late endosomes. *Virology* 378:21–33
143. Spoden G, Freitag K, Husmann M, Boller K, Sapp M, et al. 2008. Clathrin- and caveolin-independent entry of human papillomavirus type 16–involvement of tetraspanin-enriched microdomains (TEMs). *PLoS ONE* 3:e3313
144. Misinzo G, Delputte PL, Lefebvre DJ, Nauwynck HJ. 2009. Porcine circovirus 2 infection of epithelial cells is clathrin-, caveolae- and dynamin-independent, actin and Rho-GTPase-mediated, and enhanced by cholesterol depletion. *Virus Res.* 139:1–9
145. Misinzo G, Meerts P, Bublot M, Mast J, Weingartl HM, Nauwynck HJ. 2005. Binding and entry characteristics of porcine circovirus 2 in cells of the porcine monocytic line 3D4/31. *J. Gen. Virol.* 86:2057–68
146. Clement C, Tiwari V, Scanlan PM, Valyi-Nagy T, Yue BY, Shukla D. 2006. A novel role for phagocytosis-like uptake in herpes simplex virus entry. *J. Cell Biol.* 174:1009–21
147. Ghigo E, Kartenbeck J, Lien P, Pelkmans L, Capo C, et al. 2008. Ameobal pathogen mimivirus infects macrophages through phagocytosis. *PLoS Pathog.* 4:e1000087
148. Johannes L, Lamaze C. 2002. Clathrin-dependent or not: Is it still the question? *Traffic* 3:443–51
149. Robinson MS. 2004. Adaptable adaptors for coated vesicles. *Trends Cell Biol.* 14:167–74
150. Motley A, Bright NA, Seaman MN, Robinson MS. 2003. Clathrin-mediated endocytosis in AP-2-depleted cells. *J. Cell Biol.* 162:909–18
151. Wang X, Huong SM, Chiu ML, Raab-Traub N, Huang ES. 2003. Epidermal growth factor receptor is a cellular receptor for human cytomegalovirus. *Nature* 424:456–61
152. Watts C, Marsh M. 1992. Endocytosis: What goes in and how? *J. Cell Sci.* 103(Part 1):1–8
153. Swanson JA, Watts C. 1995. Macropinocytosis. *Trends Cell Biol.* 5:424–28
154. Swanson JA. 1989. Phorbol esters stimulate macropinocytosis and solute flow through macrophages. *J. Cell Sci.* 94(Part 1):135–42

155. Hewlett LJ, Prescott AR, Watts C. 1994. The coated pit and macropinocytic pathways serve distinct endosome populations. *J. Cell Biol.* 124:689–703
156. Racoosin EL, Swanson JA. 1993. Macropinosome maturation and fusion with tubular lysosomes in macrophages. *J. Cell Biol.* 121:1011–20
157. Watarai M, Derre I, Kirby J, Growney JD, Dietrich WF, Isberg RR. 2001. *Legionella pneumophila* is internalized by a macropinocytotic uptake pathway controlled by the Dot/Icm system and the mouse Lgn1 locus. *J. Exp. Med.* 194:1081–96
158. Erwig LP, Henson PM. 2008. Clearance of apoptotic cells by phagocytes. *Cell Death Differ.* 15:243–50
159. Xu W, Roos A, Schlagwein N, Woltman AM, Daha MR, van Kooten C. 2006. IL-10-producing macrophages preferentially clear early apoptotic cells. *Blood* 107:4930–37
160. Hoffmann PR, deCathelineau AM, Ogden CA, Leverrier Y, Bratton DL, et al. 2001. Phosphatidylserine (PS) induces PS receptor-mediated macropinocytosis and promotes clearance of apoptotic cells. *J. Cell Biol.* 155:649–59
161. Fabbri A, Falzano L, Travaglione S, Stringaro A, Malorni W, et al. 2002. Rho-activating *Escherichia coli* cytotoxic necrotizing factor 1: macropinocytosis of apoptotic bodies in human epithelial cells. *Int. J. Med. Microbiol.* 291:551–54
162. Francis CL, Ryan TA, Jones BD, Smith SJ, Falkow S. 1993. Ruffles induced by *Salmonella* and other stimuli direct macropinocytosis of bacteria. *Nature* 364:639–42
163. Norbury CC. 2006. Drinking a lot is good for dendritic cells. *Immunology* 117:443–51
164. Jones AT. 2007. Macropinocytosis: searching for an endocytic identity and role in the uptake of cell penetrating peptides. *J. Cell. Mol. Med.* 11:670–84
165. Donaldson JG, Porat-Shliom N, Cohen LA. 2009. Clathrin-independent endocytosis: a unique platform for cell signaling and PM remodeling. *Cell Signal.* 21:1–6
166. Amyere M, Mettlen M, Van Der Smissen P, Platek A, Payrastre B, et al. 2002. Origin, originality, functions, subversions and molecular signalling of macropinocytosis. *Int. J. Med. Microbiol.* 291:487–94
167. Sansonetti PJ. 2001. Microbes and microbial toxins: paradigms for microbial-mucosal interactions III. Shigellosis: from symptoms to molecular pathogenesis. *Am. J. Physiol. Gastrointest. Liver Physiol.* 280:G319–23
168. Kerr MC, Teasdale RD. 2009. Defining macropinocytosis. *Traffic* 10:364–71
169. Henson PM, Bratton DL, Fadok VA. 2001. Apoptotic cell removal. *Curr. Biol.* 11:R795–805
170. Meier O, Boucke K, Hammer SV, Keller S, Stidwill RP, et al. 2002. Adenovirus triggers macropinocytosis and endosomal leakage together with its clathrin-mediated uptake. *J. Cell Biol.* 158:1119–31
171. Schelhaas M, Malmstrom J, Pelkmans L, Haugstetter J, Ellgaard L, et al. 2007. Simian virus 40 depends on ER protein folding and quality control factors for entry into host cells. *Cell* 131:516–29
172. Jiang M, Abend JR, Tsai B, Imperiale MJ. 2009. Early events during BK virus entry and disassembly. *J. Virol.* 83:1350–58
173. Lilley BN, Gilbert JM, Ploegh HL, Benjamin TL. 2006. Murine polyomavirus requires the endoplasmic reticulum protein Derlin-2 to initiate infection. *J. Virol.* 80:8739–44
174. Magnuson B, Rainey EK, Benjamin T, Baryshev M, Mkrtchian S, Tsai B. 2005. ERp29 triggers a conformational change in polyomavirus to stimulate membrane binding. *Mol. Cell* 20:289–300
175. Lajoie P, Nabi IR. 2007. Regulation of raft-dependent endocytosis. *J. Cell. Mol. Med.* 11:644–53
176. Eash S, Manley K, Gasparovic M, Querbes W, Atwood WJ. 2006. The human polyomaviruses. *Cell. Mol. Life Sci.* 63:865–76
177. Low JA, Magnuson B, Tsai B, Imperiale MJ. 2006. Identification of gangliosides GD1b and GT1b as receptors for BK virus. *J. Virol.* 80:1361–66
178. Smith AE, Lilie H, Helenius A. 2003. Ganglioside-dependent cell attachment and endocytosis of murine polyomavirus-like particles. *FEBS Lett.* 555:199–203
179. Tsai B, Gilbert JM, Stehle T, Lencer W, Benjamin TL, Rapoport TA. 2003. Gangliosides are receptors for murine polyoma virus and SV40. *EMBO J.* 22:4346–55
180. Stehle T, Yan Y, Benjamin TL, Harrison SC. 1994. Structure of murine polyomavirus complexed with an oligosaccharide receptor fragment. *Nature* 369:160–63
181. Empig CJ, Goldsmith MA. 2002. Association of the caveola vesicular system with cellular entry by filoviruses. *J. Virol.* 76:5266–70

182. Nomura R, Kiyota A, Suzaki E, Kataoka K, Ohe Y, et al. 2004. Human coronavirus 229E binds to CD13 in rafts and enters the cell through caveolae. *J. Virol.* 78:8701–8
183. Werling D, Hope JC, Chaplin P, Collins RA, Taylor G, Howard CJ. 1999. Involvement of caveolae in the uptake of respiratory syncytial virus antigen by dendritic cells. *J. Leukoc. Biol.* 66:50–58
184. Liddington RC, Yan Y, Moulai J, Sahli R, Benjamin TL, Harrison SC. 1991. Structure of simian virus 40 at 3.8-Å resolution. *Nature* 354:278–84
185. Neu U, Stehle T, Atwood WJ. 2009. The *Polyomaviridae*: contributions of virus structure to our understanding of virus receptors and infectious entry. *Virology* 384:389–99
186. Ewers H, Römer W, Smith AE, Bacia K, Dmitrieff S, et al. 2009. GM1 structure determines SV40-induced membrane invagination and infection. *Nat. Cell Biol.* 12:11–18
187. Hummeler K, Tomassini N, Sokol F. 1970. Morphological aspects of the uptake of simian virus 40 by permissive cells. *J. Virol.* 6:87–93
188. Kartenbeck J, Stukenbrok H, Helenius A. 1989. Endocytosis of simian virus 40 into the endoplasmic reticulum. *J. Cell Biol.* 109:2721–29
189. Mackay RL, Consigli RA. 1976. Early events in polyoma virus infection: attachment, penetration, and nuclear entry. *J. Virol.* 19:620–36
190. Maul GG, Rovera G, Vorbrodt A, Abramczuk J. 1978. Membrane fusion as a mechanism of simian virus 40 entry into different cellular compartments. *J. Virol.* 28:936–44
191. Pelkmans L, Helenius A. 2003. Insider information: what viruses tell us about endocytosis. *Curr. Opin. Cell Biol.* 15:414–22
192. Norkin LC. 2001. Caveolae in the uptake and targeting of infectious agents and secreted toxins. *Adv. Drug Deliv. Rev.* 49:301–15
193. Parton RG, Lindsay M. 1999. Exploitation of major histocompatibility complex class I molecules and caveole by simian vius 40. *Immunol. Rev.* 168:23–31
194. Pelkmans L, Puntener D, Helenius A. 2002. Local actin polymerization and dynamin recruitment in SV40-induced internalization of caveolae. *Science* 296:535–39
195. Roy S, Luetterforst R, Harding A, Apolloni A, Etheridge M, et al. 1999. Dominant-negative caveolin inhibits H-Ras function by disrupting cholesterol-rich plasma membrane domains. *Nat. Cell Biol.* 1:98–105
196. Tagawa A, Mezzacasa A, Hayer A, Longatti A, Pelkmans L, Helenius A. 2005. Assembly and trafficking of caveolar domains in the cell: caveolae as stable, cargo-triggered, vesicular transporters. *J. Cell Biol.* 170:769–79
197. Parton RG. 1994. Ultrastructural localization of gangliosides; GM1 is concentrated in caveolae. *J. Histochem. Cytochem.* 42:155–66
198. Damm EM, Pelkmans L, Kartenbeck J, Mezzacasa A, Kurzchalia T, Helenius A. 2005. Clathrin- and caveolin-1-independent endocytosis: entry of simian virus 40 into cells devoid of caveolae. *J. Cell Biol.* 168:477–88
199. Norkin LC, Anderson HA, Wolfrom SA, Oppenheim A. 2002. Caveolar endocytosis of simian virus 40 is followed by brefeldin A-sensitive transport to the endoplasmic reticulum, where the virus disassembles. *J. Virol.* 76:5156–66
200. Thomsen P, Roepstorff K, Stahlhut M, van Deurs B. 2002. Caveolae are highly immobile plasma membrane microdomains, which are not involved in constitutive endocytic trafficking. *Mol. Biol. Cell* 13:238–50
201. Kojic LD, Joshi B, Lajoie P, Le PU, Cox ME, et al. 2007. Raft-dependent endocytosis of autocrine motility factor is phosphatidylinositol 3-kinase-dependent in breast carcinoma cells. *J. Biol. Chem.* 282:29305–13
202. Parton RG, Richards AA. 2003. Lipid rafts and caveolae as portals for endocytosis: new insights and common mechanisms. *Traffic* 4:724–38
203. Pelkmans L, Bürli T, Zerial M, Helenius A. 2004. Caveolin-stabilized membrane domains as multifunctional transport and sorting devices in endocytic membrane traffic. *Cell* 118:767–80
204. López S, Arias CF. 2004. Multistep entry of rotavirus into cells: a Versaillesque dance. *Trends Microbiol.* 12:271–78

205. Lamaze C, Dujeancourt A, Baba T, Lo CG, Benmerah A, Dautry-Varsat A. 2001. Interleukin 2 receptors and detergent-resistant membrane domains define a clathrin-independent endocytic pathway. *Mol. Cell* 7:661–71
206. Sauvonnet N, Dujeancourt A, Dautry-Varsat A. 2005. Cortactin and dynamin are required for the clathrin-independent endocytosis of γc cytokine receptor. *J. Cell Biol.* 168:155–63
207. Neuman BW, Adair BD, Burns JW, Milligan RA, Buchmeier MJ, Yeager M. 2005. Complementarity in the supramolecular design of arenaviruses and retroviruses revealed by electron cryomicroscopy and image analysis. *J. Virol.* 79:3822–30
208. Buchmeier MJ. 2002. Arenaviruses: protein structure and function. *Curr. Top. Microbiol. Immunol.* 262:159–73
209. Cao W, Henry MD, Borrow P, Yamada H, Elder JH, et al. 1998. Identification of α-dystroglycan as a receptor for lymphocytic choriomeningitis virus and Lassa fever virus. *Science* 282:2079–81
210. Borrow P, Oldstone MB. 1994. Mechanism of lymphocytic choriomeningitis virus entry into cells. *Virology* 198:1–9
211. Longworth MS, Laimins LA. 2004. Pathogenesis of human papillomaviruses in differentiating epithelia. *Microbiol. Mol. Biol. Rev.* 68:362–72
212. Joyce JG, Tung JS, Przysiecki CT, Cook JC, Lehman ED, et al. 1999. The L1 major capsid protein of human papillomavirus type 11 recombinant virus-like particles interacts with heparin and cell-surface glycosaminoglycans on human keratinocytes. *J. Biol. Chem.* 274:5810–22
213. Bienkowska-Haba M, Patel HD, Sapp M. 2009. Target cell cyclophilins facilitate human papillomavirus type 16 infection. *PLoS Pathog.* 5:e1000524
214. Day PM, Lowy DR, Schiller JT. 2008. Heparan sulfate-independent cell binding and infection with furin-precleaved papillomavirus capsids. *J. Virol.* 82:12565–68
215. Hansen CG, Nichols BJ. 2009. Molecular mechanisms of clathrin-independent endocytosis. *J. Cell Sci.* 122:1713–21
216. Kinchen JM, Ravichandran KS. 2008. Phagosome maturation: going through the acid test. *Nat. Rev. Mol. Cell Biol.* 9:781–95
217. Melendez AJ, Tay HK. 2008. Phagocytosis: a repertoire of receptors and Ca^{2+} as a key second messenger. *Biosci. Rep.* 28:287–98
218. Cherry S. 2009. What have RNAi screens taught us about viral-host interactions? *Curr. Opin. Microbiol.* 12:446–52
219. Pelkmans L. 2009. Systems biology of virus infection in mammalian cells. *Curr. Opin. Microbiol.* 12:429–31
220. Cudmore S, Cossart P, Griffiths G, Way M. 1995. Actin-based motility of vaccinia virus. *Nature* 378:636–38
221. Münter S, Way M, Frischknecht F. 2006. Signaling during pathogen infection. *Sci. STKE* 2006:re5
222. Sherer NM, Mothes W. 2008. Cytonemes and tunneling nanotubules in cell-cell communication and viral pathogenesis. *Trends Cell Biol.* 18:414–20
223. Sattentau Q. 2008. Avoiding the void: cell-to-cell spread of human viruses. *Nat. Rev. Microbiol.* 6:815–26

Cumulative Indexes

Contributing Authors, Volumes 75–79

Chapter Titles, Volumes 75–79

Mutagenesis

Chromatin and Chromosomes

Recombination and Transposition

Enzymes and Binding Proteins

RNA

Chemistry and Structure

Transcription and Gene Regulation

Splicing, Posttranscriptional Processing, and Modifications

Translation

Enzymology

Kinetics

Catalytic Mechanisms

Cofactors and Prosthetic Groups

Metalloenzymes

Biomembranes: Composition, Biology, Structure, and Function

Vesicular Trafficking and Secretion

Intracellular Targeting and Localization

Organelles and Organelle Biogenesis

Organismal Biochemistry

Development and Differentiation

Biochemical Basis of Disease

Molecular Physiology and Nutritional Biochemistry